$132.95 per copy (in United States).
Price subject to change without prior notice.

0049

REF

D1397369

NEW FOR 2009!

Line items found in this cost data that have the icon shown above are green.

The identified items fall into a broad definition of what is considered green.

Please see page ix for more information.

2009

R.S. Means Company, Inc.
Construction Publishers & Consultants
63 Smiths Lane
Kingston, MA 02364-3008
(781) 422-5000

Copyright©2008 by R.S. Means Company, Inc.
All rights reserved.

Printed in the United States of America.
ISSN 0271-5945
ISBN 978-0-87629-206-8

RSMeans
Repair & Remodeling Cost Data

30th Annual Edition

Senior Editor
Bob Mewis, CCC

Contributing Editors
Christopher Babbitt
Ted Baker
Barbara Balboni
Robert A. Bastoni
John H. Chiang, PE
Gary W. Christensen
David G. Drain, PE
Cheryl Elsmore
Robert J. Kuchta
Robert C. McNichols
Melville J. Mossman, PE
Jeannene D. Murphy
Stephen C. Plotner
Eugene R. Spencer
Marshall J. Stetson
Phillip R. Waier, PE

Editorial Advisory Board
James E. Armstrong, CPE, CEM
Senior Energy Consultant
KEMA Services, Inc.

William R. Barry, CCC
Cost Consultant

Robert F. Cox, PhD
Department Head and Professor
of Building Construction Management
Purdue University

Senior Vice President & General Manager
John Ware

Vice President of Operations
Dave Walsh

Vice President of Sales & Marketing
Sev Ritchie

Marketing Director
John M. Shea

Director of Product Development
Thomas J. Dion

Engineering Manager
Bob Mewis, CCC

Roy F. Gilley, AIA
Principal
Gilley Design Associates

Kenneth K. Humphreys, PhD, PE, CCE

Patricia L. Jackson, PE
Jackson A&E Associates, Inc.

Martin F. Joyce
Executive Vice President
Bond Brothers, Inc.

Production Manager
Michael Kokernak

Production Coordinator
Jill Goodman

Technical Support
Jonathan Forgit
Mary Lou Geary
Roger Hancock
Gary L. Hoitt
Genevieve Medeiros
Debbie Panarelli
Paula Reale-Camelio
Kathryn S. Rodriguez
Sheryl A. Rose

Book & Cover Design
Norman R. Forgit

This book is printed on recycled paper (10% PCW cover, 20% PCW text) using soy-based printing ink. This book is recyclable.

Foreword

Our Mission

Since 1942, RSMeans has been actively engaged in construction cost publishing and consulting throughout North America.

Today, over 60 years after RSMeans began, our primary objective remains the same: to provide you, the construction and facilities professional, with the most current and comprehensive construction cost data possible.

Whether you are a contractor, an owner, an architect, an engineer, a facilities manager, or anyone else who needs a reliable construction cost estimate, you'll find this publication to be a highly useful and necessary tool.

With the constant flow of new construction methods and materials today, it's difficult to find the time to look at and evaluate all the different construction cost possibilities. In addition, because labor and material costs keep changing, last year's cost information is not a reliable basis for today's estimate or budget.

That's why so many construction professionals turn to RSMeans. We keep track of the costs for you, along with a wide range of other key information, from city cost indexes . . . to productivity rates . . . to crew composition . . . to contractor's overhead and profit rates.

RSMeans performs these functions by collecting data from all facets of the industry and organizing it in a format that is instantly accessible to you. From the preliminary budget to the detailed unit price estimate, you'll find the data in this book useful for all phases of construction cost determination.

The Staff, the Organization, and Our Services

When you purchase one of RSMeans' publications, you are, in effect, hiring the services of a full-time staff of construction and engineering professionals.

Our thoroughly experienced and highly qualified staff works daily at collecting, analyzing, and disseminating comprehensive cost information for your needs. These staff members have years of practical construction experience and engineering training prior to joining the firm. As a result, you can count on them not only for the cost figures, but also for additional background reference information that will help you create a realistic estimate.

The RSMeans organization is always prepared to help you solve construction problems through its four major divisions: Construction and Cost Data Publishing, Electronic Products and Services, Consulting and Business Solutions, and Professional Development Services.

Besides a full array of construction cost estimating books, RSMeans also publishes a number of other reference works for the construction industry. Subjects include construction estimating and project and business management; special topics such as HVAC, roofing, plumbing, and hazardous waste remediation; and a library of facility management references.

In addition, you can access all of our construction cost data electronically using *Means CostWorks*® CD or on the Web using Means CostWorks.com.

What's more, you can increase your knowledge and improve your construction estimating and management performance with an RSMeans Construction Seminar or In-House Training Program. These two-day seminar programs offer unparalleled opportunities for everyone in your organization to get updated on a wide variety of construction-related issues.

RSMeans is also a worldwide provider of construction cost management and analysis services for commercial and government owners.

In short, RSMeans can provide you with the tools and expertise for constructing accurate and dependable construction estimates and budgets in a variety of ways.

Robert Snow Means Established a Tradition of Quality That Continues Today

Robert Snow Means spent years building RSMeans, making certain he always delivered a quality product.

Today, at RSMeans, we do more than talk about the quality of our data and the usefulness of our books. We stand behind all of our data, from historical cost indexes to construction materials and techniques to current costs.

If you have any questions about our products or services, please call us toll-free at 1-800-334-3509. Our customer service representatives will be happy to assist you. You can also visit our Web site at www.rsmeans.com.

Table of Contents

Related RSMeans Products and Services

The engineers at RSMeans suggest the following products and services as companion information resources to *RSMeans Repair & Remodeling Cost Data:*

Construction Cost Data Books
Building Construction Cost Data 2009
Facilities Construction Cost Data 2009
Facilities Maintenance & Repair Cost Data 2009

Reference Books
ADA Compliance Pricing Guide, 2nd Ed.
Building Security: Strategies & Costs
Designing & Building with the IBC, 2nd Ed.
Estimating Building Costs
Estimating Handbook, 2nd Ed.
Green Building: Project Planning & Estimating, 2nd Ed.
How to Estimate with Means Data and CostWorks, 3rd Ed.
Plan Reading & Material Takeoff
Project Scheduling & Management for Construction, 3rd Ed.
Repair & Remodeling Estimating Methods, 4th Ed.

Seminars and In-House Training
Repair & Remodeling Estimating
Means CostWorks® Training
Means Data for Job Order Contracting (JOC)
Plan Reading & Material Takeoff
Scheduling & Project Management

RSMeans on the Internet
Visit RSMeans at **www.rsmeans.com.** The site contains useful interactive cost and reference material. Request or download **FREE** estimating software demos. Visit our bookstore for convenient ordering and to learn more about new publications and companion products.

RSMeans Electronic Data
Get the information found in RSMeans cost books electronically on *Means CostWorks®* CD or on the Web at MeansCostworks.com

RSMeans Business Solutions
Engineers and Analysts offer research studies, benchmark analysis, predictive cost modeling, analytics, job order contracting, and real property management consultation, as well as custom-designed, Web-based dashboards and calculators that apply RCD/RSMeans extensive databases. Clients include federal government agencies, architects, construction management firms, and institutional organizations such as school systems, health care facilities, associations, and corporations.

RSMeans for Job Order Contracting (JOC)
Best practice JOC tools for cost estimating and project management to help streamline delivery processes for renovation projects. Renovation is a $147 billion market in the U.S., and includes projects in school districts, municipalities, health care facilities, colleges and universities, and corporations.
- RSMeans Engineers consult in contracting methods and conduct JOC Facility Audits
- JOCWorks™ Software (Basic, Advanced, PRO levels)
- RSMeans Job Order Contracting Cost Data for the entire U.S.

Construction Costs for Software Applications
Over 25 unit price and assemblies cost databases are available through a number of leading estimating and facilities management software providers (listed below). For more information see the "Other RSMeans Products" pages at the back of this publication.

MeansData™ is also available to federal, state, and local government agencies as multi-year, multi-seat licenses.

- 4Clicks-Solutions, LLC
- ArenaSoft Estimating
- Beck Technology
- BSD – Building Systems Design, Inc.
- CMS – Construction Management Software
- Corecon Technologies, Inc.
- CorVet Systems
- Estimating Systems, Inc.
- Maximus Asset Solutions
- Sage Timberline Office
- US Cost, Inc.
- VFA – Vanderweil Facility Advisers
- WinEstimator, Inc.

How the Book Is Built: An Overview

The Construction Specifications Institute (CSI) and Construction Specifications Canada (CSC) have produced the 2004 edition of MasterFormat, a new and updated system of titles and numbers used extensively to organize construction information.

All unit price data in the RSMeans cost data books is now arranged in the 50-division MasterFormat 2004 system.

A Powerful Construction Tool

You have in your hands one of the most powerful construction tools available today. A successful project is built on the foundation of an accurate and dependable estimate. This book will enable you to construct just such an estimate.

For the casual user the book is designed to be:

- quickly and easily understood so you can get right to your estimate.
- filled with valuable information so you can understand the necessary factors that go into the cost estimate.

For the regular user, the book is designed to be:

- a handy desk reference that can be quickly referred to for key costs.
- a comprehensive, fully reliable source of current construction costs and productivity rates so you'll be prepared to estimate any project.
- a source book for preliminary project cost, product selections, and alternate materials and methods.

To meet all of these requirements we have organized the book into the following clearly defined sections.

New: Quick Start

See our new "Quick Start" instructions on the following page.

Estimating with RSMeans Unit Price Cost Data

Please see these steps to complete an estimate using RSMeans unit price cost data.

How To Use the Book:
The Details

This section contains an in-depth explanation of how the book is arranged . . . and how you can use it to determine a reliable construction cost estimate. It includes information about how we develop our cost figures and how to completely prepare your estimate.

Unit Price Section

All unit price cost data has been divided into the 50 divisions according to the MasterFormat system of classification and numbering. For a listing of these divisions and an outline of their subdivisions, see the Unit Price Section Table of Contents.

Estimating tips are included at the beginning of each division.

Assemblies Section

The cost data in this section has been organized in an "Assemblies" format. These assemblies are the functional elements of a building and are arranged according to the 7 divisions of the UNIFORMAT II classification system. For a complete explanation of a typical "Assemblies" page, see "How To Use the Assemblies Cost Tables."

Reference Section

This section includes information on Equipment Rental Costs, Crew Listings, Historical Cost Indexes, City Cost Indexes, Location Factors, Reference Tables, Change Orders, Square Foot Costs, and a listing of Abbreviations.

Equipment Rental Costs: This section contains the average costs to rent and operate hundreds of pieces of construction equipment.

Crew Listings: This section lists all the crews referenced in the book. For the purposes of this book, a crew is composed of more than one trade classification and/or the addition of power equipment to any trade classification. Power equipment is included in the cost of the crew. Costs are shown both with the bare labor rates and with the installing contractor's overhead and profit added. For each, the total crew cost per eight-hour day and the composite cost per labor-hour are listed.

Historical Cost Indexes: These indexes provide you with data to adjust construction costs over time.

City Cost Indexes: All costs in this book are U.S. national averages. Costs vary because of the regional economy. You can adjust costs by CSI Division to over 316 locations throughout the U.S. and Canada by using the data in this section.

Location Factors: You can adjust total project costs to over 900 locations throughout the U.S. and Canada by using the data in this section.

Reference Tables: At the beginning of selected major classifications in the Unit Price and Assemblies sections are "reference numbers" shown in a shaded box. These numbers refer you to related information in the Reference Section. In this section,

you'll find reference tables, explanations, estimating information that support how we develop the unit price data, technical data, and estimating procedures.

Change Orders: This section includes information on the factors that influence the pricing of change orders.

Square Foot Costs: This section contains costs for 59 different building types that allow you to make a rough estimate for the overall cost of a project or its major components.

Abbreviations: A listing of abbreviations used throughout this book, along with the terms they represent, is included in this section.

Index

A comprehensive listing of all terms and subjects in this book will help you quickly find what you need when you are not sure where it falls in MasterFormat.

The Scope of This Book

This book is designed to be as comprehensive and as easy to use as possible. To that end we have made certain assumptions and limited its scope in two key ways:

1. We have established material prices based on a national average.
2. We have computed labor costs based on a 30-city national average of union wage rates.

For a more detailed explanation of how the cost data is developed, see "How To Use the Book: The Details."

Project Size

The material prices in Means data cost books are "contractor's prices." They are the prices that contractors can expect to pay at the lumberyards, suppliers/distributers warehouses, etc. Small orders of speciality items would be higher than the costs shown, while very large orders, such as truckload lots, would be less. The variation would depend on the size, timing, and negotiating power of the contractor. The labor costs are primarily for new construction or major renovations rather than repairs or minor alterations. **With reasonable exercise of judgment, the figures can be used for any building work.**

Absolute Essentials for a Quick Start

If you feel you are ready to use this book and don't think you will need the detailed instructions that begin on the following page, this Absolute Essentials for a Quick Start page is for you. These steps will allow you to get started estimating in a matter of minutes.

1 Scope

Think through the project that you will be estimating and identify the many individual work tasks that will need to be covered in your estimate.

2 Quantify

Determine the number of units that will be required for each work task that you identified.

3 Pricing

Locate individual Unit Price line items that match the work tasks you identified. The Unit Price Section Table of Contents that begins on page 1 and the Index in the back of the book will help you find these line items.

4 Multiply

Multiply the Total Incl O&P cost for a Unit Price line item in the book by your quantity for that item. The price you calculate will be an estimate for a completed item of work performed by a subcontractor. Keep adding line items in this manner to build your estimate.

5 Project Overhead

Include project overhead items in your estimate. These items are needed to make the job run and are typically, but not always, provided by the General Contractor. They can be found in Division 1. An alternate method of estimating project overhead costs is to apply a percentage of the total project cost.

Include rented tools not included in crews, waste, rubbish handling and cleanup.

6 Estimate Summary

Include General Contractor's markup on subcontractors, General Contractor's office overhead and profit, and sales tax on materials and equipment.

Adjust your estimate to the project's location by using the City Cost Indexes or Location Factors found in the Reference Section.

Editors' Note: We urge you to spend time reading and understanding the supporting material in the front of this book. An accurate estimate requires experience, knowledge and careful calculation. The more you know about how we at RSMeans developed the data, the more accurate your estimate will be. In addition, it is important to take into consideration the reference material in the back of the book such as Equipment Listings, Crew Listings, City Cost Indexes, Location Factors and Reference Numbers.

Estimating with RSMeans Unit Price Cost Data

Following these steps will allow you to complete an accurate estimate using RSMeans Unit Price cost data.

1 Scope out the project

- Identify the individual work tasks that will need to be covered in your estimate.
- The Unit Price data inside this book has been divided into 44 Divisions according to the CSI MasterFormat2004 – their titles are listed on the back cover of your book.
- Think through the project that you will be estimating and identify those CSI Divisions that you will need to use in your estimate.
- The Unit Price Section Table of Contents that begins on page 1 may also be helpful when scoping out your project.
- Experienced estimators find it helpful to begin an estimate with Division 2, estimating Division 1 after the full scope of the project is known.

2 Quantify

- Determine the number of units that will be required for each work task that you previously identified.
- Experienced estimators will include an allowance for waste in their quantities (waste is not included in RSMeans Unit Price line items unless so stated).

3 Price the quantities

- Use the Unit Price Table of Contents, and the Index, to locate the first individual Unit Price line item for your estimate.
- Reference Numbers indicated within a Unit Price section refer to additional information that you may find useful.
- The crew will tell you who is performing the work for that task. Crew codes are expanded in the Crew Listings in the Reference Section to include all trades and equipment that comprise the crew.
- The Daily Output is the amount of work the crew is expected to do in one day.

- The Labor-Hours value is the amount of time it will take for the average crew member to install one unit of measure.
- The abbreviated Unit designation indicates the unit of measure upon which the crew, productivity and prices are based.
- Bare Costs are shown for materials, labor, and equipment needed to complete the Unit Price line item. Bare costs do not include waste, project overhead, payroll insurance, payroll taxes, main office overhead, or profit.
- The Total Incl O&P cost is the billing rate or invoice amount of the installing contractor or subcontractor who performs the work for the Unit Price line item.

4 Multiply

- Multiply the total number of units needed for your project by the Total Incl O&P cost for the Unit Price line item.
- Be careful that the final unit of measure for your quantity of units matches the unit of measure in the Unit column in the book.
- The price you calculate will be an estimate for a completed item of work.
- Keep scoping individual tasks, determining the number of units required for those tasks, matching them up with individual Unit Price line items in the book, and multiplying quantities by Total Incl O&P costs. In this manner keep building your estimate.
- An estimate completed to this point in this manner will be priced as if a subcontractor, or set of subcontractors, will perform the work. The estimate does not yet include Project Overhead or Estimate Summary components such as General Contractor markups on subcontracted and self-performed work, General Contractor office overhead and profit, contingency, and location factor.

5 Project Overhead

- Include project overhead items from Division 1 – General Requirements.
- These are items that will be needed to make the job run. These are typically, but not always, provided by the General Contractor. They include, but are not limited to, such items as field personnel, insurance, performance bond, permits, testing, temporary utilities, field office and storage facilities, temporary scaffolding and platforms, equipment mobilization and demobilization, temporary roads and sidewalks, winter protection, temporary barricades and fencing, temporary security, temporary signs, field engineering and layout, final cleaning and commissioning.
- These items should be scoped, quantified, matched to individual Unit Price line items in Division 1, priced and added to your estimate.
- An alternate method of estimating project overhead costs is to apply a percentage of the total project cost, usually 5% to 15% with an average of 10% (see General Conditions, page ix).
- Include other project related expenses in your estimate such as:
  - Rented equipment not itemized in the Crew Listings.
  - Rubbish handling throughout the project (see 02 41 19.23).

6 Estimate Summary

- Includes sales tax on materials and equipment.
  Note: Sales tax must be added for materials in subcontracted work.
- Include the General Contractor's markup on self-performed work, usually 5% to 15% with an average of 10%.
- Include the General Contractor's markup on subcontracted work, usually 5% to 15% with an average of 10%.
- Include General Contractor's main office overhead and profit.
- RSMeans gives general guidelines on the General Contractor's main office overhead (see section 01 31 13.50 and Reference Number R013113-50).
- RSMeans gives no guidance on the General Contractor's profit.
- Markups will depend on the size of the General Contractor's operations, his projected annual revenue, the level of risk he is taking on, and on the level of competitiveness in the local area and for this project in particular.
- Include a contingency, usually 3% to 5%.

- Adjust your estimate to the project's location by using the City Cost Indexes or the Location Factors in the Reference Section.
- Look at the rules on the pages for How to Use the City Cost Indexes to see how to apply the Indexes for your location.
- When the proper Index or Factor has been identified for the project's location, convert it to a multiplier by dividing it by 100, then multiply that multiplier by your estimate total cost. Your original estimate total cost will now be adjusted up or down from the national average to a total that is appropriate for your location.

Editors' Notes:

1) *We urge you to spend time reading and understanding the supporting material in the front of this book. An accurate estimate requires experience, knowledge, and careful calculation. The more you know about how we at RSMeans developed the data, the more accurate your estimate will be. In addition, it is important to take into consideration the reference material in the back of the book such as Equipment Listings, Crew Listings, City Cost Indexes, Location Factors, and Reference Numbers.*

2) *Contractors who are bidding or are involved in JOC, DOC, SABER, or IDIQ type contracts are cautioned that workers' compensation Insurance, federal and state payroll taxes, waste, project supervision, project overhead, main office overhead, and profit are not included in bare costs. Your coefficient or multiplier must cover these costs.*

How to Use the Book: The Details

What's Behind the Numbers? The Development of Cost Data

The staff at RSMeans continuously monitors developments in the construction industry in order to ensure reliable, thorough, and up-to-date cost information.

While *overall* construction costs may vary relative to general economic conditions, price fluctuations within the industry are dependent upon many factors. Individual price variations may, in fact, be opposite to overall economic trends. Therefore, costs are continually monitored and complete updates are published yearly. Also, new items are frequently added in response to changes in materials and methods.

Costs—$ (U.S.)

All costs represent U.S. national averages and are given in U.S. dollars. The RSMeans City Cost Indexes can be used to adjust costs to a particular location. The City Cost Indexes for Canada can be used to adjust U.S. national averages to local costs in Canadian dollars. No exchange rate conversion is necessary.

G The processes or products identified by the green symbol in this publication have been determined to be environmentally responsible and/or resource-efficient solely by the RSMeans engineering staff. The inclusion of the green symbol does not represent compliance with any specific industry association or standard.

Material Costs

The RSMeans staff contacts manufacturers, dealers, distributors, and contractors all across the U.S. and Canada to determine national average material costs. If you have access to current material costs for your specific location, you may wish to make adjustments to reflect differences from the national average. Included within material costs are fasteners for a normal installation. RSMeans engineers use manufacturers' recommendations, written specifications, and/or standard construction practice for size and spacing of fasteners. Adjustments to material costs may be required for your specific application or location. Material costs do not include sales tax.

Labor Costs

Labor costs are based on the average of wage rates from 30 major U.S. cities. Rates are determined from labor union agreements or prevailing wages for construction trades for the current year. Rates, along with overhead and profit markups, are listed on the inside back cover of this book.

- If wage rates in your area vary from those used in this book, or if rate increases are expected within a given year, labor costs should be adjusted accordingly.

Labor costs reflect productivity based on actual working conditions. In addition to actual installation, these figures include time spent during a normal weekday on tasks such as, material receiving and handling, mobilization at site, site movement, breaks, and cleanup.

Productivity data is developed over an extended period so as not to be influenced by abnormal variations and reflects a typical average.

Equipment Costs

Equipment costs include not only rental, but also operating costs for equipment under normal use. The operating costs include parts and labor for routine servicing such as repair and replacement of pumps, filters, and worn lines. Normal operating expendables, such as fuel, lubricants, tires, and electricity (where applicable), are also included. Extraordinary operating expendables with highly variable wear patterns, such as diamond bits and blades, are excluded. These costs are included under materials. Equipment rental rates are obtained from industry sources throughout North America—contractors, suppliers, dealers, manufacturers, and distributors.

Equipment costs do not include operators' wages; nor do they include the cost to move equipment to a job site (mobilization) or from a job site (demobilization).

Equipment Cost/Day—The cost of power equipment required for each crew is included in the Crew Listings in the Reference Section (small tools that are considered as essential everyday tools are not listed out separately). The Crew Listings itemize specialized tools and heavy equipment along with labor trades. The daily cost of itemized equipment included in a crew is based on dividing the weekly bare rental rate by 5 (number of working days per week) and then adding the hourly operating cost times 8 (the number of hours per day). This Equipment Cost/Day is shown in the last column of the Equipment Rental Cost pages in the Reference Section.

Mobilization/Demobilization—The cost to move construction equipment from an equipment yard or rental company to the job site and back again is not included in equipment costs. Mobilization (to the site) and demobilization (from the site) costs can be found in the Unit Price Section. If a piece of equipment is already at the job site, it is not appropriate to utilize mob./demob. costs again in an estimate.

General Conditions

Cost data in this book is presented in two ways: Bare Costs and Total Cost including O&P (Overhead and Profit). General Conditions, when applicable, should also be added to the Total Cost including O&P. The costs for General Conditions are listed in Division 1 of the Unit Price Section and the Reference Section of this book. General Conditions for the *Installing Contractor* may range from 0% to 10% of the Total Cost including O&P. For the *General* or *Prime Contractor*, costs for General Conditions may range from 5% to 15% of the Total Cost including O&P, with a figure of 10% as the most typical allowance.

Overhead and Profit

Total Cost including O&P for the *Installing Contractor* is shown in the last column on both the Unit Price and the Assemblies pages of this book. This figure is the sum of the bare material cost plus 10% for profit, the bare labor cost plus total overhead and profit, and the bare equipment cost plus 10% for profit. Details for the calculation of Overhead and Profit on labor are shown on the inside back cover and in the

Reference Section of this book. (See the "How to Use the Unit Price Pages" for an example of this calculation.)

Factors Affecting Costs

Costs can vary depending upon a number of variables. Here's how we have handled the main factors affecting costs.

Quality—The prices for materials and the workmanship upon which productivity is based represent sound construction work. They are also in line with U.S. government specifications.

Overtime—We have made no allowance for overtime. If you anticipate premium time or work beyond normal working hours, be sure to make an appropriate adjustment to your labor costs.

Productivity—The productivity, daily output, and labor-hour figures for each line item are based on working an eight-hour day in daylight hours in moderate temperatures. For work that extends beyond normal work hours or is performed under adverse conditions, productivity may decrease. (See the section in "How To Use the Unit Price Pages" for more on productivity.)

Size of Project—The size, scope of work, and type of construction project will have a significant impact on cost. Economies of scale can reduce costs for large projects. Unit costs can often run higher for small projects. Costs in this book are intended for the size and type of project as previously described in "How the Book Is Built: An Overview." Costs for projects of a significantly different size or type should be adjusted accordingly.

Location—Material prices in this book are for metropolitan areas. However, in dense urban areas, traffic and site storage limitations may increase costs. Beyond a 20-mile radius of large cities, extra trucking or transportation charges may also increase the material costs slightly. On the other hand, lower wage rates may be in effect. Be sure to consider both of these factors when preparing an estimate, particularly if the job site is located in a central city or remote rural location.

In addition, highly specialized subcontract items may require travel and per-diem expenses for mechanics.

Other Factors—
- season of year
- contractor management
- weather conditions
- local union restrictions
- building code requirements
- availability of:
  - adequate energy
  - skilled labor
  - building materials
- owner's special requirements/restrictions
- safety requirements
- environmental considerations

Unpredictable Factors—General business conditions influence "in-place" costs of all items. Substitute materials and construction methods may have to be employed. These may affect the installed cost and/or life cycle costs. Such factors may be difficult to evaluate and cannot necessarily be predicted on the basis of the job's location in a particular section of the country. Thus, where these factors apply, you may find significant but unavoidable cost variations for which you will have to apply a measure of judgment to your estimate.

Rounding of Costs

In general, all unit prices in excess of $5.00 have been rounded to make them easier to use and still maintain adequate precision of the results. The rounding rules we have chosen are in the following table.

Prices from ...	Rounded to the nearest ...
$.01 to $5.00	$.01
$5.01 to $20.00	$.05
$20.01 to $100.00	$.50
$100.01 to $300.00	$1.00
$300.01 to $1,000.00	$5.00
$1,000.01 to $10,000.00	$25.00
$10,000.01 to $50,000.00	$100.00
$50,000.01 and above	$500.00

Final Checklist

Estimating can be a straightforward process provided you remember the basics. Here's a checklist of some of the steps you should remember to complete before finalizing your estimate.

Did you remember to . . .

- factor in the City Cost Index for your locale?
- take into consideration which items have been marked up and by how much?
- mark up the entire estimate sufficiently for your purposes?
- read the background information on techniques and technical matters that could impact your project time span and cost?
- include all components of your project in the final estimate?
- make use of Minimum Labor/Equipment Charges for Small Quantities (see the following page for more details)?
- double check your figures for accuracy?
- call RSMeans if you have any questions about your estimate or the data you've found in our publications?

Remember, RSMeans stands behind its publications. If you have any questions about your estimate . . . about the costs you've used from our books . . . or even about the technical aspects of the job that may affect your estimate, feel free to call the RSMeans editors at 1-800-334-3509.

Using Minimum Labor/Equipment Charges for Small Quantities

Estimating small construction or repair tasks often creates situations in which the quantity of work to be performed is very small. When this occurs, the labor and/or equipment costs to perform the work may be too low to allow for the crew to get to the job, receive instructions, find materials, get set up, perform the work, clean up, and get to the next job. In these situations, the estimator should compare the developed labor and/or equipment costs for performing the work (e.g., quantity x labor and/or equipment costs) with the *"minimum labor/equipment charge"* within that Unit Price section of the book.(These "minimum labor/equipment charge" line items appear only in the Facilities Construction Cost Data and the Repair & Remodeling Cost Data titles).

If the labor and/or equipment costs developed by the estimator are LOWER THAN the *"minimum labor/equipment charge"* listed at the bottom of specific sections of Unit Price costs, the estimator should adjust the developed costs upward to the *"minimum labor/equipment charge."* The proper use of a *"minimum labor/equipment charge"* results in having enough money in the estimate to cover the contractor's higher cost of performing a very small amount of work during a partial workday.

A *"minimum labor/equipment charge"* should be used only when the task being estimated is the only task the crew will perform at the job site that day. If, however, the crew will be able to perform other tasks at the job site that day, the use of a *"minimum labor/equipment charge"* is not appropriate.

---

08 52 Wood Windows
08 52 10 – Plain Wood Windows

08 52 10.40 Casement Window		Crew	Daily Output	Labor-Hours	Unit	Material	2009 Bare Costs Labor	Equipment	Total	Total Incl O&P
0010	**CASEMENT WINDOW**, including frame, screen and grilles									
0100	Avg. quality, bldrs. model, 2'-0" x 3'-0" H, dbl. insulated glass	G	1 Carp	10	.800	Ea.	276	32		308	360
0150	Low E glass	G		10	.800		225	32		257	300
0200	2'-0" x 4'-6" high, double insulated glass	G		9	.889		345	35.50		380.50	440
0250	Low E glass	G		9	.889		263	35.50		298.50	350
0300	2'-4" x 6'-0" high, double insulated glass	G		8	1		380	40		420	485
0350	Low E glass	G		8	1		465	40		505	580
0522	Vinyl clad, premium, double insulated glass, 2'-0" x 3'-0"	G		10	.800		266	32		298	345
0524	2'-0" x 4'-0"	G		9	.889		310	35.50		345.50	400
0525	2'-0" x 5'-0"	G		8	1		355	40		395	455
0528	2'-0" x 6'-0"	G		8	1		355	40		395	455
8100	Metal clad, deluxe, dbl. insul. glass, 2'-0" x 3'-0" high	G		10	.800		226	32		258	300
9000	Minimum labor/equipment charge		1 Carp	3	2.667	Job		107		107	176

Example:

Establish the bid price to install two casement windows. Assume installation of 2' x 4' metal clad windows with insulating glass [Unit Price line number 08 52 10.40 0200], and that this is the only task this crew will perform at the job site that day.

Solution:

Step One —Develop the Bare Labor Cost for this task:

Bare Labor Cost = 2 windows @ $35.50/each = $71.00

Step Two —Evaluate the *"minimum labor/equipment charge"* for this Unit Price section against the developed Bare Labor Cost for this task:

   "Minimum labor/equipment charge" = $107.00 (compare with $71.00)

Step Three —Choose to adjust the developed labor cost upward to the *"minimum labor/ equipment charge."*

Step Four —Develop the bid price for this task (including O&P):

   Add together the marked-up Bare Material Cost for this task and the marked-up *"minimum labor/equipment charge"* for this Unit Price section.

   2 x ($345.00 + 10%) + ($107.00 + 65.1%)

   = 2 x ($345.00 + $34.50) + ($107.00 + $69.66)

   = 2 x ($379.50) + $176.66

   = $759.00 + $176.66

   = $935.66

ANSWER: $935.66 is the correct bid price to use. This sum takes into consideration the Material Cost (with 10% for profit) for these two windows, plus the *"minimum labor/equipment charge"* (with O&P included) for this section of the Unit Price book.

Unit Price Section

Table of Contents

1

Table of Contents (cont.)

Table of Contents (cont.)

How to Use the Unit Price Pages

The following is a detailed explanation of a sample entry in the Unit Price Section. Next to each bold number below is the described item with the appropriate component of the sample entry following in parentheses. Some prices are listed as bare costs; others as costs that include overhead and profit of the installing contractor. In most cases, if the work is to be subcontracted, the general contractor will need to add an additional markup (RSMeans suggests using 10%) to the figures in the column "Total Incl. O&P."

1 Division Number/Title (03 30/Cast-In-Place Concrete)

Use the Unit Price Section Table of Contents to locate specific items. The sections are classified according to the CSI MasterFormat 2004 system.

2 Line Numbers (03 30 53.40 3920)

Each unit price line item has been assigned a unique 12-digit code based on the CSI MasterFormat classification.

```
MasterFormat Division (03)
    MasterFormat Level 2 (03 30 00)
        MasterFormat Level 3
    03  30  53.40    3920
                MasterFormat Level 4
                RSMeans 12-Digit Line Number
```

3 Description (Concrete In Place, etc.)

Each line item is described in detail. Sub-items and additional sizes are indented beneath the appropriate line items. The first line or two after the main item (in boldface) may contain descriptive information that pertains to all line items beneath this boldface listing.

4 Reference Number Information

R033053 -50 You'll see reference numbers shown in shaded boxes at the beginning of some sections. These refer to related items in the Reference Section, visually identified by a vertical gray bar on the page edges.

The relation may be an estimating procedure that should be read before estimating, or technical information.

The "R" designates the Reference Section. The numbers refer to the MasterFormat 2004 classification system.

It is strongly recommended that you review all reference numbers that appear within the section in which you are working.

Note: Not all reference numbers appear in all RSMeans publications.

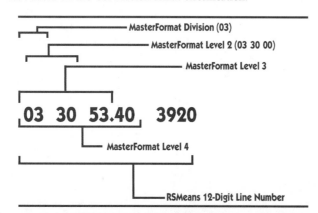

03 30 Cast-In-Place Concrete

03 30 53 – Miscellaneous Cast-In-Place Concrete

03 30 53.40 Concrete In Place		Crew	Daily Output	Labor-Hours	Unit	Material	2009 Bare Costs Labor	Equipment	Total	Total Incl O&P
3570	6' x 6' x 8" thick		14	3.429		188	135	1.85	324.85	430
3580	8' x 8' x 10" thick		8	6		395	236	3.24	634.24	830
3590	10' x 10' x 12" thick		5	9.600		655	380	5.20	1,040.20	1,350
3800	Footings, spread under 1 C.Y.	C-14C	28	4	C.Y.	209	153	.92	362.92	485
3825	1 C.Y to 5 C.Y.		43	2.605		221	99.50	.60	321.10	410
3850	Over 5 C.Y.		75	1.493		198	57	.34	255.34	310
3900	Footings, strip, 18" x 9", unreinforced	C-14L	40	2.400		120	89.50	.65	210.15	280
3920	18" x 9", reinforced	C-14C	35	3.200		157	122	.74	279.74	375
3925	20" x 10", unreinforced	C-14L	45	2.133		117	79.50	.58	197.08	260
3930	20" x 10", reinforced	C-14C	40	2.400		147	107	.64	254.64	340
3935	24" x 12", unreinforced	C-14L	55	1.745		116	65	.47	181.47	236
3940	24" x 12", reinforced	C-14C	48	2.333		147	89.50	.54	237.04	310
3945	36" x 12", unreinforced	C-14L	70	1.371		113	51	.37	164.37	208
3950	36" x 12", reinforced	C-14C	60	1.867		141	71.50	.43	212.93	273

Crew (C-14C)

The "Crew" column designates the typical trade or crew used to install the item. If an installation can be accomplished by one trade and requires no power equipment, that trade and the number of workers are listed (for example, "2 Carpenters"). If an installation requires a composite crew, a crew code designation is listed (for example, "C-14C"). You'll find full details on all composite crews in the Crew Listings.

- For a complete list of all trades utilized in this book and their abbreviations, see the inside back cover.

Crews

Crew No.	Bare Costs		Incl. Subs O & P		Cost Per Labor-Hour	
Crew C-14C	Hr.	Daily	Hr.	Daily	Bare Costs	Incl. O&P
1 Carpenter Foreman (out)	$41.95	$335.60	$69.25	$554.00	$38.25	$63.24
6 Carpenters	39.95	1917.60	65.95	3165.60		
2 Rodmen (reinf.)	44.55	712.80	75.95	1215.20		
4 Laborers	31.60	1011.20	52.15	1668.80		
1 Cement Finisher	38.30	306.40	59.90	479.20		
1 Gas Engine Vibrator		26.00		28.60	.23	.26
112 L.H., Daily Totals		$4309.60		$7111.40	$38.48	$63.49

Productivity: Daily Output (35.0)/ Labor-Hours (3.20)

The "Daily Output" represents the typical number of units the designated crew will install in a normal 8-hour day. To find out the number of days the given crew would require to complete the installation, divide your quantity by the daily output. For example:

Quantity	÷	Daily Output	=	Duration
100 C.Y.	÷	35.0/ Crew Day	=	2.86 Crew Days

The "Labor-Hours" figure represents the number of labor-hours required to install one unit of work. To find out the number of labor-hours required for your particular task, multiply the quantity of the item times the number of labor-hours shown. For example:

Quantity	x	Productivity Rate	=	Duration
100 C.Y.	x	3.20 Labor-Hours/ C.Y.	=	320 Labor-Hours

Unit (C.Y.)

The abbreviated designation indicates the unit of measure upon which the price, production, and crew are based (C.Y. = Cubic Yard). For a complete listing of abbreviations, refer to the Abbreviations Listing in the Reference Section of this book.

Bare Costs:
Mat. (Bare Material Cost) (157)

The unit material cost is the "bare" material cost with no overhead or profit included. Costs shown reflect national average material prices for January of the current year and include delivery to the job site. No sales taxes are included.

Labor (122)

The unit labor cost is derived by multiplying bare labor-hour costs for Crew C-14C by labor-hour units. The bare labor-hour cost is found in the Crew Section under C-14C. (If a trade is listed, the hourly labor cost—the wage rate—is found on the inside back cover.)

Labor-Hour Cost Crew C-14C	x	Labor-Hour Units	=	Labor
$38.25	x	3.20	=	$122.00

Equip. (Equipment) (.74)

Equipment costs for each crew are listed in the description of each crew. Tools or equipment whose value justifies purchase or ownership by a contractor are considered overhead as shown on the inside back cover. The unit equipment cost is derived by multiplying the bare equipment hourly cost by the labor-hour units.

Equipment Cost Crew C-14C	x	Labor-Hour Units	=	Equip.
$.23	x	3.20	=	$.74

Total (279.74)

The total of the bare costs is the arithmetic total of the three previous columns: mat., labor, and equip.

Material	+	Labor	+	Equip.	=	Total
$157	+	$122	+	$.74	=	$279.74

Total Costs Including O &P

This figure is the sum of the bare material cost plus 10% for profit; the bare labor cost plus total overhead and profit (per the inside back cover or, if a crew is listed, from the crew listings); and the bare equipment cost plus 10% for profit.

Material is Bare Material Cost + 10% = 157 + 15.70	=	$172.70
Labor for Crew C-14C = Labor-Hour Cost (63.24) x Labor-Hour Units (3.20)	=	$202.37
Equip. is Bare Equip. Cost + 10% = .74 + .07	=	$ .81
Total (Rounded)	=	$ 375

Estimating Tips

01 20 00 Price and Payment Procedures

- Allowances that should be added to estimates to cover contingencies and job conditions that are not included in the national average material and labor costs are shown in section 01 21.

- When estimating historic preservation projects (depending on the condition of the existing structure and the owner's requirements), a 15%–20% contingency or allowance is recommended, regardless of the stage of the drawings.

01 30 00 Administrative Requirements

- Before determining a final cost estimate, it is a good practice to review all the items listed in Subdivisions 01 31 and 01 32 to make final adjustments for items that may need customizing to specific job conditions.

- Requirements for initial and periodic submittals can represent a significant cost to the General Requirements of a job. Thoroughly check the submittal specifications when estimating a project to determine any costs that should be included.

01 40 00 Quality Requirements

- All projects will require some degree of Quality Control. This cost is not included in the unit cost of construction listed in each division. Depending upon the terms of the contract, the various costs of inspection and testing can be the responsibility of either the owner or the contractor. Be sure to include the required costs in your estimate.

01 50 00 Temporary Facilities and Controls

- Barricades, access roads, safety nets, scaffolding, security, and many more requirements for the execution of a safe project are elements of direct cost. These costs can easily be overlooked when preparing an estimate. When looking through the major classifications of this subdivision, determine which items apply to each division in your estimate.

- Construction Equipment Rental Costs can be found in the Reference Section in section 01 54 33. Operators' wages are not included in equipment rental costs.

- Equipment mobilization and demobilization costs are not included in equipment rental costs and must be considered separately in section 01 54 36.50.

- The cost of small tools provided by the installing contractor for his workers is covered in the "Overhead" column on the "Installing Contractor's Overhead and Profit" table that lists labor trades, base rates and markups and, therefore, is included in the "Total Incl. O&P" cost of any Unit Price line item. For those users who are constrained by contract terms to use only bare costs, there are two line items in section 01 54 39.70 for small tools as a percentage of bare labor cost. If some of those users are further constrained by contract terms to refrain from using line items from Division 1, you are advised to cover the cost of small tools within your coefficient or multiplier.

01 70 00 Execution and Closeout Requirements

- When preparing an estimate, thoroughly read the specifications to determine the requirements for Contract Closeout. Final cleaning, record documentation, operation and maintenance data, warranties and bonds, and spare parts and maintenance materials can all be elements of cost for the completion of a contract. Do not overlook these in your estimate.

Reference Numbers

Reference numbers are shown in shaded boxes at the beginning of some major classifications. These numbers refer to related items in the Reference Section. The reference information may be an estimating procedure, an alternate pricing method, or technical information.

Note: Not all subdivisions listed here necessarily appear in this publication.

01 11 Summary of Work

01 11 31 - Professional Consultants

01 11 31.10 Architectural Fees

	01 11 31.10 Architectural Fees		Crew	Daily Output	Labor-Hours	Unit	Material	2009 Bare Costs Labor	Equipment	Total	Total Incl O&P
0010	ARCHITECTURAL FEES	R011110-10									
0020	For new construction										
0060	Minimum					Project					4.90%
0090	Maximum										16%
0100	For alteration work, to $500,000, add to new construction fee										50%
0150	Over $500,000, add to new construction fee										25%

01 11 31.20 Construction Management Fees

	01 11 31.20 Construction Management Fees	Crew	Daily Output	Labor-Hours	Unit	Material	Labor	Equipment	Total	Total Incl O&P
0010	CONSTRUCTION MANAGEMENT FEES									
0060	For work to $100,000				Project					10%
0070	To $250,000									9%
0090	To $1,000,000									6%
0100	To $5,000,000									5%
0110	To $10,000,000									4%

01 11 31.30 Engineering Fees

	01 11 31.30 Engineering Fees		Crew	Daily Output	Labor-Hours	Unit	Material	Labor	Equipment	Total	Total Incl O&P
0010	ENGINEERING FEES	R011110-30									
0020	Educational planning consultant, minimum					Project					.50%
0100	Maximum					"					2.50%
0400	Elevator & conveying systems, minimum					Contrct					2.50%
0500	Maximum										5%
1000	Mechanical (plumbing & HVAC), minimum										4.10%
1100	Maximum										10.10%
1200	Structural, minimum					Project					1%
1300	Maximum					"					2.50%

01 11 31.75 Renderings

	01 11 31.75 Renderings	Crew	Daily Output	Labor-Hours	Unit	Material	Labor	Equipment	Total	Total Incl O&P
0010	RENDERINGS Color, matted, 20" x 30", eye level,									
0020	1 building, minimum				Ea.	1,950			1,950	2,150
0050	Average					2,775			2,775	3,075
0100	Maximum					4,450			4,450	4,900
1000	5 buildings, minimum					3,900			3,900	4,300
1100	Maximum					7,800			7,800	8,575

01 21 Allowances

01 21 16 - Contingency Allowances

01 21 16.50 Contingencies

	01 21 16.50 Contingencies	Crew	Daily Output	Labor-Hours	Unit	Material	Labor	Equipment	Total	Total Incl O&P
0010	CONTINGENCIES, Add to estimate									
0020	Conceptual stage				Project					25%
0050	Schematic stage									20%
0100	Preliminary working drawing stage (Design Dev.)									15%
0150	Final working drawing stage									8%

01 21 53 - Factors Allowance

01 21 53.50 Factors

	01 21 53.50 Factors		Crew	Daily Output	Labor-Hours	Unit	Material	Labor	Equipment	Total	Total Incl O&P
0010	FACTORS Cost adjustments	R012153-10									
0100	Add to construction costs for particular job requirements										
0500	Cut & patch to match existing construction, add, minimum					Costs	2%	3%			
0550	Maximum						5%	9%			
0800	Dust protection, add, minimum						1%	2%			
0850	Maximum						4%	11%			
1100	Equipment usage curtailment, add, minimum						1%	1%			
1150	Maximum						3%	10%			
1400	Material handling & storage limitation, add, minimum						1%	1%			

01 21 Allowances

01 21 53 – Factors Allowance

01 21 53.50 Factors

	01 21 53.50 Factors	Crew	Daily Output	Labor-Hours	Unit	Material	2009 Bare Costs Labor	Equipment	Total	Total Incl O&P
1450	Maximum				Costs	6%	7%			
1700	Protection of existing work, add, minimum					2%	2%			
1750	Maximum					5%	7%			
2000	Shift work requirements, add, minimum						5%			
2050	Maximum						30%			
2300	Temporary shoring and bracing, add, minimum					2%	5%			
2350	Maximum					5%	12%			
2400	Work inside prisons and high security areas, add, minimum						30%			
2450	Maximum						50%			

01 21 55 – Job Conditions Allowance

01 21 55.50 Job Conditions

	01 21 55.50 Job Conditions	Crew	Daily Output	Labor-Hours	Unit	Material	Labor	Equipment	Total	Total Incl O&P
0010	**JOB CONDITIONS** Modifications to applicable									
0020	cost summaries									
0100	Economic conditions, favorable, deduct				Project				2%	2%
0200	Unfavorable, add								5%	5%
0300	Hoisting conditions, favorable, deduct								2%	2%
0400	Unfavorable, add								5%	5%
0500	General Contractor management, experienced, deduct								2%	2%
0600	Inexperienced, add								10%	10%
0700	Labor availability, surplus, deduct								1%	1%
0800	Shortage, add								10%	10%
0900	Material storage area, available, deduct								1%	1%
1000	Not available, add								2%	2%
1100	Subcontractor availability, surplus, deduct								5%	5%
1200	Shortage, add								12%	12%
1300	Work space, available, deduct								2%	2%
1400	Not available, add								5%	5%

01 21 63 – Taxes

01 21 63.10 Taxes

	01 21 63.10 Taxes	Crew	Daily Output	Labor-Hours	Unit	Material	Labor	Equipment	Total	Total Incl O&P
0010	**TAXES** R012909-80									
0020	Sales tax, State, average				%	4.91%				
0050	Maximum R012909-85					7.25%				
0200	Social Security, on first $102,000 of wages						7.65%			
0300	Unemployment, combined Federal and State, minimum						.80%			
0350	Average						6.20%			
0400	Maximum						11.76%			

01 31 Project Management and Coordination

01 31 13 – Project Coordination

01 31 13.20 Field Personnel

	01 31 13.20 Field Personnel	Crew	Daily Output	Labor-Hours	Unit	Material	Labor	Equipment	Total	Total Incl O&P
0010	**FIELD PERSONNEL**									
0020	Clerk, average				Week		380		380	590
0180	Project manager, minimum						1,650		1,650	2,550
0200	Average						1,925		1,925	2,975
0220	Maximum						2,175		2,175	3,375
0240	Superintendent, minimum						1,600		1,600	2,475
0260	Average						1,775		1,775	2,750
0280	Maximum						2,025		2,025	3,125

01 31 13.30 Insurance

	01 31 13.30 Insurance	Crew	Daily Output	Labor-Hours	Unit	Material	Labor	Equipment	Total	Total Incl O&P
0010	**INSURANCE** R013113-40									

01 31 Project Management and Coordination

01 31 13 – Project Coordination

01 31 13.30 Insurance

		Crew	Daily Output	Labor-Hours	Unit	Material	2009 Bare Costs Labor	Equipment	Total	Total Incl O&P
0020	Builders risk, standard, minimum				Job					.24%
0050	Maximum	R013113-60								.64%
0200	All-risk type, minimum									.25%
0250	Maximum				▼					.62%
0400	Contractor's equipment floater, minimum				Value					.50%
0450	Maximum				"					1.50%
0600	Public liability, average				Job					2.02%
0810	Workers' compensation & employer's liability									
2000	Range of 35 trades in 50 states, excl. wrecking, min.				Payroll		2.09%			
2100	Average						15.50%			
2200	Maximum				▼		168.17%			

01 31 13.40 Main Office Expense

		Crew	Daily Output	Labor-Hours	Unit	Material	Labor	Equipment	Total	Total Incl O&P
0010	**MAIN OFFICE EXPENSE** Average for General Contractors									
0020	As a percentage of their annual volume									
0030	Annual volume to $300,000, minimum				% Vol.					20%
0040	Maximum									30%
0060	To $500,000, minimum									17%
0070	Maximum									22%
0080	To $1,000,000, minimum									16%
0090	Maximum									19%
0110	To $3,000,000, minimum									14%
0120	Maximum									16%
0130	To $5,000,000, minimum									8%
0140	Maximum				▼					10%

01 31 13.50 General Contractor's Mark-Up

		Crew	Daily Output	Labor-Hours	Unit	Material	Labor	Equipment	Total	Total Incl O&P
0010	**GENERAL CONTRACTOR'S MARK-UP** on Change Orders									
0200	Extra work, by subcontractors, add				%					10%
0250	By General Contractor, add									15%
0400	Omitted work, by subcontractors, deduct all but									5%
0450	By General Contractor, deduct all but									7.50%
0600	Overtime work, by subcontractors, add									15%
0650	By General Contractor, add				▼					10%

01 31 13.60 Installing Contractor's Main Office Overhead

		Crew	Daily Output	Labor-Hours	Unit	Material	Labor	Equipment	Total	Total Incl O&P
0010	**INSTALLING CONTRACTOR'S MAIN OFFICE OVERHEAD** R013113-50									
0020	As percent of direct costs, minimum				%					5%
0050	Average									16%
0100	Maximum				▼					30%

01 31 13.80 Overhead and Profit

		Crew	Daily Output	Labor-Hours	Unit	Material	Labor	Equipment	Total	Total Incl O&P
0010	**OVERHEAD & PROFIT** Allowance to add to items in this R013113-50									
0020	book that do not include Subs O&P, average				%					30%
0100	Allowance to add to items in this book that									
0110	do include Subs O&P, minimum				%					5%
0150	Average									10%
0200	Maximum									15%
0290	Typical, by size of project, under $50,000									40%
0310	$50,000 to $100,000									35%
0320	$100,000 to $500,000									25%
0330	$500,000 to $1,000,000				▼					20%

01 41 Regulatory Requirements

01 41 26 – Permits

01 41 26.50 Permits	Crew	Daily Output	Labor-Hours	Unit	Material	2009 Bare Costs Labor	Equipment	Total	Total Incl O&P
0010 **PERMITS**									
0020 Rule of thumb, most cities, minimum				Job					.50%
0100 Maximum				"					2%

01 45 Quality Control

01 45 23 – Testing and Inspecting Services

01 45 23.50 Testing	Crew	Daily Output	Labor-Hours	Unit	Material	2009 Bare Costs Labor	Equipment	Total	Total Incl O&P
0010 **TESTING** and Inspecting Services									
1800 Compressive test, cylinder, delivered to lab, ASTM C 39				Ea.				12	13
1900 Picked up by lab, minimum								14	15
1950 Average								18	20
2000 Maximum								27	30
2200 Compressive strength, cores (not incl. drilling), ASTM C 42								36	40
2300 Patching core holes								22	24
4730 Soil testing									
4735 Soil density, nuclear method, ASTM D2922				Ea.				35	38.67
4740 Sand cone method ASTM D1556								27	30.17
4750 Moisture content, ASTM D 2216								9	10
4780 Permeability test, double ring infiltrometer								500	550
4800 Permeability, var. or constant head, undist., ASTM D 2434								227	250
4850 Recompacted								250	275
4900 Proctor compaction, 4" standard mold, ASTM D 698								123	135
4950 6" modified mold								68	75
5100 Shear tests, triaxial, minimum								409	450
5150 Maximum								545	600
5550 Technician for inspection, per day, earthwork								210	231
5650 Bolting								268	295
5750 Roofing								244	256
5790 Welding								257	283
5820 Non-destructive metal testing, dye penetrant				Day				310	341
5840 Magnetic particle								310	341
5860 Radiography								450	495
5880 Ultrasonic								309	340
6000 Welding certification, minimum				Ea.				91	100
6100 Maximum				"				250	275
7000 Underground storage tank									
7500 Volumetric tightness test ,<=12,000 gal				Ea.				435	478
7510 <=30,000 gal				"				613	675
7600 Vadose zone (soil gas) sampling, 10-40 samples, min.				Day				1,364	1,500
7610 Maximum				"				2,273	2,500
7700 Ground water monitoring incl. drilling 3 wells, min.				Total				4,545	5,000
7710 Maximum				"				6,364	7,000
8000 X-ray concrete slabs				Ea.				182	200

01 51 Temporary Utilities

01 51 13 – Temporary Electricity

01 51 13.80 Temporary Utilities	Crew	Daily Output	Labor-Hours	Unit	Material	2009 Bare Costs Labor	2009 Bare Costs Equipment	Total	Total Incl O&P
0010 **TEMPORARY UTILITIES**									
0100 Heat, incl. fuel and operation, per week, 12 hrs. per day	1 Skwk	100	.080	CSF Flr	27	3.27		30.27	35
0200 24 hrs. per day	"	60	.133		52	5.45		57.45	66.50
0350 Lighting, incl. service lamps, wiring & outlets, minimum	1 Elec	34	.235		2.63	11.05		13.68	19.90
0360 Maximum	"	17	.471		5.70	22		27.70	40.50
0400 Power for temp lighting only, per month, min/month 6.6 KWH								.75	.83
0450 Maximum/month 23.6 KWH								2.85	3.14
0600 Power for job duration incl. elevator, etc., minimum								47	51.70
0650 Maximum				▼				110	121

01 52 Construction Facilities

01 52 13 – Field Offices and Sheds

01 52 13.20 Office and Storage Space

	Crew	Daily Output	Labor-Hours	Unit	Material	2009 Bare Costs Labor	2009 Bare Costs Equipment	Total	Total Incl O&P
0010 **OFFICE AND STORAGE SPACE**									
0020 Trailer, furnished, no hookups, 20' x 8', buy	2 Skwk	1	16	Ea.	8,200	655		8,855	10,100
0250 Rent per month					163			163	179
0300 32' x 8', buy	2 Skwk	.70	22.857		12,200	935		13,135	15,000
0350 Rent per month					200			200	220
0400 50' x 10', buy	2 Skwk	.60	26.667		23,200	1,100		24,300	27,400
0450 Rent per month					281			281	310
0500 50' x 12', buy	2 Skwk	.50	32		27,900	1,300		29,200	32,800
0550 Rent per month					375			375	410
0700 For air conditioning, rent per month, add				▼	41			41	45
0800 For delivery, add per mile				Mile	4.50			4.50	4.95
1200 Storage boxes, 20' x 8', buy	2 Skwk	1.80	8.889	Ea.	4,675	365		5,040	5,750
1250 Rent per month					72			72	79
1300 40' x 8', buy	2 Skwk	1.40	11.429		6,400	465		6,865	7,800
1350 Rent per month				▼	99			99	109

01 54 Construction Aids

01 54 16 – Temporary Hoists

01 54 16.50 Weekly Forklift Crew

	Crew	Daily Output	Labor-Hours	Unit	Material	2009 Bare Costs Labor	2009 Bare Costs Equipment	Total	Total Incl O&P
0010 **WEEKLY FORKLIFT CREW**									
0100 All-terrain forklift, 45' lift, 35' reach, 9000 lb. capacity	A-3P	.20	40	Month		1,550	2,225	3,775	4,900

01 54 19 – Temporary Cranes

01 54 19.50 Daily Crane Crews

	Crew	Daily Output	Labor-Hours	Unit	Material	2009 Bare Costs Labor	2009 Bare Costs Equipment	Total	Total Incl O&P
0010 **DAILY CRANE CREWS** for small jobs, portal to portal									
0100 12-ton truck-mounted hydraulic crane	A-3H	1	8	Day		340	1,025	1,365	1,650
0200 25-ton	A-3I	1	8			340	1,050	1,390	1,675
0300 40-ton	A-3J	1	8			340	1,050	1,390	1,700
0400 55-ton	A-3K	1	16			635	1,625	2,260	2,800
0500 80-ton	A-3L	1	16			635	1,850	2,485	3,025
0600 100-ton	A-3M	1	16	▼		635	2,525	3,160	3,800

01 54 19.60 Monthly Tower Crane Crew

	Crew	Daily Output	Labor-Hours	Unit	Material	2009 Bare Costs Labor	2009 Bare Costs Equipment	Total	Total Incl O&P
0010 **MONTHLY TOWER CRANE CREW**, excludes concrete footing									
0100 Static tower crane, 130' high, 106' jib, 6200 lb. capaccity	A-3N	.05	176	Month		7,500	22,500	30,000	36,500

01 54 Construction Aids

01 54 23 – Temporary Scaffolding and Platforms

01 54 23.60 Pump Staging

		Crew	Daily Output	Labor-Hours	Unit	Material	2009 Bare Costs Labor	Equipment	Total	Total Incl O&P
0010	**PUMP STAGING**, Aluminum R015423-20									
0200	24' long pole section, buy				Ea.	415			415	460
0300	18' long pole section, buy					325			325	355
0400	12' long pole section, buy					218			218	240
0500	6' long pole section, buy					115			115	126
0600	6' long splice joint section, buy					85.50			85.50	94
0700	Pump jack, buy					139			139	153
0900	Foldable brace, buy					56.50			56.50	62.50
1000	Workbench/back safety rail support, buy					73.50			73.50	81
1100	Scaffolding planks/workbench, 14" wide x 24' long, buy					680			680	750
1200	Plank end safety rail, buy					370			370	405
1250	Safety net, 22' long, buy					335			335	365
1300	System in place, 50' working height, per use based on 50 uses	2 Carp	84.80	.189	C.S.F.	6.40	7.55		13.95	19.50
1400	100 uses		84.80	.189		3.21	7.55		10.76	16
1500	150 uses		84.80	.189		2.15	7.55		9.70	14.80

01 54 23.70 Scaffolding

		Crew	Daily Output	Labor-Hours	Unit	Material	2009 Bare Costs Labor	Equipment	Total	Total Incl O&P
0010	**SCAFFOLDING** R015423-10									
0015	Steel tube, regular, no plank, labor only to erect & dismantle									
0090	Building exterior, wall face, 1 to 5 stories, 6'-4" x 5' frames	3 Carp	8	3	C.S.F.		120		120	198
0200	6 to 12 stories	4 Carp	8	4			160		160	264
0301	13 to 20 stories	5 Clab	8	5			158		158	261
0460	Building interior, wall face area, up to 16' high	3 Carp	12	2			80		80	132
0560	16' to 40' high		10	2.400			96		96	158
0800	Building interior floor area, up to 30' high		150	.160	C.C.F.		6.40		6.40	10.55
0900	Over 30' high	4 Carp	160	.200	"		8		8	13.20
0906	Complete system for face of walls, no plank, material only rent/mo				C.S.F.	35.50			35.50	39
0908	Interior spaces, no plank, material only rent/mo				C.C.F.	3.40			3.40	3.74
0910	Steel tubular, heavy duty shoring, buy									
0920	Frames 5' high 2' wide				Ea.	93			93	102
0925	5' high 4' wide					105			105	116
0930	6' high 2' wide					106			106	117
0935	6' high 4' wide					124			124	136
0940	Accessories									
0945	Cross braces				Ea.	18			18	19.80
0950	U-head, 8" x 8"					21.50			21.50	24
0955	J-head, 4" x 8"					15.80			15.80	17.40
0960	Base plate, 8" x 8"					17.60			17.60	19.35
0965	Leveling jack					38			38	41.50
1000	Steel tubular, regular, buy									
1100	Frames 3' high 5' wide				Ea.	71			71	78
1150	5' high 5' wide					83			83	91.50
1200	6'-4" high 5' wide					104			104	114
1350	7'-6" high 6' wide					179			179	197
1500	Accessories cross braces					20.50			20.50	22.50
1550	Guardrail post					18.60			18.60	20.50
1600	Guardrail 7' section					9.90			9.90	10.90
1650	Screw jacks & plates					27			27	29.50
1700	Sidearm brackets					43			43	47.50
1750	8" casters					35.50			35.50	39
1800	Plank 2" x 10" x 16'-0"					51			51	56
1900	Stairway section					325			325	360
1910	Stairway starter bar					36.50			36.50	40

01 54 23 – Temporary Scaffolding and Platforms

01 54 23.70 **Scaffolding**	Crew	Daily Output	Labor-Hours	Unit	Material	2009 Bare Costs Labor	Equipment	Total	Total Incl O&P
1920 Stairway inside handrail				Ea.	66			66	72.50
1930 Stairway outside handrail					98			98	108
1940 Walk-thru frame guardrail				↓	47			47	51.50
2000 Steel tubular, regular, rent/mo.									
2100 Frames 3' high 5' wide				Ea.	5			5	5.50
2150 5' high 5' wide					5			5	5.50
2200 6'-4" high 5' wide					5.05			5.05	5.55
2250 7'-6" high 6' wide					7			7	7.70
2500 Accessories, cross braces					1			1	1.10
2550 Guardrail post					1			1	1.10
2600 Guardrail 7' section					1			1	1.10
2650 Screw jacks & plates					2			2	2.20
2700 Sidearm brackets					2			2	2.20
2750 8" casters					8			8	8.80
2800 Outrigger for rolling tower					3			3	3.30
2850 Plank 2" x 10" x 16'-0"					6			6	6.60
2900 Stairway section					40			40	44
2940 Walk-thru frame guardrail				↓	2.50			2.50	2.75
3000 Steel tubular, heavy duty shoring, rent/mo.									
3250 5' high 2' & 4' wide				Ea.	5			5	5.50
3300 6' high 2' & 4' wide					5			5	5.50
3500 Accessories, cross braces					1			1	1.10
3600 U - head, 8" x 8"					1			1	1.10
3650 J - head, 4" x 8"					1			1	1.10
3700 Base plate, 8" x 8"					1			1	1.10
3750 Leveling jack					2			2	2.20
5700 Planks, 2x10x16'-0", labor only to erect & remove to 50' H	3 Carp	72	.333			13.30		13.30	22
5800 Over 50' high	4 Carp	80	.400	↓		16		16	26.50

01 54 23.75 **Scaffolding Specialties**	Crew	Daily Output	Labor-Hours	Unit	Material	2009 Bare Costs Labor	Equipment	Total	Total Incl O&P
0010 **SCAFFOLDING SPECIALTIES**									
1200 Sidewalk bridge, heavy duty steel posts & beams, including									
1210 parapet protection & waterproofing									
1220 8' to 10' wide, 2 posts	3 Carp	15	1.600	L.F.	35	64		99	145
1230 3 posts	"	10	2.400	"	54	96		150	218
1500 Sidewalk bridge using tubular steel									
1510 scaffold frames, including planking	3 Carp	45	.533	L.F.	5.60	21.50		27.10	41
1600 For 2 uses per month, deduct from all above					50%				
1700 For 1 use every 2 months, add to all above					100%				
1900 Catwalks, 20" wide, no guardrails, 7' span, buy				Ea.	145			145	160
2000 10' span, buy					203			203	223
3720 Putlog, standard, 8' span, with hangers, buy					75			75	82.50
3730 Rent per month					10			10	11
3750 12' span, buy					113			113	124
3755 Rent per month					15			15	16.50
3760 Trussed type, 16' span, buy					260			260	286
3770 Rent per month					20			20	22
3790 22' span, buy					310			310	345
3795 Rent per month					30			30	33
4000 7 step					655			655	720
4100 Rolling towers, buy, 5' wide, 7' long, 10' high					1,350			1,350	1,500
4200 For 5' high added sections, to buy, add				↓	207			207	228
4300 Complete incl. wheels, railings, outriggers,									

01 54 Construction Aids

01 54 23 – Temporary Scaffolding and Platforms

01 54 23.75 Scaffolding Specialties	Crew	Daily Output	Labor-Hours	Unit	Material	2009 Bare Costs Labor	Equipment	Total	Total Incl O&P
4350	21' high, to buy				Ea.	2,250			2,250	2,475
4400	Rent/month = 5% of purchase cost				"	7.50			7.50	8.25

01 54 23.80 Staging Aids

		Crew	Daily Output	Labor-Hours	Unit	Material	Labor	Equipment	Total	Total Incl O&P
0010	**STAGING AIDS** and fall protection equipment									
0100	Sidewall staging bracket, tubular, buy				Ea.	43			43	47.50
0110	Cost each per day, based on 250 days use				Day	.17			.17	.19
0200	Guard post, buy				Ea.	22.50			22.50	25
0210	Cost each per day, based on 250 days use				Day	.09			.09	.10
0300	End guard chains, buy per pair				Pair	31			31	34
0310	Cost per set per day, based on 250 days use				Day	.17			.17	.19
1010	Cost each per day, based on 250 days use				"	.04			.04	.04
1100	Wood bracket, buy				Ea.	16.60			16.60	18.25
1110	Cost each per day, based on 250 days use				Day	.07			.07	.07
2010	Cost per pair per day, based on 250 days use				"	.43			.43	.48
2100	Steel siderail jack, buy per pair				Pair	122			122	134
2110	Cost per pair per day, based on 250 days use				Day	.49			.49	.54
3010	Cost each per day, based on 250 days use				"	.20			.20	.22
3100	Aluminum scaffolding plank, 20" wide x 24' long, buy				Ea.	890			890	975
3110	Cost each per day, based on 250 days use				Day	3.55			3.55	3.91
4000	Nylon full body harness, lanyard and rope grab				Ea.	215			215	236
4010	Cost each per day, based on 250 days use				Day	.86			.86	.95
4100	Rope for safety line, 5/8" x 100' nylon, buy				Ea.	70			70	77
4110	Cost each per day, based on 250 days use				Day	.28			.28	.31
4200	Permanent U-Bolt roof anchor, buy				Ea.	35			35	38.50
4300	Temporary (one use) roof ridge anchor, buy				"	28.50			28.50	31
5000	Installation (setup and removal) of staging aids									
5010	Sidewall staging bracket	2 Carp	64	.250	Ea.		10		10	16.50
5020	Guard post with 2 wood rails	"	64	.250			10		10	16.50
5030	End guard chains, set	1 Carp	64	.125			4.99		4.99	8.25
5100	Roof shingling bracket		96	.083			3.33		3.33	5.50
5200	Ladder jack		64	.125			4.99		4.99	8.25
5300	Wood plank, 2x10x16'	2 Carp	80	.200			8		8	13.20
5310	Aluminum scaffold plank, 20" x 24'	"	40	.400			16		16	26.50
5410	Safety rope	1 Carp	40	.200			8		8	13.20
5420	Permanent U-Bolt roof anchor (install only)	2 Carp	40	.400			16		16	26.50
5430	Temporary roof ridge anchor (install only)	1 Carp	64	.125			4.99		4.99	8.25

01 54 26 – Temporary Swing Staging

01 54 26.50 Swing Staging

		Crew	Daily Output	Labor-Hours	Unit	Material	Labor	Equipment	Total	Total Incl O&P
0010	**SWING STAGING**, 500 lb cap., 2' wide to 24' long, hand operated									
0020	steel cable type, with 60' cables, buy				Ea.	4,050			4,050	4,450
0030	Rent per month				"	405			405	445
0600	Lightweight (not for masons) 24' long for 150' height,									
0610	manual type, buy				Ea.	4,150			4,150	4,575
0620	Rent per month					415			415	455
0700	Powered, electric or air, to 150' high, buy					13,300			13,300	14,700
0710	Rent per month					935			935	1,025
0780	To 300' high, buy					13,500			13,500	14,800
0800	Rent per month					945			945	1,025
1000	Bosun's chair or work basket 3' x 3.5', to 300' high, electric, buy					8,625			8,625	9,500
1010	Rent per month					605			605	665
2200	Move swing staging (setup and remove)	E-4	2	16	Move		725	67	792	1,425

15

01 54 Construction Aids

01 54 36 – Equipment Mobilization

01 54 36.50 Mobilization or Demob.	Crew	Daily Output	Labor-Hours	Unit	Material	2009 Bare Costs Labor	Equipment	Total	Total Incl O&P
0010 **MOBILIZATION OR DEMOB.** (One or the other, unless noted) R015433-10									
0015 Up to 25 mi haul dist (50 mi RT for mob/demob crew)									
0020 Dozer, loader, backhoe, excav., grader, paver, roller, 70 to 150 H.P.	B-34N	4	2	Ea.		64	117	181	233
0100 Above 150 HP	B-34K	3	2.667			85	261	346	430
0300 Scraper, towed type (incl. tractor), 6 C.Y. capacity		3	2.667			85	261	346	430
0400 10 C.Y.		2.50	3.200			102	315	417	515
0600 Self-propelled scraper, 15 C.Y.		2.50	3.200			102	315	417	515
0700 24 C.Y.		2	4			128	390	518	640
0900 Shovel or dragline, 3/4 C.Y.		3.60	2.222			71	218	289	355
1000 1-1/2 C.Y.		3	2.667			85	261	346	430
1100 Small equipment, placed in rear of, or towed by pickup truck	A-3A	8	1			31	16.05	47.05	68.50
1150 Equip up to 70 HP, on flatbed trailer behind pickup truck	A-3D	4	2			62	57.50	119.50	166
2000 Mob & demob truck-mounted crane up to 75 ton, driver only	1 Eqhv	3.60	2.222			94.50		94.50	149
2100 Crane, truck-mounted, over 75 ton	A-3E	2.50	6.400			238	51.50	289.50	435
2200 Crawler-mounted, up to 75 ton	A-3F	2	8			298	405	703	920
2300 Over 75 ton	A-3G	1.50	10.667			395	610	1,005	1,300
2500 For each additional 5 miles haul distance, add						10%	10%		
3000 For large pieces of equipment, allow for assembly/knockdown									
3100 For mob/demob of micro-tunneling equip, see Div. 33 05 23.19									
3200 For mob/demob of pile driving equip, see Div. 31 62 19.10									

01 54 39 – Construction Equipment

01 54 39.70 Small Tools

	Crew	Daily Output	Labor-Hours	Unit	Material	2009 Bare Costs Labor	Equipment	Total	Total Incl O&P
0010 **SMALL TOOLS** R013113-50									
0020 As % of contractor's bare labor cost for project, minimum				Total		.50%			
0100 Maximum				"		2%			

01 55 Vehicular Access and Parking

01 55 23 – Temporary Roads

01 55 23.50 Roads and Sidewalks

	Crew	Daily Output	Labor-Hours	Unit	Material	2009 Bare Costs Labor	Equipment	Total	Total Incl O&P
0010 **ROADS AND SIDEWALKS** Temporary									
0050 Roads, gravel fill, no surfacing, 4" gravel depth	B-14	715	.067	S.Y.	4.97	2.23	.41	7.61	9.55
0100 8" gravel depth	"	615	.078	"	9.95	2.59	.48	13.02	15.70

01 56 Temporary Barriers and Enclosures

01 56 13 – Temporary Air Barriers

01 56 13.60 Tarpaulins

	Crew	Daily Output	Labor-Hours	Unit	Material	2009 Bare Costs Labor	Equipment	Total	Total Incl O&P
0010 **TARPAULINS**									
0020 Cotton duck, 10 oz. to 13.13 oz. per S.Y., minimum				S.F.	.54			.54	.59
0050 Maximum					.53			.53	.58
0100 Polyvinyl coated nylon, 14 oz. to 18 oz., minimum					.48			.48	.53
0150 Maximum					.68			.68	.75
0200 Reinforced polyethylene 3 mils thick, white					.15			.15	.17
0300 4 mils thick, white, clear or black					.20			.20	.22
0400 5.5 mils thick, clear					.21			.21	.23
0500 White, fire retardant					.35			.35	.39
0600 7.5 mils, oil resistant, fire retardant					.40			.40	.44
0700 8.5 mils, black					.53			.53	.58
0720 Steel reinforced polyethylene, 4 mils thick					.53			.53	.58

01 56 Temporary Barriers and Enclosures

01 56 13 – Temporary Air Barriers

01 56 13.60 Tarpaulins

		Crew	Daily Output	Labor-Hours	Unit	Material	2009 Bare Costs Labor	Equipment	Total	Total Incl O&P
0730	Polyester reinforced w/ integral fastening system 11 mils thick				S.F.	1.07			1.07	1.18
0740	Mylar polyester, non-reinforced, 7 mils thick				↓	1.17			1.17	1.29

01 56 13.90 Winter Protection

		Crew	Daily Output	Labor-Hours	Unit	Material	2009 Bare Costs Labor	Equipment	Total	Total Incl O&P
0010	**WINTER PROTECTION**									
0100	Framing to close openings	2 Clab	750	.021	S.F.	.39	.67		1.06	1.54
0200	Tarpaulins hung over scaffolding, 8 uses, not incl. scaffolding		1500	.011		.25	.34		.59	.84
0250	Tarpaulin polyester reinf. w/ integral fastening system 11 mils thick		1600	.010		.80	.32		1.12	1.40
0300	Prefab fiberglass panels, steel frame, 8 uses	↓	1200	.013	↓	.85	.42		1.27	1.64

01 56 23 – Temporary Barricades

01 56 23.10 Barricades

		Crew	Daily Output	Labor-Hours	Unit	Material	2009 Bare Costs Labor	Equipment	Total	Total Incl O&P
0010	**BARRICADES**									
0020	5' high, 3 rail @ 2" x 8", fixed	2 Carp	20	.800	L.F.	4.99	32		36.99	58.50
0150	Movable	"	30	.533	"	4.12	21.50		25.62	39.50
0300	Stock units, 6' high, 8' wide, plain, buy				Ea.	435			435	480
0350	With reflective tape, buy				"	525			525	580
0400	Break-a-way 3" PVC pipe barricade									
0410	with 3 ea. 1' x 4' reflectorized panels, buy				Ea.	305			305	335
0500	Plywood with steel legs, 32" wide					72			72	79
0600	Telescoping Christmas tree, 9' high, 5 flags, buy					122			122	134
0800	Traffic cones, PVC, 18" high					8.40			8.40	9.25
0850	28" high					14.20			14.20	15.65
0900	Barrels, 55 gal., with flasher	1 Clab	96	.083	↓	75.50	2.63		78.13	87.50
1000	Guardrail, wooden, 3' high, 1" x 6", on 2" x 4" posts	2 Carp	200	.080	L.F.	1.02	3.20		4.22	6.40
1100	2" x 6", on 4" x 4" posts	"	165	.097		2.16	3.87		6.03	8.75
1200	Portable metal with base pads, buy					19.75			19.75	21.50
1250	Typical installation, assume 10 reuses	2 Carp	600	.027	↓	2.50	1.07		3.57	4.51
1300	Barricade tape, polyethylene, 7 mil, 3" wide x 500' long roll				Ea.	25			25	27.50

01 56 26 – Temporary Fencing

01 56 26.50 Temporary Fencing

		Crew	Daily Output	Labor-Hours	Unit	Material	2009 Bare Costs Labor	Equipment	Total	Total Incl O&P
0010	**TEMPORARY FENCING**									
0020	Chain link, 11 ga, 5' high	2 Clab	400	.040	L.F.	7.25	1.26		8.51	10.10
0100	6' high		300	.053		7.75	1.69		9.44	11.35
0200	Rented chain link, 6' high, to 1000' (up to 12 mo.)		400	.040		2.69	1.26		3.95	5.05
0250	Over 1000' (up to 12 mo.)	↓	300	.053		2.59	1.69		4.28	5.65
0350	Plywood, painted, 2" x 4" frame, 4' high	A-4	135	.178		5.05	6.80		11.85	16.70
0400	4" x 4" frame, 8' high	"	110	.218		9.85	8.35		18.20	24.50
0500	Wire mesh on 4" x 4" posts, 4' high	2 Carp	100	.160		9.85	6.40		16.25	21.50
0550	8' high	"	80	.200	↓	15.05	8		23.05	30

01 56 29 – Temporary Protective Walkways

01 56 29.50 Protection

		Crew	Daily Output	Labor-Hours	Unit	Material	2009 Bare Costs Labor	Equipment	Total	Total Incl O&P
0010	**PROTECTION**									
0020	Stair tread, 2" x 12" planks, 1 use	1 Carp	75	.107	Tread	4.16	4.26		8.42	11.60
0100	Exterior plywood, 1/2" thick, 1 use		65	.123		1.39	4.92		6.31	9.60
0200	3/4" thick, 1 use		60	.133	↓	2.49	5.35		7.84	11.55
2200	Sidewalks, 2" x 12" planks, 2 uses		350	.023	S.F.	.69	.91		1.60	2.27
2300	Exterior plywood, 2 uses, 1/2" thick		750	.011		.23	.43		.66	.95
2400	5/8" thick		650	.012		.34	.49		.83	1.18
2500	3/4" thick	↓	600	.013		.42	.53		.95	1.34

01 56 Temporary Barriers and Enclosures

01 56 32 – Temporary Security

01 56 32.50 Watchman	Crew	Daily Output	Labor-Hours	Unit	Material	2009 Bare Costs Labor	Equipment	Total	Total Incl O&P
0010 **WATCHMAN**									
0020 Service, monthly basis, uniformed person, minimum				Hr.				25	27.50
0100 Maximum								45.45	50
0200 Person and command dog, minimum								31	34
0300 Maximum				↓				54.55	60
0500 Sentry dog, leased, with job patrol (yard.dog), 1 dog				Week				290	319
0600 2 dogs				"				390	429
0800 Purchase, trained sentry dog, minimum				Ea.				1,364	1,500
0900 Maximum				"				2,727	3,000

01 58 Project Identification

01 58 13 – Temporary Project Signage

01 58 13.50 Signs

	Crew	Daily Output	Labor-Hours	Unit	Material	2009 Bare Costs Labor	Equipment	Total	Total Incl O&P
0010 **SIGNS**									
0020 High intensity reflectorized, no posts, buy				S.F.	21			21	23

01 71 Examination and Preparation

01 71 23 – Field Engineering

01 71 23.13 Construction Layout

	Crew	Daily Output	Labor-Hours	Unit	Material	2009 Bare Costs Labor	Equipment	Total	Total Incl O&P
0010 **CONSTRUCTION LAYOUT**									
1100 Crew for layout of building, trenching or pipe laying, 2 person crew	A-6	1	16	Day		635	70	705	1,100
1200 3 person crew	A-7	1	24			1,025	70	1,095	1,750
1400 Crew for roadway layout, 4 person crew	A-8	1	32	↓		1,350	70	1,420	2,250

01 74 Cleaning and Waste Management

01 74 13 – Progress Cleaning

01 74 13.20 Cleaning Up

	Crew	Daily Output	Labor-Hours	Unit	Material	2009 Bare Costs Labor	Equipment	Total	Total Incl O&P
0010 **CLEANING UP**									
0020 After job completion, allow, minimum				Job					.30%
0040 Maximum				"					1%
0052 Cleanup of floor area, continuous, per day, during const.	A-5	16	1.125	M.S.F.	1.70	35.50	3.04	40.24	63.50
0100 Final by GC at end of job	"	11.50	1.565	"	2.71	49.50	4.23	56.44	89

01 91 Commissioning

01 91 13 – General Commissioning Requirements

01 91 13.50 Commissioning

	Crew	Daily Output	Labor-Hours	Unit	Material	2009 Bare Costs Labor	Equipment	Total	Total Incl O&P
0010 **COMMISSIONING** Including documentation of design intent									
0100 performance verification, O&M, training, minimum				Project					.50%
0150 Maximum				"					.75%

01 93 09 – Facility Equipment

01 93 09.50 Moving Equipment	Crew	Daily Output	Labor-Hours	Unit	Material	2009 Bare Costs Labor	Equipment	Total	Total Incl O&P
0010 **MOVING EQUIPMENT**, Remove and reset, 100' distance,									
0020 No obstructions, no assembly or leveling unless noted									
0100 Annealing furnace, 24' overall	B-67	4	4	Ea.		161	70.50	231.50	330
0200 Annealing oven, small		14	1.143			46	20	66	94.50
0240 Very large		1	16			645	281	926	1,325
0400 Band saw, small		12	1.333			54	23.50	77.50	111
0440 Large		8	2			80.50	35	115.50	166
0500 Blue print copy machine		7	2.286			92	40	132	189
0600 Bonding mill, 6"		7	2.286			92	40	132	189
0620 12"		6	2.667			108	47	155	221
0640 18"		4	4			161	70.50	231.50	330
0660 24"		2	8			325	141	466	660
0700 Boring machine (jig)	B-68	7	3.429			140	40	180	263
0800 Bridgeport mill, standard	B-67	14	1.143			46	20	66	94.50
1000 Calibrator, 6 unit	"	14	1.143			46	20	66	94.50
1100 Comparitor, bench top	2 Clab	14	1.143			36		36	59.50
1140 Floor mounted	B-67	7	2.286			92	40	132	189
1200 Computer, desk top	2 Clab	25	.640			20		20	33.50
1300 Copy machine	"	25	.640			20		20	33.50
1500 Deflasher	B-67	14	1.143			46	20	66	94.50
1600 Degreaser, small		14	1.143			46	20	66	94.50
1640 Large 24' overall		1	16			645	281	926	1,325
1700 Desk with chair	2 Clab	25	.640			20		20	33.50
1800 Dial press	B-67	7	2.286			92	40	132	189
1900 Drafting table	2 Clab	14	1.143			36		36	59.50
2000 Drill press, bench top	"	14	1.143			36		36	59.50
2040 Floor mounted	B-67	14	1.143			46	20	66	94.50
2080 Industrial radial	"	7	2.286			92	40	132	189
2100 Dust collector, portable	2 Clab	25	.640			20		20	33.50
2140 Stationary, small	B-67	7	2.286			92	40	132	189
2180 Stationary, large	"	2	8			325	141	466	660
2300 Electric discharge machine	B-68	7	3.429			140	40	180	263
2400 Environmental chamber walls, including assembly	4 Clab	18	1.778	L.F.		56		56	92.50
2600 File cabinet	2 Clab	25	.640	Ea.		20		20	33.50
2800 Grinder/sander, pedestal mount	B-67	14	1.143			46	20	66	94.50
3000 Hack saw, power	2 Clab	24	.667			21		21	35
3100 Hydraulic press	B-67	14	1.143			46	20	66	94.50
3500 Laminar flow tables	"	14	1.143			46	20	66	94.50
3600 Lathe, bench	2 Clab	14	1.143			36		36	59.50
3640 6"	B-67	14	1.143			46	20	66	94.50
3680 10"		13	1.231			49.50	21.50	71	102
3720 12"		12	1.333			54	23.50	77.50	111
4000 Milling machine		8	2			80.50	35	115.50	166
4100 Molding press, 25 ton		5	3.200			129	56.50	185.50	265
4140 60 ton		4	4			161	70.50	231.50	330
4180 100 ton		2	8			325	141	466	660
4220 150 ton		1.50	10.667			430	188	618	880
4260 200 ton		1	16			645	281	926	1,325
4300 300 ton		.75	21.333			860	375	1,235	1,750
4700 Oil pot stand		14	1.143			46	20	66	94.50
5000 Press, 10 ton		14	1.143			46	20	66	94.50
5040 15 ton		12	1.333			54	23.50	77.50	111
5080 20 ton		10	1.600			64.50	28	92.50	132

01 93 Facility Maintenance

01 93 09 – Facility Equipment

01 93 09.50 Moving Equipment	Crew	Daily Output	Labor-Hours	Unit	Material	2009 Bare Costs Labor	2009 Bare Costs Equipment	Total	Total Incl O&P
5120	30 ton	B-67	8	2	Ea.		80.50	35	115.50	166
5160	45 ton		6	2.667			108	47	155	221
5200	60 ton		4	4			161	70.50	231.50	330
5240	75 ton		2.50	6.400			258	113	371	530
5280	100 ton		2	8			325	141	466	660
5500	Raised floor, including assembly	2 Carp	250	.064	S.F.		2.56		2.56	4.22
5600	Rolling mill, 6"	B-67	7	2.286	Ea.		92	40	132	189
5640	9"		6	2.667			108	47	155	221
5680	12"		4	4			161	70.50	231.50	330
5720	13"		3.50	4.571			184	80.50	264.50	380
5760	18"		2	8			325	141	466	660
5800	25"		1	16			645	281	926	1,325
6000	Sander, floor stand		14	1.143			46	20	66	94.50
6100	Screw machine		7	2.286			92	40	132	189
6200	Shaper, 16"		14	1.143			46	20	66	94.50
6300	Shear, power assist		4	4			161	70.50	231.50	330
6400	Slitter, 6"		14	1.143			46	20	66	94.50
6440	8"		13	1.231			49.50	21.50	71	102
6480	10"		12	1.333			54	23.50	77.50	111
6520	12"		11	1.455			58.50	25.50	84	120
6560	16"		10	1.600			64.50	28	92.50	132
6600	20"		8	2			80.50	35	115.50	166
6640	24"		6	2.667			108	47	155	221
6800	Snag and tap machine		7	2.286			92	40	132	189
6900	Solder machine (auto)		7	2.286			92	40	132	189
7000	Storage cabinet metal, small	2 Clab	36	.444			14.05		14.05	23
7040	Large		25	.640			20		20	33.50
7100	Storage rack open, small		14	1.143			36		36	59.50
7140	Large		7	2.286			72		72	119
7200	Surface bench, small	B-67	14	1.143			46	20	66	94.50
7240	Large	"	5	3.200			129	56.50	185.50	265
7300	Surface grinder, large wet	B-68	5	4.800			196	56.50	252.50	365
7500	Time check machine	2 Clab	14	1.143			36		36	59.50
8000	Welder, 30 KVA (bench)		14	1.143			36		36	59.50
8100	Work bench with chair		25	.640			20		20	33.50

Estimating Tips

02 30 00 Subsurface Investigation

In preparing estimates on structures involving earthwork or foundations, all information concerning soil characteristics should be obtained. Look particularly for hazardous waste, evidence of prior dumping of debris, and previous stream beds.

02 41 00 Demolition and Structure Moving

The costs shown for selective demolition do not include rubbish handling or disposal. These items should be estimated separately using RSMeans data or other sources.

- Historic preservation often requires that the contractor remove materials from the existing structure, rehab them, and replace them. The estimator must be aware of any related measures and precautions that must be taken when doing selective demolition and cutting and patching. Requirements may include special handling and storage, as well as security.
- In addition to Subdivision 02 41 00, you can find selective demolition items in each division. Example: Roofing demolition is in Division 7.

02 80 00 Hazardous Material Disposal/ Remediation

This subdivision includes information on hazardous waste handling, asbestos remediation, lead remediation, and mold remediation. See reference R028213-20 and R028319-60 for further guidance in using these unit price lines.

Reference Numbers

Reference numbers are shown in shaded boxes at the beginning of some major classifications. These numbers refer to related items in the Reference Section. The reference information may be an estimating procedure, an alternate pricing method, or technical information.

Note: Not all subdivisions listed here necessarily appear in this publication.

Division 2 - Existing Conditions

02 21 Surveys

02 21 13 – Site Surveys

02 21 13.09 Topographical Surveys	Crew	Daily Output	Labor-Hours	Unit	Material	2009 Bare Costs Labor	Equipment	Total	Total Incl O&P
0010 **TOPOGRAPHICAL SURVEYS**									
0020 Topographical surveying, conventional, minimum	A-7	3.30	7.273	Acre	18	315	21	354	550
0100 Maximum	A-8	.60	53.333	"	55	2,250	117	2,422	3,800

02 21 13.13 Boundary and Survey Markers

02 21 13.13 Boundary and Survey Markers	Crew	Daily Output	Labor-Hours	Unit	Material	Labor	Equipment	Total	Total Incl O&P
0010 **BOUNDARY AND SURVEY MARKERS**									
0300 Lot location and lines, large quantities, minimum	A-7	2	12	Acre	32	520	35	587	910
0320 Average	"	1.25	19.200	↓	51	830	56	937	1,450
0400 Small quantities, maximum	A-8	1	32	↓	68	1,350	70	1,488	2,325

02 32 Geotechnical Investigations

02 32 13 – Subsurface Drilling and Sampling

02 32 13.10 Boring and Exploratory Drilling

02 32 13.10 Boring and Exploratory Drilling	Crew	Daily Output	Labor-Hours	Unit	Material	Labor	Equipment	Total	Total Incl O&P
0010 **BORING AND EXPLORATORY DRILLING**									
0020 Borings, initial field stake out & determination of elevations	A-6	1	16	Day		635	70	705	1,100
0100 Drawings showing boring details				Total		255		255	372
0200 Report and recommendations from P.E.				↓		580		580	847
0300 Mobilization and demobilization, minimum	B-55	4	6	↓		188	237	425	570
0350 For over 100 miles, per added mile		450	.053	Mile		1.67	2.11	3.78	5.10
0600 Auger holes in earth, no samples, 2-1/2" diameter		78.60	.305	L.F.		9.60	12.10	21.70	29
0650 4" diameter		67.50	.356			11.15	14.05	25.20	34
0800 Cased borings in earth, with samples, 2-1/2" diameter		55.50	.432		20	13.55	17.10	50.65	63.50
0850 4" diameter	↓	32.60	.736		31.50	23	29	83.50	105
1000 Drilling in rock, "BX" core, no sampling	B-56	34.90	.458			16.20	37.50	53.70	67
1050 With casing & sampling		31.70	.505		20	17.85	41	78.85	95.50
1200 "NX" core, no sampling		25.92	.617			22	50	72	90
1250 With casing and sampling	↓	25	.640	↓	24.50	22.50	52	99	121
1400 Borings, earth, drill rig and crew with truck mounted auger	B-55	1	24	Day		755	950	1,705	2,300
1450 Rock using crawler type drill	B-56	1	16	"		565	1,300	1,865	2,325
1500 For inner city borings add, minimum									10%
1510 Maximum									20%

02 41 Demolition

02 41 13 – Selective Site Demolition

02 41 13.15 Hydrodemolition

02 41 13.15 Hydrodemolition	Crew	Daily Output	Labor-Hours	Unit	Material	Labor	Equipment	Total	Total Incl O&P
0010 **HYDRODEMOLITION**									
0015 Hydrodemolition, concrete pavement, 4000 PSI, 2" depth	B-5	500	.112	S.F.		3.88	2.36	6.24	8.90
0120 4" depth		450	.124			4.31	2.63	6.94	9.90
0130 6" depth		400	.140			4.85	2.95	7.80	11.15
0410 6000 PSI, 2" depth		410	.137			4.74	2.88	7.62	10.85
0420 4" depth		350	.160			5.55	3.38	8.93	12.70
0430 6" depth		300	.187			6.45	3.94	10.39	14.85
0510 8000 PSI, 2" depth		330	.170			5.90	3.58	9.48	13.50
0520 4" depth		280	.200			6.95	4.22	11.17	15.90
0530 6" depth	↓	240	.233	↓		8.10	4.92	13.02	18.55

02 41 13.17 Demolish, Remove Pavement and Curb

02 41 13.17 Demolish, Remove Pavement and Curb	Crew	Daily Output	Labor-Hours	Unit	Material	Labor	Equipment	Total	Total Incl O&P
0010 **DEMOLISH, REMOVE PAVEMENT AND CURB** R024119-10									
5010 Pavement removal, bituminous roads, 3" thick	B-38	690	.058	S.Y.		2.05	1.51	3.56	4.99
5050 4" to 6" thick		420	.095			3.38	2.49	5.87	8.20
5100 Bituminous driveways	↓	640	.063	↓		2.22	1.63	3.85	5.40

02 41 Demolition

02 41 13 – Selective Site Demolition

02 41 13.17 Demolish, Remove Pavement and Curb

		Crew	Daily Output	Labor-Hours	Unit	Material	Labor	Equipment	Total	Total Incl O&P
							2009 Bare Costs			
5200	Concrete to 6" thick, hydraulic hammer, mesh reinforced	B-38	255	.157	S.Y.		5.55	4.10	9.65	13.50
5300	Rod reinforced		200	.200	↓		7.10	5.20	12.30	17.20
5400	Concrete, 7" to 24" thick, plain		33	1.212	C.Y.		43	31.50	74.50	105
5500	Reinforced		24	1.667	"		59	43.50	102.50	144
5590	Minimum labor/equipment charge	↓	6	6.667	Job		236	174	410	570
5600	With hand held air equipment, bituminous, to 6" thick	B-39	1900	.025	S.F.		.84	.10	.94	1.48
5700	Concrete to 6" thick, no reinforcing		1600	.030			1	.12	1.12	1.76
5800	Mesh reinforced		1400	.034			1.14	.14	1.28	2.01
5900	Rod reinforced	↓	765	.063	↓		2.08	.25	2.33	3.68
5990	Minimum labor/equipment charge	B-38	6	6.667	Job		236	174	410	570
6000	Curbs, concrete, plain	B-6	360	.067	L.F.		2.27	.82	3.09	4.58
6100	Reinforced		275	.087			2.97	1.07	4.04	6
6200	Granite		360	.067			2.27	.82	3.09	4.58
6300	Bituminous		528	.045	↓		1.55	.56	2.11	3.12
6390	Minimum labor/equipment charge	↓	6	4	Job		136	49	185	275

02 41 13.33 Minor Site Demolition

		Crew	Daily Output	Labor-Hours	Unit	Material	Labor	Equipment	Total	Total Incl O&P
0010	**MINOR SITE DEMOLITION** R024119-10									
0015	No hauling, abandon catch basin or manhole	B-6	7	3.429	Ea.		117	42	159	235
0020	Remove existing catch basin or manhole, masonry		4	6			204	73.50	277.50	410
0030	Catch basin or manhole frames and covers, stored		13	1.846			63	22.50	85.50	127
0040	Remove and reset	↓	7	3.429			117	42	159	235
0100	Roadside delineators, remove only	B-80	175	.183			6.20	3.74	9.94	14.15
0110	Remove and reset	"	100	.320	↓		10.80	6.55	17.35	25
0400	Minimum labor/equipment charge	B-6	4	6	Job		204	73.50	277.50	410
0800	Guiderail, corrugated steel, remove only	B-80A	100	.240	L.F.		7.60	2.43	10.03	15.15
0850	Remove and reset	"	40	.600	"		18.95	6.05	25	38
0860	Guide posts, remove only	B-80B	120	.267	Ea.		8.90	1.87	10.77	16.60
0870	Remove and reset	B-55	50	.480	"		15.05	19	34.05	46
0890	Minimum labor/equipment charge	2 Clab	4	4	Job		126		126	209
0900	Hydrants, fire, remove only	B-21A	5	8	Ea.		315	120	435	625
0950	Remove and reset	"	2	20	"		780	300	1,080	1,575
0990	Minimum labor/equipment charge	2 Plum	2	8	Job		390		390	605
1000	Masonry walls, block, solid	B-5	1800	.031	C.F.		1.08	.66	1.74	2.47
1200	Brick, solid		900	.062			2.16	1.31	3.47	4.94
1400	Stone, with mortar		900	.062			2.16	1.31	3.47	4.94
1500	Dry set	↓	1500	.037	↓		1.29	.79	2.08	2.97
1600	Median barrier, precast concrete, remove and store	B-3	430	.112	L.F.		3.76	4.78	8.54	11.40
1610	Remove and reset	"	390	.123	"		4.15	5.25	9.40	12.55
1650	Minimum labor/equipment charge	A-1	4	2	Job		63	15.80	78.80	121
2900	Pipe removal, sewer/water, no excavation, 12" diameter	B-6	175	.137	L.F.		4.67	1.68	6.35	9.45
2930	15"-18" diameter		150	.160			5.45	1.96	7.41	11
2960	21"-24" diameter		120	.200			6.80	2.45	9.25	13.75
3000	27"-36" diameter		90	.267			9.10	3.26	12.36	18.35
3200	Steel, welded connections, 4" diameter		160	.150			5.10	1.84	6.94	10.30
3300	10" diameter		80	.300	↓		10.20	3.67	13.87	20.50
3390	Minimum labor/equipment charge	↓	3	8	Job		273	98	371	550
3500	Railroad track removal, ties and track	B-13	330	.170	L.F.		5.80	2.39	8.19	12.10
3600	Ballast	B-14	500	.096	C.Y.		3.19	.59	3.78	5.85
3700	Remove and re-install, ties & track using new bolts & spikes		50	.960	L.F.		32	5.90	37.90	58.50
3800	Turnouts using new bolts and spikes		1	48	Ea.		1,600	294	1,894	2,925
3890	Minimum labor/equipment charge	↓	5	9.600	Job		320	59	379	585
4000	Sidewalk removal, bituminous, 2-1/2" thick	B-6	325	.074	S.Y.		2.52	.90	3.42	5.05

23

02 41 Demolition

02 41 13 – Selective Site Demolition

02 41 13.33 Minor Site Demolition	Crew	Daily Output	Labor-Hours	Unit	Material	2009 Bare Costs Labor	Equipment	Total	Total Incl O&P
4050 Brick, set in mortar	B-6	185	.130	S.Y.		4.42	1.59	6.01	8.90
4100 Concrete, plain, 4"		160	.150			5.10	1.84	6.94	10.30
4200 Mesh reinforced		150	.160			5.45	1.96	7.41	11
4290 Minimum labor/equipment charge	B-39	12	4	Job		133	15.95	148.95	235

02 41 13.60 Selective Demolition Fencing

0010 SELECTIVE DEMOLITION FENCING R024119-10									
1600 Fencing, barbed wire, 3 strand	2 Clab	430	.037	L.F.		1.18		1.18	1.94
1650 5 strand	"	280	.057			1.81		1.81	2.98
1700 Chain link, posts & fabric, 8' to 10' high, remove only	B-6	445	.054			1.84	.66	2.50	3.71

02 41 16 – Structure Demolition

02 41 16.13 Building Demolition

0010 BUILDING DEMOLITION Large urban projects, incl. 20 mi. haul R024119-10									
0011 No foundation or dump fees, C.F. is vol. of building standing									
0020 Steel	B-8	21500	.003	C.F.		.10	.13	.23	.32
0050 Concrete		15300	.004			.15	.19	.34	.44
0080 Masonry		20100	.003			.11	.14	.25	.34
0100 Mixture of types, average		20100	.003			.11	.14	.25	.34
0500 Small bldgs, or single bldgs, no salvage included, steel	B-3	14800	.003			.11	.14	.25	.33
0600 Concrete		11300	.004			.14	.18	.32	.43
0650 Masonry		14800	.003			.11	.14	.25	.33
0700 Wood		14800	.003			.11	.14	.25	.33
1000 Single family, one story house, wood, minimum				Ea.				3,870	4,260
1020 Maximum								6,650	7,315
1200 Two family, two story house, wood, minimum								5,080	5,590
1220 Maximum								9,680	10,650
1300 Three family, three story house, wood, minimum								6,660	7,325
1320 Maximum								11,735	12,910
5000 For buildings with no interior walls, deduct								50%	

02 41 16.17 Building Demolition Footings and Foundations

0010 BUILDING DEMOLITION FOOTINGS AND FOUNDATIONS R024119-10									
0200 Floors, concrete slab on grade,									
0240 4" thick, plain concrete	B-9	500	.080	S.F.		2.56	.38	2.94	4.64
0280 Reinforced, wire mesh		470	.085			2.72	.41	3.13	4.94
0300 Rods		400	.100			3.20	.48	3.68	5.85
0400 6" thick, plain concrete		375	.107			3.41	.51	3.92	6.20
0420 Reinforced, wire mesh		340	.118			3.76	.56	4.32	6.80
0440 Rods		300	.133			4.27	.64	4.91	7.75
1000 Footings, concrete, 1' thick, 2' wide	B-5	300	.187	L.F.		6.45	3.94	10.39	14.85
1080 1'-6" thick, 2' wide		250	.224			7.75	4.73	12.48	17.80
1120 3' wide		200	.280			9.70	5.90	15.60	22.50
1140 2' thick, 3' wide		175	.320			11.10	6.75	17.85	25.50
1200 Average reinforcing, add								10%	10%
1220 Heavy reinforcing, add								20%	20%
2000 Walls, block, 4" thick	1 Clab	180	.044	S.F.		1.40		1.40	2.32
2040 6" thick		170	.047			1.49		1.49	2.45
2080 8" thick		150	.053			1.69		1.69	2.78
2100 12" thick		150	.053			1.69		1.69	2.78
2200 For horizontal reinforcing, add								10%	10%
2220 For vertical reinforcing, add								20%	20%
2400 Concrete, plain concrete, 6" thick	B-9	160	.250			8	1.20	9.20	14.50
2420 8" thick		140	.286			9.15	1.37	10.52	16.60
2440 10" thick		120	.333			10.65	1.60	12.25	19.35

02 41 Demolition

02 41 16 – Structure Demolition

02 41 16.17 Building Demolition Footings and Foundations

		Crew	Daily Output	Labor-Hours	Unit	Material	2009 Bare Costs Labor	Equipment	Total	Total Incl O&P
2500	12" thick	B-9	100	.400	S.F.		12.80	1.92	14.72	23
2600	For average reinforcing, add								10%	10%
2620	For heavy reinforcing, add				↓				20%	20%
9000	Minimum labor/equipment charge	A-1	2	4	Job		126	31.50	157.50	244

02 41 16.19 Minor Building Deconstruction

			Crew	Daily Output	Labor-Hours	Unit	Material	2009 Bare Costs Labor	Equipment	Total	Total Incl O&P
0010	**MINOR BUILDING DECONSTRUCTION**	R024119-10									
0011	For salvage, avg house										
0225	2 story, pre 1970 house, 1400 SF, labor for all salv mat, min	G	6 Clab	25	1.920	SF Flr.		60.50		60.50	100
0230	Maximum	G	"	15	3.200	"		101		101	167
0235	Salvage of carpet, tackless	G	2 Clab	2400	.007	S.F.		.21		.21	.35
0240	Wood floors, incl denailing and packaging	G	3 Clab	270	.089	"		2.81		2.81	4.64
0260	Wood doors and trim, standard	G	1 Clab	16	.500	Ea.		15.80		15.80	26
0280	Base or cove mouldings, incl denailing and packaging	G		500	.016	L.F.		.51		.51	.83
0320	Closet shelving and trim, incl denailing and packaging	G	↓	18	.444	Set		14.05		14.05	23
0340	Kitchen cabinets, uppers and lowers, prefab type	G	2 Clab	24	.667	L.F.		21		21	35
0360	Kitchen cabinets, uppers and lowers, built-in	G		12	1.333	"		42		42	69.50
0370	Bath fixt, incl toilet, tub, vanity, and med cabinet	G	↓	3	5.333	Set		169		169	278

02 41 19 – Selective Structure Demolition

02 41 19.13 Selective Building Demolition

0010	**SELECTIVE BUILDING DEMOLITION**									
0020	Costs related to selective demolition of specific building components									
0025	are included under Common Work Results (XX 05 00)									
0030	in the component's appropriate division.									

02 41 19.16 Selective Demolition, Cutout

			Crew	Daily Output	Labor-Hours	Unit	Material	2009 Bare Costs Labor	Equipment	Total	Total Incl O&P
0010	**SELECTIVE DEMOLITION, CUTOUT**	R024119-10									
0020	Concrete, elev. slab, light reinforcement, under 6 C.F.		B-9	65	.615	C.F.		19.70	2.95	22.65	35.50
0050	Light reinforcing, over 6 C.F.			75	.533	"		17.05	2.55	19.60	31
0200	Slab on grade to 6" thick, not reinforced, under 8 S.F.			85	.471	S.F.		15.05	2.25	17.30	27.50
0250	8 – 16 S.F.		↓	175	.229	"		7.30	1.09	8.39	13.25
0255	For over 16 S.F. see Div. 02 41 16.17 0400										
0600	Walls, not reinforced, under 6 C.F.		B-9	60	.667	C.F.		21.50	3.19	24.69	38.50
0650	6 – 12 C.F.		"	80	.500	"		16	2.40	18.40	29
0655	For over 12 C.F. see Div. 02 41 16.17 2500										
1000	Concrete, elevated slab, bar reinforced, under 6 C.F.		B-9	45	.889	C.F.		28.50	4.26	32.76	51.50
1050	Bar reinforced, over 6 C.F.			50	.800	"		25.50	3.83	29.33	46.50
1200	Slab on grade to 6" thick, bar reinforced, under 8 S.F.			75	.533	S.F.		17.05	2.55	19.60	31
1250	8 – 16 S.F.		↓	150	.267	"		8.55	1.28	9.83	15.50
1255	For over 16 S.F. see Div. 02 41 16.17 0440										
1400	Walls, bar reinforced, under 6 C.F.		B-9	50	.800	C.F.		25.50	3.83	29.33	46.50
1450	6 – 12 C.F.		"	70	.571	"		18.30	2.74	21.04	33
1455	For over 12 C.F. see Div. 02 41 16.17 2500 and 2600										
2000	Brick, to 4 S.F. opening, not including toothing										
2040	4" thick		B-9	30	1.333	Ea.		42.50	6.40	48.90	77.50
2060	8" thick			18	2.222			71	10.65	81.65	129
2080	12" thick			10	4			128	19.15	147.15	232
2400	Concrete block, to 4 S.F. opening, 2" thick			35	1.143			36.50	5.45	41.95	66.50
2420	4" thick			30	1.333			42.50	6.40	48.90	77.50
2440	8" thick			27	1.481			47.50	7.10	54.60	86
2460	12" thick			24	1.667			53.50	8	61.50	97
2600	Gypsum block, to 4 S.F. opening, 2" thick			80	.500			16	2.40	18.40	29
2620	4" thick			70	.571			18.30	2.74	21.04	33
2640	8" thick		↓	55	.727	↓		23.50	3.48	26.98	42.50

02 41 Demolition

02 41 19 – Selective Structure Demolition

02 41 19.16 Selective Demolition, Cutout

		Crew	Daily Output	Labor-Hours	Unit	Material	Labor	Equipment	Total	Total Incl O&P
							2009 Bare Costs			
2800	Terra cotta, to 4 S.F. opening, 4" thick	B-9	70	.571	Ea.		18.30	2.74	21.04	33
2840	8" thick		65	.615			19.70	2.95	22.65	35.50
2880	12" thick	↓	50	.800	↓		25.50	3.83	29.33	46.50
4000	For toothing masonry, see Div. 04 01 20.50									
6000	Walls, interior, not including re-framing,									
6010	openings to 5 S.F.									
6100	Drywall to 5/8" thick	1 Clab	24	.333	Ea.		10.55		10.55	17.40
6200	Paneling to 3/4" thick		20	.400			12.65		12.65	21
6300	Plaster, on gypsum lath		20	.400			12.65		12.65	21
6340	On wire lath	↓	14	.571	↓		18.05		18.05	30
7000	Wood frame, not including re-framing, openings to 5 S.F.									
7200	Floors, sheathing and flooring to 2" thick	1 Clab	5	1.600	Ea.		50.50		50.50	83.50
7310	Roofs, sheathing to 1" thick, not including roofing		6	1.333			42		42	69.50
7410	Walls, sheathing to 1" thick, not including siding		7	1.143	↓		36		36	59.50
8500	Minimum labor/equipment charge	↓	4	2	Job		63		63	104

02 41 19.18 Selective Demolition, Disposal Only

		Crew	Daily Output	Labor-Hours	Unit	Material	Labor	Equipment	Total	Total Incl O&P
0010	**SELECTIVE DEMOLITION, DISPOSAL ONLY** R024119-10									
0015	Urban bldg w/salvage value allowed									
0020	Including loading and 5 mile haul to dump									
0200	Steel frame	B-3	430	.112	C.Y.		3.76	4.78	8.54	11.40
0300	Concrete frame		365	.132			4.43	5.65	10.08	13.40
0400	Masonry construction		445	.108			3.63	4.62	8.25	11
0500	Wood frame	↓	247	.194	↓		6.55	8.30	14.85	19.80

02 41 19.19 Selective Demolition, Dump Charges

		Crew	Daily Output	Labor-Hours	Unit	Material	Labor	Equipment	Total	Total Incl O&P
0010	**SELECTIVE DEMOLITION, DUMP CHARGES** R024119-10									
0020	Dump charges, typical urban city, tipping fees only									
0100	Building construction materials				Ton	100			100	110
0200	Trees, brush, lumber					75			75	82.50
0300	Rubbish only					90			90	99
0500	Reclamation station, usual charge				↓	100			100	110

02 41 19.21 Selective Demolition, Gutting

		Crew	Daily Output	Labor-Hours	Unit	Material	Labor	Equipment	Total	Total Incl O&P
0010	**SELECTIVE DEMOLITION, GUTTING** R024119-10									
0020	Building interior, including disposal, dumpster fees not included									
0500	Residential building									
0560	Minimum	B-16	400	.080	SF Flr.		2.58	1.33	3.91	5.70
0580	Maximum	"	360	.089	"		2.86	1.48	4.34	6.35
0900	Commercial building									
1000	Minimum	B-16	350	.091	SF Flr.		2.94	1.52	4.46	6.55
1020	Maximum		250	.128	"		4.12	2.13	6.25	9.15
3000	Minimum labor/equipment charge	↓	4	8	Job		258	133	391	570

02 41 19.23 Selective Demolition, Rubbish Handling

		Crew	Daily Output	Labor-Hours	Unit	Material	Labor	Equipment	Total	Total Incl O&P
0010	**SELECTIVE DEMOLITION, RUBBISH HANDLING** R024119-10									
0020	The following are to be added to the demolition prices									
0400	Chute, circular, prefabricated steel, 18" diameter	B-1	40	.600	L.F.	47	19.35		66.35	84
0440	30" diameter	"	30	.800	"	48.50	26		74.50	96
0725	Dumpster, weekly rental, 1 dump/week, 20 C.Y. capacity (8 Tons) R024119-20				Week	775			775	852.50
0800	30 C.Y. capacity (10 Tons)					1,000			1,000	1,100
0840	40 C.Y. capacity (13 Tons)				↓	1,300			1,300	1,430
0900	Alternate pricing for dumpsters									
0910	Delivery, average for all sizes				Ea.	75			75	82.50
0920	Haul, average for all sizes					220			220	242
0930	Rent per day, average for all sizes				↓	10			10	11

02 41 Demolition

02 41 19 – Selective Structure Demolition

02 41 19.23 Selective Demolition, Rubbish Handling

		Crew	Daily Output	Labor-Hours	Unit	Material	2009 Bare Costs Labor	Equipment	Total	Total Incl O&P
0940	Rent per month, average for all sizes				Ea.	75			75	82.50
0950	Disposal fee per ton, average for all sizes				Ton	75			75	82.50
1000	Dust partition, 6 mil polyethylene, 1" x 3" frame	2 Carp	2000	.008	S.F.	.44	.32		.76	1.02
1080	2" x 4" frame	"	2000	.008	"	.27	.32		.59	.83
2000	Load, haul, and dump, 50' haul	2 Clab	24	.667	C.Y.		21		21	35
2040	100' haul		16.50	.970			30.50		30.50	50.50
2080	Over 100' haul, add per 100 L.F.		35.50	.451			14.25		14.25	23.50
2120	In elevators, per 10 floors, add		140	.114			3.61		3.61	5.95
3000	Loading & trucking, including 2 mile haul, chute loaded	B-16	45	.711			23	11.85	34.85	50.50
3040	Hand loading truck, 50' haul	"	48	.667			21.50	11.10	32.60	47.50
3080	Machine loading truck	B-17	120	.267			8.95	5.15	14.10	20.50
3120	Wheeled 50' and ramp dump loaded	2 Clab	24	.667			21		21	35
5000	Haul, per mile, up to 8 C.Y. truck	B-34B	1165	.007			.22	.46	.68	.86
5100	Over 8 C.Y. truck	"	1550	.005			.16	.34	.50	.65

02 41 19.25 Selective Demolition, Saw Cutting

		Crew	Daily Output	Labor-Hours	Unit	Material	2009 Bare Costs Labor	Equipment	Total	Total Incl O&P
0010	**SELECTIVE DEMOLITION, SAW CUTTING** R024119-10									
0015	Asphalt, up to 3" deep	B-89	1050	.015	L.F.	.40	.53	.38	1.31	1.72
0020	Each additional inch of depth	"	1800	.009		.08	.31	.22	.61	.84
1200	Masonry walls, hydraulic saw, brick, per inch of depth	B-89B	300	.053		.42	1.87	2.28	4.57	5.95
1220	Block walls, solid, per inch of depth	"	250	.064		.42	2.24	2.74	5.40	7.05
2000	Brick or masonry w/hand held saw, per inch of depth	A-1	125	.064		.34	2.02	.51	2.87	4.27
5000	Wood sheathing to 1" thick, on walls	1 Carp	200	.040			1.60		1.60	2.64
5020	On roof	"	250	.032			1.28		1.28	2.11
9000	Minimum labor/equipment charge	A-1	2	4	Job		126	31.50	157.50	244

02 41 19.27 Selective Demolition, Torch Cutting

		Crew	Daily Output	Labor-Hours	Unit	Material	2009 Bare Costs Labor	Equipment	Total	Total Incl O&P
0010	**SELECTIVE DEMOLITION, TORCH CUTTING** R024119-10									
0020	Steel, 1" thick plate	1 Clab	360	.022	L.F.	.22	.70		.92	1.40
0040	1" diameter bar	"	210	.038	Ea.		1.20		1.20	1.99
1000	Oxygen lance cutting, reinforced concrete walls									
1040	12" to 16" thick walls	1 Clab	10	.800	L.F.		25.50		25.50	41.50
1080	24" thick walls	"	6	1.333	"		42		42	69.50
1090	Minimum labor/equipment charge	E-25	2	4	Job		187	46.50	233.50	395
1100	See Div. 05 05 21.10									

02 43 Structure Moving

02 43 13 – Structure Relocation

02 43 13.13 Building Relocation

			Crew	Daily Output	Labor-Hours	Unit	Material	2009 Bare Costs Labor	Equipment	Total	Total Incl O&P
0010	**BUILDING RELOCATION**										
0011	One day move, up to 24' wide										
0020	Reset on new foundation, patch & hook-up, average move					Total					11,765
0040	Wood or steel frame bldg., based on ground floor area	G	B-4	185	.259	S.F.		8.30	2.31	10.61	16.25
0060	Masonry bldg., based on ground floor area	G	"	137	.350			11.20	3.13	14.33	22
0200	For 24' to 42' wide, add										15%
0220	For each additional day on road, add	G	B-4	1	48	Day		1,525	430	1,955	3,000
0240	Construct new basement, move building, 1 day										
0300	move, patch & hook-up, based on ground floor area	G	B-3	155	.310	S.F.	9	10.45	13.25	32.70	41.50

02 58 Snow Control

02 58 13 – Snow Fencing

02 58 13.10 Snow Fencing System		Crew	Daily Output	Labor-Hours	Unit	Material	2009 Bare Costs Labor	Equipment	Total	Total Incl O&P
0010	**SNOW FENCING SYSTEM**									
7001	Snow fence on steel posts 10' O.C., 4' high	B-1	500	.048	L.F.	2.70	1.55		4.25	5.55

02 65 Underground Storage Tank Removal

02 65 10 – Underground Tank and Contaminated Soil Removal

02 65 10.30 Removal of Underground Storage Tanks

			Crew	Daily Output	Labor-Hours	Unit	Material	2009 Bare Costs Labor	Equipment	Total	Total Incl O&P
0010	**REMOVAL OF UNDERGROUND STORAGE TANKS**	R026510-20									
0011	Petroleum storage tanks, non-leaking										
0100	Excavate & load onto trailer										
0110	3000 gal. to 5000 gal. tank	G	B-14	4	12	Ea.		400	73.50	473.50	730
0120	6000 gal to 8000 gal tank	G	B-3A	3	13.333			445	289	734	1,050
0130	9000 gal to 12000 gal tank	G	"	2	20			670	435	1,105	1,575
0190	Known leaking tank, add					%				100%	100%
0200	Remove sludge, water and remaining product from tank bottom										
0201	of tank with vacuum truck										
0300	3000 gal to 5000 gal tank	G	A-13	5	1.600	Ea.		62.50	152	214.50	266
0310	6000 gal to 8000 gal tank	G		4	2			78	190	268	330
0320	9000 gal to 12000 gal tank	G		3	2.667			104	253	357	440
0390	Dispose of sludge off-site, average					Gal.				6.60	7.25
0400	Insert inert solid CO$_2$ "dry ice" into tank										
0401	For cleaning/transporting tanks (1.5 lbs./100 gal. cap)	G	1 Clab	500	.016	Lb.	1.69	.51		2.20	2.69
0403	Insert solid carbon dioxide, 1.5 lbs./100 gal.	G	"	400	.020	"	1.69	.63		2.32	2.90
0503	Disconnect and remove piping	G	1 Plum	160	.050	L.F.		2.44		2.44	3.78
0603	Transfer liquids, 10% of volume	G	"	1600	.005	Gal.		.24		.24	.38
0703	Cut accessway into underground storage tank	G	1 Clab	5.33	1.501	Ea.		47.50		47.50	78.50
0813	Remove sludge, wash and wipe tank, 500 gal	G	1 Plum	8	1			49		49	75.50
0823	3,000 gal	G		6.67	1.199			58.50		58.50	90.50
0833	5,000 gal	G		6.15	1.301			63.50		63.50	98.50
0843	8,000 gal	G		5.33	1.501			73		73	113
0853	10,000 gal	G		4.57	1.751			85.50		85.50	132
0863	12,000 gal	G		4.21	1.900			92.50		92.50	144
1020	Haul tank to certified salvage dump, 100 miles round trip										
1023	3000 gal. to 5000 gal. tank					Ea.				725	798
1026	6000 gal. to 8000 gal. tank									860	946
1029	9,000 gal. to 12,000 gal. tank									1,160	1,276
1100	Disposal of contaminated soil to landfill										
1110	Minimum					C.Y.				130.70	144
1111	Maximum					"				392	431
1120	Disposal of contaminated soil to										
1121	bituminous concrete batch plant										
1130	Minimum					C.Y.				65.30	72
1131	Maximum					"				130.70	144
1203	Excavate, pull, & load tank, backfill hole, 8,000 gal +	G	B-12C	.50	32	Ea.		1,175	2,375	3,550	4,500
1213	Haul tank to certified dump, 100 miles rt, 8,000 gal +	G	B-34K	1	8			256	785	1,041	1,275
1223	Excavate, pull, & load tank, backfill hole, 500 gal	G	B-11C	1	16			585	294	879	1,275
1233	Excavate, pull, & load tank, backfill hole, 3-5,000 gal	G	B-11M	.50	32			1,175	690	1,865	2,625
1243	Haul tank to certified dump, 100 miles rt, 500 gal	G	B-34L	1	8			310	194	504	705
2010	Decontamination of soil on site incl poly tarp on top/bottom										
2011	Soil containment berm, and chemical treatment										
2020	Minimum	G	B-11C	100	.160	C.Y.	6.90	5.85	2.94	15.69	20
2021	Maximum	G	"	100	.160		9	5.85	2.94	17.79	22.50

02 65 Underground Storage Tank Removal

02 65 10 – Underground Tank and Contaminated Soil Removal

02 65 10.30 Removal of Underground Storage Tanks	Crew	Daily Output	Labor-Hours	Unit	Material	2009 Bare Costs Labor	Equipment	Total	Total Incl O&P
2050	Disposal of decontaminated soil, minimum				C.Y.				78.20	86
2055	Maximum				↓				160.40	176.50

02 81 Transportation and Disposal of Hazardous Materials

02 81 20 – Hazardous Waste Handling

02 81 20.10 Hazardous Waste Cleanup/Pickup/Disposal

		Crew	Daily Output	Labor-Hours	Unit	Material	2009 Bare Costs Labor	Equipment	Total	Total Incl O&P
0010	**HAZARDOUS WASTE CLEANUP/PICKUP/DISPOSAL**									
0100	For contractor rental equipment, i.e., Dozer,									
0110	Front end loader, Dump truck, etc., see 01 54 33 Reference Section									
1000	Solid pickup									
1100	55 gal. drums				Ea.				239.60	263.50
1120	Bulk material, minimum				Ton				180.20	198
1130	Maximum				"				599	659
1200	Transportation to disposal site									
1220	Truckload = 80 drums or 25 C.Y. or 18 tons									
1260	Minimum				Mile				3.72	4.09
1270	Maximum				"				6.53	7.18
3000	Liquid pickup, vacuum truck, stainless steel tank									
3100	Minimum charge, 4 hours									
3110	1 compartment, 2200 gallon				Hr.				130.70	144
3120	2 compartment, 5000 gallon				"				177.75	195.50
3400	Transportation in 6900 gallon bulk truck				Mile				7.05	7.75
3410	In teflon lined truck				"				8.17	9
5000	Heavy sludge or dry vacuumable material				Hr.				130.70	144
6000	Dumpsite disposal charge, minimum				Ton				130.70	144
6020	Maximum				"				479.20	527

02 82 Asbestos Remediation

02 82 13 – Asbestos Abatement

02 82 13.41 Asbestos Abatement Equip.

		Crew	Daily Output	Labor-Hours	Unit	Material	2009 Bare Costs Labor	Equipment	Total	Total Incl O&P
0010	**ASBESTOS ABATEMENT EQUIP.** R028213-20									
0011	Equipment and supplies, buy									
0200	Air filtration device, 2000 C.F.M.				Ea.	795			795	875
0250	Large volume air sampling pump, minimum					305			305	335
0260	Maximum					455			455	500
0300	Airless sprayer unit, 2 gun					3,125			3,125	3,425
0350	Light stand, 500 watt				↓	72.50			72.50	80
0400	Personal respirators									
0410	Negative pressure, 1/2 face, dual operation, min.				Ea.	25.50			25.50	28
0420	Maximum					31			31	34.50
0450	P.A.P.R., full face, minimum					124			124	137
0460	Maximum					150			150	165
0470	Supplied air, full face, incl. air line, minimum					157			157	172
0480	Maximum					239			239	263
0500	Personnel sampling pump, minimum					201			201	221
1500	Power panel, 20 unit, incl. G.F.I.					1,100			1,100	1,200
1600	Shower unit, including pump and filters					1,075			1,075	1,200
1700	Supplied air system (type C)					9,800			9,800	10,800
1750	Vacuum cleaner, HEPA, 16 gal., stainless steel, wet/dry				↓	675			675	745

02 82 Asbestos Remediation

02 82 13 – Asbestos Abatement

02 82 13.41 Asbestos Abatement Equip.

		Crew	Daily Output	Labor-Hours	Unit	Material	2009 Bare Costs Labor	Equipment	Total	Total Incl O&P
1760	55 gallon				Ea.	1,700			1,700	1,875
1800	Vacuum loader, 9-18 ton/hr					90,000			90,000	99,000
1900	Water atomizer unit, including 55 gal. drum					230			230	253
2000	Worker protection, whole body, foot, head cover & gloves, plastic					6.80			6.80	7.45
2500	Respirator, single use					10.90			10.90	12
2550	Cartridge for respirator					12.30			12.30	13.55
2570	Glove bag, 7 mil, 50" x 64"					9.20			9.20	10.10
2580	10 mil, 44" x 60"					74.50			74.50	82
3000	HEPA vacuum for work area, minimum					1,325			1,325	1,475
3050	Maximum					3,100			3,100	3,425
6000	Disposable polyethylene bags, 6 mil, 3 C.F.					1.24			1.24	1.36
6300	Disposable fiber drums, 3 C.F.					12			12	13.20
6400	Pressure sensitive caution labels, 3" x 5"					1.19			1.19	1.31
6450	11" x 17"					7.05			7.05	7.75
6500	Negative air machine, 1800 C.F.M.					770			770	845

02 82 13.42 Preparation of Asbestos Containment Area

		Crew	Daily Output	Labor-Hours	Unit	Material	2009 Bare Costs Labor	Equipment	Total	Total Incl O&P
0010	**PREPARATION OF ASBESTOS CONTAINMENT AREA**									
0100	Pre-cleaning, HEPA vacuum and wet wipe, flat surfaces	A-9	12000	.005	S.F.	.02	.24		.26	.40
0200	Protect carpeted area, 2 layers 6 mil poly on 3/4" plywood	"	1000	.064		1.50	2.83		4.33	6.20
0300	Separation barrier, 2" x 4" @ 16", 1/2" plywood ea. side, 8' high	2 Carp	400	.040		1.25	1.60		2.85	4.02
0310	12' high		320	.050		1.40	2		3.40	4.84
0320	16' high		200	.080		1.50	3.20		4.70	6.95
0400	Personnel decontam. chamber, 2" x 4" @ 16", 3/4" ply ea. side		280	.057		2.50	2.28		4.78	6.50
0450	Waste decontam. chamber, 2" x 4" studs @ 16", 3/4" ply ea. side		360	.044		3	1.78		4.78	6.25
0500	Cover surfaces with polyethylene sheeting									
0501	Including glue and tape									
0550	Floors, each layer, 6 mil	A-9	8000	.008	S.F.	.06	.35		.41	.63
0551	4 mil		9000	.007		.04	.31		.35	.55
0560	Walls, each layer, 6 mil		6000	.011		.06	.47		.53	.82
0561	4 mil		7000	.009		.04	.40		.44	.69
0570	For heights above 12', add						20%			
0575	For heights above 20', add						30%			
0580	For fire retardant poly, add					100%				
0590	For large open areas, deduct					10%	20%			
0600	Seal floor penetrations with foam firestop to 36 Sq. In.	2 Carp	200	.080	Ea.	1,275	3.20		1,278.20	1,400
0610	36 Sq. In. to 72 Sq. In.		125	.128		2,550	5.10		2,555.10	2,825
0615	72 Sq. In. to 144 Sq. In.		80	.200		5,125	8		5,133	5,650
0620	Wall penetrations, to 36 square inches		180	.089		1,275	3.55		1,278.55	1,400
0630	36 Sq. In. to 72 Sq. In.		100	.160		2,550	6.40		2,556.40	2,825
0640	72 Sq. In. to 144 Sq. In.		60	.267		5,125	10.65		5,135.65	5,650
0800	Caulk seams with latex	1 Carp	230	.035	L.F.	.15	1.39		1.54	2.46
0900	Set up neg. air machine, 1-2k C.F.M. /25 M.C.F. volume	1 Asbe	4.30	1.860	Ea.		82		82	133

02 82 13.43 Bulk Asbestos Removal

		Crew	Daily Output	Labor-Hours	Unit	Material	2009 Bare Costs Labor	Equipment	Total	Total Incl O&P
0010	**BULK ASBESTOS REMOVAL**									
0020	Includes disposable tools and 2 suits and 1 respirator filter/day/worker									
0100	Beams, W 10 x 19	A-9	235	.272	L.F.	.88	12.05		12.93	20.50
0110	W 12 x 22		210	.305		.99	13.45		14.44	23
0120	W 14 x 26		180	.356		1.15	15.70		16.85	27
0130	W 16 x 31		160	.400		1.29	17.65		18.94	30
0140	W 18 x 40		140	.457		1.48	20		21.48	34
0150	W 24 x 55		110	.582		1.88	25.50		27.38	43.50
0160	W 30 x 108		85	.753		2.43	33.50		35.93	56.50

02 82 Asbestos Remediation

02 82 13 – Asbestos Abatement

02 82 13.43 Bulk Asbestos Removal

		Crew	Daily Output	Labor-Hours	Unit	Material	2009 Bare Costs Labor	Equipment	Total	Total Incl O&P
0170	W 36 x 150	A-9	72	.889	L.F.	2.87	39.50		42.37	66.50
0200	Boiler insulation	↓	480	.133	S.F.	.50	5.90		6.40	10.05
0210	With metal lath, add				%				50%	
0300	Boiler breeching or flue insulation	A-9	520	.123	S.F.	.40	5.45		5.85	9.25
0310	For active boiler, add				%				100%	
0400	Duct or AHU insulation	A-10B	440	.073	S.F.	.24	3.22		3.46	5.45
0500	Duct vibration isolation joints, up to 24 Sq. In. duct	A-9	56	1.143	Ea.	3.69	50.50		54.19	85.50
0520	25 Sq. In. to 48 Sq. In. duct		48	1.333		4.31	59		63.31	99.50
0530	49 Sq. In. to 76 Sq. In. duct		40	1.600	↓	5.15	70.50		75.65	120
0600	Pipe insulation, air cell type, up to 4" diameter pipe		900	.071	L.F.	.23	3.14		3.37	5.35
0610	4" to 8" diameter pipe		800	.080		.26	3.53		3.79	6
0620	10" to 12" diameter pipe		700	.091		.30	4.04		4.34	6.90
0630	14" to 16" diameter pipe		550	.116	↓	.38	5.15		5.53	8.70
0650	Over 16" diameter pipe		650	.098	S.F.	.32	4.35		4.67	7.40
0700	With glove bag up to 3" diameter pipe		200	.320	L.F.	2.15	14.15		16.30	25.50
1000	Pipe fitting insulation up to 4" diameter pipe		320	.200	Ea.	.65	8.85		9.50	15
1100	6" to 8" diameter pipe		304	.211		.68	9.30		9.98	15.80
1110	10" to 12" diameter pipe		192	.333		1.08	14.70		15.78	25
1120	14" to 16" diameter pipe		128	.500	↓	1.62	22		23.62	37.50
1130	Over 16" diameter pipe		176	.364	S.F.	1.18	16.05		17.23	27.50
1200	With glove bag, up to 8" diameter pipe		75	.853	L.F.	6.55	37.50		44.05	68
2000	Scrape foam fireproofing from flat surface		2400	.027	S.F.	.09	1.18		1.27	1.99
2100	Irregular surfaces		1200	.053		.17	2.36		2.53	4
3000	Remove cementitious material from flat surface		1800	.036		.11	1.57		1.68	2.67
3100	Irregular surface		1000	.064		.15	2.83		2.98	4.73
4000	Scrape acoustical coating/fireproofing, from ceiling		3200	.020		.06	.88		.94	1.50
5000	Remove VAT from floor by hand	↓	2400	.027		.09	1.18		1.27	1.99
5100	By machine	A-11	4800	.013	↓	.04	.59	.01	.64	1.01
5150	For 2 layers, add				%				50%	
6000	Remove contaminated soil from crawl space by hand	A-9	400	.160	C.F.	.52	7.05		7.57	11.95
6100	With large production vacuum loader	A-12	700	.091	"	.30	4.04	1.08	5.42	8.05
7000	Radiator backing, not including radiator removal	A-9	1200	.053	S.F.	.17	2.36		2.53	4
8000	Cement-asbestos transite board	2 Asbe	1000	.016		.14	.71		.85	1.30
8100	Transite shingle siding	A-10B	750	.043		.22	1.89		2.11	3.29
8200	Shingle roofing	"	2000	.016		.08	.71		.79	1.23
8250	Built-up, no gravel, non-friable	B-2	1400	.029		.08	.91		.99	1.60
8300	Asbestos millboard	2 Asbe	1000	.016	↓	.08	.71		.79	1.23
9000	For type C (supplied air) respirator equipment, add				%					10%

02 82 13.44 Demolition In Asbestos Contaminated Area

		Crew	Daily Output	Labor-Hours	Unit	Material	2009 Bare Costs Labor	Equipment	Total	Total Incl O&P
0010	**DEMOLITION IN ASBESTOS CONTAMINATED AREA**									
0200	Ceiling, including suspension system, plaster and lath	A-9	2100	.030	S.F.	.10	1.35		1.45	2.29
0210	Finished plaster, leaving wire lath		585	.109		.35	4.83		5.18	8.20
0220	Suspended acoustical tile		3500	.018		.06	.81		.87	1.38
0230	Concealed tile grid system		3000	.021		.07	.94		1.01	1.60
0240	Metal pan grid system		1500	.043		.14	1.88		2.02	3.20
0250	Gypsum board		2500	.026	↓	.08	1.13		1.21	1.92
0260	Lighting fixtures up to 2' x 4'	↓	72	.889	Ea.	2.87	39.50		42.37	66.50
0400	Partitions, non load bearing									
0410	Plaster, lath, and studs	A-9	690	.093	S.F.	.87	4.10		4.97	7.55
0450	Gypsum board and studs	"	1390	.046	"	.15	2.03		2.18	3.45
9000	For type C (supplied air) respirator equipment, add				%					10%

02 82 Asbestos Remediation

02 82 13 – Asbestos Abatement

02 82 13.45 OSHA Testing

		Crew	Daily Output	Labor-Hours	Unit	Material	2009 Bare Costs Labor	Equipment	Total	Total Incl O&P
0010	**OSHA TESTING**									
0100	Certified technician, minimum				Day				240	264
0110	Maximum				"				320	352
0200	Personal sampling, PCM analysis, NIOSH 7400, minimum	1 Asbe	8	1	Ea.	2.75	44		46.75	74.50
0210	Maximum	"	4	2	"	3	88		91	146
0300	Industrial hygienist, minimum				Day				320	352
0310	Maximum				"				480	528
1000	Cleaned area samples	1 Asbe	8	1	Ea.	2.63	44		46.63	74.50
1100	PCM air sample analysis, NIOSH 7400, minimum		8	1		30.50	44		74.50	106
1110	Maximum	↓	4	2		3.20	88		91.20	147
1200	TEM air sample analysis, NIOSH 7402, minimum								100	125
1210	Maximum			↓					400	500

02 82 13.46 Decontamination of Asbestos Containment Area

		Crew	Daily Output	Labor-Hours	Unit	Material	2009 Bare Costs Labor	Equipment	Total	Total Incl O&P
0010	**DECONTAMINATION OF ASBESTOS CONTAINMENT AREA**									
0100	Spray exposed substrate with surfactant (bridging)									
0200	Flat surfaces	A-9	6000	.011	S.F.	.36	.47		.83	1.16
0250	Irregular surfaces		4000	.016	"	.31	.71		1.02	1.48
0300	Pipes, beams, and columns		2000	.032	L.F.	.56	1.41		1.97	2.90
1000	Spray encapsulate polyethylene sheeting		8000	.008	S.F.	.31	.35		.66	.91
1100	Roll down polyethylene sheeting		8000	.008	"		.35		.35	.57
1500	Bag polyethylene sheeting		400	.160	Ea.	.77	7.05		7.82	12.25
2000	Fine clean exposed substrate, with nylon brush		2400	.027	S.F.		1.18		1.18	1.90
2500	Wet wipe substrate		4800	.013			.59		.59	.95
2600	Vacuum surfaces, fine brush	↓	6400	.010	↓		.44		.44	.71
3000	Structural demolition									
3100	Wood stud walls	A-9	2800	.023	S.F.		1.01		1.01	1.63
3500	Window manifolds, not incl. window replacement		4200	.015			.67		.67	1.09
3600	Plywood carpet protection	↓	2000	.032	↓		1.41		1.41	2.28
4000	Remove custom decontamination facility	A-10A	8	3	Ea.	15	133		148	232
4100	Remove portable decontamination facility	3 Asbe	12	2	"	12.75	88		100.75	157
5000	HEPA vacuum, shampoo carpeting	A-9	4800	.013	S.F.	.05	.59		.64	1.01
9000	Final cleaning of protected surfaces	A-10A	8000	.003	"		.13		.13	.21

02 82 13.47 Asbestos Waste Pkg., Handling, and Disp.

		Crew	Daily Output	Labor-Hours	Unit	Material	2009 Bare Costs Labor	Equipment	Total	Total Incl O&P
0010	**ASBESTOS WASTE PACKAGING, HANDLING, AND DISPOSAL**									
0100	Collect and bag bulk material, 3 C.F. bags, by hand	A-9	400	.160	Ea.	1.24	7.05		8.29	12.75
0200	Large production vacuum loader	A-12	880	.073		.80	3.21	.86	4.87	7.05
1000	Double bag and decontaminate	A-9	960	.067		1.24	2.94		4.18	6.10
2000	Containerize bagged material in drums, per 3 C.F. drum	"	800	.080	↓	12	3.53		15.53	18.90
3000	Cart bags 50' to dumpster	2 Asbe	400	.040	↓		1.76		1.76	2.85
5000	Disposal charges, not including haul, minimum				C.Y.				50	55
5020	Maximum				"				170	187
5100	Remove refrigerant from system	1 Plum	40	.200	Lb.		9.75		9.75	15.10
9000	For type C (supplied air) respirator equipment, add				%					10%

02 82 13.48 Asbestos Encapsulation With Sealants

		Crew	Daily Output	Labor-Hours	Unit	Material	2009 Bare Costs Labor	Equipment	Total	Total Incl O&P
0010	**ASBESTOS ENCAPSULATION WITH SEALANTS**									
0100	Ceilings and walls, minimum	A-9	21000	.003	S.F.	.27	.13		.40	.52
0110	Maximum		10600	.006		.42	.27		.69	.89
0200	Columns and beams, minimum		13300	.005	↓	.27	.21		.48	.64
0210	Maximum	↓	5325	.012	↓	.47	.53		1	1.38
0300	Pipes to 12" diameter including minor repairs, minimum		800	.080	L.F.	.37	3.53		3.90	6.10
0310	Maximum	↓	400	.160	"	1.04	7.05		8.09	12.55

02 83 19 – Lead-Based Paint Remediation

02 83 19.23 Encapsulation of Lead-Based Paint	Crew	Daily Output	Labor-Hours	Unit	Material	2009 Bare Costs Labor	Equipment	Total	Total Incl O&P
0010 **ENCAPSULATION OF LEAD-BASED PAINT**									
0020 Interior, brushwork, trim, under 6"	1 Pord	240	.033	L.F.	2.46	1.17		3.63	4.58
0030 6" to 12" wide		180	.044		3.28	1.56		4.84	6.10
0040 Balustrades		300	.027		1.98	.94		2.92	3.68
0050 Pipe to 4" diameter		500	.016		1.19	.56		1.75	2.21
0060 To 8" diameter		375	.021		1.57	.75		2.32	2.93
0070 To 12" diameter		250	.032		2.36	1.13		3.49	4.40
0080 To 16" diameter		170	.047		3.47	1.66		5.13	6.45
0090 Cabinets, ornate design		200	.040	S.F.	2.97	1.41		4.38	5.50
0100 Simple design		250	.032	"	2.36	1.13		3.49	4.40
0110 Doors, 3' x 7', both sides, incl. frame & trim									
0120 Flush	1 Pord	6	1.333	Ea.	30.50	47		77.50	109
0130 French, 10-15 lite		3	2.667		6.05	94		100.05	157
0140 Panel		4	2		36.50	70.50		107	153
0150 Louvered		2.75	2.909		33.50	102		135.50	201
0160 Windows, per interior side, per 15 S.F.									
0170 1 to 6 lite	1 Pord	14	.571	Ea.	21	20		41	55
0180 7 to 10 lite		7.50	1.067		23	37.50		60.50	85.50
0190 12 lite		5.75	1.391		31	49		80	113
0200 Radiators		8	1		73.50	35		108.50	138
0210 Grilles, vents		275	.029	S.F.	2.15	1.02		3.17	4.01
0220 Walls, roller, drywall or plaster		1000	.008		.59	.28		.87	1.10
0230 With spunbonded reinforcing fabric		720	.011		.68	.39		1.07	1.37
0240 Wood		800	.010		.74	.35		1.09	1.37
0250 Ceilings, roller, drywall or plaster		900	.009		.68	.31		.99	1.25
0260 Wood		700	.011		.84	.40		1.24	1.56
0270 Exterior, brushwork, gutters and downspouts		300	.027	L.F.	1.98	.94		2.92	3.68
0280 Columns		400	.020	S.F.	1.47	.70		2.17	2.75
0290 Spray, siding		600	.013	"	.99	.47		1.46	1.84
0300 Miscellaneous									
0310 Electrical conduit, brushwork, to 2" diameter	1 Pord	500	.016	L.F.	1.19	.56		1.75	2.21
0320 Brick, block or concrete, spray		500	.016	S.F.	1.19	.56		1.75	2.21
0330 Steel, flat surfaces and tanks to 12"		500	.016		1.19	.56		1.75	2.21
0340 Beams, brushwork		400	.020		1.47	.70		2.17	2.75
0350 Trusses		400	.020		1.47	.70		2.17	2.75

02 83 19.26 Removal of Lead-Based Paint

	Crew	Daily Output	Labor-Hours	Unit	Material	2009 Bare Costs Labor	Equipment	Total	Total Incl O&P
0010 **REMOVAL OF LEAD-BASED PAINT** R028319-60									
0011 By chemicals, per application									
0050 Baseboard, to 6" wide	1 Pord	64	.125	L.F.	.66	4.40		5.06	7.80
0070 To 12" wide		32	.250	"	1.32	8.80		10.12	15.50
0200 Balustrades, one side		28	.286	S.F.	1.32	10.05		11.37	17.50
1400 Cabinets, simple design		32	.250		1.32	8.80		10.12	15.50
1420 Ornate design		25	.320		1.32	11.25		12.57	19.45
1600 Cornice, simple design		60	.133		1.32	4.69		6.01	8.95
1620 Ornate design		20	.400		5.05	14.10		19.15	28
2800 Doors, one side, flush		84	.095		1	3.35		4.35	6.45
2820 Two panel		80	.100		1.32	3.52		4.84	7.10
2840 Four panel		45	.178		1.32	6.25		7.57	11.45
2880 For trim, one side, add		64	.125	L.F.	.66	4.40		5.06	7.80
3000 Fence, picket, one side		30	.267	S.F.	1.32	9.40		10.72	16.45
3200 Grilles, one side, simple design		30	.267		1.32	9.40		10.72	16.45
3220 Ornate design		25	.320		1.32	11.25		12.57	19.45

02 83 Lead Remediation

02 83 19 – Lead-Based Paint Remediation

02 83 19.26 Removal of Lead-Based Paint

		Crew	Daily Output	Labor-Hours	Unit	Material	2009 Bare Costs Labor	Equipment	Total	Total Incl O&P
4400	Pipes, to 4" diameter	1 Pord	90	.089	L.F.	1	3.13		4.13	6.10
4420	To 8" diameter		50	.160		2	5.65		7.65	11.20
4440	To 12" diameter		36	.222		3	7.80		10.80	15.80
4460	To 16" diameter		20	.400	▼	4	14.10		18.10	27
4500	For hangers, add		40	.200	Ea.	2.54	7.05		9.59	14.05
4800	Siding		90	.089	S.F.	1.17	3.13		4.30	6.30
5000	Trusses, open		55	.145	SF Face	1.85	5.10		6.95	10.25
6200	Windows, one side only, double hung, 1/1 light, 24" x 48" high		4	2	Ea.	25.50	70.50		96	141
6220	30" x 60" high		3	2.667		34	94		128	188
6240	36" x 72" high		2.50	3.200		41	113		154	225
6280	40" x 80" high		2	4		51	141		192	281
6400	Colonial window, 6/6 light, 24" x 48" high		2	4		51	141		192	282
6420	30" x 60" high		1.50	5.333		68	188		256	375
6440	36" x 72" high		1	8		102	282		384	560
6480	40" x 80" high		1	8		102	282		384	560
6600	8/8 light, 24" x 48" high		2	4		51	141		192	282
6620	40" x 80" high		1	8		102	282		384	560
6800	12/12 light, 24" x 48" high		1	8		102	282		384	560
6820	40" x 80" high	▼	.75	10.667	▼	136	375		511	750
6840	Window frame & trim items, included in pricing above									
9000	Minimum labor/equipment charge	1 Pord	3	2.667	Job		94		94	150

02 85 Mold Remediation

02 85 16 – Mold Remediation Preparation and Containment

02 85 16.40 Mold Remediation Plans and Methods

		Crew	Daily Output	Labor-Hours	Unit	Material	2009 Bare Costs Labor	Equipment	Total	Total Incl O&P
0010	**MOLD REMEDIATION PLANS AND METHODS**									
0020	Initial inspection, average 3 bedroom home				Total				250	275
0030	Average 5 bedroom home								340	374
0040	Testing, air sample each								250	275
0050	Swab sample								150	165
0060	Tape sample								150	165
0070	Post remediation air test								250	275
0080	Mold abatement plan, average 3 bedroom home								1,300	1,430
0090	Average 5 bedroom home								1,580	1,740
0100	Packup & removal of contents, average 3 bedroom home, excl storage								10,500	11,550
0110	Average 5 bedroom home, excl storage				▼				26,400	29,000
0120	For demolition in mold contaminated areas, see Div. 02 85 33.50									
0130	For personal protection equipment, see Div. 02 82 13.41									

02 85 16.50 Preparation of Mold Containment Area

		Crew	Daily Output	Labor-Hours	Unit	Material	2009 Bare Costs Labor	Equipment	Total	Total Incl O&P
0010	**PREPARATION OF MOLD CONTAINMENT AREA**									
0100	Pre-cleaning, HEPA vacuum and wet wipe, flat surfaces	A-9	12000	.005	S.F.	.02	.24		.26	.40
0300	Separation barrier, 2" x 4" @ 16", 1/2" plywood ea. side, 8' high	2 Carp	400	.040		1.25	1.60		2.85	4.02
0310	12' high		320	.050		1.40	2		3.40	4.84
0320	16' high		200	.080		1.50	3.20		4.70	6.95
0400	Personnel decontam. chamber, 2" x 4" @ 16", 3/4" ply ea. side		280	.057		2.50	2.28		4.78	6.50
0450	Waste decontam. chamber, 2" x 4" studs @ 16", 3/4" ply each side	▼	360	.044	▼	3	1.78		4.78	6.25
0500	Cover surfaces with polyethylene sheeting									
0501	Including glue and tape									
0550	Floors, each layer, 6 mil	A-9	8000	.008	S.F.	.09	.35		.44	.67
0551	4 mil		9000	.007		.05	.31		.36	.57
0560	Walls, each layer, 6 mil	▼	6000	.011	▼	.09	.47		.56	.86

02 85 Mold Remediation

02 85 16 – Mold Remediation Preparation and Containment

02 85 16.50 Preparation of Mold Containment Area

		Crew	Daily Output	Labor-Hours	Unit	Material	2009 Bare Costs Labor	Equipment	Total	Total Incl O&P
0561	4 mil	A-9	7000	.009	S.F.	.07	.40		.47	.73
0570	For heights above 12', add						20%			
0575	For heights above 20', add						30%			
0580	For fire retardant poly, add					100%				
0590	For large open areas, deduct					10%	20%			
0600	Seal floor penetrations with foam firestop to 36 sq in	2 Carp	200	.080	Ea.	1,275	3.20		1,278.20	1,400
0610	36 sq in to 72 sq in		125	.128		2,550	5.10		2,555.10	2,825
0615	72 sq in to 144 sq in		80	.200		5,125	8		5,133	5,650
0620	Wall penetrations, to 36 square inches		180	.089		1,275	3.55		1,278.55	1,400
0630	36 sq. in. to 72 sq. in.		100	.160		2,550	6.40		2,556.40	2,825
0640	72 Sq. in. to 144 sq. in.	↓	60	.267	↓	5,125	10.65		5,135.65	5,650
0800	Caulk seams with latex caulk	1 Carp	230	.035	L.F.	.15	1.39		1.54	2.46
0900	Set up neg. air machine, 1-2k C.F.M. /25 M.C.F. volume	1 Asbe	4.30	1.860	Ea.		82		82	133

02 85 33 – Removal and Disposal of Materials with Mold

02 85 33.50 Demolition in Mold Contaminated Area

		Crew	Daily Output	Labor-Hours	Unit	Material	2009 Bare Costs Labor	Equipment	Total	Total Incl O&P
0010	**DEMOLITION IN MOLD CONTAMINATED AREA**									
0200	Ceiling, including suspension system, plaster and lath	A-9	2100	.030	S.F.	.10	1.35		1.45	2.29
0210	Finished plaster, leaving wire lath		585	.109		.35	4.83		5.18	8.20
0220	Suspended acoustical tile		3500	.018		.06	.81		.87	1.38
0230	Concealed tile grid system		3000	.021		.07	.94		1.01	1.60
0240	Metal pan grid system		1500	.043		.14	1.88		2.02	3.20
0250	Gypsum board		2500	.026		.08	1.13		1.21	1.92
0255	Plywood		2500	.026	↓	.08	1.13		1.21	1.92
0260	Lighting fixtures up to 2' x 4'	↓	72	.889	Ea.	2.87	39.50		42.37	66.50
0400	Partitions, non load bearing									
0410	Plaster, lath, and studs	A-9	690	.093	S.F.	.87	4.10		4.97	7.55
0450	Gypsum board and studs		1390	.046		.15	2.03		2.18	3.45
0465	Carpet & pad		1390	.046	↓	.15	2.03		2.18	3.45
0600	Pipe insulation, air cell type, up to 4" diameter pipe		900	.071	L.F.	.23	3.14		3.37	5.35
0610	4" to 8" diameter pipe		800	.080		.26	3.53		3.79	6
0620	10" to 12" diameter pipe		700	.091		.30	4.04		4.34	6.90
0630	14" to 16" diameter pipe		550	.116	↓	.38	5.15		5.53	8.70
0650	Over 16" diameter pipe	↓	650	.098	S.F.	.32	4.35		4.67	7.40
9000	For type C (supplied air) respirator equipment, add				%					10%

Division Notes

		CREW	DAILY OUTPUT	LABOR-HOURS	UNIT	2009 BARE COSTS				TOTAL INCL O&P
						MAT.	LABOR	EQUIP.	TOTAL	

Estimating Tips

General

- Carefully check all the plans and specifications. Concrete often appears on drawings other than structural drawings, including mechanical and electrical drawings for equipment pads. The cost of cutting and patching is often difficult to estimate. See Subdivision 03 81 for Concrete Cutting, Subdivision 02 41 19.16 for Cutout Demolition, Subdivision 03 05 05.10 for Concrete Demolition, and Subdivision 02 41 19.23 for Rubbish Handling (handling, loading and hauling of debris).

- Always obtain concrete prices from suppliers near the job site. A volume discount can often be negotiated, depending upon competition in the area. Remember to add for waste, particularly for slabs and footings on grade.

03 10 00 Concrete Forming and Accessories

- A primary cost for concrete construction is forming. Most jobs today are constructed with prefabricated forms. The selection of the forms best suited for the job and the total square feet of forms required for efficient concrete forming and placing are key elements in estimating concrete construction. Enough forms must be available for erection to make efficient use of the concrete placing equipment and crew.

- Concrete accessories for forming and placing depend upon the systems used. Study the plans and specifications to ensure that all special accessory requirements have been included in the cost estimate, such as anchor bolts, inserts, and hangers.

- Included within costs for forms-in-place are all necessary bracing and shoring.

03 20 00 Concrete Reinforcing

- Ascertain that the reinforcing steel supplier has included all accessories, cutting, bending, and an allowance for lapping, splicing, and waste. A good rule of thumb is 10% for lapping, splicing, and waste. Also, 10% waste should be allowed for welded wire fabric.

- The unit price items in the subdivision for Reinforcing In Place include the labor to install accessories such as beam and slab bolsters, high chairs, and bar ties and tie wire. The material cost for these accessories is not included; they may be obtained from the Accessories Division.

03 30 00 Cast-In-Place Concrete

- When estimating structural concrete, pay particular attention to requirements for concrete additives, curing methods, and surface treatments. Special consideration for climate, hot or cold, must be included in your estimate. Be sure to include requirements for concrete placing equipment, and concrete finishing.

- For accurate concrete estimating, the estimator must consider each of the following major components individually: forms, reinforcing steel, ready-mix concrete, placement of the concrete, and finishing of the top surface. For faster estimating, Subdivision 03 30 53.40 for Concrete-In-Place can be used; here, various items of concrete work are presented that include the costs of all five major components (unless specifically stated otherwise).

03 40 00 Precast Concrete
03 50 00 Cast Decks and Underlayment

- The cost of hauling precast concrete structural members is often an important factor. For this reason, it is important to get a quote from the nearest supplier. It may become economically feasible to set up precasting beds on the site if the hauling costs are prohibitive.

Reference Numbers

Reference numbers are shown in shaded boxes at the beginning of some major classifications. These numbers refer to related items in the Reference Section. The reference information may be an estimating procedure, an alternate pricing method, or technical information.

Note: Not all subdivisions listed here necessarily appear in this publication.

03 01 Maintenance of Concrete

03 01 30 – Maintenance of Cast-In-Place Concrete

03 01 30.62 Concrete Patching

		Daily Output	Labor-Hours	Unit	Material	2009 Bare Costs Labor	Equipment	Total	Total Incl O&P
0010	**CONCRETE PATCHING**								
0100	Floors, 1/4" thick, small areas, regular grout	Crew							
0100	Floors, 1/4" thick, small areas, regular grout	1 Cefi	170	.047	S.F.	1.07	1.80		2.87	3.99
0150	Epoxy grout	"	100	.080	"	6.85	3.06		9.91	12.30
2000	Walls, including chipping, cleaning and epoxy grout									
2100	1/4" deep	1 Cefi	65	.123	S.F.	7.50	4.71		12.21	15.60
2150	1/2" deep		50	.160		15.05	6.15		21.20	26
2200	3/4" deep		40	.200		22.50	7.65		30.15	37
9000	Minimum labor/equipment charge		4.50	1.778	Job		68		68	106

03 05 Common Work Results for Concrete

03 05 05 – Selective Concrete Demolition

03 05 05.10 Selective Demolition, Concrete

		Crew	Daily Output	Labor-Hours	Unit	Material	2009 Bare Costs Labor	Equipment	Total	Total Incl O&P
0010	**SELECTIVE DEMOLITION, CONCRETE** R024119-10									
0012	Excludes saw cutting, torch cutting, loading or hauling									
0050	Break up into small pieces, minimum reinforcing	B-9	24	1.667	C.Y.		53.50	8	61.50	97
0060	Average reinforcing		16	2.500			80	12	92	145
0070	Maximum reinforcing		8	5			160	24	184	291
0150	Remove whole pieces, up to 2 tons per piece	E-18	36	1.111	Ea.		49.50	30	79.50	122
0160	2 – 5 tons per piece		30	1.333			59	36	95	147
0170	5 – 10 tons per piece		24	1.667			74	45	119	183
0180	10 – 15 tons per piece		18	2.222			98.50	60	158.50	244
0250	Precast unit embedded in masonry, up to 1 CF	D-1	16	1			36.50		36.50	59
0260	1 – 2 CF		12	1.333			48.50		48.50	78.50
0270	2 – 5 CF		10	1.600			58		58	94
0280	5 – 10 CF		8	2			72.50		72.50	118
1910	Minimum labor/equipment charge	B-9	2	20	Job		640	96	736	1,150

03 05 13 – Basic Concrete Materials

03 05 13.20 Concrete Admixtures and Surface Treatments

		Crew	Daily Output	Labor-Hours	Unit	Material	2009 Bare Costs Labor	Equipment	Total	Total Incl O&P
0010	**CONCRETE ADMIXTURES AND SURFACE TREATMENTS**									
0040	Abrasives, aluminum oxide, over 20 tons				Lb.	1.63			1.63	1.79
0050	1 to 20 tons					1.68			1.68	1.85
0070	Under 1 ton					1.75			1.75	1.93
0100	Silicon carbide, black, over 20 tons					2.23			2.23	2.45
0110	1 to 20 tons					2.34			2.34	2.58
0120	Under 1 ton					2.44			2.44	2.68
0200	Air entraining agent, .7 to 1.5 oz. per bag, 55 gallon drum				Gal.	9.95			9.95	10.95
0220	5 gallon pail					12.50			12.50	13.75
0300	Bonding agent, acrylic latex, 250 S.F. per gallon, 5 gallon pail					18.50			18.50	20.50
0320	Epoxy resin, 80 S.F. per gallon, 3.5 gallon unit					45			45	49.50
0400	Calcium chloride, 50 lb. bags, TL lots				Ton	525			525	580
0420	Less than truckload lots				Bag	17.75			17.75	19.55
0500	Carbon black, liquid, 2 to 8 lbs. per bag of cement				Lb.	5.90			5.90	6.45
0600	Colors, integral, 2 to 10 lb. per bag of cement, minimum					2.41			2.41	2.65
0610	Average					3.40			3.40	3.74
0620	Maximum					4.80			4.80	5.30
0920	Dustproofing compound, 250 SF/gal, 5 gallon pail				Gal.	6.80			6.80	7.45
1010	Epoxy based, 125 SF/Gal, 5 gallon pail				"	45.50			45.50	50
1100	Hardeners, metallic, 55 lb. bags, natural (grey)				Lb.	.93			.93	1.02
1200	Colors					1.32			1.32	1.45
1300	Non-metallic, 55 lb. bags, natural grey					.37			.37	.41

03 05 Common Work Results for Concrete

03 05 13 – Basic Concrete Materials

03 05 13.20 Concrete Admixtures and Surface Treatments		Crew	Daily Output	Labor-Hours	Unit	Material	2009 Bare Costs Labor	Equipment	Total	Total Incl O&P
1320	Colors				Lb.	.66			.66	.72
1600	Sealer, hardener and dustproofer, epoxy-based, 125 SF/gal, 5 gallon, min				Gal.	45.50			45.50	50
1620	5 gallon pail, maximum					50			50	55
1630	Sealer, solvent-based, 250 SF/gal, 55 gallon drum					15.25			15.25	16.75
1640	5 gallon pail					17.40			17.40	19.15
1650	Sealer, water based, 350 S.F., 55 gallon drum					18.30			18.30	20
1660	5 gallon pail					21			21	23
1900	Set retarder, 100 SF/gallon, 5 gallon pail					18.35			18.35	20
2000	Waterproofing, integral 1 lb. per bag of cement				Lb.	1.38			1.38	1.52
2100	Powdered metallic, 40 lbs. per 100 S.F., minimum					1.63			1.63	1.79
2120	Maximum					2.28			2.28	2.51
3000	For integral colors, 2500 psi (5 bag mix)									
3100	Red, yellow or brown, 1.8 lb. per bag, add				C.Y.	21.50			21.50	24
3200	9.4 lb. per bag, add					113			113	125
3400	Black, 1.8 lb. per bag, add					30.50			30.50	33.50
3500	7.5 lb. per bag, add					128			128	140
3700	Green, 1.8 lb. per bag, add					43			43	47.50
3800	7.5 lb. per bag, add					180			180	198

03 05 13.85 Winter Protection

		Crew	Daily Output	Labor-Hours	Unit	Material	2009 Bare Costs Labor	Equipment	Total	Total Incl O&P
0010	**WINTER PROTECTION**									
0012	For heated ready mix, add, minimum				C.Y.	5.25			5.25	5.80
0050	Maximum				"	6.60			6.60	7.25
0100	Temporary heat to protect concrete, 24 hours, minimum	2 Clab	50	.320	M.S.F.	520	10.10		530.10	590
0150	Maximum	"	25	.640	"	680	20		700	780
0200	Temporary shelter for slab on grade, wood frame/polyethylene sheeting									
0201	Build or remove, minimum	2 Carp	10	1.600	M.S.F.	245	64		309	375
0210	Maximum	"	3	5.333	"	293	213		506	670
0710	Electrically, heated pads, 15 watts/S.F., 20 uses, minimum				S.F.	.22			.22	.24
0800	Maximum				"	.36			.36	.40

03 11 Concrete Forming

03 11 13 – Structural Cast-In-Place Concrete Forming

03 11 13.20 Forms In Place, Beams and Girders

			Crew	Daily Output	Labor-Hours	Unit	Material	2009 Bare Costs Labor	Equipment	Total	Total Incl O&P
0010	**FORMS IN PLACE, BEAMS AND GIRDERS**	R031113-40									
0500	Exterior spandrel, job-built plywood, 12" wide, 1 use	R031113-60	C-2	225	.213	SFCA	3.71	8.30		12.01	17.80
0650	4 use			310	.155		1.20	6		7.20	11.30
1000	18" wide, 1 use			250	.192		3.01	7.45		10.46	15.65
1150	4 use			315	.152		.99	5.95		6.94	10.90
1500	24" wide, 1 use			265	.181		2.75	7.05		9.80	14.70
1650	4 use			325	.148		.89	5.75		6.64	10.50
2000	Interior beam, job-built plywood, 12" wide, 1 use			300	.160		4.30	6.20		10.50	15
2150	4 use			377	.127		1.40	4.95		6.35	9.70
2500	24" wide, 1 use			320	.150		2.80	5.85		8.65	12.75
2650	4 use			395	.122		.90	4.73		5.63	8.80
3000	Encasing steel beam, hung, job-built plywood, 1 use			325	.148		3.71	5.75		9.46	13.60
3150	4 use			430	.112		1.21	4.34		5.55	8.50
9000	Minimum labor/equipment charge		2 Carp	2	8	Job		320		320	530

03 11 13.25 Forms In Place, Columns

			Crew	Daily Output	Labor-Hours	Unit	Material	2009 Bare Costs Labor	Equipment	Total	Total Incl O&P
0010	**FORMS IN PLACE, COLUMNS**	R031113-40									
0500	Round fiberglass, 4 use per mo., rent, 12" diameter		C-1	160	.200	L.F.	7.10	7.55		14.65	20.50

03 11 Concrete Forming

03 11 13 – Structural Cast-In-Place Concrete Forming

03 11 13.25 Forms In Place, Columns

		Crew	Daily Output	Labor-Hours	Unit	Material	2009 Bare Costs Labor	Equipment	Total	Total Incl O&P
0550	16" diameter R031113-60	C-1	150	.213	L.F.	8.40	8.10		16.50	22.50
0600	18" diameter		140	.229		9.45	8.65		18.10	24.50
0650	24" diameter		135	.237		11.75	8.95		20.70	28
0700	28" diameter		130	.246		13.10	9.30		22.40	30
0800	30" diameter		125	.256		13.70	9.70		23.40	31
0850	36" diameter		120	.267		18.25	10.10		28.35	36.50
1500	Round fiber tube, recycled paper, 1 use, 8" diameter G		155	.206		1.39	7.80		9.19	14.45
1550	10" diameter G		155	.206		1.98	7.80		9.78	15.10
1600	12" diameter G		150	.213		2.33	8.10		10.43	15.90
1650	14" diameter G		145	.221		3.21	8.35		11.56	17.35
1700	16" diameter G		140	.229		4	8.65		12.65	18.70
1720	18" diameter G		140	.229		4.68	8.65		13.33	19.45
1750	20" diameter G		135	.237		6.05	8.95		15	21.50
1800	24" diameter G		130	.246		7.95	9.30		17.25	24
1850	30" diameter G		125	.256		11.65	9.70		21.35	29
1900	36" diameter G		115	.278		14.90	10.55		25.45	34
1950	42" diameter G		100	.320		35	12.10		47.10	58.50
2000	48" diameter G		85	.376		41.50	14.25		55.75	69.50
2200	For seamless type, add					15%				
3000	Round, steel, 4 use per mo., rent, regular duty, 14" diameter G	C-1	145	.221	L.F.	15.15	8.35		23.50	30.50
3050	16" diameter G		125	.256		15.50	9.70		25.20	33
3100	Heavy duty, 20" diameter G		105	.305		17	11.55		28.55	38
3150	24" diameter G		85	.376		18.65	14.25		32.90	44
3200	30" diameter G		70	.457		21.50	17.30		38.80	52
3250	36" diameter G		60	.533		23	20		43	59
3300	48" diameter G		50	.640		68	24		92	115
3350	60" diameter G		45	.711		42	27		69	90.50
4500	For second and succeeding months, deduct					50%				
5000	Job-built plywood, 8" x 8" columns, 1 use	C-1	165	.194	SFCA	2.41	7.35		9.76	14.75
5050	2 use		195	.164		1.37	6.20		7.57	11.75
5100	3 use		210	.152		.96	5.75		6.71	10.55
5150	4 use		215	.149		.79	5.65		6.44	10.15
5500	12" x 12" columns, 1 use		180	.178		2.27	6.75		9.02	13.60
5550	2 use		210	.152		1.25	5.75		7	10.85
5600	3 use		220	.145		.91	5.50		6.41	10.10
5650	4 use		225	.142		.74	5.40		6.14	9.70
6000	16" x 16" columns, 1 use		185	.173		2.24	6.55		8.79	13.25
6050	2 use		215	.149		1.19	5.65		6.84	10.60
6100	3 use		230	.139		.90	5.25		6.15	9.70
6150	4 use		235	.136		.73	5.15		5.88	9.30
6500	24" x 24" columns, 1 use		190	.168		2.49	6.40		8.89	13.30
6550	2 use		216	.148		1.37	5.60		6.97	10.75
6600	3 use		230	.139		.99	5.25		6.24	9.80
6650	4 use		238	.134		.81	5.10		5.91	9.30
7000	36" x 36" columns, 1 use		200	.160		1.81	6.05		7.86	12
7050	2 use		230	.139		1.03	5.25		6.28	9.85
7100	3 use		245	.131		.73	4.94		5.67	8.95
7150	4 use		250	.128		.59	4.85		5.44	8.65
7400	Steel framed plywood, based on 50 uses of purchased									
7420	forms, and 4 uses of bracing lumber									
7500	8" x 8" column	C-1	340	.094	SFCA	4.53	3.56		8.09	10.90
7550	10" x 10"		350	.091		3.99	3.46		7.45	10.10
7600	12" x 12"		370	.086		3.38	3.27		6.65	9.10

03 11 Concrete Forming

03 11 13 – Structural Cast-In-Place Concrete Forming

03 11 13.25 Forms In Place, Columns

		Crew	Daily Output	Labor-Hours	Unit	Material	2009 Bare Costs Labor	Equipment	Total	Total Incl O&P
7650	16" x 16"	C-1	400	.080	SFCA	2.63	3.03		5.66	7.90
7700	20" x 20"		420	.076		2.35	2.88		5.23	7.35
7750	24" x 24"		440	.073		1.67	2.75		4.42	6.40
7755	30" x 30"		440	.073		2.14	2.75		4.89	6.90
7760	36" x 36"		460	.070		1.90	2.63		4.53	6.45
9000	Minimum labor/equipment charge	2 Carp	2	8	Job		320		320	530

03 11 13.35 Forms In Place, Elevated Slabs

		Crew	Daily Output	Labor-Hours	Unit	Material	2009 Bare Costs Labor	Equipment	Total	Total Incl O&P
0010	**FORMS IN PLACE, ELEVATED SLABS** R031113-40									
1000	Flat plate, job-built plywood, to 15' high, 1 use	C-2	470	.102	S.F.	4.53	3.97		8.50	11.55
1150	4 use R031113-60		560	.086		1.47	3.33		4.80	7.10
1500	15' to 20' high ceilings, 4 use		495	.097		1.67	3.77		5.44	8.10
2000	Flat slab, drop panels, job-built plywood, to 15' high, 1 use		449	.107		4.77	4.16		8.93	12.10
2150	4 use		544	.088		1.55	3.43		4.98	7.35
2250	15' to 20' high ceilings, 4 use		480	.100		5.10	3.89		8.99	12
3500	Floor slab, with 1-way joist pans, 1 use		415	.116		6.50	4.50		11	14.60
3650	4 use		500	.096		3.26	3.73		6.99	9.75
4500	With 2-way joist domes, 1 use		405	.119		6.25	4.61		10.86	14.45
4550	4 use		470	.102		3.02	3.97		6.99	9.85
5000	Box out for slab openings, over 16" deep, 1 use		190	.253	SFCA	5	9.80		14.80	21.50
5050	2 use		240	.200	"	2.75	7.80		10.55	15.90
5500	Shallow slab box outs, to 10 S.F.		42	1.143	Ea.	9.50	44.50		54	84
5550	Over 10 S.F. (use perimeter)		600	.080	L.F.	1.27	3.11		4.38	6.55
6000	Bulkhead forms for slab, with keyway, 1 use, 2 piece		500	.096		2.30	3.73		6.03	8.70
6100	3 piece (see also edge forms)		460	.104		2.36	4.06		6.42	9.30
6200	Slab bulkhead form, 4-1/2" high, exp metal, w/ keyway & stakes [G]	C-1	1200	.027		2.78	1.01		3.79	4.73
6210	5-1/2" high [G]		1100	.029		3.23	1.10		4.33	5.35
6215	7-1/2" high [G]		960	.033		4.26	1.26		5.52	6.75
6220	9-1/2" high [G]		840	.038		4.84	1.44		6.28	7.70
6500	Curb forms, wood, 6" to 12" high, on elevated slabs, 1 use		180	.178	SFCA	1.27	6.75		8.02	12.50
6550	2 use		205	.156		.70	5.90		6.60	10.50
6600	3 use		220	.145		.51	5.50		6.01	9.65
6650	4 use		225	.142		.41	5.40		5.81	9.35
7000	Edge forms to 6" high, on elevated slab, 4 use		500	.064	L.F.	.17	2.42		2.59	4.19
7500	Depressed area forms to 12" high, 4 use		300	.107		1.12	4.04		5.16	7.90
7550	12" to 24" high, 4 use		175	.183		1.52	6.90		8.42	13.10
8000	Perimeter deck and rail for elevated slabs, straight		90	.356		13.70	13.45		27.15	37
8050	Curved		65	.492		18.75	18.65		37.40	51.50
8500	Void forms, round fiber, 3" diameter [G]		450	.071		.97	2.69		3.66	5.50
8650	8" diameter [G]		375	.085		3.85	3.23		7.08	9.60
9000	Minimum labor/equipment charge	2 Carp	2	8	Job		320		320	530

03 11 13.40 Forms In Place, Equipment Foundations

		Crew	Daily Output	Labor-Hours	Unit	Material	2009 Bare Costs Labor	Equipment	Total	Total Incl O&P
0010	**FORMS IN PLACE, EQUIPMENT FOUNDATIONS** R031113-40									
0020	1 use	C-2	160	.300	SFCA	3.60	11.65		15.25	23
0050	2 use R031113-60		190	.253		1.98	9.80		11.78	18.40
0100	3 use		200	.240		1.44	9.35		10.79	17
0150	4 use		205	.234		1.17	9.10		10.27	16.35
9000	Minimum labor/equipment charge	1 Carp	3	2.667	Job		107		107	176

03 11 13.45 Forms In Place, Footings

		Crew	Daily Output	Labor-Hours	Unit	Material	2009 Bare Costs Labor	Equipment	Total	Total Incl O&P
0010	**FORMS IN PLACE, FOOTINGS** R031113-40									
0020	Continuous wall, plywood, 1 use	C-1	375	.085	SFCA	7.45	3.23		10.68	13.55
0150	4 use R031113-60	"	485	.066	"	2.42	2.50		4.92	6.80
1500	Keyway, 4 use, tapered wood, 2" x 4"	1 Carp	530	.015	L.F.	.16	.60		.76	1.17

03 11 Concrete Forming

03 11 13 – Structural Cast-In-Place Concrete Forming

03 11 13.45 Forms In Place, Footings

		Crew	Daily Output	Labor-Hours	Unit	Material	2009 Bare Costs Labor	Equipment	Total	Total Incl O&P
1550	2" x 6"	1 Carp	500	.016	L.F.	.26	.64		.90	1.35
3000	Pile cap, square or rectangular, job-built plywood, 1 use	C-1	290	.110	SFCA	2.46	4.18		6.64	9.60
3150	4 use		383	.084		.80	3.16		3.96	6.10
5000	Spread footings, job-built lumber, 1 use		305	.105		2.16	3.97		6.13	8.95
5150	4 use	↓	414	.077	↓	.70	2.93		3.63	5.60
9000	Minimum labor/equipment charge	1 Carp	3	2.667	Job		107		107	176

03 11 13.47 Forms In Place, Gas Station Forms

			Crew	Daily Output	Labor-Hours	Unit	Material	2009 Bare Costs Labor	Equipment	Total	Total Incl O&P
0010	**FORMS IN PLACE, GAS STATION FORMS**										
0050	Curb fascia, with template, 12 ga. steel, left in place, 9" high	G	1 Carp	50	.160	L.F.	15.70	6.40		22.10	28
1000	Sign or light bases, 18" diameter, 9" high	G		9	.889	Ea.	99	35.50		134.50	168
1050	30" diameter, 13" high	G		8	1	"	157	40		197	239
1990	Minimum labor/equipment charge		↓	2	4	Job		160		160	264
2000	Island forms, 10' long, 9" high, 3'- 6" wide	G	C-1	10	3.200	Ea.	440	121		561	685
2050	4' wide	G		9	3.556		455	135		590	720
2500	20' long, 9" high, 4' wide	G		6	5.333		730	202		932	1,125
2550	5' wide	G	↓	5	6.400	↓	760	242		1,002	1,225
9000	Minimum labor/equipment charge		1 Carp	3	2.667	Job		107		107	176

03 11 13.50 Forms In Place, Grade Beam

			Crew	Daily Output	Labor-Hours	Unit	Material	2009 Bare Costs Labor	Equipment	Total	Total Incl O&P
0010	**FORMS IN PLACE, GRADE BEAM**	R031113-40									
0020	Job-built plywood, 1 use		C-2	530	.091	SFCA	3.25	3.52		6.77	9.40
0150	4 use	R031113-60	"	605	.079	"	1.06	3.09		4.15	6.25
9000	Minimum labor/equipment charge		2 Carp	2	8	Job		320		320	530

03 11 13.65 Forms In Place, Slab On Grade

			Crew	Daily Output	Labor-Hours	Unit	Material	2009 Bare Costs Labor	Equipment	Total	Total Incl O&P
0010	**FORMS IN PLACE, SLAB ON GRADE**	R031113-40									
1000	Bulkhead forms w/keyway, wood, 6" high, 1 use	R031113-60	C-1	510	.063	L.F.	.84	2.38		3.22	4.84
1400	Bulkhead form for slab, 4-1/2" high, exp metal, incl keyway & stakes	G		1200	.027		2.78	1.01		3.79	4.73
1410	5-1/2" high	G		1100	.029		3.23	1.10		4.33	5.35
1420	7-1/2" high	G		960	.033		4.26	1.26		5.52	6.75
1430	9-1/2" high	G		840	.038	↓	4.84	1.44		6.28	7.70
2000	Curb forms, wood, 6" to 12" high, on grade, 1 use			215	.149	SFCA	2.83	5.65		8.48	12.40
2150	4 use			275	.116	"	.92	4.41		5.33	8.25
3000	Edge forms, wood, 4 use, on grade, to 6" high			600	.053	L.F.	.38	2.02		2.40	3.75
3050	7" to 12" high			435	.074	SFCA	.74	2.79		3.53	5.40
3500	For depressed slabs, 4 use, to 12" high			300	.107	L.F.	.56	4.04		4.60	7.25
3550	To 24" high			175	.183		.76	6.90		7.66	12.30
4000	For slab blockouts, to 12" high, 1 use			200	.160		.63	6.05		6.68	10.70
4050	To 24" high, 1 use			120	.267		.80	10.10		10.90	17.55
4100	Plastic (extruded), to 6" high, multiple use, on grade		↓	800	.040	↓	5.50	1.51		7.01	8.55
5000	Screed, 24 ga. metal key joint, see Div. 03 15 05.25										
5020	Wood, incl. wood stakes, 1" x 3"		C-1	900	.036	L.F.	.68	1.35		2.03	2.97
5050	2" x 4"			900	.036	"	.63	1.35		1.98	2.91
6000	Trench forms in floor, wood, 1 use			160	.200	SFCA	1.48	7.55		9.03	14.10
6150	4 use			185	.173	"	.48	6.55		7.03	11.35
8760	Void form, corrugated fiberboard, 6" x 12", 10' long	G	↓	240	.133	S.F.	.84	5.05		5.89	9.25
9000	Minimum labor/equipment charge		1 Carp	2	4	Job		160		160	264

03 11 13.85 Forms In Place, Walls

			Crew	Daily Output	Labor-Hours	Unit	Material	2009 Bare Costs Labor	Equipment	Total	Total Incl O&P
0010	**FORMS IN PLACE, WALLS**	R031113-10									
0100	Box out for wall openings, to 16" thick, to 10 S.F.		C-2	24	2	Ea.	21.50	78		99.50	152
0150	Over 10 S.F. (use perimeter)	R031113-40	"	280	.171	L.F.	1.78	6.65		8.43	12.95
0250	Brick shelf, 4" w, add to wall forms, use wall area abv shelf										
0260	1 use	R031113-60	C-2	240	.200	SFCA	1.87	7.80		9.67	14.90
0350	4 use		↓	300	.160	"	.75	6.20		6.95	11.05

03 11 13.85 Forms In Place, Walls		Crew	Daily Output	Labor-Hours	Unit	Material	2009 Bare Costs Labor	Equipment	Total	Total Incl O&P
0500	Bulkhead, wood with keyway, 1 use, 2 piece	C-2	265	.181	L.F.	1.68	7.05		8.73	13.50
0600	Bulkhead forms with keyway, 1 piece expanded metal, 8" wall G	C-1	1000	.032		4.26	1.21		5.47	6.70
0610	10" wall G		800	.040		4.84	1.51		6.35	7.80
0620	12" wall G		525	.061		5.80	2.31		8.11	10.20
0700	Buttress, to 8' high, 1 use	C-2	350	.137	SFCA	7.05	5.35		12.40	16.55
0850	4 use		480	.100	"	2.33	3.89		6.22	8.95
1000	Corbel or haunch, to 12" wide, add to wall forms, 1 use		150	.320	L.F.	1.95	12.45		14.40	22.50
1150	4 use		180	.267	"	.63	10.35		10.98	17.80
2000	Wall, job-built plywood, to 8' high, 1 use		370	.130	SFCA	2.62	5.05		7.67	11.25
2050	2 use		435	.110		1.61	4.29		5.90	8.85
2100	3 use		495	.097		1.17	3.77		4.94	7.55
2150	4 use		505	.095		.95	3.70		4.65	7.15
2400	Over 8' to 16' high, 1 use		280	.171		9.40	6.65		16.05	21.50
2450	2 use		345	.139		1.34	5.40		6.74	10.40
2500	3 use		375	.128		.96	4.98		5.94	9.25
2550	4 use		395	.122		.78	4.73		5.51	8.65
2700	Over 16' high, 1 use		235	.204		2.71	7.95		10.66	16.10
2750	2 use		290	.166		1.49	6.45		7.94	12.30
2800	3 use		315	.152		1.08	5.95		7.03	11
2850	4 use		330	.145		.88	5.65		6.53	10.30
3000	For architectural finish, add		1820	.026		2.72	1.03		3.75	4.68
4000	Radial, smooth curved, job-built plywood, 1 use		245	.196		2.57	7.60		10.17	15.45
4150	4 use		335	.143		.85	5.55		6.40	10.15
4200	Below grade, job-built plywood, 1 use		225	.213		4.52	8.30		12.82	18.70
4210	2 use		225	.213		2.52	8.30		10.82	16.45
4220	3 use		225	.213		1.94	8.30		10.24	15.85
4230	4 use		225	.213		1.46	8.30		9.76	15.30
4600	Retaining wall, battered, job-built plyw'd, to 8' high, 1 use		300	.160		2	6.20		8.20	12.45
4750	4 use		390	.123		.50	4.79		5.29	8.45
4900	Over 8' to 16' high, 1 use		240	.200		2.14	7.80		9.94	15.20
5050	4 use		320	.150		.69	5.85		6.54	10.40
5750	Liners for forms (add to wall forms), A.B.S. plastic									
5800	Aged wood, 4" wide, 1 use	1 Carp	256	.031	SFCA	3.24	1.25		4.49	5.60
5820	2 use		256	.031		1.78	1.25		3.03	4.02
5830	3 use		256	.031		1.30	1.25		2.55	3.49
5840	4 use		256	.031		1.05	1.25		2.30	3.22
5900	Fractured rope rib, 1 use		192	.042		4.71	1.66		6.37	7.95
5925	2 use		192	.042		2.59	1.66		4.25	5.60
5950	3 use		192	.042		1.88	1.66		3.54	4.82
6000	4 use		192	.042		1.53	1.66		3.19	4.43
6100	Ribbed, 3/4" deep x 1-1/2" O.C., 1 use		224	.036		2.72	1.43		4.15	5.35
6125	2 use		224	.036		1.50	1.43		2.93	4.01
6150	3 use		224	.036		1.09	1.43		2.52	3.56
6200	4 use		224	.036		.88	1.43		2.31	3.33
6300	Rustic brick pattern, 1 use		224	.036		3.24	1.43		4.67	5.90
6325	2 use		224	.036		1.78	1.43		3.21	4.32
6350	3 use		224	.036		1.30	1.43		2.73	3.79
6400	4 use		224	.036		1.05	1.43		2.48	3.52
6500	3/8" striated, random, 1 use		224	.036		2.12	1.43		3.55	4.69
6525	2 use		224	.036		1.17	1.43		2.60	3.64
6550	3 use		224	.036		.85	1.43		2.28	3.29
6600	4 use		224	.036		.69	1.43		2.12	3.12
7500	Lintel or sill forms, 1 use		30	.267		2.66	10.65		13.31	20.50

03 11 Concrete Forming

03 11 13 – Structural Cast-In-Place Concrete Forming

03 11 13.85 Forms In Place, Walls

		Crew	Daily Output	Labor-Hours	Unit	Material	2009 Bare Costs Labor	Equipment	Total	Total Incl O&P
7560	4 use	1 Carp	37	.216	SFCA	.86	8.65		9.51	15.20
7800	Modular prefabricated plywood, based on 20 uses of purchased									
7820	forms, and 4 uses of bracing lumber									
7860	To 8' high	C-2	800	.060	SFCA	1.47	2.33		3.80	5.45
8060	Over 8' to 16' high		600	.080		1.51	3.11		4.62	6.80
8600	Pilasters, 1 use		270	.178		3.07	6.90		9.97	14.80
8660	4 use		385	.125		1	4.85		5.85	9.10
9475	For elevated walls, add						10%			
9480	For battered walls, 1 side battered, add					10%	10%			
9485	For battered walls, 2 sides battered, add					15%	15%			
9900	Minimum labor/equipment charge	2 Carp	2	8	Job		320		320	530

03 15 Concrete Accessories

03 15 05 – Concrete Forming Accessories

03 15 05.25 Expansion Joints

			Crew	Daily Output	Labor-Hours	Unit	Material	2009 Bare Costs Labor	Equipment	Total	Total Incl O&P
0010	**EXPANSION JOINTS**										
0020	Keyed, cold, 24 ga., incl. stakes, 3-1/2" high	G	1 Carp	200	.040	L.F.	2.44	1.60		4.04	5.30
0050	4-1/2" high	G		200	.040		2.78	1.60		4.38	5.70
0100	5-1/2" high	G		195	.041		3.23	1.64		4.87	6.25
0150	7-1/2" high	G		190	.042		4.26	1.68		5.94	7.45
0160	9-1/2" high	G		185	.043		4.84	1.73		6.57	8.15
0300	Poured asphalt, plain, 1/2" x 1"		1 Clab	450	.018		.51	.56		1.07	1.49
0350	1" x 2"			400	.020		2.05	.63		2.68	3.30
0500	Neoprene, liquid, cold applied, 1/2" x 1"			450	.018		2.21	.56		2.77	3.36
0550	1" x 2"			400	.020		8.85	.63		9.48	10.75
0700	Polyurethane, poured, 2 part, 1/2" x 1"			400	.020		1.40	.63		2.03	2.58
0750	1" x 2"			350	.023		5.60	.72		6.32	7.35
0900	Rubberized asphalt, hot or cold applied, 1/2" x 1"			450	.018		.76	.56		1.32	1.77
0950	1" x 2"			400	.020		3.05	.63		3.68	4.40
1100	Hot applied, fuel resistant, 1/2" x 1"			450	.018		1.14	.56		1.70	2.18
1150	1" x 2"			400	.020		4.58	.63		5.21	6.10
2000	Premolded, bituminous fiber, 1/2" x 6"		1 Carp	375	.021		.40	.85		1.25	1.85
2050	1" x 12"			300	.027		1.91	1.07		2.98	3.86
2140	Concrete expansion joint, recycled paper and fiber, 1/2" x 6"	G		390	.021		.35	.82		1.17	1.73
2150	1/2" x 12"	G		360	.022		.69	.89		1.58	2.23
2250	Cork with resin binder, 1/2" x 6"			375	.021		1.81	.85		2.66	3.40
2300	1" x 12"			300	.027		7.55	1.07		8.62	10.05
2500	Neoprene sponge, closed cell, 1/2" x 6"			375	.021		1.88	.85		2.73	3.48
2550	1" x 12"			300	.027		7.05	1.07		8.12	9.50
2750	Polyethylene foam, 1/2" x 6"			375	.021		.59	.85		1.44	2.06
2800	1" x 12"			300	.027		1.92	1.07		2.99	3.87
3000	Polyethylene backer rod, 3/8" diameter			460	.017		.06	.69		.75	1.21
3050	3/4" diameter			460	.017		.09	.69		.78	1.25
3100	1" diameter			460	.017		.16	.69		.85	1.33
3500	Polyurethane foam, with polybutylene, 1/2" x 1/2"			475	.017		.95	.67		1.62	2.16
3550	1" x 1"			450	.018		2.40	.71		3.11	3.81
3750	Polyurethane foam, regular, closed cell, 1/2" x 6"			375	.021		.70	.85		1.55	2.18
3800	1" x 12"			300	.027		2.50	1.07		3.57	4.51
4000	Polyvinyl chloride foam, closed cell, 1/2" x 6"			375	.021		1.90	.85		2.75	3.50
4050	1" x 12"			300	.027		6.55	1.07		7.62	8.95
4250	Rubber, gray sponge, 1/2" x 6"			375	.021		1.60	.85		2.45	3.17

03 15 Concrete Accessories

03 15 05 – Concrete Forming Accessories

03 15 05.25 Expansion Joints

	03 15 05.25 Expansion Joints		Crew	Daily Output	Labor-Hours	Unit	Material	2009 Bare Costs Labor	Equipment	Total	Total Incl O&P
4300	1" x 12"		1 Carp	300	.027	L.F.	5.75	1.07		6.82	8.10
4500	Lead wool for joints					Lb.	8.80			8.80	9.70
5000	For installation in walls, add							75%			
5250	For installation in boxouts, add							25%			

03 15 05.70 Shores

	03 15 05.70 Shores		Crew	Daily Output	Labor-Hours	Unit	Material	2009 Bare Costs Labor	Equipment	Total	Total Incl O&P
0010	**SHORES**										
0020	Erect and strip, by hand, horizontal members										
0500	Aluminum joists and stringers	G	2 Carp	60	.267	Ea.		10.65		10.65	17.60
0600	Steel, adjustable beams	G		45	.356			14.20		14.20	23.50
0700	Wood joists			50	.320			12.80		12.80	21
0800	Wood stringers			30	.533			21.50		21.50	35
1000	Vertical members to 10' high	G		55	.291			11.60		11.60	19.20
1050	To 13' high	G		50	.320			12.80		12.80	21
1100	To 16' high	G		45	.356			14.20		14.20	23.50
1500	Reshoring	G		1400	.011	S.F.	.50	.46		.96	1.30
1600	Flying truss system	G	C-17D	9600	.009	SFCA		.36	.07	.43	.66
1760	Horizontal, aluminum joists, 6-1/4" high x 5' to 21' span, buy	G				L.F.	36			36	39.50
1770	Beams, 7-1/4" high x 4' to 30' span	G				"	47.50			47.50	52.50
1810	Horizontal, steel beam, adjustable, 4' to 7' span, buy	G				Ea.	330			330	360
1830	6' to 10' span	G					330			330	360
1920	9' to 15' span	G					545			545	600
1940	12' to 20' span	G					650			650	715
1970	Steel stringer, W8x10, 4' to 16' span, buy	G				L.F.	25.50			25.50	28
3000	Rent for job duration, aluminum joist @ 2' O.C., per mo	G				SF Flr.	.90			.90	.99
3050	Steel W8x10	G					.64			.64	.70
3060	Steel adjustable	G					.82			.82	.90
3500	#1 post shore, steel, 5'-7" to 9'-6" high, 10000# cap., buy	G				Ea.	410			410	450
3550	#2 post shore, 7'-3" to 12'-10" high, 7800# capacity	G					465			465	515
3600	#3 post shore, 8'-10" to 16'-1" high, 3800# capacity	G					540			540	595
5010	Frame shoring systems, steel, 12000#/leg, buy										
5040	Frame, 2' wide x 6' high	G				Ea.	106			106	117
5250	X-brace	G					18			18	19.80
5550	Base plate	G					17.60			17.60	19.35
5600	Screw jack	G					38			38	41.50
5650	U-head, 8" x 8"	G					21.50			21.50	24

03 15 05.85 Stair Tread Inserts

	03 15 05.85 Stair Tread Inserts		Crew	Daily Output	Labor-Hours	Unit	Material	2009 Bare Costs Labor	Equipment	Total	Total Incl O&P
0010	**STAIR TREAD INSERTS**										
0015	Cast iron, abrasive, 3" wide	G	1 Carp	90	.089	L.F.	10.65	3.55		14.20	17.55
0020	4" wide	G		80	.100		14.15	4		18.15	22
0040	6" wide	G		75	.107		21.50	4.26		25.76	30.50
0050	9" wide	G		70	.114		32	4.57		36.57	42.50
0100	12" wide	G		65	.123		42.50	4.92		47.42	55
0300	Cast aluminum, compared to cast iron, deduct						10%				
0500	Extruded aluminum safety tread, 3" wide	G	1 Carp	75	.107		12.10	4.26		16.36	20.50
0550	4" wide	G		75	.107		16.15	4.26		20.41	25
0600	6" wide	G		75	.107		24	4.26		28.26	33.50
0650	9" wide to resurface stairs	G		70	.114		36.50	4.57		41.07	47.50
1700	Cement fill for pan-type metal treads, plain		1 Cefi	115	.070	S.F.	1.87	2.66		4.53	6.25
1750	Non-slip		"	100	.080	"	2.06	3.06		5.12	7.05

03 15 13 – Waterstops

03 15 13.50 Waterstops

0010	**WATERSTOPS**, PVC and Rubber										

03 15 Concrete Accessories

03 15 13 – Waterstops

03 15 13.50 Waterstops		Crew	Daily Output	Labor- Hours	Unit	Material	2009 Bare Costs Labor	Equipment	Total	Total Incl O&P
0020	PVC, ribbed 3/16" thick, 4" wide	1 Carp	155	.052	L.F.	1.08	2.06		3.14	4.59
0050	6" wide		145	.055		1.80	2.20		4	5.60
0500	With center bulb, 6" wide, 3/16" thick		135	.059		1.52	2.37		3.89	5.60
0550	3/8" thick		130	.062		3.20	2.46		5.66	7.60
0600	9" wide x 3/8" thick		125	.064		4.50	2.56		7.06	9.15

03 21 Reinforcing Steel

03 21 10 – Uncoated Reinforcing Steel

03 21 10.60 Reinforcing In Place

			Crew	Daily Output	Labor- Hours	Unit	Material	2009 Bare Costs Labor	Equipment	Total	Total Incl O&P
0015	**REINFORCING IN PLACE** A615 Grade 60, incl. access. labor	R032110-10									
0030	Made from recycled materials										
0102	Beams & Girders, #3 to #7	G	4 Rodm	3200	.010	Lb.	.81	.45		1.26	1.65
0152	#8 to #18	G		5400	.006		.81	.26		1.07	1.34
0202	Columns, #3 to #7	G		3000	.011		.81	.48		1.29	1.70
0252	#8 to #18	G		4600	.007		.81	.31		1.12	1.42
0402	Elevated slabs, #4 to #7	G		5800	.006		.84	.25		1.09	1.35
0502	Footings, #4 to #7	G		4200	.008		.81	.34		1.15	1.47
0552	#8 to #18	G		7200	.004		.81	.20		1.01	1.23
0602	Slab on grade, #3 to #7	G		4200	.008		.78	.34		1.12	1.43
0702	Walls, #3 to #7	G		6000	.005		.81	.24		1.05	1.29
0752	#8 to #18	G		8000	.004		.81	.18		.99	1.19
2000	Unloading & sorting, add to above		C-5	100	.560	Ton		24.50	7.90	32.40	49
2200	Crane cost for handling, add to above, minimum			135	.415			18	5.85	23.85	36.50
2210	Average			92	.609			26.50	8.60	35.10	53.50
2220	Maximum			35	1.600			69.50	22.50	92	141
2400	Dowels, 2 feet long, deformed, #3	G	2 Rodm	520	.031	Ea.	.64	1.37		2.01	3.05
2410	#4	G		480	.033		1.14	1.48		2.62	3.78
2420	#5	G		435	.037		1.78	1.64		3.42	4.75
2430	#6	G		360	.044		2.56	1.98		4.54	6.20
2450	Longer and heavier dowels, add	G		725	.022	Lb.	.85	.98		1.83	2.62
2500	Smooth dowels, 12" long, 1/4" or 3/8" diameter	G		140	.114	Ea.	1.21	5.10		6.31	10.05
2520	5/8" diameter	G		125	.128		2.11	5.70		7.81	12
2530	3/4" diameter	G		110	.145		2.62	6.50		9.12	13.95
2600	Dowel sleeves for CIP concrete, 2-part system										
2610	Sleeve base, plastic, for 5/8" smooth dowel sleeve, fasten to edge form		1 Rodm	200	.040	Ea.	.52	1.78		2.30	3.61
2615	Sleeve, plastic, 12" long, for 5/8" smooth dowel, snap onto base			400	.020		1.03	.89		1.92	2.65
2620	Sleeve base, for 3/4" smooth dowel sleeve			175	.046		.49	2.04		2.53	4.01
2625	Sleeve, 12" long, for 3/4" smooth dowel			350	.023		1.08	1.02		2.10	2.93
2630	Sleeve base, for 1" smooth dowel sleeve			150	.053		.64	2.38		3.02	4.75
2635	Sleeve, 12" long, for 1" smooth dowel			300	.027		1.84	1.19		3.03	4.05
2700	Dowel caps, visual warning only, plastic, #3 to #8		2 Rodm	800	.020		.51	.89		1.40	2.08
2720	#8 to #18			750	.021		1.08	.95		2.03	2.81
2750	Impalement protective, plastic, #4 to #9			800	.020		1.95	.89		2.84	3.67
9000	Minimum labor/equipment charge		1 Rodm	4	2	Job		89		89	152

03 21 10.70 Glass Fiber Reinforced Polymer Bars

			Crew	Daily Output	Labor- Hours	Unit	Material	2009 Bare Costs Labor	Equipment	Total	Total Incl O&P
0010	**GLASS FIBER REINFORCED POLYMER BARS**										
0050	#2 bar, .043 lbs/ ft.					L.F.	.35			.35	.39
0100	#3 bar, .092 lbs/ ft.						.48			.48	.53
0150	#4 bar, .160 lbs/ ft.						.70			.70	.77
0200	#5 bar, .258 lbs/ ft.						.98			.98	1.08
0250	#6 bar, .372 lbs/ ft.						1.35			1.35	1.49

03 21 Reinforcing Steel

03 21 10 – Uncoated Reinforcing Steel

03 21 10.70 Glass Fiber Reinforced Polymer Bars	Crew	Daily Output	Labor-Hours	Unit	Material	2009 Bare Costs Labor	Equipment	Total	Total Incl O&P
0300	#7 bar, .497 lbs/ft.				L.F.	1.70			1.70	1.87
0350	#8 bar, .620 lbs/ft.					2.25			2.25	2.48
0400	#9 bar, .800 lbs/ft.					2.90			2.90	3.19
0450	#10 bar, 1.08 lbs/ft.					3.45			3.45	3.80
0500	For Bends, add per bend				Ea.	1			1	1.10

03 22 Welded Wire Fabric Reinforcing

03 22 05 – Uncoated Welded Wire Fabric

03 22 05.50 Welded Wire Fabric

			Crew	Daily Output	Labor-Hours	Unit	Material	Labor	Equipment	Total	Total Incl O&P
0010	**WELDED WIRE FABRIC** ASTM A185	R032205-30									
0030	Made from recycled materials										
0050	Sheets										
0100	6 x 6 - W1.4 x W1.4 (10 x 10) 21 lb. per C.S.F.	G	2 Rodm	35	.457	C.S.F.	18.05	20.50		38.55	54.50
0200	6 x 6 - W2.1 x W2.1 (8 x 8) 30 lb. per C.S.F.	G		31	.516		26.50	23		49.50	68
0300	6 x 6 - W2.9 x W2.9 (6 x 6) 42 lb. per C.S.F.	G		29	.552		32.50	24.50		57	78
0400	6 x 6 - W4 x W4 (4 x 4) 58 lb. per C.S.F.	G		27	.593		46.50	26.50		73	96
0500	4 x 4 - W1.4 x W1.4 (10 x 10) 31 lb. per C.S.F.	G		31	.516		26.50	23		49.50	68.50
0600	4 x 4 - W2.1 x W2.1 (8 x 8) 44 lb. per C.S.F.	G		29	.552		38	24.50		62.50	83.50
0650	4 x 4 - W2.9 x W2.9 (6 x 6) 61 lb. per C.S.F.	G		27	.593		50.50	26.50		77	101
0700	4 x 4 - W4 x W4 (4 x 4) 85 lb. per C.S.F.	G		25	.640		65.50	28.50		94	121
0750	Rolls										
0800	2 x 2 - #14 galv., 21 lb/C.S.F., beam & column wrap	G	2 Rodm	6.50	2.462	C.S.F.	66	110		176	260
0900	2 x 2 - #12 galv. for gunite reinforcing	G	"	6.50	2.462	"	65.50	110		175.50	260
9000	Minimum labor/equipment charge		1 Rodm	4	2	Job		89		89	152

03 24 Fibrous Reinforcing

03 24 05 – Reinforcing Fibers

03 24 05.30 Synthetic Fibers

						Unit	Material	Labor	Equipment	Total	Total Incl O&P
0010	**SYNTHETIC FIBERS**										
0100	Synthetic fibers, add to concrete					Lb.	4.43			4.43	4.87
0110	1-1/2 lb. per C.Y.					C.Y.	6.85			6.85	7.55

03 24 05.70 Steel Fibers

						Unit	Material	Labor	Equipment	Total	Total Incl O&P
0010	**STEEL FIBERS**										
0150	Steel fibers, add to concrete	G				Lb.	.70			.70	.77
0155	25 lb. per C.Y.	G				C.Y.	17.50			17.50	19.25
0160	50 lb. per C.Y.	G					35			35	38.50
0170	75 lb. per C.Y.	G					54			54	59.50
0180	100 lb. per C.Y.	G					70			70	77

03 30 53.40 Concrete In Place	Crew	Daily Output	Labor-Hours	Unit	Material	2009 Bare Costs Labor	Equipment	Total	Total Incl O&P
0010 **CONCRETE IN PLACE**									
0020 Including forms (4 uses), reinforcing steel, concrete, placement,									
0050 and finishing unless otherwise indicated									
0300 Beams, 5 kip per L.F., 10' span	C-14A	15.62	12.804	C.Y.	400	515	49	964	1,350
0350 25' span	"	18.55	10.782		430	430	41	901	1,225
0500 Chimney foundations, industrial, minimum	C-14C	32.22	3.476		166	133	.80	299.80	405
0510 Maximum	"	23.71	4.724		203	181	1.09	385.09	525
0700 Columns, square, 12" x 12", minimum reinforcing	C-14A	11.96	16.722		400	670	64	1,134	1,600
0720 Average reinforcing		10.13	19.743		720	790	75.50	1,585.50	2,175
0740 Maximum reinforcing		9.03	22.148		1,150	890	84.50	2,124.50	2,850
0800 16" x 16", minimum reinforcing		16.22	12.330		315	495	47	857	1,225
0820 Average reinforcing		12.57	15.911		625	640	61	1,326	1,800
0840 Maximum reinforcing		10.25	19.512		1,025	780	74.50	1,879.50	2,525
0900 24" x 24", minimum reinforcing		23.66	8.453		275	340	32.50	647.50	895
0920 Average reinforcing		17.71	11.293		570	455	43	1,068	1,425
0940 Maximum reinforcing		14.15	14.134		965	565	54	1,584	2,050
1000 36" x 36", minimum reinforcing		33.69	5.936		249	238	22.50	509.50	695
1020 Average reinforcing		23.32	8.576		505	345	33	883	1,150
1040 Maximum reinforcing		17.82	11.223		910	450	43	1,403	1,800
1200 16" diameter, minimum reinforcing		31.49	6.351		315	255	24.50	594.50	790
1220 Average reinforcing		19.12	10.460		635	420	40	1,095	1,450
1240 Maximum reinforcing		13.77	14.524		1,025	580	55.50	1,660.50	2,150
1300 20" diameter, minimum reinforcing		41.04	4.873		315	195	18.60	528.60	690
1320 Average reinforcing		24.05	8.316		610	335	32	977	1,250
1340 Maximum reinforcing		17.01	11.758		1,025	470	45	1,540	1,950
1400 24" diameter, minimum reinforcing		51.85	3.857		294	155	14.75	463.75	595
1420 Average reinforcing		27.06	7.391		620	296	28	944	1,200
1440 Maximum reinforcing		18.29	10.935		1,000	440	42	1,482	1,875
1500 36" diameter, minimum reinforcing		75.04	2.665		310	107	10.20	427.20	530
1520 Average reinforcing		37.49	5.335		595	214	20.50	829.50	1,025
1540 Maximum reinforcing		22.84	8.757		990	350	33.50	1,373.50	1,725
1900 Elevated slabs, flat slab with drops, 125 psf Sup. Load, 20' span	C-14B	38.45	5.410		299	216	19.85	534.85	705
1950 30' span		50.99	4.079		340	163	14.95	517.95	660
2100 Flat plate, 125 psf Sup. Load, 15' span		30.24	6.878		278	275	25	578	790
2150 25' span		49.60	4.194		300	168	15.40	483.40	625
2300 Waffle const., 30" domes, 125 psf Sup. Load, 20' span		37.07	5.611		267	225	20.50	512.50	685
2350 30' span		44.07	4.720		255	189	17.30	461.30	610
2500 One way joists, 30" pans, 125 psf Sup. Load, 15' span		27.38	7.597		297	305	28	630	855
2550 25' span		31.15	6.677		294	267	24.50	585.50	790
2700 One way beam & slab, 125 psf Sup. Load, 15' span		20.59	10.102		305	405	37	747	1,050
2750 25' span		28.36	7.334		297	294	27	618	840
2900 Two way beam & slab, 125 psf Sup. Load, 15' span		24.04	8.652		296	345	32	673	930
2950 25' span		35.87	5.799		256	232	21.50	509.50	690
3100 Elevated slabs including finish, not									
3110 including forms or reinforcing									
3150 Regular concrete, 4" slab	C-8	2613	.021	S.F.	1.36	.75	.28	2.39	3.01
3200 6" slab		2585	.022		2.01	.76	.29	3.06	3.75
3250 2-1/2" thick floor fill		2685	.021		.87	.73	.27	1.87	2.44
3300 Lightweight, 110# per C.F., 2-1/2" thick floor fill		2585	.022		1.17	.76	.29	2.22	2.83
3400 Cellular concrete, 1-5/8" fill, under 5000 S.F.		2000	.028		.78	.99	.37	2.14	2.86
3450 Over 10,000 S.F.		2200	.025		.75	.90	.34	1.99	2.63
3500 Add per floor for 3 to 6 stories high		31800	.002			.06	.02	.08	.13
3520 For 7 to 20 stories high		21200	.003			.09	.03	.12	.19

03 30 53.40 Concrete In Place	Crew	Daily Output	Labor-Hours	Unit	Material	2009 Bare Costs Labor	Equipment	Total	Total Incl O&P
3540	Equipment pad, 3' x 3' x 6" thick	C-14H	45	1.067	Ea.	58	42	.58	100.58	134
3550	4' x 4' x 6" thick		30	1.600		84	63	.86	147.86	197
3560	5' x 5' x 8" thick		18	2.667		140	105	1.44	246.44	330
3570	6' x 6' x 8" thick		14	3.429		188	135	1.85	324.85	430
3580	8' x 8' x 10" thick		8	6		395	236	3.24	634.24	830
3590	10' x 10' x 12" thick		5	9.600		655	380	5.20	1,040.20	1,350
3800	Footings, spread under 1 C.Y.	C-14C	28	4	C.Y.	209	153	.92	362.92	485
3825	1 C.Y to 5 C.Y.		43	2.605		221	99.50	.60	321.10	410
3850	Over 5 C.Y.		75	1.493		198	57	.34	255.34	310
3900	Footings, strip, 18" x 9", unreinforced	C-14L	40	2.400		120	89.50	.65	210.15	280
3920	18" x 9", reinforced	C-14C	35	3.200		157	122	.74	279.74	375
3925	20" x 10", unreinforced	C-14L	45	2.133		117	79.50	.58	197.08	260
3930	20" x 10", reinforced	C-14C	40	2.800		147	107	.64	254.64	340
3935	24" x 12", unreinforced	C-14L	55	1.745		116	65	.47	181.47	236
3940	24" x 12", reinforced	C-14C	48	2.333		147	89.50	.54	237.04	310
3945	36" x 12", unreinforced	C-14L	70	1.371		113	51	.37	164.37	208
3950	36" x 12", reinforced	C-14C	60	1.867		141	71.50	.43	212.93	273
4000	Foundation mat, under 10 C.Y.		38.67	2.896		255	111	.67	366.67	465
4050	Over 20 C.Y.		56.40	1.986		218	76	.46	294.46	365
4200	Wall, free-standing, 8" thick, 8' high	C-14D	45.83	4.364		195	173	16.65	384.65	520
4250	14' high		27.26	7.337		251	291	28	570	785
4260	12" thick, 8' high		64.32	3.109		175	124	11.90	310.90	410
4270	14' high		40.01	4.999		191	199	19.10	409.10	555
4300	15" thick, 8' high		80.02	2.499		166	99.50	9.55	275.05	355
4350	12' high		51.26	3.902		166	155	14.90	335.90	455
4500	18' high		48.85	4.094		191	163	15.65	369.65	495
4520	Handicap access ramp, railing both sides, 3' wide	C-14H	14.58	3.292	L.F.	300	130	1.78	431.78	545
4525	5' wide		12.22	3.928		310	155	2.12	467.12	595
4530	With 6" curb and rails both sides, 3' wide		8.55	5.614		310	221	3.03	534.03	710
4535	5' wide		7.31	6.566		315	259	3.55	577.55	775
4650	Slab on grade, not including finish, 4" thick	C-14E	60.75	1.449	C.Y.	127	57	.43	184.43	235
4700	6" thick	"	92	.957	"	121	37.50	.29	158.79	196
4751	Slab on grade, incl. troweled finish, not incl. forms									
4760	or reinforcing, over 10,000 S.F., 4" thick	C-14F	3425	.021	S.F.	1.35	.76	.01	2.12	2.70
4820	6" thick		3350	.021		1.97	.78	.01	2.76	3.42
4840	8" thick		3184	.023		2.70	.82	.01	3.53	4.27
4900	12" thick		2734	.026		4.04	.96	.01	5.01	6
4950	15" thick		2505	.029		5.10	1.04	.01	6.15	7.25
5000	Slab on grade, incl. textured finish, not incl. forms									
5001	or reinforcing, 4" thick	C-14G	2873	.019	S.F.	1.31	.70	.01	2.02	2.56
5010	6" thick		2590	.022		2.05	.77	.01	2.83	3.50
5020	8" thick		2320	.024		2.68	.86	.01	3.55	4.33
5200	Lift slab in place above the foundation, incl. forms,									
5210	reinforcing, concrete and columns, minimum	C-14B	2113	.098	S.F.	7.45	3.94	.36	11.75	15.10
5250	Average		1650	.126		8.15	5.05	.46	13.66	17.80
5300	Maximum		1500	.139		8.85	5.55	.51	14.91	19.45
5500	Lightweight, ready mix, including screed finish only,									
5510	not including forms or reinforcing									
5550	1:4 for structural roof decks	C-14B	260	.800	C.Y.	138	32	2.94	172.94	208
5600	1:6 for ground slab with radiant heat	C-14F	92	.783		143	28.50	.28	171.78	202
5650	1:3:2 with sand aggregate, roof deck	C-14B	260	.800		132	32	2.94	166.94	201
5700	Ground slab	C-14F	107	.673		132	24.50	.24	156.74	184
5900	Pile caps, incl. forms and reinf., sq. or rect., under 5 C.Y.	C-14C	54.14	2.069		172	79	.48	251.48	320

03 30 Cast-In-Place Concrete

03 30 53 – Miscellaneous Cast-In-Place Concrete

03 30 53.40 Concrete In Place

		Crew	Daily Output	Labor-Hours	Unit	Material	2009 Bare Costs Labor	Equipment	Total	Total Incl O&P
5950	Over 10 C.Y.	C-14C	75	1.493	C.Y.	167	57	.34	224.34	279
6000	Triangular or hexagonal, under 5 C.Y.		53	2.113		127	81	.49	208.49	275
6050	Over 10 C.Y.	↓	85	1.318		152	50.50	.30	202.80	251
6200	Retaining walls, gravity, 4' high see Div. 32 32	C-14D	66.20	3.021		153	120	11.55	284.55	380
6250	10' high		125	1.600		146	63.50	6.10	215.60	273
6300	Cantilever, level backfill loading, 8' high		70	2.857		175	113	10.90	298.90	390
6350	16' high	↓	91	2.198	↓	166	87.50	8.40	261.90	335
6800	Stairs, not including safety treads, free standing, 3'-6" wide	C-14H	83	.578	LF Nose	6.20	23	.31	29.51	44.50
6850	Cast on ground		125	.384	"	5	15.10	.21	20.31	30.50
7000	Stair landings, free standing		200	.240	S.F.	5.25	9.45	.13	14.83	21.50
7050	Cast on ground	↓	475	.101	"	4	3.98	.05	8.03	11
9000	Minimum labor/equipment charge	2 Carp	1	16	Job		640		640	1,050

03 31 Structural Concrete

03 31 05 – Normal Weight Structural Concrete

03 31 05.35 Normal Weight Concrete, Ready Mix

			Crew	Daily Output	Labor-Hours	Unit	Material	2009 Bare Costs Labor	Equipment	Total	Total Incl O&P
0010	**NORMAL WEIGHT CONCRETE, READY MIX**, delivered										
0012	Includes local aggregate, sand, Portland cement, and water										
0015	Excludes all additives and treatments	R033105-20									
0020	2000 psi					C.Y.	97			97	107
0100	2500 psi						98.50			98.50	108
0150	3000 psi						101			101	111
0200	3500 psi						104			104	114
0300	4000 psi						106			106	116
0350	4500 psi						109			109	120
0400	5000 psi						111			111	122
0411	6000 psi						127			127	139
0412	8000 psi						206			206	227
0413	10,000 psi						293			293	320
0414	12,000 psi						355			355	390
1000	For high early strength cement, add						10%				
1300	For winter concrete (hot water), add						5.25			5.25	5.80
1400	For hot weather concrete (ice), add						7.20			7.20	7.90
1410	For mid-range water reducer, add						3.95			3.95	4.35
1420	For high-range water reducer/superplasticizer, add						5.85			5.85	6.40
1430	For retarder, add						2			2	2.20
1440	For non-Chloride accelerator, add						4.75			4.75	5.25
1450	For Chloride accelerator, per 1%, add						2.75			2.75	3.03
1460	For fiber reinforcing, synthetic (1 Lb./C.Y.), add						6			6	6.60
1500	For Saturday delivery, add					↓	5.50			5.50	6.05
1510	For truck holding/waiting time past 1st hour per load, add						Hr.	92			92	101
1520	For short load (less than 4 C.Y.), add per load						Ea.	100			100	110
2000	For all lightweight aggregate, add						C.Y.	45%				

03 31 05.70 Placing Concrete

			Crew	Daily Output	Labor-Hours	Unit	Material	2009 Bare Costs Labor	Equipment	Total	Total Incl O&P
0010	**PLACING CONCRETE**	R033105-70									
0020	Includes labor and equipment to place and vibrate										
0050	Beams, elevated, small beams, pumped		C-20	60	1.067	C.Y.		36	13.15	49.15	73.50
0100	With crane and bucket		C-7	45	1.600			55	26.50	81.50	118
0200	Large beams, pumped		C-20	90	.711			24	8.80	32.80	48.50
0250	With crane and bucket		C-7	65	1.108			38	18.40	56.40	82
0400	Columns, square or round, 12" thick, pumped		C-20	60	1.067	↓		36	13.15	49.15	73.50

03 31 Structural Concrete

03 31 05 – Normal Weight Structural Concrete

03 31 05.70 Placing Concrete		Crew	Daily Output	Labor-Hours	Unit	Material	2009 Bare Costs		Total	Total Incl O&P
							Labor	Equipment		
0450	With crane and bucket	C-7	40	1.800	C.Y.		61.50	30	91.50	133
0600	18" thick, pumped	C-20	90	.711			24	8.80	32.80	48.50
0650	With crane and bucket	C-7	55	1.309			45	22	67	96.50
0800	24" thick, pumped	C-20	92	.696			23.50	8.60	32.10	48
0850	With crane and bucket	C-7	70	1.029			35	17.10	52.10	76
1000	36" thick, pumped	C-20	140	.457			15.50	5.65	21.15	31
1050	With crane and bucket	C-7	100	.720			24.50	11.95	36.45	53
1400	Elevated slabs, less than 6" thick, pumped	C-20	140	.457			15.50	5.65	21.15	31
1450	With crane and bucket	C-7	95	.758			26	12.60	38.60	56
1500	6" to 10" thick, pumped	C-20	160	.400			13.55	4.94	18.49	27.50
1550	With crane and bucket	C-7	110	.655			22.50	10.90	33.40	48.50
1600	Slabs over 10" thick, pumped	C-20	180	.356			12.05	4.39	16.44	24.50
1650	With crane and bucket	C-7	130	.554			18.95	9.20	28.15	40.50
1900	Footings, continuous, shallow, direct chute	C-6	120	.400			13.20	.43	13.63	22
1950	Pumped	C-20	150	.427			14.45	5.25	19.70	29.50
2000	With crane and bucket	C-7	90	.800			27.50	13.30	40.80	59
2100	Footings, continuous, deep, direct chute	C-6	140	.343			11.35	.37	11.72	18.90
2150	Pumped	C-20	160	.400			13.55	4.94	18.49	27.50
2200	With crane and bucket	C-7	110	.655			22.50	10.90	33.40	48.50
2400	Footings, spread, under 1 C.Y., direct chute	C-6	55	.873			29	.94	29.94	48
2450	Pumped	C-20	65	.985			33.50	12.15	45.65	68
2500	With crane and bucket	C-7	45	1.600			55	26.50	81.50	118
2600	Over 5 C.Y., direct chute	C-6	120	.400			13.20	.43	13.63	22
2650	Pumped	C-20	150	.427			14.45	5.25	19.70	29.50
2700	With crane and bucket	C-7	100	.720			24.50	11.95	36.45	53
2900	Foundation mats, over 20 C.Y., direct chute	C-6	350	.137			4.53	.15	4.68	7.55
2950	Pumped	C-20	400	.160			5.45	1.97	7.42	10.95
3000	With crane and bucket	C-7	300	.240			8.20	3.99	12.19	17.70
3200	Grade beams, direct chute	C-6	150	.320			10.60	.35	10.95	17.70
3250	Pumped	C-20	180	.356			12.05	4.39	16.44	24.50
3300	With crane and bucket	C-7	120	.600			20.50	10	30.50	44.50
3500	High rise, for more than 5 stories, pumped, add per story	C-20	2100	.030			1.03	.38	1.41	2.09
3510	With crane and bucket, add per story	C-7	2100	.034			1.17	.57	1.74	2.53
3700	Pile caps, under 5 C.Y., direct chute	C-6	90	.533			17.65	.58	18.23	29.50
3750	Pumped	C-20	110	.582			19.75	7.20	26.95	40
3800	With crane and bucket	C-7	80	.900			31	14.95	45.95	66.50
3850	Pile cap, 5 C.Y. to 10 C.Y., direct chute	C-6	175	.274			9.05	.30	9.35	15.15
3900	Pumped	C-20	200	.320			10.85	3.95	14.80	22
3950	With crane and bucket	C-7	150	.480			16.45	8	24.45	35.50
4000	Over 10 C.Y., direct chute	C-6	215	.223			7.40	.24	7.64	12.30
4050	Pumped	C-20	240	.267			9.05	3.29	12.34	18.30
4100	With crane and bucket	C-7	185	.389			13.30	6.45	19.75	28.50
4300	Slab on grade, up to 6" thick, direct chute	C-6	110	.436			14.40	.47	14.87	24
4350	Pumped	C-20	130	.492			16.70	6.10	22.80	33.50
4400	With crane and bucket	C-7	110	.655			22.50	10.90	33.40	48.50
4600	Over 6" thick, direct chute	C-6	165	.291			9.60	.31	9.91	16.05
4650	Pumped	C-20	185	.346			11.75	4.27	16.02	24
4700	With crane and bucket	C-7	145	.497			17	8.25	25.25	36.50
4900	Walls, 8" thick, direct chute	C-6	90	.533			17.65	.58	18.23	29.50
4950	Pumped	C-20	100	.640			21.50	7.90	29.40	44
5000	With crane and bucket	C-7	80	.900			31	14.95	45.95	66.50
5050	12" thick, direct chute	C-6	100	.480			15.85	.52	16.37	26.50
5100	Pumped	C-20	110	.582			19.75	7.20	26.95	40

03 31 Structural Concrete

03 31 05 – Normal Weight Structural Concrete

03 31 05.70 Placing Concrete

	Crew	Daily Output	Labor-Hours	Unit	Material	2009 Bare Costs Labor	Equipment	Total	Total Incl O&P
5200	With crane and bucket	C-7	90	.800	C.Y.		27.50	13.30	40.80	59
5300	15" thick, direct chute	C-6	105	.457			15.10	.49	15.59	25
5350	Pumped	C-20	120	.533			18.10	6.60	24.70	37
5400	With crane and bucket	C-7	95	.758			26	12.60	38.60	56
5600	Wheeled concrete dumping, add to placing costs above									
5610	Walking cart, 50' haul, add	C-18	32	.281	C.Y.		8.95	1.72	10.67	16.65
5620	150' haul, add		24	.375			11.95	2.30	14.25	22
5700	250' haul, add		18	.500			15.90	3.07	18.97	30
5800	Riding cart, 50' haul, add	C-19	80	.113			3.58	1.08	4.66	7.10
5810	150' haul, add		60	.150			4.77	1.44	6.21	9.50
5900	250' haul, add		45	.200			6.35	1.92	8.27	12.60
9000	Minimum labor/equipment charge	C-6	2	24	Job		795	26	821	1,325

03 35 Concrete Finishing

03 35 29 – Tooled Concrete Finishing

03 35 29.30 Finishing Floors

	Crew	Daily Output	Labor-Hours	Unit	Material	2009 Bare Costs Labor	Equipment	Total	Total Incl O&P
0010	**FINISHING FLOORS**									
0020	Manual screed finish	C-10	4800	.005	S.F.		.18		.18	.29
0100	Manual screed and bull float		4000	.006			.22		.22	.34
0125	Manual screed, bull float, manual float		2000	.012			.43		.43	.69
0150	Manual screed, bull float, manual float & broom finish		1850	.013			.47		.47	.74
0200	Manual screed, bull float, manual float, manual steel trowel		1265	.019			.68		.68	1.09
0250	Manual screed, bull float, machine float & trowel (walk-behind)	C-10C	1715	.014			.50	.02	.52	.83
0300	Power screed, bull float, machine float & trowel (walk-behind)	C-10D	2400	.010			.36	.05	.41	.62
0350	Power screed, bull float, machine float & trowel (ride-on)	C-10E	4000	.006			.22	.06	.28	.41
0370	Minimum labor/equipment charge	C-10	2	12	Job		435		435	690
0400	Integral topping and finish, using 1:1:2 mix, 3/16" thick	C-10B	1000	.040	S.F.	.10	1.37	.24	1.71	2.58
0450	1/2" thick		950	.042		.26	1.44	.25	1.95	2.89
0500	3/4" thick		850	.047		.39	1.61	.28	2.28	3.34
0600	1" thick		750	.053		.52	1.83	.32	2.67	3.87
0800	Granolithic topping, laid after, 1:1:1-1/2 mix, 1/2" thick		590	.068		.29	2.32	.41	3.02	4.52
0820	3/4" thick		580	.069		.43	2.36	.41	3.20	4.74
0850	1" thick		575	.070		.57	2.38	.42	3.37	4.93
0950	2" thick		500	.080		1.15	2.74	.48	4.37	6.20
9100	Minimum labor/equipment charge	C-10	2	12	Job		435		435	690

03 35 29.35 Control Joints, Saw Cut

	Crew	Daily Output	Labor-Hours	Unit	Material	2009 Bare Costs Labor	Equipment	Total	Total Incl O&P
0010	**CONTROL JOINTS, SAW CUT**									
0100	Sawcut in green concrete									
0120	1" depth	C-27	2000	.008	L.F.	.07	.31	.07	.45	.63
0140	1-1/2" depth		1800	.009		.10	.34	.08	.52	.73
0160	2" depth		1600	.010		.13	.38	.09	.60	.85
0200	Clean out control joint of debris	C-28	6000	.001			.05		.05	.08
0300	Joint sealant									
0320	Backer rod, polyethylene, 1/4" diameter	1 Cefi	460	.017	L.F.	.04	.67		.71	1.09
0340	Sealant, polyurethane									
0360	1/4" x 1/4" (308 LF/Gal)	1 Cefi	270	.030	L.F.	.18	1.13		1.31	1.96
0380	1/4" x 1/2" (154 LF/Gal)	"	255	.031	"	.35	1.20		1.55	2.27

03 35 29.60 Finishing Walls

	Crew	Daily Output	Labor-Hours	Unit	Material	2009 Bare Costs Labor	Equipment	Total	Total Incl O&P
0010	**FINISHING WALLS**									
0020	Break ties and patch voids	1 Cefi	540	.015	S.F.	.03	.57		.60	.92

03 35 Concrete Finishing

03 35 29 – Tooled Concrete Finishing

03 35 29.60 Finishing Walls

		Crew	Daily Output	Labor-Hours	Unit	Material	2009 Bare Costs Labor	2009 Bare Costs Equipment	Total	Total Incl O&P
0050	Burlap rub with grout	1 Cefi	450	.018	S.F.	.03	.68		.71	1.10
0300	Bush hammer, green concrete	B-39	1000	.048			1.59	.19	1.78	2.81
0350	Cured concrete	"	650	.074			2.45	.29	2.74	4.33
0700	Sandblast, light penetration	E-11	1100	.029		.70	1.04	.18	1.92	2.80
0750	Heavy penetration	"	375	.085	↓	1.41	3.04	.52	4.97	7.45
0850	Grind form fins flush	1 Clab	700	.011	L.F.		.36		.36	.60
9000	Minimum labor/equipment charge	C-10	2	12	Job		435		435	690

03 35 33 – Stamped Concrete Finishing

03 35 33.50 Slab Texture Stamping

		Crew	Daily Output	Labor-Hours	Unit	Material	2009 Bare Costs Labor	2009 Bare Costs Equipment	Total	Total Incl O&P
0010	**SLAB TEXTURE STAMPING**									
0020	Buy re-usable stamp, small (2 S.F.)				Ea.	82.50			82.50	91
0030	Medium (4 S.F.)					165			165	182
0120	Large (8 S.F.)				↓	330			330	365
0200	Commonly used chemicals for texture systems									
0210	Hardener, colored powder				S.F.	.53			.53	.59
0220	Release agent, colored powder					.09			.09	.10
0230	Curing & sealing compound, solvent based					.06			.06	.06
0300	Broadcasting hardener & release agent, stamping, tooling	3 Cefi	1000	.024	↓		.92		.92	1.44

03 37 Specialty Placed Concrete

03 37 13 – Shotcrete

03 37 13.30 Gunite (Dry-Mix)

		Crew	Daily Output	Labor-Hours	Unit	Material	2009 Bare Costs Labor	2009 Bare Costs Equipment	Total	Total Incl O&P
0010	**GUNITE (DRY-MIX)**									
0020	Applied in 1" layers, no mesh included	C-8	2000	.028	S.F.	.32	.99	.37	1.68	2.35
0300	Typical in place, including mesh, 2" thick, minimum	C-16	1000	.072		1.30	2.68	.74	4.72	6.65
0350	Maximum		500	.144		1.30	5.35	1.48	8.13	11.85
0500	4" thick, minimum		750	.096		1.94	3.58	.98	6.50	9.05
0550	Maximum	↓	350	.206		1.94	7.65	2.11	11.70	17
0900	Prepare old walls, no scaffolding, minimum	C-10	1000	.024			.87		.87	1.38
0950	Maximum	"	275	.087			3.15		3.15	5
1100	For high finish requirement or close tolerance, add, minimum						50%			
1150	Maximum				↓		110%			
9000	Minimum labor/equipment charge	C-10	1	24	Job		865		865	1,375

03 39 Concrete Curing

03 39 13 – Water Concrete Curing

03 39 13.50 Water Curing

		Crew	Daily Output	Labor-Hours	Unit	Material	2009 Bare Costs Labor	2009 Bare Costs Equipment	Total	Total Incl O&P
0010	**WATER CURING**									
0015	With burlap, 4 uses assumed, 7.5 oz.	2 Clab	55	.291	C.S.F.	8.65	9.20		17.85	24.50
0100	10 oz.	"	55	.291	"	15.55	9.20		24.75	32.50
9000	Minimum labor/equipment charge	1 Clab	5	1.600	Job		50.50		50.50	83.50

03 39 23 – Membrane Concrete Curing

03 39 23.13 Chemical Compound Membrane Concrete Curing

		Crew	Daily Output	Labor-Hours	Unit	Material	2009 Bare Costs Labor	2009 Bare Costs Equipment	Total	Total Incl O&P
0010	**CHEMICAL COMPOUND MEMBRANE CONCRETE CURING**									
0300	Sprayed membrane curing compound	2 Clab	95	.168	C.S.F.	5.25	5.30		10.55	14.55
0700	Curing compound, solvent based, 400 S.F./gal, 55 gallon lots				Gal.	15.25			15.25	16.75
0720	5 gallon lots					17.40			17.40	19.15
0800	Curing compound, water based, 250 S.F./gal, 55 gallon lots				↓	18.30			18.30	20

03 39 Concrete Curing

03 39 23 – Membrane Concrete Curing

03 39 23.13 Chemical Compound Membrane Concrete Curing	Crew	Daily Output	Labor-Hours	Unit	Material	2009 Bare Costs Labor	Equipment	Total	Total Incl O&P
0820	5 gallon lots				Gal.	21			21	23

03 39 23.23 Sheet Membrane Concrete Curing

0010	**SHEET MEMBRANE CONCRETE CURING**									
0200	Curing blanket, burlap/poly, 2-ply	2 Clab	70	.229	C.S.F.	15.95	7.20		23.15	29.50

03 41 Precast Structural Concrete

03 41 05 – Precast Concrete Members

03 41 05.10 Precast Beams

		Crew	Daily Output	Labor-Hours	Unit	Material	2009 Bare Costs Labor	Equipment	Total	Total Incl O&P
0010	**PRECAST BEAMS** R034105-30									
0011	L-shaped, 20' span, 12" x 20"	C-11	32	2.250	Ea.	1,750	98.50	59	1,907.50	2,175
1000	Inverted tee beams, add to above, small beams					15%				
1050	Large beams					20%				
1200	Rectangular, 20' span, 12" x 20"	C-11	32	2.250		1,300	98.50	59	1,457.50	1,675
1250	18" x 36"		24	3		1,475	131	79	1,685	1,950
1300	24" x 44"		22	3.273		1,850	143	86	2,079	2,375
1400	30' span, 12" x 36"		24	3		2,000	131	79	2,210	2,525
1450	18" x 44"		20	3.600		2,500	158	94.50	2,752.50	3,125
1500	24" x 52"		16	4.500		3,225	197	118	3,540	4,025
1600	40' span, 12" x 52"		20	3.600		3,700	158	94.50	3,952.50	4,425
1650	18" x 52"		16	4.500		4,400	197	118	4,715	5,325
1700	24" x 52"		12	6		4,850	263	158	5,271	5,975
2000	"T" shaped, 20' span, 12" x 20"		32	2.250		2,200	98.50	59	2,357.50	2,675
2050	18" x 36"		24	3		2,500	131	79	2,710	3,100
2100	24" x 44"		22	3.273		3,150	143	86	3,379	3,800
2200	30' span, 12" x 36"		24	3		3,400	131	79	3,610	4,050
2250	18" x 44"		20	3.600		4,275	158	94.50	4,527.50	5,075
2300	24" x 52"		16	4.500		5,475	197	118	5,790	6,500
2500	40' span, 12" x 52"		20	3.600		6,275	158	94.50	6,527.50	7,275
2550	18" x 52"		16	4.500		7,475	197	118	7,790	8,700
2600	24" x 52"		12	6		8,225	263	158	8,646	9,700

03 41 05.15 Precast Columns

		Crew	Daily Output	Labor-Hours	Unit	Material	2009 Bare Costs Labor	Equipment	Total	Total Incl O&P
0010	**PRECAST COLUMNS** R034105-30									
0020	Rectangular to 12' high, small columns	C-11	120	.600	L.F.	106	26.50	15.75	148.25	180
0050	Large columns	"	96	.750	"	184	33	19.70	236.70	283

03 41 13 – Precast Concrete Hollow Core Planks

03 41 13.50 Precast Slab Planks

		Crew	Daily Output	Labor-Hours	Unit	Material	2009 Bare Costs Labor	Equipment	Total	Total Incl O&P
0010	**PRECAST SLAB PLANKS** R034105-30									
0020	Prestressed roof/floor members, grouted, solid, 4" thick	C-11	2400	.030	S.F.	5.70	1.31	.79	7.80	9.50
0050	6" thick		2800	.026		6.65	1.13	.67	8.45	10.10
0100	Hollow, 8" thick		3200	.023		7.35	.99	.59	8.93	10.45
0150	10" thick		3600	.020		7.75	.88	.53	9.16	10.65
0200	12" thick		4000	.018		8.35	.79	.47	9.61	11.10

03 41 23 – Precast Concrete Stairs

03 41 23.50 Precast Stairs

		Crew	Daily Output	Labor-Hours	Unit	Material	2009 Bare Costs Labor	Equipment	Total	Total Incl O&P
0010	**PRECAST STAIRS**									
0020	Precast concrete treads on steel stringers, 3' wide	C-12	75	.640	Riser	137	25	10.25	172.25	203
0300	Front entrance, 5' wide with 48" platform, 2 risers		16	3	Flight	450	118	48	616	740
0350	5 risers		12	4		705	157	64	926	1,100
0500	6' wide, 2 risers		15	3.200		505	126	51.50	682.50	820

03 41 Precast Structural Concrete

03 41 23 – Precast Concrete Stairs

03 41 23.50 Precast Stairs	Crew	Daily Output	Labor-Hours	Unit	Material	2009 Bare Costs Labor	2009 Bare Costs Equipment	Total	Total Incl O&P
0550 5 risers	C-12	11	4.364	Flight	785	172	70	1,027	1,225
1200 Basement entrance stairs, steel bulkhead doors, minimum	B-51	22	2.182	↓	1,350	69.50	8.85	1,428.35	1,600
1250 Maximum	"	11	4.364	↓	2,250	139	17.65	2,406.65	2,725

03 45 Precast Architectural Concrete

03 45 13 – Faced Architectural Precast Concrete

03 45 13.50 Precast Wall Panels

	Crew	Daily Output	Labor-Hours	Unit	Material	2009 Bare Costs Labor	2009 Bare Costs Equipment	Total	Total Incl O&P
0010 **PRECAST WALL PANELS**									
0750 20' x 10', 6" thick, smooth gray	C-11	1400	.051	S.F.	25.50	2.25	1.35	29.10	33.50
2200 Fiberglass reinforced cement with urethane core									
2210 R20, 8' x 8', minimum	E-2	750	.075	S.F.	21.50	3.25	2.32	27.07	32.50
2220 Maximum	"	600	.093	"	24	4.06	2.90	30.96	37

03 48 Precast Concrete Specialties

03 48 43 – Precast Concrete Trim

03 48 43.40 Precast Lintels

	Crew	Daily Output	Labor-Hours	Unit	Material	2009 Bare Costs Labor	2009 Bare Costs Equipment	Total	Total Incl O&P
0010 **PRECAST LINTELS**									
0800 Precast concrete, 4" wide, 8" high, to 5' long	D-10	28	1.143	Ea.	25.50	45	21.50	92	124
0850 5'-12' long		24	1.333		82.50	52.50	25	160	203
1000 6" wide, 8" high, to 5' long		26	1.231		39.50	48.50	23	111	147
1050 5'-12' long		22	1.455		99.50	57.50	27.50	184.50	232
1200 8" wide, 8" high, to 5' long		24	1.333		47.50	52.50	25	125	165
1250 5'-12' long		20	1.600		135	63	30	228	283
1400 10" wide, 8" high U-Shape, to 14' long		18	1.778		209	70	33.50	312.50	380
1450 12" wide, 8" high U-Shape, to 19' long	↓	16	2	↓	275	79	37.50	391.50	475

03 48 43.90 Precast Window Sills

	Crew	Daily Output	Labor-Hours	Unit	Material	2009 Bare Costs Labor	2009 Bare Costs Equipment	Total	Total Incl O&P
0010 **PRECAST WINDOW SILLS**									
0600 Precast concrete, 4" tapers to 3", 9" wide	D-1	70	.229	L.F.	11.75	8.30		20.05	26.50
0650 11" wide		60	.267		15.60	9.70		25.30	33
0700 13" wide, 3 1/2" tapers to 2 1/2", 12" wall	↓	50	.320	↓	15	11.65		26.65	35.50

03 51 Cast Roof Decks

03 51 13 – Cementitious Wood Fiber Decks

03 51 13.50 Cementitious/Wood Fiber Planks

	Crew	Daily Output	Labor-Hours	Unit	Material	2009 Bare Costs Labor	2009 Bare Costs Equipment	Total	Total Incl O&P
0010 **CEMENTITIOUS/WOOD FIBER PLANKS**									
0050 Plank, beveled edge, 1" thick	2 Carp	1000	.016	S.F.	2.10	.64		2.74	3.37
0100 1-1/2" thick		975	.016		2.41	.66		3.07	3.73
0150 T & G, 2" thick		950	.017		2.69	.67		3.36	4.07
0200 2-1/2" thick		925	.017		3.02	.69		3.71	4.46
0300 3-1/2" thick		875	.018		4.94	.73		5.67	6.65
0350 4" thick	↓	850	.019		5.50	.75		6.25	7.30
1000 Bulb tee, sub-purlin and grout, 6' span, add	E-1	5000	.005		2.18	.21	.03	2.42	2.80
1100 8' span	"	4200	.006	↓	2.47	.25	.03	2.75	3.20

03 52 Lightweight Concrete Roof Insulation

03 52 16 – Lightweight Insulating Concrete

03 52 16.13 Lightweight Cellular Insulating Concrete	Crew	Daily Output	Labor-Hours	Unit	Material	2009 Bare Costs Labor	Equipment	Total	Total Incl O&P
0010 **LIGHTWEIGHT CELLULAR INSULATING CONCRETE** R035216-10									
0020 Portland cement and foaming agent [G]	C-8	50	1.120	C.Y.	118	39.50	14.75	172.25	210

03 52 16.16 Lightweight Aggregate Insulating Concrete

	Crew	Daily Output	Labor-Hours	Unit	Material	2009 Bare Costs Labor	Equipment	Total	Total Incl O&P
0010 **LIGHTWEIGHT AGGREGATE INSULATING CONCRETE** R035216-10									
0100 Poured vermiculite or perlite, field mix,									
0110 1:6 field mix [G]	C-8	50	1.120	C.Y.	119	39.50	14.75	173.25	210
0200 Ready mix, 1:6 mix, roof fill, 2" thick [G]		10000	.006	S.F.	.66	.20	.07	.93	1.12
0250 3" thick [G]		7700	.007	"	.99	.26	.10	1.35	1.61

03 54 Cast Underlayment

03 54 16 – Hydraulic Cement Underlayment

03 54 16.50 Cement Underlayment

	Crew	Daily Output	Labor-Hours	Unit	Material	2009 Bare Costs Labor	Equipment	Total	Total Incl O&P
0010 **CEMENT UNDERLAYMENT**									
2510 Underlayment, P.C based self-leveling, 4100 psi, pumped, 1/4"	C-8	20000	.003	S.F.	1.44	.10	.04	1.58	1.78
2520 1/2"		19000	.003		2.87	.10	.04	3.01	3.37
2530 3/4"		18000	.003		4.31	.11	.04	4.46	4.97
2540 1"		17000	.003		5.75	.12	.04	5.91	6.55
2550 1-1/2"		15000	.004		8.60	.13	.05	8.78	9.75
2560 Hand mix, 1/2"	C-18	4000	.002		2.87	.07	.01	2.95	3.30
2610 Topping, P.C. based self-level/dry 6100 psi, pumped, 1/4"	C-8	20000	.003		2.21	.10	.04	2.35	2.64
2620 1/2"		19000	.003		4.43	.10	.04	4.57	5.10
2630 3/4"		18000	.003		6.65	.11	.04	6.80	7.55
2660 1"		17000	.003		8.85	.12	.04	9.01	10
2670 1-1/2"		15000	.004		13.30	.13	.05	13.48	14.85
2680 Hand mix, 1/2"	C-18	4000	.002		4.43	.07	.01	4.51	5

03 63 Epoxy Grouting

03 63 05 – Grouting of Dowels and Fasteners

03 63 05.10 Epoxy Only

	Crew	Daily Output	Labor-Hours	Unit	Material	2009 Bare Costs Labor	Equipment	Total	Total Incl O&P
0010 **EPOXY ONLY**									
1500 Chemical anchoring, epoxy cartridge, excludes layout, drilling, fastener									
1530 For fastener 3/4" diam. x 6" embedment	2 Skwk	72	.222	Ea.	3.54	9.10		12.64	18.70
1535 1" diam. x 8" embedment		66	.242		5.30	9.90		15.20	22
1540 1-1/4" diam. x 10" embedment		60	.267		10.60	10.90		21.50	29.50
1545 1-3/4" diam. x 12" embedment		54	.296		17.70	12.10		29.80	39
1550 14" embedment		48	.333		21	13.60		34.60	45.50
1555 2" diam. x 12" embedment		42	.381		28.50	15.55		44.05	56.50
1560 18" embedment		32	.500	"	35.50	20.50		56	72.50

03 64 Injection Grouting

03 64 23 – Epoxy Injection Grouting

03 64 23.10 Crack Repair

		Crew	Daily Output	Labor-Hours	Unit	Material	2009 Bare Costs Labor	Equipment	Total	Total Incl O&P
0010	**CRACK REPAIR**									
0100	Epoxy injection, 1/8" wide, 12" deep	B-9	80	.500	L.F.	6.50	16	2.40	24.90	36.50
0110	1/4" wide, 12" deep		60	.667		13	21.50	3.19	37.69	53
0200	Latex injection, 1/8" wide, 12" deep		100	.400		2.28	12.80	1.92	17	25.50
0210	1/4" wide, 12" deep		75	.533		4.57	17.05	2.55	24.17	36

03 81 Concrete Cutting

03 81 13 – Flat Concrete Sawing

03 81 13.50 Concrete Floor/Slab Cutting

		Crew	Daily Output	Labor-Hours	Unit	Material	2009 Bare Costs Labor	Equipment	Total	Total Incl O&P
0010	**CONCRETE FLOOR/SLAB CUTTING**									
0400	Concrete slabs, mesh reinforcing, up to 3" deep	B-89	980	.016	L.F.	.49	.57	.41	1.47	1.91
0420	Each additional inch of depth	"	1600	.010	"	.16	.35	.25	.76	1.02

03 81 16 – Track Mounted Concrete Wall Sawing

03 81 16.50 Concrete Wall Cutting

		Crew	Daily Output	Labor-Hours	Unit	Material	2009 Bare Costs Labor	Equipment	Total	Total Incl O&P
0010	**CONCRETE WALL CUTTING**									
0800	Concrete walls, hydraulic saw, plain, per inch of depth	B-89B	250	.064	L.F.	.44	2.24	2.74	5.42	7.10
0820	Rod reinforcing, per inch of depth	"	150	.107	"	.62	3.73	4.56	8.91	11.70

03 82 Concrete Boring

03 82 13 – Concrete Core Drilling

03 82 13.10 Core Drilling

		Crew	Daily Output	Labor-Hours	Unit	Material	2009 Bare Costs Labor	Equipment	Total	Total Incl O&P
0010	**CORE DRILLING**									
0020	Reinf. conc slab, up to 6" thick, incl. bit, layout & set up									
0100	1" diameter core	B-89A	17	.941	Ea.	3.45	34	6.50	43.95	67
0150	Each added inch thick in same hole, add		1440	.011		.63	.40	.08	1.11	1.43
0300	3" diameter core		16	1		7.40	36	6.90	50.30	75.50
0350	Each added inch thick in same hole, add		720	.022		1.38	.81	.15	2.34	3.01
0500	4" diameter core		15	1.067		7.40	38.50	7.35	53.25	80
0550	Each added inch thick in same hole, add		480	.033		1.81	1.21	.23	3.25	4.22
0700	6" diameter core		14	1.143		12.20	41.50	7.85	61.55	90
0750	Each added inch thick in same hole, add		360	.044		2.17	1.61	.31	4.09	5.35
0900	8" diameter core		13	1.231		16.70	44.50	8.50	69.70	101
0950	Each added inch thick in same hole, add		288	.056		2.88	2.01	.38	5.27	6.90
1100	10" diameter core		12	1.333		22	48.50	9.20	79.70	113
1150	Each added inch thick in same hole, add		240	.067		3.62	2.42	.46	6.50	8.45
1300	12" diameter core		11	1.455		27	52.50	10	89.50	128
1350	Each added inch thick in same hole, add		206	.078		4.35	2.81	.54	7.70	10
1500	14" diameter core		10	1.600		33	58	11	102	143
1550	Each added inch thick in same hole, add		180	.089		5.85	3.22	.61	9.68	12.35
1700	18" diameter core		9	1.778		42.50	64.50	12.25	119.25	165
1750	Each added inch thick in same hole, add		144	.111		7.60	4.03	.77	12.40	15.80
1760	For horizontal holes, add to above						20%	20%		
1770	Prestressed hollow core plank, 8" thick									
1780	1" diameter core	B-89A	17.50	.914	Ea.	2.30	33	6.30	41.60	63.50
1790	Each added inch thick in same hole, add		3840	.004		.46	.15	.03	.64	.79
1800	3" diameter core		17	.941		4.81	34	6.50	45.31	68.50
1810	Each added inch thick in same hole, add		1920	.008		.83	.30	.06	1.19	1.46
1820	4" diameter core		16.50	.970		6.60	35	6.70	48.30	72
1830	Each added inch thick in same hole, add		1280	.013		1.21	.45	.09	1.75	2.16

03 82 Concrete Boring

03 82 13 – Concrete Core Drilling

03 82 13.10 Core Drilling

		Crew	Daily Output	Labor-Hours	Unit	Material	2009 Bare Costs Labor	Equipment	Total	Total Incl O&P
1840	6" diameter core	B-89A	15.50	1.032	Ea.	8.10	37.50	7.10	52.70	77.50
1850	Each added inch thick in same hole, add		960	.017		1.38	.60	.11	2.09	2.64
1860	8" diameter core		15	1.067		11.50	38.50	7.35	57.35	84.50
1870	Each added inch thick in same hole, add		768	.021		1.93	.75	.14	2.82	3.52
1880	10" diameter core		14	1.143		15.25	41.50	7.85	64.60	93.50
1890	Each added inch thick in same hole, add		640	.025		2.01	.91	.17	3.09	3.88
1900	12" diameter core		13.50	1.185		17.85	43	8.15	69	99
1910	Each added inch thick in same hole, add		548	.029		3.22	1.06	.20	4.48	5.50
1999	Drilling, core, minimum labor/equipment charge		4	4	Job		145	27.50	172.50	268
3010	Bits for core drill, diamond, premium, 1" diameter				Ea.	117			117	129
3020	3" diameter					288			288	315
3040	4" diameter					320			320	350
3050	6" diameter					510			510	565
3080	8" diameter					705			705	775
3120	12" diameter					1,050			1,050	1,150
3180	18" diameter					1,900			1,900	2,075
3240	24" diameter					2,600			2,600	2,875

03 82 16 – Concrete Drilling

03 82 16.10 Concrete Impact Drilling

		Crew	Daily Output	Labor-Hours	Unit	Material	2009 Bare Costs Labor	Equipment	Total	Total Incl O&P
0010	**CONCRETE IMPACT DRILLING**									
0050	Up to 4" deep in conc/brick floor/wall, incl. bit & layout, no anchor									
0100	Holes, 1/4" diameter	1 Carp	75	.107	Ea.	.06	4.26		4.32	7.10
0150	For each additional inch of depth, add		430	.019		.01	.74		.75	1.25
0200	3/8" diameter		63	.127		.05	5.05		5.10	8.40
0250	For each additional inch of depth, add		340	.024		.01	.94		.95	1.56
0300	1/2" diameter		50	.160		.06	6.40		6.46	10.60
0350	For each additional inch of depth, add		250	.032		.01	1.28		1.29	2.13
0400	5/8" diameter		48	.167		.08	6.65		6.73	11.10
0450	For each additional inch of depth, add		240	.033		.02	1.33		1.35	2.22
0500	3/4" diameter		45	.178		.10	7.10		7.20	11.80
0550	For each additional inch of depth, add		220	.036		.03	1.45		1.48	2.43
0600	7/8" diameter		43	.186		.12	7.45		7.57	12.40
0650	For each additional inch of depth, add		210	.038		.03	1.52		1.55	2.54
0700	1" diameter		40	.200		.14	8		8.14	13.35
0750	For each additional inch of depth, add		190	.042		.04	1.68		1.72	2.82
0800	1-1/4" diameter		38	.211		.20	8.40		8.60	14.10
0850	For each additional inch of depth, add		180	.044		.05	1.78		1.83	2.99
0900	1-1/2" diameter		35	.229		.29	9.15		9.44	15.35
0950	For each additional inch of depth, add		165	.048		.07	1.94		2.01	3.28
1000	For ceiling installations, add						40%			

Estimating Tips

04 05 00 Common Work Results for Masonry

- The terms *mortar* and *grout* are often used interchangeably, and incorrectly. Mortar is used to bed masonry units, seal the entry of air and moisture, provide architectural appearance, and allow for size variations in the units. Grout is used primarily in reinforced masonry construction and is used to bond the masonry to the reinforcing steel. Common mortar types are M(2500 psi), S(1800 psi), N(750 psi), and O(350 psi), and conform to ASTM C270. Grout is either fine or coarse and conforms to ASTM C476, and in-place strengths generally exceed 2500 psi. Mortar and grout are different components of masonry construction and are placed by entirely different methods. An estimator should be aware of their unique uses and costs.

- Waste, specifically the loss/droppings of mortar and the breakage of brick and block, is included in all masonry assemblies in this division. A factor of 25% is added for mortar and 3% for brick and concrete masonry units.

- Scaffolding or staging is not included in any of the Division 4 costs. Refer to Subdivision 01 54 23 for scaffolding and staging costs.

04 20 00 Unit Masonry

- The most common types of unit masonry are brick and concrete masonry. The major classifications of brick are building brick (ASTM C62), facing brick (ASTM C216), glazed brick, fire brick, and pavers. Many varieties of texture and appearance can exist within these classifications, and the estimator would be wise to check local custom and availability within the project area. For repair and remodeling jobs, matching the existing brick may be the most important criteria.

- Brick and concrete block are priced by the piece and then converted into a price per square foot of wall. Openings less than two square feet are generally ignored by the estimator because any savings in units used is offset by the cutting and trimming required.

- It is often difficult and expensive to find and purchase small lots of historic brick. Costs can vary widely. Many design issues affect costs, selection of mortar mix, and repairs or replacement of masonry materials. Cleaning techniques must be reflected in the estimate.

- All masonry walls, whether interior or exterior, require bracing. The cost of bracing walls during construction should be included by the estimator, and this bracing must remain in place until permanent bracing is complete. Permanent bracing of masonry walls is accomplished by masonry itself, in the form of pilasters or abutting wall corners, or by anchoring the walls to the structural frame. Accessories in the form of anchors, anchor slots, and ties are used, but their supply and installation can be by different trades. For instance, anchor slots on spandrel beams and columns are supplied and welded in place by the steel fabricator, but the ties from the slots into the masonry are installed by the bricklayer. Regardless of the installation method, the estimator must be certain that these accessories are accounted for in pricing.

Reference Numbers

Reference numbers are shown in shaded boxes at the beginning of some major classifications. These numbers refer to related items in the Reference Section. The reference information may be an estimating procedure, an alternate pricing method, or technical information.

Note: Not all subdivisions listed here necessarily appear in this publication.

04 01 20 – Maintenance of Unit Masonry

04 01 20.20 Pointing Masonry		Crew	Daily Output	Labor-Hours	Unit	Material	2009 Bare Costs Labor	Equipment	Total	Total Incl O&P
0010	**POINTING MASONRY**									
0300	Cut and repoint brick, hard mortar, running bond	1 Bric	80	.100	S.F.	.53	4.05		4.58	7.15
0320	Common bond		77	.104		.53	4.21		4.74	7.40
0360	Flemish bond		70	.114		.56	4.63		5.19	8.10
0400	English bond		65	.123		.56	4.98		5.54	8.65
0600	Soft old mortar, running bond		100	.080		.53	3.24		3.77	5.85
0620	Common bond		96	.083		.53	3.37		3.90	6.05
0640	Flemish bond		90	.089		.56	3.60		4.16	6.40
0680	English bond		82	.098		.56	3.95		4.51	7
0700	Stonework, hard mortar		140	.057	L.F.	.70	2.31		3.01	4.51
0720	Soft old mortar		160	.050	"	.70	2.03		2.73	4.05
1000	Repoint, mask and grout method, running bond		95	.084	S.F.	.70	3.41		4.11	6.25
1020	Common bond		90	.089		.70	3.60		4.30	6.55
1040	Flemish bond		86	.093		.74	3.77		4.51	6.90
1060	English bond		77	.104		.74	4.21		4.95	7.60
2000	Scrub coat, sand grout on walls, minimum		120	.067		2.84	2.70		5.54	7.50
2020	Maximum		98	.082		3.94	3.31		7.25	9.70
9000	Minimum labor/equipment charge		3	2.667	Job		108		108	175

04 01 20.50 Toothing Masonry										
0010	**TOOTHING MASONRY**									
0500	Brickwork, soft old mortar	1 Clab	40	.200	V.L.F.		6.30		6.30	10.45
0520	Hard mortar		30	.267			8.45		8.45	13.90
0700	Blockwork, soft old mortar		70	.114			3.61		3.61	5.95
0720	Hard mortar		50	.160			5.05		5.05	8.35
9000	Minimum labor/equipment charge		4	2	Job		63		63	104

04 01 30 – Unit Masonry Cleaning

04 01 30.20 Cleaning Masonry		Crew	Daily Output	Labor-Hours	Unit	Material	2009 Bare Costs Labor	Equipment	Total	Total Incl O&P
0010	**CLEANING MASONRY**									
0200	Chemical cleaning, new construction, brush and wash, minimum	D-1	1000	.016	S.F.	.05	.58		.63	.99
0220	Average		800	.020		.07	.73		.80	1.26
0240	Maximum		600	.027		.09	.97		1.06	1.67
0260	Light restoration, minimum		800	.020		.07	.73		.80	1.25
0270	Average		400	.040		.10	1.45		1.55	2.46
0280	Maximum		330	.048		.13	1.76		1.89	3
0300	Heavy restoration, minimum		600	.027		.09	.97		1.06	1.67
0310	Average		400	.040		.14	1.45		1.59	2.51
0320	Maximum		250	.064		.19	2.33		2.52	3.97
0400	High pressure water only, minimum	B-9	2000	.020			.64	.10	.74	1.17
0420	Average		1500	.027			.85	.13	.98	1.55
0440	Maximum		1000	.040			1.28	.19	1.47	2.32
0800	High pressure water and chemical, minimum		1800	.022		.08	.71	.11	.90	1.38
0820	Average		1200	.033		.12	1.07	.16	1.35	2.07
0840	Maximum		800	.050		.16	1.60	.24	2	3.08
1200	Sandblast, wet system, minimum		1750	.023		.47	.73	.11	1.31	1.85
1220	Average		1100	.036		.70	1.16	.17	2.03	2.88
1240	Maximum		700	.057		.94	1.83	.27	3.04	4.35
1400	Dry system, minimum		2500	.016		.47	.51	.08	1.06	1.45
1420	Average		1750	.023		.70	.73	.11	1.54	2.10
1440	Maximum		1000	.040		.94	1.28	.19	2.41	3.35
1800	For walnut shells, add					.50			.50	.55
1820	For corn chips, add					.50			.50	.55
2000	Steam cleaning, minimum	B-9	3000	.013			.43	.06	.49	.77

04 01 Maintenance of Masonry

04 01 30 – Unit Masonry Cleaning

04 01 30.20 Cleaning Masonry	Crew	Daily Output	Labor-Hours	Unit	Material	2009 Bare Costs Labor	Equipment	Total	Total Incl O&P
2020	Average	B-9	2500	.016	S.F.		.51	.08	.59	.93
2040	Maximum	↓	1500	.027			.85	.13	.98	1.55
4000	Add for masking doors and windows	1 Clab	800	.010	↓	.06	.32		.38	.59
4200	Add for pedestrian protection				Job					10%
9000	Minimum labor/equipment charge	D-4	2	16	"		575	63.50	638.50	995

04 05 Common Work Results for Masonry

04 05 05 – Selective Masonry Demolition

04 05 05.10 Selective Demolition

		Crew	Daily Output	Labor-Hours	Unit	Material	2009 Bare Costs Labor	Equipment	Total	Total Incl O&P
0010	**SELECTIVE DEMOLITION** R024119-10									
0200	Bond beams, 8" block with #4 bar	2 Clab	32	.500	L.F.		15.80		15.80	26
0300	Concrete block walls, unreinforced, 2" thick		1200	.013	S.F.		.42		.42	.70
0310	4" thick		1150	.014			.44		.44	.73
0320	6" thick		1100	.015			.46		.46	.76
0330	8" thick		1050	.015			.48		.48	.79
0340	10" thick		1000	.016			.51		.51	.83
0360	12" thick		950	.017			.53		.53	.88
0380	Reinforced alternate courses, 2" thick		1130	.014			.45		.45	.74
0390	4" thick		1080	.015			.47		.47	.77
0400	6" thick		1035	.015			.49		.49	.81
0410	8" thick		990	.016			.51		.51	.84
0420	10" thick		940	.017			.54		.54	.89
0430	12" thick		890	.018			.57		.57	.94
0440	Reinforced alternate courses & vertically 48" OC, 4" thick		900	.018			.56		.56	.93
0450	6" thick		850	.019			.59		.59	.98
0460	8" thick		800	.020			.63		.63	1.04
0480	10" thick		750	.021			.67		.67	1.11
0490	12" thick	↓	700	.023	↓		.72		.72	1.19
1000	Chimney, 16" x 16", soft old mortar	1 Clab	55	.145	C.F.		4.60		4.60	7.60
1020	Hard mortar		40	.200			6.30		6.30	10.45
1030	16" x 20", soft old mortar		55	.145			4.60		4.60	7.60
1040	Hard mortar		40	.200			6.30		6.30	10.45
1050	16" x 24", soft old mortar		55	.145			4.60		4.60	7.60
1060	Hard mortar		40	.200			6.30		6.30	10.45
1080	20" x 20", soft old mortar		55	.145			4.60		4.60	7.60
1100	Hard mortar		40	.200			6.30		6.30	10.45
1110	20" x 24", soft old mortar		55	.145			4.60		4.60	7.60
1120	Hard mortar		40	.200			6.30		6.30	10.45
1140	20" x 32", soft old mortar		55	.145			4.60		4.60	7.60
1160	Hard mortar		40	.200			6.30		6.30	10.45
1200	48" x 48", soft old mortar		55	.145			4.60		4.60	7.60
1220	Hard mortar	↓	40	.200	↓		6.30		6.30	10.45
1250	Metal, high temp steel jacket, 24" diameter	E-2	130	.431	V.L.F.		18.75	13.40	32.15	48.50
1260	60" diameter	"	60	.933			40.50	29	69.50	105
1280	Flue lining, up to 12" x 12"	1 Clab	200	.040			1.26		1.26	2.09
1282	Up to 24" x 24"		150	.053			1.69		1.69	2.78
2000	Columns, 8" x 8", soft old mortar		48	.167			5.25		5.25	8.70
2020	Hard mortar		40	.200			6.30		6.30	10.45
2060	16" x 16", soft old mortar		16	.500			15.80		15.80	26
2100	Hard mortar		14	.571			18.05		18.05	30
2140	24" x 24", soft old mortar	↓	8	1	↓		31.50		31.50	52

04 05 05 – Selective Masonry Demolition

04 05 05.10 Selective Demolition	Crew	Daily Output	Labor-Hours	Unit	Material	2009 Bare Costs Labor	Equipment	Total	Total Incl O&P
2160 Hard mortar	1 Clab	6	1.333	V.L.F.		42		42	69.50
2200 36" x 36", soft old mortar		4	2			63		63	104
2220 Hard mortar		3	2.667	↓		84.50		84.50	139
2230 Alternate pricing method, soft old mortar		30	.267	C.F.		8.45		8.45	13.90
2240 Hard mortar	↓	23	.348	"		11		11	18.15
3000 Copings, precast or masonry, to 8" wide									
3020 Soft old mortar	1 Clab	180	.044	L.F.		1.40		1.40	2.32
3040 Hard mortar	"	160	.050	"		1.58		1.58	2.61
3100 To 12" wide									
3120 Soft old mortar	1 Clab	160	.050	L.F.		1.58		1.58	2.61
3140 Hard mortar	"	140	.057	"		1.81		1.81	2.98
4000 Fireplace, brick, 30" x 24" opening									
4020 Soft old mortar	1 Clab	2	4	Ea.		126		126	209
4040 Hard mortar		1.25	6.400			202		202	335
4100 Stone, soft old mortar		1.50	5.333			169		169	278
4120 Hard mortar		1	8	↓		253		253	415
5000 Veneers, brick, soft old mortar		140	.057	S.F.		1.81		1.81	2.98
5020 Hard mortar		125	.064			2.02		2.02	3.34
5100 Granite and marble, 2" thick		180	.044			1.40		1.40	2.32
5120 4" thick		170	.047			1.49		1.49	2.45
5140 Stone, 4" thick		180	.044			1.40		1.40	2.32
5160 8" thick		175	.046	↓		1.44		1.44	2.38
5400 Alternate pricing method, stone, 4" thick		60	.133	C.F.		4.21		4.21 *	6.95
5420 8" thick		85	.094	"		2.97		2.97	4.91
9000 Minimum labor/equipment charge	↓	2	4	Job		126		126	209

04 05 13 – Masonry Mortaring

04 05 13.10 Cement

		Crew	Daily Output	Labor-Hours	Unit	Material	Labor	Equipment	Total	Total Incl O&P
0010	**CEMENT**									
0100	Masonry, 70 lb. bag, T.L. lots				Bag	9.05			9.05	9.95
0150	L.T.L. lots					9.60			9.60	10.55
0200	White, 70 lb. bag, T.L. lots					15.05			15.05	16.55
0250	L.T.L. lots				↓	16.05			16.05	17.65

04 05 13.20 Lime

		Crew	Daily Output	Labor-Hours	Unit	Material	Labor	Equipment	Total	Total Incl O&P
0010	**LIME**									
0020	Masons, hydrated, 50 lb. bag, T.L. lots				Bag	8.95			8.95	9.85
0050	L.T.L. lots					9.30			9.30	10.25
0200	Finish, double hydrated, 50 lb. bag, T.L. lots					9.90			9.90	10.90
0250	L.T.L. lots				↓	11			11	12.10

04 05 13.30 Mortar

		Crew	Daily Output	Labor-Hours	Unit	Material	Labor	Equipment	Total	Total Incl O&P
0010	**MORTAR**									
0100	Type M, 1:1:6 mix	1 Brhe	143	.056	C.F.	4.75	1.80		6.55	8.10
0200	Type N, 1:3 mix		143	.056		4.29	1.80		6.09	7.65
0300	Type O, 1:3 mix		143	.056		6.45	1.80		8.25	10
0400	Type PM, 1:1:6 mix, 2500 psi		143	.056		6.85	1.80		8.65	10.45
0500	Type S, 1/2:1:4 mix	↓	143	.056	↓	7.25	1.80		9.05	10.90
2000	With portland cement and lime									
2100	Type M, 1:1/4:3 mix	1 Brhe	143	.056	C.F.	8.45	1.80		10.25	12.20
2200	Type N, 1:1:6 mix, 750 psi		143	.056		6.80	1.80		8.60	10.35
2300	Type O, 1:2:9 mix (Pointing Mortar)		143	.056		7.50	1.80		9.30	11.15
2400	Type PL, 1:1/2:4 mix, 2500 psi		143	.056		7.45	1.80		9.25	11.10
2500	Type K, 1:3:12 mix, 75 psi		143	.056		6.45	1.80		8.25	10
2600	Type S, 1:1/2:4 mix, 1800 psi	↓	143	.056	↓	9.15	1.80		10.95	12.95

04 05 Common Work Results for Masonry

04 05 13 – Masonry Mortaring

04 05 13.30 Mortar		Crew	Daily Output	Labor-Hours	Unit	Material	2009 Bare Costs Labor	Equipment	Total	Total Incl O&P
2650	Pre-mixed, type S or N				C.F.	4.94			4.94	5.45
2700	Mortar for glass block	1 Brhe	143	.056	↓	10.20	1.80		12	14.10
2900	Mortar for Fire Brick, 80 lb. bag, T.L. Lots				Bag	25.50			25.50	28

04 05 16 – Masonry Grouting

04 05 16.30 Grouting		Crew	Daily Output	Labor-Hours	Unit	Material	2009 Bare Costs Labor	Equipment	Total	Total Incl O&P
0010	**GROUTING**									
0011	Bond beams & lintels, 8" deep, 6" thick, 0.15 C.F. per L.F.	D-4	1480	.022	L.F.	.64	.78	.09	1.51	2.04
0020	8" thick, 0.2 C.F. per L.F.		1400	.023		.98	.82	.09	1.89	2.50
0050	10" thick, 0.25 C.F. per L.F.		1200	.027		1.06	.96	.11	2.13	2.83
0060	12" thick, 0.3 C.F. per L.F.	↓	1040	.031	↓	1.28	1.11	.12	2.51	3.31
0200	Concrete block cores, solid, 4" thk., by hand, 0.067 C.F./S.F. of wall	D-8	1100	.036	S.F.	.28	1.35		1.63	2.50
0210	6" thick, pumped, 0.175 C.F. per S.F.	D-4	720	.044		.74	1.60	.18	2.52	3.58
0250	8" thick, pumped, 0.258 C.F. per S.F.		680	.047		1.10	1.69	.19	2.98	4.14
0300	10" thick, pumped, 0.340 C.F. per S.F.		660	.048		1.45	1.74	.19	3.38	4.60
0350	12" thick, pumped, 0.422 C.F. per S.F.		640	.050		1.79	1.80	.20	3.79	5.10
0500	Cavity walls, 2" space, pumped, 0.167 C.F./S.F. of wall		1700	.019		.71	.68	.07	1.46	1.95
0550	3" space, 0.250 C.F./S.F.		1200	.027		1.06	.96	.11	2.13	2.83
0600	4" space, 0.333 C.F. per S.F.		1150	.028		1.42	1	.11	2.53	3.29
0700	6" space, 0.500 C.F. per S.F.		800	.040	↓	2.13	1.44	.16	3.73	4.82
0800	Door frames, 3' x 7' opening, 2.5 C.F. per opening		60	.533	Opng.	10.65	19.20	2.11	31.96	45
0850	6' x 7' opening, 3.5 C.F. per opening		45	.711	"	14.90	25.50	2.82	43.22	60.50
2000	Grout, C476, for bond beams, lintels and CMU cores	↓	350	.091	C.F.	4.25	3.29	.36	7.90	10.40
9000	Minimum labor/equipment charge	1 Bric	2	4	Job		162		162	262

04 05 19 – Masonry Anchorage and Reinforcing

04 05 19.05 Anchor Bolts

04 05 19.05 Anchor Bolts		Crew	Daily Output	Labor-Hours	Unit	Material	2009 Bare Costs Labor	Equipment	Total	Total Incl O&P
0010	**ANCHOR BOLTS**									
0020	Hooked, with nut and washer, 1/2" diam., 8" long	1 Bric	200	.040	Ea.	.83	1.62		2.45	3.53
0030	12" long		190	.042		1.26	1.71		2.97	4.15
0040	5/8" diameter, 8" long		180	.044		1.13	1.80		2.93	4.15
0050	12" long		170	.047		1.24	1.91		3.15	4.45
0060	3/4" diameter, 8" long		160	.050		1.86	2.03		3.89	5.35
0070	12" long	↓	150	.053	↓	2.33	2.16		4.49	6.05

04 05 19.16 Masonry Anchors

04 05 19.16 Masonry Anchors		Crew	Daily Output	Labor-Hours	Unit	Material	2009 Bare Costs Labor	Equipment	Total	Total Incl O&P
0010	**MASONRY ANCHORS**									
0020	For brick veneer, galv., corrugated, 7/8" x 7", 22 Ga.	1 Bric	10.50	.762	C	9.80	31		40.80	61
0100	24 Ga.		10.50	.762		9.15	31		40.15	60
0150	16 Ga.		10.50	.762		24	31		55	76.50
0200	Buck anchors, galv., corrugated, 16 gauge, 2" bend, 8" x 2"		10.50	.762		134	31		165	197
0250	8" x 3"	↓	10.50	.762	↓	138	31		169	202
0300	Adjustable, rectangular, 4-1/8" wide									
0350	Anchor and tie, 3/16" wire, mill galv.									
0400	2-3/4" eye, 3-1/4" tie	1 Bric	1.05	7.619	M	415	310		725	960
0500	4-3/4" tie		1.05	7.619		465	310		775	1,000
0520	5-1/2" tie		1.05	7.619		490	310		800	1,050
0550	4-3/4" eye, 3-1/4" tie		1.05	7.619		460	310		770	1,000
0570	4-3/4" tie		1.05	7.619		525	310		835	1,075
0580	5-1/2" tie		1.05	7.619	↓	550	310		860	1,100
0670	3/16" diameter		10.50	.762	C	21	31		52	73
0680	1/4" diameter		10.50	.762		39.50	31		70.50	93.50
0850	8" long, 3/16" diameter		10.50	.762		24	31		55	76
0855	1/4" diameter	↓	10.50	.762	↓	46.50	31		77.50	101

04 05 19.16 Masonry Anchors		Crew	Daily Output	Labor-Hours	Unit	Material	2009 Bare Costs Labor	Equipment	Total	Total Incl O&P
1000	Rectangular type, galvanized, 1/4" diameter, 2" x 6"	1 Bric	10.50	.762	C	67	31		98	124
1050	4" x 6"		10.50	.762		83.50	31		114.50	142
1100	3/16" diameter, 2" x 6"		10.50	.762		37.50	31		68.50	91.50
1150	4" x 6"		10.50	.762		39	31		70	93
1200	Mesh wall tie, 1/2" mesh, hot dip galvanized									
1400	16 gauge, 12" long, 3" wide	1 Bric	9	.889	C	87.50	36		123.50	155
1420	6" wide		9	.889		130	36		166	201
1440	12" wide		8.50	.941		208	38		246	291
1500	Rigid partition anchors, plain, 8" long, 1" x 1/8"		10.50	.762		103	31		134	163
1550	1" x 1/4"		10.50	.762		170	31		201	237
1580	1-1/2" x 1/8"		10.50	.762		138	31		169	201
1600	1-1/2" x 1/4"		10.50	.762		280	31		311	360
1650	2" x 1/8"		10.50	.762		178	31		209	246
1700	2" x 1/4"		10.50	.762		325	31		356	410
2000	Column flange ties, wire, galvanized									
2300	3/16" diameter, up to 3" wide	1 Bric	10.50	.762	C	81.50	31		112.50	140
2350	To 5" wide		10.50	.762		89	31		120	148
2400	To 7" wide		10.50	.762		95	31		126	155
2600	To 9" wide		10.50	.762		103	31		134	163
2650	1/4" diameter, up to 3" wide		10.50	.762		111	31		142	172
2700	To 5" wide		10.50	.762		122	31		153	184
2800	To 7" wide		10.50	.762		132	31		163	195
2850	To 9" wide		10.50	.762		144	31		175	208
2900	For hot dip galvanized, add					35%				
4000	Channel slots, 1-3/8" x 1/2" x 8"									
4100	12 gauge, plain	1 Bric	10.50	.762	C	238	31		269	310
4150	16 gauge, galvanized	"	10.50	.762	"	136	31		167	199
4200	Channel slot anchors									
4300	16 gauge, galvanized, 1-1/4" x 3-1/2"				C	52			52	57.50
4350	1-1/4" x 5-1/2"					65			65	72
4400	1-1/4" x 7-1/2"					74.50			74.50	82
4500	1/8" plain, 1-1/4" x 3-1/2"					100			100	110
4550	1-1/4" x 5-1/2"					107			107	117
4600	1-1/4" x 7-1/2"					117			117	128
4700	For corrugation, add					54			54	59
4750	For hot dip galvanized, add					35%				
5000	Dowels									
5100	Plain, 1/4" diameter, 3" long				C	38.50			38.50	42.50
5150	4" long					43			43	47
5200	6" long					52.50			52.50	57.50
5300	3/8" diameter, 3" long					51			51	56
5350	4" long					63.50			63.50	70
5400	6" long					73			73	80.50
5500	1/2" diameter, 3" long					74.50			74.50	82
5550	4" long					88.50			88.50	97
5600	6" long					117			117	129
5700	5/8" diameter, 3" long					102			102	112
5750	4" long					126			126	138
5800	6" long					173			173	190
6000	3/4" diameter, 3" long					127			127	140
6100	4" long					163			163	179
6150	6" long					230			230	254
6300	For hot dip galvanized, add					35%				

04 05 19 – Masonry Anchorage and Reinforcing

04 05 19.26 Masonry Reinforcing Bars

		Crew	Daily Output	Labor-Hours	Unit	Material	2009 Bare Costs Labor	Equipment	Total	Total Incl O&P
0010	**MASONRY REINFORCING BARS**									
0015	Steel bars A615, placed horiz., #3 & #4 bars	1 Bric	450	.018	Lb.	.78	.72		1.50	2.01
0020	#5 & #6 bars		800	.010		.78	.41		1.19	1.51
0050	Placed vertical, #3 & #4 bars		350	.023		.78	.93		1.71	2.35
0060	#5 & #6 bars	↓	650	.012	↓	.78	.50		1.28	1.66
0500	Joint reinforcing, ladder type, mill std galvanized									
0600	9 ga. sides, 9 ga. ties, 4" wall	1 Bric	30	.267	C.L.F.	11.20	10.80		22	30
0650	6" wall		30	.267		12.75	10.80		23.55	31.50
0700	8" wall		25	.320		13.10	12.95		26.05	35.50
0750	10" wall		20	.400		13.95	16.20		30.15	41.50
0800	12" wall	↓	20	.400	↓	14.75	16.20		30.95	42
1000	Truss type									
1100	9 ga. sides, 9 ga. ties, 4" wall	1 Bric	30	.267	C.L.F.	16.35	10.80		27.15	35.50
1150	6" wall		30	.267		17.25	10.80		28.05	36.50
1200	8" wall		25	.320		17.95	12.95		30.90	41
1250	10" wall		20	.400		18.95	16.20		35.15	47
1300	12" wall		20	.400		20	16.20		36.20	48
1500	3/16" sides, 9 ga. ties, 4" wall		30	.267		21	10.80		31.80	41
1550	6" wall		30	.267		21	10.80		31.80	40.50
1600	8" wall		25	.320		21.50	12.95		34.45	44.50
1650	10" wall		20	.400		22.50	16.20		38.70	50.50
1700	12" wall		20	.400		26	16.20		42.20	54.50
2000	3/16" sides, 3/16" ties, 4" wall		30	.267		24	10.80		34.80	44
2050	6" wall		30	.267		24.50	10.80		35.30	44
2100	8" wall		25	.320		25	12.95		37.95	48.50
2150	10" wall		20	.400		26.50	16.20		42.70	55
2200	12" wall	↓	20	.400	↓	31	16.20		47.20	60
2500	Cavity truss type, galvanized									
2600	9 ga. sides, 9 ga. ties, 4" wall	1 Bric	25	.320	C.L.F.	29.50	12.95		42.45	53.50
2650	6" wall		25	.320		30	12.95		42.95	53.50
2700	8" wall		20	.400		30.50	16.20		46.70	59.50
2750	10" wall		15	.533		32	21.50		53.50	70
2800	12" wall		15	.533		33	21.50		54.50	71.50
3000	3/16" sides, 9 ga. ties, 4" wall		25	.320		38	12.95		50.95	63
3050	6" wall		25	.320		38.50	12.95		51.45	63.50
3100	8" wall		20	.400		39.50	16.20		55.70	69.50
3150	10" wall		15	.533		40	21.50		61.50	79
3200	12" wall	↓	15	.533	↓	42	21.50		63.50	81
3500	For hot dip galvanizing, add					80%				

04 05 23 – Masonry Accessories

04 05 23.13 Control Joint

		Crew	Daily Output	Labor-Hours	Unit	Material	2009 Bare Costs Labor	Equipment	Total	Total Incl O&P
0010	**CONTROL JOINT**									
0020	Rubber, for double wythe 8" minimum wall (Brick/CMU)	1 Bric	400	.020	L.F.	1.90	.81		2.71	3.40
0025	"T" shaped		320	.025		1.46	1.01		2.47	3.25
0030	Cross-shaped for CMU units		280	.029		2	1.16		3.16	4.07
0050	PVC, for double wythe 8" minimum wall (Brick/CMU)		400	.020		1.18	.81		1.99	2.61
0120	"T" shaped		320	.025		1.07	1.01		2.08	2.82
0160	Cross-shaped for CMU units	↓	280	.029	↓	1.23	1.16		2.39	3.22

04 05 23.19 Vent Box

		Crew	Daily Output	Labor-Hours	Unit	Material	2009 Bare Costs Labor	Equipment	Total	Total Incl O&P
0010	**VENT BOX**									
0020	Extruded aluminum, 4" deep, 2-3/8" x 8-1/8"	1 Bric	30	.267	Ea.	30.50	10.80		41.30	51

04 05 23 – Masonry Accessories

04 05 23.19 Vent Box

		Crew	Daily Output	Labor-Hours	Unit	Material	2009 Bare Costs Labor	Equipment	Total	Total Incl O&P
0050	5" x 8-1/8"	1 Bric	25	.320	Ea.	40	12.95		52.95	65
0100	2-1/4" x 25"		25	.320		69.50	12.95		82.45	97.50
0150	5" x 16-1/2"		22	.364		58.50	14.75		73.25	88
0200	6" x 16-1/2"		22	.364		81	14.75		95.75	113
0250	7-3/4" x 16-1/2"		20	.400		69	16.20		85.20	102
0400	For baked enamel finish, add					35%				
0500	For cast aluminum, painted, add					60%				
1000	Stainless steel ventilators, 6" x 6"	1 Bric	25	.320		126	12.95		138.95	159
1050	8" x 8"		24	.333		132	13.50		145.50	168
1100	12" x 12"		23	.348		153	14.10		167.10	191
1150	12" x 6"		24	.333		134	13.50		147.50	169
1200	Foundation block vent, galv., 1-1/4" thk, 8" high, 16" long, no damper		30	.267		24.50	10.80		35.30	44.50
1250	For damper, add					8.20			8.20	9

04 21 13 – Brick Masonry

04 21 13.13 Brick Veneer Masonry

		Crew	Daily Output	Labor-Hours	Unit	Material	2009 Bare Costs Labor	Equipment	Total	Total Incl O&P
0010	**BRICK VENEER MASONRY**, T.L. lots, excl. scaff., grout & reinforcing R042110-20									
0015	Material costs incl. 3% brick and 25% mortar waste									
2000	Standard, sel. common, 4" x 2-2/3" x 8", (6.75/S.F.) R042110-50	D-8	230	.174	S.F.	3.60	6.45		10.05	14.40
2020	Standard, red, 4" x 2-2/3" x 8", running bond (6.75/SF)		220	.182		5.65	6.75		12.40	17.20
2050	Full header every 6th course (7.88/S.F.)		185	.216		6.60	8.05		14.65	20.50
2100	English, full header every 2nd course (10.13/S.F.)		140	.286		8.50	10.60		19.10	26.50
2150	Flemish, alternate header every course (9.00/S.F.)		150	.267		7.55	9.90		17.45	24.50
2200	Flemish, alt. header every 6th course (7.13/S.F.)		205	.195		6	7.25		13.25	18.35
2250	Full headers throughout (13.50/S.F.)		105	.381		11.30	14.15		25.45	35.50
2300	Rowlock course (13.50/S.F.)		100	.400		11.30	14.85		26.15	36.50
2350	Rowlock stretcher (4.50/S.F.)		310	.129		3.80	4.79		8.59	11.95
2400	Soldier course (6.75/S.F.)		200	.200		5.65	7.45		13.10	18.25
2450	Sailor course (4.50/S.F.)		290	.138		3.80	5.15		8.95	12.50
2600	Buff or gray face, running bond, (6.75/S.F.)		220	.182		6	6.75		12.75	17.55
2700	Glazed face brick, running bond		210	.190		11.15	7.10		18.25	23.50
2750	Full header every 6th course (7.88/S.F.)		170	.235		13	8.75		21.75	28.50
3000	Jumbo, 6" x 4" x 12" running bond (3.00/S.F.)		435	.092		4.86	3.42		8.28	10.90
3050	Norman, 4" x 2-2/3" x 12" running bond, (4.5/S.F.)		320	.125		5.65	4.65		10.30	13.75
3100	Norwegian, 4" x 3-1/5" x 12" (3.75/S.F.)		375	.107		4.14	3.96		8.10	10.95
3150	Economy, 4" x 4" x 8" (4.50/S.F.)		310	.129		4.18	4.79		8.97	12.35
3200	Engineer, 4" x 3-1/5" x 8" (5.63/S.F.)		260	.154		3.46	5.70		9.16	13.05
3250	Roman, 4" x 2" x 12" (6.00/S.F.)		250	.160		5.75	5.95		11.70	15.90
3300	SCR, 6" x 2-2/3" x 12" (4.50/S.F.)		310	.129		5.25	4.79		10.04	13.50
3350	Utility, 4" x 4" x 12" (3.00/S.F.)		450	.089		4.40	3.30		7.70	10.20
3400	For cavity wall construction, add						15%			
3450	For stacked bond, add						10%			
3500	For interior veneer construction, add						15%			
3550	For curved walls, add						30%			
9000	Minimum labor/equipment charge	D-1	2	8	Job		291		291	470

04 21 13.15 Chimney

		Crew	Daily Output	Labor-Hours	Unit	Material	2009 Bare Costs Labor	Equipment	Total	Total Incl O&P
0010	**CHIMNEY**, excludes foundation, scaffolding, grout and reinforcing									
0100	Brick, 16" x 16", 8" flue	D-1	18.20	.879	V.L.F.	20.50	32		52.50	74.50
0150	16" x 20" with one 8" x 12" flue		16	1		32.50	36.50		69	95
0200	16" x 24" with two 8" x 8" flues		14	1.143		47.50	41.50		89	120

04 21 Clay Unit Masonry

04 21 13 – Brick Masonry

04 21 13.15 Chimney		Crew	Daily Output	Labor-Hours	Unit	Material	2009 Bare Costs Labor	Equipment	Total	Total Incl O&P
0250	20" x 20" with one 12" x 12" flue	D-1	13.70	1.168	V.L.F.	38.50	42.50		81	111
0300	20" x 24" with two 8" x 12" flues		12	1.333		54	48.50		102.50	138
0350	20" x 32" with two 12" x 12" flues	↓	10	1.600	↓	67.50	58		125.50	169

04 21 13.18 Columns

		Crew	Daily Output	Labor-Hours	Unit	Material	Labor	Equipment	Total	Total Incl O&P
0010	**COLUMNS**, solid, excludes scaffolding, grout and reinforcing									
0050	Brick, 8" x 8", 9 brick per VLF	D-1	56	.286	V.L.F.	7.35	10.40		17.75	25
0100	12" x 8", 13.5 brick per VLF		37	.432		11.05	15.70		26.75	37.50
0200	12" x 12", 20 brick per VLF		25	.640		16.35	23.50		39.85	55.50
0300	16" x 12", 27 brick per VLF		19	.842		22	30.50		52.50	74
0400	16" x 16", 36 brick per VLF		14	1.143		29.50	41.50		71	99.50
0500	20" x 16", 45 brick per VLF		11	1.455		37	53		90	126
0600	20" x 20", 56 brick per VLF		9	1.778		46	64.50		110.50	155
0700	24" x 20", 68 brick per VLF		7	2.286		55.50	83		138.50	195
0800	24" x 24", 81 brick per VLF		6	2.667		66	97		163	230
1000	36" x 36", 182 brick per VLF		3	5.333	↓	149	194		343	480
9000	Minimum labor/equipment charge	↓	2	8	Job		291		291	470

04 21 13.30 Oversized Brick

		Crew	Daily Output	Labor-Hours	Unit	Material	Labor	Equipment	Total	Total Incl O&P
0010	**OVERSIZED BRICK**, excludes scaffolding, grout and reinforcing									
0100	Veneer, 4" x 2.25" x 16"	D-8	387	.103	S.F.	5.05	3.84		8.89	11.75
0105	4" x 2.75" x 16"		412	.097		4.55	3.61		8.16	10.85
0110	4" x 4" x 16"		460	.087		4.48	3.23		7.71	10.20
0120	4" x 8" x 16"		533	.075		5.50	2.79		8.29	10.55
0125	Loadbearing, 6" x 4" x 16", grouted and reinforced		387	.103		8.10	3.84		11.94	15.10
0130	8" x 4" x 16", grouted and reinforced		327	.122		8.25	4.55		12.80	16.40
0135	6" x 8" x 16", grouted and reinforced		440	.091		7.45	3.38		10.83	13.65
0140	8" x 8" x 16", grouted and reinforced		400	.100		8.35	3.72		12.07	15.20
0145	Curtainwall / reinforced veneer, 6" x 4" x 16"		387	.103		12.05	3.84		15.89	19.45
0150	8" x 4" x 16"		327	.122		14.55	4.55		19.10	23.50
0155	6" x 8" x 16"		440	.091		12.15	3.38		15.53	18.80
0160	8" x 8" x 16"	↓	400	.100		14.65	3.72		18.37	22
0200	For 1 to 3 slots in face, add					25%				
0210	For 4 to 7 slots in face, add					15%				
0220	For bond beams, add					20%				
0230	For bullnose shapes, add					20%				
0240	For open end knockout, add					10%				
0250	For white or gray color group, add					10%				
0260	For 135 degree corner, add				↓	250%				

04 21 13.35 Common Building Brick

		Crew	Daily Output	Labor-Hours	Unit	Material	Labor	Equipment	Total	Total Incl O&P
0010	**COMMON BUILDING BRICK**, C62, TL lots, material only R042110-20									
0020	Standard, minimum				M	340			340	375
0050	Average (select)				"	430			430	475

04 21 13.40 Structural Brick

		Crew	Daily Output	Labor-Hours	Unit	Material	Labor	Equipment	Total	Total Incl O&P
0010	**STRUCTURAL BRICK** C652, Grade SW, incl. mortar, scaffolding not incl.									
0100	Standard unit, 4-5/8" x 2-3/4" x 9-5/8"	D-8	245	.163	S.F.	5.15	6.05		11.20	15.50
0120	Bond beam		225	.178		9.65	6.60		16.25	21.50
0140	V cut bond beam		225	.178		10.45	6.60		17.05	22
0160	Stretcher quoin, 5-5/8" x 2-3/4" x 9-5/8"		245	.163		10.10	6.05		16.15	21
0180	Corner quoin		245	.163		11.30	6.05		17.35	22.50
0200	Corner, 45 deg, 4-5/8" x 2-3/4" x 10-7/16"	↓	235	.170	↓	11.70	6.35		18.05	23

04 21 13.45 Face Brick

		Crew	Daily Output	Labor-Hours	Unit	Material	Labor	Equipment	Total	Total Incl O&P
0010	**FACE BRICK** Material Only, C216, TL lots R042110-20									

04 21 Clay Unit Masonry

04 21 13 – Brick Masonry

04 21 13.45 Face Brick		Crew	Daily Output	Labor-Hours	Unit	Material	2009 Bare Costs Labor	Equipment	Total	Total Incl O&P
0300	Standard modular, 4" x 2-2/3" x 8", minimum				M	730			730	805
0350	Maximum	R042110-50				835			835	920
0450	Economy, 4" x 4" x 8", minimum					815			815	895
0500	Maximum					1,050			1,050	1,150
0510	Economy, 4" x 4" x 12", minimum					1,200			1,200	1,325
0520	Maximum					1,550			1,550	1,725
0550	Jumbo, 6" x 4" x 12", minimum					1,475			1,475	1,625
0600	Maximum					1,875			1,875	2,075
0610	Jumbo, 8" x 4" x 12", minimum					1,475			1,475	1,625
0620	Maximum					1,875			1,875	2,075
0650	Norwegian, 4" x 3-1/5" x 12", minimum					980			980	1,075
0700	Maximum					1,275			1,275	1,400
0710	Norwegian, 6" x 3-1/5" x 12", minimum					1,425			1,425	1,575
0720	Maximum					1,875			1,875	2,050
0850	Standard glazed, plain colors, 4" x 2-2/3" x 8", minimum					1,525			1,525	1,675
0900	Maximum					1,975			1,975	2,175
1000	Deep trim shades, 4" x 2-2/3" x 8", minimum					1,600			1,600	1,750
1050	Maximum					1,900			1,900	2,100
1080	Jumbo utility, 4" x 4" x 12"					1,325			1,325	1,450
1120	4" x 8" x 8"					1,550			1,550	1,700
1140	4" x 8" x 16"					3,100			3,100	3,400
1260	Engineer, 4" x 3-1/5" x 8", minimum					510			510	560
1270	Maximum					625			625	690
1350	King, 4" x 2-3/4" x 10", minimum					525			525	575
1360	Maximum					560			560	620
1400	Norman, 4" x 2-3/4" x 12"					525			525	575
1450	Roman, 4" x 2" x 12"					525			525	575
1500	SCR, 6" x 2-2/3" x 12"					525			525	575
1550	Double, 4" x 5-1/3" x 8"					525			525	575
1600	Triple, 4" x 5-1/3" x 12"					525			525	575
1770	Standard modular, double glazed, 4" x 2-2/3" x 8"					1,775			1,775	1,950
1850	Jumbo, colored glazed ceramic, 6" x 4" x 12"					2,400			2,400	2,650
2050	Jumbo utility, glazed, 4" x 4" x 12"					4,050			4,050	4,450
2100	4" x 8" x 8"					4,725			4,725	5,200
2150	4" x 16" x 8"					6,200			6,200	6,825
2170	For less than truck load lots, add					15			15	16.50
2180	For buff or gray brick, add					16			16	17.60
3050	Used brick, minimum					395			395	430
3100	Maximum					375			375	415
3150	Add for brick to match existing work, minimum					5%				
3200	Maximum					50%				

04 21 26 – Glazed Structural Clay Tile Masonry

04 21 26.10 Structural Facing Tile		Crew	Daily Output	Labor-Hours	Unit	Material	2009 Bare Costs Labor	Equipment	Total	Total Incl O&P
0010	**STRUCTURAL FACING TILE**, std. colors, excl. scaffolding, grout, reinforcing									
0020	6T series, 5-1/3" x 12", 2.3 pieces per S.F., glazed 1 side, 2" thick	D-8	225	.178	S.F.	8.45	6.60		15.05	20
0100	4" thick		220	.182		11.05	6.75		17.80	23
0150	Glazed 2 sides		195	.205		14.15	7.60		21.75	28
0250	6" thick		210	.190		16.60	7.10		23.70	29.50
0300	Glazed 2 sides		185	.216		19.45	8.05		27.50	34.50
0400	8" thick		180	.222		22	8.25		30.25	37.50
9000	Minimum labor/equipment charge	D-1	2	8	Job		291		291	470

04 21 Clay Unit Masonry

04 21 29 – Terra Cotta Masonry

04 21 29.10 Terra Cotta Masonry Components	Crew	Daily Output	Labor-Hours	Unit	Material	2009 Bare Costs Labor	Equipment	Total	Total Incl O&P
0010 **TERRA COTTA MASONRY COMPONENTS**									
0020 Coping, split type, not glazed, 9" wide	D-1	90	.178	L.F.	10.40	6.45		16.85	22
0100 13" wide		80	.200		15.75	7.25		23	29
0200 Coping, split type, glazed, 9" wide		90	.178		17.40	6.45		23.85	29.50
0250 13" wide		80	.200		22.50	7.25		29.75	37
0500 Partition or back-up blocks, scored, in C.L. lots									
0700 Non-load bearing 12" x 12", 3" thick, special order	D-8	550	.073	S.F.	19.30	2.70		22	25.50
0850 8" thick		400	.100		11.05	3.72		14.77	18.20
1000 Load bearing, 12" x 12", 4" thick, in walls		500	.080		6.95	2.97		9.92	12.45
1400 8" thick, in walls		400	.100		11.05	3.72		14.77	18.20
9000 Minimum labor/equipment charge	D-1	2	8	Job		291		291	470

04 22 Concrete Unit Masonry

04 22 10 – Concrete Masonry Units

04 22 10.11 Autoclave Aerated Concrete Block

	Crew	Daily Output	Labor-Hours	Unit	Material	2009 Bare Costs Labor	Equipment	Total	Total Incl O&P
0010 **AUTOCLAVE AERATED CONCRETE BLOCK**, excl. scaffolding, grout & reinforcing									
0050 Solid, 4" x 12" x 24", incl mortar [G]	D-8	600	.067	S.F.	1.60	2.48		4.08	5.75
0060 6" x 12" x 24" [G]		600	.067		2.17	2.48		4.65	6.40
0070 8" x 8" x 24" [G]		575	.070		2.80	2.59		5.39	7.25
0080 10" x 12" x 24" [G]		575	.070		3.59	2.59		6.18	8.15
0090 12" x 12" x 24" [G]		550	.073		4.23	2.70		6.93	9

04 22 10.14 Concrete Block, Back-Up

	Crew	Daily Output	Labor-Hours	Unit	Material	2009 Bare Costs Labor	Equipment	Total	Total Incl O&P
0010 **CONCRETE BLOCK, BACK-UP**, C90, 2000 psi									
0020 Normal weight, 8" x 16" units, tooled joint 1 side									
0050 Not-reinforced, 2000 psi, 2" thick	D-8	475	.084	S.F.	1.36	3.13		4.49	6.55
0200 4" thick		460	.087		1.52	3.23		4.75	6.90
0300 6" thick		440	.091		2.12	3.38		5.50	7.80
0350 8" thick		400	.100		2.23	3.72		5.95	8.45
0400 10" thick		330	.121		3.03	4.50		7.53	10.65
0450 12" thick	D-9	310	.155		3.44	5.65		9.09	12.90
1000 Reinforced, alternate courses, 4" thick	D-8	450	.089		1.64	3.30		4.94	7.15
1100 6" thick		430	.093		2.25	3.46		5.71	8.10
1150 8" thick		395	.101		2.37	3.76		6.13	8.70
1200 10" thick		320	.125		3.18	4.65		7.83	11
1250 12" thick	D-9	300	.160		3.59	5.80		9.39	13.35
9000 Minimum labor/equipment charge	D-1	2	8	Job		291		291	470

04 22 10.16 Concrete Block, Bond Beam

	Crew	Daily Output	Labor-Hours	Unit	Material	2009 Bare Costs Labor	Equipment	Total	Total Incl O&P
0010 **CONCRETE BLOCK, BOND BEAM**, C90, 2000 psi									
0020 Not including grout or reinforcing									
0125 Regular block, 6" thick	D-8	584	.068	L.F.	2.21	2.55		4.76	6.55
0130 8" high, 8" thick	"	565	.071		2.51	2.63		5.14	7
0150 12" thick	D-9	510	.094		3.43	3.42		6.85	9.35
0525 Lightweight, 6" thick	D-8	592	.068		2.53	2.51		5.04	6.85
2000 Including grout and 2 #5 bars									
2100 Regular block, 8" high, 8" thick	D-8	300	.133	L.F.	5.15	4.95		10.10	13.65
2150 12" thick	D-9	250	.192	"	6.60	7		13.60	18.60
9000 Minimum labor/equipment charge	D-1	2	8	Job		291		291	470

04 22 10.19 Concrete Block, Insulation Inserts

0010 **CONCRETE BLOCK, INSULATION INSERTS**									
0100 Styrofoam, plant installed, add to block prices									

04 22 Concrete Unit Masonry

04 22 10 – Concrete Masonry Units

04 22 10.44 Glazed Concrete Block		Crew	Daily Output	Labor-Hours	Unit	Material	2009 Bare Costs Labor	Equipment	Total	Total Incl O&P
0300	8" thick	D-8	310	.129	S.F.	9.20	4.79		13.99	17.90
0350	10" thick	↓	295	.136		11.10	5.05		16.15	20.50
0400	12" thick	D-9	280	.171		11.75	6.25		18	23
0700	Double face, 8" x 16" units, 4" thick	D-8	340	.118		12.20	4.37		16.57	20.50
0750	6" thick		320	.125		14.50	4.65		19.15	23.50
0800	8" thick		300	.133	↓	15.15	4.95		20.10	24.50
1500	Cove base, 8" x 16", 2" thick		315	.127	L.F.	11	4.72		15.72	19.75
1550	4" thick		285	.140		7.15	5.20		12.35	16.30
1600	6" thick		265	.151		7.70	5.60		13.30	17.50
1650	8" thick	↓	245	.163	↓	8.10	6.05		14.15	18.70
9000	Minimum labor/equipment charge	D-1	2	8	Job		291		291	470

04 23 Glass Unit Masonry

04 23 13 – Vertical Glass Unit Masonry

04 23 13.10 Glass Block

		Crew	Daily Output	Labor-Hours	Unit	Material	2009 Bare Costs Labor	Equipment	Total	Total Incl O&P
0010	**GLASS BLOCK**									
0100	Plain, 4" thick, under 1,000 S.F., 6" x 6"	D-8	115	.348	S.F.	20.50	12.95		33.45	43.50
0150	8" x 8"		160	.250		14.05	9.30		23.35	30.50
0160	end block		160	.250		39.50	9.30		48.80	58
0170	90 deg corner		160	.250		39	9.30		48.30	58
0180	45 deg corner		160	.250		19.15	9.30		28.45	36
0200	12" x 12"		175	.229		16	8.50		24.50	31.50
0210	4" x 8"		160	.250		10.25	9.30		19.55	26.50
0220	6" x 8"		160	.250		11	9.30		20.30	27
0300	1,000 to 5,000 S.F., 6" x 6"		135	.296		20	11		31	40
0350	8" x 8"		190	.211		13.75	7.80		21.55	28
0400	12" x 12"		215	.186		15.70	6.90		22.60	28.50
0410	4" x 8"		215	.186		4.43	6.90		11.33	16.05
0420	6" x 8"	↓	215	.186	↓	4.79	6.90		11.69	16.45
0700	For solar reflective blocks, add					100%				
1000	Thinline, plain, 3-1/8" thick, under 1,000 S.F., 6" x 6"	D-8	115	.348	S.F.	15.05	12.95		28	37.50
1050	8" x 8"		160	.250		9.35	9.30		18.65	25.50
1400	For cleaning block after installation (both sides), add	↓	1000	.040	↓	.11	1.49		1.60	2.52
9000	Minimum labor/equipment charge	D-1	2	8	Job		291		291	470

04 27 Multiple-Wythe Unit Masonry

04 27 10 – Multiple-Wythe Masonry

04 27 10.10 Cornices

		Crew	Daily Output	Labor-Hours	Unit	Material	2009 Bare Costs Labor	Equipment	Total	Total Incl O&P
0010	**CORNICES**									
0110	Face bricks, 12 brick/S.F., minimum	D-1	30	.533	SF Face	5.95	19.40		25.35	38
0150	15 brick/S.F., maximum		23	.696	"	7.15	25.50		32.65	49
9000	Minimum labor/equipment charge	↓	1.50	10.667	Job		390		390	625

04 27 10.30 Brick Walls

			Crew	Daily Output	Labor-Hours	Unit	Material	2009 Bare Costs Labor	Equipment	Total	Total Incl O&P
0010	**BRICK WALLS**, including mortar, excludes scaffolding										
0800	Face brick, 4" thick wall, 6.75 brick/S.F.		D-8	215	.186	S.F.	5.60	6.90		12.50	17.35
0850	Common brick, 4" thick wall, 6.75 brick/S.F.	R042110-50		240	.167		2.90	6.20		9.10	13.20
0900	8" thick, 13.50 bricks per S.F.			135	.296		6.05	11		17.05	24.50
1000	12" thick, 20.25 bricks per S.F.			95	.421		9.10	15.65		24.75	35.50
1050	16" thick, 27.00 bricks per S.F.		↓	75	.533	↓	12.35	19.80		32.15	45.50

04 27 10 – Multiple-Wythe Masonry

04 27 10.30 Brick Walls		Crew	Daily Output	Labor-Hours	Unit	Material	2009 Bare Costs Labor	Equipment	Total	Total Incl O&P
1200	Reinforced, face brick, 4" thick wall, 6.75 brick/S.F.	D-8	210	.190	S.F.	5.85	7.10		12.95	17.90
1220	Common brick, 4" thick wall, 6.75 brick/S.F.		235	.170		3.16	6.35		9.51	13.75
1250	8" thick, 13.50 bricks per S.F.		130	.308		6.55	11.45		18	25.50
1300	12" thick, 20.25 bricks per S.F.		90	.444		9.90	16.50		26.40	37.50
1350	16" thick, 27.00 bricks per S.F.	↓	70	.571	↓	13.40	21		34.40	49
9000	Minimum labor/equipment charge	D-1	2	8	Job		291		291	470

04 41 10 – Dry Placed Stone

04 41 10.10 Rough Stone Wall

			Crew	Daily Output	Labor-Hours	Unit	Material	2009 Bare Costs Labor	Equipment	Total	Total Incl O&P
0011	**ROUGH STONE WALL**, Dry										
0100	Random fieldstone, under 18" thick	G	D-12	60	.533	C.F.	11.15	19.65		30.80	44.50
0150	Over 18" thick	G	"	63	.508	"	13.40	18.70		32.10	45.50
0500	Field stone veneer	G	D-8	120	.333	S.F.	9.60	12.40		22	30.50
0510	Valley stone veneer	G		120	.333		9.60	12.40		22	30.50
0520	River stone veneer	G	↓	120	.333	↓	9.60	12.40		22	30.50
0600	Rubble stone walls, in mortar bed, up to 18" thick	G	D-11	75	.320	C.F.	13.50	12.30		25.80	34.50
9000	Minimum labor/equipment charge		D-1	2	8	Job		291		291	470

04 43 10 – Masonry with Natural and Processed Stone

04 43 10.10 Bluestone

		Crew	Daily Output	Labor-Hours	Unit	Material	2009 Bare Costs Labor	Equipment	Total	Total Incl O&P
0010	**BLUESTONE**, cut to size									
0500	Sills, natural cleft, 10" wide to 6' long, 1-1/2" thick	D-11	70	.343	L.F.	13.35	13.15		26.50	36
0550	2" thick		63	.381		14.60	14.60		29.20	39.50
0600	Smooth finish, 1-1/2" thick		70	.343		25	13.15		38.15	49
0650	2" thick		63	.381		30	14.60		44.60	56.50
0800	Thermal finish, 1-1/2" thick		70	.343		30	13.15		43.15	54.50
0850	2" thick	↓	63	.381		35	14.60		49.60	62
1000	Stair treads, natural cleft, 12" wide, 6' long, 1-1/2" thick	D-10	115	.278		30	10.95	5.25	46.20	56.50
1050	2" thick		105	.305		30	12	5.75	47.75	58.50
1100	Smooth finish, 1-1/2" thick		115	.278		15	10.95	5.25	31.20	40
1150	2" thick		105	.305		15	12	5.75	32.75	42
1300	Thermal finish		115	.278		15	10.95	5.25	31.20	40
1350	2" thick	↓	105	.305	↓	15	12	5.75	32.75	42
9000	Minimum labor/equipment charge	D-1	2.50	6.400	Job		233		233	375

04 43 10.45 Granite

		Crew	Daily Output	Labor-Hours	Unit	Material	2009 Bare Costs Labor	Equipment	Total	Total Incl O&P
0010	**GRANITE**, cut to size									
0050	Veneer, polished face, 3/4" to 1-1/2" thick									
0150	Low price, gray, light gray, etc.	D-10	130	.246	S.F.	25.50	9.70	4.63	39.83	48.50
0180	Medium price, pink, brown, etc.		130	.246		28	9.70	4.63	42.33	51
0220	High price, red, black, etc.	↓	130	.246	↓	40	9.70	4.63	54.33	64.50
2500	Steps, copings, etc., finished on more than one surface									
2550	Minimum	D-10	50	.640	C.F.	91	25	12.05	128.05	154
2600	Maximum	"	50	.640	"	146	25	12.05	183.05	214
9000	Minimum labor/equipment charge	D-1	2	8	Job		291		291	470

04 43 10.55 Limestone

0010	**LIMESTONE**, cut to size
0020	Veneer facing panels

04 43 10 – Masonry with Natural and Processed Stone

04 43 10.55 Limestone

		Crew	Daily Output	Labor-Hours	Unit	Material	2009 Bare Costs Labor	2009 Bare Costs Equipment	Total	Total Incl O&P
0750	5" thick, 5' x 14' panels	D-10	275	.116	S.F.	52.50	4.59	2.19	59.28	67.50
1000	Sugarcube finish, 2" Thick, 3' x 5' panels		275	.116		35	4.59	2.19	41.78	48.50
1050	3" Thick, 4' x 9' panels		275	.116		38	4.59	2.19	44.78	52
1200	4" Thick, 5' x 11' panels		275	.116		41	4.59	2.19	47.78	55
1400	Sugarcube, textured finish, 4-1/2" thick, 5' x 12'		275	.116		45	4.59	2.19	51.78	59.50
1450	5" thick, 5' x 14' panels		275	.116		53.50	4.59	2.19	60.28	68.50
2000	Coping, sugarcube finish, top & 2 sides		30	1.067	C.F.	58.50	42	20	120.50	154
2100	Sills, lintels, jambs, trim, stops, sugarcube finish, average		20	1.600		58.50	63	30	151.50	199
2150	Detailed		20	1.600		58.50	63	30	151.50	199
2300	Steps, extra hard, 14" wide, 6" rise		50	.640	L.F.	35	25	12.05	72.05	92.50
3000	Quoins, plain finish, 6"x12"x12"	D-12	25	1.280	Ea.	58.50	47		105.50	141
3050	6"x16"x24"	"	25	1.280	"	78	47		125	162
9000	Minimum labor/equipment charge	D-1	2	8	Job		291		291	470

04 43 10.60 Marble

		Crew	Daily Output	Labor-Hours	Unit	Material	2009 Bare Costs Labor	2009 Bare Costs Equipment	Total	Total Incl O&P
0011	**MARBLE**, ashlar, split face, 4" + or - thick, random									
0040	Lengths 1' to 4' & heights 2" to 7-1/2", average	D-8	175	.229	S.F.	16.90	8.50		25.40	32.50
0100	Base, polished, 3/4" or 7/8" thick, polished, 6" high	D-10	65	.492	L.F.	15.35	19.40	9.25	44	58
0300	Carvings or bas relief, from templates, average		80	.400	S.F.	137	15.75	7.50	160.25	185
0350	Maximum		80	.400	"	320	15.75	7.50	343.25	385
1000	Facing, polished finish, cut to size, 3/4" to 7/8" thick									
1050	Average	D-10	130	.246	S.F.	22.50	9.70	4.63	36.83	45.50
1100	Maximum	"	130	.246	"	52.50	9.70	4.63	66.83	78
2500	Flooring, polished tiles, 12" x 12" x 3/8" thick									
2510	Thin set, average	D-11	90	.267	S.F.	11.35	10.25		21.60	29
2600	Maximum		90	.267		174	10.25		184.25	209
2700	Mortar bed, average		65	.369		10.15	14.15		24.30	34
2740	Maximum		65	.369		174	14.15		188.15	215
3500	Thresholds, 3' long, 7/8" thick, 4" to 5" wide, plain	D-12	24	1.333	Ea.	16.30	49		65.30	97.50
3550	Beveled		24	1.333	"	18.95	49		67.95	101
3700	Window stools, polished, 7/8" thick, 5" wide		85	.376	L.F.	15.20	13.85		29.05	39.50
9000	Minimum labor/equipment charge	D-1	2	8	Job		291		291	470

04 43 10.75 Sandstone or Brownstone

		Crew	Daily Output	Labor-Hours	Unit	Material	2009 Bare Costs Labor	2009 Bare Costs Equipment	Total	Total Incl O&P
0011	**SANDSTONE OR BROWNSTONE**									
0100	Sawed face veneer, 2-1/2" thick, to 2' x 4' panels	D-10	130	.246	S.F.	17.25	9.70	4.63	31.58	39.50
0150	4' thick, to 3'-6" x 8' panels		100	.320		17.25	12.60	6	35.85	46
0300	Split face, random sizes		100	.320		12.40	12.60	6	31	41
9000	Minimum labor/equipment charge	D-1	2.50	6.400	Job		233		233	375

04 43 10.80 Slate

		Crew	Daily Output	Labor-Hours	Unit	Material	2009 Bare Costs Labor	2009 Bare Costs Equipment	Total	Total Incl O&P
0010	**SLATE**									
0040	Pennsylvania - blue gray to black									
0050	Vermont - unfading green, mottled green & purple, gray & purple									
0100	Virginia, blue black									
3100	Stair landings, 1" thick, black, clear	D-1	65	.246	S.F.	20	8.95		28.95	36.50
3200	Ribbon	"	65	.246	"	22	8.95		30.95	38.50
3500	Stair treads, sand finish, 1" thick x 12" wide									
3550	Under 3 L.F.	D-10	85	.376	L.F.	22	14.85	7.10	43.95	56
3600	3 L.F. to 6 L.F.	"	120	.267	"	24	10.50	5	39.50	48.50
3700	Ribbon, sand finish, 1" thick x 12" wide									
3750	To 6 L.F.	D-10	120	.267	L.F.	20	10.50	5	35.50	44.50
4000	Stools or sills, sand finish, 1" thick, 6" wide	D-12	160	.200		11.50	7.35		18.85	24.50
4200	10" wide		90	.356		17.75	13.10		30.85	40.50
4400	2" thick, 6" wide		140	.229		18.50	8.40		26.90	34

04 43 Stone Masonry

04 43 10 – Masonry with Natural and Processed Stone

04 43 10.80 Slate		Crew	Daily Output	Labor-Hours	Unit	Material	2009 Bare Costs Labor	Equipment	Total	Total Incl O&P
4600	10" wide	D-12	90	.356	L.F.	29	13.10		42.10	53
4800	For lengths over 3', add				↓	25%				
9000	Minimum labor/equipment charge	D-1	2.50	6.400	Job		233		233	375

04 43 10.85 Window Sill		Crew	Daily Output	Labor-Hours	Unit	Material	2009 Bare Costs Labor	Equipment	Total	Total Incl O&P
0010	**WINDOW SILL**									
0020	Bluestone, thermal top, 10" wide, 1-1/2" thick	D-1	85	.188	S.F.	16.05	6.85		22.90	28.50
0050	2" thick		75	.213	"	18.75	7.75		26.50	33
0100	Cut stone, 5" x 8" plain		48	.333	L.F.	11.90	12.10		24	32.50
0200	Face brick on edge, brick, 8" wide		80	.200		2.53	7.25		9.78	14.55
0400	Marble, 9" wide, 1" thick		85	.188		8.50	6.85		15.35	20.50
0900	Slate, colored, unfading, honed, 12" wide, 1" thick		85	.188		17.10	6.85		23.95	30
0950	2" thick	↓	70	.229	↓	24	8.30		32.30	39.50
9000	Minimum labor/equipment charge	1 Bric	2	4	Job		162		162	262

04 54 Refractory Brick Masonry

04 54 10 – Refractory Brick Work

04 54 10.20 Fire Clay		Crew	Daily Output	Labor-Hours	Unit	Material	2009 Bare Costs Labor	Equipment	Total	Total Incl O&P
0010	**FIRE CLAY**									
0020	Gray, high duty, 100 lb. bag				Bag	59			59	65
0050	100 lb. drum, premixed (400 brick per drum)				Drum	73			73	80.50

04 57 Masonry Fireplaces

04 57 10 – Brick or Stone Fireplaces

04 57 10.10 Fireplace		Crew	Daily Output	Labor-Hours	Unit	Material	2009 Bare Costs Labor	Equipment	Total	Total Incl O&P
0010	**FIREPLACE**									
0100	Brick fireplace, not incl. foundations or chimneys									
0110	30" x 29" opening, incl. chamber, plain brickwork	D-1	.40	40	Ea.	535	1,450		1,985	2,950
0200	Fireplace box only (110 brick)	"	2	8	"	176	291		467	665
0300	For elaborate brickwork and details, add					35%	35%			
0400	For hearth, brick & stone, add	D-1	2	8	Ea.	197	291		488	685
0410	For steel angle, damper, cleanouts, add		4	4		138	145		283	385
0600	Plain brickwork, incl. metal circulator		.50	32	↓	1,025	1,175		2,200	3,000
0800	Face brick only, standard size, 8" x 2-2/3" x 4"	↓	.30	53.333	M	535	1,950		2,485	3,725
0900	Stone fireplace, fieldstone, add				SF Face	14.05			14.05	15.50
1000	Cut stone, add				"	15.50			15.50	17.05
9000	Minimum labor/equipment charge	D-1	2	8	Job		291		291	470

04 71 Manufactured Brick Masonry

04 71 10 – Simulated or Manufactured Brick

04 71 10.10 Simulated Brick		Crew	Daily Output	Labor-Hours	Unit	Material	2009 Bare Costs Labor	Equipment	Total	Total Incl O&P
0010	**SIMULATED BRICK**									
0020	Aluminum, baked on colors	1 Carp	200	.040	S.F.	3.53	1.60		5.13	6.50
0050	Fiberglass panels		200	.040		3.87	1.60		5.47	6.90
0100	Urethane pieces cemented in mastic		150	.053		6.70	2.13		8.83	10.85
0150	Vinyl siding panels	↓	200	.040	↓	2.67	1.60		4.27	5.60

04 72 10 – Cast Stone Masonry Features

04 72 10.10 Coping	Crew	Daily Output	Labor-Hours	Unit	Material	2009 Bare Costs Labor	Equipment	Total	Total Incl O&P
0010 **COPING**, stock units									
0050 Precast concrete, 10" wide, 4" tapers to 3-1/2", 8" wall	D-1	75	.213	L.F.	17.40	7.75		25.15	31.50
0100 12" wide, 3-1/2" tapers to 3", 10" wall		70	.229		17.40	8.30		25.70	32.50
0150 16" wide, 4" tapers to 3-1/2", 14" wall		60	.267		17.10	9.70		26.80	34.50
0300 Limestone for 12" wall, 4" thick		90	.178		15.35	6.45		21.80	27.50
0350 6" thick		80	.200		23	7.25		30.25	37.50
0500 Marble, to 4" thick, no wash, 9" wide		90	.178		23	6.45		29.45	36
0550 12" wide		80	.200		35	7.25		42.25	50.50
0700 Terra cotta, 9" wide		90	.178		5.60	6.45		12.05	16.60
0750 12" wide		80	.200		9.20	7.25		16.45	22
0800 Aluminum, for 12" wall		80	.200		13.30	7.25		20.55	26.50
9000 Minimum labor/equipment charge		2	8	Job		291		291	470

04 72 20 – Cultured Stone Veneer

04 72 20.10 Cultured Stone Veneer Components	Crew	Daily Output	Labor-Hours	Unit	Material	2009 Bare Costs Labor	Equipment	Total	Total Incl O&P
0010 **CULTURED STONE VENEER COMPONENTS**									
0110 On wood frame and sheathing substrate, random sized cobbles, corner stones	D-8	70	.571	V.L.F.	12	21		33	47.50
0120 Field stones		140	.286	S.F.	8.70	10.60		19.30	26.50
0130 Random sized flats, corner stones		70	.571	V.L.F.	11.80	21		32.80	47.50
0140 Field stones		140	.286	S.F.	9.90	10.60		20.50	28
0150 Horizontal lined ledgestones, corner stones		75	.533	V.L.F.	12	19.80		31.80	45
0160 Field stones		150	.267	S.F.	8.70	9.90		18.60	25.50
0170 Random shaped flats, corner stones		65	.615	V.L.F.	12	23		35	50
0180 Field stones		150	.267	S.F.	8.70	9.90		18.60	25.50
0190 Random shaped / textured face, corner stones		65	.615	V.L.F.	12	23		35	50
0200 Field stones		130	.308	S.F.	8.70	11.45		20.15	28
0210 Random shaped river rock, corner stones		65	.615	V.L.F.	12	23		35	50
0220 Field stones		130	.308	S.F.	8.70	11.45		20.15	28
0240 On concrete or CMU substrate, random sized cobbles, corner stones		70	.571	V.L.F.	11.25	21		32.25	47
0250 Field stones		140	.286	S.F.	8.30	10.60		18.90	26.50
0260 Random sized flats, corner stones		70	.571	V.L.F.	11	21		32	46.50
0270 Field stones		140	.286	S.F.	9.50	10.60		20.10	27.50
0280 Horizontal lined ledgestones, corner stones		75	.533	V.L.F.	11.25	19.80		31.05	44.50
0290 Field stones		150	.267	S.F.	8.30	9.90		18.20	25
0300 Random shaped flats, corner stones		70	.571	V.L.F.	11.25	21		32.25	47
0310 Field stones		140	.286	S.F.	8.30	10.60		18.90	26.50
0320 Random shaped / textured face, corner stones		65	.615	V.L.F.	11.25	23		34.25	49.50
0330 Field stones		130	.308	S.F.	8.30	11.45		19.75	27.50
0340 Random shaped river rock, corner stones		65	.615	V.L.F.	11.25	23		34.25	49.50
0350 Field stones		130	.308	S.F.	8.30	11.45		19.75	27.50
0360 Cultured stone veneer, #15 felt weather resistant barrier	1 Clab	3700	.002	Sq.	4.77	.07		4.84	5.35
0370 Expanded metal lath, diamond, 2.5 lb/sy, galvanized	1 Lath	85	.094	S.Y.	3.47	3.35		6.82	9.10
0390 Water table or window sill, 18" long	1 Bric	80	.100	Ea.	47.50	4.05		51.55	58.50

Estimating Tips

05 05 00 Common Work Results for Metals

- Nuts, bolts, washers, connection angles, and plates can add a significant amount to both the tonnage of a structural steel job and the estimated cost. As a rule of thumb, add 10% to the total weight to account for these accessories.

- Type 2 steel construction, commonly referred to as "simple construction," consists generally of field-bolted connections with lateral bracing supplied by other elements of the building, such as masonry walls or x-bracing. The estimator should be aware, however, that shop connections may be accomplished by welding or bolting. The method may be particular to the fabrication shop and may have an impact on the estimated cost.

05 12 23 Structural Steel

- Steel items can be obtained from two sources: a fabrication shop or a metals service center. Fabrication shops can fabricate items under more controlled conditions than can crews in the field. They are also more efficient and can produce items more economically. Metal service centers serve as a source of long mill shapes to both fabrication shops and contractors.

- Most line items in this structural steel subdivision, and most items in 05 50 00 Metal Fabrications, are indicated as being shop fabricated. The bare material cost for these shop fabricated items is the "Invoice Cost" from the shop and includes the mill base price of steel plus mill extras, transportation to the shop, shop drawings and detailing where warranted, shop fabrication and handling, sandblasting and a shop coat of primer paint, all necessary structural bolts, and delivery to the job site. The bare labor cost and bare equipment cost for these shop fabricated items is for field installation or erection.

- Line items in Subdivision 05 12 23.40 Lightweight Framing, and other items scattered in Division 5, are indicated as being field fabricated. The bare material cost for these field fabricated items is the "Invoice Cost" from the metals service center and includes the mill base price of steel plus mill extras, transportation to the metals service center, material handling, and delivery of long lengths of mill shapes to the job site. Material costs for structural bolts and welding rods should be added to the estimate. The bare labor cost and bare equipment cost for these items is for both field fabrication and field installation or erection, and include time for cutting, welding and drilling in the fabricated metal items. Drilling into concrete and fasteners to fasten field fabricated items to other work are not included and should be added to the estimate.

05 20 00 Steel Joist Framing

- In any given project the total weight of open web steel joists is determined by the loads to be supported and the design. However, economies can be realized in minimizing the amount of labor used to place the joists. This is done by maximizing the joist spacing, and therefore minimizing the number of joists required to be installed on the job. Certain spacings and locations may be required by the design, but in other cases maximizing the spacing and keeping it as uniform as possible will keep the costs down.

05 30 00 Steel Decking

- The takeoff and estimating of metal deck involves more than simply the area of the floor or roof and the type of deck specified or shown on the drawings. Many different sizes and types of openings may exist. Small openings for individual pipes or conduits may be drilled after the floor/roof is installed, but larger openings may require special deck lengths as well as reinforcing or structural support. The estimator should determine who will be supplying this reinforcing. Additionally, some deck terminations are part of the deck package, such as screed angles and pour stops, and others will be part of the steel contract, such as angles attached to structural members and cast-in-place angles and plates. The estimator must ensure that all pieces are accounted for in the complete estimate.

05 50 00 Metal Fabrications

- The most economical steel stairs are those that use common materials, standard details, and most importantly, a uniform and relatively simple method of field assembly. Commonly available A36 channels and plates are very good choices for the main stringers of the stairs, as are angles and tees for the carrier members. Risers and treads are usually made by specialty shops, and it is most economical to use a typical detail in as many places as possible. The stairs should be pre-assembled and shipped directly to the site. The field connections should be simple and straightforward to be accomplished efficiently, and with minimum equipment and labor.

Reference Numbers

Reference numbers are shown in shaded boxes at the beginning of some major classifications. These numbers refer to related items in the Reference Section. The reference information may be an estimating procedure, an alternate pricing method, or technical information. *Note:* Not all subdivisions listed here necessarily appear in this publication.

05 05 Common Work Results for Metals

05 05 05 – Selective Metals Demolition

05 05 05.10 Selective Demolition, Metals

		Crew	Daily Output	Labor-Hours	Unit	Material	2009 Bare Costs Labor	2009 Bare Costs Equipment	Total	Total Incl O&P
0010	**SELECTIVE DEMOLITION, METALS** R024119-10									
0015	Excludes shores, bracing, cutting, loading, hauling, dumping									
0020	Remove nuts only up to 3/4" diameter	1 Sswk	480	.017	Ea.		.75		.75	1.38
0030	7/8" to 1-1/4" diameter		240	.033			1.49		1.49	2.76
0040	1-3/8" to 2" diameter		160	.050			2.24		2.24	4.14
0060	Unbolt and remove structural bolts up to 3/4" diameter		240	.033			1.49		1.49	2.76
0070	7/8" to 2" diameter		160	.050			2.24		2.24	4.14
0140	Light weight framing members, remove whole or cut up, up to 20 lb		240	.033			1.49		1.49	2.76
0150	21 – 40 lb	2 Sswk	210	.076			3.41		3.41	6.30
0160	41 – 80 lb	3 Sswk	180	.133			5.95		5.95	11.05
0170	81 – 120 lb	4 Sswk	150	.213			9.55		9.55	17.65
0230	Structural members, remove whole or cut up, up to 500 lb	E-19	48	.500			21.50	22.50	44	63.50
0240	1/4 – 2 tons	E-18	36	1.111			49.50	30	79.50	122
0250	2 – 5 tons	E-24	30	1.067			47	26.50	73.50	113
0260	5 – 10 tons	E-20	24	2.667			117	55.50	172.50	269
0270	10 – 15 tons	E-2	18	3.111			135	96.50	231.50	345
0340	Fabricated item, remove whole or cut up, up to 20 lb	1 Sswk	96	.083			3.72		3.72	6.90
0350	21 – 40 lb	2 Sswk	84	.190			8.50		8.50	15.75
0360	41 – 80 lb	3 Sswk	72	.333			14.90		14.90	27.50
0370	81 – 120 lb	4 Sswk	60	.533			24		24	44
0380	121 – 500 lb	E-19	48	.500			21.50	22.50	44	63.50
0390	501 – 1000 lb	"	36	.667			29	30	59	84.50
2950	Minimum labor/equipment charge	B-13	2	28	Job		955	395	1,350	1,975

05 05 21 – Fastening Methods for Metal

05 05 21.10 Cutting Steel

		Crew	Daily Output	Labor-Hours	Unit	Material	2009 Bare Costs Labor	2009 Bare Costs Equipment	Total	Total Incl O&P
0010	**CUTTING STEEL**									
0020	Hand burning, incl. preparation, torch cutting & grinding, no staging									
0100	Steel to 1/2" thick	E-25	320	.025	L.F.		1.17	.29	1.46	2.48
0150	3/4" thick		260	.031			1.44	.36	1.80	3.05
0200	1" thick		200	.040			1.87	.46	2.33	3.97
9000	Minimum labor/equipment charge		2	4	Job		187	46.50	233.50	395

05 05 21.15 Drilling Steel

		Crew	Daily Output	Labor-Hours	Unit	Material	2009 Bare Costs Labor	2009 Bare Costs Equipment	Total	Total Incl O&P
0010	**DRILLING STEEL**									
1910	Drilling & layout for steel, up to 1/4" deep, no anchor									
1920	Holes, 1/4" diameter	1 Sswk	112	.071	Ea.	.08	3.19		3.27	6
1925	For each additional 1/4" depth, add		336	.024		.08	1.06		1.14	2.06
1930	3/8" diameter		104	.077		.09	3.44		3.53	6.45
1935	For each additional 1/4" depth, add		312	.026		.09	1.15		1.24	2.22
1940	1/2" diameter		96	.083		.10	3.72		3.82	7
1945	For each additional 1/4" depth, add		288	.028		.10	1.24		1.34	2.41
1950	5/8" diameter		88	.091		.13	4.06		4.19	7.70
1955	For each additional 1/4" depth, add		264	.030		.13	1.35		1.48	2.66
1960	3/4" diameter		80	.100		.15	4.47		4.62	8.45
1965	For each additional 1/4" depth, add		240	.033		.15	1.49		1.64	2.93
1970	7/8" diameter		72	.111		.18	4.97		5.15	9.40
1975	For each additional 1/4" depth, add		216	.037		.18	1.66		1.84	3.27
1980	1" diameter		64	.125		.20	5.60		5.80	10.55
1985	For each additional 1/4" depth, add		192	.042		.20	1.86		2.06	3.67
1990	For drilling up, add						40%			
2000	Minimum labor/equipment charge	1 Carp	4	2	Job		80		80	132

05 05 21.90 Welding Steel

		Crew	Daily Output	Labor-Hours	Unit	Material	2009 Bare Costs Labor	2009 Bare Costs Equipment	Total	Total Incl O&P
0010	**WELDING STEEL**, Structural R050521-20									

05 05 21 – Fastening Methods for Metal

05 05 21.90 Welding Steel		Crew	Daily Output	Labor-Hours	Unit	Material	2009 Bare Costs Labor	Equipment	Total	Total Incl O&P
0020	Field welding, 1/8" E6011, cost per welder, no oper. engr	E-14	8	1	Hr.	4.73	46.50	16.80	68.03	110
0200	With 1/2 operating engineer	E-13	8	1.500		4.73	66	16.75	87.48	141
0300	With 1 operating engineer	E-12	8	2	▼	4.73	86	16.80	107.53	172
0500	With no operating engineer, 2# weld rod per ton	E-14	8	1	Ton	4.73	46.50	16.80	68.03	110
0600	8# E6011 per ton	"	2	4		18.90	187	67	272.90	440
0800	With one operating engineer per welder, 2# E6011 per ton	E-12	8	2		4.73	86	16.80	107.53	172
0900	8# E6011 per ton	"	2	8	▼	18.90	345	67	430.90	685
1200	Continuous fillet, stick welding, incl. equipment									
1300	Single pass, 1/8" thick, 0.1#/L.F.	E-14	150	.053	L.F.	.24	2.49	.89	3.62	5.85
1400	3/16" thick, 0.2#/L.F.		75	.107		.47	4.98	1.79	7.24	11.75
1500	1/4" thick, 0.3#/L.F.		50	.160		.71	7.45	2.68	10.84	17.60
1610	5/16" thick, 0.4#/L.F.		38	.211		.95	9.85	3.53	14.33	23
1800	3 passes, 3/8" thick, 0.5#/L.F.		30	.267		1.18	12.45	4.47	18.10	29
2010	4 passes, 1/2" thick, 0.7#/L.F.		22	.364		1.66	17	6.10	24.76	40
2200	5 to 6 passes, 3/4" thick, 1.3#/L.F.		12	.667		3.07	31	11.20	45.27	73
2400	8 to 11 passes, 1" thick, 2.4#/L.F.	▼	6	1.333		5.65	62.50	22.50	90.65	146
2600	For all position welding, add, minimum						20%			
2700	Maximum						300%			
2900	For semi-automatic welding, deduct, minimum						5%			
3000	Maximum				▼		15%			
4000	Cleaning and welding plates, bars, or rods									
4010	to existing beams, columns, or trusses	E-14	12	.667	L.F.	1.18	31	11.20	43.38	71
9000	Minimum labor/equipment charge	"	4	2	Job		93.50	33.50	127	210

05 05 23 – Metal Fastenings

05 05 23.15 Chemical Anchors

		Crew	Daily Output	Labor-Hours	Unit	Material	2009 Bare Costs Labor	Equipment	Total	Total Incl O&P
0010	**CHEMICAL ANCHORS**									
0020	Includes layout & drilling									
1430	Chemical anchor, w/rod & epoxy cartridge, 3/4" diam. x 9-1/2" long	B-89A	27	.593	Ea.	7.55	21.50	4.08	33.13	48
1435	1" diameter x 11-3/4" long		24	.667		12.80	24	4.59	41.39	58.50
1440	1-1/4" diameter x 14" long		21	.762		23.50	27.50	5.25	56.25	77
1445	1-3/4" diameter x 15" long		20	.800		49.50	29	5.50	84	108
1450	18" long		17	.941		59	34	6.50	99.50	128
1455	2" diameter x 18" long		16	1		78.50	36	6.90	121.40	154
1460	24" long	▼	15	1.067	▼	102	38.50	7.35	147.85	184

05 05 23.20 Expansion Anchors

			Crew	Daily Output	Labor-Hours	Unit	Material	2009 Bare Costs Labor	Equipment	Total	Total Incl O&P
0010	**EXPANSION ANCHORS**										
0100	Anchors for concrete, brick or stone, no layout and drilling										
0200	Expansion shields, zinc, 1/4" diameter, 1-5/16" long, single	G	1 Carp	90	.089	Ea.	.28	3.55		3.83	6.15
0300	1-3/8" long, double	G		85	.094		.33	3.76		4.09	6.55
0400	3/8" diameter, 1-1/2" long, single	G		85	.094		.49	3.76		4.25	6.75
0500	2" long, double	G		80	.100		1.10	4		5.10	7.80
0600	1/2" diameter, 2-1/16" long, single	G		80	.100		1.11	4		5.11	7.80
0700	2-1/2" long, double	G		75	.107		1.40	4.26		5.66	8.60
0800	5/8" diameter, 2-5/8" long, single	G		75	.107		1.52	4.26		5.78	8.70
0900	2-3/4" long, double	G		70	.114		2.14	4.57		6.71	9.90
1000	3/4" diameter, 2-3/4" long, single	G		70	.114		1.99	4.57		6.56	9.75
1100	3-15/16" long, double	G		65	.123		3.36	4.92		8.28	11.80
1600	Self-drilling anchor, snap-off, for 3/8" diameter bolt	G		23	.348		1.46	13.90		15.36	24.50
1900	3/4" diameter bolt	G	▼	16	.500	▼	6.30	20		26.30	40
2100	Hollow wall anchors for gypsum wall board, plaster or tile										
2300	1/8" diameter, short	G	1 Carp	160	.050	Ea.	.18	2		2.18	3.50
2400	Long	G	▼	150	.053	▼	.21	2.13		2.34	3.75

05 05 23.20 Expansion Anchors

		Crew	Daily Output	Labor-Hours	Unit	Material	2009 Bare Costs Labor	Equipment	Total	Total Incl O&P
2500	3/16" diameter, short	G 1 Carp	150	.053	Ea.	.36	2.13		2.49	3.92
2600	Long	G	140	.057		.47	2.28		2.75	4.29
2700	1/4" diameter, short	G	140	.057		.45	2.28		2.73	4.27
2800	Long	G	130	.062		.49	2.46		2.95	4.60
3000	Toggle bolts, bright steel, 1/8" diameter, 2" long	G	85	.094		.17	3.76		3.93	6.40
3100	4" long	G	80	.100		.20	4		4.20	6.80
3200	3/16" diameter, 3" long	G	80	.100		.22	4		4.22	6.85
3300	6" long	G	75	.107		.31	4.26		4.57	7.40
3400	1/4" diameter, 3" long	G	75	.107		.30	4.26		4.56	7.40
3500	6" long	G	70	.114		.42	4.57		4.99	8
3600	3/8" diameter, 3" long	G	70	.114		.68	4.57		5.25	8.30
3700	6" long	G	60	.133		1.13	5.35		6.48	10.05
3800	1/2" diameter, 4" long	G	60	.133		1.58	5.35		6.93	10.55
3900	6" long	G	50	.160		1.86	6.40		8.26	12.60
4000	Nailing anchors									
4100	Nylon nailing anchor, 1/4" diameter, 1" long	1 Carp	3.20	2.500	C	9.75	100		109.75	176
4200	1-1/2" long		2.80	2.857		11.20	114		125.20	200
4300	2" long		2.40	3.333		14.25	133		147.25	236
4400	Metal nailing anchor, 1/4" diameter, 1" long	G	3.20	2.500		14.75	100		114.75	181
4500	1-1/2" long	G	2.80	2.857		19.55	114		133.55	210
4600	2" long	G	2.40	3.333		24.50	133		157.50	247
8000	Wedge anchors, not including layout or drilling									
8050	Carbon steel, 1/4" diameter, 1-3/4" long	G 1 Carp	150	.053	Ea.	.25	2.13		2.38	3.79
8100	3-1/4" long	G	140	.057		.32	2.28		2.60	4.13
8150	3/8" diameter, 2-1/4" long	G	145	.055		.32	2.20		2.52	4
8200	5" long	G	140	.057		.57	2.28		2.85	4.40
8250	1/2" diameter, 2-3/4" long	G	140	.057		.97	2.28		3.25	4.84
8300	7" long	G	125	.064		1.66	2.56		4.22	6.05
8350	5/8" diameter, 3-1/2" long	G	130	.062		1	2.46		3.46	5.15
8400	8-1/2" long	G	115	.070		2.12	2.78		4.90	6.90
8450	3/4" diameter, 4-1/4" long	G	115	.070		1.66	2.78		4.44	6.40
8500	10" long	G	95	.084		3.76	3.36		7.12	9.70
8550	1" diameter, 6" long	G	100	.080		6	3.20		9.20	11.90
8575	9" long	G	85	.094		7.80	3.76		11.56	14.80
8600	12" long	G	75	.107		8.45	4.26		12.71	16.35
8650	1-1/4" diameter, 9" long	G	70	.114		19.45	4.57		24.02	29
8700	12" long	G	60	.133		25	5.35		30.35	36.50
8750	For type 303 stainless steel, add					350%				
8800	For type 316 stainless steel, add					450%				
8950	Self-drilling concrete screw, hex washer head, 3/16" diam. x 1-3/4" long	G 1 Carp	300	.027	Ea.	.16	1.07		1.23	1.94
8960	2-1/4" long	G	250	.032		.18	1.28		1.46	2.31
8970	Phillips flat head, 3/16" diam. x 1-3/4" long	G	300	.027		.15	1.07		1.22	1.93
8980	2-1/4" long	G	250	.032		.17	1.28		1.45	2.30
9000	Minimum labor/equipment charge		4	2	Job		80		80	132

05 05 23.30 Lag Screws

		Crew	Daily Output	Labor-Hours	Unit	Material	2009 Bare Costs Labor	Equipment	Total	Total Incl O&P
0010	**LAG SCREWS**									
0020	Steel, 1/4" diameter, 2" long	G 1 Carp	200	.040	Ea.	.08	1.60		1.68	2.73
0200	1/2" diameter, 3" long	G	130	.062		.45	2.46		2.91	4.56
0300	5/8" diameter, 3" long	G	120	.067		1.12	2.66		3.78	5.65

05 05 23.40 Machinery Anchors

		Crew	Daily Output	Labor-Hours	Unit	Material	2009 Bare Costs Labor	Equipment	Total	Total Incl O&P
0010	**MACHINERY ANCHORS**, heavy duty, incl. sleeve, floating base nut,									
0020	lower stud & coupling nut, fiber plug, connecting stud, washer & nut.									

05 05 Common Work Results for Metals

05 05 23 – Metal Fastenings

05 05 23.40 Machinery Anchors

		Crew	Daily Output	Labor-Hours	Unit	Material	2009 Bare Costs Labor	Equipment	Total	Total Incl O&P
0030	For flush mounted embedment in poured concrete heavy equip. pads.									
0200	Stud & bolt, 1/2" diameter	E-16	40	.400	Ea.	67	18.30	3.36	88.66	112
0300	5/8" diameter		35	.457		74.50	21	3.84	99.34	125
0500	3/4" diameter		30	.533		86	24.50	4.47	114.97	144
0600	7/8" diameter		25	.640		93.50	29.50	5.35	128.35	163
0800	1" diameter		20	.800		98.50	36.50	6.70	141.70	183
0900	1-1/4" diameter		15	1.067		131	49	8.95	188.95	244

05 05 23.50 Powder Actuated Tools and Fasteners

		Crew	Daily Output	Labor-Hours	Unit	Material	2009 Bare Costs Labor	Equipment	Total	Total Incl O&P
0010	**POWDER ACTUATED TOOLS & FASTENERS**									
0020	Stud driver, .22 caliber, buy, minimum				Ea.	224			224	246
0100	Maximum				"	360			360	395
0300	Powder charges for above, low velocity				C	8.65			8.65	9.50
0400	Standard velocity					12.30			12.30	13.55
0600	Drive pins & studs, 1/4" & 3/8" diam., to 3" long, minimum	1 Carp	4.80	1.667		2.88	66.50		69.38	113
0700	Maximum	"	4	2		11.25	80		91.25	144
0800	Pneumatic stud driver for 1/8" diameter studs				Ea.	1,550			1,550	1,725
0900	Drive pins for above, 1/2" to 3/4" long	1 Carp	1	8	M	355	320		675	925

05 05 23.87 Weld Studs

		Crew	Daily Output	Labor-Hours	Unit	Material	2009 Bare Costs Labor	Equipment	Total	Total Incl O&P
0010	**WELD STUDS**									
0020	1/4" diameter, 2-11/16" long	E-10	1120	.014	Ea.	.32	.65	.34	1.31	1.93
0100	4-1/8" long		1080	.015		.30	.68	.35	1.33	1.96
0200	3/8" diameter, 4-1/8" long		1080	.015		.35	.68	.35	1.38	2.02
0300	6-1/8" long		1040	.015		.45	.70	.36	1.51	2.20
9000	Minimum labor/equipment charge	1 Sswk	2	4	Job		179		179	330

05 12 Structural Steel Framing

05 12 23 – Structural Steel for Buildings

05 12 23.05 Canopy Framing

		Crew	Daily Output	Labor-Hours	Unit	Material	2009 Bare Costs Labor	Equipment	Total	Total Incl O&P
0010	**CANOPY FRAMING**									
0020	6" and 8" members, shop fabricated	E-4	3000	.011	Lb.	1.80	.48	.04	2.32	2.92
9000	Minimum labor/equipment charge	1 Sswk	1	8	Job		360		360	660

05 12 23.10 Ceiling Supports

		Crew	Daily Output	Labor-Hours	Unit	Material	2009 Bare Costs Labor	Equipment	Total	Total Incl O&P
0010	**CEILING SUPPORTS**									
1000	Entrance door/folding partition supports, shop fabricated	E-4	60	.533	L.F.	30	24	2.23	56.23	80
1100	Linear accelerator door supports		14	2.286		137	103	9.60	249.60	350
1200	Lintels or shelf angles, hung, exterior hot dipped galv.		267	.120		20.50	5.40	.50	26.40	33
1250	Two coats primer paint instead of galv.		267	.120		17.75	5.40	.50	23.65	30
1400	Monitor support, ceiling hung, expansion bolted		4	8	Ea.	475	360	33.50	868.50	1,225
1450	Hung from pre-set inserts		6	5.333		510	241	22.50	773.50	1,025
1600	Motor supports for overhead doors		4	8		242	360	33.50	635.50	975
1700	Partition support for heavy folding partitions, without pocket		24	1.333	L.F.	68.50	60.50	5.60	134.60	193
1750	Supports at pocket only		12	2.667		137	121	11.15	269.15	385
2000	Rolling grilles & fire door supports		34	.941		58.50	42.50	3.94	104.94	148
2100	Spider-leg light supports, expansion bolted to ceiling slab		8	4	Ea.	195	181	16.75	392.75	570
2150	Hung from pre-set inserts		12	2.667	"	210	121	11.15	342.15	465
2400	Toilet partition support		36	.889	L.F.	68.50	40	3.72	112.22	154
2500	X-ray travel gantry support		12	2.667	"	234	121	11.15	366.15	490

05 12 23.17 Columns, Structural

0010	**COLUMNS, STRUCTURAL** R051223-10									
0015	Made from recycled materials									

05 12 23 – Structural Steel for Buildings

05 12 23.17 Columns, Structural

		Crew	Daily Output	Labor-Hours	Unit	Material	2009 Bare Costs Labor	Equipment	Total	Total Incl O&P
0020	Shop fab'd for 100-ton, 1-2 story project, bolted connections									
0800	Steel, concrete filled, extra strong pipe, 3-1/2" diameter	E-2	660	.085	L.F.	49.50	3.70	2.64	55.84	64
0830	4" diameter		780	.072		55	3.13	2.23	60.36	68.50
0890	5" diameter		1020	.055		65.50	2.39	1.71	69.60	78.50
0930	6" diameter		1200	.047		87	2.03	1.45	90.48	101
0940	8" diameter		1100	.051		87	2.22	1.58	90.80	102
1500	Steel pipe, extra strong, no concrete, 3" to 5" diameter Ⓖ		16000	.004	Lb.	1.50	.15	.11	1.76	2.04
3300	Structural tubing, square, A500GrB, 4" to 6" square, light section Ⓖ		11270	.005	"	1.50	.22	.15	1.87	2.21
8090	For projects 75 to 99 tons, add				All	10%				
8092	50 to 74 tons, add					20%				
8094	25 to 49 tons, add					30%	10%			
8096	10 to 24 tons, add					50%	25%			
8098	2 to 9 tons, add					75%	50%			
8099	Less than 2 tons, add					100%	100%			
9000	Minimum labor/equipment charge	1 Sswk	1	8	Job		360		360	660

05 12 23.20 Curb Edging

		Crew	Daily Output	Labor-Hours	Unit	Material	2009 Bare Costs Labor	Equipment	Total	Total Incl O&P
0010	**CURB EDGING**									
0020	Steel angle w/anchors, shop fabricated, on forms, 1" x 1", 0.8#/L.F. Ⓖ	E-4	350	.091	L.F.	1.76	4.13	.38	6.27	10
0300	4" x 4" angles, 8.2#/L.F. Ⓖ		275	.116	"	15.50	5.25	.49	21.24	27.50
9000	Minimum labor/equipment charge		4	8	Job		360	33.50	393.50	705

05 12 23.40 Lightweight Framing

		Crew	Daily Output	Labor-Hours	Unit	Material	2009 Bare Costs Labor	Equipment	Total	Total Incl O&P
0010	**LIGHTWEIGHT FRAMING**									
0015	Made from recycled materials									
0200	For load-bearing steel studs see Div. 05 41 13.30									
0400	Angle framing, field fabricated, 4" and larger Ⓖ	E-3	440	.055	Lb.	.87	2.47	.30	3.64	5.90
0450	Less than 4" angles Ⓖ		265	.091		.90	4.11	.51	5.52	9.15
0600	Channel framing, field fabricated, 8" and larger Ⓖ		500	.048		.90	2.18	.27	3.35	5.30
0650	Less than 8" channels Ⓖ		335	.072		.90	3.25	.40	4.55	7.45
1000	Continuous slotted channel framing system, shop fab, minimum Ⓖ	2 Sswk	2400	.007		4.65	.30		4.95	5.65
1200	Maximum Ⓖ	"	1600	.010		5.25	.45		5.70	6.65
1250	Plate & bar stock for reinforcing beams and trusses Ⓖ					1.65			1.65	1.82
1300	Cross bracing, rods, shop fabricated, 3/4" diameter Ⓖ	E-3	700	.034		1.80	1.56	.19	3.55	5.05
1310	7/8" diameter Ⓖ		850	.028		1.80	1.28	.16	3.24	4.52
1320	1" diameter Ⓖ		1000	.024		1.80	1.09	.13	3.02	4.15
1330	Angle, 5" x 5" x 3/8" Ⓖ		2800	.009		1.80	.39	.05	2.24	2.75
1350	Hanging lintels, shop fabricated, average Ⓖ		850	.028		1.80	1.28	.16	3.24	4.52
1380	Roof frames, shop fabricated, 3'-0" square, 5' span Ⓖ	E-2	4200	.013		1.80	.58	.41	2.79	3.47
1400	Tie rod, not upset, 1-1/2" to 4" diameter, with turnbuckle Ⓖ	2 Sswk	800	.020		1.95	.89		2.84	3.81
1420	No turnbuckle Ⓖ		700	.023		1.88	1.02		2.90	3.95
1500	Upset, 1-3/4" to 4" diameter, with turnbuckle Ⓖ		800	.020		1.95	.89		2.84	3.81
1520	No turnbuckle Ⓖ		700	.023		1.88	1.02		2.90	3.95
9000	Minimum labor/equipment charge		2	8	Job		360		360	660

05 12 23.45 Lintels

		Crew	Daily Output	Labor-Hours	Unit	Material	2009 Bare Costs Labor	Equipment	Total	Total Incl O&P
0010	**LINTELS**									
0015	Made from recycled materials									
0020	Plain steel angles, shop fabricated, under 500 lb. Ⓖ	1 Bric	550	.015	Lb.	1.16	.59		1.75	2.22
0100	500 to 1000 lb. Ⓖ		640	.013		1.13	.51		1.64	2.06
0200	1,000 to 2,000 lb. Ⓖ		640	.013		1.10	.51		1.61	2.02
0300	2,000 to 4,000 lb. Ⓖ		640	.013		1.07	.51		1.58	1.99
0500	For built-up angles and plates, add to above Ⓖ					.38			.38	.41
0700	For engineering, add to above					.15			.15	.17
0900	For galvanizing, add to above, under 500 lb.					.50			.50	.55

05 12 Structural Steel Framing

05 12 23 – Structural Steel for Buildings

05 12 23.45 Lintels		Crew	Daily Output	Labor-Hours	Unit	Material	2009 Bare Costs Labor	Equipment	Total	Total Incl O&P
0950	500 to 2,000 lb.				Lb.	.45			.45	.50
1000	Over 2,000 lb.					.40			.40	.44
2000	Steel angles, 3-1/2" x 3", 1/4" thick, 2'-6" long	G 1 Bric	47	.170	Ea.	16.20	6.90		23.10	29
2100	4'-6" long	G	26	.308		29	12.45		41.45	52
2500	3-1/2" x 3-1/2" x 5/16", 5'-0" long	G	18	.444		43	18		61	76.50
2600	4" x 3-1/2", 1/4" thick, 5'-0" long	G	21	.381		37	15.45		52.45	66
2700	9'-0" long	G	12	.667		67	27		94	117
2800	4" x 3-1/2" x 5/16", 7'-0" long	G	12	.667		64.50	27		91.50	115
2900	5" x 3-1/2" x 5/16", 10'-0" long	G	8	1		104	40.50		144.50	181
9000	Minimum labor/equipment charge		4	2	Job		81		81	131

05 12 23.60 Pipe Support Framing

		Crew	Daily Output	Labor-Hours	Unit	Material	Labor	Equipment	Total	Total Incl O&P
0010	**PIPE SUPPORT FRAMING**									
0020	Under 10#/L.F., shop fabricated	G E-4	3900	.008	Lb.	2.01	.37	.03	2.41	2.94
0200	10.1 to 15#/L.F.	G	4300	.007		1.98	.34	.03	2.35	2.83
0400	15.1 to 20#/L.F.	G	4800	.007		1.95	.30	.03	2.28	2.74
0600	Over 20#/L.F.	G	5400	.006		1.92	.27	.02	2.21	2.64

05 12 23.65 Plates

		Crew	Daily Output	Labor-Hours	Unit	Material	Labor	Equipment	Total	Total Incl O&P
0010	**PLATES**									
0015	Made from recycled materials									
0020	For connections & stiffener plates, shop fabricated									
0050	1/8" thick (5.1 Lb./S.F.)	G			S.F.	7.65			7.65	8.40
0100	1/4" thick (10.2 Lb./S.F.)	G				15.30			15.30	16.85
0300	3/8" thick (15.3 Lb./S.F.)	G				23			23	25.50
0400	1/2" thick (20.4 Lb./S.F.)	G				30.50			30.50	33.50
0450	3/4" thick (30.6 Lb./S.F.)	G				46			46	50.50
0500	1" thick (40.8 Lb.S.F.)	G				61			61	67.50
2000	Steel plate, warehouse prices, no shop fabrication									
2100	1/4" thick (10.2 Lb./S.F.)	G			S.F.	11.20			11.20	12.35

05 12 23.75 Structural Steel Members

		Crew	Daily Output	Labor-Hours	Unit	Material	Labor	Equipment	Total	Total Incl O&P
0010	**STRUCTURAL STEEL MEMBERS** R051223-10									
0015	Made from recycled materials									
0020	Shop fab'd for 100-ton, 1-2 story project, bolted connections									
0102	W 6 x 9 R051223-15	G E-2	600	.093	L.F.	14.85	4.06	2.90	21.81	27
0302	W 8 x 10	G	600	.093		16.50	4.06	2.90	23.46	28.50
0502	x 31	G	550	.102		51	4.43	3.17	58.60	68
0702	W 10 x 22	G	600	.093		36.50	4.06	2.90	43.46	50.50
0902	x 49	G	550	.102		81	4.43	3.17	88.60	100
1102	W 12 x 16	G	880	.064		26.50	2.77	1.98	31.25	36
1302	x 22	G	880	.064		36.50	2.77	1.98	41.25	47
1502	x 26	G	880	.064		43	2.77	1.98	47.75	54
1902	W 14 x 26	G	990	.057		43	2.46	1.76	47.22	53.50
2102	x 30	G	900	.062		49.50	2.71	1.93	54.14	61.50
2702	W 16 x 26	G	1000	.056		43	2.44	1.74	47.18	53.50
2902	x 31	G	900	.062		51	2.71	1.93	55.64	63.50
8490	For projects 75 to 99 tons, add					10%				
8492	50 to 74 tons, add					20%				
8494	25 to 49 tons, add					30%	10%			
8496	10 to 24 tons, add					50%	25%			
8498	2 to 9 tons, add					75%	50%			
8499	Less than 2 tons, add					100%	100%			
9000	Minimum labor/equipment charge	E-2	2	28	Job		1,225	870	2,095	3,125

05 12 Structural Steel Framing

05 12 23 – Structural Steel for Buildings

05 12 23.77 Structural Steel Projects

05 12 23.77 Structural Steel Projects		Crew	Daily Output	Labor-Hours	Unit	Material	2009 Bare Costs Labor	Equipment	Total	Total Incl O&P
0010	**STRUCTURAL STEEL PROJECTS**									
0015	Made from recycled materials									
0020	Shop fab'd for 100-ton, 1-2 story project, bolted connections									
0700	Offices, hospitals, etc., steel bearing, 1 to 2 stories R050521-20 **G**	E-5	10.30	7.767	Ton	3,000	345	182	3,527	4,125
1300	Industrial bldgs., 1 story, beams & girders, steel bearing **G**		12.90	6.202		3,000	273	145	3,418	3,950
1400	Masonry bearing R051223-10 **G**		10	8		3,000	355	188	3,543	4,150
1600	1 story with roof trusses, steel bearing **G**		10.60	7.547		3,550	335	177	4,062	4,700
1700	Masonry bearing R051223-15 **G**		8.30	9.639		3,550	425	226	4,201	4,925
5390	For projects 75 to 99 tons, add					10%				
5392	50 to 74 tons, add					20%				
5394	25 to 49 tons, add					30%	10%			
5396	10 to 24 tons, add					50%	25%			
5398	2 to 9 tons, add					75%	50%			
5399	Less than 2 tons, add					100%	100%			

05 15 Wire Rope Assemblies

05 15 16 – Steel Wire Rope Assemblies

05 15 16.70 Temporary Cable Safety Railing

05 15 16.70 Temporary Cable Safety Railing		Crew	Daily Output	Labor-Hours	Unit	Material	2009 Bare Costs Labor	Equipment	Total	Total Incl O&P
0010	**TEMPORARY CABLE SAFETY RAILING**, Each 100' strand incl.									
0020	2 eyebolts, 1 turnbuckle, 100' cable, 2 thimbles, 6 clips									
0025	Made from recycled materials									
0100	One strand using 1/4" cable & accessories **G**	2 Sswk	4	4	C.L.F.	194	179		373	545
0200	1/2" cable & accessories **G**	"	2	8	"	390	360		750	1,100

05 21 Steel Joist Framing

05 21 13 – Deep Longspan Steel Joist Framing

05 21 13.50 Deep Longspan Joists

05 21 13.50 Deep Longspan Joists		Crew	Daily Output	Labor-Hours	Unit	Material	2009 Bare Costs Labor	Equipment	Total	Total Incl O&P
0010	**DEEP LONGSPAN JOISTS**									
3010	DLH series, 40-ton job lots, bolted cross bridging, shop primer									
3015	Made from recycled materials									
3020	Spans to 144' (shipped in 2 pieces), minimum **G**	E-7	16	5	Ton	2,225	221	126	2,572	2,975
3040	Average **G**		13	6.154		2,400	271	155	2,826	3,300
3100	Maximum **G**		11	7.273		2,900	320	183	3,403	3,975
3500	For less than 40-ton job lots									
3502	For 30 to 39 tons, add					10%				
3504	20 to 29 tons, add					20%				
3506	10 to 19 tons, add					30%				
3507	5 to 9 tons, add					50%	25%			
3508	1 to 4 tons, add					75%	50%			
3509	Less than 1 ton, add					100%	100%			
4010	SLH series, 40-ton job lots, bolted cross bridging, shop primer									
4020	Spans to 200' (shipped in 3 pieces), minimum **G**	E-7	16	5	Ton	2,200	221	126	2,547	2,950
4040	Average **G**		13	6.154		2,475	271	155	2,901	3,375
4060	Maximum **G**		11	7.273		2,950	320	183	3,453	4,000
6100	For less than 40-ton job lots									
6102	For 30 to 39 tons, add					10%				
6104	20 to 29 tons, add					20%				
6106	10 to 19 tons, add					30%				
6107	5 to 9 tons, add					50%	25%			

05 21 Steel Joist Framing

05 21 13 – Deep Longspan Steel Joist Framing

05 21 13.50 Deep Longspan Joists		Crew	Daily Output	Labor-Hours	Unit	Material	2009 Bare Costs Labor	Equipment	Total	Total Incl O&P
6108	1 to 4 tons, add					75%	50%			
6109	Less than 1 ton, add					100%	100%			

05 21 16 – Longspan Steel Joist Framing

05 21 16.50 Longspan Joists

		Crew	Daily Output	Labor-Hours	Unit	Material	2009 Bare Costs Labor	Equipment	Total	Total Incl O&P
0010	**LONGSPAN JOISTS**									
2000	LH series, 40-ton job lots, bolted cross bridging, shop primer									
2015	Made from recycled materials									
2020	Spans to 96', minimum	G E-7	16	5	Ton	2,050	221	126	2,397	2,775
2040	Average	G	13	6.154		2,250	271	155	2,676	3,125
2080	Maximum	G	11	7.273		2,675	320	183	3,178	3,700
2600	For less than 40-ton job lots									
2602	For 30 to 39 tons, add					10%				
2604	20 to 29 tons, add					20%				
2606	10 to 19 tons, add					30%				
2607	5 to 9 tons, add					50%	25%			
2608	1 to 4 tons, add					75%	50%			
2609	Less than 1 ton, add					100%	100%			

05 21 19 – Open Web Steel Joist Framing

05 21 19.10 Open Web Joists

		Crew	Daily Output	Labor-Hours	Unit	Material	2009 Bare Costs Labor	Equipment	Total	Total Incl O&P
0010	**OPEN WEB JOISTS**									
0015	Made from recycled materials									
0020	K series, 40-ton lots, horiz. bridging, spans to 30', shop primer, minimum	G E-7	15	5.333	Ton	1,825	235	134	2,194	2,575
0050	Average	G	12	6.667		2,050	294	167	2,511	2,975
0080	Maximum	G	9	8.889		2,450	390	223	3,063	3,650
0410	Span 30' to 50', minimum	G	17	4.706		1,775	208	118	2,101	2,450
0440	Average	G	17	4.706		2,000	208	118	2,326	2,700
0460	Maximum	G	10	8		2,125	355	201	2,681	3,175
0800	For less than 40-ton job lots									
0802	For 30 to 39 tons, add					10%				
0804	20 to 29 tons, add					20%				
0806	10 to 19 tons, add					30%				
0807	5 to 9 tons, add					50%	25%			
0808	1 to 4 tons, add					75%	50%			
0809	Less than 1 ton, add					100%	100%			
1010	CS series, 40-ton job lots, horizontal bridging, shop primer									
1020	Spans to 30', minimum	G E-7	15	5.333	Ton	1,875	235	134	2,244	2,650
1040	Average	G	12	6.667		2,100	294	167	2,561	3,025
1060	Maximum	G	9	8.889		2,475	390	223	3,088	3,675
1500	For less than 40-ton job lots									
1502	For 30 to 39 tons, add					10%				
1504	20 to 29 tons, add					20%				
1506	10 to 19 tons, add					30%				
1507	5 to 9 tons, add					50%	25%			
1508	1 to 4 tons, add					75%	50%			
1509	Less than 1 ton, add					100%	100%			
6400	Individual steel bearing plate, 6" x 6" x 1/4" with J-hook	G 1 Bric	160	.050	Ea.	9	2.03		11.03	13.20
9000	Minimum labor and equipment, bar joists	F-6	1	40	Job		1,475	770	2,245	3,275

05 31 Steel Decking

05 31 13 – Steel Floor Decking

05 31 13.50 Floor Decking

			Daily Output	Labor-Hours	Unit	Material	2009 Bare Costs Labor	Equipment	Total	Total Incl O&P
0010	**FLOOR DECKING**	R053100-10								
0015	Made from recycled materials									
5200	Non-cellular composite deck, galv., 2" deep, 22 gauge	G E-4	3860	.008	S.F.	2.44	.37	.03	2.84	3.41
5300	20 gauge	G	3600	.009		2.71	.40	.04	3.15	3.76
5400	18 gauge	G	3380	.009		3.43	.43	.04	3.90	4.60
5500	16 gauge	G	3200	.010		4.30	.45	.04	4.79	5.60
5700	3" deep, galv., 22 gauge	G	3200	.010		2.66	.45	.04	3.15	3.82
5800	20 gauge	G	3000	.011		2.98	.48	.04	3.50	4.22
5900	18 gauge	G	2850	.011		3.65	.51	.05	4.21	5
6000	16 gauge	G	2700	.012		4.89	.54	.05	5.48	6.45
9000	Minimum labor/equipment charge	1 Sswk	1	8	Job		360		360	660

05 31 23 – Steel Roof Decking

05 31 23.50 Roof Decking

			Daily Output	Labor-Hours	Unit	Material	2009 Bare Costs Labor	Equipment	Total	Total Incl O&P
0010	**ROOF DECKING**									
0015	Made from recycled materials									
2100	Open type, galv., 1-1/2" deep wide rib, 22 gauge, under 50 squares	G E-4	4500	.007	S.F.	2.58	.32	.03	2.93	3.46
2400	Over 500 squares	G "	5100	.006	"	1.85	.28	.03	2.16	2.60

05 31 33 – Steel Form Decking

05 31 33.50 Form Decking

			Daily Output	Labor-Hours	Unit	Material	2009 Bare Costs Labor	Equipment	Total	Total Incl O&P
0010	**FORM DECKING**									
0015	Made from recycled materials									
6100	Slab form, steel, 28 gauge, 9/16" deep, uncoated	G E-4	4000	.008	S.F.	1.72	.36	.03	2.11	2.60
6200	Galvanized	G "	4000	.008	"	1.52	.36	.03	1.91	2.38

05 35 Raceway Decking Assemblies

05 35 13 – Steel Cellular Decking

05 35 13.50 Cellular Decking

			Daily Output	Labor-Hours	Unit	Material	2009 Bare Costs Labor	Equipment	Total	Total Incl O&P
0010	**CELLULAR DECKING**									
0015	Made from recycled materials									
0200	Cellular units, galv, 2" deep, 20-20 gauge, over 15 squares	G E-4	1460	.022	S.F.	10.20	.99	.09	11.28	13.20
0400	3" deep, galvanized, 20-20 gauge	G	1375	.023		11.25	1.05	.10	12.40	14.40
1000	4-1/2" deep, galvanized, 20-18 gauge	G	1100	.029		15.70	1.31	.12	17.13	19.80
1500	For acoustical deck, add					15%				
1900	For multi-story or congested site, add						50%			

05 41 Structural Metal Stud Framing

05 41 13 – Load-Bearing Metal Stud Framing

05 41 13.05 Bracing

			Daily Output	Labor-Hours	Unit	Material	2009 Bare Costs Labor	Equipment	Total	Total Incl O&P
0010	**BRACING**, shear wall X-bracing, per 10' x 10' bay, one face									
0015	Made of recycled materials									
0120	Metal strap, 20 ga x 4" wide	G 2 Carp	18	.889	Ea.	27	35.50		62.50	88
0130	6" wide	G	18	.889		42	35.50		77.50	105
0160	18 ga x 4" wide	G	16	1		39	40		79	109
0170	6" wide	G	16	1		57	40		97	129
0410	Continuous strap bracing, per horizontal row on both faces									
0420	Metal strap, 20 ga x 2" wide, studs 12" O.C.	G 1 Carp	7	1.143	C.L.F.	68	45.50		113.50	151
0430	16" O.C.	G	8	1		68	40		108	141
0440	24" O.C.	G	10	.800		68	32		100	128

05 41 Structural Metal Stud Framing

05 41 13 – Load-Bearing Metal Stud Framing

05 41 13.05 Bracing

		Crew	Daily Output	Labor-Hours	Unit	Material	2009 Bare Costs Labor	2009 Bare Costs Equipment	Total	Total Incl O&P
0450	18 ga x 2" wide, studs 12" O.C.	G 1 Carp	6	1.333	C.L.F.	94.50	53.50		148	192
0460	16" O.C.	G	7	1.143		94.50	45.50		140	180
0470	24" O.C.	G ↓	8	1	↓	94.50	40		134.50	170

05 41 13.10 Bridging

		Crew	Daily Output	Labor-Hours	Unit	Material	2009 Bare Costs Labor	2009 Bare Costs Equipment	Total	Total Incl O&P
0010	**BRIDGING**, solid between studs w/ 1-1/4" leg track, per stud bay									
0015	Made from recycled materials									
0200	Studs 12" O.C., 18 ga x 2-1/2" wide	G 1 Carp	125	.064	Ea.	1.16	2.56		3.72	5.50
0210	3-5/8" wide	G	120	.067		1.40	2.66		4.06	5.95
0220	4" wide	G	120	.067		1.48	2.66		4.14	6.05
0230	6" wide	G	115	.070		1.93	2.78		4.71	6.70
0240	8" wide	G	110	.073		2.45	2.91		5.36	7.50
0300	16 ga x 2-1/2" wide	G	115	.070		1.44	2.78		4.22	6.20
0310	3-5/8" wide	G	110	.073		1.76	2.91		4.67	6.75
0320	4" wide	G	110	.073		1.88	2.91		4.79	6.85
0330	6" wide	G	105	.076		2.42	3.04		5.46	7.65
0340	8" wide	G	100	.080		3.08	3.20		6.28	8.70
1200	Studs 16" O.C., 18 ga x 2-1/2" wide	G	125	.064		1.49	2.56		4.05	5.85
1210	3-5/8" wide	G	120	.067		1.80	2.66		4.46	6.40
1220	4" wide	G	120	.067		1.90	2.66		4.56	6.50
1230	6" wide	G	115	.070		2.48	2.78		5.26	7.30
1240	8" wide	G	110	.073		3.14	2.91		6.05	8.25
1300	16 ga x 2-1/2" wide	G	115	.070		1.85	2.78		4.63	6.60
1310	3-5/8" wide	G	110	.073		2.26	2.91		5.17	7.30
1320	4" wide	G	110	.073		2.41	2.91		5.32	7.45
1330	6" wide	G	105	.076		3.10	3.04		6.14	8.40
1340	8" wide	G	100	.080		3.94	3.20		7.14	9.65
2200	Studs 24" O.C., 18 ga x 2-1/2" wide	G	125	.064		2.15	2.56		4.71	6.60
2210	3-5/8" wide	G	120	.067		2.60	2.66		5.26	7.25
2220	4" wide	G	120	.067		2.75	2.66		5.41	7.40
2230	6" wide	G	115	.070		3.58	2.78		6.36	8.55
2240	8" wide	G	110	.073		4.54	2.91		7.45	9.80
2300	16 ga x 2-1/2" wide	G	115	.070		2.67	2.78		5.45	7.55
2310	3-5/8" wide	G	110	.073		3.27	2.91		6.18	8.40
2320	4" wide	G	110	.073		3.49	2.91		6.40	8.65
2330	6" wide	G	105	.076		4.49	3.04		7.53	9.95
2340	8" wide	G ↓	100	.080	↓	5.70	3.20		8.90	11.60
3000	Continuous bridging, per row									
3100	16 ga x 1-1/2" channel thru studs 12" O.C.	G 1 Carp	6	1.333	C.L.F.	65	53.50		118.50	160
3110	16" O.C.	G	7	1.143		65	45.50		110.50	147
3120	24" O.C.	G	8.80	.909		65	36.50		101.50	132
4100	2" x 2" angle x 18 ga, studs 12" O.C.	G	7	1.143		94.50	45.50		140	180
4110	16" O.C.	G	9	.889		94.50	35.50		130	163
4120	24" O.C.	G	12	.667		94.50	26.50		121	148
4200	16 ga, studs 12" O.C.	G	5	1.600		120	64		184	238
4210	16" O.C.	G	7	1.143		120	45.50		165.50	208
4220	24" O.C.	G ↓	10	.800	↓	120	32		152	185

05 41 13.25 Framing, Boxed Headers/Beams

		Crew	Daily Output	Labor-Hours	Unit	Material	2009 Bare Costs Labor	2009 Bare Costs Equipment	Total	Total Incl O&P
0010	**FRAMING, BOXED HEADERS/BEAMS**									
0015	Made from recycled materials									
0200	Double, 18 ga x 6" deep	G 2 Carp	220	.073	L.F.	6.70	2.91		9.61	12.20
0210	8" deep	G	210	.076		7.45	3.04		10.49	13.20
0220	10" deep	G ↓	200	.080	↓	9	3.20		12.20	15.25

05 41 13 – Load-Bearing Metal Stud Framing

05 41 13.25 Framing, Boxed Headers/Beams		Crew	Daily Output	Labor-Hours	Unit	Material	2009 Bare Costs Labor	Equipment	Total	Total Incl O&P
0230	12" deep	G	2 Carp	190	.084	L.F.	9.90	3.36		13.26	16.45
0300	16 ga x 8" deep	G		180	.089		8.55	3.55		12.10	15.25
0310	10" deep	G		170	.094		10.30	3.76		14.06	17.50
0320	12" deep	G		160	.100		11.20	4		15.20	18.90
0400	14 ga x 10" deep	G		140	.114		11.95	4.57		16.52	20.50
0410	12" deep	G		130	.123		13.10	4.92		18.02	22.50
1210	Triple, 18 ga x 8" deep	G		170	.094		10.80	3.76		14.56	18.10
1220	10" deep	G		165	.097		13	3.87		16.87	20.50
1230	12" deep	G		160	.100		14.35	4		18.35	22.50
1300	16 ga x 8" deep	G		145	.110		12.45	4.41		16.86	21
1310	10" deep	G		140	.114		14.85	4.57		19.42	24
1320	12" deep	G		135	.119		16.20	4.73		20.93	25.50
1400	14 ga x 10" deep	G		115	.139		16.30	5.55		21.85	27
1410	12" deep	G	▼	110	.145	▼	18.10	5.80		23.90	29.50

05 41 13.30 Framing, Stud Walls

			Crew	Daily Output	Labor-Hours	Unit	Material	2009 Bare Costs Labor	Equipment	Total	Total Incl O&P
0010	**FRAMING, STUD WALLS** w/ top & bottom track, no openings,										
0020	Headers, beams, bridging or bracing										
0025	Made from recycled materials										
4100	8' high walls, 18 ga x 2-1/2" wide, studs 12" O.C.	G	2 Carp	54	.296	L.F.	11.25	11.85		23.10	32
4110	16" O.C.	G		77	.208		9	8.30		17.30	23.50
4120	24" O.C.	G		107	.150		6.75	5.95		12.70	17.25
4130	3-5/8" wide, studs 12" O.C.	G		53	.302		13.40	12.05		25.45	34.50
4140	16" O.C.	G		76	.211		10.70	8.40		19.10	25.50
4150	24" O.C.	G		105	.152		8.05	6.10		14.15	18.90
4160	4" wide, studs 12" O.C.	G		52	.308		14	12.30		26.30	36
4170	16" O.C.	G		74	.216		11.20	8.65		19.85	26.50
4180	24" O.C.	G		103	.155		8.40	6.20		14.60	19.50
4190	6" wide, studs 12" O.C.	G		51	.314		17.85	12.55		30.40	40
4200	16" O.C.	G		73	.219		14.30	8.75		23.05	30
4210	24" O.C.	G		101	.158		10.75	6.35		17.10	22.50
4220	8" wide, studs 12" O.C.	G		50	.320		22	12.80		34.80	45
4230	16" O.C.	G		72	.222		17.50	8.90		26.40	34
4240	24" O.C.	G		100	.160		13.20	6.40		19.60	25
4300	16 ga x 2-1/2" wide, studs 12" O.C.	G		47	.340		13.25	13.60		26.85	37
4310	16" O.C.	G		68	.235		10.50	9.40		19.90	27
4320	24" O.C.	G		94	.170		7.75	6.80		14.55	19.75
4330	3-5/8" wide, studs 12" O.C.	G		46	.348		15.85	13.90		29.75	40.50
4340	16" O.C.	G		66	.242		12.55	9.70		22.25	30
4350	24" O.C.	G		92	.174		9.25	6.95		16.20	21.50
4360	4" wide, studs 12" O.C.	G		45	.356		16.75	14.20		30.95	42
4370	16" O.C.	G		65	.246		13.25	9.85		23.10	31
4380	24" O.C.	G		90	.178		9.80	7.10		16.90	22.50
4390	6" wide, studs 12" O.C.	G		44	.364		21	14.55		35.55	47
4400	16" O.C.	G		64	.250		16.70	10		26.70	35
4410	24" O.C.	G		88	.182		12.35	7.25		19.60	25.50
4420	8" wide, studs 12" O.C.	G		43	.372		26	14.85		40.85	53
4430	16" O.C.	G		63	.254		20.50	10.15		30.65	39.50
4440	24" O.C.	G		86	.186		15.25	7.45		22.70	29
5100	10' high walls, 18 ga x 2-1/2" wide, studs 12" O.C.	G		54	.296		13.50	11.85		25.35	34.50
5110	16" O.C.	G		77	.208		10.70	8.30		19	25.50
5120	24" O.C.	G		107	.150		7.85	5.95		13.80	18.50
5130	3-5/8" wide, studs 12" O.C.	G	▼	53	.302		16.05	12.05		28.10	37.50

05 41 13 – Load-Bearing Metal Stud Framing

05 41 13.30 Framing, Stud Walls		Crew	Daily Output	Labor-Hours	Unit	Material	2009 Bare Costs Labor	Equipment	Total	Total Incl O&P
5140	16" O.C.	G	2 Carp	76	.211	L.F.	12.70	8.40		21.10	28
5150	24" O.C.	G		105	.152		9.35	6.10		15.45	20.50
5160	4" wide, studs 12" O.C.	G		52	.308		16.80	12.30		29.10	39
5170	16" O.C.	G		74	.216		13.30	8.65		21.95	29
5180	24" O.C.	G		103	.155		9.80	6.20		16	21
5190	6" wide, studs 12" O.C.	G		51	.314		21.50	12.55		34.05	44
5200	16" O.C.	G		73	.219		16.95	8.75		25.70	33
5210	24" O.C.	G		101	.158		12.55	6.35		18.90	24.50
5220	8" wide, studs 12" O.C.	G		50	.320		26	12.80		38.80	49.50
5230	16" O.C.	G		72	.222		20.50	8.90		29.40	37.50
5240	24" O.C.	G		100	.160		15.35	6.40		21.75	27.50
5300	16 ga x 2-1/2" wide, studs 12" O.C.	G		47	.340		16	13.60		29.60	40
5310	16" O.C.	G		68	.235		12.55	9.40		21.95	29.50
5320	24" O.C.	G		94	.170		9.10	6.80		15.90	21.50
5330	3-5/8" wide, studs 12" O.C.	G		46	.348		19.15	13.90		33.05	44
5340	16" O.C.	G		66	.242		15.05	9.70		24.75	32.50
5350	24" O.C.	G		92	.174		10.90	6.95		17.85	23.50
5360	4" wide, studs 12" O.C.	G		45	.356		20	14.20		34.20	45.50
5370	16" O.C.	G		65	.246		15.85	9.85		25.70	33.50
5380	24" O.C.	G		90	.178		11.50	7.10		18.60	24.50
5390	6" wide, studs 12" O.C.	G		44	.364		25.50	14.55		40.05	52
5400	16" O.C.	G		64	.250		19.95	10		29.95	38.50
5410	24" O.C.	G		88	.182		14.55	7.25		21.80	28
5420	8" wide, studs 12" O.C.	G		43	.372		31	14.85		45.85	59
5430	16" O.C.	G		63	.254		24.50	10.15		34.65	44
5440	24" O.C.	G		86	.186		17.90	7.45		25.35	32
6190	12' high walls, 18 ga x 6" wide, studs 12" O.C.	G		41	.390		25	15.60		40.60	53
6200	16" O.C.	G		58	.276		19.60	11		30.60	39.50
6210	24" O.C.	G		81	.198		14.30	7.90		22.20	29
6220	8" wide, studs 12" O.C.	G		40	.400		30.50	16		46.50	60
6230	16" O.C.	G		57	.281		24	11.20		35.20	45
6240	24" O.C.	G		80	.200		17.50	8		25.50	32.50
6390	16 ga x 6" wide, studs 12" O.C.	G		35	.457		29.50	18.25		47.75	62.50
6400	16" O.C.	G		51	.314		23	12.55		35.55	46
6410	24" O.C.	G		70	.229		16.70	9.15		25.85	33.50
6420	8" wide, studs 12" O.C.	G		34	.471		36.50	18.80		55.30	71
6430	16" O.C.	G		50	.320		28.50	12.80		41.30	52.50
6440	24" O.C.	G		69	.232		20.50	9.25		29.75	38
6530	14 ga x 3-5/8" wide, studs 12" O.C.	G		34	.471		28	18.80		46.80	62
6540	16" O.C.	G		48	.333		22	13.30		35.30	46
6550	24" O.C.	G		65	.246		15.80	9.85		25.65	33.50
6560	4" wide, studs 12" O.C.	G		33	.485		29.50	19.35		48.85	64.50
6570	16" O.C.	G		47	.340		23	13.60		36.60	48
6580	24" O.C.	G		64	.250		16.65	10		26.65	35
6730	12 ga x 3-5/8" wide, studs 12" O.C.	G		31	.516		39	20.50		59.50	77
6740	16" O.C.	G		43	.372		30	14.85		44.85	57.50
6750	24" O.C.	G		59	.271		21	10.85		31.85	41.50
6760	4" wide, studs 12" O.C.	G		30	.533		41.50	21.50		63	81
6770	16" O.C.	G		42	.381		32	15.20		47.20	60.50
6780	24" O.C.	G		58	.276		22.50	11		33.50	43
7390	16' high walls, 16 ga x 6" wide, studs 12" O.C.	G		33	.485		38.50	19.35		57.85	74
7400	16" O.C.	G		48	.333		29.50	13.30		42.80	54.50
7410	24" O.C.	G		67	.239		21	9.55		30.55	39

05 41 Structural Metal Stud Framing

05 41 13 – Load-Bearing Metal Stud Framing

05 41 13.30 Framing, Stud Walls		Crew	Daily Output	Labor-Hours	Unit	Material	2009 Bare Costs Labor	Equipment	Total	Total Incl O&P
7420	8" wide, studs 12" O.C.	G	2 Carp	32	.500	L.F.	47	20		67	85
7430	16" O.C.	G		47	.340		36.50	13.60		50.10	62.50
7440	24" O.C.	G		66	.242		26	9.70		35.70	44.50
7560	14 ga x 4" wide, studs 12" O.C.	G		31	.516		38.50	20.50		59	76.50
7570	16" O.C.	G		45	.356		29.50	14.20		43.70	56
7580	24" O.C.	G		61	.262		21	10.50		31.50	40.50
7590	6" wide, studs 12" O.C.	G		30	.533		48.50	21.50		70	88.50
7600	16" O.C.	G		44	.364		37.50	14.55		52.05	65
7610	24" O.C.	G		60	.267		26.50	10.65		37.15	46.50
7760	12 ga x 4" wide, studs 12" O.C.	G		29	.552		54.50	22		76.50	96.50
7770	16" O.C.	G		40	.400		41.50	16		57.50	72.50
7780	24" O.C.	G		55	.291		29	11.60		40.60	51
7790	6" wide, studs 12" O.C.	G		28	.571		69	23		92	114
7800	16" O.C.	G		39	.410		53	16.40		69.40	85
7810	24" O.C.	G		54	.296		37	11.85		48.85	60
8590	20' high walls, 14 ga x 6" wide, studs 12" O.C.	G		29	.552		59.50	22		81.50	102
8600	16" O.C.	G		42	.381		45.50	15.20		60.70	75.50
8610	24" O.C.	G		57	.281		32	11.20		43.20	53.50
8620	8" wide, studs 12" O.C.	G		28	.571		72.50	23		95.50	118
8630	16" O.C.	G		41	.390		56	15.60		71.60	87
8640	24" O.C.	G		56	.286		39.50	11.40		50.90	62
8790	12 ga x 6" wide, studs 12" O.C.	G		27	.593		85	23.50		108.50	133
8800	16" O.C.	G		37	.432		65	17.30		82.30	100
8810	24" O.C.	G		51	.314		45	12.55		57.55	70
8820	8" wide, studs 12" O.C.	G		26	.615		103	24.50		127.50	155
8830	16" O.C.	G		36	.444		79	17.75		96.75	117
8840	24" O.C.	G		50	.320		54.50	12.80		67.30	81
9000	Minimum labor/equipment charge			4	4	Job		160		160	264

05 42 Cold-Formed Metal Joist Framing

05 42 13 – Cold-Formed Metal Floor Joist Framing

05 42 13.05 Bracing

		Crew	Daily Output	Labor-Hours	Unit	Material	2009 Bare Costs Labor	Equipment	Total	Total Incl O&P
0010	**BRACING**, continuous, per row, top & bottom										
0015	Made from recycled materials										
0120	Flat strap, 20 ga x 2" wide, joists at 12" O.C.	G	1 Carp	4.67	1.713	C.L.F.	71.50	68.50		140	192
0130	16" O.C.	G		5.33	1.501		69	60		129	175
0140	24" O.C.	G		6.66	1.201		66.50	48		114.50	152
0150	18 ga x 2" wide, joists at 12" O.C.	G		4	2		93.50	80		173.50	235
0160	16" O.C.	G		4.67	1.713		92	68.50		160.50	214
0170	24" O.C.	G		5.33	1.501		90.50	60		150.50	199

05 42 13.10 Bridging

		Crew	Daily Output	Labor-Hours	Unit	Material	2009 Bare Costs Labor	Equipment	Total	Total Incl O&P
0010	**BRIDGING**, solid between joists w/ 1-1/4" leg track, per joist bay										
0015	Made from recycled materials										
0230	Joists 12" O.C., 18 ga track x 6" wide	G	1 Carp	80	.100	Ea.	1.93	4		5.93	8.70
0240	8" wide	G		75	.107		2.45	4.26		6.71	9.75
0250	10" wide	G		70	.114		3.02	4.57		7.59	10.90
0260	12" wide	G		65	.123		3.50	4.92		8.42	11.95
0330	16 ga track x 6" wide	G		70	.114		2.42	4.57		6.99	10.20
0340	8" wide	G		65	.123		3.08	4.92		8	11.50
0350	10" wide	G		60	.133		3.78	5.35		9.13	12.95
0360	12" wide	G		55	.145		4.36	5.80		10.16	14.40

05 42 Cold-Formed Metal Joist Framing

05 42 13 – Cold-Formed Metal Floor Joist Framing

05 42 13.10 Bridging		Crew	Daily Output	Labor-Hours	Unit	Material	2009 Bare Costs Labor	Equipment	Total	Total Incl O&P
0440	14 ga track x 8" wide	G	1 Carp	60	.133	Ea.	3.87	5.35		9.22	13.05
0450	10" wide	G		55	.145		4.76	5.80		10.56	14.85
0460	12" wide	G		50	.160		5.50	6.40		11.90	16.60
0550	12 ga track x 10" wide	G		45	.178		7	7.10		14.10	19.40
0560	12" wide	G		40	.200		7.90	8		15.90	22
1230	16" O.C., 18 ga track x 6" wide	G		80	.100		2.48	4		6.48	9.30
1240	8" wide	G		75	.107		3.14	4.26		7.40	10.50
1250	10" wide	G		70	.114		3.88	4.57		8.45	11.80
1260	12" wide	G		65	.123		4.49	4.92		9.41	13.05
1330	16 ga track x 6" wide	G		70	.114		3.10	4.57		7.67	10.95
1340	8" wide	G		65	.123		3.94	4.92		8.86	12.45
1350	10" wide	G		60	.133		4.85	5.35		10.20	14.15
1360	12" wide	G		55	.145		5.60	5.80		11.40	15.75
1440	14 ga track x 8" wide	G		60	.133		4.97	5.35		10.32	14.25
1450	10" wide	G		55	.145		6.10	5.80		11.90	16.30
1460	12" wide	G		50	.160		7.05	6.40		13.45	18.30
1550	12 ga track x 10" wide	G		45	.178		9	7.10		16.10	21.50
1560	12" wide	G		40	.200		10.15	8		18.15	24.50
2230	24" O.C., 18 ga track x 6" wide	G		80	.100		3.58	4		7.58	10.55
2240	8" wide	G		75	.107		4.54	4.26		8.80	12.05
2250	10" wide	G		70	.114		5.60	4.57		10.17	13.70
2260	12" wide	G		65	.123		6.50	4.92		11.42	15.25
2330	16 ga track x 6" wide	G		70	.114		4.49	4.57		9.06	12.50
2340	8" wide	G		65	.123		5.70	4.92		10.62	14.40
2350	10" wide	G		60	.133		7	5.35		12.35	16.50
2360	12" wide	G		55	.145		8.10	5.80		13.90	18.50
2440	14 ga track x 8" wide	G		60	.133		7.20	5.35		12.55	16.70
2450	10" wide	G		55	.145		8.85	5.80		14.65	19.30
2460	12" wide	G		50	.160		10.20	6.40		16.60	22
2550	12 ga track x 10" wide	G		45	.178		13	7.10		20.10	26
2560	12" wide	G		40	.200		14.70	8		22.70	29.50

05 42 13.25 Framing, Band Joist

		Crew	Daily Output	Labor-Hours	Unit	Material	Labor	Equipment	Total	Total Incl O&P
0010	**FRAMING, BAND JOIST** (track) fastened to bearing wall									
0015	Made from recycled materials									
0220	18 ga track x 6" deep	G	2 Carp	1000	.016	L.F.	1.58	.64		2.22	2.79
0230	8" deep	G		920	.017		2	.69		2.69	3.34
0240	10" deep	G		860	.019		2.47	.74		3.21	3.94
0320	16 ga track x 6" deep	G		900	.018		1.97	.71		2.68	3.34
0330	8" deep	G		840	.019		2.51	.76		3.27	4.02
0340	10" deep	G		780	.021		3.09	.82		3.91	4.75
0350	12" deep	G		740	.022		3.56	.86		4.42	5.35
0430	14 ga track x 8" deep	G		750	.021		3.16	.85		4.01	4.89
0440	10" deep	G		720	.022		3.89	.89		4.78	5.75
0450	12" deep	G		700	.023		4.49	.91		5.40	6.45
0540	12 ga track x 10" deep	G		670	.024		5.70	.95		6.65	7.85
0550	12" deep	G		650	.025		6.45	.98		7.43	8.70

05 42 13.30 Framing, Boxed Headers/Beams

		Crew	Daily Output	Labor-Hours	Unit	Material	Labor	Equipment	Total	Total Incl O&P
0010	**FRAMING, BOXED HEADERS/BEAMS**									
0015	Made from recycled materials									
0200	Double, 18 ga x 6" deep	G	2 Carp	220	.073	L.F.	6.70	2.91		9.61	12.20
0210	8" deep	G		210	.076		7.45	3.04		10.49	13.20
0220	10" deep	G		200	.080		9	3.20		12.20	15.25

05 42 13 – Cold-Formed Metal Floor Joist Framing

05 42 13.30 Framing, Boxed Headers/Beams

			Crew	Daily Output	Labor-Hours	Unit	Material	2009 Bare Costs Labor	Equipment	Total	Total Incl O&P
0230	12" deep	G	2 Carp	190	.084	L.F.	9.90	3.36		13.26	16.45
0300	16 ga x 8" deep	G		180	.089		8.55	3.55		12.10	15.25
0310	10" deep	G		170	.094		10.30	3.76		14.06	17.50
0320	12" deep	G		160	.100		11.20	4		15.20	18.90
0400	14 ga x 10" deep	G		140	.114		11.95	4.57		16.52	20.50
0410	12" deep	G		130	.123		13.10	4.92		18.02	22.50
0500	12 ga x 10" deep	G		110	.145		15.80	5.80		21.60	27
0510	12" deep	G		100	.160		17.50	6.40		23.90	30
1210	Triple, 18 ga x 8" deep	G		170	.094		10.80	3.76		14.56	18.10
1220	10" deep	G		165	.097		13	3.87		16.87	20.50
1230	12" deep	G		160	.100		14.35	4		18.35	22.50
1300	16 ga x 8" deep	G		145	.110		12.45	4.41		16.86	21
1310	10" deep	G		140	.114		14.85	4.57		19.42	24
1320	12" deep	G		135	.119		16.20	4.73		20.93	25.50
1400	14 ga x 10" deep	G		115	.139		17.35	5.55		22.90	28.50
1410	12" deep	G		110	.145		19.15	5.80		24.95	30.50
1500	12 ga x 10" deep	G		90	.178		23	7.10		30.10	37
1510	12" deep	G		85	.188		25.50	7.50		33	41

05 42 13.40 Framing, Joists

			Crew	Daily Output	Labor-Hours	Unit	Material	2009 Bare Costs Labor	Equipment	Total	Total Incl O&P
0010	**FRAMING, JOISTS**, no band joists (track), web stiffeners, headers,										
0020	Beams, bridging or bracing										
0025	Made from recycled materials										
0030	Joists (2" flange) and fasteners, materials only										
0220	18 ga x 6" deep	G				L.F.	2.08			2.08	2.29
0230	8" deep	G					2.47			2.47	2.71
0240	10" deep	G					2.90			2.90	3.19
0320	16 ga x 6" deep	G					2.54			2.54	2.80
0330	8" deep	G					3.05			3.05	3.35
0340	10" deep	G					3.56			3.56	3.92
0350	12" deep	G					4.03			4.03	4.44
0430	14 ga x 8" deep	G					3.84			3.84	4.23
0440	10" deep	G					4.43			4.43	4.87
0450	12" deep	G					5.05			5.05	5.55
0540	12 ga x 10" deep	G					6.45			6.45	7.10
0550	12" deep	G					7.35			7.35	8.10
1010	Installation of joists to band joists, beams & headers, labor only										
1220	18 ga x 6" deep		2 Carp	110	.145	Ea.		5.80		5.80	9.60
1230	8" deep			90	.178			7.10		7.10	11.70
1240	10" deep			80	.200			8		8	13.20
1320	16 ga x 6" deep			95	.168			6.75		6.75	11.10
1330	8" deep			70	.229			9.15		9.15	15.05
1340	10" deep			60	.267			10.65		10.65	17.60
1350	12" deep			55	.291			11.60		11.60	19.20
1430	14 ga x 8" deep			65	.246			9.85		9.85	16.25
1440	10" deep			45	.356			14.20		14.20	23.50
1450	12" deep			35	.457			18.25		18.25	30
1540	12 ga x 10" deep			40	.400			16		16	26.50
1550	12" deep			30	.533			21.50		21.50	35
9000	Minimum labor/equipment charge			4	4	Job		160		160	264

05 42 13.45 Framing, Web Stiffeners

0010	**FRAMING, WEB STIFFENERS** at joist bearing, fabricated from										
0020	Stud piece (1-5/8" flange) to stiffen joist (2" flange)										

05 42 Cold-Formed Metal Joist Framing

05 42 13 — Cold-Formed Metal Floor Joist Framing

05 42 13.45 Framing, Web Stiffeners	Crew	Daily Output	Labor-Hours	Unit	Material	2009 Bare Costs Labor	Equipment	Total	Total Incl O&P
0025	Made from recycled materials									
2120	For 6" deep joist, with 18 ga x 2-1/2" stud	[G]	1 Carp	120	.067	Ea.	2.49	2.66		5.15	7.15
2130	3-5/8" stud	[G]		110	.073		2.75	2.91		5.66	7.80
2140	4" stud	[G]		105	.076		2.66	3.04		5.70	7.95
2150	6" stud	[G]		100	.080		2.92	3.20		6.12	8.50
2160	8" stud	[G]		95	.084		3	3.36		6.36	8.85
2220	8" deep joist, with 2-1/2" stud	[G]		120	.067		2.73	2.66		5.39	7.40
2230	3-5/8" stud	[G]		110	.073		2.96	2.91		5.87	8.05
2240	4" stud	[G]		105	.076		2.91	3.04		5.95	8.20
2250	6" stud	[G]		100	.080		3.20	3.20		6.40	8.80
2260	8" stud	[G]		95	.084		3.44	3.36		6.80	9.35
2320	10" deep joist, with 2-1/2" stud	[G]		110	.073		3.85	2.91		6.76	9.05
2330	3-5/8" stud	[G]		100	.080		4.23	3.20		7.43	9.95
2340	4" stud	[G]		95	.084		4.18	3.36		7.54	10.15
2350	6" stud	[G]		90	.089		4.55	3.55		8.10	10.85
2360	8" stud	[G]		85	.094		4.62	3.76		8.38	11.30
2420	12" deep joist, with 2-1/2" stud	[G]		110	.073		4.07	2.91		6.98	9.25
2430	3-5/8" stud	[G]		100	.080		4.42	3.20		7.62	10.15
2440	4" stud	[G]		95	.084		4.34	3.36		7.70	10.30
2450	6" stud	[G]		90	.089		4.78	3.55		8.33	11.10
2460	8" stud	[G]		85	.094		5.15	3.76		8.91	11.85
3130	For 6" deep joist, with 16 ga x 3-5/8" stud	[G]		100	.080		2.89	3.20		6.09	8.50
3140	4" stud	[G]		95	.084		2.87	3.36		6.23	8.70
3150	6" stud	[G]		90	.089		3.15	3.55		6.70	9.30
3160	8" stud	[G]		85	.094		3.31	3.76		7.07	9.85
3230	8" deep joist, with 3-5/8" stud	[G]		100	.080		3.21	3.20		6.41	8.85
3240	4" stud	[G]		95	.084		3.15	3.36		6.51	9
3250	6" stud	[G]		90	.089		3.49	3.55		7.04	9.70
3260	8" stud	[G]		85	.094		3.73	3.76		7.49	10.30
3330	10" deep joist, with 3-5/8" stud	[G]		85	.094		4.38	3.76		8.14	11
3340	4" stud	[G]		80	.100		4.48	4		8.48	11.50
3350	6" stud	[G]		75	.107		4.86	4.26		9.12	12.40
3360	8" stud	[G]		70	.114		5.05	4.57		9.62	13.10
3430	12" deep joist, with 3-5/8" stud	[G]		85	.094		4.79	3.76		8.55	11.45
3440	4" stud	[G]		80	.100		4.70	4		8.70	11.75
3450	6" stud	[G]		75	.107		5.20	4.26		9.46	12.80
3460	8" stud	[G]		70	.114		5.55	4.57		10.12	13.65
4230	For 8" deep joist, with 14 ga x 3-5/8" stud	[G]		90	.089		4.16	3.55		7.71	10.45
4240	4" stud	[G]		85	.094		4.24	3.76		8	10.85
4250	6" stud	[G]		80	.100		4.59	4		8.59	11.65
4260	8" stud	[G]		75	.107		4.92	4.26		9.18	12.45
4330	10" deep joist, with 3-5/8" stud	[G]		75	.107		5.85	4.26		10.11	13.50
4340	4" stud	[G]		70	.114		5.80	4.57		10.37	13.90
4350	6" stud	[G]		65	.123		6.35	4.92		11.27	15.10
4360	8" stud	[G]		60	.133		6.65	5.35		12	16.10
4430	12" deep joist, with 3-5/8" stud	[G]		75	.107		6.20	4.26		10.46	13.90
4440	4" stud	[G]		70	.114		6.30	4.57		10.87	14.50
4450	6" stud	[G]		65	.123		6.85	4.92		11.77	15.65
4460	8" stud	[G]		60	.133		7.35	5.35		12.70	16.90
5330	For 10" deep joist, with 12 ga x 3-5/8" stud	[G]		65	.123		6.15	4.92		11.07	14.90
5340	4" stud	[G]		60	.133		6.35	5.35		11.70	15.75
5350	6" stud	[G]		55	.145		7	5.80		12.80	17.30
5360	8" stud	[G]		50	.160		7.70	6.40		14.10	19

05 42 Cold-Formed Metal Joist Framing

05 42 13 – Cold-Formed Metal Floor Joist Framing

05 42 13.45 Framing, Web Stiffeners

		Crew	Daily Output	Labor-Hours	Unit	Material	2009 Bare Costs Labor	Equipment	Total	Total Incl O&P
5430	12" deep joist, with 3-5/8" stud	[G] 1 Carp	65	.123	Ea.	6.85	4.92		11.77	15.60
5440	4" stud	[G]	60	.133		6.70	5.35		12.05	16.15
5450	6" stud	[G]	55	.145		7.65	5.80		13.45	18
5460	8" stud	[G]	50	.160		8.80	6.40		15.20	20

05 42 23 – Cold-Formed Metal Roof Joist Framing

05 42 23.05 Framing, Bracing

		Crew	Daily Output	Labor-Hours	Unit	Material	2009 Bare Costs Labor	Equipment	Total	Total Incl O&P
0010	**FRAMING, BRACING**									
0015	Made from recycled materials									
0020	Continuous bracing, per row									
0100	16 ga x 1-1/2" channel thru rafters/trusses @ 16" O.C.	[G] 1 Carp	4.50	1.778	C.L.F.	65	71		136	189
0120	24" O.C.	[G]	6	1.333		65	53.50		118.50	160
0300	2" x 2" angle x 18 ga, rafters/trusses @ 16" O.C.	[G]	6	1.333		94.50	53.50		148	192
0320	24" O.C.	[G]	8	1		94.50	40		134.50	170
0400	16 ga, rafters/trusses @ 16" O.C.	[G]	4.50	1.778		120	71		191	249
0420	24" O.C.	[G]	6.50	1.231		120	49		169	213

05 42 23.10 Framing, Bridging

		Crew	Daily Output	Labor-Hours	Unit	Material	2009 Bare Costs Labor	Equipment	Total	Total Incl O&P
0010	**FRAMING, BRIDGING**									
0015	Made from recycled materials									
0020	Solid, between rafters w/ 1-1/4" leg track, per rafter bay									
1200	Rafters 16" O.C., 18 ga x 4" deep	[G] 1 Carp	60	.133	Ea.	1.90	5.35		7.25	10.90
1210	6" deep	[G]	57	.140		2.48	5.60		8.08	11.95
1220	8" deep	[G]	55	.145		3.14	5.80		8.94	13.05
1230	10" deep	[G]	52	.154		3.88	6.15		10.03	14.40
1240	12" deep	[G]	50	.160		4.49	6.40		10.89	15.50
2200	24" O.C., 18 ga x 4" deep	[G]	60	.133		2.75	5.35		8.10	11.80
2210	6" deep	[G]	57	.140		3.58	5.60		9.18	13.20
2220	8" deep	[G]	55	.145		4.54	5.80		10.34	14.60
2230	10" deep	[G]	52	.154		5.60	6.15		11.75	16.30
2240	12" deep	[G]	50	.160		6.50	6.40		12.90	17.70

05 42 23.50 Framing, Parapets

		Crew	Daily Output	Labor-Hours	Unit	Material	2009 Bare Costs Labor	Equipment	Total	Total Incl O&P
0010	**FRAMING, PARAPETS**									
0015	Made from recycled materials									
0100	3' high installed on 1st story, 18 ga x 4" wide studs, 12" O.C.	[G] 2 Carp	100	.160	L.F.	7	6.40		13.40	18.25
0110	16" O.C.	[G]	150	.107		5.95	4.26		10.21	13.60
0120	24" O.C.	[G]	200	.080		4.92	3.20		8.12	10.70
0200	6" wide studs, 12" O.C.	[G]	100	.160		9	6.40		15.40	20.50
0210	16" O.C.	[G]	150	.107		7.65	4.26		11.91	15.45
0220	24" O.C.	[G]	200	.080		6.35	3.20		9.55	12.25
1100	Installed on 2nd story, 18 ga x 4" wide studs, 12" O.C.	[G]	95	.168		7	6.75		13.75	18.80
1110	16" O.C.	[G]	145	.110		5.95	4.41		10.36	13.85
1120	24" O.C.	[G]	190	.084		4.92	3.36		8.28	10.95
1200	6" wide studs, 12" O.C.	[G]	95	.168		9	6.75		15.75	21
1210	16" O.C.	[G]	145	.110		7.65	4.41		12.06	15.70
1220	24" O.C.	[G]	190	.084		6.35	3.36		9.71	12.50
2100	Installed on gable, 18 ga x 4" wide studs, 12" O.C.	[G]	85	.188		7	7.50		14.50	20
2110	16" O.C.	[G]	130	.123		5.95	4.92		10.87	14.65
2120	24" O.C.	[G]	170	.094		4.92	3.76		8.68	11.60
2200	6" wide studs, 12" O.C.	[G]	85	.188		9	7.50		16.50	22.50
2210	16" O.C.	[G]	130	.123		7.65	4.92		12.57	16.50
2220	24" O.C.	[G]	170	.094		6.35	3.76		10.11	13.15

05 42 Cold-Formed Metal Joist Framing

05 42 23 – Cold-Formed Metal Roof Joist Framing

05 42 23.60 Framing, Roof Rafters

		Crew	Daily Output	Labor-Hours	Unit	Material	2009 Bare Costs Labor	Equipment	Total	Total Incl O&P
0010	**FRAMING, ROOF RAFTERS**									
0015	Made from recycled materials									
0100	Boxed ridge beam, double, 18 ga x 6" deep	G 2 Carp	160	.100	L.F.	6.70	4		10.70	14
0110	8" deep	G	150	.107		7.45	4.26		11.71	15.25
0120	10" deep	G	140	.114		9	4.57		13.57	17.50
0130	12" deep	G	130	.123		9.90	4.92		14.82	19
0200	16 ga x 6" deep	G	150	.107		7.60	4.26		11.86	15.40
0210	8" deep	G	140	.114		8.55	4.57		13.12	16.95
0220	10" deep	G	130	.123		10.30	4.92		15.22	19.40
0230	12" deep	G	120	.133		11.20	5.35		16.55	21
1100	Rafters, 2" flange, material only, 18 ga x 6" deep	G				2.08			2.08	2.29
1110	8" deep	G				2.47			2.47	2.71
1120	10" deep	G				2.90			2.90	3.19
1130	12" deep	G				3.37			3.37	3.71
1200	16 ga x 6" deep	G				2.54			2.54	2.80
1210	8" deep	G				3.05			3.05	3.35
1220	10" deep	G				3.56			3.56	3.92
1230	12" deep	G				4.03			4.03	4.44
2100	Installation only, ordinary rafter to 4:12 pitch, 18 ga x 6" deep	2 Carp	35	.457	Ea.		18.25		18.25	30
2110	8" deep		30	.533			21.50		21.50	35
2120	10" deep		25	.640			25.50		25.50	42
2130	12" deep		20	.800			32		32	53
2200	16 ga x 6" deep		30	.533			21.50		21.50	35
2210	8" deep		25	.640			25.50		25.50	42
2220	10" deep		20	.800			32		32	53
2230	12" deep		15	1.067			42.50		42.50	70.50
8100	Add to labor, ordinary rafters on steep roofs						25%			
8110	Dormers & complex roofs						50%			
8200	Hip & valley rafters to 4:12 pitch						25%			
8210	Steep roofs						50%			
8220	Dormers & complex roofs						75%			
8300	Hip & valley jack rafters to 4:12 pitch						50%			
8310	Steep roofs						75%			
8320	Dormers & complex roofs						100%			
9000	Minimum labor/equipment charge	2 Carp	4	4	Job		160		160	264

05 42 23.70 Framing, Soffits and Canopies

		Crew	Daily Output	Labor-Hours	Unit	Material	2009 Bare Costs Labor	Equipment	Total	Total Incl O&P
0010	**FRAMING, SOFFITS & CANOPIES**									
0015	Made from recycled materials									
0130	Continuous ledger track @ wall, studs @ 16" O.C., 18 ga x 4" wide	G 2 Carp	535	.030	L.F.	1.27	1.19		2.46	3.36
0140	6" wide	G	500	.032		1.65	1.28		2.93	3.93
0150	8" wide	G	465	.034		2.09	1.37		3.46	4.57
0160	10" wide	G	430	.037		2.59	1.49		4.08	5.30
0230	Studs @ 24" O.C., 18 ga x 4" wide	G	800	.020		1.21	.80		2.01	2.65
0240	6" wide	G	750	.021		1.58	.85		2.43	3.14
0250	8" wide	G	700	.023		2	.91		2.91	3.70
0260	10" wide	G	650	.025		2.47	.98		3.45	4.33
1000	Horizontal soffit and canopy members, material only									
1030	1-5/8" flange studs, 18 ga x 4" deep	G			L.F.	1.68			1.68	1.85
1040	6" deep	G				2.12			2.12	2.34
1050	8" deep	G				2.57			2.57	2.82
1140	2" flange joists, 18 ga x 6" deep	G				2.38			2.38	2.61
1150	8" deep	G				2.82			2.82	3.10

05 42 Cold-Formed Metal Joist Framing

05 42 23 – Cold-Formed Metal Roof Joist Framing

05 42 23.70 Framing, Soffits and Canopies	Crew	Daily Output	Labor-Hours	Unit	Material	Labor	Equipment	Total	Total Incl O&P
						2009 Bare Costs			
1160	10" deep	G				L.F.	3.31			3.31	3.64
4030	Installation only, 18 ga, 1-5/8" flange x 4" deep	2 Carp	130	.123	Ea.		4.92		4.92	8.10
4040	6" deep		110	.145			5.80		5.80	9.60
4050	8" deep		90	.178			7.10		7.10	11.70
4140	2" flange, 18 ga x 6" deep		110	.145			5.80		5.80	9.60
4150	8" deep		90	.178			7.10		7.10	11.70
4160	10" deep		80	.200			8		8	13.20
6010	Clips to attach facia to rafter tails, 2" x 2" x 18 ga angle	G	1 Carp	120	.067		1.12	2.66		3.78	5.65
6020	16 ga angle	G	"	100	.080		1.42	3.20		4.62	6.85
9000	Minimum labor/equipment charge	2 Carp	4	4	Job		160		160	264

05 44 Cold-Formed Metal Trusses

05 44 13 – Cold-Formed Metal Roof Trusses

05 44 13.60 Framing, Roof Trusses

05 44 13.60 Framing, Roof Trusses		Crew	Daily Output	Labor-Hours	Unit	Material	Labor	Equipment	Total	Total Incl O&P
0010	**FRAMING, ROOF TRUSSES**									
0015	Made from recycled materials									
0020	Fabrication of trusses on ground, Fink (W) or King Post, to 4:12 pitch									
0120	18 ga x 4" chords, 16' span	G	2 Carp	12	1.333	Ea.	78.50	53.50		132	174
0130	20' span	G		11	1.455		98	58		156	204
0140	24' span	G		11	1.455		118	58		176	225
0150	28' span	G		10	1.600		137	64		201	257
0160	32' span	G		10	1.600		157	64		221	278
0250	6" chords, 28' span	G		9	1.778		173	71		244	310
0260	32' span	G		9	1.778		198	71		269	335
0270	36' span	G		8	2		223	80		303	375
0280	40' span	G		8	2		248	80		328	405
1120	5:12 to 8:12 pitch, 18 ga x 4" chords, 16' span	G		10	1.600		89.50	64		153.50	205
1130	20' span	G		9	1.778		112	71		183	240
1140	24' span	G		9	1.778		134	71		205	265
1150	28' span	G		8	2		157	80		237	305
1160	32' span	G		8	2		179	80		259	330
1250	6" chords, 28' span	G		7	2.286		198	91.50		289.50	370
1260	32' span	G		7	2.286		227	91.50		318.50	400
1270	36' span	G		6	2.667		255	107		362	455
1280	40' span	G		6	2.667		283	107		390	485
2120	9:12 to 12:12 pitch, 18 ga x 4" chords, 16' span	G		8	2		112	80		192	255
2130	20' span	G		7	2.286		140	91.50		231.50	305
2140	24' span	G		7	2.286		168	91.50		259.50	335
2150	28' span	G		6	2.667		196	107		303	390
2160	32' span	G		6	2.667		224	107		331	420
2250	6" chords, 28' span	G		5	3.200		248	128		376	485
2260	32' span	G		5	3.200		283	128		411	520
2270	36' span	G		4	4		320	160		480	615
2280	40' span	G		4	4		355	160		515	655
4900	Minimum labor/equipment charge		4	4	Job		160		160	264
5120	Erection only of roof trusses, to 4:12 pitch, 16' span	F-6	48	.833	Ea.		31	16	47	68
5130	20' span		46	.870			32.50	16.70	49.20	71
5140	24' span		44	.909			34	17.45	51.45	74
5150	28' span		42	.952			35.50	18.30	53.80	77.50
5160	32' span		40	1			37	19.20	56.20	81.50
5170	36' span		38	1.053			39	20	59	86.50

05 44 Cold-Formed Metal Trusses

05 44 13 – Cold-Formed Metal Roof Trusses

05 44 13.60 Framing, Roof Trusses		Crew	Daily Output	Labor-Hours	Unit	Material	2009 Bare Costs		Total	Total Incl O&P
							Labor	Equipment		
5180	40' span	F-6	36	1.111	Ea.		41.50	21.50	63	91
5220	5:12 to 8:12 pitch, 16' span		42	.952			35.50	18.30	53.80	77.50
5230	20' span		40	1			37	19.20	56.20	81.50
5240	24' span		38	1.053			39	20	59	86.50
5250	28' span		36	1.111			41.50	21.50	63	91
5260	32' span		34	1.176			43.50	22.50	66	96.50
5270	36' span		32	1.250			46.50	24	70.50	103
5280	40' span		30	1.333			49.50	25.50	75	109
5320	9:12 to 12:12 pitch, 16' span		36	1.111			41.50	21.50	63	91
5330	20' span		34	1.176			43.50	22.50	66	96.50
5340	24' span		32	1.250			46.50	24	70.50	103
5350	28' span		30	1.333			49.50	25.50	75	109
5360	32' span		28	1.429			53	27.50	80.50	117
5370	36' span		26	1.538			57	29.50	86.50	126
5380	40' span		24	1.667			62	32	94	136
9000	Minimum labor/equipment charge		2	20	Job		745	385	1,130	1,650

05 51 Metal Stairs

05 51 13 – Metal Pan Stairs

05 51 13.50 Pan Stairs

			Crew	Daily Output	Labor-Hours	Unit	Material	Labor	Equipment	Total	Total Incl O&P
0010	**PAN STAIRS**, shop fabricated, steel stringers										
0015	Made from recycled materials										
0200	Cement fill metal pan, picket rail, 3'-6" wide	G	E-4	35	.914	Riser	560	41.50	3.83	605.33	700
0300	4'-0" wide	G		30	1.067		635	48	4.47	687.47	795
0350	Wall rail, both sides, 3'-6" wide	G		53	.604		430	27.50	2.53	460.03	530
1500	Landing, steel pan, conventional	G		160	.200	S.F.	75	9.05	.84	84.89	100

05 51 16 – Metal Floor Plate Stairs

05 51 16.50 Floor Plate Stairs

			Crew	Daily Output	Labor-Hours	Unit	Material	Labor	Equipment	Total	Total Incl O&P
0010	**FLOOR PLATE STAIRS**, shop fabricated, steel stringers										
0015	Made from recycled materials										
0400	Cast iron tread and pipe rail, 3'-6" wide	G	E-4	35	.914	Riser	600	41.50	3.83	645.33	740
0450	4'-0" wide	G		30	1.067		635	48	4.47	687.47	795
0475	5'-0" wide	G		28	1.143		695	51.50	4.79	751.29	860
0500	Checkered plate tread, industrial, 3'-6" wide	G		28	1.143		375	51.50	4.79	431.29	510
0550	Circular, for tanks, 3'-0" wide	G		33	.970		410	44	4.06	458.06	540
0600	For isolated stairs, add							100%			
0800	Custom steel stairs, 3'-6" wide, minimum	G	E-4	35	.914		560	41.50	3.83	605.33	700
0810	Average	G		30	1.067		750	48	4.47	802.47	920
0900	Maximum	G		20	1.600		935	72.50	6.70	1,014.20	1,175
1100	For 4' wide stairs, add							5%	5%		
1300	For 5' wide stairs, add							10%	10%		

05 51 19 – Metal Grating Stairs

05 51 19.50 Grating Stairs

			Crew	Daily Output	Labor-Hours	Unit	Material	Labor	Equipment	Total	Total Incl O&P
0010	**GRATING STAIRS**, shop fabricated, steel stringers, safety nosing on treads										
0015	Made from recycled materials										
0020	Grating tread and pipe railing, 3'-6" wide	G	E-4	35	.914	Riser	375	41.50	3.83	420.33	490
0100	4'-0" wide	G		30	1.067	"	485	48	4.47	537.47	630
9000	Minimum labor/equipment charge			2	16	Job		725	67	792	1,425

05 51 Metal Stairs

05 51 23 – Metal Fire Escapes

05 51 23.25 Fire Escapes

05 51 23.25 Fire Escapes		Crew	Daily Output	Labor-Hours	Unit	Material	2009 Bare Costs Labor	Equipment	Total	Total Incl O&P
0010	**FIRE ESCAPES**, shop fabricated									
0200	2' wide balcony, 1" x 1/4" bars 1-1/2" O.C.	G	1 Sswk	5	1.600	L.F.	65.50	71.50		137	204
0400	1st story cantilevered stair, standard	G	"	.09	88.889	Ea.	2,725	3,975		6,700	10,400
0500	Cable counterweight	G				"	2,550			2,550	2,800
0700	Platform & fixed stair, 36" x 40"	G	1 Sswk	.17	47.059	Flight	1,200	2,100		3,300	5,225
0900	For 3'-6" wide escapes, add to above						100%	150%			

05 51 23.50 Fire Escape Stairs

05 51 23.50 Fire Escape Stairs		Crew	Daily Output	Labor-Hours	Unit	Material	2009 Bare Costs Labor	Equipment	Total	Total Incl O&P
0010	**FIRE ESCAPE STAIRS**, shop fabricated									
0020	One story, disappearing, stainless steel	G	2 Sswk	20	.800	V.L.F.	223	36		259	310
0100	Portable ladder					Ea.	93			93	102
1100	Fire escape, galvanized steel, 8'-0" to 10'-4" ceiling	G	2 Carp	1	16		1,525	640		2,165	2,725
1110	10'-6" to 13'-6" ceiling	G	"	1	16		2,000	640		2,640	3,250

05 51 33 – Metal Ladders

05 51 33.13 Vertical Metal Ladders

05 51 33.13 Vertical Metal Ladders		Crew	Daily Output	Labor-Hours	Unit	Material	2009 Bare Costs Labor	Equipment	Total	Total Incl O&P
0010	**VERTICAL METAL LADDERS**, shop fabricated									
0015	Made from recycled materials									
0020	Steel, 20" wide, bolted to concrete, with cage	G	E-4	50	.640	V.L.F.	78.50	29	2.68	110.18	143
0100	Without cage	G		85	.376		41	17	1.58	59.58	78
0300	Aluminum, bolted to concrete, with cage	G		50	.640		111	29	2.68	142.68	178
0400	Without cage	G		85	.376		64.50	17	1.58	83.08	104
9000	Minimum labor/equipment charge			2	16	Job		725	67	792	1,425

05 51 33.16 Inclined Metal Ladders

05 51 33.16 Inclined Metal Ladders		Crew	Daily Output	Labor-Hours	Unit	Material	2009 Bare Costs Labor	Equipment	Total	Total Incl O&P
0010	**INCLINED METAL LADDERS**, shop fabricated									
3900	Industrial ships ladder, 3' W, grating treads, 2 line pipe rail	G	E-4	30	1.067	Riser	203	48	4.47	255.47	315
4000	Aluminum	G	"	30	1.067	"	310	48	4.47	362.47	440

05 51 33.23 Alternating Tread Ladders

05 51 33.23 Alternating Tread Ladders		Crew	Daily Output	Labor-Hours	Unit	Material	2009 Bare Costs Labor	Equipment	Total	Total Incl O&P
0010	**ALTERNATING TREAD LADDERS**, shop fabricated									
0015	Made from recycled materials									
1350	Alternating tread stair, 56/68°, steel, standard paint color	G	2 Sswk	50	.320	V.L.F.	168	14.30		182.30	212
1360	Non-standard paint color	G		50	.320		190	14.30		204.30	236
1370	Galvanized steel	G		50	.320		196	14.30		210.30	243
1380	Stainless steel	G		50	.320		282	14.30		296.30	335
1390	68°, aluminum	G		50	.320		206	14.30		220.30	254

05 52 Metal Railings

05 52 13 – Pipe and Tube Railings

05 52 13.50 Railings, Pipe

05 52 13.50 Railings, Pipe		Crew	Daily Output	Labor-Hours	Unit	Material	2009 Bare Costs Labor	Equipment	Total	Total Incl O&P
0010	**RAILINGS, PIPE**, shop fab'd, 3'-6" high, posts @ 5' O.C.									
0015	Made from recycled materials									
0020	Aluminum, 2 rail, satin finish, 1-1/4" diameter	G	E-4	160	.200	L.F.	34	9.05	.84	43.89	55
0030	Clear anodized	G		160	.200		42	9.05	.84	51.89	64
0040	Dark anodized	G		160	.200		47.50	9.05	.84	57.39	70
0080	1-1/2" diameter, satin finish	G		160	.200		41	9.05	.84	50.89	62.50
0090	Clear anodized	G		160	.200		45.50	9.05	.84	55.39	67.50
0100	Dark anodized	G		160	.200		50.50	9.05	.84	60.39	73
0140	Aluminum, 3 rail, 1-1/4" diam., satin finish	G		137	.234		52	10.55	.98	63.53	78
0150	Clear anodized	G		137	.234		65.50	10.55	.98	77.03	92.50
0160	Dark anodized	G		137	.234		72.50	10.55	.98	84.03	100
0200	1-1/2" diameter, satin finish	G		137	.234		62.50	10.55	.98	74.03	89.50

05 52 Metal Railings

05 52 13 – Pipe and Tube Railings

05 52 13.50 Railings, Pipe

			Crew	Daily Output	Labor-Hours	Unit	Material	2009 Bare Costs Labor	Equipment	Total	Total Incl O&P
0210	Clear anodized	G	E-4	137	.234	L.F.	71	10.55	.98	82.53	98.50
0220	Dark anodized	G		137	.234		78	10.55	.98	89.53	106
0500	Steel, 2 rail, on stairs, primed, 1-1/4" diameter	G		160	.200		28	9.05	.84	37.89	48.50
0520	1-1/2" diameter	G		160	.200		30.50	9.05	.84	40.39	51.50
0540	Galvanized, 1-1/4" diameter	G		160	.200		39	9.05	.84	48.89	60
0560	1-1/2" diameter	G		160	.200		43.50	9.05	.84	53.39	65.50
0580	Steel, 3 rail, primed, 1-1/4" diameter	G		137	.234		41.50	10.55	.98	53.03	66.50
0600	1-1/2" diameter	G		137	.234		44.50	10.55	.98	56.03	69
0620	Galvanized, 1-1/4" diameter	G		137	.234		58.50	10.55	.98	70.03	85
0640	1-1/2" diameter	G		137	.234		69	10.55	.98	80.53	96.50
0700	Stainless steel, 2 rail, 1-1/4" diam. #4 finish	G		137	.234		101	10.55	.98	112.53	132
0720	High polish	G		137	.234		162	10.55	.98	173.53	200
0740	Mirror polish	G		137	.234		203	10.55	.98	214.53	245
0760	Stainless steel, 3 rail, 1-1/2" diam., #4 finish	G		120	.267		152	12.05	1.12	165.17	191
0770	High polish	G		120	.267		251	12.05	1.12	264.17	300
0780	Mirror finish	G		120	.267		305	12.05	1.12	318.17	360
0900	Wall rail, alum. pipe, 1-1/4" diam., satin finish	G		213	.150		19.50	6.80	.63	26.93	35
0905	Clear anodized	G		213	.150		24	6.80	.63	31.43	39.50
0910	Dark anodized	G		213	.150		29	6.80	.63	36.43	45
0915	1-1/2" diameter, satin finish	G		213	.150		21.50	6.80	.63	28.93	37
0920	Clear anodized	G		213	.150		27	6.80	.63	34.43	43.50
0925	Dark anodized	G		213	.150		33.50	6.80	.63	40.93	50.50
0930	Steel pipe, 1-1/4" diameter, primed	G		213	.150		16.95	6.80	.63	24.38	32
0935	Galvanized	G		213	.150		24.50	6.80	.63	31.93	40.50
0940	1-1/2" diameter	G		176	.182		17.45	8.20	.76	26.41	35
0945	Galvanized	G		213	.150		24.50	6.80	.63	31.93	40.50
0955	Stainless steel pipe, 1-1/2" diam., #4 finish	G		107	.299		80.50	13.50	1.25	95.25	115
0960	High polish	G		107	.299		164	13.50	1.25	178.75	206
0965	Mirror polish	G		107	.299		193	13.50	1.25	207.75	238
2000	Aluminum pipe & picket railing, double top rail, pickets @ 4-1/2" OC										
2010	36" high, straight & level	G	2 Sswk	80	.200	L.F.	81	8.95		89.95	106
2020	Curved & level	G		60	.267		111	11.90		122.90	144
2030	Straight & sloped	G		40	.400		93	17.90		110.90	135
9000	Minimum labor/equipment charge		1 Sswk	2	4	Job		179		179	330

05 52 16 – Industrial Railings

05 52 16.50 Railings, Industrial

			Crew	Daily Output	Labor-Hours	Unit	Material	2009 Bare Costs Labor	Equipment	Total	Total Incl O&P
0010	**RAILINGS, INDUSTRIAL**, shop fab'd, 3'-6" high, posts @ 5' O.C.										
0020	2 rail, 3'-6" high, 1-1/2" pipe	G	E-4	255	.125	L.F.	31	5.65	.53	37.18	45
0200	For 4" high kick plate, 10 gauge, add	G					6.40			6.40	7.05
0500	For curved rails, add						30%	30%			
9000	Minimum labor/equipment charge		1 Sswk	2	4	Job		179		179	330

05 56 Metal Castings

05 56 13 – Metal Construction Castings

05 56 13.50 Construction Castings		Crew	Daily Output	Labor-Hours	Unit	Material	2009 Bare Costs Labor	2009 Bare Costs Equipment	Total	Total Incl O&P
0010	**CONSTRUCTION CASTINGS**									
0020	Manhole covers and frames, see Div. 33 44 13.13									
0100	Column bases, cast iron, 16" x 16", approx. 65 lb.	G	E-4	46	.696	Ea.	142	31.50	2.91	176.41	218
0200	32" x 32", approx. 256 lb.	G		23	1.391	"	530	63	5.85	598.85	705
0600	Miscellaneous C.I. castings, light sections, less than 150 lb.	G		3200	.010	Lb.	2.20	.45	.04	2.69	3.31
1300	Special low volume items	G		3200	.010	"	3.82	.45	.04	4.31	5.10

05 58 Formed Metal Fabrications

05 58 23 – Formed Columns

05 58 23.10 Aluminum Columns

05 58 23.10 Aluminum Columns		Crew	Daily Output	Labor-Hours	Unit	Material	Labor	Equipment	Total	Total Incl O&P
0010	**ALUMINUM COLUMNS**									
0015	Made from recycled materials									
0020	Aluminum, extruded, stock units, no cap or base, 6" diameter	G	E-4	240	.133	L.F.	11.65	6.05	.56	18.26	24.50
0100	8" diameter	G	"	170	.188		15	8.50	.79	24.29	33
0500	For square columns, add to column prices above						50%			

05 58 25 – Formed Lamp Posts

05 58 25.40 Lamp Posts

05 58 25.40 Lamp Posts		Crew	Daily Output	Labor-Hours	Unit	Material	Labor	Equipment	Total	Total Incl O&P
0010	**LAMP POSTS**									
0020	Aluminum, 7' high, stock units, post only	G	1 Carp	16	.500	Ea.	28.50	20		48.50	64.50
0100	Mild steel, plain	G		16	.500	"	23	20		43	58.50
9000	Minimum labor/equipment charge			4	2	Job		80		80	132

05 58 27 – Formed Guards

05 58 27.90 Window Guards

05 58 27.90 Window Guards		Crew	Daily Output	Labor-Hours	Unit	Material	Labor	Equipment	Total	Total Incl O&P
0010	**WINDOW GUARDS**, shop fabricated									
0015	Expanded metal, steel angle frame, permanent	G	E-4	350	.091	S.F.	26	4.13	.38	30.51	36.50
0025	Steel bars, 1/2" x 1/2", spaced 5" O.C.	G	"	290	.110	"	18	4.99	.46	23.45	29.50
0030	Hinge mounted, add	G				Opng.	52			52	57.50
0040	Removable type, add	G				"	33			33	36.50
0050	For galvanized guards, add					S.F.	35%			
0070	For pivoted or projected type, add						105%	40%		
0100	Mild steel, stock units, economy	G	E-4	405	.079		7.05	3.57	.33	10.95	14.70
0200	Deluxe	G		405	.079		14.50	3.57	.33	18.40	23
0400	Woven wire, stock units, 3/8" channel frame, 3' x 5' opening	G		40	.800	Opng.	191	36	3.35	230.35	281
0500	4' x 6' opening	G		38	.842		305	38	3.53	346.53	410
0800	Basket guards for above, add	G					261			261	287
1000	Swinging guards for above, add	G					89.50			89.50	98.50
9000	Minimum labor/equipment charge		1 Sswk	2	4	Job		179		179	330

05 71 Decorative Metal Stairs

05 71 13 – Fabricated Metal Spiral Stairs

05 71 13.50 Spiral Stairs

05 71 13.50 Spiral Stairs		Crew	Daily Output	Labor-Hours	Unit	Material	Labor	Equipment	Total	Total Incl O&P
0010	**SPIRAL STAIRS**, shop fabricated									
1810	Spiral aluminum, 5'-0" diameter, stock units	G	E-4	45	.711	Riser	680	32	2.98	714.98	815
1820	Custom units	G		45	.711		1,275	32	2.98	1,309.98	1,500
1900	Spiral, cast iron, 4'-0" diameter, ornamental, minimum	G		45	.711		605	32	2.98	639.98	730
1920	Maximum	G		25	1.280		825	58	5.35	888.35	1,025
2000	Spiral, steel, industrial checkered plate, 4' diameter	G		45	.711		605	32	2.98	639.98	730
2200	Stock units, 6'-0" diameter	G		40	.800		735	36	3.35	774.35	880

05 71 Decorative Metal Stairs

05 71 13 – Fabricated Metal Spiral Stairs

05 71 13.50 Spiral Stairs		Crew	Daily Output	Labor-Hours	Unit	Material	2009 Bare Costs Labor	Equipment	Total	Total Incl O&P
3110	Spiral steel, stock units, primed, flat metal tread, 3'-6" dia **G**	2 Carp	1.60	10	Flight	1,250	400		1,650	2,000
3120	4'-0" dia **G**		1.45	11.034		1,425	440		1,865	2,300
3130	4'-6" dia **G**		1.35	11.852		1,575	475		2,050	2,500
3140	5'-0" dia **G**		1.25	12.800		1,950	510		2,460	3,000
3210	Galvanized, 3'-6" dia **G**		1.60	10		2,100	400		2,500	2,975
3220	4'-0" dia **G**		1.45	11.034		2,350	440		2,790	3,300
3230	4'-6" dia **G**		1.35	11.852		2,550	475		3,025	3,575
3240	5'-0" dia **G**		1.25	12.800		3,050	510		3,560	4,200
3310	Checkered plate tread, 3'-6" dia **G**		1.45	11.034		1,450	440		1,890	2,325
3320	4'-0" dia **G**		1.35	11.852		1,650	475		2,125	2,575
3330	4'-6" dia **G**		1.25	12.800		1,800	510		2,310	2,825
3340	5'-0" dia **G**		1.15	13.913		2,200	555		2,755	3,350
3410	Galvanized, 3'-6" dia **G**		1.45	11.034		2,400	440		2,840	3,375
3420	4'-0" dia **G**		1.35	11.852		2,650	475		3,125	3,675
3430	4'-6" dia **G**		1.25	12.800		2,850	510		3,360	3,975
3440	5'-0" dia **G**		1.15	13.913		3,350	555		3,905	4,625
3510	Red oak tread on flat metal, 3'-6" dia		1.35	11.852		2,375	475		2,850	3,375
3520	4'-0" dia		1.25	12.800		2,575	510		3,085	3,675
3530	4'-6" dia		1.15	13.913		2,800	555		3,355	4,000
3540	5'-0" dia		1.05	15.238		3,250	610		3,860	4,575

05 73 Decorative Metal Railings

05 73 16 – Wire Rope Decorative Metal Railings

05 73 16.10 Cable Railings

		Crew	Daily Output	Labor-Hours	Unit	Material	Labor	Equipment	Total	Total Incl O&P
0010	**CABLE RAILINGS**, with 316 stainless steel 1 x 19 cable, 3/16" diameter									
0015	Made from recycled materials									
0100	1-3/4" diameter stainless steel posts x 42" high, cables 4" OC **G**	2 Sswk	25	.640	L.F.	43	28.50		71.50	100

05 73 23 – Ornamental Railings

05 73 23.50 Railings, Ornamental

		Crew	Daily Output	Labor-Hours	Unit	Material	Labor	Equipment	Total	Total Incl O&P
0010	**RAILINGS, ORNAMENTAL**, shop fab'd, 3'-6" high, posts @ 5' O.C.									
0020	Bronze or stainless, minimum **G**	1 Sswk	24	.333	L.F.	36.50	14.90		51.40	68
0100	Maximum **G**		9	.889		365	39.50		404.50	475
0200	Aluminum ornamental rail, minimum **G**		15	.533		36.50	24		60.50	84.50
0300	Maximum **G**		8	1		110	44.50		154.50	204
0400	Hand-forged wrought iron, minimum **G**		12	.667		115	30		145	181
0500	Maximum **G**		8	1		345	44.50		389.50	465
0600	Composite metal/wood/glass, minimum		6	1.333		220	59.50		279.50	350
0700	Maximum		5	1.600		440	71.50		511.50	615
9000	Minimum labor/equipment charge		2	4	Job		179		179	330

Estimating Tips

06 05 00 Common Work Results for Wood, Plastics, and Composites

- Common to any wood-framed structure are the accessory connector items such as screws, nails, adhesives, hangers, connector plates, straps, angles, and hold-downs. For typical wood-framed buildings, such as residential projects, the aggregate total for these items can be significant, especially in areas where seismic loading is a concern. For floor and wall framing, the material cost is based on 10 to 25 lbs. per MBF. Hold-downs, hangers, and other connectors should be taken off by the piece.

06 10 00 Carpentry

- Lumber is a traded commodity and therefore sensitive to supply and demand in the marketplace. Even in "budgetary" estimating of wood-framed projects, it is advisable to call local suppliers for the latest market pricing.

- Common quantity units for wood-framed projects are "thousand board feet" (MBF). A board foot is a volume of wood, 1" x 1' x 1', or 144 cubic inches. Board-foot quantities are generally calculated using nominal material dimensions—dressed sizes are ignored. Board foot per lineal foot of any stick of lumber can be calculated by dividing the nominal cross-sectional area by 12. As an example, 2,000 lineal feet of 2 x 12 equates to 4 MBF by dividing the nominal area, 2 x 12, by 12, which equals 2, and multiplying by 2,000 to give 4,000 board feet. This simple rule applies to all nominal dimensioned lumber.

- Waste is an issue of concern at the quantity takeoff for any area of construction. Framing lumber is sold in even foot lengths, i.e., 10', 12', 14', 16', and depending on spans, wall heights, and the grade of lumber, waste is inevitable. A rule of thumb for lumber waste is 5%–10% depending on material quality and the complexity of the framing.

- Wood in various forms and shapes is used in many projects, even where the main structural framing is steel, concrete, or masonry. Plywood as a back-up partition material and 2x boards used as blocking and cant strips around roof edges are two common examples. The estimator should ensure that the costs of all wood materials are included in the final estimate.

06 20 00 Finish Carpentry

- It is necessary to consider the grade of workmanship when estimating labor costs for erecting millwork and interior finish. In practice, there are three grades: premium, custom, and economy. The RSMeans daily output for base and case moldings is in the range of 200 to 250 L.F. per carpenter per day. This is appropriate for most average custom-grade projects. For premium projects, an adjustment to productivity of 25%–50% should be made, depending on the complexity of the job.

Reference Numbers

Reference numbers are shown in shaded boxes at the beginning of some major classifications. These numbers refer to related items in the Reference Section. The reference information may be an estimating procedure, an alternate pricing method, or technical information.

Note: Not all subdivisions listed here necessarily appear in this publication.

06 05 Common Work Results for Wood, Plastics and Composites

06 05 05 – Selective Wood and Plastics Demolition

06 05 05.10 Selective Demolition Wood Framing	Crew	Daily Output	Labor-Hours	Unit	Material	2009 Bare Costs Labor	Equipment	Total	Total Incl O&P
0010 **SELECTIVE DEMOLITION WOOD FRAMING** R024119-10									
0100 Timber connector, nailed, small	1 Clab	96	.083	Ea.		2.63		2.63	4.35
0110 Medium		60	.133			4.21		4.21	6.95
0120 Large		48	.167			5.25		5.25	8.70
0130 Bolted, small		48	.167			5.25		5.25	8.70
0140 Medium		32	.250			7.90		7.90	13.05
0150 Large		24	.333			10.55		10.55	17.40
2958 Beams, 2" x 6"	2 Clab	1100	.015	L.F.		.46		.46	.76
2960 2" x 8"		825	.019			.61		.61	1.01
2965 2" x 10"		665	.024			.76		.76	1.25
2970 2" x 12"		550	.029			.92		.92	1.52
2972 2" x 14"		470	.034			1.08		1.08	1.78
2975 4" x 8"	B-1	413	.058			1.88		1.88	3.09
2980 4" x 10"		330	.073			2.35		2.35	3.87
2985 4" x 12"		275	.087			2.82		2.82	4.65
3000 6" x 8"		275	.087			2.82		2.82	4.65
3040 6" x 10"		220	.109			3.52		3.52	5.80
3080 6" x 12"		185	.130			4.19		4.19	6.90
3120 8" x 12"		140	.171			5.55		5.55	9.15
3160 10" x 12"		110	.218			7.05		7.05	11.60
3162 Alternate pricing method		1.10	21.818	M.B.F.		705		705	1,150
3170 Blocking, in 16" OC wall framing, 2" x 4"	1 Clab	600	.013	L.F.		.42		.42	.70
3172 2" x 6"		400	.020			.63		.63	1.04
3174 In 24" OC wall framing, 2" x 4"		600	.013			.42		.42	.70
3176 2" x 6"		400	.020			.63		.63	1.04
3178 Alt method, wood blocking removal from wood framing		.40	20	M.B.F.		630		630	1,050
3179 Wood blocking removal from steel framing		.36	22.222	"		700		700	1,150
3180 Bracing, let in, 1" x 3", studs 16" OC		1050	.008	L.F.		.24		.24	.40
3181 Studs 24" OC		1080	.007			.23		.23	.39
3182 1" x 4", studs 16" OC		1050	.008			.24		.24	.40
3183 Studs 24" OC		1080	.007			.23		.23	.39
3184 1" x 6", studs 16" OC		1050	.008			.24		.24	.40
3185 Studs 24" OC		1080	.007			.23		.23	.39
3186 2" x 3", studs 16" OC		800	.010			.32		.32	.52
3187 Studs 24" OC		830	.010			.30		.30	.50
3188 2" x 4", studs 16" OC		800	.010			.32		.32	.52
3189 Studs 24" OC		830	.010			.30		.30	.50
3190 2" x 6", studs 16" OC		800	.010			.32		.32	.52
3191 Studs 24" OC		830	.010			.30		.30	.50
3192 2" x 8", studs 16" OC		800	.010			.32		.32	.52
3193 Studs 24" OC		830	.010			.30		.30	.50
3194 "T" shaped metal bracing, studs at 16" OC		1060	.008			.24		.24	.39
3195 Studs at 24" OC		1200	.007			.21		.21	.35
3196 Metal straps, studs at 16" OC		1200	.007			.21		.21	.35
3197 Studs at 24" OC		1240	.006			.20		.20	.34
3200 Columns, round, 8' to 14' tall	2 Clab	40	.200	Ea.		6.30		6.30	10.45
3202 Dimensional lumber sizes		1.10	14.545	M.B.F.		460		460	760
3250 Blocking, between joists	1 Clab	320	.025	Ea.		.79		.79	1.30
3252 Bridging, metal strap, between joists		320	.025	Pr.		.79		.79	1.30
3254 Wood, between joists		320	.025	"		.79		.79	1.30
3260 Door buck, studs, header & access., 8' high 2" x 4" wall, 3' wide		32	.250	Ea.		7.90		7.90	13.05
3261 4' wide		32	.250			7.90		7.90	13.05
3262 5' wide		32	.250			7.90		7.90	13.05

06 05 05 – Selective Wood and Plastics Demolition

06 05 05.10 Selective Demolition Wood Framing		Crew	Daily Output	Labor-Hours	Unit	Material	Labor	2009 Bare Costs Equipment	Total	Total Incl O&P
3263	6' wide	1 Clab	32	.250	Ea.		7.90		7.90	13.05
3264	8' wide		30	.267			8.45		8.45	13.90
3265	10' wide		30	.267			8.45		8.45	13.90
3266	12' wide		30	.267			8.45		8.45	13.90
3267	2" x 6" wall, 3' wide		32	.250			7.90		7.90	13.05
3268	4' wide		32	.250			7.90		7.90	13.05
3269	5' wide		32	.250			7.90		7.90	13.05
3270	6' wide		32	.250			7.90		7.90	13.05
3271	8' wide		30	.267			8.45		8.45	13.90
3272	10' wide		30	.267			8.45		8.45	13.90
3273	12' wide		30	.267			8.45		8.45	13.90
3274	Window buck, studs, header & access, 8' high 2" x 4" wall, 2' wide		24	.333			10.55		10.55	17.40
3275	3' wide		24	.333			10.55		10.55	17.40
3276	4' wide		24	.333			10.55		10.55	17.40
3277	5' wide		24	.333			10.55		10.55	17.40
3278	6' wide		24	.333			10.55		10.55	17.40
3279	7' wide		24	.333			10.55		10.55	17.40
3280	8' wide		22	.364			11.50		11.50	18.95
3281	10' wide		22	.364			11.50		11.50	18.95
3282	12' wide		22	.364			11.50		11.50	18.95
3283	2" x 6" wall, 2' wide		24	.333			10.55		10.55	17.40
3284	3' wide		24	.333			10.55		10.55	17.40
3285	4' wide		24	.333			10.55		10.55	17.40
3286	5' wide		24	.333			10.55		10.55	17.40
3287	6' wide		24	.333			10.55		10.55	17.40
3288	7' wide		24	.333			10.55		10.55	17.40
3289	8' wide		22	.364			11.50		11.50	18.95
3290	10' wide		22	.364			11.50		11.50	18.95
3291	12' wide		22	.364			11.50		11.50	18.95
3400	Fascia boards, 1" x 6"		500	.016	L.F.		.51		.51	.83
3440	1" x 8"		450	.018			.56		.56	.93
3480	1" x 10"		400	.020			.63		.63	1.04
3490	2" x 6"		450	.018			.56		.56	.93
3500	2" x 8"		400	.020			.63		.63	1.04
3510	2" x 10"		350	.023			.72		.72	1.19
3610	Furring, on wood walls or ceiling		4000	.002	S.F.		.06		.06	.10
3620	On masonry or concrete walls or ceiling		1200	.007	"		.21		.21	.35
3800	Headers over openings, 2 @ 2" x 6"		110	.073	L.F.		2.30		2.30	3.79
3840	2 @ 2" x 8"		100	.080			2.53		2.53	4.17
3880	2 @ 2" x 10"		90	.089			2.81		2.81	4.64
3885	Alternate pricing method		.26	30.651	M.B.F.		970		970	1,600
3920	Joists, 1" x 4"		1250	.006	L.F.		.20		.20	.33
3930	1" x 6"		1135	.007			.22		.22	.37
3940	1" x 8"		1000	.008			.25		.25	.42
3950	1" x 10"		895	.009			.28		.28	.47
3960	1" x 12"		765	.010			.33		.33	.55
4200	2" x 4"	2 Clab	1000	.016			.51		.51	.83
4230	2" x 6"		970	.016			.52		.52	.86
4240	2" x 8"		940	.017			.54		.54	.89
4250	2" x 10"		910	.018			.56		.56	.92
4280	2" x 12"		880	.018			.57		.57	.95
4281	2" x 14"		850	.019			.59		.59	.98
4282	Composite joists, 9-1/2"		960	.017			.53		.53	.87

06 05 05.10 Selective Demolition Wood Framing		Crew	Daily Output	Labor-Hours	Unit	Material	2009 Bare Costs Labor	Equipment	Total	Total Incl O&P
4283	11-7/8"	2 Clab	930	.017	L.F.		.54		.54	.90
4284	14"		897	.018			.56		.56	.93
4285	16"		865	.019			.58		.58	.96
4290	Wood joists, alternate pricing method		1.50	10.667	M.B.F.		335		335	555
4500	Open web joist, 12" deep		500	.032	L.F.		1.01		1.01	1.67
4505	14" deep		475	.034			1.06		1.06	1.76
4510	16" deep		450	.036			1.12		1.12	1.85
4520	18" deep		425	.038			1.19		1.19	1.96
4530	24" deep		400	.040			1.26		1.26	2.09
4550	Ledger strips, 1" x 2"	1 Clab	1200	.007			.21		.21	.35
4560	1" x 3"		1200	.007			.21		.21	.35
4570	1" x 4"		1200	.007			.21		.21	.35
4580	2" x 2"		1100	.007			.23		.23	.38
4590	2" x 4"		1000	.008			.25		.25	.42
4600	2" x 6"		1000	.008			.25		.25	.42
4601	2" x 8" or 2" x 10"		800	.010			.32		.32	.52
4602	4" x 6"		600	.013			.42		.42	.70
4604	4" x 8"		450	.018			.56		.56	.93
5400	Posts, 4" x 4"	2 Clab	800	.020			.63		.63	1.04
5405	4" x 6"		550	.029			.92		.92	1.52
5410	4" x 8"		440	.036			1.15		1.15	1.90
5425	4" x 10"		390	.041			1.30		1.30	2.14
5430	4" x 12"		350	.046			1.44		1.44	2.38
5440	6" x 6"		400	.040			1.26		1.26	2.09
5445	6" x 8"		350	.046			1.44		1.44	2.38
5450	6" x 10"		320	.050			1.58		1.58	2.61
5455	6" x 12"		290	.055			1.74		1.74	2.88
5480	8" x 8"		300	.053			1.69		1.69	2.78
5500	10" x 10"		240	.067			2.11		2.11	3.48
5660	Tongue and groove floor planks		2	8	M.B.F.		253		253	415
5750	Rafters, ordinary, 16" OC, 2" x 4"		880	.018	S.F.		.57		.57	.95
5755	2" x 6"		840	.019			.60		.60	.99
5760	2" x 8"		820	.020			.62		.62	1.02
5770	2" x 10"		820	.020			.62		.62	1.02
5780	2" x 12"		810	.020			.62		.62	1.03
5785	24" OC, 2" x 4"		1170	.014			.43		.43	.71
5786	2" x 6"		1117	.014			.45		.45	.75
5787	2" x 8"		1091	.015			.46		.46	.77
5788	2" x 10"		1091	.015			.46		.46	.77
5789	2" x 12"		1077	.015			.47		.47	.77
5795	Rafters, ordinary, 2" x 4" (alternate method)		862	.019	L.F.		.59		.59	.97
5800	2" x 6" (alternate method)		850	.019			.59		.59	.98
5840	2" x 8" (alternate method)		837	.019			.60		.60	1
5855	2" x 10" (alternate method)		825	.019			.61		.61	1.01
5865	2" x 12" (alternate method)		812	.020			.62		.62	1.03
5870	Sill plate, 2" x 4"	1 Clab	1170	.007			.22		.22	.36
5871	2" x 6"		780	.010			.32		.32	.54
5872	2" x 8"		586	.014			.43		.43	.71
5873	Alternate pricing method		.78	10.256	M.B.F.		325		325	535
5885	Ridge board, 1" x 4"	2 Clab	900	.018	L.F.		.56		.56	.93
5886	1" x 6"		875	.018			.58		.58	.95
5887	1" x 8"		850	.019			.59		.59	.98
5888	1" x 10"		825	.019			.61		.61	1.01

06 05 05 — Selective Wood and Plastics Demolition

06 05 05.10 Selective Demolition Wood Framing		Crew	Daily Output	Labor-Hours	Unit	Material	2009 Bare Costs Labor	Equipment	Total	Total Incl O&P
5889	1" x 12"	2 Clab	800	.020	L.F.		.63		.63	1.04
5890	2" x 4"		900	.018			.56		.56	.93
5892	2" x 6"		875	.018			.58		.58	.95
5894	2" x 8"		850	.019			.59		.59	.98
5896	2" x 10"		825	.019			.61		.61	1.01
5898	2" x 12"		800	.020			.63		.63	1.04
5900	Hip & valley rafters, 2" x 6"		500	.032			1.01		1.01	1.67
5940	2" x 8"		420	.038			1.20		1.20	1.99
6050	Rafter tie, 1" x 4"		1250	.013			.40		.40	.67
6052	1" x 6"		1135	.014			.45		.45	.74
6054	2" x 4"		1000	.016			.51		.51	.83
6056	2" x 6"		970	.016			.52		.52	.86
6070	Sleepers, on concrete, 1" x 2"	1 Clab	4700	.002			.05		.05	.09
6075	1" x 3"		4000	.002			.06		.06	.10
6080	2" x 4"		3000	.003			.08		.08	.14
6085	2" x 6"		2600	.003			.10		.10	.16
6086	Sheathing from roof, 5/16"	2 Clab	1600	.010	S.F.		.32		.32	.52
6088	3/8"		1525	.010			.33		.33	.55
6090	1/2"		1400	.011			.36		.36	.60
6092	5/8"		1300	.012			.39		.39	.64
6094	3/4"		1200	.013			.42		.42	.70
6096	Board sheathing from roof		1400	.011			.36		.36	.60
6100	Sheathing, from walls, 1/4"		1200	.013			.42		.42	.70
6110	5/16"		1175	.014			.43		.43	.71
6120	3/8"		1150	.014			.44		.44	.73
6130	1/2"		1125	.014			.45		.45	.74
6140	5/8"		1100	.015			.46		.46	.76
6150	3/4"		1075	.015			.47		.47	.78
6152	Board sheathing from walls		1500	.011			.34		.34	.56
6158	Subfloor, with boards		1050	.015			.48		.48	.79
6160	Plywood, 1/2" thick		768	.021			.66		.66	1.09
6162	5/8" thick		760	.021			.67		.67	1.10
6164	3/4" thick		750	.021			.67		.67	1.11
6165	1-1/8" thick		720	.022			.70		.70	1.16
6166	Underlayment, particle board, 3/8" thick	1 Clab	780	.010			.32		.32	.54
6168	1/2" thick		768	.010			.33		.33	.54
6170	5/8" thick		760	.011			.33		.33	.55
6172	3/4" thick		750	.011			.34		.34	.56
6200	Stairs and stringers, minimum	2 Clab	40	.400	Riser		12.65		12.65	21
6240	Maximum	"	26	.615	"		19.45		19.45	32
6300	Components, tread	1 Clab	110	.073	Ea.		2.30		2.30	3.79
6320	Riser		80	.100	"		3.16		3.16	5.20
6390	Stringer, 2" x 10"		260	.031	L.F.		.97		.97	1.60
6400	2" x 12"		260	.031			.97		.97	1.60
6410	3" x 10"		250	.032			1.01		1.01	1.67
6420	3" x 12"		250	.032			1.01		1.01	1.67
6590	Wood studs, 2" x 3"	2 Clab	3076	.005			.16		.16	.27
6600	2" x 4"		2000	.008			.25		.25	.42
6640	2" x 6"		1600	.010			.32		.32	.52
6720	Wall framing, including studs, plates and blocking, 2" x 4"	1 Clab	600	.013	S.F.		.42		.42	.70
6740	2" x 6"		480	.017	"		.53		.53	.87
6750	Headers, 2" x 4"		1125	.007	L.F.		.22		.22	.37
6755	2" x 6"		1125	.007			.22		.22	.37

06 05 05 – Selective Wood and Plastics Demolition

06 05 05.10 Selective Demolition Wood Framing		Crew	Daily Output	Labor-Hours	Unit	Material	2009 Bare Costs Labor	2009 Bare Costs Equipment	Total	Total Incl O&P
6760	2" x 8"	1 Clab	1050	.008	L.F.		.24		.24	.40
6765	2" x 10"		1050	.008			.24		.24	.40
6770	2" x 12"		1000	.008			.25		.25	.42
6780	4" x 10"		525	.015			.48		.48	.79
6785	4" x 12"		500	.016			.51		.51	.83
6790	6" x 8"		560	.014			.45		.45	.75
6795	6" x 10"		525	.015			.48		.48	.79
6797	6" x 12"		500	.016			.51		.51	.83
8000	Soffit, T & G wood		520	.015	S.F.		.49		.49	.80
8010	Hardboard, vinyl or aluminum		640	.013			.40		.40	.65
8030	Plywood	2 Carp	315	.051			2.03		2.03	3.35
9000	Minimum labor/equipment charge	1 Clab	4	2	Job		63		63	104

06 05 05.20 Selective Demolition Millwork and Trim

06 05 05.20		Crew	Daily Output	Labor-Hours	Unit	Material	Labor	Equipment	Total	Total Incl O&P
0010	SELECTIVE DEMOLITION MILLWORK AND TRIM R024119-10									
1000	Cabinets, wood, base cabinets, per L.F.	2 Clab	80	.200	L.F.		6.30		6.30	10.45
1020	Wall cabinets, per L.F.	"	80	.200	"		6.30		6.30	10.45
1060	Remove and reset, base cabinets	2 Carp	18	.889	Ea.		35.50		35.50	58.50
1070	Wall cabinets	"	20	.800	"		32		32	53
1100	Steel, painted, base cabinets	2 Clab	60	.267	L.F.		8.45		8.45	13.90
1120	Wall cabinets		60	.267	"		8.45		8.45	13.90
1200	Casework, large area		320	.050	S.F.		1.58		1.58	2.61
1220	Selective		200	.080	"		2.53		2.53	4.17
1500	Counter top, minimum		200	.080	L.F.		2.53		2.53	4.17
1510	Maximum		120	.133			4.21		4.21	6.95
1550	Remove and reset, minimum	2 Carp	50	.320			12.80		12.80	21
1560	Maximum	"	40	.400			16		16	26.50
2000	Paneling, 4' x 8' sheets	2 Clab	2000	.008	S.F.		.25		.25	.42
2100	Boards, 1" x 4"		700	.023			.72		.72	1.19
2120	1" x 6"		750	.021			.67		.67	1.11
2140	1" x 8"		800	.020			.63		.63	1.04
3000	Trim, baseboard, to 6" wide		1200	.013	L.F.		.42		.42	.70
3040	Greater than 6" and up to 12" wide		1000	.016			.51		.51	.83
3080	Remove and reset, minimum	2 Carp	400	.040			1.60		1.60	2.64
3090	Maximum	"	300	.053			2.13		2.13	3.52
3100	Ceiling trim	2 Clab	1000	.016			.51		.51	.83
3120	Chair rail		1200	.013			.42		.42	.70
3140	Railings with balusters		240	.067			2.11		2.11	3.48
3160	Wainscoting		700	.023	S.F.		.72		.72	1.19
9000	Minimum labor/equipment charge	1 Clab	4	2	Job		63		63	104

06 05 23 – Wood, Plastic, and Composite Fastenings

06 05 23.10 Nails

06 05 23.10		Crew	Daily Output	Labor-Hours	Unit	Material	Labor	Equipment	Total	Total Incl O&P
0010	NAILS, material only, based upon 50# box purchase									
0020	Copper nails, plain				Lb.	10.20			10.20	11.20
0400	Stainless steel, plain					6.65			6.65	7.30
0500	Box, 3d to 20d, bright					.91			.91	1
0520	Galvanized					1.67			1.67	1.84
0600	Common, 3d to 60d, plain					1.11			1.11	1.22
0700	Galvanized					1.53			1.53	1.68
0800	Aluminum					4.45			4.45	4.90
1000	Annular or spiral thread, 4d to 60d, plain					1.82			1.82	2
1200	Galvanized					1.99			1.99	2.19
1400	Drywall nails, plain					.79			.79	.87

06 05 23.10 Nails

		Crew	Daily Output	Labor-Hours	Unit	Material	2009 Bare Costs Labor	Equipment	Total	Total Incl O&P
1600	Galvanized				Lb.	1.59			1.59	1.75
1800	Finish nails, 4d to 10d, plain					1.11			1.11	1.22
2000	Galvanized					1.75			1.75	1.93
2100	Aluminum					4.10			4.10	4.51
2300	Flooring nails, hardened steel, 2d to 10d, plain					2.04			2.04	2.24
2400	Galvanized					3.88			3.88	4.27
2500	Gypsum lath nails, 1-1/8", 13 ga. flathead, blued					2.64			2.64	2.90
2600	Masonry nails, hardened steel, 3/4" to 3" long, plain					1.97			1.97	2.17
2700	Galvanized					2.78			2.78	3.06
2900	Roofing nails, threaded, galvanized					1.38			1.38	1.52
3100	Aluminum					4.80			4.80	5.30
3300	Compressed lead head, threaded, galvanized					2.50			2.50	2.75
3600	Siding nails, plain shank, galvanized					1.50			1.50	1.65
3800	Aluminum					4.11			4.11	4.52
5000	Add to prices above for cement coating					.10			.10	.11
5200	Zinc or tin plating					.13			.13	.14
5500	Vinyl coated sinkers, 8d to 16d					.60			.60	.66

06 05 23.20 Pneumatic Nails

		Crew	Daily Output	Labor-Hours	Unit	Material	2009 Bare Costs Labor	Equipment	Total	Total Incl O&P
0010	**PNEUMATIC NAILS**									
0020	Framing, per carton of 5000, 2"				Ea.	40			40	44
0100	2-3/8"					45			45	49.50
0200	Per carton of 4000, 3"					40			40	44
0300	3-1/4"					40			40	44
0400	Per carton of 5000, 2-3/8", galv.					60			60	66
0500	Per carton of 4000, 3", galv.					65			65	71.50
0600	3-1/4", galv.					82			82	90
0700	Roofing, per carton of 7200, 1"					35.50			35.50	39
0800	1-1/4"					35			35	38.50
0900	1-1/2"					38.50			38.50	42.50
1000	1-3/4"					47.50			47.50	52.50

06 05 23.50 Wood Screws

		Crew	Daily Output	Labor-Hours	Unit	Material	2009 Bare Costs Labor	Equipment	Total	Total Incl O&P
0010	**WOOD SCREWS**									
0020	#8 x 1" long, steel				C	5.10			5.10	5.60
0100	Brass					11.50			11.50	12.65
0600	#10, 2" long, steel					5.95			5.95	6.55
0700	Brass					23.50			23.50	26
1500	#12, 3" long, steel					11.45			11.45	12.60

06 05 23.60 Timber Connectors

		Crew	Daily Output	Labor-Hours	Unit	Material	2009 Bare Costs Labor	Equipment	Total	Total Incl O&P
0010	**TIMBER CONNECTORS**									
0100	Connector plates, steel, with bolts, straight	2 Carp	75	.213	Ea.	25.50	8.50		34	42
0110	Tee, 7 gauge		50	.320		29.50	12.80		42.30	53
0120	T- Strap, 14 gauge, 12" x 8" x 2"		50	.320		29.50	12.80		42.30	53
0150	Anchor plates, 7 ga, 9" x 7"		75	.213		25.50	8.50		34	42
0200	Bolts, machine, sq. hd. with nut & washer, 1/2" diameter, 4" long	1 Carp	140	.057		.79	2.28		3.07	4.64
0300	7-1/2" long		130	.062		1.37	2.46		3.83	5.55
0500	3/4" diameter, 7-1/2" long		130	.062		4.12	2.46		6.58	8.60
0610	Machine bolts, w/ nut, washer, 3/4" diam., 15" L, HD's & beam hangers		95	.084		7.60	3.36		10.96	13.90
0800	Drilling bolt holes in timber, 1/2" diameter		450	.018	Inch		.71		.71	1.17
0900	1" diameter		350	.023	"		.91		.91	1.51
1100	Framing anchor, angle, 3" x 3" x 1-1/2", 12 ga		175	.046	Ea.	2.20	1.83		4.03	5.45
1150	Framing anchors, 18 gauge, 4-1/2" x 2-3/4"		175	.046		2.20	1.83		4.03	5.45
1160	Framing anchors, 18 gauge, 4-1/2" x 3"		175	.046		2.20	1.83		4.03	5.45

06 05 23.60 Timber Connectors

		Crew	Daily Output	Labor-Hours	Unit	Material	2009 Bare Costs Labor	Equipment	Total	Total Incl O&P
1170	Clip anchors plates, 18 gauge, 12" x 1-1/8"	1 Carp	175	.046	Ea.	2.20	1.83		4.03	5.45
1250	Holdowns, 3 gauge base, 10 gauge body		8	1		21	40		61	89
1260	Holdowns, 7 gauge 11-1/16" x 3-1/4"		8	1		21	40		61	89
1270	Holdowns, 7 gauge 14-3/8" x 3-1/8"		8	1		21	40		61	89
1275	Holdowns, 12 gauge 8" x 2-1/2"		8	1		21	40		61	89
1300	Joist and beam hangers, 18 ga. galv., for 2" x 4" joist		175	.046		.61	1.83		2.44	3.68
1400	2" x 6" to 2" x 10" joist		165	.048		1.17	1.94		3.11	4.49
1600	16 ga. galv., 3" x 6" to 3" x 10" joist		160	.050		2.49	2		4.49	6.05
1700	3" x 10" to 3" x 14" joist		160	.050		4.14	2		6.14	7.85
1800	4" x 6" to 4" x 10" joist		155	.052		2.58	2.06		4.64	6.25
1900	4" x 10" to 4" x 14" joist		155	.052		4.23	2.06		6.29	8.05
2000	Two-2" x 6" to two-2" x 10" joists		150	.053		3.52	2.13		5.65	7.40
2100	Two-2" x 10" to two-2" x 14" joists		150	.053		3.94	2.13		6.07	7.85
2300	3/16" thick, 6" x 8" joist		145	.055		54.50	2.20		56.70	63.50
2400	6" x 10" joist		140	.057		57	2.28		59.28	66.50
2500	6" x 12" joist		135	.059		59.50	2.37		61.87	69
2700	1/4" thick, 6" x 14" joist		130	.062		61.50	2.46		63.96	72
2900	Plywood clips, extruded aluminum H clip, for 3/4" panels					.20			.20	.22
3000	Galvanized 18 ga. back-up clip					.16			.16	.18
3200	Post framing, 16 ga. galv. for 4" x 4" base, 2 piece	1 Carp	130	.062		13.80	2.46		16.26	19.20
3300	Cap		130	.062		19.15	2.46		21.61	25
3500	Rafter anchors, 18 ga. galv., 1-1/2" wide, 5-1/4" long		145	.055		.41	2.20		2.61	4.09
3600	10-3/4" long		145	.055		1.23	2.20		3.43	4.99
3800	Shear plates, 2-5/8" diameter		120	.067		2	2.66		4.66	6.60
3900	4" diameter		115	.070		4.75	2.78		7.53	9.85
4000	Sill anchors, embedded in concrete or block, 25-1/2" long		115	.070		10.70	2.78		13.48	16.35
4100	Spike grids, 3" x 6"		120	.067		.81	2.66		3.47	5.30
4400	Split rings, 2-1/2" diameter		120	.067		1.65	2.66		4.31	6.20
4500	4" diameter		110	.073		2.50	2.91		5.41	7.55
4550	Tie plate, 20 gauge, 7" x 3 1/8"		110	.073		2.50	2.91		5.41	7.55
4560	Tie plate, 20 gauge, 5" x 4 1/8"		110	.073		2.50	2.91		5.41	7.55
4575	Twist straps, 18 gauge, 12" x 1 1/4"		110	.073		2.50	2.91		5.41	7.55
4580	Twist straps, 18 gauge, 16" x 1 1/4"		110	.073		2.50	2.91		5.41	7.55
4600	Strap ties, 20 ga., 2-1/16" wide, 12 13/16" long		180	.044		.80	1.78		2.58	3.81
4700	Strap ties, 16 ga., 1-3/8" wide, 12" long		180	.044		.80	1.78		2.58	3.81
4800	21-5/8" x 1-1/4"		160	.050		2.49	2		4.49	6.05
5000	Toothed rings, 2-5/8" or 4" diameter		90	.089		1.49	3.55		5.04	7.50
5200	Truss plates, nailed, 20 gauge, up to 32' span		17	.471	Truss	10.85	18.80		29.65	43
5400	Washers, 2" x 2" x 1/8"				Ea.	.34			.34	.37
5500	3" x 3" x 3/16"				"	.89			.89	.98
9000	Minimum labor/equipment charge	1 Carp	4	2	Job		80		80	132

06 05 23.70 Rough Hardware

0010	**ROUGH HARDWARE**, average percent of carpentry material									
0020	Minimum					.50%				
0200	Maximum					1.50%				

06 05 23.80 Metal Bracing

		Crew	Daily Output	Labor-Hours	Unit	Material	Labor	Equipment	Total	Total Incl O&P
0010	**METAL BRACING**									
0302	Let-in, "T" shaped, 22 ga. galv. steel, studs at 16" O.C.	1 Carp	580	.014	L.F.	.75	.55		1.30	1.74
0402	Studs at 24" O.C.		600	.013		.75	.53		1.28	1.71
0502	16 ga. galv. steel straps, studs at 16" O.C.		600	.013		.97	.53		1.50	1.95
0602	Studs at 24" O.C.		620	.013		.97	.52		1.49	1.92

06 11 Wood Framing

06 11 10 – Framing with Dimensional, Engineered or Composite Lumber

06 11 10.02 Blocking

		Crew	Daily Output	Labor-Hours	Unit	Material	2009 Bare Costs Labor	Equipment	Total	Total Incl O&P
0010	**BLOCKING**									
2600	Miscellaneous, to wood construction									
2620	2" x 4"	1 Carp	.17	47.059	M.B.F.	435	1,875		2,310	3,575
2625	Pneumatic nailed		.21	38.095		435	1,525		1,960	2,975
2660	2" x 8"		.27	29.630		530	1,175		1,705	2,525
2665	Pneumatic nailed		.33	24.242		530	970		1,500	2,175
2720	To steel construction									
2740	2" x 4"	1 Carp	.14	57.143	M.B.F.	435	2,275		2,710	4,250
2780	2" x 8"		.21	38.095	"	530	1,525		2,055	3,075
9000	Minimum labor/equipment charge		4	2	Job		80		80	132

06 11 10.04 Wood Bracing

		Crew	Daily Output	Labor-Hours	Unit	Material	2009 Bare Costs Labor	Equipment	Total	Total Incl O&P
0012	**BRACING** Let-in, with 1" x 6" boards, studs @ 16" O.C.	1Carp	150	.053		.64	2.13		2.77	4.22
0202	Studs @ 24" O.C.	"	230	.035		.64	1.39		2.03	2.99

06 11 10.06 Bridging

		Crew	Daily Output	Labor-Hours	Unit	Material	2009 Bare Costs Labor	Equipment	Total	Total Incl O&P
0012	**BRIDGING** Wood, for joists 16" O.C., 1" x 3"	1 Carp	130	.062		.51	2.46		2.97	4.62
0017	Pneumatic nailed		170	.047		.51	1.88		2.39	3.66
0102	2" x 3" bridging		130	.062		.47	2.46		2.93	4.58
0107	Pneumatic nailed		170	.047		.47	1.88		2.35	3.62
0302	Steel, galvanized, 18 ga., for 2" x 10" joists at 12" O.C.		130	.062		1.58	2.46		4.04	5.80
0402	24" O.C.		140	.057		1.69	2.28		3.97	5.65
0902	Compression type, 16" O.C., 2" x 8" joists		200	.040		1.64	1.60		3.24	4.44
1002	2" x 12" joists		200	.040		1.64	1.60		3.24	4.44

06 11 10.10 Beam and Girder Framing

		Crew	Daily Output	Labor-Hours	Unit	Material	2009 Bare Costs Labor	Equipment	Total	Total Incl O&P
0010	**BEAM AND GIRDER FRAMING** R061110-30									
1000	Single, 2" x 6"	2 Carp	700	.023	L.F.	.52	.91		1.43	2.08
1005	Pneumatic nailed		812	.020		.52	.79		1.31	1.87
1020	2" x 8"		650	.025		.71	.98		1.69	2.40
1025	Pneumatic nailed		754	.021		.71	.85		1.56	2.18
1040	2" x 10"		600	.027		.99	1.07		2.06	2.85
1045	Pneumatic nailed		696	.023		.99	.92		1.91	2.61
1060	2" x 12"		550	.029		1.39	1.16		2.55	3.44
1065	Pneumatic nailed		638	.025		1.39	1		2.39	3.17
1080	2" x 14"		500	.032		2.14	1.28		3.42	4.46
1085	Pneumatic nailed		580	.028		2.14	1.10		3.24	4.17
1100	3" x 8"		550	.029		2.10	1.16		3.26	4.23
1120	3" x 10"		500	.032		2.94	1.28		4.22	5.35
1140	3" x 12"		450	.036		3.54	1.42		4.96	6.25
1160	3" x 14"		400	.040		4.48	1.60		6.08	7.55
1170	4" x 6"	F-3	1100	.036		2.34	1.47	.70	4.51	5.75
1180	4" x 8"		1000	.040		2.79	1.62	.77	5.18	6.55
1200	4" x 10"		950	.042		3.68	1.70	.81	6.19	7.75
1220	4" x 12"		900	.044		4.74	1.80	.85	7.39	9.10
1240	4" x 14"		850	.047		5.60	1.90	.90	8.40	10.25
2000	Double, 2" x 6"	2 Carp	625	.026		1.04	1.02		2.06	2.84
2005	Pneumatic nailed		725	.022		1.04	.88		1.92	2.61
2020	2" x 8"		575	.028		1.42	1.11		2.53	3.40
2025	Pneumatic nailed		667	.024		1.42	.96		2.38	3.14
2040	2" x 10"		550	.029		1.98	1.16		3.14	4.10
2045	Pneumatic nailed		638	.025		1.98	1		2.98	3.83
2060	2" x 12"		525	.030		2.77	1.22		3.99	5.05
2065	Pneumatic nailed		610	.026		2.77	1.05		3.82	4.78
2080	2" x 14"		475	.034		4.28	1.35		5.63	6.90

06 11 Wood Framing

06 11 10 – Framing with Dimensional, Engineered or Composite Lumber

06 11 10.10 Beam and Girder Framing

		Crew	Daily Output	Labor-Hours	Unit	Material	2009 Bare Costs Labor	Equipment	Total	Total Incl O&P
2085	Pneumatic nailed	2 Carp	551	.029	L.F.	4.28	1.16		5.44	6.60
3000	Triple, 2" x 6"		550	.029		1.56	1.16		2.72	3.64
3005	Pneumatic nailed		638	.025		1.56	1		2.56	3.37
3020	2" x 8"		525	.030		2.12	1.22		3.34	4.34
3025	Pneumatic nailed		609	.026		2.12	1.05		3.17	4.06
3040	2" x 10"		500	.032		2.98	1.28		4.26	5.40
3045	Pneumatic nailed		580	.028		2.98	1.10		4.08	5.10
3060	2" x 12"		475	.034		4.16	1.35		5.51	6.80
3065	Pneumatic nailed		551	.029		4.16	1.16		5.32	6.50
3080	2" x 14"		450	.036		6.40	1.42		7.82	9.40
3085	Pneumatic nailed		522	.031		6.40	1.22		7.62	9.05
9000	Minimum labor/equipment charge	1 Carp	2	4	Job		160		160	264

06 11 10.12 Ceiling Framing

		Crew	Daily Output	Labor-Hours	Unit	Material	2009 Bare Costs Labor	Equipment	Total	Total Incl O&P
0010	**CEILING FRAMING**									
6000	Suspended, 2" x 3"	2 Carp	1000	.016	L.F.	.32	.64		.96	1.41
6050	2" x 4"		900	.018		.29	.71		1	1.49
6100	2" x 6"		800	.020		.52	.80		1.32	1.89
6150	2" x 8"		650	.025		.71	.98		1.69	2.40
9000	Minimum labor/equipment charge	1 Carp	4	2	Job		80		80	132

06 11 10.14 Posts and Columns

		Crew	Daily Output	Labor-Hours	Unit	Material	2009 Bare Costs Labor	Equipment	Total	Total Incl O&P
0010	**POSTS AND COLUMNS**									
0100	4" x 4"	2 Carp	390	.041	L.F.	1.56	1.64		3.20	4.43
0150	4" x 6"		275	.058		2.34	2.32		4.66	6.40
0200	4" x 8"		220	.073		2.79	2.91		5.70	7.85
0250	6" x 6"		215	.074		4.87	2.97		7.84	10.25
0300	6" x 8"		175	.091		7.60	3.65		11.25	14.40
0350	6" x 10"		150	.107		10.90	4.26		15.16	19.05
9000	Minimum labor/equipment charge	1 Carp	2	4	Job		160		160	264

06 11 10.18 Joist Framing

		Crew	Daily Output	Labor-Hours	Unit	Material	2009 Bare Costs Labor	Equipment	Total	Total Incl O&P
0010	**JOIST FRAMING** R061110-30									
2002	Joists, 2" x 4"	2 Carp	1250	.013	L.F.	.29	.51		.80	1.16
2007	Pneumatic nailed		1438	.011		.29	.44		.73	1.05
2100	2" x 6"		1250	.013		.52	.51		1.03	1.41
2105	Pneumatic nailed		1438	.011		.52	.44		.96	1.30
2152	2" x 8"		1100	.015		.71	.58		1.29	1.74
2157	Pneumatic nailed		1265	.013		.71	.51		1.22	1.61
2202	2" x 10"		900	.018		.99	.71		1.70	2.26
2207	Pneumatic nailed		1035	.015		.99	.62		1.61	2.11
2252	2" x 12"		875	.018		1.39	.73		2.12	2.73
2257	Pneumatic nailed		1006	.016		1.39	.64		2.03	2.57
2302	2" x 14"		770	.021		2.14	.83		2.97	3.72
2307	Pneumatic nailed		886	.018		2.14	.72		2.86	3.54
2352	3" x 6"		925	.017		1.52	.69		2.21	2.82
2402	3" x 10"		780	.021		2.94	.82		3.76	4.58
2452	3" x 12"		600	.027		3.54	1.07		4.61	5.65
2502	4" x 6"		800	.020		2.34	.80		3.14	3.89
2552	4" x 10"		600	.027		3.68	1.07		4.75	5.80
2602	4" x 12"		450	.036		4.74	1.42		6.16	7.55
2607	Sister joist, 2" x 6"		800	.020		.52	.80		1.32	1.89
2608	Pneumatic nailed		960	.017		.52	.67		1.19	1.67
2612	2" x 8"		640	.025		.71	1		1.71	2.43
2613	Pneumatic nailed		768	.021		.71	.83		1.54	2.15

06 11 10.18 Joist Framing

		Crew	Daily Output	Labor-Hours	Unit	Material	Labor	Equipment	Total	Total Incl O&P
2617	2" x 10"	2 Carp	535	.030	L.F.	.99	1.19		2.18	3.06
2618	Pneumatic nailed		642	.025		.99	1		1.99	2.73
2622	2" x 12"		455	.035		1.39	1.40		2.79	3.84
2627	Pneumatic nailed		546	.029		1.39	1.17		2.56	3.45
3000	Composite wood joist 9-1/2" deep		.90	17.778	M.L.F.	2,000	710		2,710	3,375
3010	11-1/2" deep		.88	18.182		2,175	725		2,900	3,575
3020	14" deep		.82	19.512		2,800	780		3,580	4,350
3030	16" deep		.78	20.513		3,250	820		4,070	4,925
4000	Open web joist 12" deep		.88	18.182		3,050	725		3,775	4,550
4010	14" deep		.82	19.512		3,275	780		4,055	4,875
4020	16" deep		.78	20.513		3,250	820		4,070	4,925
4030	18" deep		.74	21.622		3,500	865		4,365	5,275
6000	Composite rim joist, 1-1/4" x 9-1/2"		.90	17.778		1,825	710		2,535	3,200
6010	1-1/4" x 11-1/2"		.88	18.182		2,275	725		3,000	3,700
6020	1-1/4" x 14-1/2"		.82	19.512		2,625	780		3,405	4,150
6030	1-1/4" x 16-1/2"		.78	20.513		3,100	820		3,920	4,775
9000	Minimum labor/equipment charge	1 Carp	4	2	Job		80		80	132

06 11 10.24 Miscellaneous Framing

		Crew	Daily Output	Labor-Hours	Unit	Material	Labor	Equipment	Total	Total Incl O&P
0010	**MISCELLANEOUS FRAMING**									
2002	Firestops, 2" x 4"	2 Carp	780	.021	L.F.	.29	.82		1.11	1.67
2007	Pneumatic nailed		952	.017		.29	.67		.96	1.43
2102	2" x 6"		600	.027		.52	1.07		1.59	2.33
2107	Pneumatic nailed		732	.022		.52	.87		1.39	2.01
5002	Nailers, treated, wood construction, 2" x 4"		800	.020		.44	.80		1.24	1.81
5007	Pneumatic nailed		960	.017		.44	.67		1.11	1.59
5102	2" x 6"		750	.021		.69	.85		1.54	2.16
5107	Pneumatic nailed		900	.018		.69	.71		1.40	1.92
5122	2" x 8"		700	.023		1.05	.91		1.96	2.66
5127	Pneumatic nailed		840	.019		1.05	.76		1.81	2.41
5202	Steel construction, 2" x 4"		750	.021		.44	.85		1.29	1.90
5222	2" x 6"		700	.023		.69	.91		1.60	2.26
5242	2" x 8"		650	.025		1.05	.98		2.03	2.77
7002	Rough bucks, treated, for doors or windows, 2" x 6"		400	.040		.69	1.60		2.29	3.39
7007	Pneumatic nailed		480	.033		.69	1.33		2.02	2.95
7102	2" x 8"		380	.042		1.05	1.68		2.73	3.93
7107	Pneumatic nailed		456	.035		1.05	1.40		2.45	3.46
8000	Stair stringers, 2" x 10"		130	.123		.99	4.92		5.91	9.20
8100	2" x 12"		130	.123		1.39	4.92		6.31	9.60
8150	3" x 10"		125	.128		2.94	5.10		8.04	11.70
8200	3" x 12"		125	.128		3.54	5.10		8.64	12.35
8870	Laminated structural lumber, 1-1/4" x 11-1/2"		130	.123		2.27	4.92		7.19	10.60
8880	1-1/4" x 14-1/2"		130	.123		2.62	4.92		7.54	11
9000	Minimum labor/equipment charge	1 Carp	4	2	Job		80		80	132

06 11 10.26 Partitions

		Crew	Daily Output	Labor-Hours	Unit	Material	Labor	Equipment	Total	Total Incl O&P
0010	**PARTITIONS**									
0020	Single bottom and double top plate, no waste, std. & better lumber									
0182	2" x 4" studs, 8' high, studs 12" O.C.	2 Carp	80	.200	L.F.	3.51	8		11.51	17.05
0187	12" O.C., pneumatic nailed		96	.167		3.51	6.65		10.16	14.85
0202	16" O.C.		100	.160		2.87	6.40		9.27	13.70
0207	16" O.C., pneumatic nailed		120	.133		2.87	5.35		8.22	11.95
0302	24" O.C.		125	.128		2.23	5.10		7.33	10.90
0307	24" O.C., pneumatic nailed		150	.107		2.23	4.26		6.49	9.50

06 11 Wood Framing

06 11 10 – Framing with Dimensional, Engineered or Composite Lumber

06 11 10.26 Partitions

		Crew	Daily Output	Labor-Hours	Unit	Material	2009 Bare Costs Labor	Equipment	Total	Total Incl O&P
0382	10' high, studs 12" O.C.	2 Carp	80	.200	L.F.	4.15	8		12.15	17.75
0387	12" O.C., pneumatic nailed		96	.167		4.15	6.65		10.80	15.55
0402	16" O.C.		100	.160		3.35	6.40		9.75	14.25
0407	16" O.C., pneumatic nailed		120	.133		3.35	5.35		8.70	12.50
0502	24" O.C.		125	.128		2.55	5.10		7.65	11.25
0507	24" O.C., pneumatic nailed		150	.107		2.55	4.26		6.81	9.85
0582	12' high, studs 12" O.C.		65	.246		4.78	9.85		14.63	21.50
0587	12" O.C., pneumatic nailed		78	.205		4.78	8.20		12.98	18.80
0602	16" O.C.		80	.200		3.83	8		11.83	17.40
0607	16" O.C., pneumatic nailed		96	.167		3.83	6.65		10.48	15.20
0701	24" O.C.	↓	100	.160	↓	2.87	6.40		9.27	13.70
0702										
0706	24" O.C., pneumatic nailed	2 Carp	120	.133	L.F.	2.87	5.35		8.22	11.95
0782	2" x 6" studs, 8' high, studs 12" O.C.		70	.229		6.30	9.15		15.45	22
0787	12" O.C., pneumatic nailed		84	.190		6.30	7.60		13.90	19.50
0802	16" O.C.		90	.178		5.15	7.10		12.25	17.35
0807	16" O.C., pneumatic nailed		108	.148		5.15	5.90		11.05	15.40
0902	24" O.C.		115	.139		4.01	5.55		9.56	13.60
0907	24" O.C., pneumatic nailed		138	.116		4.01	4.63		8.64	12.05
0982	10' high, studs 12" O.C.		70	.229		7.45	9.15		16.60	23.50
0987	12" O.C., pneumatic nailed		84	.190		7.45	7.60		15.05	21
1002	16" O.C.		90	.178		6	7.10		13.10	18.30
1007	16" O.C., pneumatic nailed		108	.148		6	5.90		11.90	16.35
1102	24" O.C.		115	.139		4.58	5.55		10.13	14.25
1107	24" O.C., pneumatic nailed		138	.116		4.58	4.63		9.21	12.70
1182	12' high, studs 12" O.C.		55	.291		8.60	11.60		20.20	28.50
1187	12" O.C., pneumatic nailed		66	.242		8.60	9.70		18.30	25.50
1202	16" O.C.		70	.229		6.90	9.15		16.05	22.50
1207	16" O.C., pneumatic nailed		84	.190		6.90	7.60		14.50	20
1302	24" O.C.		90	.178		5.15	7.10		12.25	17.35
1307	24" O.C., pneumatic nailed		108	.148		5.15	5.90		11.05	15.40
1402	For horizontal blocking, 2" x 4", add		600	.027		.32	1.07		1.39	2.11
1502	2" x 6", add		600	.027		.57	1.07		1.64	2.39
1600	For openings, add	↓	250	.064	↓		2.56		2.56	4.22
1702	Headers for above openings, material only, add				B.F.	.58			.58	.64
9000	Minimum labor/equipment charge	1 Carp	4	2	Job		80		80	132

06 11 10.28 Porch or Deck Framing

		Crew	Daily Output	Labor-Hours	Unit	Material	2009 Bare Costs Labor	Equipment	Total	Total Incl O&P
0010	**PORCH OR DECK FRAMING**									
0100	Treated lumber, posts or columns, 4" x 4"	2 Carp	390	.041	L.F.	1.52	1.64		3.16	4.38
0110	4" x 6"		275	.058		2.24	2.32		4.56	6.30
0120	4" x 8"		220	.073		3.21	2.91		6.12	8.35
0130	Girder, single, 4" x 4"		675	.024		1.52	.95		2.47	3.23
0140	4" x 6"		600	.027		2.24	1.07		3.31	4.22
0150	4" x 8"		525	.030		3.21	1.22		4.43	5.55
0160	Double, 2" x 4"		625	.026		.91	1.02		1.93	2.69
0170	2" x 6"		600	.027		1.42	1.07		2.49	3.32
0180	2" x 8"		575	.028		2.16	1.11		3.27	4.22
0190	2" x 10"		550	.029		2.63	1.16		3.79	4.81
0200	2" x 12"		525	.030		4.15	1.22		5.37	6.55
0210	Triple, 2" x 4"		575	.028		1.37	1.11		2.48	3.35
0220	2" x 6"		550	.029		2.13	1.16		3.29	4.26
0230	2" x 8"	↓	525	.030	↓	3.24	1.22		4.46	5.55

06 11 Wood Framing

06 11 10 – Framing with Dimensional, Engineered or Composite Lumber

06 11 10.28 Porch or Deck Framing		Crew	Daily Output	Labor-Hours	Unit	Material	2009 Bare Costs Labor	Equipment	Total	Total Incl O&P
0240	2" x 10"	2 Carp	500	.032	L.F.	3.95	1.28		5.23	6.45
0250	2" x 12"		475	.034		6.20	1.35		7.55	9.05
0260	Ledger, bolted 4' O.C., 2" x 4"		400	.040		.55	1.60		2.15	3.25
0270	2" x 6"		395	.041		.80	1.62		2.42	3.55
0280	2" x 8"		390	.041		1.16	1.64		2.80	3.99
0300	2" x 12"		380	.042		2.14	1.68		3.82	5.15
0310	Joists, 2" x 4"		1250	.013		.46	.51		.97	1.34
0320	2" x 6"		1250	.013		.71	.51		1.22	1.62
0330	2" x 8"		1100	.015		1.08	.58		1.66	2.15
0340	2" x 10"		900	.018		1.32	.71		2.03	2.62
0350	2" x 12"		875	.018		1.58	.73		2.31	2.95
0440	Balusters, square, 2" x 2"		660	.024		.34	.97		1.31	1.97
0450	Turned, 2" x 2"		420	.038		.45	1.52		1.97	3
0460	Stair stringer, 2" x 10"		130	.123		1.32	4.92		6.24	9.55
0470	2" x 12"		130	.123		1.58	4.92		6.50	9.85
0480	Stair treads, 1" x 4"		140	.114		.52	4.57		5.09	8.15
0490	2" x 4"		140	.114		.46	4.57		5.03	8.05
0500	2" x 6"		160	.100		.69	4		4.69	7.35
0510	5/4" x 6"		160	.100	↓	1.10	4		5.10	7.80
0520	Turned handrail post, 4" x 4"		64	.250	Ea.	31.50	10		41.50	51.50
0530	Lattice panel, 4' x 8'		1600	.010	S.F.	.68	.40		1.08	1.41
0540	Cedar, posts or columns, 4" x 4"		390	.041	L.F.	3.21	1.64		4.85	6.25
0550	4" x 6"		275	.058		4.86	2.32		7.18	9.20
0560	4" x 8"		220	.073		7.30	2.91		10.21	12.80
0800	Decking, 1" x 4"		550	.029		1.76	1.16		2.92	3.85
0810	2" x 4"		600	.027		3.63	1.07		4.70	5.75
0820	2" x 6"		640	.025		6.95	1		7.95	9.30
0830	5/4" x 6"		640	.025		4.60	1		5.60	6.70
0840	Railings and trim, 1" x 4"		600	.027		1.76	1.07		2.83	3.69
0860	2" x 4"		600	.027		3.63	1.07		4.70	5.75
0870	2" x 6"		600	.027		6.95	1.07		8.02	9.40
0920	Stair treads, 1" x 4"		140	.114		1.76	4.57		6.33	9.50
0930	2" x 4"		140	.114		3.63	4.57		8.20	11.55
0940	2" x 6"		160	.100		6.95	4		10.95	14.25
0950	5/4" x 6"		160	.100		4.60	4		8.60	11.65
0980	Redwood, posts or columns, 4" x 4"		390	.041		6.45	1.64		8.09	9.80
0990	4" x 6"		275	.058		11.20	2.32		13.52	16.15
1000	4" x 8"		220	.073		20.50	2.91		23.41	27.50
1280	Railings and trim, 1" x 4"		600	.027		1.67	1.07		2.74	3.60
1310	2" x 6"		600	.027		4.93	1.07		6	7.15
1420	Alternative decking, wood / plastic composite, 5/4" x 6" G		640	.025		3.01	1		4.01	4.97
1430	Vinyl, 1-1/2" x 5-1/2"		640	.025		4.15	1		5.15	6.20
1440	1" x 4" square edge fir		550	.029		1.57	1.16		2.73	3.64
1450	1" x 4" tongue and groove fir		450	.036		1.41	1.42		2.83	3.90
1460	1" x 4" mahogany		550	.029		2.51	1.16		3.67	4.68
1462	5/4" x 6" PVC	↓	550	.029	↓	2.72	1.16		3.88	4.91
1470	Accessories, joist hangers, 2" x 4"	1 Carp	160	.050	Ea.	.61	2		2.61	3.97
1480	2" x 6" through 2" x 12"	"	150	.053		1.17	2.13		3.30	4.81
1530	Post footing, incl excav, backfill, tube form & concrete, 4' deep, 8" dia	F-7	12	2.667		10.60	95.50		106.10	169
1540	10" diameter		11	2.909		16	104		120	190
1550	12" diameter	↓	10	3.200	↓	21.50	115		136.50	213

115

06 11 Wood Framing

06 11 10 – Framing with Dimensional, Engineered or Composite Lumber

06 11 10.30 Roof Framing		Crew	Daily Output	Labor-Hours	Unit	Material	2009 Bare Costs Labor	Equipment	Total	Total Incl O&P
0010	**ROOF FRAMING** R061110-30									
2000	Fascia boards, 2" x 8"	2 Carp	225	.071	L.F.	.71	2.84		3.55	5.45
2100	2" x 10"		180	.089		.99	3.55		4.54	6.95
5000	Rafters, to 4 in 12 pitch, 2" x 6", ordinary		1000	.016		.52	.64		1.16	1.63
5060	2" x 8", ordinary		950	.017		.71	.67		1.38	1.89
5250	Composite rafter, 9-1/2" deep		575	.028		2	1.11		3.11	4.04
5260	11-1/2" deep		575	.028		2.17	1.11		3.28	4.22
5300	Hip and valley rafters, 2" x 6", ordinary		760	.021		.52	.84		1.36	1.96
5360	2" x 8", ordinary		720	.022		.71	.89		1.60	2.25
5540	Hip and valley jacks, 2" x 6", ordinary		600	.027		.52	1.07		1.59	2.33
5600	2" x 8", ordinary		490	.033		.71	1.30		2.01	2.93
5761	For slopes steeper than 4 in 12, add						30%			
5770	For dormers or complex roofs, add						50%			
5780	Rafter tie, 1" x 4", #3	2 Carp	800	.020	L.F.	.44	.80		1.24	1.80
5800	Ridge board, #2 or better, 1" x 6"		600	.027		.64	1.07		1.71	2.46
5820	1" x 8"		550	.029		.84	1.16		2	2.84
5840	1" x 10"		500	.032		1.23	1.28		2.51	3.46
5860	2" x 6"		500	.032		.52	1.28		1.80	2.68
5880	2" x 8"		450	.036		.71	1.42		2.13	3.13
5900	2" x 10"		400	.040		.99	1.60		2.59	3.73
5920	Roof cants, split, 4" x 4"		650	.025		1.56	.98		2.54	3.34
5940	6" x 6"		600	.027		4.87	1.07		5.94	7.10
5960	Roof curbs, untreated, 2" x 6"		520	.031		.52	1.23		1.75	2.60
5980	2" x 12"		400	.040		1.39	1.60		2.99	4.16
6000	Sister rafters, 2" x 6"		800	.020		.52	.80		1.32	1.89
6020	2" x 8"		640	.025		.71	1		1.71	2.43
6040	2" x 10"		535	.030		.99	1.19		2.18	3.06
6060	2" x 12"		455	.035		1.39	1.40		2.79	3.84
9000	Minimum labor/equipment charge	1 Carp	4	2	Job		80		80	132

06 11 10.32 Sill and Ledger Framing

		Crew	Daily Output	Labor-Hours	Unit	Material	Labor	Equipment	Total	Total Incl O&P
0010	**SILL AND LEDGER FRAMING**									
2002	Ledgers, nailed, 2" x 4"	2 Carp	755	.021	L.F.	.29	.85		1.14	1.72
2052	2" x 6"		600	.027		.52	1.07		1.59	2.33
2102	Bolted, not including bolts, 3" x 6"		325	.049		1.52	1.97		3.49	4.93
2152	3" x 12"		233	.069		3.54	2.74		6.28	8.40
2602	Mud sills, redwood, construction grade, 2" x 4"		895	.018		2.27	.71		2.98	3.67
2622	2" x 6"		780	.021		3.40	.82		4.22	5.10
4002	Sills, 2" x 4"		600	.027		.29	1.07		1.36	2.08
4052	2" x 6"		550	.029		.52	1.16		1.68	2.49
4082	2" x 8"		500	.032		.71	1.28		1.99	2.89
4202	Treated, 2" x 4"		550	.029		.44	1.16		1.60	2.41
4222	2" x 6"		500	.032		.69	1.28		1.97	2.86
4242	2" x 8"		450	.036		1.05	1.42		2.47	3.50
4402	4" x 4"		450	.036		1.49	1.42		2.91	3.99
4422	4" x 6"		350	.046		2.19	1.83		4.02	5.40
4462	4" x 8"		300	.053		3.14	2.13		5.27	7
4481	4" x 10"		260	.062		4.10	2.46		6.56	8.55
9000	Minimum labor/equipment charge	1 Carp	4	2	Job		80		80	132

06 11 10.34 Sleepers

		Crew	Daily Output	Labor-Hours	Unit	Material	Labor	Equipment	Total	Total Incl O&P
0010	**SLEEPERS**									
0100	On concrete, treated, 1" x 2"	2 Carp	2350	.007	L.F.	.38	.27		.65	.87
0150	1" x 3"		2000	.008		.52	.32		.84	1.10

06 11 Wood Framing

06 11 10 – Framing with Dimensional, Engineered or Composite Lumber

06 11 10.34 Sleepers		Crew	Daily Output	Labor-Hours	Unit	Material	2009 Bare Costs Labor	2009 Bare Costs Equipment	Total	Total Incl O&P
0200	2" x 4"	2 Carp	1500	.011	L.F.	.44	.43		.87	1.19
0250	2" x 6"	↓	1300	.012	↓	.69	.49		1.18	1.56
9000	Minimum labor/equipment charge	1 Carp	4	2	Job		80		80	132

06 11 10.36 Soffit and Canopy Framing

0010	**SOFFIT AND CANOPY FRAMING**									
1002	Canopy or soffit framing , 1" x 4"	2 Carp	900	.018	L.F.	.44	.71		1.15	1.65
1042	1" x 8"		750	.021		.84	.85		1.69	2.33
1102	2" x 4"		620	.026		.29	1.03		1.32	2.02
1142	2" x 8"		500	.032		.71	1.28		1.99	2.89
1202	3" x 4"		500	.032		.93	1.28		2.21	3.13
1242	3" x 10"	↓	300	.053	↓	2.94	2.13		5.07	6.75
9000	Minimum labor/equipment charge	1 Carp	4	2	Job		80		80	132

06 11 10.38 Treated Lumber Framing Material

0010	**TREATED LUMBER FRAMING MATERIAL**									
0100	2" x 4"				M.B.F.	660			660	730
0110	2" x 6"					685			685	755
0120	2" x 8"					785			785	865
0130	2" x 10"					765			765	845
0140	2" x 12"					1,025			1,025	1,125
0200	4" x 4"					1,125			1,125	1,225
0210	4" x 6"					1,100			1,100	1,200
0220	4" x 8"				↓	1,175			1,175	1,300

06 11 10.40 Wall Framing

0010	**WALL FRAMING** R061110-30									
0100	Door buck, studs, header, access, 8' H, 2"x4" wall, 3' W	1 Carp	32	.250	Ea.	12.65	10		22.65	30.50
0110	4' wide		32	.250		13.70	10		23.70	31.50
0120	5' wide		32	.250		16.70	10		26.70	35
0130	6' wide		32	.250		18.10	10		28.10	36.50
0140	8' wide		30	.267		25.50	10.65		36.15	45.50
0150	10' wide		30	.267		37.50	10.65		48.15	59
0160	12' wide		30	.267		61.50	10.65		72.15	85.50
0170	2" x 6" wall, 3' wide		32	.250		20	10		30	38.50
0180	4' wide		32	.250		21	10		31	39.50
0190	5' wide		32	.250		24	10		34	43
0200	6' wide		32	.250		25.50	10		35.50	44.50
0210	8' wide		30	.267		33	10.65		43.65	54
0220	10' wide		30	.267		45	10.65		55.65	67
0230	12' wide		30	.267		69	10.65		79.65	93.50
0240	Window buck, studs, header & access, 8' high 2" x 4" wall, 2' wide		24	.333		12.90	13.30		26.20	36
0250	3' wide		24	.333		15.55	13.30		28.85	39
0260	4' wide		24	.333		17.20	13.30		30.50	41
0270	5' wide		24	.333		20	13.30		33.30	44
0280	6' wide		24	.333		22.50	13.30		35.80	46.50
0300	8' wide		22	.364		31.50	14.55		46.05	59
0310	10' wide		22	.364		44.50	14.55		59.05	73
0320	12' wide		22	.364		70.50	14.55		85.05	102
0330	2" x 6" wall, 2' wide		24	.333		22	13.30		35.30	46.50
0340	3' wide		24	.333		25.50	13.30		38.80	50
0350	4' wide		24	.333		27	13.30		40.30	52
0360	5' wide		24	.333		30.50	13.30		43.80	55.50
0370	6' wide		24	.333		33.50	13.30		46.80	58.50
0380	7' wide	↓	24	.333	↓	40.50	13.30		53.80	66.50

06 11 Wood Framing

06 11 10 – Framing with Dimensional, Engineered or Composite Lumber

06 11 10.40 Wall Framing

		Crew	Daily Output	Labor-Hours	Unit	Material	2009 Bare Costs Labor	Equipment	Total	Total Incl O&P
0390	8' wide	1 Carp	22	.364	Ea.	44	14.55		58.55	72.50
0400	10' wide		22	.364		57.50	14.55		72.05	87.50
0410	12' wide		22	.364		84.50	14.55		99.05	117
2002	Headers over openings, 2" x 6"	2 Carp	360	.044	L.F.	.52	1.78		2.30	3.50
2007	2" x 6", pneumatic nailed		432	.037		.52	1.48		2	3.01
2052	2" x 8"		340	.047		.71	1.88		2.59	3.88
2057	2" x 8", pneumatic nailed		408	.039		.71	1.57		2.28	3.37
2100	2" x 10"		320	.050		.99	2		2.99	4.39
2105	2" x 10", pneumatic nailed		384	.042		.99	1.66		2.65	3.84
2152	2" x 12"		300	.053		1.39	2.13		3.52	5.05
2157	2" x 12", pneumatic nailed		360	.044		1.39	1.78		3.17	4.45
2202	4" x 12"		190	.084		4.74	3.36		8.10	10.75
2207	4" x 12", pneumatic nailed		228	.070		4.74	2.80		7.54	9.85
2252	6" x 12"		140	.114		9.15	4.57		13.72	17.65
5002	Plates, untreated, 2" x 3"		850	.019		.32	.75		1.07	1.59
5007	2" x 3", pneumatic nailed		1020	.016		.32	.63		.95	1.38
5022	2" x 4"		800	.020		.29	.80		1.09	1.64
5027	2" x 4", pneumatic nailed		960	.017		.29	.67		.96	1.42
5041	2" x 6"		750	.021		.52	.85		1.37	1.98
5046	2" x 6", pneumatic nailed		900	.018		.52	.71		1.23	1.74
5122	Studs, 8' high wall, 2" x 3"		1200	.013		.32	.53		.85	1.23
5127	2" x 3", pneumatic nailed		1440	.011		.32	.44		.76	1.08
5142	2" x 4"		1100	.015		.29	.58		.87	1.28
5147	2" x 4", pneumatic nailed		1320	.012		.29	.48		.77	1.12
5162	2" x 6"		1000	.016		.52	.64		1.16	1.63
5167	2" x 6", pneumatic nailed		1200	.013		.52	.53		1.05	1.45
5182	3" x 4"		800	.020		.93	.80		1.73	2.34
5187	3" x 4", pneumatic nailed		960	.017		.93	.67		1.60	2.12
8200	For 12' high walls, deduct						5%			
8220	For stub wall, 6' high, add						20%			
8240	3' high, add						40%			
8250	For second story & above, add						5%			
8300	For dormer & gable, add						15%			
9000	Minimum labor/equipment charge	1 Carp	4	2	Job		80		80	132

06 11 10.42 Furring

		Crew	Daily Output	Labor-Hours	Unit	Material	2009 Bare Costs Labor	Equipment	Total	Total Incl O&P
0010	**FURRING**									
0012	Wood strips, 1" x 2", on walls, on wood	1 Carp	550	.015	L.F.	.25	.58		.83	1.24
0017	Pneumatic nailed		710	.011		.25	.45		.70	1.02
0300	On masonry		495	.016		.25	.65		.90	1.35
0400	On concrete		260	.031		.25	1.23		1.48	2.31
0600	1" x 3", on walls, on wood		550	.015		.34	.58		.92	1.34
0607	Pneumatic nailed		710	.011		.34	.45		.79	1.12
0700	On masonry		495	.016		.34	.65		.99	1.45
0800	On concrete		260	.031		.34	1.23		1.57	2.41
0850	On ceilings, on wood		350	.023		.34	.91		1.25	1.89
0857	Pneumatic nailed		450	.018		.34	.71		1.05	1.55
0900	On masonry		320	.025		.34	1		1.34	2.03
0950	On concrete		210	.038		.34	1.52		1.86	2.89
9000	Minimum labor/equipment charge		4	2	Job		80		80	132

06 11 10.44 Grounds

		Crew	Daily Output	Labor-Hours	Unit	Material	2009 Bare Costs Labor	Equipment	Total	Total Incl O&P
0010	**GROUNDS**									
0020	For casework, 1" x 2" wood strips, on wood	1 Carp	330	.024	L.F.	.25	.97		1.22	1.88

06 11 Wood Framing

06 11 10 – Framing with Dimensional, Engineered or Composite Lumber

06 11 10.44 Grounds

		Crew	Daily Output	Labor-Hours	Unit	Material	2009 Bare Costs Labor	Equipment	Total	Total Incl O&P
0100	On masonry	1 Carp	285	.028	L.F.	.25	1.12		1.37	2.13
0200	On concrete		250	.032		.25	1.28		1.53	2.39
0400	For plaster, 3/4" deep, on wood		450	.018		.25	.71		.96	1.45
0500	On masonry		225	.036		.25	1.42		1.67	2.63
0600	On concrete		175	.046		.25	1.83		2.08	3.29
0700	On metal lath		200	.040		.25	1.60		1.85	2.92
9000	Minimum labor/equipment charge		4	2	Job		80		80	132

06 12 Structural Panels

06 12 10 – Structural Insulated Panels

06 12 10.10 OSB Faced Panels

			Crew	Daily Output	Labor-Hours	Unit	Material	2009 Bare Costs Labor	Equipment	Total	Total Incl O&P
0010	**OSB FACED PANELS**										
0100	Structural insul. panels, 7/16" OSB both faces, EPS insul, 3-5/8" T	G	F-3	2075	.019	S.F.	2.84	.78	.37	3.99	4.81
0110	5-5/8" thick	G		1725	.023		3.37	.94	.45	4.76	5.75
0120	7-3/8" thick	G		1425	.028		3.84	1.14	.54	5.52	6.65
0130	9-3/8" thick	G		1125	.036		4.40	1.44	.68	6.52	7.95
0140	7/16" OSB one face, EPS insul, 3-5/8" thick	G		2175	.018		1.65	.74	.35	2.74	3.43
0150	5-5/8" thick	G		1825	.022		2.06	.89	.42	3.37	4.18
0160	7-3/8" thick	G		1525	.026		2.40	1.06	.50	3.96	4.93
0170	9-3/8" thick	G		1225	.033		2.81	1.32	.63	4.76	5.95
0190	7/16" OSB - 1/2" GWB faces , EPS insul, 3-5/8" T	G		2075	.019		3.03	.78	.37	4.18	5
0200	5-5/8" thick	G		1725	.023		3.34	.94	.45	4.73	5.70
0210	7-3/8" thick	G		1425	.028		3.65	1.14	.54	5.33	6.45
0220	9-3/8" thick	G		1125	.036		4	1.44	.68	6.12	7.50
0240	7/16" OSB - 1/2" MRGWB faces , EPS insul, 3-5/8" T	G		2075	.019		3.03	.78	.37	4.18	5
0250	5-5/8" thick	G		1725	.023		3.34	.94	.45	4.73	5.70
0260	7-3/8" thick	G		1425	.028		3.65	1.14	.54	5.33	6.45
0270	9-3/8" thick	G		1125	.036		4	1.44	.68	6.12	7.50
0300	For 1/2" GWB added to OSB skin, add	G					.60			.60	.66
0310	For 1/2" MRGWB added to OSB skin, add	G					.75			.75	.83
0320	For one T1-11 skin, add to OSB-OSB	G					.74			.74	.81
0330	For one 19/32" CDX skin, add to OSB-OSB	G					.69			.69	.76
0500	Structural insulated panel, 7/16" OSB both sides, straw core										
0510	4-3/8" T, walls (w/sill, splines, plates)	G	F-6	2400	.017	S.F.	7.60	.62	.32	8.54	9.70
0520	Floors (w/ splines)	G		2400	.017		7.60	.62	.32	8.54	9.70
0530	Roof (w/ splines)	G		2400	.017		7.60	.62	.32	8.54	9.70
0550	7-7/8" T, walls (w/sill, splines, plates)	G		2400	.017		10.95	.62	.32	11.89	13.40
0560	Floors (w/ splines)	G		2400	.017		10.95	.62	.32	11.89	13.40
0570	Roof (w/ splines)	G		2400	.017		10.95	.62	.32	11.89	13.40

06 13 Heavy Timber

06 13 23 – Heavy Timber Construction

06 13 23.10 Heavy Framing

	06 13 23.10 Heavy Framing	Crew	Daily Output	Labor-Hours	Unit	Material	2009 Bare Costs Labor	2009 Bare Costs Equipment	Total	Total Incl O&P
0010	**HEAVY FRAMING**									
0020	Beams, single 6" x 10"	2 Carp	1.10	14.545	M.B.F.	1,825	580		2,405	2,950
0100	Single 8" x 16"		1.20	13.333	"	2,900	535		3,435	4,075
0202	Built from 2" lumber, multiple 2" x 14"		900	.018	B.F.	.92	.71		1.63	2.18
0212	Built from 3" lumber, multiple 3" x 6"		700	.023		1.02	.91		1.93	2.63
0222	Multiple 3" x 8"		800	.020		1.05	.80		1.85	2.48
0232	Multiple 3" x 10"		900	.018		1.18	.71		1.89	2.46
0242	Multiple 3" x 12"		1000	.016		1.18	.64		1.82	2.36
0252	Built from 4" lumber, multiple 4" x 6"		800	.020		1.17	.80		1.97	2.61
0262	Multiple 4" x 8"		900	.018		1.05	.71		1.76	2.32
0272	Multiple 4" x 10"	▼	1000	.016	▼	1.10	.64		1.74	2.27
0281										
0282	Multiple 4" x 12"	2 Carp	1100	.015	B.F.	1.18	.58		1.76	2.26
0292	Columns, structural grade, 1500f, 4" x 4"		450	.036	L.F.	1.75	1.42		3.17	4.28
0302	6" x 6"		225	.071		4.85	2.84		7.69	10.05
0402	8" x 8"		240	.067		10.75	2.66		13.41	16.25
0502	10" x 10"		90	.178		19	7.10		26.10	32.50
0602	12" x 12"		70	.229	▼	28	9.15		37.15	45.50
0802	Floor planks, 2" thick, T & G, 2" x 6"		1050	.015	B.F.	1.54	.61		2.15	2.71
0902	2" x 10"		1100	.015		1.54	.58		2.12	2.66
1102	3" thick, 3" x 6"		1050	.015		1.30	.61		1.91	2.44
1202	3" x 10"		1100	.015		1.30	.58		1.88	2.39
1402	Girders, structural grade, 12" x 12"		800	.020		2.32	.80		3.12	3.88
1502	10" x 16"		1000	.016		2.28	.64		2.92	3.57
2302	Roof purlins, 4" thick, structural grade		1050	.015		1.46	.61		2.07	2.62
2502	Roof trusses, add timber connectors, Division 06 05 23.60	▼	450	.036	▼	1.34	1.42		2.76	3.83
9000	Minimum labor/equipment charge	1 Carp	2	4	Job		160		160	264

06 15 Wood Decking

06 15 16 – Wood Roof Decking

06 15 16.10 Solid Wood Roof Decking

	06 15 16.10 Solid Wood Roof Decking	Crew	Daily Output	Labor-Hours	Unit	Material	Labor	Equipment	Total	Total Incl O&P
0010	**SOLID WOOD ROOF DECKING**									
0400	Cedar planks, 3" thick	2 Carp	320	.050	S.F.	6.20	2		8.20	10.15
0500	4" thick		250	.064		8.35	2.56		10.91	13.40
0702	Douglas fir, 3" thick		320	.050		2.33	2		4.33	5.85
0802	4" thick		250	.064		3.12	2.56		5.68	7.65
1002	Hemlock, 3" thick		320	.050		2.33	2		4.33	5.85
1102	4" thick		250	.064		3.11	2.56		5.67	7.65
1302	Western white spruce, 3" thick		320	.050		2.24	2		4.24	5.75
1402	4" thick	▼	250	.064	▼	2.99	2.56		5.55	7.50
9000	Minimum labor/equipment charge	1 Carp	2	4	Job		160		160	264

06 15 23 – Laminated Wood Decking

06 15 23.10 Laminated Roof Deck

	06 15 23.10 Laminated Roof Deck	Crew	Daily Output	Labor-Hours	Unit	Material	Labor	Equipment	Total	Total Incl O&P
0010	**LAMINATED ROOF DECK**									
0020	Pine or hemlock, 3" thick	2 Carp	425	.038	S.F.	3.36	1.50		4.86	6.20
0100	4" thick		325	.049		4.47	1.97		6.44	8.15
0300	Cedar, 3" thick		425	.038		4.11	1.50		5.61	7
0400	4" thick		325	.049		5.25	1.97		7.22	9
0600	Fir, 3" thick		425	.038		3.43	1.50		4.93	6.25
0700	4" thick	▼	325	.049	▼	4.29	1.97		6.26	7.95

06 15 Wood Decking

06 15 23 - Laminated Wood Decking

06 15 23.10 Laminated Roof Deck		Crew	Daily Output	Labor-Hours	Unit	Material	2009 Bare Costs Labor	2009 Bare Costs Equipment	Total	Total Incl O&P
9000	Minimum labor/equipment charge	1 Carp	3	2.667	Job		107		107	176

06 16 Sheathing

06 16 23 - Subflooring

06 16 23.10 Subfloor

		Crew	Daily Output	Labor-Hours	Unit	Material	2009 Bare Costs Labor	2009 Bare Costs Equipment	Total	Total Incl O&P
0010	**SUBFLOOR** R061636-20									
0011	Plywood, CDX, 1/2" thick	2 Carp	1500	.011	SF Flr.	.46	.43		.89	1.21
0015	Pneumatic nailed		1860	.009		.46	.34		.80	1.08
0102	5/8" thick		1350	.012		.67	.47		1.14	1.52
0107	Pneumatic nailed		1674	.010		.67	.38		1.05	1.37
0202	3/4" thick		1250	.013		.83	.51		1.34	1.75
0207	Pneumatic nailed		1550	.010		.83	.41		1.24	1.59
0302	1-1/8" thick, 2-4-1 including underlayment		1050	.015		1.63	.61		2.24	2.80
0440	With boards, 1" x 6", S4S, laid regular		900	.018		1.40	.71		2.11	2.70
0452	1" x 8", laid regular		1000	.016		1.34	.64		1.98	2.53
0462	Laid diagonal		850	.019		1.34	.75		2.09	2.71
0502	1" x 10", laid regular		1100	.015		1.54	.58		2.12	2.66
0602	Laid diagonal		900	.018		1.54	.71		2.25	2.87
8990	Subfloor adhesive, 3/8" bead	1 Carp	2300	.003	L.F.	.09	.14		.23	.33
9000	Minimum labor/equipment charge	"	4	2	Job		80		80	132

06 16 26 - Underlayment

06 16 26.10 Wood Product Underlayment

		Crew	Daily Output	Labor-Hours	Unit	Material	2009 Bare Costs Labor	2009 Bare Costs Equipment	Total	Total Incl O&P
0010	**WOOD PRODUCT UNDERLAYMENT** R061636-20									
0030	Plywood, underlayment grade, 3/8" thick	2 Carp	1500	.011	SF Flr.	1.03	.43		1.46	1.83
0080	Pneumatic nailed		1860	.009		1.03	.34		1.37	1.70
0102	1/2" thick		1450	.011		1	.44		1.44	1.83
0107	Pneumatic nailed		1798	.009		1	.36		1.36	1.69
0202	5/8" thick		1400	.011		1.29	.46		1.75	2.17
0207	Pneumatic nailed		1736	.009		1.29	.37		1.66	2.03
0302	3/4" thick		1300	.012		1.33	.49		1.82	2.27
0306	Pneumatic nailed		1612	.010		1.33	.40		1.73	2.11
0502	Particle board, 3/8" thick		1500	.011		.41	.43		.84	1.15
0507	Pneumatic nailed		1860	.009		.41	.34		.75	1.02
0602	1/2" thick		1450	.011		.45	.44		.89	1.23
0607	Pneumatic nailed		1798	.009		.45	.36		.81	1.09
0802	5/8" thick		1400	.011		.54	.46		1	1.34
0807	Pneumatic nailed		1736	.009		.54	.37		.91	1.20
0902	3/4" thick		1300	.012		.65	.49		1.14	1.53
0907	Pneumatic nailed		1612	.010		.65	.40		1.05	1.37
0955	Particleboard, 100% recycled straw/wheat, 4' x 8' x 1/4" [G]		1450	.011	S.F.	.26	.44		.70	1.02
0960	4' x 8' x 3/8" [G]		1450	.011		.39	.44		.83	1.16
0965	4' x 8' x 1/2" [G]		1350	.012		.53	.47		1	1.36
0970	4' x 8' x 5/8" [G]		1300	.012		.63	.49		1.12	1.50
0975	4' x 8' x 3/4" [G]		1250	.013		.73	.51		1.24	1.64
0980	4' x 8' x 1" [G]		1150	.014		.98	.56		1.54	2
0985	4' x 8' x 1-1/4" [G]		1100	.015		1.15	.58		1.73	2.23
1102	Hardboard, underlayment grade, 4' x 4', .215" thick		1500	.011	SF Flr.	.50	.43		.93	1.25
9000	Minimum labor/equipment charge	1 Carp	4	2	Job		80		80	132

06 16 Sheathing

06 16 36 – Wood Panel Product Sheathing

06 16 36.10 Sheathing		Crew	Daily Output	Labor-Hours	Unit	Material	2009 Bare Costs Labor	Equipment	Total	Total Incl O&P
0010	**SHEATHING**, plywood on roofs	R061110-30								
0012	Plywood on roofs, CDX									
0032	5/16" thick	2 Carp	1600	.010	S.F.	.53	.40		.93	1.24
0037	Pneumatic nailed	R061636-20		1952	.008		.53	.33		.86	1.12
0052	3/8" thick		1525	.010		.45	.42		.87	1.18
0057	Pneumatic nailed		1860	.009		.45	.34		.79	1.06
0102	1/2" thick		1400	.011		.46	.46		.92	1.26
0103	Pneumatic nailed		1708	.009		.46	.37		.83	1.13
0202	5/8" thick		1300	.012		.67	.49		1.16	1.55
0207	Pneumatic nailed		1586	.010		.67	.40		1.07	1.41
0302	3/4" thick		1200	.013		.83	.53		1.36	1.79
0307	Pneumatic nailed		1464	.011		.83	.44		1.27	1.63
0502	Plywood on walls with exterior CDX, 3/8" thick		1200	.013		.45	.53		.98	1.37
0507	Pneumatic nailed		1488	.011		.45	.43		.88	1.20
0602	1/2" thick		1125	.014		.46	.57		1.03	1.45
0607	Pneumatic nailed		1395	.011		.46	.46		.92	1.27
0702	5/8" thick		1050	.015		.67	.61		1.28	1.75
0707	Pneumatic nailed		1302	.012		.67	.49		1.16	1.55
0802	3/4" thick		975	.016		.83	.66		1.49	1.99
0807	Pneumatic nailed		1209	.013		.83	.53		1.36	1.78
0840	Oriented strand board, 7/16" thick	G	1400	.011		.22	.46		.68	.99
0845	Pneumatic nailed	G	1736	.009		.22	.37		.59	.85
0846	1/2" thick	G	1325	.012		.22	.48		.70	1.04
0847	Pneumatic nailed	G	1643	.010		.22	.39		.61	.88
0852	5/8" thick	G	1250	.013		.35	.51		.86	1.23
0857	Pneumatic nailed	G	1550	.010		.35	.41		.76	1.07
1000	For shear wall construction, add						20%			
1200	For structural 1 exterior plywood, add				S.F.	10%				
1402	With boards, on roof 1" x 6" boards, laid horizontal	2 Carp	725	.022		1.40	.88		2.28	2.99
1502	Laid diagonal		650	.025		1.40	.98		2.38	3.15
1702	1" x 8" boards, laid horizontal		875	.018		1.34	.73		2.07	2.68
1802	Laid diagonal		725	.022		1.34	.88		2.22	2.93
2000	For steep roofs, add						40%			
2200	For dormers, hips and valleys, add					5%	50%			
2402	Boards on walls, 1" x 6" boards, laid regular	2 Carp	650	.025		1.40	.98		2.38	3.15
2502	Laid diagonal		585	.027		1.40	1.09		2.49	3.33
2702	1" x 8" boards, laid regular		765	.021		1.34	.84		2.18	2.85
2802	Laid diagonal		650	.025		1.34	.98		2.32	3.09
2852	Gypsum, weatherproof, 1/2" thick		1050	.015		.61	.61		1.22	1.68
2900	With embedded glass mats		1100	.015		.49	.58		1.07	1.50
3000	Wood fiber, regular, no vapor barrier, 1/2" thick		1200	.013		.56	.53		1.09	1.50
3100	5/8" thick		1200	.013		.71	.53		1.24	1.66
3300	No vapor barrier, in colors, 1/2" thick		1200	.013		.75	.53		1.28	1.71
3400	5/8" thick		1200	.013		.80	.53		1.33	1.76
3600	With vapor barrier one side, white, 1/2" thick		1200	.013		.55	.53		1.08	1.49
3700	Vapor barrier 2 sides, 1/2" thick		1200	.013		.77	.53		1.30	1.73
3800	Asphalt impregnated, 25/32" thick		1200	.013		.28	.53		.81	1.19
3850	Intermediate, 1/2" thick		1200	.013		.18	.53		.71	1.08
9000	Minimum labor/equipment charge	1 Carp	2	4	Job		160		160	264

06 17 33 – Wood I-Joists

	06 17 33.10 Wood and Composite I-Joists	Crew	Daily Output	Labor-Hours	Unit	Material	2009 Bare Costs Labor	Equipment	Total	Total Incl O&P
0010	**WOOD AND COMPOSITE I-JOISTS**									
0100	Plywood webs, incl. bridging & blocking, panels 24" O.C.									
1200	15' to 24' span, 50 psf live load	F-5	2400	.013	SF Flr.	2.28	.54		2.82	3.40
1300	55 psf live load		2250	.014		2.47	.58		3.05	3.67
1400	24' to 30' span, 45 psf live load		2600	.012		3.18	.50		3.68	4.32
1500	55 psf live load		2400	.013		3.70	.54		4.24	4.96
1600	Tubular steel open webs, 45 psf, 24" O.C., 40' span	F-3	6250	.006		2.24	.26	.12	2.62	3.02
1700	55' span		7750	.005		2.17	.21	.10	2.48	2.84
1800	70' span		9250	.004		2.82	.17	.08	3.07	3.48
1900	85 psf live load, 26' span		2300	.017		2.63	.70	.33	3.66	4.41

06 17 53 – Shop-Fabricated Wood Trusses

06 17 53.10 Roof Trusses

		Crew	Daily Output	Labor-Hours	Unit	Material	2009 Bare Costs Labor	Equipment	Total	Total Incl O&P
0010	**ROOF TRUSSES**									
0100	Fink (W) or King post type, 2'-0" O.C.									
0200	Metal plate connected, 4 in 12 slope									
0210	24' to 29' span	F-3	3000	.013	SF Flr.	2.06	.54	.26	2.86	3.43
0300	30' to 43' span		3000	.013		2.28	.54	.26	3.08	3.67
0400	44' to 60' span		3000	.013		2.52	.54	.26	3.32	3.93
0700	Glued and nailed, add					50%				

06 18 13 – Glued-Laminated Beams

06 18 13.20 Laminated Framing

		Crew	Daily Output	Labor-Hours	Unit	Material	2009 Bare Costs Labor	Equipment	Total	Total Incl O&P
0010	**LAMINATED FRAMING**									
0020	30 lb., short term live load, 15 lb. dead load									
0200	Straight roof beams, 20' clear span, beams 8' O.C.	F-3	2560	.016	SF Flr.	1.98	.63	.30	2.91	3.54
0300	Beams 16' O.C.		3200	.013		1.43	.51	.24	2.18	2.66
0500	40' clear span, beams 8' O.C.		3200	.013		3.78	.51	.24	4.53	5.25
0600	Beams 16' O.C.		3840	.010		3.09	.42	.20	3.71	4.31
0800	60' clear span, beams 8' O.C.	F-4	2880	.017		6.50	.66	.39	7.55	8.65
0900	Beams 16' O.C.	"	3840	.013		4.84	.50	.29	5.63	6.45
1100	Tudor arches, 30' to 40' clear span, frames 8' O.C.	F-3	1680	.024		8.45	.96	.46	9.87	11.40
1200	Frames 16' O.C.	"	2240	.018		6.65	.72	.34	7.71	8.85
1400	50' to 60' clear span, frames 8' O.C.	F-4	2200	.022		9.15	.87	.51	10.53	12
1500	Frames 16' O.C.		2640	.018		7.80	.72	.43	8.95	10.20
1700	Radial arches, 60' clear span, frames 8' O.C.		1920	.025		8.55	1	.59	10.14	11.65
1800	Frames 16' O.C.		2880	.017		6.55	.66	.39	7.60	8.75
2000	100' clear span, frames 8' O.C.		1600	.030		8.85	1.20	.70	10.75	12.45
2100	Frames 16' O.C.		2400	.020		7.80	.80	.47	9.07	10.35
2300	120' clear span, frames 8' O.C.		1440	.033		11.75	1.33	.78	13.86	15.95
2400	Frames 16' O.C.		1920	.025		10.75	1	.59	12.34	14.05
2600	Bowstring trusses, 20' O.C., 40' clear span	F-3	2400	.017		5.30	.67	.32	6.29	7.30
2700	60' clear span	F-4	3600	.013		4.76	.53	.31	5.60	6.45
2800	100' clear span		4000	.012		6.75	.48	.28	7.51	8.50
2900	120' clear span		3600	.013		7.25	.53	.31	8.09	9.15
3100	For premium appearance, add to S.F. prices					5%				
3300	For industrial type, deduct					15%				
3500	For stain and varnish, add					5%				
3900	For 3/4" laminations, add to straight					25%				
4100	Add to curved					15%				

06 18 Glued-Laminated Construction

06 18 13 – Glued-Laminated Beams

06 18 13.20 Laminated Framing

06 18 13.20 Laminated Framing	Crew	Daily Output	Labor-Hours	Unit	Material	2009 Bare Costs Labor	Equipment	Total	Total Incl O&P
4300	Alternate pricing method: (use nominal footage of									
4310	components). Straight beams, camber less than 6"	F-3	3.50	11.429	M.B.F.	2,925	465	220	3,610	4,225
4400	Columns, including hardware		2	20		3,150	810	385	4,345	5,225
4600	Curved members, radius over 32'		2.50	16		3,225	650	310	4,185	4,950
4700	Radius 10' to 32'		3	13.333		3,200	540	256	3,996	4,675
4900	For complicated shapes, add maximum					100%				
5100	For pressure treating, add to straight					35%				
5200	Add to curved					45%				
6000	Laminated veneer members, southern pine or western species									
6050	1-3/4" wide x 5-1/2" deep	2 Carp	480	.033	L.F.	2.99	1.33		4.32	5.50
6100	9-1/2" deep		480	.033		4.79	1.33		6.12	7.45
6150	14" deep		450	.036		7	1.42		8.42	10.05
6200	18" deep		450	.036		9.05	1.42		10.47	12.35
6300	Parallel strand members, southern pine or western species									
6350	1-3/4" wide x 9-1/4" deep	2 Carp	480	.033	L.F.	4.79	1.33		6.12	7.45
6400	11-1/4" deep		450	.036		5.95	1.42		7.37	8.85
6450	14" deep		400	.040		7	1.60		8.60	10.35
6500	3-1/2" wide x 9-1/4" deep		480	.033		15.85	1.33		17.18	19.65
6550	11-1/4" deep		450	.036		19.15	1.42		20.57	23.50
6600	14" deep		400	.040		23.50	1.60		25.10	28
6650	7" wide x 9-1/4" deep		450	.036		33	1.42		34.42	39
6700	11-1/4" deep		420	.038		39.50	1.52		41.02	46
6750	14" deep		400	.040		46.50	1.60		48.10	54
9000	Minimum labor/equipment charge	F-3	2.50	16	Job		650	310	960	1,400

06 22 Millwork

06 22 13 – Standard Pattern Wood Trim

06 22 13.15 Moldings, Base

		Crew	Daily Output	Labor-Hours	Unit	Material	Labor	Equipment	Total	Total Incl O&P
0010	**MOLDINGS, BASE**									
0500	Base, stock pine, 9/16" x 3-1/2"	1 Carp	240	.033	L.F.	2.22	1.33		3.55	4.64
0550	9/16" x 4-1/2"		200	.040		2.47	1.60		4.07	5.35
0561	Base shoe, oak, 3/4" x 1"		240	.033		1.13	1.33		2.46	3.44
9000	Minimum labor/equipment charge		4	2	Job		80		80	132

06 22 13.30 Moldings, Casings

		Crew	Daily Output	Labor-Hours	Unit	Material	Labor	Equipment	Total	Total Incl O&P
0010	**MOLDINGS, CASINGS**									
0090	Apron, stock pine, 5/8" x 2"	1 Carp	250	.032	L.F.	1.24	1.28		2.52	3.47
.0110	5/8" x 3-1/2"		220	.036		1.88	1.45		3.33	4.47
0300	Band, stock pine, 11/16" x 1-1/8"		270	.030		.82	1.18		2	2.85
0350	11/16" x 1-3/4"		250	.032		1.10	1.28		2.38	3.32
0700	Casing, stock pine, 11/16" x 2-1/2"		240	.033		1.33	1.33		2.66	3.66
0750	11/16" x 3-1/2"		215	.037		1.89	1.49		3.38	4.53
0760	Door & window casing, exterior, 1-1/4" x 2"		200	.040		2.14	1.60		3.74	4.99
0770	Finger jointed, 1-1/4" x 2"		200	.040		.99	1.60		2.59	3.73
9000	Minimum labor/equipment charge		4	2	Job		80		80	132

06 22 13.35 Moldings, Ceilings

		Crew	Daily Output	Labor-Hours	Unit	Material	Labor	Equipment	Total	Total Incl O&P
0010	**MOLDINGS, CEILINGS**									
0600	Bed, stock pine, 9/16" x 1-3/4"	1 Carp	270	.030	L.F.	.97	1.18		2.15	3.02
0650	9/16" x 2"		240	.033		1.23	1.33		2.56	3.55
1200	Cornice molding, stock pine, 9/16" x 1-3/4"		330	.024		1.02	.97		1.99	2.72
1300	9/16" x 2-1/4"		300	.027		1.24	1.07		2.31	3.12

06 22 Millwork

06 22 13 – Standard Pattern Wood Trim

06 22 13.35 Moldings, Ceilings

		Crew	Daily Output	Labor-Hours	Unit	Material	2009 Bare Costs Labor	Equipment	Total	Total Incl O&P
2400	Cove scotia, stock pine, 9/16" x 1-3/4"	1 Carp	270	.030	L.F.	.93	1.18		2.11	2.97
2500	11/16" x 2-3/4"		255	.031		1.35	1.25		2.60	3.56
2600	Crown, stock pine, 9/16" x 3-5/8"		250	.032		1.92	1.28		3.20	4.22
2700	11/16" x 4-5/8"		220	.036	↓	3.22	1.45		4.67	5.95
9000	Minimum labor/equipment charge	↓	4	2	Job		80		80	132

06 22 13.40 Moldings, Exterior

		Crew	Daily Output	Labor-Hours	Unit	Material	2009 Bare Costs Labor	Equipment	Total	Total Incl O&P
0010	**MOLDINGS, EXTERIOR**									
1500	Cornice, boards, pine, 1" x 2"	1 Carp	330	.024	L.F.	.27	.97		1.24	1.90
1700	1" x 6"		250	.032		1.15	1.28		2.43	3.38
2000	1" x 12"		180	.044		2.31	1.78		4.09	5.45
2200	Three piece, built-up, pine, minimum		80	.100		2.15	4		6.15	8.95
2300	Maximum		65	.123		5.45	4.92		10.37	14.05
3000	Corner board, sterling pine, 1" x 4"		200	.040		.69	1.60		2.29	3.40
3100	1" x 6"		200	.040		1.05	1.60		2.65	3.80
3350	Fascia, sterling pine, 1" x 6"		250	.032		1.05	1.28		2.33	3.27
3370	1" x 8"		225	.036		1.43	1.42		2.85	3.92
3400	Trim, back band, 11/16" x 1-1/16"		250	.032		.89	1.28		2.17	3.09
3500	Casing, 11/16" x 4-1/4"		250	.032		2.19	1.28		3.47	4.52
3600	Crown, 11/16" x 4-1/4"		250	.032		3.65	1.28		4.93	6.15
3700	1" x 4" and 1" x 6" railing w/ balusters 4" O.C.		22	.364		22.50	14.55		37.05	49
3800	Insect screen framing stock, 1-1/16" x 1-3/4"		395	.020		2.03	.81		2.84	3.57
4100	Verge board, sterling pine, 1" x 4"		200	.040		.69	1.60		2.29	3.40
4200	1" x 6"		200	.040		1.05	1.60		2.65	3.80
4300	2" x 6"		165	.048		1.98	1.94		3.92	5.40
4400	2" x 8"	↓	165	.048	↓	2.63	1.94		4.57	6.10
4700	For redwood trim, add					200%				
9000	Minimum labor/equipment charge	1 Carp	4	2	Job		80		80	132

06 22 13.45 Moldings, Trim

		Crew	Daily Output	Labor-Hours	Unit	Material	2009 Bare Costs Labor	Equipment	Total	Total Incl O&P
0010	**MOLDINGS, TRIM**									
0200	Astragal, stock pine, 11/16" x 1-3/4"	1 Carp	255	.031	L.F.	1.19	1.25		2.44	3.38
0250	1-5/16" x 2-3/16"		240	.033		2.57	1.33		3.90	5.05
0800	Chair rail, stock pine, 5/8" x 2-1/2"		270	.030		1.45	1.18		2.63	3.55
0900	5/8" x 3-1/2"		240	.033		2.27	1.33		3.60	4.70
1000	Closet pole, stock pine, 1-1/8" diameter		200	.040		1.01	1.60		2.61	3.75
1100	Fir, 1-5/8" diameter		200	.040		1.67	1.60		3.27	4.48
3300	Half round, stock pine, 1/4" x 1/2"		270	.030		.24	1.18		1.42	2.21
3350	1/2" x 1"	↓	255	.031	↓	.61	1.25		1.86	2.74
3400	Handrail, fir, single piece, stock, hardware not included									
3450	1-1/2" x 1-3/4"	1 Carp	80	.100	L.F.	1.85	4		5.85	8.65
3470	Pine, 1-1/2" x 1-3/4"		80	.100		1.85	4		5.85	8.65
3500	1-1/2" x 2-1/2"		76	.105		2.30	4.21		6.51	9.50
3600	Lattice, stock pine, 1/4" x 1-1/8"		270	.030		.42	1.18		1.60	2.41
3700	1/4" x 1-3/4"		250	.032		.66	1.28		1.94	2.84
3800	Miscellaneous, custom, pine, 1" x 1"		270	.030		.34	1.18		1.52	2.32
3900	1" x 3"		240	.033		.69	1.33		2.02	2.96
4100	Birch or oak, nominal 1" x 1"		240	.033		.53	1.33		1.86	2.78
4200	Nominal 1" x 3"		215	.037		1.81	1.49		3.30	4.44
4400	Walnut, nominal 1" x 1"		215	.037		.87	1.49		2.36	3.41
4500	Nominal 1" x 3"		200	.040		2.61	1.60		4.21	5.50
4700	Teak, nominal 1" x 1"		215	.037		1.23	1.49		2.72	3.80
4800	Nominal 1" x 3"		200	.040		3.52	1.60		5.12	6.50
4900	Quarter round, stock pine, 1/4" x 1/4"	↓	275	.029	↓	.22	1.16		1.38	2.16

06 22 13 - Standard Pattern Wood Trim

06 22 13.45 Moldings, Trim

		Crew	Daily Output	Labor-Hours	Unit	Material	2009 Bare Costs Labor	Equipment	Total	Total Incl O&P
4950	3/4" x 3/4"	1 Carp	255	.031	L.F.	.59	1.25		1.84	2.72
5600	Wainscot moldings, 1-1/8" x 9/16", 2' high, minimum		76	.105	S.F.	11.05	4.21		15.26	19.10
5700	Maximum		65	.123	"	20	4.92		24.92	30.50
9000	Minimum labor/equipment charge		4	2	Job		80		80	132

06 22 13.50 Moldings, Window and Door

		Crew	Daily Output	Labor-Hours	Unit	Material	2009 Bare Costs Labor	Equipment	Total	Total Incl O&P
0010	**MOLDINGS, WINDOW AND DOOR**									
2800	Door moldings, stock, decorative, 1-1/8" wide, plain	1 Carp	17	.471	Set	45	18.80		63.80	80.50
2900	Detailed		17	.471	"	87.50	18.80		106.30	128
2960	Clear pine door jamb, no stops, 11/16" x 4-9/16"		240	.033	L.F.	5.05	1.33		6.38	7.75
3150	Door trim set, 1 head and 2 sides, pine, 2-1/2 wide		5.90	1.356	Opng.	22.50	54		76.50	115
3170	3-1/2" wide		5.30	1.509	"	32	60.50		92.50	135
3250	Glass beads, stock pine, 3/8" x 1/2"		275	.029	L.F.	.31	1.16		1.47	2.26
3270	3/8" x 7/8"		270	.030		.41	1.18		1.59	2.40
4850	Parting bead, stock pine, 3/8" x 3/4"		275	.029		.36	1.16		1.52	2.32
4870	1/2" x 3/4"		255	.031		.45	1.25		1.70	2.57
5000	Stool caps, stock pine, 11/16" x 3-1/2"		200	.040		2.27	1.60		3.87	5.15
5100	1-1/16" x 3-1/4"		150	.053		3.59	2.13		5.72	7.45
5300	Threshold, oak, 3' long, inside, 5/8" x 3-5/8"		32	.250	Ea.	9	10		19	26.50
5400	Outside, 1-1/2" x 7-5/8"		16	.500	"	37	20		57	73.50
5900	Window trim sets, including casings, header, stops,									
5910	stool and apron, 2-1/2" wide, minimum	1 Carp	13	.615	Opng.	32.50	24.50		57	76
5950	Average		10	.800		38	32		70	95
6000	Maximum		6	1.333		62.50	53.50		116	157
9000	Minimum labor/equipment charge		4	2	Job		80		80	132

06 22 13.60 Moldings, Soffits

		Crew	Daily Output	Labor-Hours	Unit	Material	2009 Bare Costs Labor	Equipment	Total	Total Incl O&P
0010	**MOLDINGS, SOFFITS** R061110-30									
0200	Soffits, pine, 1" x 4"	2 Carp	420	.038	L.F.	.44	1.52		1.96	2.99
0210	1" x 6"		420	.038		.64	1.52		2.16	3.21
0220	1" x 8"		420	.038		.84	1.52		2.36	3.43
0230	1" x 10"		400	.040		1.23	1.60		2.83	3.99
0240	1" x 12"		400	.040		1.60	1.60		3.20	4.40
0250	STK cedar, 1" x 4"		420	.038		.65	1.52		2.17	3.23
0260	1" x 6"		420	.038		1.09	1.52		2.61	3.71
0270	1" x 8"		420	.038		1.52	1.52		3.04	4.18
0280	1" x 10"		400	.040		2.09	1.60		3.69	4.94
0290	1" x 12"		400	.040		2.70	1.60		4.30	5.60
1000	Exterior AC plywood, 1/4" thick		420	.038	S.F.	.85	1.52		2.37	3.45
1050	3/8" thick		420	.038		1.03	1.52		2.55	3.64
1100	1/2" thick		420	.038		1	1.52		2.52	3.61
1150	Polyvinyl chloride, white, solid	1 Carp	230	.035		1.09	1.39		2.48	3.49
1160	Perforated	"	230	.035		1.09	1.39		2.48	3.49
9000	Minimum labor/equipment charge	2 Carp	5	3.200	Job		128		128	211

06 25 Prefinished Paneling

06 25 13 – Prefinished Hardboard Paneling

06 25 13.10 Paneling, Hardboard		Crew	Daily Output	Labor-Hours	Unit	Material	2009 Bare Costs Labor	Equipment	Total	Total Incl O&P
0010	**PANELING, HARDBOARD**									
0050	Not incl. furring or trim, hardboard, tempered, 1/8" thick [G]	2 Carp	500	.032	S.F.	.38	1.28		1.66	2.53
0100	1/4" thick [G]		500	.032		.53	1.28		1.81	2.69
0300	Tempered pegboard, 1/8" thick [G]		500	.032		.47	1.28		1.75	2.63
0400	1/4" thick [G]		500	.032		.65	1.28		1.93	2.83
0600	Untempered hardboard, natural finish, 1/8" thick [G]		500	.032		.39	1.28		1.67	2.54
0700	1/4" thick [G]		500	.032		.42	1.28		1.70	2.57
0900	Untempered pegboard, 1/8" thick [G]		500	.032		.38	1.28		1.66	2.53
1000	1/4" thick [G]		500	.032		.42	1.28		1.70	2.57
1200	Plastic faced hardboard, 1/8" thick [G]		500	.032		.64	1.28		1.92	2.81
1300	1/4" thick [G]		500	.032		.85	1.28		2.13	3.05
1500	Plastic faced pegboard, 1/8" thick [G]		500	.032		.61	1.28		1.89	2.78
1600	1/4" thick [G]		500	.032		.75	1.28		2.03	2.94
1800	Wood grained, plain or grooved, 1/4" thick, minimum [G]		500	.032		.57	1.28		1.85	2.74
1900	Maximum [G]		425	.038		1.19	1.50		2.69	3.79
2100	Moldings for hardboard, wood or aluminum, minimum		500	.032	L.F.	.38	1.28		1.66	2.53
2200	Maximum		425	.038	"	1.08	1.50		2.58	3.67
9000	Minimum labor/equipment charge	1 Carp	2	4	Job		160		160	264

06 25 16 – Prefinished Plywood Paneling

06 25 16.10 Paneling, Plywood		Crew	Daily Output	Labor-Hours	Unit	Material	2009 Bare Costs Labor	Equipment	Total	Total Incl O&P
0010	**PANELING, PLYWOOD** R061636-20									
2400	Plywood, prefinished, 1/4" thick, 4' x 8' sheets									
2410	with vertical grooves. Birch faced, minimum	2 Carp	500	.032	S.F.	.89	1.28		2.17	3.09
2420	Average		420	.038		1.35	1.52		2.87	4
2430	Maximum		350	.046		1.97	1.83		3.80	5.20
2600	Mahogany, African		400	.040		2.51	1.60		4.11	5.40
2700	Philippine (Lauan)		500	.032		1.08	1.28		2.36	3.30
2900	Oak or Cherry, minimum		500	.032		2.10	1.28		3.38	4.42
3000	Maximum		400	.040		3.22	1.60		4.82	6.20
3200	Rosewood		320	.050		4.58	2		6.58	8.35
3400	Teak		400	.040		3.22	1.60		4.82	6.20
3600	Chestnut		375	.043		4.77	1.70		6.47	8.05
3800	Pecan		400	.040		2.06	1.60		3.66	4.91
3900	Walnut, minimum		500	.032		2.75	1.28		4.03	5.15
3950	Maximum		400	.040		5.20	1.60		6.80	8.40
4000	Plywood, prefinished, 3/4" thick, stock grades, minimum		320	.050		1.24	2		3.24	4.66
4100	Maximum		224	.071		5.40	2.85		8.25	10.60
4300	Architectural grade, minimum		224	.071		3.98	2.85		6.83	9.10
4400	Maximum		160	.100		6.10	4		10.10	13.30
4600	Plywood, "A" face, birch, V.C., 1/2" thick, natural		450	.036		1.89	1.42		3.31	4.43
4700	Select		450	.036		2.06	1.42		3.48	4.62
4900	Veneer core, 3/4" thick, natural		320	.050		1.99	2		3.99	5.50
5000	Select		320	.050		2.23	2		4.23	5.75
5200	Lumber core, 3/4" thick, natural		320	.050		2.99	2		4.99	6.60
5500	Plywood, knotty pine, 1/4" thick, A2 grade		450	.036		1.63	1.42		3.05	4.14
5600	A3 grade		450	.036		2.06	1.42		3.48	4.62
5800	3/4" thick, veneer core, A2 grade		320	.050		2.11	2		4.11	5.60
5900	A3 grade		320	.050		2.38	2		4.38	5.90
6100	Aromatic cedar, 1/4" thick, plywood		400	.040		2.08	1.60		3.68	4.93
6200	1/4" thick, particle board		400	.040		1.01	1.60		2.61	3.75
9000	Minimum labor/equipment charge	1 Carp	2	4	Job		160		160	264

06 25 Prefinished Paneling

06 25 26 – Panel System

06 25 26.10 Panel Systems

06 25 26.10 Panel Systems	Crew	Daily Output	Labor-Hours	Unit	Material	2009 Bare Costs Labor	Equipment	Total	Total Incl O&P
0010 **PANEL SYSTEMS**									
0100 Raised panel, eng. wood core w/ wood veneer, std., paint grade	2 Carp	300	.053	S.F.	11.90	2.13		14.03	16.60
0110 Oak veneer		300	.053		19.70	2.13		21.83	25
0120 Maple veneer		300	.053		25	2.13		27.13	31
0130 Cherry veneer		300	.053		32	2.13		34.13	38.50
0300 Class I fire rated, paint grade		300	.053		14.25	2.13		16.38	19.15
0310 Oak veneer		300	.053		23.50	2.13		25.63	29.50
0320 Maple veneer		300	.053		30	2.13		32.13	36.50
0330 Cherry veneer		300	.053		38.50	2.13		40.63	45.50
0510 Beadboard, 5/8" MDF, standard, primed		300	.053		8.40	2.13		10.53	12.75
0520 Oak veneer, unfinished		300	.053		14.60	2.13		16.73	19.55
0530 Maple veneer, unfinished		300	.053		17.25	2.13		19.38	22.50
0610 Rustic paneling, 5/8" MDF, standard, maple veneer, unfinished		300	.053		22	2.13		24.13	28

06 26 Board Paneling

06 26 13 – Profile Board Paneling

06 26 13.10 Paneling, Boards

06 26 13.10 Paneling, Boards	Crew	Daily Output	Labor-Hours	Unit	Material	2009 Bare Costs Labor	Equipment	Total	Total Incl O&P
0010 **PANELING, BOARDS**									
6400 Wood board paneling, 3/4" thick, knotty pine	2 Carp	300	.053	S.F.	1.44	2.13		3.57	5.10
6500 Rough sawn cedar		300	.053		1.85	2.13		3.98	5.55
6700 Redwood, clear, 1" x 4" boards		300	.053		4.30	2.13		6.43	8.25
6900 Aromatic cedar, closet lining, boards		275	.058		3.32	2.32		5.64	7.50
9000 Minimum labor/equipment charge	1 Carp	2	4	Job		160		160	264

06 43 Wood Stairs and Railings

06 43 13 – Wood Stairs

06 43 13.20 Prefabricated Wood Stairs

06 43 13.20 Prefabricated Wood Stairs	Crew	Daily Output	Labor-Hours	Unit	Material	2009 Bare Costs Labor	Equipment	Total	Total Incl O&P
0010 **PREFABRICATED WOOD STAIRS**									
0100 Box stairs, prefabricated, 3'-0" wide									
0110 Oak treads, up to 14 risers	2 Carp	39	.410	Riser	85	16.40		101.40	121
0600 With pine treads for carpet, up to 14 risers	"	39	.410	"	55	16.40		71.40	87.50
1100 For 4' wide stairs, add				Flight	25%				
1550 Stairs, prefabricated stair handrail with balusters	1 Carp	30	.267	L.F.	75	10.65		85.65	99.50
1700 Basement stairs, prefabricated, pine treads									
1710 Pine risers, 3' wide, up to 14 risers	2 Carp	52	.308	Riser	55	12.30		67.30	81
4000 Residential, wood, oak treads, prefabricated		1.50	10.667	Flight	1,100	425		1,525	1,900
4200 Built in place		.44	36.364	"	1,575	1,450		3,025	4,125
4400 Spiral, oak, 4'-6" diameter, unfinished, prefabricated,									
4500 incl. railing, 9' high	2 Carp	1.50	10.667	Flight	3,625	425		4,050	4,675
9000 Minimum labor/equipment charge	"	3	5.333	Job		213		213	350

06 43 13.40 Wood Stair Parts

06 43 13.40 Wood Stair Parts	Crew	Daily Output	Labor-Hours	Unit	Material	2009 Bare Costs Labor	Equipment	Total	Total Incl O&P
0010 **WOOD STAIR PARTS**									
0020 Pin top balusters, 1-1/4", oak, 34"	1 Carp	96	.083	Ea.	9.15	3.33		12.48	15.55
0030 36"		96	.083		9.60	3.33		12.93	16.05
0040 42"		96	.083		11.65	3.33		14.98	18.30
0050 Poplar, 34"		96	.083		6.10	3.33		9.43	12.25
0060 36"		96	.083		6.50	3.33		9.83	12.65
0070 42"		96	.083		14.20	3.33		17.53	21

06 43 13 – Wood Stairs

06 43 13.40 Wood Stair Parts	Crew	Daily Output	Labor-Hours	Unit	Material	2009 Bare Costs Labor	Equipment	Total	Total Incl O&P
0080	Maple, 31"	1 Carp	96	.083	Ea.	11.65	3.33		14.98	18.30
0090	34"		96	.083		12.95	3.33		16.28	19.75
0100	36"		96	.083		13.50	3.33		16.83	20.50
0110	39"		96	.083		15.45	3.33		18.78	22.50
0120	41"		96	.083		17.35	3.33		20.68	24.50
0130	Primed, 31"		96	.083		4.03	3.33		7.36	9.95
0140	34"		96	.083		4.33	3.33		7.66	10.25
0150	36"		96	.083		4.51	3.33		7.84	10.45
0160	39"		96	.083		5.05	3.33		8.38	11.05
0170	41"		96	.083		5.55	3.33		8.88	11.65
0180	Box top balusters, 1-1/4", oak, 34"		60	.133		9.40	5.35		14.75	19.15
0190	36"		60	.133		10.25	5.35		15.60	20
0200	42"		60	.133		11.45	5.35		16.80	21.50
0210	Poplar, 34"		60	.133		6	5.35		11.35	15.40
0220	36"		60	.133		6.75	5.35		12.10	16.20
0230	42"		60	.133		7.50	5.35		12.85	17.05
0240	Maple, 31"		60	.133		13.45	5.35		18.80	23.50
0250	34"		60	.133		14.15	5.35		19.50	24.50
0260	36"		60	.133		16.35	5.35		21.70	27
0270	39"		60	.133		16.60	5.35		21.95	27
0280	41"		60	.133		17.15	5.35		22.50	27.50
0290	Primed, 31"		60	.133		4.46	5.35		9.81	13.70
0300	34"		60	.133		4.79	5.35		10.14	14.05
0310	36"		60	.133		4.97	5.35		10.32	14.25
0320	39"		60	.133		5.60	5.35		10.95	15
0330	41"		60	.133	▼	6.25	5.35		11.60	15.70
0340	Square balusters, cut from lineal stock, pine, 1-1/16" x 1-1/16"		180	.044	L.F.	1.32	1.78		3.10	4.38
0350	1-5/16" x 1-5/16"		180	.044		1.85	1.78		3.63	4.97
0360	1-5/8" x 1-5/8"		180	.044	▼	3.03	1.78		4.81	6.25
0370	Turned newel, oak, 3-1/2" square, 50" high		8	1	Ea.	119	40		159	196
0380	65" high		8	1		154	40		194	236
0390	Poplar, 3-1/2" square, 50" high		8	1		99	40		139	175
0400	65" high		8	1		128	40		168	206
0410	Maple, 3-1/2" square, 50" high		8	1		138	40		178	217
0420	65" high		8	1		160	40		200	241
0430	Square newel, oak, 3-1/2" square, 50" high		8	1		226	40		266	315
0440	65" high		8	1		242	40		282	330
0450	Poplar, 3-1/2" square, 50" high		8	1		215	40		255	300
0460	65" high		8	1		220	40		260	310
0470	Maple, 3" square, 50" high		8	1		243	40		283	335
0480	65" high		8	1	▼	265	40		305	360
0490	Railings, oak, minimum		96	.083	L.F.	7.50	3.33		10.83	13.75
0500	Average		96	.083		9.20	3.33		12.53	15.60
0510	Maximum		96	.083		9.35	3.33		12.68	15.80
0520	Maple, minimum		96	.083		8.80	3.33		12.13	15.20
0530	Average		96	.083		11	3.33		14.33	17.60
0540	Maximum		96	.083		13.20	3.33		16.53	20
0550	Oak, for bending rail, minimum		48	.167		21	6.65		27.65	34
0560	Average		48	.167		30	6.65		36.65	44
0570	Maximum		48	.167		32	6.65		38.65	46.50
0580	Maple, for bending rail, minimum		48	.167		32	6.65		38.65	46.50
0590	Average		48	.167		35	6.65		41.65	49.50
0600	Maximum	▼	48	.167	▼	37.50	6.65		44.15	52

06 43 13.40 Wood Stair Parts		Crew	Daily Output	Labor-Hours	Unit	Material	2009 Bare Costs		Total	Total Incl O&P
							Labor	Equipment		
0610	Risers, oak, 3/4" x 8", 36" long	1 Carp	80	.100	Ea.	22	4		26	30.50
0620	42" long		80	.100		25.50	4		29.50	34.50
0630	48" long		80	.100		29.50	4		33.50	38.50
0640	54" long		80	.100		33	4		37	42.50
0650	60" long		80	.100		36.50	4		40.50	47
0660	72" long		80	.100		44	4		48	55
0670	Poplar, 3/4" x 8", 36" long		80	.100		14.35	4		18.35	22.50
0680	42" long		80	.100		16.75	4		20.75	25
0690	48" long		80	.100		19.15	4		23.15	27.50
0700	54" long		80	.100		21.50	4		25.50	30
0710	60" long		80	.100		24	4		28	33
0720	72" long		80	.100		28.50	4		32.50	38
0730	Pine, 1" x 8", 36" long		80	.100		2.51	4		6.51	9.35
0740	42" long		80	.100		2.93	4		6.93	9.80
0750	48" long		80	.100		3.34	4		7.34	10.30
0760	54" long		80	.100		3.76	4		7.76	10.75
0770	60" long		80	.100		4.18	4		8.18	11.20
0780	72" long		80	.100		5	4		9	12.10
0790	Treads, oak, no returns, 1-1/32" x 11-1/2" x 36" long		32	.250		43.50	10		53.50	64.50
0800	42" long		32	.250		50.50	10		60.50	72.50
0810	48" long		32	.250		58	10		68	80
0820	54" long		32	.250		65	10		75	88
0830	60" long		32	.250		72.50	10		82.50	96
0840	72" long		32	.250		87	10		97	112
0850	Mitred return one end, 1-1/32" x 11-1/2" x 36" long		24	.333		65	13.30		78.30	93.50
0860	42" long		24	.333		76	13.30		89.30	106
0870	48" long		24	.333		86.50	13.30		99.80	118
0880	54" long		24	.333		97.50	13.30		110.80	129
0890	60" long		24	.333		108	13.30		121.30	141
0900	72" long		24	.333		130	13.30		143.30	165
0910	Mitred return two ends, 1-1/32" x 11-1/2" x 36" long		12	.667		81	26.50		107.50	134
0920	42" long		12	.667		95	26.50		121.50	148
0930	48" long		12	.667		108	26.50		134.50	163
0940	54" long		12	.667		122	26.50		148.50	178
0950	60" long		12	.667		135	26.50		161.50	193
0960	72" long		12	.667		162	26.50		188.50	223
0970	Starting step, oak, 48", bullnose		8	1		163	40		203	246
0980	Double end bullnose		8	1		193	40		233	278
0990	Half circle		8	1		216	40		256	305
1010	Double end half circle		8	1		286	40		326	380
1030	Skirt board, pine, 1" x 10"		55	.145	L.F.	1.23	5.80		7.03	10.95
1040	1" x 12"		52	.154	"	1.60	6.15		7.75	11.90
1050	Oak landing tread, 1-1/16" thick		54	.148	S.F.	8.25	5.90		14.15	18.85
1060	Oak cove molding		96	.083	L.F.	2.75	3.33		6.08	8.55
1070	Oak stringer molding		96	.083		2.75	3.33		6.08	8.55
1080	Oak shoe molding		96	.083		2.75	3.33		6.08	8.55
1090	Rail bolt, 5/16" x 3-1/2"		48	.167	Ea.	1.65	6.65		8.30	12.80
1100	5/16" x 4-1/2"		48	.167		2.20	6.65		8.85	13.40
1120	Newel post anchor		16	.500		27.50	20		47.50	63.50
1130	Tapered plug, 1/2"		240	.033		.28	1.33		1.61	2.51
1140	1"		240	.033		.39	1.33		1.72	2.63
1150	Rail bolt kit		24	.333		11	13.30		24.30	34
9000	Minimum labor/equipment charge		3	2.667	Job		107		107	176

06 43 Wood Stairs and Railings

06 43 16 – Wood Railings

06 43 16.10 Wood Handrails and Railings	Crew	Daily Output	Labor-Hours	Unit	Material	2009 Bare Costs Labor	Equipment	Total	Total Incl O&P
0010 **WOOD HANDRAILS AND RAILINGS**									
0020 Custom design, architectural grade, hardwood, minimum	1 Carp	38	.211	L.F.	9.50	8.40		17.90	24.50
0100 Maximum		30	.267		51.50	10.65		62.15	74
0300 Stock interior railing with spindles 4" O.C., 4' long		40	.200		31.50	8		39.50	48
0400 8' long		48	.167		31.50	6.65		38.15	46
9000 Minimum labor/equipment charge		3	2.667	Job		107		107	176

06 44 Ornamental Woodwork

06 44 19 – Wood Grilles

06 44 19.10 Grilles

	Crew	Daily Output	Labor-Hours	Unit	Material	2009 Bare Costs Labor	Equipment	Total	Total Incl O&P
0010 **GRILLES** and panels, hardwood, sanded									
0020 2' x 4' to 4' x 8', custom designs, unfinished, minimum	1 Carp	38	.211	S.F.	14.65	8.40		23.05	30
0050 Average		30	.267		31.50	10.65		42.15	52.50
0100 Maximum		19	.421		49	16.80		65.80	81.50
0300 As above, but prefinished, minimum		38	.211		14.65	8.40		23.05	30
0400 Maximum		19	.421		55	16.80		71.80	88.50
9000 Minimum labor/equipment charge		2	4	Job		160		160	264

06 44 33 – Wood Mantels

06 44 33.10 Fireplace Mantels

	Crew	Daily Output	Labor-Hours	Unit	Material	2009 Bare Costs Labor	Equipment	Total	Total Incl O&P
0010 **FIREPLACE MANTELS**									
0015 6" molding, 6' x 3'-6" opening, minimum	1 Carp	5	1.600	Opng.	228	64		292	355
0100 Maximum		5	1.600		410	64		474	560
0300 Prefabricated pine, colonial type, stock, deluxe		2	4		1,250	160		1,410	1,675
0400 Economy		3	2.667		530	107		637	760
9000 Minimum labor/equipment charge		3	2.667	Job		107		107	176

06 44 33.20 Fireplace Mantel Beam

	Crew	Daily Output	Labor-Hours	Unit	Material	2009 Bare Costs Labor	Equipment	Total	Total Incl O&P
0010 **FIREPLACE MANTEL BEAM**									
0020 Rough texture wood, 4" x 8"	1 Carp	36	.222	L.F.	6	8.90		14.90	21.50
0100 4" x 10"		35	.229	"	7	9.15		16.15	23
0300 Laminated hardwood, 2-1/4" x 10-1/2" wide, 6' long		5	1.600	Ea.	110	64		174	227
0400 8' long		5	1.600	"	150	64		214	271
0600 Brackets for above, rough sawn		12	.667	Pr.	10	26.50		36.50	55
0700 Laminated		12	.667	"	15	26.50		41.50	60.50
9000 Minimum labor/equipment charge		4	2	Job		80		80	132

06 44 39 – Wood Posts and Columns

06 44 39.10 Decorative Beams

	Crew	Daily Output	Labor-Hours	Unit	Material	2009 Bare Costs Labor	Equipment	Total	Total Incl O&P
0010 **DECORATIVE BEAMS**									
0020 Rough sawn cedar, non-load bearing, 4" x 4"	2 Carp	180	.089	L.F.	1.47	3.55		5.02	7.45
0100 4" x 6"		170	.094		2.21	3.76		5.97	8.65
0200 4" x 8"		160	.100		2.95	4		6.95	9.85
0300 4" x 10"		150	.107		3.68	4.26		7.94	11.10
0400 4" x 12"		140	.114		4.42	4.57		8.99	12.40
0500 8" x 8"		130	.123		7.50	4.92		12.42	16.35
9000 Minimum labor/equipment charge	1 Carp	3	2.667	Job		107		107	176

06 44 39.20 Columns

	Crew	Daily Output	Labor-Hours	Unit	Material	2009 Bare Costs Labor	Equipment	Total	Total Incl O&P
0010 **COLUMNS**									
0050 Aluminum, round colonial, 6" diameter	2 Carp	80	.200	V.L.F.	16	8		24	31
0100 8" diameter		62.25	.257		17	10.25		27.25	35.50
0200 10" diameter		55	.291		19.10	11.60		30.70	40

06 44 Ornamental Woodwork

06 44 39 – Wood Posts and Columns

06 44 39.20 Columns

	06 44 39.20 Columns	Crew	Daily Output	Labor-Hours	Unit	Material	2009 Bare Costs Labor	Equipment	Total	Total Incl O&P
0250	Fir, stock units, hollow round, 6" diameter	2 Carp	80	.200	V.L.F.	17.20	8		25.20	32
0300	8" diameter		80	.200		20.50	8		28.50	35.50
0350	10" diameter		70	.229		25.50	9.15		34.65	43.50
0400	Solid turned, to 8' high, 3-1/2" diameter		80	.200		8.40	8		16.40	22.50
0500	4-1/2" diameter		75	.213		12.80	8.50		21.30	28
0600	5-1/2" diameter		70	.229		16.40	9.15		25.55	33
0800	Square columns, built-up, 5" x 5"		65	.246		10.20	9.85		20.05	27.50
0900	Solid, 3-1/2" x 3-1/2"		130	.123		6.70	4.92		11.62	15.45
1600	Hemlock, tapered, T & G, 12" diam, 10' high		100	.160		31.50	6.40		37.90	45
1700	16' high		65	.246		46.50	9.85		56.35	67.50
1900	10' high, 14" diameter		100	.160		78.50	6.40		84.90	97
2000	18' high		65	.246		63.50	9.85		73.35	86.50
2200	18" diameter, 12' high		65	.246		114	9.85		123.85	142
2300	20' high		50	.320		80	12.80		92.80	109
2500	20" diameter, 14' high		40	.400		129	16		145	169
2600	20' high		35	.457		120	18.25		138.25	162
2800	For flat pilasters, deduct					33%				
3000	For splitting into halves, add				Ea.	116			116	128
4000	Rough sawn cedar posts, 4" x 4"	2 Carp	250	.064	V.L.F.	6.30	2.56		8.86	11.15
4100	4" x 6"		235	.068		15	2.72		17.72	21
4200	6" x 6"		220	.073		26.50	2.91		29.41	34
4300	8" x 8"		200	.080		60	3.20		63.20	71.50
9000	Minimum labor/equipment charge	1 Carp	3	2.667	Job		107		107	176

06 48 Wood Frames

06 48 13 – Exterior Wood Door Frames

06 48 13.10 Exterior Wood Door Frames and Accessories

	06 48 13.10 Exterior Wood Door Frames and Accessories	Crew	Daily Output	Labor-Hours	Unit	Material	2009 Bare Costs Labor	Equipment	Total	Total Incl O&P
0010	**EXTERIOR WOOD DOOR FRAMES AND ACCESSORIES**									
0400	Exterior frame, incl. ext. trim, pine, 5/4 x 4-9/16" deep	2 Carp	375	.043	L.F.	5.55	1.70		7.25	8.90
0420	5-3/16" deep		375	.043		11.30	1.70		13	15.25
0440	6-9/16" deep		375	.043		8.50	1.70		10.20	12.15
0600	Oak, 5/4 x 4-9/16" deep		350	.046		12	1.83		13.83	16.20
0620	5-3/16" deep		350	.046		13.50	1.83		15.33	17.85
0640	6-9/16" deep		350	.046		15.30	1.83		17.13	19.85
0800	Walnut, 5/4 x 4-9/16" deep		350	.046		14	1.83		15.83	18.40
0820	5-3/16" deep		350	.046		18	1.83		19.83	23
0840	6-9/16" deep		350	.046		20	1.83		21.83	25
1000	Sills, 8/4 x 8" deep, oak, no horns		100	.160		19	6.40		25.40	31.50
1020	2" horns		100	.160		19	6.40		25.40	31.50
1040	3" horns		100	.160		19.95	6.40		26.35	32.50
1100	8/4 x 10" deep, oak, no horns		90	.178		21.50	7.10		28.60	35.50
1120	2" horns		90	.178		19.80	7.10		26.90	33.50
1140	3" horns		90	.178		20.50	7.10		27.60	34.50
2000	Exterior, colonial, frame & trim, 3' opng., in-swing, minimum		22	.727	Ea.	315	29		344	400
2010	Average		21	.762		470	30.50		500.50	565
2020	Maximum		20	.800		1,075	32		1,107	1,225
2100	5'-4" opening, in-swing, minimum		17	.941		360	37.50		397.50	455
2120	Maximum		15	1.067		1,075	42.50		1,117.50	1,250
2140	Out-swing, minimum		17	.941		370	37.50		407.50	465
2160	Maximum		15	1.067		1,100	42.50		1,142.50	1,300
2400	6'-0" opening, in-swing, minimum		16	1		345	40		385	445

06 48 Wood Frames

06 48 13 – Exterior Wood Door Frames

06 48 13.10 Exterior Wood Door Frames and Accessories	Crew	Daily Output	Labor-Hours	Unit	Material	2009 Bare Costs Labor	Equipment	Total	Total Incl O&P
2420	Maximum	2 Carp	10	1.600	Ea.	1,100	64		1,164	1,325
2460	Out-swing, minimum		16	1		375	40		415	480
2480	Maximum		10	1.600		1,300	64		1,364	1,525
2600	For two sidelights, add, minimum		30	.533	Opng.	355	21.50		376.50	425
2620	Maximum		20	.800	"	1,125	32		1,157	1,300
2700	Custom birch frame, 3'-0" opening		16	1	Ea.	205	40		245	292
2750	6'-0" opening		16	1		305	40		345	400
2900	Exterior, modern, plain trim, 3' opng., in-swing, minimum		26	.615		35.50	24.50		60	79.50
2920	Average		24	.667		42	26.50		68.50	90
2940	Maximum		22	.727		50.50	29		79.50	104

06 48 16 – Interior Wood Door Frames

06 48 16.10 Interior Wood Door Jamb and Frames

		Crew	Daily Output	Labor-Hours	Unit	Material	2009 Bare Costs Labor	Equipment	Total	Total Incl O&P
0010	**INTERIOR WOOD DOOR JAMB AND FRAMES**									
3000	Interior frame, pine, 11/16" x 3-5/8" deep	2 Carp	375	.043	L.F.	5.50	1.70		7.20	8.85
3020	4-9/16" deep		375	.043		6.45	1.70		8.15	9.90
3200	Oak, 11/16" x 3-5/8" deep		350	.046		4.35	1.83		6.18	7.80
3220	4-9/16" deep		350	.046		4.50	1.83		6.33	7.95
3240	5-3/16" deep		350	.046		4.58	1.83		6.41	8.05
3400	Walnut, 11/16" x 3-5/8" deep		350	.046		7.30	1.83		9.13	11.05
3420	4-9/16" deep		350	.046		7.50	1.83		9.33	11.25
3440	5-3/16" deep		350	.046		7.90	1.83		9.73	11.65
3600	Pocket door frame		16	1	Ea.	67	40		107	140
3800	Threshold, oak, 5/8" x 3-5/8" deep		200	.080	L.F.	3.83	3.20		7.03	9.50
3820	4-5/8" deep		190	.084		4.58	3.36		7.94	10.60
3840	5-5/8" deep		180	.089		7.15	3.55		10.70	13.70
9000	Minimum labor/equipment charge	1 Carp	4	2	Job		80		80	132

06 49 Wood Screens and Exterior Wood Shutters

06 49 19 – Exterior Wood Shutters

06 49 19.10 Shutters, Exterior

		Crew	Daily Output	Labor-Hours	Unit	Material	2009 Bare Costs Labor	Equipment	Total	Total Incl O&P
0010	**SHUTTERS, EXTERIOR**									
0012	Aluminum, louvered, 1'-4" wide, 3'-0" long	1 Carp	10	.800	Pr.	55.50	32		87.50	114
0400	6'-8" long		9	.889		112	35.50		147.50	182
1000	Pine, louvered, primed, each 1'-2" wide, 3'-3" long		10	.800		96.50	32		128.50	159
1100	4'-7" long		10	.800		130	32		162	196
1250	Each 1'-4" wide, 3'-0" long		10	.800		98.50	32		130.50	161
1350	5'-3" long		10	.800		148	32		180	216
1500	Each 1'-6" wide, 3'-3" long		10	.800		102	32		134	166
1600	4'-7" long		10	.800		144	32		176	211
1610	Door blinds, 6'-9" long 1'-3" wide		9	.889		180	35.50		215.50	257
1615	1'-6" wide		9	.889		198	35.50		233.50	277
1620	Hemlock, louvered, 1'-2" wide, 5'-7" long		10	.800		158	32		190	227
1630	Each 1'-4" wide, 2'-2" long		10	.800		98.50	32		130.50	161
1640	3'-0" long		10	.800		98.50	32		130.50	161
1650	3'-3" long		10	.800		106	32		138	170
1660	3'-11" long		10	.800		121	32		153	186
1670	4'-3" long		10	.800		118	32		150	183
1680	5'-3" long		10	.800		147	32		179	214
1690	5'-11" long		10	.800		166	32		198	236
1700	Door blinds, 6'-9" long, each 1'-3" wide		9	.889		167	35.50		202.50	243

06 49 Wood Screens and Exterior Wood Shutters

06 49 19 – Exterior Wood Shutters

06 49 19.10 Shutters, Exterior		Crew	Daily Output	Labor-Hours	Unit	Material	2009 Bare Costs Labor	Equipment	Total	Total Incl O&P
1710	1'-6" wide	1 Carp	9	.889	Pr.	180	35.50		215.50	257
1720	Hemlock, solid raised panel, each 1'-4" wide, 3'-3" long		10	.800		159	32		191	228
1730	3'-11" long		10	.800		186	32		218	258
1740	4'-3" long		10	.800		202	32		234	275
1750	4'-7" long		10	.800		217	32		249	292
1760	4'-11" long		10	.800		237	32		269	315
1770	5'-11" long		10	.800		269	32		301	350
1800	Door blinds, 6'-9" long, each 1'-3" wide		9	.889		305	35.50		340.50	395
1900	1'-6" wide		9	.889		330	35.50		365.50	425
2500	Polystyrene, solid raised panel, each 1'-4" wide, 3'-3" long		10	.800		40.50	32		72.50	97.50
2700	4'-7" long		10	.800		51	32		83	109
4500	Polystyrene, louvered, each 1'-2" wide, 3'-3" long		10	.800		27.50	32		59.50	83.50
4600	4'-7" long		10	.800		34	32		66	90.50
4750	5'-3" long		10	.800		39	32		71	96
4850	6'-8" long		9	.889		46.50	35.50		82	110
6000	Vinyl, louvered, each 1'-2" x 4'-7" long		10	.800		34	32		66	90.50
6200	Each 1'-4" x 6'-8" long		9	.889		47.50	35.50		83	111
9000	Minimum labor/equipment charge		4	2	Job		80		80	132

06 51 Structural Plastic Shapes and Plates

06 51 13 – Plastic Lumber

06 51 13.10 Recycled Plastic Lumber

			Crew	Daily Output	Labor-Hours	Unit	Material	2009 Bare Costs Labor	Equipment	Total	Total Incl O&P
0010	**RECYCLED PLASTIC LUMBER**										
4000	Sheeting, recycled plastic, black or white, 4' x 8' x 1/8"	G	2 Carp	1100	.015	S.F.	1.39	.58		1.97	2.49
4010	4' x 8' x 3/16"	G		1100	.015		2.10	.58		2.68	3.27
4020	4' x 8' x 1/4"	G		950	.017		2.53	.67		3.20	3.89
4030	4' x 8' x 3/8"	G		950	.017		3.53	.67		4.20	4.99
4040	4' x 8' x 1/2"	G		900	.018		4.56	.71		5.27	6.15
4050	4' x 8' x 5/8"	G		900	.018		6	.71		6.71	7.75
4060	4' x 8' x 3/4"	G		850	.019		6.30	.75		7.05	8.20
4070	Add for colors					Ea.	5%			5%	6%
8500	100% recycled plastic, var colors, NLB, 2" x 2"	G				L.F.	2.62			2.62	2.88
8510	2" x 4"	G					2.80			2.80	3.08
8520	2" x 6"	G					4.37			4.37	4.81
8530	2" x 8"	G					6.05			6.05	6.65
8540	2" x 10"	G					8.90			8.90	9.80
8550	5/4" x 4"	G					2.20			2.20	2.42
8560	5/4" x 6"	G					3.08			3.08	3.39
8570	1" x 6"	G					2.08			2.08	2.29
8580	1/2" x 8"	G					2.30			2.30	2.53
8590	2" x 10" T & G	G					6.35			6.35	7
8600	3" x 10" T & G	G					10.15			10.15	11.15
8610	Add for premium colors						20%			20%	22%
8620	Mailbox post, recycled plastic, black/green , 4"x 4"x 6' H, incl. box	G	1 Clab	3	2.667	Ea.	58	84.50		142.50	203

06 63 Plastic Railings

06 63 10 – Plastic (PVC) Railings

06 63 10.10 Plastic Railings	Crew	Daily Output	Labor-Hours	Unit	Material	2009 Bare Costs Labor	Equipment	Total	Total Incl O&P
0010 **PLASTIC RAILINGS**									
0100 Horizontal PVC handrail with balusters, 3-1/2" wide, 36" high	1 Carp	96	.083	L.F.	23.50	3.33		26.83	31
0150 42" high		96	.083		26.50	3.33		29.83	34.50
0200 Angled PVC handrail with balusters, 3-1/2" wide, 36" high		72	.111		27	4.44		31.44	37
0250 42" high		72	.111		30	4.44		34.44	40.50
0300 Post sleeve for 4 x 4 post		96	.083		12	3.33		15.33	18.70
0400 Post cap for 4 x 4 post, flat profile		48	.167	Ea.	11.35	6.65		18	23.50
0450 Newel post style profile		48	.167		21	6.65		27.65	34
0500 Raised corbeled profile		48	.167		22	6.65		28.65	35.50
0550 Post base trim for 4 x 4 post		96	.083		12.80	3.33		16.13	19.55

06 65 Plastic Simulated Wood Trim

06 65 10 – PVC Trim

06 65 10.10 PVC Trim, Exterior	Crew	Daily Output	Labor-Hours	Unit	Material	2009 Bare Costs Labor	Equipment	Total	Total Incl O&P
0010 **PVC TRIM, EXTERIOR**									
0100 Cornerboards, 5/4" x 6" x 6"	1 Carp	240	.033	L.F.	5.70	1.33		7.03	8.50
0110 Door / window casing, 1" x 4"		200	.040		1.28	1.60		2.88	4.05
0120 1" x 6"		200	.040		2.02	1.60		3.62	4.86
0130 1" x 8"		195	.041		2.66	1.64		4.30	5.65
0140 1" x 10"		195	.041		3.39	1.64		5.03	6.45
0150 1" x 12"		190	.042		4.12	1.68		5.80	7.30
0160 5/4" x 4"		195	.041		1.63	1.64		3.27	4.50
0170 5/4" x 6"		195	.041		2.57	1.64		4.21	5.55
0180 5/4" x 8"		190	.042		3.38	1.68		5.06	6.50
0190 5/4" x 10"		190	.042		4.32	1.68		6	7.55
0200 5/4" x 12"		185	.043		5.25	1.73		6.98	8.65
0210 Fascia, 1" x 4"		250	.032		1.28	1.28		2.56	3.52
0220 1" x 6"		250	.032		2.02	1.28		3.30	4.33
0230 1" x 8"		225	.036		2.66	1.42		4.08	5.30
0240 1" x 10"		225	.036		3.39	1.42		4.81	6.10
0250 1" x 12"		200	.040		4.12	1.60		5.72	7.15
0260 5/4" x 4"		240	.033		1.63	1.33		2.96	3.99
0270 5/4" x 6"		240	.033		2.57	1.33		3.90	5.05
0280 5/4" x 8"		215	.037		3.38	1.49		4.87	6.15
0290 5/4" x 10"		215	.037		4.32	1.49		5.81	7.20
0300 5/4" x 12"		190	.042		5.25	1.68		6.93	8.60
0310 Frieze, 1" x 4"		250	.032		1.28	1.28		2.56	3.52
0320 1" x 6"		250	.032		2.02	1.28		3.30	4.33
0330 1" x 8"		225	.036		2.66	1.42		4.08	5.30
0340 1" x 10"		225	.036		3.39	1.42		4.81	6.10
0350 1" x 12"		200	.040		4.12	1.60		5.72	7.15
0360 5/4" x 4"		240	.033		1.63	1.33		2.96	3.99
0370 5/4" x 6"		240	.033		2.57	1.33		3.90	5.05
0380 5/4" x 8"		215	.037		3.38	1.49		4.87	6.15
0390 5/4" x 10"		215	.037		4.32	1.49		5.81	7.20
0400 5/4" x 12"		190	.042		5.25	1.68		6.93	8.60
0410 Rake, 1" x 4"		200	.040		1.28	1.60		2.88	4.05
0420 1" x 6"		200	.040		2.02	1.60		3.62	4.86
0430 1" x 8"		190	.042		2.66	1.68		4.34	5.70
0440 1" x 10"		190	.042		3.39	1.68		5.07	6.50
0450 1" x 12"		180	.044		4.12	1.78		5.90	7.45

06 65 Plastic Simulated Wood Trim

06 65 10 – PVC Trim

06 65 10.10 PVC Trim, Exterior		Crew	Daily Output	Labor-Hours	Unit	Material	2009 Bare Costs Labor	Equipment	Total	Total Incl O&P
0460	5/4" x 4"	1 Carp	195	.041	L.F.	1.63	1.64		3.27	4.50
0470	5/4" x 6"		195	.041		2.57	1.64		4.21	5.55
0480	5/4" x 8"		185	.043		3.38	1.73		5.11	6.55
0490	5/4" x 10"		185	.043		4.32	1.73		6.05	7.60
0500	5/4" x 12"		175	.046		5.25	1.83		7.08	8.80
0510	Rake trim, 1" x 4"		225	.036		1.28	1.42		2.70	3.76
0520	1" x 6"		225	.036		2.02	1.42		3.44	4.57
0560	5/4" x 4"		220	.036		1.63	1.45		3.08	4.19
0570	5/4" x 6"		220	.036		2.57	1.45		4.02	5.25
0610	Soffit, 1" x 4"	2 Carp	420	.038		1.28	1.52		2.80	3.92
0620	1" x 6"		420	.038		2.02	1.52		3.54	4.73
0630	1" x 8"		420	.038		2.66	1.52		4.18	5.45
0640	1" x 10"		400	.040		3.39	1.60		4.99	6.35
0650	1" x 12"		400	.040		4.12	1.60		5.72	7.15
0660	5/4" x 4"		410	.039		1.63	1.56		3.19	4.36
0670	5/4" x 6"		410	.039		2.57	1.56		4.13	5.40
0680	5/4" x 8"		410	.039		3.38	1.56		4.94	6.30
0690	5/4" x 10"		390	.041		4.32	1.64		5.96	7.45
0700	5/4" x 12"		390	.041		5.25	1.64		6.89	8.50

Estimating Tips

07 10 00 Dampproofing and Waterproofing

- Be sure of the job specifications before pricing this subdivision. The difference in cost between waterproofing and dampproofing can be great. Waterproofing will hold back standing water. Dampproofing prevents the transmission of water vapor. Also included in this section are vapor retarding membranes.

07 20 00 Thermal Protection

- Insulation and fireproofing products are measured by area, thickness, volume or R-value. Specifications may give only what the specific R-value should be in a certain situation. The estimator may need to choose the type of insulation to meet that R-value.

07 30 00 Steep Slope Roofing
07 40 00 Roofing and Siding Panels

- Many roofing and siding products are bought and sold by the square. One square is equal to an area that measures 100 square feet.

This simple change in unit of measure could create a large error if the estimator is not observant. Accessories necessary for a complete installation must be figured into any calculations for both material and labor.

07 50 00 Membrane Roofing
07 60 00 Flashing and Sheet Metal
07 70 00 Roofing and Wall Specialties and Accessories

- The items in these subdivisions compose a roofing system. No one component completes the installation, and all must be estimated. Built-up or single-ply membrane roofing systems are made up of many products and installation trades. Wood blocking at roof perimeters or penetrations, parapet coverings, reglets, roof drains, gutters, downspouts, sheet metal flashing, skylights, smoke vents, and roof hatches all need to be considered along with the roofing material. Several different installation trades will need to work together on the roofing system. Inherent difficulties in the scheduling and coordination of various trades must be accounted for when estimating labor costs.

07 90 00 Joint Protection

- To complete the weather-tight shell, the sealants and caulkings must be estimated. Where different materials meet—at expansion joints, at flashing penetrations, and at hundreds of other locations throughout a construction project—they provide another line of defense against water penetration. Often, an entire system is based on the proper location and placement of caulking or sealants. The detailed drawings that are included as part of a set of architectural plans show typical locations for these materials. When caulking or sealants are shown at typical locations, this means the estimator must include them for all the locations where this detail is applicable. Be careful to keep different types of sealants separate, and remember to consider backer rods and primers if necessary.

Reference Numbers

Reference numbers are shown in shaded boxes at the beginning of some major classifications. These numbers refer to related items in the Reference Section. The reference information may be an estimating procedure, an alternate pricing method, or technical information.

Note: Not all subdivisions listed here necessarily appear in this publication.

07 01 Operation and Maint. of Thermal and Moisture Protection

07 01 50 – Maintenance of Membrane Roofing

07 01 50.10 Roof Coatings

		Crew	Daily Output	Labor-Hours	Unit	Material	2009 Bare Costs Labor	Equipment	Total	Total Incl O&P
0010	**ROOF COATINGS**									
0012	Asphalt, brush grade, material only				Gal.	8.25			8.25	9.10
0200	Asphalt base, fibered aluminum coating G					12.20			12.20	13.45
0300	Asphalt primer, 5 gallon					5.95			5.95	6.55
0600	Coal tar pitch, 200 lb. barrels				Ton	840			840	925
0700	Tar roof cement, 5 gal. lots				Gal.	8.40			8.40	9.25
0800	Glass fibered roof & patching cement, 5 gallon				"	7.35			7.35	8.05
0900	Reinforcing glass membrane, 450 S.F./roll				Ea.	48.50			48.50	53
1000	Neoprene roof coating, 5 gal, 2 gal/sq				Gal.	24.50			24.50	26.50
1100	Roof patch & flashing cement, 5 gallon					11.70			11.70	12.85
1200	Roof resaturant, glass fibered, 3 gal/sq					7.75			7.75	8.55
1300	Mineral rubber, 3 gal/sq					4.99			4.99	5.50

07 05 Common Work Results for Thermal and Moisture Protection

07 05 05 – Selective Demolition

07 05 05.10 Selective Demo., Thermal and Moist. Protection

			Crew	Daily Output	Labor-Hours	Unit	Material	2009 Bare Costs Labor	Equipment	Total	Total Incl O&P
0010	**SELECTIVE DEMOLITION, THERMAL AND MOISTURE PROTECTION**										
0020	Caulking / sealant, to 1" x 1" joint	R024119-10	1 Clab	600	.013	L.F.		.42		.42	.70
0120	Downspouts, including hangers			350	.023	"		.72		.72	1.19
0220	Flashing, sheet metal			290	.028	S.F.		.87		.87	1.44
0420	Gutters, aluminum or wood, edge hung			240	.033	L.F.		1.05		1.05	1.74
0520	Built-in			100	.080	"		2.53		2.53	4.17
0620	Insulation, air / vapor barrier			3500	.002	S.F.		.07		.07	.12
0670	Batts or blankets			1400	.006	C.F.		.18		.18	.30
0720	Foamed or sprayed in place		2 Clab	1000	.016	B.F.		.51		.51	.83
0770	Loose fitting		1 Clab	3000	.003	C.F.		.08		.08	.14
0870	Rigid board			3450	.002	B.F.		.07		.07	.12
1120	Roll roofing, cold adhesive			12	.667	Sq.		21		21	35
1170	Roof accessories, adjustable metal chimney flashing			9	.889	Ea.		28		28	46.50
1325	Plumbing vent flashing			32	.250	"		7.90		7.90	13.05
1375	Ridge vent strip, aluminum			310	.026	L.F.		.82		.82	1.35
1620	Skylight to 10 S.F.			8	1	Ea.		31.50		31.50	52
2120	Roof edge, aluminum soffit and fascia			570	.014	L.F.		.44		.44	.73
2170	Concrete coping, up to 12" wide		2 Clab	160	.100			3.16		3.16	5.20
2220	Drip edge		1 Clab	1650	.005			.15		.15	.25
2270	Gravel stop			1650	.005			.15		.15	.25
2370	Sheet metal coping, up to 12" wide			240	.033			1.05		1.05	1.74
2470	Roof insulation board, over 2" thick		B-2	7800	.005	B.F.		.16		.16	.27
2520	Up to 2" thick		"	3900	.010	S.F.		.33		.33	.54
2620	Roof ventilation, louvered gable vent		1 Clab	16	.500	Ea.		15.80		15.80	26
2675	Rafter vents			960	.008	"		.26		.26	.43
2720	Soffit vent and/or fascia vent			575	.014	L.F.		.44		.44	.73
2775	Soffit vent strip, aluminum, 3" to 4" wide			160	.050			1.58		1.58	2.61
2820	Roofing accessories, shingle moulding, to 1" x 4"			1600	.005			.16		.16	.26
2870	Cant strip		B-2	2000	.020			.64		.64	1.06
2920	Concrete block walkway		1 Clab	230	.035			1.10		1.10	1.81
3070	Roofing, felt paper, 15#			70	.114	Sq.		3.61		3.61	5.95
3125	#30 felt			30	.267	"		8.45		8.45	13.90
3170	Asphalt shingles, 1 layer		B-2	3500	.011	S.F.		.37		.37	.60
3370	Modified bitumen			26	1.538	Sq.		49		49	81.50
3420	Built-up, no gravel, 3 ply			25	1.600			51		51	84.50

138

07 05 Common Work Results for Thermal and Moisture Protection

07 05 05 – Selective Demolition

07 05 05.10 Selective Demo., Thermal and Moist. Protection	Crew	Daily Output	Labor-Hours	Unit	Material	2009 Bare Costs Labor	Equipment	Total	Total Incl O&P
3470	4 ply	B-2	21	1.905	Sq.		61		61	101
3620	5 ply		1600	.025	S.F.		.80		.80	1.32
3720	5 ply, with gravel		890	.045			1.44		1.44	2.37
3725	Gravel removal, minimum		5000	.008			.26		.26	.42
3730	Maximum		2000	.020			.64		.64	1.06
3870	Fiberglass sheet		1200	.033			1.07		1.07	1.76
4120	Slate shingles		1900	.021	▼		.67		.67	1.11
4170	Ridge shingles, clay or slate		2000	.020	L.F.		.64		.64	1.06
4320	Single ply membrane, attached at seams		52	.769	Sq.		24.50		24.50	40.50
4370	Ballasted		75	.533			17.05		17.05	28
4420	Fully adhered	▼	39	1.026	▼		33		33	54
4550	Roof hatch, 2'-6" x 3'-0"	1 Clab	10	.800	Ea.		25.50		25.50	41.50
4670	Wood shingles	B-2	2200	.018	S.F.		.58		.58	.96
4820	Sheet metal roofing	"	2150	.019			.60		.60	.98
4970	Siding, horizontal wood clapboards	1 Clab	380	.021			.67		.67	1.10
5025	Exterior insulation finish system	"	120	.067			2.11		2.11	3.48
5070	Tempered hardboard, remove and reset	1 Carp	380	.021			.84		.84	1.39
5120	Tempered hardboard sheet siding	"	375	.021	▼		.85		.85	1.41
5170	Metal, corner strips	1 Clab	850	.009	L.F.		.30		.30	.49
5225	Horizontal strips		444	.018	S.F.		.57		.57	.94
5320	Vertical strips		400	.020			.63		.63	1.04
5520	Wood shingles		350	.023			.72		.72	1.19
5620	Stucco siding		360	.022			.70		.70	1.16
5670	Textured plywood		725	.011			.35		.35	.58
5720	Vinyl siding		510	.016	▼		.50		.50	.82
5770	Corner strips		900	.009	L.F.		.28		.28	.46
5870	Wood, boards, vertical	▼	400	.020	S.F.		.63		.63	1.04
5920	Waterproofing, protection / drain board	2 Clab	3900	.004	B.F.		.13		.13	.21
5970	Over 1/2" thick		1750	.009	S.F.		.29		.29	.48
6020	To 1/2" thick	▼	2000	.008	"		.25		.25	.42
9000	Minimum labor/equipment charge	1 Clab	2	4	Job		126		126	209

07 11 Dampproofing

07 11 13 – Bituminous Dampproofing

07 11 13.10 Bituminous Asphalt Coating

		Crew	Daily Output	Labor-Hours	Unit	Material	2009 Bare Costs Labor	Equipment	Total	Total Incl O&P
0010	**BITUMINOUS ASPHALT COATING**									
0030	Brushed on, below grade, 1 coat	1 Rofc	665	.012	S.F.	.16	.41		.57	.92
0100	2 coat		500	.016		.32	.55		.87	1.34
0300	Sprayed on, below grade, 1 coat, 25.6 S.F./gal.		830	.010		.16	.33		.49	.77
0400	2 coat, 20.5 S.F./gal.		500	.016		.32	.55		.87	1.33
0600	Troweled on, asphalt with fibers, 1/16" thick		500	.016		.37	.55		.92	1.39
0700	1/8" thick		400	.020		.65	.69		1.34	1.94
1000	1/2" thick		350	.023	▼	2.12	.78		2.90	3.73
9000	Minimum labor/equipment charge	▼	3	2.667	Job		91.50		91.50	163

07 12 Built-up Bituminous Waterproofing

07 12 13 – Built-Up Asphalt Waterproofing

07 12 13.20 Membrane Waterproofing	Crew	Daily Output	Labor-Hours	Unit	Material	2009 Bare Costs Labor	Equipment	Total	Total Incl O&P
0010 **MEMBRANE WATERPROOFING**									
0012 On slabs, 1 ply, felt, mopped	G-1	3000	.019	S.F.	.34	.60	.15	1.09	1.61
0015 On walls, 1 ply, felt, mopped		3000	.019		.34	.60	.15	1.09	1.61
0100 On slabs, 1 ply, glass fiber fabric, mopped		2100	.027		.36	.85	.21	1.42	2.14
0105 On walls, 1 ply, glass fiber fabric, mopped		2100	.027		.36	.85	.21	1.42	2.14
0300 On slabs, 2 ply, felt, mopped		2500	.022		.69	.72	.18	1.59	2.23
0305 On walls, 2 ply, felt, mopped		2500	.022		.69	.72	.18	1.59	2.23
0400 On slabs, 2 ply, glass fiber fabric, mopped		1650	.034		.79	1.09	.27	2.15	3.10
0405 On walls, 2 ply, glass fiber fabric, mopped		1650	.034		.79	1.09	.27	2.15	3.10
0600 On slabs, 3 ply, felt, mopped		2100	.027		1.03	.85	.21	2.09	2.89
0605 On walls, 3 ply, felt, mopped		2100	.027		1.03	.85	.21	2.09	2.89
0700 On slabs, 3 ply, glass fiber fabric, mopped		1550	.036		1.07	1.16	.28	2.51	3.55
0705 On walls, 3 ply, glass fiber fabric, mopped	▼	1550	.036		1.07	1.16	.28	2.51	3.55
0710 Asphaltic hardboard protection board, 1/8" thick	2 Rofc	500	.032		.40	1.10		1.50	2.40
0715 1/4" thick		450	.036		.68	1.22		1.90	2.92
1000 1/4" EPS membrane protection board		3500	.005		.21	.16		.37	.51
1050 3/8" thick		3500	.005		.23	.16		.39	.53
1060 1/2" thick		3500	.005	▼	.26	.16		.42	.57
1070 Fiberglass fabric, black, 20/10 mesh		116	.138	Sq.	18	4.72		22.72	28.50
1080 White, 20/10 mesh		116	.138	"	18	4.72		22.72	28.50
9000 Minimum labor/equipment charge	▼	2	8	Job		274		274	490

07 13 Sheet Waterproofing

07 13 53 – Elastomeric Sheet Waterproofing

07 13 53.10 Elastomeric Sheet Waterproofing and Access.

	Crew	Daily Output	Labor-Hours	Unit	Material	2009 Bare Costs Labor	Equipment	Total	Total Incl O&P
0010 **ELASTOMERIC SHEET WATERPROOFING AND ACCESS.**									
0090 EPDM, plain, 45 mils thick	2 Rofc	580	.028	S.F.	1.19	.95		2.14	3
0100 60 mils thick		570	.028		1.25	.96		2.21	3.09
0300 Nylon reinforced sheets, 45 mils thick		580	.028		1.16	.95		2.11	2.96
0400 60 mils thick	▼	570	.028	▼	1.47	.96		2.43	3.33
0600 Vulcanizing splicing tape for above, 2" wide				C.L.F.	43.50			43.50	47.50
0700 4" wide				"	99.50			99.50	109
0900 Adhesive, bonding, 60 SF per gal				Gal.	16.50			16.50	18.15
1000 Splicing, 75 SF per gal				"	32			32	35
1200 Neoprene sheets, plain, 45 mils thick	2 Rofc	580	.028	S.F.	1.68	.95		2.63	3.54
1300 60 mils thick		570	.028		2.86	.96		3.82	4.87
1500 Nylon reinforced, 45 mils thick		580	.028		1.75	.95		2.70	3.62
1600 60 mils thick		570	.028		2.46	.96		3.42	4.43
1800 120 mils thick	▼	500	.032	▼	4	1.10		5.10	6.35
1900 Adhesive, splicing, 150 S.F. per gal. per coat				Gal.	17.50			17.50	19.25
2100 Fiberglass reinforced, fluid applied, 1/8" thick	2 Rofc	500	.032	S.F.	1.88	1.10		2.98	4.03
2200 Polyethylene and rubberized asphalt sheets, 1/8" thick		550	.029		.73	1		1.73	2.58
2210 Asphaltic hardboard protection board, 1/8" thick		500	.032		.40	1.10		1.50	2.40
2220 1/4" thick		450	.036		.68	1.22		1.90	2.92
2400 Polyvinyl chloride sheets, plain, 10 mils thick		580	.028		.17	.95		1.12	1.88
2500 20 mils thick		570	.028		.28	.96		1.24	2.03
2700 30 mils thick	▼	560	.029	▼	.39	.98		1.37	2.18
3000 Adhesives, trowel grade, 40-100 SF per gal				Gal.	27			27	29.50
3100 Brush grade, 100-250 SF per gal.				"	27			27	29.50
3300 Bitumen modified polyurethane, fluid applied, 55 mils thick	2 Rofc	665	.024	S.F.	.68	.82		1.50	2.22
3600 Vinyl plastic, sprayed on, 25 to 40 mils thick	▼	475	.034	"	1.20	1.15		2.35	3.38

07 13 Sheet Waterproofing

07 13 53 – Elastomeric Sheet Waterproofing

07 13 53.10 Elastomeric Sheet Waterproofing and Access.	Crew	Daily Output	Labor-Hours	Unit	Material	2009 Bare Costs Labor	Equipment	Total	Total Incl O&P
9000	Minimum labor/equipment charge	2 Rofc	2	8	Job		274		274	490

07 16 Cementitious and Reactive Waterproofing

07 16 16 – Crystalline Waterproofing

07 16 16.20 Cementitious Waterproofing

		Crew	Daily Output	Labor-Hours	Unit	Material	Labor	Equipment	Total	Total Incl O&P
0010	**CEMENTITIOUS WATERPROOFING**									
0020	1/8" application, sprayed on	G-2A	1000	.024	S.F.	1.62	.73	.61	2.96	3.72
0030	2 coat, cementitious/metallic slurry, troweled, 1/4" thick	1 Cefi	2.48	3.226	C.S.F.	24	124		148	220
0040	3 coat, 3/8" thick		1.84	4.348		42.50	167		209.50	305
0050	4 coat, 1/2" thick		1.20	6.667		56.50	255		311.50	460

07 19 Water Repellents

07 19 19 – Silicone Water Repellents

07 19 19.10 Silicone Based Water Repellents

		Crew	Daily Output	Labor-Hours	Unit	Material	Labor	Equipment	Total	Total Incl O&P
0010	**SILICONE BASED WATER REPELLENTS**									
0020	Water base liquid, roller applied	2 Rofc	7000	.002	S.F.	.65	.08		.73	.86
0200	Silicone or stearate, sprayed on CMU, 1 coat	1 Rofc	4000	.002		.36	.07		.43	.52
0300	2 coats		3000	.003		.73	.09		.82	.96
9000	Minimum labor/equipment charge		3	2.667	Job		91.50		91.50	163

07 21 Thermal Insulation

07 21 13 – Board Insulation

07 21 13.10 Rigid Insulation

			Crew	Daily Output	Labor-Hours	Unit	Material	Labor	Equipment	Total	Total Incl O&P
0010	**RIGID INSULATION**, for walls										
0040	Fiberglass, 1.5#/CF, unfaced, 1" thick, R4.1	G	1 Carp	1000	.008	S.F.	.44	.32		.76	1.01
0060	1-1/2" thick, R6.2	G		1000	.008		.64	.32		.96	1.23
0080	2" thick, R8.3	G		1000	.008		.69	.32		1.01	1.29
0120	3" thick, R12.4	G		800	.010		.80	.40		1.20	1.54
0370	3#/CF, unfaced, 1" thick, R4.3	G		1000	.008		.49	.32		.81	1.07
0390	1-1/2" thick, R6.5	G		1000	.008		.95	.32		1.27	1.58
0400	2" thick, R8.7	G		890	.009		1.15	.36		1.51	1.86
0420	2-1/2" thick, R10.9	G		800	.010		1.01	.40		1.41	1.77
0440	3" thick, R13	G		800	.010		1.01	.40		1.41	1.77
0520	Foil faced, 1" thick, R4.3	G		1000	.008		.92	.32		1.24	1.54
0540	1-1/2" thick, R6.5	G		1000	.008		1.36	.32		1.68	2.03
0560	2" thick, R8.7	G		890	.009		1.71	.36		2.07	2.47
0580	2-1/2" thick, R10.9	G		800	.010		2.02	.40		2.42	2.88
0600	3" thick, R13	G		800	.010		2.19	.40		2.59	3.07
0670	6#/CF, unfaced, 1" thick, R4.3	G		1000	.008		.98	.32		1.30	1.61
0690	1-1/2" thick, R6.5	G		890	.009		1.50	.36		1.86	2.24
0700	2" thick, R8.7	G		800	.010		2.12	.40		2.52	2.99
0721	2-1/2" thick, R10.9	G		800	.010		2.32	.40		2.72	3.21
0741	3" thick, R13	G		730	.011		2.78	.44		3.22	3.78
0821	Foil faced, 1" thick, R4.3	G		1000	.008		1.38	.32		1.70	2.05
0840	1-1/2" thick, R6.5	G		890	.009		1.98	.36		2.34	2.77
0850	2" thick, R8.7	G		800	.010		2.59	.40		2.99	3.51
0880	2-1/2" thick, R10.9	G		800	.010		3.11	.40		3.51	4.08

07 21 13 – Board Insulation

07 21 13.10 Rigid Insulation		Crew	Daily Output	Labor-Hours	Unit	Material	2009 Bare Costs Labor	Equipment	Total	Total Incl O&P
0900	3" thick, R13	G 1 Carp	730	.011	S.F.	3.72	.44		4.16	4.81
1500	Foamglass, 1-1/2" thick, R4.5	G	800	.010		1.37	.40		1.77	2.17
1550	3" thick, R9	G	730	.011		3.29	.44		3.73	4.34
1600	Isocyanurate, 4' x 8' sheet, foil faced, both sides									
1610	1/2" thick, R3.9	G 1 Carp	800	.010	S.F.	.30	.40		.70	.99
1620	5/8" thick, R4.5	G	800	.010		.51	.40		.91	1.22
1630	3/4" thick, R5.4	G	800	.010		.35	.40		.75	1.05
1640	1" thick, R7.2	G	800	.010		.55	.40		.95	1.27
1650	1-1/2" thick, R10.8	G	730	.011		.64	.44		1.08	1.42
1660	2" thick, R14.4	G	730	.011		.81	.44		1.25	1.61
1670	3" thick, R21.6	G	730	.011		1.90	.44		2.34	2.81
1680	4" thick, R28.8	G	730	.011		2.14	.44		2.58	3.07
1700	Perlite, 1" thick, R2.77	G	800	.010		.30	.40		.70	.99
1750	2" thick, R5.55	G	730	.011		.60	.44		1.04	1.38
1900	Extruded polystyrene, 25 PSI compressive strength, 1" thick, R5	G	800	.010		.51	.40		.91	1.22
1940	2" thick R10	G	730	.011		1.04	.44		1.48	1.86
1960	3" thick, R15	G	730	.011		1.40	.44		1.84	2.26
2100	Expanded polystyrene, 1" thick, R3.85	G	800	.010		.24	.40		.64	.92
2120	2" thick, R7.69	G	730	.011		.61	.44		1.05	1.39
2140	3" thick, R11.49	G	730	.011		.78	.44		1.22	1.58
9000	Minimum labor/equipment charge		4	2	Job		80		80	132

07 21 13.13 Foam Board Insulation

0010	**FOAM BOARD INSULATION**									
0600	Polystyrene, expanded, 1" thick, R4	G 1 Carp	680	.012	S.F.	.31	.47		.78	1.12
0700	2" thick, R8	G	675	.012	"	.62	.47		1.09	1.46
9000	Minimum labor/equipment charge		4	2	Job		80		80	132

07 21 16 – Blanket Insulation

07 21 16.10 Blanket Insulation for Floors

0010	**BLANKET INSULATION FOR FLOORS**									
0020	Including spring type wire fasteners									
2000	Fiberglass, blankets or batts, paper or foil backing									
2100	1 side, 3-1/2" thick, R11	G 1 Carp	700	.011	S.F.	.42	.46		.88	1.21
2150	6" thick, R19	G	600	.013		.52	.53		1.05	1.45
2200	8-1/2" thick, R30	G	550	.015		.74	.58		1.32	1.77
9000	Minimum labor/equipment charge		4	2	Job		80		80	132

07 21 16.20 Blanket Insulation for Walls

0010	**BLANKET INSULATION FOR WALLS**									
0020	Kraft faced fiberglass, 3-1/2" thick, R11, 15" wide	G 1 Carp	1350	.006	S.F.	.34	.24		.58	.76
0030	23" wide	G	1600	.005		.34	.20		.54	.70
0060	R13, 11" wide	G	1150	.007		.34	.28		.62	.83
0080	15" wide	G	1350	.006		.34	.24		.58	.76
0100	23" wide	G	1600	.005		.34	.20		.54	.70
0110	R-15, 11" wide	G	1150	.007		.41	.28		.69	.91
0120	15" wide	G	1350	.006		.41	.24		.65	.84
0130	23" wide	G	1600	.005		.41	.20		.61	.78
0140	6" thick, R19, 11" wide	G	1150	.007		.44	.28		.72	.94
0160	15" wide	G	1350	.006		.44	.24		.68	.87
0180	23" wide	G	1600	.005		.44	.20		.64	.81
0182	R21, 11" wide	G	1150	.007		.50	.28		.78	1.01
0184	15" wide	G	1350	.006		.50	.24		.74	.94
0186	23" wide	G	1600	.005		.50	.20		.70	.88
0188	9" thick, R-30, 11" wide	G	985	.008		.66	.32		.98	1.27

07 21 Thermal Insulation

07 21 16 – Blanket Insulation

07 21 16.20 Blanket Insulation for Walls		Crew	Daily Output	Labor-Hours	Unit	Material	2009 Bare Costs Labor	Equipment	Total	Total Incl O&P
0200	15" wide	G	1 Carp	1150	.007	S.F.	.66	.28		.94	1.19
0220	23" wide	G		1350	.006		.66	.24		.90	1.12
0230	12" thick, R38, 11" wide	G		985	.008		.95	.32		1.27	1.59
0240	15" wide	G		1150	.007		.95	.28		1.23	1.51
0260	23" wide	G		1350	.006		.95	.24		1.19	1.44
0410	Foil faced fiberglass, 3-1/2" thick, R13, 11" wide	G		1150	.007		.57	.28		.85	1.09
0420	15" wide	G		1350	.006		.57	.24		.81	1.02
0440	23" wide	G		1600	.005		.57	.20		.77	.96
0442	R15, 11" wide	G		1150	.007		.56	.28		.84	1.08
0444	15" wide	G		1350	.006		.56	.24		.80	1.01
0446	23" wide	G		1600	.005		.56	.20		.76	.95
0448	6" thick, R19, 11" wide	G		1150	.007		.77	.28		1.05	1.31
0460	15" wide	G		1350	.006		.77	.24		1.01	1.24
0480	23" wide	G		1600	.005		.77	.20		.97	1.18
0482	R-21, 11" wide	G		1150	.007		.82	.28		1.10	1.36
0484	15" wide	G		1350	.006		.82	.24		1.06	1.29
0486	23" wide	G		1600	.005		.82	.20		1.02	1.23
0488	9" thick, R-30, 11" wide	G		985	.008		.89	.32		1.21	1.52
0500	9" thick, R30, 15" wide	G		1150	.007		.89	.28		1.17	1.44
0550	23" wide	G		1350	.006		.89	.24		1.13	1.37
0560	12" thick, R-38, 11" wide	G		985	.008		.90	.32		1.22	1.53
0570	15" wide	G		1150	.007		.90	.28		1.18	1.45
0580	23" wide	G		1350	.006		.90	.24		1.14	1.38
0620	Unfaced fiberglass, 3-1/2" thick, R-13, 11" wide	G		1150	.007		.29	.28		.57	.78
0820	15" wide	G		1350	.006		.29	.24		.53	.71
0830	23" wide	G		1600	.005		.29	.20		.49	.65
0832	R15, 11" wide	G		1150	.007		.37	.28		.65	.87
0836	23" wide	G		1600	.005		.37	.20		.57	.74
0838	6" thick, R19, 11" wide	G		1150	.007		.42	.28		.70	.92
0860	15" wide	G		1150	.007		.42	.28		.70	.92
0880	23" wide	G		1350	.006		.42	.24		.66	.85
0882	R-21, 11" wide	G		1150	.007		.46	.28		.74	.97
0886	15" wide	G		1350	.006		.46	.24		.70	.90
0888	23" wide	G		1600	.005		.46	.20		.66	.84
0890	9" thick, R30, 11" wide	G		985	.008		.75	.32		1.07	1.37
0900	15" wide	G		1150	.007		.75	.28		1.03	1.29
0920	23" wide	G		1350	.006		.75	.24		.99	1.22
0930	12" thick, R38, 11" wide	G		985	.008		1.28	.32		1.60	1.95
0940	15" wide	G		1000	.008		1.28	.32		1.60	1.94
0960	23" wide	G		1150	.007		1.28	.28		1.56	1.87
1300	Mineral fiber batts, kraft faced		1 Carp								
1320	3-1/2" thick, R12	G		1600	.005	S.F.	.38	.20		.58	.75
1340	6" thick, R19	G		1600	.005		.48	.20		.68	.86
1380	10" thick, R30	G		1350	.006		.73	.24		.97	1.19
1850	Friction fit wire insulation supports, 16" O.C.			960	.008	Ea.	.08	.33		.41	.64
9000	Minimum labor/equipment charge			4	2	Job		80		80	132

07 21 23 – Loose-Fill Insulation

07 21 23.10 Poured Loose-Fill Insulation

		Crew	Daily Output	Labor-Hours	Unit	Material	2009 Bare Costs Labor	Equipment	Total	Total Incl O&P
0010	**POURED LOOSE-FILL INSULATION**										
0020	Cellulose fiber, R3.8 per inch	G	1 Carp	200	.040	C.F.	.60	1.60		2.20	3.30
0040	Ceramic type (perlite), R3.2 per inch	G		200	.040		1.75	1.60		3.35	4.57
0080	Fiberglass wool, R4 per inch	G		200	.040		.46	1.60		2.06	3.15

07 21 Thermal Insulation

07 21 23 – Loose-Fill Insulation

07 21 23.10 Poured Loose-Fill Insulation

			Crew	Daily Output	Labor-Hours	Unit	Material	2009 Bare Costs Labor	Equipment	Total	Total Incl O&P
0100	Mineral wool, R3 per inch	G	1 Carp	200	.040	C.F.	.39	1.60		1.99	3.07
0300	Polystyrene, R4 per inch	G		200	.040		3.09	1.60		4.69	6.05
0400	Vermiculite or perlite, R2.7 per inch	G		200	.040		1.75	1.60		3.35	4.57
9000	Minimum labor/equipment charge			4	2	Job		80		80	132

07 21 23.20 Masonry Loose-Fill Insulation

			Crew	Daily Output	Labor-Hours	Unit	Material	2009 Bare Costs Labor	Equipment	Total	Total Incl O&P
0010	**MASONRY LOOSE-FILL INSULATION**, vermiculite or perlite										
0100	In cores of concrete block, 4" thick wall, .115 CF/SF	G	D-1	4800	.003	S.F.	.20	.12		.32	.42
0200	6" thick wall, .175 CF/SF	G		3000	.005		.31	.19		.50	.65
0300	8" thick wall, .258 CF/SF	G		2400	.007		.45	.24		.69	.89
0400	10" thick wall, .340 CF/SF	G		1850	.009		.60	.31		.91	1.16
0500	12" thick wall, .422 CF/SF	G		1200	.013		.74	.48		1.22	1.59
0550	For sand fill, deduct from above	G					70%				
0600	Poured cavity wall, vermiculite or perlite, water repellant	G	D-1	250	.064	C.F.	1.75	2.33		4.08	5.70
0700	Foamed in place, urethane in 2-5/8" cavity	G	G-2A	1035	.023	S.F.	.42	.71	.58	1.71	2.33
0800	For each 1" added thickness, add	G	"	2372	.010	"	.12	.31	.26	.69	.94

07 21 26 – Blown Insulation

07 21 26.10 Blown Insulation

			Crew	Daily Output	Labor-Hours	Unit	Material	2009 Bare Costs Labor	Equipment	Total	Total Incl O&P
0010	**BLOWN INSULATION** Ceilings, with open access										
0020	Cellulose, 3-1/2" thick, R13	G	G-4	5000	.005	S.F.	.19	.15	.07	.41	.54
0030	5-3/16" thick, R19	G		3800	.006		.28	.20	.09	.57	.75
0050	6-1/2" thick, R22	G		3000	.008		.36	.26	.11	.73	.95
0100	8-11/16" thick, R30	G		2600	.009		.48	.30	.13	.91	1.16
1000	Fiberglass, 5.5" thick, R11	G		3800	.006		.19	.20	.09	.48	.65
1050	6" thick, R12	G		3000	.008		.23	.26	.11	.60	.80
1100	8.8" thick, R19	G		2200	.011		.33	.35	.15	.83	1.11
1200	10" thick, R22	G		1800	.013		.38	.43	.19	1	1.33
1300	11.5" thick, R26	G		1500	.016		.46	.52	.22	1.20	1.60
1350	13" thick, R30	G		1400	.017		.50	.55	.24	1.29	1.72
1450	16" thick, R38	G		1145	.021		.61	.68	.29	1.58	2.11
1500	20" thick, R49	G		920	.026		.77	.84	.36	1.97	2.63
2000	Mineral wool, 4" thick, R12	G		3500	.007		.23	.22	.10	.55	.72
2050	6" thick, R17	G		2500	.010		.26	.31	.13	.70	.95
2100	9" thick, R23	G		1750	.014		.34	.44	.19	.97	1.31
2500	Wall installation, incl. drilling & patching from outside, two 1"										
2510	diam. holes @ 16" O.C., top & mid-point of wall, add to above										
2700	For masonry	G	G-4	415	.058	S.F.	.06	1.87	.80	2.73	4.03
2800	For wood siding	G		840	.029		.06	.92	.40	1.38	2.03
2900	For stucco/plaster	G		665	.036		.06	1.16	.50	1.72	2.54
9000	Minimum labor/equipment charge			4	6	Job		194	83.50	277.50	410

07 21 27 – Reflective Insulation

07 21 27.10 Reflective Insulation Options

			Crew	Daily Output	Labor-Hours	Unit	Material	2009 Bare Costs Labor	Equipment	Total	Total Incl O&P
0010	**REFLECTIVE INSULATION OPTIONS**										
0020	Aluminum foil on reinforced scrim	G	1 Carp	19	.421	C.S.F.	14.20	16.80		31	43.50
0100	Reinforced with woven polyolefin	G		19	.421		17.20	16.80		34	47
0500	With single bubble air space, R8.8	G		15	.533		25.50	21.50		47	63
0600	With double bubble air space, R9.8	G		15	.533		26.50	21.50		48	64
9000	Minimum labor/equipment charge			4	2	Job		80		80	132

07 21 29 – Sprayed Insulation

07 21 29.10 Sprayed-On Insulation

			Crew	Daily Output	Labor-Hours	Unit	Material	2009 Bare Costs Labor	Equipment	Total	Total Incl O&P
0010	**SPRAYED-ON INSULATION**										
0020	Fibrous/cementitious, finished wall, 1" thick, R3.7	G	G-2	2050	.012	S.F.	.26	.39	.06	.71	.99

07 21 Thermal Insulation

07 21 29 – Sprayed Insulation

07 21 29.10 Sprayed-On Insulation		Crew	Daily Output	Labor-Hours	Unit	Material	2009 Bare Costs Labor	2009 Bare Costs Equipment	Total	Total Incl O&P
0100	Attic, 5.2" thick, R19	G	G-2	1550	.015	S.F.	.42	.52	.08	1.02	1.39
0300	Closed cell, spray polyurethane foam, 2 pounds per cubic foot density										
0310	1" thick	G	G-2A	6000	.004	S.F.	.70	.12	.10	.92	1.09
0320	2" thick	G		3000	.008		1.40	.24	.20	1.84	2.18
0330	3" thick	G		2000	.012		2.11	.36	.30	2.77	3.28
0335	3-1/2" thick	G		1715	.014		2.45	.43	.35	3.23	3.82
0340	4" thick	G		1500	.016		2.80	.49	.40	3.69	4.37
0350	5" thick	G		1200	.020		3.50	.61	.50	4.61	5.45
0355	5-1/2" thick	G		1090	.022		3.85	.67	.56	5.08	6
0360	6" thick	G		1000	.024		4.22	.73	.61	5.56	6.60
9000	Minimum labor/equipment charge		G-2	2	12	Job		400	63.50	463.50	720

07 22 Roof and Deck Insulation

07 22 16 – Roof Board Insulation

07 22 16.10 Roof Deck Insulation		Crew	Daily Output	Labor-Hours	Unit	Material	2009 Bare Costs Labor	2009 Bare Costs Equipment	Total	Total Incl O&P
0010	**ROOF DECK INSULATION**										
0020	Fiberboard low density, 1/2" thick R1.39	G	1 Rofc	1000	.008	S.F.	.24	.27		.51	.75
0030	1" thick R2.78	G		800	.010		.42	.34		.76	1.07
0080	1 1/2" thick R4.17	G		800	.010		.64	.34		.98	1.31
0100	2" thick R5.56	G		800	.010		.85	.34		1.19	1.55
0110	Fiberboard high density, 1/2" thick R1.3	G		1000	.008		.22	.27		.49	.73
0120	1" thick R2.5	G		800	.010		.44	.34		.78	1.09
0130	1-1/2" thick R3.8	G		800	.010		.66	.34		1	1.34
0200	Fiberglass, 3/4" thick R2.78	G		1000	.008		.56	.27		.83	1.11
0400	15/16" thick R3.70	G		1000	.008		.74	.27		1.01	1.30
0460	1-1/16" thick R4.17	G		1000	.008		.93	.27		1.20	1.51
0600	1-5/16" thick R5.26	G		1000	.008		1.26	.27		1.53	1.88
0650	2-1/16" thick R8.33	G		800	.010		1.35	.34		1.69	2.10
0700	2-7/16" thick R10	G		800	.010		1.55	.34		1.89	2.32
1500	Foamglass, 1-1/2" thick R4.5	G		800	.010		1.35	.34		1.69	2.10
1530	3" thick R9	G		700	.011		2.75	.39		3.14	3.73
1600	Tapered for drainage	G		600	.013	B.F.	1.16	.46		1.62	2.10
1650	Perlite, 1/2" thick R1.32	G		1050	.008	S.F.	.34	.26		.60	.84
1655	3/4" thick R2.08	G		800	.010		.37	.34		.71	1.02
1660	1" thick R2.78	G		800	.010		.46	.34		.80	1.12
1670	1-1/2" thick R4.17	G		800	.010		.48	.34		.82	1.14
1680	2" thick R5.56	G		700	.011		.80	.39		1.19	1.58
1685	2-1/2" thick R6.67	G		700	.011		.95	.39		1.34	1.75
1690	Tapered for drainage	G		800	.010	B.F.	.73	.34		1.07	1.41
1700	Polyisocyanurate, 2#/CF density, 3/4" thick, R5.1	G		1500	.005	S.F.	.44	.18		.62	.81
1705	1" thick R7.14	G		1400	.006		.50	.20		.70	.90
1715	1-1/2" thick R10.87	G		1250	.006		.63	.22		.85	1.08
1725	2" thick R14.29	G		1100	.007		.81	.25		1.06	1.33
1735	2-1/2" thick R16.67	G		1050	.008		1.01	.26		1.27	1.58
1745	3" thick R21.74	G		1000	.008		1.26	.27		1.53	1.88
1755	3-1/2" thick R25	G		1000	.008		1.93	.27		2.20	2.61
1765	Tapered for drainage	G		1400	.006	B.F.	1.93	.20		2.13	2.47
1900	Extruded Polystyrene										
1910	15 PSI compressive strength, 1" thick, R5	G	1 Rofc	1500	.005	S.F.	.49	.18		.67	.87
1920	2" thick, R10	G		1250	.006		.62	.22		.84	1.07
1930	3" thick R15	G		1000	.008		1.27	.27		1.54	1.89

145

07 22 Roof and Deck Insulation

07 22 16 – Roof Board Insulation

07 22 16.10 Roof Deck Insulation		Crew	Daily Output	Labor-Hours	Unit	Material	2009 Bare Costs Labor	Equipment	Total	Total Incl O&P
1932	4" thick R20	G	1 Rofc	1000	.008	S.F.	1.64	.27		1.91	2.29
1934	Tapered for drainage	G		1500	.005	B.F.	.53	.18		.71	.91
1940	25 PSI compressive strength, 1" thick R5	G		1500	.005	S.F.	.67	.18		.85	1.07
1942	2" thick R10	G		1250	.006		1.27	.22		1.49	1.79
1944	3" thick R15	G		1000	.008		1.93	.27		2.20	2.61
1946	4" thick R20	G		1000	.008	↓	2.72	.27		2.99	3.48
1948	Tapered for drainage	G		1500	.005	B.F.	.56	.18		.74	.95
1950	40 psi compressive strength, 1" thick R5	G		1500	.005	S.F.	.51	.18		.69	.89
1952	2" thick R10	G		1250	.006		.96	.22		1.18	1.45
1954	3" thick R15	G		1000	.008		1.40	.27		1.67	2.03
1956	4" thick R20	G		1000	.008	↓	1.88	.27		2.15	2.56
1958	Tapered for drainage	G		1400	.006	B.F.	.71	.20		.91	1.13
1960	60 PSI compressive strength, 1" thick R5	G		1450	.006	S.F.	.59	.19		.78	.99
1962	2" thick R10	G		1200	.007		1.05	.23		1.28	1.57
1964	3" thick R15	G		975	.008		1.55	.28		1.83	2.21
1966	4" thick R20	G		950	.008	↓	2.17	.29		2.46	2.90
1968	Tapered for drainage	G		1400	.006	B.F.	.85	.20		1.05	1.29
2010	Expanded polystyrene, 1#/CF density, 3/4" thick R2.89	G		1500	.005	S.F.	.31	.18		.49	.67
2020	1" thick R3.85	G		1500	.005		.31	.18		.49	.67
2100	2" thick R7.69	G		1250	.006		.62	.22		.84	1.07
2110	3" thick R11.49	G		1250	.006		.91	.22		1.13	1.39
2120	4" thick R15.38	G		1200	.007		.98	.23		1.21	1.49
2130	5" thick R19.23	G		1150	.007		1.05	.24		1.29	1.59
2140	6" thick R23.26	G		1150	.007	↓	1.23	.24		1.47	1.78
2150	Tapered for drainage	G	↓	1500	.005	B.F.	.49	.18		.67	.87
2400	Composites with 2" EPS										
2410	1" fiberboard	G	1 Rofc	950	.008	S.F.	1.07	.29		1.36	1.69
2420	7/16" oriented strand board	G		800	.010		1.26	.34		1.60	2
2430	1/2" plywood	G		800	.010		1.37	.34		1.71	2.12
2440	1" perlite	G	↓	800	.010	↓	1.12	.34		1.46	1.84
2450	Composites with 1-1/2" polyisocyanurate										
2460	1" fiberboard	G	1 Rofc	800	.010	S.F.	1.46	.34		1.80	2.22
2470	1" perlite	G		850	.009		1.53	.32		1.85	2.26
2480	7/16" oriented strand board	G		800	.010	↓	1.77	.34		2.11	2.56
9000	Minimum labor/equipment charge		↓	3.25	2.462	Job		84.50		84.50	151

07 24 Exterior Insulation and Finish Systems

07 24 13 – Polymer Based Exterior Insulation and Finish Systems

07 24 13.10 Exterior Insulation and Finish Systems

			Crew	Daily Output	Labor-Hours	Unit	Material	2009 Bare Costs Labor	Equipment	Total	Total Incl O&P
0010	**EXTERIOR INSULATION AND FINISH SYSTEMS**										
0095	Field applied, 1" EPS insulation	G	J-1	295	.136	S.F.	2.18	4.69	.43	7.30	10.40
0100	With 1/2" cement board sheathing	G		220	.182		2.90	6.30	.58	9.78	13.90
0105	2" EPS insulation	G		295	.136		2.55	4.69	.43	7.67	10.85
0110	With 1/2" cement board sheathing	G		220	.182		3.27	6.30	.58	10.15	14.35
0115	3" EPS insulation	G		295	.136		2.72	4.69	.43	7.84	11
0120	With 1/2" cement board sheathing	G		220	.182		3.44	6.30	.58	10.32	14.50
0125	4" EPS insulation	G		295	.136		3.16	4.69	.43	8.28	11.50
0130	With 1/2" cement board sheathing	G		220	.182		4.60	6.30	.58	11.48	15.80
0140	Premium finish add			1265	.032		.31	1.09	.10	1.50	2.21
0150	Heavy duty reinforcement add		↓	914	.044	↓	1.08	1.51	.14	2.73	3.78
0160	2.5#/S.Y. metal lath substrate add		1 Lath	75	.107	S.Y.	2.37	3.79		6.16	8.60

07 24 Exterior Insulation and Finish Systems

07 24 13 – Polymer Based Exterior Insulation and Finish Systems

07 24 13.10 Exterior Insulation and Finish Systems	Crew	Daily Output	Labor-Hours	Unit	Material	2009 Bare Costs Labor	Equipment	Total	Total Incl O&P
0170	3.4#/S.Y. metal lath substrate add	1 Lath	75	.107	S.Y.	2.58	3.79		6.37	8.85
0180	Color or texture change,	J-1	1265	.032	S.F.	.83	1.09	.10	2.02	2.78
0190	With substrate leveling base coat	1 Plas	530	.015		.83	.55		1.38	1.79
0210	With substrate sealing base coat	1 Pord	1224	.007		.08	.23		.31	.46
0370	V groove shape in panel face				L.F.	.60			.60	.66
0380	U groove shape in panel face				"	.79			.79	.87
0433	Crack repair, acrylic rubber, fluid applied, 20 mils thick	1 Plas	350	.023	S.F.	1.43	.83		2.26	2.90
0437	50 mils thick, reinforced	"	200	.040	"	2.62	1.45		4.07	5.20
0440	For higher than one story, add						25%			

07 26 Vapor Retarders

07 26 10 – Vapor Retarders Made of Natural and Synthetic Materials

07 26 10.10 Vapor Retarders

			Crew	Daily Output	Labor-Hours	Unit	Material	2009 Bare Costs Labor	Equipment	Total	Total Incl O&P
0010	**VAPOR RETARDERS**										
0020	Aluminum and kraft laminated, foil 1 side	G	1 Carp	37	.216	Sq.	5.20	8.65		13.85	20
0100	Foil 2 sides	G		37	.216		8.70	8.65		17.35	24
0400	Asphalt felt sheathing paper, 15#			37	.216		4.77	8.65		13.42	19.50
0450	Housewrap, exterior, spun bonded polypropylene										
0470	Small roll	G	1 Carp	3800	.002	S.F.	.23	.08		.31	.39
0480	Large roll	G	"	4000	.002	"	.12	.08		.20	.26
0500	Material only, 3' x 111.1' roll	G				Ea.	75			75	82.50
0520	9' x 111.1' roll	G				"	120			120	132
0600	Polyethylene vapor barrier, standard, .002" thick	G	1 Carp	37	.216	Sq.	1.09	8.65		9.74	15.45
0700	.004" thick	G		37	.216		3.54	8.65		12.19	18.15
0900	.006" thick	G		37	.216		5.55	8.65		14.20	20.50
1200	.010" thick	G		37	.216		6.70	8.65		15.35	21.50
1300	Clear reinforced, fire retardant, .008" thick	G		37	.216		10.15	8.65		18.80	25.50
1350	Cross laminated type, .003" thick	G		37	.216		7.10	8.65		15.75	22
1400	.004" thick	G		37	.216		7.80	8.65		16.45	23
1500	Red rosin paper, 5 sq rolls, 4 lbs. per square			37	.216		2.19	8.65		10.84	16.65
1600	5 lbs. per square			37	.216		2.85	8.65		11.50	17.40
1800	Reinf. waterproof, .002" polyethylene backing, 1 side			37	.216		5.65	8.65		14.30	20.50
1900	2 sides			37	.216		7.45	8.65		16.10	22.50
2100	Asphalt felt roof deck vapor barrier, class 1 metal decks		1 Rofc	37	.216		18.65	7.40		26.05	33.50
2200	For all other decks		"	37	.216		13.85	7.40		21.25	28.50
2400	Waterproofed kraft with sisal or fiberglass fibers, minimum		1 Carp	37	.216		6.10	8.65		14.75	21
2500	Maximum			37	.216		15.10	8.65		23.75	31
9950	Minimum labor/equipment charge			4	2	Job		80		80	132

07 31 Shingles and Shakes

07 31 13 – Asphalt Shingles

07 31 13.10 Asphalt Roof Shingles

		Crew	Daily Output	Labor-Hours	Unit	Material	2009 Bare Costs Labor	Equipment	Total	Total Incl O&P
0010	**ASPHALT ROOF SHINGLES**									
0100	Standard strip shingles									
0150	Inorganic, class A, 210-235 lb/sq	1 Rofc	5.50	1.455	Sq.	50	50		100	144
0155	Pneumatic nailed		7	1.143		50	39		89	125
0200	Organic, class C, 235-240 lb/sq		5	1.600		51	55		106	154
0205	Pneumatic nailed		6.25	1.280		51	44		95	135
0250	Standard, laminated multi-layered shingles									

07 31 Shingles and Shakes

07 31 13 – Asphalt Shingles

07 31 13.10 Asphalt Roof Shingles

		Crew	Daily Output	Labor-Hours	Unit	Material	2009 Bare Costs Labor	2009 Bare Costs Equipment	Total	Total Incl O&P
0300	Class A, 240-260 lb/sq	1 Rofc	4.50	1.778	Sq.	61.50	61		122.50	177
0305	Pneumatic nailed		5.63	1.422		61.50	48.50		110	155
0350	Class C, 260-300 lb/square, 4 bundles/square		4	2		67	68.50		135.50	196
0355	Pneumatic nailed		5	1.600		67	55		122	172
0400	Premium, laminated multi-layered shingles									
0450	Class A, 260-300 lb, 4 bundles/sq	1 Rofc	3.50	2.286	Sq.	80	78.50		158.50	228
0455	Pneumatic nailed		4.37	1.831		80	62.50		142.50	200
0500	Class C, 300-385 lb/square, 5 bundles/square		3	2.667		106	91.50		197.50	280
0505	Pneumatic nailed		3.75	2.133		106	73		179	247
0800	#15 felt underlayment		64	.125		4.77	4.28		9.05	12.90
0825	#30 felt underlayment		58	.138		9.55	4.72		14.27	18.95
0850	Self adhering polyethylene and rubberized asphalt underlayment		22	.364		55.50	12.45		67.95	83
0900	Ridge shingles		330	.024	L.F.	1.49	.83		2.32	3.12
0905	Pneumatic nailed		412.50	.019	"	1.49	.66		2.15	2.83
1000	For steep roofs (7 to 12 pitch or greater), add						50%			
9000	Minimum labor/equipment charge	1 Rofc	3	2.667	Job		91.50		91.50	163

07 31 16 – Metal Shingles

07 31 16.10 Aluminum Shingles

		Crew	Daily Output	Labor-Hours	Unit	Material	2009 Bare Costs Labor	2009 Bare Costs Equipment	Total	Total Incl O&P
0010	**ALUMINUM SHINGLES**									
0020	Mill finish, .019 thick	1 Carp	5	1.600	Sq.	214	64		278	340
0100	.020" thick	"	5	1.600		208	64		272	335
0300	For colors, add					16			16	17.60
0600	Ridge cap, .024" thick	1 Carp	170	.047	L.F.	2.33	1.88		4.21	5.65
0700	End wall flashing, .024" thick		170	.047		1.80	1.88		3.68	5.10
0900	Valley section, .024" thick		170	.047		3.09	1.88		4.97	6.50
1000	Starter strip, .024" thick		400	.020		1.56	.80		2.36	3.04
1200	Side wall flashing, .024" thick		170	.047		1.81	1.88		3.69	5.10
9000	Minimum labor/equipment charge		3	2.667	Job		107		107	176

07 31 16.20 Steel Shingles

		Crew	Daily Output	Labor-Hours	Unit	Material	2009 Bare Costs Labor	2009 Bare Costs Equipment	Total	Total Incl O&P
0010	**STEEL SHINGLES**									
0012	Galvanized, 26 gauge	1 Rots	2.20	3.636	Sq.	232	124		356	475
0200	24 gauge	"	2.20	3.636		244	124		368	490
0300	For colored galvanized shingles, add					55			55	60.50
0500	For 1" factory applied polystyrene insulation, add					40			40	44
9000	Minimum labor/equipment charge	1 Rots	3	2.667	Job		91		91	162

07 31 19 – Mineral-Fiber Cement Shingles

07 31 19.10 Fiber Cement Shingles

		Crew	Daily Output	Labor-Hours	Unit	Material	2009 Bare Costs Labor	2009 Bare Costs Equipment	Total	Total Incl O&P
0010	**FIBER CEMENT SHINGLES**									
0012	Field shingles, 16" x 9.35", 500 lb per square	1 Rofc	2.20	3.636	Sq.	335	125		460	590
0200	Shakes, 16" x 9.35", 550 lb per square		2.20	3.636	"	305	125		430	555
0300	Hip & ridge, 4.75 x 14"		1	8	C.L.F.	825	274		1,099	1,400
0400	Hexagonal, 16" x 16"		3	2.667	Sq.	227	91.50		318.50	410
0500	Square, 16" x 16"		3	2.667		203	91.50		294.50	385
2000	For steep roofs (7/12 pitch or greater), add						50%			
9000	Minimum labor/equipment charge	1 Rofc	3	2.667	Job		91.50		91.50	163

07 31 26 – Slate Shingles

07 31 26.10 Slate Roof Shingles

			Crew	Daily Output	Labor-Hours	Unit	Material	2009 Bare Costs Labor	2009 Bare Costs Equipment	Total	Total Incl O&P
0010	**SLATE ROOF SHINGLES**										
0100	Buckingham Virginia black, 3/16" - 1/4" thick	G	1 Rots	1.75	4.571	Sq.	475	156		631	805
0200	1/4" thick	G		1.75	4.571		475	156		631	805
0900	Pennsylvania black, Bangor, #1 clear	G		1.75	4.571		490	156		646	820

07 31 Shingles and Shakes

07 31 26 – Slate Shingles

07 31 26.10 Slate Roof Shingles

07 31 26.10 Slate Roof Shingles		Crew	Daily Output	Labor-Hours	Unit	Material	2009 Bare Costs Labor	Equipment	Total	Total Incl O&P
1200	Vermont, unfading, green, mottled green	G	1 Rots	1.75	4.571	Sq.	480	156		636	805
1300	Semi-weathering green & gray	G		1.75	4.571		350	156		506	660
1400	Purple	G		1.75	4.571		425	156		581	745
1500	Black or gray	G		1.75	4.571		460	156		616	785
2500	Slate roof repair, extensive replacement			1	8		590	273		863	1,125
2600	Repair individual pieces, scattered			19	.421	Ea.	5.90	14.35		20.25	32
9000	Minimum labor/equipment charge			3	2.667	Job		91		91	162

07 31 29 – Wood Shingles and Shakes

07 31 29.13 Wood Shingles

0010	WOOD SHINGLES	R061110-30									
0012	16" No. 1 red cedar shingles, 5" exposure, on roof		1 Carp	2.50	3.200	Sq.	330	128		458	575
0015	Pneumatic nailed			3.25	2.462		330	98.50		428.50	525
0200	7-1/2" exposure, on walls			2.05	3.902		221	156		377	500
0205	Pneumatic nailed			2.67	2.996		221	120		341	440
0300	18" No. 1 red cedar perfections, 5-1/2" exposure, on roof			2.75	2.909		330	116		446	550
0305	Pneumatic nailed			3.57	2.241		330	89.50		419.50	510
0600	Resquared, and rebutted, 5-1/2" exposure, on roof			3	2.667		365	107		472	580
0605	Pneumatic nailed			3.90	2.051		365	82		447	540
0900	7-1/2" exposure, on walls			2.45	3.265		269	130		399	510
0905	Pneumatic nailed			3.18	2.516		269	101		370	460
1000	Add to above for fire retardant shingles, 16" long						63			63	69.50
1050	18" long						63			63	69.50
1060	Preformed ridge shingles		1 Carp	400	.020	L.F.	3.50	.80		4.30	5.15
2000	White cedar shingles, 16" long, extras, 5" exposure, on roof			2.40	3.333	Sq.	178	133		311	415
2005	Pneumatic nailed			3.12	2.564		178	102		280	365
2050	5" exposure on walls			2	4		178	160		338	460
2055	Pneumatic nailed			2.60	3.077		178	123		301	400
2100	7-1/2" exposure, on walls			2	4		127	160		287	405
2105	Pneumatic nailed			2.60	3.077		127	123		250	345
2150	"B" grade, 5" exposure on walls			2	4		150	160		310	430
2155	Pneumatic nailed			2.60	3.077		150	123		273	370
2300	For 15# organic felt underlayment on roof, 1 layer, add			64	.125		4.77	4.99		9.76	13.50
2400	2 layers, add			32	.250		9.55	10		19.55	27
2600	For steep roofs (7/12 pitch or greater), add to above							50%			
2700	Panelized systems, No.1 cedar shingles on 5/16" CDX plywood										
2800	On walls, 8' strips, 7" or 14" exposure		2 Carp	700	.023	S.F.	5.55	.91		6.46	7.60
3500	On roofs, 8' strips, 7" or 14" exposure		1 Carp	3	2.667	Sq.	555	107		662	785
3505	Pneumatic nailed			4	2	"	555	80		635	740
9000	Minimum labor/equipment charge			3	2.667	Job		107		107	176

07 31 29.16 Wood Shakes

0010	WOOD SHAKES										
1100	Hand-split red cedar shakes, 1/2" thick x 24" long, 10" exp. on roof		1 Carp	2.50	3.200	Sq.	252	128		380	490
1105	Pneumatic nailed			3.25	2.462		252	98.50		350.50	440
1110	3/4" thick x 24" long, 10" exp. on roof			2.25	3.556		252	142		394	510
1115	Pneumatic nailed			2.92	2.740		252	109		361	460
1200	1/2" thick, 18" long, 8-1/2" exp. on roof			2	4		239	160		399	525
1205	Pneumatic nailed			2.60	3.077		239	123		362	465
1210	3/4" thick x 18" long, 8 1/2" exp. on roof			1.80	4.444		239	178		417	555
1215	Pneumatic nailed			2.34	3.419		239	137		376	490
1255	10" exp. on walls			2	4		231	160		391	520
1260	10" exposure on walls, pneumatic nailed			2.60	3.077		231	123		354	455
1700	Add to above for fire retardant shakes, 24" long						63			63	69.50

07 31 Shingles and Shakes

07 31 29 – Wood Shingles and Shakes

07 31 29.16 Wood Shakes		Crew	Daily Output	Labor-Hours	Unit	Material	2009 Bare Costs Labor	Equipment	Total	Total Incl O&P
1800	18" long				Sq.	63			63	69.50
1810	Ridge shakes	1 Carp	350	.023	L.F.	3.50	.91		4.41	5.35

07 32 Roof Tiles

07 32 13 – Clay Roof Tiles

07 32 13.10 Clay Tiles

			Crew	Daily Output	Labor-Hours	Unit	Material	2009 Bare Costs Labor	Equipment	Total	Total Incl O&P
0010	**CLAY TILES**										
0200	Lanai tile or Classic tile, 158 pc per sq	G	1 Rots	1.65	4.848	Sq.	500	165		665	845
0300	Americana, 158 pc per sq, most colors	G		1.65	4.848		660	165		825	1,025
0350	Green, gray or brown	G		1.65	4.848		625	165		790	980
0400	Blue	G		1.65	4.848		625	165		790	980
0600	Spanish tile, 171 pc per sq, red	G		1.80	4.444		325	152		477	625
0800	Buff, green, gray, brown	G		1.80	4.444		600	152		752	930
0900	Glazed white			1.80	4.444		665	152		817	1,000
1100	Mission tile, 192 pc per sq, machine scored finish, red	G		1.15	6.957		740	237		977	1,250
1700	French tile, 133 pc per sq, smooth finish, red	G		1.35	5.926		675	202		877	1,100
1750	Blue or green	G		1.35	5.926		870	202		1,072	1,325
1800	Norman black 317 pc per sq	G		1	8		1,050	273		1,323	1,625
2200	Williamsburg tile, 158 pc per sq, aged cedar	G		1.35	5.926		730	202		932	1,150
2250	Gray or green	G		1.35	5.926		595	202		797	1,025
3000	For steep roofs (7/12 pitch or greater), add to above							50%			
3010	Clay tile, #15 felt underlayment		1 Rofc	64	.125		4.77	4.28		9.05	12.90
3020	Clay tile, #30 felt underlayment			58	.138		9.55	4.72		14.27	18.95
3040	Clay tile, polyethylene and rubberized asph. underlayment			22	.364		55.50	12.45		67.95	83
9000	Minimum labor/equipment charge		1 Rots	3	2.667	Job		91		91	162

07 32 16 – Concrete Roof Tiles

07 32 16.10 Concrete Tiles

		Crew	Daily Output	Labor-Hours	Unit	Material	2009 Bare Costs Labor	Equipment	Total	Total Incl O&P
0010	**CONCRETE TILES**									
0020	Corrugated, 13" x 16-1/2", 90 per sq, 950 lb per sq									
0050	Earthtone colors, nailed to wood deck	1 Rots	1.35	5.926	Sq.	96.50	202		298.50	465
0150	Blues		1.35	5.926		97.50	202		299.50	465
0200	Greens		1.35	5.926		97.50	202		299.50	465
0250	Premium colors		1.35	5.926		214	202		416	595
0500	Shakes, 13" x 16-1/2", 90 per sq, 950 lb per sq									
0600	All colors, nailed to wood deck	1 Rots	1.50	5.333	Sq.	252	182		434	605
1500	Accessory pieces, ridge & hip, 10" x 16-1/2", 8 lbs. each				Ea.	3.40			3.40	3.74
1700	Rake, 6-1/2" x 16-3/4", 9 lbs. each					3.40			3.40	3.74
1800	Mansard hip, 10" x 16-1/2", 9.2 lbs. each					3.40			3.40	3.74
1900	Hip starter, 10" x 16-1/2", 10.5 lbs. each					10.70			10.70	11.80
2000	3 or 4 way apex, 10" each side, 11.5 lbs. each					12.35			12.35	13.60
9000	Minimum labor/equipment charge	1 Rots	3	2.667	Job		91		91	162

07 32 19 – Metal Roof Tiles

07 32 19.10 Metal Roof Tiles

		Crew	Daily Output	Labor-Hours	Unit	Material	2009 Bare Costs Labor	Equipment	Total	Total Incl O&P
0010	**METAL ROOF TILES**									
0020	Accessories included, .032" thick aluminum, mission tile	1 Carp	2.50	3.200	Sq.	815	128		943	1,100
0200	Spanish tiles		3	2.667	"	560	107		667	790
9000	Minimum labor/equipment charge		3	2.667	Job		107		107	176

07 33 63 – Vegetated Roofing

07 33 63.10 Green Roof Systems		Crew	Daily Output	Labor-Hours	Unit	Material	2009 Bare Costs Labor	Equipment	Total	Total Incl O&P
0010	**GREEN ROOF SYSTEMS**									
0020	Soil mixture for green roof 30% sand, 55% gravel, 15% soil									
0100	Hoist and spread soil mixture 4 inch depth up to five stories tall roof [G]	B-13B	4000	.014	S.F.	.25	.48	.28	1.01	1.37
0150	6 inch depth [G]		2667	.021		.38	.72	.42	1.52	2.05
0200	8 inch depth [G]		2000	.028		.50	.96	.56	2.02	2.74
0250	10 inch depth [G]		1600	.035		.63	1.20	.70	2.53	3.40
0300	12 inch depth [G]		1335	.042		.76	1.43	.84	3.03	4.09
0350	Mobilization 55 ton crane to site [G]	1 Eqhv	3.60	2.222	Ea.		94.50		94.50	149
0355	Hoisting cost to five stories per day (Avg. 28 picks per day) [G]	B-13B	1	56	Day		1,925	1,125	3,050	4,350
0360	Mobilization or demobilization, 100 ton crane to site driver & escort [G]	A-3E	2.50	6.400	Ea.		238	51.50	289.50	435
0365	Hoisting cost six to ten stories per day (Avg. 21 picks per day) [G]	B-13C	1	56	Day		1,925	1,600	3,525	4,875
0370	Hoist and spread soil mixture 4 inch depth six to ten stories tall roof [G]		4000	.014	S.F.	.25	.48	.40	1.13	1.50
0375	6 inch depth [G]		2667	.021		.38	.72	.60	1.70	2.25
0380	8 inch depth [G]		2000	.028		.50	.96	.81	2.27	3.01
0385	10 inch depth [G]		1600	.035		.63	1.20	1.01	2.84	3.74
0390	12 inch depth [G]		1335	.042		.76	1.43	1.21	3.40	4.49
0400	Green roof edging treated lumber 4" x4" no hoisting included [G]	2 Carp	400	.040	L.F.	1.52	1.60		3.12	4.31
0410	4" x 6" [G]		400	.040		2.24	1.60		3.84	5.10
0420	4" x 8" [G]		360	.044		3.21	1.78		4.99	6.45
0430	4" x 6" double stacked [G]		300	.053		4.47	2.13		6.60	8.45
0500	Green roof edging redwood lumber 4" x4" no hoisting included [G]		400	.040		6.45	1.60		8.05	9.75
0510	4" x 6" [G]		400	.040		11.20	1.60		12.80	14.95
0520	4" x 8" [G]		360	.044		20.50	1.78		22.28	25.50
0530	4" x 6" double stacked [G]		300	.053		22.50	2.13		24.63	28
0600	Planting sedum, light soil, potted, 2-1/4" diameter, two per SF [G]	1 Clab	420	.019	S.F.	7	.60		7.60	8.70
0610	one per SF [G]	"	840	.010		3.50	.30		3.80	4.35
0630	Planting sedum mat per SF including shipping (4000 SF Minimum) [G]	4 Clab	4000	.008		8.40	.25		8.65	9.65
0640	Installation sedum mat system (no soil required) per SF (4000 SF minimum) [G]	"	4000	.008		11.70	.25		11.95	13.25
0645	Note: pricing of sedum mats shipped in full truck loads (4000-5000 SF)									

07 41 Roof Panels

07 41 13 – Metal Roof Panels

07 41 13.10 Aluminum Roof Panels

		Crew	Daily Output	Labor-Hours	Unit	Material	2009 Bare Costs Labor	Equipment	Total	Total Incl O&P
0010	**ALUMINUM ROOF PANELS**									
0020	Corrugated or ribbed, .0155" thick, natural	G-3	1200	.027	S.F.	.87	1.05		1.92	2.66
0300	Painted		1200	.027		1.27	1.05		2.32	3.10
0400	Corrugated, .018" thick, on steel frame, natural finish		1200	.027		1.15	1.05		2.20	2.97
0600	Painted		1200	.027		1.41	1.05		2.46	3.25
0700	Corrugated, on steel frame, natural, .024" thick		1200	.027		1.64	1.05		2.69	3.50
0800	Painted		1200	.027		1.99	1.05		3.04	3.89
0900	.032" thick, natural		1200	.027		2.11	1.05		3.16	4.02
1200	Painted		1200	.027		2.70	1.05		3.75	4.67
9000	Minimum labor/equipment charge	1 Rofc	3	2.667	Job		91.50		91.50	163

07 41 13.20 Steel Roofing Panels

		Crew	Daily Output	Labor-Hours	Unit	Material	2009 Bare Costs Labor	Equipment	Total	Total Incl O&P
0010	**STEEL ROOFING PANELS**									
0012	Corrugated or ribbed, on steel framing, 30 ga galv	G-3	1100	.029	S.F.	1.68	1.15		2.83	3.70
0100	28 ga		1050	.030		1.79	1.20		2.99	3.91
0300	26 ga		1000	.032		1.85	1.26		3.11	4.07
0400	24 ga		950	.034		2.91	1.33		4.24	5.35
0600	Colored, 28 ga		1050	.030		1.68	1.20		2.88	3.79
0700	26 ga		1000	.032		1.99	1.26		3.25	4.22

07 41 Roof Panels

07 41 13 – Metal Roof Panels

07 41 13.20 Steel Roofing Panels

		Crew	Daily Output	Labor-Hours	Unit	Material	2009 Bare Costs Labor	Equipment	Total	Total Incl O&P
0710	Flat profile, 1-3/4" standing seams, 10" wide, standard finish, 26 ga	G-3	1000	.032	S.F.	3.89	1.26		5.15	6.30
0715	24 ga		950	.034		4.51	1.33		5.84	7.10
0720	22 ga		900	.036		5.55	1.40		6.95	8.40
0725	Zinc aluminum alloy finish, 26 ga		1000	.032		3.05	1.26		4.31	5.40
0730	24 ga		950	.034		3.64	1.33		4.97	6.15
0735	22 ga		900	.036		4.17	1.40		5.57	6.85
0740	12" wide, standard finish, 26 ga		1000	.032		3.88	1.26		5.14	6.30
0745	24 ga		950	.034		5.10	1.33		6.43	7.75
0750	Zinc aluminum alloy finish, 26 ga		1000	.032		4.41	1.26		5.67	6.90
0755	24 ga		950	.034		3.63	1.33		4.96	6.15
0840	Flat profile, 1" x 3/8" batten, 12" wide, standard finish, 26 ga		1000	.032		3.42	1.26		4.68	5.80
0845	24 ga		950	.034		4.01	1.33		5.34	6.55
0850	22 ga		900	.036		4.80	1.40		6.20	7.55
0855	Zinc aluminum alloy finish, 26 ga		1000	.032		3.28	1.26		4.54	5.65
0860	24 ga		950	.034		3.66	1.33		4.99	6.15
0865	22 ga		900	.036		4.23	1.40		5.63	6.90
0870	16-1/2" wide, standard finish, 24 ga		950	.034		3.95	1.33		5.28	6.50
0875	22 ga		900	.036		4.42	1.40		5.82	7.10
0880	Zinc aluminum alloy finish, 24 ga		950	.034		3.45	1.33		4.78	5.95
0885	22 ga		900	.036		3.85	1.40		5.25	6.50
0890	Flat profile, 2" x 2" batten, 12" wide, standard finish, 26 ga		1000	.032		3.93	1.26		5.19	6.35
0895	24 ga		950	.034		4.69	1.33		6.02	7.30
0900	22 ga		900	.036		5.75	1.40		7.15	8.60
0905	Zinc aluminum alloy finish, 26 ga		1000	.032		3.66	1.26		4.92	6.05
0910	24 ga		950	.034		4.18	1.33		5.51	6.75
0915	22 ga		900	.036		4.87	1.40		6.27	7.60
0920	16-1/2" wide, standard finish, 24 ga		950	.034		4.32	1.33		5.65	6.90
0925	22 ga		900	.036		5.05	1.40		6.45	7.80
0930	Zinc aluminum alloy finish, 24 ga		950	.034		3.91	1.33		5.24	6.45
0935	22 ga		900	.036		4.45	1.40		5.85	7.15
9000	Minimum labor/equipment charge	1 Rofc	2	4	Job		137		137	245

07 41 33 – Plastic Roof Panels

07 41 33.10 Fiberglass Panels

		Crew	Daily Output	Labor-Hours	Unit	Material	2009 Bare Costs Labor	Equipment	Total	Total Incl O&P
0010	**FIBERGLASS PANELS**									
0012	Corrugated panels, roofing, 8 oz per S.F.	G-3	1000	.032	S.F.	1.58	1.26		2.84	3.77
0100	12 oz per S.F.		1000	.032		3.43	1.26		4.69	5.80
0300	Corrugated siding, 6 oz per S.F.		880	.036		1.58	1.43		3.01	4.05
0400	8 oz per S.F.		880	.036		1.58	1.43		3.01	4.05
0500	Fire retardant		880	.036		3.21	1.43		4.64	5.85
0600	12 oz. siding, textured		880	.036		3.32	1.43		4.75	5.95
0700	Fire retardant		880	.036		4.39	1.43		5.82	7.15
0900	Flat panels, 6 oz per S.F., clear or colors		880	.036		1.83	1.43		3.26	4.32
1100	Fire retardant, class A		880	.036		3.18	1.43		4.61	5.80
1300	8 oz per S.F., clear or colors		880	.036		2.38	1.43		3.81	4.93
9000	Minimum labor/equipment charge	1 Rofc	2	4	Job		137		137	245

07 42 Wall Panels

07 42 13 – Metal Wall Panels

	07 42 13.10 Mansard Panels	Crew	Daily Output	Labor-Hours	Unit	Material	2009 Bare Costs Labor	Equipment	Total	Total Incl O&P
0010	**MANSARD PANELS**									
0600	Aluminum, stock units, straight surfaces	1 Shee	115	.070	S.F.	3.04	3.28		6.32	8.55
0700	Concave or convex surfaces		75	.107	"	3.34	5.05		8.39	11.65
0800	For framing, to 5' high, add		115	.070	L.F.	3.34	3.28		6.62	8.85
0900	Soffits, to 1' wide		125	.064	S.F.	1.66	3.02		4.68	6.65
9000	Minimum labor/equipment charge		2.50	3.200	Job		151		151	240

07 42 13.20 Aluminum Siding Panels

		Crew	Daily Output	Labor-Hours	Unit	Material	2009 Bare Costs Labor	Equipment	Total	Total Incl O&P
0010	**ALUMINUM SIDING PANELS**									
0012	Corrugated, on steel framing, .019 thick, natural finish	G-3	775	.041	S.F.	1.35	1.63		2.98	4.12
0100	Painted		775	.041		1.45	1.63		3.08	4.23
0400	Farm type, .021" thick on steel frame, natural		775	.041		1.36	1.63		2.99	4.13
0600	Painted		775	.041		1.45	1.63		3.08	4.23
0700	Industrial type, corrugated, on steel, .024" thick, mill		775	.041		1.89	1.63		3.52	4.71
0900	Painted		775	.041		2.04	1.63		3.67	4.87
1000	.032" thick, mill		775	.041		2.17	1.63		3.80	5
1200	Painted		775	.041		2.65	1.63		4.28	5.55
1300	V-Beam, on steel frame, .032" thick, mill		775	.041		2.45	1.63		4.08	5.35
1500	Painted		775	.041		2.81	1.63		4.44	5.70
1600	.040" thick, mill		775	.041		2.98	1.63		4.61	5.90
1800	Painted		775	.041		3.49	1.63		5.12	6.45
3800	Horizontal, colored clapboard, 8" wide, plain	2 Carp	515	.031		1.57	1.24		2.81	3.78
3810	Insulated		515	.031		1.54	1.24		2.78	3.74
3830	8" embossed, painted		515	.031		1.71	1.24		2.95	3.93
3840	Insulated		515	.031		1.90	1.24		3.14	4.14
3860	12" painted, smooth		600	.027		1.59	1.07		2.66	3.51
3870	Insulated		600	.027		1.84	1.07		2.91	3.78
3890	12" embossed, painted		600	.027		1.77	1.07		2.84	3.71
3900	Insulated		515	.031		1.98	1.24		3.22	4.23
4000	Vertical board & batten, colored, non-insulated		515	.031		1.53	1.24		2.77	3.73
4200	For simulated wood design, add					.10			.10	.11
4300	Corners for above, outside	2 Carp	515	.031	V.L.F.	2.36	1.24		3.60	4.65
4500	Inside corners	"	515	.031	"	1.39	1.24		2.63	3.58
9000	Minimum labor/equipment charge	1 Carp	3	2.667	Job		107		107	176

07 42 13.30 Steel Siding

		Crew	Daily Output	Labor-Hours	Unit	Material	2009 Bare Costs Labor	Equipment	Total	Total Incl O&P
0010	**STEEL SIDING**									
0020	Beveled, vinyl coated, 8" wide	1 Carp	265	.030	S.F.	1.81	1.21		3.02	3.98
0050	10" wide	"	275	.029		1.94	1.16		3.10	4.05
0080	Galv, corrugated or ribbed, on steel frame, 30 gauge	G-3	800	.040		1.21	1.58		2.79	3.87
0100	28 gauge		795	.040		1.27	1.59		2.86	3.96
0300	26 gauge		790	.041		1.78	1.60		3.38	4.54
0400	24 gauge		785	.041		1.79	1.61		3.40	4.56
0600	22 gauge		770	.042		2.06	1.64		3.70	4.91
0700	Colored, corrugated/ribbed, on steel frame, 10 yr fnsh, 28 ga.		800	.040		1.88	1.58		3.46	4.61
0900	26 gauge		795	.040		1.96	1.59		3.55	4.72
1000	24 gauge		790	.041		2.29	1.60		3.89	5.10
1020	20 gauge		785	.041		2.90	1.61		4.51	5.80
9000	Minimum labor/equipment charge	1 Carp	3	2.667	Job		107		107	176

07 46 Siding

07 46 23 – Wood Siding

07 46 23.10 Wood Board Siding

		Crew	Daily Output	Labor-Hours	Unit	Material	2009 Bare Costs Labor	2009 Bare Costs Equipment	Total	Total Incl O&P
0010	**WOOD BOARD SIDING**									
3200	Wood, cedar bevel, A grade, 1/2" x 6"	1 Carp	295	.027	S.F.	4.62	1.08		5.70	6.90
3300	1/2" x 8"		330	.024		4.97	.97		5.94	7.05
3500	3/4" x 10", clear grade		375	.021		6.70	.85		7.55	8.75
3600	"B" grade		375	.021		6.25	.85		7.10	8.30
3800	Cedar, rough sawn, 1" x 4", A grade, natural		220	.036		3.77	1.45		5.22	6.55
3900	Stained		220	.036		4.17	1.45		5.62	7
4100	1" x 12", board & batten, #3 & Btr., natural		420	.019		3.94	.76		4.70	5.60
4200	Stained		420	.019		4.20	.76		4.96	5.90
4400	1" x 8" channel siding, #3 & Btr., natural		330	.024		2.34	.97		3.31	4.17
4500	Stained		330	.024		2.50	.97		3.47	4.35
4700	Redwood, clear, beveled, vertical grain, 1/2" x 4"		220	.036		3.10	1.45		4.55	5.80
4750	1/2" x 6"		295	.027		3.10	1.08		4.18	5.20
4800	1/2" x 8"		330	.024		3.24	.97		4.21	5.15
5000	3/4" x 10"		375	.021		3.32	.85		4.17	5.05
5200	Channel siding, 1" x 10", B grade		375	.021		3.32	.85		4.17	5.05
5250	Redwood, T&G boards, B grade, 1" x 4"		220	.036		3.10	1.45		4.55	5.80
5270	1" x 8"		330	.024		3.24	.97		4.21	5.15
5400	White pine, rough sawn, 1" x 8", natural		330	.024		2.20	.97		3.17	4.02
5500	Stained		330	.024		2.60	.97		3.57	4.46
9000	Minimum labor/equipment charge		2	4	Job		160		160	264

07 46 29 – Plywood Siding

07 46 29.10 Plywood Siding Options

		Crew	Daily Output	Labor-Hours	Unit	Material	2009 Bare Costs Labor	2009 Bare Costs Equipment	Total	Total Incl O&P
0010	**PLYWOOD SIDING OPTIONS**									
0900	Plywood, medium density overlaid, 3/8" thick	2 Carp	750	.021	S.F.	.96	.85		1.81	2.47
1000	1/2" thick		700	.023		1.27	.91		2.18	2.91
1100	3/4" thick		650	.025		1.31	.98		2.29	3.06
1600	Texture 1-11, cedar, 5/8" thick, natural		675	.024		2.45	.95		3.40	4.26
1700	Factory stained		675	.024		1.96	.95		2.91	3.72
1900	Texture 1-11, fir, 5/8" thick, natural		675	.024		1.13	.95		2.08	2.80
2000	Factory stained		675	.024		1.73	.95		2.68	3.46
2050	Texture 1-11, S.Y.P., 5/8" thick, natural		675	.024		1.21	.95		2.16	2.89
2100	Factory stained		675	.024		1.25	.95		2.20	2.94
2200	Rough sawn cedar, 3/8" thick, natural		675	.024		1.21	.95		2.16	2.89
2300	Factory stained		675	.024		.97	.95		1.92	2.63
2500	Rough sawn fir, 3/8" thick, natural		675	.024		.77	.95		1.72	2.41
2600	Factory stained		675	.024		.97	.95		1.92	2.63
2800	Redwood, textured siding, 5/8" thick		675	.024		1.94	.95		2.89	3.69
3000	Polyvinyl chloride coated, 3/8" thick		750	.021		1.12	.85		1.97	2.64
9000	Minimum labor/equipment charge	1 Carp	2	4	Job		160		160	264

07 46 33 – Plastic Siding

07 46 33.10 Vinyl Siding

		Crew	Daily Output	Labor-Hours	Unit	Material	2009 Bare Costs Labor	2009 Bare Costs Equipment	Total	Total Incl O&P
0010	**VINYL SIDING**									
3995	Clapboard profile, woodgrain texture, .048 thick, double 4	2 Carp	495	.032	S.F.	.92	1.29		2.21	3.14
4000	Double 5		550	.029		.92	1.16		2.08	2.93
4005	Single 8		495	.032		.96	1.29		2.25	3.19
4010	Single 10		550	.029		1.15	1.16		2.31	3.19
4015	.044 thick, double 4		495	.032		.84	1.29		2.13	3.05
4020	Double 5		550	.029		.84	1.16		2	2.84
4025	.042 thick, double 4		495	.032		.84	1.29		2.13	3.05
4030	Double 5		550	.029		.84	1.16		2	2.84
4035	Cross sawn texture, .040 thick, double 4		495	.032		.66	1.29		1.95	2.86

07 46 Siding

07 46 33 – Plastic Siding

07 46 33.10 Vinyl Siding

		Crew	Daily Output	Labor-Hours	Unit	Material	2009 Bare Costs Labor	2009 Bare Costs Equipment	Total	Total Incl O&P
4040	Double 5	2 Carp	550	.029	S.F.	.66	1.16		1.82	2.65
4045	Smooth texture, .042 thick, double 4		495	.032		.75	1.29		2.04	2.96
4050	Double 5		550	.029		.75	1.16		1.91	2.75
4055	Single 8		495	.032		.96	1.29		2.25	3.19
4060	Cedar texture, .044 thick, double 4		495	.032		.92	1.29		2.21	3.14
4065	Double 6		600	.027		.92	1.07		1.99	2.77
4070	Dutch lap profile, woodgrain texture, .048 thick, double 5		550	.029		.92	1.16		2.08	2.93
4075	.044 thick, double 4.5		525	.030		.92	1.22		2.14	3.02
4080	.042 thick, double 4.5		525	.030		.75	1.22		1.97	2.84
4085	.040 thick, double 4.5		525	.030		.66	1.22		1.88	2.74
4090	Shingle profile, random grooves, double 7		400	.040		2.57	1.60		4.17	5.45
4095	Triple 5		400	.040		2.57	1.60		4.17	5.45
4100	Shake profile, 10" wide		400	.040		2.19	1.60		3.79	5.05
4105	Vertical pattern, .046 thick, double 5		550	.029		1.22	1.16		2.38	3.26
4110	.044 thick, triple 3		550	.029		1.37	1.16		2.53	3.43
4115	.040 thick, triple 4		550	.029		1.22	1.16		2.38	3.26
4120	.040 thick, triple 2.66		550	.029		1.47	1.16		2.63	3.54
4125	Insulation, fan folded extruded polystyrene, 1/4"		2000	.008		.25	.32		.57	.81
4130	3/8"		2000	.008	▼	.27	.32		.59	.83
4135	Accessories, J channel, 5/8" pocket		700	.023	L.F.	.40	.91		1.31	1.94
4140	3/4" pocket		695	.023		.44	.92		1.36	2
4145	1-1/4" pocket		680	.024		.56	.94		1.50	2.16
4150	Flexible, 3/4" pocket		600	.027		1.57	1.07		2.64	3.48
4155	Under sill finish trim		500	.032		.39	1.28		1.67	2.53
4160	Vinyl starter strip		700	.023		.31	.91		1.22	1.85
4165	Aluminum starter strip		700	.023		.22	.91		1.13	1.75
4170	Window casing, 2-1/2" wide, 3/4" pocket		510	.031		.99	1.25		2.24	3.15
4175	Outside corner, woodgrain finish, 4" face, 3/4" pocket		700	.023		1.65	.91		2.56	3.33
4180	5/8" pocket		700	.023		1.67	.91		2.58	3.35
4185	Smooth finish, 4" face, 3/4" pocket		700	.023		1.73	.91		2.64	3.41
4190	7/8" pocket		690	.023		1.78	.93		2.71	3.49
4195	1-1/4" pocket		700	.023		1.16	.91		2.07	2.79
4200	Soffit and fascia, 1' overhang, solid		120	.133		3.31	5.35		8.66	12.45
4205	Vented		120	.133		3.31	5.35		8.66	12.45
4210	2' overhang, solid		110	.145		4.52	5.80		10.32	14.55
4215	Vented	▼	110	.145	▼	4.52	5.80		10.32	14.55
4220	Colors for siding and soffits, add				S.F.	.16			.16	.18
4225	Colors for accessories and trim, add				L.F.	.32			.32	.35
9000	Minimum labor/equipment charge	1 Carp	3	2.667	Job		107		107	176

07 46 46 – Mineral-Fiber Cement Siding

07 46 46.10 Fiber Cement Siding

		Crew	Daily Output	Labor-Hours	Unit	Material	2009 Bare Costs Labor	2009 Bare Costs Equipment	Total	Total Incl O&P
0010	**FIBER CEMENT SIDING**									
0020	Lap siding, 5/16" thick, 6" wide, 4-3/4" exposure, smooth texture	2 Carp	415	.039	S.F.	1.34	1.54		2.88	4.01
0025	Woodgrain texture		415	.039		1.34	1.54		2.88	4.01
0030	7-1/2" wide, 6-1/4" exposure, smooth texture		425	.038		1.36	1.50		2.86	3.98
0035	Woodgrain texture		425	.038		1.36	1.50		2.86	3.98
0040	8" wide, 6-3/4" exposure, smooth texture		425	.038		1.51	1.50		3.01	4.14
0045	Roughsawn texture		425	.038		1.51	1.50		3.01	4.14
0050	9-1/2" wide, 8-1/4" exposure, smooth texture		440	.036		1.45	1.45		2.90	4
0055	Woodgrain texture		440	.036		1.45	1.45		2.90	4
0060	12" wide, 10-3/8" exposure, smooth texture		455	.035		1.34	1.40		2.74	3.79
0065	Woodgrain texture	▼	455	.035	▼	1.34	1.40		2.74	3.79

07 46 Siding

07 46 46 – Mineral-Fiber Cement Siding

07 46 46.10 Fiber Cement Siding	Crew	Daily Output	Labor-Hours	Unit	Material	2009 Bare Costs Labor	Equipment	Total	Total Incl O&P
0070	Panel siding, 5/16" thick, smooth texture	2 Carp	750	.021	S.F.	1.12	.85		1.97	2.65
0075	Stucco texture		750	.021		1.12	.85		1.97	2.65
0080	Grooved woodgrain texture		750	.021		1.12	.85		1.97	2.65
0085	V - grooved woodgrain texture		750	.021		1.12	.85		1.97	2.65
0090	Wood starter strip		400	.040	L.F.	.52	1.60		2.12	3.21

07 46 73 – Soffit

07 46 73.10 Soffit Options

		Crew	Daily Output	Labor-Hours	Unit	Material	Labor	Equipment	Total	Total Incl O&P
0010	**SOFFIT OPTIONS**									
0012	Aluminum, residential, .020" thick	1 Carp	210	.038	S.F.	1.62	1.52		3.14	4.29
0100	Baked enamel on steel, 16 or 18 gauge		105	.076		5.35	3.04		8.39	10.90
0300	Polyvinyl chloride, white, solid		230	.035		1.09	1.39		2.48	3.49
0400	Perforated		230	.035		1.09	1.39		2.48	3.49
0500	For colors, add					.13			.13	.14
9000	Minimum labor/equipment charge	1 Carp	3	2.667	Job		107		107	176

07 51 Built-Up Bituminous Roofing

07 51 13 – Built-Up Asphalt Roofing

07 51 13.10 Built-Up Roofing Components

		Crew	Daily Output	Labor-Hours	Unit	Material	Labor	Equipment	Total	Total Incl O&P
0010	**BUILT-UP ROOFING COMPONENTS**									
0012	Asphalt saturated felt, #30, 2 square per roll	1 Rofc	58	.138	Sq.	9.55	4.72		14.27	18.95
0200	#15, 4 sq per roll, plain or perforated, not mopped		58	.138		4.77	4.72		9.49	13.70
0300	Roll roofing, smooth, #65		15	.533		9.30	18.25		27.55	42.50
0500	#90		15	.533		32	18.25		50.25	67.50
0520	Mineralized		15	.533		33.50	18.25		51.75	69
0540	D.C. (Double coverage), 19" selvage edge		10	.800		49.50	27.50		77	104
0580	Adhesive (lap cement)				Gal.	4.50			4.50	4.95
0600	Steep, flat or dead level asphalt, 10 ton lots, bulk				Ton	445			445	490
0800	Packaged				"	755			755	835
9000	Minimum labor/equipment charge	1 Rofc	4	2	Job		68.50		68.50	122

07 51 13.13 Cold-Applied Built-Up Asphalt Roofing

		Crew	Daily Output	Labor-Hours	Unit	Material	Labor	Equipment	Total	Total Incl O&P
0010	**COLD-APPLIED BUILT-UP ASPHALT ROOFING**									
0020	3 ply system, installation only (components listed below)	G-5	50	.800	Sq.		25	3.46	28.46	48.50
0100	Spunbond poly. fabric, 1.35 oz/SY, 36"W, 10.8 Sq/roll				Ea.	177			177	195
0200	49" wide, 14.6 Sq./roll					244			244	268
0300	2.10 oz./S.Y., 36" wide, 10.8 Sq./roll					266			266	293
0400	49" wide, 14.6 Sq./roll					360			360	395
0500	Base & finish coat, 3 gal./Sq., 5 gal./can				Gal.	4.70			4.70	5.15
0600	Coating, ceramic granules, 1/2 Sq./bag				Ea.	16.60			16.60	18.25
0700	Aluminum, 2 gal./Sq.				Gal.	13.35			13.35	14.70
0800	Emulsion, fibered or non-fibered, 4 gal./Sq.				"	6.15			6.15	6.75

07 51 13.20 Built-Up Roofing Systems

		Crew	Daily Output	Labor-Hours	Unit	Material	Labor	Equipment	Total	Total Incl O&P
0010	**BUILT-UP ROOFING SYSTEMS**									
0120	Asphalt flood coat with gravel/slag surfacing, not including									
0140	Insulation, flashing or wood nailers									
0200	Asphalt base sheet, 3 plies #15 asphalt felt, mopped	G-1	22	2.545	Sq.	84	81.50	20	185.50	260
0350	On nailable decks		21	2.667		88.50	85.50	21	195	273
0500	4 plies #15 asphalt felt, mopped		20	2.800		117	89.50	22	228.50	310
0550	On nailable decks		19	2.947		104	94.50	23	221.50	310
0700	Coated glass base sheet, 2 plies glass (type IV), mopped		22	2.545		85	81.50	20	186.50	261
0850	3 plies glass, mopped		20	2.800		102	89.50	22	213.50	296

07 51 13 – Built-Up Asphalt Roofing

07 51 13.20 Built-Up Roofing Systems

		Crew	Daily Output	Labor-Hours	Unit	Material	2009 Bare Costs Labor	Equipment	Total	Total Incl O&P
0950	On nailable decks	G-1	19	2.947	Sq.	95	94.50	23	212.50	299
1100	4 plies glass fiber felt (type IV), mopped		20	2.800		125	89.50	22	236.50	320
1150	On nailable decks		19	2.947		112	94.50	23	229.50	315
1200	Coated & saturated base sheet, 3 plies #15 asph. felt, mopped		20	2.800		93.50	89.50	22	205	287
1250	On nailable decks		19	2.947		87	94.50	23	204.50	290
1300	4 plies #15 asphalt felt, mopped		22	2.545		109	81.50	20	210.50	287
2000	Asphalt flood coat, smooth surface									
2200	Asphalt base sheet & 3 plies #15 asphalt felt, mopped	G-1	24	2.333	Sq.	89.50	74.50	18.35	182.35	252
2400	On nailable decks		23	2.435		83	78	19.15	180.15	252
2600	4 plies #15 asphalt felt, mopped		24	2.333		105	74.50	18.35	197.85	268
2700	On nailable decks		23	2.435		98.50	78	19.15	195.65	268
2900	Coated glass fiber base sheet, mopped, and 2 plies of									
2910	glass fiber felt (type IV)	G-1	25	2.240	Sq.	80	71.50	17.60	169.10	235
3100	On nailable decks		24	2.333		75	74.50	18.35	167.85	236
3200	3 plies, mopped		23	2.435		96.50	78	19.15	193.65	266
3300	On nailable decks		22	2.545		90	81.50	20	191.50	266
3800	4 plies glass fiber felt (type IV), mopped		23	2.435		113	78	19.15	210.15	284
3900	On nailable decks		22	2.545		106	81.50	20	207.50	284
4000	Coated & saturated base sheet, 3 plies #15 asph. felt, mopped		24	2.333		88.50	74.50	18.35	181.35	250
4200	On nailable decks		23	2.435		82	78	19.15	179.15	250
4300	4 plies #15 organic felt, mopped		22	2.545		104	81.50	20	205.50	281
4500	Coal tar pitch with gravel/slag surfacing									
4600	4 plies #15 tarred felt, mopped	G-1	21	2.667	Sq.	156	85.50	21	262.50	345
4800	3 plies glass fiber felt (type IV), mopped	"	19	2.947	"	127	94.50	23	244.50	335
5000	Coated glass fiber base sheet, and 2 plies of									
5010	glass fiber felt, (type IV), mopped	G-1	19	2.947	Sq.	127	94.50	23	244.50	335
5300	On nailable decks		18	3.111		113	99.50	24.50	237	330
5600	4 plies glass fiber felt (type IV), mopped		21	2.667		177	85.50	21	283.50	370
5800	On nailable decks		20	2.800		163	89.50	22	274.50	365

07 51 13.30 Cants

		Crew	Daily Output	Labor-Hours	Unit	Material	Labor	Equipment	Total	Total Incl O&P
0010	**CANTS**									
0012	Lumber, treated, 4" x 4" cut diagonally	1 Rofc	325	.025	L.F.	1.52	.84		2.36	3.18
0100	Foamglass		325	.025		2.23	.84		3.07	3.96
0300	Mineral or fiber, trapezoidal, 1"x 4" x 48"		325	.025		.19	.84		1.03	1.72
0400	1-1/2" x 5-5/8" x 48"		325	.025		.31	.84		1.15	1.85
9000	Minimum labor/equipment charge		4	2	Job		68.50		68.50	122

07 51 13.40 Felts

		Crew	Daily Output	Labor-Hours	Unit	Material	Labor	Equipment	Total	Total Incl O&P
0010	**FELTS**									
0012	Glass fibered roofing felt, #15, not mopped	1 Rofc	58	.138	Sq.	6	4.72		10.72	15.05
0300	Base sheet, #45, channel vented		58	.138		25.50	4.72		30.22	36.50
0400	#50, coated		58	.138		13.80	4.72		18.52	23.50
0500	Cap, mineral surfaced		58	.138		25.50	4.72		30.22	36.50
0600	Flashing membrane, #65		16	.500		39	17.15		56.15	73.50
0800	Coal tar fibered, #15, no mopping		58	.138		11.60	4.72		16.32	21
0900	Asphalt felt, #15, 4 sq per roll, no mopping		58	.138		4.77	4.72		9.49	13.70
1100	#30, 2 sq per roll		58	.138		9.55	4.72		14.27	18.95
1200	Double coated, #33		58	.138		9.50	4.72		14.22	18.90
1400	#40, base sheet		58	.138		10.80	4.72		15.52	20.50
1450	Coated and saturated		58	.138		9.45	4.72		14.17	18.85
1500	Tarred felt, organic, #15, 4 sq rolls		58	.138		13.90	4.72		18.62	24
1550	#30, 2 sq roll		58	.138		24.50	4.72		29.22	35.50
1700	Add for mopping above felts, per ply, asphalt, 24 lb per sq	G-1	192	.292		9.10	9.35	2.29	20.74	29

07 51 Built-Up Bituminous Roofing

07 51 13 – Built-Up Asphalt Roofing

07 51 13.40 Felts

		Crew	Daily Output	Labor-Hours	Unit	Material	2009 Bare Costs Labor	Equipment	Total	Total Incl O&P
1800	Coal tar mopping, 30 lb per sq	G-1	186	.301	Sq.	12.60	9.65	2.37	24.62	33.50
1900	Flood coat, with asphalt, 60 lb per sq		60	.933		22.50	30	7.35	59.85	86.50
2000	With coal tar, 75 lb per sq	↓	56	1	↓	31.50	32	7.85	71.35	100
9000	Minimum labor/equipment charge	1 Rofc	4	2	Job		68.50		68.50	122

07 51 13.50 Walkways for Built-Up Roofs

		Crew	Daily Output	Labor-Hours	Unit	Material	2009 Bare Costs Labor	Equipment	Total	Total Incl O&P
0010	**WALKWAYS FOR BUILT-UP ROOFS**									
0020	Asphalt impregnated, 3' x 6' x 1/2" thick	1 Rofc	400	.020	S.F.	1.55	.69		2.24	2.93
0100	3' x 3' x 3/4" thick		400	.020	"	2.93	.69		3.62	4.44
0600	100% recycled rubber, 3' x 4' x 3/8" [G]		400	.020	L.F.	4.35	.69		5.04	6
0610	3' x 4' x 1/2" [G]		400	.020		4.65	.69		5.34	6.30
0620	3' x 4' x 3/4" [G]		400	.020	↓	5.80	.69		6.49	7.60
9000	Minimum labor/equipment charge	↓	2.75	2.909	Job		99.50		99.50	178

07 52 Modified Bituminous Membrane Roofing

07 52 13 – Atactic-Polypropylene-Modified Bituminous Membrane Roofing

07 52 13.10 APP Modified Bituminous Membrane

		Crew	Daily Output	Labor-Hours	Unit	Material	2009 Bare Costs Labor	Equipment	Total	Total Incl O&P
0010	**APP MODIFIED BITUMINOUS MEMBRANE** R075213-30									
0020	Base sheet, #15 glass fiber felt, nailed to deck	1 Rofc	58	.138	Sq.	6.75	4.72		11.47	15.90
0030	Spot mopped to deck	G-1	295	.190		10.55	6.05	1.49	18.09	24
0040	Fully mopped to deck	"	192	.292		15.10	9.35	2.29	26.74	36
0050	#15 organic felt, nailed to deck	1 Rofc	58	.138		5.55	4.72	∘	10.27	14.55
0060	Spot mopped to deck	G-1	295	.190		9.30	6.05	1.49	16.84	22.50
0070	Fully mopped to deck	"	192	.292	↓	13.85	9.35	2.29	25.49	34.50
2100	APP mod., smooth surf. cap sheet, poly. reinf., torched, 160 mils	G-5	2100	.019	S.F.	.48	.59	.08	1.15	1.68
2150	170 mils		2100	.019		.55	.59	.08	1.22	1.76
2200	Granule surface cap sheet, poly. reinf., torched, 180 mils		2000	.020		.59	.62	.09	1.30	1.86
2250	Smooth surface flashing, torched, 160 mils		1260	.032		.48	.99	.14	1.61	2.44
2300	170 mils		1260	.032		.55	.99	.14	1.68	2.52
2350	Granule surface flashing, torched, 180 mils	↓	1260	.032		.59	.99	.14	1.72	2.56
2400	Fibrated aluminum coating	1 Rofc	3800	.002	↓	.12	.07		.19	.26

07 52 16 – Styrene-Butadiene-Styrene Modified Bituminous Membrane Roofing

07 52 16.10 SBS Modified Bituminous Membrane

		Crew	Daily Output	Labor-Hours	Unit	Material	2009 Bare Costs Labor	Equipment	Total	Total Incl O&P
0010	**SBS MODIFIED BITUMINOUS MEMBRANE**									
0080	SBS modified, granule surf cap sheet, polyester rein., mopped									
1500	Glass fiber reinforced, mopped, 160 mils	G-1	2000	.028	S.F.	.49	.90	.22	1.61	2.38
1600	Smooth surface cap sheet, mopped, 145 mils		2100	.027		.49	.85	.21	1.55	2.29
1700	Smooth surface flashing, 145 mils		1260	.044		.49	1.42	.35	2.26	3.46
1800	150 mils		1260	.044		.48	1.42	.35	2.25	3.45
1900	Granular surface flashing, 150 mils		1260	.044		.53	1.42	.35	2.30	3.50
2000	160 mils	↓	1260	.044	↓	.77	1.42	.35	2.54	3.77

07 53 Elastomeric Membrane Roofing

07 53 16 – Chlorosulfonate Polyethylene Roofing

07 53 16.10 Chlorosulfonated Polyethylene Roofing

	Crew	Daily Output	Labor-Hours	Unit	Material	2009 Bare Costs Labor	Equipment	Total	Total Incl O&P
0010 **CHLOROSULFONATED POLYETHYLENE ROOFING**									
0800 Chlorosulfonated polyethylene-hypalon (CSPE), 45 mils,									
0900 0.29 P.S.F., fully adhered	G-5	26	1.538	Sq.	185	48	6.65	239.65	297
1200 Mechanically attached	"	35	1.143	"	186	35.50	4.95	226.45	274

07 53 23 – Ethylene-Propylene-Diene-Monomer Roofing

07 53 23.20 Ethylene-Propylene-Diene-Monomer Roofing

	Crew	Daily Output	Labor-Hours	Unit	Material	2009 Bare Costs Labor	Equipment	Total	Total Incl O&P
0010 **ETHYLENE-PROPYLENE-DIENE-MONOMER ROOFING (E.P.D.M.)**									
3500 Ethylene propylene diene monomer (EPDM), 45 mils, 0.28 P.S.F.									
3600 Loose-laid & ballasted with stone (10 P.S.F.)	G-5	51	.784	Sq.	74	24.50	3.40	101.90	128
3700 Mechanically attached		35	1.143		64.50	35.50	4.95	104.95	140
3800 Fully adhered with adhesive	↓	26	1.538	↓	93	48	6.65	147.65	195
4500 60 mils, 0.40 P.S.F.									
4600 Loose-laid & ballasted with stone (10 P.S.F.)	G-5	51	.784	Sq.	92.50	24.50	3.40	120.40	149
4700 Mechanically attached		35	1.143		82	35.50	4.95	122.45	159
4800 Fully adhered with adhesive	↓	26	1.538		111	48	6.65	165.65	215
4810 45 mil, .28 PSF, membrane only					37			37	40.50
4820 60 mil, .40 PSF, membrane only					53			53	58.50
4850 Seam tape for membrane, 3" x 100' roll				Ea.	43.50			43.50	47.50
4900 Batten strips, 10' sections					3.45			3.45	3.80
4910 Cover tape for batten strips, 6" x 100' roll				↓	135			135	149
4930 Plate anchors				M	65			65	71.50
4970 Adhesive for fully adhered systems, 60 S.F./gal.				Gal.	17.65			17.65	19.40

07 53 29 – Polyisobutylene Roofing

07 53 29.10 Polyisobutylene Roofing

	Crew	Daily Output	Labor-Hours	Unit	Material	2009 Bare Costs Labor	Equipment	Total	Total Incl O&P
0010 **POLYISOBUTYLENE ROOFING**									
7500 Polyisobutylene (PIB), 100 mils, 0.57 P.S.F.									
7600 Loose-laid & ballasted with stone/gravel (10 P.S.F.)	G-5	51	.784	Sq.	181	24.50	3.40	208.90	246
7700 Partially adhered with adhesive		35	1.143		227	35.50	4.95	267.45	320
7800 Hot asphalt attachment		35	1.143		217	35.50	4.95	257.45	305
7900 Fully adhered with contact cement	↓	26	1.538	↓	234	48	6.65	288.65	350

07 54 Thermoplastic Membrane Roofing

07 54 19 – Polyvinyl-Chloride Roofing

07 54 19.10 Polyvinyl-Chloride Roofing (P.V.C.)

	Crew	Daily Output	Labor-Hours	Unit	Material	2009 Bare Costs Labor	Equipment	Total	Total Incl O&P
0010 **POLYVINYL-CHLORIDE ROOFING (P.V.C.)**									
8200 Heat welded seams									
8700 Reinforced, 48 mils, 0.33 P.S.F.									
8750 Loose-laid & ballasted with stone/gravel (12 P.S.F.)	G-5	51	.784	Sq.	116	24.50	3.40	143.90	175
8800 Mechanically attached	↓	35	1.143	↓	106	35.50	4.95	146.45	186
8850 Fully adhered with adhesive	↓	26	1.538	↓	155	48	6.65	209.65	264
8860 Reinforced, 60 mils, .40 P.S.F.									
8870 Loose-laid & ballasted with stone/gravel (12 P.S.F.)	G-5	51	.784	Sq.	118	24.50	3.40	145.90	177
8880 Mechanically attached	↓	35	1.143		108	35.50	4.95	148.45	188
8890 Fully adhered with adhesive	↓	26	1.538	↓	157	48	6.65	211.65	266

07 54 23 – Thermoplastic Polyolefin Roofing

07 54 23.10 Thermoplastic Polyolefin Roofing (T.P.O)

	Crew	Daily Output	Labor-Hours	Unit	Material	2009 Bare Costs Labor	Equipment	Total	Total Incl O&P
0010 **THERMOPLASTIC POLYOLEFIN ROOFING (T.P.O.)**									
0100 45 mils, loose laid & ballasted with stone(1/2 ton/sq.)	G-5	51	.784	Sq.	98.50	24.50	3.40	126.40	156
0120 Fully adhered	↓	25	1.600	↓	144	49.50	6.95	200.45	255

07 54 Thermoplastic Membrane Roofing

07 54 23 – Thermoplastic Polyolefin Roofing

07 54 23.10 Thermoplastic Polyolefin Roofing (T.P.O)	Crew	Daily Output	Labor-Hours	Unit	Material	2009 Bare Costs Labor	Equipment	Total	Total Incl O&P
0140	Mechanically attached	G-5	34	1.176	Sq.	72.50	36.50	5.10	114.10	151
0160	Self adhered		35	1.143		81.50	35.50	4.95	121.95	158
0180	60 mil membrane, heat welded seams, ballasted		50	.800		116	25	3.46	144.46	176
0200	Fully adhered		25	1.600		161	49.50	6.95	217.45	274
0220	Mechanically attached		34	1.176		106	36.50	5.10	147.60	188
0240	Self adhered	▼	35	1.143	▼	100	35.50	4.95	140.45	179

07 56 Fluid-Applied Roofing

07 56 10 – Fluid-Applied Roofing Elastomers

07 56 10.10 Elastomeric Roofing	Crew	Daily Output	Labor-Hours	Unit	Material	2009 Bare Costs Labor	Equipment	Total	Total Incl O&P
0010	**ELASTOMERIC ROOFING**									
0110	Acrylic rubber, fluid applied, 20 mils thick	G-5	2000	.020	S.F.	2.27	.62	.09	2.98	3.71
0120	50 mils, reinforced		1200	.033		3.52	1.04	.14	4.70	5.90
0130	For walking surface, add	▼	900	.044		1.07	1.38	.19	2.64	3.86
0300	Hypalon neoprene, fluid applied, 20 mil thick, not-reinforced	G-1	1135	.049		2.57	1.58	.39	4.54	6.10
0600	Non-woven polyester, reinforced		960	.058		2.61	1.87	.46	4.94	6.70
0700	5 coat neoprene deck, 60 mil thick, under 10,000 SF		325	.172		5.75	5.50	1.35	12.60	17.65
0900	Over 10,000 SF		625	.090		5.35	2.87	.70	8.92	11.75
1300	Vinyl plastic traffic deck, sprayed, 2 to 4 mils thick		625	.090		1.68	2.87	.70	5.25	7.70
1500	Vinyl and neoprene membrane traffic deck	▼	1550	.036	▼	1.77	1.16	.28	3.21	4.32
9000	Minimum labor/equipment charge	1 Rofc	2	4	Job		137		137	245

07 57 Coated Foamed Roofing

07 57 13 – Sprayed Polyurethane Foam Roofing

07 57 13.10 Sprayed Polyurethane Foam Roofing (S.P.F.)	Crew	Daily Output	Labor-Hours	Unit	Material	2009 Bare Costs Labor	Equipment	Total	Total Incl O&P
0010	**SPRAYED POLYURETHANE FOAM ROOFING (S.P.F.)**									
0100	Primer for metal substrate (when required)	G-2A	3000	.008	S.F.	.42	.24	.20	.86	1.10
0200	Primer for non-metal substrate (when required)		3000	.008		.16	.24	.20	.60	.82
0300	Closed cell spray, polyureathane foam, 3 lbs per CF density, 1", R6.7		15000	.002		.70	.05	.04	.79	.93
0400	2", R13.4		13125	.002		1.40	.06	.05	1.51	1.69
0500	3", R20.1		11485	.002		2.10	.06	.05	2.21	2.48
0700	Spray-on silicone coating	▼	2500	.010		.92	.29	.24	1.45	1.79
0800	Warranty 5-20 year manufacturer's									.15
0900	Warranty 20 year, no dollar limit				▼					.20

07 58 Roll Roofing

07 58 10 – Asphalt Roll Roofing

07 58 10.10 Roll Roofing	Crew	Daily Output	Labor-Hours	Unit	Material	2009 Bare Costs Labor	Equipment	Total	Total Incl O&P
0010	**ROLL ROOFING**									
0100	Asphalt, mineral surface									
0200	1 ply #15 organic felt, 1 ply mineral surfaced									
0300	Selvage roofing, lap 19", nailed & mopped	G-1	27	2.074	Sq.	67.50	66.50	16.30	150.30	210
0400	3 plies glass fiber felt (type IV), 1 ply mineral surfaced									
0500	Selvage roofing, lapped 19", mopped	G-1	25	2.240	Sq.	104	71.50	17.60	193.10	261
0600	Coated glass fiber base sheet, 2 plies of glass fiber									
0700	Felt (type IV), 1 ply mineral surfaced selvage									
0800	Roofing, lapped 19", mopped	G-1	25	2.240	Sq.	112	71.50	17.60	201.10	270

07 58 Roll Roofing

07 58 10 - Asphalt Roll Roofing

07 58 10.10 Roll Roofing	Crew	Daily Output	Labor-Hours	Unit	Material	2009 Bare Costs Labor	2009 Bare Costs Equipment	Total	Total Incl O&P
0900 On nailable decks	G-1	24	2.333	Sq.	102	74.50	18.35	194.85	266
1000 3 plies glass fiber felt (type III), 1 ply mineral surfaced									
1100 Selvage roofing, lapped 19", mopped	G-1	25	2.240	Sq.	104	71.50	17.60	193.10	261

07 61 Sheet Metal Roofing

07 61 13 - Standing Seam Sheet Metal Roofing

07 61 13.10 Standing Seam Sheet Metal Roofing, Field Fab.

	Crew	Daily Output	Labor-Hours	Unit	Material	2009 Bare Costs Labor	2009 Bare Costs Equipment	Total	Total Incl O&P
0010 **STANDING SEAM SHEET METAL ROOFING, FIELD FABRICATED**									
0400 Copper standing seam roofing, over 10 squares, 16 oz, 125 lb per sq	1 Shee	1.30	6.154	Sq.	1,075	290		1,365	1,625
0600 18 oz, 140 lb per sq		1.20	6.667		1,200	315		1,515	1,800
0700 20 oz, 150 lb per sq		1.10	7.273		1,300	345		1,645	2,000
1200 For abnormal conditions or small areas, add					25%	100%			
1300 For lead-coated copper, add					25%				
1400 Lead standing seam roofing, 5 lb. per S.F.	1 Shee	1.30	6.154		565	290		855	1,075
1500 Zinc standing seam roofing, .020" thick		1.20	6.667		1,050	315		1,365	1,675
1510 .027" thick		1.15	6.957		1,500	330		1,830	2,200
1520 .032" thick		1.10	7.273		1,800	345		2,145	2,525
1530 .040" thick		1.05	7.619		2,150	360		2,510	2,950
9000 Minimum labor/equipment charge		2	4	Job		189		189	300

07 61 16 - Batten Seam Sheet Metal Roofing

07 61 16.10 Batten Seam Sheet Metal Roofing, Field Fabricated

	Crew	Daily Output	Labor-Hours	Unit	Material	2009 Bare Costs Labor	2009 Bare Costs Equipment	Total	Total Incl O&P
0009 **BATTEN SEAM SHEET METAL ROOFING, FIELD FABRICATED**									
0012 Copper batten seam roofing, over 10 sq, 16 oz, 130 lb per sq	1 Shee	1.10	7.273	Sq.	1,075	345		1,420	1,725
0020 Lead batten seam roofing, 5 lb. per SF		1.20	6.667		565	315		880	1,125
0100 Zinc / copper alloy batten seam roofing, .020 thick		1.20	6.667		1,050	315		1,365	1,675
0200 Copper roofing, batten seam, over 10 sq, 18 oz, 145 lb per sq		1	8		1,200	380		1,580	1,900
0300 20 oz, 160 lb per sq		1	8		1,300	380		1,680	2,050
0500 Stainless steel batten seam roofing, type 304, 28 gauge		1.20	6.667		685	315		1,000	1,250
0600 26 gauge		1.15	6.957		635	330		965	1,225
0800 Zinc, copper alloy roofing, batten seam, .027" thick		1.15	6.957		1,500	330		1,830	2,200
0900 .032" thick		1.10	7.273		1,800	345		2,145	2,525
1000 .040" thick		1.05	7.619		2,150	360		2,510	2,950
9000 Minimum labor/equipment charge		2	4	Job		189		189	300

07 61 19 - Flat Seam Sheet Metal Roofing

07 61 19.10 Flat Seam Sheet Metal Roofing, Field Fabricated

	Crew	Daily Output	Labor-Hours	Unit	Material	2009 Bare Costs Labor	2009 Bare Costs Equipment	Total	Total Incl O&P
0010 **FLAT SEAM SHEET METAL ROOFING, FIELD FABRICATED**									
0900 Copper flat seam roofing, over 10 squares, 16 oz, 115 lb per sq	1 Shee	1.20	6.667	Sq.	1,075	315		1,390	1,675
0950 18 oz, 130 lb/sq		1.15	6.957		1,200	330		1,530	1,825
1000 20 oz, 145 lb per sq		1.10	7.273		1,300	345		1,645	2,000
1008 Zinc flat seam roofing, .020" thick		1.20	6.667		1,050	315		1,365	1,675
1010 .027" thick		1.15	6.957		1,500	330		1,830	2,200
1020 .032" thick		1.12	7.143		1,800	335		2,135	2,500
1100 Lead flat seam roofing, 5 lb per SF		1.30	6.154		565	290		855	1,075
9000 Minimum labor/equipment charge		2.75	2.909	Job		137		137	218

07 62 Sheet Metal Flashing and Trim

07 62 10 – Sheet Metal Trim

07 62 10.10 Sheet Metal Cladding		Crew	Daily Output	Labor-Hours	Unit	Material	2009 Bare Costs Labor	Equipment	Total	Total Incl O&P
0010	**SHEET METAL CLADDING**									
0100	Aluminum, up to 6 bends, .032" thick, window casing	1 Carp	180	.044	S.F.	.87	1.78		2.65	3.89
0200	Window sill		72	.111	L.F.	.87	4.44		5.31	8.30
0300	Door casing		180	.044	S.F.	.87	1.78		2.65	3.89
0400	Fascia		250	.032		.87	1.28		2.15	3.07
0500	Rake trim		225	.036		.87	1.42		2.29	3.31
0700	.024" thick, window casing		180	.044		1.55	1.78		3.33	4.64
0800	Window sill		72	.111	L.F.	1.55	4.44		5.99	9.05
0900	Door casing		180	.044	S.F.	1.55	1.78		3.33	4.64
1000	Fascia		250	.032		1.55	1.28		2.83	3.82
1100	Rake trim		225	.036		1.55	1.42		2.97	4.06
1200	Vinyl coated aluminum, up to 6 bends, window casing		180	.044		1.03	1.78		2.81	4.06
1300	Window sill		72	.111	L.F.	1.03	4.44		5.47	8.50
1400	Door casing		180	.044	S.F.	1.03	1.78		2.81	4.06
1500	Fascia		250	.032		1.03	1.28		2.31	3.24
1600	Rake trim		225	.036		1.03	1.42		2.45	3.48

07 65 Flexible Flashing

07 65 10 – Sheet Metal Flashing

07 65 10.10 Sheet Metal Flashing and Counter Flashing		Crew	Daily Output	Labor-Hours	Unit	Material	2009 Bare Costs Labor	Equipment	Total	Total Incl O&P
0010	**SHEET METAL FLASHING AND COUNTER FLASHING**									
0011	Including up to 4 bends									
0020	Aluminum, mill finish, .013" thick	1 Rofc	145	.055	S.F.	.67	1.89		2.56	4.11
0030	.016" thick		145	.055		.79	1.89		2.68	4.24
0060	.019" thick		145	.055		1.02	1.89		2.91	4.49
0100	.032" thick		145	.055		1.48	1.89		3.37	5
0200	.040" thick		145	.055		2.21	1.89		4.10	5.80
0300	.050" thick		145	.055		2.52	1.89		4.41	6.15
0325	Mill finish 5" x 7" step flashing, .016" thick		1920	.004	Ea.	.16	.14		.30	.44
0350	Mill finish 12" x 12" step flashing, .016" thick		1600	.005	"	.61	.17		.78	.98
0400	Painted finish, add				S.F.	.37			.37	.41
1000	Mastic-coated 2 sides, .005" thick	1 Rofc	330	.024		1.67	.83		2.50	3.32
1100	.016" thick		330	.024		1.84	.83		2.67	3.50
1600	Copper, 16 oz, sheets, under 1000 lbs.		115	.070		6.80	2.38		9.18	11.70
1700	Over 4000 lbs.		155	.052		8	1.77		9.77	11.95
1900	20 oz sheets, under 1000 lbs.		110	.073		7.90	2.49		10.39	13.15
2000	Over 4000 lbs.		145	.055		8	1.89		9.89	12.15
2200	24 oz sheets, under 1000 lbs.		105	.076		9.60	2.61		12.21	15.25
2300	Over 4000 lbs.		135	.059		9.75	2.03		11.78	14.30
2500	32 oz sheets, under 1000 lbs.		100	.080		12.80	2.74		15.54	19
2600	Over 4000 lbs.		130	.062		20	2.11		22.11	26
2700	W shape for valleys, 16 oz, 24" wide		100	.080	L.F.	14.65	2.74		17.39	21
5800	Lead, 2.5 lb. per SF, up to 12" wide		135	.059	S.F.	4.14	2.03		6.17	8.15
5900	Over 12" wide		135	.059		4.14	2.03		6.17	8.15
8650	Copper, 16 oz		100	.080		5.25	2.74		7.99	10.65
8900	Stainless steel sheets, 32 ga, .010" thick		155	.052		3.79	1.77		5.56	7.35
9000	28 ga, .015" thick		155	.052		4.70	1.77		6.47	8.30
9100	26 ga, .018" thick		155	.052		5.70	1.77		7.47	9.40
9200	24 ga, .025" thick		155	.052		7.40	1.77		9.17	11.30
9290	For mechanically keyed flashing, add					40%				
9320	Steel sheets, galvanized, 20 gauge	1 Rofc	130	.062	S.F.	1.19	2.11		3.30	5.05

07 65 Flexible Flashing

07 65 10 – Sheet Metal Flashing

07 65 10.10 Sheet Metal Flashing and Counter Flashing

	07 65 10.10 Sheet Metal Flashing and Counter Flashing	Crew	Daily Output	Labor-Hours	Unit	Material	2009 Bare Costs Labor	Equipment	Total	Total Incl O&P
9340	30 gauge	1 Rofc	160	.050	S.F.	.50	1.71		2.21	3.61
9400	Terne coated stainless steel, .015" thick, 28 ga		155	.052		7.35	1.77		9.12	11.25
9500	.018" thick, 26 ga		155	.052		8.30	1.77		10.07	12.25
9600	Zinc and copper alloy (brass), .020" thick		155	.052		5.15	1.77		6.92	8.80
9700	.027" thick		155	.052		6.90	1.77		8.67	10.75
9800	.032" thick		155	.052		8.05	1.77		9.82	12
9900	.040" thick		155	.052		9.85	1.77		11.62	13.95
9950	Minimum labor/equipment charge		3	2.667	Job		91.50		91.50	163

07 65 12 – Fabric and Mastic Flashings

07 65 12.10 Fabric and Mastic Flashing and Counter Flashing

		Crew	Daily Output	Labor-Hours	Unit	Material	2009 Bare Costs Labor	Equipment	Total	Total Incl O&P
0010	**FABRIC AND MASTIC FLASHING AND COUNTER FLASHING**									
1300	Asphalt flashing cement, 5 gallon				Gal.	5.80			5.80	6.40
4900	Fabric, asphalt-saturated cotton, specification grade	1 Rofc	35	.229	S.Y.	2.37	7.85		10.22	16.60
5000	Utility grade		35	.229		1.50	7.85		9.35	15.65
5200	Open-mesh fabric, saturated, 40 oz per S.Y.		35	.229		1.67	7.85		9.52	15.85
5300	Close-mesh fabric, saturated, 17 oz per S.Y.		35	.229		1.75	7.85		9.60	15.95
5500	Fiberglass, resin-coated		35	.229		1.43	7.85		9.28	15.55
5600	Asphalt-coated, 40 oz per S.Y.		35	.229		9.75	7.85		17.60	24.50
8500	Shower pan, bituminous membrane, 7 oz		155	.052	S.F.	1.83	1.77		3.60	5.15

07 65 13 – Laminated Sheet Flashing

07 65 13.10 Laminated Sheet Flashing

		Crew	Daily Output	Labor-Hours	Unit	Material	2009 Bare Costs Labor	Equipment	Total	Total Incl O&P
0010	**LAMINATED SHEET FLASHING**, Including up to 4 bends									
0500	Fabric-backed 2 sides, .004" thick	1 Rofc	330	.024	S.F.	1.41	.83		2.24	3.03
0700	.005" thick		330	.024		1.67	.83		2.50	3.32
0750	Mastic-backed, self adhesive		460	.017		3.46	.60		4.06	4.87
0800	Mastic-coated 2 sides, .004" thick		330	.024		1.41	.83		2.24	3.03
2800	Copper, paperbacked 1 side, 2 oz		330	.024		1.42	.83		2.25	3.04
2900	3 oz		330	.024		1.85	.83		2.68	3.52
3100	Paperbacked 2 sides, 2 oz		330	.024		1.43	.83		2.26	3.05
3150	3 oz		330	.024		1.84	.83		2.67	3.50
3200	5 oz		330	.024		2.76	.83		3.59	4.52
3250	7 oz		330	.024		4.50	.83		5.33	6.45
3400	Mastic-backed 2 sides, copper, 2 oz		330	.024		1.69	.83		2.52	3.34
3500	3 oz		330	.024		2.08	.83		2.91	3.77
3700	5 oz		330	.024		3.05	.83		3.88	4.84
3800	Fabric-backed 2 sides, copper, 2 oz		330	.024		1.79	.83		2.62	3.45
4000	3 oz		330	.024		2.31	.83		3.14	4.02
4100	5 oz		330	.024		3.14	.83		3.97	4.93
4300	Copper-clad stainless steel, .015" thick, under 500 lbs.		115	.070		5.25	2.38		7.63	10
4400	Over 2000 lbs.		155	.052		5.05	1.77		6.82	8.70
4600	.018" thick, under 500 lbs.		100	.080		6.90	2.74		9.64	12.50
4700	Over 2000 lbs.		145	.055		5.05	1.89		6.94	8.90
6100	Lead-coated copper, fabric-backed, 2 oz		330	.024		2.42	.83		3.25	4.14
6200	5 oz		330	.024		2.78	.83		3.61	4.54
6400	Mastic-backed 2 sides, 2 oz		330	.024		1.89	.83		2.72	3.56
6500	5 oz		330	.024		2.34	.83		3.17	4.05
6700	Paperbacked 1 side, 2 oz		330	.024		1.63	.83		2.46	3.27
6800	3 oz		330	.024		1.92	.83		2.75	3.59
7000	Paperbacked 2 sides, 2 oz		330	.024		1.69	.83		2.52	3.34
7100	5 oz		330	.024		2.75	.83		3.58	4.51
8550	3 ply copper and fabric, 3 oz		155	.052		2.73	1.77		4.50	6.15
8600	7 oz		155	.052		5.70	1.77		7.47	9.40

07 65 Flexible Flashing

07 65 13 – Laminated Sheet Flashing

07 65 13.10 Laminated Sheet Flashing	Crew	Daily Output	Labor-Hours	Unit	Material	2009 Bare Costs Labor	Equipment	Total	Total Incl O&P
8700	Lead on copper and fabric, 5 oz	1 Rofc	155	.052	S.F.	2.78	1.77		4.55	6.20
8800	7 oz		155	.052		4.94	1.77		6.71	8.60
9300	Stainless steel, paperbacked 2 sides, .005" thick		330	.024		3.34	.83		4.17	5.15

07 65 19 – Plastic Sheet Flashing

07 65 19.10 Plastic Sheet Flashing and Counter Flashing

		Crew	Daily Output	Labor-Hours	Unit	Material	2009 Bare Costs Labor	Equipment	Total	Total Incl O&P
0010	PLASTIC SHEET FLASHING AND COUNTER FLASHING									
7300	Polyvinyl chloride, black, .010" thick	1 Rofc	285	.028	S.F.	.26	.96		1.22	2.01
7400	.020" thick		285	.028		.36	.96		1.32	2.12
7600	.030" thick		285	.028		.47	.96		1.43	2.24
7700	.056" thick		285	.028		1.11	.96		2.07	2.94
7900	Black or white for exposed roofs, .060" thick		285	.028		2.45	.96		3.41	4.42
8060	PVC tape, 5" x 45 mils, for joint covers, 100 L.F./roll				Ea.	120			120	132

07 65 23 – Rubber Sheet Flashing

07 65 23.10 Rubber Sheet Flashing and Counterflashing

		Crew	Daily Output	Labor-Hours	Unit	Material	2009 Bare Costs Labor	Equipment	Total	Total Incl O&P
0010	RUBBER SHEET FLASHING AND COUNTERFLASHING									
8100	Rubber, butyl, 1/32" thick	1 Rofc	285	.028	S.F.	.96	.96		1.92	2.78
8200	1/16" thick		285	.028		1.43	.96		2.39	3.29
8300	Neoprene, cured, 1/16" thick		285	.028		2.03	.96		2.99	3.95
8400	1/8" thick		285	.028		4.09	.96		5.05	6.20

07 71 Roof Specialties

07 71 19 – Manufactured Gravel Stops and Fascias

07 71 19.10 Gravel Stop

		Crew	Daily Output	Labor-Hours	Unit	Material	2009 Bare Costs Labor	Equipment	Total	Total Incl O&P
0010	GRAVEL STOP									
0020	Aluminum, .050" thick, 4" face height, mill finish	1 Shee	145	.055	L.F.	5.95	2.60		8.55	10.70
0080	Duranodic finish		145	.055		5.75	2.60		8.35	10.45
0100	Painted		145	.055		6.65	2.60		9.25	11.45
0300	6" face height		135	.059		6.40	2.80		9.20	11.45
0350	Duranodic finish		135	.059		6.75	2.80		9.55	11.90
0400	Painted		135	.059		7.85	2.80		10.65	13.10
0600	8" face height		125	.064		8	3.02		11.02	13.60
0650	Duranodic finish		125	.064		7.75	3.02		10.77	13.35
0700	Painted		125	.064		7.90	3.02		10.92	13.50
0900	12" face height, .080 thick, 2 piece		100	.080		10.60	3.78		14.38	17.70
0950	Duranodic finish		100	.080		9.75	3.78		13.53	16.75
1000	Painted		100	.080		11.55	3.78		15.33	18.70
1500	Polyvinyl chloride, 6" face height		135	.059		4.32	2.80		7.12	9.20
1600	9" face height		125	.064		5.10	3.02		8.12	10.40
1800	Stainless steel, 24 ga., 6" face height		135	.059		11.15	2.80		13.95	16.70
1900	12" face height		100	.080		23.50	3.78		27.28	31.50
2100	20 ga., 6" face height		135	.059		12.60	2.80		15.40	18.30
2200	12" face height		100	.080		26.50	3.78		30.28	35.50
9000	Minimum labor/equipment charge		3.50	2.286	Job		108		108	171

07 71 19.30 Fascia

		Crew	Daily Output	Labor-Hours	Unit	Material	2009 Bare Costs Labor	Equipment	Total	Total Incl O&P
0010	FASCIA									
0100	Aluminum, reverse board and batten, .032" thick, colored, no furring incl	1 Shee	145	.055	S.F.	5.75	2.60		8.35	10.45
0200	Residential type, aluminum	1 Carp	200	.040	L.F.	1.51	1.60		3.11	4.30
0220	Vinyl	"	200	.040	"	1.16	1.60		2.76	3.92
0300	Steel, galv and enameled, stock, no furring, long panels	1 Shee	145	.055	S.F.	3.76	2.60		6.36	8.30
0600	Short panels		115	.070	"	5	3.28		8.28	10.70

07 71 Roof Specialties

07 71 19 – Manufactured Gravel Stops and Fascias

07 71 19.30 Fascia		Crew	Daily Output	Labor-Hours	Unit	Material	2009 Bare Costs Labor	2009 Bare Costs Equipment	Total	Total Incl O&P
9000	Minimum labor/equipment charge	1 Shee	4	2	Job		94.50		94.50	150

07 71 23 – Manufactured Gutters and Downspouts

07 71 23.10 Downspouts

		Crew	Daily Output	Labor-Hours	Unit	Material	2009 Bare Costs Labor	2009 Bare Costs Equipment	Total	Total Incl O&P
0010	**DOWNSPOUTS**									
0020	Aluminum 2" x 3", .020" thick, embossed	1 Shee	190	.042	L.F.	1.04	1.99		3.03	4.30
0100	Enameled		190	.042		1.59	1.99		3.58	4.91
0300	Enameled, .024" thick, 2" x 3"		180	.044		1.84	2.10		3.94	5.35
0400	3" x 4"		140	.057		2.55	2.70		5.25	7.10
0600	Round, corrugated aluminum, 3" diameter, .020" thick		190	.042		1.42	1.99		3.41	4.72
0700	4" diameter, .025" thick		140	.057		2.39	2.70		5.09	6.90
0900	Wire strainer, round, 2" diameter		155	.052	Ea.	3.32	2.44		5.76	7.50
1000	4" diameter		155	.052		6.10	2.44		8.54	10.60
1200	Rectangular, perforated, 2" x 3"		145	.055		2.30	2.60		4.90	6.65
1300	3" x 4"		145	.055		3.32	2.60		5.92	7.80
1500	Copper, round, 16 oz., stock, 2" diameter		190	.042	L.F.	9	1.99		10.99	13.05
1600	3" diameter		190	.042		8.60	1.99		10.59	12.60
1800	4" diameter		145	.055		9	2.60		11.60	14.05
1900	5" diameter		130	.062		14.75	2.90		17.65	21
2100	Rectangular, corrugated copper, stock, 2" x 3"		190	.042		7.15	1.99		9.14	11.05
2200	3" x 4"		145	.055		9.30	2.60		11.90	14.35
2400	Rectangular, plain copper, stock, 2" x 3"		190	.042		7	1.99		8.99	10.85
2500	3" x 4"		145	.055		9.10	2.60		11.70	14.15
2700	Wire strainers, rectangular, 2" x 3"		145	.055	Ea.	5.50	2.60		8.10	10.20
2800	3" x 4"		145	.055		6.60	2.60		9.20	11.40
3000	Round, 2" diameter		145	.055		4.40	2.60		7	9
3100	3" diameter		145	.055		6.10	2.60		8.70	10.90
3300	4" diameter		145	.055		11.30	2.60		13.90	16.55
3400	5" diameter		115	.070		16.60	3.28		19.88	23.50
3600	Lead-coated copper, round, stock, 2" diameter		190	.042	L.F.	11.45	1.99		13.44	15.75
3700	3" diameter		190	.042		12.60	1.99		14.59	17
3900	4" diameter		145	.055		14.70	2.60		17.30	20.50
4000	5" diameter, corrugated		130	.062		23.50	2.90		26.40	30
4200	6" diameter, corrugated		105	.076		24	3.60		27.60	32
4300	Rectangular, corrugated, stock, 2" x 3"		190	.042		12.75	1.99		14.74	17.20
4500	Plain, stock, 2" x 3"		190	.042		9.75	1.99		11.74	13.90
4600	3" x 4"		145	.055		10	2.60		12.60	15.15
4800	Steel, galvanized, round, corrugated, 2" or 3" diameter, 28 gauge		190	.042		1.74	1.99		3.73	5.05
4900	4" diameter, 28 gauge		145	.055		2.48	2.60		5.08	6.85
5100	5" diameter, 28 gauge		130	.062		3.68	2.90		6.58	8.65
5200	26 gauge		130	.062		2.97	2.90		5.87	7.90
5400	6" diameter, 28 gauge		105	.076		4.83	3.60		8.43	11
5500	26 gauge		105	.076		6.10	3.60		9.70	12.45
5700	Rectangular, corrugated, 28 gauge, 2" x 3"		190	.042		1.60	1.99		3.59	4.92
5800	3" x 4"		145	.055		1.60	2.60		4.20	5.90
6000	Rectangular, plain, 28 gauge, galvanized, 2" x 3"		190	.042		1.74	1.99		3.73	5.05
6100	3" x 4"		145	.055		2.51	2.60		5.11	6.90
6300	Epoxy painted, 24 gauge, corrugated, 2" x 3"		190	.042		1.90	1.99		3.89	5.25
6400	3" x 4"		145	.055		2.12	2.60		4.72	6.45
6600	Wire strainers, rectangular, 2" x 3"		145	.055	Ea.	5.90	2.60		8.50	10.65
6700	3" x 4"		145	.055		11.75	2.60		14.35	17.10
6900	Round strainers, 2" or 3" diameter		145	.055		2.84	2.60		5.44	7.25
7000	4" diameter		145	.055		11.75	2.60		14.35	17.10

07 71 23 – Manufactured Gutters and Downspouts

07 71 23.10 Downspouts

		Crew	Daily Output	Labor-Hours	Unit	Material	2009 Bare Costs Labor	Equipment	Total	Total Incl O&P
9000	Minimum labor/equipment charge	1 Shee	4	2	Job		94.50		94.50	150

07 71 23.20 Downspout Elbows

		Crew	Daily Output	Labor-Hours	Unit	Material	2009 Bare Costs Labor	Equipment	Total	Total Incl O&P
0010	**DOWNSPOUT ELBOWS**									
0020	Aluminum, 2" x 3", embossed	1 Shee	100	.080	Ea.	1.17	3.78		4.95	7.30
0100	Enameled		100	.080		1.17	3.78		4.95	7.30
0200	3" x 4", .025" thick, embossed		100	.080		3.99	3.78		7.77	10.40
0300	Enameled		100	.080		5.15	3.78		8.93	11.65
0400	Round corrugated, 3", embossed, .020" thick		100	.080		2.38	3.78		6.16	8.60
0500	4", .025" thick		100	.080		6.50	3.78		10.28	13.15
0600	Copper, 16 oz. round, 2" diameter		100	.080		13.40	3.78		17.18	21
0700	3" diameter		100	.080		7.70	3.78		11.48	14.45
0800	4" diameter		100	.080		12.85	3.78		16.63	20
1000	2" x 3" corrugated		100	.080		7.10	3.78		10.88	13.80
1100	3" x 4" corrugated		100	.080		12	3.78		15.78	19.20
9000	Minimum labor/equipment charge		4	2	Job		94.50		94.50	150

07 71 23.30 Gutters

		Crew	Daily Output	Labor-Hours	Unit	Material	2009 Bare Costs Labor	Equipment	Total	Total Incl O&P
0010	**GUTTERS**									
0012	Aluminum, stock units, 5" K type, .027" thick, plain	1 Shee	120	.067	L.F.	2.32	3.15		5.47	7.55
0100	Enameled		120	.067		2.45	3.15		5.60	7.70
0300	5" K type type, .032" thick, plain		120	.067		3.27	3.15		6.42	8.60
0400	Enameled		120	.067		2.92	3.15		6.07	8.20
0700	Copper, half round, 16 oz, stock units, 4" wide		120	.067		6.25	3.15		9.40	11.90
0900	5" wide		120	.067		6.30	3.15		9.45	11.95
1000	6" wide		115	.070		8.90	3.28		12.18	15
1200	K type, 16 oz, stock, 4" wide		120	.067		6.15	3.15		9.30	11.75
1300	5" wide		120	.067		7.30	3.15		10.45	13.05
1500	Lead coated copper, half round, stock, 4" wide		120	.067		10.50	3.15		13.65	16.55
1600	6" wide		115	.070		21	3.28		24.28	28
1800	K type, stock, 4" wide		120	.067		11.35	3.15		14.50	17.50
1900	5" wide		120	.067		14.25	3.15		17.40	20.50
2100	Stainless steel, half round or box, stock, 4" wide		120	.067		5.90	3.15		9.05	11.45
2200	5" wide		120	.067		6.35	3.15		9.50	11.95
2400	Steel, galv, half round or box, 28 ga, 5" wide, plain		120	.067		1.46	3.15		4.61	6.60
2500	Enameled		120	.067		1.51	3.15		4.66	6.65
2700	26 ga, stock, 5" wide		120	.067		1.59	3.15		4.74	6.75
2800	6" wide		120	.067		2.30	3.15		5.45	7.55
3000	Vinyl, O.G., 4" wide	1 Carp	110	.073		1.14	2.91		4.05	6.05
3100	5" wide		110	.073		1.35	2.91		4.26	6.30
3200	4" half round, stock units		110	.073		.91	2.91		3.82	5.80
3250	Joint connectors				Ea.	1.82			1.82	2
3300	Wood, clear treated cedar, fir or hemlock, 3" x 4"	1 Carp	100	.080	L.F.	7	3.20		10.20	13
3400	4" x 5"	"	100	.080	"	8	3.20		11.20	14.10
9000	Minimum labor/equipment charge	1 Shee	3.75	2.133	Job		101		101	160

07 71 23.35 Gutter Guard

		Crew	Daily Output	Labor-Hours	Unit	Material	2009 Bare Costs Labor	Equipment	Total	Total Incl O&P
0010	**GUTTER GUARD**									
0020	6" wide strip, aluminum mesh	1 Carp	500	.016	L.F.	2.18	.64		2.82	3.46
0100	Vinyl mesh		500	.016	"	1.45	.64		2.09	2.66
9000	Minimum labor/equipment charge		4	2	Job		80		80	132

07 71 26 – Reglets

07 71 26.10 Reglets and Accessories

0010	**REGLETS AND ACCESSORIES**									

07 71 Roof Specialties

07 71 26 – Reglets

07 71 26.10 Reglets and Accessories		Crew	Daily Output	Labor-Hours	Unit	Material	2009 Bare Costs Labor	Equipment	Total	Total Incl O&P
0020	Aluminum, .025" thick, in concrete parapet	1 Carp	225	.036	L.F.	1.27	1.42		2.69	3.75
0100	Copper, 10 oz.		225	.036		2.41	1.42		3.83	5
0300	16 oz.		225	.036		4.71	1.42		6.13	7.55
0400	Galvanized steel, 24 gauge		225	.036		.86	1.42		2.28	3.30
0600	Stainless steel, .020" thick		225	.036		2.90	1.42		4.32	5.55
0700	Zinc and copper alloy, 20 oz.		225	.036		2.41	1.42		3.83	5
0900	Counter flashing for above, 12" wide, .032" aluminum	1 Shee	150	.053		1.50	2.52		4.02	5.65
1000	Copper, 10 oz.		150	.053		4.55	2.52		7.07	9
1200	16 oz.		150	.053		5.25	2.52		7.77	9.75
1300	Galvanized steel, .020" thick		150	.053		.85	2.52		3.37	4.94
1500	Stainless steel, .020" thick		150	.053		5	2.52		7.52	9.50
1600	Zinc and copper alloy, 20 oz.		150	.053		4.39	2.52		6.91	8.85
9000	Minimum labor/equipment charge	1 Carp	3	2.667	Job		107		107	176

07 71 29 – Manufactured Roof Expansion Joints

07 71 29.10 Expansion Joints		Crew	Daily Output	Labor-Hours	Unit	Material	2009 Bare Costs Labor	Equipment	Total	Total Incl O&P
0010	**EXPANSION JOINTS**									
0300	Butyl or neoprene center with foam insulation, metal flanges									
0400	Aluminum, .032" thick for openings to 2-1/2"	1 Rofc	165	.048	L.F.	11.30	1.66		12.96	15.40
0600	For joint openings to 3-1/2"		165	.048		13.20	1.66		14.86	17.50
0610	For joint openings to 5"		165	.048		15.90	1.66		17.56	20.50
0620	For joint openings to 8"		165	.048		25	1.66		26.66	30.50
0700	Copper, 16 oz. for openings to 2-1/2"		165	.048		18.40	1.66		20.06	23
0900	For joint openings to 3-1/2"		165	.048		21	1.66		22.66	26
0910	For joint openings to 5"		165	.048		25	1.66		26.66	30
0920	For joint openings to 8"		165	.048		37	1.66		38.66	44
1000	Galvanized steel, 26 ga. for openings to 2-1/2"		165	.048		9.30	1.66		10.96	13.15
1200	For joint openings to 3-1/2"		165	.048		11	1.66		12.66	15.05
1210	For joint openings to 5"		165	.048		13.75	1.66		15.41	18.10
1220	For joint openings to 8"		165	.048		23.50	1.66		25.16	29
1300	Lead-coated copper, 16 oz. for openings to 2-1/2"		165	.048		32.50	1.66		34.16	39
1500	For joint openings to 3-1/2"		165	.048		38.50	1.66		40.16	45.50
1600	Stainless steel, .018", for openings to 2-1/2"		165	.048		14.15	1.66		15.81	18.50
1800	For joint openings to 3-1/2"		165	.048		16.10	1.66		17.76	20.50
1810	For joint openings to 5"		165	.048		19.95	1.66		21.61	25
1820	For joint openings to 8"		165	.048		30.50	1.66		32.16	36.50
1900	Neoprene, double-seal type with thick center, 4-1/2" wide		125	.064		13.20	2.19		15.39	18.45
1950	Polyethylene bellows, with galv steel flat flanges		100	.080		5.10	2.74		7.84	10.55
1960	With galvanized angle flanges		100	.080		5.65	2.74		8.39	11.10
2000	Roof joint with extruded aluminum cover, 2"	1 Shee	115	.070		35.50	3.28		38.78	44
2100	Roof joint, plastic curbs, foam center, standard	1 Rofc	100	.080		12.90	2.74		15.64	19.10
2200	Large		100	.080		17.25	2.74		19.99	24
2300	Transitions, regular, minimum		10	.800	Ea.	111	27.50		138.50	172
2350	Maximum		4	2		142	68.50		210.50	278
2400	Large, minimum		9	.889		163	30.50		193.50	235
2450	Maximum		3	2.667		171	91.50		262.50	350
2500	Roof to wall joint with extruded aluminum cover	1 Shee	115	.070	L.F.	30	3.28		33.28	38
2650										
2700	Wall joint, closed cell foam on PVC cover, 9" wide	1 Rofc	125	.064	L.F.	4.60	2.19		6.79	8.95
2800	12" wide	"	115	.070	"	5.20	2.38		7.58	9.95
9000	Minimum labor/equipment charge	1 Shee	3	2.667	Job		126		126	200

07 71 Roof Specialties

07 71 43 – Drip Edge

07 71 43.10 Drip Edge, Rake Edge, Ice Belts	Crew	Daily Output	Labor-Hours	Unit	Material	2009 Bare Costs Labor	Equipment	Total	Total Incl O&P
0010 **DRIP EDGE, RAKE EDGE, ICE BELTS**									
0020 Aluminum, .016" thick, 5" wide, mill finish	1 Carp	400	.020	L.F.	.42	.80		1.22	1.78
0100 White finish		400	.020		.46	.80		1.26	1.83
0200 8" wide, mill finish		400	.020		.56	.80		1.36	1.94
0300 Ice belt, 28" wide, mill finish		100	.080		3.80	3.20		7	9.50
0310 Vented, mill finish		400	.020		2.63	.80		3.43	4.21
0320 Painted finish		400	.020		2.72	.80		3.52	4.31
0400 Galvanized, 5" wide		400	.020		.54	.80		1.34	1.91
0500 8" wide, mill finish		400	.020		.78	.80		1.58	2.18
0510 Rake edge, aluminum, 1-1/2" x 1-1/2"		400	.020		.24	.80		1.04	1.58
0520 3-1/2" x 1-1/2"		400	.020		.46	.80		1.26	1.83
9000 Minimum labor/equipment charge		4	2	Job		80		80	132

07 72 Roof Accessories

07 72 23 – Relief Vents

07 72 23.10 Roof Vents

	Crew	Daily Output	Labor-Hours	Unit	Material	Labor	Equipment	Total	Total Incl O&P
0010 **ROOF VENTS**									
0020 Mushroom shape, for built-up roofs, aluminum	1 Rofc	30	.267	Ea.	68.50	9.15		77.65	92
0100 PVC, 6" high		30	.267	"	28.50	9.15		37.65	48
9000 Minimum labor/equipment charge		2.75	2.909	Job		99.50		99.50	178

07 72 23.20 Vents

	Crew	Daily Output	Labor-Hours	Unit	Material	Labor	Equipment	Total	Total Incl O&P
0010 **VENTS**									
0100 Soffit or eave, aluminum, mill finish, strips, 2-1/2" wide	1 Carp	200	.040	L.F.	.38	1.60		1.98	3.06
0200 3" wide		200	.040		.40	1.60		2	3.08
0300 Enamel finish, 3" wide		200	.040		.50	1.60		2.10	3.19
0400 Mill finish, rectangular, 4" x 16"		72	.111	Ea.	1.43	4.44		5.87	8.90
0500 8" x 16"		72	.111		1.84	4.44		6.28	9.35
2420 Roof ventilator	Q-9	16	1		48	42.50		90.50	121

07 72 26 – Ridge Vents

07 72 26.10 Ridge Vents and Accessories

	Crew	Daily Output	Labor-Hours	Unit	Material	Labor	Equipment	Total	Total Incl O&P
0010 **RIDGE VENTS AND ACCESSORIES**									
0100 Aluminum strips, mill finish	1 Rofc	160	.050	L.F.	2.50	1.71		4.21	5.80
0150 Painted finish		160	.050	"	3.61	1.71		5.32	7.05
0200 Connectors		48	.167	Ea.	3.50	5.70		9.20	14.05
0300 End caps		48	.167	"	1.30	5.70		7	11.65
0400 Galvanized strips		160	.050	L.F.	2.14	1.71		3.85	5.40
0430 Molded polyethylene, shingles not included		160	.050	"	3.05	1.71		4.76	6.40
0440 End plugs		48	.167	Ea.	1.30	5.70		7	11.65
0450 Flexible roll, shingles not included		160	.050	L.F.	2.48	1.71		4.19	5.80
2300 Ridge vent strip, mill finish	1 Shee	155	.052	"	2.76	2.44		5.20	6.90

07 72 33 – Roof Hatches

07 72 33.10 Roof Hatch Options

	Crew	Daily Output	Labor-Hours	Unit	Material	Labor	Equipment	Total	Total Incl O&P
0010 **ROOF HATCH OPTIONS**									
0500 2'-6" x 3', aluminum curb and cover	G-3	10	3.200	Ea.	985	126		1,111	1,275
0520 Galvanized steel curb and aluminum cover		10	3.200		595	126		721	860
0540 Galvanized steel curb and cover		10	3.200		500	126		626	755
0600 2'-6" x 4'-6", aluminum curb and cover		9	3.556		850	140		990	1,150
0800 Galvanized steel curb and aluminum cover		9	3.556		600	140		740	885
0900 Galvanized steel curb and cover		9	3.556		825	140		965	1,125

07 72 Roof Accessories

07 72 33 – Roof Hatches

07 72 33.10 Roof Hatch Options

	07 72 33.10 Roof Hatch Options	Crew	Daily Output	Labor-Hours	Unit	Material	2009 Bare Costs Labor	Equipment	Total	Total Incl O&P
1100	4' x 4' aluminum curb and cover	G-3	8	4	Ea.	1,675	158		1,833	2,075
1120	Galvanized steel curb and aluminum cover		8	4		1,675	158		1,833	2,075
1140	Galvanized steel curb and cover		8	4		1,600	158		1,758	2,025
1200	2'-6" x 8'-0", aluminum curb and cover		6.60	4.848		1,675	191		1,866	2,125
1400	Galvanized steel curb and aluminum cover		6.60	4.848		1,600	191		1,791	2,050
1500	Galvanized steel curb and cover	↓	6.60	4.848		1,200	191		1,391	1,625
1800	For plexiglass panels, 2'-6" x 3'-0", add to above				↓	400			400	440
9000	Minimum labor/equipment charge	2 Carp	2	8	Job		320		320	530

07 72 36 – Smoke Vents

07 72 36.10 Smoke Hatches

		Crew	Daily Output	Labor-Hours	Unit	Material	2009 Bare Costs Labor	Equipment	Total	Total Incl O&P
0010	**SMOKE HATCHES**									
0200	For 3'-0" long, add to roof hatches from Division 07 72 33.10				Ea.	25%	5%			
0250	For 4'-0" long, add to roof hatches from Division 07 72 33.10					20%	5%			
0300	For 8'-0" long, add to roof hatches from Division 07 72 33.10				↓	10%	5%			

07 72 36.20 Smoke Vent Options

		Crew	Daily Output	Labor-Hours	Unit	Material	2009 Bare Costs Labor	Equipment	Total	Total Incl O&P
0010	**SMOKE VENT OPTIONS**									
0100	4' x 4' aluminum cover and frame	G-3	13	2.462	Ea.	1,900	97		1,997	2,250
0200	Galvanized steel cover and frame		13	2.462		1,675	97		1,772	1,975
0300	4' x 8' aluminum cover and frame		8	4		2,600	158		2,758	3,100
0400	Galvanized steel cover and frame	↓	8	4	↓	2,200	158		2,358	2,675
9000	Minimum labor/equipment charge	2 Carp	2	8	Job		320		320	530

07 72 53 – Snow Guards

07 72 53.10 Snow Guard Options

		Crew	Daily Output	Labor-Hours	Unit	Material	2009 Bare Costs Labor	Equipment	Total	Total Incl O&P
0010	**SNOW GUARD OPTIONS**									
0100	Slate & asphalt shingle roofs	1 Rofc	160	.050	Ea.	9.40	1.71		11.11	13.40
0200	Standing seam metal roofs		48	.167		14.60	5.70		20.30	26.50
0300	Surface mount for metal roofs		48	.167	↓	7.70	5.70		13.40	18.65
0400	Double rail pipe type, including pipe	↓	130	.062	L.F.	22.50	2.11		24.61	29

07 72 73 – Pitch Pockets

07 72 73.10 Pitch Pockets, Variable Sizes

		Crew	Daily Output	Labor-Hours	Unit	Material	2009 Bare Costs Labor	Equipment	Total	Total Incl O&P
0010	**PITCH POCKETS, VARIABLE SIZES**									
0100	Adjustable, 4" to 7", welded corners, 4" deep	1 Rofc	48	.167	Ea.	12.90	5.70		18.60	24.50
0200	Side extenders, 6"	"	240	.033	"	1.97	1.14		3.11	4.21

07 72 80 – Vents

07 72 80.30 Vent Options

		Crew	Daily Output	Labor-Hours	Unit	Material	2009 Bare Costs Labor	Equipment	Total	Total Incl O&P
0010	**VENT OPTIONS**									
0020	Plastic, for insulated decks, 1 per M.S.F., minimum	1 Rofc	40	.200	Ea.	13.50	6.85		20.35	27
0100	Maximum		20	.400		31	13.70		44.70	59
0300	Aluminum	↓	30	.267		13.50	9.15		22.65	31
0800	Polystyrene baffles, 12" wide for 16" O.C. rafter spacing	1 Carp	90	.089		.40	3.55		3.95	6.30
0900	For 24" O.C. rafter spacing		110	.073	↓	.60	2.91		3.51	5.45
9000	Minimum labor/equipment charge	↓	3	2.667	Job		107		107	176

07 81 Applied Fireproofing

07 81 16 – Cementitious Fireproofing

07 81 16.10 Sprayed Cementitious Fireproofing	Crew	Daily Output	Labor-Hours	Unit	Material	2009 Bare Costs Labor	Equipment	Total	Total Incl O&P
0010 **SPRAYED CEMENTITIOUS FIREPROOFING**									
0050 Not incl tamping or canvas protection									
0100 1" thick, on flat plate steel	G-2	3000	.008	S.F.	.53	.27	.04	.84	1.06
0200 Flat decking		2400	.010		.53	.33	.05	.91	1.18
0400 Beams		1500	.016		.53	.53	.08	1.14	1.54
0500 Corrugated or fluted decks		1250	.019		.79	.64	.10	1.53	2.02
0700 Columns, 1-1/8" thick		1100	.022		.59	.73	.12	1.44	1.96
0800 2-3/16" thick		700	.034		1.13	1.14	.18	2.45	3.29
0850 For tamping, add						10%			
0900 For canvas protection, add	G-2	5000	.005	S.F.	.07	.16	.03	.26	.37
9000 Minimum labor/equipment charge	"	3	8	Job		267	42	309	480

07 84 Firestopping

07 84 13 – Penetration Firestopping

07 84 13.10 Firestopping

	Crew	Daily Output	Labor-Hours	Unit	Material	2009 Bare Costs Labor	Equipment	Total	Total Incl O&P
0010 **FIRESTOPPING** R078413-30									
0100 Metallic piping, non insulated									
0110 Through walls, 2" diameter	1 Carp	16	.500	Ea.	12.40	20		32.40	46.50
0120 4" diameter		14	.571		18.90	23		41.90	58.50
0130 6" diameter		12	.667		25.50	26.50		52	72
0140 12" diameter		10	.800		45	32		77	103
0150 Through floors, 2" diameter		32	.250		7.50	10		17.50	25
0160 4" diameter		28	.286		10.80	11.40		22.20	31
0170 6" diameter		24	.333		14.20	13.30		27.50	37.50
0180 12" diameter		20	.400		24	16		40	53
0190 Metallic piping, insulated									
0200 Through walls, 2" diameter	1 Carp	16	.500	Ea.	17.55	20		37.55	52.50
0210 4" diameter		14	.571		24	23		47	64
0220 6" diameter		12	.667		30.50	26.50		57	77.50
0230 12" diameter		10	.800		50	32		82	108
0240 Through floors, 2" diameter		32	.250		12.70	10		22.70	30.50
0250 4" diameter		28	.286		15.95	11.40		27.35	36.50
0260 6" diameter		24	.333		19.35	13.30		32.65	43.50
0270 12" diameter		20	.400		24	16		40	53
0280 Non metallic piping, non insulated									
0290 Through walls, 2" diameter	1 Carp	12	.667	Ea.	51	26.50		77.50	100
0300 4" diameter		10	.800		64	32		96	124
0310 6" diameter		8	1		89	40		129	164
0330 Through floors, 2" diameter		16	.500		40	20		60	77
0340 4" diameter		6	1.333		49.50	53.50		103	143
0350 6" diameter		6	1.333		59.50	53.50		113	154
0370 Ductwork, insulated & non insulated, round									
0380 Through walls, 6" diameter	1 Carp	12	.667	Ea.	26	26.50		52.50	72.50
0390 12" diameter		10	.800		51.50	32		83.50	110
0400 18" diameter		8	1		84	40		124	159
0410 Through floors, 6" diameter		16	.500		14.20	20		34.20	48.50
0420 12" diameter		14	.571		26	23		49	66
0430 18" diameter		12	.667		45	26.50		71.50	93.50
0440 Ductwork, insulated & non insulated, rectangular									
0450 With stiffener/closure angle, through walls, 6" x 12"	1 Carp	8	1	Ea.	21.50	40		61.50	89.50
0460 12" x 24"		6	1.333		28.50	53.50		82	120

07 84 Firestopping

07 84 13 – Penetration Firestopping

07 84 13.10 Firestopping	Crew	Daily Output	Labor-Hours	Unit	Material	2009 Bare Costs Labor	Equipment	Total	Total Incl O&P
0470 24" x 48"	1 Carp	4	2	Ea.	81.50	80		161.50	222
0480 With stiffener/closure angle, through floors, 6" x 12"		10	.800		11.65	32		43.65	66
0490 12" x 24"		8	1		21	40		61	89
0500 24" x 48"		6	1.333		41	53.50		94.50	133
0510 Multi trade openings									
0520 Through walls, 6" x 12"	1 Carp	2	4	Ea.	45	160		205	315
0530 12" x 24"	"	1	8		182	320		502	730
0540 24" x 48"	2 Carp	1	16		730	640		1,370	1,850
0550 48" x 96"	"	.75	21.333		2,925	850		3,775	4,625
0560 Through floors, 6" x 12"	1 Carp	2	4		45	160		205	315
0570 12" x 24"	"	1	8		182	320		502	730
0580 24" x 48"	2 Carp	.75	21.333		730	850		1,580	2,200
0590 48" x 96"	"	.50	32		2,925	1,275		4,200	5,325
0600 Structural penetrations, through walls									
0610 Steel beams, W8 x 10	1 Carp	8	1	Ea.	28.50	40		68.50	97.50
0620 W12 x 14		6	1.333		45	53.50		98.50	138
0630 W21 x 44		5	1.600		90.50	64		154.50	206
0640 W36 x 135		3	2.667		220	107		327	420
0650 Bar joists, 18" deep		6	1.333		41.50	53.50		95	134
0660 24" deep		6	1.333		51.50	53.50		105	145
0670 36" deep		5	1.600		77.50	64		141.50	192
0680 48" deep		4	2		90.50	80		170.50	232
0690 Construction joints, floor slab at exterior wall									
0700 Precast, brick, block or drywall exterior									
0710 2" wide joint	1 Carp	125	.064	L.F.	6.45	2.56		9.01	11.30
0720 4" wide joint	"	75	.107	"	12.90	4.26		17.16	21.50
0730 Metal panel, glass or curtain wall exterior									
0740 2" wide joint	1 Carp	40	.200	L.F.	15.30	8		23.30	30
0750 4" wide joint	"	25	.320	"	21	12.80		33.80	44
0760 Floor slab to drywall partition									
0770 Flat joint	1 Carp	100	.080	L.F.	6.35	3.20		9.55	12.25
0780 Fluted joint		50	.160		12.90	6.40		19.30	25
0790 Etched fluted joint		75	.107		8.40	4.26		12.66	16.30
0800 Floor slab to concrete/masonry partition									
0810 Flat joint	1 Carp	75	.107	L.F.	14.20	4.26		18.46	22.50
0820 Fluted joint	"	50	.160	"	16.80	6.40		23.20	29
0830 Concrete/CMU wall joints									
0840 1" wide	1 Carp	100	.080	L.F.	7.75	3.20		10.95	13.85
0850 2" wide		75	.107		14.20	4.26		18.46	22.50
0860 4" wide		50	.160		27	6.40		33.40	40.50
0870 Concrete/CMU floor joints									
0880 1" wide	1 Carp	200	.040	L.F.	3.88	1.60		5.48	6.90
0890 2" wide		150	.053		7.10	2.13		9.23	11.30
0900 4" wide		100	.080		13.55	3.20		16.75	20

07 92 10 – Caulking and Sealants

07 92 10.10 Caulking and Sealant Options	Crew	Daily Output	Labor-Hours	Unit	Material	2009 Bare Costs Labor	Equipment	Total	Total Incl O&P
0010 **CAULKING AND SEALANT OPTIONS**									
0020 Acoustical sealant, elastomeric, cartridges				Ea.	4.60			4.60	5.05
0032 Backer rod, polyethylene, 1/4" diameter	1 Bric	460	.017	L.F.	.04	.70		.74	1.19
0052 1/2" diameter		460	.017		.07	.70		.77	1.22
0072 3/4" diameter		460	.017		.09	.70		.79	1.24
0092 1" diameter		460	.017		.16	.70		.86	1.32
0100 Acrylic latex caulk, white									
0200 11 fl. oz cartridge				Ea.	2.14			2.14	2.35
0500 1/4" x 1/2"	1 Bric	248	.032	L.F.	.17	1.31		1.48	2.30
0600 1/2" x 1/2"		250	.032		.35	1.30		1.65	2.48
0800 3/4" x 3/4"		230	.035		.79	1.41		2.20	3.15
0900 3/4" x 1"		200	.040		1.05	1.62		2.67	3.77
1000 1" x 1"		180	.044		1.31	1.80		3.11	4.35
1400 Butyl based, bulk				Gal.	26			26	28.50
1500 Cartridges				"	31.50			31.50	34.50
1700 Bulk, in place 1/4" x 1/2", 154 L.F./gal.	1 Bric	230	.035	L.F.	.17	1.41		1.58	2.47
1800 1/2" x 1/2", 77 L.F./gal.	"	180	.044	"	.34	1.80		2.14	3.28
2000 Latex acrylic based, bulk				Gal.	27			27	30
2100 Cartridges				"	33.50			33.50	36.50
2200 Bulk in place, 1/4" x 1/2", 154 L.F./gal.	1 Bric	230	.035	L.F.	.18	1.41		1.59	2.47
2250									
2300 Polysulfide compounds, 1 component, bulk				Gal.	51			51	56
2400 Cartridges				"	54.50			54.50	60
2600 1 or 2 component, in place, 1/4" x 1/4", 308 L.F./gal.	1 Bric	145	.055	L.F.	.17	2.23		2.40	3.79
2700 1/2" x 1/4", 154 L.F./gal.		135	.059		.33	2.40		2.73	4.25
2900 3/4" x 3/8", 68 L.F./gal.		130	.062		.75	2.49		3.24	4.86
3000 1" x 1/2", 38 L.F./gal.		130	.062		1.35	2.49		3.84	5.50
3200 Polyurethane, 1 or 2 component				Gal.	54			54	59.50
3300 Cartridges				"	56			56	61.50
3500 Bulk, in place, 1/4" x 1/4"	1 Bric	150	.053	L.F.	.18	2.16		2.34	3.68
3600 1/2" x 1/4"		145	.055		.35	2.23		2.58	4
3800 3/4" x 3/8", 68 L.F./gal.		130	.062		.79	2.49		3.28	4.90
3900 1" x 1/2"		110	.073		1.40	2.95		4.35	6.30
4100 Silicone rubber, bulk				Gal.	41			41	45
4200 Cartridges				"	41			41	45
4300 Bulk in place, 1/4" x 1/2", 154 L.F./gal.	1 Bric	235	.034	L.F.	.27	1.38		1.65	2.52
4350									
4400 Neoprene gaskets, closed cell, adhesive, 1/8" x 3/8"	1 Bric	240	.033	L.F.	.26	1.35		1.61	2.47
4500 1/4" x 3/4"		215	.037		.58	1.51		2.09	3.08
4700 1/2" x 1"		200	.040		1.30	1.62		2.92	4.05
4800 3/4" x 1-1/2"		165	.048		1.44	1.96		3.40	4.76
5500 Resin epoxy coating, 2 component, heavy duty				Gal.	30.50			30.50	33.50
5802 Tapes, sealant, P.V.C. foam adhesive, 1/16" x 1/4"				L.F.	.06			.06	.06
5902 1/16" x 1/2"					.08			.08	.09
5952 1/16" x 1"					.14			.14	.15
6002 1/8" x 1/2"					.09			.09	.10
6200 Urethane foam, 2 component, handy pack, 1 C.F.				Ea.	34			34	37
6300 50.0 C.F. pack				C.F.	17.20			17.20	18.95
9000 Minimum labor/equipment charge	1 Bric	4	2	Job		81		81	131

07 92 13 – Elastomeric Joint Sealants

07 92 13.10 Masonry Joint Sealants

0010 **MASONRY JOINT SEALANTS**, 1/2" x 1/2" joint									

07 92 Joint Sealants

07 92 13 – Elastomeric Joint Sealants

07 92 13.10 Masonry Joint Sealants

		Crew	Daily Output	Labor-Hours	Unit	Material	2009 Bare Costs Labor	Equipment	Total	Total Incl O&P
0050	Re-caulk only, oil base	1 Bric	225	.036	L.F.	.45	1.44		1.89	2.82
0100	Acrylic latex		205	.039		.22	1.58		1.80	2.80
0200	Polyurethane		200	.040		.43	1.62		2.05	3.10
0300	Silicone		195	.041		.58	1.66		2.24	3.33
1000	Cut out and re-caulk, oil base		145	.055		.45	2.23		2.68	4.10
1050	Acrylic latex		130	.062		.22	2.49		2.71	4.27
1100	Polyurethane		125	.064		.43	2.59		3.02	4.67
1150	Silicone		120	.067		.58	2.70		3.28	5
9000	Minimum labor/equipment charge		4	2	Job		81		81	131

07 95 Expansion Control

07 95 13 – Expansion Joint Cover Assemblies

07 95 13.50 Expansion Joint Assemblies

		Crew	Daily Output	Labor-Hours	Unit	Material	2009 Bare Costs Labor	Equipment	Total	Total Incl O&P
0010	**EXPANSION JOINT ASSEMBLIES**									
0200	Floor cover assemblies, 1" space, aluminum	1 Sswk	38	.211	L.F.	21.50	9.40		30.90	41
0300	Bronze		38	.211		43.50	9.40		52.90	65.50
0500	2" space, aluminum		38	.211		26	9.40		35.40	46
0600	Bronze		38	.211		47	9.40		56.40	69
0800	Wall and ceiling assemblies, 1" space, aluminum		38	.211		12.80	9.40		22.20	31.50
0900	Bronze		38	.211		40.50	9.40		49.90	62.50
1100	2" space, aluminum		38	.211		21.50	9.40		30.90	41.50
1200	Bronze		38	.211		50	9.40		59.40	72.50
1400	Floor to wall assemblies, 1" space, aluminum		38	.211		19.55	9.40		28.95	39
1500	Bronze or stainless		38	.211		49.50	9.40		58.90	72
1700	Gym floor angle covers, aluminum, 3" x 3" angle		46	.174		16.90	7.75		24.65	33
1800	3" x 4" angle		46	.174		19.95	7.75		27.70	36.50
2000	Roof closures, aluminum, flat roof, low profile, 1" space		57	.140		39	6.25		45.25	54.50
2100	High profile		57	.140		48	6.25		54.25	64
2300	Roof to wall, low profile, 1" space		57	.140		21.50	6.25		27.75	35.50
2400	High profile		57	.140		28	6.25		34.25	42
9000	Minimum labor/equipment charge		2	4	Job		179		179	330

Division Notes

		CREW	DAILY OUTPUT	LABOR-HOURS	UNIT	2009 BARE COSTS				TOTAL INCL O&P
						MAT.	LABOR	EQUIP.	TOTAL	

Estimating Tips

08 10 00 Doors and Frames

All exterior doors should be addressed for their energy conservation.

- Most metal doors and frames look alike, but there may be significant differences among them. When estimating these items, be sure to choose the line item that most closely compares to the specification or door schedule requirements regarding:
  - type of metal
  - metal gauge
  - door core material
  - fire rating
  - finish
- Wood and plastic doors vary considerably in price. The primary determinant is the veneer material. Lauan, birch, and oak are the most common veneers. Other variables include the following:
  - hollow or solid core
  - fire rating
  - flush or raised panel
  - finish

08 30 00 Specialty Doors and Frames

- There are many varieties of special doors, and they are usually priced per each. Add frames, hardware, or operators required for a complete installation.

08 40 00 Entrances, Storefronts, and Curtain Walls

- Glazed curtain walls consist of the metal tube framing and the glazing material. The cost data in this subdivision is presented for the metal tube framing alone or the composite wall. If your estimate requires a detailed takeoff of the framing, be sure to add the glazing cost.

08 50 00 Windows

- Most metal windows are delivered preglazed. However, some metal windows are priced without glass. Refer to 08 80 00 Glazing for glass pricing. The grade C indicates commercial grade windows, usually ASTM C-35.
- All wood windows are priced preglazed. The two glazing options priced are single pane float glass and insulating glass 1/2" thick. Add the cost of screens and grills if required.

08 70 00 Hardware

- Hardware costs add considerably to the cost of a door. The most efficient method to determine the hardware requirements for a project is to review the door schedule.

- Door hinges are priced by the pair, with most doors requiring 1-1/2 pairs per door. The hinge prices do not include installation labor because it is included in door installation. Hinges are classified according to the frequency of use.

08 80 00 Glazing

- Different openings require different types of glass. The three most common types are:
  - float
  - tempered
  - insulating
- Most exterior windows are glazed with insulating glass. Entrance doors and window walls, where the glass is less than 18" from the floor, are generally glazed with tempered glass. Interior windows and some residential windows are glazed with float glass.
- Coastal communities are starting to address the use of impact-resistant glass.
- The insulation or 'u' value is a strong consideration.

Reference Numbers

Reference numbers are shown in shaded boxes at the beginning of some major classifications. These numbers refer to related items in the Reference Section. The reference information may be an estimating procedure, an alternate pricing method, or technical information.

Note: Not all subdivisions listed here necessarily appear in this publication.

Division 8 - Openings

08 01 53.81 Solid Vinyl Replacement Windows		Crew	Daily Output	Labor-Hours	Unit	Material	2009 Bare Costs Labor	Equipment	Total	Total Incl O&P
0010	**SOLID VINYL REPLACEMENT WINDOWS** R085313-20									
0020	Double hung, insulated glass, up to 83 united inches G	2 Carp	8	2	Ea.	330	80		410	490
0040	84 to 93 G		8	2		360	80		440	525
0060	94 to 101 G		6	2.667		360	107		467	570
0080	102 to 111 G		6	2.667		380	107		487	590
0100	112 to 120 G		6	2.667		410	107		517	630
0120	For each united inch over 120 , add G		800	.020	Inch	5.20	.80		6	7
0140	Casement windows, one operating sash , 42 to 60 united inches G		8	2	Ea.	191	80		271	340
0160	61 to 70 G		8	2		217	80		297	370
0180	71 to 80 G		8	2		236	80		316	390
0200	81 to 96 G		8	2		252	80		332	410
0220	Two operating sash, 58 to 78 united inches G		8	2		380	80		460	550
0240	79 to 88 G		8	2		405	80		485	580
0260	89 to 98 G		8	2		445	80		525	620
0280	99 to 108 G		6	2.667		465	107		572	690
0300	109 to 121 G		6	2.667		500	107		607	730
0320	Three operating sash, 73 to 108 united inches G		8	2		600	80		680	790
0340	109 to 118 G		8	2		635	80		715	830
0360	119 to 128 G		6	2.667		655	107		762	895
0380	129 to 138 G		6	2.667		700	107		807	945
0400	139 to 156 G		6	2.667		740	107		847	990
0420	Four operating sash, 98 to 118 united inches G		8	2		865	80		945	1,075
0440	119 to 128 G		8	2		930	80		1,010	1,150
0460	129 to 138 G		6	2.667		985	107		1,092	1,250
0480	139 to 148 G		6	2.667		1,050	107		1,157	1,325
0500	149 to 168 G		6	2.667		1,100	107		1,207	1,400
0520	169 to 178 G		6	2.667		1,200	107		1,307	1,500
0540	For venting unit to fixed unit, deduct G					16			16	17.60
0560	Fixed picture window, up to 63 united inches G	2 Carp	8	2		136	80		216	282
0580	64 to 83 G		8	2		160	80		240	310
0600	84 to 101 G		8	2		207	80		287	360
0620	For each united inch over 101, add G		900	.018	Inch	2.29	.71		3	3.69
0640	Picture window opt., low E glazing, up to 101 united inches G				Ea.	19.50			19.50	21.50
0660	102 to 124 G					26			26	28.50
0680	124 and over G					38			38	42
0700	Options, low E glazing, up to w/gas 101 united inches G					31			31	34
0720	102 to 124 G					31			31	34
0740	124 and over G					31			31	34
0760	Muntins, between glazing, square, per lite					3			3	3.30
0780	Diamond shape, per full or partial diamond					4			4	4.40
0800	Cellulose fiber insulation, poured into sash balance cavity G	1 Carp	36	.222	C.F.	.60	8.90		9.50	15.30
0820	Silicone caulking at perimeter G	"	800	.010	L.F.	.13	.40		.53	.81

08 05 05.10 Selective Demolition Doors

		Crew	Daily Output	Labor-Hours	Unit	Material	2009 Bare Costs Labor	2009 Bare Costs Equipment	Total	Total Incl O&P
0010	**SELECTIVE DEMOLITION DOORS** R024119-10									
0200	Doors, exterior, 1-3/4" thick, single, 3' x 7' high	1 Clab	16	.500	Ea.		15.80		15.80	26
0220	Double, 6' x 7' high		12	.667			21		21	35
0500	Interior, 1-3/8" thick, single, 3' x 7' high		20	.400			12.65		12.65	21
0520	Double, 6' x 7' high		16	.500			15.80		15.80	26
0700	Bi-folding, 3' x 6'-8" high		20	.400			12.65		12.65	21
0720	6' x 6'-8" high		18	.444			14.05		14.05	23
0900	Bi-passing, 3' x 6'-8" high		16	.500			15.80		15.80	26
0940	6' x 6'-8" high		14	.571			18.05		18.05	30
1500	Remove and reset, minimum	1 Carp	8	1			40		40	66
1520	Maximum		6	1.333			53.50		53.50	88
2000	Frames, including trim, metal		8	1			40		40	66
2200	Wood	2 Carp	32	.500			20		20	33
2201	Alternate pricing method	1 Carp	200	.040	L.F.		1.60		1.60	2.64
3000	Special doors, counter doors	2 Carp	6	2.667	Ea.		107		107	176
3100	Double acting		10	1.600			64		64	106
3200	Floor door (trap type), or access type		8	2			80		80	132
3300	Glass, sliding, including frames		12	1.333			53.50		53.50	88
3400	Overhead, commercial, 12' x 12' high		4	4			160		160	264
3440	up to 20' x 16' high		3	5.333			213		213	350
3445	up to 35' x 30' high		1	16			640		640	1,050
3500	Residential, 9' x 7' high		8	2			80		80	132
3540	16' x 7' high		7	2.286			91.50		91.50	151
3600	Remove and reset, minimum		4	4			160		160	264
3620	Maximum		2.50	6.400			256		256	420
3700	Roll-up grille		5	3.200			128		128	211
3800	Revolving door		2	8			320		320	530
3900	Storefront swing door		3	5.333			213		213	350
7100	Remove double swing pneumatic doors, openers and sensors	2 Skwk	.50	32	Opng.		1,300		1,300	2,125
7570	Remove shock absorbing door	2 Sswk	1.90	8.421	"		375		375	695
9000	Minimum labor/equipment charge	1 Carp	4	2	Job		80		80	132

08 05 05.20 Selective Demolition of Windows

		Crew	Daily Output	Labor-Hours	Unit	Material	2009 Bare Costs Labor	2009 Bare Costs Equipment	Total	Total Incl O&P
0010	**SELECTIVE DEMOLITION OF WINDOWS** R024119-10									
0200	Aluminum, including trim, to 12 S.F.	1 Clab	16	.500	Ea.		15.80		15.80	26
0240	To 25 S.F.		11	.727			23		23	38
0280	To 50 S.F.		5	1.600			50.50		50.50	83.50
0320	Storm windows/screens, to 12 S.F.		27	.296			9.35		9.35	15.45
0360	To 25 S.F.		21	.381			12.05		12.05	19.85
0400	To 50 S.F.		16	.500			15.80		15.80	26
0600	Glass, minimum		200	.040	S.F.		1.26		1.26	2.09
0620	Maximum		150	.053	"		1.69		1.69	2.78
1000	Steel, including trim, to 12 S.F.		13	.615	Ea.		19.45		19.45	32
1020	To 25 S.F.		9	.889			28		28	46.50
1040	To 50 S.F.		4	2			63		63	104
2000	Wood, including trim, to 12 S.F.		22	.364			11.50		11.50	18.95
2020	To 25 S.F.		18	.444			14.05		14.05	23
2060	To 50 S.F.		13	.615			19.45		19.45	32
2065	To 180 S.F.		8	1			31.50		31.50	52
4410	Remove skylight, plstc domes, flush/curb mtd	G-3	395	.081	S.F.		3.19		3.19	5.15
5020	Remove and reset window, minimum	1 Carp	6	1.333	Ea.		53.50		53.50	88
5040	Average		4	2			80		80	132
5080	Maximum		2	4			160		160	264

08 05 Common Work Results for Openings

08 05 05 – Selective Windows and Doors Demolition

08 05 05.20 Selective Demolition of Windows	Crew	Daily Output	Labor-Hours	Unit	Material	2009 Bare Costs Labor	Equipment	Total	Total Incl O&P
9000 Minimum labor/equipment charge	1 Clab	4	2	Job		63		63	104

08 11 Metal Doors and Frames

08 11 16 – Aluminum Doors and Frames

08 11 16.10 Entrance Doors and Frames

		Crew	Daily Output	Labor-Hours	Unit	Material	2009 Bare Costs Labor	Equipment	Total	Total Incl O&P
0010	**ENTRANCE DOORS AND FRAMES** Aluminum, narrow stile									
0011	Including standard hardware, clear finish, no glass									
0020	3'-0" x 7'-0" opening	2 Sswk	2	8	Ea.	670	360		1,030	1,400
0030	3'-6" x 7'-0" opening		2	8		685	360		1,045	1,400
0100	3'-0" x 10'-0" opening, 3' high transom		1.80	8.889		1,175	395		1,570	2,025
0200	3'-6" x 10'-0" opening, 3' high transom		1.80	8.889		1,175	395		1,570	2,025
0280	5'-0" x 7'-0" opening		2	8		1,050	360		1,410	1,825
0300	6'-0" x 7'-0" opening		1.30	12.308		1,000	550		1,550	2,125
0400	6'-0" x 10'-0" opening, 3' high transom		1.10	14.545	Pr.	1,250	650		1,900	2,575
0420	7'-0" x 7'-0" opening		1	16	"	1,075	715		1,790	2,500
0520	3'-0" x 7'-0" opening, wide stile		2	8	Ea.	885	360		1,245	1,625
0540	3'-6" x 7'-0" opening		2	8		960	360		1,320	1,700
0560	5'-0" x 7'-0" opening		2	8		1,400	360		1,760	2,175
0580	6'-0" x 7'-0" opening		1.30	12.308	Pr.	1,425	550		1,975	2,600
0600	7'-0" x 7'-0" opening		1	16	"	1,525	715		2,240	3,000
1100	For full vision doors, with 1/2" glass, add				Leaf	55%				
1200	For non-standard size, add					67%				
1300	Light bronze finish, add					36%				
1400	Dark bronze finish, add					18%				
1500	For black finish, add					36%				
1600	Concealed panic device, add					1,050			1,050	1,175
1700	Electric striker release, add				Opng.	275			275	305
1800	Floor check, add				Leaf	810			810	890
1900	Concealed closer, add				"	540			540	595
2000	Flush 3' x 7' Insulated, 12"x 12" lite, clear finish	2 Sswk	2	8	Ea.	1,150	360		1,510	1,925
9000	Minimum labor/equipment charge	2 Carp	4	4	Job		160		160	264

08 11 63 – Metal Screen and Storm Doors and Frames

08 11 63.23 Aluminum Screen and Storm Doors and Frames

		Crew	Daily Output	Labor-Hours	Unit	Material	2009 Bare Costs Labor	Equipment	Total	Total Incl O&P
0010	**ALUMINUM SCREEN AND STORM DOORS AND FRAMES**									
0020	Combination storm and screen									
0400	Clear anodic coating, 6'-8" x 2'-6" wide	2 Carp	15	1.067	Ea.	172	42.50		214.50	260
0420	2'-8" wide		14	1.143		156	45.50		201.50	247
0440	3'-0" wide		14	1.143		156	45.50		201.50	247
0500	For 7' door height, add					5%				
1000	Mill finish, 6'-8" x 2'-6" wide	2 Carp	15	1.067	Ea.	225	42.50		267.50	320
1020	2'-8" wide		14	1.143		225	45.50		270.50	325
1040	3'-0" wide		14	1.143		246	45.50		291.50	345
1100	For 7'-0" door, add					5%				
1500	White painted, 6'-8" x 2'-6" wide	2 Carp	15	1.067		305	42.50		347.50	405
1520	2'-8" wide		14	1.143		260	45.50		305.50	360
1540	3'-0" wide		14	1.143		305	45.50		350.50	410
1600	For 7'-0" door, add					5%				
2000	Wood door & screen, see Div. 08 14 33.20									
9000	Minimum labor/equipment charge	1 Carp	4	2	Job		80		80	132

08 12 13 – Hollow Metal Frames

08 12 13.13 Standard Hollow Metal Frames	Crew	Daily Output	Labor-Hours	Unit	Material	2009 Bare Costs Labor	Equipment	Total	Total Incl O&P
0010 **STANDARD HOLLOW METAL FRAMES**									
0020　16 ga., up to 5-3/4" jamb depth									
0025　　6'-8" high, 3'-0" wide, single	2 Carp	16	1	Ea.	123	40		163	202
0028　　　3'-6" wide, single		16	1		116	40		156	193
0030　　　4'-0" wide, single		16	1		116	40		156	193
0040　　　6'-0" wide, double		14	1.143		153	45.50		198.50	245
0045　　　8'-0" wide, double		14	1.143		159	45.50		204.50	251
0100　　7'-0" high, 3'-0" wide, single		16	1		110	40		150	187
0110　　　3'-6" wide, single		16	1		119	40		159	197
0112　　　4'-0" wide, single		16	1		151	40		191	233
0140　　　6'-0" wide, double		14	1.143		160	45.50		205.50	252
0145　　　8'-0" wide, double		14	1.143		150	45.50		195.50	241
1000　16 ga., up to 4-7/8" deep, 7'-0" H, 3'-0" W, single		16	1		130	40		170	209
1140　　　6'-0" wide, double		14	1.143		158	45.50		203.50	249
2800　14 ga., up to 3-7/8" deep, 7'-0" high, 3'-0" wide, single		16	1		134	40		174	213
2840　　　6'-0" wide, double		14	1.143		161	45.50		206.50	253
3000　14 ga., up to 5-3/4" deep, 6'-8" high, 3'-0" wide, single		16	1		140	40		180	220
3002　　　3'-6" wide, single		16	1		140	40		180	220
3005　　　4'-0" wide, single		16	1		140	40		180	220
3600　　up to 5-3/4" jamb depth, 7'-0" high, 4'-0" wide, single		15	1.067		111	42.50		153.50	193
3620　　　6'-0" wide, double		12	1.333		174	53.50		227.50	279
3640　　　8'-0" wide, double		12	1.333		164	53.50		217.50	268
3700　　8'-0" high, 4'-0" wide, single		15	1.067		146	42.50		188.50	231
3740　　　8'-0" wide, double		12	1.333		181	53.50		234.50	287
4000　6-3/4" deep, 7'-0" high, 4'-0" wide, single		15	1.067		172	42.50		214.50	260
4020　　　6'-0" wide, double		12	1.333		204	53.50		257.50	310
4040　　　8'-0", wide double		12	1.333		222	53.50		275.50	330
4100　　8'-0" high, 4'-0" wide, single		15	1.067		174	42.50		216.50	262
4140　　　8'-0" wide, double		12	1.333		202	53.50		255.50	310
4400　8-3/4" deep, 7'-0" high, 4'-0" wide, single		15	1.067		163	42.50		205.50	250
4440　　　8'-0" wide, double		12	1.333		231	53.50		284.50	345
4500　　8'-0" high, 4'-0" wide, single		15	1.067		174	42.50		216.50	262
4540　　　8'-0" wide, double	▼	12	1.333		203	53.50		256.50	310
4900　For welded frames, add					47.50			47.50	52.50
5400　14 ga., "B" label, up to 5-3/4" deep, 7'-0" high, 4'-0" wide, single	2 Carp	15	1.067		170	42.50		212.50	258
5440　　　8'-0" wide, double		12	1.333		208	53.50		261.50	315
5800　　6-3/4" deep, 7'-0" high, 4'-0" wide, single		15	1.067		150	42.50		192.50	236
5840　　　8'-0" wide, double		12	1.333		249	53.50		302.50	360
6200　　8-3/4" deep, 7'-0" high, 4'-0" wide, single		15	1.067		220	42.50		262.50	315
6240　　　8'-0" wide, double	▼	12	1.333	▼	245	53.50		298.50	360
6300　For "A" label use same price as "B" label									
6400　For baked enamel finish, add					30%	15%			
6500　For galvanizing, add					15%				
6600　For hospital stop, add				Ea.	282			282	310
7900　Transom lite frames, fixed, add	2 Carp	155	.103	S.F.	45.50	4.12		49.62	57
8000　　Movable, add	"	130	.123	"	55	4.92		59.92	68.50
9000　Minimum labor/equipment charge	1 Carp	4	2	Job		80		80	132

08 12 13.25 Channel Metal Frames

	Crew	Daily Output	Labor-Hours	Unit	Material	Labor	Equipment	Total	Total Incl O&P
0010 **CHANNEL METAL FRAMES**									
0020　Steel channels with anchors and bar stops									
0100　　6" channel @ 8.2#/L.F., 3' x 7' door, weighs 150#	E-4	13	2.462	Ea.	270	111	10.30	391.30	515
0200　　8" channel @ 11.5#/L.F., 6' x 8' door, weighs 275#	↓	9	3.556	↓	495	161	14.90	670.90	860

08 12 Metal Frames

08 12 13 – Hollow Metal Frames

08 12 13.25 Channel Metal Frames		Crew	Daily Output	Labor-Hours	Unit	Material	2009 Bare Costs Labor	Equipment	Total	Total Incl O&P
0300	8' x 12' door, weighs 400#	E-4	6.50	4.923	Ea.	720	223	20.50	963.50	1,225
0800	For frames without bar stops, light sections, deduct					15%				
0900	Heavy sections, deduct					10%				
9000	Minimum labor/equipment charge	E-4	4	8	Job		360	33.50	393.50	705

08 13 Metal Doors

08 13 13 – Hollow Metal Doors

08 13 13.13 Standard Hollow Metal Doors

0010	STANDARD HOLLOW METAL DOORS R081313-20	Crew	Daily Output	Labor-Hours	Unit	Material	2009 Bare Costs Labor	Equipment	Total	Total Incl O&P
0015	Flush, full panel, hollow core									
0020	1-3/8" thick, 20 ga., 2'-0"x 6'-8"	2 Carp	20	.800	Ea.	267	32		299	345
0040	2'-8" x 6'-8"		18	.889		274	35.50		309.50	360
0060	3'-0" x 6'-8"		17	.941		280	37.50		317.50	365
0100	3'-0" x 7'-0"		17	.941		295	37.50		332.50	385
0120	For vision lite, add					92.50			92.50	102
0140	For narrow lite, add					101			101	111
0320	Half glass, 20 ga., 2'-0" x 6'-8"	2 Carp	20	.800		395	32		427	490
0340	2'-8" x 6'-8"		18	.889		405	35.50		440.50	505
0360	3'-0" x 6'-8"		17	.941		410	37.50		447.50	515
0400	3'-0" x 7'-0"		17	.941		420	37.50		457.50	520
0410	1-3/8" thick, 18 ga., 2'-0"x 6'-8"		20	.800		320	32		352	405
0420	3'-0" x 6'-8"		17	.941		325	37.50		362.50	415
0425	3'-0" x 7'-0"		17	.941		330	37.50		367.50	425
0450	For vision lite, add					92.50			92.50	102
0452	For narrow lite, add					101			101	111
0460	Half glass, 18 ga., 2'-0" x 6'-8"	2 Carp	20	.800		450	32		482	550
0465	2'-8" x 6'-8"		18	.889		465	35.50		500.50	570
0470	3'-0" x 6'-8"		17	.941		455	37.50		492.50	560
0475	3'-0" x 7'-0"		17	.941		460	37.50		497.50	565
0500	Hollow core, 1-3/4" thick, full panel, 20 ga., 2'-8" x 6'-8"		18	.889		305	35.50		340.50	395
0520	3'-0" x 6'-8"		17	.941		280	37.50		317.50	365
0640	3'-0" x 7'-0"		17	.941		325	37.50		362.50	420
0680	4'-0" x 7'-0"		15	1.067		465	42.50		507.50	580
0700	4'-0" x 8'-0"		13	1.231		530	49		579	665
1000	18 ga., 2'-8" x 6'-8"		17	.941		310	37.50		347.50	400
1020	3'-0" x 6'-8"		16	1		299	40		339	395
1120	3'-0" x 7'-0"		17	.941		370	37.50		407.50	465
1180	4'-0" x 7'-0"		14	1.143		450	45.50		495.50	570
1200	4'-0" x 8'-0"		17	.941		535	37.50		572.50	645
1212	For vision lite, add					92.50			92.50	102
1214	For narrow lite, add					101			101	111
1230	Half glass, 20 ga., 2'-8" x 6'-8"	2 Carp	20	.800		430	32		462	530
1240	3'-0" x 6'-8"		18	.889		420	35.50		455.50	525
1260	3'-0" x 7'-0"		18	.889		425	35.50		460.50	525
1320	18 ga., 2'-8" x 6'-8"		18	.889		460	35.50		495.50	565
1340	3'-0" x 6'-8"		17	.941		455	37.50		492.50	560
1360	3'-0" x 7'-0"		17	.941		465	37.50		502.50	570
1380	4'-0" x 7'-0"		15	1.067		580	42.50		622.50	705
1400	4'-0" x 8'-0"		14	1.143		660	45.50		705.50	805
1720	Insulated, 1-3/4" thick, full panel, 18 ga., 3'-0" x 6'-8"		15	1.067		400	42.50		442.50	515
1740	2'-8" x 7'-0"		16	1		420	40		460	525

08 13 Metal Doors

08 13 13 – Hollow Metal Doors

08 13 13.13 Standard Hollow Metal Doors		Crew	Daily Output	Labor-Hours	Unit	Material	2009 Bare Costs Labor	Equipment	Total	Total Incl O&P
1760	3'-0" x 7'-0"	2 Carp	15	1.067	Ea.	415	42.50		457.50	525
1800	4'-0" x 8'-0"	↓	13	1.231		605	49		654	745
1805	For vision lite, add					92.50			92.50	102
1810	For narrow lite, add					101			101	111
1820	Half glass, 18 ga., 3'-0" x 6'-8"	2 Carp	16	1		535	40		575	655
1840	2'-8" x 7'-0"		17	.941		545	37.50		582.50	660
1860	3'-0" x 7'-0"		16	1		580	40		620	705
1900	4'-0" x 8'-0"	↓	14	1.143		655	45.50		700.50	795
2000	For bottom louver, add				↓	162			162	178
2020	For baked enamel finish, add					30%	15%			
2040	For galvanizing, add					15%				
9000	Minimum labor/equipment charge	1 Carp	4	2	Job		80		80	132

08 13 13.15 Metal Fire Doors

08 13 13.15 Metal Fire Doors		Crew	Daily Output	Labor-Hours	Unit	Material	2009 Bare Costs Labor	Equipment	Total	Total Incl O&P
0010	**METAL FIRE DOORS** R081313-20									
0015	Steel, flush, "B" label, 90 minute									
0020	Full panel, 20 ga., 2'-0" x 6'-8"	2 Carp	20	.800	Ea.	340	32		372	430
0040	2'-8" x 6'-8"		18	.889		345	35.50		380.50	440
0060	3'-0" x 6'-8"		17	.941		345	37.50		382.50	440
0080	3'-0" x 7'-0"		17	.941		360	37.50		397.50	455
0140	18 ga., 3'-0" x 6'-8"		16	1		395	40		435	500
0160	2'-8" x 7'-0"		17	.941		415	37.50		452.50	515
0180	3'-0" x 7'-0"		16	1		395	40		435	500
0200	4'-0" x 7'-0"	↓	15	1.067	↓	520	42.50		562.50	640
0220	For "A" label, 3 hour, 18 ga., use same price as "B" label									
0240	For vision lite, add				Ea.	112			112	124
0520	Flush, "B" label 90 min., composite, 20 ga., 2'-0" x 6'-8"	2 Carp	18	.889		430	35.50		465.50	535
0540	2'-8" x 6'-8"		17	.941		435	37.50		472.50	540
0560	3'-0" x 6'-8"		16	1		440	40		480	545
0580	3'-0" x 7'-0"		16	1		450	40		490	560
0640	Flush, "A" label 3 hour, composite, 18 ga., 3'-0" x 6'-8"		15	1.067		495	42.50		537.50	615
0660	2'-8" x 7'-0"		16	1		515	40		555	635
0680	3'-0" x 7'-0"		15	1.067		505	42.50		547.50	625
0700	4'-0" x 7'-0"	↓	14	1.143	↓	610	45.50		655.50	750
9000	Minimum labor/equipment charge	1 Carp	4	2	Job		80		80	132

08 13 13.20 Residential Steel Doors

08 13 13.20 Residential Steel Doors		Crew	Daily Output	Labor-Hours	Unit	Material	2009 Bare Costs Labor	Equipment	Total	Total Incl O&P
0010	**RESIDENTIAL STEEL DOORS**									
0020	Prehung, insulated, exterior									
0030	Embossed, full panel, 2'-8" x 6'-8" **G**	2 Carp	17	.941	Ea.	277	37.50		314.50	365
0040	3'-0" x 6'-8" **G**		15	1.067		249	42.50		291.50	345
0060	3'-0" x 7'-0" **G**		15	1.067		291	42.50		333.50	390
0070	5'-4" x 6'-8", double **G**		8	2		575	80		655	765
0220	Half glass, 2'-8" x 6'-8"		17	.941		278	37.50		315.50	365
0240	3'-0" x 6'-8"		16	1		278	40		318	370
0260	3'-0" x 7'-0"		16	1		335	40		375	435
0270	5'-4" x 6'-8", double		8	2		570	80		650	760
0720	Raised plastic face, full panel, 2'-8" x 6'-8"		16	1		278	40		318	370
0740	3'-0" x 6'-8"		15	1.067		280	42.50		322.50	380
0760	3'-0" x 7'-0"		15	1.067		284	42.50		326.50	380
0780	5'-4" x 6'-8", double		8	2		525	80		605	710
0820	Half glass, 2'-8" x 6'-8"		17	.941		310	37.50		347.50	405
0840	3'-0" x 6'-8"		16	1		315	40		355	410
0860	3'-0" x 7'-0"	↓	16	1	↓	345	40		385	445

08 13 Metal Doors

08 13 13 – Hollow Metal Doors

08 13 13.20 Residential Steel Doors	Crew	Daily Output	Labor-Hours	Unit	Material	2009 Bare Costs Labor	Equipment	Total	Total Incl O&P
0880	5'-4" x 6'-8", double	2 Carp	8	2	Ea.	670	80		750	865
1320	Flush face, full panel, 2'-8" x 6'-8"		16	1		230	40		270	320
1340	3'-0" x 6'-8"		15	1.067		230	42.50		272.50	325
1360	3'-0" x 7'-0"		15	1.067		290	42.50		332.50	390
1380	5'-4" x 6'-8", double		8	2		470	80		550	645
1420	Half glass, 2'-8" x 6'-8"		17	.941		289	37.50		326.50	380
1440	3'-0" x 6'-8"		16	1		289	40		329	385
1460	3'-0" x 7'-0"		16	1		335	40		375	435
1480	5'-4" x 6'-8", double		8	2		555	80		635	740
1500	Sidelight, full lite, 1'-0" x 6'-8" with grille					228			228	251
1510	1'-0" x 6'-8", low e					248			248	273
1520	1'-0" x 6'-8", half lite					254			254	280
1530	1'-0" x 6'-8", half lite, low e					262			262	289
2300	Interior, residential, closet, bi-fold, 6'-8" x 2'-0" wide	2 Carp	16	1		148	40		188	229
2330	3'-0" wide		16	1		165	40		205	248
2360	4'-0" wide		15	1.067		251	42.50		293.50	345
2400	5'-0" wide		14	1.143		290	45.50		335.50	395
2420	6'-0" wide		13	1.231		325	49		374	440
9000	Minimum labor/equipment charge	1 Carp	4	2	Job		80		80	132

08 13 16 – Aluminum Doors

08 13 16.10 Commercial Aluminum Doors

08 13 16.10 Commercial Aluminum Doors	Crew	Daily Output	Labor-Hours	Unit	Material	2009 Bare Costs Labor	Equipment	Total	Total Incl O&P
0010	**COMMERCIAL ALUMINUM DOORS**, no glazing									
0020	Incl. hinges, push/pull, deadlock, cyl., threshold									
1000	Narrow stile, no glazing, standard hardware, 3'-0" x 7'-0", single	2 Carp	3	5.333	Ea.	590	213		803	995
1200	Pair of 3'-0" x 7'-0"		1.70	9.412	Pr.	1,175	375		1,550	1,900
1500	3'-6" x 7'-0", single		3	5.333	Ea.	850	213		1,063	1,275
2100	Medium stile, 3'-0" x 7'-0", single		3	5.333	"	720	213		933	1,150
2200	Pair of 3'-0" x 7'-0"		1.70	9.412	Pr.	1,400	375		1,775	2,175
2300	3'-6" x 7'-0", single		3	5.333	Ea.	930	213		1,143	1,375
5000	Flush panel doors, pair of 2'-6" x 7'-0"	2 Sswk	2	8	Pr.	1,275	360		1,635	2,050
5050	3'-0" x 7'-0", single		2.50	6.400	Ea.	640	286		926	1,225
5100	Pair of 3'-0" x 7'-0"		2	8	Pr.	1,275	360		1,635	2,050
5150	3'-6" x 7'-0", single		2.50	6.400	Ea.	760	286		1,046	1,375

08 14 Wood Doors

08 14 13 – Carved Wood Doors

08 14 13.10 Types of Wood Doors, Carved

08 14 13.10 Types of Wood Doors, Carved	Crew	Daily Output	Labor-Hours	Unit	Material	2009 Bare Costs Labor	Equipment	Total	Total Incl O&P
0010	**TYPES OF WOOD DOORS, CARVED**									
3000	Solid wood, 1-3/4" thick stile and rail									
3020	Mahogany, 3'-0" x 7'-0", minimum	2 Carp	14	1.143	Ea.	850	45.50		895.50	1,000
3030	Maximum		10	1.600		1,350	64		1,414	1,575
3040	3'-6" x 8'-0", minimum		10	1.600		980	64		1,044	1,175
3050	Maximum		8	2		1,800	80		1,880	2,100
3100	Pine, 3'-0" x 7'-0", minimum		14	1.143		415	45.50		460.50	530
3110	Maximum		10	1.600		705	64		769	880
3120	3'-6" x 8'-0", minimum		10	1.600		750	64		814	930
3130	Maximum		8	2		1,200	80		1,280	1,450
3200	Red oak, 3'-0" x 7'-0", minimum		14	1.143		1,600	45.50		1,645.50	1,850
3210	Maximum		10	1.600		1,925	64		1,989	2,225
3220	3'-6" x 8'-0", minimum		10	1.600		1,775	64		1,839	2,050

08 14 Wood Doors

08 14 13 – Carved Wood Doors

08 14 13.10 Types of Wood Doors, Carved

		Crew	Daily Output	Labor-Hours	Unit	Material	2009 Bare Costs Labor	Equipment	Total	Total Incl O&P
3230	Maximum	2 Carp	8	2	Ea.	3,150	80		3,230	3,600
4000	Hand carved door, mahogany									
4020	3'-0" x 7'-0", minimum	2 Carp	14	1.143	Ea.	1,600	45.50		1,645.50	1,825
4030	Maximum		11	1.455		3,125	58		3,183	3,550
4040	3'-6" x 8'-0", minimum		10	1.600		2,000	64		2,064	2,300
4050	Maximum		8	2		3,075	80		3,155	3,500
4200	Rose wood, 3'-0" x 7'-0", minimum		14	1.143		4,850	45.50		4,895.50	5,425
4210	Maximum		11	1.455		13,300	58		13,358	14,700
4220	3'-6" x 8'-0", minimum		10	1.600		5,475	64		5,539	6,125
4280	For 6'-8" high door, deduct from 7'-0" door					34			34	37.50
4400	For custom finish, add					370			370	405
4600	Side light, mahogany, 7'-0" x 1'-6" wide, minimum	2 Carp	18	.889		895	35.50		930.50	1,050
4610	Maximum		14	1.143		2,625	45.50		2,670.50	2,975
4620	8'-0" x 1'-6" wide, minimum		14	1.143		1,675	45.50		1,720.50	1,925
4630	Maximum		10	1.600		1,900	64		1,964	2,200
4640	Side light, oak, 7'-0" x 1'-6" wide, minimum		18	.889		1,025	35.50		1,060.50	1,175
4650	Maximum		14	1.143		1,825	45.50		1,870.50	2,100
4660	8'-0" x 1'-6" wide, minimum		14	1.143		970	45.50		1,015.50	1,150
4670	Maximum		10	1.600		1,825	64		1,889	2,125

08 14 16 – Flush Wood Doors

08 14 16.09 Smooth Wood Doors

		Crew	Daily Output	Labor-Hours	Unit	Material	2009 Bare Costs Labor	Equipment	Total	Total Incl O&P
0010	**SMOOTH WOOD DOORS**									
0015	Flush, int., 1-3/8", 7 ply, hollow core,									
0020	Lauan face, 2'-0" x 6'-8"	2 Carp	17	.941	Ea.	32.50	37.50		70	97.50
0080	3'-0" x 6'-8"		17	.941		50	37.50		87.50	117
0100	4'-0" x 6'-8"		16	1		81	40		121	155
0120	Birch face, 2'-0" x 6'-8"		17	.941		44.50	37.50		82	111
0140	2'-6" x 6'-8"		17	.941		81	37.50		118.50	151
0180	3'-0" x 6'-8"		17	.941		90	37.50		127.50	161
0200	4'-0" x 6'-8"		16	1		110	40		150	187
0220	Oak face, 2'-0" x 6'-8"		17	.941		84.50	37.50		122	155
0240	2'-6" x 6'-8"		17	.941		90	37.50		127.50	161
0280	3'-0" x 6'-8"		17	.941		96	37.50		133.50	168
0300	4'-0" x 6'-8"		16	1		122	40		162	200
0320	Walnut face, 2'-0" x 6'-8"		17	.941		170	37.50		207.50	249
0340	2'-6" x 6'-8"		17	.941		174	37.50		211.50	253
0380	3'-0" x 6'-8"		17	.941		180	37.50		217.50	260
0400	4'-0" x 6'-8"		16	1		204	40		244	291
0430	For 7'-0" high, add					15.20			15.20	16.70
0440	For 8'-0" high, add					28			28	31
0480	For prefinishing, clear, add					34			34	37.50
0500	For prefinishing, stain, add					45			45	49.50
1320	M.D. overlay on hardboard, 2'-0" x 6'-8"	2 Carp	17	.941		94	37.50		131.50	165
1340	2'-6" x 6'-8"		17	.941		94	37.50		131.50	165
1380	3'-0" x 6'-8"		17	.941		112	37.50		149.50	185
1400	4'-0" x 6'-8"		16	1		154	40		194	235
1420	For 7'-0" high, add					8.50			8.50	9.35
1440	For 8'-0" high, add					23			23	25.50
1720	H.P. plastic laminate, 2'-0" x 6'-8"	2 Carp	16	1		224	40		264	310
1740	2'-6" x 6'-8"		16	1		224	40		264	310
1780	3'-0" x 6'-8"		15	1.067		268	42.50		310.50	365
1800	4'-0" x 6'-8"		14	1.143		355	45.50		400.50	470

08 14 16 – Flush Wood Doors

08 14 16.09 Smooth Wood Doors		Crew	Daily Output	Labor-Hours	Unit	Material	2009 Bare Costs Labor	Equipment	Total	Total Incl O&P
1820	For 7'-0" high, add				Ea.	9			9	9.90
1840	For 8'-0" high, add					23			23	25.50
2020	5 ply particle core, lauan face, 2'-6" x 6'-8"	2 Carp	15	1.067		82	42.50		124.50	161
2040	3'-0" x 6'-8"		14	1.143		85.50	45.50		131	170
2080	3'-0" x 7'-0"		13	1.231		92.50	49		141.50	183
2100	4'-0" x 7'-0"		12	1.333		110	53.50		163.50	210
2120	Birch face, 2'-6" x 6'-8"		15	1.067		93.50	42.50		136	174
2140	3'-0" x 6'-8"		14	1.143		102	45.50		147.50	189
2180	3'-0" x 7'-0"		13	1.231		105	49		154	196
2200	4'-0" x 7'-0"		12	1.333		128	53.50		181.50	228
2220	Oak face, 2'-6" x 6'-8"		15	1.067		103	42.50		145.50	185
2240	3'-0" x 6'-8"		14	1.143		114	45.50		159.50	201
2280	3'-0" x 7'-0"		13	1.231		117	49		166	209
2300	4'-0" x 7'-0"		12	1.333		143	53.50		196.50	245
2320	Walnut face, 2'-0" x 6'-8"		15	1.067		114	42.50		156.50	197
2340	2'-6" x 6'-8"		14	1.143		131	45.50		176.50	220
2380	3'-0" x 6'-8"		13	1.231		147	49		196	243
2400	4'-0" x 6'-8"	▼	12	1.333		192	53.50		245.50	299
2440	For 8'-0" high, add					28			28	31
2460	For 8'-0" high walnut, add					15.35			15.35	16.90
2480	For solid wood core, add					34.50			34.50	38
2720	For prefinishing, clear, add					22.50			22.50	24.50
2740	For prefinishing, stain, add					50			50	55
3320	M.D. overlay on hardboard, 2'-6" x 6'-8"	2 Carp	14	1.143		104	45.50		149.50	190
3340	3'-0" x 6'-8"		13	1.231		109	49		158	200
3380	3'-0" x 7'-0"		12	1.333		111	53.50		164.50	210
3400	4'-0" x 7'-0"	▼	10	1.600		136	64		200	255
3440	For 8'-0" height, add					30.50			30.50	33.50
3460	For solid wood core, add					39.50			39.50	43.50
3720	H.P. plastic laminate, 2'-6" x 6'-8"	2 Carp	13	1.231		152	49		201	248
3740	3'-0" x 6'-8"		12	1.333		172	53.50		225.50	278
3780	3'-0" x 7'-0"		11	1.455		179	58		237	292
3800	4'-0" x 7'-0"	▼	8	2		217	80		297	370
3840	For 8'-0" height, add					30.50			30.50	33.50
3860	For solid wood core, add					37			37	40.50
4000	Exterior, flush, solid wood stave core, birch, 1-3/4" x 7'-0" x 2'-6"	2 Carp	15	1.067		162	42.50		204.50	249
4020	2'-8" wide		15	1.067		170	42.50		212.50	258
4040	3'-0" wide		14	1.143		185	45.50		230.50	280
4100	Oak faced 1-3/4" x 7'-0" x 2'-6" wide		15	1.067		179	42.50		221.50	268
4120	2'-8" wide		15	1.067		192	42.50		234.50	282
4140	3'-0" wide	▼	14	1.143		204	45.50		249.50	300
4160	Walnut faced, 1-3/4" x 6'-8" x 3'-0" wide	1 Carp	17	.471		274	18.80		292.80	330
4180	3'-6" wide	"	17	.471		375	18.80		393.80	440
4200	Walnut faced, 1-3/4" x 7'-0" x 2'-6" wide	2 Carp	15	1.067		263	42.50		305.50	360
4220	2'-8" wide		15	1.067		275	42.50		317.50	375
4240	3'-0" wide	▼	14	1.143		287	45.50		332.50	390
4300	For 6'-8" high door, deduct from 7'-0" door				▼	15			15	16.50
9000	Minimum labor/equipment charge	1 Carp	4	2	Job		80		80	132

08 14 16.10 Wood Doors Decorator

		Crew	Daily Output	Labor-Hours	Unit	Material	Labor	Equipment	Total	Total Incl O&P
0010	**WOOD DOORS DECORATOR**									
0040	7 ply hollow core lauan face, 2'-6" x 6'-8"	2 Carp	17	.941	Ea.	36	37.50		73.50	102

08 14 16 – Flush Wood Doors

08 14 16.20 Wood Fire Doors

		Crew	Daily Output	Labor-Hours	Unit	Material	2009 Bare Costs Labor	Equipment	Total	Total Incl O&P
0010	**WOOD FIRE DOORS**									
0020	Particle core, 7 face plys, "B" label,									
0040	1 hour, birch face, 1-3/4" x 2'-6" x 6'-8"	2 Carp	14	1.143	Ea.	320	45.50		365.50	425
0080	3'-0" x 6'-8"		13	1.231		330	49		379	445
0090	3'-0" x 7'-0"		12	1.333		345	53.50		398.50	470
0100	4'-0" x 7'-0"		12	1.333		500	53.50		553.50	640
0140	Oak face, 2'-6" x 6'-8"		14	1.143		310	45.50		355.50	420
0180	3'-0" x 6'-8"		13	1.231		325	49		374	435
0190	3'-0" x 7'-0"		12	1.333		340	53.50		393.50	460
0200	4'-0" x 7'-0"		12	1.333		440	53.50		493.50	575
0240	Walnut face, 2'-6" x 6'-8"		14	1.143		410	45.50		455.50	525
0280	3'-0" x 6'-8"		13	1.231		420	49		469	540
0290	3'-0" x 7'-0"		12	1.333		440	53.50		493.50	570
0300	4'-0" x 7'-0"		12	1.333		590	53.50		643.50	740
0440	M.D. overlay on hardboard, 2'-6" x 6'-8"		15	1.067		274	42.50		316.50	370
0480	3'-0" x 6'-8"		14	1.143		285	45.50		330.50	390
0490	3'-0" x 7'-0"		13	1.231		300	49		349	410
0500	4'-0" x 7'-0"		12	1.333		370	53.50		423.50	495
0740	90 minutes, birch face, 1-3/4" x 2'-6" x 6'-8"		14	1.143		310	45.50		355.50	415
0780	3'-0" x 6'-8"		13	1.231		305	49		354	415
0790	3'-0" x 7'-0"		12	1.333		340	53.50		393.50	465
0800	4'-0" x 7'-0"		12	1.333		415	53.50		468.50	545
0840	Oak face, 2'-6" x 6'-8"		14	1.143		282	45.50		327.50	385
0880	3'-0" x 6'-8"		13	1.231		292	49		341	400
0890	3'-0" x 7'-0"		12	1.333		305	53.50		358.50	425
0900	4'-0" x 7'-0"		12	1.333		425	53.50		478.50	555
0940	Walnut face, 2'-6" x 6'-8"		14	1.143		385	45.50		430.50	500
0980	3'-0" x 6'-8"		13	1.231		395	49		444	515
0990	3'-0" x 7'-0"		12	1.333		415	53.50		468.50	545
1000	4'-0" x 7'-0"		12	1.333		595	53.50		648.50	745
1140	M.D. overlay on hardboard, 2'-6" x 6'-8"		15	1.067		305	42.50		347.50	405
1180	3'-0" x 6'-8"		14	1.143		315	45.50		360.50	420
1190	3'-0" x 7'-0"		13	1.231		325	49		374	440
1200	4'-0" x 7'-0"		12	1.333		445	53.50		498.50	580
1240	For 8'-0" height, add					60			60	66
1260	For 8'-0" height walnut, add					75			75	82.50
2200	Custom architectural "B" label, flush, 1-3/4" thick, birch,									
2210	Solid core									
2220	2'-6" x 7'-0"	2 Carp	15	1.067	Ea.	289	42.50		331.50	390
2260	3'-0" x 7'-0"		14	1.143		300	45.50		345.50	405
2300	4'-0" x 7'-0"		13	1.231		395	49		444	515
2420	4'-0" x 8'-0"		11	1.455		375	58		433	510
2480	For oak veneer, add					50%				
2500	For walnut veneer, add					75%				
9000	Minimum labor/equipment charge	1 Carp	4	2	Job		80		80	132

08 14 23 – Clad Wood Doors

08 14 23.13 Metal-Faced Wood Doors

		Crew	Daily Output	Labor-Hours	Unit	Material	2009 Bare Costs Labor	Equipment	Total	Total Incl O&P
0010	**METAL-FACED WOOD DOORS**									
0020	Interior, flush type, 3' x 7'	2 Carp	4.30	3.721	Opng.	233	149		382	500
9000	Minimum labor/equipment charge	1 Carp	2	4	Job		160		160	264

08 14 23 – Clad Wood Doors

08 14 23.20 Tin Clad Wood Doors	Crew	Daily Output	Labor-Hours	Unit	Material	2009 Bare Costs Labor	Equipment	Total	Total Incl O&P
0010 **TIN CLAD WOOD DOORS**									
0020 3 ply, 6' x 7', double sliding, doors only	2 Carp	1	16	Opng.	1,825	640		2,465	3,075
1000 For electric operator, add	1 Elec	2	4	"	3,250	188		3,438	3,875
9000 Minimum labor/equipment charge	2 Carp	1	16	Job		640		640	1,050

08 14 33 – Stile and Rail Wood Doors

08 14 33.10 Wood Doors Paneled

08 14 33.10 Wood Doors Paneled	Crew	Daily Output	Labor-Hours	Unit	Material	2009 Bare Costs Labor	Equipment	Total	Total Incl O&P
0010 **WOOD DOORS PANELED**									
0020 Interior, six panel, hollow core, 1-3/8" thick									
0040 Molded hardboard, 2'-0" x 6'-8"	2 Carp	17	.941	Ea.	55	37.50		92.50	123
0060 2'-6" x 6'-8"		17	.941		59.50	37.50		97	128
0070 2'-8" x 6'-8"		17	.941		62	37.50		99.50	131
0080 3'-0" x 6'-8"		17	.941		65.50	37.50		103	134
0140 Embossed print, molded hardboard, 2'-0" x 6'-8"		17	.941		59.50	37.50		97	128
0160 2'-6" x 6'-8"		17	.941		59.50	37.50		97	128
0180 3'-0" x 6'-8"		17	.941		65.50	37.50		103	134
0540 Six panel, solid, 1-3/8" thick, pine, 2'-0" x 6'-8"		15	1.067		144	42.50		186.50	229
0560 2'-6" x 6'-8"		14	1.143		162	45.50		207.50	254
0580 3'-0" x 6'-8"		13	1.231		186	49		235	286
1020 Two panel, bored rail, solid, 1-3/8" thick, pine, 1'-6" x 6'-8"		16	1		263	40		303	355
1040 2'-0" x 6'-8"		15	1.067		345	42.50		387.50	450
1060 2'-6" x 6'-8"		14	1.143		395	45.50		440.50	510
1340 Two panel, solid, 1-3/8" thick, fir, 2'-0" x 6'-8"		15	1.067		144	42.50		186.50	229
1360 2'-6" x 6'-8"		14	1.143		162	45.50		207.50	254
1380 3'-0" x 6'-8"		13	1.231		395	49		444	515
1740 Five panel, solid, 1-3/8" thick, fir, 2'-0" x 6'-8"		15	1.067		258	42.50		300.50	355
1760 2'-6" x 6'-8"		14	1.143		415	45.50		460.50	530
1780 3'-0" x 6'-8"		13	1.231		415	49		464	535
9000 Minimum labor/equipment charge	1 Carp	4	2	Job		80		80	132

08 14 33.20 Wood Doors Residential

08 14 33.20 Wood Doors Residential	Crew	Daily Output	Labor-Hours	Unit	Material	2009 Bare Costs Labor	Equipment	Total	Total Incl O&P
0010 **WOOD DOORS RESIDENTIAL**									
0200 Exterior, combination storm & screen, pine									
0260 2'-8" wide	2 Carp	10	1.600	Ea.	276	64		340	410
0280 3'-0" wide		9	1.778		282	71		353	425
0300 7'-1" x 3'-0" wide		9	1.778		325	71		396	470
0400 Full lite, 6'-9" x 2'-6" wide		11	1.455		296	58		354	420
0420 2'-8" wide		10	1.600		296	64		360	430
0440 3'-0" wide		9	1.778		300	71		371	445
0500 7'-1" x 3'-0" wide		9	1.778		325	71		396	470
0700 Dutch door, pine, 1-3/4" x 6'-8" x 2'-8" wide, minimum		12	1.333		675	53.50		728.50	835
0720 Maximum		10	1.600		825	64		889	1,025
0800 3'-0" wide, minimum		12	1.333		690	53.50		743.50	845
0820 Maximum		10	1.600		900	64		964	1,100
1000 Entrance door, colonial, 1-3/4" x 6'-8" x 2'-8" wide		16	1		450	40		490	560
1020 6 panel pine, 3'-0" wide		15	1.067		430	42.50		472.50	545
1100 8 panel pine, 2'-8" wide		16	1		570	40		610	690
1120 3'-0" wide		15	1.067		555	42.50		597.50	680
1200 For tempered safety glass lites, (min of 2) add					65.50			65.50	72
1300 Flush, birch, solid core, 1-3/4" x 6'-8" x 2'-8" wide	2 Carp	16	1		102	40		142	179
1320 3'-0" wide		15	1.067		114	42.50		156.50	196
1350 7'-0" x 2'-8" wide		16	1		112	40		152	189
1360 3'-0" wide		15	1.067		119	42.50		161.50	202

08 14 33.20 Wood Doors Residential	Crew	Daily Output	Labor-Hours	Unit	Material	2009 Bare Costs Labor	Equipment	Total	Total Incl O&P
1380	For tempered safety glass lites, add				Ea.	97.50			97.50	107
2700	Interior, closet, bi-fold, w/hardware, no frame or trim incl.									
2720	Flush, birch, 6'-6" or 6'-8" x 2'-6" wide	2 Carp	13	1.231	Ea.	51	49		100	137
2740	3'-0" wide		13	1.231		55	49		104	141
2760	4'-0" wide		12	1.333		97.50	53.50		151	195
2780	5'-0" wide		11	1.455		100	58		158	206
2800	6'-0" wide		10	1.600		109	64		173	226
2820	Flush, hardboard, primed, 6'-8" x 2'-6" wide		13	1.231		46.50	49		95.50	132
2840	3'-0" wide		13	1.231		51	49		100	137
2860	4'-0" wide		12	1.333		81	53.50		134.50	177
2880	5'-0" wide		11	1.455		92.50	58		150.50	197
2900	6'-0" wide		10	1.600		113	64		177	231
3000	Raised panel pine, 6'-6" or 6'-8" x 2'-6" wide		13	1.231		162	49		211	260
3020	3'-0" wide		13	1.231		254	49		303	360
3040	4'-0" wide		12	1.333		305	53.50		358.50	425
3060	5'-0" wide		11	1.455		365	58		423	495
3080	6'-0" wide		10	1.600		400	64		464	545
3200	Louvered, pine 6'-6" or 6'-8" x 2'-6" wide		13	1.231		106	49		155	197
3220	3'-0" wide		13	1.231		165	49		214	262
3240	4'-0" wide		12	1.333		198	53.50		251.50	305
3260	5'-0" wide		11	1.455		224	58		282	345
3280	6'-0" wide		10	1.600		248	64		312	380
4400	Bi-passing closet, incl. hardware and frame, no trim incl.									
4420	Flush, lauan, 6'-8" x 4'-0" wide	2 Carp	12	1.333	Opng.	169	53.50		222.50	273
4440	5'-0" wide		11	1.455		184	58		242	298
4460	6'-0" wide		10	1.600		198	64		262	325
4600	Flush, birch, 6'-8" x 4'-0" wide		12	1.333		213	53.50		266.50	320
4620	5'-0" wide		11	1.455		209	58		267	325
4640	6'-0" wide		10	1.600		256	64		320	385
4800	Louvered, pine, 6'-8" x 4'-0" wide		12	1.333		420	53.50		473.50	550
4820	5'-0" wide		11	1.455		390	58		448	520
4840	6'-0" wide		10	1.600		515	64		579	670
5000	Paneled, pine, 6'-8" x 4'-0" wide		12	1.333		495	53.50		548.50	635
5020	5'-0" wide		11	1.455		405	58		463	540
5040	6'-0" wide		10	1.600		590	64		654	755
6100	Folding accordion, closet, including track and frame									
6120	Vinyl, 2 layer, stock see Div. 10 22 26.13	2 Carp	10	1.600	Ea.	51.50	64		115.50	163
6200	Rigid PVC	"	10	1.600	"	32	64		96	141
6220	For custom partition, add					25%	10%			
7310	Passage doors, flush, no frame included									
7320	Hardboard, hollow core, 1-3/8" x 6'-8" x 1'-6" wide	2 Carp	18	.889	Ea.	42.50	35.50		78	105
7330	2'-0" wide		18	.889		42	35.50		77.50	105
7340	2'-6" wide		18	.889		46.50	35.50		82	110
7350	2'-8" wide		18	.889		48	35.50		83.50	112
7360	3'-0" wide		17	.941		51	37.50		88.50	118
7420	Lauan, hollow core, 1-3/8" x 6'-8" x 1'-6" wide		18	.889		30.50	35.50		66	92
7440	2'-0" wide		18	.889		28	35.50		63.50	89
7450	2'-4" wide		18	.889		31.50	35.50		67	93
7460	2'-6" wide		18	.889		31.50	35.50		67	93
7480	2'-8" wide		18	.889		33	35.50		68.50	95
7500	3'-0" wide		17	.941		34.50	37.50		72	100
7700	Birch, hollow core, 1-3/8" x 6'-8" x 1'-6" wide		18	.889		37	35.50		72.50	99.50
7720	2'-0" wide		18	.889		41	35.50		76.50	104

187

08 14 Wood Doors

08 14 33 – Stile and Rail Wood Doors

08 14 33.20 Wood Doors Residential

		Crew	Daily Output	Labor-Hours	Unit	Material	2009 Bare Costs Labor	Equipment	Total	Total Incl O&P
7740	2'-6" wide	2 Carp	18	.889	Ea.	49	35.50		84.50	112
7760	2'-8" wide		18	.889		49.50	35.50		85	113
7780	3'-0" wide		17	.941		51	37.50		88.50	118
8000	Pine louvered, 1-3/8" x 6'-8" x 1'-6" wide		19	.842		101	33.50		134.50	167
8020	2'-0" wide		18	.889		128	35.50		163.50	199
8040	2'-6" wide		18	.889		139	35.50		174.50	212
8060	2'-8" wide		18	.889		146	35.50		181.50	220
8080	3'-0" wide		17	.941		157	37.50		194.50	234
8300	Pine paneled, 1-3/8" x 6'-8" x 1'-6" wide		19	.842		111	33.50		144.50	178
8320	2'-0" wide		18	.889		128	35.50		163.50	199
8330	2'-4" wide		18	.889		150	35.50		185.50	224
8340	2'-6" wide		18	.889		158	35.50		193.50	232
8360	2'-8" wide		18	.889		161	35.50		196.50	236
8380	3'-0" wide		17	.941		165	37.50		202.50	244
8804	Pocket door,6 panel pine, 2'-6" x 6'-8"		10.50	1.524		218	61		279	340
8814	2'-8" x 6'-8"		10.50	1.524		230	61		291	355
8824	3'-0" x 6'-8"		10.50	1.524		239	61		300	365
9900	Minimum labor/equipment charge	1 Carp	4	2	Job		80		80	132

08 14 40 – Interior Cafe Doors

08 14 40.10 Cafe Style Doors

		Crew	Daily Output	Labor-Hours	Unit	Material	2009 Bare Costs Labor	Equipment	Total	Total Incl O&P
0010	**CAFE STYLE DOORS**									
6520	Interior cafe doors, 2'-6" opening, stock, panel pine	2 Carp	16	1	Ea.	188	40		228	272
6540	3'-0" opening	"	16	1	"	195	40		235	281
6550	Louvered pine									
6560	2'-6" opening	2 Carp	16	1	Ea.	164	40		204	246
8000	3'-0" opening		16	1		165	40		205	248
8010	2'-6" opening, hardwood		16	1		283	40		323	375
8020	3'-0" opening		16	1		315	40		355	410
9000	Minimum labor/equipment charge	1 Carp	4	2	Job		80		80	132

08 16 Composite Doors

08 16 13 – Fiberglass Doors

08 16 13.10 Entrance Doors, Fiberous Glass

		Crew	Daily Output	Labor-Hours	Unit	Material	2009 Bare Costs Labor	Equipment	Total	Total Incl O&P
0010	**ENTRANCE DOORS, FIBEROUS GLASS**									
0020	Exterior, fiberglass, door, 2'-8" wide x 6'-8" high	2 Carp	15	1.067	Ea.	248	42.50		290.50	345
0040	3'-0" wide x 6'-8" high		15	1.067		248	42.50		290.50	345
0060	3'-0" wide x 7'-0" high		15	1.067		445	42.50		487.50	560
0080	3'-0" wide x 6'-8" high, with two lites		15	1.067		278	42.50		320.50	375
0100	3'-0" wide x 8'-0" high, with two lites		15	1.067		450	42.50		492.50	565
0110	Half glass, 3'-0" wide x 6'-8" high		15	1.067		287	42.50		329.50	385
0120	3'-0" wide x 6'-8" high, low e		15	1.067		315	42.50		357.50	420
0130	3'-0" wide x 8'-0" high		15	1.067		570	42.50		612.50	695
0140	3'-0" wide x 8'-0" high, low e		15	1.067		630	42.50		672.50	760
0150	Side lights, 1'-0" wide x 6'-8" high,					232			232	255
0160	1'-0" wide x 6'-8" high, low e					246			246	271
0180	1'-0" wide x 6'-8" high, full glass					275			275	305
0190	1'-0" wide x 6'-8" high, low e					284			284	315

08 17 23 – Integrated Wood Door Opening Assemblies

08 17 23.10 Pre-Hung Doors	Crew	Daily Output	Labor-Hours	Unit	Material	2009 Bare Costs Labor	Equipment	Total	Total Incl O&P
0010 **PRE-HUNG DOORS**									
0300 Exterior, wood, comb. storm & screen, 6'-9" x 2'-6" wide	2 Carp	15	1.067	Ea.	289	42.50		331.50	390
0320 2'-8" wide		15	1.067		289	42.50		331.50	390
0340 3'-0" wide		15	1.067		296	42.50		338.50	395
0360 For 7'-0" high door, add					22.50			22.50	25
1600 Entrance door, flush, birch, solid core									
1620 4-5/8" solid jamb, 1-3/4" x 6'-8" x 2'-8" wide	2 Carp	16	1	Ea.	285	40		325	380
1640 3'-0" wide		16	1		350	40		390	450
1642 5-5/8" jamb		16	1		325	40		365	420
1680 For 7'-0" high door, add					20.50			20.50	22.50
2000 Entrance door, colonial, 6 panel pine									
2020 4-5/8" solid jamb, 1-3/4" x 6'-8" x 2'-8" wide	2 Carp	16	1	Ea.	530	40		570	645
2040 3'-0" wide	"	16	1		530	40		570	645
2060 For 7'-0" high door, add					52.50			52.50	58
2200 For 5-5/8" solid jamb, add					40.50			40.50	44.50
2270 Pine french door, 8 panels/leaf, 6'-8" x 5'-0" wide, including frame	2 Carp	7	2.286	Pr.	1,250	91.50		1,341.50	1,525
4000 Interior, passage door, 4-5/8" solid jamb									
4400 Lauan, flush, solid core, 1-3/8" x 6'-8" x 2'-6" wide	2 Carp	17	.941	Ea.	182	37.50		219.50	262
4420 2'-8" wide		17	.941		182	37.50		219.50	262
4440 3'-0" wide		16	1		195	40		235	281
4600 Hollow core, 1-3/8" x 6'-8" x 2'-6" wide		17	.941		123	37.50		160.50	197
4620 2'-8" wide		17	.941		123	37.50		160.50	197
4640 3'-0" wide		16	1		124	40		164	202
4700 For 7'-0" high door, add					24			24	26.50
5000 Birch, flush, solid core, 1-3/8" x 6'-8" x 2'-6" wide	2 Carp	17	.941		170	37.50		207.50	249
5020 2'-8" wide		17	.941		195	37.50		232.50	277
5040 3'-0" wide		16	1		205	40		245	292
5200 Hollow core, 1-3/8" x 6'-8" x 2'-6" wide		17	.941		141	37.50		178.50	217
5220 2'-8" wide		17	.941		147	37.50		184.50	223
5240 3'-0" wide		16	1		148	40		188	228
5280 For 7'-0" high door, add					21			21	23
5500 Hardboard paneled, 1-3/8" x 6'-8" x 2'-6" wide	2 Carp	17	.941		142	37.50		179.50	218
5520 2'-8" wide		17	.941		149	37.50		186.50	225
5540 3'-0" wide		16	1		146	40		186	227
6000 Pine paneled, 1-3/8" x 6'-8" x 2'-6" wide		17	.941		247	37.50		284.50	335
6020 2'-8" wide		17	.941		267	37.50		304.50	355
6040 3'-0" wide		16	1		271	40		311	365
6500 For 5-5/8" solid jamb, add					14.60			14.60	16.05
6520 For split jamb, deduct					16.45			16.45	18.10
9000 Minimum labor/equipment charge	1 Carp	4	2	Job		80		80	132

08 31 Access Doors and Panels

08 31 13 – Access Doors and Frames

08 31 13.10 Types of Framed Access Doors	Crew	Daily Output	Labor-Hours	Unit	Material	2009 Bare Costs Labor	Equipment	Total	Total Incl O&P
0010 **TYPES OF FRAMED ACCESS DOORS**									
1000 Fire rated door with lock									
1100 Metal, 12" x 12"	1 Carp	10	.800	Ea.	157	32		189	226
1150 18" x 18"		9	.889		205	35.50		240.50	285
1200 24" x 24"		9	.889		325	35.50		360.50	420
1250 24" x 36"		8	1		335	40		375	430
1300 24" x 48"		8	1		410	40		450	520
1350 36" x 36"		7.50	1.067		495	42.50		537.50	615
1400 48" x 48"		7.50	1.067		635	42.50		677.50	770
1600 Stainless steel, 12" x 12"		10	.800		281	32		313	365
1650 18" x 18"		9	.889		410	35.50		445.50	510
1700 24" x 24"		9	.889		500	35.50		535.50	610
1750 24" x 36"	▼	8	1	▼	640	40		680	770
2000 Flush door for finishing									
2100 Metal 8" x 8"	1 Carp	10	.800	Ea.	45.50	32		77.50	103
2150 12" x 12"	"	10	.800	"	50.50	32		82.50	109
3000 Recessed door for acoustic tile									
3100 Metal, 12" x 12"	1 Carp	4.50	1.778	Ea.	69.50	71		140.50	193
3150 12" x 24"		4.50	1.778		90.50	71		161.50	217
3200 24" x 24"		4	2		121	80		201	265
3250 24" x 36"	▼	4	2	▼	156	80		236	305
4000 Recessed door for drywall									
4100 Metal 12" x 12"	1 Carp	6	1.333	Ea.	77.50	53.50		131	173
4150 12" x 24"		5.50	1.455		115	58		173	222
4200 24" x 36"	▼	5	1.600	▼	183	64		247	305
6000 Standard door									
6100 Metal, 8" x 8"	1 Carp	10	.800	Ea.	40	32		72	97
6150 12" x 12"		10	.800		45.50	32		77.50	103
6200 18" x 18"		9	.889		62.50	35.50		98	128
6250 24" x 24"		9	.889		81.50	35.50		117	148
6300 24" x 36"		8	1		120	40		160	198
6350 36" x 36"		8	1		147	40		187	227
6500 Stainless steel, 8" x 8"		10	.800		80	32		112	141
6550 12" x 12"		10	.800		107	32		139	170
6600 18" x 18"		9	.889		197	35.50		232.50	276
6650 24" x 24"	▼	9	.889		258	35.50		293.50	345
7010 Aluminum cover, Ceiling hatches, 2'-6" x 2'-6", single leaf, st fr	G-3	11	2.909	▼	540	115		655	780
9000 Minimum labor/equipment charge	1 Carp	4	2	Job		80		80	132

08 31 13.20 Bulkhead/Cellar Doors

	Crew	Daily Output	Labor-Hours	Unit	Material	Labor	Equipment	Total	Total Incl O&P
0010 **BULKHEAD/CELLAR DOORS**									
0020 Steel, not incl. sides, 44" x 62"	1 Carp	5.50	1.455	Ea.	350	58		408	480
0100 52" x 73"		5.10	1.569		365	62.50		427.50	505
0500 With sides and foundation plates, 57" x 45" x 24"		4.70	1.702		650	68		718	825
0600 42" x 49" x 51"		4.30	1.860	▼	755	74.50		829.50	955
9000 Minimum labor/equipment charge		2	4	Job		160		160	264

08 31 13.30 Commercial Floor Doors

	Crew	Daily Output	Labor-Hours	Unit	Material	Labor	Equipment	Total	Total Incl O&P
0010 **COMMERCIAL FLOOR DOORS**									
0020 Aluminum tile, steel frame, one leaf, 2' x 2' opng.	2 Sswk	3.50	4.571	Opng.	775	204		979	1,225
0050 3'-6" x 3'-6" opening		3.50	4.571		1,100	204		1,304	1,575
0500 Double leaf, 4' x 4' opening		3	5.333		1,575	238		1,813	2,175
0550 5' x 5' opening		3	5.333	▼	2,800	238		3,038	3,525
9000 Minimum labor/equipment charge	▼	2	8	Job		360		360	660

190

08 31 Access Doors and Panels

08 31 13 – Access Doors and Frames

08 31 13.35 Industrial Floor Doors		Crew	Daily Output	Labor-Hours	Unit	Material	2009 Bare Costs Labor	Equipment	Total	Total Incl O&P
0010	**INDUSTRIAL FLOOR DOORS**									
0020	Steel 300 psf L.L., single leaf, 2' x 2', 175#	2 Sswk	6	2.667	Opng.	630	119		749	915
0050	3' x 3' opening, 300#		5.50	2.909		870	130		1,000	1,200
0300	Double leaf, 4' x 4' opening, 455#		5	3.200		1,575	143		1,718	2,000
0350	5' x 5' opening, 645#		4.50	3.556		2,400	159		2,559	2,925
1000	Aluminum, 300 psf L.L., single leaf, 2' x 2', 60#		6	2.667		625	119		744	905
1050	3' x 3' opening, 100#		5.50	2.909		970	130		1,100	1,325
1500	Double leaf, 4' x 4' opening, 160#		5	3.200		1,575	143		1,718	2,000
1550	5' x 5' opening, 235#		4.50	3.556		2,400	159		2,559	2,950
9000	Minimum labor/equipment charge		2	8	Job		360		360	660

08 32 Sliding Glass Doors

08 32 19 – Sliding Wood-Framed Glass Doors

08 32 19.10 Sliding Wood Doors

		Crew	Daily Output	Labor-Hours	Unit	Material	2009 Bare Costs Labor	Equipment	Total	Total Incl O&P
0010	**SLIDING WOOD DOORS**									
0020	Wood, 5/8" tempered insul. glass, 6' wide, premium	2 Carp	4	4	Ea.	1,200	160		1,360	1,600
0100	Economy		4	4		850	160		1,010	1,200
0150	8' wide, wood, premium		3	5.333		1,425	213		1,638	1,925
0200	Economy		3	5.333		935	213		1,148	1,375
0250	12' wide, wood, vinyl clad		2.50	6.400		2,900	256		3,156	3,600
0300	Economy		2.50	6.400		2,150	256		2,406	2,775
9000	Minimum labor/equipment charge	1 Carp	2	4	Job		160		160	264

08 32 19.15 Sliding Glass Vinyl-Clad Wood Doors

			Crew	Daily Output	Labor-Hours	Unit	Material	2009 Bare Costs Labor	Equipment	Total	Total Incl O&P
0010	**SLIDING GLASS VINYL-CLAD WOOD DOORS**										
0012	Vinyl clad, 1" insul. glass, 6'-0" x 6'-10" high	G	2 Carp	4	4	Opng.	1,350	160		1,510	1,750
0030	6'-0" x 8'-0" high	G		4	4	Ea.	1,950	160		2,110	2,425
0100	8'-0" x 6'-10" high	G		4	4	Opng.	2,000	160		2,160	2,475
0500	3 leaf, 9'-0" x 6'-10" high	G		3	5.333		2,400	213		2,613	2,975
0600	12'-0" x 6'-10" high	G		3	5.333		3,175	213		3,388	3,850
9000	Minimum labor/equipment charge		1 Carp	4	2	Job		80		80	132

08 33 Coiling Doors and Grilles

08 33 13 – Coiling Counter Doors

08 33 13.10 Counter Doors, Coiling Type

		Crew	Daily Output	Labor-Hours	Unit	Material	2009 Bare Costs Labor	Equipment	Total	Total Incl O&P
0010	**COUNTER DOORS, COILING TYPE**									
0020	Manual, incl. frm and hdwe, galv. stl., 4' roll-up, 6' long	2 Carp	2	8	Opng.	1,125	320		1,445	1,775
0300	Galvanized steel, UL label		1.80	8.889		1,200	355		1,555	1,900
0600	Stainless steel, 4' high roll-up, 6' long		2	8		2,050	320		2,370	2,775
0700	10' long		1.80	8.889		2,450	355		2,805	3,275
2000	Aluminum, 4' high, 4' long		2.20	7.273		1,350	291		1,641	1,975
2020	6' long		2	8		1,500	320		1,820	2,175
2040	8' long		1.90	8.421		1,725	335		2,060	2,450
2060	10' long		1.80	8.889		1,825	355		2,180	2,600
2080	14' long		1.40	11.429		2,675	455		3,130	3,675
2100	6' high, 4' long		2	8		1,525	320		1,845	2,200
2120	6' long		1.60	10		1,550	400		1,950	2,350
2140	10' long		1.40	11.429		2,125	455		2,580	3,100
9000	Minimum labor/equipment charge	1 Carp	2	4	Job		160		160	264

08 33 Coiling Doors and Grilles

08 33 16 – Coiling Counter Grilles

08 33 16.10 Coiling Grilles	Crew	Daily Output	Labor-Hours	Unit	Material	2009 Bare Costs Labor	Equipment	Total	Total Incl O&P
0010 **COILING GRILLES**									
2020 Aluminum, manual operated, mill finish	2 Sswk	82	.195	S.F.	27.50	8.70		36.20	46
2040 Bronze anodized		82	.195	"	43.50	8.70		52.20	64
2060 Steel, manual operated, 10' x 10' high		1	16	Opng.	2,450	715		3,165	4,025
2080 15' x 8' high	↓	.80	20	"	2,850	895		3,745	4,775
3000 For safety edge bottom bar, electric, add				L.F.	49			49	54
8000 For motor operation, add	2 Sswk	5	3.200	Opng.	1,250	143		1,393	1,650
9000 Minimum labor/equipment charge	"	1	16	Job		715		715	1,325

08 33 23 – Overhead Coiling Doors

08 33 23.10 Rolling Service Doors

08 33 23.10 Rolling Service Doors	Crew	Daily Output	Labor-Hours	Unit	Material	2009 Bare Costs Labor	Equipment	Total	Total Incl O&P
0010 **ROLLING SERVICE DOORS** Steel, manual, 20 ga., incl. hardware									
0050 8' x 8' high	2 Sswk	1.60	10	Ea.	1,025	445		1,470	1,950
0130 12' x 12' high, standard		1.20	13.333		1,700	595		2,295	2,975
0160 10' x 20' high, standard		.50	32		2,300	1,425		3,725	5,175
2000 Class A fire doors, manual, 20 ga., 8' x 8' high		1.40	11.429		1,375	510		1,885	2,475
2100 10' x 10' high		1.10	14.545		1,900	650		2,550	3,275
2200 20' x 10' high		.80	20		3,900	895		4,795	5,950
2300 12' x 12' high		1	16		2,975	715		3,690	4,600
2400 20' x 12' high		.80	20		4,225	895		5,120	6,300
2500 14' x 14' high		.60	26.667		3,275	1,200		4,475	5,800
2600 20' x 16' high		.50	32		5,275	1,425		6,700	8,450
2700 10' x 20' high	↓	.40	40	↓	4,075	1,800		5,875	7,800
3000 For 18 ga. doors, add				S.F.	1.27			1.27	1.40
3300 For enamel finish, add				"	. 1.52			1.52	1.67
3600 For safety edge bottom bar, pneumatic, add				L.F.	19.80			19.80	22
4000 For weatherstripping, extruded rubber, jambs, add					12.80			12.80	14.10
4100 Hood, add					9			9	9.90
4200 Sill, add				↓	5.15			5.15	5.65
4500 Motor operators, to 14' x 14' opening	2 Sswk	5	3.200	Ea.	1,075	143		1,218	1,450
4700 For fire door, additional fusible link, add				"	23.50			23.50	25.50
9000 Minimum labor/equipment charge	2 Sswk	1	16	Job		715		715	1,325

08 34 Special Function Doors

08 34 13 – Cold Storage Doors

08 34 13.10 Doors for Cold Area Storage

08 34 13.10 Doors for Cold Area Storage	Crew	Daily Output	Labor-Hours	Unit	Material	2009 Bare Costs Labor	Equipment	Total	Total Incl O&P
0010 **DOORS FOR COLD AREA STORAGE**									
0020 Single, 20 ga. galvanized steel									
0300 Horizontal sliding, 5' x 7', manual operation, 3.5" thick	2 Carp	2	8	Ea.	3,100	320		3,420	3,925
0400 4" thick		2	8		3,800	320		4,120	4,700
0500 6" thick		2	8		3,100	320		3,420	3,950
0800 5' x 7', power operation, 2" thick		1.90	8.421		5,150	335		5,485	6,225
0900 4" thick		1.90	8.421		5,250	335		5,585	6,325
1000 6" thick		1.90	8.421		5,975	335		6,310	7,125
1300 9' x 10', manual operation, 2" insulation		1.70	9.412		4,175	375		4,550	5,200
1400 4" insulation		1.70	9.412		4,300	375		4,675	5,350
1500 6" insulation		1.70	9.412		5,175	375		5,550	6,325
1800 Power operation, 2" insulation		1.60	10		7,175	400		7,575	8,550
1900 4" insulation		1.60	10		7,350	400		7,750	8,725
2000 6" insulation	↓	1.70	9.412	↓	8,325	375		8,700	9,775
2300 For stainless steel face, add					22%				

192

08 34 Special Function Doors

08 34 13 – Cold Storage Doors

08 34 13.10 Doors for Cold Area Storage

		Crew	Daily Output	Labor-Hours	Unit	Material	2009 Bare Costs Labor	Equipment	Total	Total Incl O&P
3000	Hinged, lightweight, 3' x 7'-0", galvanized 1 face, 2" thick	2 Carp	2	8	Ea.	1,350	320		1,670	2,000
3050	4" thick		1.90	8.421		1,675	335		2,010	2,375
3300	Aluminum doors, 3' x 7'-0", 4" thick		1.90	8.421		1,225	335		1,560	1,900
3350	6" thick		1.40	11.429		2,200	455		2,655	3,150
3600	Stainless steel, 3' x 7'-0", 4" thick		1.90	8.421		1,575	335		1,910	2,275
3650	6" thick		1.40	11.429		2,650	455		3,105	3,675
3900	Painted, 3' x 7'-0", 4" thick		1.90	8.421		1,125	335		1,460	1,800
3950	6" thick		1.40	11.429		2,150	455		2,605	3,125
5000	Bi-parting, electric operated									
5010	6' x 8' opening, galv. faces, 4" thick for cooler	2 Carp	.80	20	Opng.	6,750	800		7,550	8,750
5050	For freezer, 4" thick		.80	20		7,425	800		8,225	9,500
5300	For door buck framing and door protection, add		2.50	6.400		550	256		806	1,025
6000	Galvanized batten door, galvanized hinges, 4' x 7'		2	8		1,650	320		1,970	2,350
6050	6' x 8'		1.80	8.889		2,275	355		2,630	3,075
6500	Fire door, 3 hr., 6' x 8', single slide		.80	20		7,800	800		8,600	9,900
6550	Double, bi-parting		.70	22.857		12,000	915		12,915	14,700
9000	Minimum labor/equipment charge	1 Carp	2	4	Job		160		160	264

08 34 36 – Darkroom Doors

08 34 36.10 Various Types of Darkroom Doors

		Crew	Daily Output	Labor-Hours	Unit	Material	2009 Bare Costs Labor	Equipment	Total	Total Incl O&P
0010	**VARIOUS TYPES OF DARKROOM DOORS**									
0015	Revolving, standard, 2 way, 36" diameter	2 Carp	3.10	5.161	Opng.	2,075	206		2,281	2,625
0020	41" diameter		3.10	5.161		2,250	206		2,456	2,825
0050	3 way, 51" diameter		1.40	11.429		2,800	455		3,255	3,825
1000	4 way, 49" diameter		1.40	11.429		3,325	455		3,780	4,400
2000	Hinged safety, 2 way, 41" diameter		2.30	6.957		2,625	278		2,903	3,350
2500	3 way, 51" diameter		1.40	11.429		3,325	455		3,780	4,400
3000	Pop out safety, 2 way, 41" diameter		3.10	5.161		3,275	206		3,481	3,950
4000	3 way, 51" diameter		1.40	11.429		3,250	455		3,705	4,325
5000	Wheelchair-type, pop out, 51" diameter		1.40	11.429		3,300	455		3,755	4,400
5020	72" diameter		.90	17.778		6,775	710		7,485	8,625

08 34 73 – Sound Control Door Assemblies

08 34 73.10 Acoustical Doors

		Crew	Daily Output	Labor-Hours	Unit	Material	2009 Bare Costs Labor	Equipment	Total	Total Incl O&P
0010	**ACOUSTICAL DOORS**									
0020	Including framed seals, 3' x 7', wood, 27 STC rating	2 Carp	1.50	10.667	Ea.	400	425		825	1,150
0100	Steel, 40 STC rating		1.50	10.667		2,800	425		3,225	3,775
0200	45 STC rating		1.50	10.667		3,300	425		3,725	4,325
0300	48 STC rating		1.50	10.667		3,800	425		4,225	4,875
0400	52 STC rating		1.50	10.667		4,000	425		4,425	5,100
9000	Minimum labor/equipment charge	1 Carp	4	2	Job		80		80	132

08 36 Panel Doors

08 36 13 – Sectional Doors

08 36 13.10 Overhead Commercial Doors

		Crew	Daily Output	Labor-Hours	Unit	Material	2009 Bare Costs Labor	Equipment	Total	Total Incl O&P
0010	**OVERHEAD COMMERCIAL DOORS**									
1000	Stock, sectional, heavy duty, wood, 1-3/4" thick, 8' x 8' high	2 Carp	2	8	Ea.	755	320		1,075	1,350
1200	12' x 12' high		1.50	10.667		1,625	425		2,050	2,500
1300	Chain hoist, 14' x 14' high		1.30	12.308		2,550	490		3,040	3,600
1600	20' x 16' high	↓	.65	24.615	↓	5,950	985		6,935	8,175
2100	For medium duty custom door, deduct					5%	5%			
2150	For medium duty stock doors, deduct					10%	5%			
2300	Fiberglass and aluminum, heavy duty, sectional, 12' x 12' high	2 Carp	1.50	10.667	Ea.	2,550	425		2,975	3,500
2450	Chain hoist, 20' x 20' high	"	.50	32		6,450	1,275		7,725	9,200
2900	For electric trolley operator, 1/3 H.P., to 12' x 12', add	1 Carp	2	4		925	160		1,085	1,300
2950	Over 12' x 12', 1/2 H.P., add	"	1	8	↓	1,100	320		1,420	1,750
9000	Minimum labor/equipment charge	2 Carp	1.50	10.667	Job		425		425	705

08 36 13.20 Residential Garage Doors

		Crew	Daily Output	Labor-Hours	Unit	Material	2009 Bare Costs Labor	Equipment	Total	Total Incl O&P
0010	**RESIDENTIAL GARAGE DOORS**									
0050	Hinged, wood, custom, double door, 9' x 7'	2 Carp	4	4	Ea.	540	160		700	855
0070	16' x 7'		3	5.333		955	213		1,168	1,400
0200	Overhead, sectional, incl. hardware, fiberglass, 9' x 7', standard		5.28	3.030		715	121		836	985
0220	Deluxe		5.28	3.030		900	121		1,021	1,200
0300	16' x 7', standard		6	2.667		1,550	107		1,657	1,875
0320	Deluxe		6	2.667		1,925	107		2,032	2,275
0500	Hardboard, 9' x 7', standard		8	2		540	80		620	725
0520	Deluxe		8	2		765	80		845	970
0600	16' x 7', standard		6	2.667		1,125	107		1,232	1,400
0620	Deluxe		6	2.667		1,300	107		1,407	1,600
0700	Metal, 9' x 7', standard		5.28	3.030		555	121		676	815
0720	Deluxe		8	2		805	80		885	1,025
0800	16' x 7', standard		3	5.333		705	213		918	1,125
0820	Deluxe		6	2.667		1,375	107		1,482	1,675
0900	Wood, 9' x 7', standard		8	2		705	80		785	905
0920	Deluxe		8	2		2,025	80		2,105	2,350
1000	16' x 7', standard		6	2.667		1,425	107		1,532	1,725
1020	Deluxe	↓	6	2.667		2,950	107		3,057	3,425
1800	Door hardware, sectional	1 Carp	4	2		310	80		390	470
1810	Door tracks only		4	2		139	80		219	285
1820	One side only	↓	7	1.143		100	45.50		145.50	186
3000	Swing-up, including hardware, fiberglass, 9' x 7', standard	2 Carp	8	2		970	80		1,050	1,200
3020	Deluxe		8	2		790	80		870	1,000
3100	16' x 7', standard		6	2.667		1,225	107		1,332	1,525
3120	Deluxe		6	2.667		980	107		1,087	1,250
3200	Hardboard, 9' x 7', standard		8	2		345	80		425	510
3220	Deluxe		8	2		455	80		535	635
3300	16' x 7', standard		6	2.667		480	107		587	705
3320	Deluxe		6	2.667		715	107		822	960
3400	Metal, 9' x 7', standard		8	2		380	80		460	550
3420	Deluxe		8	2		800	80		880	1,000
3500	16' x 7', standard		6	2.667		600	107		707	835
3520	Deluxe		6	2.667		960	107		1,067	1,225
3600	Wood, 9' x 7', standard		8	2		450	80		530	625
3620	Deluxe		8	2		850	80		930	1,075
3700	16' x 7', standard		6	2.667		725	107		832	975
3720	Deluxe	↓	6	2.667		1,025	107		1,132	1,300
3900	Door hardware only, swing up	1 Carp	4	2	↓	142	80		222	288

08 36 Panel Doors

08 36 13 – Sectional Doors

08 36 13.20 Residential Garage Doors

		Crew	Daily Output	Labor-Hours	Unit	Material	2009 Bare Costs Labor	Equipment	Total	Total Incl O&P
3920	One side only	1 Carp	7	1.143	Ea.	69	45.50		114.50	152
4000	For electric operator, economy, add		8	1		405	40		445	510
4100	Deluxe, including remote control	↓	8	1	↓	595	40		635	715
4500	For transmitter/receiver control , add to operator				Total	97			97	107
4600	Transmitters, additional				"	45.50			45.50	50
6000	Replace section, on sectional door, fiberglass, 9' x 7'	1 Carp	4	2	Ea.	440	80		520	610
6020	16' x 7'		3.50	2.286		505	91.50		596.50	705
6200	Hardboard, 9' x 7'		4	2		110	80		190	253
6220	16' x 7'		3.50	2.286		211	91.50		302.50	385
6300	Metal, 9' x 7'		4	2		179	80		259	330
6320	16' x 7'		3.50	2.286		294	91.50		385.50	475
6500	Wood, 9' x 7'		4	2		109	80		189	252
6520	16' x 7'		3.50	2.286	↓	211	91.50		302.50	385
9000	Minimum labor/equipment charge	↓	2.50	3.200	Job		128		128	211

08 36 19 – Multi-Leaf Vertical Lift Doors

08 36 19.10 Sectional Vertical Lift Doors

		Crew	Daily Output	Labor-Hours	Unit	Material	2009 Bare Costs Labor	Equipment	Total	Total Incl O&P
0010	**SECTIONAL VERTICAL LIFT DOORS**									
0020	Motorized, 14 ga. stl, incl., frm and ctrl pnl									
0050	16' x 16' high	L-10	.50	48	Ea.	21,000	2,150	1,525	24,675	28,600
0100	10' x 20' high		1.30	18.462		29,000	825	590	30,415	34,000
0120	15' x 20' high		1.30	18.462		36,000	825	590	37,415	41,700
0140	20' x 20' high		1	24		41,600	1,075	770	43,445	48,400
0160	25' x 20' high		1	24		45,900	1,075	770	47,745	53,000
0170	32' x 24' high		.75	32		44,700	1,425	1,025	47,150	53,000
0180	20' x 25' high		1	24		47,600	1,075	770	49,445	55,000
0200	25' x 25' high		.70	34.286		53,500	1,525	1,100	56,125	63,000
0220	25' x 30' high		.70	34.286		62,000	1,525	1,100	64,625	72,500
0240	30' x 30' high		.70	34.286		71,500	1,525	1,100	74,125	82,500
0260	35' x 30' high	↓	.70	34.286	↓	78,000	1,525	1,100	80,625	89,500

08 38 Traffic Doors

08 38 19 – Rigid Traffic Doors

08 38 19.20 Double Acting Swing Doors

		Crew	Daily Output	Labor-Hours	Unit	Material	2009 Bare Costs Labor	Equipment	Total	Total Incl O&P
0010	**DOUBLE ACTING SWING DOORS**									
0020	Including frame, closer, hardware and vision panel									
1000	.063" aluminum, 7'-0" high, 4'-0" wide	2 Carp	4.20	3.810	Pr.	1,900	152		2,052	2,350
1025	6'-0" wide		4	4		2,100	160		2,260	2,575
1050	6'-8" wide	↓	4	4	↓	2,300	160		2,460	2,800
2000	Solid core wood, 3/4" thick, metal frame, stainless steel									
2010	base plate, 7' high opening, 4' wide	2 Carp	4	4	Pr.	2,200	160		2,360	2,700
2050	7' wide		3.80	4.211	"	2,500	168		2,668	3,025
9000	Minimum labor/equipment charge	↓	2	8	Job		320		320	530

08 38 19.30 Shock Absorbing Doors

		Crew	Daily Output	Labor-Hours	Unit	Material	2009 Bare Costs Labor	Equipment	Total	Total Incl O&P
0010	**SHOCK ABSORBING DOORS**									
0020	Rigid, no frame, 1-1/2" thick, 5' x 7'	2 Sswk	1.90	8.421	Opng.	1,400	375		1,775	2,250
0100	8' x 8'		1.80	8.889		2,000	395		2,395	2,925
0500	Flexible, no frame, insulated, .16" thick, economy, 5' x 7'		2	8		1,750	360		2,110	2,575
0600	Deluxe		1.90	8.421		2,625	375		3,000	3,575
1000	8' x 8' opening, economy		2	8		2,750	360		3,110	3,675
1100	Deluxe	↓	1.90	8.421	↓	3,500	375		3,875	4,550

08 38 Traffic Doors

08 38 19 – Rigid Traffic Doors

08 38 19.30 Shock Absorbing Doors	Crew	Daily Output	Labor-Hours	Unit	Material	Labor	Equipment	Total	Total Incl O&P
9000 Minimum labor/equipment charge	2 Sswk	2	8	Job		360		360	660

08 41 Entrances and Storefronts

08 41 13 – Aluminum-Framed Entrances and Storefronts

08 41 13.20 Tube Framing

		Crew	Daily Output	Labor-Hours	Unit	Material	Labor	Equipment	Total	Total Incl O&P
0010	**TUBE FRAMING,** For window walls and store fronts, aluminum stock									
0050	Plain tube frame, mill finish, 1-3/4" x 1-3/4"	2 Glaz	103	.155	L.F.	8.70	6		14.70	19.25
0150	1-3/4" x 4"		98	.163		11.85	6.30		18.15	23
0200	1-3/4" x 4-1/2"		95	.168		13.80	6.50		20.30	25.50
0250	2" x 6"		89	.180		19.55	6.95		26.50	32.50
0350	4" x 4"		87	.184		19.65	7.10		26.75	33
0400	4-1/2" x 4-1/2"		85	.188		25	7.25		32.25	39
0450	Glass bead		240	.067		2.55	2.57		5.12	6.95
1000	Flush tube frame, mill finish, 1/4" glass, 1-3/4" x 4", open header		80	.200		11.45	7.70		19.15	25
1050	Open sill		82	.195		9	7.55		16.55	22
1100	Closed back header		83	.193		16.05	7.45		23.50	29.50
1150	Closed back sill		85	.188		15.20	7.25		22.45	28.50
1160	Tube fmg, spandrel cover both sides, alum 1" wide	1 Sswk	85	.094	S.F.	86	4.21		90.21	102
1170	Tube fmg, spandrel cover both sides, alum 2" wide	"	85	.094	"	32.50	4.21		36.71	44
1200	Vertical mullion, one piece	2 Glaz	75	.213	L.F.	16.90	8.25		25.15	32
1250	Two piece		73	.219		18.10	8.45		26.55	33.50
1300	90° or 180° vertical corner post		75	.213		27.50	8.25		35.75	44
1400	1-3/4" x 4-1/2", open header		80	.200		13.90	7.70		21.60	28
1450	Open sill		82	.195		11.60	7.55		19.15	25
1500	Closed back header		83	.193		16.85	7.45		24.30	30.50
1550	Closed back sill		85	.188		16.35	7.25		23.60	29.50
1600	Vertical mullion, one piece		75	.213		18.30	8.25		26.55	33.50
1650	Two piece		73	.219		19.30	8.45		27.75	34.50
1700	90° or 180° vertical corner post		75	.213		19.70	8.25		27.95	35
2000	Flush tube frame, mil fin.,ins. glass w/thml brk, 2" x 4-1/2", open header		75	.213		14.50	8.25		22.75	29
2050	Open sill		77	.208		12.25	8		20.25	26.50
2100	Closed back header		78	.205		13.75	7.90		21.65	28
2150	Closed back sill		80	.200		14.70	7.70		22.40	28.50
2200	Vertical mullion, one piece		70	.229		15.25	8.80		24.05	31
2250	Two piece		68	.235		16.45	9.10		25.55	32.50
2300	90° or 180° vertical corner post		70	.229		15.70	8.80		24.50	31.50
5000	Flush tube frame, mill fin., thermal brk., 2-1/4"x 4-1/2", open header		74	.216		15.15	8.35		23.50	30
5050	Open sill		75	.213		13.45	8.25		21.70	28
5100	Vertical mullion, one piece		69	.232		16.50	8.95		25.45	32.50
5150	Two piece		67	.239		19.15	9.20		28.35	36
5200	90° or 180° vertical corner post		69	.232		16.85	8.95		25.80	33
6980	Door stop (snap in)		380	.042		2.85	1.63		4.48	5.75
7000	For joints, 90°, clip type, add				Ea.	23			23	25
7050	Screw spline joint, add					20.50			20.50	22.50
7100	For joint other than 90°, add					43			43	47
8000	For bronze anodized aluminum, add					15%				
8020	For black finish, add					27%				
8050	For stainless steel materials, add					350%				
8100	For monumental grade, add					52%				
8150	For steel stiffener, add	2 Glaz	200	.080	L.F.	10.40	3.09		13.49	16.45
8200	For 2 to 5 stories, add per story				Story		6%			

196

08 41 Entrances and Storefronts

08 41 13 – Aluminum-Framed Entrances and Storefronts

08 41 13.20 Tube Framing	Crew	Daily Output	Labor-Hours	Unit	Material	2009 Bare Costs Labor	Equipment	Total	Total Incl O&P
9000	Minimum labor/equipment charge	2 Glaz	2	8	Job		310		310	500

08 41 19 – Stainless-Steel-Framed Entrances and Storefronts

08 41 19.10 Stainless-Steel and Glass Entrance Unit

		Crew	Daily Output	Labor-Hours	Unit	Material	Labor	Equipment	Total	Total Incl O&P
0010	**STAINLESS-STEEL AND GLASS ENTRANCE UNIT**, narrow stiles									
0020	3' x 7' opening, including hardware, minimum	2 Sswk	1.60	10	Opng.	6,000	445		6,445	7,425
0050	Average		1.40	11.429		6,500	510		7,010	8,100
0100	Maximum		1.20	13.333		6,900	595		7,495	8,700
1000	For solid bronze entrance units, statuary finish, add					62%				
1100	Without statuary finish, add					45%				
2000	Balanced doors, 3' x 7', economy	2 Sswk	.90	17.778	Ea.	8,100	795		8,895	10,400
2100	Premium		.70	22.857	"	14,000	1,025		15,025	17,300
9000	Minimum labor/equipment charge		2	8	Job		360		360	660

08 41 26 – All-Glass Entrances and Storefronts

08 41 26.10 Window Walls Aluminum, Stock

		Crew	Daily Output	Labor-Hours	Unit	Material	Labor	Equipment	Total	Total Incl O&P
0010	**WINDOW WALLS ALUMINUM, STOCK**, including glazing									
0020	Minimum	H-2	160	.150	S.F.	39.50	5.45		44.95	52.50
0050	Average		140	.171		54	6.20		60.20	69.50
0100	Maximum		110	.218		152	7.90		159.90	180
0500	For translucent sandwich wall systems, see Div. 07 41 33.10									
0850	Cost of the above walls depends on material,									
0860	finish, repetition, and size of units.									
0870	The larger the opening, the lower the S.F. cost									
1200	Double glazed acoustical window wall for airports,									
1220	including 1" thick glass with 2" x 4-1/2" tube frame	H-2	40	.600	S.F.	99	22		121	145

08 42 Entrances

08 42 26 – All-Glass Entrances

08 42 26.10 Swinging Glass Doors

		Crew	Daily Output	Labor-Hours	Unit	Material	Labor	Equipment	Total	Total Incl O&P
0010	**SWINGING GLASS DOORS**									
0020	Including hardware, 1/2" thick, tempered, 3' x 7' opening	2 Glaz	2	8	Opng.	1,950	310		2,260	2,650
0100	6' x 7' opening		1.40	11.429	"	3,800	440		4,240	4,875
9000	Minimum labor/equipment charge		2	8	Job		310		310	500

08 42 29 – Automatic Entrances

08 42 29.23 Sliding Automatic Entrances

		Crew	Daily Output	Labor-Hours	Unit	Material	Labor	Equipment	Total	Total Incl O&P
0010	**SLIDING AUTOMATIC ENTRANCES** 12' x 7'-6" opng., 5' x 7' door, 2 way traffic									
0020	Mat or electronic activated, panic pushout, incl. operator & hardware,									
0030	not including glass or glazing	2 Glaz	.70	22.857	Opng.	7,700	880		8,580	9,900
9000	Minimum labor/equipment charge	"	.70	22.857	Job		880		880	1,425

08 42 36 – Balanced Door Entrances

08 42 36.10 Balanced Entrance Doors

		Crew	Daily Output	Labor-Hours	Unit	Material	Labor	Equipment	Total	Total Incl O&P
0010	**BALANCED ENTRANCE DOORS**									
0020	Hardware & frame, alum. & glass, 3' x 7', econ.	2 Sswk	.90	17.778	Ea.	5,800	795		6,595	7,850
0150	Premium		.70	22.857	"	7,275	1,025		8,300	9,900
9000	Minimum labor/equipment charge		1	16	Job		715		715	1,325

08 43 Storefronts

08 43 13 – Aluminum-Framed Storefronts

08 43 13.10 Aluminum-Framed Entrance Doors	Crew	Daily Output	Labor-Hours	Unit	Material	2009 Bare Costs Labor	2009 Bare Costs Equipment	Total	Total Incl O&P
0010 **ALUMINUM-FRAMED ENTRANCE DOORS** (Frame Only)									
0020 Entrance, 3' x 7' opening, clear anodized finish	2 Sswk	7	2.286	Opng.	420	102		522	650
0040 Bronze finish		7	2.286		450	102		552	685
0500 6' x 7' opening, clear finish		6	2.667		340	119		459	595
0520 Bronze finish		6	2.667		480	119		599	750
1000 With 3' high transoms, 3' x 10' opening, clear finish		6.50	2.462		430	110		540	680
1050 Bronze finish		6.50	2.462		465	110		575	720
1100 Black finish		6.50	2.462		525	110		635	785
1500 With 3' high transoms, 6' x 10' opening, clear finish		5.50	2.909		515	130		645	805
1550 Bronze finish		5.50	2.909		555	130		685	850
1600 Black finish		5.50	2.909		635	130		765	940
9000 Minimum labor/equipment charge		4	4	Job		179		179	330

08 43 13.20 Storefront Systems

	Crew	Daily Output	Labor-Hours	Unit	Material	Labor	Equipment	Total	Total Incl O&P
0010 **STOREFRONT SYSTEMS**, aluminum frame clear 3/8" plate glass									
0020 incl. 3' x 7' door with hardware (400 sq. ft. max. wall)									
0500 Wall height to 12' high, commercial grade	2 Glaz	150	.107	S.F.	18.70	4.12		22.82	27
0600 Institutional grade		130	.123		23	4.75		27.75	33
0700 Monumental grade		115	.139		35.50	5.35		40.85	47.50
1000 6' x 7' door with hardware, commercial grade		135	.119		18.90	4.57		23.47	28.50
1100 Institutional grade		115	.139		26	5.35		31.35	37
1200 Monumental grade		100	.160		48	6.20		54.20	62.50
1500 For bronze anodized finish, add					15%				
1600 For black anodized finish, add					35%				
1700 For stainless steel framing, add to monumental					76%				
9000 Minimum labor/equipment charge	2 Glaz	1	16	Job		620		620	995

08 43 29 – Sliding Storefronts

08 43 29.10 Sliding Panels

	Crew	Daily Output	Labor-Hours	Unit	Material	Labor	Equipment	Total	Total Incl O&P
0010 **SLIDING PANELS**									
0020 Mall fronts, aluminum & glass, 15' x 9' high	2 Glaz	1.30	12.308	Opng.	3,025	475		3,500	4,100
0100 24' x 9' high		.70	22.857		4,400	880		5,280	6,250
0200 48' x 9' high, with fixed panels		.90	17.778		8,200	685		8,885	10,100
0500 For bronze finish, add					17%				
9000 Minimum labor/equipment charge	2 Glaz	1	16	Job		620		620	995

08 45 Translucent Wall and Roof Assemblies

08 45 10 – Translucent Roof Assemblies

08 45 10.10 Skyroofs

	Crew	Daily Output	Labor-Hours	Unit	Material	Labor	Equipment	Total	Total Incl O&P
0010 **SKYROOFS**, Translucent panels, 2-3/4" thick									
0020 Under 500 S.F.	G-3	395	.081	SF Hor.	31.50	3.19		34.69	39.50
0100 Over 5000 S.F.		465	.069		29.50	2.71		32.21	36.50
0300 Continuous vaulted, semi-circular, to 8' wide, double glazed [G]		145	.221		49	8.70		57.70	68
0400 Single glazed		160	.200		33.50	7.90		41.40	49
0600 To 20' wide, single glazed		175	.183		37	7.20		44.20	52
0700 Over 20' wide, single glazed		200	.160		42.50	6.30		48.80	57
0900 Motorized opening type, single glazed, 1/3 opening		145	.221		44	8.70		52.70	62
1000 Full opening		130	.246		50	9.70		59.70	70.50
1200 Pyramid type units, self-supporting, to 30' clear opening,									
1300 square or circular, single glazed, minimum	G-3	200	.160	SF Hor.	25.50	6.30		31.80	38
1310 Average		165	.194		34.50	7.65		42.15	50.50
1400 Maximum		130	.246		50	9.70		59.70	70.50

08 45 Translucent Wall and Roof Assemblies

08 45 10 – Translucent Roof Assemblies

08 45 10.10 Skyroofs		Crew	Daily Output	Labor-Hours	Unit	Material	2009 Bare Costs Labor	Equipment	Total	Total Incl O&P
1500	Grid type, 4' to 10' modules, single glass glazed, minimum	G-3	200	.160	SF Hor.	33.50	6.30		39.80	47
1550	Maximum		128	.250		52.50	9.85		62.35	73.50
1600	Preformed acrylic, minimum		300	.107		40	4.20		44.20	51
1650	Maximum		175	.183		55	7.20		62.20	72
9000	Minimum labor/equipment charge	2 Carp	8	2	Job		80		80	132

08 51 Metal Windows

08 51 13 – Aluminum Windows

08 51 13.10 Aluminum Sash

		Crew	Daily Output	Labor-Hours	Unit	Material	2009 Bare Costs Labor	Equipment	Total	Total Incl O&P
0010	**ALUMINUM SASH**									
0020	Stock, grade C, glaze & trim not incl., casement	2 Sswk	200	.080	S.F.	36	3.58		39.58	46
0050	Double hung		200	.080		36.50	3.58		40.08	46.50
0100	Fixed casement		200	.080		15.65	3.58		19.23	24
0150	Picture window		200	.080		16.65	3.58		20.23	25
0200	Projected window		200	.080		33	3.58		36.58	42.50
0250	Single hung		200	.080		15.80	3.58		19.38	24
0300	Sliding		200	.080		20.50	3.58		24.08	29
1000	Mullions for above, tubular		240	.067	L.F.	5.45	2.98		8.43	11.45
2950	Single glazing for above, add	2 Glaz	200	.080	S.F.	6.80	3.09		9.89	12.50
3000	Double glazing for above, add		200	.080		11.35	3.09		14.44	17.50
3100	Triple glazing for above, add		85	.188		13.30	7.25		20.55	26.50
9000	Minimum labor/equipment charge	1 Sswk	2	4	Job		179		179	330

08 51 13.20 Aluminum Windows

		Crew	Daily Output	Labor-Hours	Unit	Material	2009 Bare Costs Labor	Equipment	Total	Total Incl O&P
0010	**ALUMINUM WINDOWS**, incl. frame and glazing, commercial grade									
1000	Stock units, casement, 3'-1" x 3'-2" opening	2 Sswk	10	1.600	Ea.	355	71.50		426.50	520
1050	Add for storms					115			115	126
1600	Projected, with screen, 3'-1" x 3'-2" opening	2 Sswk	10	1.600		335	71.50		406.50	500
1700	Add for storms					112			112	123
2000	4'-5" x 5'-3" opening	2 Sswk	8	2		380	89.50		469.50	585
2100	Add for storms					120			120	132
2500	Enamel finish windows, 3'-1" x 3'-2"	2 Sswk	10	1.600		340	71.50		411.50	505
2600	4'-5" x 5'-3"		8	2		385	89.50		474.50	590
3000	Single hung, 2' x 3' opening, enameled, standard glazed		10	1.600		198	71.50		269.50	350
3100	Insulating glass		10	1.600		240	71.50		311.50	395
3300	2'-8" x 6'-8" opening, standard glazed		8	2		350	89.50		439.50	550
3400	Insulating glass		8	2		450	89.50		539.50	660
3700	3'-4" x 5'-0" opening, standard glazed		9	1.778		286	79.50		365.50	460
3800	Insulating glass		9	1.778		315	79.50		394.50	495
4000	Sliding aluminum, 3' x 2' opening, standard glazed		10	1.600		209	71.50		280.50	360
4100	Insulating glass		10	1.600		224	71.50		295.50	380
4300	5' x 3' opening, standard glazed		9	1.778		320	79.50		399.50	495
4400	Insulating glass		9	1.778		370	79.50		449.50	555
4600	8' x 4' opening, standard glazed		6	2.667		335	119		454	590
4700	Insulating glass		6	2.667		545	119		664	815
5000	9' x 5' opening, standard glazed		4	4		510	179		689	890
5100	Insulating glass		4	4		820	179		999	1,225
5500	Sliding, with thermal barrier and screen, 6' x 4', 2 track		8	2		695	89.50		784.50	930
5700	4 track		8	2		880	89.50		969.50	1,125
6000	For above units with bronze finish, add					12%				
6200	For installation in concrete openings, add					6%				
6400										

08 51 Metal Windows

08 51 23 – Steel Windows

08 51 23.10 Steel Sash

		Crew	Daily Output	Labor-Hours	Unit	Material	2009 Bare Costs Labor	Equipment	Total	Total Incl O&P
0010	**STEEL SASH** Custom units, glazing and trim not included									
0100	Casement, 100% vented	2 Sswk	200	.080	S.F.	48	3.58		51.58	59
0200	50% vented		200	.080		44	3.58		47.58	55
0300	Fixed		200	.080		30	3.58		33.58	39.50
1000	Projected, commercial, 40% vented		200	.080		48	3.58		51.58	59.50
1100	Intermediate, 50% vented		200	.080		54	3.58		57.58	65.50
1500	Industrial, horizontally pivoted		200	.080		49.50	3.58		53.08	61
1600	Fixed		200	.080		28.50	3.58		32.08	38
2000	Industrial security sash, 50% vented		200	.080		53.50	3.58		57.08	65.50
2100	Fixed		200	.080		43.50	3.58		47.08	54.50
2500	Picture window		200	.080		28	3.58		31.58	37.50
3000	Double hung		200	.080		56	3.58		59.58	68
5000	Mullions for above, open interior face		240	.067	L.F.	9.80	2.98		12.78	16.25
5100	With interior cover		240	.067	"	16.35	2.98		19.33	23.50
5200	Single glazing for above, add	2 Glaz	200	.080	S.F.	6.65	3.09		9.74	12.30
6000	Double glazing for above, add		200	.080		11.90	3.09		14.99	18.10
6100	Triple glazing for above, add		85	.188		11.85	7.25		19.10	25
9000	Minimum labor/equipment charge	1 Sswk	2	4	Job		179		179	330

08 51 23.20 Steel Windows

		Crew	Daily Output	Labor-Hours	Unit	Material	2009 Bare Costs Labor	Equipment	Total	Total Incl O&P
0010	**STEEL WINDOWS** Stock, including frame, trim and insul. glass R085123-10									
1000	Custom units, double hung, 2'-8" x 4'-6" opening	2 Sswk	12	1.333	Ea.	680	59.50		739.50	860
1100	2'-4" x 3'-9" opening		12	1.333		565	59.50		624.50	730
1500	Commercial projected, 3'-9" x 5'-5" opening		10	1.600		1,200	71.50		1,271.50	1,425
1600	6'-9" x 4'-1" opening		7	2.286		1,575	102		1,677	1,925
2000	Intermediate projected, 2'-9" x 4'-1" opening		12	1.333		670	59.50		729.50	845
2100	4'-1" x 5'-5" opening		10	1.600		1,350	71.50		1,421.50	1,600
9000	Minimum labor/equipment charge	1 Sswk	3	2.667	Job		119		119	221

08 51 66 – Metal Window Screens

08 51 66.10 Screens

		Crew	Daily Output	Labor-Hours	Unit	Material	2009 Bare Costs Labor	Equipment	Total	Total Incl O&P
0010	**SCREENS**									
0020	For metal sash, aluminum or bronze mesh, flat screen	2 Sswk	1200	.013	S.F.	4.10	.60		4.70	5.60
0500	Wicket screen, inside window		1000	.016		6.30	.72		7.02	8.25
0800	Security screen, aluminum frame with stainless steel cloth		1200	.013		22.50	.60		23.10	26
0900	Steel grate, painted, on steel frame		1600	.010		12.40	.45		12.85	14.50
1000	For solar louvers, add		160	.100		23.50	4.47		27.97	34
4000	See Div. 05 58 27.90									

08 52 Wood Windows

08 52 10 – Plain Wood Windows

08 52 10.20 Awning Window

		Crew	Daily Output	Labor-Hours	Unit	Material	2009 Bare Costs Labor	Equipment	Total	Total Incl O&P
0010	**AWNING WINDOW**, Including frame, screens and grilles									
0100	Average quality, builders model, 34" x 22", double insulated glass	1 Carp	10	.800	Ea.	249	32		281	325
0200	Low E glass		10	.800		258	32		290	335
0300	40" x 28", double insulated glass		9	.889		310	35.50		345.50	400
0400	Low E Glass		9	.889		330	35.50		365.50	425
0500	48" x 36", double insulated glass		8	1		450	40		490	560
0600	Low E glass		8	1		475	40		515	590
1000	Vinyl clad, 34" x 22"		10	.800		259	32		291	340
1100	40" x 22"		10	.800		283	32		315	365
1200	36" x 28"		9	.889		300	35.50		335.50	390

08 52 Wood Windows

08 52 10 – Plain Wood Windows

08 52 10.20 Awning Window

		Crew	Daily Output	Labor-Hours	Unit	Material	2009 Bare Costs Labor	Equipment	Total	Total Incl O&P
1300	36" x 36"	1 Carp	9	.889	Ea.	335	35.50		370.50	430
1400	48" x 28"		8	1		360	40		400	465
1500	60" x 36"		8	1		505	40		545	620
2200	Metal clad, 36" x 25"		9	.889		269	35.50		304.50	355
2300	40" x 30"		9	.889		335	35.50		370.50	430
2400	48" x 28"		8	1		345	40		385	445
2500	60" x 36"		8	1		365	40		405	470
4000	Impact windows, minimum, add					60%				
4010	Impact windows, maximum, add					160%				
9000	Minimum labor/equipment charge	1 Carp	4	2	Job		80		80	132

08 52 10.30 Wood Windows

		Crew	Daily Output	Labor-Hours	Unit	Material	2009 Bare Costs Labor	Equipment	Total	Total Incl O&P
0010	**WOOD WINDOWS**, double hung									
0020	Including frame, double insulated glass, screens and grilles									
0040	Double hung, 2'-2" x 3'-4" high	2 Carp	15	1.067	Ea.	194	42.50		236.50	284
0060	2'-2" x 4'-4"		14	1.143		217	45.50		262.50	315
0080	2'-6" x 3'-4"		13	1.231		203	49		252	305
0100	2'-6" x 4'-0"		12	1.333		220	53.50		273.50	330
0120	2'-6" x 4'-8"		12	1.333		242	53.50		295.50	355
0140	2'-10" x 3'-4"		10	1.600		213	64		277	340
0160	2'-10" x 4'-0"		10	1.600		238	64		302	365
0180	3'-7" x 3'-4"		9	1.778		242	71		313	385
0200	3'-7" x 5'-4"		9	1.778		310	71		381	455
0220	3'-10" x 5'-4"		8	2		490	80		570	665
6000	Replacement sash, double hung, double glazing, to 12 S.F.	1 Carp	64	.125	S.F.	24	4.99		28.99	35
6100	12 S.F. to 20 S.F.		94	.085	"	19.05	3.40		22.45	26.50
7000	Sash, single lite, 2'-0" x 2'-0" high		20	.400	Ea.	45	16		61	76
7050	2'-6" x 2'-0" high		19	.421		56.50	16.80		73.30	90
7100	2'-6" x 2'-6" high		18	.444		67.50	17.75		85.25	104
7150	3'-0" x 2'-0" high		17	.471		88	18.80		106.80	128

08 52 10.40 Casement Window

			Crew	Daily Output	Labor-Hours	Unit	Material	2009 Bare Costs Labor	Equipment	Total	Total Incl O&P
0010	**CASEMENT WINDOW**, including frame, screen and grilles										
0100	Avg. quality, bldrs. model, 2'-0" x 3'-0" H, dbl. insulated glass	G	1 Carp	10	.800	Ea.	276	32		308	360
0150	Low E glass	G		10	.800		225	32		257	300
0200	2'-0" x 4'-6" high, double insulated glass	G		9	.889		345	35.50		380.50	440
0250	Low E glass	G		9	.889		263	35.50		298.50	350
0300	2'-4" x 6'-0" high, double insulated glass	G		8	1		380	40		420	485
0350	Low E glass	G		8	1		465	40		505	580
0522	Vinyl clad, premium, double insulated glass, 2'-0" x 3'-0"	G		10	.800		266	32		298	345
0524	2'-0" x 4'-0"	G		9	.889		310	35.50		345.50	400
0525	2'-0" x 5'-0"	G		8	1		355	40		395	455
0528	2'-0" x 6'-0"	G		8	1		355	40		395	455
8100	Metal clad, deluxe, dbl. insul. glass, 2'-0" x 3'-0" high	G		10	.800		226	32		258	300
8120	2'-0" x 4'-0" high	G		9	.889		273	35.50		308.50	360
8140	2'-0" x 5'-0" high	G		8	1		310	40		350	405
8160	2'-0" x 6'-0" high	G		8	1		355	40		395	455
8190	For installation, add per leaf							15%			
8200	For multiple leaf units, deduct for stationary sash										
8220	2' high					Ea.	22			22	24
8240	4'-6" high						25			25	28
8260	6' high						33.50			33.50	37
8300	Impact windows, minimum, add						60%				
8310	Impact windows, maximum, add						160%				

08 52 Wood Windows

08 52 10 – Plain Wood Windows

	08 52 10.40 Casement Window		Crew	Daily Output	Labor-Hours	Unit	Material	2009 Bare Costs Labor	Equipment	Total	Total Incl O&P
9000	Minimum labor/equipment charge		1 Carp	3	2.667	Job		107		107	176

08 52 10.50 Double Hung

			Crew	Daily Output	Labor-Hours	Unit	Material	Labor	Equipment	Total	Total Incl O&P
0010	**DOUBLE HUNG**, Including frame, screens and grilles										
0100	Avg. quality, bldrs. model, 2'-0" x 3'-0" high, dbl insul. glass	G	1 Carp	10	.800	Ea.	184	32		216	255
0150	Low E glass	G		10	.800		208	32		240	282
0200	3'-0" x 4'-0" high, double insulated glass	G		9	.889		257	35.50		292.50	340
0250	Low E glass	G		9	.889		275	35.50		310.50	365
0300	4'-0" x 4'-6" high, double insulated glass	G		8	1		315	40		355	410
0350	Low E glass	G		8	1		340	40		380	440
1000	Vinyl clad, premium, double insulated glass, 2'-6" x 3'-0"	G		10	.800		225	32		257	300
1100	3'-0" x 3'-6"	G		10	.800		261	32		293	340
1200	3'-0" x 4'-0"	G		9	.889		370	35.50		405.50	465
1300	3'-0" x 4'-6"	G		9	.889		340	35.50		375.50	430
1400	3'-0" x 5'-0"	G		8	1		315	40		355	410
1500	3'-6" x 6'-0"	G		8	1		370	40		410	470
2000	Metal clad, deluxe, dbl. insul. glass, 2'-6" x 3'-0" high	G		10	.800		262	32		294	340
2100	3'-0" x 3'-6" high	G		10	.800		299	32		331	385
2200	3'-0" x 4'-0" high	G		9	.889		315	35.50		350.50	405
2300	3'-0" x 4'-6" high	G		9	.889		330	35.50		365.50	425
2400	3'-0" x 5'-0" high	G		8	1		360	40		400	460
2500	3'-6" x 6'-0" high	G		8	1		435	40		475	540
8000	Impact windows, minimum, add						60%				
8010	Impact windows, maximum, add						160%				
9000	Minimum labor/equipment charge		1 Carp	3	2.667	Job		107		107	176

08 52 10.55 Picture Window

			Crew	Daily Output	Labor-Hours	Unit	Material	Labor	Equipment	Total	Total Incl O&P
0010	**PICTURE WINDOW**, Including frame and grilles										
0100	Average quality, bldrs. model, 3'-6" x 4'-0" high, dbl insulated glass		2 Carp	12	1.333	Ea.	420	53.50		473.50	555
0150	Low E glass			12	1.333		465	53.50		518.50	600
0200	4'-0" x 4'-6" high, double insulated glass			11	1.455		475	58		533	620
0250	Low E glass			11	1.455		500	58		558	645
0300	5'-0" x 4'-0" high, double insulated glass			11	1.455		555	58		613	705
0350	Low E glass			11	1.455		575	58		633	730
0400	6'-0" x 4'-6" high, double insulated glass			10	1.600		605	64		669	770
0450	Low E glass			10	1.600		610	64		674	780

08 52 10.65 Wood Sash

			Crew	Daily Output	Labor-Hours	Unit	Material	Labor	Equipment	Total	Total Incl O&P
0010	**WOOD SASH**, Including glazing but not trim										
0050	Custom, 5'-0" x 4'-0", 1" dbl. glazed, 3/16" thick lites		2 Carp	3.20	5	Ea.	167	200		367	515
0100	1/4" thick lites			5	3.200		172	128		300	400
0200	1" thick, triple glazed			5	3.200		395	128		523	640
0300	7'-0" x 4'-6" high, 1" double glazed, 3/16" thick lites			4.30	3.721		400	149		549	685
0400	1/4" thick lites			4.30	3.721		450	149		599	740
0500	1" thick, triple glazed			4.30	3.721		515	149		664	810
0600	8'-6" x 5'-0" high, 1" double glazed, 3/16" thick lites			3.50	4.571		540	183		723	890
0700	1/4" thick lites			3.50	4.571		590	183		773	950
0800	1" thick, triple glazed			3.50	4.571		595	183		778	955
0900	Window frames only, based on perimeter length					L.F.	3.68			3.68	4.05
1200	Window sill, stock, per lineal foot						7.65			7.65	8.40
1250	Casing, stock						3.07			3.07	3.38

08 52 10.70 Sliding Windows

			Crew	Daily Output	Labor-Hours	Unit	Material	Labor	Equipment	Total	Total Incl O&P
0010	**SLIDING WINDOWS**										
0100	Average quality, bldrs. model, 3'-0" x 3'-0" high, double insulated	G	1 Carp	10	.800	Ea.	244	32		276	320
0120	Low E glass	G		10	.800		264	32		296	345

08 52 Wood Windows

08 52 10 – Plain Wood Windows

08 52 10.70 Sliding Windows		Crew	Daily Output	Labor-Hours	Unit	Material	2009 Bare Costs Labor	Equipment	Total	Total Incl O&P
0200	4'-0" x 3'-6" high, double insulated	G	1 Carp	9	.889	Ea.	282	35.50		317.50	370
0220	Low E glass	G		9	.889		305	35.50		340.50	395
0300	6'-0" x 5'-0" high, double insulated	G		8	1		420	40		460	525
0320	Low E glass	G		8	1		460	40		500	570
9000	Minimum labor/equipment charge			3	2.667	Job		107		107	176

08 52 13 – Metal-Clad Wood Windows

08 52 13.10 Awning Windows, Metal-Clad

		Crew	Daily Output	Labor-Hours	Unit	Material	2009 Bare Costs Labor	Equipment	Total	Total Incl O&P
0010	**AWNING WINDOWS, METAL-CLAD**									
2000	Metal clad, awning deluxe, double insulated glass, 34" x 22"	1 Carp	10	.800	Ea.	242	32		274	320
2100	40" x 22"	"	10	.800	"	285	32		317	370

08 52 13.35 Picture and Sliding Windows Metal-Clad

		Crew	Daily Output	Labor-Hours	Unit	Material	2009 Bare Costs Labor	Equipment	Total	Total Incl O&P
0010	**PICTURE AND SLIDING WINDOWS METAL-CLAD**									
2000	Metal clad, dlx picture, dbl. insul. glass, 4'-0" x 4'-0" high	2 Carp	12	1.333	Ea.	360	53.50		413.50	490
2100	4'-0" x 6'-0" high		11	1.455		535	58		593	685
2200	5'-0" x 6'-0" high		10	1.600		590	64		654	750
2300	6'-0" x 6'-0" high		10	1.600		675	64		739	850
2400	Metal clad, dlx sliding, double insulated glass, 3'-0" x 3'-0" high	G	1 Carp	10	.800		320	32		352	410
2420	4'-0" x 3'-6" high	G		9	.889		395	35.50		430.50	495
2440	5'-0" x 4'-0" high	G		9	.889		475	35.50		510.50	580
2460	6'-0" x 5'-0" high	G		8	1		700	40		740	835
9000	Minimum labor/equipment charge		2 Carp	2.75	5.818	Job		232		232	385

08 52 16 – Plastic-Clad Wood Windows

08 52 16.10 Bow Window

		Crew	Daily Output	Labor-Hours	Unit	Material	2009 Bare Costs Labor	Equipment	Total	Total Incl O&P
0010	**BOW WINDOW** Including frames, screens, and grilles									
0020	End panels operable									
1000	Bow type, casement, wood, bldrs mdl, 8' x 5' dbl insltd glass, 4 panel	2 Carp	10	1.600	Ea.	1,500	64		1,564	1,750
1050	Low E glass		10	1.600		1,300	64		1,364	1,550
1100	10'-0" x 5'-0", double insulated glass, 6 panels		6	2.667		1,350	107		1,457	1,675
1200	Low E glass, 6 panels		6	2.667		1,450	107		1,557	1,775
1300	Vinyl clad, bldrs model, double insulated glass, 6'-0" x 4'-0", 3 panel		10	1.600		1,475	64		1,539	1,725
1340	9'-0" x 4'-0", 4 panel		8	2		1,450	80		1,530	1,725
1380	10'-0" x 6'-0", 5 panels		7	2.286		2,250	91.50		2,341.50	2,625
1420	12'-0" x 6'-0", 6 panels		6	2.667		2,925	107		3,032	3,400
1600	Metal clad, casement, bldrs mdl, 6'-0" x 4'-0", dbl insltd gls, 3 panels		10	1.600		920	64		984	1,100
1640	9'-0" x 4'-0", 4 panels		8	2		1,250	80		1,330	1,500
1680	10'-0" x 5'-0", 5 panels		7	2.286		1,725	91.50		1,816.50	2,050
1720	12'-0" x 6'-0", 6 panels		6	2.667		2,400	107		2,507	2,825
2000	Bay window, builders model, 8' x 5' dbl insul glass,		10	1.600		1,825	64		1,889	2,125
2050	Low E glass,		10	1.600		2,225	64		2,289	2,550
2100	12'-0" x 6'-0", double insulated glass, 6 panels		6	2.667		2,300	107		2,407	2,700
2200	Low E glass		6	2.667		2,175	107		2,282	2,575
2300	Vinyl clad, premium, double insulated glass, 8'-0" x 5'-0"		10	1.600		1,300	64		1,364	1,525
2340	10'-0" x 5'-0"		8	2		1,850	80		1,930	2,150
2380	10'-0" x 6'-0"		7	2.286		1,925	91.50		2,016.50	2,275
2420	12'-0" x 6'-0"		6	2.667		2,300	107		2,407	2,700
2600	Metal clad, deluxe, dbl insul. glass, 8'-0" x 5'-0" high, 4 panels		10	1.600		1,625	64		1,689	1,875
2640	10'-0" x 5'-0" high, 5 panels		8	2		1,750	80		1,830	2,050
2680	10'-0" x 6'-0" high, 5 panels		7	2.286		2,050	91.50		2,141.50	2,425
2720	12'-0" x 6'-0" high, 6 panels		6	2.667		2,850	107		2,957	3,325
3000	Double hung, bldrs. model, bay, 8' x 4' high, dbl insulated glass		10	1.600		1,300	64		1,364	1,525
3050	Low E glass		10	1.600		1,375	64		1,439	1,625

08 52 Wood Windows

08 52 16 – Plastic-Clad Wood Windows

08 52 16.10 Bow Window

		Crew	Daily Output	Labor-Hours	Unit	Material	2009 Bare Costs Labor	Equipment	Total	Total Incl O&P
3100	9'-0" x 5'-0" high, double insulated glass	2 Carp	6	2.667	Ea.	1,400	107		1,507	1,700
3200	Low E glass		6	2.667		1,475	107		1,582	1,800
3300	Vinyl clad, premium, double insulated glass, 7'-0" x 4'-6"		10	1.600		1,325	64		1,389	1,575
3340	8'-0" x 4'-6"		8	2		1,375	80		1,455	1,625
3380	8'-0" x 5'-0"		7	2.286		1,425	91.50		1,516.50	1,725
3420	9'-0" x 5'-0"		6	2.667		1,475	107		1,582	1,800
3600	Metal clad, deluxe, dbl insul. glass, 7'-0" x 4'-0" high		10	1.600		1,225	64		1,289	1,450
3640	8'-0" x 4'-0" high		8	2		1,275	80		1,355	1,525
3680	8'-0" x 5'-0" high		7	2.286		1,325	91.50		1,416.50	1,600
3720	9'-0" x 5'-0" high		6	2.667		1,400	107		1,507	1,725
9000	Minimum labor/equipment charge		2.50	6.400	Job		256		256	420

08 52 16.20 Half Round, Vinyl Clad

		Crew	Daily Output	Labor-Hours	Unit	Material	2009 Bare Costs Labor	Equipment	Total	Total Incl O&P
0010	**HALF ROUND, VINYL CLAD**, double insulated glass, incl. grille									
0800	14" height x 24" base	2 Carp	9	1.778	Ea.	405	71		476	560
1040	15" height x 25" base		8	2		375	80		455	545
1060	16" height x 28" base		7	2.286		405	91.50		496.50	595
1080	17" height x 29" base		7	2.286		420	91.50		511.50	610
2000	19" height x 33" base	1 Carp	6	1.333		450	53.50		503.50	585
2100	20" height x 35" base		6	1.333		450	53.50		503.50	585
2200	21" height x 37" base		6	1.333		475	53.50		528.50	615
2250	23" height x 41" base	2 Carp	6	2.667		520	107		627	745
2300	26" height x 48" base		6	2.667		535	107		642	760
2350	30" height x 56" base		6	2.667		615	107		722	850
3000	36" height x 67"base	1 Carp	4	2		1,050	80		1,130	1,275
3040	38" height x 71" base	2 Carp	5	3.200		975	128		1,103	1,275
3050	40" height x 75" base	"	5	3.200		1,275	128		1,403	1,600
5000	Elliptical, 71" x 16"	1 Carp	11	.727		975	29		1,004	1,125
5100	95" x 21"	"	10	.800		1,375	32		1,407	1,575

08 52 16.40 Transom Windows

		Crew	Daily Output	Labor-Hours	Unit	Material	2009 Bare Costs Labor	Equipment	Total	Total Incl O&P
0010	**TRANSOM WINDOWS**									
0050	Vinyl clad, premium, double insulated glass, 32" x 8"	1 Carp	16	.500	Ea.	173	20		193	223
0100	36" x 8"		16	.500		184	20		204	235
0110	36" x 12"		16	.500		196	20		216	249
2000	Custom sizes, up to 350 sq. in.		12	.667		240	26.50		266.50	310
2100	351 to 750 sq. in.		12	.667		305	26.50		331.50	380
2200	751 to 1150 sq. in.		11	.727		375	29		404	465
2300	1151 to 1450 sq. in.		11	.727		430	29		459	525
2400	1451 to 1850 sq. in.	2 Carp	12	1.333		495	53.50		548.50	635
2500	1851 to 2250 sq. in.		12	1.333		565	53.50		618.50	715
2600	2251 to 2650 sq. in.		11	1.455		635	58		693	795
2700	2651 to 3050 sq. in.		11	1.455		640	58		698	800
2800	3051 to 3450 sq. in.		11	1.455		705	58		763	870
2900	3451 to 3850 sq. in.		10	1.600		760	64		824	940
3000	3851 to 4250 sq. in.		10	1.600		815	64		879	1,000
3100	4251 to 4650 sq. in.		10	1.600		830	64		894	1,025
3200	4651 to 5050 sq. in.		10	1.600		880	64		944	1,075
3300	5051 to 5450 sq. in.		9	1.778		950	71		1,021	1,175
3400	5451 to 5850 sq. in.		9	1.778		920	71		991	1,125
3600	6251 to 6650 sq. in.		8	2		1,100	80		1,180	1,350
3700	6651 to 7050 sq. in.		8	2		1,175	80		1,255	1,400

08 52 16.45 Trapezoid Windows

		Crew	Daily Output	Labor-Hours	Unit	Material	2009 Bare Costs Labor	Equipment	Total	Total Incl O&P
0010	**TRAPEZOID WINDOWS**									

08 52 Wood Windows

08 52 16 – Plastic-Clad Wood Windows

08 52 16.45 Trapezoid Windows

		Crew	Daily Output	Labor-Hours	Unit	Material	2009 Bare Costs Labor	Equipment	Total	Total Incl O&P
0900	20" base x 44" leg x 53" leg	2 Carp	13	1.231	Ea.	375	49		424	495
1000	24" base x 90" leg x 102" leg		8	2		640	80		720	835
3000	36" base x 40" leg x 22" leg		12	1.333		410	53.50		463.50	540
3010	36" base x 44" leg x 25" leg		13	1.231		430	49		479	555
3050	36" base x 26" leg x 48" leg		9	1.778		440	71		511	600
3100	36" base x 42" legs, 50" peak		9	1.778		520	71		591	685
3200	36" base x 60" leg x 81" leg		11	1.455		680	58		738	845
4320	44" base x 23" leg x 56" leg		11	1.455		535	58		593	685
4350	44" base x 59" leg x 92" leg		10	1.600		800	64		864	985
4500	46" base x 15" leg x 46" leg		8	2		405	80		485	575
4550	46" base x 16" leg x 48" leg		8	2		430	80		510	605
4600	46" base x 50" leg x 80" leg		7	2.286		650	91.50		741.50	865
6600	66" base x 12" leg x 42" leg		8	2		550	80		630	735
6650	66" base x 12" legs, 28" peak		9	1.778		440	71		511	600
6700	68" base x 23" legs, 31" peak		8	2		550	80		630	735

08 52 16.70 Vinyl Clad, Premium, Dbl. Insulated Glass

			Crew	Daily Output	Labor-Hours	Unit	Material	2009 Bare Costs Labor	Equipment	Total	Total Incl O&P
0010	**VINYL CLAD, PREMIUM, DBL. INSULATED GLASS**										
1000	Sliding, 3'-0" x 3'-0"	G	1 Carp	10	.800	Ea.	575	32		607	685
1050	4'-0" x 3'-6"	G		9	.889		670	35.50		705.50	795
1100	5'-0" x 4'-0"	G		9	.889		860	35.50		895.50	1,000
1150	6'-0" x 5'-0"	G		8	1		1,075	40		1,115	1,275

08 52 50 – Window Accessories

08 52 50.10 Window Grille or Muntin

		Crew	Daily Output	Labor-Hours	Unit	Material	2009 Bare Costs Labor	Equipment	Total	Total Incl O&P
0010	**WINDOW GRILLE OR MUNTIN**, snap in type									
0020	Standard pattern interior grilles									
2000	Wood, awning window, glass size 28" x 16" high	1 Carp	30	.267	Ea.	22.50	10.65		33.15	42.50
2060	44" x 24" high		32	.250		32.50	10		42.50	52
2100	Casement, glass size, 20" x 36" high		30	.267		27.50	10.65		38.15	48
2180	20" x 56" high		32	.250		39.50	10		49.50	60
2200	Double hung, glass size, 16" x 24" high		24	.333	Set	48	13.30		61.30	74.50
2280	32" x 32" high		34	.235	"	128	9.40		137.40	157
2500	Picture, glass size, 48" x 48" high		30	.267	Ea.	112	10.65		122.65	141
2580	60" x 68" high		28	.286	"	160	11.40		171.40	195
2600	Sliding, glass size, 14" x 36" high		24	.333	Set	25.50	13.30		38.80	50
2680	36" x 36" high		22	.364	"	39	14.55		53.55	67
9000	Minimum labor/equipment charge		5	1.600	Job		64		64	106

08 52 66 – Wood Window Screens

08 52 66.10 Wood Screens

		Crew	Daily Output	Labor-Hours	Unit	Material	2009 Bare Costs Labor	Equipment	Total	Total Incl O&P
0010	**WOOD SCREENS**									
0020	Over 3 S.F., 3/4" frames	2 Carp	375	.043	S.F.	4.28	1.70		5.98	7.50
0100	1-1/8" frames		375	.043		7.60	1.70		9.30	11.15
0200	Rescreen wood frame		500	.032		1.09	1.28		2.37	3.31
9000	Minimum labor/equipment charge	1 Carp	4	2	Job		80		80	132

08 52 69 – Wood Storm Windows

08 52 69.10 Storm Windows

		Crew	Daily Output	Labor-Hours	Unit	Material	2009 Bare Costs Labor	Equipment	Total	Total Incl O&P
0010	**STORM WINDOWS**, aluminum residential									
0300	Basement, mill finish, incl. fiberglass screen									
0320	1'-10" x 1'-0" high	2 Carp	30	.533	Ea.	32	21.50		53.50	70
0340	2'-9" x 1'-6" high		30	.533		34	21.50		55.50	72.50
0360	3'-4" x 2'-0" high		30	.533		42	21.50		63.50	81
1600	Double-hung, combination, storm & screen									

08 52 Wood Windows

08 52 69 – Wood Storm Windows

08 52 69.10 Storm Windows	Crew	Daily Output	Labor-Hours	Unit	Material	2009 Bare Costs Labor	Equipment	Total	Total Incl O&P
1700	Custom, clear anodic coating, 2'-0" x 3'-5" high	2 Carp	30	.533	Ea.	82	21.50		103.50	125
1720	2'-6" x 5'-0" high		28	.571		108	23		131	157
1740	4'-0" x 6'-0" high		25	.640		226	25.50		251.50	291
1800	White painted, 2'-0" x 3'-5" high		30	.533		96	21.50		117.50	141
1820	2'-6" x 5'-0" high		28	.571		154	23		177	207
1840	4'-0" x 6'-0" high		25	.640		273	25.50		298.50	340
2000	Average quality, clear anodic coating, 2'-0" x 3'-5" high		30	.533		82	21.50		103.50	125
2020	2'-6" x 5'-0" high		28	.571		102	23		125	150
2040	4'-0" x 6'-0" high		25	.640		120	25.50		145.50	174
2400	White painted, 2'-0" x 3'-5" high		30	.533		81	21.50		102.50	124
2420	2'-6" x 5'-0" high		28	.571		88	23		111	135
2440	4'-0" x 6'-0" high		25	.640		96	25.50		121.50	148
2600	Mill finish, 2'-0" x 3'-5" high		30	.533		73	21.50		94.50	116
2620	2'-6" x 5'-0" high		28	.571		82	23		105	128
2640	4'-0" x 6'-8" high	↓	25	.640	↓	92	25.50		117.50	143
4000	Picture window, storm, 1 lite, white or bronze finish									
4020	4'-6" x 4'-6" high	2 Carp	25	.640	Ea.	122	25.50		147.50	176
4040	5'-8" x 4'-6" high		20	.800		138	32		170	205
4400	Mill finish, 4'-6" x 4'-6" high		25	.640		122	25.50		147.50	176
4420	5'-8" x 4'-6" high	↓	20	.800	↓	138	32		170	205
4600	3 lite, white or bronze finish									
4620	4'-6" x 4'-6" high	2 Carp	25	.640	Ea.	148	25.50		173.50	205
4640	5'-8" x 4'-6" high		20	.800		164	32		196	233
4800	Mill finish, 4'-6" x 4'-6" high		25	.640		131	25.50		156.50	186
4820	5'-8" x 4'-6" high	↓	20	.800	↓	148	32		180	216
6000	Sliding window, storm, 2 lite, white or bronze finish									
6020	3'-4" x 2'-7" high	2 Carp	28	.571	Ea.	110	23		133	159
6040	4'-4" x 3'-3" high		25	.640		150	25.50		175.50	207
6060	5'-4" x 6'-0" high	↓	20	.800	↓	240	32		272	315
8000	PVC framed									
8100	Double-hung, combination, storm & screen									
8120	2'-6" x 3'-5"	2 Carp	15	1.067	Ea.	93.50	42.50		136	174
8140	4' x 6'	"	12.50	1.280	"	128	51		179	226
8600	Single lite picture storm									
8620	4'-6" x 4'-6"	2 Carp	12.50	1.280	Ea.	144	51		195	243
8640	5'-8" x 4'-6"	"	10	1.600	"	153	64		217	274
9000	Magnetic interior storm window									
9100	3/16" plate glass	1 Glaz	107	.075	S.F.	4.95	2.89		7.84	10.10
9410	Minimum labor/equipment charge	1 Carp	4	2	Job		80		80	132

08 53 Plastic Windows

08 53 13 – Vinyl Windows

08 53 13.20 Vinyl Single Hung Windows

0010	VINYL SINGLE HUNG WINDOWS		Crew	Daily Output	Labor-Hours	Unit	Material	Labor	Equipment	Total	Total Incl O&P
0100	Grids, low E, J fin, ext. jambs, 21" x 53"	G	2 Carp	18	.889	Ea.	151	35.50		186.50	225
0110	21" x 57"	G		17	.941		155	37.50		192.50	232
0120	21" x 65"	G		16	1		161	40		201	243
0130	25" x 41"	G		20	.800		142	32		174	209
0140	25" x 49"	G		18	.889		157	35.50		192.50	232
0150	25" x 57"	G		17	.941		161	37.50		198.50	239
0160	25" x 65"	G	↓	16	1	↓	168	40		208	251

08 53 Plastic Windows

08 53 13 – Vinyl Windows

08 53 13.20 Vinyl Single Hung Windows		Crew	Daily Output	Labor-Hours	Unit	Material	Labor	2009 Bare Costs Equipment	Total	Total Incl O&P
0170	29" x 41"	G	2 Carp	18	.889	Ea.	151	35.50		186.50	226
0180	29" x 53"	G		18	.889		162	35.50		197.50	238
0190	29" x 57"	G		17	.941		166	37.50		203.50	245
0200	29" x 65"	G		16	1		173	40		213	256
0210	33" x 41"	G		20	.800		157	32		189	225
0220	33" x 53"	G		18	.889		169	35.50		204.50	245
0230	33" x 57"	G		17	.941		173	37.50		210.50	252
0240	33" x 65"	G		16	1		180	40		220	264
0250	37" x 41"	G		20	.800		165	32		197	235
0260	37" x 53"	G		18	.889		178	35.50		213.50	254
0270	37" x 57"	G		17	.941		181	37.50		218.50	261
0280	37" x 65"	G		16	1		189	40		229	274

08 53 13.30 Vinyl Double Hung Windows		Crew	Daily Output	Labor-Hours	Unit	Material	Labor	2009 Bare Costs Equipment	Total	Total Incl O&P
0010	**VINYL DOUBLE HUNG WINDOWS**										
0100	Grids, low E, J fin, ext. jambs, 21" x 53"	G	2 Carp	18	.889	Ea.	173	35.50		208.50	249
0102	21" x 37"	G		18	.889		154	35.50		189.50	228
0104	21" x 41"	G		18	.889		157	35.50		192.50	232
0106	21" x 49"	G		18	.889		165	35.50		200.50	241
0110	21" x 57"	G		17	.941		176	37.50		213.50	256
0120	21" x 65"	G		16	1		184	40		224	268
0128	25" x 37"	G		20	.800		162	32		194	232
0130	25" x 41"	G		20	.800		166	32		198	235
0140	25" x 49"	G		18	.889		171	35.50		206.50	247
0145	25" x 53"	G		18	.889		178	35.50		213.50	255
0150	25" x 57"	G		17	.941		179	37.50		216.50	259
0160	25" x 65"	G		16	1		190	40		230	275
0162	25" x 69"	G		16	1		198	40		238	284
0164	25" x 77"	G		16	1		209	40		249	296
0168	29" x 37"	G		18	.889		168	35.50		203.50	243
0170	29" x 41"	G		18	.889		171	35.50		206.50	247
0172	29" x 49"	G		18	.889		180	35.50		215.50	257
0180	29" x 53"	G		18	.889		184	35.50		219.50	261
0190	29" x 57"	G		17	.941		188	37.50		225.50	268
0200	29" x 65"	G		16	1		195	40		235	281
0202	29" x 69"	G		16	1		202	40		242	289
0205	29" x 77"	G		16	1		214	40		254	300
0208	33" x 37"	G		20	.800		172	32		204	242
0210	33" x 41"	G		20	.800		176	32		208	247
0215	33" x 49"	G		20	.800		186	32		218	257
0220	33" x 53"	G		18	.889		190	35.50		225.50	268
0230	33" x 57"	G		17	.941		194	37.50		231.50	275
0240	33" x 65"	G		16	1		199	40		239	285
0242	33" x 69"	G		16	1		211	40		251	298
0246	33" x 77"	G		16	1		222	40		262	310
0250	37" x 41"	G		20	.800		180	32		212	251
0255	37" x 49"	G		20	.800		190	32		222	262
0260	37" x 53"	G		18	.889		197	35.50		232.50	276
0270	37" x 57"	G		17	.941		202	37.50		239.50	284
0280	37" x 65"	G		16	1		207	40		247	294
0282	37" x 69"	G		16	1		212	40		252	299
0286	37" x 77"	G		16	1		231	40		271	320
0300	Solid vinyl, average quality, double insulated glass, 2'-0" x 3'-0"	G	1 Carp	10	.800		279	32		311	360

08 53 Plastic Windows

08 53 13 – Vinyl Windows

08 53 13.30 Vinyl Double Hung Windows		Crew	Daily Output	Labor-Hours	Unit	Material	2009 Bare Costs Labor	Equipment	Total	Total Incl O&P
0310	3'-0" x 4'-0"	G	1 Carp	9	.889	Ea.	175	35.50		210.50	252
0320	4'-0" x 4'-6"	G		8	1		286	40		326	380
0330	Premium, double insulated glass, 2'-6" x 3'-0"	G		10	.800		194	32		226	266
0340	3'-0" x 3'-6"	G		9	.889		225	35.50		260.50	305
0350	3'-0" x 4'-0"	G		9	.889		238	35.50		273.50	320
0360	3'-0" x 4'-6"	G		9	.889		243	35.50		278.50	325
0370	3'-0" x 5'-0"	G		8	1		261	40		301	355
0380	3'-6" x 6'-0"	G		8	1		300	40		340	395

08 53 13.40 Vinyl Casement Windows		Crew	Daily Output	Labor-Hours	Unit	Material	2009 Bare Costs Labor	Equipment	Total	Total Incl O&P
0010	**VINYL CASEMENT WINDOWS**										
0100	Grids, low E, J fin, ext. jambs, 1 lt, 21" x 41"	G	2 Carp	20	.800	Ea.	225	32		257	300
0110	21" x 47"	G		20	.800		245	32		277	325
0120	21" x 53"	G		20	.800		265	32		297	345
0128	24" x 35"	G		19	.842		216	33.50		249.50	294
0130	24" x 41"	G		19	.842		235	33.50		268.50	315
0140	24" x 47"	G		19	.842		255	33.50		288.50	335
0150	24" x 53"	G		19	.842		275	33.50		308.50	355
0158	28" x 35"	G		19	.842		230	33.50		263.50	310
0160	28" x 41"	G		19	.842		249	33.50		282.50	330
0170	28" x 47"	G		19	.842		269	33.50		302.50	350
0180	28" x 53"	G		19	.842		297	33.50		330.50	380
0184	28" x 59"	G		19	.842		300	33.50		333.50	390
0188	Two lites, 33" x 35"	G		18	.889		370	35.50		405.50	465
0190	33" x 41"	G		18	.889		395	35.50		430.50	495
0200	33" x 47"	G		18	.889		425	35.50		460.50	530
0210	33" x 53"	G		18	.889		455	35.50		490.50	560
0212	33" x 59"	G		18	.889		485	35.50		520.50	590
0215	33" x 72"	G		18	.889		500	35.50		535.50	610
0220	41" x 41"	G		18	.889		435	35.50		470.50	535
0230	41" x 47"	G		18	.889		465	35.50		500.50	570
0240	41" x 53"	G		17	.941		490	37.50		527.50	600
0242	41" x 59"	G		17	.941		520	37.50		557.50	630
0246	41" x 72"	G		17	.941		545	37.50		582.50	655
0250	47" x 41"	G		17	.941		440	37.50		477.50	540
0260	47" x 47"	G		17	.941		465	37.50		502.50	575
0270	47" x 53"	G		17	.941		495	37.50		532.50	605
0272	47" x 59"	G		17	.941		540	37.50		577.50	655
0280	56" x 41"	G		15	1.067		470	42.50		512.50	585
0290	56" x 47"	G		15	1.067		495	42.50		537.50	615
0300	56" x 53"	G		15	1.067		540	42.50		582.50	665
0302	56" x 59"	G		15	1.067		565	42.50		607.50	690
0310	56" x 72"	G		15	1.067		610	42.50		652.50	745
0340	Solid vinyl, premium, double insulated glass, 2'-0" x 3'-0" high	G	1 Carp	10	.800		252	32		284	330
0360	2'-0" x 4'-0" high	G		9	.889		280	35.50		315.50	370
0380	2'-0" x 5'-0" high	G		8	1		276	40		316	370

08 53 13.50 Vinyl Picture Windows		Crew	Daily Output	Labor-Hours	Unit	Material	2009 Bare Costs Labor	Equipment	Total	Total Incl O&P
0010	**VINYL PICTURE WINDOWS**										
0100	Grids, low E, J fin, ext. jambs, 33" x 47"		2 Carp	12	1.333	Ea.	221	53.50		274.50	330
0110	35" x 71"			12	1.333		230	53.50		283.50	340
0120	41" x 47"			12	1.333		252	53.50		305.50	365
0130	41" x 71"			12	1.333		273	53.50		326.50	390
0140	47" x 47"			12	1.333		285	53.50		338.50	405

08 53 Plastic Windows

08 53 13 – Vinyl Windows

08 53 13.50 Vinyl Picture Windows

		Crew	Daily Output	Labor-Hours	Unit	Material	2009 Bare Costs Labor	Equipment	Total	Total Incl O&P
0150	47" x 71"	2 Carp	11	1.455	Ea.	299	58		357	425
0160	53" x 47"		11	1.455		280	58		338	405
0170	53" x 71"		11	1.455		293	58		351	420
0180	59" x 47"		11	1.455		320	58		378	445
0190	59" x 71"		11	1.455		340	58		398	470
0200	71" x 47"		10	1.600		350	64		414	495
0210	71" x 71"	↓	10	1.600	↓	370	64		434	515

08 53 13.60 Vinyl Half Round Windows

		Crew	Daily Output	Labor-Hours	Unit	Material	2009 Bare Costs Labor	Equipment	Total	Total Incl O&P
0010	**VINYL HALF ROUND WINDOWS**, Including grille, j fin low E, ext. jambs									
0100	10" height x 20" base	2 Carp	8	2	Ea.	475	80		555	650
0110	15" height x 30" base		8	2		415	80		495	590
0120	17" height x 34" base		7	2.286		265	91.50		356.50	440
0130	19" height x 38" base		7	2.286		286	91.50		377.50	465
0140	19" height x 33" base	↓	7	2.286		555	91.50		646.50	765
0150	24" height x 48" base	1 Carp	6	1.333		297	53.50		350.50	415
0160	25" height x 50" base	"	6	1.333		755	53.50		808.50	920
0170	30" height x 60" base	2 Carp	6	2.667	↓	620	107		727	860

08 54 Composite Windows

08 54 13 – Fiberglass Windows

08 54 13.10 Fiberglass Single Hung Windows

			Crew	Daily Output	Labor-Hours	Unit	Material	2009 Bare Costs Labor	Equipment	Total	Total Incl O&P
0010	**FIBERGLASS SINGLE HUNG WINDOWS**										
0100	Grids, low E, 18" x 24"	G	2 Carp	18	.889	Ea.	255	35.50		290.50	340
0110	18" x 40"	G		17	.941		275	37.50		312.50	365
0130	24" x 40"	G		20	.800		295	32		327	380
0230	36" x 36"	G		17	.941		290	37.50		327.50	380
0250	36" x 48"	G		20	.800		330	32		362	420
0260	36" x 60"	G		18	.889		350	35.50		385.50	445
0280	36" x 72"	G		16	1		360	40		400	460
0290	48" x 40"	G	↓	16	1	↓	360	40		400	460

08 56 Special Function Windows

08 56 63 – Detention Windows

08 56 63.13 Visitor Cubicle Windows

		Crew	Daily Output	Labor-Hours	Unit	Material	2009 Bare Costs Labor	Equipment	Total	Total Incl O&P
0010	**VISITOR CUBICLE WINDOWS**									
4000	Visitor cubicle, vision panel, no intercom	E-4	2	16	Ea.	2,925	725	67	3,717	4,650

08 62 Unit Skylights

08 62 13 – Domed Unit Skylights

08 62 13.20 Skylights

			Crew	Daily Output	Labor-Hours	Unit	Material	2009 Bare Costs Labor	Equipment	Total	Total Incl O&P
0010	**SKYLIGHTS**, Plastic domes, flush or curb mounted ten or										
0100	more units										
0300	Nominal size under 10 S.F., double	G	G-3	130	.246	S.F.	26.50	9.70		36.20	44.50
0400	Single			160	.200		23.50	7.90		31.40	38
0600	10 S.F. to 20 S.F., double	G		315	.102		22.50	4		26.50	31
0700	Single			395	.081		23	3.19		26.19	30
0900	20 S.F. to 30 S.F., double	G	↓	395	.081	↓	21	3.19		24.19	28

08 62 Unit Skylights

08 62 13 – Domed Unit Skylights

08 62 13.20 Skylights

		Crew	Daily Output	Labor-Hours	Unit	Material	2009 Bare Costs Labor	Equipment	Total	Total Incl O&P
1000	Single	G-3	465	.069	S.F.	19.05	2.71		21.76	25.50
1200	30 S.F. to 65 S.F., double [G]		465	.069		21	2.71		23.71	27.50
1300	Single		610	.052		16.50	2.07		18.57	21.50
1500	For insulated 4" curbs, double, add					27%				
1600	Single, add					30%				
1800	For integral insulated 9" curbs, double, add					30%				
1900	Single, add					40%				
2120	Ventilating insulated plexiglass dome with									
2130	curb mounting, 36" x 36" [G]	G-3	12	2.667	Ea.	380	105		485	590
2150	52" x 52" [G]		12	2.667		570	105		675	800
2160	28" x 52" [G]		10	3.200		455	126		581	705
2170	36" x 52" [G]		10	3.200		495	126		621	750
2180	For electric opening system, add [G]					292			292	320
2200	Field fabricated, factory type, aluminum and wire glass [G]	G-3	120	.267	S.F.	15.10	10.50		25.60	33.50
2300	Insulated safety glass with aluminum frame [G]		160	.200		88	7.90		95.90	110
2400	Sandwich panels, fiberglass, for walls, 1-9/16" thick, to 250 SF [G]		200	.160		16	6.30	.	22.30	28
2500	250 SF and up [G]		265	.121		14.30	4.76		19.06	23.50
2700	As above, but for roofs, 2-3/4" thick, to 250 SF [G]		295	.108		23	4.27		27.27	32.50
2800	250 SF and up [G]		330	.097		18.90	3.82		22.72	27

08 71 Door Hardware

08 71 13 – Automatic Door Operators

08 71 13.10 Automatic Openers Commercial

		Crew	Daily Output	Labor-Hours	Unit	Material	2009 Bare Costs Labor	Equipment	Total	Total Incl O&P
0010	**AUTOMATIC OPENERS COMMERCIAL**									
0020	Pneumatic, incl opener, motion sens, control box, tubing, compressor									
0050	For single swing door, per opening	2 Skwk	.80	20	Ea.	4,125	815		4,940	5,875
0100	Pair, per opening		.50	32	Opng.	6,875	1,300		8,175	9,700
1000	For single sliding door, per opening		.60	26.667		4,575	1,100		5,675	6,800
1300	Bi-parting pair		.50	32		6,900	1,300		8,200	9,725
1420	Electronic door opener incl motion sens, 12V control box, motor									
1450	For single swing door, per opening	2 Skwk	.80	20	Opng.	3,400	815		4,215	5,050
1500	Pair, per opening		.50	32		5,550	1,300		6,850	8,225
1600	For single sliding door, per opening		.60	26.667		3,675	1,100		4,775	5,825
1700	Bi-parting pair		.50	32		5,625	1,300		6,925	8,325
1750	Handicap actuator buttons, 2, including 12V DC wiring, add	1 Carp	1.50	5.333	Pr.	415	213		628	805

08 71 20 – Hardware

08 71 20.15 Hardware

		Crew	Daily Output	Labor-Hours	Unit	Material	2009 Bare Costs Labor	Equipment	Total	Total Incl O&P
0009	**HARDWARE**									
0010	Average percentage for hardware, total job cost									
0050	Maximum									4%
0500	Total hardware for building, average distribution					85%	15%			
1000	Door hardware, apartment, interior				Door	140			140	154
1500	Hospital bedroom, minimum					310			310	340
2000	Maximum					700			700	770
2250	School, single exterior, incl. lever, not incl. panic device					445			445	490
2500	Single interior, regular use, no lever included					315			315	345
2600	Heavy use, incl. lever and closer					550			550	605
2850	Stairway, single interior					790			790	870
3100	Double exterior, with panic device				Pr.	1,075			1,075	1,200
3600	Toilet, public, single interior				Door	170			170	187

08 71 Door Hardware

08 71 20 – Hardware

08 71 20.15 Hardware		Crew	Daily Output	Labor-Hours	Unit	Material	2009 Bare Costs Labor	Equipment	Total	Total Incl O&P
6020	Add for door holder, electro-magnetic	1 Elec	4	2	Ea.	86	94		180	240

08 71 20.30 Door Closers

		Crew	Daily Output	Labor-Hours	Unit	Material	Labor	Equipment	Total	Total Incl O&P
0010	**DOOR CLOSERS**									
0020	Adjustable backcheck, 3 way mount, all sizes, regular arm	1 Carp	6	1.333	Ea.	156	53.50		209.50	260
0040	Hold open arm		6	1.333		168	53.50		221.50	273
0100	Fusible link		6.50	1.231		130	49		179	224
0200	Non sized, regular arm		6	1.333		147	53.50		200.50	250
0240	Hold open arm		6	1.333		191	53.50		244.50	299
0400	4 way mount, non sized, regular arm		6	1.333		208	53.50		261.50	315
0440	Hold open arm		6	1.333		214	53.50		267.50	325
2000	Backcheck and adjustable power, hinge face mount									
2010	All sizes, regular arm	1 Carp	6.50	1.231	Ea.	187	49		236	287
2040	Hold open arm		6.50	1.231		235	49		284	340
2400	Top jamb mount, all sizes, regular arm		6	1.333		184	53.50		237.50	290
2440	Hold open arm		6	1.333		203	53.50		256.50	310
2800	Top face mount, all sizes, regular arm		6.50	1.231		184	49		233	283
2840	Hold open arm		6.50	1.231		203	49		252	305
4000	Backcheck, overhead concealed, all sizes, regular arm		5.50	1.455		200	58		258	315
4040	Concealed arm		5	1.600		217	64		281	345
4400	Compact overhead, concealed, all sizes, regular arm		5.50	1.455		340	58		398	465
4440	Concealed arm		5	1.600		375	64		439	520
4800	Concealed in door, all sizes, regular arm		5.50	1.455		129	58		187	238
4840	Concealed arm		5	1.600		139	64		203	259
4900	Floor concealed, all sizes, single acting		2.20	3.636		164	145		309	420
4940	Double acting		2.20	3.636		210	145		355	470
5000	For cast aluminum cylinder, deduct					17.90			17.90	19.70
5040	For delayed action, add					31			31	34
5080	For fusible link arm, add					13.20			13.20	14.50
5120	For shock absorbing arm, add					39			39	43
5160	For spring power adjustment, add					30			30	33
6000	Closer-holder, hinge face mount, all sizes, exposed arm	1 Carp	6.50	1.231		140	49		189	235
7000	Electronic closer-holder, hinge facemount, concealed arm		5	1.600		218	64		282	345
7400	With built-in detector		5	1.600		655	64		719	825
9000	Minimum labor/equipment charge		4	2	Job		80		80	132

08 71 20.31 Door Closers

		Crew	Daily Output	Labor-Hours	Unit	Material	Labor	Equipment	Total	Total Incl O&P
0010	**DOOR CLOSERS**									
0015	Door closer, rack and pinion	1 Carp	6.50	1.231	Ea.	148	49		197	243

08 71 20.35 Panic Devices

		Crew	Daily Output	Labor-Hours	Unit	Material	Labor	Equipment	Total	Total Incl O&P
0010	**PANIC DEVICES**									
0015	For rim locks, single door exit only	1 Carp	6	1.333	Ea.	475	53.50		528.50	615
0020	Outside key and pull		5	1.600		520	64		584	675
0200	Bar and vertical rod, exit only		5	1.600		700	64		764	870
0210	Outside key and pull		4	2		775	80		855	985
0400	Bar and concealed rod		4	2		570	80		650	755
0600	Touch bar, exit only		6	1.333		485	53.50		538.50	625
0610	Outside key and pull		5	1.600		555	64		619	715
0700	Touch bar and vertical rod, exit only		5	1.600		630	64		694	795
0710	Outside key and pull		4	2		730	80		810	935
1000	Mortise, bar, exit only		4	2		515	80		595	695
1600	Touch bar, exit only		4	2		590	80		670	780
2000	Narrow stile, rim mounted, bar, exit only		6	1.333		620	53.50		673.50	770
2010	Outside key and pull		5	1.600		675	64		739	845

08 71 Door Hardware

08 71 20 – Hardware

08 71 20.35 Panic Devices

		Crew	Daily Output	Labor-Hours	Unit	Material	2009 Bare Costs Labor	Equipment	Total	Total Incl O&P
2200	Bar and vertical rod, exit only	1 Carp	5	1.600	Ea.	635	64		699	805
2210	Outside key and pull		4	2		635	80		715	830
2400	Bar and concealed rod, exit only		3	2.667		745	107		852	990
3000	Mortise, bar, exit only		4	2		570	80		650	755
3600	Touch bar, exit only		4	2		825	80		905	1,050
4000	Double doors, exit only		2	4	Pr.	740	160		900	1,075
4500	Exit & entrance		2	4	"	845	160		1,005	1,200
9000	Minimum labor/equipment charge		2.50	3.200	Job		128		128	211

08 71 20.40 Lockset

		Crew	Daily Output	Labor-Hours	Unit	Material	2009 Bare Costs Labor	Equipment	Total	Total Incl O&P
0010	**LOCKSET**, Standard duty									
0020	Non-keyed, passage	1 Carp	12	.667	Ea.	49	26.50		75.50	97.50
0100	Privacy		12	.667		61	26.50		87.50	111
0400	Keyed, single cylinder function		10	.800		83.50	32		115.50	145
0420	Hotel		8	1		120	40		160	198
0500	Lever handled, keyed, single cylinder function		10	.800		148	32		180	216
1000	Heavy duty with sectional trim, non-keyed, passages		12	.667		138	26.50		164.50	196
1100	Privacy		12	.667		173	26.50		199.50	234
1400	Keyed, single cylinder function		10	.800		210	32		242	284
1420	Hotel		8	1		310	40		350	405
1600	Communicating		10	.800		235	32		267	310
1690	For re-core cylinder, add					34			34	37.50
1700	Residential, interior door, minimum	1 Carp	16	.500		17	20		37	51.50
1720	Maximum		8	1		43.50	40		83.50	114
1800	Exterior, minimum		14	.571		36.50	23		59.50	77.50
1810	Average		8	1		69	40		109	142
1820	Maximum		8	1		156	40		196	238
9000	Minimum labor/equipment charge		6	1.333	Job		53.50		53.50	88

08 71 20.41 Dead Locks

		Crew	Daily Output	Labor-Hours	Unit	Material	2009 Bare Costs Labor	Equipment	Total	Total Incl O&P
0010	**DEAD LOCKS**									
0011	Mortise heavy duty outside key (security item)	1 Carp	9	.889	Ea.	150	35.50		185.50	224
0020	Double cylinder		9	.889		162	35.50		197.50	237
0100	Medium duty, outside key		10	.800		110	32		142	174
0110	Double cylinder		10	.800		142	32		174	209
1000	Tubular, standard duty, outside key		10	.800		61	32		93	120
1010	Double cylinder		10	.800		79	32		111	140
1200	Night latch, outside key		10	.800		78.50	32		110.50	140

08 71 20.42 Mortise Locksets

		Crew	Daily Output	Labor-Hours	Unit	Material	2009 Bare Costs Labor	Equipment	Total	Total Incl O&P
0010	**MORTISE LOCKSETS**, Comm., wrought knobs & full escutcheon trim									
0020	Non-keyed, passage, minimum	1 Carp	9	.889	Ea.	175	35.50		210.50	252
0030	Maximum		8	1		288	40		328	380
0040	Privacy, minimum		9	.889		190	35.50		225.50	268
0050	Maximum		8	1		310	40		350	405
0100	Keyed, office/entrance/apartment, minimum		8	1		220	40		260	310
0110	Maximum		7	1.143		370	45.50		415.50	480
0120	Single cylinder, typical, minimum		8	1		189	40		229	274
0130	Maximum		7	1.143		350	45.50		395.50	460
0200	Hotel, minimum		7	1.143		220	45.50		265.50	320
0210	Maximum		6	1.333		365	53.50		418.50	490
0300	Communication, double cylinder, minimum		8	1		215	40		255	305
0310	Maximum		7	1.143		286	45.50		331.50	390
1000	Wrought knobs and sectional trim, non-keyed, passage, minimum		10	.800		119	32		151	184
1010	Maximum		9	.889		230	35.50		265.50	310

08 71 Door Hardware

08 71 20.42 Mortise Locksets

		Crew	Daily Output	Labor-Hours	Unit	Material	2009 Bare Costs Labor	2009 Bare Costs Equipment	Total	Total Incl O&P
1040	Privacy, minimum	1 Carp	10	.800	Ea.	136	32		168	203
1050	Maximum		9	.889		250	35.50		285.50	335
1100	Keyed, entrance, office/apartment, minimum		9	.889		206	35.50		241.50	286
1110	Maximum		8	1		295	40		335	390
1120	Single cylinder, typical, minimum		9	.889		200	35.50		235.50	279
1130	Maximum	↓	8	1	↓	288	40		328	380
2000	Cast knobs and full escutcheon trim									
2010	Non-keyed, passage, minimum	1 Carp	9	.889	Ea.	249	35.50		284.50	335
2020	Maximum		8	1		400	40		440	505
2040	Privacy, minimum		9	.889		305	35.50		340.50	395
2050	Maximum		8	1		440	40		480	545
2120	Keyed, single cylinder, typical, minimum		8	1		310	40		350	405
2130	Maximum		7	1.143		480	45.50		525.50	605
2200	Hotel, minimum		7	1.143		320	45.50		365.50	425
2210	Maximum		6	1.333		570	53.50		623.50	715
3000	Cast knob and sectional trim, non-keyed, passage, minimum		10	.800		196	32		228	269
3010	Maximum		10	.800		380	32		412	475
3040	Privacy, minimum		10	.800		220	32		252	295
3050	Maximum		10	.800		390	32		422	485
3100	Keyed, office/entrance/apartment, minimum		9	.889		250	35.50		285.50	335
3110	Maximum		9	.889		400	35.50		435.50	500
3120	Single cylinder, typical, minimum		9	.889		250	35.50		285.50	335
3130	Maximum	↓	9	.889		490	35.50		525.50	600
3190	For re-core cylinder, add					33			33	36.50
3800	Cipher lockset w/key pad (security item)	1 Carp	13	.615		775	24.50		799.50	890
3900	Cipher lockset, with dial for swinging doors (security item)		13	.615		1,675	24.50		1,699.50	1,900
3920	with dial for swinging doors & drill resistant plate (security item)		12	.667		2,025	26.50		2,051.50	2,275
3950	Cipher lockset with dial for safe/vault door (security item)	↓	12	.667	↓	1,300	26.50		1,326.50	1,475
3980	Keyless, pushbutton type									
4000	Residential/light commercial, deadbolt, standard	1 Carp	9	.889	Ea.	114	35.50		149.50	184
4010	Heavy duty		9	.889		134	35.50		169.50	206
4020	Industrial, heavy duty, with deadbolt		9	.889		271	35.50		306.50	355
4030	Key override		9	.889		300	35.50		335.50	395
4040	Lever activated handle		9	.889		320	35.50		355.50	415
4050	Key override		9	.889		360	35.50		395.50	455
4060	Double sided pushbutton type		8	1		660	40		700	795
4070	Key override	↓	8	1	↓	655	40		695	785

08 71 20.45 Peepholes

		Crew	Daily Output	Labor-Hours	Unit	Material	2009 Bare Costs Labor	2009 Bare Costs Equipment	Total	Total Incl O&P
0010	**PEEPHOLES**									
2010	Peephole	1 Carp	32	.250	Ea.	15.80	10		25.80	34

08 71 20.50 Door Stops

		Crew	Daily Output	Labor-Hours	Unit	Material	2009 Bare Costs Labor	2009 Bare Costs Equipment	Total	Total Incl O&P
0010	**DOOR STOPS**									
0020	Holder & bumper, floor or wall	1 Carp	32	.250	Ea.	33.50	10		43.50	53.50
1300	Wall bumper, 4" diameter, with rubber pad, aluminum		32	.250		10.25	10		20.25	28
1600	Door bumper, floor type, aluminum		32	.250		5.85	10		15.85	23
1900	Plunger type, door mounted		32	.250	↓	27	10		37	46
9000	Minimum labor/equipment charge	↓	6	1.333	Job		53.50		53.50	88

08 71 20.55 Push-Pull Plates

		Crew	Daily Output	Labor-Hours	Unit	Material	2009 Bare Costs Labor	2009 Bare Costs Equipment	Total	Total Incl O&P
0010	**PUSH-PULL PLATES**									
0100	Push plate, .050 thick, 4" x 16", aluminum	1 Carp	12	.667	Ea.	8.65	26.50		35.15	53.50
0500	Bronze		12	.667		20.50	26.50		47	66.50
1500	Pull handle and push bar, aluminum	↓	11	.727	↓	126	29		155	187

213

08 71 Door Hardware

08 71 20 – Hardware

08 71 20.55 Push-Pull Plates

		Crew	Daily Output	Labor-Hours	Unit	Material	2009 Bare Costs Labor	Equipment	Total	Total Incl O&P
2000	Bronze	1 Carp	10	.800	Ea.	163	32		195	233
3000	Push plate both sides, aluminum		14	.571		17.75	23		40.75	57
3500	Bronze		13	.615		41.50	24.50		66	86
4000	Door pull, designer style, cast aluminum, minimum		12	.667		67.50	26.50		94	119
5000	Maximum		8	1		365	40		405	465
6000	Cast bronze, minimum		12	.667		85.50	26.50		112	138
7000	Maximum		8	1		400	40		440	505
8000	Walnut, minimum		12	.667		65.50	26.50		92	116
9000	Maximum		8	1		365	40		405	465
9800	Minimum labor/equipment charge		5	1.600	Job		64		64	106

08 71 20.60 Entrance Locks

		Crew	Daily Output	Labor-Hours	Unit	Material	2009 Bare Costs Labor	Equipment	Total	Total Incl O&P
0010	**ENTRANCE LOCKS**									
0015	Cylinder, grip handle deadlocking latch	1 Carp	9	.889	Ea.	127	35.50		162.50	199
0020	Deadbolt		8	1		154	40		194	235
0100	Push and pull plate, dead bolt		8	1		146	40		186	227
0900	For handicapped lever, add					161			161	177

08 71 20.65 Thresholds

		Crew	Daily Output	Labor-Hours	Unit	Material	2009 Bare Costs Labor	Equipment	Total	Total Incl O&P
0010	**THRESHOLDS**									
0011	Threshold 3' long saddles aluminum	1 Carp	48	.167	L.F.	4.02	6.65		10.67	15.40
0100	Aluminum, 8" wide, 1/2" thick		12	.667	Ea.	38.50	26.50		65	86
0500	Bronze		60	.133	L.F.	38.50	5.35		43.85	51.50
0600	Bronze, panic threshold, 5" wide, 1/2" thick		12	.667	Ea.	65.50	26.50		92	116
0700	Rubber, 1/2" thick, 5-1/2" wide		20	.400		37.50	16		53.50	68
0800	2-3/4" wide		20	.400		16	16		32	44
9000	Minimum labor/equipment charge		4	2	Job		80		80	132

08 71 20.75 Door Hardware Accessories

		Crew	Daily Output	Labor-Hours	Unit	Material	2009 Bare Costs Labor	Equipment	Total	Total Incl O&P
0010	**DOOR HARDWARE ACCESSORIES**									
0050	Door closing coordinator, 36" (for paired openings up to 56")	1 Carp	8	1	Ea.	89.50	40		129.50	165
0060	48" (for paired openings up to 84")		8	1		96	40		136	171
0070	56" (for paired openings up to 96")		8	1		105	40		145	182
1000	Knockers, brass, standard		16	.500		39.50	20		59.50	76.50
1100	Deluxe		10	.800		123	32		155	188
4100	Deluxe		18	.444		44	17.75		61.75	78
4500	Rubber door silencers		540	.015		.26	.59		.85	1.27
9000	Minimum labor/equipment charge		6	1.333	Job		53.50		53.50	88

08 71 20.80 Hasps

		Crew	Daily Output	Labor-Hours	Unit	Material	2009 Bare Costs Labor	Equipment	Total	Total Incl O&P
0010	**HASPS**, steel assembly									
0015	3"	1 Carp	26	.308	Ea.	3.93	12.30		16.23	25
0020	4-1/2"		13	.615		5.05	24.50		29.55	46
0040	6"		12.50	.640		8.40	25.50		33.90	51.50

08 71 20.90 Hinges

		Crew	Daily Output	Labor-Hours	Unit	Material	2009 Bare Costs Labor	Equipment	Total	Total Incl O&P
0010	**HINGES** R087120-10									
0012	Full mortise, avg. freq., steel base, USP, 4-1/2" x 4-1/2"				Pr.	24.50			24.50	27
0100	5" x 5", USP					41			41	45
0200	6" x 6", USP					87			87	95.50
0400	Brass base, 4-1/2" x 4-1/2", US10					50			50	55
0500	5" x 5", US10					73.50			73.50	81
0600	6" x 6", US10					125			125	137
0800	Stainless steel base, 4-1/2" x 4-1/2", US32					74.50			74.50	82
0900	For non removable pin, add(security item)				Ea.	4.24			4.24	4.66
0910	For floating pin, driven tips, add					3.09			3.09	3.40

08 71 20.90 Hinges

		Daily Output	Labor-Hours	Unit	Material	2009 Bare Costs Labor	Equipment	Total	Total Incl O&P
0930	For hospital type tip on pin, add			Ea.	13.35			13.35	14.70
0940	For steeple type tip on pin, add			↓	11.70			11.70	12.85
0950	Full mortise, high frequency, steel base, 3-1/2" x 3-1/2", US26D			Pr.	25.50			25.50	28.50
1000	4-1/2" x 4-1/2", USP				58.50			58.50	64.50
1100	5" x 5", USP				54.50			54.50	60
1200	6" x 6", USP				134			134	147
1400	Brass base, 3-1/2" x 3-1/2", US4				45			45	49.50
1430	4-1/2" x 4-1/2", US10				78.50			78.50	86.50
1500	5" x 5", US10				118			118	130
1600	6" x 6", US10				170			170	187
1800	Stainless steel base, 4-1/2" x 4-1/2", US32				125			125	137
1810	5" x 4-1/2", US32			↓	175			175	193
1930	For hospital type tip on pin, add			Ea.	11.65			11.65	12.80
1950	Full mortise, low frequency, steel base, 3-1/2" x 3-1/2", US26D			Pr.	10.55			10.55	11.60
2000	4-1/2" x 4-1/2", USP				12.05			12.05	13.25
2100	5" x 5", USP				30			30	33
2200	6" x 6", USP				59.50			59.50	65.50
2300	4-1/2" x 4-1/2", US3				17.40			17.40	19.15
2310	5" x 5", US3				43			43	47.50
2400	Brass bass, 4-1/2" x 4-1/2", US10				42			42	46
2500	5" x 5", US10				63.50			63.50	70
2800	Stainless steel base, 4-1/2" x 4-1/2", US32			↓	72			72	79.50

08 71 20.91 Special Hinges

		Daily Output	Labor-Hours	Unit	Material	2009 Bare Costs Labor	Equipment	Total	Total Incl O&P
0010	**SPECIAL HINGES**								
0015	Paumelle, high frequency								
0020	Steel base, 6" x 4-1/2", US10			Pr.	139			139	153
0100	Bronze base, 5" x 4-1/2", US10				174			174	192
0200	Paumelle, average frequency, steel base, 4-1/2" x 3-1/2", US10				94			94	104
0400	Olive knuckle, low frequency, brass base, 6" x 4-1/2", US10			↓	159			159	175
1000	Electric hinge with concealed conductor, average frequency								
1010	Steel base, 4-1/2" x 4-1/2", US26D			Pr.	310			310	340
1100	Bronze base, 4-1/2" x 4-1/2", US26D			"	315			315	345
1200	Electric hinge with concealed conductor, high frequency								
1210	Steel base, 4-1/2" x 4-1/2", US26D			Pr.	231			231	255
1600	Double weight, 800 lb., steel base, removable pin, 5" x 6", USP				128			128	141
1700	Steel base-welded pin, 5" x 6", USP				158			158	174
1800	Triple weight, 2000 lb., steel base, welded pin, 5" x 6", USP				147			147	162
2000	Pivot reinf., high frequency, steel base, 7-3/4" door plate, USP				177			177	195
2200	Bronze base, 7-3/4" door plate, US10			↓	222			222	245
3000	Swing clear, full mortise, full or half surface, high frequency,								
3010	Steel base, 5" high, USP			Pr.	154			154	169
3200	Swing clear, full mortise, average frequency								
3210	Steel base, 4-1/2" high, USP			Pr.	122			122	135
4000	Wide throw, average frequency, steel base, 4-1/2" x 6", USP				93.50			93.50	103
4200	High frequency, steel base, 4-1/2" x 6", USP				143			143	157
4440	US26D			↓	86			86	94.50
4600	Spring hinge, single acting, 6" flange, steel			Ea.	52.50			52.50	57.50
4700	Brass				92			92	101
4900	Double acting, 6" flange, steel				94			94	103
4950	Brass			↓	153			153	168
5000	T-strap, galvanized, 4"			Pr.	18.50			18.50	20.50
5010	6"			↓	29			29	32

08 71 Door Hardware

08 71 20 – Hardware

08 71 20.91 Special Hinges

		Crew	Daily Output	Labor-Hours	Unit	Material	2009 Bare Costs Labor	Equipment	Total	Total Incl O&P
5020	8"				Pr.	44			44	48.50
9000	Continuous hinge, steel, full mortise, heavy duty	2 Carp	64	.250	L.F.	12.40	10		22.40	30

08 71 20.95 Kick Plates

		Crew	Daily Output	Labor-Hours	Unit	Material	2009 Bare Costs Labor	Equipment	Total	Total Incl O&P
0010	**KICK PLATES**									
0020	Stainless steel, 6" high, for 3' door	1 Carp	15	.533	Ea.	31	21.50		52.50	69
0500	Bronze 6" high, for 3' door		15	.533		43.50	21.50		65	83
2000	Aluminum, .050, with 3 beveled edges, 10" x 28"		15	.533		22	21.50		43.50	59
2010	10" x 30"		15	.533		23.50	21.50		45	61
2020	10" x 34"		15	.533		24.50	21.50		46	62
2040	10" x 38"		15	.533		26.50	21.50		48	64.50
9000	Minimum labor/equipment charge		6	1.333	Job		53.50		53.50	88

08 71 21 – Astragals

08 71 21.10 Exterior Mouldings, Astragals

		Crew	Daily Output	Labor-Hours	Unit	Material	2009 Bare Costs Labor	Equipment	Total	Total Incl O&P
0010	**EXTERIOR MOULDINGS, ASTRAGALS**									
0400	One piece, overlapping cadmium plated steel, flat, 3/16" x 2"	1 Carp	90	.089	L.F.	3.40	3.55		6.95	9.60
0600	Prime coated steel, flat, 1/8" x 3"		90	.089		4.65	3.55		8.20	10.95
0800	Stainless steel, flat, 3/32" x 1-5/8"		90	.089		17.80	3.55		21.35	25.50
1000	Aluminum, flat, 1/8" x 2"		90	.089		3.45	3.55		7	9.65
1200	Nail on, "T" extrusion		120	.067		.78	2.66		3.44	5.25
1300	Vinyl bulb insert		105	.076		1.25	3.04		4.29	6.40
1600	Screw on, "T" extrusion		90	.089		5.05	3.55		8.60	11.40
1700	Vinyl insert		75	.107		3.40	4.26		7.66	10.80
2000	"L" extrusion, neoprene bulbs		75	.107		1.95	4.26		6.21	9.20
2100	Neoprene sponge insert		75	.107		5.95	4.26		10.21	13.60
2200	Magnetic		75	.107		9.80	4.26		14.06	17.85
2400	Spring hinged security seal, with cam		75	.107		6.30	4.26		10.56	14
2600	Spring loaded locking bolt, vinyl insert		45	.178		8.60	7.10		15.70	21
2800	Neoprene sponge strip, "Z" shaped, aluminum		60	.133		4.10	5.35		9.45	13.30
2900	Solid neoprene strip, nail on aluminum strip		90	.089		3.40	3.55		6.95	9.60
3000	One piece stile protection									
3020	Neoprene fabric loop, nail on aluminum strips	1 Carp	60	.133	L.F.	.65	5.35		6	9.50
3110	Flush mounted aluminum extrusion, 1/2" x 1-1/4"		60	.133		3.26	5.35		8.61	12.40
3140	3/4" x 1-3/8"		60	.133		4.15	5.35		9.50	13.35
3160	1-1/8" x 1-3/4"		60	.133		6.60	5.35		11.95	16.05
3300	Mortise, 9/16" x 3/4"		60	.133		3.50	5.35		8.85	12.65
3320	13/16" x 1-3/8"		60	.133		3.80	5.35		9.15	13
3600	Spring bronze strip, nail on type		105	.076		3	3.04		6.04	8.30
3620	Screw on, with retainer		75	.107		2.45	4.26		6.71	9.75
3800	Flexible stainless steel housing, pile insert, 1/2" door		105	.076		6.75	3.04		9.79	12.45
3820	3/4" door		105	.076		7.60	3.04		10.64	13.35
4000	Extruded aluminum retainer, flush mount, pile insert		105	.076		2.36	3.04		5.40	7.60
4080	Mortise, felt insert		90	.089		4.20	3.55		7.75	10.45
4160	Mortise with spring, pile insert		90	.089		3.30	3.55		6.85	9.50
4400	Rigid vinyl retainer, mortise, pile insert		105	.076		2.35	3.04		5.39	7.60
4600	Wool pile filler strip, aluminum backing		105	.076		2.36	3.04		5.40	7.60
5000	Two piece overlapping astragal, extruded aluminum retainer									
5010	Pile insert	1 Carp	60	.133	L.F.	3.06	5.35		8.41	12.15
5020	Vinyl bulb insert		60	.133		1.98	5.35		7.33	11
5040	Vinyl flap insert		60	.133		6.15	5.35		11.50	15.60
5060	Solid neoprene flap insert		60	.133		6.10	5.35		11.45	15.50
5080	Hypalon rubber flap insert		60	.133		6.25	5.35		11.60	15.65
5090	Snap on cover, pile insert		60	.133		7.10	5.35		12.45	16.60

08 71 Door Hardware

08 71 21 – Astragals

08 71 21.10 Exterior Mouldings, Astragals	Crew	Daily Output	Labor-Hours	Unit	Material	2009 Bare Costs Labor	Equipment	Total	Total Incl O&P
5400	Magnetic aluminum, surface mounted	1 Carp	60	.133	L.F.	24	5.35		29.35	35.50
5500	Interlocking aluminum, 5/8" x 1" neoprene bulb insert		45	.178		3.80	7.10		10.90	15.90
5600	Adjustable aluminum, 9/16" x 21/32", pile insert		45	.178		18.15	7.10		25.25	31.50
5790	For vinyl bulb, deduct					.50			.50	.55
5800	Magnetic, adjustable, 9/16" x 21/32"	1 Carp	45	.178		23.50	7.10		30.60	37
6000	Two piece stile protection									
6010	Cloth backed rubber loop, 1" gap, nail on aluminum strips	1 Carp	45	.178	L.F.	3.90	7.10		11	16
6040	Screw on aluminum strips		45	.178		6.10	7.10		13.20	18.40
6100	1-1/2" gap, screw on aluminum extrusion		45	.178		5.45	7.10		12.55	17.70
6240	Vinyl fabric loop, slotted aluminum extrusion, 1" gap		45	.178		1.90	7.10		9	13.80
6300	1-1/4" gap		45	.178		5.75	7.10		12.85	18.05

08 71 25 – Weatherstripping

08 71 25.10 Mechanical Seals, Weatherstripping

		Crew	Daily Output	Labor-Hours	Unit	Material	Labor	Equipment	Total	Total Incl O&P
0010	**MECHANICAL SEALS, WEATHERSTRIPPING**									
1000	Doors, wood frame, interlocking, for 3' x 7' door, zinc	1 Carp	3	2.667	Opng.	15.50	107		122.50	193
1100	Bronze		3	2.667		24	107		131	203
1300	6' x 7' opening, zinc		2	4		16.90	160		176.90	283
1400	Bronze		2	4		32	160		192	299
1700	Wood frame, spring type, bronze									
1800	3' x 7' door	1 Carp	7.60	1.053	Opng.	21.50	42		63.50	93
1900	6' x 7' door	"	7	1.143	"	26.50	45.50		72	105
2200	Metal frame, spring type, bronze									
2300	3' x 7' door	1 Carp	3	2.667	Opng.	35.50	107		142.50	215
2400	6' x 7' door	"	2.50	3.200	"	46	128		174	262
2500	For stainless steel, spring type, add					133%				
2700	Metal frame, extruded sections, 3' x 7' door, aluminum	1 Carp	2	4	Opng.	45	160		205	315
2800	Bronze		2	4		114	160		274	390
3100	6' x 7' door, aluminum		1.20	6.667		57.50	266		323.50	505
3200	Bronze		1.20	6.667		134	266		400	585
3500	Threshold weatherstripping									
3650	Door sweep, flush mounted, aluminum	1 Carp	25	.320	Ea.	13.30	12.80		26.10	35.50
3700	Vinyl		25	.320		15.70	12.80		28.50	38.50
5000	Garage door bottom weatherstrip, 12' aluminum, clear		14	.571		21.50	23		44.50	61
5010	Bronze		14	.571		81.50	23		104.50	127
5050	Bottom protection, Rubber		14	.571		22	23		45	62
5100	Threshold		14	.571		88.50	23		111.50	135
9000	Minimum labor/equipment charge		3	2.667	Job		107		107	176

08 74 Access Control Hardware

08 74 13 – Card Key Access Control Hardware

08 74 13.50 Card Key Access

		Crew	Daily Output	Labor-Hours	Unit	Material	Labor	Equipment	Total	Total Incl O&P
0010	**CARD KEY ACCESS**									
0020	Card type, 1 time zone, minimum				Ea.	615			615	675
0040	Maximum					1,200			1,200	1,300
0060	3 time zones, minimum					1,400			1,400	1,550
0080	Maximum					1,900			1,900	2,100
0100	System with printer, and control console, 3 zones				Total	9,150			9,150	10,100
0120	6 zones				"	12,000			12,000	13,200
0140	For each door, minimum, add				Ea.	1,400			1,400	1,550
0160	Maximum, add				"	2,000			2,000	2,200

217

08 74 Access Control Hardware

08 74 19 – Biometric Identity Access Control Hardware

08 74 19.50 Biometric Identity Access

08 74 19.50 Biometric Identity Access	Crew	Daily Output	Labor-Hours	Unit	Material	2009 Bare Costs Labor	Equipment	Total	Total Incl O&P
0010 **BIOMETRIC IDENTITY ACCESS**									
0220　Hand geometry scanner, mem of 512 users, excl striker/powr	1 Elec	3	2.667	Ea.	1,900	125		2,025	2,300
0230　　Memory upgrade for, adds 9,700 user profiles		8	1		249	47		296	345
0240　　　Adds 32,500 user profiles		8	1		550	47		597	680
0250　Prison type, memory of 256 users, excl striker, power		3	2.667		2,450	125		2,575	2,900
0260　　Memory upgrade for, adds 3,300 user profiles		8	1		199	47		246	292
0270　　　Adds 9,700 user profiles		8	1		400	47		447	515
0280　　　Adds 27,900 user profiles		8	1		550	47		597	680
0290　All weather, mem of 512 users, excl striker/pwr		3	2.667		3,675	125		3,800	4,225
0300　Facial & fingerprint scanner, combination unit, excl striker/power		3	2.667		4,200	125		4,325	4,825
0310　Access for, for initial setup, excl striker/power		3	2.667		1,000	125		1,125	1,300

08 75 Window Hardware

08 75 10 – Window Handles and Latches

08 75 10.10 Handles and Latches

08 75 10.10 Handles and Latches	Crew	Daily Output	Labor-Hours	Unit	Material	2009 Bare Costs Labor	Equipment	Total	Total Incl O&P
0010 **HANDLES AND LATCHES**									
1000　Handles, surface mounted, aluminum	1 Carp	24	.333	Ea.	3.40	13.30		16.70	25.50
1020　　Brass		24	.333		5.75	13.30		19.05	28.50
1040　　Chrome		24	.333		5.65	13.30		18.95	28
1500　　Recessed, aluminum		12	.667		1.95	26.50		28.45	46
1520　　Brass		12	.667		2.05	26.50		28.55	46.50
1540　　Chrome		12	.667		2	26.50		28.50	46
2000　Latches, aluminum		20	.400		1.96	16		17.96	28.50
2020　　Brass		20	.400		2.21	16		18.21	29
2040　　Chrome		20	.400		2.11	16		18.11	29
9000　Minimum labor/equipment charge		6	1.333	Job		53.50		53.50	88

08 75 30 – Weatherstripping

08 75 30.10 Mechanical Weather Seals

08 75 30.10 Mechanical Weather Seals	Crew	Daily Output	Labor-Hours	Unit	Material	2009 Bare Costs Labor	Equipment	Total	Total Incl O&P
0010 **MECHANICAL WEATHER SEALS**, Window, double hung, 3' X 5'									
0020　　Zinc	1 Carp	7.20	1.111	Opng.	12.30	44.50		56.80	87
0100　　Bronze		7.20	1.111		25.50	44.50		70	102
0200　　Vinyl V strip		7	1.143		4.55	45.50		50.05	80.50
0500　As above but heavy duty, zinc		4.60	1.739		17.30	69.50		86.80	134
0600　　Bronze		4.60	1.739		29	69.50		98.50	147
9000　Minimum labor/equipment charge	1 Clab	4.60	1.739	Job		55		55	90.50

08 79 Hardware Accessories

08 79 20 – Door Accessories

08 79 20.10 Door Hardware Accessories

08 79 20.10 Door Hardware Accessories	Crew	Daily Output	Labor-Hours	Unit	Material	2009 Bare Costs Labor	Equipment	Total	Total Incl O&P
0010 **DOOR HARDWARE ACCESSORIES**									
0140　Door bolt, surface, 4"	1 Carp	32	.250	Ea.	8.65	10		18.65	26
0160　Door latch	"	12	.667	"	8.05	26.50		34.55	53
0200　Sliding closet door									
0220　　Track and hanger, single	1 Carp	10	.800	Ea.	49.50	32		81.50	108
0240　　　Double		8	1		72.50	40		112.50	146
0260　　Door guide, single		48	.167		23.50	6.65		30.15	37
0280　　　Double		48	.167		32	6.65		38.65	46.50
0600　Deadbolt and lock cover plate, brass or stainless steel		30	.267		26.50	10.65		37.15	47

218

08 79 Hardware Accessories

08 79 20 – Door Accessories

08 79 20.10 Door Hardware Accessories	Crew	Daily Output	Labor-Hours	Unit	Material	2009 Bare Costs Labor	Equipment	Total	Total Incl O&P
0620	Hole cover plate, brass or chrome	1 Carp	35	.229	Ea.	6.95	9.15		16.10	22.50
2240	Mortise lockset, passage, lever handle		9	.889		194	35.50		229.50	272
4000	Security chain, standard		18	.444		7.40	17.75		25.15	37.50

08 81 Glass Glazing

08 81 10 – Float Glass

08 81 10.10 Various Types and Thickness of Float Glass

		Crew	Daily Output	Labor-Hours	Unit	Material	2009 Bare Costs Labor	Equipment	Total	Total Incl O&P
0010	**VARIOUS TYPES AND THICKNESS OF FLOAT GLASS**									
0020	3/16" Plain	2 Glaz	130	.123	S.F.	4.48	4.75		9.23	12.60
0200	Tempered, clear		130	.123		6.05	4.75		10.80	14.30
0300	Tinted		130	.123		7.05	4.75		11.80	15.40
0600	1/4" thick, clear, plain		120	.133		5.25	5.15		10.40	14.10
0700	Tinted		120	.133		7.40	5.15		12.55	16.45
0800	Tempered, clear		120	.133		7.45	5.15		12.60	16.50
0900	Tinted		120	.133		9.75	5.15		14.90	19
1600	3/8" thick, clear, plain		75	.213		9.10	8.25		17.35	23.50
1700	Tinted		75	.213		14.80	8.25		23.05	29.50
1800	Tempered, clear		75	.213		15.10	8.25		23.35	30
1900	Tinted		75	.213		16.85	8.25		25.10	32
2200	1/2" thick, clear, plain		55	.291		18.15	11.25		29.40	38
2300	Tinted		55	.291		26	11.25		37.25	46.50
2400	Tempered, clear		55	.291		22	11.25		33.25	42.50
2500	Tinted		55	.291		26	11.25		37.25	46.50
2800	5/8" thick, clear, plain		45	.356		26	13.70		39.70	50.50
2900	Tempered, clear		45	.356		29.50	13.70		43.20	54.50
3200	3/4" thick, clear, plain		35	.457		33.50	17.65		51.15	65
3300	Tempered, clear		35	.457		39	17.65		56.65	71.50
3600	1" thick, clear, plain		30	.533		55.50	20.50		76	94
8900	For low emissivity coating for 3/16" & 1/4" only, add to above					16%				
9000	Minimum labor/equipment charge	1 Glaz	2	4	Job		154		154	249

08 81 20 – Vision Panels

08 81 20.10 Full Vision

		Crew	Daily Output	Labor-Hours	Unit	Material	2009 Bare Costs Labor	Equipment	Total	Total Incl O&P
0010	**FULL VISION**, window system with 3/4" glass mullions									
0020	Up to 10' high	H-2	130	.185	S.F.	57.50	6.70		64.20	74
0100	10' to 20' high, minimum		110	.218		61	7.90		68.90	80
0150	Average		100	.240		65.50	8.70		74.20	86.50
0200	Maximum		80	.300		74	10.90		84.90	98.50
9000	Minimum labor/equipment charge	1 Glaz	2	4	Job		154		154	249

08 81 25 – Glazing Variables

08 81 25.10 Applications of Glazing

		Crew	Daily Output	Labor-Hours	Unit	Material	2009 Bare Costs Labor	Equipment	Total	Total Incl O&P
0010	**APPLICATIONS OF GLAZING** R088110-10									
0500	For high rise glazing, exterior, add per S.F. per story				S.F.					.28
0600	For glass replacement, add				"		100%			
0700	For gasket settings, add				L.F.	5.55			5.55	6.10
0900	For sloped glazing, add				S.F.		25%			
2000	Fabrication, polished edges, 1/4" thick				Inch	.48			.48	.53
2100	1/2" thick					1.21			1.21	1.33
2500	Mitered edges, 1/4" thick					1.21			1.21	1.33
2600	1/2" thick					1.96			1.96	2.16

08 81 Glass Glazing

08 81 30 – Insulating Glass

08 81 30.10 Reduce Heat Transfer Glass

	08 81 30.10 Reduce Heat Transfer Glass		Crew	Daily Output	Labor-Hours	Unit	Material	2009 Bare Costs Labor	Equipment	Total	Total Incl O&P
0010	**REDUCE HEAT TRANSFER GLASS**, 2 lites 1/8" float, 1/2" thk under 15 S.F.										
0020	Clear R088110-10	G	2 Glaz	95	.168	S.F.	8.95	6.50		15.45	20.50
0100	Tinted	G		95	.168		12.55	6.50		19.05	24.50
0200	2 lites 3/16" float, for 5/8" thk unit, 15 to 30 S.F., clear	G		90	.178		12.95	6.85		19.80	25.50
0400	1" thk, dbl. glazed, 1/4" float, 30-70 S.F., clear	G		75	.213		15.20	8.25		23.45	30
0500	Tinted	G		75	.213		22	8.25		30.25	38
2000	Both lites, light & heat reflective	G		85	.188		29.50	7.25		36.75	44
2500	Heat reflective, film inside, 1" thick unit, clear	G		85	.188		26	7.25		33.25	40
2600	Tinted	G		85	.188		28	7.25		35.25	42.50
3000	Film on weatherside, clear, 1/2" thick unit	G		95	.168		18.50	6.50		25	31
3100	5/8" thick unit	G		90	.178		18.10	6.85		24.95	31
3200	1" thick unit	G		85	.188		25.50	7.25		32.75	39.50
3350	Minimum	G	1 Glaz	50	.160		11.70	6.20		17.90	23
3360	Maximum	G	"	25	.320		12.30	12.35		24.65	33.50
3370	Reflective or tinted, add	G					3.25			3.25	3.58
9000	Minimum labor/equipment charge		1 Glaz	2	4	Job		154		154	249

08 81 40 – Plate Glass

08 81 40.10 Plate Glass

	08 81 40.10 Plate Glass	Crew	Daily Output	Labor-Hours	Unit	Material	2009 Bare Costs Labor	Equipment	Total	Total Incl O&P
0010	**PLATE GLASS** Twin ground, polished,									
0020	3/16" thick	2 Glaz	100	.160	S.F.	4.73	6.20		10.93	15.15
0100	1/4" thick		94	.170		6.50	6.55		13.05	17.75
0200	3/8" thick		60	.267		11.15	10.30		21.45	29
0300	1/2" thick		40	.400		21.50	15.45		36.95	48.50

08 81 55 – Window Glass

08 81 55.10 Sheet Glass

	08 81 55.10 Sheet Glass	Crew	Daily Output	Labor-Hours	Unit	Material	2009 Bare Costs Labor	Equipment	Total	Total Incl O&P
0010	**SHEET GLASS** (window), clear float, stops, putty bed									
0015	1/8" thick, clear float	2 Glaz	480	.033	S.F.	4.27	1.29		5.56	6.75
0500	3/16" thick, clear		480	.033		5.25	1.29		6.54	7.85
0600	Tinted		480	.033		7.15	1.29		8.44	9.90
0700	Tempered		480	.033		8.65	1.29		9.94	11.55
2000	Replace broken window lite, 1/8" glass (9 S.F. maximum)	1 Glaz	48	.167		4.40	6.45		10.85	15.20
2100	1/4" plate (16 S.F. maximum)	"	48	.167		5.30	6.45		11.75	16.20
9000	Minimum labor/equipment charge	2 Glaz	5	3.200	Job		124		124	199

08 81 65 – Wire Glass

08 81 65.10 Glass Reinforced With Wire

	08 81 65.10 Glass Reinforced With Wire	Crew	Daily Output	Labor-Hours	Unit	Material	2009 Bare Costs Labor	Equipment	Total	Total Incl O&P
0010	**GLASS REINFORCED WITH WIRE**									
0012	1/4" thick rough obscure	2 Glaz	135	.119	S.F.	13.75	4.57		18.32	22.50
1000	Polished wire, 1/4" thick, diamond, clear		135	.119		18.75	4.57		23.32	28
1500	Pinstripe, obscure		135	.119		21	4.57		25.57	30.50

08 83 Mirrors

08 83 13 – Mirrored Glass Glazing

08 83 13.10 Mirrors		Crew	Daily Output	Labor-Hours	Unit	Material	2009 Bare Costs Labor	Equipment	Total	Total Incl O&P
0010	**MIRRORS**, No frames, wall type, 1/4" plate glass, polished edge									
0100	Up to 5 S.F.	2 Glaz	125	.128	S.F.	9.10	4.94		14.04	17.95
0200	Over 5 S.F.		160	.100		8.80	3.86		12.66	15.90
0500	Door type, 1/4" plate glass, up to 12 S.F.		160	.100		7.85	3.86		11.71	14.85
1000	Float glass, up to 10 S.F., 1/8" thick		160	.100		5.05	3.86		8.91	11.80
1100	3/16" thick		150	.107		6.15	4.12		10.27	13.45
1500	12" x 12" wall tiles, square edge, clear		195	.082		1.78	3.17		4.95	7.05
1600	Veined		195	.082		4.69	3.17		7.86	10.25
2000	1/4" thick, stock sizes, one way transparent		125	.128		17.70	4.94		22.64	27.50
2010	Bathroom, unframed, laminated		160	.100		13.20	3.86		17.06	20.50
2500	Tempered		160	.100		16.65	3.86		20.51	24.50

08 83 13.15 Reflective Glass

08 83 13.15 Reflective Glass			Crew	Daily Output	Labor-Hours	Unit	Material	2009 Bare Costs Labor	Equipment	Total	Total Incl O&P
0010	**REFLECTIVE GLASS**										
0100	1/4" float with fused metallic oxide fixed	G	2 Glaz	115	.139	S.F.	15.40	5.35		20.75	25.50
0500	1/4" float glass with reflective applied coating	G	"	115	.139	"	12.85	5.35		18.20	23

08 84 Plastic Glazing

08 84 10 – Plexiglass Glazing

08 84 10.10 Plexiglass Acrylic

08 84 10.10 Plexiglass Acrylic		Crew	Daily Output	Labor-Hours	Unit	Material	2009 Bare Costs Labor	Equipment	Total	Total Incl O&P
0010	**PLEXIGLASS ACRYLIC**, clear, masked,									
0020	1/8" thick, cut sheets	2 Glaz	170	.094	S.F.	4.69	3.63		8.32	11
0200	Full sheets		195	.082		2.45	3.17		5.62	7.80
0500	1/4" thick, cut sheets		165	.097		8.30	3.74		12.04	15.15
0600	Full sheets		185	.086		4.50	3.34		7.84	10.35
0900	3/8" thick, cut sheets		155	.103		15.15	3.98		19.13	23
1000	Full sheets		180	.089		8.15	3.43		11.58	14.55
1300	1/2" thick, cut sheets		135	.119		17.45	4.57		22.02	26.50
1400	Full sheets		150	.107		16.90	4.12		21.02	25.50
1700	3/4" thick, cut sheets		115	.139		62	5.35		67.35	76.50
1800	Full sheets		130	.123		36	4.75		40.75	47
2100	1" thick, cut sheets		105	.152		70	5.90		75.90	86.50
2200	Full sheets		125	.128		43	4.94		47.94	55.50
3000	Colored, 1/8" thick, cut sheets		170	.094		14.40	3.63		18.03	21.50
3200	Full sheets		195	.082		9.30	3.17		12.47	15.35
3500	1/4" thick, cut sheets		165	.097		16.10	3.74		19.84	24
3600	Full sheets		185	.086		11	3.34		14.34	17.50
4000	Mirrors, untinted, cut sheets, 1/8" thick		185	.086		6.75	3.34		10.09	12.85
4200	1/4" thick		180	.089		10.90	3.43		14.33	17.55

08 84 20 – Polycarbonate

08 84 20.10 Thermoplastic

08 84 20.10 Thermoplastic		Crew	Daily Output	Labor-Hours	Unit	Material	2009 Bare Costs Labor	Equipment	Total	Total Incl O&P
0010	**THERMOPLASTIC**, clear, masked, cut sheets									
0020	1/8" thick	2 Glaz	170	.094	S.F.	9.25	3.63		12.88	16
0500	3/16" thick		165	.097		10.55	3.74		14.29	17.65
1000	1/4" thick		155	.103		11.90	3.98		15.88	19.45
1500	3/8" thick		150	.107		21	4.12		25.12	29.50
9000	Minimum labor/equipment charge	1 Glaz	2	4	Job		154		154	249

08 85 Glazing Accessories

08 85 10 – Miscellaneous Glazing Accessories

08 85 10.10 Glazing Gaskets	Crew	Daily Output	Labor-Hours	Unit	Material	2009 Bare Costs Labor	Equipment	Total	Total Incl O&P
0010 **GLAZING GASKETS,** Neoprene for glass tongued mullion									
0015 1/4" glass	2 Glaz	200	.080	L.F.	.63	3.09		3.72	5.65
0020 3/8"		200	.080		.80	3.09		3.89	5.85
0040 1/2"		200	.080		.89	3.09		3.98	5.95
0060 3/4"		180	.089		1.32	3.43		4.75	7
0080 1"		180	.089		1.07	3.43		4.50	6.75
1000 Glazing compound, wood	1 Glaz	58	.138		.80	5.30		6.10	9.45
1005 Glazing compound, per window, up to 30 L.F.					.80			.80	.87
1006 Glazing compound, per window, up to 30 L.F. E.A.				Ea.	24			24	26
1020 Metal	1 Glaz	58	.138	L.F.	.80	5.30		6.10	9.45

08 87 Glazing Surface Films

08 87 13 – Solar Control Films

08 87 13.10 Solar Films On Glass

		Crew	Daily Output	Labor-Hours	Unit	Material	2009 Bare Costs Labor	Equipment	Total	Total Incl O&P
0010	**SOLAR FILMS ON GLASS** (glass not included)									
2000	Minimum	G 2 Glaz	180	.089	S.F.	6.20	3.43		9.63	12.40
2050	Maximum	G "	225	.071	"	14.40	2.74		17.14	20

08 87 53 – Security Films

08 87 53.10 Security Film

		Crew	Daily Output	Labor-Hours	Unit	Material	2009 Bare Costs Labor	Equipment	Total	Total Incl O&P
0010	**SECURITY FILM,** clear, 32000psi tensile strength, adhered to glass R088110-10									
0100	.002" thick, daylight installation	H-2	950	.025	S.F.	1.45	.92		2.37	3.09
0150	.004" thick, daylight installation		800	.030		1.60	1.09		2.69	3.53
0200	.006" thick, daylight installation		700	.034		1.70	1.24		2.94	3.89
0210	Install for anchorage		600	.040		1.89	1.45		3.34	4.43
0400	.007" thick, daylight installation		600	.040		1.80	1.45		3.25	4.33
0410	Install for anchorage		500	.048		2	1.74		3.74	5
0500	.008" thick, daylight installation		500	.048		2.10	1.74		3.84	5.15
0510	Install for anchorage		500	.048		2.33	1.74		4.07	5.40
0600	.015" thick, daylight installation		400	.060		3.50	2.18		5.68	7.40
0610	Install for anchorage		400	.060		2.33	2.18		4.51	6.10
0900	Security film anchorage, mechanical attachment and cover plate	H-3	370	.043	L.F.	8.70	1.49		10.19	12
0950	Security film anchorage, wet glaze structural caulking	1 Glaz	225	.036	"	.83	1.37		2.20	3.12
1000	Adhered security film removal	1 Clab	275	.029	S.F.		.92		.92	1.52

08 88 Special Function Glazing

08 88 56 – Ballistics-Resistant Glazing

08 88 56.10 Laminated Glass

		Crew	Daily Output	Labor-Hours	Unit	Material	2009 Bare Costs Labor	Equipment	Total	Total Incl O&P
0010	**LAMINATED GLASS**									
0020	Clear float .03" vinyl 1/4"	2 Glaz	90	.178	S.F.	11.10	6.85		17.95	23.50
0100	3/8" thick		78	.205		20	7.90		27.90	35.50
0200	.06" vinyl, 1/2" thick		65	.246		23.50	9.50		33	41.50
1000	5/8" thick		90	.178		28	6.85		34.85	41.50
2000	Bullet-resisting, 1-3/16" thick, to 15 S.F.		16	1		69.50	38.50		108	139
2100	Over 15 S.F.		16	1		88	38.50		126.50	159
2500	2-1/4" thick, to 15 S.F.		12	1.333		91	51.50		142.50	183
2600	Over 15 S.F.		12	1.333		81	51.50		132.50	172
2700	Level 2 (.357 magnum)		12	1.333		60	51.50		111.50	149
2750	Level 3 (.44 magnum)		12	1.333		65	51.50		116.50	155
2800	Level 4 (AK-47)		12	1.333		85	51.50		136.50	177

08 88 Special Function Glazing

08 88 56 – Ballistics-Resistant Glazing

08 88 56.10 Laminated Glass

	08 88 56.10 Laminated Glass	Crew	Daily Output	Labor-Hours	Unit	Material	2009 Bare Costs Labor	Equipment	Total	Total Incl O&P
2850	Level 5 (M-16)	2 Glaz	12	1.333	S.F.	95	51.50		146.50	188
2900	Level (7.62 Armor Piercing)	↓	12	1.333	↓	112	51.50		163.50	206

08 91 Louvers

08 91 19 – Fixed Louvers

08 91 19.10 Aluminum Louvers

		Crew	Daily Output	Labor-Hours	Unit	Material	2009 Bare Costs Labor	Equipment	Total	Total Incl O&P
0010	**ALUMINUM LOUVERS**									
0020	Aluminum with screen, residential, 8" x 8"	1 Carp	38	.211	Ea.	10.50	8.40		18.90	25.50
0100	12" x 12"		38	.211		11.85	8.40		20.25	27
0200	12" x 18"		35	.229		13.50	9.15		22.65	30
0250	14" x 24"		30	.267		17	10.65		27.65	36.50
0300	18" x 24"		27	.296		24.50	11.85		36.35	46.50
0500	24" x 30"		24	.333		34	13.30		47.30	59.50
0700	Triangle, adjustable, small		20	.400		29	16		45	58.50
0800	Large		15	.533		50	21.50		71.50	90.50
2100	Midget, aluminum, 3/4" deep, 1" diameter		85	.094		.77	3.76		4.53	7.05
2150	3" diameter		60	.133		1.62	5.35		6.97	10.60
2200	4" diameter		50	.160		3.07	6.40		9.47	13.95
2250	6" diameter	↓	30	.267	↓	3.63	10.65		14.28	21.50

08 95 Vents

08 95 13 – Soffit Vents

08 95 13.10 Wall Louvers

		Crew	Daily Output	Labor-Hours	Unit	Material	2009 Bare Costs Labor	Equipment	Total	Total Incl O&P
0010	**WALL LOUVERS**									
2330	Soffit vent, continuous, 3" wide, aluminum, mill finish	1 Carp	200	.040	L.F.	.53	1.60		2.13	3.22
2340	Baked enamel finish		200	.040	"	.48	1.60		2.08	3.17
2400	Under eaves vent, aluminum, mill finish, 16" x 4"		48	.167	Ea.	1.29	6.65		7.94	12.40
2500	16" x 8"	↓	48	.167	"	1.49	6.65		8.14	12.65

08 95 16 – Wall Vents

08 95 16.10 Louvers

		Crew	Daily Output	Labor-Hours	Unit	Material	2009 Bare Costs Labor	Equipment	Total	Total Incl O&P
0010	**LOUVERS**									
0020	Redwood, 2'-0" diameter, full circle	1 Carp	16	.500	Ea.	159	20		179	207
0100	Half circle		16	.500		152	20		172	200
0200	Octagonal		16	.500		121	20		141	166
0300	Triangular, 5/12 pitch, 5'-0" at base		16	.500		257	20		277	315
7000	Vinyl gable vent, 8" x 8"		38	.211		11.20	8.40		19.60	26.50
7020	12" x 12"		38	.211		23.50	8.40		31.90	40
7080	12" x 18"		35	.229		30	9.15		39.15	48
7200	18" x 24"		30	.267	↓	36.50	10.65		47.15	58
9000	Minimum labor/equipment charge	↓	3.50	2.286	Job		91.50		91.50	151

Division Notes

	CREW	DAILY OUTPUT	LABOR-HOURS	UNIT	2009 BARE COSTS				TOTAL INCL O&P
					MAT.	LABOR	EQUIP.	TOTAL	

Estimating Tips

General

- Room Finish Schedule: A complete set of plans should contain a room finish schedule. If one is not available, it would be well worth the time and effort to obtain one.

09 20 00 Plaster and Gypsum Board

- Lath is estimated by the square yard plus a 5% allowance for waste. Furring, channels, and accessories are measured by the linear foot. An extra foot should be allowed for each accessory miter or stop.

- Plaster is also estimated by the square yard. Deductions for openings vary by preference, from zero deduction to 50% of all openings over 2 feet in width. The estimator should allow one extra square foot for each linear foot of horizontal interior or exterior angle located below the ceiling level. Also, double the areas of small radius work.

- Drywall accessories, studs, track, and acoustical caulking are all measured by the linear foot. Drywall taping is figured by the square foot. Gypsum wallboard is estimated by the square foot. No material deductions should be made for door or window openings under 32 S.F.

09 60 00 Flooring

- Tile and terrazzo areas are taken off on a square foot basis. Trim and base materials are measured by the linear foot. Accent tiles are listed per each. Two basic methods of installation are used. Mud set is approximately 30% more expensive than thin set. In terrazzo work, be sure to include the linear footage of embedded decorative strips, grounds, machine rubbing, and power cleanup.

- Wood flooring is available in strip, parquet, or block configuration. The latter two types are set in adhesives with quantities estimated by the square foot. The laying pattern will influence labor costs and material waste. In addition to the material and labor for laying wood floors, the estimator must make allowances for sanding and finishing these areas unless the flooring is prefinished.

- Sheet flooring is measured by the square yard. Roll widths vary, so consideration should be given to use the most economical width, as waste must be figured into the total quantity. Consider also the installation methods available, direct glue down or stretched.

09 70 00 Wall Finishes

- Wall coverings are estimated by the square foot. The area to be covered is measured, length by height of wall above baseboards, to calculate the square footage of each wall. This figure is divided by the number of square feet in the single roll which is being used. Deduct, in full, the areas of openings such as doors and windows. Where a pattern match is required allow 25%–30% waste.

09 80 00 Acoustic Treatment

- Acoustical systems fall into several categories. The takeoff of these materials should be by the square foot of area with a 5% allowance for waste. Do not forget about scaffolding, if applicable, when estimating these systems.

09 90 00 Painting and Coating

- A major portion of the work in painting involves surface preparation. Be sure to include cleaning, sanding, filling, and masking costs in the estimate.

- Protection of adjacent surfaces is not included in painting costs. When considering the method of paint application, an important factor is the amount of protection and masking required. These must be estimated separately and may be the determining factor in choosing the method of application.

Reference Numbers

Reference numbers are shown in shaded boxes at the beginning of some major classifications. These numbers refer to related items in the Reference Section. The reference information may be an estimating procedure, an alternate pricing method, or technical information.

Note: Not all subdivisions listed here necessarily appear in this publication.

09 01 Maintenance of Finishes

09 01 70 – Maintenance of Wall Finishes

09 01 70.10 Gypsum Wallboard Repairs

		Crew	Daily Output	Labor-Hours	Unit	Material	2009 Bare Costs Labor	Equipment	Total	Total Incl O&P
0010	**GYPSUM WALLBOARD REPAIRS**									
0100	Fill and sand, pin / nail holes	1 Carp	960	.008	Ea.		.33		.33	.55
0110	Screw head pops		480	.017			.67		.67	1.10
0120	Dents, up to 2" square		48	.167		.01	6.65		6.66	11
0130	2" to 4" square		24	.333		.03	13.30		13.33	22
0140	Cut square, patch, sand and finish, holes, up to 2" square		12	.667		.04	26.50		26.54	44
0150	2" to 4" square		11	.727		.10	29		29.10	48
0160	4" to 8" square		10	.800		.25	32		32.25	53.50
0170	8" to 12" square		8	1		.48	40		40.48	66.50
0180	12" to 32" square		6	1.333		3.04	53.50		56.54	91.50
0210	16" by 48"		5	1.600		2.31	64		66.31	109
0220	32" by 48"		4	2		4.13	80		84.13	137
0230	48" square		3.50	2.286		5.85	91.50		97.35	157
0240	60" square		3.20	2.500		9.80	100		109.80	176
0500	Skim coat surface with joint compound		1600	.005	S.F.	.03	.20		.23	.36
9000	Minimum labor/equipment charge		2	4	Job		160		160	264

09 05 Common Work Results for Finishes

09 05 05 – Selective Finishes Demolition

09 05 05.10 Selective Demolition, Ceilings

		Crew	Daily Output	Labor-Hours	Unit	Material	2009 Bare Costs Labor	Equipment	Total	Total Incl O&P
0010	**SELECTIVE DEMOLITION, CEILINGS** R024119-10									
0200	Ceiling, drywall, furred and nailed or screwed	2 Clab	800	.020	S.F.		.63		.63	1.04
0220	On metal frame		760	.021			.67		.67	1.10
0240	On suspension system, including system		720	.022			.70		.70	1.16
1000	Plaster, lime and horse hair, on wood lath, incl. lath		700	.023			.72		.72	1.19
1020	On metal lath		570	.028			.89		.89	1.46
1100	Gypsum, on gypsum lath		720	.022			.70		.70	1.16
1120	On metal lath		500	.032			1.01		1.01	1.67
1200	Suspended ceiling, mineral fiber, 2' x 2' or 2' x 4'		1500	.011			.34		.34	.56
1250	On suspension system, incl. system		1200	.013			.42		.42	.70
1500	Tile, wood fiber, 12" x 12", glued		900	.018			.56		.56	.93
1540	Stapled		1500	.011			.34		.34	.56
1580	On suspension system, incl. system		760	.021			.67		.67	1.10
2000	Wood, tongue and groove, 1" x 4"		1000	.016			.51		.51	.83
2040	1" x 8"		1100	.015			.46		.46	.76
2400	Plywood or wood fiberboard, 4' x 8' sheets		1200	.013			.42		.42	.70
9000	Minimum labor/equipment charge	1 Clab	2	4	Job		126		126	209

09 05 05.20 Selective Demolition, Flooring

		Crew	Daily Output	Labor-Hours	Unit	Material	2009 Bare Costs Labor	Equipment	Total	Total Incl O&P
0010	**SELECTIVE DEMOLITION, FLOORING** R024119-10									
0200	Brick with mortar	2 Clab	475	.034	S.F.		1.06		1.06	1.76
0400	Carpet, bonded, including surface scraping		2000	.008			.25		.25	.42
0480	Tackless		9000	.002			.06		.06	.09
0550	Carpet tile, releasable adhesive		5000	.003			.10		.10	.17
0560	Permanent adhesive		1850	.009			.27		.27	.45
0600	Composition, acrylic or epoxy		400	.040			1.26		1.26	2.09
0700	Concrete, scarify skin	A-1A	225	.036			1.45	.97	2.42	3.43
0800	Resilient, sheet goods	2 Clab	1400	.011			.36		.36	.60
0820	For gym floors	"	900	.018			.56		.56	.93
0850	Vinyl or rubber cove base	1 Clab	1000	.008	L.F.		.25		.25	.42
0860	Vinyl or rubber cove base, molded corner	"	1000	.008	Ea.		.25		.25	.42
0870	For glued and caulked installation, add to labor						50%			

226

09 05 05 – Selective Finishes Demolition

09 05 05.20 Selective Demolition, Flooring

		Crew	Daily Output	Labor-Hours	Unit	Material	2009 Bare Costs Labor	Equipment	Total	Total Incl O&P
0900	Vinyl composition tile, 12" x 12"	2 Clab	1000	.016	S.F.		.51		.51	.83
2000	Tile, ceramic, thin set		675	.024			.75		.75	1.24
2020	Mud set		625	.026			.81		.81	1.34
2200	Marble, slate, thin set		675	.024			.75		.75	1.24
2220	Mud set		625	.026			.81		.81	1.34
2600	Terrazzo, thin set		450	.036			1.12		1.12	1.85
2620	Mud set		425	.038			1.19		1.19	1.96
2640	Terrazzo, cast in place		300	.053			1.69		1.69	2.78
3000	Wood, block, on end	1 Carp	400	.020			.80		.80	1.32
3200	Parquet		450	.018			.71		.71	1.17
3400	Strip flooring, interior, 2-1/4" x 25/32" thick		325	.025			.98		.98	1.62
3500	Exterior, porch flooring, 1" x 4"		220	.036			1.45		1.45	2.40
3800	Subfloor, tongue and groove, 1" x 6"		325	.025			.98		.98	1.62
3820	1" x 8"		430	.019			.74		.74	1.23
3840	1" x 10"		520	.015			.61		.61	1.01
4000	Plywood, nailed		600	.013			.53		.53	.88
4100	Glued and nailed		400	.020			.80		.80	1.32
4200	Hardboard, 1/4" thick		760	.011			.42		.42	.69
8000	Remove flooring, bead blast, minimum	A-1A	1000	.008			.33	.22	.55	.77
8100	Maximum		400	.020			.82	.54	1.36	1.93
8150	Mastic only		1500	.005			.22	.15	.37	.51
9000	Minimum labor/equipment charge	1 Clab	4	2	Job		63		63	104

09 05 05.30 Selective Demolition, Walls and Partitions

		Crew	Daily Output	Labor-Hours	Unit	Material	2009 Bare Costs Labor	Equipment	Total	Total Incl O&P
0010	**SELECTIVE DEMOLITION, WALLS AND PARTITIONS** R024119-10									
0020	Walls, concrete, reinforced	B-39	120	.400	C.F.		13.25	1.60	14.85	23.50
0025	Plain	"	160	.300			9.95	1.20	11.15	17.60
0100	Brick, 4" to 12" thick	B-9	220	.182			5.80	.87	6.67	10.55
0200	Concrete block, 4" thick		1000	.040	S.F.		1.28	.19	1.47	2.32
0280	8" thick		810	.049			1.58	.24	1.82	2.87
0300	Exterior stucco 1" thick over mesh		3200	.013			.40	.06	.46	.73
1000	Drywall, nailed or screwed	1 Clab	1000	.008			.25		.25	.42
1010	2 layers		400	.020			.63		.63	1.04
1020	Glued and nailed		900	.009			.28		.28	.46
1500	Fiberboard, nailed		900	.009			.28		.28	.46
1520	Glued and nailed		800	.010			.32		.32	.52
1568	Plenum barrier, sheet lead		300	.027			.84		.84	1.39
2000	Movable walls, metal, 5' high		300	.027			.84		.84	1.39
2020	8' high		400	.020			.63		.63	1.04
2200	Metal or wood studs, finish 2 sides, fiberboard	B-1	520	.046			1.49		1.49	2.46
2250	Lath and plaster		260	.092			2.98		2.98	4.92
2300	Plasterboard (drywall)		520	.046			1.49		1.49	2.46
2350	Plywood		450	.053			1.72		1.72	2.84
2800	Paneling, 4' x 8' sheets	1 Clab	475	.017			.53		.53	.88
3000	Plaster, lime and horsehair, on wood lath		400	.020			.63		.63	1.04
3020	On metal lath		335	.024			.75		.75	1.25
3400	Gypsum or perlite, on gypsum lath		410	.020			.62		.62	1.02
3420	On metal lath		300	.027			.84		.84	1.39
3450	Plaster, interior gypsum, acoustic, or cement		60	.133	S.Y.		4.21		4.21	6.95
3500	Stucco, on masonry		145	.055			1.74		1.74	2.88
3510	Commercial 3-coat		80	.100			3.16		3.16	5.20
3520	Interior stucco		25	.320			10.10		10.10	16.70
3760	Tile, ceramic, on walls, thin set		300	.027	S.F.		.84		.84	1.39

09 05 Common Work Results for Finishes

09 05 05 – Selective Finishes Demolition

09 05 05.30 Selective Demolition, Walls and Partitions

		Crew	Daily Output	Labor-Hours	Unit	Material	2009 Bare Costs Labor	Equipment	Total	Total Incl O&P
3765	Mud set	1 Clab	250	.032	S.F.		1.01		1.01	1.67
3800	Toilet partitions, slate or marble		5	1.600	Ea.		50.50		50.50	83.50
3820	Metal or plastic	↓	8	1	"		31.50		31.50	52
5000	Wallcovering, vinyl	1 Pape	700	.011	S.F.		.41		.41	.65
5010	With release agent		1500	.005			.19		.19	.30
5025	Wallpaper, 2 layers or less, by hand		250	.032			1.13		1.13	1.81
5035	3 layers or more		165	.048			1.72		1.72	2.75
5040	Designer	↓	480	.017	↓		.59		.59	.94
9000	Minimum labor/equipment charge	1 Clab	4	2	Job		63		63	104

09 21 Plaster and Gypsum Board Assemblies

09 21 16 – Gypsum Board Assemblies

09 21 16.23 Gypsum Board Shaft Wall Assemblies

		Crew	Daily Output	Labor-Hours	Unit	Material	2009 Bare Costs Labor	Equipment	Total	Total Incl O&P
0010	**GYPSUM BOARD SHAFT WALL ASSEMBLIES**									
0030	1" thick coreboard wall liner on shaft side									
0040	2-hour assembly with double layer									
0060	5/8" fire rated gypsum board on room side	2 Carp	220	.073	S.F.	1.56	2.91		4.47	6.50
0100	3-hour assembly with triple layer									
0300	5/8" fire rated gypsum board on room side	2 Carp	180	.089	S.F.	1.90	3.55		5.45	7.95
0400	4-hour assembly, 1" coreboard, 5/8" fire rated gypsum board									
0600	and 3/4" galv. metal furring channels, 24" O.C., with									
0700	Double layer 5/8" fire rated gypsum board on room side	2 Carp	110	.145	S.F.	1.77	5.80		7.57	11.55
0900	For taping & finishing, add per side	1 Carp	1050	.008	"	.04	.30		.34	.55
1000	For insulation, see Div. 07 21									
5200	For work over 8' high, add	2 Carp	3060	.005	S.F.		.21		.21	.34
5300	For distribution cost over 3 stories high, add per story	"	6100	.003	"		.10		.10	.17

09 21 16.33 Partition Wall

		Crew	Daily Output	Labor-Hours	Unit	Material	2009 Bare Costs Labor	Equipment	Total	Total Incl O&P
0010	**PARTITION WALL** Stud wall, 8' to 12' high									
0050	1/2", interior, gypsum board, std, tape & finish 2 sides									
0500	Installed on and incl. 2" x 4" wood studs, 16" O.C.	2 Carp	310	.052	S.F.	1.03	2.06		3.09	4.53
1000	Metal studs, NLB, 25 ga., 16" O.C., 3-5/8" wide		350	.046		1.16	1.83		2.99	4.28
1200	6" wide		330	.048		1.39	1.94		3.33	4.72
1400	Water resistant, on 2" x 4" wood studs, 16" O.C.		310	.052		1.37	2.06		3.43	4.91
1600	Metal studs, NLB, 25 ga., 16" O.C., 3-5/8" wide		350	.046		1.50	1.83		3.33	4.65
1800	6" wide		330	.048		1.73	1.94		3.67	5.10
2000	Fire res., 2 layers, 1-1/2 hr., on 2" x 4" wood studs, 16" O.C.		210	.076		1.69	3.04		4.73	6.85
2200	Metal studs, NLB, 25 ga., 16" O.C., 3-5/8" wide		250	.064		1.82	2.56		4.38	6.20
2400	6" wide		230	.070		2.05	2.78		4.83	6.85
2600	Fire & water res., 2 layers, 1-1/2 hr., 2" x 4" studs, 16" O.C.		210	.076		1.69	3.04		4.73	6.85
2800	Metal studs, NLB, 25 ga., 16" O.C., 3-5/8" wide		250	.064		1.82	2.56		4.38	6.20
3000	6" wide	↓	230	.070	↓	2.05	2.78		4.83	6.85
3200	5/8", interior, gypsum board, std, tape & finish 2 sides									
3400	Installed on and including 2" x 4" wood studs, 16" O.C.	2 Carp	300	.053	S.F.	1.09	2.13		3.22	4.72
3600	24" O.C.		330	.048		1.02	1.94		2.96	4.32
3800	Metal studs, NLB, 25 ga., 16" O.C., 3-5/8" wide		340	.047		1.22	1.88		3.10	4.44
4000	6" wide		320	.050		1.45	2		3.45	4.89
4200	24" O.C., 3-5/8" wide		360	.044		1.10	1.78		2.88	4.14
4400	6" wide		340	.047		1.27	1.88		3.15	4.49
4800	Water resistant, on 2" x 4" wood studs, 16" O.C.		300	.053		1.35	2.13		3.48	5
5000	24" O.C.		330	.048		1.28	1.94		3.22	4.61
5200	Metal studs, NLB, 25 ga. 16" O.C., 3-5/8" wide	↓	340	.047	↓	1.48	1.88		3.36	4.72

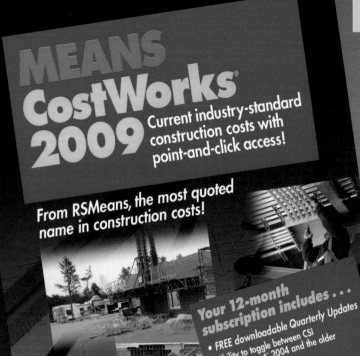

09 21 Plaster and Gypsum Board Assemblies

09 21 16 – Gypsum Board Assemblies

09 21 16.33 Partition Wall		Crew	Daily Output	Labor-Hours	Unit	Material	2009 Bare Costs Labor	Equipment	Total	Total Incl O&P
5400	6" wide	2 Carp	320	.050	S.F.	1.71	2		3.71	5.20
5600	24" O.C., 3-5/8" wide		360	.044		1.36	1.78		3.14	4.42
5800	6" wide		340	.047		1.53	1.88		3.41	4.78
6000	Fire resistant, 2 layers, 2 hr., on 2" x 4" wood studs, 16" O.C.		205	.078		1.78	3.12		4.90	7.10
6200	24" O.C.		235	.068		1.78	2.72		4.50	6.45
6400	Metal studs, NLB, 25 ga., 16" O.C., 3-5/8" wide		245	.065		2	2.61		4.61	6.50
6600	6" wide		225	.071		2.21	2.84		5.05	7.10
6800	24" O.C., 3-5/8" wide		265	.060		1.86	2.41		4.27	6
7000	6" wide		245	.065		2.03	2.61		4.64	6.55
7200	Fire & water resistant, 2 layers, 2 hr., 2" x 4" studs, 16" O.C.		205	.078		1.85	3.12		4.97	7.20
7400	24" O.C.		235	.068		1.78	2.72		4.50	6.45
7600	Metal studs, NLB, 25 ga., 16" O.C., 3-5/8" wide		245	.065		1.98	2.61		4.59	6.50
7800	6" wide		225	.071		2.21	2.84		5.05	7.10
8000	24" O.C., 3-5/8" wide		265	.060		1.86	2.41		4.27	6
8200	6" wide		245	.065		2.03	2.61		4.64	6.55
8600	1/2" blueboard, mesh tape both sides									
8620	Installed on and including 2" x 4" wood studs, 16" O.C.	2 Carp	300	.053	S.F.	1.09	2.13		3.22	4.72
8640	Metal studs, NLB, 25 ga., 16" O.C., 3-5/8" wide		340	.047		1.22	1.88		3.10	4.44
8660	6" wide		320	.050		1.45	2		3.45	4.89
9000	Exterior, 1/2" gypsum sheathing, 1/2" gypsum finished, interior,									
9100	including foil faced insulation, metal studs, 20 ga.									
9200	16" O.C., 3-5/8" wide	2 Carp	290	.055	S.F.	2.17	2.20		4.37	6.05
9400	6" wide	"	270	.059	"	2.42	2.37		4.79	6.55

09 22 Supports for Plaster and Gypsum Board

09 22 03 – Fastening Methods for Finishes

09 22 03.20 Drilling Plaster/Drywall

		Crew	Daily Output	Labor-Hours	Unit	Material	Labor	Equipment	Total	Total Incl O&P
0010	**DRILLING PLASTER/DRYWALL**									
1100	Drilling & layout for drywall/plaster walls, up to 1" deep, no anchor									
1200	Holes, 1/4" diameter	1 Carp	150	.053	Ea.	.01	2.13		2.14	3.53
1300	3/8" diameter		140	.057		.01	2.28		2.29	3.78
1400	1/2" diameter		130	.062		.01	2.46		2.47	4.07
1500	3/4" diameter		120	.067		.01	2.66		2.67	4.41
1600	1" diameter		110	.073		.02	2.91		2.93	4.82
1700	1-1/4" diameter		100	.080		.03	3.20		3.23	5.35
1800	1-1/2" diameter		90	.089		.04	3.55		3.59	5.90
1900	For ceiling installations, add						40%			

09 22 13 – Metal Furring

09 22 13.13 Metal Channel Furring

		Crew	Daily Output	Labor-Hours	Unit	Material	Labor	Equipment	Total	Total Incl O&P
0010	**METAL CHANNEL FURRING**									
0030	Beams and columns, 7/8" channels, galvanized, 12" O.C.	1 Lath	155	.052	S.F.	.38	1.83		2.21	3.31
0050	16" O.C.		170	.047		.31	1.67		1.98	2.98
0070	24" O.C.		185	.043		.21	1.54		1.75	2.66
0100	Ceilings, on steel, 7/8" channels, galvanized, 12" O.C.		210	.038		.34	1.35		1.69	2.52
0300	16" O.C.		290	.028		.31	.98		1.29	1.89
0400	24" O.C.		420	.019		.21	.68		.89	1.30
0600	1-5/8" channels, galvanized, 12" O.C.		190	.042		.46	1.50		1.96	2.87
0700	16" O.C.		260	.031		.41	1.09		1.50	2.19
0900	24" O.C.		390	.021		.28	.73		1.01	1.45
0930	7/8" channels with sound isolation clips, 12" O.C.		120	.067		1.53	2.37		3.90	5.40

09 22 13 – Metal Furring

09 22 13.13 Metal Channel Furring	Crew	Daily Output	Labor-Hours	Unit	Material	2009 Bare Costs Labor	Equipment	Total	Total Incl O&P
0940	16" O.C.	1 Lath	100	.080	S.F.	2.12	2.84		4.96	6.80
0950	24" O.C.		165	.048		1.39	1.72		3.11	4.25
0960	1-5/8" channels, galvanized, 12" O.C.		110	.073		1.64	2.59		4.23	5.90
0970	16" O.C.		100	.080		2.22	2.84		5.06	6.95
0980	24" O.C.		155	.052		1.46	1.83		3.29	4.51
1000	Walls, 7/8" channels, galvanized, 12" O.C.		235	.034		.34	1.21		1.55	2.29
1200	16" O.C.		265	.030		.31	1.07		1.38	2.04
1300	24" O.C.		350	.023		.21	.81		1.02	1.51
1500	1-5/8" channels, galvanized, 12" O.C.		210	.038		.46	1.35		1.81	2.65
1600	16" O.C.		240	.033		.41	1.18		1.59	2.33
1800	24" O.C.		305	.026		.28	.93		1.21	1.77
1920	7/8" channels with sound isolation clips, 12" O.C.		125	.064		1.53	2.28		3.81	5.25
1940	16" O.C.		100	.080		2.12	2.84		4.96	6.80
1950	24" O.C.		150	.053		1.39	1.90		3.29	4.52
1960	1-5/8" channels, galvanized, 12" O.C.		115	.070		1.64	2.47		4.11	5.70
1970	16" O.C.		95	.084		2.22	2.99		5.21	7.20
1980	24" O.C.		140	.057		1.46	2.03		3.49	4.82
9000	Minimum labor/equipment charge		4	2	Job		71		71	112

09 22 16 – Non-Structural Metal Framing

09 22 16.13 Metal Studs and Track

	09 22 16.13 Metal Studs and Track	Crew	Daily Output	Labor-Hours	Unit	Material	2009 Bare Costs Labor	Equipment	Total	Total Incl O&P
0010	**METAL STUDS AND TRACK**									
1600	Non-load bearing, galv, 8' high, 25 ga. 1-5/8" wide, 16" O.C.	1 Carp	619	.013	S.F.	.33	.52		.85	1.21
1610	24" O.C.		950	.008		.25	.34		.59	.83
1620	2-1/2" wide, 16" O.C.		613	.013		.42	.52		.94	1.32
1630	24" O.C.		938	.009		.31	.34		.65	.90
1640	3-5/8" wide, 16" O.C.		600	.013		.48	.53		1.01	1.40
1650	24" O.C.		925	.009		.36	.35		.71	.96
1660	4" wide, 16" O.C.		594	.013		.51	.54		1.05	1.45
1670	24" O.C.		925	.009		.38	.35		.73	.99
1680	6" wide, 16" O.C.		588	.014		.72	.54		1.26	1.69
1690	24" O.C.		906	.009		.54	.35		.89	1.17
1700	20 ga. studs, 1-5/8" wide, 16" O.C.		494	.016		.53	.65		1.18	1.65
1710	24" O.C.		763	.010		.40	.42		.82	1.13
1720	2-1/2" wide, 16" O.C.		488	.016		.62	.65		1.27	1.76
1730	24" O.C.		750	.011		.47	.43		.90	1.21
1740	3-5/8" wide, 16" O.C.		481	.017		.67	.66		1.33	1.84
1750	24" O.C.		738	.011		.50	.43		.93	1.26
1760	4" wide, 16" O.C.		475	.017		.80	.67		1.47	1.98
1770	24" O.C.		738	.011		.60	.43		1.03	1.37
1780	6" wide, 16" O.C.		469	.017		.94	.68		1.62	2.16
1790	24" O.C.		725	.011		.70	.44		1.14	1.50
2000	Non-load bearing, galv, 10' high, 25 ga. 1-5/8" wide, 16" O.C.		495	.016		.31	.65		.96	1.41
2100	24" O.C.		760	.011		.23	.42		.65	.94
2200	2-1/2" wide, 16" O.C.		490	.016		.39	.65		1.04	1.51
2250	24" O.C.		750	.011		.29	.43		.72	1.02
2300	3-5/8" wide, 16" O.C.		480	.017		.45	.67		1.12	1.60
2350	24" O.C.		740	.011		.33	.43		.76	1.07
2400	4" wide, 16" O.C.		475	.017		.49	.67		1.16	1.64
2450	24" O.C.		740	.011		.36	.43		.79	1.10
2500	6" wide, 16" O.C.		470	.017		.68	.68		1.36	1.87
2550	24" O.C.		725	.011		.50	.44		.94	1.28
2600	20 ga. studs, 1-5/8" wide, 16" O.C.		395	.020		.50	.81		1.31	1.89

09 22 Supports for Plaster and Gypsum Board

09 22 16 – Non-Structural Metal Framing

09 22 16.13 Metal Studs and Track

		Crew	Daily Output	Labor-Hours	Unit	Material	2009 Bare Costs Labor	Equipment	Total	Total Incl O&P
2650	24" O.C.	1 Carp	610	.013	S.F.	.37	.52		.89	1.27
2700	2-1/2" wide, 16" O.C.		390	.021		.59	.82		1.41	2
2750	24" O.C.		600	.013		.43	.53		.96	1.36
2800	3-5/8" wide, 16" OC		385	.021		.64	.83		1.47	2.07
2850	24" O.C.		590	.014		.47	.54		1.01	1.41
2900	4" wide, 16" O.C.		380	.021		.75	.84		1.59	2.22
2950	24" O.C.		590	.014		.56	.54		1.10	1.50
3000	6" wide, 16" O.C.		375	.021		.89	.85		1.74	2.39
3050	24" O.C.		580	.014		.65	.55		1.20	1.63
3060	Non-load bearing, galv, 12' high, 25 ga. 1-5/8" wide, 16" O.C.		413	.019		.30	.77		1.07	1.61
3070	24" O.C.		633	.013		.22	.51		.73	1.07
3080	2-1/2" wide, 16" O.C.		408	.020		.38	.78		1.16	1.71
3090	24" O.C.		625	.013		.27	.51		.78	1.14
3100	3-5/8" wide, 16" O.C.		400	.020		.43	.80		1.23	1.79
3110	24" O.C.		617	.013		.31	.52		.83	1.20
3120	4" wide, 16" O.C.		396	.020		.47	.81		1.28	1.84
3130	24" O.C.		617	.013		.34	.52		.86	1.23
3140	6" wide, 16" O.C.		392	.020		.65	.82		1.47	2.07
3150	24" O.C.		604	.013		.47	.53		1	1.39
3160	20 ga. studs, 1-5/8" wide, 16" O.C.		329	.024		.48	.97		1.45	2.13
3170	24" O.C.		508	.016		.35	.63		.98	1.42
3180	2-1/2" wide, 16" O.C.		325	.025		.56	.98		1.54	2.24
3190	24" O.C.		500	.016		.41	.64		1.05	1.51
3200	3-5/8" wide, 16" O.C.		321	.025		.61	1		1.61	2.31
3210	24" O.C.		492	.016		.44	.65		1.09	1.56
3220	4" wide, 16" O.C.		317	.025		.72	1.01		1.73	2.45
3230	24" O.C.		492	.016		.52	.65		1.17	1.65
3240	6" wide, 16" O.C.		313	.026		.85	1.02		1.87	2.63
3250	24" O.C.	▼	483	.017	▼	.62	.66		1.28	1.77
5000	Load bearing studs, see Div. 05 41 13.30									
9000	Minimum labor/equipment charge	1 Carp	4	2	Job		80		80	132

09 22 26 – Suspension Systems

09 22 26.13 Ceiling Suspension Systems

		Crew	Daily Output	Labor-Hours	Unit	Material	2009 Bare Costs Labor	Equipment	Total	Total Incl O&P
0010	**CEILING SUSPENSION SYSTEMS** For gypsum board or plaster									
8000	Suspended ceilings, including carriers									
8200	1-1/2" carriers, 24" O.C. with:									
8300	7/8" channels, 16" O.C.	1 Lath	215	.037	S.F.	.58	1.32		1.90	2.73
8320	24" O.C.		275	.029		.48	1.03		1.51	2.16
8400	1-5/8" channels, 16" O.C.		205	.039		.69	1.39		2.08	2.95
8420	24" O.C.	▼	265	.030	▼	.55	1.07		1.62	2.31
8600	2" carriers, 24" O.C. with:									
8700	7/8" channels, 16" O.C.	1 Lath	205	.039	S.F.	.56	1.39		1.95	2.80
8720	24" O.C.		265	.030		.46	1.07		1.53	2.20
8800	1-5/8" channels, 16" O.C.		195	.041		.66	1.46		2.12	3.03
8820	24" O.C.	▼	255	.031	▼	.53	1.12		1.65	2.34

09 22 36 – Lath

09 22 36.13 Gypsum Lath

		Crew	Daily Output	Labor-Hours	Unit	Material	2009 Bare Costs Labor	Equipment	Total	Total Incl O&P
0010	**GYPSUM LATH** R092000-50									
0020	Plain or perforated, nailed, 3/8" thick	1 Lath	85	.094	S.Y.	5.50	3.35		8.85	11.35
0100	1/2" thick		80	.100		5.85	3.56		9.41	12.05
0300	Clipped to steel studs, 3/8" thick		75	.107		5.50	3.79		9.29	12.05
0400	1/2" thick	▼	70	.114	▼	5.85	4.06		9.91	12.85

09 22 36 – Lath

09 22 36.13 Gypsum Lath

		Crew	Daily Output	Labor-Hours	Unit	Material	2009 Bare Costs Labor	Equipment	Total	Total Incl O&P
1500	For ceiling installations, add	1 Lath	216	.037	S.Y.		1.32		1.32	2.08
1600	For columns and beams, add		170	.047	↓		1.67		1.67	2.64
9000	Minimum labor/equipment charge	↓	4.25	1.882	Job		67		67	106

09 22 36.23 Metal Lath

		Crew	Daily Output	Labor-Hours	Unit	Material	2009 Bare Costs Labor	Equipment	Total	Total Incl O&P
0010	**METAL LATH** R092000-50									
0020	Diamond, expanded, 2.5 lb. per S.Y., painted				S.Y.	3.32			3.32	3.65
0100	Galvanized					3.47			3.47	3.82
0300	3.4 lb. per S.Y., painted					3.99			3.99	4.39
0400	Galvanized					4.09			4.09	4.50
0600	For 15# asphalt sheathing paper, add					.43			.43	.47
0900	Flat rib, 1/8" high, 2.75 lb., painted					3.62			3.62	3.98
1000	Foil backed					3.56			3.56	3.92
1200	3.4 lb. per S.Y., painted					4.61			4.61	5.05
1300	Galvanized					4.23			4.23	4.65
1500	For 15# asphalt sheathing paper, add					.43			.43	.47
1800	High rib, 3/8" high, 3.4 lb. per S.Y., painted					5.55			5.55	6.15
1900	Galvanized				↓	5.55			5.55	6.15
2400	3/4" high, painted, .60 lb. per S.F.				S.F.	.58			.58	.64
2500	.75 lb. per S.F.				"	1.25			1.25	1.38
2800	Stucco mesh, painted, 3.6 lb.				S.Y.	4.30			4.30	4.73
3000	K-lath, perforated, absorbent paper, regular					4.43			4.43	4.87
3100	Heavy duty					5.25			5.25	5.75
3300	Waterproof, heavy duty, grade B backing					5.10			5.10	5.65
3400	Fire resistant backing					5.65			5.65	6.25
3600	2.5 lb. diamond painted, on wood framing, on walls	1 Lath	85	.094		3.32	3.35		6.67	8.95
3700	On ceilings		75	.107		3.32	3.79		7.11	9.65
3900	3.4 lb. diamond painted, on wood framing, on walls		80	.100		4.61	3.56		8.17	10.65
4000	On ceilings		70	.114		4.61	4.06		8.67	11.45
4200	3.4 lb. diamond painted, wired to steel framing		75	.107		4.61	3.79		8.40	11.05
4300	On ceilings		60	.133		4.61	4.74		9.35	12.55
4600	Cornices, wired to steel		35	.229		4.61	8.15		12.76	17.90
4800	Screwed to steel studs, 2.5 lb.		80	.100		3.32	3.56		6.88	9.25
4900	3.4 lb.		75	.107		3.99	3.79		7.78	10.40
5100	Rib lath, painted, wired to steel, on walls, 2.5 lb.		75	.107		3.62	3.79		7.41	10
5200	3.4 lb.		70	.114		5.55	4.06		9.61	12.55
5400	4.0 lb.	↓	65	.123		5.70	4.38		10.08	13.20
5500	For self-furring lath, add					.10			.10	.11
5700	Suspended ceiling system, incl. 3.4 lb. diamond lath, painted	1 Lath	15	.533	↓	4.23	18.95		23.18	34.50
5800	Galvanized	"	15	.533		7.50	18.95		26.45	38.50
6000	Hollow metal stud partitions, 3.4 lb. painted lath both sides									
6010	Non-load bearing, 25 ga., w/rib lath 2-1/2" studs, 12" O.C.	1 Lath	20.30	.394	S.Y.	15.65	14		29.65	39
6300	16" O.C.		21.10	.379		14.70	13.50		28.20	37.50
6350	24" O.C.		22.70	.352		13.75	12.55		26.30	35
6400	3-5/8" studs, 16" O.C.		19.50	.410		15.20	14.60		29.80	39.50
6600	24" O.C.		20.40	.392		14.10	13.95		28.05	37.50
6700	4" studs, 16" O.C.		20.40	.392		15.50	13.95		29.45	39
6900	24" O.C.		21.60	.370		14.35	13.15		27.50	37
7000	6" studs, 16" O.C.		19.50	.410		17.25	14.60		31.85	42
7100	24" O.C.		21.10	.379		15.65	13.50		29.15	38.50
7200	L.B. partitions, 16 ga., w/rib lath, 2-1/2" studs, 16" O.C.		20	.400		16.95	14.20		31.15	41
7300	3-5/8" studs, 16 ga.		19.70	.406		19.20	14.45		33.65	44
7500	4" studs, 16 ga.	↓	19.50	.410	↓	20	14.60		34.60	45

09 22 Supports for Plaster and Gypsum Board

09 22 36 – Lath

09 22 36.23 Metal Lath	Crew	Daily Output	Labor-Hours	Unit	Material	2009 Bare Costs Labor	Equipment	Total	Total Incl O&P
7600 6" studs, 16 ga.	1 Lath	18.70	.428	S.Y.	23.50	15.20		38.70	50
9000 Minimum labor/equipment charge	↓	4.25	1.882	Job		67		67	106

09 23 Gypsum Plastering

09 23 13 – Acoustical Gypsum Plastering

09 23 13.10 Perlite or Vermiculite Plaster

	Crew	Daily Output	Labor-Hours	Unit	Material	Labor	Equipment	Total	Total Incl O&P
0010 **PERLITE OR VERMICULITE PLASTER** R092000-50									
0020 In 100 lb. bags, under 200 bags				Bag	14.40			14.40	15.85
0300 2 coats, no lath included, on walls	J-1	92	.435	S.Y.	4.05	15.05	1.38	20.48	30
0400 On ceilings		79	.506		4.05	17.50	1.61	23.16	34
0900 3 coats, no lath included, on walls		74	.541		6.30	18.70	1.71	26.71	39
1000 On ceilings	↓	63	.635		6.30	22	2.01	30.31	44.50
1700 For irregular or curved surfaces, add to above						30%			
1800 For columns and beams, add to above						50%			
1900 For soffits, add to ceiling prices				↓		40%			
9000 Minimum labor/equipment charge	1 Plas	1	8	Job		289		289	465

09 23 20 – Gypsum Plaster

09 23 20.10 Gypsum Plaster On Walls and Ceilings

	Crew	Daily Output	Labor-Hours	Unit	Material	Labor	Equipment	Total	Total Incl O&P
0010 **GYPSUM PLASTER ON WALLS AND CEILINGS** R092000-50									
0020 80# bag, less than 1 ton				Bag	15.20			15.20	16.70
0300 2 coats, no lath included, on walls	J-1	105	.381	S.Y.	3.77	13.20	1.21	18.18	26.50
0400 On ceilings		92	.435		3.77	15.05	1.38	20.20	29.50
0900 3 coats, no lath included, on walls		87	.460		5.40	15.90	1.46	22.76	33
1000 On ceilings	↓	78	.513	↓	5.40	17.75	1.63	24.78	36
1600 For irregular or curved surfaces, add						30%			
1800 For columns & beams, add						50%			
9000 Minimum labor/equipment charge	1 Plas	1	8	Job		289		289	465

09 24 Portland Cement Plastering

09 24 23 – Portland Cement Stucco

09 24 23.40 Stucco

	Crew	Daily Output	Labor-Hours	Unit	Material	Labor	Equipment	Total	Total Incl O&P
0010 **STUCCO** R092000-50									
0015 3 coats 1" thick, float finish, with mesh, on wood frame	J-2	63	.762	S.Y.	6.70	26.50	2.01	35.21	52
0100 On masonry construction, no mesh incl.	J-1	67	.597		2.41	20.50	1.89	24.80	37.50
0300 For trowel finish, add	1 Plas	170	.047			1.70		1.70	2.74
0400 For 3/4" thick, on masonry, deduct	J-1	880	.045		.62	1.57	.14	2.33	3.37
0600 For coloring and special finish, add, minimum		685	.058		.40	2.02	.19	2.61	3.89
0700 Maximum	↓	200	.200		1.39	6.90	.63	8.92	13.40
0900 For soffits, add	J-2	155	.310		2.15	10.75	.82	13.72	20.50
1000 Exterior stucco, with bonding agent, 3 coats, on walls, no mesh incl.	J-1	200	.200		3.61	6.90	.63	11.14	15.80
1200 Ceilings		180	.222		3.61	7.70	.70	12.01	17.10
1300 Beams		80	.500		3.61	17.30	1.59	22.50	33.50
1500 Columns	↓	100	.400	↓	3.61	13.85	1.27	18.73	28
1550 Minimum labor/equipment charge	1 Plas	1	8	Job		289		289	465
1600 Mesh, painted, nailed to wood, 1.8 lb.	1 Lath	60	.133	S.Y.	5.75	4.74		10.49	13.85
1800 3.6 lb.		55	.145		4.30	5.15		9.45	12.90
1900 Wired to steel, painted, 1.8 lb.		53	.151		5.75	5.35		11.10	14.85
2100 3.6 lb.		50	.160		4.30	5.70		10	13.75
9000 Minimum labor/equipment charge	↓	4	2	Job		71		71	112

233

09 26 Veneer Plastering

09 26 13 – Gypsum Veneer Plastering

09 26 13.20 Blueboard

		Crew	Daily Output	Labor-Hours	Unit	Material	2009 Bare Costs Labor	Equipment	Total	Total Incl O&P
0010	**BLUEBOARD** For use with thin coat									
0100	plaster application (see Div. 09 26 13.80)									
1000	3/8" thick, on walls or ceilings, standard, no finish included	2 Carp	1900	.008	S.F.	.28	.34		.62	.87
1100	With thin coat plaster finish		875	.018		.36	.73		1.09	1.61
1400	On beams, columns, or soffits, standard, no finish included		675	.024		.32	.95		1.27	1.91
1450	With thin coat plaster finish		475	.034		.40	1.35		1.75	2.66
3000	1/2" thick, on walls or ceilings, standard, no finish included		1900	.008		.33	.34		.67	.92
3100	With thin coat plaster finish		875	.018		.41	.73		1.14	1.66
3300	Fire resistant, no finish included		1900	.008		.33	.34		.67	.92
3400	With thin coat plaster finish		875	.018		.41	.73		1.14	1.66
3450	On beams, columns, or soffits, standard, no finish included		675	.024		.38	.95		1.33	1.98
3500	With thin coat plaster finish		475	.034		.46	1.35		1.81	2.73
3700	Fire resistant, no finish included		675	.024		.38	.95		1.33	1.98
3800	With thin coat plaster finish		475	.034		.46	1.35		1.81	2.73
5000	5/8" thick, on walls or ceilings, fire resistant, no finish included		1900	.008		.34	.34		.68	.93
5100	With thin coat plaster finish		875	.018		.42	.73		1.15	1.67
5500	On beams, columns, or soffits, no finish included		675	.024		.39	.95		1.34	1.99
5600	With thin coat plaster finish		475	.034		.47	1.35		1.82	2.74
6000	For high ceilings, over 8' high, add		3060	.005			.21		.21	.34
6500	For over 3 stories high, add per story		6100	.003			.10		.10	.17
9000	Minimum labor/equipment charge	1 Carp	2	4	Job		160		160	264

09 26 13.80 Thin Coat Plaster

			Crew	Daily Output	Labor-Hours	Unit	Material	2009 Bare Costs Labor	Equipment	Total	Total Incl O&P
0010	**THIN COAT PLASTER**	R092000-50									
0012	1 coat veneer, not incl. lath		J-1	3600	.011	S.F.	.08	.38	.04	.50	.75
1000	In 50 lb. bags					Bag	11.10			11.10	12.20

09 28 Backing Boards and Underlayments

09 28 13 – Cementitious Backing Boards

09 28 13.10 Cementitious Backerboard

		Crew	Daily Output	Labor-Hours	Unit	Material	2009 Bare Costs Labor	Equipment	Total	Total Incl O&P
0010	**CEMENTITIOUS BACKERBOARD**									
0070	Cementitious backerboard, on floor, 3' x 4' x 1/2" sheets	2 Carp	525	.030	S.F.	.79	1.22		2.01	2.87
0080	3' x 5' x 1/2" sheets		525	.030		.78	1.22		2	2.86
0090	3' x 6' x 1/2" sheets		525	.030		.72	1.22		1.94	2.80
0100	3' x 4' x 5/8" sheets		525	.030		1.04	1.22		2.26	3.15
0110	3' x 5' x 5/8" sheets		525	.030		1.03	1.22		2.25	3.14
0120	3' x 6' x 5/8" sheets		525	.030		.96	1.22		2.18	3.07
0150	On wall, 3' x 4' x 1/2" sheets		350	.046		.79	1.83		2.62	3.87
0160	3' x 5' x 1/2" sheets		350	.046		.78	1.83		2.61	3.86
0170	3' x 6' x 1/2" sheets		350	.046		.72	1.83		2.55	3.80
0180	3' x 4' x 5/8" sheets		350	.046		1.04	1.83		2.87	4.15
0190	3' x 5' x 5/8" sheets		350	.046		1.03	1.83		2.86	4.14
0200	3' x 6' x 5/8" sheets		350	.046		.96	1.83		2.79	4.07
0250	On counter, 3' x 4' x 1/2" sheets		180	.089		.79	3.55		4.34	6.70
0260	3' x 5' x 1/2" sheets		180	.089		.78	3.55		4.33	6.70
0270	3' x 6' x 1/2" sheets		180	.089		.72	3.55		4.27	6.65
0300	3' x 4' x 5/8" sheets		180	.089		1.04	3.55		4.59	7
0310	3' x 5' x 5/8" sheets		180	.089		1.03	3.55		4.58	7
0320	3' x 6' x 5/8" sheets		180	.089		.96	3.55		4.51	6.90

09 29 Gypsum Board

09 29 10 – Gypsum Board Panels

09 29 10.10 Gypsum Board Ceilings

		Crew	Daily Output	Labor-Hours	Unit	Material	2009 Bare Costs Labor	Equipment	Total	Total Incl O&P
0010	**GYPSUM BOARD CEILINGS**, fire rated, finished									
0100	Screwed to grid, channel or joists, 1/2" thick	2 Carp	765	.021	S.F.	.32	.84		1.16	1.73
0150	Mold resistant		765	.021		.54	.84		1.38	1.97
0200	5/8" thick		765	.021		.36	.84		1.20	1.78
0250	Mold resistant		765	.021		.66	.84		1.50	2.11
0300	Over 8' high, 1/2" thick		615	.026		.32	1.04		1.36	2.07
0350	Mold resistant		615	.026		.54	1.04		1.58	2.31
0400	5/8" thick		615	.026		.36	1.04		1.40	2.12
0450	Mold resistant		615	.026		.66	1.04		1.70	2.45
0600	Grid suspension system, direct hung									
0700	1-1/2" C.R.C., with 7/8" hi hat furring channel, 16" O.C.	2 Carp	1025	.016	S.F.	1.57	.62		2.19	2.76
0800	24" O.C.		1300	.012		1.43	.49		1.92	2.38
0900	3-5/8" C.R.C., with 7/8" hi hat furring channel, 16" O.C.		1025	.016		1.66	.62		2.28	2.86
1000	24" O.C.		1300	.012		1.44	.49		1.93	2.39

09 29 10.30 Gypsum Board

		Crew	Daily Output	Labor-Hours	Unit	Material	2009 Bare Costs Labor	Equipment	Total	Total Incl O&P
0010	**GYPSUM BOARD** on walls & ceilings R092910-10									
0100	Nailed or screwed to studs unless otherwise noted									
0150	3/8" thick, on walls, standard, no finish included	2 Carp	2000	.008	S.F.	.31	.32		.63	.87
0200	On ceilings, standard, no finish included		1800	.009		.31	.36		.67	.93
0250	On beams, columns, or soffits, no finish included		675	.024		.31	.95		1.26	1.90
0300	1/2" thick, on walls, standard, no finish included		2000	.008		.31	.32		.63	.87
0350	Taped and finished (level 4 finish)		965	.017		.35	.66		1.01	1.48
0390	With compound skim coat (level 5 finish)		775	.021		.40	.83		1.23	1.80
0400	Fire resistant, no finish included		2000	.008		.32	.32		.64	.88
0450	Taped and finished (level 4 finish)		965	.017		.36	.66		1.02	1.49
0490	With compound skim coat (level 5 finish)		775	.021		.41	.83		1.24	1.81
0500	Water resistant, no finish included		2000	.008		.48	.32		.80	1.06
0550	Taped and finished (level 4 finish)		965	.017		.52	.66		1.18	1.66
0590	With compound skim coat (level 5 finish)		775	.021		.57	.83		1.40	1.98
0600	Prefinished, vinyl, clipped to studs		900	.018		.83	.71		1.54	2.08
0700	Mold resistant, no finish included		2000	.008		.49	.32		.81	1.07
0710	Taped and finished (level 4 finish)		965	.017		.53	.66		1.19	1.68
0720	With compound skim coat (level 5 finish)		775	.021		.58	.83		1.41	1.99
1000	On ceilings, standard, no finish included		1800	.009		.31	.36		.67	.93
1050	Taped and finished (level 4 finish)		765	.021		.35	.84		1.19	1.77
1090	With compound skim coat (level 5 finish)		610	.026		.40	1.05		1.45	2.17
1100	Fire resistant, no finish included		1800	.009		.32	.36		.68	.94
1150	Taped and finished (level 4 finish)		765	.021		.36	.84		1.20	1.78
1195	With compound skim coat (level 5 finish)		610	.026		.41	1.05		1.46	2.18
1200	Water resistant, no finish included		1800	.009		.48	.36		.84	1.12
1250	Taped and finished (level 4 finish)		765	.021		.52	.84		1.36	1.95
1290	With compound skim coat (level 5 finish)		610	.026		.57	1.05		1.62	2.35
1310	Mold resistant, no finish included		1800	.009		.49	.36		.85	1.13
1320	Taped and finished (level 4 finish)		765	.021		.53	.84		1.37	1.97
1330	With compound skim coat (level 5 finish)		610	.026		.58	1.05		1.63	2.36
1500	On beams, columns, or soffits, standard, no finish included		675	.024		.36	.95		1.31	1.95
1550	Taped and finished (level 4 finish)		475	.034		.35	1.35		1.70	2.61
1590	With compound skim coat (level 5 finish)		540	.030		.40	1.18		1.58	2.39
1600	Fire resistant, no finish included		675	.024		.32	.95		1.27	1.91
1650	Taped and finished (level 4 finish)		475	.034		.36	1.35		1.71	2.62
1690	With compound skim coat (level 5 finish)		540	.030		.41	1.18		1.59	2.40
1700	Water resistant, no finish included		675	.024		.55	.95		1.50	2.17

235

09 29 Gypsum Board

09 29 10 – Gypsum Board Panels

09 29 10.30 Gypsum Board		Crew	Daily Output	Labor-Hours	Unit	Material	2009 Bare Costs Labor	Equipment	Total	Total Incl O&P
1750	Taped and finished (level 4 finish)	2 Carp	475	.034	S.F.	.52	1.35		1.87	2.79
1790	With compound skim coat (level 5 finish)		540	.030		.57	1.18		1.75	2.57
1800	Mold resistant, no finish included		675	.024		.56	.95		1.51	2.18
1810	Taped and finished (level 4 finish)		475	.034		.53	1.35		1.88	2.81
1820	With compound skim coat (level 5 finish)		540	.030		.58	1.18		1.76	2.58
2000	5/8" thick, on walls, standard, no finish included		2000	.008		.34	.32		.66	.90
2050	Taped and finished (level 4 finish)		965	.017		.38	.66		1.04	1.51
2090	With compound skim coat (level 5 finish)		775	.021		.43	.83		1.26	1.83
2100	Fire resistant, no finish included		2000	.008		.36	.32		.68	.93
2150	Taped and finished (level 4 finish)		965	.017		.40	.66		1.06	1.53
2195	With compound skim coat (level 5 finish)		775	.021		.45	.83		1.28	1.85
2200	Water resistant, no finish included		2000	.008		.47	.32		.79	1.05
2250	Taped and finished (level 4 finish)		965	.017		.51	.66		1.17	1.65
2290	With compound skim coat (level 5 finish)		775	.021		.56	.83		1.39	1.97
2300	Prefinished, vinyl, clipped to studs		900	.018		1.02	.71		1.73	2.29
2510	Mold resistant, no finish included		2000	.008		.60	.32		.92	1.19
2520	Taped and finished (level 4 finish)		965	.017		.64	.66		1.30	1.80
2530	With compound skim coat (level 5 finish)		775	.021		.69	.83		1.52	2.12
3000	On ceilings, standard, no finish included		1800	.009		.34	.36		.70	.96
3050	Taped and finished (level 4 finish)		765	.021		.38	.84		1.22	1.80
3090	With compound skim coat (level 5 finish)		615	.026		.43	1.04		1.47	2.19
3100	Fire resistant, no finish included		1800	.009		.36	.36		.72	.99
3150	Taped and finished (level 4 finish)		765	.021		.40	.84		1.24	1.82
3190	With compound skim coat (level 5 finish)		615	.026		.45	1.04		1.49	2.21
3200	Water resistant, no finish included		1800	.009		.47	.36		.83	1.11
3250	Taped and finished (level 4 finish)		765	.021		.51	.84		1.35	1.94
3290	With compound skim coat (level 5 finish)		615	.026		.56	1.04		1.60	2.33
3300	Mold resistant, no finish included		1800	.009		.60	.36		.96	1.25
3310	Taped and finished (level 4 finish)		765	.021		.64	.84		1.48	2.09
3320	With compound skim coat (level 5 finish)		615	.026		.69	1.04		1.73	2.48
3500	On beams, columns, or soffits, no finish included		675	.024		.39	.95		1.34	1.99
3550	Taped and finished (level 4 finish)		475	.034		.44	1.35		1.79	2.70
3590	With compound skim coat (level 5 finish)		380	.042		.49	1.68		2.17	3.32
3600	Fire resistant, no finish included		675	.024		.41	.95		1.36	2.02
3650	Taped and finished (level 4 finish)		475	.034		.46	1.35		1.81	2.73
3690	With compound skim coat (level 5 finish)		380	.042		.45	1.68		2.13	3.27
3700	Water resistant, no finish included		675	.024		.54	.95		1.49	2.15
3750	Taped and finished (level 4 finish)		475	.034		.59	1.35		1.94	2.87
3790	With compound skim coat (level 5 finish)		380	.042		.56	1.68		2.24	3.39
3800	Mold resistant, no finish included		675	.024		.69	.95		1.64	2.32
3810	Taped and finished (level 4 finish)		475	.034		.74	1.35		2.09	3.03
3820	With compound skim coat (level 5 finish)		380	.042		.69	1.68		2.37	3.54
4000	Fireproofing, beams or columns, 2 layers, 1/2" thick, incl finish		330	.048		.68	1.94		2.62	3.95
4010	Mold resistant		330	.048		1.02	1.94		2.96	4.32
4050	5/8" thick		300	.053		.80	2.13		2.93	4.41
4060	Mold resistant		300	.053		1.28	2.13		3.41	4.93
4100	3 layers, 1/2" thick		225	.071		1	2.84		3.84	5.80
4110	Mold resistant		225	.071		1.51	2.84		4.35	6.35
4150	5/8" thick		210	.076		1.21	3.04		4.25	6.35
4160	Mold resistant		210	.076		1.93	3.04		4.97	7.10
5050	For 1" thick coreboard on columns		480	.033		.88	1.33		2.21	3.17
5100	For foil-backed board, add					.14			.14	.15
5200	For work over 8' high, add	2 Carp	3060	.005			.21		.21	.34

09 29 Gypsum Board

09 29 10 – Gypsum Board Panels

09 29 10.30 Gypsum Board		Crew	Daily Output	Labor-Hours	Unit	Material	2009 Bare Costs Labor	Equipment	Total	Total Incl O&P
5270	For textured spray, add	2 Lath	1600	.010	S.F.	.08	.36		.44	.65
5300	For distribution cost over 3 stories high, add per story	2 Carp	6100	.003	↓		.10		.10	.17
5350	For finishing inner corners, add		950	.017	L.F.	.09	.67		.76	1.21
5355	For finishing outer corners, add	↓	1250	.013	↓	.21	.51		.72	1.07
5500	For acoustical sealant, add per bead	1 Carp	500	.016	↓	.04	.64		.68	1.11
5550	Sealant, 1 quart tube				Ea.	6.85			6.85	7.55
5600	Sound deadening board, 1/4" gypsum	2 Carp	1800	.009	S.F.	.37	.36		.73	1
5650	1/2" wood fiber	"	1800	.009	"	.28	.36		.64	.90
9000	Minimum labor/equipment charge	1 Carp	2	4	Job		160		160	264

09 29 15 – Gypsum Board Accessories

09 29 15.10 Accessories, Gypsum Board

		Crew	Daily Output	Labor-Hours	Unit	Material	2009 Bare Costs Labor	Equipment	Total	Total Incl O&P
0010	**ACCESSORIES, GYPSUM BOARD**									
0020	Casing bead, galvanized steel	1 Carp	2.90	2.759	C.L.F.	22	110		132	207
0100	Vinyl		3	2.667		22.50	107		129.50	201
0300	Corner bead, galvanized steel, 1" x 1"		4	2		13.95	80		93.95	147
0400	1-1/4" x 1-1/4"		3.50	2.286		18.15	91.50		109.65	171
0600	Vinyl		4	2		17.80	80		97.80	152
0900	Furring channel, galv. steel, 7/8" deep, standard		2.60	3.077		27.50	123		150.50	234
1000	Resilient		2.55	3.137		26.50	125		151.50	237
1100	J trim, galvanized steel, 1/2" wide		3	2.667		22.50	107		129.50	201
1120	5/8" wide		2.95	2.712		24.50	108		132.50	206
1500	Z stud, galvanized steel, 1-1/2" wide		2.60	3.077	↓	37	123		160	244
9000	Minimum labor/equipment charge	↓	3	2.667	Job		107		107	176

09 30 Tiling

09 30 13 – Ceramic Tiling

09 30 13.10 Ceramic Tile

		Crew	Daily Output	Labor-Hours	Unit	Material	2009 Bare Costs Labor	Equipment	Total	Total Incl O&P
0010	**CERAMIC TILE**									
0600	Cove base, 4-1/4" x 4-1/4" high, mud set	D-7	91	.176	L.F.	3.93	6		9.93	13.70
0700	Thin set		128	.125		4.07	4.26		8.33	11.15
0900	6" x 4-1/4" high, mud set		100	.160		3.67	5.45		9.12	12.60
1000	Thin set		137	.117		3.47	3.98		7.45	10
1200	Sanitary cove base, 6" x 4-1/4" high, mud set		93	.172		4	5.85		9.85	13.55
1300	Thin set		124	.129		3.84	4.40		8.24	11.10
1500	6" x 6" high, mud set		84	.190		4.94	6.50		11.44	15.60
1600	Thin set		117	.137		4.74	4.66		9.40	12.50
2400	Bullnose trim, 4-1/4" x 4-1/4", mud set		82	.195		3.68	6.65		10.33	14.45
2500	Thin set		128	.125		3.30	4.26		7.56	10.25
2700	6" x 4-1/4" bullnose trim, mud set		84	.190		2.88	6.50		9.38	13.30
2800	Thin set		124	.129	↓	2.74	4.40		7.14	9.90
3000	Floors, natural clay, random or uniform, thin set, color group 1		183	.087	S.F.	4.34	2.98		7.32	9.45
3100	Color group 2		183	.087		4.66	2.98		7.64	9.80
3300	Porcelain type, 1 color, color group 2, 1" x 1"		183	.087		4.71	2.98		7.69	9.85
3310	2" x 2" or 2" x 1", thin set	↓	190	.084		5.40	2.87		8.27	10.45
3350	For random blend, 2 colors, add					.88			.88	.97
3360	4 colors, add					1.24			1.24	1.36
4300	Specialty tile, 4-1/4" x 4-1/4" x 1/2", decorator finish	D-7	183	.087		10	2.98		12.98	15.65
4500	Add for epoxy grout, 1/16" joint, 1" x 1" tile		800	.020		.62	.68		1.30	1.75
4600	2" x 2" tile	↓	820	.020	↓	.59	.66		1.25	1.69
4800	Pregrouted sheets, walls, 4-1/4" x 4-1/4", 6" x 4-1/4"									

09 30 13 – Ceramic Tiling

09 30 13.10 Ceramic Tile

		Crew	Daily Output	Labor-Hours	Unit	Material	2009 Bare Costs Labor	Equipment	Total	Total Incl O&P
4810	and 8-1/2" x 4-1/4", 4 S.F. sheets, silicone grout	D-7	240	.067	S.F.	4.73	2.27		7	8.75
5100	Floors, unglazed, 2 S.F. sheets,									
5110	Urethane adhesive	D-7	180	.089	S.F.	4.71	3.03		7.74	9.95
5400	Walls, interior, thin set, 4-1/4" x 4-1/4" tile		190	.084		2.49	2.87		5.36	7.25
5500	6" x 4-1/4" tile		190	.084		2.86	2.87		5.73	7.65
5700	8-1/2" x 4-1/4" tile		190	.084		3.93	2.87		6.80	8.80
5800	6" x 6" tile		200	.080		3.34	2.73		6.07	7.95
5810	8" x 8" tile		225	.071		4.36	2.42		6.78	8.60
5820	12" x 12" tile		300	.053		3.62	1.82		5.44	6.80
5830	16" x 16" tile		500	.032		3.90	1.09		4.99	6
6000	Decorated wall tile, 4-1/4" x 4-1/4", minimum		270	.059		3.66	2.02		5.68	7.20
6100	Maximum		180	.089		42	3.03		45.03	50.50
6600	Crystalline glazed, 4-1/4" x 4-1/4", mud set, plain		100	.160		4.43	5.45		9.88	13.40
6700	4-1/4" x 4-1/4", scored tile		100	.160		5.25	5.45		10.70	14.30
6900	6"x 6" plain		93	.172		5.65	5.85		11.50	15.35
7000	For epoxy grout, 1/16" joints, 4-1/4" tile, add		800	.020		.39	.68		1.07	1.50
7200	For tile set in dry mortar, add		1735	.009			.31		.31	.49
7300	For tile set in Portland cement mortar, add		290	.055		.27	1.88		2.15	3.24
9300	Ceramic tiles, recycled glass, standard colors, 2" x 2" thru 6" x 6" G		190	.084		16.70	2.87		19.57	23
9310	6" x 6" G		200	.080		16.70	2.73		19.43	22.50
9320	8" x 8" G		225	.071		18.10	2.42		20.52	23.50
9330	12"x12" G		300	.053		18.10	1.82		19.92	22.50
9340	Earthtones, 2" x 2" to 4" x 8" G		190	.084		21	2.87		23.87	27.50
9350	6" x 6" G		200	.080		21	2.73		23.73	27.50
9360	8" x 8" G		225	.071		22.50	2.42		24.92	28.50
9370	12" x 12" G		300	.053		22.50	1.82		24.32	27.50
9380	Deep colors, 2" x 2" to 4" x 8" G		190	.084		30	2.87		32.87	37.50
9390	6" x 6" G		200	.080		30	2.73		32.73	37.50
9400	8" x 8" G		225	.071		31.50	2.42		33.92	39
9410	12" x1 2" G		300	.053		31.50	1.82		33.32	38
9500	Minimum labor/equipment charge		3.25	4.923	Job		168		168	262

09 30 16 – Quarry Tiling

09 30 16.10 Quarry Tile

		Crew	Daily Output	Labor-Hours	Unit	Material	2009 Bare Costs Labor	Equipment	Total	Total Incl O&P
0010	**QUARRY TILE**									
0100	Base, cove or sanitary, mud set, to 5" high, 1/2" thick	D-7	110	.145	L.F.	5.60	4.96		10.56	13.90
0300	Bullnose trim, red, mud set, 6" x 6" x 1/2" thick		120	.133		4.36	4.54		8.90	11.90
0400	4" x 4" x 1/2" thick		110	.145		4.79	4.96		9.75	13
0600	4" x 8" x 1/2" thick, using 8" as edge		130	.123		4.25	4.19		8.44	11.20
0700	Floors, mud set, 1,000 S.F. lots, red, 4" x 4" x 1/2" thick		120	.133	S.F.	7.50	4.54		12.04	15.35
0900	6" x 6" x 1/2" thick		140	.114		7.05	3.90		10.95	13.85
1000	4" x 8" x 1/2" thick		130	.123		7.50	4.19		11.69	14.80
1300	For waxed coating, add					.67			.67	.74
1500	For non-standard colors, add					.40			.40	.44
1600	For abrasive surface, add					.47			.47	.52
1800	Brown tile, imported, 6" x 6" x 3/4"	D-7	120	.133		8.45	4.54		12.99	16.35
1900	8" x 8" x 1"		110	.145		9.10	4.96		14.06	17.75
2100	For thin set mortar application, deduct		700	.023			.78		.78	1.22
2700	Stair tread, 6" x 6" x 3/4", plain		50	.320		5.25	10.90		16.15	23
2800	Abrasive		47	.340		5.30	11.60		16.90	24
3000	Wainscot, 6" x 6" x 1/2", thin set, red		105	.152		4.24	5.20		9.44	12.75
3100	Non-standard colors		105	.152		4.69	5.20		9.89	13.25
3300	Window sill, 6" wide, 3/4" thick		90	.178	L.F.	5.25	6.05		11.30	15.30

09 30 Tiling

09 30 16 – Quarry Tiling

09 30 16.10 Quarry Tile

		Crew	Daily Output	Labor-Hours	Unit	Material	2009 Bare Costs Labor	Equipment	Total	Total Incl O&P
3400	Corners	D-7	80	.200	Ea.	5.85	6.80		12.65	17.10
9000	Minimum labor/equipment charge	↓	3.25	4.923	Job		168		168	262

09 30 29 – Metal Tiling

09 30 29.10 Metal Tile

		Crew	Daily Output	Labor-Hours	Unit	Material	2009 Bare Costs Labor	Equipment	Total	Total Incl O&P
0010	**METAL TILE** 4' x 4' sheet, 24 ga., tile pattern, nailed									
0200	Stainless steel	2 Carp	512	.031	S.F.	24.50	1.25		25.75	29
0400	Aluminized steel	"	512	.031	"	13.15	1.25		14.40	16.50
9000	Minimum labor/equipment charge	1 Carp	4	2	Job		80		80	132

09 51 Acoustical Ceilings

09 51 23 – Acoustical Tile Ceilings

09 51 23.10 Suspended Acoustic Ceiling Tiles

		Crew	Daily Output	Labor-Hours	Unit	Material	2009 Bare Costs Labor	Equipment	Total	Total Incl O&P
0010	**SUSPENDED ACOUSTIC CEILING TILES**, not including									
0100	suspension system									
0300	Fiberglass boards, film faced, 2' x 2' or 2' x 4', 5/8" thick	1 Carp	625	.013	S.F.	.73	.51		1.24	1.64
0400	3/4" thick		600	.013		1.69	.53		2.22	2.74
0500	3" thick, thermal, R11		450	.018		1.53	.71		2.24	2.85
0600	Glass cloth faced fiberglass, 3/4" thick		500	.016		3.04	.64		3.68	4.40
0700	1" thick		485	.016		2.16	.66		2.82	3.47
0820	1-1/2" thick, nubby face		475	.017		2.66	.67		3.33	4.04
1110	Mineral fiber tile, lay-in, 2' x 2' or 2' x 4', 5/8" thick, fine texture		625	.013		.57	.51		1.08	1.47
1115	Rough textured		625	.013		1.26	.51		1.77	2.23
1125	3/4" thick, fine textured		600	.013		1.56	.53		2.09	2.60
1130	Rough textured		600	.013		1.88	.53		2.41	2.95
1135	Fissured		600	.013		2.45	.53		2.98	3.58
1150	Tegular, 5/8" thick, fine textured		470	.017		1.45	.68		2.13	2.72
1155	Rough textured		470	.017		1.90	.68		2.58	3.21
1165	3/4" thick, fine textured		450	.018		2.08	.71		2.79	3.46
1170	Rough textured		450	.018		2.35	.71		3.06	3.76
1175	Fissured	↓	450	.018		3.63	.71		4.34	5.15
1180	For aluminum face, add					6.05			6.05	6.65
1185	For plastic film face, add					.94			.94	1.03
1190	For fire rating, add					.45			.45	.50
1300	Mirror faced panels, 15/16" thick, 2' x 2'	1 Carp	500	.016		10.45	.64		11.09	12.55
1900	Eggcrate, acrylic, 1/2" x 1/2" x 1/2" cubes		500	.016		1.82	.64		2.46	3.06
2100	Polystyrene eggcrate, 3/8" x 3/8" x 1/2" cubes		510	.016		1.53	.63		2.16	2.71
2200	1/2" x 1/2" x 1/2" cubes		500	.016		2.04	.64		2.68	3.30
2400	Luminous panels, prismatic, acrylic		400	.020		2.22	.80		3.02	3.76
2500	Polystyrene		400	.020		1.14	.80		1.94	2.57
2700	Flat white acrylic		400	.020		3.86	.80		4.66	5.55
2800	Polystyrene		400	.020		2.65	.80		3.45	4.24
3000	Drop pan, white, acrylic		400	.020		5.65	.80		6.45	7.55
3100	Polystyrene		400	.020		4.73	.80		5.53	6.50
3600	Perforated aluminum sheets, .024" thick, corrugated, painted		490	.016		2.25	.65		2.90	3.56
3700	Plain		500	.016		3.76	.64		4.40	5.20
3720	Mineral fiber, 24" x 24" or 48", reveal edge, painted, 5/8" thick		600	.013		1.38	.53		1.91	2.40
3740	3/4" thick		575	.014		2.24	.56		2.80	3.38
5020	66 – 78% recycled content, 3/4" thick [G]		600	.013		1.82	.53		2.35	2.88
5040	Mylar, 42% recycled content, 3/4" thick [G]		600	.013	↓	4.28	.53		4.81	5.60
9000	Minimum labor/equipment charge	↓	4	2	Job		80		80	132

09 51 Acoustical Ceilings

09 51 23 – Acoustical Tile Ceilings

09 51 23.30 Suspended Ceilings, Complete

	09 51 23.30 Suspended Ceilings, Complete	Crew	Daily Output	Labor-Hours	Unit	Material	2009 Bare Costs Labor	Equipment	Total	Total Incl O&P
0010	**SUSPENDED CEILINGS, COMPLETE** Including standard									
0100	suspension system but not incl. 1-1/2" carrier channels									
0600	Fiberglass ceiling board, 2' x 4' x 5/8", plain faced	1 Carp	500	.016	S.F.	1.36	.64		2	2.56
0700	Offices, 2' x 4' x 3/4"		380	.021		2.32	.84		3.16	3.94
0800	Mineral fiber, on 15/16" T bar susp. 2' x 2' x 3/4" lay-in board		345	.023		2.35	.93		3.28	4.11
0810	2' x 4' x 5/8" tile		380	.021		1.20	.84		2.04	2.71
0820	Tegular, 2' x 2' x 5/8" tile on 9/16" grid		250	.032		2.40	1.28		3.68	4.74
0830	2' x 4' x 3/4" tile		275	.029		2.87	1.16		4.03	5.10
0900	Luminous panels, prismatic, acrylic		255	.031		2.85	1.25		4.10	5.20
1200	Metal pan with acoustic pad, steel		75	.107		4.16	4.26		8.42	11.65
1300	Painted aluminum		75	.107		2.88	4.26		7.14	10.20
1500	Aluminum, degreased finish		75	.107		4.80	4.26		9.06	12.35
1600	Stainless steel		75	.107		9.05	4.26		13.31	17
1800	Tile, Z bar suspension, 5/8" mineral fiber tile		150	.053		2.16	2.13		4.29	5.90
1900	3/4" mineral fiber tile	▼	150	.053	▼	2.31	2.13		4.44	6.05
2402	For strip lighting, see Div. 26 51 13.50									
2500	For rooms under 500 S.F., add				S.F.		25%			
9000	Minimum labor/equipment charge	1 Carp	2	4	Job		160		160	264

09 51 53 – Direct-Applied Acoustical Ceilings

09 51 53.10 Ceiling Tile

		Crew	Daily Output	Labor-Hours	Unit	Material	2009 Bare Costs Labor	Equipment	Total	Total Incl O&P
0010	**CEILING TILE**, Stapled or cemented									
0100	12" x 12" or 12" x 24", not including furring									
0600	Mineral fiber, vinyl coated, 5/8" thick	1 Carp	300	.027	S.F.	1.85	1.07		2.92	3.80
0700	3/4" thick		300	.027		1.87	1.07		2.94	3.82
0900	Fire rated, 3/4" thick, plain faced		300	.027		1.40	1.07		2.47	3.30
1000	Plastic coated face		300	.027		1.18	1.07		2.25	3.06
1200	Aluminum faced, 5/8" thick, plain		300	.027		1.51	1.07		2.58	3.42
3000	Wood fiber tile, 1/2" thick		400	.020		.92	.80		1.72	2.33
3100	3/4" thick	▼	400	.020		1.17	.80		1.97	2.61
3300	For flameproofing, add					.10			.10	.11
3400	For sculptured 3 dimensional, add					.29			.29	.32
3900	For ceiling primer, add					.13			.13	.14
4000	For ceiling cement, add				▼	.37			.37	.41
9000	Minimum labor/equipment charge	1 Carp	4	2	Job		80		80	132

09 53 Acoustical Ceiling Suspension Assemblies

09 53 23 – Metal Acoustical Ceiling Suspension Assemblies

09 53 23.30 Ceiling Suspension Systems

		Crew	Daily Output	Labor-Hours	Unit	Material	2009 Bare Costs Labor	Equipment	Total	Total Incl O&P
0010	**CEILING SUSPENSION SYSTEMS** for boards and tile									
0050	Class A suspension system, 15/16" T bar, 2' x 4' grid	1 Carp	800	.010	S.F.	.63	.40		1.03	1.35
0300	2' x 2' grid		650	.012		.79	.49		1.28	1.67
0310	25% recycled steel, 2' x 4' grid [G]		800	.010		.73	.40		1.13	1.46
0320	2' x 2' grid [G]	▼	650	.012		.92	.49		1.41	1.82
0350	For 9/16" grid, add					.16			.16	.18
0360	For fire rated grid, add					.09			.09	.10
0370	For colored grid, add					.21			.21	.23
0400	Concealed Z bar suspension system, 12" module	1 Carp	520	.015		.65	.61		1.26	1.73
0600	1-1/2" carrier channels, 4' O.C., add	"	470	.017	▼	.15	.68		.83	1.29
0700	Carrier channels for ceilings with									
0900	recessed lighting fixtures, add	1 Carp	460	.017	S.F.	.28	.69		.97	1.45

09 53 Acoustical Ceiling Suspension Assemblies

09 53 23 – Metal Acoustical Ceiling Suspension Assemblies

09 53 23.30 Ceiling Suspension Systems	Crew	Daily Output	Labor-Hours	Unit	Material	2009 Bare Costs Labor	2009 Bare Costs Equipment	Total	Total Incl O&P
9000	Minimum labor/equipment charge	1 Carp	5	1.600	Job		64		64	106

09 63 Masonry Flooring

09 63 13 – Brick Flooring

09 63 13.10 Miscellaneous Brick Flooring

		Crew	Daily Output	Labor-Hours	Unit	Material	Labor	Equipment	Total	Total Incl O&P
0010	**MISCELLANEOUS BRICK FLOORING**									
0020	Acid proof shales, red, 8" x 3-3/4" x 1-1/4" thick	D-7	.43	37.209	M	820	1,275		2,095	2,875
0050	2-1/4" thick	D-1	.40	40		950	1,450		2,400	3,400
0200	Acid proof clay brick, 8" x 3-3/4" x 2-1/4" thick		.40	40		870	1,450		2,320	3,300
0250	9" x 4-1/2" x 3"		95	.168	S.F.	3.93	6.10		10.03	14.20
0260	Cast ceramic, pressed, 4" x 8" x 1/2", unglazed	D-7	100	.160		5.95	5.45		11.40	15.10
0270	Glazed		100	.160		7.95	5.45		13.40	17.25
0280	Hand molded flooring, 4" x 8" x 3/4", unglazed		95	.168		7.85	5.75		13.60	17.65
0290	Glazed		95	.168		9.85	5.75		15.60	19.85
0300	8" hexagonal, 3/4" thick, unglazed		85	.188		8.60	6.40		15	19.50
0310	Glazed		85	.188		15.55	6.40		21.95	27
0400	Heavy duty industrial, cement mortar bed, 2" thick, not incl. brick	D-1	80	.200		.79	7.25		8.04	12.60
0450	Acid proof joints, 1/4" wide	"	65	.246		1.37	8.95		10.32	15.95
0500	Pavers, 8" x 4", 1" to 1-1/4" thick, red	D-7	95	.168		3.47	5.75		9.22	12.80
0510	Ironspot	"	95	.168		4.89	5.75		10.64	14.40
0540	1-3/8" to 1-3/4" thick, red	D-1	95	.168		3.34	6.10		9.44	13.55
0560	Ironspot		95	.168		4.84	6.10		10.94	15.20
0580	2-1/4" thick, red		90	.178		3.40	6.45		9.85	14.20
0590	Ironspot		90	.178		5.30	6.45		11.75	16.25
0700	Paver, adobe brick, 6" x 12", 1/2" joint		42	.381		.90	13.85		14.75	23.50
0710	Mexican red, 12" x 12"	1 Tilf	48	.167		1.61	6.35		7.96	11.70
0720	Saltillo, 12" x 12"	"	48	.167		1.13	6.35		7.48	11.20
0800	For sidewalks and patios with pavers, see Div. 32 14 16.10									
0870	For epoxy joints, add	D-1	600	.027	S.F.	2.59	.97		3.56	4.42
0880	For Furan underlayment, add	"	600	.027		2.14	.97		3.11	3.92
0890	For waxed surface, steam cleaned, add	A-1H	1000	.008		.19	.25	.05	.49	.69
9000	Minimum labor/equipment charge	1 Bric	2	4	Job		162		162	262

09 63 40 – Stone Flooring

09 63 40.10 Marble

		Crew	Daily Output	Labor-Hours	Unit	Material	Labor	Equipment	Total	Total Incl O&P
0010	**MARBLE**									
0020	Thin gauge tile, 12" x 6", 3/8", white Carara	D-7	60	.267	S.F.	11.15	9.10		20.25	26.50
0100	Travertine		60	.267		12.20	9.10		21.30	27.50
0200	12" x 12" x 3/8", thin set, floors		60	.267		6.05	9.10		15.15	21
0300	On walls		52	.308		9.35	10.50		19.85	26.50
1000	Marble threshold, 4" wide x 36" long x 5/8" thick, white		60	.267	Ea.	6.05	9.10		15.15	21
9000	Minimum labor/equipment charge		3	5.333	Job		182		182	284

09 63 40.20 Slate Tile

		Crew	Daily Output	Labor-Hours	Unit	Material	Labor	Equipment	Total	Total Incl O&P
0010	**SLATE TILE**									
0020	Vermont, 6" x 6" x 1/4" thick, thin set	D-7	180	.089	S.F.	6.80	3.03		9.83	12.20
9000	Minimum labor/equipment charge	"	3	5.333	Job		182		182	284

09 64 Wood Flooring

09 64 16 – Wood Block Flooring

09 64 16.10 End Grain Block Flooring

		Crew	Daily Output	Labor-Hours	Unit	Material	2009 Bare Costs Labor	Equipment	Total	Total Incl O&P
0010	**END GRAIN BLOCK FLOORING**									
0020	End grain flooring, coated, 2" thick	1 Carp	295	.027	S.F.	3.29	1.08		4.37	5.40
0400	Natural finish, 1" thick, fir		125	.064		3.40	2.56		5.96	7.95
0600	1-1/2" thick, pine		125	.064		3.34	2.56		5.90	7.90
0700	2" thick, pine		125	.064		4.64	2.56		7.20	9.30
9000	Minimum labor/equipment charge		2	4	Job		160		160	264

09 64 23 – Wood Parquet Flooring

09 64 23.10 Wood Parquet

		Crew	Daily Output	Labor-Hours	Unit	Material	2009 Bare Costs Labor	Equipment	Total	Total Incl O&P
0010	**WOOD PARQUET** flooring									
5200	Parquetry, standard, 5/16" thick, not incl. finish, oak, minimum	1 Carp	160	.050	S.F.	4.41	2		6.41	8.15
5300	Maximum		100	.080		4.95	3.20		8.15	10.75
5500	Teak, minimum		160	.050		4.79	2		6.79	8.55
5600	Maximum		100	.080		8.35	3.20		11.55	14.50
5650	13/16" thick, select grade oak, minimum		160	.050		9.20	2		11.20	13.45
5700	Maximum		100	.080		14	3.20		17.20	20.50
5800	Custom parquetry, including finish, minimum		100	.080		15.45	3.20		18.65	22.50
5900	Maximum		50	.160		20.50	6.40		26.90	33
6700	Parquetry, prefinished white oak, 5/16" thick, minimum		160	.050		3.61	2		5.61	7.25
6800	Maximum		100	.080		7.30	3.20		10.50	13.35
7000	Walnut or teak, parquetry, minimum		160	.050		5.35	2		7.35	9.20
7100	Maximum		100	.080		9.35	3.20		12.55	15.60
7200	Acrylic wood parquet blocks, 12" x 12" x 5/16",									
7210	Irradiated, set in epoxy	1 Carp	160	.050	S.F.	7.80	2		9.80	11.90

09 64 29 – Wood Strip and Plank Flooring

09 64 29.10 Wood

		Crew	Daily Output	Labor-Hours	Unit	Material	2009 Bare Costs Labor	Equipment	Total	Total Incl O&P
0010	**WOOD** R061110-30									
0020	Fir, vertical grain, 1" x 4", not incl. finish, grade B & better	1 Carp	255	.031	S.F.	2.74	1.25		3.99	5.10
0100	C grade & better		255	.031		2.58	1.25		3.83	4.91
0300	Flat grain, 1" x 4", not incl. finish, B & better		255	.031		3.13	1.25		4.38	5.50
0400	C & better		255	.031		3.01	1.25		4.26	5.40
4000	Maple, strip, 25/32" x 2-1/4", not incl. finish, select		170	.047		5.25	1.88		7.13	8.90
4100	#2 & better		170	.047		3.32	1.88		5.20	6.75
4300	33/32" x 3-1/4", not incl. finish, #1 grade		170	.047		4.19	1.88		6.07	7.70
4400	#2 & better		170	.047		3.73	1.88		5.61	7.20
4600	Oak, white or red, 25/32" x 2-1/4", not incl. finish									
4700	#1 common	1 Carp	170	.047	S.F.	3.18	1.88		5.06	6.60
4900	Select quartered, 2-1/4" wide		170	.047		2.94	1.88		4.82	6.35
5000	Clear		170	.047		4.01	1.88		5.89	7.50
6100	Prefinished, white oak, prime grade, 2-1/4" wide		170	.047		6.40	1.88		8.28	10.15
6200	3-1/4" wide		185	.043		8.20	1.73		9.93	11.85
6400	Ranch plank		145	.055		7.90	2.20		10.10	12.35
6500	Hardwood blocks, 9" x 9", 25/32" thick		160	.050		5.75	2		7.75	9.65
7400	Yellow pine, 3/4" x 3-1/8", T & G, C & better, not incl. finish		200	.040		2.35	1.60		3.95	5.25
7500	Refinish wood floor, sand, 2 cts poly, wax, soft wood, min.	1 Clab	400	.020		.78	.63		1.41	1.90
7600	Hard wood, max		130	.062		1.16	1.94		3.10	4.49
7800	Sanding and finishing, 2 coats polyurethane		295	.027		.78	.86		1.64	2.27
7900	Subfloor and underlayment, see Div. 06 16									
8015	Transition molding, 2 1/4" wide, 5' long	1 Carp	19.20	.417	Ea.	15	16.65		31.65	44
8300	Floating floor, wood composition strip, complete	1 Clab	133	.060	S.F.	4.23	1.90		6.13	7.80
8310	Components, T & G wood composite strips					3.66			3.66	4.03
8320	Film					.15			.15	.16
8330	Foam					.30			.30	.33

09 64 Wood Flooring

09 64 29 – Wood Strip and Plank Flooring

09 64 29.10 Wood		Crew	Daily Output	Labor-Hours	Unit	Material	2009 Bare Costs Labor	Equipment	Total	Total Incl O&P
8340	Adhesive				S.F.	.17			.17	.18
8350	Installation kit				↓	.17			.17	.19
8360	Trim, 2" wide x 3' long				L.F.	2.49			2.49	2.74
8370	Reducer moulding				"	4.29			4.29	4.72
8600	Flooring, wood, bamboo strips, unfinished, 5/8" x 4" x 3' [G]	1 Carp	255	.031	S.F.	4.44	1.25		5.69	6.95
8610	5/8" x 4" x 4' [G]		275	.029		4.61	1.16		5.77	6.95
8620	5/8" x 4" x 6' [G]		295	.027		5.05	1.08		6.13	7.35
8630	Finished, 5/8" x 4" x 3' [G]		255	.031		4.89	1.25		6.14	7.45
8640	5/8" x 4" x 4' [G]		275	.029		5.10	1.16		6.26	7.55
8650	5/8" x 4" x 6' [G]		295	.027	↓	5.15	1.08		6.23	7.50
8660	Stair treads, unfinished, 1-1/16" x 11-1/2" x 4' [G]		18	.444	Ea.	42.50	17.75		60.25	76.50
8670	Finished, 1-1/16" x 11-1/2" x 4' [G]		18	.444		65	17.75		82.75	101
8680	Stair risers, unfinished, 5/8" x 7-1/2" x 4' [G]		18	.444		15.75	17.75		33.50	47
8690	Finished, 5/8" x 7-1/2" x 4' [G]		18	.444		30	17.75		47.75	62.50
8700	Stair nosing, unfinished, 6' long [G]		16	.500		35	20		55	71.50
8710	Finished, 6' long [G]		16	.500	↓	44	20		64	81
9000	Minimum labor/equipment charge	↓	2	4	Job		160		160	264

09 65 Resilient Flooring

09 65 10 – Resilient Tile Underlayment

09 65 10.10 Latex Underlayment

		Crew	Daily Output	Labor-Hours	Unit	Material	2009 Bare Costs Labor	Equipment	Total	Total Incl O&P
0010	**LATEX UNDERLAYMENT**									
3600	Latex underlayment, 1/8" thk., cementitious for resilient flooring	1 Tilf	160	.050	S.F.	1.72	1.91		3.63	4.87
4000	Liquid, fortified				Gal.	38.50			38.50	42

09 65 13 – Resilient Base and Accessories

09 65 13.13 Resilient Base

		Crew	Daily Output	Labor-Hours	Unit	Material	2009 Bare Costs Labor	Equipment	Total	Total Incl O&P
0010	**RESILIENT BASE**									
0800	Base, cove, rubber or vinyl									
1100	Standard colors, 0.080" thick, 2-1/2" high	1 Tilf	315	.025	L.F.	.75	.97		1.72	2.34
1150	4" high		315	.025		.93	.97		1.90	2.53
1200	6" high		315	.025		1.25	.97		2.22	2.89
1450	1/8" thick, 2-1/2" high		315	.025		.75	.97		1.72	2.34
1500	4" high		315	.025		.71	.97		1.68	2.29
1550	6" high		315	.025	↓	1.30	.97		2.27	2.94
1600	Corners, 2-1/2" high		315	.025	Ea.	1.45	.97		2.42	3.11
1630	4" high		315	.025		2.75	.97		3.72	4.54
1660	6" high	↓	315	.025	↓	2.85	.97		3.82	4.65

09 65 13.23 Resilient Stair Treads and Risers

		Crew	Daily Output	Labor-Hours	Unit	Material	2009 Bare Costs Labor	Equipment	Total	Total Incl O&P
0010	**RESILIENT STAIR TREADS AND RISERS**									
0300	Rubber, molded tread, 12" wide, 5/16" thick, black	1 Tilf	115	.070	L.F.	11.75	2.65		14.40	17.10
0400	Colors		115	.070		12.25	2.65		14.90	17.65
0600	1/4" thick, black		115	.070		11.50	2.65		14.15	16.80
0700	Colors		115	.070		12.25	2.65		14.90	17.65
0900	Grip strip safety tread, colors, 5/16" thick		115	.070		16	2.65		18.65	22
1000	3/16" thick		120	.067	↓	11.15	2.54		13.69	16.25
1200	Landings, smooth sheet rubber, 1/8" thick		120	.067	S.F.	5.75	2.54		8.29	10.30
1300	3/16" thick		120	.067	"	7.25	2.54		9.79	11.95
1500	Nosings, 3" wide, 3/16" thick, black		140	.057	L.F.	3.15	2.18		5.33	6.90
1600	Colors		140	.057		3.11	2.18		5.29	6.85
1800	Risers, 7" high, 1/8" thick, flat	↓	250	.032		6.15	1.22		7.37	8.70

09 65 Resilient Flooring

09 65 13 – Resilient Base and Accessories

09 65 13.23 Resilient Stair Treads and Risers

		Crew	Daily Output	Labor-Hours	Unit	Material	2009 Bare Costs Labor	2009 Bare Costs Equipment	Total	Total Incl O&P
1900	Coved	1 Tilf	250	.032	L.F.	5.05	1.22		6.27	7.50
2100	Vinyl, molded tread, 12" wide, colors, 1/8" thick		115	.070		5.65	2.65		8.30	10.35
2200	1/4" thick		115	.070		7.50	2.65		10.15	12.40
2300	Landing material, 1/8" thick		200	.040	S.F.	5.65	1.52		7.17	8.60
2400	Riser, 7" high, 1/8" thick, coved		175	.046	L.F.	3.25	1.74		4.99	6.30
2500	Tread and riser combined, 1/8" thick		80	.100	"	8.85	3.81		12.66	15.70
9000	Minimum labor/equipment charge		3	2.667	Job		102		102	159

09 65 16 – Resilient Sheet Flooring

09 65 16.10 Rubber and Vinyl Sheet Flooring

		Crew	Daily Output	Labor-Hours	Unit	Material	2009 Bare Costs Labor	2009 Bare Costs Equipment	Total	Total Incl O&P
0010	**RUBBER AND VINYL SHEET FLOORING**									
5500	Linoleum, sheet goods	1 Tilf	360	.022	S.F.	4.55	.85		5.40	6.30
5900	Rubber, sheet goods, 36" wide, 1/8" thick		120	.067		6.25	2.54		8.79	10.85
5950	3/16" thick		100	.080		8.50	3.05		11.55	14.10
6000	1/4" thick		90	.089		10	3.39		13.39	16.30
8000	Vinyl sheet goods, backed, .065" thick, minimum		250	.032		3.50	1.22		4.72	5.75
8050	Maximum		200	.040		3.39	1.52		4.91	6.10
8100	.080" thick, minimum		230	.035		3.50	1.33		4.83	5.90
8150	Maximum		200	.040		4.95	1.52		6.47	7.85
8200	.125" thick, minimum		230	.035		4	1.33		5.33	6.45
8250	Maximum		200	.040		5.95	1.52		7.47	8.95
8700	Adhesive cement, 1 gallon per 200 to 300 S.F.				Gal.	24.50			24.50	26.50
8800	Asphalt primer, 1 gallon per 300 S.F.					11.65			11.65	12.80
8900	Emulsion, 1 gallon per 140 S.F.					15			15	16.50

09 65 19 – Resilient Tile Flooring

09 65 19.10 Miscellaneous Resilient Tile Flooring

			Crew	Daily Output	Labor-Hours	Unit	Material	2009 Bare Costs Labor	2009 Bare Costs Equipment	Total	Total Incl O&P
0010	**MISCELLANEOUS RESILIENT TILE FLOORING**										
2200	Cork tile, standard finish, 1/8" thick	G	1 Tilf	315	.025	S.F.	5.75	.97		6.72	7.85
2250	3/16" thick	G		315	.025		5.75	.97		6.72	7.85
2300	5/16" thick	G		315	.025		7.15	.97		8.12	9.35
2350	1/2" thick	G		315	.025		8.45	.97		9.42	10.80
2500	Urethane finish, 1/8" thick	G		315	.025		6.85	.97		7.82	9.05
2550	3/16" thick	G		315	.025		7.20	.97		8.17	9.40
2600	5/16" thick	G		315	.025		9.20	.97		10.17	11.65
2650	1/2" thick	G		315	.025		12.40	.97		13.37	15.15
6050	Rubber tile, marbleized colors, 12" x 12", 1/8" thick			400	.020		6.20	.76		6.96	8
6100	3/16" thick			400	.020		9.25	.76		10.01	11.40
6300	Special tile, plain colors, 1/8" thick			400	.020		9.25	.76		10.01	11.40
6350	3/16" thick			400	.020		9.50	.76		10.26	11.65
7000	Vinyl composition tile, 12" x 12", 1/16" thick			500	.016		.87	.61		1.48	1.91
7050	Embossed			500	.016		2.07	.61		2.68	3.23
7100	Marbleized			500	.016		2.07	.61		2.68	3.23
7150	Solid			500	.016		2.66	.61		3.27	3.88
7200	3/32" thick, embossed			500	.016		1.30	.61		1.91	2.38
7250	Marbleized			500	.016		2.37	.61		2.98	3.56
7300	Solid			500	.016		2.20	.61		2.81	3.37
7350	1/8" thick, marbleized			500	.016		1.70	.61		2.31	2.82
7400	Solid			500	.016		2.14	.61		2.75	3.30
7450	Conductive			500	.016		5.35	.61		5.96	6.85
7500	Vinyl tile, 12" x 12", .050" thick, minimum			500	.016		2.95	.61		3.56	4.20
7550	Maximum			500	.016		5.90	.61		6.51	7.45
7600	1/8" thick, minimum			500	.016		4.20	.61		4.81	5.55
7650	Solid colors			500	.016		5.85	.61		6.46	7.40

09 65 Resilient Flooring

09 65 19 – Resilient Tile Flooring

09 65 19.10 Miscellaneous Resilient Tile Flooring	Crew	Daily Output	Labor-Hours	Unit	Material	2009 Bare Costs Labor	Equipment	Total	Total Incl O&P
7700	Marbleized or Travertine pattern	1 Tilf	500	.016	S.F.	5.65	.61		6.26	7.15
7750	Florentine pattern		500	.016		5.70	.61		6.31	7.25
7800	Maximum		500	.016		12.35	.61		12.96	14.55
9500	Minimum labor/equipment charge		4	2	Job		76		76	119

09 65 33 – Conductive Resilient Flooring

09 65 33.10 Conductive Rubber and Vinyl Flooring

		Crew	Daily Output	Labor-Hours	Unit	Material	Labor	Equipment	Total	Total Incl O&P
0010	**CONDUCTIVE RUBBER AND VINYL FLOORING**									
1700	Conductive flooring, rubber tile, 1/8" thick	1 Tilf	315	.025	S.F.	5	.97		5.97	7
1800	Homogeneous vinyl tile, 1/8" thick	"	315	.025	"	6.95	.97		7.92	9.15

09 66 Terrazzo Flooring

09 66 13 – Portland Cement Terrazzo Flooring

09 66 13.10 Portland Cement Terrazzo

		Crew	Daily Output	Labor-Hours	Unit	Material	Labor	Equipment	Total	Total Incl O&P
0010	**PORTLAND CEMENT TERRAZZO**, cast-in-place									
0020	Cove base, 6" high, 16ga. zinc	1 Mstz	20	.400	L.F.	3.32	15.10		18.42	27
0100	Curb, 6" high and 6" wide		6	1.333		5.25	50.50		55.75	84.50
0300	Divider strip for floors, 14 ga., 1-1/4" deep, zinc		375	.021		1.21	.80		2.01	2.59
0400	Brass		375	.021		2.23	.80		3.03	3.71
0600	Heavy top strip 1/4" thick, 1-1/4" deep, zinc		300	.027		1.76	1.01		2.77	3.51
1200	For thin set floors, 16 ga., 1/2" x 1/2", zinc		350	.023		.75	.86		1.61	2.18
1500	Floor, bonded to concrete, 1-3/4" thick, gray cement	J-3	75	.213	S.F.	2.87	7.30	3.38	13.55	18.30
1600	White cement, mud set		75	.213		3.27	7.30	3.38	13.95	18.75
1800	Not bonded, 3" total thickness, gray cement		70	.229		3.59	7.85	3.62	15.06	20
1900	White cement, mud set		70	.229		3.92	7.85	3.62	15.39	20.50
9000	Minimum labor/equipment charge	1 Mstz	1	8	Job		300		300	470

09 66 13.30 Terrazzo, Precast

		Crew	Daily Output	Labor-Hours	Unit	Material	Labor	Equipment	Total	Total Incl O&P
0010	**TERRAZZO, PRECAST**									
0020	Base, 6" high, straight	1 Mstz	70	.114	L.F.	10.40	4.31		14.71	18.20
0100	Cove		60	.133		12.15	5.05		17.20	21
0300	8" high, straight		60	.133		10.85	5.05		15.90	19.75
0400	Cove		50	.160		15.95	6.05		22	27
0600	For white cement, add					.43			.43	.47
0700	For 16 ga. zinc toe strip, add					1.60			1.60	1.76
0900	Curbs, 4" x 4" high	1 Mstz	40	.200		30.50	7.55		38.05	45.50
1000	8" x 8" high	"	30	.267		34.50	10.05		44.55	53.50
1200	Floor tiles, non-slip, 1" thick, 12" x 12"	D-1	60	.267	S.F.	17.70	9.70		27.40	35
1300	1-1/4" thick, 12" x 12"		60	.267		19.60	9.70		29.30	37
1500	16" x 16"		50	.320		21.50	11.65		33.15	42.50
1600	1-1/2" thick, 16" x 16"		45	.356		19.45	12.90		32.35	42.50
1800	For Venetian terrazzo, add					5.85			5.85	6.40
1900	For white cement, add					.55			.55	.61
2400	Stair treads, 1-1/2" thick, non-slip, three line pattern	2 Mstz	70	.229	L.F.	39.50	8.60		48.10	57
2500	Nosing and two lines		70	.229		39.50	8.60		48.10	57
2700	2" thick treads, straight		60	.267		43	10.05		53.05	62.50
2800	Curved		50	.320		56	12.05		68.05	80.50
3000	Stair risers, 1" thick, to 6" high, straight sections		60	.267		9.80	10.05		19.85	26.50
3100	Cove		50	.320		13.90	12.05		25.95	34
3300	Curved, 1" thick, to 6" high, vertical		48	.333		19.05	12.55		31.60	40.50
3400	Cove		38	.421		36	15.85		51.85	64.50
3600	Stair tread and riser, single piece, straight, minimum		60	.267		50	10.05		60.05	70.50

09 66 Terrazzo Flooring

09 66 13 – Portland Cement Terrazzo Flooring

09 66 13.30 Terrazzo, Precast		Crew	Daily Output	Labor-Hours	Unit	Material	2009 Bare Costs Labor	Equipment	Total	Total Incl O&P
3700	Maximum	2 Mstz	40	.400	L.F.	65	15.10		80.10	94.50
3900	Curved tread and riser, minimum		40	.400		70	15.10		85.10	101
4000	Maximum		32	.500		88.50	18.85		107.35	127
4200	Stair stringers, notched, 1" thick		25	.640		28.50	24		52.50	69
4300	2" thick		22	.727		34	27.50		61.50	80
4500	Stair landings, structural, non-slip, 1-1/2" thick		85	.188	S.F.	31	7.10		38.10	45.50
4600	3" thick		75	.213		44.50	8.05		52.55	61
4800	Wainscot, 12" x 12" x 1" tiles	1 Mstz	12	.667		6.50	25		31.50	46.50
4900	16" x 16" x 1-1/2" tiles	"	8	1		13.70	37.50		51.20	74
9500	Minimum labor/equipment charge	1 Tilf	2	4	Job		152		152	238

09 66 16 – Terrazzo Floor Tile

09 66 16.10 Tile or Terrazzo Base										
0010	**TILE OR TERRAZZO BASE**									
0020	Scratch coat only	1 Mstz	150	.053	S.F.	.41	2.01		2.42	3.59
0500	Scratch and brown coat only		75	.107	"	.78	4.02		4.80	7.15
9000	Minimum labor/equipment charge		1	8	Job		300		300	470

09 67 Fluid-Applied Flooring

09 67 20 – Epoxy-Marble Chip Flooring

09 67 20.13 Elastomeric Liquid Flooring

0010	**ELASTOMERIC LIQUID FLOORING**									
0020	Cementitious acrylic, 1/4" thick	C-6	520	.092	S.F.	1.45	3.05	.10	4.60	6.70
0200	Methyl methacrylate, 1/4" thick	C-8A	3000	.016	"	5.40	.55		5.95	6.85

09 67 20.16 Epoxy Terrazzo

0010	**EPOXY TERRAZZO**									
1800	Epoxy terrazzo, 1/4" thick, chemical resistant, minimum	J-3	200	.080	S.F.	5.25	2.74	1.27	9.26	11.50
1900	Maximum	"	150	.107	"	7.70	3.66	1.69	13.05	16.05

09 67 20.19 Polyacrylate Terrazzo

0010	**POLYACRYLATE TERRAZZO**									

09 67 20.26 Quartz Flooring

0010	**QUARTZ FLOORING**									
0600	Epoxy, with colored quartz chips, broadcast, minimum	C-6	675	.071	S.F.	2.49	2.35	.08	4.92	6.65
0700	Maximum		490	.098		3.02	3.24	.11	6.37	8.75
0900	Trowelled, minimum		560	.086		3.21	2.83	.09	6.13	8.25
1000	Maximum		480	.100		4.69	3.31	.11	8.11	10.65
1200	Heavy duty epoxy topping, 1/4" thick,									
1300	500 to 1,000 S.F.	C-6	420	.114	S.F.	4.67	3.78	.12	8.57	11.45
1500	1,000 to 2,000 S.F.		450	.107		3.87	3.53	.12	7.52	10.15
1600	Over 10,000 S.F.		480	.100		3.62	3.31	.11	7.04	9.50

09 68 Carpeting

09 68 05 – Carpet Accessories

09 68 05.11 Flooring Transition Strip

		Crew	Daily Output	Labor-Hours	Unit	Material	2009 Bare Costs Labor	Equipment	Total	Total Incl O&P
0010	**FLOORING TRANSITION STRIP**									
0107	Clamp down brass divider, 12' strip, vinyl to carpet	1 Tilf	31.25	.256	Ea.	26.50	9.75		36.25	44.50
0117	Vinyl to hard surface	"	31.25	.256	"	26.50	9.75		36.25	44.50

09 68 10 – Carpet Pad

09 68 10.10 Commercial Grade Carpet Pad

		Crew	Daily Output	Labor-Hours	Unit	Material	Labor	Equipment	Total	Total Incl O&P
0010	**COMMERCIAL GRADE CARPET PAD**									
9000	Sponge rubber pad, minimum	1 Tilf	150	.053	S.Y.	4.20	2.03		6.23	7.80
9100	Maximum		150	.053		10	2.03		12.03	14.20
9200	Felt pad, minimum		150	.053		4.03	2.03		6.06	7.60
9300	Maximum		150	.053		7.05	2.03		9.08	10.95
9400	Bonded urethane pad, minimum		150	.053		4.57	2.03		6.60	8.25
9500	Maximum		150	.053		7.80	2.03		9.83	11.80
9600	Prime urethane pad, minimum		150	.053		2.62	2.03		4.65	6.05
9700	Maximum		150	.053		4.84	2.03		6.87	8.50

09 68 13 – Tile Carpeting

09 68 13.10 Carpet Tile

		Crew	Daily Output	Labor-Hours	Unit	Material	Labor	Equipment	Total	Total Incl O&P
0010	**CARPET TILE**									
0100	Tufted nylon, 18" x 18", hard back, 20 oz.	1 Tilf	150	.053	S.Y.	22	2.03		24.03	27.50
0110	26 oz.		150	.053		38	2.03		40.03	44.50
0200	Cushion back, 20 oz.		150	.053		28	2.03		30.03	33.50
0210	26 oz.		150	.053		43.50	2.03		45.53	51
1100	Tufted, 24" x 24", 24 oz. nylon		80	.100		29.50	3.81		33.31	38
1180	35 oz.		80	.100		34	3.81		37.81	43.50
5060	42 oz.		80	.100		45.50	3.81		49.31	56

09 68 16 – Sheet Carpeting

09 68 16.10 Sheet Carpet

		Crew	Daily Output	Labor-Hours	Unit	Material	Labor	Equipment	Total	Total Incl O&P
0010	**SHEET CARPET**									
0700	Nylon, level loop, 26 oz., light to medium traffic	1 Tilf	75	.107	S.Y.	26	4.06		30.06	35
0720	28 oz., light to medium traffic		75	.107		27	4.06		31.06	36
0900	32 oz., medium traffic		75	.107		32.50	4.06		36.56	42
1100	40 oz., medium to heavy traffic		75	.107		48	4.06		52.06	59.50
2920	Nylon plush, 30 oz., medium traffic		57	.140		24	5.35		29.35	35
3000	36 oz., medium traffic		75	.107		32	4.06		36.06	41.50
3100	42 oz., medium to heavy traffic		70	.114		36	4.35		40.35	46.50
3200	46 oz., medium to heavy traffic		70	.114		42	4.35		46.35	53
3300	54 oz., heavy traffic		70	.114		46	4.35		50.35	57.50
4500	50 oz., medium to heavy traffic		75	.107		110	4.06		114.06	127
4700	Patterned, 32 oz., medium to heavy traffic		70	.114		109	4.35		113.35	127
4900	48 oz., heavy traffic		70	.114		112	4.35		116.35	130
5000	For less than full roll (approx. 1500 S.F.), add					25%				
5100	For small rooms, less than 12' wide, add						25%			
5200	For large open areas (no cuts), deduct						25%			
5600	For bound carpet baseboard, add	1 Tilf	300	.027	L.F.	2.25	1.02		3.27	4.07
5610	For stairs, not incl. price of carpet, add	"	30	.267	Riser		10.15		10.15	15.90
5620	For borders and patterns, add to labor						18%			
8950	For tackless, stretched installation, add padding to above									
9850	For "branded" fiber, add				S.Y.	25%				
9900	Carpet cleaning machine, rent				Day				34.50	37.75
9910	Minimum labor/equipment charge	1 Tilf	3	2.667	Job		102		102	159

09 68 Carpeting

09 68 20 – Athletic Carpet

09 68 20.10 Indoor Athletic Carpet

	Crew	Daily Output	Labor-Hours	Unit	Material	2009 Bare Costs Labor	2009 Bare Costs Equipment	Total	Total Incl O&P
0010 **INDOOR ATHLETIC CARPET**									
3700 Polyethylene, in rolls, no base incl., landscape surfaces	1 Tilf	275	.029	S.F.	3.05	1.11		4.16	5.10
3800 Nylon action surface, 1/8" thick		275	.029		3.20	1.11		4.31	5.25
3900 1/4" thick		275	.029		4.61	1.11		5.72	6.80
4000 3/8" thick		275	.029		5.80	1.11		6.91	8.10
5500 Polyvinyl chloride, sheet goods for gyms, 1/4" thick		80	.100		7.50	3.81		11.31	14.20
5600 3/8" thick		60	.133		8.50	5.10		13.60	17.30

09 69 Access Flooring

09 69 13 – Rigid-Grid Access Flooring

09 69 13.10 Access Floors

	Crew	Daily Output	Labor-Hours	Unit	Material	2009 Bare Costs Labor	2009 Bare Costs Equipment	Total	Total Incl O&P
0010 **ACCESS FLOORS**									
0015 Access floor package including panel, pedestal, stringers & laminate cover									
0100 Computer room, greater than 6,000 S.F.	4 Carp	750	.043	S.F.	8.65	1.70		10.35	12.30
0110 Less than 6,000 S.F.	2 Carp	375	.043		9.95	1.70		11.65	13.75
0120 Office, greater than 6,000 S.F.	4 Carp	1050	.030		4.56	1.22		5.78	7
0250 Panels, particle board or steel, 1250# load, no covering, under 6,000 S.F.	2 Carp	600	.027		4.24	1.07		5.31	6.40
0300 Over 6,000 S.F.		640	.025		3.72	1		4.72	5.75
0400 Aluminum, 24" panels		500	.032		31.50	1.28		32.78	36.50
0600 For carpet covering, add					8.55			8.55	9.40
0700 For vinyl floor covering, add					6.70			6.70	7.35
0900 For high pressure laminate covering, add					5.45			5.45	6
0910 For snap on stringer system, add	2 Carp	1000	.016		1.44	.64		2.08	2.64
0950 Office applications, steel or concrete panels,									
0960 no covering, over 6,000 S.F.	2 Carp	960	.017	S.F.	10.10	.67		10.77	12.20
1050 Pedestals, 6" to 12"	"	85	.188	Ea.	7.95	7.50		15.45	21

09 72 Wall Coverings

09 72 23 – Wallpapering

09 72 23.10 Wallpaper

	Crew	Daily Output	Labor-Hours	Unit	Material	2009 Bare Costs Labor	2009 Bare Costs Equipment	Total	Total Incl O&P
0010 **WALLPAPER** including sizing; add 10-30 percent waste @ takeoff									
0050 Aluminum foil	1 Pape	275	.029	S.F.	.97	1.03		2	2.72
0100 Copper sheets, .025" thick, vinyl backing		240	.033		5.20	1.18		6.38	7.60
0300 Phenolic backing		240	.033		6.75	1.18		7.93	9.30
0600 Cork tiles, light or dark, 12" x 12" x 3/16"		240	.033		4.25	1.18		5.43	6.55
0700 5/16" thick		235	.034		3.62	1.21		4.83	5.90
0900 1/4" basketweave		240	.033		5.55	1.18		6.73	8.05
1000 1/2" natural, non-directional pattern		240	.033		6.70	1.18		7.88	9.25
1100 3/4" natural, non-directional pattern		240	.033		10.90	1.18		12.08	13.90
1200 Granular surface, 12" x 36", 1/2" thick		385	.021		1.20	.74		1.94	2.50
1300 1" thick		370	.022		1.55	.77		2.32	2.93
1500 Polyurethane coated, 12" x 12" x 3/16" thick		240	.033		3.76	1.18		4.94	6.05
1600 5/16" thick		235	.034		5.35	1.21		6.56	7.85
1800 Cork wallpaper, paperbacked, natural		480	.017		1.90	.59		2.49	3.03
1900 Colors		480	.017		2.65	.59		3.24	3.86
2100 Flexible wood veneer, 1/32" thick, plain woods		100	.080		2.25	2.84		5.09	7
2200 Exotic woods		95	.084		3.41	2.99		6.40	8.50
2400 Gypsum-based, fabric-backed, fire resistant									
2500 for masonry walls, minimum, 21 oz./S.Y.	1 Pape	800	.010	S.F.	.80	.35		1.15	1.45

09 72 Wall Coverings

09 72 23 – Wallpapering

09 72 23.10 Wallpaper

		Crew	Daily Output	Labor-Hours	Unit	Material	2009 Bare Costs Labor	Equipment	Total	Total Incl O&P
2600	Average	1 Pape	720	.011	S.F.	1.17	.39		1.56	1.92
2700	Maximum (small quantities)	↓	640	.013		1.30	.44		1.74	2.14
2750	Acrylic, modified, semi-rigid PVC, .028" thick	2 Carp	330	.048		1.08	1.94		3.02	4.39
2800	.040" thick	"	320	.050		1.42	2		3.42	4.86
3000	Vinyl wall covering, fabric-backed, lightweight (12-15 oz./S.Y.)	1 Pape	640	.013		.65	.44		1.09	1.43
3300	Medium weight, type 2 (20-24 oz./S.Y.)	↓	480	.017		.77	.59		1.36	1.79
3400	Heavy weight, type 3 (28 oz./S.Y.)	↓	435	.018	↓	1.59	.65		2.24	2.79
3600	Adhesive, 5 gal. lots (18SY/Gal.)				Gal.	10.15			10.15	11.15
3700	Wallpaper, average workmanship, solid pattern, low cost paper	1 Pape	640	.013	S.F.	.35	.44		.79	1.10
3900	basic patterns (matching required), avg. cost paper		535	.015		.63	.53		1.16	1.54
4000	Paper at $85 per double roll, quality workmanship		435	.018		2.35	.65		3	3.63
4200	Grass cloths with lining paper, minimum		400	.020		.74	.71		1.45	1.94
4300	Maximum		350	.023		2.37	.81		3.18	3.91
5990	Wallpaper removal, 1 layer, minimum		800	.010		.06	.35		.41	.64
6000	Wallpaper removal, 3 layer, maximum		400	.020	↓	.13	.71		.84	1.27
9000	Minimum labor/equipment charge	↓	2	4	Job		142		142	227

09 77 Special Wall Surfacing

09 77 33 – Fiberglass Reinforced Panels

09 77 33.10 Fiberglass Reinforced Plastic Panels

		Crew	Daily Output	Labor-Hours	Unit	Material	2009 Bare Costs Labor	Equipment	Total	Total Incl O&P
0010	**FIBERGLASS REINFORCED PLASTIC PANELS**, .090" thick									
0020	On walls, adhesive mounted, embossed surface	2 Carp	640	.025	S.F.	1.23	1		2.23	3
0030	Smooth surface		640	.025		1.51	1		2.51	3.31
0040	Fire rated, embossed surface		640	.025		2.19	1		3.19	4.06
0050	Nylon rivet mounted, on drywall, embossed surface		480	.033		1.15	1.33		2.48	3.47
0060	Smooth surface		480	.033		1.43	1.33		2.76	3.77
0070	Fire rated, embossed surface		480	.033		2.11	1.33		3.44	4.52
0080	On masonry, embossed surface		320	.050		1.15	2		3.15	4.57
0090	Smooth surface		320	.050		1.43	2		3.43	4.87
0100	Fire rated, embossed surface		320	.050		2.11	2		4.11	5.60
0110	Nylon rivet and adhesive mounted, on drywall, embossed surface		240	.067		1.30	2.66		3.96	5.85
0120	Smooth surface		240	.067		1.58	2.66		4.24	6.15
0130	Fire rated, embossed surface		240	.067		2.26	2.66		4.92	6.90
0140	On masonry, embossed surface		190	.084		1.30	3.36		4.66	7
0150	Smooth surface		190	.084		1.58	3.36		4.94	7.30
0160	Fire rated, embossed surface	↓	190	.084	↓	2.26	3.36		5.62	8.05
0170	For moldings add	1 Carp	250	.032	L.F.	.28	1.28		1.56	2.42
0180	On ceilings, for lay in grid system, embossed surface		400	.020	S.F.	1.23	.80		2.03	2.67
0190	Smooth surface		400	.020		1.51	.80		2.31	2.98
0200	Fire rated, embossed surface	↓	400	.020	↓	2.19	.80		2.99	3.73

09 77 43 – Panel Systems

09 77 43.20 Slatwall Panels and Accessories

		Crew	Daily Output	Labor-Hours	Unit	Material	2009 Bare Costs Labor	Equipment	Total	Total Incl O&P
0010	**SLATWALL PANELS AND ACCESSORIES**									
0100	Slatwall panel, 4' x 8' x 3/4" T, MDF, paint grade	1 Carp	500	.016	S.F.	1.60	.64		2.24	2.82
0110	Melamine finish		500	.016		2.28	.64		2.92	3.57
0120	High pressure plastic laminate finish	↓	500	.016		3.70	.64		4.34	5.15
0130	Aluminum channel inserts, add				↓	3.57			3.57	3.93
0200	Accessories, corner forms, 8' L				L.F.	4.89			4.89	5.40
0210	T-connector, 8' L					5.95			5.95	6.55
0220	J-mold, 8' L				↓	1.59			1.59	1.75

09 77 43 – Panel Systems

09 77 43.20 Slatwall Panels and Accessories	Crew	Daily Output	Labor-Hours	Unit	Material	2009 Bare Costs Labor	Equipment	Total	Total Incl O&P
0230	Edge cap, 8' L				L.F.	1.16			1.16	1.28
0240	Finish end cap, 8' L				↓	3.66			3.66	4.03
0300	Display hook, metal, 4" L				Ea.	1.09			1.09	1.20
0310	6" L					1.19			1.19	1.31
0320	8" L					1.27			1.27	1.40
0330	10" L					1.43			1.43	1.57
0340	12" L					1.58			1.58	1.74
0350	Acrylic, 4" L					.91			.91	1
0360	6" L					1.05			1.05	1.16
0370	8" L					1.10			1.10	1.21
0380	10" L					1.21			1.21	1.33
0400	Waterfall hanger, metal, 12" - 16"					6.35			6.35	6.95
0410	Acrylic					10.60			10.60	11.65
0500	Shelf bracket, metal, 8"					6.70			6.70	7.35
0510	10"					6.90			6.90	7.60
0520	12"					7.15			7.15	7.90
0530	14"					7.60			7.60	8.35
0540	16"					9.05			9.05	10
0550	Acrylic, 8"					3.41			3.41	3.75
0560	10"					3.82			3.82	4.20
0570	12"					4.23			4.23	4.65
0580	14"					4.79			4.79	5.25
0600	Shelf, acrylic, 12" x 16" x 1/4"					27			27	29.50
0610	12" x 24" x 1/4"				↓	45.50			45.50	50

09 81 16 – Acoustic Blanket Insulation

09 81 16.10 Sound Attenuation Blanket	Crew	Daily Output	Labor-Hours	Unit	Material	2009 Bare Costs Labor	Equipment	Total	Total Incl O&P
0010	**SOUND ATTENUATION BLANKET**									
0020	Blanket, 1" thick	1 Carp	925	.009	S.F.	.25	.35		.60	.85
0500	1-1/2" thick		920	.009		.25	.35		.60	.85
1000	2" thick		915	.009		.34	.35		.69	.95
1500	3" thick	↓	910	.009		.50	.35		.85	1.13
3400	Urethane plastic foam, open cell, on wall, 2" thick	2 Carp	2050	.008		3.05	.31		3.36	3.87
3500	3" thick		1550	.010		4.05	.41		4.46	5.15
3600	4" thick		1050	.015		5.70	.61		6.31	7.25
3700	On ceiling, 2" thick		1700	.009		3.04	.38		3.42	3.96
3800	3" thick		1300	.012		4.05	.49		4.54	5.25
3900	4" thick	↓	900	.018	↓	5.70	.71		6.41	7.40
4000	Nylon matting 0.4" thick, with carbon black spinerette									
4010	plus polyester fabric, on floor	D-7	4000	.004	S.F.	2.26	.14		2.40	2.70
4200	Fiberglass reinf. backer board underlayment, 7/16" thick, on floor	"	800	.020	"	1.90	.68		2.58	3.16
9000	Minimum labor/equipment charge	1 Carp	5	1.600	Job		64		64	106

09 84 Acoustic Room Components

09 84 13 – Fixed Sound-Absorptive Panels

09 84 13.10 Fixed Panels

		Crew	Daily Output	Labor-Hours	Unit	Material	2009 Bare Costs Labor	Equipment	Total	Total Incl O&P
0010	**FIXED PANELS** Perforated steel facing, painted with									
0100	Fiberglass or mineral filler, no backs, 2-1/4" thick, modular									
0200	space units, ceiling or wall hung, white or colored	1 Carp	100	.080	S.F.	10	3.20		13.20	16.30
0300	Fiberboard sound deadening panels, 1/2" thick	"	600	.013	"	.32	.53		.85	1.23
0500	Fiberglass panels, 4' x 8' x 1" thick, with									
0600	glass cloth face for walls, cemented	1 Carp	155	.052	S.F.	7	2.06		9.06	11.10
0700	1-1/2" thick, dacron covered, inner aluminum frame,									
0710	wall mounted	1 Carp	300	.027	S.F.	8.75	1.07		9.82	11.35
0900	Mineral fiberboard panels, fabric covered, 30"x 108",									
1000	3/4" thick, concealed spline, wall mounted	1 Carp	150	.053	S.F.	6.25	2.13		8.38	10.40
9000	Minimum labor/equipment charge	"	4	2	Job		80		80	132

09 91 Painting

09 91 03 – Paint Restoration

09 91 03.20 Sanding

		Crew	Daily Output	Labor-Hours	Unit	Material	2009 Bare Costs Labor	Equipment	Total	Total Incl O&P
0010	**SANDING** and puttying interior trim, compared to									
0100	Painting 1 coat, on quality work				L.F.		100%			
0300	Medium work						50%			
0400	Industrial grade				↓		25%			
0500	Surface protection, placement and removal									
0510	Basic drop cloths	1 Pord	6400	.001	S.F.		.04		.04	.07
0520	Masking with paper		800	.010		.06	.35		.41	.63
0530	Volume cover up (using plastic sheathing, or building paper)	↓	16000	.001	↓		.02		.02	.03

09 91 03.30 Exterior Surface Preparation

		Crew	Daily Output	Labor-Hours	Unit	Material	2009 Bare Costs Labor	Equipment	Total	Total Incl O&P
0010	**EXTERIOR SURFACE PREPARATION**									
0015	Doors, per side, not incl. frames or trim									
0020	Scrape & sand									
0030	Wood, flush	1 Pord	616	.013	S.F.		.46		.46	.73
0040	Wood, detail		496	.016			.57		.57	.91
0050	Wood, louvered		280	.029			1.01		1.01	1.61
0060	Wood, overhead	↓	616	.013	↓		.46		.46	.73
0070	Wire brush									
0080	Metal, flush	1 Pord	640	.013	S.F.		.44		.44	.70
0090	Metal, detail		520	.015			.54		.54	.87
0100	Metal, louvered		360	.022			.78		.78	1.25
0110	Metal or fibr., overhead		640	.013			.44		.44	.70
0120	Metal, roll up		560	.014			.50		.50	.80
0130	Metal, bulkhead	↓	640	.013	↓		.44		.44	.70
0140	Power wash, based on 2500 lb. operating pressure									
0150	Metal, flush	A-1H	2240	.004	S.F.		.11	.02	.13	.22
0160	Metal, detail		2120	.004			.12	.02	.14	.23
0170	Metal, louvered		2000	.004			.13	.03	.16	.24
0180	Metal or fibr., overhead		2400	.003			.11	.02	.13	.19
0190	Metal, roll up		2400	.003			.11	.02	.13	.19
0200	Metal, bulkhead	↓	2200	.004	↓		.12	.02	.14	.22
0400	Windows, per side, not incl. trim									
0410	Scrape & sand									
0420	Wood, 1-2 lite	1 Pord	320	.025	S.F.		.88		.88	1.41
0430	Wood, 3-6 lite		280	.029			1.01		1.01	1.61
0440	Wood, 7-10 lite		240	.033			1.17		1.17	1.87
0450	Wood, 12 lite	↓	200	.040	↓		1.41		1.41	2.25

251

09 91 Painting

09 91 03 – Paint Restoration

09 91 03.30 Exterior Surface Preparation

		Crew	Daily Output	Labor-Hours	Unit	Material	2009 Bare Costs Labor	Equipment	Total	Total Incl O&P
0460	Wood, Bay / Bow	1 Pord	320	.025	S.F.		.88		.88	1.41
0470	Wire brush									
0480	Metal, 1-2 lite	1 Pord	480	.017	S.F.		.59		.59	.94
0490	Metal, 3-6 lite		400	.020			.70		.70	1.13
0500	Metal, Bay / Bow	↓	480	.017	↓		.59		.59	.94
0510	Power wash, based on 2500 lb. operating pressure									
0520	1-2 lite	A-1H	4400	.002	S.F.		.06	.01	.07	.10
0530	3-6 lite		4320	.002			.06	.01	.07	.11
0540	7-10 lite		4240	.002			.06	.01	.07	.11
0550	12 lite		4160	.002			.06	.01	.07	.11
0560	Bay / Bow	↓	4400	.002	↓		.06	.01	.07	.10
0600	Siding, scrape and sand, light=10-30%, med.=30-70%									
0610	Heavy=70-100% of surface to sand									
0650	Texture 1-11, light	1 Pord	480	.017	S.F.		.59		.59	.94
0660	Med.		440	.018			.64		.64	1.02
0670	Heavy		360	.022			.78		.78	1.25
0680	Wood shingles, shakes, light		440	.018			.64		.64	1.02
0690	Med.		360	.022			.78		.78	1.25
0700	Heavy		280	.029			1.01		1.01	1.61
0710	Clapboard, light		520	.015			.54		.54	.87
0720	Med.		480	.017			.59		.59	.94
0730	Heavy	↓	400	.020	↓		.70		.70	1.13
0740	Wire brush									
0750	Aluminum, light	1 Pord	600	.013	S.F.		.47		.47	.75
0760	Med.		520	.015			.54		.54	.87
0770	Heavy	↓	440	.018	↓		.64		.64	1.02
0780	Pressure wash, based on 2500 lb. operating pressure									
0790	Stucco	A-1H	3080	.003	S.F.		.08	.02	.10	.16
0800	Aluminum or vinyl		3200	.003			.08	.02	.10	.15
0810	Siding, masonry, brick & block	↓	2400	.003	↓		.11	.02	.13	.19
1300	Miscellaneous, wire brush									
1310	Metal, pedestrian gate	1 Pord	100	.080	S.F.		2.82		2.82	4.50
8000	For chemical washing, see Div. 04 01 30									
8010	For steam cleaning, see Div. 04 01 30.20									

09 91 03.40 Interior Surface Preparation

		Crew	Daily Output	Labor-Hours	Unit	Material	2009 Bare Costs Labor	Equipment	Total	Total Incl O&P
0010	**INTERIOR SURFACE PREPARATION**									
0020	Doors, per side, not incl. frames or trim									
0030	Scrape & sand									
0040	Wood, flush	1 Pord	616	.013	S.F.		.46		.46	.73
0050	Wood, detail		496	.016			.57		.57	.91
0060	Wood, louvered	↓	280	.029	↓		1.01		1.01	1.61
0070	Wire brush									
0080	Metal, flush	1 Pord	640	.013	S.F.		.44		.44	.70
0090	Metal, detail		520	.015			.54		.54	.87
0100	Metal, louvered	↓	360	.022	↓		.78		.78	1.25
0110	Hand wash									
0120	Wood, flush	1 Pord	2160	.004	S.F.		.13		.13	.21
0130	Wood, detailed		2000	.004			.14		.14	.23
0140	Wood, louvered		1360	.006			.21		.21	.33
0150	Metal, flush		2160	.004			.13		.13	.21
0160	Metal, detail		2000	.004			.14		.14	.23
0170	Metal, louvered	↓	1360	.006	↓		.21		.21	.33

09 91 Painting

09 91 03 – Paint Restoration

09 91 03.40 Interior Surface Preparation

		Crew	Daily Output	Labor-Hours	Unit	Material	2009 Bare Costs Labor	Equipment	Total	Total Incl O&P
0400	Windows, per side, not incl. trim									
0410	Scrape & sand									
0420	Wood, 1-2 lite	1 Pord	360	.022	S.F.		.78		.78	1.25
0430	Wood, 3-6 lite		320	.025			.88		.88	1.41
0440	Wood, 7-10 lite		280	.029			1.01		1.01	1.61
0450	Wood, 12 lite		240	.033			1.17		1.17	1.87
0460	Wood, Bay / Bow		360	.022			.78		.78	1.25
0470	Wire brush									
0480	Metal, 1-2 lite	1 Pord	520	.015	S.F.		.54		.54	.87
0490	Metal, 3-6 lite		440	.018			.64		.64	1.02
0500	Metal, Bay / Bow		520	.015			.54		.54	.87
0600	Walls, sanding, light=10-30%, medium - 30-70%,									
0610	heavy=70-100% of surface to sand									
0650	Walls, sand									
0660	Drywall, gypsum, plaster, light	1 Pord	3077	.003	S.F.		.09		.09	.15
0670	Drywall, gypsum, plaster, med.		2160	.004			.13		.13	.21
0680	Drywall, gypsum, plaster, heavy		923	.009			.31		.31	.49
0690	Wood, T&G, light		2400	.003			.12		.12	.19
0700	Wood, T&G, med.		1600	.005			.18		.18	.28
0710	Wood, T&G, heavy		800	.010			.35		.35	.56
0720	Walls, wash									
0730	Drywall, gypsum, plaster	1 Pord	3200	.003	S.F.		.09		.09	.14
0740	Wood, T&G		3200	.003			.09		.09	.14
0750	Masonry, brick & block, smooth		2800	.003			.10		.10	.16
0760	Masonry, brick & block, coarse		2000	.004			.14		.14	.23
8000	For chemical washing, see Div. 04 01 30									
8010	For steam cleaning, see Div. 04 01 30.20									

09 91 03.41 Scrape After Fire Damage

		Crew	Daily Output	Labor-Hours	Unit	Material	2009 Bare Costs Labor	Equipment	Total	Total Incl O&P
0010	**SCRAPE AFTER FIRE DAMAGE**									
0050	Boards, 1" x 4"	1 Pord	336	.024	L.F.		.84		.84	1.34
0060	1" x 6"		260	.031			1.08		1.08	1.73
0070	1" x 8"		207	.039			1.36		1.36	2.17
0080	1" x 10"		174	.046			1.62		1.62	2.59
0500	Framing, 2" x 4"		265	.030			1.06		1.06	1.70
0510	2" x 6"		221	.036			1.27		1.27	2.04
0520	2" x 8"		190	.042			1.48		1.48	2.37
0530	2" x 10"		165	.048			1.71		1.71	2.73
0540	2" x 12"		144	.056			1.96		1.96	3.13
1000	Heavy framing, 3" x 4"		226	.035			1.25		1.25	1.99
1010	4" x 4"		210	.038			1.34		1.34	2.14
1020	4" x 6"		191	.042			1.47		1.47	2.36
1030	4" x 8"		165	.048			1.71		1.71	2.73
1040	4" x 10"		144	.056			1.96		1.96	3.13
1060	4" x 12"		131	.061			2.15		2.15	3.44
2900	For sealing, minimum		825	.010	S.F.	.14	.34		.48	.70
2920	Maximum		460	.017	"	.28	.61		.89	1.29
3000	For sandblasting, see Div. 04 01 30.20									
3020										
9000	Minimum labor/equipment charge	1 Pord	3	2.667	Job		94		94	150

09 91 13 – Exterior Painting

09 91 13.30 Fences

0010	**FENCES**									

09 91 13.30 Fences

		Crew	Daily Output	Labor-Hours	Unit	Material	2009 Bare Costs Labor	Equipment	Total	Total Incl O&P
0100	Chain link or wire metal, one side, water base									
0110	Roll & brush, first coat	1 Pord	960	.008	S.F.	.05	.29		.34	.53
0120	Second coat		1280	.006		.05	.22		.27	.41
0130	Spray, first coat		2275	.004		.05	.12		.17	.26
0140	Second coat	↓	2600	.003	↓	.05	.11		.16	.23
0150	Picket, water base									
0160	Roll & brush, first coat	1 Pord	865	.009	S.F.	.06	.33		.39	.58
0170	Second coat		1050	.008		.06	.27		.33	.49
0180	Spray, first coat		2275	.004		.06	.12		.18	.26
0190	Second coat	↓	2600	.003	↓	.06	.11		.17	.23
0200	Stockade, water base									
0210	Roll & brush, first coat	1 Pord	1040	.008	S.F.	.06	.27		.33	.49
0220	Second coat		1200	.007		.06	.23		.29	.44
0230	Spray, first coat		2275	.004		.06	.12		.18	.26
0240	Second coat	↓	2600	.003	↓	.06	.11		.17	.23

09 91 13.42 Miscellaneous, Exterior

		Crew	Daily Output	Labor-Hours	Unit	Material	2009 Bare Costs Labor	Equipment	Total	Total Incl O&P
0010	**MISCELLANEOUS, EXTERIOR**									
0100	Railing, ext., decorative wood, incl. cap & baluster									
0110	Newels & spindles @ 12" O.C.									
0120	Brushwork, stain, sand, seal & varnish									
0130	First coat	1 Pord	90	.089	L.F.	.61	3.13		3.74	5.65
0140	Second coat	"	120	.067	"	.61	2.35		2.96	4.42
0150	Rough sawn wood, 42" high, 2" x 2" verticals, 6" O.C.									
0160	Brushwork, stain, each coat	1 Pord	90	.089	L.F.	.18	3.13		3.31	5.20
0170	Wrought iron, 1" rail, 1/2" sq. verticals									
0180	Brushwork, zinc chromate, 60" high, bars 6" O.C.									
0190	Primer	1 Pord	130	.062	L.F.	.60	2.17		2.77	4.12
0200	Finish coat		130	.062		.19	2.17		2.36	3.67
0210	Additional coat	↓	190	.042	↓	.22	1.48		1.70	2.61
0220	Shutters or blinds, single panel, 2' x 4', paint all sides									
0230	Brushwork, primer	1 Pord	20	.400	Ea.	.59	14.10		14.69	23
0240	Finish coat, exterior latex		20	.400		.44	14.10		14.54	23
0250	Primer & 1 coat, exterior latex		13	.615		.89	21.50		22.39	35.50
0260	Spray, primer		35	.229		.85	8.05		8.90	13.80
0270	Finish coat, exterior latex		35	.229		.93	8.05		8.98	13.85
0280	Primer & 1 coat, exterior latex	↓	20	.400	↓	.91	14.10		15.01	23.50
0290	For louvered shutters, add				S.F.	10%				
0300	Stair stringers, exterior, metal									
0310	Roll & brush, zinc chromate, to 14", each coat	1 Pord	320	.025	L.F.	.06	.88		.94	1.48
0320	Rough sawn wood, 4" x 12"									
0330	Roll & brush, exterior latex, each coat	1 Pord	215	.037	L.F.	.06	1.31		1.37	2.16
0340	Trellis/lattice, 2" x 2" @ 3" O.C. with 2" x 8" supports									
0350	Spray, latex, per side, each coat	1 Pord	475	.017	S.F.	.06	.59		.65	1.02
0450	Decking, ext., sealer, alkyd, brushwork, sealer coat		1140	.007		.09	.25		.34	.49
0460	1st coat		1140	.007		.07	.25		.32	.47
0470	2nd coat		1300	.006		.05	.22		.27	.41
0500	Paint, alkyd, brushwork, primer coat		1140	.007		.07	.25		.32	.47
0510	1st coat		1140	.007		.07	.25		.32	.47
0520	2nd coat		1300	.006		.05	.22		.27	.41
0600	Sand paint, alkyd, brushwork, 1 coat	↓	150	.053	↓	.11	1.88		1.99	3.12

09 91 13.60 Siding Exterior

		Crew	Daily Output	Labor-Hours	Unit	Material	2009 Bare Costs Labor	Equipment	Total	Total Incl O&P
0010	**SIDING EXTERIOR**, Alkyd (oil base)									

09 91 Painting

09 91 13 – Exterior Painting

09 91 13.60 Siding Exterior

		Crew	Daily Output	Labor-Hours	Unit	Material	2009 Bare Costs Labor	Equipment	Total	Total Incl O&P
0450	Steel siding, oil base, paint 1 coat, brushwork	2 Pord	2015	.008	S.F.	.06	.28		.34	.51
0500	Spray		4550	.004		.09	.12		.21	.30
0800	Paint 2 coats, brushwork		1300	.012		.11	.43		.54	.82
1000	Spray		2750	.006		.17	.20		.37	.52
1200	Stucco, rough, oil base, paint 2 coats, brushwork		1300	.012		.11	.43		.54	.82
1400	Roller		1625	.010		.12	.35		.47	.68
1600	Spray		2925	.005		.13	.19		.32	.45
1800	Texture 1-11 or clapboard, oil base, primer coat, brushwork		1300	.012		.09	.43		.52	.79
2000	Spray		4550	.004		.09	.12		.21	.30
2400	Paint 2 coats, brushwork		810	.020		.17	.70		.87	1.29
2600	Spray		2600	.006		.19	.22		.41	.55
3400	Stain 2 coats, brushwork		950	.017		.12	.59		.71	1.09
4000	Spray		3050	.005		.14	.18		.32	.45
4200	Wood shingles, oil base primer coat, brushwork		1300	.012		.08	.43		.51	.78
4400	Spray		3900	.004		.08	.14		.22	.32
5000	Paint 2 coats, brushwork		810	.020		.14	.70		.84	1.26
5200	Spray		2275	.007		.13	.25		.38	.55
6500	Stain 2 coats, brushwork		950	.017		.12	.59		.71	1.09
7000	Spray		2660	.006		.17	.21		.38	.53
8000	For latex paint, deduct					10%				
8100	For work over 12' H, from pipe scaffolding, add						15%			
8200	For work over 12' H, from extension ladder, add						25%			
8300	For work over 12' H, from swing staging, add						35%			

09 91 13.62 Siding, Misc.

		Crew	Daily Output	Labor-Hours	Unit	Material	2009 Bare Costs Labor	Equipment	Total	Total Incl O&P
0010	**SIDING, MISC.**, latex paint									
0100	Aluminum siding									
0110	Brushwork, primer	2 Pord	2275	.007	S.F.	.05	.25		.30	.46
0120	Finish coat, exterior latex		2275	.007		.04	.25		.29	.45
0130	Primer & 1 coat exterior latex		1300	.012		.10	.43		.53	.80
0140	Primer & 2 coats exterior latex		975	.016		.15	.58		.73	1.08
0150	Mineral fiber shingles									
0160	Brushwork, primer	2 Pord	1495	.011	S.F.	.09	.38		.47	.70
0170	Finish coat, industrial enamel		1495	.011		.13	.38		.51	.74
0180	Primer & 1 coat enamel		810	.020		.22	.70		.92	1.35
0190	Primer & 2 coats enamel		540	.030		.35	1.04		1.39	2.06
0200	Roll, primer		1625	.010		.10	.35		.45	.66
0210	Finish coat, industrial enamel		1625	.010		.14	.35		.49	.71
0220	Primer & 1 coat enamel		975	.016		.24	.58		.82	1.19
0230	Primer & 2 coats enamel		650	.025		.38	.87		1.25	1.80
0240	Spray, primer		3900	.004		.08	.14		.22	.32
0250	Finish coat, industrial enamel		3900	.004		.12	.14		.26	.36
0260	Primer & 1 coat enamel		2275	.007		.20	.25		.45	.61
0270	Primer & 2 coats enamel		1625	.010		.31	.35		.66	.89
0280	Waterproof sealer, first coat		4485	.004		.08	.13		.21	.29
0290	Second coat		5235	.003		.07	.11		.18	.25
0300	Rough wood incl. shingles, shakes or rough sawn siding									
0310	Brushwork, primer	2 Pord	1280	.013	S.F.	.12	.44		.56	.83
0320	Finish coat, exterior latex		1280	.013		.07	.44		.51	.78
0330	Primer & 1 coat exterior latex		960	.017		.19	.59		.78	1.15
0340	Primer & 2 coats exterior latex		700	.023		.27	.80		1.07	1.58
0350	Roll, primer		2925	.005		.16	.19		.35	.48
0360	Finish coat, exterior latex		2925	.005		.09	.19		.28	.41

09 91 Painting

09 91 13 – Exterior Painting

09 91 13.62 Siding, Misc.		Crew	Daily Output	Labor-Hours	Unit	Material	2009 Bare Costs Labor	Equipment	Total	Total Incl O&P
0370	Primer & 1 coat exterior latex	2 Pord	1790	.009	S.F.	.25	.31		.56	.77
0380	Primer & 2 coats exterior latex		1300	.012		.34	.43		.77	1.06
0390	Spray, primer		3900	.004		.13	.14		.27	.38
0400	Finish coat, exterior latex		3900	.004		.07	.14		.21	.31
0410	Primer & 1 coat exterior latex		2600	.006		.20	.22		.42	.57
0420	Primer & 2 coats exterior latex		2080	.008		.27	.27		.54	.73
0430	Waterproof sealer, first coat		4485	.004		.14	.13		.27	.35
0440	Second coat		4485	.004		.08	.13		.21	.29
0450	Smooth wood incl. butt, T&G, beveled, drop or B&B siding									
0460	Brushwork, primer	2 Pord	2325	.007	S.F.	.09	.24		.33	.48
0470	Finish coat, exterior latex		1280	.013		.07	.44		.51	.78
0480	Primer & 1 coat exterior latex		800	.020		.16	.70		.86	1.31
0490	Primer & 2 coats exterior latex		630	.025		.23	.89		1.12	1.69
0500	Roll, primer		2275	.007		.09	.25		.34	.50
0510	Finish coat, exterior latex		2275	.007		.08	.25		.33	.49
0520	Primer & 1 coat exterior latex		1300	.012		.18	.43		.61	.88
0530	Primer & 2 coats exterior latex		975	.016		.26	.58		.84	1.20
0540	Spray, primer		4550	.004		.07	.12		.19	.28
0550	Finish coat, exterior latex		4550	.004		.07	.12		.19	.28
0560	Primer & 1 coat exterior latex		2600	.006		.14	.22		.36	.51
0570	Primer & 2 coats exterior latex		1950	.008		.21	.29		.50	.69
0580	Waterproof sealer, first coat		5230	.003		.08	.11		.19	.26
0590	Second coat		5980	.003		.08	.09		.17	.24
0600	For oil base paint, add					10%				

09 91 13.70 Doors and Windows, Exterior

09 91 13.70 Doors and Windows, Exterior		Crew	Daily Output	Labor-Hours	Unit	Material	2009 Bare Costs Labor	Equipment	Total	Total Incl O&P
0010	**DOORS AND WINDOWS, EXTERIOR**									
0100	Door frames & trim, only									
0110	Brushwork, primer	1 Pord	512	.016	L.F.	.05	.55		.60	.94
0120	Finish coat, exterior latex		512	.016		.05	.55		.60	.94
0130	Primer & 1 coat, exterior latex		300	.027		.11	.94		1.05	1.62
0135	2 coats, exterior latex, both sides		15	.533	Ea.	4.87	18.75		23.62	35.50
0140	Primer & 2 coats, exterior latex		265	.030	L.F.	.16	1.06		1.22	1.88
0150	Doors, flush, both sides, incl. frame & trim									
0160	Roll & brush, primer	1 Pord	10	.800	Ea.	4.05	28		32.05	49.50
0170	Finish coat, exterior latex		10	.800		4.15	28		32.15	49.50
0180	Primer & 1 coat, exterior latex		7	1.143		8.20	40		48.20	73.50
0190	Primer & 2 coats, exterior latex		5	1.600		12.35	56.50		68.85	104
0200	Brushwork, stain, sealer & 2 coats polyurethane		4	2		21	70.50		91.50	136
0210	Doors, French, both sides, 10-15 lite, incl. frame & trim									
0220	Brushwork, primer	1 Pord	6	1.333	Ea.	2.03	47		49.03	77
0230	Finish coat, exterior latex		6	1.333		2.07	47		49.07	77.50
0240	Primer & 1 coat, exterior latex		3	2.667		4.10	94		98.10	155
0250	Primer & 2 coats, exterior latex		2	4		6.05	141		147.05	232
0260	Brushwork, stain, sealer & 2 coats polyurethane		2.50	3.200		7.70	113		120.70	189
0270	Doors, louvered, both sides, incl. frame & trim									
0280	Brushwork, primer	1 Pord	7	1.143	Ea.	4.05	40		44.05	69
0290	Finish coat, exterior latex		7	1.143		4.15	40		44.15	69
0300	Primer & 1 coat, exterior latex		4	2		8.20	70.50		78.70	122
0310	Primer & 2 coats, exterior latex		3	2.667		12.10	94		106.10	163
0320	Brushwork, stain, sealer & 2 coats polyurethane		4.50	1.778		21	62.50		83.50	123
0330	Doors, panel, both sides, incl. frame & trim									
0340	Roll & brush, primer	1 Pord	6	1.333	Ea.	4.05	47		51.05	79.50

09 91 Painting

09 91 13 – Exterior Painting

09 91 13.70 Doors and Windows, Exterior

		Crew	Daily Output	Labor-Hours	Unit	Material	2009 Bare Costs Labor	Equipment	Total	Total Incl O&P
0350	Finish coat, exterior latex	1 Pord	6	1.333	Ea.	4.15	47		51.15	79.50
0360	Primer & 1 coat, exterior latex		3	2.667		8.20	94		102.20	159
0370	Primer & 2 coats, exterior latex		2.50	3.200		12.10	113		125.10	193
0380	Brushwork, stain, sealer & 2 coats polyurethane		3	2.667		21	94		115	173
0400	Windows, per ext. side, based on 15 SF									
0410	1 to 6 lite									
0420	Brushwork, primer	1 Pord	13	.615	Ea.	.80	21.50		22.30	35.50
0430	Finish coat, exterior latex		13	.615		.82	21.50		22.32	35.50
0440	Primer & 1 coat, exterior latex		8	1		1.62	35		36.62	58.50
0450	Primer & 2 coats, exterior latex		6	1.333		2.39	47		49.39	77.50
0460	Stain, sealer & 1 coat varnish		7	1.143		3.04	40		43.04	68
0470	7 to 10 lite									
0480	Brushwork, primer	1 Pord	11	.727	Ea.	.80	25.50		26.30	42
0490	Finish coat, exterior latex		11	.727		.82	25.50		26.32	42
0500	Primer & 1 coat, exterior latex		7	1.143		1.62	40		41.62	66.50
0510	Primer & 2 coats, exterior latex		5	1.600		2.39	56.50		58.89	92.50
0520	Stain, sealer & 1 coat varnish		6	1.333		3.04	47		50.04	78.50
0530	12 lite									
0540	Brushwork, primer	1 Pord	10	.800	Ea.	.80	28		28.80	46
0550	Finish coat, exterior latex		10	.800		.82	28		28.82	46
0560	Primer & 1 coat, exterior latex		6	1.333		1.62	47		48.62	77
0570	Primer & 2 coats, exterior latex		5	1.600		2.39	56.50		58.89	92.50
0580	Stain, sealer & 1 coat varnish		6	1.333		3.06	47		50.06	78.50
0590	For oil base paint, add					10%				

09 91 13.80 Trim, Exterior

		Crew	Daily Output	Labor-Hours	Unit	Material	2009 Bare Costs Labor	Equipment	Total	Total Incl O&P
0010	**TRIM, EXTERIOR**									
0100	Door frames & trim (see Doors, interior or exterior)									
0110	Fascia, latex paint, one coat coverage									
0120	1" x 4", brushwork	1 Pord	640	.013	L.F.	.02	.44		.46	.72
0130	Roll		1280	.006		.02	.22		.24	.37
0140	Spray		2080	.004		.01	.14		.15	.24
0150	1" x 6" to 1" x 10", brushwork		640	.013		.06	.44		.50	.76
0160	Roll		1230	.007		.06	.23		.29	.44
0170	Spray		2100	.004		.05	.13		.18	.26
0180	1" x 12", brushwork		640	.013		.06	.44		.50	.76
0190	Roll		1050	.008		.06	.27		.33	.50
0200	Spray		2200	.004		.05	.13		.18	.25
0210	Gutters & downspouts, metal, zinc chromate paint									
0220	Brushwork, gutters, 5", first coat	1 Pord	640	.013	L.F.	.07	.44		.51	.77
0230	Second coat		960	.008		.06	.29		.35	.54
0240	Third coat		1280	.006		.05	.22		.27	.41
0250	Downspouts, 4", first coat		640	.013		.07	.44		.51	.77
0260	Second coat		960	.008		.06	.29		.35	.54
0270	Third coat		1280	.006		.05	.22		.27	.41
0280	Gutters & downspouts, wood									
0290	Brushwork, gutters, 5", primer	1 Pord	640	.013	L.F.	.05	.44		.49	.76
0300	Finish coat, exterior latex		640	.013		.05	.44		.49	.75
0310	Primer & 1 coat exterior latex		400	.020		.11	.70		.81	1.25
0320	Primer & 2 coats exterior latex		325	.025		.16	.87		1.03	1.56
0330	Downspouts, 4", primer		640	.013		.05	.44		.49	.76
0340	Finish coat, exterior latex		640	.013		.05	.44		.49	.75
0350	Primer & 1 coat exterior latex		400	.020		.11	.70		.81	1.25

09 91 Painting

09 91 13 – Exterior Painting

09 91 13.80 Trim, Exterior

		Crew	Daily Output	Labor-Hours	Unit	Material	2009 Bare Costs Labor	Equipment	Total	Total Incl O&P
0360	Primer & 2 coats exterior latex	1 Pord	325	.025	L.F.	.08	.87		.95	1.47
0370	Molding, exterior, up to 14" wide									
0380	Brushwork, primer	1 Pord	640	.013	L.F.	.06	.44		.50	.77
0390	Finish coat, exterior latex		640	.013		.06	.44		.50	.76
0400	Primer & 1 coat exterior latex		400	.020		.13	.70		.83	1.27
0410	Primer & 2 coats exterior latex		315	.025		.13	.89		1.02	1.57
0420	Stain & fill		1050	.008		.07	.27		.34	.51
0430	Shellac		1850	.004		.08	.15		.23	.33
0440	Varnish		1275	.006		.09	.22		.31	.45

09 91 13.90 Walls, Masonry (CMU), Exterior

		Crew	Daily Output	Labor-Hours	Unit	Material	2009 Bare Costs Labor	Equipment	Total	Total Incl O&P
0350	**WALLS, MASONRY (CMU), EXTERIOR**									
0360	Concrete masonry units (CMU), smooth surface									
0370	Brushwork, latex, first coat	1 Pord	640	.013	S.F.	.04	.44		.48	.75
0380	Second coat		960	.008		.04	.29		.33	.51
0390	Waterproof sealer, first coat		736	.011		.23	.38		.61	.87
0400	Second coat		1104	.007		.23	.26		.49	.67
0410	Roll, latex, paint, first coat		1465	.005		.05	.19		.24	.37
0420	Second coat		1790	.004		.04	.16		.20	.29
0430	Waterproof sealer, first coat		1680	.005		.23	.17		.40	.53
0440	Second coat		2060	.004		.23	.14		.37	.48
0450	Spray, latex, paint, first coat		1950	.004		.04	.14		.18	.28
0460	Second coat		2600	.003		.03	.11		.14	.21
0470	Waterproof sealer, first coat		2245	.004		.23	.13		.36	.46
0480	Second coat		2990	.003		.23	.09		.32	.41
0490	Concrete masonry unit (CMU), porous									
0500	Brushwork, latex, first coat	1 Pord	640	.013	S.F.	.09	.44		.53	.80
0510	Second coat		960	.008		.04	.29		.33	.52
0520	Waterproof sealer, first coat		736	.011		.23	.38		.61	.87
0530	Second coat		1104	.007		.23	.26		.49	.67
0540	Roll latex, first coat		1465	.005		.07	.19		.26	.38
0550	Second coat		1790	.004		.04	.16		.20	.30
0560	Waterproof sealer, first coat		1680	.005		.23	.17		.40	.53
0570	Second coat		2060	.004		.23	.14		.37	.48
0580	Spray latex, first coat		1950	.004		.05	.14		.19	.28
0590	Second coat		2600	.003		.03	.11		.14	.21
0600	Waterproof sealer, first coat		2245	.004		.23	.13		.36	.46
0610	Second coat		2990	.003		.23	.09		.32	.41

09 91 23 – Interior Painting

09 91 23.20 Cabinets and Casework

		Crew	Daily Output	Labor-Hours	Unit	Material	2009 Bare Costs Labor	Equipment	Total	Total Incl O&P
0010	**CABINETS AND CASEWORK**									
1000	Primer coat, oil base, brushwork	1 Pord	650	.012	S.F.	.05	.43		.48	.75
2000	Paint, oil base, brushwork, 1 coat		650	.012		.06	.43		.49	.76
3000	Stain, brushwork, wipe off		650	.012		.06	.43		.49	.76
4000	Shellac, 1 coat, brushwork		650	.012		.07	.43		.50	.77
4500	Varnish, 3 coats, brushwork, sand after 1st coat		325	.025		.23	.87		1.10	1.63
5000	For latex paint, deduct					10%				

09 91 23.33 Doors and Windows, Interior Alkyd (Oil Base)

		Crew	Daily Output	Labor-Hours	Unit	Material	2009 Bare Costs Labor	Equipment	Total	Total Incl O&P
0010	**DOORS AND WINDOWS, INTERIOR ALKYD (OIL BASE)**									
0500	Flush door & frame, 3' x 7', oil, primer, brushwork	1 Pord	10	.800	Ea.	2.34	28		30.34	47.50
1000	Paint, 1 coat		10	.800		2.25	28		30.25	47.50
1400	Stain, brushwork, wipe off		18	.444		1.29	15.65		16.94	26.50
1600	Shellac, 1 coat, brushwork		25	.320		1.48	11.25		12.73	19.60

09 91 23.33 Doors and Windows, Interior Alkyd (Oil Base)

		Crew	Daily Output	Labor-Hours	Unit	Material	2009 Bare Costs Labor	2009 Bare Costs Equipment	Total	Total Incl O&P
1800	Varnish, 3 coats, brushwork, sand after 1st coat	1 Pord	9	.889	Ea.	4.77	31.50		36.27	55.50
2000	Panel door & frame, 3' x 7', oil, primer, brushwork		6	1.333		1.96	47		48.96	77
2200	Paint, 1 coat		6	1.333		2.25	47		49.25	77.50
2600	Stain, brushwork, panel door, 3' x 7', not incl. frame		16	.500		1.29	17.60		18.89	29.50
2800	Shellac, 1 coat, brushwork		22	.364		1.48	12.80		14.28	22
3000	Varnish, 3 coats, brushwork, sand after 1st coat		7.50	1.067		4.77	37.50		42.27	65.50
4400	Windows, including frame and trim, per side									
4600	Colonial type, 6/6 lites, 2' x 3', oil, primer, brushwork	1 Pord	14	.571	Ea.	.31	20		20.31	32.50
5800	Paint, 1 coat		14	.571		.36	20		20.36	32.50
6200	3' x 5' opening, 6/6 lites, primer coat, brushwork		12	.667		.78	23.50		24.28	38.50
6400	Paint, 1 coat		12	.667		.89	23.50		24.39	38.50
6800	4' x 8' opening, 6/6 lites, primer coat, brushwork		8	1		1.65	35		36.65	58.50
7000	Paint, 1 coat		8	1		1.90	35		36.90	58.50
8000	Single lite type, 2' x 3', oil base, primer coat, brushwork		33	.242		.31	8.55		8.86	14
8200	Paint, 1 coat		33	.242		.36	8.55		8.91	14.05
8600	3' x 5' opening, primer coat, brushwork		20	.400		.78	14.10		14.88	23.50
8800	Paint, 1 coat		20	.400		.89	14.10		14.99	23.50
9200	4' x 8' opening, primer coat, brushwork		14	.571		1.65	20		21.65	34
9400	Paint, 1 coat		14	.571		1.90	20		21.90	34

09 91 23.35 Doors and Windows, Interior Latex

		Crew	Daily Output	Labor-Hours	Unit	Material	2009 Bare Costs Labor	2009 Bare Costs Equipment	Total	Total Incl O&P
0010	**DOORS & WINDOWS, INTERIOR LATEX**									
0100	Doors, flush, both sides, incl. frame & trim									
0110	Roll & brush, primer	1 Pord	10	.800	Ea.	3.64	28		31.64	49
0120	Finish coat, latex		10	.800		4.21	28		32.21	49.50
0130	Primer & 1 coat latex		7	1.143		7.85	40		47.85	73
0140	Primer & 2 coats latex		5	1.600		11.80	56.50		68.30	103
0160	Spray, both sides, primer		20	.400		3.83	14.10		17.93	26.50
0170	Finish coat, latex		20	.400		4.42	14.10		18.52	27.50
0180	Primer & 1 coat latex		11	.727		8.30	25.50		33.80	50
0190	Primer & 2 coats latex		8	1		12.50	35		47.50	70.50
0200	Doors, French, both sides, 10-15 lite, incl. frame & trim									
0210	Roll & brush, primer	1 Pord	6	1.333	Ea.	1.82	47		48.82	77
0220	Finish coat, latex		6	1.333		2.10	47		49.10	77.50
0230	Primer & 1 coat latex		3	2.667		3.92	94		97.92	154
0240	Primer & 2 coats latex		2	4		5.90	141		146.90	232
0260	Doors, louvered, both sides, incl. frame & trim									
0270	Roll & brush, primer	1 Pord	7	1.143	Ea.	3.64	40		43.64	68.50
0280	Finish coat, latex		7	1.143		4.21	40		44.21	69
0290	Primer & 1 coat, latex		4	2		7.65	70.50		78.15	121
0300	Primer & 2 coats, latex		3	2.667		12.05	94		106.05	163
0320	Spray, both sides, primer		20	.400		3.83	14.10		17.93	26.50
0330	Finish coat, latex		20	.400		4.42	14.10		18.52	27.50
0340	Primer & 1 coat, latex		11	.727		8.30	25.50		33.80	50
0350	Primer & 2 coats, latex		8	1		12.75	35		47.75	70.50
0360	Doors, panel, both sides, incl. frame & trim									
0370	Roll & brush, primer	1 Pord	6	1.333	Ea.	3.83	47		50.83	79
0380	Finish coat, latex		6	1.333		4.21	47		51.21	79.50
0390	Primer & 1 coat, latex		3	2.667		7.85	94		101.85	159
0400	Primer & 2 coats, latex		2.50	3.200		12.05	113		125.05	193
0420	Spray, both sides, primer		10	.800		3.83	28		31.83	49
0430	Finish coat, latex		10	.800		4.42	28		32.42	50
0440	Primer & 1 coat, latex		5	1.600		8.30	56.50		64.80	99

09 91 23.35 Doors and Windows, Interior Latex

		Crew	Daily Output	Labor-Hours	Unit	Material	2009 Bare Costs Labor	Equipment	Total	Total Incl O&P
0450	Primer & 2 coats, latex	1 Pord	4	2	Ea.	12.75	70.50		83.25	127
0460	Windows, per interior side, based on 15 SF									
0470	1 to 6 lite									
0480	Brushwork, primer	1 Pord	13	.615	Ea.	.72	21.50		22.22	35.50
0490	Finish coat, enamel		13	.615		.83	21.50		22.33	35.50
0500	Primer & 1 coat enamel		8	1		1.55	35		36.55	58
0510	Primer & 2 coats enamel		6	1.333		2.38	47		49.38	77.50
0530	7 to 10 lite									
0540	Brushwork, primer	1 Pord	11	.727	Ea.	.72	25.50		26.22	42
0550	Finish coat, enamel		11	.727		.83	25.50		26.33	42
0560	Primer & 1 coat enamel		7	1.143		1.55	40		41.55	66
0570	Primer & 2 coats enamel		5	1.600		2.38	56.50		58.88	92.50
0590	12 lite									
0600	Brushwork, primer	1 Pord	10	.800	Ea.	.72	28		28.72	46
0610	Finish coat, enamel		10	.800		.83	28		28.83	46
0620	Primer & 1 coat enamel		6	1.333		1.55	47		48.55	76.50
0630	Primer & 2 coats enamel		5	1.600		2.38	56.50		58.88	92.50
0650	For oil base paint, add					10%				

09 91 23.39 Doors and Windows, Interior Latex, Zero Voc

			Crew	Daily Output	Labor-Hours	Unit	Material	2009 Bare Costs Labor	Equipment	Total	Total Incl O&P
0010	**DOORS & WINDOWS, INTERIOR LATEX, ZERO VOC**										
0100	Doors flush, both sides, incl. frame & trim										
0110	Roll & brush, primer	G	1 Pord	10	.800	Ea.	4.75	28		32.75	50.50
0120	Finish coat, latex	G		10	.800		4.82	28		32.82	50.50
0130	Primer & 1 coat latex	G		7	1.143		9.55	40		49.55	75
0140	Primer & 2 coats latex	G		5	1.600		14.10	56.50		70.60	106
0160	Spray, both sides, primer	G		20	.400		5	14.10		19.10	28
0170	Finish coat, latex	G		20	.400		5.05	14.10		19.15	28
0180	Primer & 1 coat latex	G		11	.727		10.10	25.50		35.60	52
0190	Primer & 2 coats latex	G		8	1		14.95	35		49.95	73
0200	Doors, French, both sides, 10-15 lite, incl. frame & trim										
0210	Roll & brush, primer	G	1 Pord	6	1.333	Ea.	2.38	47		49.38	77.50
0220	Finish coat, latex	G		6	1.333		2.41	47		49.41	77.50
0230	Primer & 1 coat latex	G		3	2.667		4.79	94		98.79	155
0240	Primer & 2 coats latex	G		2	4		7.05	141		148.05	233
0360	Doors, panel, both sides, incl. frame & trim										
0370	Roll & brush, primer	G	1 Pord	6	1.333	Ea.	5	47		52	80.50
0380	Finish coat, latex	G		6	1.333		4.82	47		51.82	80.50
0390	Primer & 1 coat, latex	G		3	2.667		9.55	94		103.55	161
0400	Primer & 2 coats, latex	G		2.50	3.200		14.40	113		127.40	196
0420	Spray, both sides, primer	G		10	.800		5	28		33	50.50
0430	Finish coat, latex	G		10	.800		5.05	28		33.05	50.50
0440	Primer & 1 coat, latex	G		5	1.600		10.10	56.50		66.60	101
0450	Primer & 2 coats, latex	G		4	2		15.25	70.50		85.75	130
0460	Windows, per interior side, based on 15 SF										
0470	1 to 6 lite										
0480	Brushwork, primer	G	1 Pord	13	.615	Ea.	.94	21.50		22.44	35.50
0490	Finish coat, enamel	G		13	.615		.95	21.50		22.45	35.50
0500	Primer & 1 coat enamel	G		8	1		1.89	35		36.89	58.50
0510	Primer & 2 coats enamel	G		6	1.333		2.84	47		49.84	78

09 91 23.40 Floors, Interior

0010	**FLOORS, INTERIOR**									
0100	Concrete paint, latex									

09 91 Painting

09 91 23 – Interior Painting

09 91 23.40 Floors, Interior

		Crew	Daily Output	Labor-Hours	Unit	Material	2009 Bare Costs Labor	Equipment	Total	Total Incl O&P
0110	Brushwork									
0120	1st coat	1 Pord	975	.008	S.F.	.14	.29		.43	.62
0130	2nd coat		1150	.007		.10	.25		.35	.50
0140	3rd coat	↓	1300	.006	↓	.08	.22		.30	.43
0150	Roll									
0160	1st coat	1 Pord	2600	.003	S.F.	.19	.11		.30	.38
0170	2nd coat		3250	.002		.12	.09		.21	.27
0180	3rd coat	↓	3900	.002	↓	.09	.07		.16	.21
0190	Spray									
0200	1st coat	1 Pord	2600	.003	S.F.	.17	.11		.28	.35
0210	2nd coat		3250	.002		.09	.09		.18	.24
0220	3rd coat	↓	3900	.002	↓	.07	.07		.14	.20
0300	Acid stain and sealer									
0310	Stain, one coat	1 Pord	650	.012	S.F.	.11	.43		.54	.81
0320	Two coats		570	.014		.22	.49		.71	1.03
0330	Acrylic sealer, one coat		2600	.003		.15	.11		.26	.34
0340	Two coats	↓	1400	.006	↓	.30	.20		.50	.65

09 91 23.52 Miscellaneous, Interior

		Crew	Daily Output	Labor-Hours	Unit	Material	2009 Bare Costs Labor	Equipment	Total	Total Incl O&P
0010	**MISCELLANEOUS, INTERIOR**									
2400	Floors, conc./wood, oil base, primer/sealer coat, brushwork	2 Pord	1950	.008	S.F.	.06	.29		.35	.52
2450	Roller		5200	.003		.06	.11		.17	.23
2600	Spray		6000	.003		.06	.09		.15	.21
2650	Paint 1 coat, brushwork		1950	.008		.08	.29		.37	.55
2800	Roller		5200	.003		.09	.11		.20	.27
2850	Spray		6000	.003		.09	.09		.18	.25
3000	Stain, wood floor, brushwork, 1 coat		4550	.004		.06	.12		.18	.27
3200	Roller		5200	.003		.06	.11		.17	.24
3250	Spray		6000	.003		.06	.09		.15	.22
3400	Varnish, wood floor, brushwork		4550	.004		.08	.12		.20	.28
3450	Roller		5200	.003		.08	.11		.19	.26
3600	Spray	↓	6000	.003		.08	.09		.17	.24
3800	Grilles, per side, oil base, primer coat, brushwork	1 Pord	520	.015		.10	.54		.64	.98
3850	Spray		1140	.007		.11	.25		.36	.51
3920	Paint 2 coats, brushwork		325	.025		.23	.87		1.10	1.63
3940	Spray	↓	650	.012	↓	.26	.43		.69	.98
5000	Pipe, 1 – 4" diameter, primer or sealer coat, oil base, brushwork	2 Pord	1250	.013	L.F.	.06	.45		.51	.78
5100	Spray		2165	.007		.06	.26		.32	.48
5350	Paint 2 coats, brushwork		775	.021		.11	.73		.84	1.28
5400	Spray		1240	.013		.12	.45		.57	.87
6300	13" - 16" diameter, primer or sealer coat, brushwork		310	.052		.24	1.82		2.06	3.16
6450	Spray		540	.030		.26	1.04		1.30	1.95
6500	Paint 2 coats, brushwork		195	.082		.45	2.89		3.34	5.10
6550	Spray	↓	310	.052	↓	.50	1.82		2.32	3.45
7000	Trim, wood, incl. puttying, under 6" wide									
7200	Primer coat, oil base, brushwork	1 Pord	650	.012	L.F.	.03	.43		.46	.72
7250	Paint, 1 coat, brushwork		650	.012		.03	.43		.46	.72
7450	3 coats		325	.025		.09	.87		.96	1.47
7500	Over 6" wide, primer coat, brushwork		650	.012		.05	.43		.48	.75
7550	Paint, 1 coat, brushwork		650	.012		.06	.43		.49	.76
7650	3 coats		325	.025	↓	.17	.87		1.04	1.57
8000	Cornice, simple design, primer coat, oil base, brushwork		650	.012	S.F.	.05	.43		.48	.75
8250	Paint, 1 coat	↓	650	.012	↓	.06	.43		.49	.76

09 91 Painting

09 91 23 – Interior Painting

09 91 23.52 Miscellaneous, Interior		Crew	Daily Output	Labor-Hours	Unit	Material	2009 Bare Costs Labor	Equipment	Total	Total Incl O&P
8350	Ornate design, primer coat	1 Pord	350	.023	S.F.	.05	.80		.85	1.35
8400	Paint, 1 coat		350	.023		.06	.80		.86	1.36
8600	Balustrades, primer coat, oil base, brushwork	↓	520	.015	↓	.05	.54		.59	.93
8800										
8900	Trusses and wood frames, primer coat, oil base, brushwork	1 Pord	800	.010	S.F.	.05	.35		.40	.62
8950	Spray		1200	.007		.05	.23		.28	.44
9220	Paint 2 coats, brushwork		500	.016		.12	.56		.68	1.03
9240	Spray		600	.013		.13	.47		.60	.89
9260	Stain, brushwork, wipe off		600	.013		.06	.47		.53	.82
9280	Varnish, 3 coats, brushwork	↓	275	.029		.23	1.02		1.25	1.89
9350	For latex paint, deduct				↓	10%				

09 91 23.72 Walls and Ceilings, Interior

0010	**WALLS AND CEILINGS, INTERIOR**									
0100	Concrete, drywall or plaster, oil base, primer or sealer coat									
0200	Smooth finish, brushwork	1 Pord	1150	.007	S.F.	.05	.25		.30	.45
0240	Roller		1350	.006		.05	.21		.26	.38
0280	Spray		2750	.003		.04	.10		.14	.20
0300	Sand finish, brushwork		975	.008		.05	.29		.34	.51
0340	Roller		1150	.007		.05	.25		.30	.45
0380	Spray		2275	.004		.04	.12		.16	.25
0800	Paint 2 coats, smooth finish, brushwork		680	.012		.11	.41		.52	.79
0840	Roller		800	.010		.12	.35		.47	.69
0880	Spray		1625	.005		.10	.17		.27	.39
0900	Sand finish, brushwork		605	.013		.11	.47		.58	.86
0940	Roller		1020	.008		.12	.28		.40	.57
0980	Spray		1700	.005		.10	.17		.27	.37
1600	Glaze coating, 2 coats, spray, clear		1200	.007		.49	.23		.72	.92
1640	Multicolor	↓	1200	.007	↓	.95	.23		1.18	1.43
1660	Painting walls, complete, including surface prep, primer &									
1670	2 coats finish, on drywall or plaster, with roller	1 Pord	325	.025	S.F.	.18	.87		1.05	1.58
1700	For latex paint, deduct					10%				
1800	For ceiling installations, add				↓		25%			
2000	Masonry or concrete block, oil base, primer or sealer coat									
2100	Smooth finish, brushwork	1 Pord	1224	.007	S.F.	.06	.23		.29	.44
2180	Spray		2400	.003		.09	.12		.21	.29
2200	Sand finish, brushwork		1089	.007		.10	.26		.36	.52
2280	Spray		2400	.003		.09	.12		.21	.29
2800	Paint 2 coats, smooth finish, brushwork		756	.011		.20	.37		.57	.82
2880	Spray		1360	.006		.18	.21		.39	.53
2900	Sand finish, brushwork		672	.012		.20	.42		.62	.89
2980	Spray		1360	.006		.18	.21		.39	.53
3600	Glaze coating, 3 coats, spray, clear		900	.009		.70	.31		1.01	1.27
3620	Multicolor		900	.009		1.10	.31		1.41	1.71
4000	Block filler, 1 coat, brushwork		425	.019		.13	.66		.79	1.21
4100	Silicone, water repellent, 2 coats, spray	↓	2000	.004		.30	.14		.44	.56
4120	For latex paint, deduct					10%				
8200	For work 8 – 15' H, add						10%			
8300	For work over 15' H, add						20%			
8400	For light textured surfaces, add						10%			
8410	Heavy textured, add				↓		25%			

09 91 23.74 Walls and Ceilings, Interior, Zero VOC Latex

0010	**WALLS AND CEILINGS, INTERIOR, ZERO VOC LATEX**									

09 91 Painting

09 91 23 – Interior Painting

09 91 23.74 Walls and Ceilings, Interior, Zero VOC Latex

		Crew	Daily Output	Labor-Hours	Unit	Material	2009 Bare Costs Labor	Equipment	Total	Total Incl O&P
0100	Concrete, dry wall or plaster, latex, primer or sealer coat									
0200	Smooth finish, brushwork	G	1 Pord	1150	.007	S.F.	.06	.25		.31	.46
0240	Roller	G		1350	.006		.06	.21		.27	.39
0280	Spray	G		2750	.003		.05	.10		.15	.21
0300	Sand finish, brushwork	G		975	.008		.06	.29		.35	.52
0340	Roller	G		1150	.007		.07	.25		.32	.46
0380	Spray	G		2275	.004		.05	.12		.17	.26
0800	Paint 2 coats, smooth finish, brushwork	G		680	.012		.12	.41		.53	.79
0840	Roller	G		800	.010		.12	.35		.47	.69
0880	Spray	G		1625	.005		.10	.17		.27	.39
0900	Sand finish, brushwork	G		605	.013		.11	.47		.58	.87
0940	Roller	G		1020	.008		.12	.28		.40	.57
0980	Spray	G	▼	1700	.005		.10	.17		.27	.37
1800	For ceiling installations, add	G						25%			
8200	For work 8 – 15' H, add							10%			
8300	For work over 15' H, add					▼		20%			

09 91 23.75 Dry Fall Painting

		Crew	Daily Output	Labor-Hours	Unit	Material	2009 Bare Costs Labor	Equipment	Total	Total Incl O&P
0010	**DRY FALL PAINTING**										
0100	Walls										
0200	Wallboard and smooth plaster, one coat, brush		1 Pord	910	.009	S.F.	.05	.31		.36	.54
0210	Roll			1560	.005		.05	.18		.23	.34
0220	Spray			2600	.003		.05	.11		.16	.22
0230	Two coats, brush			520	.015		.10	.54		.64	.98
0240	Roll			877	.009		.10	.32		.42	.62
0250	Spray			1560	.005		.10	.18		.28	.40
0260	Concrete or textured plaster, one coat, brush			747	.011		.05	.38		.43	.65
0270	Roll			1300	.006		.05	.22		.27	.40
0280	Spray			1560	.005		.05	.18		.23	.34
0290	Two coats, brush			422	.019		.10	.67		.77	1.18
0300	Roll			747	.011		.10	.38		.48	.71
0310	Spray			1300	.006		.10	.22		.32	.46
0320	Concrete block, one coat, brush			747	.011		.05	.38		.43	.65
0330	Roll			1300	.006		.05	.22		.27	.40
0340	Spray			1560	.005		.05	.18		.23	.34
0350	Two coats, brush			422	.019		.10	.67		.77	1.18
0360	Roll			747	.011		.10	.38		.48	.71
0370	Spray			1300	.006		.10	.22		.32	.46
0380	Wood, one coat, brush			747	.011		.05	.38		.43	.65
0390	Roll			1300	.006		.05	.22		.27	.40
0400	Spray			877	.009		.05	.32		.37	.56
0410	Two coats, brush			487	.016		.10	.58		.68	1.03
0420	Roll		▼	747	.011		.10	.38		.48	.71
0430	Spray		▼	650	.012	▼	.10	.43		.53	.80
0440	Ceilings										
0450	Wallboard and smooth plaster, one coat, brush		1 Pord	600	.013	S.F.	.05	.47		.52	.80
0460	Roll			1040	.008		.05	.27		.32	.48
0470	Spray			1560	.005		.05	.18		.23	.34
0480	Two coats, brush			341	.023		.10	.83		.93	1.43
0490	Roll			650	.012		.10	.43		.53	.80
0500	Spray			1300	.006		.10	.22		.32	.46
0510	Concrete or textured plaster, one coat, brush			487	.016		.05	.58		.63	.97
0520	Roll		▼	877	.009	▼	.05	.32		.37	.56

09 91 Painting

09 91 23 – Interior Painting

09 91 23.75 Dry Fall Painting

		Crew	Daily Output	Labor-Hours	Unit	Material	2009 Bare Costs Labor	Equipment	Total	Total Incl O&P
0530	Spray	1 Pord	1560	.005	S.F.	.05	.18		.23	.34
0540	Two coats, brush		276	.029		.10	1.02		1.12	1.74
0550	Roll		520	.015		.10	.54		.64	.98
0560	Spray		1300	.006		.10	.22		.32	.46
0570	Structural steel, bar joists or metal deck, one coat, spray		1560	.005		.05	.18		.23	.34
0580	Two coats, spray	↓	1040	.008	↓	.10	.27		.37	.54

09 93 Staining and Transparent Finishing

09 93 23 – Interior Staining and Finishing

09 93 23.10 Varnish

		Crew	Daily Output	Labor-Hours	Unit	Material	2009 Bare Costs Labor	Equipment	Total	Total Incl O&P
0010	**VARNISH**									
0012	1 coat + sealer, on wood trim, brush, no sanding included	1 Pord	400	.020	S.F.	.07	.70		.77	1.21
0100	Hardwood floors, 2 coats, no sanding included, roller		1890	.004	"	.15	.15		.30	.41
9000	Minimum labor/equipment charge	↓	4	2	Job		70.50		70.50	113

09 96 High-Performance Coatings

09 96 23 – Graffiti-Resistant Coatings

09 96 23.10 Graffiti Resistant Treatments

		Crew	Daily Output	Labor-Hours	Unit	Material	2009 Bare Costs Labor	Equipment	Total	Total Incl O&P
0010	**GRAFFITI RESISTANT TREATMENTS**, sprayed on walls									
0100	Non-sacrificial, permanent non-stick coating, clear, on metals	1 Pord	2000	.004	S.F.	1.59	.14		1.73	1.98
0200	Concrete		2000	.004		1.81	.14		1.95	2.22
0300	Concrete block		2000	.004		2.34	.14		2.48	2.81
0400	Brick		2000	.004		2.66	.14		2.80	3.15
0500	Stone		2000	.004		2.66	.14		2.80	3.15
0600	Unpainted wood		2000	.004		3.06	.14		3.20	3.60
2000	Semi-permanent cross linking polymer primer, on metals		2000	.004		.40	.14		.54	.67
2100	Concrete		2000	.004		.48	.14		.62	.76
2200	Concrete block		2000	.004		.60	.14		.74	.89
2300	Brick		2000	.004		.48	.14		.62	.76
2400	Stone		2000	.004		.48	.14		.62	.76
2500	Unpainted wood		2000	.004		.67	.14		.81	.96
3000	Top coat, on metals		2000	.004		.43	.14		.57	.70
3100	Concrete		2000	.004		.49	.14		.63	.76
3200	Concrete block		2000	.004		.68	.14		.82	.98
3300	Brick		2000	.004		.57	.14		.71	.85
3400	Stone		2000	.004		.57	.14		.71	.85
3500	Unpainted wood		2000	.004		.68	.14		.82	.98
5000	Sacrificial, water based, on metal		2000	.004		.21	.14		.35	.46
5100	Concrete		2000	.004		.21	.14		.35	.46
5200	Concrete block		2000	.004		.21	.14		.35	.46
5300	Brick		2000	.004		.21	.14		.35	.46
5400	Stone		2000	.004		.21	.14		.35	.46
5500	Unpainted wood	↓	2000	.004	↓	.21	.14		.35	.46
8000	Cleaner for use after treatment									
8100	Towels or wipes, per package of 30				Ea.	.80			.80	.88
8200	Aerosol spray, 24 oz. can				"	19			19	21

09 96 56 – Epoxy Coatings

09 96 56.20 Wall Coatings

0010	**WALL COATINGS**									

09 96 High-Performance Coatings

09 96 56 – Epoxy Coatings

09 96 56.20 Wall Coatings		Crew	Daily Output	Labor-Hours	Unit	Material	2009 Bare Costs Labor	Equipment	Total	Total Incl O&P
0100	Acrylic glazed coatings, minimum	1 Pord	525	.015	S.F.	.28	.54		.82	1.17
0200	Maximum		305	.026		.60	.92		1.52	2.14
0300	Epoxy coatings, minimum		525	.015		.37	.54		.91	1.27
0400	Maximum		170	.047		1.12	1.66		2.78	3.88
2400	Sprayed perlite or vermiculite, 1/16" thick, minimum		2935	.003		.24	.10		.34	.41
2500	Maximum		640	.013		.68	.44		1.12	1.45
2700	Vinyl plastic wall coating, minimum		735	.011		.30	.38		.68	.94
2800	Maximum		240	.033		.75	1.17		1.92	2.70
3000	Urethane on smooth surface, 2 coats, minimum		1135	.007		.24	.25		.49	.66
3100	Maximum		665	.012		.53	.42		.95	1.26

Division Notes

		CREW	DAILY OUTPUT	LABOR-HOURS	UNIT	2009 BARE COSTS				TOTAL INCL O&P
						MAT.	LABOR	EQUIP.	TOTAL	

Estimating Tips

General

- The items in this division are usually priced per square foot or each.
- Many items in Division 10 require some type of support system or special anchors that are not usually furnished with the item. The required anchors must be added to the estimate in the appropriate division.
- Some items in Division 10, such as lockers, may require assembly before installation. Verify the amount of assembly required. Assembly can often exceed installation time.

10 20 00 Interior Specialties

- Support angles and blocking are not included in the installation of toilet compartments, shower/dressing compartments, or cubicles. Appropriate line items from Divisions 5 or 6 may need to be added to support the installations.
- Toilet partitions are priced by the stall. A stall consists of a side wall, pilaster, and door with hardware. Toilet tissue holders and grab bars are extra.
- The required acoustical rating of a folding partition can have a significant impact on costs. Verify the sound transmission coefficient rating of the panel priced to the specification requirements.

- Grab bar installation does not include supplemental blocking or backing to support the required load. When grab bars are installed at an existing facility, provisions must be made to attach the grab bars to solid structure.

Reference Numbers

Reference numbers are shown in shaded boxes at the beginning of some major classifications. These numbers refer to related items in the Reference Section. The reference information may be an estimating procedure, an alternate pricing method, or technical information.

Note: Not all subdivisions listed here necessarily appear in this publication.

Division 10 - Specialties

10 11 Visual Display Surfaces

10 11 13 – Chalkboards

10 11 13.13 Fixed Chalkboards

10 11 13.13 Fixed Chalkboards		Crew	Daily Output	Labor-Hours	Unit	Material	2009 Bare Costs Labor	Equipment	Total	Total Incl O&P
0010	**FIXED CHALKBOARDS** Porcelain enamel steel									
3900	Wall hung									
4000	Aluminum frame and chalktrough									
4300	3' x 5'	2 Carp	15	1.067	Ea.	242	42.50		284.50	335
4600	4' x 12'	"	13	1.231	"	505	49		554	635
4700	Wood frame and chalktrough									
4800	3' x 4'	2 Carp	16	1	Ea.	169	40		209	252
5300	4' x 8'	"	13	1.231	"	340	49		389	450
5400	Liquid chalk, white porcelain enamel, wall hung									
5420	Deluxe units, aluminum trim and chalktrough									
5450	4' x 4'	2 Carp	16	1	Ea.	202	40		242	288
5550	4' x 12'	"	12	1.333	"	470	53.50		523.50	605
5700	Wood trim and chalktrough									
5900	4' x 4'	2 Carp	16	1	Ea.	460	40		500	575
6200	4' x 8'		14	1.143	"	635	45.50		680.50	775
9000	Minimum labor/equipment charge	↓	3	5.333	Job		213		213	350

10 11 13.43 Portable Chalkboards

10 11 13.43 Portable Chalkboards		Crew	Daily Output	Labor-Hours	Unit	Material	2009 Bare Costs Labor	Equipment	Total	Total Incl O&P
0010	**PORTABLE CHALKBOARDS**									
0100	Freestanding, reversible									
0120	Economy, wood frame, 4' x 6'									
0140	Chalkboard both sides				Ea.	565			565	620
0160	Chalkboard one side, cork other side				"	485			485	535
0200	Standard, lightweight satin finished aluminum, 4' x 6'									
0220	Chalkboard both sides				Ea.	650			650	715
0240	Chalkboard one side, cork other side				"	580			580	640
0300	Deluxe, heavy duty extruded aluminum, 4' x 6'									
0320	Chalkboard both sides				Ea.	1,050			1,050	1,150
0340	Chalkboard one side, cork other side				"	940			940	1,025

10 11 23 – Tackboards

10 11 23.10 Fixed Tackboards

10 11 23.10 Fixed Tackboards		Crew	Daily Output	Labor-Hours	Unit	Material	2009 Bare Costs Labor	Equipment	Total	Total Incl O&P
0010	**FIXED TACKBOARDS**									
2120	Prefabricated, 1/4" cork, 3' x 5' with aluminum frame	2 Carp	16	1	Ea.	109	40		149	186
2140	Wood frame	"	16	1	"	139	40		179	219
2300	Glass enclosed cabinets, alum., cork panel, hinged doors									
2600	4' x 7', 3 door	2 Carp	10	1.600	Ea.	1,500	64		1,564	1,750
9000	Minimum labor/equipment charge	"	4	4	Job		160		160	264

10 13 Directories

10 13 10 – Building Directories

10 13 10.10 Directory Boards

10 13 10.10 Directory Boards		Crew	Daily Output	Labor-Hours	Unit	Material	2009 Bare Costs Labor	Equipment	Total	Total Incl O&P
0010	**DIRECTORY BOARDS**									
0050	Plastic, glass covered, 30" x 20"	2 Carp	3	5.333	Ea.	242	213		455	615
0100	36" x 48"		2	8		875	320		1,195	1,500
0900	Outdoor, weatherproof, black plastic, 36" x 24"		2	8		740	320		1,060	1,350
1000	36" x 36"	↓	1.50	10.667	↓	850	425		1,275	1,650
9000	Minimum labor/equipment charge	1 Carp	1	8	Job		320		320	530

268

10 14 Signage

10 14 19 – Dimensional Letter Signage

10 14 19.10 Exterior Signs

		Crew	Daily Output	Labor-Hours	Unit	Material	2009 Bare Costs Labor	Equipment	Total	Total Incl O&P
0010	**EXTERIOR SIGNS**									
3900	Plaques, custom, 20" x 30", for up to 450 letters, cast aluminum	2 Carp	4	4	Ea.	1,050	160		1,210	1,425
4000	Cast bronze		4	4		1,475	160		1,635	1,900
4200	30" x 36", up to 900 letters cast aluminum		3	5.333		1,875	213		2,088	2,400
4300	Cast bronze	↓	3	5.333		3,475	213		3,688	4,175
5100	Exit signs, 24 ga. alum., 14" x 12" surface mounted	1 Carp	30	.267		40.50	10.65		51.15	62
5200	10" x 7"	"	20	.400		24	16		40	53
6400	Replacement sign faces, 6" or 8"	1 Clab	50	.160	↓	47	5.05		52.05	60
9000	Minimum labor/equipment charge	1 Carp	4	2	Job		80		80	132

10 14 23 – Panel Signage

10 14 23.13 Engraved Interior Panel Signage

		Crew	Daily Output	Labor-Hours	Unit	Material	2009 Bare Costs Labor	Equipment	Total	Total Incl O&P
0010	**ENGRAVED INTERIOR PANEL SIGNAGE**									
1010	Flexible door sign, adhesive back, w/Braille, 5/8" letters, 4" x 4"	1 Clab	32	.250	Ea.	26.50	7.90		34.40	42.50
1050	6" x 6"		32	.250		39.50	7.90		47.40	56.50
1100	8" x 2"		32	.250		27.50	7.90		35.40	43
1150	8" x 4"		32	.250		35	7.90		42.90	51.50
1200	8" x 8"		32	.250		48	7.90		55.90	66
1250	12" x 2"		32	.250		35	7.90		42.90	51.50
1300	12" x 6"		32	.250		50	7.90		57.90	68
1350	12" x 12"		32	.250		99	7.90		106.90	122
1500	Graphic symbols, 2" x 2"		32	.250		12	7.90		19.90	26.50
1550	6" x 6"		32	.250		30	7.90		37.90	46
1600	8" x 8"	↓	32	.250	↓	30	7.90		37.90	46

10 14 53 – Traffic Signage

10 14 53.20 Traffic Signs

		Crew	Daily Output	Labor-Hours	Unit	Material	2009 Bare Costs Labor	Equipment	Total	Total Incl O&P
0010	**TRAFFIC SIGNS**									
0012	Stock, 24" x 24", no posts, .080" alum. reflectorized	B-80	70	.457	Ea.	73.50	15.45	9.35	98.30	116
0100	High intensity		70	.457		73.50	15.45	9.35	98.30	116
0300	30" x 30", reflectorized		70	.457		150	15.45	9.35	174.80	200
0400	High intensity		70	.457		150	15.45	9.35	174.80	200
0600	Guide and directional signs, 12" x 18", reflectorized		70	.457		48	15.45	9.35	72.80	88
0700	High intensity		70	.457		45.50	15.45	9.35	70.30	86
0900	18" x 24", stock signs, reflectorized		70	.457		53.50	15.45	9.35	78.30	94.50
1000	High intensity		70	.457		53.50	15.45	9.35	78.30	94.50
1200	24" x 24", stock signs, reflectorized		70	.457		66.50	15.45	9.35	91.30	109
1300	High intensity		70	.457		66.50	15.45	9.35	91.30	109
1500	Add to above for steel posts, galvanized, 10'-0" upright, bolted		200	.160		24	5.40	3.27	32.67	39
1600	12'-0" upright, bolted		140	.229	↓	32	7.75	4.68	44.43	52.50
1800	Highway road signs, aluminum, over 20 S. F., reflectorized		350	.091	S.F.	30.50	3.09	1.87	35.46	40.50
2000	High intensity		350	.091		30.50	3.09	1.87	35.46	40.50
2200	Highway, suspended over road, 80 S.F. min., reflectorized		165	.194		30.50	6.55	3.97	41.02	48.50
2300	High intensity	↓	165	.194	↓	30.50	6.55	3.97	41.02	48.50

10 17 Telephone Specialties

10 17 16 – Telephone Enclosures

10 17 16.10 Commercial Telephone Enclosures

		Crew	Daily Output	Labor-Hours	Unit	Material	2009 Bare Costs Labor	Equipment	Total	Total Incl O&P
0010	**COMMERCIAL TELEPHONE ENCLOSURES**									
0300	Shelf type, wall hung, minimum	2 Carp	5	3.200	Ea.	1,300	128		1,428	1,625
0400	Maximum		5	3.200		2,725	128		2,853	3,175
0600	Booth type, painted steel, indoor or outdoor, minimum		1.50	10.667		3,475	425		3,900	4,525
0700	Maximum (stainless steel)		1.50	10.667		11,600	425		12,025	13,400
1900	Outdoor, drive-up type, wall mounted		4	4		930	160		1,090	1,300
2000	Post mounted, stainless steel posts		3	5.333		1,425	213		1,638	1,925
9000	Minimum labor/equipment charge		4	4	Job		160		160	264

10 21 Compartments and Cubicles

10 21 13 – Toilet Compartments

10 21 13.13 Metal Toilet Compartments

		Crew	Daily Output	Labor-Hours	Unit	Material	2009 Bare Costs Labor	Equipment	Total	Total Incl O&P
0010	**METAL TOILET COMPARTMENTS**									
0110	Cubicles, ceiling hung									
0200	Powder coated steel	2 Carp	4	4	Ea.	465	160		625	780
0500	Stainless steel	"	4	4		1,775	160		1,935	2,225
0600	For handicap units, incl. 52" grab bars, add					390			390	430
0900	Floor and ceiling anchored									
1000	Powder coated steel	2 Carp	5	3.200	Ea.	485	128		613	745
1300	Stainless steel	"	5	3.200		1,700	128		1,828	2,050
1400	For handicap units, incl. 52" grab bars, add					276			276	305
1610	Floor anchored									
1700	Powder coated steel	2 Carp	7	2.286	Ea.	545	91.50		636.50	745
2000	Stainless steel	"	7	2.286		1,525	91.50		1,616.50	1,825
2100	For handicap units, incl. 52" grab bars, add					273			273	300
2200	For juvenile units, deduct					39.50			39.50	43.50
2450	Floor anchored, headrail braced									
2500	Powder coated steel	2 Carp	6	2.667	Ea.	445	107		552	660
2804	Stainless steel	"	4.60	3.478		1,125	139		1,264	1,475
2900	For handicap units, incl. 52" grab bars, add					330			330	365
3000	Wall hung partitions, powder coated steel	2 Carp	7	2.286		595	91.50		686.50	805
3300	Stainless steel	"	7	2.286		1,575	91.50		1,666.50	1,875
3400	For handicap units, incl. 52" grab bars, add					330			330	365
4000	Screens, entrance, floor mounted, 58" high, 48" wide									
4200	Powder coated steel	2 Carp	15	1.067	Ea.	233	42.50		275.50	325
4500	Stainless steel	"	15	1.067	"	850	42.50		892.50	1,000
4650	Urinal screen, 18" wide									
4704	Powder coated steel	2 Carp	6.15	2.602	Ea.	184	104		288	375
5004	Stainless steel	"	6.15	2.602	"	580	104		684	810
5100	Floor mounted, head rail braced									
5300	Powder coated steel	2 Carp	8	2	Ea.	199	80		279	350
5600	Stainless steel	"	8	2	"	555	80		635	745
5750	Pilaster, flush									
5800	Powder coated steel	2 Carp	10	1.600	Ea.	259	64		323	390
6100	Stainless steel		10	1.600		655	64		719	825
6300	Post braced, powder coated steel		10	1.600		182	64		246	305
6600	Stainless steel		10	1.600		485	64		549	640
6700	Wall hung, bracket supported									
6800	Powder coated steel	2 Carp	10	1.600	Ea.	289	64		353	425
7100	Stainless steel		10	1.600		315	64		379	455
7400	Flange supported, powder coated steel		10	1.600		95	64		159	210

10 21 Compartments and Cubicles

10 21 13 – Toilet Compartments

10 21 13.13 Metal Toilet Compartments

		Crew	Daily Output	Labor-Hours	Unit	Material	2009 Bare Costs Labor	2009 Bare Costs Equipment	Total	Total Incl O&P
7700	Stainless steel	2 Carp	10	1.600	Ea.	276	64		340	410
7800	Wedge type, powder coated steel		10	1.600		128	64		192	247
8100	Stainless steel	↓	10	1.600	↓	540	64		604	700
9000	Minimum labor/equipment charge	1 Carp	2.50	3.200	Job		128		128	211

10 21 13.14 Metal Toilet Compartment Components

		Crew	Daily Output	Labor-Hours	Unit	Material	2009 Bare Costs Labor	2009 Bare Costs Equipment	Total	Total Incl O&P
0010	**METAL TOILET COMPARTMENT COMPONENTS**									
0100	Pilasters									
0110	Overhead braced, powder coated steel, 7" wide x 82" high	2 Carp	22.20	.721	Ea.	76.50	29		105.50	132
0120	Stainless steel		22.20	.721		206	29		235	275
0130	Floor braced, powder coated steel, 7" wide x 70" high		23.30	.687		121	27.50		148.50	179
0140	Stainless steel		23.30	.687		310	27.50		337.50	385
0150	Ceiling hung, powder coated steel, 7" wide x 83" high		13.30	1.203		129	48		177	222
0160	Stainless steel		13.30	1.203		330	48		378	445
0170	Wall hung, powder coated steel, 3" wide x 58" high		18.90	.847		115	34		149	183
0180	Stainless steel	↓	18.90	.847	↓	201	34		235	277
0200	Panels									
0210	Powder coated steel, 31" wide x 58" high	2 Carp	18.90	.847	Ea.	134	34		168	203
0220	Stainless steel		18.90	.847		440	34		474	540
0230	Powder coated steel, 53" wide x 58" high		18.90	.847		171	34		205	244
0240	Stainless steel		18.90	.847		590	34		624	705
0250	Powder coated steel, 63" wide x 58" high		18.90	.847		206	34		240	282
0260	Stainless steel	↓	18.90	.847	↓	725	34		759	855
0300	Doors									
0310	Powder coated steel, 24" wide x 58" high	2 Carp	14.10	1.135	Ea.	147	45.50		192.50	237
0320	Stainless steel		14.10	1.135		415	45.50		460.50	530
0330	Powder coated steel, 26" wide x 58" high		14.10	1.135		149	45.50		194.50	239
0340	Stainless steel		14.10	1.135		420	45.50		465.50	540
0350	Powder coated steel, 28" wide x 58" high		14.10	1.135		158	45.50		203.50	249
0360	Stainless steel		14.10	1.135		445	45.50		490.50	565
0370	Powder coated steel, 36" wide x 58" high		14.10	1.135		182	45.50		227.50	276
0380	Stainless steel	↓	14.10	1.135	↓	515	45.50		560.50	645
0400	Headrails									
0410	For powder coated steel, 62" long	2 Carp	65	.246	Ea.	27.50	9.85		37.35	46.50
0420	Stainless steel		65	.246		27.50	9.85		37.35	46.50
0430	For powder coated steel, 84" long		50	.320		35	12.80		47.80	59.50
0440	Stainless steel		50	.320		35	12.80		47.80	59.50
0450	For powder coated steel, 120" long		30	.533		50.50	21.50		72	90.50
0460	Stainless steel	↓	30	.533	↓	50.50	21.50		72	90.50
9000	Minimum labor/equipment charge	1 Carp	4	2	Job		80		80	132

10 21 13.16 Plastic-Laminate-Clad Toilet Compartments

		Crew	Daily Output	Labor-Hours	Unit	Material	2009 Bare Costs Labor	2009 Bare Costs Equipment	Total	Total Incl O&P
0010	**PLASTIC-LAMINATE-CLAD TOILET COMPARTMENTS**									
0110	Cubicles, ceiling hung									
0300	Plastic laminate on particle board ♿	2 Carp	4	4	Ea.	605	160		765	930
0600	For handicap units, incl. 52" grab bars, add				"	390			390	430
0900	Floor and ceiling anchored									
1100	Plastic laminate on particle board ♿	2 Carp	5	3.200	Ea.	615	128		743	885
1400	For handicap units, incl. 52" grab bars, add				"	276			276	305
1610	Floor mounted									
1800	Plastic laminate on particle board	2 Carp	7	2.286	Ea.	545	91.50		636.50	750
2450	Floor mounted, headrail braced									
2600	Plastic laminate on particle board ♿	2 Carp	6	2.667	Ea.	575	107		682	810
3400	For handicap units, incl. 52" grab bars, add				↓	330			330	365

10 21 13 – Toilet Compartments

10 21 13.16 Plastic-Laminate-Clad Toilet Compartments

		Crew	Daily Output	Labor-Hours	Unit	Material	2009 Bare Costs Labor	Equipment	Total	Total Incl O&P
4300	Entrance screen, floor mtd., plas. lam., 58" high, 48" wide	2 Carp	15	1.067	Ea.	480	42.50		522.50	600
4800	Urinal screen, 18" wide, ceiling braced, plastic laminate		8	2		180	80		260	330
5400	Floor mounted, headrail braced		8	2		365	80		445	530
5900	Pilaster, flush, plastic laminate		10	1.600		400	64		464	545
6400	Post braced, plastic laminate		10	1.600		288	64		352	420
6700	Wall hung, bracket supported									
6900	Plastic laminate on particle board	2 Carp	10	1.600	Ea.	98	64		162	214
7450	Flange supported									
7500	Plastic laminate on particle board	2 Carp	10	1.600	Ea.	203	64		267	330
9000	Minimum labor/equipment charge	1 Carp	2.50	3.200	Job		128		128	211

10 21 13.17 Plastic-Laminate Clad Toilet Compartment Components

		Crew	Daily Output	Labor-Hours	Unit	Material	2009 Bare Costs Labor	Equipment	Total	Total Incl O&P
0010	**PLASTIC-LAMINATE CLAD TOILET COMPARTMENT COMPONENTS**									
0100	Pilasters									
0110	Overhead braced, 7" wide x 82" high	2 Carp	22.20	.721	Ea.	104	29		133	163
0130	Floor anchored, 7" wide x 70" high		23.30	.687		106	27.50		133.50	163
0150	Ceiling hung, 7" wide x 83" high		13.30	1.203		105	48		153	195
0180	Wall hung, 3" wide x 58" high		18.90	.847		115	34		149	183
0200	Panels									
0210	31" wide x 58" high	2 Carp	18.90	.847	Ea.	144	34		178	214
0230	51" wide x 58" high		18.90	.847		203	34		237	280
0250	63" wide x 58" high		18.90	.847		241	34		275	320
0300	Doors									
0310	24" wide x 58" high	2 Carp	14.10	1.135	Ea.	142	45.50		187.50	231
0330	26" wide x 58" high		14.10	1.135		147	45.50		192.50	236
0350	28" wide x 58" high		14.10	1.135		152	45.50		197.50	243
0370	36" wide x 58" high		14.10	1.135		181	45.50		226.50	274
0400	Headrails									
0410	62" long	2 Carp	65	.246	Ea.	31	9.85		40.85	50.50
0430	84" long		60	.267		35	10.65		45.65	56
0450	120" long		30	.533		50.50	21.50		72	90.50
9000	Minimum labor/equipment charge	1 Carp	4	2	Job		80		80	132

10 21 13.19 Plastic Toilet Compartments

		Crew	Daily Output	Labor-Hours	Unit	Material	2009 Bare Costs Labor	Equipment	Total	Total Incl O&P
0010	**PLASTIC TOILET COMPARTMENTS**									
0110	Cubicles, ceiling hung									
0250	Phenolic	2 Carp	4	4	Ea.	965	160		1,125	1,350
0600	For handicap units, incl. 52" grab bars, add				"	390			390	430
0900	Floor and ceiling anchored									
1050	Phenolic	2 Carp	5	3.200	Ea.	1,025	128		1,153	1,325
1400	For handicap units, incl. 52" grab bars, add				"	276			276	305
1610	Floor mounted									
1750	Phenolic	2 Carp	7	2.286	Ea.	1,275	91.50		1,366.50	1,550
2100	For handicap units, incl. 52" grab bars, add					273			273	300
2200	For juvenile units, deduct					39.50			39.50	43.50
2450	Floor mounted, headrail braced									
2550	Phenolic	2 Carp	6	2.667	Ea.	815	107		922	1,075
9000	Minimum labor/equipment charge	1 Carp	2.50	3.200	Job		128		128	211

10 21 13.20 Plastic Toilet Compartment Components

		Crew	Daily Output	Labor-Hours	Unit	Material	2009 Bare Costs Labor	Equipment	Total	Total Incl O&P
0010	**PLASTIC TOILET COMPARTMENT COMPONENTS**									
0100	Pilasters									
0110	Overhead braced, polymer plastic, 7" wide x 82" high	2 Carp	22.20	.721	Ea.	167	29		196	232
0120	Phenolic		22.20	.721		179	29		208	245
0130	Floor braced, polymer plastic, 7" wide x 70" high		23.30	.687		157	27.50		184.50	218

10 21 13 – Toilet Compartments

10 21 13.20 Plastic Toilet Compartment Components

		Crew	Daily Output	Labor-Hours	Unit	Material	2009 Bare Costs Labor	Equipment	Total	Total Incl O&P
0140	Phenolic	2 Carp	23.30	.687	Ea.	166	27.50		193.50	229
0150	Ceiling hung, polymer plastic, 7" wide x 83" high		13.30	1.203		141	48		189	235
0160	Phenolic		13.30	1.203		188	48		236	286
0180	Wall hung, phenolic, 3" wide x 58" high		18.90	.847		122	34		156	190
0200	Panels									
0210	Polymer plastic, 31" high x 55" high	2 Carp	18.90	.847	Ea.	360	34		394	455
0220	Phenolic, 31" wide x 58" high		18.90	.847		279	34		313	360
0230	Polymer plastic, 51" wide x 55" high		18.90	.847		550	34		584	665
0240	Phenolic, 51" wide x 58" high		18.90	.847		420	34		454	520
0250	Polymer plastic, 63" wide x 55" high		18.90	.847		670	34		704	790
0260	Phenolic, 63" wide x 58" high		18.90	.847		505	34		539	610
0300	Doors									
0310	Polymer plastic, 24" wide x 55" high	2 Carp	14.10	1.135	Ea.	290	45.50		335.50	395
0320	Phenolic, 24" wide x 58" high		14.10	1.135		315	45.50		360.50	425
0330	polymer plastic, 26" high x 55" high		14.10	1.135		310	45.50		355.50	415
0340	Phenolic, 26" wide x 58" high		14.10	1.135		335	45.50		380.50	445
0350	Polymer plastic, 26" wide x 55" high		14.10	1.135		325	45.50		370.50	435
0360	Phenolic, 28" wide x 58" high		14.10	1.135		355	45.50		400.50	465
0370	Polymer plasst, 36" wide x 55" high		14.10	1.135		410	45.50		455.50	525
0380	Phenolic, 36" wide x 58" high		14.10	1.135		440	45.50		485.50	560
0400	Headrails									
0410	For polymer plastic, 62" long	2 Carp	65	.246	Ea.	40.50	9.85		50.35	61
0420	Phenolic		65	.246		27.50	9.85		37.35	47
0430	For polymer plastic, 84" long		50	.320		42.50	12.80		55.30	68
0440	Phenolic		50	.320		32.50	12.80		45.30	56.50
0450	For polymer plastic, 120" long		30	.533		60	21.50		81.50	101
0460	Phenolic		30	.533		42.50	21.50		64	82
9000	Minimum labor/equipment charge	1 Carp	4	2	Job		80		80	132

10 21 13.40 Stone Toilet Compartments

		Crew	Daily Output	Labor-Hours	Unit	Material	Labor	Equipment	Total	Total Incl O&P
0010	**STONE TOILET COMPARTMENTS**									
0100	Cubicles, ceiling hung, marble	2 Marb	2	8	Ea.	1,675	310		1,985	2,325
0600	For handicap units, incl. 52" grab bars, add					390			390	430
0800	Floor & ceiling anchored, marble	2 Marb	2.50	6.400		1,825	247		2,072	2,400
1400	For handicap units, incl. 52" grab bars, add					276			276	305
1600	Floor mounted, marble	2 Marb	3	5.333		1,075	206		1,281	1,500
2400	Floor mounted, headrail braced, marble	"	3	5.333		1,025	206		1,231	1,450
2900	For handicap units, incl. 52" grab bars, add					330			330	365
4100	Entrance screen, floor mounted marble, 58" high, 48" wide	2 Marb	9	1.778		675	68.50		743.50	855
4600	Urinal screen, 18" wide, ceiling braced, marble	D-1	6	2.667		675	97		772	900
5100	Floor mounted, head rail braced									
5200	Marble	D-1	6	2.667	Ea.	575	97		672	785
5700	Pilaster, flush, marble		9	1.778		750	64.50		814.50	930
6200	Post braced, marble		9	1.778		745	64.50		809.50	925
9000	Minimum labor/equipment charge	1 Carp	2.50	3.200	Job		128		128	211

10 21 16 – Shower and Dressing Compartments

10 21 16.10 Partitions, Shower

		Crew	Daily Output	Labor-Hours	Unit	Material	Labor	Equipment	Total	Total Incl O&P
0010	**PARTITIONS, SHOWER** Floor mounted, no plumbing									
0100	Cabinet, incl. base, no door, painted steel, 1" thick walls	2 Shee	5	3.200	Ea.	905	151		1,056	1,250
0300	With door, fiberglass		4.50	3.556		705	168		873	1,050
0600	Galvanized and painted steel, 1" thick walls		5	3.200		955	151		1,106	1,300
0800	Stall, 1" thick wall, no base, enameled steel		5	3.200		980	151		1,131	1,325
1500	Circular fiberglass, cabinet 36" diameter,		4	4		725	189		914	1,100

10 21 Compartments and Cubicles

10 21 16 – Shower and Dressing Compartments

10 21 16.10 Partitions, Shower

	10 21 16.10 Partitions, Shower	Crew	Daily Output	Labor-Hours	Unit	Material	2009 Bare Costs Labor	Equipment	Total	Total Incl O&P
1700	One piece, 36" diameter, less door	2 Shee	4	4	Ea.	610	189		799	970
1800	With door	↓	3.50	4.571		1,000	216		1,216	1,450
4100	Shower doors, economy plastic, 24" wide	1 Shee	9	.889		128	42		170	207
4200	Tempered glass door, economy		8	1		169	47		216	261
4700	Deluxe, tempered glass, chrome on brass frame, minimum		8	1		223	47		270	320
4800	Maximum		1	8	↓	880	380		1,260	1,575
9990	Minimum labor/equipment charge	↓	2.50	3.200	Job		151		151	240

10 22 Partitions

10 22 16 – Folding Gates

10 22 16.10 Security Gates

	10 22 16.10 Security Gates	Crew	Daily Output	Labor-Hours	Unit	Material	2009 Bare Costs Labor	Equipment	Total	Total Incl O&P
0010	**SECURITY GATES** For roll up type, see Div. 08 33 13.10									
0300	Scissors type folding gate, ptd. steel, single, 6-1/2' high, 5-1/2'wide	2 Sswk	4	4	Opng.	180	179		359	530
0350	6-1/2' wide		4	4		187	179		366	535
0400	7-1/2' wide		4	4		185	179		364	535
0600	Double gate, 8' high, 8' wide		2.50	6.400		300	286		586	860
0650	10' wide		2.50	6.400		310	286		596	870
0700	12' wide		2	8		455	360		815	1,150
0750	14' wide		2	8		540	360		900	1,250
0900	Door gate, folding steel, 4' wide, 61" high		4	4		94.50	179		273.50	435
1000	71" high		4	4		98	179		277	440
1200	81" high		4	4		106	179		285	445
1300	Window gates, 2' to 4' wide, 31" high		4	4		55	179		234	390
1500	55" high		3.75	4.267		86	191		277	450
1600	79" high	↓	3.50	4.571	↓	99.50	204		303.50	490

10 22 19 – Demountable Partitions

10 22 19.43 Demountable Composite Partitions

	10 22 19.43 Demountable Composite Partitions	Crew	Daily Output	Labor-Hours	Unit	Material	2009 Bare Costs Labor	Equipment	Total	Total Incl O&P
0010	**DEMOUNTABLE COMPOSITE PARTITIONS**, add for doors									
0100	Do not deduct door openings from total L.F.									
0900	Demountable gypsum system on 2" to 2-1/2"									
1000	steel studs, 9' high, 3" to 3-3/4" thick									
1200	Vinyl clad gypsum	2 Carp	48	.333	L.F.	56.50	13.30		69.80	84.50
1300	Fabric clad gypsum		44	.364		141	14.55		155.55	179
1500	Steel clad gypsum	↓	40	.400	↓	153	16		169	195
1600	1.75 system, aluminum framing, vinyl clad hardboard,									
1800	paper honeycomb core panel, 1-3/4" to 2-1/2" thick									
1900	9' high	2 Carp	48	.333	L.F.	95	13.30		108.30	127
2100	7' high		60	.267		85.50	10.65		96.15	112
2200	5' high	↓	80	.200	↓	72	8		80	92.50
2250	Unitized gypsum system									
2300	Unitized panel, 9' high, 2" to 2-1/2" thick									
2350	Vinyl clad gypsum	2 Carp	48	.333	L.F.	123	13.30		136.30	157
2400	Fabric clad gypsum	"	44	.364	"	202	14.55		216.55	246
2500	Unitized mineral fiber system									
2510	Unitized panel, 9' high, 2-1/4" thick, aluminum frame									
2550	Vinyl clad mineral fiber	2 Carp	48	.333	L.F.	122	13.30		135.30	156
2600	Fabric clad mineral fiber	"	44	.364	"	181	14.55		195.55	224
2800	Movable steel walls, modular system									
2900	Unitized panels, 9' high, 48" wide									
3100	Baked enamel, pre-finished	2 Carp	60	.267	L.F.	138	10.65		148.65	170

10 22 Partitions

10 22 19 – Demountable Partitions

10 22 19.43 Demountable Composite Partitions	Crew	Daily Output	Labor-Hours	Unit	Material	2009 Bare Costs Labor	Equipment	Total	Total Incl O&P
3200 Fabric clad steel	2 Carp	56	.286	L.F.	199	11.40		210.40	238
5500 For acoustical partitions, add, minimum				S.F.	2.22			2.22	2.44
5550 Maximum				"	10.35			10.35	11.40
5700 For doors, see Div. 08 11 & 08 16									
5800 For door hardware, see Div. 08 71									
6100 In-plant modular office system, w/prehung hollow core door									
6200 3" thick polystyrene core panels									
6250 12' x 12', 2 wall	2 Clab	3.80	4.211	Ea.	3,125	133		3,258	3,675
6300 4 wall		1.90	8.421		4,400	266		4,666	5,275
6350 16' x 16', 2 wall		3.60	4.444		4,600	140		4,740	5,275
6400 4 wall		1.80	8.889		6,200	281		6,481	7,275
9000 Minimum labor/equipment charge	2 Carp	3	5.333	Job		213		213	350

10 22 23 – Portable Partitions, Screens, and Panels

10 22 23.13 Wall Screens

	Crew	Daily Output	Labor-Hours	Unit	Material	Labor	Equipment	Total	Total Incl O&P
0010 **WALL SCREENS**, divider panels, free standing, fiber core									
0020 Fabric face straight									
0100 3'-0" long, 4'-0" high	2 Carp	100	.160	L.F.	106	6.40		112.40	127
0200 5'-0" high		90	.178		111	7.10		118.10	134
0500 6'-0" high		75	.213		114	8.50		122.50	139
0900 5'-0" long, 4'-0" high		175	.091		76.50	3.65		80.15	90.50
1000 5'-0" high		150	.107		88	4.26		92.26	104
1500 6"-0" high		125	.128		91	5.10		96.10	108
1600 6'-0" long, 5'-0" high		162	.099		76	3.95		79.95	90
3200 Economical panels, fabric face, 4'-0" long, 5'-0" high		132	.121		42.50	4.84		47.34	54.50
3250 6'-0" high		112	.143		45	5.70		50.70	58.50
3300 5'-0" long, 5'-0" high		150	.107		36	4.26		40.26	46.50
3350 6'-0" high		125	.128		38	5.10		43.10	50
3450 Acoustical panels, 60 to 90 NRC, 3'-0" long, 5'-0" high		90	.178		79.50	7.10		86.60	99
3550 6'-0" high		75	.213		89.50	8.50		98	113
3600 5'-0" long, 5'-0" high		150	.107		63	4.26		67.26	76.50
3650 6'-0" high		125	.128		61	5.10		66.10	75.50
3700 6'-0" long, 5'-0" high		162	.099		56	3.95		59.95	68
3750 6'-0" high		138	.116		56	4.63		60.63	69
3800 Economy acoustical panels, 40 N.R.C., 4'-0" long, 5'-0" high		132	.121		42.50	4.84		47.34	54.50
3850 6'-0" high		112	.143		45	5.70		50.70	58.50
3900 5'-0" long, 6'-0" high		125	.128		38	5.10		43.10	50
3950 6'-0" long, 5'-0" high		162	.099		30	3.95		33.95	39.50
9000 Minimum labor/equipment charge		3	5.333	Job		213		213	350

10 22 26 – Operable Partitions

10 22 26.13 Accordion Folding Partitions

	Crew	Daily Output	Labor-Hours	Unit	Material	Labor	Equipment	Total	Total Incl O&P
0010 **ACCORDION FOLDING PARTITIONS**									
0100 Vinyl covered, over 150 S.F., frame not included									
0300 Residential, 1.25 lb. per S.F., 8' maximum height	2 Carp	300	.053	S.F.	19.35	2.13		21.48	25
0400 Commercial, 1.75 lb. per S.F., 8' maximum height		225	.071		22	2.84		24.84	29
0900 Acoustical, 3 lb. per S.F., 17' maximum height		100	.160		25.50	6.40		31.90	38.50
1200 5 lb. per S.F., 20' maximum height		95	.168		35.50	6.75		42.25	50
1500 Vinyl clad wood or steel, electric operation, 5.0 psf		160	.100		48.50	4		52.50	59.50
1900 Wood, non-acoustic, birch or mahogany, to 10' high		300	.053		26	2.13		28.13	32
9000 Minimum labor/equipment charge		4	4	Job		160		160	264

10 22 26.33 Folding Panel Partitions

0010 **FOLDING PANEL PARTITIONS**, acoustic, wood									

10 22 Partitions

10 22 26 - Operable Partitions

10 22 26.33 Folding Panel Partitions

		Crew	Daily Output	Labor-Hours	Unit	Material	2009 Bare Costs Labor	Equipment	Total	Total Incl O&P
0100	Vinyl faced, to 18' high, 6 psf, minimum	2 Carp	60	.267	S.F.	49	10.65		59.65	71.50
0150	Average		45	.356		58.50	14.20		72.70	88
0200	Maximum		30	.533		75.50	21.50		97	119
0400	Formica or hardwood finish, minimum		60	.267		50.50	10.65		61.15	73
0500	Maximum		30	.533		54	21.50		75.50	94.50
0600	Wood, low acoustical type, 4.5 psf, to 14' high		50	.320		37	12.80		49.80	61.50
9000	Minimum labor/equipment charge		4	4	Job		160		160	264

10 26 Wall and Door Protection

10 26 13 - Corner Guards

10 26 13.10 Metal Corner Guards

		Crew	Daily Output	Labor-Hours	Unit	Material	2009 Bare Costs Labor	Equipment	Total	Total Incl O&P
0010	**METAL CORNER GUARDS**									
0020	Steel angle w/anchors, 1" x 1" x 1/4", 1.5#/L.F.	2 Carp	160	.100	L.F.	5.75	4		9.75	12.95
0100	2" x 2" x 1/4" angles, 3.2#/L.F.	"	150	.107	"	10.45	4.26		14.71	18.55
9000	Minimum labor/equipment charge	1 Carp	2	4	Job		160		160	264

10 28 Toilet, Bath, and Laundry Accessories

10 28 13 - Toilet Accessories

10 28 13.13 Commercial Toilet Accessories

		Crew	Daily Output	Labor-Hours	Unit	Material	2009 Bare Costs Labor	Equipment	Total	Total Incl O&P
0010	**COMMERCIAL TOILET ACCESSORIES**									
0200	Curtain rod, stainless steel, 5' long, 1" diameter	1 Carp	13	.615	Ea.	41	24.50		65.50	86
0300	1-1/4" diameter		13	.615		36.50	24.50		61	80.50
0400	Diaper changing station, horizontal, wall mounted, plastic		10	.800		280	32		312	360
0500	Dispenser units, combined soap & towel dispensers,									
0510	mirror and shelf, flush mounted	1 Carp	10	.800	Ea.	295	32		327	380
0600	Towel dispenser and waste receptacle,									
0610	18 gallon capacity	1 Carp	10	.800	Ea.	395	32		427	490
0800	Grab bar, straight, 1-1/4" diameter, stainless steel, 18" long		24	.333		29.50	13.30		42.80	54.50
0900	24" long		23	.348		32	13.90		45.90	58
1000	30" long		22	.364		34.50	14.55		49.05	62
1100	36" long		20	.400		36	16		52	66
1105	42" long		20	.400		42	16		58	72.50
1200	1-1/2" diameter, 24" long		23	.348		34.50	13.90		48.40	61
1300	36" long		20	.400		43.50	16		59.50	74.50
1310	42" long		18	.444		47	17.75		64.75	81
1500	Tub bar, 1-1/4" diameter, 24" x 36"		14	.571		109	23		132	157
1600	Plus vertical arm		12	.667		122	26.50		148.50	178
1900	End tub bar, 1" diameter, 90° angle, 16" x 32"		12	.667		153	26.50		179.50	212
2010	Tub/shower/toilet, 2-wall, 36" x 24"		12	.667		140	26.50		166.50	198
2300	Hand dryer, surface mounted, electric, 115 volt, 20 amp		4	2		380	80		460	545
2400	230 volt, 10 amp		4	2		815	80		895	1,025
2600	Hat and coat strip, stainless steel, 4 hook, 36" long		24	.333		54	13.30		67.30	81.50
2700	6 hook, 60" long		20	.400		59	16		75	91.50
3000	Mirror, with stainless steel 3/4" square frame, 18" x 24"		20	.400		54.50	16		70.50	86.50
3100	36" x 24"		15	.533		137	21.50		158.50	186
3200	48" x 24"		10	.800		173	32		205	243
3300	72" x 24"		6	1.333		330	53.50		383.50	455
3500	With 5" stainless steel shelf, 18" x 24"		20	.400		227	16		243	276
3600	36" x 24"		15	.533		300	21.50		321.50	365

10 28 Toilet, Bath, and Laundry Accessories

10 28 13 – Toilet Accessories

10 28 13.13 Commercial Toilet Accessories

	Crew	Daily Output	Labor-Hours	Unit	Material	2009 Bare Costs Labor	Equipment	Total	Total Incl O&P
3700 48" x 24"	1 Carp	10	.800	Ea.	335	32		367	425
3800 72" x 24"		6	1.333		375	53.50		428.50	505
4100 Mop holder strip, stainless steel, 5 holders, 48" long		20	.400		80.50	16		96.50	115
4200 Napkin/tampon dispenser, recessed		15	.533		820	21.50		841.50	940
4300 Robe hook, single, regular		36	.222		15.50	8.90		24.40	31.50
4400 Heavy duty, concealed mounting		36	.222		15.55	8.90		24.45	32
4600 Soap dispenser, chrome, surface mounted, liquid		20	.400		47.50	16		63.50	78.50
4700 Powder		20	.400		46.50	16		62.50	78
5000 Recessed stainless steel, liquid		10	.800		140	32		172	207
5100 Powder		10	.800		213	32		245	287
5300 Soap tank, stainless steel, 1 gallon		10	.800		190	32		222	262
5400 5 gallon		5	1.600		262	64		326	395
5600 Shelf, stainless steel, 5" wide, 18 ga., 24" long		24	.333		63.50	13.30		76.80	91.50
5700 48" long		16	.500		100	20		120	143
5800 8" wide shelf, 18 ga., 24" long		22	.364		77.50	14.55		92.05	109
5900 48" long		14	.571		124	23		147	175
6000 Toilet seat cover dispenser, stainless steel, recessed		20	.400		143	16		159	184
6050 Surface mounted		15	.533		34.50	21.50		56	73
6100 Toilet tissue dispenser, surface mounted, SS, single roll		30	.267		19.45	10.65		30.10	39
6200 Double roll		24	.333		24	13.30		37.30	48
6290 Toilet seat	1 Plum	40	.200		21.50	9.75		31.25	38.50
6400 Towel bar, stainless steel, 18" long	1 Carp	23	.348		39.50	13.90		53.40	66.50
6500 30" long		21	.381		101	15.20		116.20	136
6700 Towel dispenser, stainless steel, surface mounted		16	.500		42	20		62	79
6800 Flush mounted, recessed		10	.800		270	32		302	350
7000 Towel holder, hotel type, 2 guest size		20	.400		54.50	16		70.50	86.50
7200 Towel shelf, stainless steel, 24" long, 8" wide		20	.400		93	16		109	129
7400 Tumbler holder, tumbler only		30	.267		38.50	10.65		49.15	60
7500 Soap, tumbler & toothbrush		30	.267		25	10.65		35.65	45
7700 Wall urn ash receiver, surface mount, 11" long		12	.667		124	26.50		150.50	180
8000 Waste receptacles, stainless steel, with top, 13 gallon		10	.800		330	32		362	415
8100 36 gallon		8	1		515	40		555	630
9000 Minimum labor/equipment charge		5	1.600	Job		64		64	106

10 28 16 – Bath Accessories

10 28 16.20 Medicine Cabinets

	Crew	Daily Output	Labor-Hours	Unit	Material	2009 Bare Costs Labor	Equipment	Total	Total Incl O&P
0010 **MEDICINE CABINETS**									
0020 With mirror, st. st. frame, 16" x 22", unlighted	1 Carp	14	.571	Ea.	81.50	23		104.50	127
0100 Wood frame		14	.571		113	23		136	162
0300 Sliding mirror doors, 20" x 16" x 4-3/4", unlighted		7	1.143		101	45.50		146.50	187
0400 24" x 19" x 8-1/2", lighted		5	1.600		159	64		223	281
0600 Triple door, 30" x 32", unlighted, plywood body		7	1.143		239	45.50		284.50	340
0700 Steel body		7	1.143		315	45.50		360.50	420
0900 Oak door, wood body, beveled mirror, single door		7	1.143		154	45.50		199.50	245
1000 Double door		6	1.333		375	53.50		428.50	500
1200 Hotel cabinets, stainless, with lower shelf, unlighted		10	.800		202	32		234	275
1300 Lighted		5	1.600		300	64		364	435
9000 Minimum labor/equipment charge		4	2	Job		80		80	132

277

10 31 Manufactured Fireplaces

10 31 13 – Manufactured Fireplace Chimneys

10 31 13.10 Fireplace Chimneys

		Crew	Daily Output	Labor-Hours	Unit	Material	2009 Bare Costs Labor	Equipment	Total	Total Incl O&P
0010	**FIREPLACE CHIMNEYS**									
0500	Chimney dbl. wall, all stainless, over 8'-6", 7" diam., add to fireplace	1 Carp	33	.242	V.L.F.	70	9.70		79.70	93
0600	10" diameter, add to fireplace		32	.250		93	10		103	119
0700	12" diameter, add to fireplace		31	.258		111	10.30		121.30	139
0800	14" diameter, add to fireplace		30	.267		180	10.65		190.65	216
1000	Simulated brick chimney top, 4' high, 16" x 16"		10	.800	Ea.	244	32		276	320
1100	24" x 24"		7	1.143	"	455	45.50		500.50	575

10 31 13.20 Chimney Accessories

		Crew	Daily Output	Labor-Hours	Unit	Material	2009 Bare Costs Labor	Equipment	Total	Total Incl O&P
0010	**CHIMNEY ACCESSORIES**									
0020	Chimney screens, galv., 13" x 13" flue	1 Bric	8	1	Ea.	52	40.50		92.50	123
0050	24" x 24" flue		5	1.600		121	65		186	238
0200	Stainless steel, 13" x 13" flue		8	1		320	40.50		360.50	415
0250	20" x 20" flue		5	1.600		435	65		500	585
2400	Squirrel and bird screens, galvanized, 8" x 8" flue		16	.500		45.50	20.50		66	83
2450	13" x 13" flue		12	.667		49	27		76	97.50
9000	Minimum labor/equipment charge		3.50	2.286	Job		92.50		92.50	150

10 31 16 – Manufactured Fireplace Forms

10 31 16.10 Fireplace Forms

		Crew	Daily Output	Labor-Hours	Unit	Material	2009 Bare Costs Labor	Equipment	Total	Total Incl O&P
0010	**FIREPLACE FORMS**									
1800	Fireplace forms, no accessories, 32" opening	1 Bric	3	2.667	Ea.	620	108		728	855
1900	36" opening		2.50	3.200		785	130		915	1,075
2000	40" opening		2	4		1,050	162		1,212	1,400
2100	78" opening		1.50	5.333		1,525	216		1,741	2,025

10 31 23 – Prefabricated Fireplaces

10 31 23.10 Fireplace, Prefabricated

		Crew	Daily Output	Labor-Hours	Unit	Material	2009 Bare Costs Labor	Equipment	Total	Total Incl O&P
0010	**FIREPLACE, PREFABRICATED**, free standing or wall hung									
0100	With hood & screen, minimum	1 Carp	1.30	6.154	Ea.	1,175	246		1,421	1,675
0150	Average		1	8		1,625	320		1,945	2,325
0200	Maximum		.90	8.889		3,125	355		3,480	4,000
1500	Simulated logs, gas fired, 40,000 BTU, 2' long, minimum		7	1.143	Set	400	45.50		445.50	515
1600	Maximum		6	1.333		940	53.50		993.50	1,125
1700	Electric, 1,500 BTU, 1'-6" long, minimum		7	1.143		166	45.50		211.50	259
1800	11,500 BTU, maximum		6	1.333		370	53.50		423.50	500
2000	Fireplace, built-in, 36" hearth, radiant		1.30	6.154	Ea.	560	246		806	1,025
2100	Recirculating, small fan		1	8		875	320		1,195	1,500
2150	Large fan		.90	8.889		1,825	355		2,180	2,575
2200	42" hearth, radiant		1.20	6.667		790	266		1,056	1,300
2300	Recirculating, small fan		.90	8.889		1,025	355		1,380	1,700
2350	Large fan		.80	10		1,825	400		2,225	2,650
2400	48" hearth, radiant		1.10	7.273		1,700	291		1,991	2,350
2500	Recirculating, small fan		.80	10		2,125	400		2,525	2,975
2550	Large fan		.70	11.429		3,275	455		3,730	4,350
3000	See through, including doors		.80	10		2,775	400		3,175	3,700
3200	Corner (2 wall)		1	8		2,775	320		3,095	3,600

10 32 Fireplace Specialties

10 32 13 – Fireplace Dampers

10 32 13.10 Dampers

		Crew	Daily Output	Labor-Hours	Unit	Material	2009 Bare Costs Labor	Equipment	Total	Total Incl O&P
0010	**DAMPERS**									
0800	Damper, rotary control, steel, 30" opening	1 Bric	6	1.333	Ea.	77.50	54		131.50	173
0850	Cast iron, 30" opening		6	1.333		86	54		140	182
0880	36" opening		6	1.333		93.50	54		147.50	191
0900	48" opening		6	1.333		133	54		187	235
0920	60" opening		6	1.333		299	54		353	420
1000	84" opening, special order		5	1.600		770	65		835	950
1050	96" opening, special order		4	2		780	81		861	990
1200	Steel plate, poker control, 60" opening		8	1		273	40.50		313.50	365
1250	84" opening, special order		5	1.600		495	65		560	650
1400	"Universal" type, chain operated, 32" x 20" opening		8	1		212	40.50		252.50	299
1450	48" x 24" opening		5	1.600		315	65		380	455

10 32 23 – Fireplace Doors

10 32 23.10 Doors

		Crew	Daily Output	Labor-Hours	Unit	Material	2009 Bare Costs Labor	Equipment	Total	Total Incl O&P
0010	**DOORS**									
0400	Cleanout doors and frames, cast iron, 8" x 8"	1 Bric	12	.667	Ea.	36.50	27		63.50	83.50
0450	12" x 12"		10	.800		44	32.50		76.50	101
0500	18" x 24"		8	1		127	40.50		167.50	206
0550	Cast iron frame, steel door, 24" x 30"		5	1.600		275	65		340	405
1600	Dutch Oven door and frame, cast iron, 12" x 15" opening		13	.615		111	25		136	163
1650	Copper plated, 12" x 15" opening		13	.615		218	25		243	280

10 35 Stoves

10 35 13 – Heating Stoves

10 35 13.10 Woodburning Stoves

		Crew	Daily Output	Labor-Hours	Unit	Material	2009 Bare Costs Labor	Equipment	Total	Total Incl O&P
0010	**WOODBURNING STOVES**									
0015	Cast iron, minimum	2 Carp	1.30	12.308	Ea.	1,125	490		1,615	2,050
0020	Average		1	16		1,525	640		2,165	2,725
0030	Maximum		.80	20		2,750	800		3,550	4,350
0050	For gas log lighter, add					37.50			37.50	41.50

10 44 Fire Protection Specialties

10 44 13 – Fire Extinguisher Cabinets

10 44 13.53 Fire Equipment Cabinets

		Crew	Daily Output	Labor-Hours	Unit	Material	2009 Bare Costs Labor	Equipment	Total	Total Incl O&P
0010	**FIRE EQUIPMENT CABINETS**, not equipped, 20 ga. steel box									
0040	recessed, D.S. glass in door, box size given									
1000	Portable extinguisher, single, 8" x 12" x 27", alum. door & frame	Q-12	8	2	Ea.	124	86.50		210.50	271
1100	Steel door and frame	"	8	2	"	86.50	86.50		173	230
3000	Hose rack assy., 1-1/2" valve & 100' hose, 24" x 40" x 5-1/2"									
3200	Steel door and frame	Q-12	6	2.667	Ea.	272	116		388	480
4000	Hose rack assy., 2-1/2" x 1-1/2" valve, 100' hose, 24" x 40" x 8"									
4200	Steel door and frame	Q-12	6	2.667	Ea.	208	116		324	410
5000	Hose rack assy., 2-1/2" x 1-1/2" valve, 100' hose									
5010	and extinguisher, 30" x 40" x 8"									
5200	Steel door and frame	Q-12	5	3.200	Ea.	208	139		347	445

10 44 16 – Fire Extinguishers

10 44 16.13 Portable Fire Extinguishers

0010	**PORTABLE FIRE EXTINGUISHERS**									

10 44 Fire Protection Specialties

10 44 16 – Fire Extinguishers

10 44 16.13 Portable Fire Extinguishers

	10 44 16.13 Portable Fire Extinguishers	Crew	Daily Output	Labor-Hours	Unit	Material	2009 Bare Costs Labor	Equipment	Total	Total Incl O&P
0120	CO₂, portable with swivel horn, 5 lb.				Ea.	159			159	175
0140	With hose and "H" horn, 10 lb.				"	215			215	237
1000	Dry chemical, pressurized									
1040	Standard type, portable, painted, 2-1/2 lb.				Ea.	35			35	38.50
1080	10 lb.					91			91	100
1100	20 lb.					135			135	149
1120	30 lb.					295			295	325
2000	ABC all purpose type, portable, 2-1/2 lb.					34			34	37
2080	9-1/2 lb.					75			75	82.50

10 51 Lockers

10 51 13 – Metal Lockers

10 51 13.10 Lockers

		Crew	Daily Output	Labor-Hours	Unit	Material	2009 Bare Costs Labor	Equipment	Total	Total Incl O&P
0011	**LOCKERS** Steel, baked enamel									
0110	Single tier box locker, 12" x 15" x 72"	1 Shee	8	1	Ea.	183	47		230	276
0120	18" x 15" x 72"		8	1		205	47		252	300
0130	12" x 18" x 72"		8	1		198	47		245	293
0140	18" x 18" x 72"		8	1		212	47		259	310
0410	Double tier, 12" x 15" x 36"		21	.381		199	18		217	248
0420	18" x 15" x 36"		21	.381		251	18		269	305
0430	12" x 18" x 36"		21	.381		201	18		219	250
0440	18" x 18" x 36"		21	.381		263	18		281	320
0500	Two person, 18" x 15" x 72"		8	1		253	47		300	355
0510	18" x 18" x 72"		8	1		271	47		318	375
0520	Duplex, 15" x 15" x 72"		8	1		299	47		346	405
0530	15" x 21" x 72"		8	1		325	47		372	430
1100	Wire meshed wardrobe, floor. mtd., open front varsity type		7.50	1.067		217	50.50		267.50	320
2400	16-person locker unit with clothing rack									
2500	72 wide x 15" deep x 72" high	1 Shee	8	1	Ea.	450	47		497	570
2550	18" deep		8	1	"	470	47		517	590
3250	Rack w/ 24 wire mesh baskets		1.50	5.333	Set	345	252		597	780
3260	30 baskets		1.25	6.400		267	300		567	775
3270	36 baskets		.95	8.421		350	395		745	1,025
3280	42 baskets		.80	10		385	470		855	1,175
3600	For hanger rods, add				Ea.	1.86			1.86	2.05
9000	Minimum labor/equipment charge	1 Shee	2.50	3.200	Job		151		151	240

10 55 Postal Specialties

10 55 23 – Mail Boxes

10 55 23.10 Commercial Mail Boxes

		Crew	Daily Output	Labor-Hours	Unit	Material	2009 Bare Costs Labor	Equipment	Total	Total Incl O&P
0010	**COMMERCIAL MAIL BOXES**									
0020	Horiz., key lock, 5"H x 6"W x 15"D, alum., rear load	1 Carp	34	.235	Ea.	35	9.40		44.40	54
0100	Front loading		34	.235		36	9.40		45.40	55
0200	Double, 5"H x 12"W x 15"D, rear loading		26	.308		71.50	12.30		83.80	99.50
0300	Front loading		26	.308		66	12.30		78.30	93
0500	Quadruple, 10"H x 12"W x 15"D, rear loading		20	.400		91	16		107	127
0600	Front loading		20	.400		108	16		124	146
1600	Vault type, horizontal, for apartments, 4" x 5"		34	.235		41	9.40		50.40	61
1700	Alphabetical directories, 120 names		10	.800		126	32		158	191

10 55 Postal Specialties

10 55 23 – Mail Boxes

	10 55 23.10 Commercial Mail Boxes	Crew	Daily Output	Labor-Hours	Unit	Material	2009 Bare Costs Labor	Equipment	Total	Total Incl O&P
1830	Lobby collection boxes, aluminum	2 Shee	5	3.200	Ea.	2,025	151		2,176	2,475
1840	Bronze or stainless	"	4.50	3.556		2,500	168		2,668	3,025
1900	Letter slot, residential	1 Carp	20	.400		86	16		102	121
2000	Post office type		8	1		110	40		150	187
9000	Minimum labor/equipment charge		5	1.600	Job		64		64	106

10 55 91 – Mail Chutes

	10 55 91.10 Commercial Mail Chutes	Crew	Daily Output	Labor-Hours	Unit	Material	2009 Bare Costs Labor	Equipment	Total	Total Incl O&P
0010	**COMMERCIAL MAIL CHUTES**									
0020	Aluminum & glass, 14-1/4" wide, 4-5/8" deep	2 Shee	4	4	Floor	705	189		894	1,075
0100	8-5/8" deep	"	3.80	4.211	"	855	199		1,054	1,250

10 56 Storage Assemblies

10 56 13 – Metal Storage Shelving

	10 56 13.10 Shelving	Crew	Daily Output	Labor-Hours	Unit	Material	2009 Bare Costs Labor	Equipment	Total	Total Incl O&P
0010	**SHELVING**									
0020	Metal, industrial, cross-braced, 3' wide, 12" deep	1 Sswk	175	.046	SF Shlf	6.55	2.04		8.59	11
0100	24" deep		330	.024		4.70	1.08		5.78	7.15
2200	Wide span, 1600 lb. capacity per shelf, 6' wide, 24" deep		380	.021		7	.94		7.94	9.45
2400	36" deep		440	.018		8.30	.81		9.11	10.60
9000	Minimum labor/equipment charge	1 Carp	4	2	Job		80		80	132

10 57 Wardrobe and Closet Specialties

10 57 23 – Closet and Utility Shelving

	10 57 23.19 Wood Closet and Utility Shelving	Crew	Daily Output	Labor-Hours	Unit	Material	2009 Bare Costs Labor	Equipment	Total	Total Incl O&P
0010	**WOOD CLOSET AND UTILITY SHELVING**									
0020	Pine, clear grade, no edge band, 1" x 8"	1 Carp	115	.070	L.F.	1.98	2.78		4.76	6.75
0100	1" x 10"		110	.073		2.22	2.91		5.13	7.25
0200	1" x 12"		105	.076		2.64	3.04		5.68	7.90
0600	Plywood, 3/4" thick with lumber edge, 12" wide		75	.107		1.67	4.26		5.93	8.90
0700	24" wide		70	.114		3	4.57		7.57	10.85
0900	Bookcase, clear grade pine, shelves 12" O.C., 8" deep, /SF shelf		70	.114	S.F.	6.45	4.57		11.02	14.65
1000	12" deep shelves		65	.123	"	8.60	4.92		13.52	17.55
1200	Adjustable closet rod and shelf, 12" wide, 3' long		20	.400	Ea.	8.50	16		24.50	36
1300	8' long		15	.533	"	24	21.50		45.50	61.50
1500	Prefinished shelves with supports, stock, 8" wide		75	.107	L.F.	3.80	4.26		8.06	11.25
1600	10" wide		70	.114	"	4.21	4.57		8.78	12.20
9000	Minimum labor/equipment charge		4	2	Job		80		80	132

10 73 Protective Covers

10 73 13 – Awnings

10 73 13.10 Awnings, Fabric	Crew	Daily Output	Labor-Hours	Unit	Material	2009 Bare Costs Labor	Equipment	Total	Total Incl O&P
0010 **AWNINGS, FABRIC**									
0020 Including acrylic canvas and frame, standard design									
0100 Door and window, slope, 3' high, 4' wide	1 Carp	4.50	1.778	Ea.	690	71		761	875
0110 6' wide		3.50	2.286		890	91.50		981.50	1,125
0120 8' wide		3	2.667		1,100	107		1,207	1,375
0200 Quarter round convex, 4' wide		3	2.667		1,075	107		1,182	1,350
0210 6' wide		2.25	3.556		1,400	142		1,542	1,750
0220 8' wide		1.80	4.444		1,700	178		1,878	2,175
0300 Dome, 4' wide		7.50	1.067		415	42.50		457.50	525
0310 6' wide		3.50	2.286		935	91.50		1,026.50	1,175
0320 8' wide		2	4		1,650	160		1,810	2,100
0350 Elongated dome, 4' wide		1.33	6.015		1,550	240		1,790	2,100
0360 6' wide		1.11	7.207		1,850	288		2,138	2,525
0370 8' wide		1	8		2,175	320		2,495	2,925
1000 Entry or walkway, peak, 12' long, 4' wide	2 Carp	.90	17.778		4,925	710		5,635	6,575
1010 6' wide		.60	26.667		7,575	1,075		8,650	10,100
1020 8' wide		.40	40		10,500	1,600		12,100	14,200
1100 Radius with dome end, 4' wide		1.10	14.545		3,725	580		4,305	5,050
1110 6' wide		.70	22.857		6,000	915		6,915	8,100
1120 8' wide		.50	32		8,525	1,275		9,800	11,500
2000 Retractable lateral arm awning, manual									
2010 To 12' wide, 8'-6" projection	2 Carp	1.70	9.412	Ea.	1,100	375		1,475	1,850
2020 To 14' wide, 8'-6" projection		1.10	14.545		1,300	580		1,880	2,375
2030 To 19' wide, 8'-6" projection		.85	18.824		1,750	750		2,500	3,175
2040 To 24' wide, 8'-6" projection		.67	23.881		2,225	955		3,180	4,025
2050 Motor for above, add	1 Carp	2.67	3		955	120		1,075	1,250
3000 Patio/deck canopy with frame									
3010 12' wide, 12' projection	2 Carp	2	8	Ea.	1,575	320		1,895	2,250
3020 16' wide, 14' projection	"	1.20	13.333		2,450	535		2,985	3,550
9000 For fire retardant canvas, add					7%				
9010 For lettering or graphics, add					35%				
9020 For painted or coated acrylic canvas, deduct					8%				
9030 For translucent or opaque vinyl canvas, add					10%				
9040 For 6 or more units, deduct					20%	15%			

10 73 16 – Canopies

10 73 16.10 Canopies, Residential	Crew	Daily Output	Labor-Hours	Unit	Material	2009 Bare Costs Labor	Equipment	Total	Total Incl O&P
0010 **CANOPIES, RESIDENTIAL** Prefabricated									
0500 Carport, free standing, baked enamel, alum., .032", 40 psf									
0520 16' x 8', 4 posts	2 Carp	3	5.333	Ea.	4,075	213		4,288	4,825
0600 20' x 10', 6 posts		2	8		4,250	320		4,570	5,200
0605 30' x 10', 8 posts		2	8		6,375	320		6,695	7,525
1000 Door canopies, extruded alum., .032", 42" projection, 4' wide	1 Carp	8	1		285	40		325	380
1020 6' wide	"	6	1.333		360	53.50		413.50	485
1040 8' wide	2 Carp	9	1.778		455	71		526	615
1060 10' wide		7	2.286		670	91.50		761.50	885
1080 12' wide		5	3.200		775	128		903	1,050
1200 54" projection, 4' wide	1 Carp	8	1		335	40		375	435
1220 6' wide	"	6	1.333		420	53.50		473.50	550
1240 8' wide	2 Carp	9	1.778		485	71		556	650
1260 10' wide		7	2.286		725	91.50		816.50	945
1280 12' wide		5	3.200		775	128		903	1,050
1300 Painted, add					20%				

10 73 Protective Covers

10 73 16 – Canopies

10 73 16.10 Canopies, Residential

		Crew	Daily Output	Labor-Hours	Unit	Material	2009 Bare Costs Labor	2009 Bare Costs Equipment	Total	Total Incl O&P
1310	Bronze anodized, add				Ea.	50%				
3000	Window awnings, aluminum, window 3' high, 4' wide	1 Carp	10	.800		325	32		357	415
3020	6' wide	"	8	1		330	40		370	430
3040	9' wide	2 Carp	9	1.778		310	71		381	455
3060	12' wide	"	5	3.200		360	128		488	605
3100	Window, 4' high, 4' wide	1 Carp	10	.800		162	32		194	231
3120	6' wide	"	8	1		230	40		270	320
3140	9' wide	2 Carp	9	1.778		385	71		456	535
3160	12' wide	"	5	3.200		410	128		538	665
3200	Window, 6' high, 4' wide	1 Carp	10	.800		241	32		273	320
3220	6' wide	"	8	1		370	40		410	470
3240	9' wide	2 Carp	9	1.778		875	71		946	1,075
3260	12' wide	"	5	3.200		1,350	128		1,478	1,675
3400	Roll-up aluminum, 2'-6" wide	1 Carp	14	.571		185	23		208	242
3420	3' wide		12	.667		196	26.50		222.50	259
3440	4' wide		10	.800		226	32		258	300
3460	6' wide		8	1		278	40		318	370
3480	9' wide	2 Carp	9	1.778		360	71		431	510
3500	12' wide	"	5	3.200		465	128		593	720
3600	Window awnings, canvas, 24" drop, 3' wide	1 Carp	30	.267	L.F.	52.50	10.65		63.15	75.50
3620	4' wide		40	.200		45.50	8		53.50	63
3700	30" drop, 3' wide		30	.267		79.50	10.65		90.15	105
3720	4' wide		40	.200		66	8		74	86
3740	5' wide		45	.178		58.50	7.10		65.60	76
3760	6' wide		48	.167		53	6.65		59.65	69.50
3780	8' wide		48	.167		43.50	6.65		50.15	58.50
3800	10' wide		50	.160		41	6.40		47.40	56
3900	Repair canvas, minimum		8	1	Ea.		40		40	66
3920	Maximum		4	2	"		80		80	132
9000	Minimum labor/equipment charge		3	2.667	Job		107		107	176

10 73 16.20 Metal Canopies

		Crew	Daily Output	Labor-Hours	Unit	Material	2009 Bare Costs Labor	2009 Bare Costs Equipment	Total	Total Incl O&P
0010	**METAL CANOPIES**									
0020	Wall hung, .032", aluminum, prefinished, 8' x 10'	K-2	1.30	18.462	Ea.	2,150	755	187	3,092	3,900
0300	8' x 20'		1.10	21.818		4,275	890	221	5,386	6,550
1000	12' x 20'		1	24		5,475	980	243	6,698	8,050
1050										
2300	Aluminum entrance canopies, flat soffit, .032"									
2500	3'-6" x 4'-0", clear anodized	2 Carp	4	4	Ea.	900	160		1,060	1,250
9000	Minimum labor/equipment charge	"	2	8	Job		320		320	530

10 74 Manufactured Exterior Specialties

10 74 23 – Cupolas

10 74 23.10 Wood Cupolas

		Crew	Daily Output	Labor-Hours	Unit	Material	2009 Bare Costs Labor	2009 Bare Costs Equipment	Total	Total Incl O&P
0010	**WOOD CUPOLAS**									
0020	Stock units, pine, painted, 18" sq., 28" high, alum. roof	1 Carp	4.10	1.951	Ea.	162	78		240	305
0100	Copper roof		3.80	2.105		189	84		273	345
0300	23" square, 33" high, aluminum roof		3.70	2.162		355	86.50		441.50	535
0400	Copper roof		3.30	2.424		480	97		577	685
0600	30" square, 37" high, aluminum roof		3.70	2.162		495	86.50		581.50	690
0700	Copper roof		3.30	2.424		535	97		632	750
0900	Hexagonal, 31" wide, 46" high, copper roof		4	2		660	80		740	855

10 74 Manufactured Exterior Specialties

10 74 23 – Cupolas

10 74 23.10 Wood Cupolas

		Crew	Daily Output	Labor-Hours	Unit	Material	2009 Bare Costs Labor	2009 Bare Costs Equipment	Total	Total Incl O&P
1000	36" wide, 50" high, copper roof	1 Carp	3.50	2.286	Ea.	1,150	91.50		1,241.50	1,425
1200	For deluxe stock units, add to above					25%				
1400	For custom built units, add to above					50%	50%			
9000	Minimum labor/equipment charge	1 Carp	2.75	2.909	Job		116		116	192

10 74 29 – Steeples

10 74 29.10 Prefabricated Steeples

		Crew	Daily Output	Labor-Hours	Unit	Material	2009 Bare Costs Labor	2009 Bare Costs Equipment	Total	Total Incl O&P
0010	**PREFABRICATED STEEPLES**									
4000	Steeples, translucent fiberglass, 30" square, 15' high	F-3	2	20	Ea.	5,450	810	385	6,645	7,750
4150	25' high	"	1.80	22.222		6,350	900	425	7,675	8,950
4600	Aluminum, baked finish, 16" square, 14' high					4,200			4,200	4,625
4640	35' high, 8' base					27,400			27,400	30,200
4680	152' high, custom					495,000			495,000	544,500

10 74 33 – Weathervanes

10 74 33.10 Residential Weathervanes

		Crew	Daily Output	Labor-Hours	Unit	Material	2009 Bare Costs Labor	2009 Bare Costs Equipment	Total	Total Incl O&P
0010	**RESIDENTIAL WEATHERVANES**									
0020	Residential types, minimum	1 Carp	8	1	Ea.	183	40		223	268
0100	Maximum		2	4	"	1,750	160		1,910	2,200
9000	Minimum labor/equipment charge		4	2	Job		80		80	132

Estimating Tips

General

- The items in this division are usually priced per square foot or each. Many of these items are purchased by the owner for installation by the contractor. Check the specifications for responsibilities and include time for receiving, storage, installation, and mechanical and electrical hookups in the appropriate divisions.

- Many items in Division 11 require some type of support system that is not usually furnished with the item. Examples of these systems include blocking for the attachment of casework and support angles for ceiling-hung projection screens. The required blocking or supports must be added to the estimate in the appropriate division.

- Some items in Division 11 may require assembly or electrical hookups. Verify the amount of assembly required or the need for a hard electrical connection and add the appropriate costs.

Reference Numbers

Reference numbers are shown in shaded boxes at the beginning of some major classifications. These numbers refer to related items in the Reference Section. The reference information may be an estimating procedure, an alternate pricing method, or technical information.

Note: Not all subdivisions listed here necessarily appear in this publication.

11 11 Vehicle Service Equipment

11 11 13 – Compressed-Air Vehicle Service Equipment

11 11 13.10 Compressed Air Equipment	Crew	Daily Output	Labor-Hours	Unit	Material	2009 Bare Costs Labor	Equipment	Total	Total Incl O&P
0010 COMPRESSED AIR EQUIPMENT									
0030 Compressors, electric, 1-1/2 H.P., standard controls	L-4	1.50	16	Ea.	370	595		965	1,375
0550 Dual controls		1.50	16		695	595		1,290	1,750
0600 5 H.P., 115/230 volt, standard controls		1	24		1,900	895		2,795	3,550
0650 Dual controls	↓	1	24	↓	2,025	895		2,920	3,700

11 11 19 – Vehicle Lubrication Equipment

11 11 19.10 Lubrication Equipment

	Crew	Daily Output	Labor-Hours	Unit	Material	Labor	Equipment	Total	Total Incl O&P
0010 LUBRICATION EQUIPMENT									
3000 Lube equipment, 3 reel type, with pumps, not including piping	L-4	.50	48	Set	7,900	1,800		9,700	11,600

11 11 33 – Vehicle Spray Painting Equipment

11 11 33.10 Spray Painting Equipment

	Crew	Daily Output	Labor-Hours	Unit	Material	Labor	Equipment	Total	Total Incl O&P
0010 SPRAY PAINTING EQUIPMENT									
4000 Spray painting booth, 26' long, complete	L-4	.40	60	Ea.	15,400	2,250		17,650	20,600

11 12 Parking Control Equipment

11 12 13 – Parking Key and Card Control Units

11 12 13.10 Parking Control Units

	Crew	Daily Output	Labor-Hours	Unit	Material	Labor	Equipment	Total	Total Incl O&P
0010 PARKING CONTROL UNITS									
5100 Card reader	1 Elec	2	4	Ea.	1,825	188		2,013	2,300
5120 Proximity with customer display	2 Elec	1	16		5,500	750		6,250	7,200
6000 Parking control software, minimum	1 Elec	.50	16		21,900	750		22,650	25,300
6020 Maximum	"	.20	40	↓	96,000	1,875		97,875	108,500

11 12 16 – Parking Ticket Dispensers

11 12 16.10 Ticket Dispensers

	Crew	Daily Output	Labor-Hours	Unit	Material	Labor	Equipment	Total	Total Incl O&P
0010 TICKET DISPENSERS									
5900 Ticket spitter with time/date stamp, standard	2 Elec	2	8	Ea.	6,275	375		6,650	7,475
5920 Mag stripe encoding	"	2	8	"	18,600	375		18,975	21,000

11 12 26 – Parking Fee Collection Equipment

11 12 26.13 Parking Fee Coin Collection Equipment

	Crew	Daily Output	Labor-Hours	Unit	Material	Labor	Equipment	Total	Total Incl O&P
0010 PARKING FEE COIN COLLECTION EQUIPMENT									
5200 Cashier booth, average	B-22	1	30	Ea.	9,575	1,100	194	10,869	12,500
5300 Collector station, pay on foot	2 Elec	.20	80		109,500	3,750		113,250	126,500
5320 Credit card only	"	.50	32	↓	20,200	1,500		21,700	24,600

11 12 26.23 Fee Equipment

	Crew	Daily Output	Labor-Hours	Unit	Material	Labor	Equipment	Total	Total Incl O&P
0010 FEE EQUIPMENT									
5600 Fee computer	1 Elec	1.50	5.333	Ea.	13,300	251		13,551	15,000

11 12 33 – Parking Gates

11 12 33.13 Parking Gate Equipment

	Crew	Daily Output	Labor-Hours	Unit	Material	Labor	Equipment	Total	Total Incl O&P
0010 PARKING GATE EQUIPMENT									
5000 Barrier gate with programmable controller	2 Elec	3	5.333	Ea.	3,300	251		3,551	4,000
5020 Industrial		3	5.333		4,525	251		4,776	5,350
5500 Exit verifier	↓	1	16		17,400	750		18,150	20,400
5700 Full sign, 4" letters	1 Elec	2	4		1,225	188		1,413	1,625
5800 Inductive loop	2 Elec	4	4		169	188		357	475
5950 Vehicle detector, microprocessor based	1 Elec	3	2.667	↓	385	125		510	620

11 13 Loading Dock Equipment

11 13 13 – Loading Dock Bumpers

11 13 13.10 Dock Bumpers

		Crew	Daily Output	Labor-Hours	Unit	Material	2009 Bare Costs Labor	Equipment	Total	Total Incl O&P
0010	**DOCK BUMPERS** Bolts not included									
0012	2" x 6" to 4" x 8", average	1 Carp	300	.027	B.F.	.63	1.07		1.70	2.45
0050	Bumpers, rubber blocks 4-1/2" thick, 10" high, 14" long		26	.308	Ea.	45.50	12.30		57.80	70.50
0200	24" long		22	.364		86.50	14.55		101.05	119
0300	36" long		17	.471		82.50	18.80		101.30	122
0500	12" high, 14" long		25	.320		91	12.80		103.80	121

11 13 16 – Loading Dock Seals and Shelters

11 13 16.10 Dock Seals and Shelters

		Crew	Daily Output	Labor-Hours	Unit	Material	2009 Bare Costs Labor	Equipment	Total	Total Incl O&P
0010	**DOCK SEALS AND SHELTERS**									
6200	Shelters, fabric, for truck or train, scissor arms, minimum	1 Carp	1	8	Ea.	1,475	320		1,795	2,150
6300	Maximum	"	.50	16	"	1,925	640		2,565	3,150

11 13 19 – Stationary Loading Dock Equipment

11 13 19.10 Dock Equipment

		Crew	Daily Output	Labor-Hours	Unit	Material	2009 Bare Costs Labor	Equipment	Total	Total Incl O&P
0010	**DOCK EQUIPMENT**									
2200	Dock boards, heavy duty, 60" x 60", aluminum, 5,000 lb. capacity				Ea.	1,150			1,150	1,250
4200	Platform lifter, 6' x 6', portable, 3,000 lb. capacity				"	8,300			8,300	9,150

11 16 Vault Equipment

11 16 16 – Safes

11 16 16.10 Commercial Safes

		Crew	Daily Output	Labor-Hours	Unit	Material	2009 Bare Costs Labor	Equipment	Total	Total Incl O&P
0010	**COMMERCIAL SAFES**									
0800	Money, "B" label, 9" x 14" x 14"				Ea.	525			525	580
0900	Tool resistive, 24" x 24" x 20"					3,575			3,575	3,925
1050	Tool and torch resistive, 24" x 24" x 20"					9,150			9,150	10,100

11 17 Teller and Service Equipment

11 17 13 – Teller Equipment Systems

11 17 13.10 Bank Equipment

		Crew	Daily Output	Labor-Hours	Unit	Material	2009 Bare Costs Labor	Equipment	Total	Total Incl O&P
0010	**BANK EQUIPMENT**									
0020	Alarm system, police	2 Elec	1.60	10	Ea.	4,675	470		5,145	5,875
0400	Bullet resistant teller window, 44" x 60"	1 Glaz	.60	13.333		3,225	515		3,740	4,375
0500	48" x 60"	"	.60	13.333		4,100	515		4,615	5,350
3000	Counters for banks, frontal only	2 Carp	1	16	Station	1,750	640		2,390	2,975
3100	Complete with steel undercounter		.50	32	"	3,400	1,275		4,675	5,850
5400	Partitions, bullet-resistant, 1-3/16" glass, 8' high		10	1.600	L.F.	189	64		253	315
5600	Pass thru, bullet-res. window, painted steel, 24" x 36"	2 Sswk	1.60	10	Ea.	2,250	445		2,695	3,300

11 19 Detention Equipment

11 19 30 - Detention Cell Equipment

11 19 30.10 Cell Equipment	Crew	Daily Output	Labor-Hours	Unit	Material	2009 Bare Costs Labor	Equipment	Total	Total Incl O&P
0010 **CELL EQUIPMENT**									
3000 Toilet apparatus including wash basin, average	L-8	1.50	13.333	Ea.	3,350	555		3,905	4,600

11 21 Mercantile and Service Equipment

11 21 13 - Cash Registers and Checking Equipment

11 21 13.10 Checkout Counter

11 21 13.10	Crew	Daily Output	Labor-Hours	Unit	Material	2009 Bare Costs Labor	Equipment	Total	Total Incl O&P
0010 **CHECKOUT COUNTER**									
0020 Supermarket conveyor, single belt	2 Clab	10	1.600	Ea.	2,475	50.50		2,525.50	2,800
0100 Double belt, power take-away		9	1.778		4,175	56		4,231	4,700
0400 Double belt, power take-away, incl. side scanning		7	2.286		5,200	72		5,272	5,825
0800 Warehouse or bulk type		6	2.667		6,075	84.50		6,159.50	6,825
1000 Scanning system, 2 lanes, w/registers, scan gun & memory				System	15,200			15,200	16,700
1100 10 lanes, single processor, full scan, with scales				"	144,500			144,500	158,500
2000 Register, restaurant, minimum				Ea.	615			615	675
2100 Maximum					2,700			2,700	2,975
2150 Store, minimum					615			615	675
2200 Maximum					2,700			2,700	2,975

11 21 53 - Barber and Beauty Shop Equipment

11 21 53.10 Barber Equipment

11 21 53.10	Crew	Daily Output	Labor-Hours	Unit	Material	2009 Bare Costs Labor	Equipment	Total	Total Incl O&P
0010 **BARBER EQUIPMENT**									
0020 Chair, hydraulic, movable, minimum	1 Carp	24	.333	Ea.	405	13.30		418.30	465
0050 Maximum	"	16	.500		2,750	20		2,770	3,050
0200 Wall hung styling station with mirrors, minimum	L-2	8	2		355	70.50		425.50	505
0300 Maximum	"	4	4		1,825	141		1,966	2,250
0500 Sink, hair washing basin, rough plumbing not incl.	1 Plum	8	1		395	49		444	505
1000 Sterilizer, liquid solution for tools					128			128	140
1100 Total equipment, rule of thumb, per chair, minimum	L-8	1	20		1,525	835		2,360	3,025
1150 Maximum	"	1	20		4,075	835		4,910	5,850

11 23 Commercial Laundry and Dry Cleaning Equipment

11 23 13 - Dry Cleaning Equipment

11 23 13.13 Dry Cleaners

11 23 13.13	Crew	Daily Output	Labor-Hours	Unit	Material	2009 Bare Costs Labor	Equipment	Total	Total Incl O&P
0010 **DRY CLEANERS** Not incl. rough-in									
2000 Dry cleaners, electric, 20 lb. capacity	L-1	.20	80	Ea.	33,300	3,825		37,125	42,500
2050 25 lb. capacity		.17	94.118		45,100	4,500		49,600	56,500
2100 30 lb. capacity		.15	106		47,600	5,100		52,700	60,500
2150 60 lb. capacity		.09	177		73,000	8,500		81,500	93,000

11 23 16 - Drying and Conditioning Equipment

11 23 16.13 Dryers

11 23 16.13	Crew	Daily Output	Labor-Hours	Unit	Material	2009 Bare Costs Labor	Equipment	Total	Total Incl O&P
0010 **DRYERS**, Not including rough-in									
1500 Industrial, 30 lb. capacity	1 Plum	2	4	Ea.	2,975	195		3,170	3,575
1600 50 lb. capacity	"	1.70	4.706		3,200	229		3,429	3,875
4700 Lint collector, ductwork not included, 8,000 to 10,000 C.F.M.	Q-10	.30	80		8,575	3,525		12,100	15,000

11 23 19 - Finishing Equipment

11 23 19.13 Folders and Spreaders

11 23 19.13	Crew	Daily Output	Labor-Hours	Unit	Material	2009 Bare Costs Labor	Equipment	Total	Total Incl O&P
0010 **FOLDERS AND SPREADERS**									
3500 Folders, blankets & sheets, minimum	1 Elec	.17	47.059	Ea.	30,600	2,200		32,800	37,100
3700 King size with automatic stacker		.10	80		55,500	3,750		59,250	67,000

11 23 Commercial Laundry and Dry Cleaning Equipment

11 23 19 – Finishing Equipment

11 23 19.13 Folders and Spreaders

		Crew	Daily Output	Labor-Hours	Unit	Material	2009 Bare Costs Labor	2009 Bare Costs Equipment	Total	Total Incl O&P
3800	For conveyor delivery, add	1 Elec	.45	17.778	Ea.	6,000	835		6,835	7,875

11 23 23 – Commercial Ironing Equipment

11 23 23.13 Irons and Pressers

0010	**IRONS AND PRESSERS**									
4500	Ironers, institutional, 110", single roll	1 Elec	.20	40	Ea.	29,700	1,875		31,575	35,600

11 23 26 – Commercial Washers and Extractors

11 23 26.13 Washers and Extractors

0010	**WASHERS AND EXTRACTORS**, not including rough-in									
6000	Combination washer/extractor, 20 lb. capacity	L-6	1.50	8	Ea.	5,350	385		5,735	6,500
6100	30 lb. capacity		.80	15		8,500	725		9,225	10,500
6200	50 lb. capacity		.68	17.647		9,925	850		10,775	12,200
6300	75 lb. capacity		.30	40		18,600	1,925		20,525	23,500
6350	125 lb. capacity		.16	75		24,900	3,625		28,525	33,000

11 23 33 – Coin-Operated Laundry Equipment

11 23 33.13 Coin Operated Washers and Dryers

0010	**COIN OPERATED WASHERS AND DRYERS**									
0990	Dryer, gas fired									
1000	Commercial, 30 lb. capacity, coin operated, single	1 Plum	3	2.667	Ea.	3,100	130		3,230	3,600
1100	Double stacked	"	2	4	"	6,275	195		6,470	7,200
5290	Clothes washer									
5300	Commercial, coin operated, average	1 Plum	3	2.667	Ea.	1,125	130		1,255	1,450

11 24 Maintenance Equipment

11 24 19 – Vacuum Cleaning Systems

11 24 19.10 Vacuum Cleaning

0010	**VACUUM CLEANING**									
0020	Central, 3 inlet, residential	1 Skwk	.90	8.889	Total	925	365		1,290	1,625
0200	Commercial		.70	11.429		1,050	465		1,515	1,925
0400	5 inlet system, residential		.50	16		1,225	655		1,880	2,425
0600	7 inlet system, commercial		.40	20		1,400	815		2,215	2,850
4010	Rule of thumb: First 1200 S.F., installed								1,150	1,275
4020	For each additional S.F., add				S.F.					.21

11 26 Unit Kitchens

11 26 13 – Metal Unit Kitchens

11 26 13.10 Commercial Unit Kitchens

0010	**COMMERCIAL UNIT KITCHENS**									
1500	Combination range, refrigerator and sink, 30" wide, minimum	L-1	2	8	Ea.	900	385		1,285	1,575
1550	Maximum	"	1	16	"	3,175	765		3,940	4,675
1640	Combination range, refrigerator, sink, microwave									
1660	Oven and ice maker	L-1	.80	20	Ea.	3,950	960		4,910	5,825

11 27 Photographic Processing Equipment

11 27 13 – Darkroom Processing Equipment

11 27 13.10 Darkroom Equipment

11 27 13.10 Darkroom Equipment	Crew	Daily Output	Labor-Hours	Unit	Material	2009 Bare Costs Labor	Equipment	Total	Total Incl O&P
0010 **DARKROOM EQUIPMENT**									
0020 Developing sink, 5" deep, 24" x 48"	Q-1	2	8	Ea.	5,025	350		5,375	6,075
0050 48" x 52"		1.70	9.412		5,100	415		5,515	6,250
0200 10" deep, 24" x 48"		1.70	9.412		6,475	415		6,890	7,775
0250 24" x 108"		1.50	10.667		8,525	470		8,995	10,100
0500 Dryers, dehumidified filtered air, 36" x 25" x 68" high	L-7	6	4.667		4,800	180		4,980	5,575
0550 48" x 25" x 68" high		5	5.600		9,000	216		9,216	10,300
2000 Processors, automatic, color print, minimum		4	7		13,100	270		13,370	14,800
2050 Maximum		.60	46.667		20,600	1,800		22,400	25,600
2300 Black and white print, minimum		2	14		9,950	540		10,490	11,800
2350 Maximum		.80	35		59,500	1,350		60,850	67,500
2600 Manual processor, 16" x 20" maximum print size		2	14		9,175	540		9,715	11,000
2650 20" x 24" maximum print size		1	28		8,725	1,075		9,800	11,400
3000 Viewing lights, 20" x 24"		6	4.667		380	180		560	710
3100 20" x 24" with color correction		6	4.667		535	180		715	885
3500 Washers, round, minimum sheet 11" x 14"	Q-1	2	8		3,075	350		3,425	3,925
3550 Maximum sheet 20" x 24"		1	16		3,475	700		4,175	4,925
3800 Square, minimum sheet 20" x 24"		1	16		3,125	700		3,825	4,525
3900 Maximum sheet 50" x 56"		.80	20		4,850	880		5,730	6,675
4500 Combination tank sink, tray sink, washers, with									
4510 Dry side tables, average	Q-1	.45	35.556	Ea.	9,600	1,550		11,150	13,000

11 31 Residential Appliances

11 31 13 – Residential Kitchen Appliances

11 31 13.13 Cooking Equipment

11 31 13.13 Cooking Equipment	Crew	Daily Output	Labor-Hours	Unit	Material	2009 Bare Costs Labor	Equipment	Total	Total Incl O&P
0010 **COOKING EQUIPMENT**									
0020 Cooking range, 30" free standing, 1 oven, minimum	2 Clab	10	1.600	Ea.	320	50.50		370.50	435
0050 Maximum	"	4	4		1,775	126		1,901	2,150
0350 Built-in, 30" wide, 1 oven, minimum	1 Elec	6	1.333		690	62.50		752.50	855
0400 Maximum	2 Carp	2	8		1,375	320		1,695	2,050
0900 Countertop cooktops, 4 burner, standard, minimum	1 Elec	6	1.333		244	62.50		306.50	365
0950 Maximum		3	2.667		585	125		710	840
1250 Microwave oven, minimum		4	2		104	94		198	259
1300 Maximum		2	4		450	188		638	785
5380 Oven, built in, standard		4	2		440	94		534	625
5390 Deluxe		2	4		1,900	188		2,088	2,400

11 31 13.23 Refrigeration Equipment

11 31 13.23 Refrigeration Equipment	Crew	Daily Output	Labor-Hours	Unit	Material	2009 Bare Costs Labor	Equipment	Total	Total Incl O&P
0010 **REFRIGERATION EQUIPMENT**									
5450 Refrigerator, no frost, 6 C.F.	2 Clab	15	1.067	Ea.	238	33.50		271.50	315
5500 Refrigerator, no frost, 10 C.F. to 12 C.F. minimum		10	1.600		445	50.50		495.50	575
5600 Maximum		6	2.667		520	84.50		604.50	715
6790 Energy-star qualified, 18 CF, minimum G	2 Carp	4	4		525	160		685	845
6795 Maximum G		2	8		1,050	320		1,370	1,675
6797 21.7 CF, minimum G		4	4		875	160		1,035	1,225
6799 Maximum G		4	4		1,150	160		1,310	1,550

11 31 13.33 Kitchen Cleaning Equipment

11 31 13.33 Kitchen Cleaning Equipment	Crew	Daily Output	Labor-Hours	Unit	Material	2009 Bare Costs Labor	Equipment	Total	Total Incl O&P
0010 **KITCHEN CLEANING EQUIPMENT**									
2750 Dishwasher, built-in, 2 cycles, minimum	L-1	4	4	Ea.	206	192		398	525
2800 Maximum		2	8		310	385		695	930
3100 Energy-star qualified, minimum G		4	4		315	192		507	640

11 31 Residential Appliances

11 31 13 – Residential Kitchen Appliances

11 31 13.33 Kitchen Cleaning Equipment

		Crew	Daily Output	Labor-Hours	Unit	Material	2009 Bare Costs Labor	Equipment	Total	Total Incl O&P
3110	Maximum	L-1	2	8	Ea.	1,000	385		1,385	1,700

11 31 13.43 Waste Disposal Equipment

		Crew	Daily Output	Labor-Hours	Unit	Material	Labor	Equipment	Total	Total Incl O&P
0010	**WASTE DISPOSAL EQUIPMENT**									
1750	Compactor, residential size, 4 to 1 compaction, minimum	1 Carp	5	1.600	Ea.	485	64		549	635
1800	Maximum	"	3	2.667		515	107		622	740
3300	Garbage disposal, sink type, minimum	L-1	10	1.600		66.50	76.50		143	192
3350	Maximum	"	10	1.600		192	76.50		268.50	330

11 31 13.53 Kitchen Ventilation Equipment

		Crew	Daily Output	Labor-Hours	Unit	Material	Labor	Equipment	Total	Total Incl O&P
0010	**KITCHEN VENTILATION EQUIPMENT**									
4150	Hood for range, 2 speed, vented, 30" wide, minimum	L-3	5	3.200	Ea.	52.50	139		191.50	281
4200	Maximum		3	5.333		810	232		1,042	1,275
4300	42" wide, minimum		5	3.200		250	139		389	500
4330	Custom		5	3.200		1,400	139		1,539	1,775
4350	Maximum		3	5.333		1,725	232		1,957	2,275
4400	Ventless hood, 2 speed, 30" wide, minimum	1 Elec	8	1		64.50	47		111.50	144
4450	Maximum		7	1.143		650	53.50		703.50	800
4510	36" wide, minimum		7	1.143		65	53.50		118.50	154
4550	Maximum		6	1.333		610	62.50		672.50	765
4580	42" wide, minimum		6	1.333		199	62.50		261.50	315
4600	Maximum		5	1.600		350	75		425	500
4650	For vented 1 speed, deduct from maximum					44			44	48.50
4700										

11 31 23 – Residential Laundry Appliances

11 31 23.13 Washers

		Crew	Daily Output	Labor-Hours	Unit	Material	Labor	Equipment	Total	Total Incl O&P
0010	**WASHERS**									
5000	Residential, 4 cycle, average	1 Plum	3	2.667	Ea.	800	130		930	1,075
6750	Energy star qualified, front loading, minimum [G]		3	2.667		510	130		640	760
6760	Maximum [G]		1	8		1,600	390		1,990	2,350
6764	Top loading, minimum [G]		3	2.667		485	130		615	735
6766	Maximum [G]		3	2.667		990	130		1,120	1,300

11 31 23.23 Dryers

		Crew	Daily Output	Labor-Hours	Unit	Material	Labor	Equipment	Total	Total Incl O&P
0010	**DRYERS**									
0500	Gas fired residential, 16 lb. capacity, average	1 Plum	3	2.667	Ea.	625	130		755	890
6770	Electric, front loading, energy-star qualified, minimum [G]	L-2	3	5.333		255	187		442	590
6780	Maximum [G]	"	2	8		1,425	281		1,706	2,025

11 31 33 – Miscellaneous Residential Appliances

11 31 33.13 Sump Pumps

		Crew	Daily Output	Labor-Hours	Unit	Material	Labor	Equipment	Total	Total Incl O&P
0010	**SUMP PUMPS**									
6400	Cellar drainer, pedestal, 1/3 H.P., molded PVC base	1 Plum	3	2.667	Ea.	119	130		249	335
6450	Solid brass	"	2	4	"	263	195		458	590
6460	Sump pump, see also Div. 22 14 29.16									

11 33 Retractable Stairs

11 33 10 – Disappearing Stairs

11 33 10.10 Disappearing Stairway

		Crew	Daily Output	Labor-Hours	Unit	Material	2009 Bare Costs Labor	Equipment	Total	Total Incl O&P
0010	**DISAPPEARING STAIRWAY** No trim included									
0020	One piece, yellow pine, 8'-0" ceiling	2 Carp	4	4	Ea.	288	160		448	580
0030	9'-0" ceiling		4	4		330	160		490	630
0040	10'-0" ceiling		3	5.333		300	213		513	680
0050	11'-0" ceiling		3	5.333		545	213		758	950
0060	12'-0" ceiling		3	5.333		545	213		758	950
0100	Custom grade, pine, 8'-6" ceiling, minimum	1 Carp	4	2		202	80		282	355
0150	Average		3.50	2.286		215	91.50		306.50	385
0200	Maximum		3	2.667		275	107		382	480
0250										
0500	Heavy duty, pivoted, from 7'-7" to 12'-10" floor to floor	1 Carp	3	2.667	Ea.	650	107		757	890
0600	16'-0" ceiling		2	4		1,325	160		1,485	1,725
0800	Economy folding, pine, 8'-6" ceiling		4	2		195	80		275	345
0900	9'-6" ceiling		4	2		225	80		305	380
1100	Automatic electric, aluminum, floor to floor height, 8' to 9'	2 Carp	1	16		8,250	640		8,890	10,100
1400	11' to 12'		.90	17.778		8,875	710		9,585	11,000
1700	14' to 15'		.70	22.857		9,475	915		10,390	11,900
9000	Minimum labor/equipment charge	1 Carp	2	4	Job		160		160	264

11 41 Food Storage Equipment

11 41 13 – Refrigerated Food Storage Cases

11 41 13.10 Refrigerated Food Cases

		Crew	Daily Output	Labor-Hours	Unit	Material	2009 Bare Costs Labor	Equipment	Total	Total Incl O&P
0010	**REFRIGERATED FOOD CASES**									
0030	Dairy, multi-deck, 12' long	Q-5	3	5.333	Ea.	9,925	237		10,162	11,300
0100	For rear sliding doors, add					1,425			1,425	1,550
0200	Delicatessen case, service deli, 12' long, single deck	Q-5	3.90	4.103		6,675	182		6,857	7,625
0300	Multi-deck, 18 S.F. shelf display		3	5.333		7,425	237		7,662	8,550
0400	Freezer, self-contained, chest-type, 30 C.F.		3.90	4.103		6,700	182		6,882	7,650
0500	Glass door, upright, 78 C.F.		3.30	4.848		9,275	215		9,490	10,500
0600	Frozen food, chest type, 12' long		3.30	4.848		6,700	215		6,915	7,700
0700	Glass door, reach-in, 5 door		3	5.333		12,900	237		13,137	14,500
0800	Island case, 12' long, single deck		3.30	4.848		7,600	215		7,815	8,675
0900	Multi-deck		3	5.333		16,100	237		16,337	18,100
1000	Meat case, 12' long, single deck		3.30	4.848		5,525	215		5,740	6,400
1050	Multi-deck		3.10	5.161		9,500	229		9,729	10,800
1100	Produce, 12' long, single deck		3.30	4.848		7,300	215		7,515	8,350
1200	Multi-deck		3.10	5.161		7,975	229		8,204	9,125

11 41 13.20 Refrigerated Food Storage Equipment

		Crew	Daily Output	Labor-Hours	Unit	Material	2009 Bare Costs Labor	Equipment	Total	Total Incl O&P
0010	**REFRIGERATED FOOD STORAGE EQUIPMENT**									
2350	Cooler, reach-in, beverage, 6' long	Q-1	6	2.667	Ea.	4,300	117		4,417	4,925
4300	Freezers, reach-in, 44 C.F.		4	4		3,150	176		3,326	3,725
4500	68 C.F.		3	5.333		4,350	234		4,584	5,150
4600	Freezer, pre-fab, 8' x 8' w/refrigeration	2 Carp	.45	35.556		10,200	1,425		11,625	13,600
4620	8' x 12'		.35	45.714		12,100	1,825		13,925	16,300
4640	8' x 16'		.25	64		15,000	2,550		17,550	20,700
4660	8' x 20'		.17	94.118		18,100	3,750		21,850	26,100
4680	Reach-in, 1 compartment	Q-1	4	4		1,550	176		1,726	1,975
4700	2 compartment		3	5.333		4,100	234		4,334	4,875
8300	Refrigerators, reach-in type, 44 C.F.		5	3.200		6,775	140		6,915	7,675
8310	With glass doors, 68 C.F.		4	4		6,625	176		6,801	7,575
8320	Refrigerator, reach-in, 1 compartment	R-18	7.80	3.333		1,950	123		2,073	2,350

11 41 Food Storage Equipment

11 41 13 – Refrigerated Food Storage Cases

11 41 13.20 Refrigerated Food Storage Equipment

		Crew	Daily Output	Labor-Hours	Unit	Material	2009 Bare Costs Labor	Equipment	Total	Total Incl O&P
8330	2 compartment	R-18	6.20	4.194	Ea.	2,900	154		3,054	3,425
8340	3 compartment	↓	5.60	4.643		3,925	171		4,096	4,575
8350	Pre-fab, with refrigeration, 8' x 8'	2 Carp	.45	35.556		5,375	1,425		6,800	8,250
8360	8' x 12'		.35	45.714		7,125	1,825		8,950	10,900
8370	8' x 16'		.25	64		10,400	2,550		12,950	15,600
8380	8' x 20'	↓	.17	94.118		13,100	3,750		16,850	20,600
8390	Pass-thru/roll-in, 1 compartment	R-18	7.80	3.333		3,550	123		3,673	4,100
8400	2 compartment		6.24	4.167		4,975	153		5,128	5,700
8410	3 compartment	↓	5.60	4.643		5,950	171		6,121	6,825
8420	Walk-in, alum, door & floor only, no refrig, 6' x 6' x 7'-6"	2 Carp	1.40	11.429		8,150	455		8,605	9,700
8430	10' x 6' x 7'-6"		.55	29.091		11,700	1,150		12,850	14,700
8440	12' x 14' x 7'-6"		.25	64		16,300	2,550		18,850	22,100
8450	12' x 20' x 7'-6"	↓	.17	94.118		20,000	3,750		23,750	28,200
8460	Refrigerated cabinets, mobile					3,425			3,425	3,775
8470	Refrigerator/freezer, reach-in, 1 compartment	R-18	5.60	4.643		4,425	171		4,596	5,150
8480	2 compartment	"	4.80	5.417	↓	6,250	199		6,449	7,200

11 41 13.30 Wine Cellar

		Crew	Daily Output	Labor-Hours	Unit	Material	2009 Bare Costs Labor	Equipment	Total	Total Incl O&P
0010	**WINE CELLAR**, refrigerated, Redwood interior, carpeted, walk-in type									
0020	6'-8" high, including racks									
0200	80" W x 48" D for 900 bottles	2 Carp	1.50	10.667	Ea.	3,350	425		3,775	4,375
0250	80" W x 72" D for 1300 bottles		1.33	12.030		4,425	480		4,905	5,650
0300	80" W x 94" D for 1900 bottles		1.17	13.675		5,725	545		6,270	7,200
0400	80" W x 124" D for 2500 bottles	↓	1	16	↓	6,925	640		7,565	8,675

11 41 33 – Food Storage Shelving

11 41 33.20 Metal Food Storage Shelving

		Crew	Daily Output	Labor-Hours	Unit	Material	2009 Bare Costs Labor	Equipment	Total	Total Incl O&P
0010	**METAL FOOD STORAGE SHELVING**									
8600	Stainless steel shelving, louvered 4-tier, 20" x 3'	1 Clab	6	1.333	Ea.	1,175	42		1,217	1,350
8605	20" x 4'		6	1.333		1,675	42		1,717	1,900
8610	20" x 6'		6	1.333		2,425	42		2,467	2,725
8615	24" x 3'		6	1.333		1,525	42		1,567	1,750
8620	24" x 4'		6	1.333		1,775	42		1,817	2,025
8625	24" x 6'		6	1.333		2,575	42		2,617	2,925
8630	Flat 4-tier, 20" x 3'		6	1.333		980	42		1,022	1,150
8635	20" x 4'		6	1.333		1,225	42		1,267	1,425
8640	20" x 5'		6	1.333		1,350	42		1,392	1,575
8645	24" x 3'		6	1.333		1,075	42		1,117	1,250
8650	24" x 4'		6	1.333		1,875	42		1,917	2,150
8655	24" x 6'		6	1.333		2,250	42		2,292	2,550
8700	Galvanized shelving, louvered 4-tier, 20" x 3'		6	1.333		595	42		637	720
8705	20" x 4'		6	1.333		660	42		702	800
8710	20" x 6'		6	1.333		700	42		742	840
8715	24" x 3'		6	1.333		525	42		567	645
8720	24" x 4'		6	1.333		710	42		752	850
8725	24" x 6'		6	1.333		1,025	42		1,067	1,200
8730	Flat 4-tier, 20" x 3'		6	1.333		425	42		467	540
8735	20" x 4'		6	1.333		480	42		522	600
8740	20" x 6'		6	1.333		835	42		877	985
8745	24" x 3'		6	1.333		430	42		472	545
8750	24" x 4'		6	1.333		615	42		657	750
8755	24" x 6'		6	1.333		865	42		907	1,025
8760	Stainless steel dunnage rack, 24" x 3'		8	1		395	31.50		426.50	485
8765	24" x 4'	↓	8	1	↓	775	31.50		806.50	905

11 41 Food Storage Equipment

11 41 33 – Food Storage Shelving

11 41 33.20 Metal Food Storage Shelving	Crew	Daily Output	Labor-Hours	Unit	Material	2009 Bare Costs Labor	Equipment	Total	Total Incl O&P
8770	Galvanized dunnage rack, 24" x 3'	1 Clab	8	1	Ea.	162	31.50		193.50	230
8775	24" x 4'	↓	8	1	↓	181	31.50		212.50	251

11 42 Food Preparation Equipment

11 42 10 – Commercial Food Preparation Equipment

11 42 10.10 Choppers, Mixers and Misc. Equipment

		Crew	Daily Output	Labor-Hours	Unit	Material	2009 Bare Costs Labor	Equipment	Total	Total Incl O&P
0010	**CHOPPERS, MIXERS AND MISC. EQUIPMENT**									
1700	Choppers, 5 pounds	R-18	7	3.714	Ea.	1,625	137		1,762	2,000
1720	16 pounds		5	5.200		1,975	191		2,166	2,475
1740	35 to 40 pounds	↓	4	6.500		2,975	239		3,214	3,650
1840	Coffee brewer, 5 burners	1 Plum	3	2.667		1,175	130		1,305	1,500
1850	Coffee urn, twin 6 gallon urns		2	4		2,525	195		2,720	3,075
1860	Single, 3 gallon	↓	3	2.667		1,800	130		1,930	2,200
3000	Fast food equipment, total package, minimum	6 Skwk	.08	600		172,000	24,500		196,500	229,500
3100	Maximum	"	.07	685		234,500	28,000		262,500	303,500
3800	Food mixers, 20 quarts	L-7	7	4		2,700	154		2,854	3,200
3850	40 quarts		5.40	5.185		6,950	200		7,150	7,975
3900	60 quarts		5	5.600		8,925	216		9,141	10,200
4040	80 quarts		3.90	7.179		11,000	277		11,277	12,600
4080	130 quarts		2.20	12.727		16,600	490		17,090	19,000
4100	Floor type, 20 quarts		15	1.867		3,125	72		3,197	3,550
4120	60 quarts		14	2		7,725	77		7,802	8,625
4140	80 quarts		12	2.333		9,575	90		9,665	10,600
4160	140 quarts	↓	8.60	3.256		18,200	126		18,326	20,200
6700	Peelers, small	R-18	8	3.250		1,450	119		1,569	1,800
6720	Large	"	6	4.333		4,375	159		4,534	5,075
6800	Pulper/extractor, close coupled, 5 HP	1 Plum	1.90	4.211		3,325	205		3,530	4,000
8580	Slicer with table	R-18	9	2.889	↓	4,675	106		4,781	5,300

11 43 Food Delivery Carts and Conveyors

11 43 13 – Food Delivery Carts

11 43 13.10 Mobile Carts, Racks and Trays

		Crew	Daily Output	Labor-Hours	Unit	Material	2009 Bare Costs Labor	Equipment	Total	Total Incl O&P
0010	**MOBILE CARTS, RACKS AND TRAYS**									
1650	Cabinet, heated, 1 compartment, reach-in	R-18	5.60	4.643	Ea.	3,250	171		3,421	3,850
1655	Pass-thru roll-in		5.60	4.643		5,125	171		5,296	5,900
1660	2 compartment, reach-in	↓	4.80	5.417		6,750	199		6,949	7,750
1670	Mobile					3,000			3,000	3,300
6850	Mobile rack w/pan slide					1,450			1,450	1,600
9180	Tray and silver dispenser, mobile	1 Clab	16	.500	↓	775	15.80		790.80	875

11 44 Food Cooking Equipment

11 44 13 – Commercial Ranges

11 44 13.10 Cooking Equipment	Crew	Daily Output	Labor-Hours	Unit	Material	2009 Bare Costs Labor	Equipment	Total	Total Incl O&P
0010 **COOKING EQUIPMENT**									
0020 Bake oven, gas, one section	Q-1	8	2	Ea.	4,300	88		4,388	4,875
0300 Two sections	↓	7	2.286		10,100	100		10,200	11,400
0600 Three sections	↓	6	2.667		13,200	117		13,317	14,700
0900 Electric convection, single deck	L-7	4	7		5,400	270		5,670	6,375
1300 Broiler, without oven, standard	Q-1	8	2		3,550	88		3,638	4,025
1550 Infrared	L-7	4	7		9,250	270		9,520	10,600
4750 Fryer, with twin baskets, modular model	Q-1	7	2.286		1,450	100		1,550	1,750
5000 Floor model, on 6" legs	"	5	3.200		2,400	140		2,540	2,850
5100 Extra single basket, large					99			99	109
5300 Griddle, SS, 24" plate, w/4" legs, elec, 208V, 3 phase, 3' long	Q-1	7	2.286		1,125	100		1,225	1,375
5550 4' long	"	6	2.667		1,450	117		1,567	1,775
6200 Iced tea brewer	1 Plum	3.44	2.326		670	113		783	910
6350 Kettle, w/steam jacket, tilting, w/positive lock, SS, 20 gallons	L-7	7	4		6,675	154		6,829	7,600
6600 60 gallons	"	6	4.667		8,750	180		8,930	9,900
6900 Range, restaurant type, 6 burners and 1 standard oven, 36" wide	Q-1	7	2.286		2,325	100		2,425	2,700
6950 Convection	↓	7	2.286		4,700	100		4,800	5,325
7150 2 standard ovens, 24" griddle, 60" wide		6	2.667		7,450	117		7,567	8,375
7200 1 standard, 1 convection oven		6	2.667		6,700	117		6,817	7,550
7450 Heavy duty, single 34" standard oven, open top		5	3.200		4,850	140		4,990	5,550
7500 Convection oven		5	3.200		3,425	140		3,565	4,000
7700 Griddle top		6	2.667		2,625	117		2,742	3,075
7750 Convection oven	↓	6	2.667		5,950	117		6,067	6,700
8850 Steamer, electric 27 KW	L-7	7	4		7,375	154		7,529	8,375
9100 Electric, 10 KW or gas 100,000 BTU	"	5	5.600		5,350	216		5,566	6,225
9150 Toaster, conveyor type, 16-22 slices per minute					1,100			1,100	1,200
9160 Pop-up, 2 slot				↓	535			535	585
9200 For deluxe models of above equipment, add					75%				
9400 Rule of thumb: Equipment cost based									
9410 on kitchen work area									
9420 Office buildings, minimum	L-7	77	.364	S.F.	76.50	14.05		90.55	108
9450 Maximum	↓	58	.483		129	18.60		147.60	173
9550 Public eating facilities, minimum		77	.364		100	14.05		114.05	134
9600 Maximum		46	.609		163	23.50		186.50	219
9750 Hospitals, minimum		58	.483		103	18.60		121.60	145
9800 Maximum	↓	39	.718	↓	190	27.50		217.50	254

11 46 Food Dispensing Equipment

11 46 16 – Service Line Equipment

11 46 16.10 Commercial Food Dispensing Equipment

	Crew	Daily Output	Labor-Hours	Unit	Material	2009 Bare Costs Labor	Equipment	Total	Total Incl O&P
0010 **COMMERCIAL FOOD DISPENSING EQUIPMENT**									
1050 Butter pat dispenser	1 Clab	13	.615	Ea.	865	19.45		884.45	985
1100 Bread dispenser, counter top		13	.615		825	19.45		844.45	940
1900 Cup and glass dispenser, drop in		4	2		1,200	63		1,263	1,425
1920 Disposable cup, drop in		16	.500		405	15.80		420.80	470
2650 Dish dispenser, drop in, 12"		11	.727		1,850	23		1,873	2,075
2660 Mobile	↓	10	.800		2,225	25.50		2,250.50	2,500
3300 Food warmer, counter, 1.2 KW					670			670	735
3550 1.6 KW					1,850			1,850	2,025
3600 Well, hot food, built-in, rectangular, 12" x 20"	R-30	10	2.600		590	98		688	800
3610 Circular, 7 qt	↓	10	2.600	↓	330	98		428	515

11 46 Food Dispensing Equipment

11 46 16 – Service Line Equipment

11 46 16.10 Commercial Food Dispensing Equipment

	11 46 16.10 Commercial Food Dispensing Equipment	Crew	Daily Output	Labor-Hours	Unit	Material	2009 Bare Costs Labor	Equipment	Total	Total Incl O&P
3620	Refrigerated, 2 compartments	R-30	10	2.600	Ea.	2,725	98		2,823	3,125
3630	3 compartments		9	2.889		3,000	109		3,109	3,475
3640	4 compartments		8	3.250		3,700	122		3,822	4,275
4720	Frost cold plate	↓	9	2.889		15,600	109		15,709	17,300
5700	Hot chocolate dispenser	1 Plum	4	2		1,025	97.50		1,122.50	1,275
6250	Jet spray dispenser	R-18	4.50	5.778		3,000	212		3,212	3,650
6300	Juice dispenser, concentrate	"	4.50	5.778		2,100	212		2,312	2,650
6690	Milk dispenser, bulk, 2 flavor	R-30	8	3.250		1,425	122		1,547	1,775
6695	3 flavor	"	8	3.250	↓	1,950	122		2,072	2,350
8800	Serving counter, straight	1 Carp	40	.200	L.F.	745	8		753	835
8820	Curved section	"	30	.267	"	935	10.65		945.65	1,050
8825	Solid surface, see Div. 12 36 61.16									
8830	Soft serve ice cream machine, medium	R-18	11	2.364	Ea.	9,850	87		9,937	10,900
8840	Large	"	9	2.889	"	17,300	106		17,406	19,200

11 47 Ice Machines

11 47 10 – Commercial Ice Machines

11 47 10.10 Commercial Ice Equipment

		Crew	Daily Output	Labor-Hours	Unit	Material	2009 Bare Costs Labor	Equipment	Total	Total Incl O&P
0010	**COMMERCIAL ICE EQUIPMENT**									
5800	Ice cube maker, 50 pounds per day	Q-1	6	2.667	Ea.	1,425	117		1,542	1,750
5900	250 pounds per day		1.20	13.333		2,400	585		2,985	3,550
6050	500 pounds per day		4	4		3,000	176		3,176	3,575
6060	With bin		1.20	13.333		2,775	585		3,360	3,950
6090	1000 pounds per day, with bin		1	16		4,775	700		5,475	6,350
6100	Ice flakers, 300 pounds per day		1.60	10		3,050	440		3,490	4,025
6120	600 pounds per day		.95	16.842		3,875	740		4,615	5,400
6130	1000 pounds per day	↓	.75	21.333		4,325	935		5,260	6,200
6140	2000 pounds per day		.65	24.615		18,200	1,075		19,275	21,800
6160	Ice storage bin, 500 pound capacity	Q-5	1	16		940	710		1,650	2,125
6180	1000 pound	"	.56	28.571	↓	1,800	1,275		3,075	3,950

11 48 Cleaning and Disposal Equipment

11 48 13 – Commercial Dishwashers

11 48 13.10 Dishwashers

		Crew	Daily Output	Labor-Hours	Unit	Material	2009 Bare Costs Labor	Equipment	Total	Total Incl O&P
0010	**DISHWASHERS**									
2700	Dishwasher, commercial, rack type									
2720	10 to 12 racks per hour	Q-1	3.20	5	Ea.	4,200	219		4,419	4,975
2750	Semi-automatic 38 to 50 racks per hour	"	1.30	12.308		8,350	540		8,890	10,000
2800	Automatic, 190 to 230 racks per hour	L-6	.35	34.286		12,400	1,650		14,050	16,300
2820	235 to 275 racks per hour	↓	.25	48		27,400	2,300		29,700	33,800
2840	8,750 to 12,500 dishes per hour		.10	120	↓	48,000	5,775		53,775	62,000
2950	Dishwasher hood, canopy type	L-3A	10	1.200	L.F.	770	52.50		822.50	930
2960	Pant leg type	"	2.50	4.800	Ea.	7,375	210		7,585	8,450
5200	Garbage disposal 1.5 HP, 100 GPH	L-1	4.80	3.333		1,825	160		1,985	2,250
5210	3 HP, 120 GPH		4.60	3.478		2,300	167		2,467	2,775
5220	5 HP, 250 GPH		4.50	3.556		3,550	170		3,720	4,175
6750	Pot sink, 3 compartment	1 Plum	7.25	1.103	L.F.	845	54		899	1,025
6760	Pot washer, small		1.60	5	Ea.	8,900	244		9,144	10,200
6770	Large	↓	1.20	6.667	↓	14,600	325		14,925	16,600

11 48 Cleaning and Disposal Equipment

11 48 13 - Commercial Dishwashers

11 48 13.10 Dishwashers

		Crew	Daily Output	Labor-Hours	Unit	Material	2009 Bare Costs Labor	Equipment	Total	Total Incl O&P
9170	Trash compactor, small, up to 125 lb. compacted weight	L-4	4	6	Ea.	20,800	224		21,024	23,300
9171	Trash Compactor for Furniture Row		12	2		16,800	74.50		16,874.50	18,600
9175	Large, up to 175 lb. compacted weight		3	8		25,400	299		25,699	28,400

11 52 Audio-Visual Equipment

11 52 13 - Projection Screens

11 52 13.10 Projection Screens, Wall or Ceiling Hung

		Crew	Daily Output	Labor-Hours	Unit	Material	2009 Bare Costs Labor	Equipment	Total	Total Incl O&P
0010	**PROJECTION SCREENS, WALL OR CEILING HUNG**, matte white									
0100	Manually operated, economy	2 Carp	500	.032	S.F.	5.85	1.28		7.13	8.55
0300	Intermediate		450	.036		6.80	1.42		8.22	9.85
0400	Deluxe		400	.040		9.45	1.60		11.05	13
9000	Minimum labor/equipment charge		3	5.333	Job		213		213	350

11 52 16 - Projectors

11 52 16.10 Movie Equipment

		Crew	Daily Output	Labor-Hours	Unit	Material	2009 Bare Costs Labor	Equipment	Total	Total Incl O&P
0011	**MOVIE EQUIPMENT**									
0020	Changeover, minimum				Ea.	465			465	510
3000	Projection screens, rigid, in wall, acrylic, 1/4" thick	2 Glaz	195	.082	S.F.	42	3.17		45.17	51
3100	1/2" thick	"	130	.123	"	48.50	4.75		53.25	61
3700	Sound systems, incl. amplifier, mono, minimum	1 Elec	.90	8.889	Ea.	3,300	420		3,720	4,275
3800	Dolby/Super Sound, maximum		.40	20		18,100	940		19,040	21,400
4100	Dual system, 2 channel, front surround, minimum		.70	11.429		4,625	535		5,160	5,925
4200	Dolby/Super Sound, 4 channel, maximum		.40	20		16,500	940		17,440	19,700
5700	Seating, painted steel, upholstered, minimum	2 Carp	35	.457		132	18.25		150.25	175
5800	Maximum	"	28	.571		420	23		443	505

11 53 Laboratory Equipment

11 53 33 - Emergency Safety Appliances

11 53 33.13 Emergency Equipment

		Crew	Daily Output	Labor-Hours	Unit	Material	2009 Bare Costs Labor	Equipment	Total	Total Incl O&P
0010	**EMERGENCY EQUIPMENT**									
1400	Safety equipment, eye wash, hand held				Ea.	405			405	445
1450	Deluge shower				"	725			725	800

11 53 43 - Service Fittings and Accessories

11 53 43.13 Fittings

		Crew	Daily Output	Labor-Hours	Unit	Material	2009 Bare Costs Labor	Equipment	Total	Total Incl O&P
0010	**FITTINGS**									
1600	Sink, one piece plastic, flask wash, hose, free standing	1 Plum	1.60	5	Ea.	1,800	244		2,044	2,350
1610	Epoxy resin sink, 25" x 16" x 10"	"	2	4	"	198	195		393	520
1630	Steel table, open underneath				L.F.	410			410	450
1640	Doors underneath				"	605			605	665
1900	Utility cabinet, stainless steel				Ea.	635			635	700
1950	Utility table, acid resistant top with drawers	2 Carp	30	.533	L.F.	147	21.50		168.50	196
8000	Alternate pricing method: as percent of lab furniture									
8050	Installation, not incl. plumbing & duct work				% Furn.				20%	22%
8100	Plumbing, final connections, simple system								9.09%	10%
8110	Moderately complex system								13.64%	15%
8120	Complex system								18.18%	20%
8150	Electrical, simple system								9.09%	10%
8160	Moderately complex system								18.18%	20%
8170	Complex system								31.90%	35%

11 57 Vocational Shop Equipment

11 57 10 – Shop Equipment

11 57 10.10 Vocational School Shop Equipment	Crew	Daily Output	Labor-Hours	Unit	Material	2009 Bare Costs Labor	Equipment	Total	Total Incl O&P
0010 **VOCATIONAL SCHOOL SHOP EQUIPMENT**									
0020 Benches, work, wood, average	2 Carp	5	3.200	Ea.	505	128		633	765
0100 Metal, average		5	3.200		370	128		498	615
0400 Combination belt & disc sander, 6"		4	4		1,275	160		1,435	1,675
0700 Drill press, floor mounted, 12", 1/2 H.P.		4	4		400	160		560	700
0800 Dust collector, not incl. ductwork, 6" diameter	1 Shee	1.10	7.273		3,000	345		3,345	3,850
1000 Grinders, double wheel, 1/2 H.P.	2 Carp	5	3.200		245	128		373	480
1300 Jointer, 4", 3/4 H.P.		4	4		1,250	160		1,410	1,650
1600 Kilns, 16 C.F., to 2000°		4	4		2,250	160		2,410	2,750
1900 Lathe, woodworking, 10", 1/2 H.P.		4	4		765	160		925	1,100
2200 Planer, 13" x 6"		4	4		1,250	160		1,410	1,650
2500 Potter's wheel, motorized		4	4		1,125	160		1,285	1,500
2800 Saws, band, 14", 3/4 H.P.		4	4		850	160		1,010	1,200
3100 Metal cutting band saw, 14"		4	4		2,000	160		2,160	2,475
3400 Radial arm saw, 10", 2 H.P.		4	4		975	160		1,135	1,350
3700 Scroll saw, 24"		4	4		575	160		735	900
4000 Table saw, 10", 3 H.P.		4	4		1,925	160		2,085	2,375
4300 Welder AC arc, 30 amp capacity		4	4		1,950	160		2,110	2,400

11 61 Theater and Stage Equipment

11 61 23 – Folding and Portable Stages

11 61 23.10 Portable Stages

	Crew	Daily Output	Labor-Hours	Unit	Material	2009 Bare Costs Labor	Equipment	Total	Total Incl O&P
0010 **PORTABLE STAGES**									
5000 Stages, portable with steps, folding legs, stock, 8" high				SF Stg.	25.50			25.50	28.50
5100 16" high					23.50			23.50	26
5200 32" high					35.50			35.50	39
5300 40" high					59.50			59.50	65

11 61 33 – Rigging Systems and Controls

11 61 33.10 Controls

	Crew	Daily Output	Labor-Hours	Unit	Material	2009 Bare Costs Labor	Equipment	Total	Total Incl O&P
0010 **CONTROLS**									
0050 Control boards with dimmers and breakers, minimum	1 Elec	1	8	Ea.	11,300	375		11,675	13,100
0150 Maximum	"	.20	40	"	119,000	1,875		120,875	134,000
0160									

11 61 43 – Stage Curtains

11 61 43.10 Curtains

	Crew	Daily Output	Labor-Hours	Unit	Material	2009 Bare Costs Labor	Equipment	Total	Total Incl O&P
0010 **CURTAINS**									
0500 Curtain track, straight, light duty	2 Carp	20	.800	L.F.	25.50	32		57.50	81
0700 Curved sections		12	1.333	"	164	53.50		217.50	268
1000 Curtains, velour, medium weight		600	.027	S.F.	7.90	1.07		8.97	10.45

11 66 Athletic Equipment

11 66 13 – Exercise Equipment

11 66 13.10 Physical Training Equipment

	11 66 13.10 Physical Training Equipment	Crew	Daily Output	Labor-Hours	Unit	Material	2009 Bare Costs Labor	Equipment	Total	Total Incl O&P
0010	**PHYSICAL TRAINING EQUIPMENT**									
0020	Abdominal rack, 2 board capacity				Ea.	470			470	515
0050	Abdominal board, upholstered					525			525	575
0200	Bicycle trainer, minimum					815			815	895
0300	Deluxe, electric					4,150			4,150	4,575
0400	Barbell set, chrome plated steel, 25 lbs.					244			244	269
0420	100 lbs.					370			370	405
0450	200 lbs.					715			715	785
0500	Weight plates, cast iron, per lb.				Lb.	5			5	5.50
0520	Storage rack, 10 station				Ea.	790			790	870
0600	Circuit training apparatus, 12 machines minimum	2 Clab	1.25	12.800	Set	27,600	405		28,005	31,000
0700	Average		1	16		33,800	505		34,305	38,000
0800	Maximum		.75	21.333		40,100	675		40,775	45,200
0820	Dumbbell set, cast iron, with rack and 5 pair					570			570	630
0900	Squat racks	2 Clab	5	3.200	Ea.	765	101		866	1,000
4150	Exercise equipment, bicycle trainer					595			595	655
4180	Chinning bar, adjustable, wall mounted	1 Carp	5	1.600		279	64		343	410
4200	Exercise ladder, 16' x 1'-7", suspended	L-2	3	5.333		1,125	187		1,312	1,550
4210	High bar, floor plate attached	1 Carp	4	2		1,100	80		1,180	1,350
4240	Parallel bars, adjustable		4	2		2,475	80		2,555	2,825
4270	Uneven parallel bars, adjustable		4	2		2,675	80		2,755	3,050
4280	Wall mounted, adjustable	L-2	1.50	10.667	Set	835	375		1,210	1,525
4300	Rope, ceiling mounted, 18' long	1 Carp	3.66	2.186	Ea.	220	87.50		307.50	385
4330	Side horse, vaulting		5	1.600		1,700	64		1,764	1,975
4360	Treadmill, motorized, deluxe, training type		5	1.600		3,500	64		3,564	3,950
4390	Weight lifting multi-station, minimum	2 Clab	1	16		395	505		900	1,275

11 66 23 – Gymnasium Equipment

11 66 23.13 Basketball Equipment

	11 66 23.13 Basketball Equipment	Crew	Daily Output	Labor-Hours	Unit	Material	2009 Bare Costs Labor	Equipment	Total	Total Incl O&P
0010	**BASKETBALL EQUIPMENT**									
1000	Backstops, wall mtd., 6' extended, fixed, minimum	L-2	1	16	Ea.	1,150	560		1,710	2,175
1100	Maximum		1	16		1,675	560		2,235	2,775
1200	Swing up, minimum		1	16		1,275	560		1,835	2,325
1250	Maximum		1	16		5,350	560		5,910	6,825
1300	Portable, manual, heavy duty, spring operated		1.90	8.421		11,400	296		11,696	13,000
1400	Ceiling suspended, stationary, minimum		.78	20.513		2,850	720		3,570	4,350
1450	Fold up, with accessories, maximum		.40	40		6,100	1,400		7,500	9,050
1600	For electrically operated, add	1 Elec	1	8		2,000	375		2,375	2,775
5800	Wall pads, 1-1/2" thick	2 Carp	640	.025	S.F.	6.80	1		7.80	9.15

11 66 23.19 Boxing Ring

	11 66 23.19 Boxing Ring	Crew	Daily Output	Labor-Hours	Unit	Material	2009 Bare Costs Labor	Equipment	Total	Total Incl O&P
0010	**BOXING RING**									
4100	Elevated, 22' x 22'	L-4	.10	240	Ea.	6,700	8,950		15,650	22,000
4110	For cellular plastic foam padding, add		.10	240		850	8,950		9,800	15,500
4120	Floor level, including posts and ropes only, 20' x 20'		.80	30		2,375	1,125		3,500	4,425
4130	Canvas, 30' x 30'		5	4.800		1,200	179		1,379	1,600

11 66 23.47 Gym Mats

	11 66 23.47 Gym Mats	Crew	Daily Output	Labor-Hours	Unit	Material	2009 Bare Costs Labor	Equipment	Total	Total Incl O&P
0010	**GYM MATS**									
5500	2" thick, naugahyde covered				S.F.	3.59			3.59	3.95
5600	Vinyl/nylon covered					6.20			6.20	6.85
6000	Wrestling mats, 1" thick, heavy duty					5.50			5.50	6.05

11 66 Athletic Equipment

11 66 43 – Interior Scoreboards

11 66 43.10 Scoreboards	Crew	Daily Output	Labor-Hours	Unit	Material	2009 Bare Costs Labor	2009 Bare Costs Equipment	Total	Total Incl O&P
0010 **SCOREBOARDS**									
7000 Baseball, minimum	R-3	1.30	15.385	Ea.	3,225	710	99.50	4,034.50	4,750
7200 Maximum		.05	400		15,600	18,500	2,575	36,675	48,700
7300 Football, minimum		.86	23.256		3,975	1,075	150	5,200	6,225
7400 Maximum		.20	100		12,600	4,625	645	17,870	21,800
7500 Basketball (one side), minimum		2.07	9.662		2,225	445	62.50	2,732.50	3,200
7600 Maximum		.30	66.667		5,550	3,075	430	9,055	11,400
7700 Hockey-basketball (four sides), minimum		.25	80		5,675	3,700	515	9,890	12,500
7800 Maximum		.15	133		5,600	6,175	860	12,635	16,600

11 66 53 – Gymnasium Dividers

11 66 53.10 Divider Curtains

	Crew	Daily Output	Labor-Hours	Unit	Material	2009 Bare Costs Labor	2009 Bare Costs Equipment	Total	Total Incl O&P
0010 **DIVIDER CURTAINS**									
4500 Gym divider curtain, mesh top, vinyl bottom, manual	L-4	500	.048	S.F.	8.05	1.79		9.84	11.80
4700 Electric roll up	L-7	400	.070	"	11.75	2.70		14.45	17.35

11 82 Solid Waste Handling Equipment

11 82 26 – Waste Compactors and Destructors

11 82 26.10 Compactors

	Crew	Daily Output	Labor-Hours	Unit	Material	2009 Bare Costs Labor	2009 Bare Costs Equipment	Total	Total Incl O&P
0010 **COMPACTORS**									
0020 Compactors, 115 volt, 250#/hr., chute fed	L-4	1	24	Ea.	10,900	895		11,795	13,500
0100 Hand fed		2.40	10		7,975	375		8,350	9,375
1000 Heavy duty industrial compactor, 0.5 C.Y. capacity		1	24		7,100	895		7,995	9,275
1050 1.0 C.Y. capacity		1	24		10,800	895		11,695	13,400
1400 For handling hazardous waste materials, 55 gallon drum packer, std.					16,800			16,800	18,500
1410 55 gallon drum packer w/HEPA filter					20,900			20,900	23,000
1420 55 gallon drum packer w/charcoal & HEPA filter					28,000			28,000	30,800
1430 All of the above made explosion proof, add					12,800			12,800	14,000

11 91 Religious Equipment

11 91 13 – Baptistries

11 91 13.10 Baptistry

	Crew	Daily Output	Labor-Hours	Unit	Material	2009 Bare Costs Labor	2009 Bare Costs Equipment	Total	Total Incl O&P
0010 **BAPTISTRY**									
0150 Fiberglass, 3'-6" deep, x 13'-7" long,									
0160 steps at both ends, incl. plumbing, minimum	L-8	1	20	Ea.	3,825	835		4,660	5,550
0200 Maximum	"	.70	28.571		6,250	1,200		7,450	8,825
0250 Add for filter, heater and lights					1,375			1,375	1,500

11 91 23 – Sanctuary Equipment

11 91 23.10 Sanctuary Furnishings

	Crew	Daily Output	Labor-Hours	Unit	Material	2009 Bare Costs Labor	2009 Bare Costs Equipment	Total	Total Incl O&P
0010 **SANCTUARY FURNISHINGS**									
0020 Altar, wood, custom design, plain	1 Carp	1.40	5.714	Ea.	1,925	228		2,153	2,475
0050 Deluxe		.20	40	"	9,225	1,600		10,825	12,900
5000 Wall cross, aluminum, extruded, 2" x 2" section		34	.235	L.F.	132	9.40		141.40	161
5150 4" x 4" section		29	.276		190	11		201	227
5300 Bronze, extruded, 1" x 2" section		31	.258		260	10.30		270.30	305
5350 2-1/2" x 2-1/2" section		34	.235		395	9.40		404.40	450

Estimating Tips

General

- The items in this division are usually priced per square foot or each. Most of these items are purchased by the owner and placed by the supplier.

  Do not assume the items in Division 12 will be purchased and installed by the supplier. Check the specifications for responsibilities and include receiving, storage, installation, and mechanical and electrical hookups in the appropriate divisions.

- Some items in this division require some type of support system that is not usually furnished with the item. Examples of these systems include blocking for the attachment of casework and heavy drapery rods. The required blocking must be added to the estimate in the appropriate division.

Reference Numbers

Reference numbers are shown in shaded boxes at the beginning of some major classifications. These numbers refer to related items in the Reference Section. The reference information may be an estimating procedure, an alternate pricing method, or technical information.

Note: Not all subdivisions listed here necessarily appear in this publication.

12 21 Window Blinds

12 21 13 – Horizontal Louver Blinds

12 21 13.13 Metal Horizontal Louver Blinds	Crew	Daily Output	Labor-Hours	Unit	Material	2009 Bare Costs Labor	Equipment	Total	Total Incl O&P
0010 **METAL HORIZONTAL LOUVER BLINDS**									
0020 Horizontal, 1" aluminum slats, solid color, stock	1 Carp	590	.014	S.F.	4.75	.54		5.29	6.15

12 23 Interior Shutters

12 23 10 – Wood Interior Shutters

12 23 10.10 Wood Interior Shutters

	Crew	Daily Output	Labor-Hours	Unit	Material	Labor	Equipment	Total	Total Incl O&P
0010 **WOOD INTERIOR SHUTTERS**, louvered									
0200 Two panel, 27" wide, 36" high	1 Carp	5	1.600	Set	114	64		178	231
0300 33" wide, 36" high		5	1.600		149	64		213	270
0500 47" wide, 36" high		5	1.600		199	64		263	325
1000 Four panel, 27" wide, 36" high		5	1.600		134	64		198	254
1100 33" wide, 36" high		5	1.600		165	64		229	287
1300 47" wide, 36" high		5	1.600		234	64		298	365

12 23 10.13 Wood Panels

	Crew	Daily Output	Labor-Hours	Unit	Material	Labor	Equipment	Total	Total Incl O&P
0010 **WOOD PANELS**									
3000 Wood folding panels with movable louvers, 7" x 20" each	1 Carp	17	.471	Pr.	46	18.80		64.80	81.50
4000 Fixed louver type, stock units, 8" x 20" each		17	.471		79.50	18.80		98.30	119
4450 18" x 40" each		17	.471		168	18.80		186.80	216

12 24 Window Shades

12 24 13 – Roller Window Shades

12 24 13.10 Shades

	Crew	Daily Output	Labor-Hours	Unit	Material	Labor	Equipment	Total	Total Incl O&P
0010 **SHADES**									
0020 Basswood, roll-up, stain finish, 3/8" slats	1 Carp	300	.027	S.F.	13.65	1.07		14.72	16.80
0030 Double layered, heat reflective		685	.012		8.55	.47		9.02	10.20
0200 7/8" slats		300	.027		12.90	1.07		13.97	15.95
0900 Mylar, single layer, non-heat reflective		685	.012		6.05	.47		6.52	7.45
1000 Double layered, heat reflective		685	.012		9.10	.47		9.57	10.75
1100 Triple layered, heat reflective		685	.012		10.55	.47		11.02	12.40
5000 Thermal, roll up, R-4		44	.182		11.05	7.25		18.30	24
5030 R-10.7		44	.182		12.70	7.25		19.95	26
5050 Magnetic clips, set of 20				Set	23.50			23.50	26

12 32 Manufactured Wood Casework

12 32 16 – Manufactured Plastic-Laminate-Clad Casework

12 32 16.20 Plastic Laminate Casework Doors

	Crew	Daily Output	Labor-Hours	Unit	Material	Labor	Equipment	Total	Total Incl O&P
0010 **PLASTIC LAMINATE CASEWORK DOORS**									
1000 For casework frames, see Div. 12 32 23.15									
1100 For casework hardware, see Div. 12 32 23.35									
6000 Plastic laminate on particle board									
6100 12" wide, 18" high	1 Carp	25	.320	Ea.	10.50	12.80		23.30	32.50
6140 30" high		23	.348		17.50	13.90		31.40	42.50
6500 18" wide, 18" high		24	.333		15.75	13.30		29.05	39.50
6600 30" high		22	.364		26.50	14.55		41.05	53

12 32 16.25 Plastic Laminate Drawer Fronts

	Crew	Daily Output	Labor-Hours	Unit	Material	Labor	Equipment	Total	Total Incl O&P
0010 **PLASTIC LAMINATE DRAWER FRONTS**									
2800 Plastic laminate on particle board front									

12 32 16 – Manufactured Plastic-Laminate-Clad Casework

12 32 16.25 Plastic Laminate Drawer Fronts	Crew	Daily Output	Labor-Hours	Unit	Material	2009 Bare Costs Labor	Equipment	Total	Total Incl O&P
3000	4" high, 12" wide	1 Carp	17	.471	Ea.	4.62	18.80		23.42	36
3200	18" wide	"	16	.500	"	6.95	20		26.95	40.50

12 32 23 – Hardwood Casework

12 32 23.10 Manufactured Wood Casework, Stock Units

		Crew	Daily Output	Labor-Hours	Unit	Material	Labor	Equipment	Total	Total Incl O&P
0010	**MANUFACTURED WOOD CASEWORK, STOCK UNITS**									
0300	Built-in drawer units, pine, 18" deep, 32" high, unfinished									
0400	Minimum	2 Carp	53	.302	L.F.	135	12.05		147.05	169
0500	Maximum	"	40	.400	"	165	16		181	209
0700	Kitchen base cabinets, hardwood, not incl. counter tops,									
0710	24" deep, 35" high, prefinished									
0800	One top drawer, one door below, 12" wide	2 Carp	24.80	.645	Ea.	227	26		253	293
0820	15" wide		24	.667		272	26.50		298.50	345
0840	18" wide		23.30	.687		297	27.50		324.50	370
0860	21" wide		22.70	.705		310	28		338	385
0880	24" wide		22.30	.717		360	28.50		388.50	450
1000	Four drawers, 12" wide		24.80	.645		335	26		361	410
1020	15" wide		24	.667		365	26.50		391.50	445
1040	18" wide		23.30	.687		385	27.50		412.50	465
1060	24" wide		22.30	.717		410	28.50		438.50	500
1200	Two top drawers, two doors below, 27" wide		22	.727		370	29		399	460
1220	30" wide		21.40	.748		370	30		400	460
1240	33" wide		20.90	.766		375	30.50		405.50	465
1260	36" wide		20.30	.788		390	31.50		421.50	480
1280	42" wide		19.80	.808		420	32.50		452.50	515
1300	48" wide		18.90	.847		450	34		484	550
1500	Range or sink base, two doors below, 30" wide		21.40	.748		300	30		330	380
1520	33" wide		20.90	.766		320	30.50		350.50	405
1540	36" wide		20.30	.788		340	31.50		371.50	420
1560	42" wide		19.80	.808		360	32.50		392.50	455
1580	48" wide		18.90	.847		380	34		414	470
1800	For sink front units, deduct					60			60	66
2000	Corner base cabinets, 36" wide, standard	2 Carp	18	.889		630	35.50		665.50	750
2100	Lazy Susan with revolving door	"	16.50	.970		540	38.50		578.50	660
4000	Kitchen wall cabinets, hardwood, 12" deep with two doors									
4050	12" high, 30" wide	2 Carp	24.80	.645	Ea.	197	26		223	260
4100	36" wide		24	.667		228	26.50		254.50	295
4400	15" high, 30" wide		24	.667		198	26.50		224.50	262
4420	33" wide		23.30	.687		223	27.50		250.50	291
4440	36" wide		22.70	.705		243	28		271	315
4450	42" wide		22.70	.705		284	28		312	355
4700	24" high, 30" wide		23.30	.687		310	27.50		337.50	385
4720	36" wide		22.70	.705		345	28		373	425
4740	42" wide		22.30	.717		335	28.50		363.50	415
5000	30" high, one door, 12" wide		22	.727		191	29		220	259
5020	15" wide		21.40	.748		210	30		240	281
5040	18" wide		20.90	.766		234	30.50		264.50	310
5060	24" wide		20.30	.788		271	31.50		302.50	350
5300	Two doors, 27" wide		19.80	.808		289	32.50		321.50	375
5320	30" wide		19.30	.829		320	33		353	405
5340	36" wide		18.80	.851		360	34		394	450
5360	42" wide		18.50	.865		385	34.50		419.50	480
5380	48" wide		18.40	.870		400	34.50		434.50	500

12 32 Manufactured Wood Casework

12 32 23 – Hardwood Casework

12 32 23.10 Manufactured Wood Casework, Stock Units	Crew	Daily Output	Labor-Hours	Unit	Material	2009 Bare Costs Labor	2009 Bare Costs Equipment	Total	Total Incl O&P
6000	Corner wall, 30" high, 24" wide	2 Carp	18	.889	Ea.	209	35.50		244.50	289
6050	30" wide		17.20	.930		260	37		297	350
6100	36" wide		16.50	.970		290	38.50		328.50	385
6500	Revolving Lazy Susan		15.20	1.053		390	42		432	500
7000	Broom cabinet, 84" high, 24" deep, 18" wide		10	1.600		515	64		579	670
7500	Oven cabinets, 84" high, 24" deep, 27" wide		8	2		785	80		865	990
7750	Valance board trim		396	.040	L.F.	12.25	1.61		13.86	16.15
9000	For deluxe models of all cabinets, add					40%				
9500	For custom built in place, add					25%	10%			
9550	Rule of thumb, kitchen cabinets not including									
9560	appliances & counter top, minimum	2 Carp	30	.533	L.F.	130	21.50		151.50	178
9600	Maximum	"	25	.640	"	310	25.50		335.50	385
9700	Minimum labor/equipment charge	1 Carp	3	2.667	Job		107		107	176

12 32 23.15 Manufactured Wood Casework Frames

		Crew	Daily Output	Labor-Hours	Unit	Material	Labor	Equipment	Total	Total Incl O&P
0010	**MANUFACTURED WOOD CASEWORK FRAMES**									
0050	Base cabinets, counter storage, 36" high									
0100	One bay, 18" wide	1 Carp	2.70	2.963	Ea.	131	118		249	340
0400	Two bay, 36" wide		2.20	3.636		200	145		345	460
1100	Three bay, 54" wide		1.50	5.333		238	213		451	610
2800	Bookcases, one bay, 7' high, 18" wide		2.40	3.333		154	133		287	390
3500	Two bay, 36" wide		1.60	5		224	200		424	575
4100	Three bay, 54" wide		1.20	6.667		370	266		636	845
5100	Coat racks, one bay, 7' high, 24" wide		4.50	1.778		154	71		225	286
5300	Two bay, 48" wide		2.75	2.909		214	116		330	430
5800	Three bay, 72" wide		2.10	3.810		315	152		467	595
6100	Wall mounted cabinet, one bay, 24" high, 18" wide		3.60	2.222		85	89		174	241
6800	Two bay, 36" wide		2.20	3.636		124	145		269	375
7400	Three bay, 54" wide		1.70	4.706		154	188		342	480
8400	30" high, one bay, 18" wide		3.60	2.222		92	89		181	248
9000	Two bay, 36" wide		2.15	3.721		122	149		271	380
9400	Three bay, 54" wide		1.60	5		153	200		353	500
9800	Wardrobe, 7' high, single, 24" wide		2.70	2.963		170	118		288	380
9880	Partition & adjustable shelves, 48" wide		1.70	4.706		216	188		404	550
9950	Partition, adjustable shelves & drawers, 48" wide		1.40	5.714		325	228		553	730
9970	Minimum labor/equipment charge		4	2	Job		80		80	132

12 32 23.20 Manufactured Hardwood Casework Doors

		Crew	Daily Output	Labor-Hours	Unit	Material	Labor	Equipment	Total	Total Incl O&P
0010	**MANUFACTURED HARDWOOD CASEWORK DOORS**									
2000	Glass panel, hardwood frame									
2200	12" wide, 18" high	1 Carp	34	.235	Ea.	18.75	9.40		28.15	36
2600	30" high		32	.250		31.50	10		41.50	51
4450	18" wide, 18" high		32	.250		28	10		38	47.50
4550	30" high		29	.276		47	11		58	69.50
5000	Hardwood, raised panel									
5100	12" wide, 18" high	1 Carp	16	.500	Ea.	27	20		47	62.50
5200	30" high		15	.533		45	21.50		66.50	84.50
5500	18" wide, 18" high		15	.533		40.50	21.50		62	79.50
5600	30" high		14	.571		67.50	23		90.50	112
9000	Minimum labor/equipment charge		4	2	Job		80		80	132

12 32 23.25 Manufactured Wood Casework Drawer Fronts

		Crew	Daily Output	Labor-Hours	Unit	Material	Labor	Equipment	Total	Total Incl O&P
0010	**MANUFACTURED WOOD CASEWORK DRAWER FRONTS**									
0100	Solid hardwood front									
1000	4" high, 12" wide	1 Carp	17	.471	Ea.	2.66	18.80		21.46	34

12 32 Manufactured Wood Casework

12 32 23 – Hardwood Casework

12 32 23.25 Manufactured Wood Casework Drawer Fronts

		Crew	Daily Output	Labor-Hours	Unit	Material	2009 Bare Costs Labor	Equipment	Total	Total Incl O&P
1200	18" wide	1 Carp	16	.500	Ea.	4	20		24	37.50
9000	Minimum labor/equipment charge	↓	4	2	Job		80		80	132

12 32 23.30 Manufactured Wood Casework Vanities

		Crew	Daily Output	Labor-Hours	Unit	Material	2009 Bare Costs Labor	Equipment	Total	Total Incl O&P
0010	**MANUFACTURED WOOD CASEWORK VANITIES**									
8000	Vanity bases, 2 doors, 30" high, 21" deep, 24" wide	2 Carp	20	.800	Ea.	275	32		307	360
8050	30" wide		16	1		315	40		355	415
8100	36" wide		13.33	1.200		345	48		393	460
8150	48" wide	↓	11.43	1.400		425	56		481	565
9000	For deluxe models of all vanities, add to above					40%				
9500	For custom built in place, add to above				↓	25%	10%			

12 32 23.35 Manufactured Wood Casework Hardware

		Crew	Daily Output	Labor-Hours	Unit	Material	2009 Bare Costs Labor	Equipment	Total	Total Incl O&P
0010	**MANUFACTURED WOOD CASEWORK HARDWARE**									
1000	Catches, minimum	1 Carp	235	.034	Ea.	1	1.36		2.36	3.34
1040	Maximum	"	80	.100	"	6.25	4		10.25	13.45
2000	Door/drawer pulls, handles									
2200	Handles and pulls, projecting, metal, minimum	1 Carp	160	.050	Ea.	4.43	2		6.43	8.15
2240	Maximum		68	.118		9.40	4.70		14.10	18.05
2300	Wood, minimum		160	.050		4.67	2		6.67	8.45
2340	Maximum		68	.118		8.55	4.70		13.25	17.20
2600	Flush, metal, minimum		160	.050		4.67	2		6.67	8.45
2640	Maximum		68	.118	↓	8.55	4.70		13.25	17.20
3000	Drawer tracks/glides, minimum		48	.167	Pr.	7.90	6.65		14.55	19.70
3040	Maximum		24	.333		23	13.30		36.30	47.50
4000	Cabinet hinges, minimum		160	.050		2.69	2		4.69	6.25
4040	Maximum	↓	68	.118	↓	10.15	4.70		14.85	18.95

12 34 Manufactured Plastic Casework

12 34 16 – Manufactured Solid-Plastic Casework

12 34 16.10 Outdoor Casework

		Crew	Daily Output	Labor-Hours	Unit	Material	2009 Bare Costs Labor	Equipment	Total	Total Incl O&P
0010	**OUTDOOR CASEWORK**									
0020	Cabinet, base, sink/range, 36"	2 Carp	20.30	.788	Ea.	1,675	31.50		1,706.50	1,900
0100	Base, 36"		20.30	.788		2,325	31.50		2,356.50	2,600
0200	Filler strip, 1" x 30"		158	.101		31	4.05		35.05	40.50
0210	Filler strip, 2" x 30"	↓	158	.101	↓	41	4.05		45.05	51.50

12 35 Specialty Casework

12 35 53 – Laboratory Casework

12 35 53.13 Metal Laboratory Casework

		Crew	Daily Output	Labor-Hours	Unit	Material	2009 Bare Costs Labor	Equipment	Total	Total Incl O&P
0010	**METAL LABORATORY CASEWORK**									
0020	Cabinets, base, door units, metal	2 Carp	18	.889	L.F.	183	35.50		218.50	260
0300	Drawer units		18	.889		410	35.50		445.50	510
0700	Tall storage cabinets, open, 7' high		20	.800		390	32		422	485
0900	With glazed doors		20	.800		470	32		502	570
1300	Wall cabinets, metal, 12-1/2" deep, open		20	.800		129	32		161	195
1500	With doors	↓	20	.800	↓	268	32		300	345

12 36 Countertops

12 36 16 – Metal Countertops

12 36 16.10 Stainless Steel Countertops	Crew	Daily Output	Labor-Hours	Unit	Material	2009 Bare Costs Labor	Equipment	Total	Total Incl O&P
0010 STAINLESS STEEL COUNTERTOPS									
3200 Stainless steel, custom	1 Carp	24	.333	S.F.	136	13.30		149.30	172

12 36 19 – Wood Countertops

12 36 19.10 Maple Countertops

	Crew	Daily Output	Labor-Hours	Unit	Material	2009 Bare Costs Labor	Equipment	Total	Total Incl O&P
0010 MAPLE COUNTERTOPS									
2900 Solid, laminated, 1-1/2" thick, no splash	1 Carp	28	.286	L.F.	59.50	11.40		70.90	84.50
3000 With square splash		28	.286	"	70.50	11.40		81.90	96.50
3400 Recessed cutting block with trim, 16" x 20" x 1"		8	1	Ea.	69	40		109	142

12 36 23 – Plastic Countertops

12 36 23.13 Plastic-Laminate-Clad Countertops

	Crew	Daily Output	Labor-Hours	Unit	Material	2009 Bare Costs Labor	Equipment	Total	Total Incl O&P
0010 PLASTIC-LAMINATE-CLAD COUNTERTOPS									
0020 Stock, 24" wide w/ backsplash, minimum	1 Carp	30	.267	L.F.	8.65	10.65		19.30	27
0100 Maximum		25	.320		18	12.80		30.80	41
0300 Custom plastic, 7/8" thick, aluminum molding, no splash		30	.267		19.85	10.65		30.50	39.50
0400 Cove splash		30	.267		26	10.65		36.65	46
0600 1-1/4" thick, no splash		28	.286		29.50	11.40		40.90	51.50
0700 Square splash		28	.286		28.50	11.40		39.90	50
0900 Square edge, plastic face, 7/8" thick, no splash		30	.267		25	10.65		35.65	45
1000 With splash		30	.267		32	10.65		42.65	52.50
1200 For stainless channel edge, 7/8" thick, add					2.62			2.62	2.88
1300 1-1/4" thick, add					3.07			3.07	3.38
1500 For solid color suede finish, add					2.55			2.55	2.81
1700 For end splash, add				Ea.	17			17	18.70
1900 For cut outs, standard, add, minimum	1 Carp	32	.250		3.40	10		13.40	20
2000 Maximum		8	1		5.65	40		45.65	72.50
2100 Postformed, including backsplash and front edge		30	.267	L.F.	10.20	10.65		20.85	29
2110 Mitred, add		12	.667	Ea.		26.50		26.50	44
2200 Built-in place, 25" wide, plastic laminate		25	.320	L.F.	13.60	12.80		26.40	36
9000 Minimum labor/equipment charge		3.75	2.133	Job		85		85	141

12 36 33 – Tile Countertops

12 36 33.10 Ceramic Tile Countertops

	Crew	Daily Output	Labor-Hours	Unit	Material	2009 Bare Costs Labor	Equipment	Total	Total Incl O&P
0010 CERAMIC TILE COUNTERTOPS									
2300 Ceramic tile mosaic	1 Carp	25	.320	L.F.	29.50	12.80		42.30	53.50

12 36 40 – Stone Countertops

12 36 40.10 Natural Stone Countertops

	Crew	Daily Output	Labor-Hours	Unit	Material	2009 Bare Costs Labor	Equipment	Total	Total Incl O&P
0010 NATURAL STONE COUNTERTOPS									
2500 Marble, stock, with splash, 1/2" thick, minimum	1 Bric	17	.471	L.F.	36.50	19.05		55.55	71
2700 3/4" thick, maximum	"	13	.615	"	91	25		116	141

12 36 53 – Laboratory Countertops

12 36 53.10 Laboratory Countertops and Sinks

	Crew	Daily Output	Labor-Hours	Unit	Material	2009 Bare Costs Labor	Equipment	Total	Total Incl O&P
0010 LABORATORY COUNTERTOPS AND SINKS									
0020 Countertops, not incl. base cabinets, acidproof, minimum	2 Carp	82	.195	S.F.	31.50	7.80		39.30	47.50
0030 Maximum		70	.229		41	9.15		50.15	60.50
0040 Stainless steel		82	.195		95.50	7.80		103.30	118

12 36 61 – Simulated Stone Countertops

12 36 61.16 Solid Surface Countertops

	Crew	Daily Output	Labor-Hours	Unit	Material	2009 Bare Costs Labor	Equipment	Total	Total Incl O&P
0010 SOLID SURFACE COUNTERTOPS, Acrylic polymer									
0020 Pricing for orders of 100 L.F. or greater									
0100 25" wide, solid colors	2 Carp	28	.571	L.F.	44.50	23		67.50	86.50

12 36 61.16 Solid Surface Countertops

		Crew	Daily Output	Labor-Hours	Unit	Material	2009 Bare Costs Labor	Equipment	Total	Total Incl O&P
0200	Patterned colors	2 Carp	28	.571	L.F.	56.50	23		79.50	99.50
0300	Premium patterned colors		28	.571		70.50	23		93.50	116
0400	With silicone attached 4" backsplash, solid colors		27	.593		49	23.50		72.50	93
0500	Patterned colors		27	.593		62	23.50		85.50	107
0600	Premium patterned colors		27	.593		77	23.50		100.50	124
0700	With hard seam attached 4" backsplash, solid colors		23	.696		49	28		77	100
0800	Patterned colors		23	.696		62	28		90	114
0900	Premium patterned colors		23	.696		77	28		105	131
1000	Pricing for order of 51 – 99 L.F.									
1100	25" wide, solid colors	2 Carp	24	.667	L.F.	51.50	26.50		78	101
1200	Patterned colors		24	.667		65	26.50		91.50	116
1300	Premium patterned colors		24	.667		81.50	26.50		108	134
1400	With silicone attached 4" backsplash, solid colors		23	.696		56.50	28		84.50	108
1500	Patterned colors		23	.696		71.50	28		99.50	125
1600	Premium patterned colors		23	.696		89	28		117	144
1700	With hard seam attached 4" backsplash, solid colors		20	.800		56.50	32		88.50	115
1800	Patterned colors		20	.800		71.50	32		103.50	132
1900	Premium patterned colors		20	.800		89	32		121	151
2000	Pricing for order of 1 – 50 L.F.									
2100	25" wide, solid colors	2 Carp	20	.800	L.F.	60	32		92	119
2200	Patterned colors		20	.800		76.50	32		108.50	137
2300	Premium patterned colors		20	.800		95.50	32		127.50	158
2400	With silicone attached 4" backsplash, solid colors		19	.842		66	33.50		99.50	128
2500	Patterned colors		19	.842		83.50	33.50		117	148
2600	Premium patterned colors		19	.842		104	33.50		137.50	171
2700	With hard seam attached 4" backsplash, solid colors		15	1.067		66	42.50		108.50	143
2800	Patterned colors		15	1.067		83.50	42.50		126	163
2900	Premium patterned colors		15	1.067		104	42.50		146.50	186
3000	Sinks, pricing for order of 100 or greater units									
3100	Single bowl, hard seamed, solid colors, 13" x 17"	1 Carp	3	2.667	Ea.	300	107		407	505
3200	10" x 15"		7	1.143		139	45.50		184.50	229
3300	Cutouts for sinks		8	1			40		40	66
3400	Sinks, pricing for order of 51 – 99 units									
3500	Single bowl, hard seamed, solid colors, 13" x 17"	1 Carp	2.55	3.137	Ea.	345	125		470	585
3600	10" x 15"		6	1.333		160	53.50		213.50	264
3700	Cutouts for sinks		7	1.143			45.50		45.50	75.50
3800	Sinks, pricing for order of 1 – 50 units									
3900	Single bowl, hard seamed, solid colors, 13" x 17"	1 Carp	2	4	Ea.	405	160		565	710
4000	10" x 15"		4.55	1.758		188	70		258	325
4100	Cutouts for sinks		5.25	1.524			61		61	101
4200	Cooktop cutouts, pricing for 100 or greater units		4	2		22	80		102	157
4300	51 – 99 units		3.40	2.353		25.50	94		119.50	183
4400	1 – 50 units		3	2.667		30	107		137	209

12 36 61.17 Solid Surface Vanity Tops

		Crew	Daily Output	Labor-Hours	Unit	Material	2009 Bare Costs Labor	Equipment	Total	Total Incl O&P
0010	**SOLID SURFACE VANITY TOPS**									
0015	Solid surface, center bowl, 17" x 19"	1 Carp	12	.667	Ea.	189	26.50		215.50	252
0020	19" x 25"		12	.667		229	26.50		255.50	296
0030	19" x 31"		12	.667		278	26.50		304.50	350
0040	19" x 37"		12	.667		325	26.50		351.50	400
0050	22" x 25"		10	.800		202	32		234	275
0060	22" x 31"		10	.800		236	32		268	310
0070	22" x 37"		10	.800		274	32		306	355

12 36 Countertops

12 36 61 – Simulated Stone Countertops

12 36 61.17 Solid Surface Vanity Tops

		Crew	Daily Output	Labor-Hours	Unit	Material	2009 Bare Costs Labor	2009 Bare Costs Equipment	Total	Total Incl O&P
0080	22" x 43"	1 Carp	10	.800	Ea.	315	32		347	400
0090	22" x 49"		10	.800		345	32		377	435
0110	22" x 55"		8	1		395	40		435	500
0120	22" x 61"		8	1		450	40		490	560
0220	Double bowl, 22" x 61"		8	1		505	40		545	625
0230	Double bowl, 22" x 73"		8	1		705	40		745	840
0240	For aggregate colors, add					35%				
0250	For faucets and fittings, see Div. 22 41 39.10									

12 36 61.19 Engineered Stone Countertops

		Crew	Daily Output	Labor-Hours	Unit	Material	2009 Bare Costs Labor	2009 Bare Costs Equipment	Total	Total Incl O&P
0010	**ENGINEERED STONE COUNTERTOPS**									
0100	25" wide, 4" backsplash, color group A, minimum	2 Carp	15	1.067	L.F.	17	42.50		59.50	89
0110	Maximum		15	1.067		42	42.50		84.50	117
0120	Color group B, minimum		15	1.067		22	42.50		64.50	94.50
0130	Maximum		15	1.067		49	42.50		91.50	125
0140	Color group C, minimum		15	1.067		29.50	42.50		72	103
0150	Maximum		15	1.067		60	42.50		102.50	137
0160	Color group D, minimum		15	1.067		37	42.50		79.50	111
0170	Maximum		15	1.067		70	42.50		112.50	148

12 48 Rugs and Mats

12 48 13 – Entrance Floor Mats and Frames

12 48 13.13 Entrance Floor Mats

			Crew	Daily Output	Labor-Hours	Unit	Material	2009 Bare Costs Labor	2009 Bare Costs Equipment	Total	Total Incl O&P
0010	**ENTRANCE FLOOR MATS**										
2000	Recycled rubber tire tile, 12" x 12" x 3/8" thick	G	1 Clab	125	.064	S.F.	8.40	2.02		10.42	12.60
2510	Natural cocoa fiber, 1/2" thick	G		125	.064		7.95	2.02		9.97	12.10
2520	3/4" thick	G		125	.064		6.25	2.02		8.27	10.20
2530	1" thick	G		125	.064		6.90	2.02		8.92	10.95

12 51 Office Furniture

12 51 16 – Case Goods

12 51 16.13 Metal Case Goods

		Crew	Daily Output	Labor-Hours	Unit	Material	2009 Bare Costs Labor	2009 Bare Costs Equipment	Total	Total Incl O&P
0010	**METAL CASE GOODS**									
0020	Desks, 29" high, double pedestal, 30" x 60", metal, minimum				Ea.	450			450	495
0030	Maximum				"	1,125			1,125	1,225

12 54 Hospitality Furniture

12 54 13 – Hotel and Motel Furniture

12 54 13.10 Hotel Furniture

		Crew	Daily Output	Labor-Hours	Unit	Material	2009 Bare Costs Labor	2009 Bare Costs Equipment	Total	Total Incl O&P
0010	**HOTEL FURNITURE**									
0020	Standard quality set, minimum				Room	2,325			2,325	2,550
0200	Maximum				"	8,300			8,300	9,125

12 54 16 – Restaurant Furniture

12 54 16.20 Furniture, Restaurant

		Crew	Daily Output	Labor-Hours	Unit	Material	2009 Bare Costs Labor	2009 Bare Costs Equipment	Total	Total Incl O&P
0010	**FURNITURE, RESTAURANT**									
0020	Bars, built-in, front bar	1 Carp	5	1.600	L.F.	238	64		302	370
0200	Back bar		5	1.600	"	173	64		237	296
0500	Booth unit, molded plastic, stub wall and 2 seats, minimum		2	4	Set	320	160		480	620

12 54 Hospitality Furniture

12 54 16 – Restaurant Furniture

12 54 16.20 Furniture, Restaurant

		Crew	Daily Output	Labor-Hours	Unit	Material	2009 Bare Costs Labor	Equipment	Total	Total Incl O&P
0600	Maximum	1 Carp	1.50	5.333	Set	1,100	213		1,313	1,575
0800	Booth seat, upholstered, foursome, single (end) minimum		5	1.600	Ea.	590	64		654	755
0900	Maximum		4	2		795	80		875	1,000
1000	Foursome, double, minimum		4	2		975	80		1,055	1,200
1100	Maximum		3	2.667		1,375	107		1,482	1,675
1300	Circle booth, upholstered, 1/4 circle, minimum		3	2.667		1,075	107		1,182	1,350
1400	Maximum		2	4		1,475	160		1,635	1,900
1500	3/4 circle, minimum		1.50	5.333		4,200	213		4,413	4,975
1600	Maximum	↓	1	8	↓	5,075	320		5,395	6,125

12 55 Detention Furniture

12 55 13 – Detention Bunks

12 55 13.13 Cots

		Crew	Daily Output	Labor-Hours	Unit	Material	2009 Bare Costs Labor	Equipment	Total	Total Incl O&P
0010	**COTS**									
2500	Bolted, single, painted steel	E-4	20	1.600	Ea.	330	72.50	6.70	409.20	505
2700	Stainless steel	"	20	1.600	"	960	72.50	6.70	1,039.20	1,200

12 56 Institutional Furniture

12 56 43 – Dormitory Furniture

12 56 43.10 Dormitory Furnishings

		Crew	Daily Output	Labor-Hours	Unit	Material	2009 Bare Costs Labor	Equipment	Total	Total Incl O&P
0010	**DORMITORY FURNISHINGS**									
0300	Bunkable bed, twin, minimum				Ea.	270			270	297
0320	Maximum				"	650			650	715

12 56 51 – Library Furniture

12 56 51.10 Library Furnishings

		Crew	Daily Output	Labor-Hours	Unit	Material	2009 Bare Costs Labor	Equipment	Total	Total Incl O&P
0010	**LIBRARY FURNISHINGS**									
1710	Carrels, hardwood, 36" x 24", minimum	1 Carp	5	1.600	Ea.	510	64		574	665
1720	Maximum		4	2		785	80		865	995
1730	Metal, minimum		5	1.600		241	64		305	370
1740	Maximum		4	2	↓	575	80		655	765
6010	Bookshelf, mtl, 90" high, 10" shelf, dbl face		11.50	.696	L.F.	118	28		146	176
6020	Single face	↓	12	.667	"	113	26.50		139.50	168
6050	For 8" shelving, subtract from above					10%				
6070	For 42" high with countertop, subtract from above					20%				

12 56 70 – Healthcare Furniture

12 56 70.10 Furniture, Hospital

		Crew	Daily Output	Labor-Hours	Unit	Material	2009 Bare Costs Labor	Equipment	Total	Total Incl O&P
0010	**FURNITURE, HOSPITAL**									
0020	Beds, manual, minimum				Ea.	730			730	800
0100	Maximum				"	2,500			2,500	2,750
1100	Patient wall systems, not incl. plumbing, minimum				Room	1,025			1,025	1,125
1200	Maximum				"	1,900			1,900	2,100

12 63 Stadium and Arena Seating

12 63 13 – Stadium and Arena Bench Seating

12 63 13.13 Bleachers

12 63 13.13 Bleachers		Crew	Daily Output	Labor-Hours	Unit	Material	2009 Bare Costs Labor	Equipment	Total	Total Incl O&P
0010	**BLEACHERS**									
3000	Telescoping, manual to 15 tier, minimum	F-5	65	.492	Seat	76	19.90		95.90	117
3100	Maximum		60	.533		114	21.50		135.50	162
3300	16 to 20 tier, minimum		60	.533		183	21.50		204.50	237
3400	Maximum		55	.582		228	23.50		251.50	290
3600	21 to 30 tier, minimum		50	.640		190	26		216	252
3700	Maximum		40	.800		228	32.50		260.50	305
3900	For integral power operation, add, minimum	2 Elec	300	.053		38	2.51		40.51	46
4000	Maximum	"	250	.064		61	3.01		64.01	71.50
5000	Benches, folding, in wall, 14' table, 2 benches	L-4	2	12	Set	645	450		1,095	1,450

12 67 Pews and Benches

12 67 13 – Pews

12 67 13.13 Sanctuary Pews

12 67 13.13 Sanctuary Pews		Crew	Daily Output	Labor-Hours	Unit	Material	2009 Bare Costs Labor	Equipment	Total	Total Incl O&P
0010	**SANCTUARY PEWS**									
1500	Bench type, hardwood, minimum	1 Carp	20	.400	L.F.	74	16		90	108
1550	Maximum	"	15	.533		147	21.50		168.50	196
1570	For kneeler, add					18.15			18.15	19.95

12 93 Site Furnishings

12 93 23 – Trash and Litter Receptors

12 93 23.10 Trash Receptacles

12 93 23.10 Trash Receptacles			Crew	Daily Output	Labor-Hours	Unit	Material	2009 Bare Costs Labor	Equipment	Total	Total Incl O&P
0010	**TRASH RECEPTACLES**										
0500	Recycled plastic, var colors, round, 32 Gal, 28" x 38" H	G	2 Clab	5	3.200	Ea.	290	101		391	485
0510	32 Gal, 31" x 32" H	G	"	5	3.200	"	360	101		461	560

12 93 33 – Manufactured Planters

12 93 33.10 Planters

12 93 33.10 Planters		Crew	Daily Output	Labor-Hours	Unit	Material	2009 Bare Costs Labor	Equipment	Total	Total Incl O&P
0010	**PLANTERS**									
0012	Concrete, sandblasted, precast, 48" diameter, 24" high	2 Clab	15	1.067	Ea.	635	33.50		668.50	755
0100	Fluted, precast, 7' diameter, 36" high		10	1.600		1,075	50.50		1,125.50	1,250
0300	Fiberglass, circular, 36" diameter, 24" high		15	1.067		470	33.50		503.50	575
0400	60" diameter, 24" high		10	1.600		800	50.50		850.50	960
9000	Minimum labor/equipment charge	1 Clab	2	4	Job		126		126	209

12 93 43 – Site Seating and Tables

12 93 43.13 Site Seating

12 93 43.13 Site Seating		Crew	Daily Output	Labor-Hours	Unit	Material	2009 Bare Costs Labor	Equipment	Total	Total Incl O&P
0010	**SITE SEATING**									
0012	Seating, benches, park, precast conc, w/backs, wood rails, 4' long	2 Clab	5	3.200	Ea.	430	101		531	640
0100	8' long		4	4		860	126		986	1,150
0300	Fiberglass, without back, one piece, 4' long		10	1.600		565	50.50		615.50	710
0400	8' long		7	2.286		1,175	72		1,247	1,400
0500	Steel barstock pedestals w/backs, 2" x 3" wood rails, 4' long		10	1.600		1,050	50.50		1,100.50	1,225
0510	8' long		7	2.286		1,250	72		1,322	1,500
0520	3" x 8" wood plank, 4' long		10	1.600		1,050	50.50		1,100.50	1,250
0530	8' long		7	2.286		1,100	72		1,172	1,350
0540	Backless, 4" x 4" wood plank, 4' square		10	1.600		1,025	50.50		1,075.50	1,200
0550	8' long		7	2.286		975	72		1,047	1,200
0600	Aluminum pedestals, with backs, aluminum slats, 8' long		8	2		335	63		398	475
0610	15' long		5	3.200		500	101		601	715

12 93 Site Furnishings

12 93 43 – Site Seating and Tables

12 93 43.13 Site Seating		Crew	Daily Output	Labor-Hours	Unit	Material	2009 Bare Costs Labor	Equipment	Total	Total Incl O&P
0620	Portable, aluminum slats, 8' long	2 Clab	8	2	Ea.	335	63		398	475
0630	15' long		5	3.200		655	101		756	885
0800	Cast iron pedestals, back & arms, wood slats, 4' long		8	2		370	63		433	515
0820	8' long		5	3.200		1,075	101		1,176	1,375
0840	Backless, wood slats, 4' long		8	2		645	63		708	815
0860	8' long		5	3.200		815	101		916	1,075
1700	Steel frame, fir seat, 10' long		10	1.600		239	50.50		289.50	345
9000	Minimum labor/equipment charge		2	8	Job		253		253	415

Division Notes

		CREW	DAILY OUTPUT	LABOR-HOURS	UNIT	2009 BARE COSTS				TOTAL INCL O&P
						MAT.	LABOR	EQUIP.	TOTAL	

Estimating Tips

General

- The items and systems in this division are usually estimated, purchased, supplied, and installed as a unit by one or more subcontractors. The estimator must ensure that all parties are operating from the same set of specifications and assumptions, and that all necessary items are estimated and will be provided. Many times the complex items and systems are covered, but the more common ones, such as excavation or a crane, are overlooked for the very reason that everyone assumes nobody could miss them. The estimator should be the central focus and be able to ensure that all systems are complete.

- Another area where problems can develop in this division is at the interface between systems. The estimator must ensure, for instance, that anchor bolts, nuts, and washers are estimated and included for the air-supported structures and pre-engineered buildings to be bolted to their foundations. Utility supply is a common area where essential items or pieces of equipment can be missed or overlooked due to the fact that each subcontractor may feel it is another's responsibility. The estimator should also be aware of certain items which may be supplied as part of a package but installed by others, and ensure that the installing contractor's estimate includes the cost of installation. Conversely, the estimator must also ensure that items are not costed by two different subcontractors, resulting in an inflated overall estimate.

13 30 00 Special Structures

- The foundations and floor slab, as well as rough mechanical and electrical, should be estimated, as this work is required for the assembly and erection of the structure. Generally, as noted in the book, the pre-engineered building comes as a shell and additional features, such as windows and doors, must be included by the estimator. Here again, the estimator must have a clear understanding of the scope of each portion of the work and all the necessary interfaces.

Reference Numbers

Reference numbers are shown in shaded boxes at the beginning of some major classifications. These numbers refer to related items in the Reference Section. The reference information may be an estimating procedure, an alternate pricing method, or technical information.

Note: Not all subdivisions listed here necessarily appear in this publication.

Division 13 – Special Construction

13 05 05 – Selective Special Construction Demolition

13 05 05.10 Selective Demolition, Air Supported Structures		Crew	Daily Output	Labor-Hours	Unit	Material	2009 Bare Costs Labor	Equipment	Total	Total Incl O&P
0010	**SELECTIVE DEMOLITION, AIR SUPPORTED STRUCTURES**									
0020	Tank covers, scrim, dbl layer, vinyl poly w/ hdw, blower & controls									
0050	Round and rectangular R024119-10	B-2	9000	.004	S.F.		.14		.14	.23
0100	Warehouse structures									
0120	Poly/vinyl fabric, 28 oz, incl tension cables & inflation system	4 Clab	9000	.004	SF Flr.		.11		.11	.19
0150	Reinforced vinyl, 12 oz., 3000 S.F.	"	5000	.006			.20		.20	.33
0200	12,000 to 24,000 S.F.	8 Clab	20000	.003			.10		.10	.17
0250	Tedlar vinyl fabric, 28 oz. w/liner, to 3000 S.F.	4 Clab	5000	.006			.20		.20	.33
0300	12,000 to 24,000 S.F.	8 Clab	20000	.003	↓		.10		.10	.17
0350	Greenhouse/shelter, woven polyethylene with liner									
0400	3000 S.F.	4 Clab	5000	.006	SF Flr.		.20		.20	.33
0450	12,000 to 24,000 S.F.	8 Clab	20000	.003			.10		.10	.17
0500	Tennis/gymnasium, poly/vinyl fabric, 28 oz., incl thermal liner	4 Clab	9000	.004			.11		.11	.19
0600	Stadium/convention center, teflon coated fiberglass, incl thermal liner	9 Clab	40000	.002	↓		.06		.06	.09
0700	Doors, air lock, 15' long, 10' x 10'	2 Carp	1.50	10.667	Ea.		425		425	705
0720	15' x 15'		.80	20			800		800	1,325
0750	Revolving personnel door, 6' dia. x 6'-6" high	↓	1.50	10.667	↓		425		425	705

13 05 05.20 Selective Demolition, Garden Houses

		Crew	Daily Output	Labor-Hours	Unit	Material	Labor	Equipment	Total	Total Incl O&P
0010	**SELECTIVE DEMOLITION, GARDEN HOUSES**									
0020	Prefab, wood, excl foundation, average	2 Clab	400	.040	SF Flr.		1.26		1.26	2.09

13 05 05.25 Selective Demolition, Geodesic Domes

		Crew	Daily Output	Labor-Hours	Unit	Material	Labor	Equipment	Total	Total Incl O&P
0010	**SELECTIVE DEMOLITION, GEODESIC DOMES**									
0050	Shell only, interlocking plywood panels, 30' diameter	F-5	3.20	10	Ea.		405		405	670
0060	34' diameter		2.30	13.913			565		565	930
0070	39' diameter		2	16			645		645	1,075
0080	45' diameter		2.20	14.545			590		590	970
0090	55' diameter		2	16			645		645	1,075
0100	60' diameter		2	16			645		645	1,075
0110	65' diameter	↓	1.60	20	↓		810		810	1,325

13 05 05.30 Selective Demolition, Greenhouses

		Crew	Daily Output	Labor-Hours	Unit	Material	Labor	Equipment	Total	Total Incl O&P
0010	**SELECTIVE DEMOLITION, GREENHOUSES**									
0020	Resi-type, free standing, excl foundations, 9' long x 8' wide	2 Clab	160	.100	SF Flr.		3.16		3.16	5.20
0030	9' long x 11' wide		170	.094			2.97		2.97	4.91
0040	9' long x 14' wide		220	.073			2.30		2.30	3.79
0050	9' long x 17' wide		320	.050			1.58		1.58	2.61
0060	Lean-to type, 4' wide		64	.250			7.90		7.90	13.05
0070	7' wide		120	.133	↓		4.21		4.21	6.95
0080	Geodesic hemisphere, 1/8" plexiglass glazing, 8' dia		4	4	Ea.		126		126	209
0090	24' dia		.80	20			630		630	1,050
0100	48' dia	↓	.40	40	↓		1,275		1,275	2,075

13 05 05.35 Selective Demolition, Hangars

		Crew	Daily Output	Labor-Hours	Unit	Material	Labor	Equipment	Total	Total Incl O&P
0010	**SELECTIVE DEMOLITION, HANGARS**									
0020	T type hangars, prefab, steel , galv roof & walls, incl doors, excl fndtn	E-2	2550	.022	SF Flr.		.96	.68	1.64	2.45
0030	Circular type, prefab, steel frame, plastic skin, incl foundation, 80' dia	"	.50	112	Total		4,875	3,475	8,350	12,500

13 05 05.45 Selective Demolition, Lightning Protection

		Crew	Daily Output	Labor-Hours	Unit	Material	Labor	Equipment	Total	Total Incl O&P
0010	**SELECTIVE DEMOLITION, LIGHTNING PROTECTION**									
0020	Air terminal & base, copper, 3/8" diam. x 10", to 75' h	1 Clab	16	.500	Ea.		15.80		15.80	26
0030	1/2" diam. x 12", over 75' h		16	.500			15.80		15.80	26
0050	Aluminum, 1/2" diam. x 12", to 75' h		16	.500			15.80		15.80	26
0060	5/8" diam. x 12", over 75' h		16	.500	↓		15.80		15.80	26
0070	Cable, copper, 220 lb per thousand feet, to 75' high	↓	640	.013	L.F.		.40		.40	.65

13 05 Common Work Results for Special Construction

13 05 05 – Selective Special Construction Demolition

13 05 05.45 Selective Demolition, Lightning Protection

		Crew	Daily Output	Labor-Hours	Unit	Material	2009 Bare Costs Labor	Equipment	Total	Total Incl O&P
0080	375 lb per thousand feet, over 75' high	1 Clab	460	.017	L.F.		.55		.55	.91
0090	Aluminum, 101 lb per thousand feet, to 75' high		560	.014			.45		.45	.75
0100	199 lb per thousand feet, over 75' high		480	.017			.53		.53	.87
0110	Arrester, 175 V AC, to ground		16	.500	Ea.		15.80		15.80	26
0120	650 V AC, to ground		13	.615	"		19.45		19.45	32

13 05 05.50 Selective Demolition, Pre-Engineered Steel Buildings

		Crew	Daily Output	Labor-Hours	Unit	Material	2009 Bare Costs Labor	Equipment	Total	Total Incl O&P
0010	**SELECTIVE DEMOLITION, PRE-ENGINEERED STEEL BUILDINGS**									
0500	Pre-engd steel bldgs, rigid frame, clear span & multi post, excl salvage									
0550	3,500 to 7,500 S.F	L-11	1350	.024	SF Flr.		.86	1.39	2.25	2.97
0600	7,501 to 12,500 S.F		2000	.016			.58	.94	1.52	2
0650	12,500 S.F or greater		2200	.015			.53	.85	1.38	1.82
0700	Pre-engd steel building components									
0710	Entrance canopy, including frame 4' x 4'	E-24	8	4	Ea.		175	98.50	273.50	425
0720	4' x 8'	"	7	4.571			201	113	314	485
0730	H.M doors, self framing, single leaf	2 Skwk	8	2			81.50		81.50	133
0740	Double leaf		5	3.200			131		131	213
0760	Gutter, eave type		600	.027	L.F.		1.09		1.09	1.77
0770	Sash, single slide, double slide or fixed		24	.667	Ea.		27		27	44.50
0780	Skylight, fiberglass, to 30 S.F.		16	1			41		41	66.50
0785	Roof vents, circular, 12" to 24" diameter		12	1.333			54.50		54.50	88.50
0790	Continuous, 10' long		8	2			81.50		81.50	133
0900	Shelters, aluminum frame									
0910	Acrylic glazing, 3' x 9' x 8' high	2 Skwk	2	8	Ea.		325		325	530
0920	9' x 12' x 8' high	"	1.50	10.667	"		435		435	710

13 05 05.60 Selective Demolition, Silos

		Crew	Daily Output	Labor-Hours	Unit	Material	2009 Bare Costs Labor	Equipment	Total	Total Incl O&P
0010	**SELECTIVE DEMOLITION, SILOS**									
0020	Conc stave, indstrl, conical/sloping bott, excl fndtn, 12' dia, 35' h	D-8	.22	181	Ea.		6,750		6,750	10,900
0030	16' dia, 45' h		.16	250			9,300		9,300	15,000
0040	25' dia, 75' h		.10	400			14,900		14,900	24,000
0050	Steel, factory fabricated, 30,000 gal cap, painted or epoxy lined	L-5	2	28			1,250	395	1,645	2,700

13 05 05.65 Selective Demolition, Sound Control

		Crew	Daily Output	Labor-Hours	Unit	Material	2009 Bare Costs Labor	Equipment	Total	Total Incl O&P
0010	**SELECTIVE DEMOLITION, SOUND CONTROL**									
0120	Acoustical enclosure, 4" thick walls & ceiling panels, 8 lbs/SF	3 Carp	144	.167	SF Surf		6.65		6.65	11
0130	10.5 lbs/SF		128	.188			7.50		7.50	12.35
0140	Reverb chamber, parallel walls, 4" thick		120	.200			8		8	13.20
0150	Skewed walls, parallel roof, 4" thick		110	.218			8.70		8.70	14.40
0160	Skewed walls/roof, 4" layer/air space		96	.250			10		10	16.50
0170	Sound-absorbing panels, painted metal, 2'-6" x 8', under 1,000 SF		430	.056			2.23		2.23	3.68
0180	Over 1,000 SF		480	.050			2		2	3.30
0190	Flexible transparent curtain, clear	3 Shee	430	.056			2.63		2.63	4.19
0192	50% clear, 50% foam		430	.056			2.63		2.63	4.19
0194	25% clear, 75% foam		430	.056			2.63		2.63	4.19
0196	100% foam		430	.056			2.63		2.63	4.19
0200	Audio-masking sys, incl speakers, amplfr, signal gnrtr, clng mntd, 5,000 SF									
0205	Ceiling mounted, 5,000 SF	2 Elec	4800	.003	S.F.		.16		.16	.24
0210	10,000 SF		5600	.003			.13		.13	.21
0220	Plenum mounted, 5,000 SF		7600	.002			.10		.10	.15
0230	10,000 SF		8800	.002			.09		.09	.13

13 05 05.70 Selective Demolition, Special Purpose Rooms

		Crew	Daily Output	Labor-Hours	Unit	Material	2009 Bare Costs Labor	Equipment	Total	Total Incl O&P
0010	**SELECTIVE DEMOLITION, SPECIAL PURPOSE ROOMS**									
0100	Audiometric rooms, under 500 S.F. surface	4 Carp	200	.160	SF Surf		6.40		6.40	10.55
0110	Over 500 S.F. surface	"	240	.133	"		5.35		5.35	8.80

315

13 05 05 – Selective Special Construction Demolition

13 05 05.70 Selective Demolition, Special Purpose Rooms

		Crew	Daily Output	Labor-Hours	Unit	Material	2009 Bare Costs Labor	2009 Bare Costs Equipment	Total	Total Incl O&P
0200	Clean rooms, 12' x 12' soft wall, class 100	1 Carp	.30	26.667	Ea.		1,075		1,075	1,750
0210	Class 1000		.30	26.667			1,075		1,075	1,750
0220	Class 10,000		.35	22.857			915		915	1,500
0230	Class 100,000		.35	22.857			915		915	1,500
0300	Darkrooms, shell complete, 8' high	2 Carp	220	.073	SF Flr.		2.91		2.91	4.80
0310	12' high		110	.145	"		5.80		5.80	9.60
0350	Darkrooms doors, mini-cylindrical, revolving		4	4	Ea.		160		160	264
0400	Music room, practice modular		140	.114	SF Surf		4.57		4.57	7.55
0500	Refrigeration structures and finishes									
0510	Wall finish, 2 coat portland cement plaster, 1/2" thick	1 Clab	200	.040	S.F.		1.26		1.26	2.09
0520	Fiberglass panels, 1/8" thick		400	.020			.63		.63	1.04
0530	Ceiling finish, polystyrene plastic, 1" to 2" thick		500	.016			.51		.51	.83
0540	4" thick		450	.018			.56		.56	.93
0550	Refrigerator, prefab aluminum walk-in, 7'-6" high, 6' x 6' OD	2 Carp	100	.160	SF Flr.		6.40		6.40	10.55
0560	10' x 10' OD		160	.100			4		4	6.60
0570	Over 150 S.F.		200	.080			3.20		3.20	5.30
0600	Sauna, prefabricated, including heater & controls, 7' high, to 30 S.F.		120	.133			5.35		5.35	8.80
0610	To 40 S.F.		140	.114			4.57		4.57	7.55
0620	To 60 S.F.		175	.091			3.65		3.65	6.05
0630	To 100 S.F.		220	.073			2.91		2.91	4.80
0640	To 130 S.F.		250	.064			2.56		2.56	4.22

13 05 05.75 Selective Demolition, Storage Tanks

		Crew	Daily Output	Labor-Hours	Unit	Material	2009 Bare Costs Labor	2009 Bare Costs Equipment	Total	Total Incl O&P
0010	**SELECTIVE DEMOLITION, STORAGE TANKS**									
0500	Steel tank, single wall, above ground, not incl fdn, pumps or piping									
0510	Single wall, 275 gallon	Q-1	3	5.333	Ea.		234		234	365
0520	550 thru 2,000 gallon	B-34P	2	12			485	268	753	1,075
0530	5,000 thru 10,000 gallon	B-34Q	2	12			490	610	1,100	1,450
0540	15,000 thru 30,000 gallon	B-34N	2	4			128	233	361	465
0600	Steel tank, double wall, above ground not incl fdn, pumps & piping									
0620	500 thru 2,000 gallon	B-34P	2	12	Ea.		485	268	753	1,075

13 05 05.85 Selective Demolition, Swimming Pool Equip

		Crew	Daily Output	Labor-Hours	Unit	Material	2009 Bare Costs Labor	2009 Bare Costs Equipment	Total	Total Incl O&P
0010	**SELECTIVE DEMOLITION, SWIMMING POOL EQUIP**									
0020	Diving stand, stainless steel, 3 meter	2 Clab	3	5.333	Ea.		169		169	278
0030	1 meter		5	3.200			101		101	167
0040	Diving board, 16' long, aluminum		5.40	2.963			93.50		93.50	155
0050	Fiberglass		5.40	2.963			93.50		93.50	155
0070	Ladders, heavy duty, stainless steel, 2 tread		14	1.143			36		36	59.50
0080	4 tread		12	1.333			42		42	69.50
0090	Lifeguard chair, stainless steel, fixed		5	3.200			101		101	167
0100	Slide, tubular, fiberglass, aluminum handrails & ladder, 5', straight		4	4			126		126	209
0110	8', curved		6	2.667			84.50		84.50	139
0120	10', curved		3	5.333			169		169	278
0130	12' straight, with platform		2.50	6.400			202		202	335
0140	Removable access ramp, stainless steel		4	4			126		126	209
0150	Removable stairs, stainless steel, collapsible		4	4			126		126	209

13 05 05.90 Selective Demolition, Tension Structures

		Crew	Daily Output	Labor-Hours	Unit	Material	2009 Bare Costs Labor	2009 Bare Costs Equipment	Total	Total Incl O&P
0010	**SELECTIVE DEMOLITION, TENSION STRUCTURES**									
0020	Tension structure, steel/alum frame, fabric shell, 60' clear span, 6,000 SF	B-41	2000	.022	SF Flr.		.72	.13	.85	1.33
0030	12,000 SF		2200	.020			.65	.12	.77	1.20
0040	80' clear span, 20,800 SF		2440	.018			.59	.11	.70	1.09
0050	100' clear span, 10,000 SF	L-5	4350	.013			.58	.18	.76	1.24
0060	26,000 SF	"	4600	.012			.54	.17	.71	1.18

13 05 Common Work Results for Special Construction

13 05 05 – Selective Special Construction Demolition

13 05 05.95 Selective Demo, X-Ray/Radio Freq Protection	Crew	Daily Output	Labor-Hours	Unit	Material	2009 Bare Costs Labor	Equipment	Total	Total Incl O&P
0010 **SELECTIVE DEMO, X-RAY/RADIO FREQ PROTECTION**									
0020 Shielding lead, lined door frame, excl hdwe, 1/16" thick	1 Clab	4.80	1.667	Ea.		52.50		52.50	87
0030 Lead sheets, 1/16" thick	2 Clab	270	.059	S.F.		1.87		1.87	3.09
0050 Lead shielding, 1/4" thick		270	.059			1.87		1.87	3.09
0060 1/2" thick		240	.067			2.11		2.11	3.48
0070 Lead glass, 1/4" thick, 2.0 mm LE, 12" x 16"	2 Glaz	26	.615	Ea.		24		24	38.50
0080 24" x 36"		16	1			38.50		38.50	62
0090 36" x 60"		4	4			154		154	249
0100 Lead glass window frame, with 1/16" lead & voice passage, 36" x 60"		4	4			154		154	249
0110 Lead glass window frame, 24" x 36"		16	1			38.50		38.50	62
0120 Lead gypsum board, 5/8" thick with 1/16" lead	2 Clab	320	.050	S.F.		1.58		1.58	2.61
0130 1/8" lead		280	.057			1.81		1.81	2.98
0140 1/32" lead		400	.040			1.26		1.26	2.09
0150 Butt joints, 1/8" lead or thicker, 2" x 7' long batten strip		480	.033	Ea.		1.05		1.05	1.74
0160 X-ray protection, average radiography room, up to 300 SF, 1/16" lead, min		.50	32	Total		1,000		1,000	1,675
0170 Maximum		.30	53.333			1,675		1,675	2,775
0180 Deep therapy X-ray room, 250 KV cap, up to 300 SF, 1/4" lead, min		.20	80			2,525		2,525	4,175
0190 Maximum		.12	133			4,225		4,225	6,950
0880 Radio frequency shielding, prefab or screen-type copper or steel, minimum		360	.044	SF Surf		1.40		1.40	2.32
0890 Average		310	.052			1.63		1.63	2.69
0895 Maximum		290	.055			1.74		1.74	2.88

13 11 Swimming Pools

13 11 13 – Below-Grade Swimming Pools

13 11 13.50 Swimming Pools

13 11 13.50 Swimming Pools	Crew	Daily Output	Labor-Hours	Unit	Material	2009 Bare Costs Labor	Equipment	Total	Total Incl O&P
0011 **SWIMMING POOLS**, Outdoor, incl. equip. & houses, minimum				SF Surf				32.73	36
0300 Maximum								61.82	68
0400 Residential, incl. equipment, permanent type, minimum								10.91	12
0700 Maximum								22.73	25
0900 Municipal, including equipment only, over 5000 S.F., minimum								22.73	25
1000 Maximum								46.36	51
1300 Motel or apt., incl. equipment only, under 5000 S.F., minimum								20	22
1400 Maximum								30.91	34

13 12 Fountains

13 12 13 – Exterior Fountains

13 12 13.10 Yard Fountains

13 12 13.10 Yard Fountains	Crew	Daily Output	Labor-Hours	Unit	Material	2009 Bare Costs Labor	Equipment	Total	Total Incl O&P
0010 **YARD FOUNTAINS**									
0100 Outdoor fountain, 48" high with bowl and figures	2 Clab	2	8	Ea.	250	253		503	690

13 17 Tubs and Pools

13 17 13 – Redwood Hot Tub System

13 17 13.10 Redwood Hot Tub System	Crew	Daily Output	Labor-Hours	Unit	Material	2009 Bare Costs Labor	Equipment	Total	Total Incl O&P
0010 **REDWOOD HOT TUB SYSTEM**									
7050 4' diameter x 4' deep	Q-1	1	16	Ea.	2,025	700		2,725	3,325
7150 6' diameter x 4' deep		.80	20		3,125	880		4,005	4,800
7200 8' diameter x 4' deep		.80	20		4,350	880		5,230	6,125

13 17 33 – Whirlpool Tubs

13 17 33.10 Whirlpool Bath	Crew	Daily Output	Labor-Hours	Unit	Material	2009 Bare Costs Labor	Equipment	Total	Total Incl O&P
0010 **WHIRLPOOL BATH**									
6000 Whirlpool, bath with vented overflow, molded fiberglass									
6100 66" x 48" x 24"	Q-1	1	16	Ea.	3,050	700		3,750	4,475
6400 72" x 36" x 24"		1	16		3,000	700		3,700	4,400
6500 60" x 30" x 21"		1	16		2,575	700		3,275	3,925
6600 72" x 42" x 22"		1	16		4,025	700		4,725	5,550
6700 83" x 65"		.30	53.333		5,250	2,350		7,600	9,400
6710 For color add					10%				
6711 For designer colors and trim add					25%				

13 21 Controlled Environment Rooms

13 21 26 – Cold Storage Rooms

13 21 26.50 Refrigeration	Crew	Daily Output	Labor-Hours	Unit	Material	2009 Bare Costs Labor	Equipment	Total	Total Incl O&P
0010 **REFRIGERATION**									
0020 Curbs, 12" high, 4" thick, concrete	2 Carp	58	.276	L.F.	4.14	11		15.14	23
6300 Rule of thumb for complete units, w/o doors & refrigeration, cooler		146	.110	SF Flr.	120	4.38		124.38	139
6400 Freezer		109.60	.146	"	142	5.85		147.85	167

13 24 Special Activity Rooms

13 24 16 – Saunas

13 24 16.50 Saunas and Heaters	Crew	Daily Output	Labor-Hours	Unit	Material	2009 Bare Costs Labor	Equipment	Total	Total Incl O&P
0010 **SAUNAS AND HEATERS**									
0020 Prefabricated, incl. heater & controls, 7' high, 6' x 4', C/C	L-7	2.20	12.727	Ea.	4,625	490		5,115	5,900
1700 Door only, cedar, 2'x6', with tempered insulated glass window	2 Carp	3.40	4.706		610	188		798	980
1800 Prehung, incl. jambs, pulls & hardware	"	12	1.333		615	53.50		668.50	765
2500 Heaters only (incl. above), wall mounted, to 200 C.F.					605			605	665
4480 For additional equipment, see Div. 11 66 13.10									

13 24 26 – Steam Baths

13 24 26.50 Steam Baths and Components	Crew	Daily Output	Labor-Hours	Unit	Material	2009 Bare Costs Labor	Equipment	Total	Total Incl O&P
0010 **STEAM BATHS AND COMPONENTS**									
0020 Heater, timer & head, single, to 140 C.F.	1 Plum	1.20	6.667	Ea.	1,400	325		1,725	2,050
0500 To 300 C.F.	"	1.10	7.273		1,625	355		1,980	2,325
2000 Multiple, motels, apts., 2 baths, w/ blow-down assm., 500 C.F.	Q-1	1.30	12.308		5,900	540		6,440	7,350
2500 4 baths	"	.70	22.857		7,450	1,000		8,450	9,750

13 28 Athletic and Recreational Special Construction

13 28 33 - Athletic and Recreational Court Walls

13 28 33.50 Sport Court

13 28 33.50 Sport Court		Crew	Daily Output	Labor-Hours	Unit	Material	2009 Bare Costs Labor	Equipment	Total	Total Incl O&P
0010	**SPORT COURT**									
0020	Floors, No. 2 & better maple, 25/32" thick				SF Flr.				6.05	6.65
0300	Squash, regulation court in existing building, minimum				Court	35,000			35,000	38,500
0400	Maximum				"	39,000			39,000	42,900
0450	Rule of thumb for components:									
0470	Walls	3 Carp	.15	160	Court	10,500	6,400		16,900	22,200
0500	Floor	"	.25	96		8,300	3,825		12,125	15,500
0550	Lighting	2 Elec	.60	26.667	↓	2,000	1,250		3,250	4,125

13 34 Fabricated Engineered Structures

13 34 13 - Glazed Structures

13 34 13.13 Greenhouses

		Crew	Daily Output	Labor-Hours	Unit	Material	2009 Bare Costs Labor	Equipment	Total	Total Incl O&P
0010	**GREENHOUSES**, Shell only, stock units, not incl. 2' stub walls,									
0020	foundation, floors, heat or compartments									
0300	Residential type, free standing, 8'-6" long x 7'-6" wide	2 Carp	59	.271	SF Flr.	47.50	10.85		58.35	70.50
0400	10'-6" wide		85	.188		37	7.50		44.50	53.50
0600	13'-6" wide		108	.148		35	5.90		40.90	48
0700	17'-0" wide		160	.100		35	4		39	45
0900	Lean-to type, 3'-10" wide		34	.471		38.50	18.80		57.30	73.50
1000	6'-10" wide		58	.276		37.50	11		48.50	59
1050	8'-0" wide	↓	60	.267	↓	45.50	10.65		56.15	67.50
1060										
1100	Wall mounted, to existing window, 3' x 3'	1 Carp	4	2	Ea.	445	80		525	620
1120	4' x 5'	"	3	2.667	"	665	107		772	910
3900	For cooling, add, minimum				SF Flr.	2.81			2.81	3.09
4000	Maximum					6.95			6.95	7.65
4200	For heaters, 13.6 MBH, add					5.35			5.35	5.85
4300	60 MBH, add				↓	2			2	2.20
4500	For benches, 2' x 3'-6", add				SF Hor.	24			24	26.50
4600	3' x 10', add				S.F.	12.95			12.95	14.25
4800	For controls, add, minimum				Total	2,400			2,400	2,650
4900	Maximum				"	14,300			14,300	15,700
5100	For humidification equipment, add				M.C.F.	6.15			6.15	6.75
5200	For vinyl shading, add				S.F.	1.29			1.29	1.42

13 34 13.19 Swimming Pool Enclosures

		Crew	Daily Output	Labor-Hours	Unit	Material	2009 Bare Costs Labor	Equipment	Total	Total Incl O&P
0010	**SWIMMING POOL ENCLOSURES** Translucent, free standing									
0020	not including foundations, heat or light									
0200	Economy, minimum	2 Carp	200	.080	SF Hor.	14.80	3.20		18	21.50
0300	Maximum		100	.160		41.50	6.40		47.90	56
0400	Deluxe, minimum		100	.160		48.50	6.40		54.90	64
0600	Maximum	↓	70	.229	↓	225	9.15		234.15	262

13 34 23 - Fabricated Structures

13 34 23.16 Fabricated Control Booths

		Crew	Daily Output	Labor-Hours	Unit	Material	2009 Bare Costs Labor	Equipment	Total	Total Incl O&P
0010	**FABRICATED CONTROL BOOTHS**									
0100	Guard House, prefab conc w/bullet resistant doors & windows, roof & wiring									
0110	8' x 8', Level III	L-10	1	24	Ea.	22,000	1,075	770	23,845	26,900
0120	8' x 8', Level IV	"	1	24	"	27,500	1,075	770	29,345	33,000

13 34 23.45 Kiosks

		Crew	Daily Output	Labor-Hours	Unit	Material	2009 Bare Costs Labor	Equipment	Total	Total Incl O&P
0010	**KIOSKS**									
0020	Round, 5' diameter, 8' high, 1/4" fiberglass wall				Ea.	5,800			5,800	6,375

13 34 Fabricated Engineered Structures

13 34 23 – Fabricated Structures

13 34 23.45 Kiosks		Crew	Daily Output	Labor-Hours	Unit	Material	2009 Bare Costs Labor	Equipment	Total	Total Incl O&P
0100	1" insulated double wall, fiberglass				Ea.	6,600			6,600	7,250
0500	Rectangular, 5' x 9', 7'-6" high, 1/4" fiberglass wall					8,450			8,450	9,300
0600	1" insulated double wall, fiberglass				↓	10,000			10,000	11,000

13 34 63 – Natural Fiber Construction

13 34 63.50 Straw Bale Construction

		Crew	Daily Output	Labor-Hours	Unit	Material	Labor	Equipment	Total	Total Incl O&P
0010	**STRAW BALE CONSTRUCTION**									
2000	Straw bales wall, incl labor & material complete, minimum				SF Flr.					130
2010	Maximum				"					160
2020	Straw bales in walls w/ modified post and beam frame [G]	2 Carp	320	.050	S.F.	4.28	2		6.28	8

13 42 Building Modules

13 42 63 – Detention Cell Modules

13 42 63.16 Steel Detention Cell Modules

		Crew	Daily Output	Labor-Hours	Unit	Material	Labor	Equipment	Total	Total Incl O&P
0010	**STEEL DETENTION CELL MODULES**									
2000	Cells, prefab., 5' to 6' wide, 7' to 8' high, 7' to 8' deep,									
2010	bar front, cot, not incl. plumbing	E-4	1.50	21.333	Ea.	9,650	965	89.50	10,704.50	12,500

13 48 Sound, Vibration, and Seismic Control

13 48 13 – Manufactured Sound and Vibration Control Components

13 48 13.50 Audio Masking

		Crew	Daily Output	Labor-Hours	Unit	Material	Labor	Equipment	Total	Total Incl O&P
0010	**AUDIO MASKING**, acoustical enclosure, 4" thick wall and ceiling									
0020	8# per S.F., up to 12' span	3 Carp	72	.333	SF Surf	32	13.30		45.30	57.50
0300	Better quality panels, 10.5# per S.F.		64	.375		36.50	15		51.50	64.50
0400	Reverb-chamber, 4" thick, parallel walls	↓	60	.400	↓	45.50	16		61.50	76.50

13 49 Radiation Protection

13 49 13 – Lead Sheet

13 49 13.50 Lead Sheets

		Crew	Daily Output	Labor-Hours	Unit	Material	Labor	Equipment	Total	Total Incl O&P
0010	**LEAD SHEETS**									
0300	Lead sheets, 1/16" thick	2 Lath	135	.119	S.F.	7.15	4.21		11.36	14.55
0400	1/8" thick		120	.133		14.45	4.74		19.19	23.50
0500	Lead shielding, 1/4" thick		135	.119		34	4.21		38.21	44
0550	1/2" thick	↓	120	.133	↓	59.50	4.74		64.24	73
1200	X-ray protection, average radiography or fluoroscopy									
1210	room, up to 300 S.F. floor, 1/16" lead, minimum	2 Lath	.25	64	Total	8,075	2,275		10,350	12,500
1500	Maximum, 7'-0" walls	"	.15	106	"	9,700	3,800		13,500	16,700
1600	Deep therapy X-ray room, 250 kV capacity,									
1800	up to 300 S.F. floor, 1/4" lead, minimum	2 Lath	.08	200	Total	22,600	7,100		29,700	36,000
1900	Maximum, 7'-0" walls		.06	266	"	27,800	9,475		37,275	45,600
1999	Minimum labor/equipment charge	↓	4.50	3.556	Job		126		126	200

13 49 21 – Lead Glazing

13 49 21.50 Lead Glazing

		Crew	Daily Output	Labor-Hours	Unit	Material	Labor	Equipment	Total	Total Incl O&P
0010	**LEAD GLAZING**									
0600	Lead glass, 1/4" thick, 2.0 mm LE, 12" x 16"	2 Glaz	13	1.231	Ea.	265	47.50		312.50	370
0700	24" x 36"	"	8	2	"	980	77		1,057	1,200
2000	X-ray viewing panels, clear lead plastic									

13 49 21 – Lead Glazing

	13 49 21.50 Lead Glazing	Crew	Daily Output	Labor-Hours	Unit	Material	2009 Bare Costs Labor	Equipment	Total	Total Incl O&P
2010	7 mm thick, 0.3 mm LE, 2.3 lbs/S.F.	H-3	139	.115	S.F.	143	3.97		146.97	164
2020	12 mm thick, 0.5 mm LE, 3.9 lbs/S.F.		82	.195		194	6.70		200.70	224
2030	18 mm thick, 0.8 mm LE, 5.9 lbs/S.F.		54	.296		210	10.20		220.20	248
2040	22 mm thick, 1.0 mm LE, 7.2 lbs/S.F.		44	.364		215	12.55		227.55	257
2050	35 mm thick, 1.5 mm LE, 11.5 lbs/S.F.		28	.571		241	19.70		260.70	297
2060	46 mm thick, 2.0 mm LE, 15.0 lbs/S.F.	↓	21	.762	↓	315	26.50		341.50	395
2090	For panels 12 S.F. to 48 S.F., add crating charge				Ea.					50

13 49 23 – Modular Shielding Partitions

	13 49 23.50 Modular Shielding Partitions	Crew	Daily Output	Labor-Hours	Unit	Material	2009 Bare Costs Labor	Equipment	Total	Total Incl O&P
0010	**MODULAR SHIELDING PARTITIONS**									
4000	X-ray barriers, modular, panels mounted within framework for									
4002	attaching to floor, wall or ceiling, upper portion is clear lead									
4005	plastic window panels 48"H, lower portion is opaque leaded									
4008	steel panels 36"H, structural supports not incl.									
4010	1-section barrier, 36"W x 84"H overall									
4020	0.5 mm LE panels	H-3	6.40	2.500	Ea.	4,525	86		4,611	5,125
4030	0.8 mm LE panels		6.40	2.500		4,825	86		4,911	5,450
4040	1.0 mm LE panels		5.33	3.002		4,900	103		5,003	5,575
4050	1.5 mm LE panels	↓	5.33	3.002	↓	5,225	103		5,328	5,925
4060	2-section barrier, 72"W x 84"H overall									
4070	0.5 mm LE panels	H-3	4	4	Ea.	9,250	138		9,388	10,400
4080	0.8 mm LE panels		4	4		9,775	138		9,913	11,000
4090	1.0 mm LE panels		3.56	4.494		9,975	155		10,130	11,300
5000	1.5 mm LE panels	↓	3.20	5	↓	11,300	172		11,472	12,800
5010	3-section barrier, 108"W x 84"H overall									
5020	0.5 mm LE panels	H-3	3.20	5	Ea.	13,900	172		14,072	15,600
5030	0.8 mm LE panels		3.20	5		14,700	172		14,872	16,400
5040	1.0 mm LE panels		2.67	5.993		14,900	206		15,106	16,700
5050	1.5 mm LE panels	↓	2.46	6.504	↓	16,200	224		16,424	18,200
7000	X-ray barriers, mobile, mounted within framework w/casters on									
7005	bottom, clear lead plastic window panels on upper portion,									
7010	opaque on lower, 30"W x 75"H overall, incl. framework									
7020	24"H upper w/0.5 mm LE, 48"H lower w/0.8 mm LE	1 Carp	16	.500	Ea.	2,400	20		2,420	2,675
7030	48"W x 75"H overall, incl. framework									
7040	36"H upper w/0.5 mm LE, 36"H lower w/0.8 mm LE	1 Carp	16	.500	Ea.	4,425	20		4,445	4,875
7050	36"H upper w/1.0 mm LE, 36"H lower w/1.5 mm LE	"	16	.500	"	5,250	20		5,270	5,800
7060	72"W x 75"H overall, incl. framework									
7070	36"H upper w/0.5 mm LE, 36"H lower w/0.8 mm LE	1 Carp	16	.500	Ea.	5,225	20		5,245	5,775
7080	36"H upper w/1.0 mm LE, 36"H lower w/1.5 mm LE	"	16	.500	"	6,550	20		6,570	7,225

Division Notes

		CREW	DAILY OUTPUT	LABOR-HOURS	UNIT	2009 BARE COSTS				TOTAL INCL O&P
						MAT.	LABOR	EQUIP.	TOTAL	

Estimating Tips

General

- Many products in Division 14 will require some type of support or blocking for installation not included with the item itself. Examples are supports for conveyors or tube systems, attachment points for lifts, and footings for hoists or cranes. Add these supports in the appropriate division.

14 10 00 Dumbwaiters
14 20 00 Elevators

- Dumbwaiters and elevators are estimated and purchased in a method similar to buying a car. The manufacturer has a base unit with standard features. Added to this base unit price will be whatever options the owner or specifications require. Increased load capacity, additional vertical travel, additional stops, higher speed, and cab finish options are items to be considered. When developing an estimate for dumbwaiters and elevators, remember that some items needed by the installers may have to be included as part of the general contract.

Examples are:
- — shaftway
- — rail support brackets
- — machine room
- — electrical supply
- — sill angles
- — electrical connections
- — pits
- — roof penthouses
- — pit ladders

Check the job specifications and drawings before pricing.

- Installation of elevators and handicapped lifts in historic structures can require significant additional costs. The associated structural requirements may involve cutting into and repairing finishes, mouldings, flooring, etc. The estimator must account for these special conditions.

14 30 00 Escalators and Moving Walks

- Escalators and moving walks are specialty items installed by specialty contractors. There are numerous options associated with these items. For specific options, contact a manufacturer or contractor. In a method similar to estimating dumbwaiters and elevators, you should verify the extent of general contract work and add items as necessary.

14 40 00 Lifts
14 90 00 Other Conveying Equipment

- Products such as correspondence lifts, chutes, and pneumatic tube systems, as well as other items specified in this subdivision, may require trained installers. The general contractor might not have any choice as to who will perform the installation or when it will be performed. Long lead times are often required for these products, making early decisions in scheduling necessary.

Reference Numbers

Reference numbers are shown in shaded boxes at the beginning of some major classifications. These numbers refer to related items in the Reference Section. The reference information may be an estimating procedure, an alternate pricing method, or technical information.

Note: Not all subdivisions listed here necessarily appear in this publication.

14 11 Manual Dumbwaiters

14 11 10 – Hand Operated Dumbwaiters

14 11 10.20 Manual Dumbwaiters		Crew	Daily Output	Labor-Hours	Unit	Material	2009 Bare Costs		Total	Total Incl O&P
							Labor	Equipment		
0010	**MANUAL DUMBWAITERS**									
0020	2 stop, minimum	2 Elev	.75	21.333	Ea.	2,925	1,200		4,125	5,075
0100	Maximum		.50	32	"	6,650	1,800		8,450	10,100
0300	For each additional stop, add	↓	.75	21.333	Stop	1,050	1,200		2,250	3,025

14 12 Electric Dumbwaiters

14 12 10 – Dumbwaiters

14 12 10.10 Electric Dumbwaiters

		Crew	Daily Output	Labor-Hours	Unit	Material	Labor	Equipment	Total	Total Incl O&P
0010	**ELECTRIC DUMBWAITERS**									
0020	2 stop, minimum	2 Elev	.13	123	Ea.	7,375	6,975		14,350	18,800
0100	Maximum		.11	145	"	22,200	8,225		30,425	37,100
0600	For each additional stop, add	↓	.54	29.630	Stop	3,250	1,675		4,925	6,150
0750	Correspondence lift, 1 floor, 2 stop, 45 lb capacity	2 Elec	.20	80	Ea.	8,350	3,750		12,100	15,000

14 21 Electric Traction Elevators

14 21 13 – Electric Traction Freight Elevators

14 21 13.10 Electric Traction Freight Elevators and Options

		Crew	Daily Output	Labor-Hours	Unit	Material	Labor	Equipment	Total	Total Incl O&P
0010	**ELECTRIC TRACTION FREIGHT ELEVATORS AND OPTIONS** R142000-10									
0425	Electric freight, base unit, 4000 lb, 200 fpm, 4 stop, std. fin.	2 Elev	.05	320	Ea.	99,500	18,100		117,600	137,500
0450	For 5000 lb capacity, add R142000-40					5,600			5,600	6,150
0500	For 6000 lb capacity, add					17,400			17,400	19,200
0525	For 7000 lb capacity, add					21,500			21,500	23,600
0550	For 8000 lb capacity, add					29,600			29,600	32,600
0575	For 10000 lb capacity, add					32,500			32,500	35,800
0600	For 12000 lb capacity, add					39,400			39,400	43,300
0625	For 16000 lb capacity, add					50,000			50,000	55,000
0650	For 20000 lb capacity, add					59,500			59,500	65,500
0675	For increased speed, 250 fpm, add					22,500			22,500	24,700
0700	300 fpm, geared electric, add					28,400			28,400	31,300
0725	350 fpm, geared electric, add					34,300			34,300	37,700
0750	400 fpm, geared electric, add					39,900			39,900	43,800
0775	500 fpm, gearless electric, add					47,300			47,300	52,000
0800	600 fpm, gearless electric, add					57,500			57,500	63,000
0825	700 fpm, gearless electric, add					68,500			68,500	75,500
0850	800 fpm, gearless electric, add					79,000			79,000	87,000
0875	For class "B" loading, add					6,525			6,525	7,175
0900	For class "C-1" loading, add					9,075			9,075	9,975
0925	For class "C-2" loading, add					10,300			10,300	11,400
0950	For class "C-3" loading, add					13,000			13,000	14,400
0975	For travel over 40 V.L.F., add	2 Elev	7.25	2.207	V.L.F.	540	125		665	785
1000	For number of stops over 4, add	"	.27	59.259	Stop	4,150	3,350		7,500	9,725

14 21 23 – Electric Traction Passenger Elevators

14 21 23.10 Electric Traction Passenger Elevators and Options

		Crew	Daily Output	Labor-Hours	Unit	Material	Labor	Equipment	Total	Total Incl O&P
0010	**ELECTRIC TRACTION PASSENGER ELEVATORS AND OPTIONS**									
1625	Electric pass., base unit, 2000 lb, 200 fpm, 4 stop, std. fin.	2 Elev	.05	320	Ea.	91,500	18,100		109,600	128,500
1650	For 2500 lb capacity, add					3,800			3,800	4,200
1675	For 3000 lb capacity, add					4,550			4,550	5,000
1700	For 3500 lb capacity, add					7,675			7,675	8,425
1725	For 4000 lb capacity, add					7,100			7,100	7,800

14 21 Electric Traction Elevators

14 21 23 – Electric Traction Passenger Elevators

14 21 23.10 Electric Traction Passenger Elevators and Options	Crew	Daily Output	Labor-Hours	Unit	Material	2009 Bare Costs Labor	Equipment	Total	Total Incl O&P
1750	For 4500 lb capacity, add				Ea.	8,975			8,975	9,875
1775	For 5000 lb capacity, add					10,800			10,800	11,900
1800	For increased speed, 250 fpm, geared electric, add					7,000			7,000	7,700
1825	300 fpm, geared electric, add					10,600			10,600	11,700
1850	350 fpm, geared electric, add					12,100			12,100	13,300
1875	400 fpm, geared electric, add					15,400			15,400	17,000
1900	500 fpm, gearless electric, add					32,600			32,600	35,900
1925	600 fpm, gearless electric, add					36,100			36,100	39,700
1950	700 fpm, gearless electric, add					41,000			41,000	45,100
1975	800 fpm, gearless electric, add					46,800			46,800	51,500
2000	For travel over 40 V.L.F., add	2 Elev	7.25	2.207	V.L.F.	650	125		775	905
2025	For number of stops over 4, add		.27	59.259	Stop	7,825	3,350		11,175	13,800
2400	Electric hospital, base unit, 4000 lb, 200 fpm, 4 stop, std fin.		.05	320	Ea.	79,000	18,100		97,100	115,000
2425	For 4500 lb capacity, add					5,650			5,650	6,200
2450	For 5000 lb capacity, add					6,775			6,775	7,475
2475	For increased speed, 250 fpm, geared electric, add					7,125			7,125	7,850
2500	300 fpm, geared electric, add					10,600			10,600	11,700
2525	350 fpm, geared electric, add					12,200			12,200	13,400
2550	400 fpm, geared electric, add					15,400			15,400	17,000
2575	500 fpm, gearless electric, add					31,700			31,700	34,900
2600	600 fpm, gearless electric, add					36,100			36,100	39,700
2625	700 fpm, gearless electric, add					40,900			40,900	45,000
2650	800 fpm, gearless electric, add					46,800			46,800	51,500
2675	For travel over 40 V.L.F., add	2 Elev	7.25	2.207	V.L.F.	665	125		790	925
2700	For number of stops over 4, add	"	.27	59.259	Stop	4,150	3,350		7,500	9,725

14 21 33 – Electric Traction Residential Elevators

14 21 33.20 Residential Elevators

		Crew	Daily Output	Labor-Hours	Unit	Material	Labor	Equipment	Total	Total Incl O&P
0010	**RESIDENTIAL ELEVATORS**									
7000	Residential, cab type, 1 floor, 2 stop, minimum	2 Elev	.20	80	Ea.	11,000	4,525		15,525	19,100
7100	Maximum		.10	160		18,600	9,050		27,650	34,400
7200	2 floor, 3 stop, minimum		.12	133		16,400	7,550		23,950	29,600
7300	Maximum		.06	266		26,700	15,100		41,800	52,500

14 24 Hydraulic Elevators

14 24 13 – Hydraulic Freight Elevators

14 24 13.10 Hydraulic Freight Elevators and Options

		Crew	Daily Output	Labor-Hours	Unit	Material	Labor	Equipment	Total	Total Incl O&P
0010	**HYDRAULIC FREIGHT ELEVATORS AND OPTIONS**									
1025	Hydraulic freight, base unit, 2000 lb, 50 fpm, 2 stop, std. fin.	2 Elev	.10	160	Ea.	82,000	9,050		91,050	104,000
1050	For 2500 lb capacity, add					5,025			5,025	5,525
1075	For 3000 lb capacity, add					8,850			8,850	9,750
1100	For 3500 lb capacity, add					16,000			16,000	17,600
1125	For 4000 lb capacity, add					17,100			17,100	18,800
1150	For 4500 lb capacity, add					20,000			20,000	22,000
1175	For 5000 lb capacity, add					25,300			25,300	27,800
1200	For 6000 lb capacity, add					31,500			31,500	34,600
1225	For 7000 lb capacity, add					37,900			37,900	41,700
1250	For 8000 lb capacity, add					41,700			41,700	45,900
1275	For 10000 lb capacity, add					52,500			52,500	57,500
1300	For 12000 lb capacity, add					68,500			68,500	75,500
1325	For 16000 lb capacity, add					94,000			94,000	103,000

14 24 Hydraulic Elevators

14 24 13 – Hydraulic Freight Elevators

14 24 13.10 Hydraulic Freight Elevators and Options		Crew	Daily Output	Labor-Hours	Unit	Material	2009 Bare Costs Labor	Equipment	Total	Total Incl O&P
1350	For 20000 lb capacity, add				Ea.	114,000			114,000	125,500
1375	For increased speed, 100 fpm, add					2,150			2,150	2,375
1400	125 fpm, add					4,650			4,650	5,100
1425	150 fpm, add					5,525			5,525	6,075
1450	175 fpm, add					9,400			9,400	10,400
1475	For class "B" loading, add					6,500			6,500	7,150
1500	For class "C-1" loading, add					9,025			9,025	9,925
1525	For class "C-2" loading, add					10,300			10,300	11,300
1550	For class "C-3" loading, add					13,000			13,000	14,300
1575	For travel over 20 V.L.F., add	2 Elev	7.25	2.207	V.L.F.	1,675	125		1,800	2,050
1600	For number of stops over 2, add	"	.27	59.259	Stop	4,575	3,350		7,925	10,200

14 24 23 – Hydraulic Passenger Elevators

14 24 23.10 Hydraulic Passenger Elevators and Options		Crew	Daily Output	Labor-Hours	Unit	Material	2009 Bare Costs Labor	Equipment	Total	Total Incl O&P
0010	**HYDRAULIC PASSENGER ELEVATORS AND OPTIONS**									
2050	Hyd. pass., base unit, 1500 lb, 100 fpm, 2 stop, std. fin.	2 Elev	.10	160	Ea.	39,900	9,050		48,950	57,500
2075	For 2000 lb capacity, add					2,025			2,025	2,250
2100	For 2500 lb. capacity, add					3,225			3,225	3,550
2125	For 3000 lb capacity, add					4,275			4,275	4,700
2150	For 3500 lb capacity, add					6,225			6,225	6,850
2175	For 4000 lb capacity, add					7,600			7,600	8,350
2200	For 4500 lb capacity, add					9,000			9,000	9,900
2225	For 5000 lb capacity, add					11,900			11,900	13,100
2250	For increased speed, 125 fpm, add					2,975			2,975	3,275
2275	150 fpm, add					3,775			3,775	4,150
2300	175 fpm, add					5,650			5,650	6,225
2325	200 fpm, add					7,850			7,850	8,625
2350	For travel over 12 V.L.F., add	2 Elev	7.25	2.207	V.L.F.	890	125		1,015	1,175
2375	For number of stops over 2, add		.27	59.259	Stop	2,575	3,350		5,925	7,975
2725	Hydraulic hospital, base unit, 4000 lb, 100 fpm, 2 stop, std. fin.		.10	160	Ea.	62,000	9,050		71,050	82,000
2750	For 4000 lb capacity, add					7,150			7,150	7,850
2775	For 4500 lb capacity, add					6,900			6,900	7,600
2800	For 5000 lb capacity, add					9,150			9,150	10,100
2825	For increased speed, 125 fpm, add					3,450			3,450	3,775
2850	150 fpm, add					4,150			4,150	4,550
2875	175 fpm, add					6,000			6,000	6,600
2900	200 fpm, add					7,275			7,275	8,000
2925	For travel over 12 V.L.F., add	2 Elev	7.25	2.207	V.L.F.	745	125		870	1,000
2950	For number of stops over 2, add	"	.27	59.259	Stop	4,200	3,350		7,550	9,775

14 27 Custom Elevator Cabs

14 27 13 – Custom Elevator Cab Finishes

14 27 13.10 Cab Finishes		Crew	Daily Output	Labor-Hours	Unit	Material	2009 Bare Costs Labor	Equipment	Total	Total Incl O&P
0010	**CAB FINISHES**									
3325	Passenger elevator cab finishes (based on 3500 lb cab size)									
3350	Acrylic panel ceiling				Ea.	780			780	860
3375	Aluminum eggcrate ceiling					1,225			1,225	1,350
3400	Stainless steel doors					2,975			2,975	3,250
3425	Carpet flooring					650			650	715
3450	Epoxy flooring					480			480	530
3475	Quarry tile flooring					825			825	910

14 27 Custom Elevator Cabs

14 27 13 – Custom Elevator Cab Finishes

14 27 13.10 Cab Finishes

		Crew	Daily Output	Labor-Hours	Unit	Material	2009 Bare Costs Labor	Equipment	Total	Total Incl O&P
3500	Slate flooring				Ea.	1,525			1,525	1,675
3525	Textured rubber flooring					200			200	220
3550	Stainless steel walls					5,175			5,175	5,700
3575	Stainless steel returns at door					1,200			1,200	1,325
4450	Hospital elevator cab finishes (based on 3500 lb cab size)									
4475	Aluminum eggcrate ceiling				Ea.	1,200			1,200	1,325
4500	Stainless steel doors					2,675			2,675	2,950
4525	Epoxy flooring					375			375	415
4550	Quarry tile flooring					845			845	930
4575	Textured rubber flooring					720			720	790
4600	Stainless steel walls					5,175			5,175	5,700
4625	Stainless steel returns at door					1,200			1,200	1,325

14 28 Elevator Equipment and Controls

14 28 10 – Elevator Equipment and Control Options

14 28 10.10 Elevator Controls and Doors

		Crew	Daily Output	Labor-Hours	Unit	Material	2009 Bare Costs Labor	Equipment	Total	Total Incl O&P
0010	**ELEVATOR CONTROLS AND DOORS**									
2975	Passenger elevator options									
3000	2 car group automatic controls	2 Elev	.66	24.242	Ea.	6,650	1,375		8,025	9,400
3025	3 car group automatic controls		.44	36.364		13,000	2,050		15,050	17,500
3050	4 car group automatic controls		.33	48.485		26,500	2,750		29,250	33,300
3075	5 car group automatic controls		.26	61.538		48,100	3,475		51,575	58,500
3100	6 car group automatic controls		.22	72.727		98,500	4,125		102,625	115,000
3125	Intercom service		3	5.333		1,375	300		1,675	2,000
3150	Duplex car selective collective		.66	24.242		12,200	1,375		13,575	15,600
3175	Center opening 1 speed doors		2	8		1,850	455		2,305	2,750
3200	Center opening 2 speed doors		2	8		2,550	455		3,005	3,500
3225	Rear opening doors (opposite front)		2	8		3,125	455		3,580	4,150
3250	Side opening 2 speed doors		2	8		4,900	455		5,355	6,075
3275	Automatic emergency power switching		.66	24.242		1,200	1,375		2,575	3,425
3300	Manual emergency power switching		8	2		500	113		613	725
3625	Hall finishes, stainless steel doors					2,000			2,000	2,200
3650	Stainless steel frames					1,450			1,450	1,600
3675	12 month maintenance contract								3,600	3,960
3700	Signal devices, hall lanterns	2 Elev	8	2		540	113		653	770
3725	Position indicators, up to 3		9.40	1.702		650	96.50		746.50	860
3750	Position indicators, per each over 3		32	.500		620	28.50		648.50	730
3775	High speed heavy duty door opener					4,200			4,200	4,625
3800	Variable voltage, O.H. gearless machine, min.	2 Elev	.16	100		70,500	5,650		76,150	86,000
3815	Maximum		.07	228		105,500	12,900		118,400	136,000
3825	Basement installed geared machine		.33	48.485		72,000	2,750		74,750	83,500
3850	Freight elevator options									
3875	Doors, bi-parting	2 Elev	.66	24.242	Ea.	9,975	1,375		11,350	13,100
3900	Power operated door and gate	"	.66	24.242		23,400	1,375		24,775	27,900
3925	Finishes, steel plate floor					2,475			2,475	2,725
3950	14 ga. 1/4" x 4' steel plate walls					1,900			1,900	2,100
3975	12 month maintenance contract								3,600	3,960
4000	Signal devices, hall lanterns	2 Elev	8	2		540	113		653	770
4025	Position indicators, up to 3		9.40	1.702		650	96.50		746.50	860
4050	Position indicators, per each over 3		32	.500		620	28.50		648.50	730
4075	Variable voltage basement installed geared machine		.66	24.242		19,600	1,375		20,975	23,700

14 28 Elevator Equipment and Controls

14 28 10 – Elevator Equipment and Control Options

14 28 10.10 Elevator Controls and Doors	Crew	Daily Output	Labor-Hours	Unit	Material	2009 Bare Costs Labor	Equipment	Total	Total Incl O&P
4100	Hospital elevator options									
4125	2 car group automatic controls	2 Elev	.66	24.242	Ea.	6,650	1,375		8,025	9,400
4150	3 car group automatic controls		.44	36.364		13,000	2,050		15,050	17,500
4175	4 car group automatic controls		.33	48.485		26,500	2,750		29,250	33,300
4200	5 car group automatic controls		.26	61.538		48,100	3,475		51,575	58,500
4225	6 car group automatic controls		.22	72.727		98,500	4,125		102,625	115,000
4250	Intercom service		3	5.333		1,375	300		1,675	2,000
4275	Duplex car selective collective		.66	24.242		12,200	1,375		13,575	15,600
4300	Center opening 1 speed doors		2	8		1,850	455		2,305	2,750
4325	Center opening 2 speed doors		2	8		2,525	455		2,980	3,500
4350	Rear opening doors (opposite front)		2	8		3,125	455		3,580	4,150
4375	Side opening 2 speed doors		2	8		3,925	455		4,380	5,025
4400	Automatic emergency power switching		.66	24.242		1,200	1,375		2,575	3,425
4425	Manual emergency power switching	↓	8	2		500	113		613	725
4675	Hall finishes, stainless steel doors					2,000			2,000	2,200
4700	Stainless steel frames					1,450			1,450	1,600
4725	12 month maintenance contract								3,600	3,960
4750	Signal devices, hall lanterns	2 Elev	8	2		540	113		653	770
4775	Position indicators, up to 3		9.40	1.702		650	96.50		746.50	860
4800	Position indicators, per each over 3	↓	32	.500		620	28.50		648.50	730
4825	High speed heavy duty door opener					4,200			4,200	4,625
4850	Variable voltage, O.H. gearless machine, min.	2 Elev	.16	100		70,500	5,650		76,150	86,000
4865	Maximum		.07	228		105,500	12,900		118,400	136,000
4875	Basement installed geared machine	↓	.33	48.485	↓	26,900	2,750		29,650	33,700
5000	Drilling for piston, casing included, 18" diameter	B-48	80	.700	V.L.F.	45	24.50	36	105.50	129

14 42 Wheelchair Lifts

14 42 13 – Inclined Wheelchair Lifts

14 42 13.10 Inclined Wheelchair Lifts and Stairclimbers

0010	**INCLINED WHEELCHAIR LIFTS AND STAIRCLIMBERS**	Crew	Daily Output	Labor-Hours	Unit	Material	Labor	Equipment	Total	Total Incl O&P
7700	Stair climber (chair lift), single seat, minimum	2 Elev	1	16	Ea.	5,325	905		6,230	7,275
7800	Maximum		.20	80		7,350	4,525		11,875	15,100
8700	Stair lift, minimum		1	16		14,500	905		15,405	17,300
8900	Maximum	↓	.20	80	↓	22,900	4,525		27,425	32,200

14 42 16 – Vertical Wheelchair Lifts

14 42 16.10 Wheelchair Lifts

0010	**WHEELCHAIR LIFTS**	Crew	Daily Output	Labor-Hours	Unit	Material	Labor	Equipment	Total	Total Incl O&P
8000	Wheelchair lift, minimum	2 Elev	1	16	Ea.	7,325	905		8,230	9,450
8500	Maximum	"	.50	32	"	17,300	1,800		19,100	21,800

14 45 Vehicle Lifts

14 45 10 - Hydraulic Vehicle Lifts

14 45 10.10 Hydraulic Lifts	Crew	Daily Output	Labor-Hours	Unit	Material	2009 Bare Costs Labor	Equipment	Total	Total Incl O&P
0010 **HYDRAULIC LIFTS**									
0200 Double post, 9000 lb frame	L-4	1.15	20.870	Ea.	6,000	780		6,780	7,875
0500 Four post, 50000 lb frame		.18	133		54,000	4,975		58,975	67,000
2200 Hoists, single post, 8,000# capacity, swivel arms		.40	60		4,975	2,250		7,225	9,125
2400 Two posts, adjustable frames, 11,000# capacity		.25	96		6,400	3,575		9,975	12,900
2500 24,000# capacity		.15	160		8,550	5,975		14,525	19,200
2700 7,500# capacity, frame supports		.50	48		7,125	1,800		8,925	10,800
2800 Four post, roll on ramp		.50	48		6,400	1,800		8,200	9,975
2810 Hydraulic lifts, above ground, 2 post, clear floor, 6000 lb cap		2.67	8.989		7,025	335		7,360	8,275
2815 9000 lb capacity		2.29	10.480		16,700	390		17,090	18,900
2820 15,000 lb capacity		2	12		18,800	450		19,250	21,400
2825 30,000 lb capacity		1.60	15		41,600	560		42,160	46,700
2830 4 post, ramp style, 25,000 lb capacity		2	12		16,300	450		16,750	18,600
2835 35,000 lb capacity		1	24		76,000	895		76,895	85,000
2840 50,000 lb capacity		1	24		85,000	895		85,895	95,000
2845 75,000 lb capacity		1	24		98,500	895		99,395	110,000
2850 For drive thru tracks, add, minimum					1,050			1,050	1,150
2855 Maximum					1,800			1,800	1,975
2860 Ramp extensions, 3' (set of 2)					860			860	945
2865 Rolling jack platform					3,000			3,000	3,300
2870 Elec/hyd jacking beam					8,000			8,000	8,800
2880 Scissor lift, portable, 6000 lb capacity					7,850			7,850	8,625

14 51 Correspondence and Parcel Lifts

14 51 10 - Electric Correspondence and Parcel Lifts

14 51 10.10 Correspondence Lifts	Crew	Daily Output	Labor-Hours	Unit	Material	2009 Bare Costs Labor	Equipment	Total	Total Incl O&P
0010 **CORRESPONDENCE LIFTS**									
0020 1 floor, 2 stop, 25 lb capacity, electric	2 Elev	.20	80	Ea.	5,975	4,525		10,500	13,600
0100 Hand, 5 lb capacity	"	.20	80	"	2,250	4,525		6,775	9,475

14 91 Facility Chutes

14 91 33 - Laundry and Linen Chutes

14 91 33.10 Chutes	Crew	Daily Output	Labor-Hours	Unit	Material	2009 Bare Costs Labor	Equipment	Total	Total Incl O&P
0011 **CHUTES**, linen, trash or refuse									
0050 Aluminized steel, 16 ga., 18" diameter	2 Shee	3.50	4.571	Floor	1,100	216		1,316	1,575
0100 24" diameter		3.20	5		1,175	236		1,411	1,650
0400 Galvanized steel, 16 ga., 18" diameter		3.50	4.571		1,000	216		1,216	1,450
0500 24" diameter		3.20	5		1,125	236		1,361	1,625
0800 Stainless steel, 18" diameter		3.50	4.571		2,350	216		2,566	2,925
0900 24" diameter		3.20	5		2,550	236		2,786	3,175

14 91 82 - Trash Chutes

14 91 82.10 Trash Chutes and Accessories	Crew	Daily Output	Labor-Hours	Unit	Material	2009 Bare Costs Labor	Equipment	Total	Total Incl O&P
0010 **TRASH CHUTES AND ACCESSORIES**									
9000 Minimum labor/equipment charge	1 Shee	1	8	Job		380		380	600

14 92 Pneumatic Tube Systems

14 92 10 – Conventional, Automatic and Computer Controlled Pneumatic Tube Systems

14 92 10.10 Pneumatic Tube Systems	Crew	Daily Output	Labor-Hours	Unit	Material	2009 Bare Costs Labor	Equipment	Total	Total Incl O&P
0010 **PNEUMATIC TUBE SYSTEMS**									
0020 100' long, single tube, 2 stations, stock									
0100 3" diameter	2 Stpi	.12	133	Total	6,600	6,575		13,175	17,500
0300 4" diameter	"	.09	177	"	7,450	8,775		16,225	21,800
0400 Twin tube, two stations or more, conventional system									
0700 3" round	2 Stpi	46	.348	L.F.	15.25	17.15		32.40	43.50
1050 Add for blower		2	8	System	5,200	395		5,595	6,325
1110 Plus for each round station, add		7.50	2.133	Ea.	585	105		690	810
1200 Alternate pricing method: base cost, minimum		.75	21.333	Total	6,375	1,050		7,425	8,650
1300 Maximum		.25	64	"	12,700	3,150		15,850	18,800
1500 Plus total system length, add, minimum		93.40	.171	L.F.	8.20	8.45		16.65	22
1600 Maximum		37.60	.426	"	24.50	21		45.50	59

Estimating Tips

Pipe for fire protection and all uses is located in Subdivision 22 11 13.

The labor adjustment factors listed in Subdivision 22 01 02.20 also apply to Division 21.

Many, but not all, areas require backflow protection in the fire system. It is advisable to check local building codes for specific requirements.

For your reference, the following is a list of the most applicable Fire Codes and Standards which may be purchased from the NFPA, 1 Batterymarch Park, Quincy, MA 02169-7471.

NFPA 1: Uniform Fire Code

NFPA 10: Portable Fire Extinguishers

NFPA 11: Low-, Medium-, and High-Expansion Foam

NFPA 12: Carbon Dioxide Extinguishing Systems (Also companion 12A)

NFPA 13: Installation of Sprinkler Systems (Also companion 13D, 13E, and 13R)

NFPA 14: Installation of Standpipe and Hose Systems

NFPA 15: Water Spray Fixed Systems for Fire Protection

NFPA 16: Installation of Foam-Water Sprinkler and Foam-Water Spray Systems

NFPA 17: Dry Chemical Extinguishing Systems (Also companion 17A)

NFPA 18: Wetting Agents

NFPA 20: Installation of Stationary Pumps for Fire Protection

NFPA 22: Water Tanks for Private Fire Protection

NFPA 24: Installation of Private Fire Service Mains and their Appurtenances

NFPA 25: Inspection, Testing and Maintenance of Water-Based Fire Protection

Reference Numbers

Reference numbers are shown in shaded boxes at the beginning of some major classifications. These numbers refer to related items in the Reference Section. The reference information may be an estimating procedure, an alternate pricing method, or technical information.

Note: Not all subdivisions listed here necessarily appear in this publication.

Note: **Trade Service,** *in part, has been used as a reference source for some of the material prices used in Division 21.*

21 05 Common Work Results for Fire Suppression

21 05 23 – General-Duty Valves for Water-Based Fire-Suppression Piping

21 05 23.50 General-Duty Valves for Water-Based Fire Supp.	Crew	Daily Output	Labor-Hours	Unit	Material	2009 Bare Costs Labor	Equipment	Total	Total Incl O&P
0010 **GENERAL-DUTY VALVES FOR WATER-BASED FIRE SUPPRESSION PIPING**									
6500 Check, swing, C.I. body, brass fittings, auto. ball drip									
6520 4" size	Q-12	3	5.333	Ea.	255	231		486	640
9990 Minimum labor/equipment charge	1 Spri	3	2.667	Job		128		128	199

21 11 Facility Fire-Suppression Water-Service Piping

21 11 16 – Facility Fire Hydrants

21 11 16.50 Fire Hydrants for Buildings

	Crew	Daily Output	Labor-Hours	Unit	Material	Labor	Equipment	Total	Total Incl O&P
0010 **FIRE HYDRANTS FOR BUILDINGS**									
3750 Hydrants, wall, w/caps, single, flush, polished brass									
3800 2-1/2" x 2-1/2"	Q-12	5	3.200	Ea.	175	139		314	410
4350 Double, projecting, polished brass									
4400 2-1/2" x 2-1/2" x 4"	Q-12	5	3.200	Ea.	209	139		348	445

21 11 19 – Fire-Department Connections

21 11 19.50 Connections for the Fire-Department

	Crew	Daily Output	Labor-Hours	Unit	Material	Labor	Equipment	Total	Total Incl O&P
0010 **CONNECTIONS FOR THE FIRE-DEPARTMENT**									
7140 Standpipe connections, wall, w/plugs & chains									
7280 Double, flush, polished brass									
7300 2-1/2" x 2-1/2" x 4"	Q-12	5	3.200	Ea.	425	139		564	685
9990 Minimum labor/equipment charge	1 Spri	4	2	Job		96.50		96.50	149

21 12 Fire-Suppression Standpipes

21 12 13 – Fire-Suppression Hoses and Nozzles

21 12 13.50 Fire Hoses and Nozzles

	Crew	Daily Output	Labor-Hours	Unit	Material	Labor	Equipment	Total	Total Incl O&P
0010 **FIRE HOSES AND NOZZLES**									
2200 Hose, less couplings									
2260 Synthetic jacket, lined, 300 lb. test, 1-1/2" diameter	Q-12	2600	.006	L.F.	2.39	.27		2.66	3.04
2280 2-1/2" diameter	"	2200	.007	"	3.98	.32		4.30	4.87
5600 Nozzles, brass									
5620 Adjustable fog, 3/4" booster line				Ea.	88.50			88.50	97.50
5640 1-1/2" leader line					63.50			63.50	70
5660 2-1/2" direct connection				↓	168			168	185
9900 Minimum labor/equipment charge	1 Plum	2	4	Job		195		195	300

21 12 19 – Fire-Suppression Hose Racks

21 12 19.50 Fire Hose Racks

	Crew	Daily Output	Labor-Hours	Unit	Material	Labor	Equipment	Total	Total Incl O&P
0010 **FIRE HOSE RACKS**									
2600 Hose rack, swinging, for 1-1/2" diameter hose,									
2620 Enameled steel, 50' & 75' lengths of hose	Q-12	20	.800	Ea.	48.50	34.50		83	107

21 12 23 – Fire-Suppression Hose Valves

21 12 23.70 Fire Hose Valves

	Crew	Daily Output	Labor-Hours	Unit	Material	Labor	Equipment	Total	Total Incl O&P
0010 **FIRE HOSE VALVES**									
3000 Gate, hose, wheel handle, N.R.S., rough brass, 1-1/2"	1 Spri	12	.667	Ea.	111	32		143	172
3040 2-1/2", 300 lb.	"	7	1.143	"	156	55		211	257

21 13 13 – Wet-Pipe Sprinkler Systems

21 13 13.50 Wet-Pipe Sprinkler System Components

	Crew	Daily Output	Labor-Hours	Unit	Material	2009 Bare Costs Labor	Equipment	Total	Total Incl O&P
0010 **WET-PIPE SPRINKLER SYSTEM COMPONENTS**									
1220 Water motor, complete with gong	1 Spri	4	2	Ea.	305	96.50		401.50	485
1800 Firecycle system, controls, includes panel,									
1820 batteries, solenoid valves and pressure switches	Q-13	1	32	Ea.	10,800	1,475		12,275	14,200
1860 Detector	1 Spri	16	.500	"	525	24		549	615
1900 Flexible sprinkler head connectors									
1910 Braided stainless steel hose with mounting bracket									
1920 1/2" and 1" outlet size									
1930 24" length	1 Spri	32	.250	Ea.	24	12.05		36.05	45
1940 36" length		30	.267		28	12.85		40.85	50.50
1950 48" length		26	.308		31.50	14.80		46.30	57.50
1960 60" length		22	.364		35.50	17.50		53	66
1970 72" length		18	.444		39	21.50		60.50	76
1982 May replace hard-pipe armovers									
1984 For wet, pre-action, deluge or dry pipe systems									
2600 Sprinkler heads, not including supply piping									
3700 Standard spray, pendent or upright, brass, 135° to 286° F									
3740 1/2" NPT, 1/2" orifice	1 Spri	16	.500	Ea.	7.35	24		31.35	45.50
3790 For riser & feeder piping, see Div. 22 11 13.00									

21 13 16 – Dry-Pipe Sprinkler Systems

21 13 16.50 Dry-Pipe Sprinkler System Components

	Crew	Daily Output	Labor-Hours	Unit	Material	2009 Bare Costs Labor	Equipment	Total	Total Incl O&P
0010 **DRY-PIPE SPRINKLER SYSTEM COMPONENTS**									
0800 Air compressor for dry pipe system, automatic, complete									
0820 280 gal. system capacity, 3/4 HP	1 Spri	1.30	6.154	Ea.	700	296		996	1,225
2600 Sprinkler heads, not including supply piping									
2640 Dry, pendent, 1/2" orifice, 3/4" or 1" NPT									
2660 1/2" to 6" length	1 Spri	14	.571	Ea.	90.50	27.50		118	142
2700 15-1/4" to 18" length	"	14	.571		106	27.50		133.50	160
8200 Dry pipe valve, incl. trim and gauges, 3" size	Q-12	2	8		1,850	345		2,195	2,575
8220 4" size	"	1	16		2,000	695		2,695	3,300

21 13 39 – Foam-Water Systems

21 13 39.50 Foam-Water System Components

	Crew	Daily Output	Labor-Hours	Unit	Material	2009 Bare Costs Labor	Equipment	Total	Total Incl O&P
0010 **FOAM-WATER SYSTEM COMPONENTS**									
2600 Sprinkler heads, not including supply piping									
3600 Foam-water, pendent or upright, 1/2" NPT	1 Spri	12	.667	Ea.	130	32		162	192

Division Notes

	CREW	DAILY OUTPUT	LABOR-HOURS	UNIT	2009 BARE COSTS				TOTAL INCL O&P
					MAT.	LABOR	EQUIP.	TOTAL	

Estimating Tips

22 10 00 Plumbing Piping and Pumps

This subdivision is primarily basic pipe and related materials. The pipe may be used by any of the mechanical disciplines, i.e., plumbing, fire protection, heating, and air conditioning.

- The labor adjustment factors listed in Subdivision 22 01 02.20 apply throughout Divisions 21, 22, and 23. CAUTION: the correct percentage may vary for the same items. For example, the percentage add for the basic pipe installation should be based on the maximum height that the craftsman must install for that particular section. If the pipe is to be located 14' above the floor but it is suspended on threaded rod from beams, the bottom flange of which is 18' high (4' rods), then the height is actually 18' and the add is 20%. The pipe coverer, however, does not have to go above the 14', and so his or her add should be 10%.

- Most pipe is priced first as straight pipe with a joint (coupling, weld, etc.) every 10' and a hanger usually every 10'.

There are exceptions with hanger spacing such as for cast iron pipe (5') and plastic pipe (3 per 10'). Following each type of pipe there are several lines listing sizes and the amount to be subtracted to delete couplings and hangers. This is for pipe that is to be buried or supported together on trapeze hangers. The reason that the couplings are deleted is that these runs are usually long, and frequently longer lengths of pipe are used. By deleting the couplings, the estimator is expected to look up and add back the correct reduced number of couplings.

- When preparing an estimate, it may be necessary to approximate the fittings. Fittings usually run between 25% and 50% of the cost of the pipe. The lower percentage is for simpler runs, and the higher number is for complex areas, such as mechanical rooms.

- For historic restoration projects, the systems must be as invisible as possible, and pathways must be sought for pipes, conduit, and ductwork. While installations in accessible spaces (such as basements and attics) are relatively straightforward to estimate, labor costs may be more difficult to determine when delivery systems must be concealed.

22 40 00 Plumbing Fixtures

- Plumbing fixture costs usually require two lines: the fixture itself and its "rough-in, supply, and waste."

- In the Assemblies Section (Plumbing D2010) for the desired fixture, the System Components Group at the center of the page shows the fixture on the first line. The rest of the list (fittings, pipe, tubing, etc.) will total up to what we refer to in the Unit Price section as "Rough-in, supply, waste, and vent." Note that for most fixtures we allow a nominal 5' of tubing to reach from the fixture to a main or riser.

- Remember that gas- and oil-fired units need venting.

Reference Numbers

Reference numbers are shown in shaded boxes at the beginning of some major classifications. These numbers refer to related items in the Reference Section. The reference information may be an estimating procedure, an alternate pricing method, or technical information.

Note: Not all subdivisions listed here necessarily appear in this publication.

Note: **Trade Service,** *in part, has been used as a reference source for some of the material prices used in Division 22.*

22 01 Operation and Maintenance of Plumbing

22 01 02 – Labor Adjustments

22 01 02.20 Labor Adjustment Factors		Crew	Daily Output	Labor-Hours	Unit	Material	2009 Bare Costs Labor	Equipment	Total	Total Incl O&P
0010	**LABOR ADJUSTMENT FACTORS**, (For Div. 21,22 and 23)									
1000	Add to labor for elevated installation (Above floor level)									
1080	10' to 14.5' high						10%			
1100	15' to 19.5' high						20%			
1120	20' to 24.5' high						25%			
1140	25' to 29.5' high						35%			
1160	30' to 34.5' high						40%			
1180	35' to 39.5' high						50%			
1200	40' and higher						55%			
2000	Add to labor for crawl space									
2100	3' high						40%			
2140	4' high						30%			
3000	Add to labor for multi-story building									
3100	Add per floor for floors 3 thru 19						2%			
3140	Add per floor for floors 20 and up						4%			
4000	Add to labor for working in existing occupied buildings									
4100	Hospital						35%			
4140	Office building						25%			
4180	School						20%			
4220	Factory or warehouse						15%			
4260	Multi dwelling						15%			
5000	Add to labor, miscellaneous									
5100	Cramped shaft						35%			
5140	Congested area						15%			
5180	Excessive heat or cold						30%			
9000	Labor factors, The above are reasonable suggestions, however									
9010	each project should be evaluated for its own peculiarities.									
9100	Other factors to be considered are:									
9140	Movement of material and equipment through finished areas									
9180	Equipment room									
9220	Attic space									
9260	No service road									
9300	Poor unloading/storage area									
9340	Congested site area/heavy traffic									

22 05 Common Work Results for Plumbing

22 05 05 – Selective Plumbing Demolition

22 05 05.10 Plumbing Demolition

22 05 05.10 Plumbing Demolition		Crew	Daily Output	Labor-Hours	Unit	Material	2009 Bare Costs Labor	Equipment	Total	Total Incl O&P
0010	**PLUMBING DEMOLITION** R220105-10									
1020	Fixtures, including 10' piping									
1100	Bath tubs, cast iron R024119-10	1 Plum	4	2	Ea.		97.50		97.50	151
1120	Fiberglass		6	1.333			65		65	101
1140	Steel		5	1.600			78		78	121
1200	Lavatory, wall hung		10	.800			39		39	60.50
1220	Counter top		8	1			49		49	75.50
1300	Sink, single compartment		8	1			49		49	75.50
1320	Double compartment		7	1.143			55.50		55.50	86.50
1400	Water closet, floor mounted		8	1			49		49	75.50
1420	Wall mounted		7	1.143			55.50		55.50	86.50
1500	Urinal, floor mounted		4	2			97.50		97.50	151
1520	Wall mounted		7	1.143			55.50		55.50	86.50

22 05 05 – Selective Plumbing Demolition

22 05 05.10 Plumbing Demolition	Crew	Daily Output	Labor-Hours	Unit	Material	2009 Bare Costs Labor	2009 Bare Costs Equipment	Total	Total Incl O&P
1600 Water fountains, free standing	1 Plum	8	1	Ea.		49		49	75.50
1620 Wall or deck mounted		6	1.333	↓		65		65	101
2000 Piping, metal, up thru 1-1/2" diameter		200	.040	L.F.		1.95		1.95	3.02
2050 2" thru 3-1/2" diameter	↓	150	.053			2.60		2.60	4.03
2100 4" thru 6" diameter	2 Plum	100	.160			7.80		7.80	12.10
2150 8" thru 14" diameter	"	60	.267			13		13	20
2153 16" thru 20" diameter	Q-18	70	.343			15.80	.80	16.60	25.50
2155 24" thru 26" diameter		55	.436			20	1.01	21.01	32
2156 30" thru 36" diameter	↓	40	.600	↓		27.50	1.39	28.89	44.50
2250 Water heater, 40 gal.	1 Plum	6	1.333	Ea.		65		65	101
6000 Remove and reset fixtures, minimum		6	1.333			65		65	101
6100 Maximum		4	2	↓		97.50		97.50	151
9000 Minimum labor/equipment charge	↓	2	4	Job		195		195	300

22 05 23 – General-Duty Valves for Plumbing Piping

22 05 23.20 Valves, Bronze

	Crew	Daily Output	Labor-Hours	Unit	Material	Labor	Equipment	Total	Total Incl O&P
0010 **VALVES, BRONZE**									
1020 Angle, 150 lb., rising stem, threaded									
1070 3/4"	1 Plum	20	.400	Ea.	123	19.50		142.50	165
1080 1"		19	.421		177	20.50		197.50	227
1100 1-1/2"		13	.615		296	30		326	370
1110 2"	↓	11	.727	↓	480	35.50		515.50	580
1380 Ball, 150 psi, threaded									
1460 3/4"	1 Plum	20	.400	Ea.	16.95	19.50		36.45	48.50
1750 Check, swing, class 150, regrinding disc, threaded									
1850 1/2"	1 Plum	24	.333	Ea.	44.50	16.25		60.75	73.50
1860 3/4"		20	.400		64	19.50		83.50	100
1870 1"		19	.421		91.50	20.50		112	133
1890 1-1/2"		13	.615		154	30		184	216
1900 2"	↓	11	.727	↓	226	35.50		261.50	305
2850 Gate, N.R.S., soldered, 125 psi									
2920 1/2"	1 Plum	24	.333	Ea.	31	16.25		47.25	59
2940 3/4"		20	.400		35.50	19.50		55	69
2950 1"		19	.421		50.50	20.50		71	87.50
2970 1-1/2"		13	.615		85.50	30		115.50	141
2980 2"	↓	11	.727	↓	122	35.50		157.50	189
4850 Globe, class 150, rising stem, threaded									
4950 1/2"	1 Plum	24	.333	Ea.	67	16.25		83.25	98.50
4960 3/4"	"	20	.400	"	90	19.50		109.50	129
5600 Relief, pressure & temperature, self-closing, ASME, threaded									
5640 3/4"	1 Plum	28	.286	Ea.	118	13.95		131.95	152
5650 1"		24	.333		180	16.25		196.25	224
5660 1-1/4"		20	.400		345	19.50		364.50	410
5670 1-1/2"		18	.444		660	21.50		681.50	760
5680 2"	↓	16	.500	↓	720	24.50		744.50	830
6400 Pressure, water, ASME, threaded									
6440 3/4"	1 Plum	28	.286	Ea.	70	13.95		83.95	98.50
6450 1"		24	.333		149	16.25		165.25	189
6470 1-1/2"	↓	18	.444	↓	325	21.50		346.50	395
6900 Reducing, water pressure									
6940 1/2"	1 Plum	24	.333	Ea.	214	16.25		230.25	261
6960 1"	"	19	.421	"	330	20.50		350.50	395
8350 Tempering, water, sweat connections									

22 05 23 – General-Duty Valves for Plumbing Piping

22 05 23.20 Valves, Bronze

		Crew	Daily Output	Labor-Hours	Unit	Material	2009 Bare Costs Labor	Equipment	Total	Total Incl O&P
8400	1/2"	1 Plum	24	.333	Ea.	71	16.25		87.25	103
8440	3/4"	"	20	.400	"	87	19.50		106.50	126
8650	Threaded connections									
8700	1/2"	1 Plum	24	.333	Ea.	87	16.25		103.25	121
8740	3/4"		20	.400	"	370	19.50		389.50	440
9000	Minimum labor/equipment charge		4	2	Job		97.50		97.50	151

22 05 23.60 Valves, Plastic

		Crew	Daily Output	Labor-Hours	Unit	Material	2009 Bare Costs Labor	Equipment	Total	Total Incl O&P
0010	**VALVES, PLASTIC**									
0020										
1150	Ball, PVC, socket or threaded, single union									
1280	2"	1 Plum	17	.471	Ea.	68.50	23		91.50	111
1650	CPVC, socket or threaded, single union									
1770	2"	1 Plum	17	.471	Ea.	128	23		151	177
3150	Ball check, PVC, socket or threaded									
3290	2"	1 Plum	17	.471	Ea.	103	23		126	149
9000	Minimum labor/equipment charge	"	3.50	2.286	Job		111		111	173

22 05 76 – Facility Drainage Piping Cleanouts

22 05 76.10 Cleanouts

		Crew	Daily Output	Labor-Hours	Unit	Material	2009 Bare Costs Labor	Equipment	Total	Total Incl O&P
0010	**CLEANOUTS**									
0060	Floor type									
0080	Round or square, scoriated nickel bronze top									
0100	2" pipe size	1 Plum	10	.800	Ea.	202	39		241	283
0140	4" pipe size	"	6	1.333	"	285	65		350	415
0980	Round top, recessed for terrazzo									
1000	2" pipe size	1 Plum	9	.889	Ea.	202	43.50		245.50	289
1100	4" pipe size	"	4	2		285	97.50		382.50	465
1120	5" pipe size	Q-1	6	2.667		465	117		582	690
9000	Minimum labor/equipment charge	1 Plum	3	2.667	Job		130		130	202

22 05 76.20 Cleanout Tees

		Crew	Daily Output	Labor-Hours	Unit	Material	2009 Bare Costs Labor	Equipment	Total	Total Incl O&P
0010	**CLEANOUT TEES**									
0100	Cast iron, B&S, with countersunk plug									
0220	3" pipe size	1 Plum	3.60	2.222	Ea.	205	108		313	395
0240	4" pipe size	"	3.30	2.424	"	256	118		374	465
0500	For round smooth access cover, same price									
4000	Plastic, tees and adapters. Add plugs									
4010	ABS, DWV									
4020	Cleanout tee, 1-1/2" pipe size	1 Plum	15	.533	Ea.	6.60	26		32.60	48
9000	Minimum labor/equipment charge	"	2.75	2.909	Job		142		142	220

22 07 Plumbing Insulation

22 07 16 – Plumbing Equipment Insulation

22 07 16.10 Insulation for Plumbing Equipment

			Crew	Daily Output	Labor-Hours	Unit	Material	2009 Bare Costs Labor	Equipment	Total	Total Incl O&P
0010	**INSULATION FOR PLUMBING EQUIPMENT**										
2900	Domestic water heater wrap kit										
2920	1-1/2" with vinyl jacket, 20-60 gal.	G	1 Plum	8	1	Ea.	16.05	49		65.05	93
2925	50 to 80 gallons	G	"	8	1	"	19.15	49		68.15	96.50

22 07 19 – Plumbing Piping Insulation

22 07 19.10 Piping Insulation

0010	**PIPING INSULATION**

22 07 19.10 Piping Insulation		Crew	Daily Output	Labor-Hours	Unit	Material	2009 Bare Costs Labor	Equipment	Total	Total Incl O&P
0100	Rule of thumb, as a percentage of total mechanical costs				Job				10%	
0110	Insulation req'd is based on the surface size/area to be covered									
2930	Insulated protectors, (ADA)									
2935	For exposed piping under sinks or lavatories ♿									
2940	Vinyl coated foam, velcro tabs									
2945	P Trap, 1-1/4" or 1-1/2"	1 Plum	32	.250	Ea.	18.50	12.20		30.70	39.50
2960	Valve and supply cover									
2965	1/2", 3/8", and 7/16" pipe size	1 Plum	32	.250	Ea.	17.60	12.20		29.80	38.50
2970	Extension drain cover									
2975	1-1/4", or 1-1/2" pipe size	1 Plum	32	.250	Ea.	19.30	12.20		31.50	40.50
2980	Tailpiece offset (wheelchair) ♿									
2985	1-1/4" pipe size	1 Plum	32	.250	Ea.	21.50	12.20		33.70	42.50
4000	Pipe covering (price copper tube one size less than IPS)									
6120	2" wall, 1/2" iron pipe size **G**	Q-14	145	.110	L.F.	3.89	4.38		8.27	11.40
6210	4" iron pipe size **G**	"	125	.128	"	7.10	5.10		12.20	16
6600	Fiberglass, with all service jacket									
6840	1" wall, 1/2" iron pipe size **G**	Q-14	240	.067	L.F.	.89	2.65		3.54	5.25
6860	3/4" iron pipe size **G**		230	.070		.97	2.76		3.73	5.55
6870	1" iron pipe size **G**		220	.073		1.04	2.89		3.93	5.80
6880	1-1/4" iron pipe size **G**		210	.076		1.12	3.02		4.14	6.10
6890	1-1/2" iron pipe size **G**		210	.076		1.21	3.02		4.23	6.20
6900	2" iron pipe size **G**		200	.080		1.31	3.18		4.49	6.60
7080	1-1/2" wall, 1/2" iron pipe size **G**		230	.070		1.67	2.76		4.43	6.30
7140	2" iron pipe size **G**		190	.084		2.26	3.34		5.60	7.90
7160	3" iron pipe size **G**	↓	170	.094	↓	2.54	3.74		6.28	8.85
7879	Rubber tubing, flexible closed cell foam									
8100	1/2" wall, 1/4" iron pipe size **G**	1 Asbe	90	.089	L.F.	.44	3.92		4.36	6.85
8120	3/8" iron pipe size **G**		90	.089		.49	3.92		4.41	6.90
8130	1/2" iron pipe size **G**		89	.090		.54	3.96		4.50	7
8140	3/4" iron pipe size **G**		89	.090		.60	3.96		4.56	7.05
8150	1" iron pipe size **G**		88	.091		.67	4.01		4.68	7.25
8170	1-1/2" iron pipe size **G**		87	.092		.94	4.06		5	7.60
8180	2" iron pipe size **G**		86	.093		1.22	4.10		5.32	8
8200	3" iron pipe size **G**		85	.094		1.72	4.15		5.87	8.60
8220	4" iron pipe size **G**		80	.100		2.55	4.41		6.96	9.95
8300	3/4" wall, 1/4" iron pipe size **G**		90	.089		.68	3.92		4.60	7.10
8330	1/2" iron pipe size **G**		89	.090		.90	3.96		4.86	7.40
8340	3/4" iron pipe size **G**		89	.090		1.09	3.96		5.05	7.60
8350	1" iron pipe size **G**		88	.091		1.25	4.01		5.26	7.90
8360	1-1/4" iron pipe size **G**		87	.092		1.68	4.06		5.74	8.40
8370	1-1/2" iron pipe size **G**		87	.092		1.90	4.06		5.96	8.65
8380	2" iron pipe size **G**		86	.093		2.23	4.10		6.33	9.10
8444	1" wall, 1/2" iron pipe size **G**		86	.093		1.71	4.10		5.81	8.55
8445	3/4" iron pipe size **G**		84	.095		2.07	4.20		6.27	9.10
8446	1" iron pipe size **G**		84	.095		2.41	4.20		6.61	9.45
8447	1-1/4" iron pipe size **G**		82	.098		2.72	4.30		7.02	9.95
8448	1-1/2" iron pipe size **G**		82	.098		3.16	4.30		7.46	10.45
8449	2" iron pipe size **G**		80	.100		4.23	4.41		8.64	11.80
8450	2-1/2" iron pipe size **G**	↓	80	.100	↓	5.50	4.41		9.91	13.20
8456	Rubber insulation tape, 1/8" x 2" x 30' **G**				Ea.	11.40			11.40	12.50
9600	Minimum labor/equipment charge	1 Plum	4	2	Job		97.50		97.50	151

22 11 13.14 Pipe, Brass

		Crew	Daily Output	Labor-Hours	Unit	Material	2009 Bare Costs Labor	Equipment	Total	Total Incl O&P
0010	**PIPE, BRASS**, Plain end									
0900	Field threaded, coupling & clevis hanger assembly 10' O.C.									
0920	Regular weight									
1120	1/2" diameter	1 Plum	48	.167	L.F.	4.62	8.15		12.77	17.70
1140	3/4" diameter		46	.174		6.30	8.50		14.80	20
1160	1" diameter		43	.186		9.30	9.05		18.35	24.50
1180	1-1/4" diameter	Q-1	72	.222		14.30	9.75		24.05	31
1200	1-1/2" diameter		65	.246		17.10	10.80		27.90	35.50
1220	2" diameter		53	.302		24	13.25		37.25	47
9000	Minimum labor/equipment charge	1 Plum	4	2	Job		97.50		97.50	151

22 11 13.16 Pipe Fittings, Brass

		Crew	Daily Output	Labor-Hours	Unit	Material	2009 Bare Costs Labor	Equipment	Total	Total Incl O&P
0010	**PIPE FITTINGS, BRASS**, Rough bronze, threaded.									
0020										
1000	Standard wt., 90° Elbow									
1100	1/2"	1 Plum	12	.667	Ea.	12.90	32.50		45.40	64.50
1120	3/4"		11	.727		17.25	35.50		52.75	74
1140	1"		10	.800		28	39		67	91.50
1160	1-1/4"	Q-1	17	.941		45.50	41.50		87	114
1180	1-1/2"	"	16	1		56	44		100	130
1500	45° Elbow, 1/8"	1 Plum	13	.615		15.85	30		45.85	64
1580	1/2"		12	.667		15.85	32.50		48.35	68
1600	3/4"		11	.727		22.50	35.50		58	80
1620	1"		10	.800		38.50	39		77.50	103
1640	1-1/4"	Q-1	17	.941		61	41.50		102.50	131
1660	1-1/2"	"	16	1		77	44		121	153
2000	Tee, 1/8"	1 Plum	9	.889		15.15	43.50		58.65	83.50
2080	1/2"		8	1		15.15	49		64.15	92
2100	3/4"		7	1.143		21.50	55.50		77	110
2120	1"		6	1.333		39	65		104	144
2140	1-1/4"	Q-1	10	1.600		67	70		137	183
2160	1-1/2"	"	9	1.778		75.50	78		153.50	204
2500	Coupling, 1/8"	1 Plum	26	.308		10.80	15		25.80	35.50
2580	1/2"		15	.533		10.80	26		36.80	52.50
2600	3/4"		14	.571		15.15	28		43.15	59.50
2620	1"		13	.615		26	30		56	75
2640	1-1/4"	Q-1	22	.727		43	32		75	97
2660	1-1/2"		20	.800		56	35		91	116
2680	2"		18	.889		93	39		132	163
9000	Minimum labor/equipment charge	1 Plum	4	2	Job		97.50		97.50	151

22 11 13.23 Pipe, Copper

		Crew	Daily Output	Labor-Hours	Unit	Material	2009 Bare Costs Labor	Equipment	Total	Total Incl O&P
0010	**PIPE, COPPER**, Solder joints									
1000	Type K tubing, couplings & clevis hanger assemblies 10' O.C.									
1100	1/4" diameter	1 Plum	84	.095	L.F.	3.43	4.64		8.07	10.95
1200	1" diameter		66	.121		12.10	5.90		18	22.50
1260	2" diameter		40	.200		30.50	9.75		40.25	48.50
2000	Type L tubing, couplings & clevis hanger assemblies 10' O.C.									
2100	1/4" diameter	1 Plum	88	.091	L.F.	2.23	4.43		6.66	9.30
2120	3/8" diameter		84	.095		3.26	4.64		7.90	10.80
2140	1/2" diameter		81	.099		3.83	4.82		8.65	11.65
2160	5/8" diameter		79	.101		5.45	4.94		10.39	13.65
2180	3/4" diameter		76	.105		6	5.15		11.15	14.55
2200	1" diameter		68	.118		9.05	5.75		14.80	18.85

22 11 13 – Facility Water Distribution Piping

22 11 13.23 Pipe, Copper

		Crew	Daily Output	Labor-Hours	Unit	Material	2009 Bare Costs Labor	Equipment	Total	Total Incl O&P
2220	1-1/4" diameter	1 Plum	58	.138	L.F.	12.50	6.70		19.20	24
2240	1-1/2" diameter		52	.154		16.10	7.50		23.60	29.50
2260	2" diameter		42	.190		25	9.30		34.30	42
2280	2-1/2" diameter	Q-1	62	.258		38	11.30		49.30	59.50
2300	3" diameter		56	.286		51.50	12.55		64.05	76.50
2320	3-1/2" diameter		43	.372		70	16.35		86.35	103
2340	4" diameter		39	.410		87	18		105	124
2360	5" diameter		34	.471		186	20.50		206.50	236
2380	6" diameter	Q-2	40	.600		256	27.50		283.50	325
2410	For other than full hard temper, add					21%				
2590	For silver solder, add						15%			
3000	Type M tubing, couplings & clevis hanger assemblies 10' O.C.									
3140	1/2" diameter	1 Plum	84	.095	L.F.	2.79	4.64		7.43	10.25
3180	3/4" diameter		78	.103		4.41	5		9.41	12.60
3200	1" diameter		70	.114		6.95	5.55		12.50	16.25
4000	Type DWV tubing, couplings & clevis hanger assemblies 10' O.C.									
4100	1-1/4" diameter	1 Plum	60	.133	L.F.	10.90	6.50		17.40	22
4120	1-1/2" diameter		54	.148		13.70	7.20		20.90	26.50
4140	2" diameter		44	.182		18.45	8.85		27.30	34.50
4160	3" diameter	Q-1	58	.276		32.50	12.10		44.60	55
4180	4" diameter		40	.400		57.50	17.55		75.05	90.50
4200	5" diameter		36	.444		165	19.50		184.50	211
4220	6" diameter	Q-2	42	.571		235	26		261	300
9000	Minimum labor/equipment charge	1 Plum	4	2	Job		97.50		97.50	151

22 11 13.25 Pipe Fittings, Copper

		Crew	Daily Output	Labor-Hours	Unit	Material	2009 Bare Costs Labor	Equipment	Total	Total Incl O&P
0010	**PIPE FITTINGS, COPPER**, Wrought unless otherwise noted									
0040	Solder joints, copper x copper									
0070	90° elbow, 1/4"	1 Plum	22	.364	Ea.	3.89	17.75		21.64	32
0100	1/2"		20	.400		1.23	19.50		20.73	31.50
0120	3/4"		19	.421		2.76	20.50		23.26	35
0130	1"		16	.500		6.80	24.50		31.30	45.50
0250	45° elbow, 1/4"		22	.364		7.30	17.75		25.05	35.50
0270	3/8"		22	.364		5.90	17.75		23.65	34
0280	1/2"		20	.400		2.26	19.50		21.76	32.50
0290	5/8"		19	.421		11.30	20.50		31.80	44.50
0300	3/4"		19	.421		3.96	20.50		24.46	36.50
0310	1"		16	.500		9.95	24.50		34.45	49
0320	1-1/4"		15	.533		13.45	26		39.45	55.50
0330	1-1/2"		13	.615		16.20	30		46.20	64.50
0340	2"		11	.727		27	35.50		62.50	85
0350	2-1/2"	Q-1	13	1.231		57.50	54		111.50	148
0360	3"		13	1.231		85.50	54		139.50	178
0370	3-1/2"		10	1.600		144	70		214	267
0380	4"		9	1.778		182	78		260	320
0390	5"		6	2.667		705	117		822	955
0400	6"	Q-2	9	2.667		1,125	121		1,246	1,425
0450	Tee, 1/4"	1 Plum	14	.571		8.15	28		36.15	52
0470	3/8"		14	.571		6.25	28		34.25	50
0480	1/2"		13	.615		2.11	30		32.11	49
0490	5/8"		12	.667		13.55	32.50		46.05	65.50
0500	3/4"		12	.667		5.10	32.50		37.60	56
0510	1"		10	.800		15.70	39		54.70	78

22 11 13.25 Pipe Fittings, Copper		Crew	Daily Output	Labor-Hours	Unit	Material	2009 Bare Costs Labor	Equipment	Total	Total Incl O&P
0520	1-1/4"	1 Plum	9	.889	Ea.	21.50	43.50		65	90.50
0530	1-1/2"		8	1		33	49		82	112
0540	2"	↓	7	1.143		51.50	55.50		107	143
0550	2-1/2"	Q-1	8	2		104	88		192	250
0560	3"		7	2.286		159	100		259	330
0570	3-1/2"		6	2.667		460	117		577	685
0580	4"		5	3.200		385	140		525	640
0590	5"	↓	4	4		1,250	176		1,426	1,650
0600	6"	Q-2	6	4		1,725	182		1,907	2,175
0612	Tee, reducing on the outlet, 1/4"	1 Plum	15	.533		13.15	26		39.15	55
0613	3/8"		15	.533		12.35	26		38.35	54
0614	1/2"		14	.571		10.85	28		38.85	55
0615	5/8"		13	.615		19.40	30		49.40	68
0616	3/4"		12	.667		7	32.50		39.50	58
0617	1"		11	.727		15.75	35.50		51.25	72.50
0618	1-1/4"		10	.800		29.50	39		68.50	93
0619	1-1/2"		9	.889		31.50	43.50		75	102
0620	2"	↓	8	1		44.50	49		93.50	125
0621	2-1/2"	Q-1	9	1.778		119	78		197	251
0622	3"		8	2		133	88		221	283
0623	4"		6	2.667		245	117		362	450
0624	5"	↓	5	3.200		1,250	140		1,390	1,600
0625	6"	Q-2	7	3.429		1,625	156		1,781	2,050
0626	8"	"	6	4		6,625	182		6,807	7,575
0630	Tee, reducing on the run, 1/4"	1 Plum	15	.533		16.45	26		42.45	58.50
0631	3/8"		15	.533		22	26		48	64.50
0632	1/2"		14	.571		14.90	28		42.90	59.50
0633	5/8"		13	.615		22	30		52	70.50
0634	3/4"		12	.667		6	32.50		38.50	57
0635	1"		11	.727		18.70	35.50		54.20	75.50
0636	1-1/4"		10	.800		29.50	39		68.50	93
0637	1-1/2"		9	.889		49.50	43.50		93	122
0638	2"	↓	8	1		72	49		121	155
0639	2-1/2"	Q-1	9	1.778		157	78		235	293
0640	3"		8	2		231	88		319	390
0641	4"		6	2.667		510	117		627	745
0642	5"	↓	5	3.200		1,200	140		1,340	1,525
0643	6"	Q-2	7	3.429		1,825	156		1,981	2,250
0644	8"	"	6	4		6,525	182		6,707	7,450
0650	Coupling, 1/4"	1 Plum	24	.333		.90	16.25		17.15	26
0670	3/8"		24	.333		1.19	16.25		17.44	26.50
0680	1/2"		22	.364		.94	17.75		18.69	28.50
0690	5/8"		21	.381		2.81	18.55		21.36	32
0700	3/4"		21	.381		1.87	18.55		20.42	31
0710	1"		18	.444		3.73	21.50		25.23	37.50
0715	1-1/4"		17	.471		6.65	23		29.65	43
0716	1-1/2"		15	.533		8.75	26		34.75	50
0718	2"	↓	13	.615		14.65	30		44.65	62.50
0721	2-1/2"	Q-1	15	1.067		31	47		78	107
0722	3"		13	1.231		47	54		101	136
0724	3-1/2"		8	2		89.50	88		177.50	235
0726	4"		7	2.286		98.50	100		198.50	264
0728	5"	↓	6	2.667	↓	242	117		359	445

22 11 Facility Water Distribution

22 11 13 – Facility Water Distribution Piping

22 11 13.25 Pipe Fittings, Copper

		Crew	Daily Output	Labor-Hours	Unit	Material	2009 Bare Costs Labor	Equipment	Total	Total Incl O&P
0731	6"	Q-2	8	3	Ea.	400	137		537	650
2000	DWV, solder joints, copper x copper									
2030	90° Elbow, 1-1/4"	1 Plum	13	.615	Ea.	11.90	30		41.90	59.50
2050	1-1/2"		12	.667		17.85	32.50		50.35	70
2070	2"	↓	10	.800		23.50	39		62.50	86
2090	3"	Q-1	10	1.600		57.50	70		127.50	173
2100	4"	"	9	1.778		281	78		359	430
2250	Tee, Sanitary, 1-1/4"	1 Plum	9	.889		23.50	43.50		67	93
2270	1-1/2"		8	1		29.50	49		78.50	108
2290	2"	↓	7	1.143		34	55.50		89.50	124
2310	3"	Q-1	7	2.286		126	100		226	295
2330	4"	"	6	2.667		320	117		437	535
2400	Coupling, 1-1/4"	1 Plum	14	.571		5.55	28		33.55	49
2420	1-1/2"		13	.615		6.90	30		36.90	54
2440	2"	↓	11	.727		9.60	35.50		45.10	65.50
2460	3"	Q-1	11	1.455		18.65	64		82.65	120
2480	4"	"	10	1.600	↓	59.50	70		129.50	175
9000	Minimum labor/equipment charge	1 Plum	4	2	Job		97.50		97.50	151

22 11 13.44 Pipe, Steel

		Crew	Daily Output	Labor-Hours	Unit	Material	2009 Bare Costs Labor	Equipment	Total	Total Incl O&P
0010	**PIPE, STEEL** R221113-50									
0020	All pipe sizes are to Spec. A-53 unless noted otherwise									
0050	Schedule 40, threaded, with couplings, and clevis hanger									
0060	assemblies sized for covering, 10' O.C.									
0540	Black, 1/4" diameter	1 Plum	66	.121	L.F.	2.17	5.90		8.07	11.55
0560	1/2" diameter		63	.127		2.23	6.20		8.43	12.05
0570	3/4" diameter		61	.131		2.64	6.40		9.04	12.80
0580	1" diameter	↓	53	.151		3.87	7.35		11.22	15.65
0590	1-1/4" diameter	Q-1	89	.180		4.89	7.90		12.79	17.65
0600	1-1/2" diameter		80	.200		5.75	8.80		14.55	19.95
0610	2" diameter		64	.250		7.70	10.95		18.65	25.50
0620	2-1/2" diameter		50	.320		12.05	14.05		26.10	35.50
0630	3" diameter		43	.372		15.50	16.35		31.85	42.50
0640	3-1/2" diameter		40	.400		25.50	17.55		43.05	55
0650	4" diameter	↓	36	.444	↓	23	19.50		42.50	55
1280	All pipe sizes are to Spec. A-53 unless noted otherwise									
1281	Schedule 40, threaded, with couplings and clevis hanger									
1282	assemblies sized for covering, 10' O. C.									
1290	Galvanized, 1/4" diameter	1 Plum	66	.121	L.F.	2.36	5.90		8.26	11.75
1310	1/2" diameter		63	.127		3.11	6.20		9.31	13
1320	3/4" diameter		61	.131		3.76	6.40		10.16	14.05
1330	1" diameter	↓	53	.151		5.10	7.35		12.45	17.05
1350	1-1/2" diameter	Q-1	80	.200		7.80	8.80		16.60	22
1360	2" diameter		64	.250		10.35	10.95		21.30	28.50
1370	2-1/2" diameter		50	.320		17.60	14.05		31.65	41.50
1380	3" diameter		43	.372		22	16.35		38.35	49.50
1400	4" diameter	↓	36	.444	↓	32.50	19.50		52	65.50
2000	Welded, sch. 40, on yoke & roll hanger assys, sized for covering, 10' O.C.									
2040	Black, 1" diameter	Q-15	93	.172	L.F.	4.30	7.55	.60	12.45	17.10
2070	2" diameter		61	.262		7.95	11.50	.91	20.36	27.50
2090	3" diameter		43	.372		15.55	16.35	1.29	33.19	44
2110	4" diameter		37	.432		18.10	19	1.50	38.60	51
2120	5" diameter	↓	32	.500	↓	24.50	22	1.74	48.24	63

343

22 11 13.44 Pipe, Steel		Crew	Daily Output	Labor-Hours	Unit	Material	2009 Bare Costs Labor	2009 Bare Costs Equipment	Total	Total Incl O&P
2130	6" diameter	Q-16	36	.667	L.F.	31	30.50	1.55	63.05	82.50
9990	Minimum labor/equipment charge	1 Plum	3	2.667	Job		130		130	202

22 11 13.45 Pipe Fittings, Steel, Threaded

	22 11 13.45 Pipe Fittings, Steel, Threaded	Crew	Daily Output	Labor-Hours	Unit	Material	Labor	Equipment	Total	Total Incl O&P
0010	**PIPE FITTINGS, STEEL, THREADED** R221113-50									
0020	Cast Iron									
0040	Standard weight, black									
0060	90° Elbow, straight									
0070	1/4"	1 Plum	16	.500	Ea.	5.80	24.50		30.30	44.50
0080	3/8"		16	.500		8.40	24.50		32.90	47.50
0090	1/2"		15	.533		3.69	26		29.69	44.50
0100	3/4"		14	.571		3.84	28		31.84	47
0110	1"		13	.615		4.54	30		34.54	51.50
0130	1-1/2"	Q-1	20	.800		8.90	35		43.90	64.50
0140	2"		18	.889		13.95	39		52.95	76
0150	2-1/2"		14	1.143		33.50	50		83.50	115
0160	3"		10	1.600		54.50	70		124.50	169
0180	4"		6	2.667		101	117		218	293
0500	Tee, straight									
0510	1/4"	1 Plum	10	.800	Ea.	9.10	39		48.10	70.50
0520	3/8"		10	.800		8.85	39		47.85	70
0540	3/4"		9	.889		6.70	43.50		50.20	74.50
0550	1"		8	1		5.95	49		54.95	82
0570	1-1/2"	Q-1	13	1.231		14.15	54		68.15	99.50
0580	2"		11	1.455		19.70	64		83.70	121
0590	2-1/2"		9	1.778		51	78		129	177
0600	3"		6	2.667		79	117		196	268
0620	4"		4	4		153	176		329	440
0700	Standard weight, galvanized cast iron									
0720	90° Elbow, straight									
0730	1/4"	1 Plum	16	.500	Ea.	9.10	24.50		33.60	48
0740	3/8"		16	.500		9.10	24.50		33.60	48
0750	1/2"		15	.533		6.90	26		32.90	48
0760	3/4"		14	.571		10.20	28		38.20	54
0770	1"		13	.615		11.75	30		41.75	59.50
0790	1-1/2"	Q-1	20	.800		25	35		60	82
0800	2"		18	.889		37	39		76	102
0810	2-1/2"		14	1.143		76	50		126	162
0820	3"		10	1.600		116	70		186	236
0840	4"		6	2.667		212	117		329	415
1100	Tee, straight									
1110	1/4"	1 Plum	10	.800	Ea.	16.85	39		55.85	79
1120	3/8"		10	.800		11.60	39		50.60	73.50
1140	3/4"		9	.889		14.65	43.50		58.15	83
1150	1"		8	1		15.90	49		64.90	93
1170	1-1/2"	Q-1	13	1.231		37	54		91	125
1180	2"		11	1.455		46	64		110	150
1190	2-1/2"		9	1.778		99	78		177	230
1200	3"		6	2.667		256	117		373	460
1220	4"		4	4		294	176		470	595
5000	Malleable iron, 150 lb.									
5020	Black									
5040	90° elbow, straight									

22 11 Facility Water Distribution

22 11 13 – Facility Water Distribution Piping

22 11 13.45 Pipe Fittings, Steel, Threaded

		Crew	Daily Output	Labor-Hours	Unit	Material	2009 Bare Costs Labor	Equipment	Total	Total Incl O&P
5060	1/4"	1 Plum	16	.500	Ea.	2.94	24.50		27.44	41
5070	3/8"		16	.500		2.94	24.50		27.44	41
5090	3/4"		14	.571		3.37	28		31.37	46.50
5100	1"		13	.615		4.27	30		34.27	51
5120	1-1/2"	Q-1	20	.800		9.25	35		44.25	64.50
5130	2"		18	.889		16.70	39		55.70	79
5140	2-1/2"		14	1.143		35.50	50		85.50	117
5150	3"		10	1.600		52	70		122	167
5170	4"		6	2.667		112	117		229	305
5450	Tee, straight									
5470	1/4"	1 Plum	10	.800	Ea.	4.27	39		43.27	65
5480	3/8"		10	.800		4.27	39		43.27	65
5500	3/4"		9	.889		3.91	43.50		47.41	71.50
5510	1"		8	1		6.70	49		55.70	83
5520	1-1/4"	Q-1	14	1.143		10.85	50		60.85	90
5530	1-1/2"		13	1.231		13.50	54		67.50	99
5540	2"		11	1.455		23	64		87	125
5550	2-1/2"		9	1.778		49.50	78		127.50	176
5560	3"		6	2.667		73	117		190	262
5570	3-1/2"		5	3.200		164	140		304	400
5580	4"		4	4		176	176		352	465
5650	Coupling									
5670	1/4"	1 Plum	19	.421	Ea.	3.65	20.50		24.15	36
5680	3/8"		19	.421		3.65	20.50		24.15	36
5690	1/2"		19	.421		2.81	20.50		23.31	35
5700	3/4"		18	.444		3.30	21.50		24.80	37
5710	1"		15	.533		4.93	26		30.93	46
5730	1-1/2"	Q-1	24	.667		8.65	29.50		38.15	55
5740	2"		21	.762		12.80	33.50		46.30	66
5750	2-1/2"		18	.889		35.50	39		74.50	99.50
5760	3"		14	1.143		48	50		98	131
5780	4"		10	1.600		96	70		166	215
6000	For galvanized elbows, tees, and couplings add					20%				
9990	Minimum labor/equipment charge	1 Plum	4	2	Job		97.50		97.50	151

22 11 13.47 Pipe Fittings, Steel

		Crew	Daily Output	Labor-Hours	Unit	Material	2009 Bare Costs Labor	Equipment	Total	Total Incl O&P
0010	**PIPE FITTINGS, STEEL**, Flanged, Welded & Special									
3000	Weld joint, butt, carbon steel, standard weight									
3040	90° elbow, long radius									
3050	1/2" pipe size	Q-15	16	1	Ea.	27.50	44	3.48	74.98	102
3060	3/4" pipe size		16	1		27.50	44	3.48	74.98	102
3070	1" pipe size		16	1		12.80	44	3.48	60.28	86
3100	2" pipe size		10	1.600		14.40	70	5.55	89.95	131
3120	3" pipe size		7	2.286		20.50	100	7.95	128.45	187
3130	4" pipe size		5	3.200		34	140	11.15	185.15	268
3136	5" pipe size		4	4		75	176	13.90	264.90	370
3140	6" pipe size	Q-16	5	4.800		74.50	218	11.15	303.65	435
3350	Tee, straight									
3360	1/2" pipe size	Q-15	10	1.600	Ea.	68.50	70	5.55	144.05	191
3370	3/4" pipe size		10	1.600		68.50	70	5.55	144.05	191
3380	1" pipe size		10	1.600		34	70	5.55	109.55	153
3410	2" pipe size		6	2.667		34	117	9.30	160.30	229
3430	3" pipe size		4	4		51.50	176	13.90	241.40	345

22 11 13 – Facility Water Distribution Piping

22 11 13.47 Pipe Fittings, Steel

		Crew	Daily Output	Labor-Hours	Unit	Material	2009 Bare Costs Labor	Equipment	Total	Total Incl O&P
3440	4" pipe size	Q-15	3	5.333	Ea.	72.50	234	18.55	325.05	465
3446	5" pipe size	↓	2.50	6.400		120	281	22.50	423.50	590
3450	6" pipe size	Q-16	3	8	↓	130	365	18.55	513.55	730
9990	Minimum labor/equipment charge	Q-15	3	5.333	Job		234	18.55	252.55	385

22 11 13.48 Pipe, Fittings and Valves, Steel, Grooved-Joint

		Crew	Daily Output	Labor-Hours	Unit	Material	2009 Bare Costs Labor	Equipment	Total	Total Incl O&P
0010	**PIPE, FITTINGS AND VALVES, STEEL, GROOVED-JOINT**									
0012	Fittings are ductile iron. Steel fittings noted.									
0020	Pipe includes coupling & clevis type hanger assemblies, 10' O.C.									
1000	Schedule 40, black									
1040	3/4" diameter	1 Plum	71	.113	L.F.	3.53	5.50		9.03	12.40
1050	1" diameter		63	.127		4.23	6.20		10.43	14.25
1060	1-1/4" diameter		58	.138		5.45	6.70		12.15	16.45
1070	1-1/2" diameter		51	.157		6.30	7.65		13.95	18.80
1080	2" diameter	↓	40	.200		8.05	9.75		17.80	24
1090	2-1/2" diameter	Q-1	57	.281		10.05	12.30		22.35	30
1100	3" diameter		50	.320		16.10	14.05		30.15	39.50
1110	4" diameter		45	.356		19.10	15.60		34.70	45
1120	5" diameter	↓	37	.432	↓	26.50	19		45.50	58.50
4000	Elbow, 90° or 45°, painted									
4030	3/4" diameter	1 Plum	50	.160	Ea.	36.50	7.80		44.30	52
4040	1" diameter		50	.160		19.35	7.80		27.15	33.50
4050	1-1/4" diameter		40	.200		19.35	9.75		29.10	36.50
4060	1-1/2" diameter		33	.242		19.35	11.80		31.15	40
4070	2" diameter	↓	25	.320		19.35	15.60		34.95	45.50
4080	2-1/2" diameter	Q-1	40	.400		19.35	17.55		36.90	48.50
4100	4" diameter		25	.640		37.50	28		65.50	84.50
4110	5" diameter	↓	20	.800		90	35		125	154
4250	For galvanized elbows, add				↓	26%				
4690	Tee, painted									
4700	3/4" diameter	1 Plum	38	.211	Ea.	39	10.25		49.25	59
4740	1" diameter		33	.242		30	11.80		41.80	51.50
4750	1-1/4" diameter		27	.296		30	14.45		44.45	55.50
4760	1-1/2" diameter		22	.364		30	17.75		47.75	60.50
4770	2" diameter	↓	17	.471		30	23		53	68.50
4780	2-1/2" diameter	Q-1	27	.593		30	26		56	73.50
4800	4" diameter		17	.941		63.50	41.50		105	134
4810	5" diameter	↓	13	1.231	↓	148	54		202	247
4900	For galvanized tees, add					24%				
9990	Minimum labor/equipment charge	1 Plum	4	2	Job		97.50		97.50	151

22 11 13.74 Pipe, Plastic

		Crew	Daily Output	Labor-Hours	Unit	Material	2009 Bare Costs Labor	Equipment	Total	Total Incl O&P
0010	**PIPE, PLASTIC**									
1800	PVC, couplings 10' O.C., clevis hanger assy's, 3 per 10'									
1820	Schedule 40									
1860	1/2" diameter	1 Plum	54	.148	L.F.	1.43	7.20		8.63	12.75
1870	3/4" diameter		51	.157		1.67	7.65		9.32	13.70
1880	1" diameter		46	.174		1.96	8.50		10.46	15.30
1890	1-1/4" diameter		42	.190		2.49	9.30		11.79	17.15
1900	1-1/2" diameter	↓	36	.222		2.73	10.85		13.58	19.80
1910	2" diameter	Q-1	59	.271		3.33	11.90		15.23	22
1920	2-1/2" diameter		56	.286		4.93	12.55		17.48	25
1930	3" diameter		53	.302		6.60	13.25		19.85	28
1940	4" diameter	↓	48	.333		9.15	14.65		23.80	32.50

22 11 13 – Facility Water Distribution Piping

22 11 13.74 Pipe, Plastic		Crew	Daily Output	Labor-Hours	Unit	Material	2009 Bare Costs Labor	Equipment	Total	Total Incl O&P
1950	5" diameter	Q-1	43	.372	L.F.	12.85	16.35		29.20	39.50
1960	6" diameter	↓	39	.410	↓	16.95	18		34.95	46.50
4100	DWV type, schedule 40, couplings 10' O.C., clevis hanger assy's, 3 per 10'									
4120	ABS									
4140	1-1/4" diameter	1 Plum	42	.190	L.F.	1.58	9.30		10.88	16.15
4150	1-1/2" diameter	"	36	.222		1.72	10.85		12.57	18.70
4160	2" diameter	Q-1	59	.271	↓	2.21	11.90		14.11	21
4400	PVC									
4410	1-1/4" diameter	1 Plum	42	.190	L.F.	1.85	9.30		11.15	16.45
4460	2" diameter	Q-1	59	.271		2.20	11.90		14.10	21
4470	3" diameter		53	.302		4.21	13.25		17.46	25
4480	4" diameter		48	.333		5.75	14.65		20.40	29
4490	6" diameter	↓	39	.410	↓	10.55	18		28.55	39.50
5360	CPVC, couplings 10' O.C., clevis hanger assy's, 3 per 10'									
5380	Schedule 40									
5460	1/2" diameter	1 Plum	54	.148	L.F.	2.84	7.20		10.04	14.30
5470	3/4" diameter		51	.157		3.37	7.65		11.02	15.55
5480	1" diameter		46	.174		4.13	8.50		12.63	17.70
5490	1-1/4" diameter		42	.190		5.05	9.30		14.35	19.95
5500	1-1/2" diameter	↓	36	.222		5.70	10.85		16.55	23
5510	2" diameter	Q-1	59	.271		7.05	11.90		18.95	26
5520	2-1/2" diameter		56	.286		11.50	12.55		24.05	32
5530	3" diameter	↓	53	.302	↓	14.20	13.25		27.45	36
5560										
9900	Minimum labor/equipment charge	1 Plum	4	2	Job		97.50		97.50	151

22 11 13.76 Pipe Fittings, Plastic										
0010	**PIPE FITTINGS, PLASTIC**									
2700	PVC (white), schedule 40, socket joints									
2760	90° elbow, 1/2"	1 Plum	33.30	.240	Ea.	.41	11.70		12.11	18.60
2810	2"	Q-1	36.40	.440	"	2.42	19.30		21.72	32.50
4500	DWV, ABS, non pressure, socket joints									
4540	1/4 Bend, 1-1/4"	1 Plum	20.20	.396	Ea.	2.78	19.30		22.08	33
4560	1-1/2"	"	18.20	.440	↓	1.96	21.50		23.46	35
4570	2"	Q-1	33.10	.483		3.04	21		24.04	36.50
4800	Tee, sanitary									
4820	1-1/4"	1 Plum	13.50	.593	Ea.	3.31	29		32.31	48.50
4830	1-1/2"	"	12.10	.661	↓	2.67	32		34.67	53
4840	2"	Q-1	20	.800	↓	4.12	35		39.12	59
5000	DWV, PVC, schedule 40, socket joints									
5040	1/4 bend, 1-1/4"	1 Plum	20.20	.396	Ea.	5.70	19.30		25	36.50
5060	1-1/2"	"	18.20	.440		1.90	21.50		23.40	35
5070	2"	Q-1	33.10	.483		2.89	21		23.89	36
5080	3"		20.80	.769		7.30	34		41.30	60.50
5090	4"		16.50	.970		13.75	42.50		56.25	81
5100	6"	↓	10.10	1.584		63.50	69.50		133	178
5105	8"	Q-2	9.30	2.581		92.50	117		209.50	284
5110	1/4 bend, long sweep, 1-1/2"	1 Plum	18.20	.440		3.41	21.50		24.91	37
5112	2"	Q-1	33.10	.483		3.68	21		24.68	37
5114	3"		20.80	.769		8.45	34		42.45	62
5116	4"	↓	16.50	.970		16.15	42.50		58.65	84
5215	8"	Q-2	9.30	2.581		69.50	117		186.50	259
5250	Tee, sanitary 1-1/4"	1 Plum	13.50	.593	↓	4.10	29		33.10	49.50

22 11 13.76 Pipe Fittings, Plastic		Crew	Daily Output	Labor-Hours	Unit	Material	2009 Bare Costs Labor	Equipment	Total	Total Incl O&P
5254	1-1/2"	1 Plum	12.10	.661	Ea.	2.57	32		34.57	53
5255	2"	Q-1	20	.800		3.78	35		38.78	58.50
5256	3"		13.90	1.151		9.70	50.50		60.20	89
5257	4"		11	1.455		16.85	64		80.85	118
5259	6"		6.70	2.388		73.50	105		178.50	244
5261	8"	Q-2	6.20	3.871		184	176		360	475
5264	2" x 1-1/2"	Q-1	22	.727		3.34	32		35.34	53
5266	3" x 1-1/2"		15.50	1.032		6.15	45.50		51.65	77.50
5268	4" x 3"		12.10	1.322		21.50	58		79.50	114
5271	6" x 4"		6.90	2.319		79	102		181	245
5314	Combination Y & 1/8 bend, 1-1/2"	1 Plum	12.10	.661		5.60	32		37.60	56
5315	2"	Q-1	20	.800		7.40	35		42.40	62.50
5317	3"		13.90	1.151		17.90	50.50		68.40	98
5318	4"		11	1.455		32	64		96	134
5324	Combination Y & 1/8 bend, reducing									
5325	2" x 2" x 1-1/2"	Q-1	22	.727	Ea.	8	32		40	58.50
5327	3" x 3" x 1-1/2"		15.50	1.032		13.85	45.50		59.35	86
5328	3" x 3" x 2"		15.30	1.046		10.65	46		56.65	82.50
5329	4" x 4" x 2"		12.20	1.311		17.75	57.50		75.25	109
5331	Wye, 1-1/4"	1 Plum	13.50	.593		5.25	29		34.25	51
5332	1-1/2"	"	12.10	.661		3.93	32		35.93	54.50
5333	2"	Q-1	20	.800		4.60	35		39.60	59.50
5334	3"		13.90	1.151		11.85	50.50		62.35	91.50
5335	4"		11	1.455		22	64		86	123
5336	6"		6.70	2.388		68	105		173	238
5337	8"	Q-2	6.20	3.871		185	176		361	475
5341	2" x 1-1/2"	Q-1	22	.727		5.65	32		37.65	56
5342	3" x 1-1/2"		15.50	1.032		7.95	45.50		53.45	79
5343	4" x 3"		12.10	1.322		17.50	58		75.50	109
5344	6" x 4"		6.90	2.319		50	102		152	213
5345	8" x 6"	Q-2	6.40	3.750		151	171		322	430
5347	Double wye, 1-1/2"	1 Plum	9.10	.879		8.90	43		51.90	76.50
5348	2"	Q-1	16.60	.964		9.55	42.50		52.05	76
5349	3"		10.40	1.538		24.50	67.50		92	132
5350	4"		8.25	1.939		50	85		135	187
5354	2" x 1-1/2"		16.80	.952		8.75	42		50.75	74.50
5355	3" x 2"		10.60	1.509		18.35	66		84.35	123
5356	4" x 3"		8.45	1.893		39.50	83		122.50	173
5357	6" x 4"		7.25	2.207		100	97		197	260
5410	Reducer bushing, 2" x 1-1/4"		36.50	.438		2.51	19.25		21.76	33
5412	3" x 1-1/2"		27.30	.586		6.05	25.50		31.55	46.50
5414	4" x 2"		18.20	.879		11.25	38.50		49.75	72.50
5416	6" x 4"		11.10	1.441		30	63.50		93.50	131
5418	8" x 6"	Q-2	10.20	2.353		73	107		180	247
5500	CPVC, Schedule 80, threaded joints									
5540	90° Elbow, 1/4"	1 Plum	32	.250	Ea.	10.20	12.20		22.40	30
5560	1/2"		30.30	.264		5.90	12.85		18.75	26.50
5570	3/4"		26	.308		8.85	15		23.85	33.50
5580	1"		22.70	.352		12.40	17.20		29.60	40
5590	1-1/4"		20.20	.396		24	19.30		43.30	56.50
5600	1-1/2"		18.20	.440		26	21.50		47.50	61.50
5610	2"	Q-1	33.10	.483		34.50	21		55.50	71
5620	2-1/2"		24.20	.661		107	29		136	163

22 11 Facility Water Distribution

22 11 13 – Facility Water Distribution Piping

22 11 13.76 Pipe Fittings, Plastic

		Crew	Daily Output	Labor-Hours	Unit	Material	2009 Bare Costs Labor	2009 Bare Costs Equipment	Total	Total Incl O&P
5630	3"	Q-1	20.80	.769	Ea.	115	34		149	179
6000	Coupling, 1/4"	1 Plum	32	.250		13	12.20		25.20	33
6020	1/2"		30.30	.264		13	12.85		25.85	34.50
6030	3/4"		26	.308		15.60	15		30.60	40.50
6040	1"		22.70	.352		17.25	17.20		34.45	45.50
6050	1-1/4"		20.20	.396		21	19.30		40.30	53
6060	1-1/2"		18.20	.440		22.50	21.50		44	57.50
6070	2"	Q-1	33.10	.483		26.50	21		47.50	62
6080	2-1/2"		24.20	.661		47	29		76	97
6090	3"		20.80	.769		55	34		89	113
9900	Minimum labor/equipment charge	1 Plum	4	2	Job		97.50		97.50	151

22 11 19 – Domestic Water Piping Specialties

22 11 19.10 Flexible Connectors

		Crew	Daily Output	Labor-Hours	Unit	Material	2009 Bare Costs Labor	2009 Bare Costs Equipment	Total	Total Incl O&P
0010	**FLEXIBLE CONNECTORS**, Corrugated, 7/8" O.D., 1/2" I.D.									
0050	Gas, seamless brass, steel fittings									
0200	12" long	1 Plum	36	.222	Ea.	15.10	10.85		25.95	33.50
0220	18" long		36	.222		18.35	10.85		29.20	37
0240	24" long		34	.235		22	11.45		33.45	42.50
0260	30" long		34	.235		24	11.45		35.45	44.50
9000	Minimum labor/equipment charge		4	2	Job		97.50		97.50	151

22 11 19.18 Mixing Valve

		Crew	Daily Output	Labor-Hours	Unit	Material	2009 Bare Costs Labor	2009 Bare Costs Equipment	Total	Total Incl O&P
0010	**MIXING VALVE**, Automatic, water tempering.									
0040	1/2" size	1 Stpi	19	.421	Ea.	435	21		456	510
0050	3/4" size		18	.444	"	435	22		457	515
9000	Minimum labor/equipment charge		5	1.600	Job		79		79	122

22 11 19.38 Water Supply Meters

		Crew	Daily Output	Labor-Hours	Unit	Material	2009 Bare Costs Labor	2009 Bare Costs Equipment	Total	Total Incl O&P
0010	**WATER SUPPLY METERS**									
2000	Domestic/commercial, bronze									
2020	Threaded									
2060	5/8" diameter, to 20 GPM	1 Plum	16	.500	Ea.	42	24.50		66.50	84
2080	3/4" diameter, to 30 GPM		14	.571		76.50	28		104.50	127
2100	1" diameter, to 50 GPM		12	.667		116	32.50		148.50	179
2300	Threaded/flanged									
2340	1-1/2" diameter, to 100 GPM	1 Plum	8	1	Ea.	284	49		333	385
9000	Minimum labor/equipment charge	"	3.25	2.462	Job		120		120	186

22 11 19.42 Backflow Preventers

		Crew	Daily Output	Labor-Hours	Unit	Material	2009 Bare Costs Labor	2009 Bare Costs Equipment	Total	Total Incl O&P
0010	**BACKFLOW PREVENTERS**, Includes valves									
0020	and four test cocks, corrosion resistant, automatic operation									
1000	Double check principle									
1010	Threaded, with ball valves									
1020	3/4" pipe size	1 Plum	16	.500	Ea.	223	24.50		247.50	283
1030	1" pipe size		14	.571		257	28		285	325
1040	1-1/2" pipe size		10	.800		670	39		709	800
1050	2" pipe size		7	1.143		800	55.50		855.50	965
1080	Threaded, with gate valves									
1100	3/4" pipe size	1 Plum	16	.500	Ea.	860	24.50		884.50	985
1120	1" pipe size		14	.571		865	28		893	1,000
1140	1-1/2" pipe size		10	.800		1,125	39		1,164	1,275
1160	2" pipe size		7	1.143		1,375	55.50		1,430.50	1,575
4000	Reduced pressure principle									
4100	Threaded, bronze, valves are ball									

22 11 Facility Water Distribution

22 11 19 – Domestic Water Piping Specialties

22 11 19.42 Backflow Preventers

		Crew	Daily Output	Labor-Hours	Unit	Material	2009 Bare Costs Labor	Equipment	Total	Total Incl O&P
4120	3/4" pipe size	1 Plum	16	.500	Ea.	330	24.50		354.50	405
4140	1" pipe size		14	.571		355	28		383	435
4160	1-1/2" pipe size		10	.800		670	39		709	795
4180	2" pipe size		7	1.143		750	55.50		805.50	910
5000	Flanged, bronze, valves are OS&Y									
5060	2-1/2" pipe size	Q-1	5	3.200	Ea.	3,200	140		3,340	3,750
5080	3" pipe size		4.50	3.556		3,350	156		3,506	3,950
5100	4" pipe size		3	5.333		4,225	234		4,459	5,025
5120	6" pipe size	Q-2	3	8		6,100	365		6,465	7,300
5600	Flanged, iron, valves are OS&Y									
5660	2-1/2" pipe size	Q-1	5	3.200	Ea.	1,750	140		1,890	2,150
5680	3" pipe size		4.50	3.556		2,700	156		2,856	3,200
5700	4" pipe size		3	5.333		3,450	234		3,684	4,175
5720	6" pipe size	Q-2	3	8		4,875	365		5,240	5,950
5800	Rebuild 4" diameter reduced pressure BFP	1 Plum	4	2		290	97.50		387.50	470
5810	6" diameter		2.66	3.008		305	147		452	565
5820	8" diameter		2	4		380	195		575	720
5830	10" diameter		1.60	5		445	244		689	870
9010	Minimum labor/equipment charge		2	4	Job		195		195	300

22 11 19.50 Vacuum Breakers

		Crew	Daily Output	Labor-Hours	Unit	Material	2009 Bare Costs Labor	Equipment	Total	Total Incl O&P
0010	**VACUUM BREAKERS**									
0013	See also backflow preventers Div. 22 11 19.42									
1000	Anti-siphon continuous pressure type									
1010	Max. 150 PSI - 210° F									
1020	Bronze body									
1030	1/2" size	1 Stpi	24	.333	Ea.	143	16.45		159.45	183
1040	3/4" size		20	.400		143	19.75		162.75	188
1050	1" size		19	.421		148	21		169	195
1060	1-1/4" size		15	.533		291	26.50		317.50	360
1070	1-1/2" size		13	.615		360	30.50		390.50	440
1080	2" size		11	.727		370	36		406	460
1200	Max. 125 PSI with atmospheric vent									
1210	Brass, in-line construction									
1220	1/4" size	1 Stpi	24	.333	Ea.	57.50	16.45		73.95	89
1230	3/8" size	"	24	.333		57.50	16.45		73.95	89
1260	For polished chrome finish, add					13%				
2000	Anti-siphon, non-continuous pressure type									
2010	Hot or cold water 125 PSI - 210° F									
2020	Bronze body									
2030	1/4" size	1 Stpi	24	.333	Ea.	36	16.45		52.45	65
2040	3/8" size		24	.333		36	16.45		52.45	65
2050	1/2" size		24	.333		41	16.45		57.45	70.50
2060	3/4" size		20	.400		48.50	19.75		68.25	84
2070	1" size		19	.421		75.50	21		96.50	115
2080	1-1/4" size		15	.533		133	26.50		159.50	187
2090	1-1/2" size		13	.615		137	30.50		167.50	198
2100	2" size		11	.727		243	36		279	325
2110	2-1/2" size		8	1		695	49.50		744.50	840
2120	3" size		6	1.333		925	66		991	1,125
2150	For polished chrome finish, add					50%				

22 11 19.54 Water Hammer Arresters/Shock Absorbers

0010	**WATER HAMMER ARRESTERS/SHOCK ABSORBERS**									

22 11 Facility Water Distribution

22 11 19 – Domestic Water Piping Specialties

22 11 19.54 Water Hammer Arresters/Shock Absorbers		Crew	Daily Output	Labor-Hours	Unit	Material	2009 Bare Costs Labor	2009 Bare Costs Equipment	Total	Total Incl O&P
0490	Copper									
0500	3/4" male I.P.S. For 1 to 11 fixtures	1 Plum	12	.667	Ea.	17	32.50		49.50	69
0600	1" male I.P.S., For 12 to 32 fixtures		8	1		41.50	49		90.50	122
0700	1-1/4" male I.P.S. For 33 to 60 fixtures		8	1		44	49		93	124
0800	1-1/2" male I.P.S. For 61 to 113 fixtures		8	1		63	49		112	145
0900	2" male I.P.S.For 114 to 154 fixtures		8	1		92	49		141	177
1000	2-1/2" male I.P.S. For 155 to 330 fixtures		4	2		285	97.50		382.50	465
9000	Minimum labor/equipment charge		3.50	2.286	Job		111		111	173

22 11 19.64 Hydrants

		Crew	Daily Output	Labor-Hours	Unit	Material	2009 Bare Costs Labor	2009 Bare Costs Equipment	Total	Total Incl O&P
0010	**HYDRANTS**									
0050	Wall type, moderate climate, bronze, encased									
0200	3/4" IPS connection	1 Plum	16	.500	Ea.	535	24.50		559.50	630
1000	Non-freeze, bronze, exposed									
1100	3/4" IPS connection, 4" to 9" thick wall	1 Plum	14	.571	Ea.	365	28		393	445
1120	10" to 14" thick wall	"	12	.667		475	32.50		507.50	570
1280	For anti-siphon type, add					95			95	105
9000	Minimum labor/equipment charge	1 Plum	3	2.667	Job		130		130	202

22 13 Facility Sanitary Sewerage

22 13 16 – Sanitary Waste and Vent Piping

22 13 16.20 Pipe, Cast Iron

		Crew	Daily Output	Labor-Hours	Unit	Material	2009 Bare Costs Labor	2009 Bare Costs Equipment	Total	Total Incl O&P
0010	**PIPE, CAST IRON**, Soil, on clevis hanger assemblies, 5' O.C. R221113-50									
0020	Single hub, service wt., lead & oakum joints 10' O.C.									
2120	2" diameter	Q-1	63	.254	L.F.	6.35	11.15		17.50	24.50
2140	3" diameter		60	.267		8.90	11.70		20.60	28
2160	4" diameter		55	.291		11.50	12.75		24.25	32.50
2180	5" diameter	Q-2	76	.316		15.85	14.35		30.20	40
2200	6" diameter	"	73	.329		19.50	14.95		34.45	44.50
2220	8" diameter	Q-3	59	.542		30.50	25		55.50	72.50
4000	No hub, couplings 10' O.C.									
4100	1-1/2" diameter	Q-1	71	.225	L.F.	6.90	9.90		16.80	23
4120	2" diameter		67	.239		7.10	10.50		17.60	24
4140	3" diameter		64	.250		9.80	10.95		20.75	28
4160	4" diameter		58	.276		12.70	12.10		24.80	32.50
4180	5" diameter	Q-2	83	.289		18.05	13.15		31.20	40.50
9000	Minimum labor/equipment charge	1 Plum	4	2	Job		97.50		97.50	151

22 13 16.30 Pipe Fittings, Cast Iron

		Crew	Daily Output	Labor-Hours	Unit	Material	2009 Bare Costs Labor	2009 Bare Costs Equipment	Total	Total Incl O&P
0010	**PIPE FITTINGS, CAST IRON**, Soil R221113-50									
0040	Hub and spigot, service weight, lead & oakum joints									
0080	1/4 bend, 2"	Q-1	16	1	Ea.	13.70	44		57.70	83
0120	3"		14	1.143		18.40	50		68.40	98.50
0140	4"		13	1.231		28.50	54		82.50	116
0160	5"	Q-2	18	1.333		40.50	60.50		101	139
0180	6"	"	17	1.412		50	64		114	155
0200	8"	Q-3	11	2.909		151	135		286	375
0340	1/8 bend, 2"	Q-1	16	1		9.80	44		53.80	79
0350	3"		14	1.143		15.35	50		65.35	95
0360	4"		13	1.231		22.50	54		76.50	109
0380	5"	Q-2	18	1.333		31	60.50		91.50	128
0400	6"	"	17	1.412		38	64		102	142

22 13 16.30 Pipe Fittings, Cast Iron		Crew	Daily Output	Labor-Hours	Unit	Material	2009 Bare Costs Labor	Equipment	Total	Total Incl O&P
0420	8"	Q-3	11	2.909	Ea.	113	135		248	335
0500	Sanitary tee, 2"	Q-1	10	1.600		19.25	70		89.25	130
0540	3"		9	1.778		31	78		109	155
0620	4"	↓	8	2		38	88		126	178
0700	5"	Q-2	12	2		75.50	91		166.50	224
0800	6"	"	11	2.182		86	99.50		185.50	249
0880	8"	Q-3	7	4.571		227	212		439	580
0954	10" x 6"	"	8	4	↓	330	186		516	650
5990	No hub									
6000	Cplg. & labor required at joints not incl. in fitting									
6010	price. Add 1 coupling per joint for installed price									
6020	1/4 Bend, 1-1/2"				Ea.	7.70			7.70	8.45
6060	2"					8.45			8.45	9.30
6080	3"					11.60			11.60	12.80
6120	4"					16.80			16.80	18.50
6184	1/4 Bend, long sweep, 1-1/2"					18.20			18.20	20
6186	2"					18.20			18.20	20
6188	3"					21.50			21.50	23.50
6189	4"					34.50			34.50	38
6190	5"					63.50			63.50	70
6191	6"					77.50			77.50	85
6192	8"					187			187	206
6193	10"					335			335	370
6200	1/8 Bend, 1-1/2"					6.45			6.45	7.10
6210	2"					6.45			6.45	7.10
6212	3"					9.75			9.75	10.70
6214	4"					12.30			12.30	13.55
6380	Sanitary Tee, tapped, 1-1/2"					14.20			14.20	15.65
6382	2" x 1-1/2"					12.80			12.80	14.10
6384	2"					14.35			14.35	15.80
6386	3" x 2"					20			20	22
6388	3"					37			37	40.50
6390	4" x 1-1/2"					17.70			17.70	19.50
6392	4" x 2"					20			20	22
6393	4"					20			20	22
6394	6" x 1-1/2"					44			44	48.50
6396	6" x 2"					45			45	49.50
6459	Sanitary Tee, 1-1/2"					10.70			10.70	11.80
6460	2"					11.60			11.60	12.80
6470	3"					14.20			14.20	15.65
6472	4"				↓	22			22	24
8000	Coupling, standard (by CISPI Mfrs.)									
8020	1-1/2"	Q-1	48	.333	Ea.	14.45	14.65		29.10	38.50
8040	2"		44	.364		14.45	15.95		30.40	41
8080	3"		38	.421		17.25	18.50		35.75	47.50
8120	4"	↓	33	.485	↓	20.50	21.50		42	55.50
9000	Minimum labor/equipment charge	1 Plum	4	2	Job		97.50		97.50	151

22 13 16.60 Traps

		Crew	Daily Output	Labor-Hours	Unit	Material	2009 Bare Costs Labor	Equipment	Total	Total Incl O&P
0010	**TRAPS**									
0030	Cast iron, service weight									
0050	Running P trap, without vent									
1100	2"	Q-1	16	1	Ea.	99	44		143	177

22 13 Facility Sanitary Sewerage

22 13 16 – Sanitary Waste and Vent Piping

22 13 16.60 Traps

		Crew	Daily Output	Labor-Hours	Unit	Material	2009 Bare Costs Labor	2009 Bare Costs Equipment	Total	Total Incl O&P
1150	4"	Q-1	13	1.231	Ea.	99	54		153	193
1160	6"	Q-2	17	1.412		430	64		494	575
3000	P trap, B&S, 2" pipe size	Q-1	16	1		23.50	44		67.50	94
3040	3" pipe size	"	14	1.143		35	50		85	117
3900										
4700	Copper, drainage, drum trap									
4840	3" x 6" swivel, 1-1/2" pipe size	1 Plum	16	.500	Ea.	155	24.50		179.50	208
5100	P trap, standard pattern									
5200	1-1/4" pipe size	1 Plum	18	.444	Ea.	72	21.50		93.50	113
5240	1-1/2" pipe size		17	.471		69.50	23		92.50	112
5260	2" pipe size		15	.533		107	26		133	159
5280	3" pipe size		11	.727		258	35.50		293.50	340
9000	Minimum labor/equipment charge		3	2.667	Job		130		130	202

22 13 16.80 Vent Flashing and Caps

		Crew	Daily Output	Labor-Hours	Unit	Material	2009 Bare Costs Labor	2009 Bare Costs Equipment	Total	Total Incl O&P
0010	**VENT FLASHING AND CAPS**									
0120	Vent caps									
0140	Cast iron									
0180	2-1/2" - 3-5/8" pipe	1 Plum	21	.381	Ea.	40	18.55		58.55	73
0190	4" - 4-1/8" pipe	"	19	.421	"	48	20.50		68.50	85
0900	Vent flashing									
1000	Aluminum with lead ring									
1050	3" pipe	1 Plum	17	.471	Ea.	11.70	23		34.70	48.50
1060	4" pipe	"	16	.500	"	14.15	24.50		38.65	53.50
1350	Copper with neoprene ring									
1440	2" pipe	1 Plum	18	.444	Ea.	21	21.50		42.50	56.50
1450	3" pipe		17	.471		24.50	23		47.50	62.50
1460	4" pipe		16	.500		27	24.50		51.50	68
9000	Minimum labor/equipment charge		4	2	Job		97.50		97.50	151

22 13 19 – Sanitary Waste Piping Specialties

22 13 19.13 Sanitary Drains

		Crew	Daily Output	Labor-Hours	Unit	Material	2009 Bare Costs Labor	2009 Bare Costs Equipment	Total	Total Incl O&P
0010	**SANITARY DRAINS**									
0400	Deck, auto park, C.I., 13" top									
0440	3", 4", 5", and 6" pipe size	Q-1	8	2	Ea.	1,000	88		1,088	1,225
0480	For galvanized body, add				"	545			545	600
0500										
2000	Floor, medium duty, C.I., deep flange, 7" dia top									
2040	2" and 3" pipe size	Q-1	12	1.333	Ea.	141	58.50		199.50	247
2080	For galvanized body, add					67			67	74
2120	For polished bronze top, add					102			102	112
2500	Heavy duty, cleanout & trap w/bucket, C.I., 15" top									
2540	2", 3", and 4" pipe size	Q-1	6	2.667	Ea.	4,625	117		4,742	5,250
2560	For galvanized body, add					1,175			1,175	1,300
2580	For polished bronze top, add					510			510	560

22 13 19.14 Floor Receptors

		Crew	Daily Output	Labor-Hours	Unit	Material	2009 Bare Costs Labor	2009 Bare Costs Equipment	Total	Total Incl O&P
0010	**FLOOR RECEPTORS**, For connection to 2", 3" & 4" diameter pipe									
0200	12-1/2" square top, 25 sq in open area	Q-1	10	1.600	Ea.	680	70		750	860
0300	For grate with 4" diam. x 3-3/4" high funnel, add					137			137	150
0400	For grate with 6" diameter x 6" high funnel, add					202			202	222
0700	For acid-resisting bucket, add					192			192	211
0900	For stainless steel mesh bucket liner, add					152			152	167
2000	12-5/8" diameter top, 40 sq. in. open area	Q-1	10	1.600		540	70		610	705
2100	For options, add same prices as square top									

22 13 Facility Sanitary Sewerage

22 13 19 – Sanitary Waste Piping Specialties

22 13 19.14 Floor Receptors

		Crew	Daily Output	Labor-Hours	Unit	Material	2009 Bare Costs Labor	Equipment	Total	Total Incl O&P
3000	8" x 4" rectangular top, 7.5 sq. in. open area	Q-1	14	1.143	Ea.	525	50		575	655
3100	For trap primer connections, add				"	61.50			61.50	68
9000	Minimum labor/equipment charge	Q-1	3	5.333	Job		234		234	365

22 13 23 – Sanitary Waste Interceptors

22 13 23.10 Interceptors

		Crew	Daily Output	Labor-Hours	Unit	Material	2009 Bare Costs Labor	Equipment	Total	Total Incl O&P
0010	**INTERCEPTORS**									
0150	Grease, cast iron, 4 GPM, 8 lb. fat capacity	1 Plum	4	2	Ea.	835	97.50		932.50	1,075
0200	7 GPM, 14 lb. fat capacity		4	2		1,150	97.50		1,247.50	1,425
1000	10 GPM, 20 lb. fat capacity		4	2		1,375	97.50		1,472.50	1,650
1160	100 GPM, 200 lb. fat capacity	Q-1	2	8		10,900	350		11,250	12,500
1240	300 GPM, 600 lb. fat capacity	"	1	16		22,800	700		23,500	26,100
9000	Minimum labor/equipment charge	1 Plum	3	2.667	Job		130		130	202

22 13 29 – Sanitary Sewerage Pumps

22 13 29.14 Sewage Ejector Pumps

		Crew	Daily Output	Labor-Hours	Unit	Material	2009 Bare Costs Labor	Equipment	Total	Total Incl O&P
0010	**SEWAGE EJECTOR PUMPS**, With operating and level controls									
0100	Simplex system incl. tank, cover, pump 15' head									
0500	37 gal PE tank, 12 GPM, 1/2 HP, 2" discharge	Q-1	3.20	5	Ea.	440	219		659	825
0510	3" discharge		3.10	5.161		480	226		706	875
0530	87 GPM, .7 HP, 2" discharge		3.20	5		670	219		889	1,075
0540	3" discharge		3.10	5.161		725	226		951	1,150
0600	45 gal. coated stl tank, 12 GPM, 1/2 HP, 2" discharge		3	5.333		780	234		1,014	1,225
0610	3" discharge		2.90	5.517		810	242		1,052	1,275
0630	87 GPM, .7 HP, 2" discharge		3	5.333		1,000	234		1,234	1,475
0640	3" discharge		2.90	5.517		1,050	242		1,292	1,525
0660	134 GPM, 1 HP, 2" discharge		2.80	5.714		1,075	251		1,326	1,600
0680	3" discharge		2.70	5.926		1,150	260		1,410	1,650
0700	70 gal. PE tank, 12 GPM, 1/2 HP, 2" discharge		2.60	6.154		850	270		1,120	1,350
0710	3" discharge		2.40	6.667		910	293		1,203	1,450
0730	87 GPM, 0.7 HP, 2" discharge		2.50	6.400		1,100	281		1,381	1,625
0740	3" discharge		2.30	6.957		1,150	305		1,455	1,750
0760	134 GPM, 1 HP, 2" discharge		2.20	7.273		1,200	320		1,520	1,800
0770	3" discharge		2	8		1,275	350		1,625	1,950
9000	Minimum labor/equipment charge		2.50	6.400	Job		281		281	435

22 14 Facility Storm Drainage

22 14 26 – Facility Storm Drains

22 14 26.13 Roof Drains

		Crew	Daily Output	Labor-Hours	Unit	Material	2009 Bare Costs Labor	Equipment	Total	Total Incl O&P
0010	**ROOF DRAINS**									
0140	Cornice, C.I., 45° or 90° outlet									
0200	3" and 4" pipe size	Q-1	12	1.333	Ea.	231	58.50		289.50	345
0260	For galvanized body, add					53			53	58.50
0280	For polished bronze dome, add					39			39	43
3860	Roof, flat metal deck, C.I. body, 12" C.I. dome									
3890	3" pipe size	Q-1	14	1.143	Ea.	277	50		327	385
3900	4" pipe size	"	13	1.231	"	277	54		331	390
4620	Main, all aluminum, 12" low profile dome									
4640	2", 3" and 4" pipe size	Q-1	14	1.143	Ea.	305	50		355	415
9000	Minimum labor/equipment charge	1 Plum	4	2	Job		97.50		97.50	151

22 14 Facility Storm Drainage

22 14 29 – Sump Pumps

22 14 29.16 Submersible Sump Pumps

22 14 29.16 Submersible Sump Pumps	Crew	Daily Output	Labor-Hours	Unit	Material	2009 Bare Costs Labor	Equipment	Total	Total Incl O&P
0010 SUBMERSIBLE SUMP PUMPS									
7000 Sump pump, automatic									
7100 Plastic, 1-1/4" discharge, 1/4 HP	1 Plum	6	1.333	Ea.	136	65		201	250
7140 1/3 HP		5	1.600		163	78		241	300
7180 1-1/2" discharge, 1/2 HP		4	2		223	97.50		320.50	395
7500 Cast iron, 1-1/4" discharge, 1/4 HP		6	1.333		156	65		221	272
7540 1/3 HP		6	1.333		184	65		249	305
7560 1/2 HP		5	1.600		222	78		300	365
9000 Minimum labor/equipment charge		4	2	Job		97.50		97.50	151

22 15 General Service Compressed-Air Systems

22 15 19 – General Service Packaged Air Compressors and Receivers

22 15 19.10 Air Compressors

22 15 19.10 Air Compressors	Crew	Daily Output	Labor-Hours	Unit	Material	2009 Bare Costs Labor	Equipment	Total	Total Incl O&P
0010 AIR COMPRESSORS									
5250 Air, reciprocating air cooled, splash lubricated, tank mounted									
5300 Single stage, 1 phase, 140 psi									
5303 1/2 HP, 30 gal tank	1 Stpi	3	2.667	Ea.	1,250	132		1,382	1,575
5305 3/4 HP, 30 gal tank		2.60	3.077		1,275	152		1,427	1,625
5307 1 HP, 30 gal tank		2.20	3.636		1,625	179		1,804	2,075

22 31 Domestic Water Softeners

22 31 13 – Residential Domestic Water Softeners

22 31 13.10 Residential Water Softeners

22 31 13.10 Residential Water Softeners	Crew	Daily Output	Labor-Hours	Unit	Material	2009 Bare Costs Labor	Equipment	Total	Total Incl O&P
0010 RESIDENTIAL WATER SOFTENERS									
7350 Water softener, automatic, to 30 grains per gallon	2 Plum	5	3.200	Ea.	410	156		566	690
7400 To 100 grains per gallon	"	4	4	"	680	195		875	1,050

22 33 Electric Domestic Water Heaters

22 33 13 – Instantaneous Electric Domestic Water Heaters

22 33 13.20 Instantaneous Elec. Point-Of-use Water Heaters

22 33 13.20 Instantaneous Elec. Point-Of-use Water Heaters	Crew	Daily Output	Labor-Hours	Unit	Material	2009 Bare Costs Labor	Equipment	Total	Total Incl O&P
0010 INSTANTANEOUS ELECTRIC POINT-OF-USE WATER HEATERS									
8965 Point of use, electric, glass lined									
8969 Energy saver									
8970 2.5 gal. single element	1 Plum	2.80	2.857	Ea.	261	139		400	505
8971 4 gal. single element		2.80	2.857		276	139		415	520
8974 6 gal. single element		2.50	3.200		295	156		451	565
8975 10 gal. single element		2.50	3.200		315	156		471	585
8976 15 gal. single element		2.40	3.333		330	163		493	610
8977 20 gal. single element		2.40	3.333		360	163		523	645
8978 30 gal. single element		2.30	3.478		400	170		570	705
8979 40 gal. single element		2.20	3.636		510	177		687	840
8988 Commercial (ASHRAE energy std. 90)									
8989 6 gallon	1 Plum	2.50	3.200	Ea.	560	156		716	860
8990 10 gallon		2.50	3.200		600	156		756	900
8991 15 gallon		2.40	3.333		635	163		798	950
8992 20 gallon		2.40	3.333		680	163		843	1,000
8993 30 gallon		2.30	3.478		1,425	170		1,595	1,850

22 33 Electric Domestic Water Heaters

22 33 13 – Instantaneous Electric Domestic Water Heaters

22 33 13.20 Instantaneous Elec. Point-Of-use Water Heaters	Crew	Daily Output	Labor-Hours	Unit	Material	2009 Bare Costs Labor	Equipment	Total	Total Incl O&P
8995	Under the sink, copper, w/bracket									
8996	2.5 gallon	1 Plum	4	2	Ea.	249	97.50		346.50	425
9000	Minimum labor/equipment charge	"	1.75	4.571	Job		223		223	345

22 33 30 – Residential, Electric Domestic Water Heaters

22 33 30.13 Residential, Small-Capacity Elec. Water Heaters

		Crew	Daily Output	Labor-Hours	Unit	Material	Labor	Equipment	Total	Total Incl O&P
0010	**RESIDENTIAL, SMALL-CAPACITY ELECTRIC DOMESTIC WATER HEATERS**									
0050										
1000	Residential, electric, glass lined tank, 5 yr, 10 gal., single element	1 Plum	2.30	3.478	Ea.	315	170		485	610
1060	30 gallon, double element		2.20	3.636		465	177		642	790
1080	40 gallon, double element		2	4		500	195		695	845
1100	52 gallon, double element		2	4		560	195		755	915
1120	66 gallon, double element		1.80	4.444		755	217		972	1,175
1140	80 gallon, double element		1.60	5		840	244		1,084	1,300

22 33 33 – Light-Commercial Electric Domestic Water Heaters

22 33 33.10 Commercial Electric Water Heaters

		Crew	Daily Output	Labor-Hours	Unit	Material	Labor	Equipment	Total	Total Incl O&P
0010	**COMMERCIAL ELECTRIC WATER HEATERS**									
4000	Commercial, 100° rise. NOTE: for each size tank, a range of									
4010	heaters between the ones shown are available									
4020	Electric									
4100	5 gal., 3 kW, 12 GPH, 208V	1 Plum	2	4	Ea.	2,675	195		2,870	3,250
4160	50 gal., 36 kW, 148 GPH, 208V	"	1.80	4.444		6,325	217		6,542	7,275
4480	400 gal., 210 kW, 860 GPH, 480V	Q-1	1	16		37,600	700		38,300	42,500

22 34 Fuel-Fired Domestic Water Heaters

22 34 30 – Residential Gas Domestic Water Heaters

22 34 30.13 Residential, Atmos, Gas Domestic Wtr Heaters

		Crew	Daily Output	Labor-Hours	Unit	Material	Labor	Equipment	Total	Total Incl O&P
0010	**RESIDENTIAL, ATMOSPHERIC, GAS DOMESTIC WATER HEATERS**									
2000	Gas fired, foam lined tank, 10 yr, vent not incl.,									
2040	30 gallon	1 Plum	2	4	Ea.	630	195		825	990
2060	40 gallon		1.90	4.211		895	205		1,100	1,300
2100	75 gallon		1.50	5.333		1,225	260		1,485	1,750

22 34 36 – Commercial Gas Domestic Water Heaters

22 34 36.13 Commercial, Atmos., Gas Domestic Water Htrs.

		Crew	Daily Output	Labor-Hours	Unit	Material	Labor	Equipment	Total	Total Incl O&P
0010	**COMMERCIAL, ATMOSPHERIC, GAS DOMESTIC WATER HEATERS**									
6000	Gas fired, flush jacket, std. controls, vent not incl.									
6040	75 MBH input, 73 GPH	1 Plum	1.40	5.714	Ea.	1,725	279		2,004	2,325
6060	98 MBH input, 95 GPH		1.40	5.714		3,975	279		4,254	4,800
6180	200 MBH input, 192 GPH		.60	13.333		6,275	650		6,925	7,925

22 34 46 – Oil-Fired Domestic Water Heaters

22 34 46.10 Residential Oil-Fired Water Heaters

		Crew	Daily Output	Labor-Hours	Unit	Material	Labor	Equipment	Total	Total Incl O&P
0010	**RESIDENTIAL OIL-FIRED WATER HEATERS**									
3000	Oil fired, glass lined tank, 5 yr, vent not included, 30 gallon	1 Plum	2	4	Ea.	665	195		860	1,025
3040	50 gallon	"	1.80	4.444	"	1,175	217		1,392	1,600

22 34 46.20 Commercial Oil-Fired Water Heaters

		Crew	Daily Output	Labor-Hours	Unit	Material	Labor	Equipment	Total	Total Incl O&P
0010	**COMMERCIAL OIL-FIRED WATER HEATERS**									
8000	Oil fired, glass lined, UL listed, std. controls, vent not incl.									
8060	140 gal., 140 MBH input, 134 GPH	Q-1	2.13	7.512	Ea.	14,200	330		14,530	16,100
8080	140 gal., 199 MBH input, 191 GPH		2	8		14,600	350		14,950	16,600

22 34 Fuel-Fired Domestic Water Heaters

22 34 46 – Oil-Fired Domestic Water Heaters

22 34 46.20 Commercial Oil-Fired Water Heaters	Crew	Daily Output	Labor-Hours	Unit	Material	2009 Bare Costs Labor	Equipment	Total	Total Incl O&P
8160	140 gal., 540 MBH input, 519 GPH	Q-1	.96	16.667	Ea.	19,900	730		20,630	23,000
8280	201 gal., 1250 MBH input, 1200 GPH	Q-2	1.22	19.672	↓	31,100	895		31,995	35,600

22 41 Residential Plumbing Fixtures

22 41 13 – Residential Water Closets, Urinals, and Bidets

22 41 13.40 Water Closets

		Crew	Daily Output	Labor-Hours	Unit	Material	Labor	Equipment	Total	Total Incl O&P
0010	**WATER CLOSETS** R224000-40									
0032	For automatic flush, see Div. 22 42 39.10 0972									
0150	Tank type, vitreous china, incl. seat, supply pipe w/stop									
0200	Wall hung									
0400	Two piece, close coupled	Q-1	5.30	3.019	Ea.	555	132		687	815
0960	For rough-in, supply, waste, vent and carrier		2.73	5.861		630	257		887	1,100
1000	Floor mounted, one piece		5.30	3.019		530	132		662	790
1100	Two piece, close coupled	↓	5.30	3.019	↓	184	132		316	410
1960	For color, add					30%				
1980	For rough-in, supply, waste and vent	Q-1	3.05	5.246	Ea.	275	230		505	655

22 41 16 – Residential Lavatories and Sinks

22 41 16.10 Lavatories

		Crew	Daily Output	Labor-Hours	Unit	Material	Labor	Equipment	Total	Total Incl O&P
0010	**LAVATORIES**, With trim, white unless noted otherwise R224000-40									
0020										
0500	Vanity top, porcelain enamel on cast iron									
0600	20" x 18"	Q-1	6.40	2.500	Ea.	231	110		341	425
0640	33" x 19" oval	"	6.40	2.500	"	505	110		615	725
0860	For color, add					25%				
1000	Cultured marble, 19" x 17", single bowl	Q-1	6.40	2.500	Ea.	169	110		279	355
1120	25" x 22", single bowl		6.40	2.500		185	110		295	375
1900	Stainless steel, self-rimming, 25" x 22", single bowl, ledge		6.40	2.500		330	110		440	530
1960	17" x 22", single bowl		6.40	2.500		320	110		430	520
2600	Steel, enameled, 20" x 17", single bowl		5.80	2.759		159	121		280	365
2900	Vitreous china, 20" x 16", single bowl		5.40	2.963		260	130		390	485
2960	20" x 17", single bowl		5.40	2.963		176	130		306	395
3580	Rough-in, supply, waste and vent for all above lavatories	↓	2.30	6.957	↓	224	305		529	720
4000	Wall hung									
4040	Porcelain enamel on cast iron, 16" x 14", single bowl	Q-1	8	2	Ea.	410	88		498	585
4180	20" x 18", single bowl	"	8	2	"	297	88		385	460
4580	For color, add					30%				
6000	Vitreous china, 18" x 15", single bowl with backsplash	Q-1	7	2.286	Ea.	231	100		331	410
6960	Rough-in, supply, waste and vent for above lavatories	"	1.66	9.639	"	365	425		790	1,050
9000	Minimum labor/equipment charge	1 Plum	3	2.667	Job		130		130	202

22 41 16.30 Sinks

		Crew	Daily Output	Labor-Hours	Unit	Material	Labor	Equipment	Total	Total Incl O&P
0010	**SINKS**, With faucets and drain									
2000	Kitchen, counter top style, P.E. on C.I., 24" x 21" single bowl	Q-1	5.60	2.857	Ea.	272	125		397	495
2100	31" x 22" single bowl		5.60	2.857		278	125		403	500
2200	32" x 21" double bowl		4.80	3.333		370	146		516	635
3000	Stainless steel, self rimming, 19" x 18" single bowl		5.60	2.857		510	125		635	755
3100	25" x 22" single bowl		5.60	2.857		570	125		695	820
3200	33" x 22" double bowl		4.80	3.333		830	146		976	1,125
3300	43" x 22" double bowl		4.80	3.333		965	146		1,111	1,275
4000	Steel, enameled, with ledge, 24" x 21" single bowl		5.60	2.857		145	125		270	355
4100	32" x 21" double bowl	↓	4.80	3.333	↓	166	146		312	410

22 41 Residential Plumbing Fixtures

22 41 16 – Residential Lavatories and Sinks

22 41 16.30 Sinks

		Crew	Daily Output	Labor-Hours	Unit	Material	2009 Bare Costs Labor	Equipment	Total	Total Incl O&P
4960	For color sinks except stainless steel, add				Ea.	10%				
4980	For rough-in, supply, waste and vent, counter top sinks	Q-1	2.14	7.477		266	330		596	805
5790	For rough-in, supply, waste & vent, sinks	"	1.85	8.649	↓	266	380		646	885

22 41 19 – Residential Bathtubs

22 41 19.10 Baths

		Crew	Daily Output	Labor-Hours	Unit	Material	2009 Bare Costs Labor	Equipment	Total	Total Incl O&P
0010	**BATHS** R224000-40									
0100	Tubs, recessed porcelain enamel on cast iron, with trim									
0180	48" x 42"	Q-1	4	4	Ea.	1,800	176		1,976	2,250
0220	72" x 36"	"	3	5.333	"	1,850	234		2,084	2,425
0300	Mat bottom									
0380	5' long	Q-1	4.40	3.636	Ea.	810	160		970	1,125
0560	Corner 48" x 44"		4.40	3.636		1,800	160		1,960	2,225
2000	Enameled formed steel, 4'-6" long	↓	5.80	2.759	↓	365	121		486	590
4600	Module tub & showerwall surround, molded fiberglass									
4610	5' long x 34" wide x 76" high	Q-1	4	4	Ea.	560	176		736	885
4750	Handicap with 1-1/2" OD grab bar, antiskid bottom									
4760	60" x 32-3/4" x 72" high	Q-1	4	4	Ea.	670	176		846	1,000
4770	60" x 30" x 71" high with molded seat		3.50	4.571		655	201		856	1,025
9600	Rough-in, supply, waste and vent, for all above tubs, add		2.07	7.729	↓	300	340		640	855
9900	Minimum labor/equipment charge	↓	3	5.333	Job		234		234	365

22 41 23 – Residential Shower Receptors and Basins

22 41 23.20 Showers

		Crew	Daily Output	Labor-Hours	Unit	Material	2009 Bare Costs Labor	Equipment	Total	Total Incl O&P
0011	**SHOWERS**, Stall, with drain only									
0020										
1520	32" square	Q-1	5	3.200	Ea.	365	140		505	620
1530	36" square		4.80	3.333		870	146		1,016	1,175
1540	Terrazzo receptor, 32" square		5	3.200		740	140		880	1,025
1580	36" corner angle		4.80	3.333		1,275	146		1,421	1,625
3000	Fiberglass, one piece, with 3 walls, 32" x 32" square		5.50	2.909		460	128		588	710
3100	36" x 36" square		5.50	2.909		525	128		653	775
4200	Rough-in, supply, waste and vent for above showers	↓	2.05	7.805	↓	395	340		735	965

22 41 23.40 Shower System Components

		Crew	Daily Output	Labor-Hours	Unit	Material	2009 Bare Costs Labor	Equipment	Total	Total Incl O&P
0010	**SHOWER SYSTEM COMPONENTS**									
5500	Head, water economizer [G]	1 Plum	24	.333	Ea.	49.50	16.25		65.75	79.50

22 41 36 – Residential Laundry Trays

22 41 36.10 Laundry Sinks

		Crew	Daily Output	Labor-Hours	Unit	Material	2009 Bare Costs Labor	Equipment	Total	Total Incl O&P
0010	**LAUNDRY SINKS**, With trim									
0020	Porcelain enamel on cast iron, black iron frame									
0050	24" x 21", single compartment	Q-1	6	2.667	Ea.	415	117		532	635
0100	26" x 21", single compartment	"	6	2.667	"	405	117		522	625
2000	Molded stone, on wall hanger or legs									
2020	22" x 23", single compartment	Q-1	6	2.667	Ea.	143	117		260	340
2100	45" x 21", double compartment	"	5	3.200	"	213	140		353	455
3000	Plastic, on wall hanger or legs									
3020	18" x 23", single compartment	Q-1	6.50	2.462	Ea.	99.50	108		207.50	278
3100	20" x 24", single compartment		6.50	2.462		133	108		241	315
3200	36" x 23", double compartment		5.50	2.909		158	128		286	370
3300	40" x 24", double compartment		5.50	2.909		236	128		364	455
5000	Stainless steel, counter top, 22" x 17" single compartment		6	2.667		58	117		175	245
5200	33" x 22", double compartment		5	3.200		70	140		210	295
9600	Rough-in, supply, waste and vent, for all laundry sinks	↓	2.14	7.477	↓	266	330		596	805

22 41 Residential Plumbing Fixtures

22 41 36 – Residential Laundry Trays

		Crew	Daily Output	Labor-Hours	Unit	Material	2009 Bare Costs Labor	Equipment	Total	Total Incl O&P
22 41 36.10	**Laundry Sinks**									
9810	Minimum labor/equipment charge	1 Plum	3	2.667	Job		130		130	202

22 41 39 – Residential Faucets, Supplies and Trim

22 41 39.10 Faucets and Fittings

		Crew	Daily Output	Labor-Hours	Unit	Material	2009 Bare Costs Labor	Equipment	Total	Total Incl O&P
0010	**FAUCETS AND FITTINGS**									
0150	Bath, faucets, diverter spout combination, sweat	1 Plum	8	1	Ea.	104	49		153	190
0200	For integral stops, IPS unions, add					109			109	120
0420	Bath, press-bal mix valve w/diverter, spout, shower hd, arm/flange	1 Plum	8	1		137	49		186	227
0500	Drain, central lift, 1-1/2" IPS male		20	.400		40.50	19.50		60	74.50
0600	Trip lever, 1-1/2" IPS male		20	.400		41	19.50		60.50	75
0810	Bidet									
0812	Fitting, over the rim, swivel spray/pop-up drain	1 Plum	8	1	Ea.	175	49		224	268
1000	Kitchen sink faucets, top mount, cast spout		10	.800		55	39		94	121
1100	For spray, add		24	.333		16.15	16.25		32.40	43
1200	Wall type, swing tube spout		10	.800		74.50	39		113.50	142
1300	Single control lever handle									
1310	With pull out spray									
1320	Polished chrome	1 Plum	10	.800	Ea.	192	39		231	272
1330	Polished brass		10	.800		230	39		269	315
1340	White		10	.800		211	39		250	293
1348	With spray thru escutcheon									
1350	Polished chrome	1 Plum	10	.800	Ea.	152	39		191	228
1360	Polished brass		10	.800		183	39		222	262
1370	White		10	.800		164	39		203	241
2000	Laundry faucets, shelf type, IPS or copper unions		12	.667		45.50	32.50		78	101
2100	Lavatory faucet, centerset, without drain		10	.800		44.50	39		83.50	110
2120	With pop-up drain		6.66	1.201		61.50	58.50		120	159
2210	Porcelain cross handles and pop-up drain									
2220	Polished chrome	1 Plum	6.66	1.201	Ea.	126	58.50		184.50	230
2230	Polished brass	"	6.66	1.201	"	189	58.50		247.50	299
2260	Single lever handle and pop-up drain									
2270	Black nickel	1 Plum	6.66	1.201	Ea.	220	58.50		278.50	335
2280	Polished brass		6.66	1.201		220	58.50		278.50	335
2290	Polished chrome		6.66	1.201		169	58.50		227.50	277
2800	Self-closing, center set		10	.800		131	39		170	205
2810	Automatic sensor and operator, with faucet head [G]		6.15	1.301		355	63.50		418.50	490
4000	Shower by-pass valve with union		18	.444		68.50	21.50		90	109
4100	Shower arm with flange and head		22	.364		69	17.75		86.75	104
4140	Shower, hand held, pin mount, massage action, chrome		22	.364		69	17.75		86.75	104
4142	Polished brass		22	.364		97	17.75		114.75	135
4144	Shower, hand held, wall mtd, adj. spray, 2 wall mounts, chrome		20	.400		119	19.50		138.50	160
4146	Polished brass		20	.400		242	19.50		261.50	296
4148	Shower, hand held head, bar mounted 24", adj. spray, chrome		20	.400		158	19.50		177.50	203
4150	Polished brass		20	.400		385	19.50		404.50	450
4200	Shower thermostatic mixing valve, concealed		8	1		320	49		369	425
4204	For inlet strainer, check, and stops, add					45			45	49.50
4220	Shower pressure balancing mixing valve,									
4230	With shower head, arm, flange and diverter tub spout									
4240	Chrome	1 Plum	6.14	1.303	Ea.	170	63.50		233.50	286
4250	Polished brass		6.14	1.303		238	63.50		301.50	360
4260	Satin		6.14	1.303		238	63.50		301.50	360
4270	Polished chrome/brass		6.14	1.303		196	63.50		259.50	315
5000	Sillcock, compact, brass, IPS or copper to hose		24	.333		8.05	16.25		24.30	34

359

22 41 Residential Plumbing Fixtures

22 41 39 – Residential Faucets, Supplies and Trim

22 41 39.10 Faucets and Fittings

	22 41 39.10 Faucets and Fittings	Crew	Daily Output	Labor-Hours	Unit	Material	2009 Bare Costs Labor	Equipment	Total	Total Incl O&P
6000	Stop and waste valves, bronze									
6100	Angle, solder end 1/2"	1 Plum	24	.333	Ea.	7.05	16.25		23.30	33
6110	3/4"		20	.400		7.75	19.50		27.25	38.50
6300	Straightway, solder end 3/8"		24	.333		6.15	16.25		22.40	32
6310	1/2"		24	.333		2.81	16.25		19.06	28
6330	1"		19	.421		19.95	20.50		40.45	54
6400	Straightway, threaded 3/8"		24	.333		7.15	16.25		23.40	33
6410	1/2"		24	.333		3.82	16.25		20.07	29
6420	3/4"		20	.400		6.35	19.50		25.85	37
6430	1"		19	.421		12.60	20.50		33.10	46
7800	Water closet, wax gasket		96	.083		1.42	4.06		5.48	7.85
7820	Gasket toilet tank to bowl	↓	32	.250	↓	1.24	12.20		13.44	20.50
8000	Water supply stops, polished chrome plate									
8200	Angle, 3/8"	1 Plum	24	.333	Ea.	7.35	16.25		23.60	33
8300	1/2"		22	.364		7.35	17.75		25.10	35.50
8400	Straight, 3/8"		26	.308		7.75	15		22.75	32
8500	1/2"		24	.333		7.75	16.25		24	33.50
8600	Water closet, angle, w/flex riser, 3/8"		24	.333	↓	28	16.25		44.25	56
9000	Minimum labor/equipment charge	↓	4	2	Job		97.50		97.50	151

22 42 Commercial Plumbing Fixtures

22 42 13 – Commercial Water Closets, Urinals, and Bidets

22 42 13.30 Urinals

		Crew	Daily Output	Labor-Hours	Unit	Material	2009 Bare Costs Labor	Equipment	Total	Total Incl O&P
0010	**URINALS**									
3000	Wall hung, vitreous china, with hanger & self-closing valve									
3100	Siphon jet type	Q-1	3	5.333	Ea.	248	234		482	640
3120	Blowout type		3	5.333		385	234		619	790
3300	Rough-in, supply, waste & vent		2.83	5.654		240	248		488	650
5000	Stall type, vitreous china, includes valve		2.50	6.400		575	281		856	1,075
6980	Rough-in, supply, waste and vent		1.99	8.040	↓	350	355		705	930
9000	Minimum labor/equipment charge	↓	4	4	Job		176		176	272

22 42 13.40 Water Closets

		Crew	Daily Output	Labor-Hours	Unit	Material	2009 Bare Costs Labor	Equipment	Total	Total Incl O&P
0010	**WATER CLOSETS**									
3000	Bowl only, with flush valve, seat									
3100	Wall hung	Q-1	5.80	2.759	Ea.	284	121		405	505
3200	For rough-in, supply, waste and vent, single WC		2.56	6.250	"	680	274		954	1,175
9000	Minimum labor/equipment charge	↓	4	4	Job		176		176	272

22 42 16 – Commercial Lavatories and Sinks

22 42 16.10 Handwasher-Dryer Module

		Crew	Daily Output	Labor-Hours	Unit	Material	2009 Bare Costs Labor	Equipment	Total	Total Incl O&P
0010	**HANDWASHER-DRYER MODULE**									
0030	Wall mounted									
0040	With electric dryer									
0050	Sensor operated	Q-1	8	2	Ea.	3,650	88		3,738	4,125
0110	Sensor operated (ADA)		8	2		3,650	88		3,738	4,150
0140	Sensor operated (ADA), surface mounted	↓	8	2	↓	4,400	88		4,488	4,975
0150	With paper towels									
0180	Sensor operated	Q-1	8	2	Ea.	3,550	88		3,638	4,025

22 42 16.14 Lavatories

0010	**LAVATORIES**, With trim, white unless noted otherwise									
0020	Commercial lavatories same as residential. See Div. 22 41 16.10									

22 42 Commercial Plumbing Fixtures

22 42 16 – Commercial Lavatories and Sinks

22 42 16.40 Service Sinks

	22 42 16.40 Service Sinks	Crew	Daily Output	Labor-Hours	Unit	Material	2009 Bare Costs Labor	Equipment	Total	Total Incl O&P
0010	**SERVICE SINKS**									
6650	Service, floor, corner, P.E. on C.I., 28" x 28"	Q-1	4.40	3.636	Ea.	625	160		785	935
6790	For rough-in, supply, waste & vent, floor service sinks		1.64	9.756	"	705	430		1,135	1,450
9000	Minimum labor/equipment charge		4	4	Job		176		176	272

22 42 23 – Commercial Shower Receptors and Basins

22 42 23.30 Group Showers

		Crew	Daily Output	Labor-Hours	Unit	Material	Labor	Equipment	Total	Total Incl O&P
0010	**GROUP SHOWERS**									
9000	Minimum labor/equipment charge	Q-1	4	4	Job		176		176	272

22 42 33 – Wash Fountains

22 42 33.20 Commercial Wash Fountains

		Crew	Daily Output	Labor-Hours	Unit	Material	Labor	Equipment	Total	Total Incl O&P
0010	**COMMERCIAL WASH FOUNTAINS**									
1900	Group, foot control									
2000	Precast terrazzo, circular, 36" diam., 5 or 6 persons	Q-2	3	8	Ea.	4,075	365		4,440	5,050
2100	54" diameter for 8 or 10 persons		2.50	9.600		5,050	435		5,485	6,250
2400	Semi-circular, 36" diam. for 3 persons		3	8		3,750	365		4,115	4,700
2500	54" diam. for 4 or 5 persons		2.50	9.600		4,525	435		4,960	5,650
5610	Group, infrared control, barrier free									
5614	Precast terrazzo									
5620	Semi-circular 36" diam. for 3 persons	Q-2	3	8	Ea.	5,725	365		6,090	6,850
5630	46" diam. for 4 persons		2.80	8.571		6,225	390		6,615	7,450
5640	Circular, 54" diam. for 8 persons, button control		2.50	9.600		7,475	435		7,910	8,900
5700	Rough-in, supply, waste and vent for above wash fountains	Q-1	1.82	8.791		410	385		795	1,050
9000	Minimum labor/equipment charge	Q-2	3	8	Job		365		365	565

22 42 39 – Commercial Faucets, Supplies, and Trim

22 42 39.10 Faucets and Fittings

		Crew	Daily Output	Labor-Hours	Unit	Material	Labor	Equipment	Total	Total Incl O&P
0010	**FAUCETS AND FITTINGS**									
0840	Flush valves, with vacuum breaker									
0850	Water closet									
0860	Exposed, rear spud	1 Plum	8	1	Ea.	130	49		179	219
0870	Top spud		8	1		120	49		169	208
0880	Concealed, rear spud		8	1		171	49		220	265
0890	Top spud		8	1		141	49		190	231
0900	Wall hung		8	1		151	49		200	242
0920	Urinal									
0930	Exposed, stall	1 Plum	8	1	Ea.	120	49		169	208
0940	Wall, (washout)		8	1		120	49		169	208
0950	Pedestal, top spud		8	1		122	49		171	211
0960	Concealed, stall		8	1		138	49		187	227
0970	Wall (washout)		8	1		148	49		197	239
0971	Automatic flush sensor and operator for									
0972	urinals or water closets [G]	1 Plum	8	1	Ea.	405	49		454	520
3000	Service sink faucet, cast spout, pail hook, hose end	"	14	.571	"	80	28		108	131

22 42 39.30 Carriers and Supports

		Crew	Daily Output	Labor-Hours	Unit	Material	Labor	Equipment	Total	Total Incl O&P
0010	**CARRIERS AND SUPPORTS**, For plumbing fixtures									
0020										
3000	Lavatory, concealed arm									
3050	Floor mounted, single									
3100	High back fixture	1 Plum	6	1.333	Ea.	350	65		415	485
3200	Flat slab fixture		6	1.333		296	65		361	425
3220	Paraplegic		6	1.333		355	65		420	490

22 42 Commercial Plumbing Fixtures

22 42 39 – Commercial Faucets, Supplies, and Trim

22 42 39.30 Carriers and Supports	Crew	Daily Output	Labor-Hours	Unit	Material	2009 Bare Costs Labor	Equipment	Total	Total Incl O&P
6980	Water closet, siphon jet									
7000	Horizontal, adjustable, caulk								
7040	Single, 4" pipe size	1 Plum	5.33	1.501	Ea.	500	73		573	665
7060	5" pipe size		5.33	1.501		855	73		928	1,050
7100	Double, 4" pipe size		5	1.600		930	78		1,008	1,150
7120	5" pipe size	↓	5	1.600	↓	1,125	78		1,203	1,375
8200	Water closet, residential								
8220	Vertical centerline, floor mount								
8240	Single, 3" caulk, 2" or 3" vent	1 Plum	6	1.333	Ea.	405	65		470	545
8260	4" caulk, 2" or 4" vent		6	1.333	"	520	65		585	675
9990	Minimum labor/equipment charge	↓	3.50	2.286	Job		111		111	173

22 43 Healthcare Plumbing Fixtures

22 43 39 – Healthcare Faucets, Supplies, and Trim

22 43 39.10 Faucets and Fittings

		Crew	Daily Output	Labor-Hours	Unit	Material	2009 Bare Costs Labor	Equipment	Total	Total Incl O&P
0010	**FAUCETS AND FITTINGS**									
2850	Medical, bedpan cleanser, with pedal valve,	1 Plum	12	.667	Ea.	600	32.50		632.50	710
2860	With screwdriver stop valve		12	.667		310	32.50		342.50	395
2870	With self-closing spray valve		12	.667		234	32.50		266.50	310
2900	Faucet, gooseneck spout, wrist handles, grid drain		10	.800		124	39		163	198
2940	Mixing valve, knee action, screwdriver stops	↓	4	2	↓	485	97.50		582.50	685

22 45 Emergency Plumbing Fixtures

22 45 13 – Emergency Showers

22 45 13.10 Emergency Showers

		Crew	Daily Output	Labor-Hours	Unit	Material	2009 Bare Costs Labor	Equipment	Total	Total Incl O&P
0010	**EMERGENCY SHOWERS**, Rough-in not included									
0020										
5000	Shower, single head, drench, ball valve, pull, freestanding	Q-1	4	4	Ea.	320	176		496	625
5200	Horizontal or vertical supply		4	4		460	176		636	775
6000	Multi-nozzle, eye/face wash combination		4	4		590	176		766	920
6400	Multi-nozzle, 12 spray, shower only		4	4		1,600	176		1,776	2,025
6600	For freeze-proof, add		6	2.667	↓	335	117		452	545
9000	Minimum labor/equipment charge	↓	3	5.333	Job		234		234	365

22 45 16 – Eyewash Equipment

22 45 16.10 Eyewash Safety Equipment

		Crew	Daily Output	Labor-Hours	Unit	Material	2009 Bare Costs Labor	Equipment	Total	Total Incl O&P
0010	**EYEWASH SAFETY EQUIPMENT**, Rough-in not included									
1000	Eye wash fountain									
1400	Plastic bowl, pedestal mounted	Q-1	4	4	Ea.	232	176		408	525

22 47 Drinking Fountains and Water Coolers

22 47 13 – Drinking Fountains

22 47 13.10 Drinking Water Fountains

	22 47 13.10 Drinking Water Fountains	Crew	Daily Output	Labor-Hours	Unit	Material	2009 Bare Costs Labor	2009 Bare Costs Equipment	Total	Total Incl O&P
0010	**DRINKING WATER FOUNTAINS,** For connection to cold water supply									
1000	Wall mounted, non-recessed									
2700	Stainless steel, single bubbler, no back	1 Plum	4	2	Ea.	1,200	97.50		1,297.50	1,475
2740	With back		4	2		2,000	97.50		2,097.50	2,350
2780	Dual handle & wheelchair projection type		4	2		625	97.50		722.50	835
2820	Dual level for handicapped type		3.20	2.500		1,275	122		1,397	1,600
3980	For rough-in, supply and waste, add		2.21	3.620		171	176		347	465
4000	Wall mounted, semi-recessed									
4200	Poly-marble, single bubbler	1 Plum	4	2	Ea.	780	97.50		877.50	1,000
4600	Stainless steel, satin finish, single bubbler	"	4	2	"	905	97.50		1,002.50	1,150
6000	Wall mounted, fully recessed									
6400	Poly-marble, single bubbler	1 Plum	4	2	Ea.	890	97.50		987.50	1,125
6800	Stainless steel, single bubbler		4	2		875	97.50		972.50	1,100
7580	For rough-in, supply and waste, add		1.83	4.372		171	213		384	520
7590										
7600	Floor mounted, pedestal type									
8600	Enameled iron, heavy duty service, 2 bubblers	1 Plum	2	4	Ea.	1,275	195		1,470	1,700
8880	For freeze-proof valve system, add		2	4		385	195		580	725
8900	For rough-in, supply and waste, add		1.83	4.372		171	213		384	520
9000	Minimum labor/equipment charge		2	4	Job		195		195	300

22 47 16 – Pressure Water Coolers

22 47 16.10 Electric Water Coolers

	22 47 16.10 Electric Water Coolers	Crew	Daily Output	Labor-Hours	Unit	Material	2009 Bare Costs Labor	2009 Bare Costs Equipment	Total	Total Incl O&P
0010	**ELECTRIC WATER COOLERS**									
0100	Wall mounted, non-recessed									
0140	4 GPH	Q-1	4	4	Ea.	575	176		751	900
0160	8 GPH, barrier free, sensor operated		4	4		725	176		901	1,075
1000	Dual height, 8.2 GPH		3.80	4.211		930	185		1,115	1,300
1040	14.3 GPH		3.80	4.211		985	185		1,170	1,350
3300	Semi-recessed, 8.1 GPH		4	4		965	176		1,141	1,325
4600	Floor mounted, flush-to-wall									
4640	4 GPH	1 Plum	3	2.667	Ea.	600	130		730	860
4980	For stainless steel cabinet, add					128			128	141
5000	Dual height, 8.2 GPH	1 Plum	2	4		1,000	195		1,195	1,400
9000	Minimum labor/equipment charge	"	2	4	Job		195		195	300

22 66 Chemical-Waste Systems for Lab. and Healthcare Facilities

22 66 53 – Laboratory Chemical-Waste and Vent Piping

22 66 53.30 Glass Pipe

	22 66 53.30 Glass Pipe	Crew	Daily Output	Labor-Hours	Unit	Material	2009 Bare Costs Labor	2009 Bare Costs Equipment	Total	Total Incl O&P
0010	**GLASS PIPE,** Borosilicate, couplings & clevis hanger assemblies, 10' O.C.									
0020	Drainage									
1100	1-1/2" diameter	Q-1	52	.308	L.F.	12.55	13.50		26.05	35
1120	2" diameter		44	.364		16.65	15.95		32.60	43.50
1140	3" diameter		39	.410		22.50	18		40.50	52.50
1160	4" diameter		30	.533		40.50	23.50		64	81
1180	6" diameter		26	.615		74	27		101	124
9000	Minimum labor/equipment charge	1 Plum	4	2	Job		97.50		97.50	151

22 66 53.40 Pipe Fittings, Glass

	22 66 53.40 Pipe Fittings, Glass									
0010	**PIPE FITTINGS, GLASS**									
0020	Drainage, beaded ends									
0040	Coupling & labor required at joints not incl. in fitting									

22 66 53 – Laboratory Chemical-Waste and Vent Piping

22 66 53.40 Pipe Fittings, Glass

		Crew	Daily Output	Labor-Hours	Unit	Material	Labor	Equipment	Total	Total Incl O&P
0050	price. Add 1 per joint for installed price									
0070	90° Bend or sweep, 1-1/2"				Ea.	27.50			27.50	30.50
0090	2"					35			35	38.50
0100	3"					57.50			57.50	63.50
0110	4"					92.50			92.50	102
0120	6" (sweep only)					281			281	310
0350	Tee, single sanitary, 1-1/2"					45			45	49
0370	2"					45			45	49
0380	3"					67			67	73.50
0390	4"					120			120	132
0400	6"					320			320	355
0500	Coupling, stainless steel, TFE seal ring									
0520	1-1/2"	Q-1	32	.500	Ea.	19.35	22		41.35	55.50
0530	2"		30	.533		24.50	23.50		48	63.50
0540	3"		25	.640		33	28		61	80
0550	4"		23	.696		56.50	30.50		87	110
0560	6"		20	.800		127	35		162	195
9000	Minimum labor/equipment charge	1 Plum	4	2	Job		97.50		97.50	151

22 66 53.60 Corrosion Resistant Pipe

		Crew	Daily Output	Labor-Hours	Unit	Material	Labor	Equipment	Total	Total Incl O&P
0010	**CORROSION RESISTANT PIPE**, No couplings or hangers									
0020	Iron alloy, drain, mechanical joint									
1000	1-1/2" diameter	Q-1	70	.229	L.F.	34.50	10.05		44.55	53.50
1100	2" diameter		66	.242		35.50	10.65		46.15	55.50
1120	3" diameter		60	.267		45.50	11.70		57.20	68
1140	4" diameter		52	.308		58.50	13.50		72	85.50
2980	Plastic, epoxy, fiberglass filament wound, B&S joint									
3000	2" diameter	Q-1	62	.258	L.F.	9.40	11.30		20.70	28
3100	3" diameter		51	.314		11.10	13.75		24.85	34
3120	4" diameter		45	.356		16.10	15.60		31.70	41.50
3160	8" diameter	Q-2	38	.632		35.50	28.50		64	83.50
3200	12" diameter	"	28	.857		59	39		98	126
9800	Minimum labor/equipment charge	1 Plum	4	2	Job		97.50		97.50	151

22 66 53.70 Pipe Fittings, Corrosion Resistant

		Crew	Daily Output	Labor-Hours	Unit	Material	Labor	Equipment	Total	Total Incl O&P
0010	**PIPE FITTINGS, CORROSION RESISTANT**									
0030	Iron alloy									
0050	Mechanical joint									
0060	1/4 Bend, 1-1/2"	Q-1	12	1.333	Ea.	47	58.50		105.50	142
0080	2"		10	1.600		76.50	70		146.50	193
0090	3"		9	1.778		92	78		170	222
0100	4"		8	2		105	88		193	252
0160	Tee and Y, sanitary, straight									
0170	1-1/2"	Q-1	8	2	Ea.	51	88		139	192
0180	2"		7	2.286		68	100		168	231
0190	3"		6	2.667		105	117		222	297
0200	4"		5	3.200		194	140		334	430
0360	Coupling, 1-1/2"		14	1.143		35.50	50		85.50	118
0380	2"		12	1.333		41	58.50		99.50	136
0390	3"		11	1.455		42.50	64		106.50	146
0400	4"		10	1.600		48.50	70		118.50	163
3000	Epoxy, filament wound									
3030	Quick-lock joint									
3040	90° Elbow, 2"	Q-1	28	.571	Ea.	79.50	25		104.50	127

22 66 53 – Laboratory Chemical-Waste and Vent Piping

22 66 53.70 Pipe Fittings, Corrosion Resistant		Crew	Daily Output	Labor-Hours	Unit	Material	2009 Bare Costs Labor	Equipment	Total	Total Incl O&P
3060	3"	Q-1	16	1	Ea.	91.50	44		135.50	169
3070	4"		13	1.231		125	54		179	222
3190	Tee, 2"		19	.842		190	37		227	267
3200	3"		11	1.455		230	64		294	350
3210	4"		9	1.778		276	78		354	425
9000	Minimum labor/equipment charge	1 Plum	4	2	Job		97.50		97.50	151

Division Notes

	CREW	DAILY OUTPUT	LABOR-HOURS	UNIT	2009 BARE COSTS				TOTAL INCL O&P
					MAT.	LABOR	EQUIP.	TOTAL	

Estimating Tips

The labor adjustment factors listed in Subdivision 22 01 02.20 also apply to Division 23.

23 10 00 Facility Fuel Systems

- The prices in this subdivision for above- and below-ground storage tanks do not include foundations or hold-down slabs. The estimator should refer to Divisions 3 and 31 for foundation system pricing. In addition to the foundations, required tank accessories, such as tank gauges, leak detection devices, and additional manholes and piping, must be added to the tank prices.

23 50 00 Central Heating Equipment

- When estimating the cost of an HVAC system, check to see who is responsible for providing and installing the temperature control system. It is possible to overlook controls, assuming that they would be included in the electrical estimate.

- When looking up a boiler, be careful on specified capacity. Some manufacturers rate their products on output while others use input.

- Include HVAC insulation for pipe, boiler, and duct (wrap and liner).

- Be careful when looking up mechanical items to get the correct pressure rating and connection type (thread, weld, flange).

23 70 00 Central HVAC Equipment

- Combination heating and cooling units are sized by the air conditioning requirements. (See Reference No. R236000-20 for preliminary sizing guide.)

- A ton of air conditioning is nominally 400 CFM.

- Rectangular duct is taken off by the linear foot for each size, but its cost is usually estimated by the pound. Remember that SMACNA standards now base duct on internal pressure.

- Prefabricated duct is estimated and purchased like pipe: straight sections and fittings.

- Note that cranes or other lifting equipment are not included on any lines in Division 23. For example, if a crane is required to lift a heavy piece of pipe into place high above a gym floor, or to put a rooftop unit on the roof of a four-story building, etc., it must be added. Due to the potential for extreme variation—from nothing additional required to a major crane or helicopter—we feel that including a nominal amount for "lifting contingency" would be useless and detract from the accuracy of the estimate. When using equipment rental cost data from RSMeans, do not forget to include the cost of the operator(s).

Reference Numbers

Reference numbers are shown in shaded boxes at the beginning of some major classifications. These numbers refer to related items in the Reference Section. The reference information may be an estimating procedure, an alternate pricing method, or technical information.

Note: Not all subdivisions listed here necessarily appear in this publication.

Note: **Trade Service,** *in part, has been used as a reference source for some of the material prices used in Division 23.*

23 05 05 – Selective HVAC Demolition

23 05 05.10 HVAC Demolition

		Crew	Daily Output	Labor-Hours	Unit	Material	2009 Bare Costs Labor	2009 Bare Costs Equipment	Total	Total Incl O&P
0010	**HVAC DEMOLITION** R220105-10									
0100	Air conditioner, split unit, 3 ton	Q-5	2	8	Ea.		355		355	550
0150	Package unit, 3 ton R024119-10	Q-6	3	8	"		370		370	570
0298	Boilers									
0300	Electric, up thru 148 kW	Q-19	2	12	Ea.		545		545	840
0310	150 thru 518 kW	"	1	24			1,075		1,075	1,675
0320	550 thru 2000 kW	Q-21	.40	80			3,700		3,700	5,725
0330	2070 kW and up	"	.30	106			4,950		4,950	7,650
0340	Gas and/or oil, up thru 150 MBH	Q-7	2.20	14.545			685		685	1,050
0350	160 thru 2000 MBH		.80	40			1,875		1,875	2,925
0360	2100 thru 4500 MBH		.50	64			3,000		3,000	4,675
0370	4600 thru 7000 MBH		.30	106			5,025		5,025	7,775
0380	7100 thru 12,000 MBH		.16	200			9,400		9,400	14,600
0390	12,200 thru 25,000 MBH		.12	266			12,500		12,500	19,400
1000	Ductwork, 4" high, 8" wide	1 Clab	200	.040	L.F.		1.26		1.26	2.09
1020	10" wide		190	.042			1.33		1.33	2.20
1040	14" wide		180	.044			1.40		1.40	2.32
1100	6" high, 8" wide		165	.048			1.53		1.53	2.53
1120	12" wide		150	.053			1.69		1.69	2.78
1140	18" wide		135	.059			1.87		1.87	3.09
1200	10" high, 12" wide		125	.064			2.02		2.02	3.34
1220	18" wide		115	.070			2.20		2.20	3.63
1240	24" wide		110	.073			2.30		2.30	3.79
1300	12"-14" high, 16"-18" wide		85	.094			2.97		2.97	4.91
1320	24" wide		75	.107			3.37		3.37	5.55
1340	48" wide		71	.113			3.56		3.56	5.90
1400	18" high, 24" wide		67	.119			3.77		3.77	6.25
1420	36" wide		63	.127			4.01		4.01	6.60
1440	48" wide		59	.136			4.28		4.28	7.05
1500	30" high, 36" wide		56	.143			4.51		4.51	7.45
1520	48" wide		53	.151			4.77		4.77	7.85
1540	72" wide		50	.160			5.05		5.05	8.35
1550	Duct heater, electric strip	1 Elec	8	1	Ea.		47		47	72.50
1850	Minimum labor/equipment charge	1 Clab	3	2.667	Job		84.50		84.50	139
2200	Furnace, electric	Q-20	2	10	Ea.		435		435	685
2300	Gas or oil, under 120 MBH	Q-9	4	4			170		170	270
2340	Over 120 MBH	"	3	5.333			227		227	360
2800	Heat pump, package unit, 3 ton	Q-5	2.40	6.667			296		296	460
2840	Split unit, 3 ton		2	8			355		355	550
3000	Mechanical equipment, light items. Unit is weight, not cooling.		.90	17.778	Ton		790		790	1,225
3600	Heavy items		1.10	14.545	"		645		645	1,000
9000	Minimum labor/equipment charge	Q-6	3	8	Job		370		370	570

23 05 23 – General-Duty Valves for HVAC Piping

23 05 23.30 Valves, Iron Body

		Crew	Daily Output	Labor-Hours	Unit	Material	2009 Bare Costs Labor	2009 Bare Costs Equipment	Total	Total Incl O&P
0010	**VALVES, IRON BODY**									
1020	Butterfly, wafer type, gear actuator, 200 lb.									
1030	2"	1 Plum	14	.571	Ea.	155	28		183	213
1060	4"	Q-1	5	3.200	"	188	140		328	425
1650	Gate, 125 lb., N.R.S.									
2150	Flanged									
2240	2-1/2"	Q-1	5	3.200	Ea.	510	140		650	780
2260	3"		4.50	3.556		570	156		726	865

23 05 Common Work Results for HVAC

23 05 23 – General-Duty Valves for HVAC Piping

23 05 23.30 Valves, Iron Body

	23 05 23.30 Valves, Iron Body	Crew	Daily Output	Labor-Hours	Unit	Material	2009 Bare Costs Labor	Equipment	Total	Total Incl O&P
2280	4"	Q-1	3	5.333	Ea.	815	234		1,049	1,250
3550	OS&Y, 125 lb., flanged									
3680	4"	Q-1	3	5.333	Ea.	510	234		744	925
3690	5"	Q-2	3.40	7.059		840	320		1,160	1,425
3700	6"	"	3	8	↓	840	365		1,205	1,500
9000	Minimum labor/equipment charge	1 Plum	3	2.667	Job		130		130	202

23 07 HVAC Insulation

23 07 13 – Duct Insulation

23 07 13.10 Duct Thermal Insulation

			Crew	Daily Output	Labor-Hours	Unit	Material	2009 Bare Costs Labor	Equipment	Total	Total Incl O&P
0010	**DUCT THERMAL INSULATION**										
0100	Rule of thumb, as a percentage of total mechanical costs					Job				10%	
0110	Insulation req'd is based on the surface size/area to be covered										
3000	Ductwork										
3020	Blanket type, fiberglass, flexible										
3140	FSK vapor barrier wrap, .75 lb. density										
3160	1" thick	G	Q-14	350	.046	S.F.	.17	1.81		1.98	3.12
3170	1-1/2" thick	G		320	.050		.20	1.99		2.19	3.43
3180	2" thick	G		300	.053		.26	2.12		2.38	3.71
3190	3" thick	G		260	.062		.39	2.44		2.83	4.38
3200	4" thick	G		242	.066		.56	2.63		3.19	4.86
3212	Vinyl Jacket, .75 lb. density, 1-1/2" thick	G	↓	320	.050	↓	.20	1.99		2.19	3.43
9600	Minimum labor/equipment charge		1 Stpi	4	2	Job		98.50		98.50	153

23 07 16 – HVAC Equipment Insulation

23 07 16.10 HVAC Equipment Thermal Insulation

			Crew	Daily Output	Labor-Hours	Unit	Material	2009 Bare Costs Labor	Equipment	Total	Total Incl O&P
0010	**HVAC EQUIPMENT THERMAL INSULATION**										
0100	Rule of thumb, as a percentage of total mechanical costs					Job				10%	
0110	Insulation req'd is based on the surface size/area to be covered										

23 09 Instrumentation and Control for HVAC

23 09 33 – Electric and Electronic Control System for HVAC

23 09 33.10 Electronic Control Systems

			Crew	Daily Output	Labor-Hours	Unit	Material	2009 Bare Costs Labor	Equipment	Total	Total Incl O&P
0010	**ELECTRONIC CONTROL SYSTEMS**										
9000	Minimum labor/equipment charge		1 Plum	8	1	Job		49		49	75.50

23 09 53 – Pneumatic and Electric Control System for HVAC

23 09 53.10 Control Components

			Crew	Daily Output	Labor-Hours	Unit	Material	2009 Bare Costs Labor	Equipment	Total	Total Incl O&P
0010	**CONTROL COMPONENTS**										
5000	Thermostats										
5030	Manual		1 Shee	8	1	Ea.	27	47		74	105
5040	1 set back, electric, timed	G		8	1		95	47		142	179
5050	2 set back, electric, timed	G		8	1		209	47		256	305
5200	24 hour, automatic, clock	G	↓	8	1	↓	115	47		162	202
6000	Valves, motorized zone										
6100	Sweat connections, 1/2" C x C		1 Stpi	20	.400	Ea.	140	19.75		159.75	185
6110	3/4" C x C			20	.400		141	19.75		160.75	186
6120	1" C x C		↓	19	.421	↓	183	21		204	234
9000	Minimum labor/equipment charge		1 Plum	4	2	Job		97.50		97.50	151

23 12 Facility Fuel Pumps

23 12 16 – Facility Gasoline Dispensing Pumps

23 12 16.20 Fuel Dispensing Equipment	Crew	Daily Output	Labor-Hours	Unit	Material	2009 Bare Costs Labor	Equipment	Total	Total Incl O&P
0010 **FUEL DISPENSING EQUIPMENT**									
1100 Product dispenser with vapor recovery for 6 nozzles, installed, not									
1110 including piping to storage tanks				Ea.	20,800			20,800	22,900

23 21 Hydronic Piping and Pumps

23 21 20 – Hydronic HVAC Piping Specialties

23 21 20.34 Dielectric Unions

	Crew	Daily Output	Labor-Hours	Unit	Material	Labor	Equipment	Total	Total Incl O&P
0010 **DIELECTRIC UNIONS**, Standard gaskets for water and air.									
0020 250 psi maximum pressure									
0280 Female IPT to sweat, straight									
0340 3/4" pipe size	1 Plum	20	.400	Ea.	4.79	19.50		24.29	35.50
0780 Female IPT to female IPT, straight									
0800 1/2" pipe size	1 Plum	24	.333	Ea.	8.80	16.25		25.05	34.50
0840 3/4" pipe size		20	.400	"	9.90	19.50		29.40	41
9000 Minimum labor/equipment charge		4	2	Job		97.50		97.50	151

23 21 20.46 Expansion Tanks

	Crew	Daily Output	Labor-Hours	Unit	Material	Labor	Equipment	Total	Total Incl O&P
0010 **EXPANSION TANKS**									
2000 Steel, liquid expansion, ASME, painted, 15 gallon capacity	Q-5	17	.941	Ea.	440	42		482	550
2040 30 gallon capacity		12	1.333		495	59		554	635
2080 60 gallon capacity		8	2		695	89		784	905
2120 100 gallon capacity		6	2.667		1,000	118		1,118	1,275
3000 Steel ASME expansion, rubber diaphragm, 19 gal. cap. accept.		12	1.333		1,925	59		1,984	2,225
3020 31 gallon capacity		8	2		2,150	89		2,239	2,525
3040 61 gallon capacity		6	2.667		3,025	118		3,143	3,500
3080 119 gallon capacity		4	4		3,250	178		3,428	3,850
9000 Minimum labor/equipment charge		4	4	Job		178		178	276

23 21 20.70 Steam Traps

	Crew	Daily Output	Labor-Hours	Unit	Material	Labor	Equipment	Total	Total Incl O&P
0010 **STEAM TRAPS**									
0030 Cast iron body, threaded									
0040 Inverted bucket									
0050 1/2" pipe size	1 Stpi	12	.667	Ea.	128	33		161	191
0100 1" pipe size		9	.889		340	44		384	445
0120 1-1/4" pipe size		8	1		515	49.50		564.50	645
1000 Float & thermostatic, 15 psi									
1010 3/4" pipe size	1 Stpi	16	.500	Ea.	103	24.50		127.50	152
1020 1" pipe size		15	.533		124	26.50		150.50	177
1040 1-1/2" pipe size		9	.889		218	44		262	310
1060 2" pipe size		6	1.333		400	66		466	540
9000 Minimum labor/equipment charge		4	2	Job		98.50		98.50	153

23 21 20.76 Strainers, Y Type, Bronze Body

	Crew	Daily Output	Labor-Hours	Unit	Material	Labor	Equipment	Total	Total Incl O&P
0010 **STRAINERS, Y TYPE, BRONZE BODY**									
0050 Screwed, 150 lb., 1/4" pipe size	1 Stpi	24	.333	Ea.	14.55	16.45		31	41.50
0100 1/2" pipe size		20	.400		19.20	19.75		38.95	51.50
0120 3/4" pipe size		19	.421		21.50	21		42.50	56
0140 1" pipe size		17	.471		23	23		46	61
0160 1-1/2" pipe size		14	.571		49.50	28		77.50	98
0180 2" pipe size		13	.615		66	30.50		96.50	120
0182 3" pipe size		12	.667		590	33		623	700
0220 3" pipe size	Q-5	16	1		695	44.50		739.50	835
0240 4" pipe size	"	15	1.067		1,575	47.50		1,622.50	1,825

23 21 Hydronic Piping and Pumps

23 21 20 – Hydronic HVAC Piping Specialties

23 21 20.76 Strainers, Y Type, Bronze Body

		Crew	Daily Output	Labor-Hours	Unit	Material	2009 Bare Costs Labor	Equipment	Total	Total Incl O&P
1000	Flanged, 150 lb., 1-1/2" pipe size	1 Stpi	11	.727	Ea.	365	36		401	455
1020	2" pipe size	"	8	1		440	49.50		489.50	560
1030	2-1/2" pipe size	Q-5	5	3.200		650	142		792	935
1040	3" pipe size		4.50	3.556		805	158		963	1,125
1060	4" pipe size		3	5.333		1,225	237		1,462	1,725
1100	6" pipe size	Q-6	3	8		2,325	370		2,695	3,125
1106	8" pipe size	"	2.60	9.231		2,550	425		2,975	3,475
1500	For 300 lb rating, add					40%				
9000	Minimum labor/equipment charge	1 Stpi	3.75	2.133	Job		105		105	163

23 21 23 – Hydronic Pumps

23 21 23.13 In-Line Centrifugal Hydronic Pumps

		Crew	Daily Output	Labor-Hours	Unit	Material	2009 Bare Costs Labor	Equipment	Total	Total Incl O&P
0010	**IN-LINE CENTRIFUGAL HYDRONIC PUMPS**									
0600	Bronze, sweat connections, 1/40 HP, in line									
0640	3/4" size	Q-1	16	1	Ea.	172	44		216	257
1000	Flange connection, 3/4" to 1-1/2" size									
1040	1/12 HP	Q-1	6	2.667	Ea.	445	117		562	670
1060	1/8 HP		6	2.667		750	117		867	1,000
1100	1/3 HP		6	2.667		855	117		972	1,125
1140	2" size, 1/6 HP		5	3.200		1,100	140		1,240	1,425
1180	2-1/2" size, 1/4 HP		5	3.200		1,375	140		1,515	1,725
1220	3" size, 1/4 HP		4	4		1,425	176		1,601	1,850
9000	Minimum labor/equipment charge		3.25	4.923	Job		216		216	335

23 31 HVAC Ducts and Casings

23 31 13 – Metal Ducts

23 31 13.13 Rectangular Metal Ducts

		Crew	Daily Output	Labor-Hours	Unit	Material	2009 Bare Costs Labor	Equipment	Total	Total Incl O&P
0010	**RECTANGULAR METAL DUCTS** R233100-20									
0020	Fabricated rectangular, includes fittings, joints, supports,									
0100	Aluminum, alloy 3003-H14, under 100 lb.	Q-10	75	.320	Lb.	3.72	14.10		17.82	26.50
0110	100 to 500 lb.		80	.300		2.48	13.20		15.68	23.50
0120	500 to 1,000 lb.		95	.253		2.31	11.15		13.46	20
0500	Galvanized steel, under 200 lb.		235	.102		1.12	4.50		5.62	8.40
0520	200 to 500 lb.		245	.098		.99	4.32		5.31	7.95
0540	500 to 1,000 lb.		255	.094		.97	4.15		5.12	7.65
0560	1,000 to 2,000 lb.		265	.091		.97	3.99		4.96	7.40
0580	Over 5,000 lb.		285	.084		.93	3.71		4.64	6.90
0590										
1000	Stainless steel, type 304, under 100 lb.	Q-10	165	.145	Lb.	4.35	6.40		10.75	15
1020	100 to 500 lb.		175	.137		4.02	6.05		10.07	14
1030	500 to 1,000 lb.		190	.126		3.68	5.55		9.23	12.90

23 31 13.16 Round and Flat-Oval Spiral Ducts

		Crew	Daily Output	Labor-Hours	Unit	Material	2009 Bare Costs Labor	Equipment	Total	Total Incl O&P
0010	**ROUND AND FLAT-OVAL SPIRAL DUCTS**									
9990	Minimum labor/equipment charge	1 Shee	3	2.667	Job		126		126	200

23 31 13.19 Metal Duct Fittings

		Crew	Daily Output	Labor-Hours	Unit	Material	2009 Bare Costs Labor	Equipment	Total	Total Incl O&P
0010	**METAL DUCT FITTINGS**									
0050	Air extractors, 12" x 4"	1 Shee	24	.333	Ea.	15.65	15.75		31.40	42
0100	8" x 6"		22	.364		15.65	17.15		32.80	44.50
0200	20" x 8"		16	.500		35.50	23.50		59	76.50
0240	18" x 10"		14	.571		34.50	27		61.50	81
0280	24" x 12"		10	.800		48	38		86	113

23 31 HVAC Ducts and Casings

23 31 16 – Nonmetal Ducts

23 31 16.13 Fibrous-Glass Ducts

	23 31 16.13 Fibrous-Glass Ducts	Crew	Daily Output	Labor-Hours	Unit	Material	2009 Bare Costs Labor	2009 Bare Costs Equipment	Total	Total Incl O&P
0010	**FIBROUS-GLASS DUCTS** R233100-20									
9990	Minimum labor/equipment charge	1 Shee	3	2.667	Job		126		126	200

23 31 16.19 PVC Ducts

0010	**PVC DUCTS**									
9990	Minimum labor/equipment charge	1 Shee	3	2.667	Job		126		126	200

23 33 Air Duct Accessories

23 33 13 – Dampers

23 33 13.13 Volume-Control Dampers

	23 33 13.13 Volume-Control Dampers	Crew	Daily Output	Labor-Hours	Unit	Material	2009 Bare Costs Labor	2009 Bare Costs Equipment	Total	Total Incl O&P
0010	**VOLUME-CONTROL DAMPERS**									
5990	Multi-blade dampers, opposed blade, 8" x 6"	1 Shee	24	.333	Ea.	20	15.75		35.75	47
5994	8" x 8"		22	.364		21	17.15		38.15	50.50
5996	10" x 10"		21	.381		24.50	18		42.50	55.50
6000	12" x 12"		21	.381		27.50	18		45.50	58.50
6020	12" x 18"		18	.444		36.50	21		57.50	74
6030	14" x 10"		20	.400		26.50	18.90		45.40	59.50
6031	14" x 14"		17	.471		32.50	22		54.50	71
6033	16" x 12"		17	.471		32.50	22		54.50	71
6035	16" x 16"		16	.500		40.50	23.50		64	82
6037	18" x 16"		15	.533		44.50	25		69.50	89
6038	18" x 18"		15	.533		48	25		73	93
6070	20" x 16"		14	.571		48	27		75	96
6072	20" x 20"		13	.615		57.50	29		86.50	110
6074	22" x 18"		14	.571		57.50	27		84.50	107
6076	24" x 16"		11	.727		57	34.50		91.50	117
6078	24" x 20"		8	1		67	47		114	149
6080	24" x 24"		8	1		78.50	47		125.50	162
6110	26" x 26"		6	1.333		86.50	63		149.50	195
6133	30" x 30"	Q-9	6.60	2.424		124	103		227	300
6135	32" x 32"		6.40	2.500		144	106		250	325
6180	48" x 36"		5.60	2.857		237	121		358	455
7500	Variable volume modulating motorized damper, incl. elect. mtr.									
7504	8" x 6"	1 Shee	15	.533	Ea.	109	25		134	160
7506	10" x 6"		14	.571		109	27		136	163
7510	10" x 10"		13	.615		112	29		141	170
7520	12" x 12"		12	.667		117	31.50		148.50	179
7522	12" x 16"		11	.727		118	34.50		152.50	185
7524	16" x 10"		12	.667		115	31.50		146.50	177
7526	16" x 14"		10	.800		120	38		158	191
7528	16" x 18"		9	.889		125	42		167	204
7542	18" x 18"		8	1		127	47		174	215
7544	20" x 14"		8	1		130	47		177	218
7546	20" x 18"		7	1.143		136	54		190	236
7560	24" x 12"		8	1		132	47		179	220
7562	24" x 18"		7	1.143		143	54		197	243
7564	24" x 24"		6	1.333		148	63		211	262
7568	28" x 10"		7	1.143		132	54		186	231
7590	30" x 14"		5	1.600		140	75.50		215.50	274
7600	30" x 18"		4	2		180	94.50		274.50	350
7610	30" x 24"		3.80	2.105		239	99.50		338.50	420

RSMeans Business Solutions

Leaders in Cost Engineering for the Design, Construction, and Facilities Management Industry...

Market Analysis and Third Party Validation – RSMeans conducts research studies from numerous perspectives, including escalation costs in local markets, cost estimates for legal damages, and trend research for industry associations, as well as product adoption and productivity studies for building product manufacturers. Means is often called upon to produce "third party validation" construction cost estimates for owners, architects, and boards of directors.

Predictive Cost Models for Facility Owners – From early planning, cost modeling is driven by close attention to standards, codes, methods, and business processes that impact cost and operational effectiveness. Predictive cost models offer cost control and standardization for facility owners, particularly in repetitive building environments. RSMeans custom models are based upon actual architectural plans and historical budgets, with cost data and location factors adjusted to regions anywhere in North America.

Custom Cost Databases and Web-Based Dashboards – Custom applications of the RSMeans cost databases use the building blocks of construction cost engineering, including a standard code of accounts, activity descriptions, material, labor, and equipment costs. RSMeans has in-house technical expertise for rapid deployment of cost databases, online cost calculators, and web-based analytic dashboards.

Job Order Contracting and FacilitiesWorks™ – A delivery method for construction renovation to reduce cycle times and overhead costs. RSMeans offers consulting in JOC management practices, FacilitiesWorks™ software for renovation cost estimating and project management, and on-site training programs. JOC is available to government agencies, colleges and universities, school districts, municipalities, and health care systems.

Life Cycle Cost Analysis and Sustainability – Building upon its extensive facility cost databases, RSMeans offers life cycle cost studies and sustainability models to analyze total cost of ownership (TCO) for high performance buildings. Analyses include capital costs, life cycle and operations costs, as well as studies specific to integrated building automation, energy efficiency and security systems. Cost savings realized in operations, maintenance and replacement are analyzed enterprise-wide.

CORPORATE • HEALTH CARE • EDUCATION • GOVERNMENT • MUNICIPAL • RETAIL
• ASSOCIATIONS • BUILDING PRODUCT MANUFACTURERS • LEGAL

For more information contact RSMeans Business Solutions at 781-422-5101 or visit our website at www.rsmeans.com or email consulting@rsmeans.com

RSMeans

23 33 Air Duct Accessories

23 33 13 – Dampers

23 33 13.13 Volume-Control Dampers

		Crew	Daily Output	Labor-Hours	Unit	Material	2009 Bare Costs Labor	Equipment	Total	Total Incl O&P
7700	For thermostat, add	1 Shee	8	1	Ea.	40.50	47		87.50	120
8000	Multi-blade dampers, parallel blade									
8100	8" x 8"	1 Shee	24	.333	Ea.	71	15.75		86.75	103
8140	16" x 10"		20	.400		91.50	18.90		110.40	130
8160	18" x 12"		18	.444		104	21		125	148
8220	28" x 16"	↓	10	.800	↓	130	38		168	203

23 33 13.16 Fire Dampers

		Crew	Daily Output	Labor-Hours	Unit	Material	2009 Bare Costs Labor	Equipment	Total	Total Incl O&P
0010	**FIRE DAMPERS**									
3000	Fire damper, curtain type, 1-1/2 hr rated, vertical, 6" x 6"	1 Shee	24	.333	Ea.	21	15.75		36.75	48
3020	8" x 6"		22	.364		21	17.15		38.15	50.50
3240	16" x 14"		18	.444		28.50	21		49.50	65
3400	24" x 20"	↓	8	1	↓	40.50	47		87.50	120

23 33 19 – Duct Silencers

23 33 19.10 Duct Silencers

		Crew	Daily Output	Labor-Hours	Unit	Material	2009 Bare Costs Labor	Equipment	Total	Total Incl O&P
0009	**DUCT SILENCERS**									
0010	Silencers, noise control for air flow, duct				MCFM	54			54	59.50

23 33 23 – Turning Vanes

23 33 23.13 Air Turning Vanes

		Crew	Daily Output	Labor-Hours	Unit	Material	2009 Bare Costs Labor	Equipment	Total	Total Incl O&P
0010	**AIR TURNING VANES**									
9900	Minimum labor/equipment charge	1 Shee	4	2	Job		94.50		94.50	150

23 33 46 – Flexible Ducts

23 33 46.10 Flexible Air Ducts

		Crew	Daily Output	Labor-Hours	Unit	Material	2009 Bare Costs Labor	Equipment	Total	Total Incl O&P
0010	**FLEXIBLE AIR DUCTS** R233100-20									
1280	Add to labor for elevated installation									
1282	of prefabricated (purchased) ductwork									
1283	10' to 15' high						10%			
1284	15' to 20' high						20%			
1285	20' to 25' high						25%			
1286	25' to 30' high						35%			
1287	30' to 35' high						40%			
1288	35' to 40' high						50%			
1289	Over 40' high						55%			
1300	Flexible, coated fiberglass fabric on corr. resist. metal helix									
1400	pressure to 12" (WG) UL-181									
1500	Non-insulated, 3" diameter	Q-9	400	.040	L.F.	.99	1.70		2.69	3.79
1540	5" diameter		320	.050		1.21	2.12		3.33	4.71
1560	6" diameter		280	.057		1.49	2.43		3.92	5.50
1580	7" diameter		240	.067		1.76	2.83		4.59	6.45
1600	8" diameter		200	.080		2.04	3.40		5.44	7.65
1640	10" diameter		160	.100		2.53	4.25		6.78	9.55
1660	12" diameter		120	.133		3.08	5.65		8.73	12.40
1900	Insulated, 1" thick, PE jacket, 3" diameter G		380	.042		2.09	1.79		3.88	5.15
1910	4" diameter G		340	.047		2.09	2		4.09	5.50
1920	5" diameter G		300	.053		2.09	2.27		4.36	5.90
1940	6" diameter G		260	.062		2.31	2.61		4.92	6.70
1960	7" diameter G		220	.073		2.70	3.09		5.79	7.90
1980	8" diameter G		180	.089		2.92	3.78		6.70	9.20
2020	10" diameter G		140	.114		3.41	4.86		8.27	11.45
2040	12" diameter G		100	.160		4.24	6.80		11.04	15.45
2060	14" diameter G	↓	80	.200	↓	5.10	8.50		13.60	19.15
9990	Minimum labor/equipment charge	1 Shee	3	2.667	Job		126		126	200

23 33 53 – Duct Liners

23 33 53.10 Duct Liner Board		Crew	Daily Output	Labor-Hours	Unit	Material	2009 Bare Costs Labor	Equipment	Total	Total Incl O&P
0010	**DUCT LINER BOARD**									
3490	Board type, fiberglass liner, 3 lb. density									
3500	Fire resistant, black pigmented, 1 side									
3520	1" thick	G	Q-14	150	.107	S.F.	.73	4.23		4.96	7.65
3540	1-1/2" thick	G	"	130	.123	"	.92	4.89		5.81	8.90
3940	Board type, non-fibrous foam									
3950	Temperature, bacteria and fungi resistant									
3960	1" thick	G	Q-14	150	.107	S.F.	2.25	4.23		6.48	9.35
3970	1-1/2" thick	G		130	.123		3	4.89		7.89	11.20
3980	2" thick	G	↓	120	.133	↓	3.60	5.30		8.90	12.50

23 34 HVAC Fans

23 34 13 – Axial HVAC Fans

23 34 13.10 Axial Flow HVAC Fans

23 34 13.10 Axial Flow HVAC Fans		Crew	Daily Output	Labor-Hours	Unit	Material	Labor	Equipment	Total	Total Incl O&P
0010	**AXIAL FLOW HVAC FANS**									
0020	Air conditioning and process air handling									
0030	Axial flow, compact, low sound, 2.5" S.P.									
0050	3,800 CFM, 5 HP	Q-20	3.40	5.882	Ea.	4,375	255		4,630	5,200
0120	15,600 CFM, 10 HP	"	1.60	12.500	"	7,650	540		8,190	9,275

23 34 14 – Blower HVAC Fans

23 34 14.10 Blower Type HVAC Fans

23 34 14.10 Blower Type HVAC Fans		Crew	Daily Output	Labor-Hours	Unit	Material	Labor	Equipment	Total	Total Incl O&P
0010	**BLOWER TYPE HVAC FANS**									
2500	Ceiling fan, right angle, extra quiet, 0.10" S.P.									
2540	210 CFM	Q-20	19	1.053	Ea.	270	45.50		315.50	370
2580	885 CFM		16	1.250		675	54		729	825
2620	2,960 CFM	↓	11	1.818		1,250	79		1,329	1,500
2640	For wall or roof cap, add	1 Shee	16	.500	↓	228	23.50		251.50	289

23 34 16 – Centrifugal HVAC Fans

23 34 16.10 Centrifugal Type HVAC Fans

23 34 16.10 Centrifugal Type HVAC Fans		Crew	Daily Output	Labor-Hours	Unit	Material	Labor	Equipment	Total	Total Incl O&P
0010	**CENTRIFUGAL TYPE HVAC FANS**									
0200	In-line centrifugal, supply/exhaust booster									
0220	aluminum wheel/hub, disconnect switch, 1/4" S.P.									
0240	500 CFM, 10" diameter connection	Q-20	3	6.667	Ea.	1,125	289		1,414	1,675
0280	1,520 CFM, 16" diameter connection		2	10		1,300	435		1,735	2,100
0320	3,480 CFM, 20" diameter connection		.80	25		1,675	1,075		2,750	3,525
0326	5,080 CFM, 20" diameter connection	↓	.75	26.667	↓	1,800	1,150		2,950	3,825
4500	Corrosive fume resistant, plastic									
4600	roof ventilators, centrifugal, V belt drive, motor									
4620	1/4" S.P., 250 CFM, 1/4 HP	Q-20	6	3.333	Ea.	3,350	145		3,495	3,925
4640	895 CFM, 1/3 HP		5	4		3,650	174		3,824	4,300
4660	1630 CFM, 1/2 HP		4	5		4,325	217		4,542	5,100
4680	2240 CFM, 1 HP	↓	3	6.667	↓	4,500	289		4,789	5,400
5000	Utility set, centrifugal, V belt drive, motor									
5020	1/4" S.P., 1200 CFM, 1/4 HP	Q-20	6	3.333	Ea.	4,075	145		4,220	4,700
5040	1520 CFM, 1/3 HP		5	4		4,075	174		4,249	4,750
5060	1850 CFM, 1/2 HP		4	5		4,100	217		4,317	4,850
5080	2180 CFM, 3/4 HP		3	6.667		4,150	289		4,439	5,000
7100	Direct drive, 320 CFM, 11" sq. damper		7	2.857		510	124		634	760
7120	600 CFM, 11" sq. damper	↓	6	3.333	↓	520	145		665	800

23 34 HVAC Fans

23 34 16 – Centrifugal HVAC Fans

23 34 16.10 Centrifugal Type HVAC Fans	Crew	Daily Output	Labor-Hours	Unit	Material	2009 Bare Costs Labor	2009 Bare Costs Equipment	Total	Total Incl O&P
9900	Minimum labor/equipment charge	1 Elec	4	2	Job		94		94	145

23 34 23 – HVAC Power Ventilators

23 34 23.10 HVAC Power Circulators and Ventilators

		Crew	Daily Output	Labor-Hours	Unit	Material	Labor	Equipment	Total	Total Incl O&P
0010	**HVAC POWER CIRCULATORS AND VENTILATORS**									
6650	Residential, bath exhaust, grille, back draft damper									
6660	50 CFM	Q-20	24	.833	Ea.	40.50	36		76.50	102
6670	110 CFM		22	.909		66.50	39.50		106	135
6680	Light combination, squirrel cage, 100 watt, 70 CFM		24	.833		73.50	36		109.50	138
6700	Light/heater combination, ceiling mounted									
6710	70 CFM, 1450 watt	Q-20	24	.833	Ea.	89	36		125	155
6800	Heater combination, recessed, 70 CFM		24	.833		42	36		78	104
6820	With 2 infrared bulbs		23	.870		64	37.50		101.50	130
6900	Kitchen exhaust, grille, complete, 160 CFM		22	.909		76	39.50		115.50	146
6910	180 CFM		20	1		66.50	43.50		110	142
6920	270 CFM		18	1.111		117	48		165	204
6940	Residential roof jacks and wall caps									
6944	Wall cap with back draft damper									
6946	3" & 4" dia. round duct	1 Shee	11	.727	Ea.	16.15	34.50		50.65	72.50
6948	6" dia. round duct	"	11	.727	"	39	34.50		73.50	97.50
6958	Roof jack with bird screen and back draft damper									
6960	3" & 4" dia. round duct	1 Shee	11	.727	Ea.	15	34.50		49.50	71
6962	3-1/4" x 10" rectangular duct	"	10	.800	"	28.50	38		66.50	91.50
6980	Transition									
6982	3-1/4" x 10" to 6" dia. round	1 Shee	20	.400	Ea.	18.55	18.90		37.45	50.50
8020	Attic, roof type									
8030	Aluminum dome, damper & curb									
8040	6" diameter, 300 CFM	1 Elec	16	.500	Ea.	380	23.50		403.50	455
8050	7" diameter, 450 CFM		15	.533		415	25		440	495
8060	9" diameter, 900 CFM		14	.571		455	27		482	540
8080	12" diameter, 1000 CFM (gravity)		10	.800		470	37.50		507.50	580
8090	16" diameter, 1500 CFM (gravity)		9	.889		570	42		612	690
8100	20" diameter, 2500 CFM (gravity)		8	1		695	47		742	840
8110	26" diameter, 4000 CFM (gravity)		7	1.143		845	53.50		898.50	1,025
8120	32" diameter, 6500 CFM (gravity)		6	1.333		1,150	62.50		1,212.50	1,375
8130	38" diameter, 8000 CFM (gravity)		5	1.600		1,725	75		1,800	2,025
8140	50" diameter, 13,000 CFM (gravity)		4	2		2,500	94		2,594	2,900
8160	Plastic, ABS dome									
8180	1050 CFM	1 Elec	14	.571	Ea.	139	27		166	194
8200	1600 CFM	"	12	.667	"	208	31.50		239.50	276
8240	Attic, wall type, with shutter, one speed									
8250	12" diameter, 1000 CFM	1 Elec	14	.571	Ea.	300	27		327	370
8260	14" diameter, 1500 CFM		12	.667		325	31.50		356.50	405
8270	16" diameter, 2000 CFM		9	.889		365	42		407	470
8290	Whole house, wall type, with shutter, one speed									
8300	30" diameter, 4800 CFM	1 Elec	7	1.143	Ea.	785	53.50		838.50	950
8310	36" diameter, 7000 CFM		6	1.333		855	62.50		917.50	1,025
8320	42" diameter, 10,000 CFM		5	1.600		960	75		1,035	1,175
8330	48" diameter, 16,000 CFM		4	2		1,200	94		1,294	1,450
8340	For two speed, add					71.50			71.50	79
8350	Whole house, lay-down type, with shutter, one speed									
8360	30" diameter, 4500 CFM	1 Elec	8	1	Ea.	835	47		882	995
8370	36" diameter, 6500 CFM		7	1.143		900	53.50		953.50	1,075

23 34 HVAC Fans

23 34 23 – HVAC Power Ventilators

23 34 23.10 HVAC Power Circulators and Ventilators	Crew	Daily Output	Labor-Hours	Unit	Material	2009 Bare Costs Labor	Equipment	Total	Total Incl O&P
8380	42" diameter, 9000 CFM	1 Elec	6	1.333	Ea.	990	62.50		1,052.50	1,175
8390	48" diameter, 12,000 CFM	↓	5	1.600		1,125	75		1,200	1,350
8440	For two speed, add					54			54	59.50
8450	For 12 hour timer switch, add	1 Elec	32	.250	↓	54	11.75		65.75	77.50

23 36 Air Terminal Units

23 36 13 – Constant-Air-Volume Units

23 36 13.10 Constant Volume Mixing Boxes

		Crew	Daily Output	Labor-Hours	Unit	Material	Labor	Equipment	Total	Total Incl O&P
0010	**CONSTANT VOLUME MIXING BOXES**									
5180	Mixing box, includes electric or pneumatic motor									
5200	Constant volume, 150 to 270 CFM	Q-9	12	1.333	Ea.	620	56.50		676.50	770
5210	270 to 600 CFM	"	11	1.455	"	640	62		702	800

23 36 16 – Variable-Air-Volume Units

23 36 16.10 Variable Volume Mixing Boxes

		Crew	Daily Output	Labor-Hours	Unit	Material	Labor	Equipment	Total	Total Incl O&P
0010	**VARIABLE VOLUME MIXING BOXES**									
5180	Mixing box, includes electric or pneumatic motor									
5500	VAV Cool only, pneumatic, pressure independent 300 to 600 CFM	Q-9	11	1.455	Ea.	400	62		462	540
5510	500 to 1000 CFM		9	1.778		410	75.50		485.50	570
5520	800 to 1600 CFM		9	1.778		425	75.50		500.50	585
5530	1100 to 2000 CFM		8	2		435	85		520	610
5540	1500 to 3000 CFM		7	2.286		460	97		557	665
5550	2000 to 4000 CFM	↓	6	2.667		465	113		578	695
5560	For electric, w/thermostat, pressure dependent, add				↓	23			23	25.50

23 37 Air Outlets and Inlets

23 37 13 – Diffusers, Registers, and Grilles

23 37 13.10 Diffusers

		Crew	Daily Output	Labor-Hours	Unit	Material	Labor	Equipment	Total	Total Incl O&P
0010	**DIFFUSERS**, Aluminum, opposed blade damper unless noted									
0100	Ceiling, linear, also for sidewall									
0120	2" wide	1 Shee	32	.250	L.F.	39.50	11.80		51.30	62.50
0160	4" wide		26	.308		51.50	14.50		66	80
0180	6" wide		24	.333		63.50	15.75		79.25	95
0200	8" wide		22	.364	↓	75	17.15		92.15	110
0500	Perforated, 24" x 24" lay-in panel size, 6" x 6"		16	.500	Ea.	56	23.50		79.50	99
0520	8" x 8"		15	.533		57	25		82	103
0530	9" x 9"		14	.571		61	27		88	111
0540	10" x 10"		14	.571		63	27		90	113
0560	12" x 12"		12	.667		64	31.50		95.50	121
0590	16" x 16"		11	.727		106	34.50		140.50	172
0600	18" x 18"		10	.800		112	38		150	183
0610	20" x 20"		10	.800		129	38		167	202
0620	24" x 24"		9	.889		161	42		203	244
1000	Rectangular, 1 to 4 way blow, 6" x 6"		16	.500		45	23.50		68.50	87
1010	8" x 8"		15	.533		54	25		79	99.50
1014	9" x 9"		15	.533		59.50	25		84.50	106
1016	10" x 10"		15	.533		64	25		89	111
1020	12" x 6"		15	.533		65.50	25		90.50	112
1040	12" x 9"		14	.571		71.50	27		98.50	122
1060	12" x 12"	↓	12	.667	↓	72	31.50		103.50	129

23 37 Air Outlets and Inlets

23 37 13 – Diffusers, Registers, and Grilles

23 37 13.10 Diffusers

		Crew	Daily Output	Labor-Hours	Unit	Material	2009 Bare Costs Labor	Equipment	Total	Total Incl O&P
1070	14" x 6"	1 Shee	13	.615	Ea.	71	29		100	124
1074	14" x 14"		12	.667		101	31.50		132.50	161
1150	18" x 18"		9	.889		126	42		168	206
1170	24" x 12"		10	.800		134	38		172	207
2000	T bar mounting, 24" x 24" lay-in frame, 6" x 6"		16	.500		42.50	23.50		66	84.50
2020	8" x 8"		14	.571		44.50	27		71.50	91.50
2040	12" x 12"		12	.667		52	31.50		83.50	108
2060	16" x 16"		11	.727		70	34.50		104.50	132
2080	18" x 18"		10	.800		78	38		116	146
6000	For steel diffusers instead of aluminum, deduct					10%				
9000	Minimum labor/equipment charge	1 Shee	4	2	Job		94.50		94.50	150

23 37 13.30 Grilles

		Crew	Daily Output	Labor-Hours	Unit	Material	2009 Bare Costs Labor	Equipment	Total	Total Incl O&P
0010	**GRILLES**									
0020	Aluminum									
1000	Air return, 6" x 6"	1 Shee	26	.308	Ea.	17.05	14.50		31.55	42
1020	10" x 6"		24	.333		17.05	15.75		32.80	44
1080	16" x 8"		22	.364		21.50	17.15		38.65	51
1100	12" x 12"		22	.364		23	17.15		40.15	53
1120	24" x 12"		18	.444		32	21		53	68.50
1300	48" x 24"		12	.667		118	31.50		149.50	180
9000	Minimum labor/equipment charge		4	2	Job		94.50		94.50	150

23 37 13.60 Registers

		Crew	Daily Output	Labor-Hours	Unit	Material	2009 Bare Costs Labor	Equipment	Total	Total Incl O&P
0010	**REGISTERS**									
0020										
0980	Air supply									
1000	Ceiling/wall, O.B. damper, anodized aluminum									
1010	One or two way deflection, adj. curved face bars									
1140	14" x 8"	1 Shee	17	.471	Ea.	48.50	22		70.50	89
3000	Baseboard, hand adj. damper, enameled steel									
3012	8" x 6"	1 Shee	26	.308	Ea.	12.85	14.50		27.35	37
3020	10" x 6"		24	.333		14	15.75		29.75	40.50
3040	12" x 5"		23	.348		15.20	16.40		31.60	42.50
3060	12" x 6"		23	.348		15.20	16.40		31.60	42.50
3080	12" x 8"		22	.364		22	17.15		39.15	51.50
3100	14" x 6"		20	.400		16.45	18.90		35.35	48
9000	Minimum labor/equipment charge		4	2	Job		94.50		94.50	150

23 37 23 – HVAC Gravity Ventilators

23 37 23.10 HVAC Gravity Air Ventilators

		Crew	Daily Output	Labor-Hours	Unit	Material	2009 Bare Costs Labor	Equipment	Total	Total Incl O&P
0010	**HVAC GRAVITY AIR VENTILATORS**, Incl. base and damper									
9000	Minimum labor/equipment charge	1 Plum	2	4	Job		195		195	300

23 38 Ventilation Hoods

23 38 13 – Commercial-Kitchen Hoods

23 38 13.10 Hood and Ventilation Equipment

23 38 13.10 Hood and Ventilation Equipment	Crew	Daily Output	Labor-Hours	Unit	Material	2009 Bare Costs Labor	Equipment	Total	Total Incl O&P
0010 **HOOD AND VENTILATION EQUIPMENT**									
2970 Exhaust hood, sst, gutter on all sides, 4' x 4' x 2'	1 Carp	1.80	4.444	Ea.	4,050	178		4,228	4,750
2980 4' x 4' x 7'	"	1.60	5		6,350	200		6,550	7,325
7950 Hood fire protection system, minimum	Q-1	3	5.333		4,100	234		4,334	4,875
8050 Maximum	"	1	16	↓	31,600	700		32,300	35,900

23 41 Particulate Air Filtration

23 41 13 – Panel Air Filters

23 41 13.10 Panel Type Air Filters

23 41 13.10 Panel Type Air Filters				Unit	Material	Labor	Equipment	Total	Total Incl O&P
0010 **PANEL TYPE AIR FILTERS**									
0020									
2950 Mechanical media filtration units									
3000 High efficiency type, with frame, non-supported [G]				MCFM	45			45	49.50
3100 Supported type [G]				"	60			60	66
5500 Throwaway glass or paper media type				Ea.	3.35			3.35	3.69

23 41 16 – Renewable-Media Air Filters

23 41 16.10 Disposable Media Air Filters

23 41 16.10 Disposable Media Air Filters				Unit	Material	Labor	Equipment	Total	Total Incl O&P
0010 **DISPOSABLE MEDIA AIR FILTERS**									
5000 Renewable disposable roll				MCFM	250			250	275

23 41 19 – Washable Air Filters

23 41 19.10 Permanent Air Filters

23 41 19.10 Permanent Air Filters				Unit	Material	Labor	Equipment	Total	Total Incl O&P
0010 **PERMANENT AIR FILTERS**									
4500 Permanent washable [G]				MCFM	20			20	22

23 41 23 – Extended Surface Filters

23 41 23.10 Expanded Surface Filters

23 41 23.10 Expanded Surface Filters				Unit	Material	Labor	Equipment	Total	Total Incl O&P
0010 **EXPANDED SURFACE FILTERS**									
4000 Medium efficiency, extended surface [G]				MCFM	5.50			5.50	6.05

23 42 Gas-Phase Air Filtration

23 42 13 – Activated-Carbon Air Filtration

23 42 13.10 Charcoal Type Air Filtration

23 42 13.10 Charcoal Type Air Filtration	Crew	Daily Output	Labor-Hours	Unit	Material	Labor	Equipment	Total	Total Incl O&P
0010 **CHARCOAL TYPE AIR FILTRATION**									
0050 Activated charcoal type, full flow				MCFM	600			600	660
0060 Full flow, impregnated media 12" deep					225			225	248
0070 HEPA filter & frame for field erection					300			300	330
0080 HEPA filter-diffuser, ceiling install.				↓	275			275	305

23 42 16 – Chemically-Impregnated Adsorption Air Filtration

23 42 16.10 Chemical Adsorption Air Filtration

23 42 16.10 Chemical Adsorption Air Filtration	Crew	Daily Output	Labor-Hours	Unit	Material	Labor	Equipment	Total	Total Incl O&P
0010 **CHEMICAL ADSORPTION AIR FILTRATION**									
9000 Minimum labor/equipment charge	1 Shee	2.25	3.556	Job		168		168	267

23 43 Electronic Air Cleaners

23 43 13 – Washable Electronic Air Cleaners

23 43 13.10 Electronic Air Cleaners	Crew	Daily Output	Labor-Hours	Unit	Material	2009 Bare Costs Labor	Equipment	Total	Total Incl O&P
0010 **ELECTRONIC AIR CLEANERS**									
2000 Electronic air cleaner, duct mounted									
2150 400 – 1000 CFM	1 Shee	2.30	3.478	Ea.	1,000	164		1,164	1,350
2200 1000 – 1400 CFM		2.20	3.636		1,225	172		1,397	1,625
2250 1400 – 2000 CFM	↓	2.10	3.810	↓	1,400	180		1,580	1,825

23 51 Breechings, Chimneys, and Stacks

23 51 13 – Draft Control Devices

23 51 13.16 Vent Dampers

	Crew	Daily Output	Labor-Hours	Unit	Material	2009 Bare Costs Labor	Equipment	Total	Total Incl O&P
0010 **VENT DAMPERS**									
5000 Vent damper, bi-metal, gas, 3" diameter	Q-9	24	.667	Ea.	38	28.50		66.50	87
5010 4" diameter		24	.667		38	28.50		66.50	87
5020 5" diameter		23	.696		38	29.50		67.50	89
5030 6" diameter		22	.727		38	31		69	91
5040 7" diameter		21	.762		40.50	32.50		73	96
5050 8" diameter	↓	20	.800	↓	44	34		78	103
9000 Minimum labor/equipment charge	1 Shee	4	2	Job		94.50		94.50	150

23 51 13.19 Barometric Dampers

	Crew	Daily Output	Labor-Hours	Unit	Material	2009 Bare Costs Labor	Equipment	Total	Total Incl O&P
0010 **BAROMETRIC DAMPERS**									
1000 Barometric, gas fired system only, 6" size for 5" and 6" pipes	1 Shee	20	.400	Ea.	54	18.90		72.90	89.50
1040 8" size, for 7" and 8" pipes	"	18	.444	"	74.50	21		95.50	116
2000 All fuel, oil, oil/gas, coal									
2020 10" for 9" and 10" pipes	1 Shee	15	.533	Ea.	125	25		150	178
3260 For thermal switch for above, add	"	24	.333	"	52.50	15.75		68.25	83

23 51 23 – Gas Vents

23 51 23.10 Gas Chimney Vents

	Crew	Daily Output	Labor-Hours	Unit	Material	2009 Bare Costs Labor	Equipment	Total	Total Incl O&P
0010 **GAS CHIMNEY VENTS,** Prefab metal, U.L. listed									
0020 Gas, double wall, galvanized steel									
0080 3" diameter	Q-9	72	.222	V.L.F.	5.35	9.45		14.80	21
0100 4" diameter		68	.235		6.70	10		16.70	23.50
0120 5" diameter		64	.250		7.80	10.60		18.40	25.50
0140 6" diameter		60	.267		9.10	11.35		20.45	28
0160 7" diameter		56	.286		13.35	12.15		25.50	34
0180 8" diameter		52	.308		14.85	13.05		27.90	37.50
0200 10" diameter		48	.333		34	14.15		48.15	59.50
0220 12" diameter		44	.364		45	15.45		60.45	74
0260 16" diameter	↓	40	.400		103	17		120	140
0300 20" diameter	Q-10	36	.667		139	29.50		168.50	200
0340 24" diameter	"	32	.750	↓	245	33		278	320

23 51 26 – All-Fuel Vent Chimneys

23 51 26.30 All-Fuel Vent Chimneys, Double Wall, St. Stl.

	Crew	Daily Output	Labor-Hours	Unit	Material	2009 Bare Costs Labor	Equipment	Total	Total Incl O&P
0010 **ALL-FUEL VENT CHIMNEYS, DOUBLE WALL, STAINLESS STEEL**									
7800 All fuel, double wall, stainless steel, 6" diameter	Q-9	60	.267	V.L.F.	56.50	11.35		67.85	80
7802 7" diameter		56	.286		73	12.15		85.15	100
7804 8" diameter		52	.308		86.50	13.05		99.55	116
7806 10" diameter		48	.333		125	14.15		139.15	160
7808 12" diameter		44	.364		167	15.45		182.45	209
7810 14" diameter	↓	42	.381	↓	219	16.20		235.20	267
8000 All fuel, double wall, stainless steel fittings									
8010 Roof support 6" diameter	Q-9	30	.533	Ea.	103	22.50		125.50	150

23 51 Breechings, Chimneys, and Stacks

23 51 26 – All-Fuel Vent Chimneys

23 51 26.30 All-Fuel Vent Chimneys, Double Wall, St. Stl.	Crew	Daily Output	Labor-Hours	Unit	Material	2009 Bare Costs Labor	2009 Bare Costs Equipment	Total	Total Incl O&P
8020	7" diameter	Q-9	28	.571	Ea.	117	24.50		141.50	167
8030	8" diameter		26	.615		130	26		156	185
8040	10" diameter		24	.667		150	28.50		178.50	210
8050	12" diameter		22	.727		181	31		212	248
8060	14" diameter		21	.762		229	32.50		261.50	305
8100	Elbow 15°, 6" diameter		30	.533		165	22.50		187.50	218
8120	7" diameter		28	.571		186	24.50		210.50	244
8140	8" diameter		26	.615		212	26		238	275
8160	10" diameter		24	.667		278	28.50		306.50	350
8180	12" diameter		22	.727		335	31		366	420
8200	14" diameter		21	.762		405	32.50		437.50	495
8300	Insulated tee with insulated tee cap, 6" diameter		30	.533		156	22.50		178.50	208
8340	7" diameter		28	.571		205	24.50		229.50	264
8360	8" diameter		26	.615		231	26		257	296
8380	10" diameter		24	.667		325	28.50		353.50	405
8400	12" diameter		22	.727		455	31		486	550
8420	14" diameter		21	.762		600	32.50		632.50	710
8500	Joist shield, 6" diameter		30	.533		47.50	22.50		70	88.50
8510	7" diameter		28	.571		51.50	24.50		76	95.50
8520	8" diameter		26	.615		63.50	26		89.50	112
8530	10" diameter		24	.667		86	28.50		114.50	140
8540	12" diameter		22	.727		107	31		138	167
8550	14" diameter		21	.762		133	32.50		165.50	198
8600	Round top, 6" diameter		30	.533		53	22.50		75.50	94
8620	7" diameter		28	.571		72	24.50		96.50	118
8640	8" diameter		26	.615		97	26		123	149
8660	10" diameter		24	.667		180	28.50		208.50	243
8680	12" diameter		22	.727		249	31		280	320
8700	14" diameter		21	.762		330	32.50		362.50	415
8800	Adjustable roof flashing, 6" diameter		30	.533		63	22.50		85.50	105
8820	7" diameter		28	.571		72	24.50		96.50	118
8840	8" diameter		26	.615		78	26		104	128
8860	10" diameter		24	.667		100	28.50		128.50	155
8880	12" diameter		22	.727		130	31		161	191
8900	14" diameter		21	.762		162	32.50		194.50	230

23 51 33 – Insulated Sectional Chimneys

23 51 33.10 Prefabricated Insulated Sectional Chimneys

		Crew	Daily Output	Labor-Hours	Unit	Material	Labor	Equipment	Total	Total Incl O&P
0010	**PREFABRICATED INSULATED SECTIONAL CHIMNEYS**									
9990	Minimum labor/equipment charge	Q-9	3	5.333	Job		227		227	360

23 52 Heating Boilers

23 52 13 – Electric Boilers

23 52 13.10 Electric Boilers, ASME

		Crew	Daily Output	Labor-Hours	Unit	Material	Labor	Equipment	Total	Total Incl O&P
0010	**ELECTRIC BOILERS, ASME**, Standard controls and trim.									
1000	Steam, 6 KW, 20.5 MBH	Q-19	1.20	20	Ea.	3,375	905		4,280	5,125
1160	60 KW, 205 MBH		1	24		5,800	1,075		6,875	8,050
1300	296 KW, 1010 MBH		.45	53.333		13,200	2,425		15,625	18,200
1400	592 KW, 2020 MBH	Q-21	.34	94.118		25,000	4,350		29,350	34,300
1540	1110 KW, 3788 MBH	"	.19	168		35,000	7,800		42,800	50,500
2000	Hot water, 7.5 KW, 25.6 MBH	Q-19	1.30	18.462		3,575	835		4,410	5,250

23 52 Heating Boilers

23 52 13 – Electric Boilers

23 52 13.10 Electric Boilers, ASME

		Crew	Daily Output	Labor-Hours	Unit	Material	2009 Bare Costs Labor	Equipment	Total	Total Incl O&P
2040	30 KW, 102 MBH	Q-19	1.20	20	Ea.	3,825	905		4,730	5,600
2060	45 KW, 164 MBH		1.20	20		4,275	905		5,180	6,100
2070	60 KW, 205 MBH		1.20	20		4,525	905		5,430	6,375
2080	75 KW, 256 MBH		1.10	21.818		4,850	990		5,840	6,875
2100	90 KW, 307 MBH		1.10	21.818		5,225	990		6,215	7,275
9000	Minimum labor/equipment charge	Q-20	1	20	Job		870		870	1,375

23 52 23 – Cast-Iron Boilers

23 52 23.20 Gas-Fired Boilers

		Crew	Daily Output	Labor-Hours	Unit	Material	2009 Bare Costs Labor	Equipment	Total	Total Incl O&P
0010	**GAS-FIRED BOILERS,** Natural or propane, standard controls, packaged.									
1000	Cast iron, with insulated jacket									
2000	Steam, gross output, 81 MBH	Q-7	1.40	22.857	Ea.	1,850	1,075		2,925	3,725
2060	163 MBH		.90	35.556		2,700	1,675		4,375	5,575
2200	440 MBH		.51	62.500		5,275	2,950		8,225	10,400
2320	1,875 MBH		.30	106		17,800	5,025		22,825	27,300
2400	3570 MBH		.18	181		25,600	8,550		34,150	41,500
2540	6,970 MBH		.10	320		83,000	15,000		98,000	115,000
2800										
3000	Hot water, gross output, 80 MBH	Q-7	1.46	21.918	Ea.	1,700	1,025		2,725	3,475
3020	100 MBH		1.35	23.704		1,950	1,125		3,075	3,875
3040	122 MBH		1.10	29.091		2,100	1,375		3,475	4,450
3060	163 MBH		1	32		2,550	1,500		4,050	5,125
3200	440 MBH		.58	54.983		5,100	2,575		7,675	9,600
3320	2,000 MBH		.26	125		21,000	5,875		26,875	32,200
3540	6,970 MBH		.09	359		89,500	16,900		106,400	124,500
7000	For tankless water heater, add					10%				
7050	For additional zone valves up to 312 MBH add					145			145	159
9900	Minimum labor/equipment charge	Q-6	1	24	Job		1,100		1,100	1,725

23 52 23.30 Gas/Oil Fired Boilers

		Crew	Daily Output	Labor-Hours	Unit	Material	2009 Bare Costs Labor	Equipment	Total	Total Incl O&P
0010	**GAS/OIL FIRED BOILERS,** Combination with burners and controls, packaged.									
1000	Cast iron with insulated jacket									
2000	Steam, gross output, 720 MBH	Q-7	.43	74.074	Ea.	13,400	3,475		16,875	20,200
2140	2,700 MBH		.19	165		26,600	7,800		34,400	41,400
2380	6,970 MBH		.09	372		92,000	17,500		109,500	128,500
2900	Hot water, gross output									
2910	200 MBH	Q-6	.62	39.024	Ea.	7,750	1,800		9,550	11,300
2920	300 MBH		.49	49.080		7,750	2,250		10,000	12,000
2930	400 MBH		.41	57.971		9,075	2,675		11,750	14,100
2940	500 MBH		.36	67.039		9,775	3,100		12,875	15,600
3000	584 MBH	Q-7	.44	72.072		13,200	3,400		16,600	19,800
3060	1,460 MBH		.28	113		18,700	5,325		24,025	28,800
3300	13,500 MBH, 403.3 BHP		.04	727		140,000	34,200		174,200	207,000

23 52 23.40 Oil-Fired Boilers

		Crew	Daily Output	Labor-Hours	Unit	Material	2009 Bare Costs Labor	Equipment	Total	Total Incl O&P
0010	**OIL-FIRED BOILERS,** Standard controls, flame retention burner, packaged									
1000	Cast iron, with insulated flush jacket									
2000	Steam, gross output, 109 MBH	Q-7	1.20	26.667	Ea.	1,875	1,250		3,125	4,025
2060	207 MBH		.90	35.556		2,575	1,675		4,250	5,425
2180	1,084 MBH		.38	85.106		8,725	4,000		12,725	15,800
2200	1,360 MBH		.33	98.160		10,100	4,625		14,725	18,300
2240	2,175 MBH		.24	133		14,700	6,300		21,000	25,900
2340	4,360 MBH		.15	214		26,300	10,100		36,400	44,600
2460	6,970 MBH		.09	363		84,000	17,100		101,100	119,000
3000	Hot water, same price as steam									

23 52 Heating Boilers

23 52 26 – Steel Boilers

23 52 26.40 Oil-Fired Boilers

	Crew	Daily Output	Labor-Hours	Unit	Material	2009 Bare Costs Labor	Equipment	Total	Total Incl O&P
0010 **OIL-FIRED BOILERS**, Standard controls, flame retention burner									
5000 Steel, with insulated flush jacket									
7000 Hot water, gross output, 103 MBH	Q-6	1.60	15	Ea.	1,675	690		2,365	2,925
7020 122 MBH		1.45	16.506		1,775	760		2,535	3,125
7040 137 MBH		1.36	17.595		1,900	810		2,710	3,325
7060 168 MBH		1.30	18.405		2,000	850		2,850	3,500
7080 225 MBH		1.22	19.704		2,625	910		3,535	4,300
7100 315 MBH		.96	25.105		5,425	1,150		6,575	7,775
7120 420 MBH		.70	34.483	↓	5,700	1,600		7,300	8,750
9000 Minimum labor/equipment charge	↓	1.75	13.714	Job		630		630	980

23 52 88 – Burners

23 52 88.10 Replacement Type Burners

	Crew	Daily Output	Labor-Hours	Unit	Material	2009 Bare Costs Labor	Equipment	Total	Total Incl O&P
0010 **REPLACEMENT TYPE BURNERS**									
0990 Residential, conversion, gas fired, LP or natural									
1000 Gun type, atmospheric input 50 to 225 MBH	Q-1	2.50	6.400	Ea.	645	281		926	1,150
1020 100 to 400 MBH		2	8		1,075	350		1,425	1,725
1040 300 to 1000 MBH	↓	1.70	9.412	↓	3,300	415		3,715	4,275
3000 Flame retention oil fired assembly, input									
3040 2.0 to 5.0 GPH	Q-1	2	8	Ea.	330	350		680	910

23 54 Furnaces

23 54 13 – Electric-Resistance Furnaces

23 54 13.10 Electric Furnaces

	Crew	Daily Output	Labor-Hours	Unit	Material	2009 Bare Costs Labor	Equipment	Total	Total Incl O&P
0010 **ELECTRIC FURNACES**, Hot air, blowers, std. controls									
0011 not including gas, oil or flue piping									
1000 Electric, UL listed									
1020 10.2 MBH	Q-20	5	4	Ea.	360	174		534	670
1100 34.1 MBH	"	4.40	4.545	"	465	197		662	820

23 54 16 – Fuel-Fired Furnaces

23 54 16.13 Gas-Fired Furnaces

	Crew	Daily Output	Labor-Hours	Unit	Material	2009 Bare Costs Labor	Equipment	Total	Total Incl O&P
0010 **GAS-FIRED FURNACES**									
3000 Gas, AGA certified, upflow, direct drive models									
3020 45 MBH input	Q-9	4	4	Ea.	575	170		745	905
3040 60 MBH input		3.80	4.211		615	179		794	960
3060 75 MBH input		3.60	4.444		685	189		874	1,050
3100 100 MBH input		3.20	5		695	212		907	1,100
3120 125 MBH input		3	5.333		700	227		927	1,125
3130 150 MBH input		2.80	5.714		705	243		948	1,150
3140 200 MBH input	↓	2.60	6.154	↓	2,425	261		2,686	3,075

23 54 16.16 Oil-Fired Furnaces

	Crew	Daily Output	Labor-Hours	Unit	Material	2009 Bare Costs Labor	Equipment	Total	Total Incl O&P
0010 **OIL-FIRED FURNACES**									
6000 Oil, UL listed, atomizing gun type burner									
6020 56 MBH output	Q-9	3.60	4.444	Ea.	1,850	189		2,039	2,325
6040 95 MBH output		3.40	4.706		1,875	200		2,075	2,400
6060 134 MBH output		3.20	5		1,950	212		2,162	2,500
6080 151 MBH output		3	5.333		2,275	227		2,502	2,850
6100 200 MBH input		2.60	6.154	↓	2,325	261		2,586	2,975
9000 Minimum labor/equipment charge	↓	2.75	5.818	Job		247		247	395

23 54 24.10 Furnace Components and Combinations	Crew	Daily Output	Labor-Hours	Unit	Material	2009 Bare Costs Labor	Equipment	Total	Total Incl O&P
0010 **FURNACE COMPONENTS AND COMBINATIONS**									
0080 Coils, A/C evaporator, for gas or oil furnaces									
0090 Add-on, with holding charge									
0100 Upflow									
0120 1-1/2 ton cooling	Q-5	4	4	Ea.	158	178		336	450
0130 2 ton cooling		3.70	4.324		188	192		380	505
0140 3 ton cooling		3.30	4.848		238	215		453	595
0150 4 ton cooling		3	5.333		325	237		562	725
0160 5 ton cooling		2.70	5.926		385	263		648	835
0300 Downflow									
0330 2-1/2 ton cooling	Q-5	3	5.333	Ea.	232	237		469	620
0340 3-1/2 ton cooling		2.60	6.154		276	273		549	730
0350 5 ton cooling		2.20	7.273		385	325		710	925
0600 Horizontal									
0630 2 ton cooling	Q-5	3.90	4.103	Ea.	156	182		338	455
0640 3 ton cooling		3.50	4.571		184	203		387	515
0650 4 ton cooling		3.20	5		224	222		446	590
0660 5 ton cooling		2.90	5.517		224	245		469	625
2000 Cased evaporator coils for air handlers									
2100 1-1/2 ton cooling	Q-5	4.40	3.636	Ea.	244	162		406	520
2110 2 ton cooling		4.10	3.902		248	173		421	540
2120 2-1/2 ton cooling		3.90	4.103		259	182		441	570
2130 3 ton cooling		3.70	4.324		286	192		478	615
2140 3-1/2 ton cooling		3.50	4.571		299	203		502	645
2150 4 ton cooling		3.20	5		360	222		582	740
2160 5 ton cooling		2.90	5.517		415	245		660	835
3010 Air handler, modular									
3100 With cased evaporator cooling coil									
3120 1-1/2 ton cooling	Q-5	3.80	4.211	Ea.	775	187		962	1,150
3130 2 ton cooling		3.50	4.571		865	203		1,068	1,275
3140 2-1/2 ton cooling		3.30	4.848		885	215		1,100	1,300
3150 3 ton cooling		3.10	5.161		945	229		1,174	1,375
3160 3-1/2 ton cooling		2.90	5.517		1,000	245		1,245	1,475
3170 4 ton cooling		2.50	6.400		1,125	284		1,409	1,700
3180 5 ton cooling		2.10	7.619		1,275	340		1,615	1,925
3500 With no cooling coil									
3520 1-1/2 ton coil size	Q-5	12	1.333	Ea.	575	59		634	720
3530 2 ton coil size		10	1.600		605	71		676	775
3540 2-1/2 ton coil size		10	1.600		605	71		676	775
3554 3 ton coil size		9	1.778		685	79		764	875
3560 3-1/2 ton coil size		9	1.778		745	79		824	940
3570 4 ton coil size		8.50	1.882		860	83.50		943.50	1,075
3580 5 ton coil size		8	2		1,000	89		1,089	1,250
4000 With heater									
4120 5 kW, 17.1 MBH	Q-5	16	1	Ea.	585	44.50		629.50	715
4130 7.5 kW, 25.6 MBH		15.60	1.026		590	45.50		635.50	720
4140 10 kW, 34.2 MBH		15.20	1.053		825	47		872	980
4150 12.5 KW, 42.7 MBH		14.80	1.081		910	48		958	1,075

23 55 Fuel-Fired Heaters

23 55 13 – Fuel-Fired Duct Heaters

23 55 13.16 Gas-Fired Duct Heaters		Crew	Daily Output	Labor-Hours	Unit	Material	2009 Bare Costs Labor	Equipment	Total	Total Incl O&P
0010	**GAS-FIRED DUCT HEATERS**, Includes burner, controls, stainless steel									
0020	heat exchanger. Gas fired, electric ignition									
1000	Outdoor installation, with vent cap									
1120	225 MBH output	Q-5	2.50	6.400	Ea.	5,225	284		5,509	6,200
1160	375 MBH output		1.60	10		6,575	445		7,020	7,925
1180	450 MBH output	↓	1.40	11.429	↓	6,875	510		7,385	8,350

23 55 33 – Fuel-Fired Unit Heaters

23 55 33.16 Gas-Fired Unit Heaters

		Crew	Daily Output	Labor-Hours	Unit	Material	2009 Bare Costs Labor	Equipment	Total	Total Incl O&P
0010	**GAS-FIRED UNIT HEATERS**, Cabinet, grilles, fan, ctrls., burner, no piping									
0022	thermostat, no piping. For flue see Div. 23 51 23.10									
1000	Gas fired, floor mounted									
1100	60 MBH output	Q-5	10	1.600	Ea.	805	71		876	995
1180	180 MBH output		6	2.667		1,275	118		1,393	1,575
2000	Suspension mounted, propeller fan, 20 MBH output		8.50	1.882		985	83.50		1,068.50	1,200
2040	60 MBH output		7	2.286		1,200	102		1,302	1,475
2100	130 MBH output		5	3.200		1,500	142		1,642	1,875
2240	320 MBH output	↓	2	8		2,625	355		2,980	3,450
2500	For powered venter and adapter, add					355			355	390
5000	Wall furnace, 17.5 MBH output	Q-5	6	2.667		615	118		733	865
5020	24 MBH output		5	3.200		640	142		782	925
5040	35 MBH output		4	4	↓	770	178		948	1,125
9000	Minimum labor/equipment charge	↓	3.50	4.571	Job		203		203	315

23 56 Solar Energy Heating Equipment

23 56 16 – Packaged Solar Heating Equipment

23 56 16.40 Solar Heating Systems

			Crew	Daily Output	Labor-Hours	Unit	Material	2009 Bare Costs Labor	Equipment	Total	Total Incl O&P
0010	**SOLAR HEATING SYSTEMS**										
0020	System/Package prices, not including connecting										
0030	pipe, insulation, or special heating/plumbing fixtures										
0500	Hot water, standard package, low temperature										
0580	2 collectors, circulator, fittings, 120 gal. tank	G	Q-1	.40	40	Ea.	1,825	1,750		3,575	4,750

23 56 19 – Solar Heating Components

23 56 19.50 Solar Heating Ancillary

			Crew	Daily Output	Labor-Hours	Unit	Material	2009 Bare Costs Labor	Equipment	Total	Total Incl O&P
0010	**SOLAR HEATING ANCILLARY**										
2250	Controller, liquid temperature	G	1 Plum	5	1.600	Ea.	96	78		174	226
2300	Circulators, air										
2310	Blowers										
2520	Space & DHW system, less duct work	G	Q-9	.50	32	Ea.	1,475	1,350		2,825	3,775
2580	8" diameter, 150 CFM	G		16	1		39.50	42.50		82	111
2660	Shutter/damper	G		12	1.333		49	56.50		105.50	144
2670	Shutter motor	G	↓	16	1		119	42.50		161.50	199
2800	Circulators, liquid, 1/25 HP, 5.3 GPM	G	Q-1	14	1.143		227	50		277	330
2820	1/20 HP, 17 GPM	G	"	12	1.333	↓	154	58.50		212.50	261
3000	Collector panels, air with aluminum absorber plate										
3010	Wall or roof mount										
3040	Flat black, plastic glazing										
3080	4' x 8'	G	Q-9	6	2.667	Ea.	635	113		748	875
3300	Collector panels, liquid with copper absorber plate										
3320	Black chrome, tempered glass glazing										
3330	Alum. frame, 4' x 8', 5/32" single glazing	G	Q-1	9.50	1.684	Ea.	665	74		739	845

23 56 19.50 Solar Heating Ancillary		Crew	Daily Output	Labor-Hours	Unit	Material	2009 Bare Costs Labor	2009 Bare Costs Equipment	Total	Total Incl O&P
3450	Flat black, alum. frame, 3.5' x 7.5'	G Q-1	9	1.778	Ea.	560	78		638	740
3500	4' x 8'	G	5.50	2.909		690	128		818	955
3600	Liquid, full wetted, plastic, alum. frame, 3' x 10'	G	5	3.200		234	140		374	475
3650	Collector panel mounting, flat roof or ground rack	G	7	2.286		217	100		317	395
3670	Roof clamps	G	70	.229	Set	2.35	10.05		12.40	18.15
3700	Roof strap, teflon	G 1 Plum	205	.039	L.F.	19.95	1.90		21.85	25
3900	Differential controller with two sensors									
3930	Thermostat, hard wired	G 1 Plum	8	1	Ea.	79.50	49		128.50	163
4050	Pool valve system	G "	2.50	3.200	"	264	156		420	535
4150	Sensors									
4220	Freeze prevention	G 1 Plum	32	.250	Ea.	22	12.20		34.20	43
4300	Heat exchanger									
4315	includes coil, blower, circulator									
4316	and controller for DHW and space hot air									
4380	70 MBH	G Q-1	3.50	4.571	Ea.	335	201		536	675
4400	80 MBH	G "	3	5.333	"	445	234		679	855
4580	Fluid to fluid package includes two circulating pumps									
4590	expansion tank, check valve, relief valve									
4600	controller, high temperature cutoff and sensors	G Q-1	2.50	6.400	Ea.	695	281		976	1,200
4650	Heat transfer fluid									
4700	Propylene glycol, inhibited anti-freeze	G 1 Plum	28	.286	Gal.	13.85	13.95		27.80	37
4800	Solar storage tanks, knocked down									
5020	7' x 10'-6" = 459 C.F./3000 gallons	G Q-10	.80	30	Ea.	14,700	1,325		16,025	18,300
5210	7' x 14' = 613 C.F./4000 gallons	G	.60	40		17,700	1,750		19,450	22,300
5230	10'-6" x 14' = 919 C.F./6000 gallons	G	.40	60		21,500	2,650		24,150	27,800
5250	14' x 17'-6" = 1531 C.F./10,000 gallons	G Q-11	.30	106		28,100	4,800		32,900	38,600
7000	Solar control valves and vents									
7070	Air eliminator, automatic 3/4" size	G 1 Plum	32	.250	Ea.	28.50	12.20		40.70	50.50
7090	Air vent, automatic, 1/8" fitting	G	32	.250		12.65	12.20		24.85	33
7100	Manual, 1/8" NPT	G	32	.250		2.78	12.20		14.98	22
7120	Backflow preventer, 1/2" pipe size	G	16	.500		58.50	24.50		83	102
7130	3/4" pipe size	G	16	.500		101	24.50		125.50	149
7150	Balancing valve, 3/4" pipe size	G	20	.400		40.50	19.50		60	75
7180	Draindown valve, 1/2" copper tube	G	9	.889		202	43.50		245.50	289
7200	Flow control valve, 1/2" pipe size	G	22	.364		102	17.75		119.75	140
7220	Expansion tank, up to 5 gal.	G	32	.250		61.50	12.20		73.70	86.50
7400	Pressure gauge, 2" dial	G	32	.250		22	12.20		34.20	43.50
7450	Relief valve, temp. and pressure 3/4" pipe size	G	30	.267		15.15	13		28.15	36.50
7500	Solenoid valve, normally closed									
7750	Vacuum relief valve, 3/4" pipe size	G 1 Plum	32	.250	Ea.	26.50	12.20		38.70	48
7800	Thermometers									
8250	Water storage tank with heat exchanger and electric element									
8270	66 gal. with 2" x 2 lb. density insulation	G 1 Plum	1.60	5	Ea.	925	244		1,169	1,400
8300	80 gal. with 2" x 2 lb. density insulation	G "	1.60	5	"	1,050	244		1,294	1,525
8500	Water storage module, plastic									
8600	Tubular, 12" diameter, 4' high	G 1 Carp	48	.167	Ea.	103	6.65		109.65	124
8610	12" diameter, 8' high	G "	40	.200		159	8		167	188
8650	Cap, 12" diameter	G				19			19	21
9000	Minimum labor/equipment charge	G 1 Plum	2	4	Job		195		195	300

23 61 Refrigerant Compressors

23 61 16 – Reciprocating Refrigerant Compressors

23 61 16.10 Reciprocating Compressors	Crew	Daily Output	Labor-Hours	Unit	Material	2009 Bare Costs Labor	Equipment	Total	Total Incl O&P
0010 **RECIPROCATING COMPRESSORS**									
0990 Refrigeration, recip. hermetic, switches & protective devices									
1000 10 ton	Q-5	1	16	Ea.	8,150	710		8,860	10,100
1100 20 ton	Q-6	.72	33.333		12,300	1,525		13,825	15,900
1400 50 ton	"	.20	120		16,800	5,525		22,325	27,100
1600 130 ton	Q-7	.21	152	↓	21,200	7,175		28,375	34,400

23 62 Packaged Compressor and Condenser Units

23 62 13 – Packaged Air-Cooled Refrigerant Compressor and Condenser Units

23 62 13.10 Packaged Air-Cooled Refrig. Condensing Units	Crew	Daily Output	Labor-Hours	Unit	Material	2009 Bare Costs Labor	Equipment	Total	Total Incl O&P
0010 **PACKAGED AIR-COOLED REFRIGERANT CONDENSING UNITS**									
0020 Condensing unit									
0030 Air cooled, compressor, standard controls									
0050 1.5 ton	Q-5	2.50	6.400	Ea.	1,125	284		1,409	1,700
0200 2.5 ton	↓	1.70	9.412	↓	1,250	420		1,670	2,025
0300 3 ton		1.30	12.308		1,300	545		1,845	2,275
0400 4 ton		.90	17.778		1,700	790		2,490	3,075
0500 5 ton	↓	.60	26.667	↓	2,050	1,175		3,225	4,075

23 64 Packaged Water Chillers

23 64 19 – Reciprocating Water Chillers

23 64 19.10 Reciprocating Type Water Chillers

23 64 19.10 Reciprocating Type Water Chillers	Crew	Daily Output	Labor-Hours	Unit	Material	2009 Bare Costs Labor	Equipment	Total	Total Incl O&P
0010 **RECIPROCATING TYPE WATER CHILLERS**, With standard controls.									
0494 Water chillers, integral air cooled condenser									
0980 Water cooled, multiple compressor, semi-hermetic, tower not incl.									
1100 50 ton cooling	Q-7	.28	113	Ea.	31,300	5,350		36,650	42,700
1160 100 ton cooling	↓	.18	179	↓	56,000	8,450		64,450	74,500
1200 140 ton cooling	↓	.16	202	↓	74,500	9,525		84,025	97,000

23 64 23 – Scroll Water Chillers

23 64 23.10 Scroll Water Chillers

23 64 23.10 Scroll Water Chillers	Crew	Daily Output	Labor-Hours	Unit	Material	2009 Bare Costs Labor	Equipment	Total	Total Incl O&P
0010 **SCROLL WATER CHILLERS**, With standard controls.									
0490 Packaged w/integral air cooled condenser, 15 ton cool	Q-7	.37	86.486	Ea.	16,500	4,075		20,575	24,400
0500 20 ton cooling	↓	.34	94.118	↓	21,800	4,425		26,225	30,800
0520 40 ton cooling	↓	.30	108	↓	29,500	5,075		34,575	40,400
0680 Scroll water cooled, single compressor, hermetic, tower not incl.									
0760 10 ton cooling	Q-6	.36	67.039	Ea.	6,125	3,100		9,225	11,500

23 64 26 – Rotary-Screw Water Chillers

23 64 26.10 Rotary-Screw Type Water Chillers

23 64 26.10 Rotary-Screw Type Water Chillers	Crew	Daily Output	Labor-Hours	Unit	Material	2009 Bare Costs Labor	Equipment	Total	Total Incl O&P
0010 **ROTARY-SCREW TYPE WATER CHILLERS**, With standard controls									
0110 Screw, liquid chiller, air cooled, insulated evaporator									
0120 130 ton	Q-7	.14	228	Ea.	79,500	10,700		90,200	104,000
0124 160 ton	"	.13	246	"	98,000	11,600		109,600	125,500

23 64 33 – Direct Expansion Water Chillers

23 64 33.10 Direct Expansion Type Water Chillers

23 64 33.10 Direct Expansion Type Water Chillers	Crew	Daily Output	Labor-Hours	Unit	Material	2009 Bare Costs Labor	Equipment	Total	Total Incl O&P
0010 **DIRECT EXPANSION TYPE WATER CHILLERS**, With standard controls									
9000 Minimum labor/equipment charge	Q-6	1	24	Job		1,100		1,100	1,725

23 65 Cooling Towers

23 65 13 – Forced-Draft Cooling Towers

23 65 13.10 Forced-Draft Type Cooling Towers	Crew	Daily Output	Labor-Hours	Unit	Material	2009 Bare Costs Labor	Equipment	Total	Total Incl O&P
0010 **FORCED-DRAFT TYPE COOLING TOWERS**, Packaged units									
0070 Galvanized steel									
0080 Induced draft, crossflow									
0100 Vertical, belt drive, 61 tons	Q-6	90	.267	TonAC	137	12.30		149.30	170
0150 100 ton		100	.240		150	11.05		161.05	182
0200 115 ton		109	.220		130	10.15		140.15	159
0250 131 ton		120	.200		169	9.20		178.20	200
0260 162 ton		132	.182		137	8.40		145.40	163
1000 For higher capacities, use multiples									
1500 Induced air, double flow									
1900 Vertical, gear drive, 167 ton	Q-6	126	.190	TonAC	133	8.80		141.80	160
2000 297 ton		129	.186		83	8.55		91.55	104
2100 582 ton		132	.182		67	8.40		75.40	86.50
2200 1016 ton		150	.160		60	7.35		67.35	77.50
3500 For pumps and piping, add		38	.632		68.50	29		97.50	121
4000 For absorption systems, add					75%	75%			
5000 Fiberglass									
5010 Draw thru									
5100 60 ton	Q-6	1.50	16	Ea.	3,650	735		4,385	5,175
5120 125 ton		.99	24.242		7,475	1,125		8,600	9,925
5140 300 ton		.43	55.814		17,500	2,575		20,075	23,200
5160 600 ton		.22	109		31,400	5,025		36,425	42,300
5180 1000 ton		.15	160		54,000	7,375		61,375	71,000
6000 Stainless steel									
6010 Induced draft, crossflow, horizontal, belt drive									
6100 57 ton	Q-6	1.50	16	Ea.	20,400	735		21,135	23,700
6120 91 ton		.99	24.242		24,700	1,125		25,825	28,800
6140 111 ton		.43	55.814		37,100	2,575		39,675	44,800
6160 126 ton		.22	109		37,100	5,025		42,125	48,600
6170 Induced draft, crossflow, vertical, gear drive									
6172 167 ton	Q-6	.75	32	Ea.	47,100	1,475		48,575	54,500
6174 297 ton		.43	55.814		52,500	2,575		55,075	61,500
6176 582 ton		.23	104		83,000	4,800		87,800	98,500
6180 1016 ton		.15	160		130,000	7,375		137,375	154,500
9000 Minimum labor/equipment charge		1	24	Job		1,100		1,100	1,725

23 72 Air-to-Air Energy Recovery Equipment

23 72 13 – Heat-Wheel Air-to-Air Energy-Recovery Equipment

23 72 13.10 Heat-Wheel Air-to-air Energy Recovery Equip.

	Crew	Daily Output	Labor-Hours	Unit	Material	2009 Bare Costs Labor	Equipment	Total	Total Incl O&P
0010 **HEAT-WHEEL AIR-TO-AIR ENERGY RECOVERY EQUIPMENT**									
0100 Air to air									
9000 Minimum labor/equipment charge	1 Shee	4	2	Job		94.50		94.50	150

23 73 Indoor Central-Station Air-Handling Units

23 73 39 – Indoor, Direct Gas-Fired Heating and Ventilating Units

23 73 39.10 Make-Up Air Unit	Crew	Daily Output	Labor-Hours	Unit	Material	2009 Bare Costs Labor	2009 Bare Costs Equipment	Total	Total Incl O&P
0010 **MAKE-UP AIR UNIT**									
0020 Indoor suspension, natural/LP gas, direct fired,									
0032 standard control. For flue see Div. 23 51 23.10									
0040 70° F temperature rise, MBH is input									
0100 2000 CFM, 168 MBH	Q-6	3	8	Ea.	9,050	370		9,420	10,500
0220 12,000 CFM, 1005 MBH	"	1	24		14,200	1,100		15,300	17,300
0300 24,000 CFM, 2007 MBH	Q-7	1	32		18,500	1,500		20,000	22,600
0400 50,000 CFM, 4180 MBH	"	.80	40	↓	25,400	1,875		27,275	30,800

23 74 Packaged Outdoor HVAC Equipment

23 74 23 – Packaged, Outdoor, Heating-Only Makeup-Air Units

23 74 23.16 Pkgd.,Ind.-Fired,Outdoor, Heat-Only Makeup-Air

	Crew	Daily Output	Labor-Hours	Unit	Material	Labor	Equipment	Total	Total Incl O&P
0010 **PACKAGED, IND.-FIRED, OUTDOOR, HEAT-ONLY MAKEUP-AIR UNITS**									
9000 Minimum labor/equipment charge	Q-6	2.75	8.727	Job		400		400	625

23 74 33 – Packaged, Outdoor, Heating and Cooling Makeup Air-Conditioners

23 74 33.10 Roof Top Air Conditioners

	Crew	Daily Output	Labor-Hours	Unit	Material	Labor	Equipment	Total	Total Incl O&P
0010 **ROOF TOP AIR CONDITIONERS**, Standard controls, curb, economizer.									
0020									
1000 Single zone, electric cool, gas heat									
1090 2 ton cooling, 55 MBH heating	Q-5	.93	17.204	Ea.	3,100	765		3,865	4,575
1100 3 ton cooling, 60 MBH heating		.70	22.857		3,200	1,025		4,225	5,100
1120 4 ton cooling, 95 MBH heating	↓	.61	26.403		3,750	1,175		4,925	5,950
1160 10 ton cooling, 200 MBH heating	Q-6	.67	35.982		8,300	1,650		9,950	11,700
1220 30 ton cooling, 540 MBH heating	Q-7	.47	68.376		25,800	3,225		29,025	33,300
1240 40 ton cooling, 675 MBH heating	"	.35	91.168	↓	32,900	4,275		37,175	42,900
2000 Multizone, electric cool, gas heat, economizer									
2120 20 ton cooling, 360 MBH heating	Q-7	.53	60.038	Ea.	57,500	2,825		60,325	68,000
2180 30 ton cooling, 540 MBH heating		.37	85.562		88,500	4,025		92,525	104,000
2210 50 ton cooling, 540 MBH heating		.23	142		125,500	6,675		132,175	148,500
2220 70 ton cooling, 1500 MBH heating		.16	198		136,000	9,350		145,350	164,000
2240 80 ton cooling, 1500 MBH heating		.14	228		155,500	10,700		166,200	187,500
2280 105 ton cooling, 1500 MBH heating	↓	.11	290		179,500	13,700		193,200	218,500
2400 For hot water heat coil, deduct					5%				
2500 For steam heat coil, deduct					2%				
2600 For electric heat, deduct				↓	3%	5%			
9000 Minimum labor/equipment charge	Q-5	1.50	10.667	Job		475		475	735

23 81 Decentralized Unitary HVAC Equipment

23 81 19 – Self-Contained Air-Conditioners

23 81 19.10 Window Unit Air Conditioners

	Crew	Daily Output	Labor-Hours	Unit	Material	Labor	Equipment	Total	Total Incl O&P
0010 **WINDOW UNIT AIR CONDITIONERS**									
4500 10,000 BTUH	1 Carp	6	1.333	Ea.	395	53.50		448.50	525
4520 12,000 BTUH	L-2	8	2	"	510	70.50		580.50	675
9000 Minimum labor/equipment charge	1 Carp	2	4	Job		160		160	264

23 81 19.20 Self-Contained Single Package

	Crew	Daily Output	Labor-Hours	Unit	Material	Labor	Equipment	Total	Total Incl O&P
0010 **SELF-CONTAINED SINGLE PACKAGE**									
0100 Air cooled, for free blow or duct, not incl. remote condenser									
0200 3 ton cooling	Q-5	1	16	Ea.	3,200	710		3,910	4,625

23 81 Decentralized Unitary HVAC Equipment

23 81 19 – Self-Contained Air-Conditioners

23 81 19.20 Self-Contained Single Package

		Crew	Daily Output	Labor-Hours	Unit	Material	2009 Bare Costs Labor	Equipment	Total	Total Incl O&P
0210	4 ton cooling	Q-5	.80	20	Ea.	3,500	890		4,390	5,225
0240	10 ton cooling	Q-7	1	32		6,600	1,500		8,100	9,575
0280	30 ton cooling	"	.80	40	↓	19,100	1,875		20,975	23,900
1000	Water cooled for free blow or duct, not including tower									
1010	Constant volume									
1100	3 ton cooling	Q-6	1	24	Ea.	3,075	1,100		4,175	5,125
1120	5 ton cooling	"	1	24		4,025	1,100		5,125	6,150
1140	10 ton cooling	Q-7	.90	35.556		7,900	1,675		9,575	11,300
1180	30 ton cooling	"	.70	45.714		34,200	2,150		36,350	40,900
1240	60 ton cooling	Q-8	.30	106	↓	63,000	5,025	186	68,211	77,000
9000	Minimum labor/equipment charge	Q-5	1	16	Job		710		710	1,100

23 81 23 – Computer-Room Air-Conditioners

23 81 23.10 Computer Room Units

		Crew	Daily Output	Labor-Hours	Unit	Material	2009 Bare Costs Labor	Equipment	Total	Total Incl O&P
0010	**COMPUTER ROOM UNITS**									
1000	Air cooled, includes remote condenser but not									
1020	interconnecting tubing or refrigerant									
1160	6 ton	Q-5	.30	53.333	Ea.	29,500	2,375		31,875	36,200
1240	10 ton	"	.25	64		32,800	2,850		35,650	40,500
1300	20 ton	Q-6	.26	92.308		43,200	4,250		47,450	54,000
1320	22 ton	"	.24	100	↓	43,800	4,600		48,400	55,500

23 81 26 – Split-System Air-Conditioners

23 81 26.10 Split Ductless Systems

		Crew	Daily Output	Labor-Hours	Unit	Material	2009 Bare Costs Labor	Equipment	Total	Total Incl O&P
0010	**SPLIT DUCTLESS SYSTEMS**									
0100	Cooling only, single zone									
0110	Wall mount									
0120	3/4 ton cooling	Q-5	2	8	Ea.	1,175	355		1,530	1,825
0130	1 ton cooling		1.80	8.889		1,300	395		1,695	2,025
0140	1-1/2 ton cooling		1.60	10		1,800	445		2,245	2,675
0150	2 ton cooling	↓	1.40	11.429	↓	2,125	510		2,635	3,125
1000	Ceiling mount									
1020	2 ton cooling	Q-5	1.40	11.429	Ea.	1,450	510		1,960	2,350
1030	3 ton cooling	"	1.20	13.333	"	3,900	590		4,490	5,225
2000	T-Bar mount									
2010	2 ton cooling	Q-5	1.40	11.429	Ea.	3,400	510		3,910	4,525
2020	3 ton cooling		1.20	13.333		3,800	590		4,390	5,100
2030	3-1/2 ton cooling	↓	1.10	14.545	↓	4,200	645		4,845	5,625
3000	Multizone									
3010	Wall mount									
3020	2 @ 3/4 ton cooling	Q-5	1.80	8.889	Ea.	1,475	395		1,870	2,225
5000	Cooling / Heating									
5010	Wall mount									
5110	1 ton cooling	Q-5	1.70	9.412	Ea.	1,100	420		1,520	1,850
5120	1-1/2 ton cooling	"	1.50	10.667	"	1,750	475		2,225	2,650
5300	Ceiling mount									
5310	3 ton cooling	Q-5	1	16	Ea.	4,125	710		4,835	5,650
7000	Accessories for all split ductless systems									
7010	Add for ambient frost control	Q-5	8	2	Ea.	130	89		219	281
7020	Add for tube / wiring kit									
7030	15' kit	Q-5	32	.500	Ea.	38.50	22		60.50	77
7040	35' kit	"	24	.667	"	125	29.50		154.50	183

23 81 Decentralized Unitary HVAC Equipment

23 81 43 – Air-Source Unitary Heat Pumps

23 81 43.10 Air-Source Heat Pumps	Crew	Daily Output	Labor-Hours	Unit	Material	2009 Bare Costs Labor	2009 Bare Costs Equipment	Total	Total Incl O&P
0010 **AIR-SOURCE HEAT PUMPS**, Not including interconnecting tubing.									
1000 Air to air, split system, not including curbs, pads, fan coil and ductwork									
1012 Outside condensing unit only, for fan coil see Div. 23 82 19.10									
1020 2 ton cooling, 8.5 MBH heat @ 0° F	Q-5	2	8	Ea.	1,650	355		2,005	2,375
1040 3 ton cooling, 13 MBH heat @ 0° F		1.20	13.333		2,100	590		2,690	3,225
1080 7.5 ton cooling, 33 MBH heat @ 0° F	↓	.45	35.556		5,050	1,575		6,625	8,000
1120 15 ton cooling, 64 MBH heat @ 0° F	Q-6	.50	48		10,200	2,200		12,400	14,600
1130 20 ton cooling, 85 MBH heat @ 0° F		.35	68.571		15,300	3,150		18,450	21,700
1140 25 ton cooling, 119 MBH heat @ 0° F	↓	.25	96	↓	18,400	4,425		22,825	27,200
1500 Single package, not including curbs, pads, or plenums									
1520 2 ton cooling, 6.5 MBH heat @ 0° F	Q-5	1.50	10.667	Ea.	2,575	475		3,050	3,550
1560 3 ton cooling, 10 MBH heat @ 0° F	"	1.20	13.333	"	3,050	590		3,640	4,275

23 81 46 – Water-Source Unitary Heat Pumps

23 81 46.10 Water Source Heat Pumps	Crew	Daily Output	Labor-Hours	Unit	Material	2009 Bare Costs Labor	2009 Bare Costs Equipment	Total	Total Incl O&P
0010 **WATER SOURCE HEAT PUMPS**, Not incl. connecting tubing or water source									
2000 Water source to air, single package									
2100 1 ton cooling, 13 MBH heat @ 75° F	Q-5	2	8	Ea.	1,300	355		1,655	1,975
2140 2 ton cooling, 19 MBH heat @ 75° F		1.70	9.412		1,550	420		1,970	2,350
2220 5 ton cooling, 29 MBH heat @ 75° F	↓	.90	17.778	↓	2,450	790		3,240	3,925
9000 Minimum labor/equipment charge	↓	1.75	9.143	Job		405		405	630

23 82 Convection Heating and Cooling Units

23 82 16 – Air Coils

23 82 16.10 Flanged Coils

	Crew	Daily Output	Labor-Hours	Unit	Material	2009 Bare Costs Labor	2009 Bare Costs Equipment	Total	Total Incl O&P
0010 **FLANGED COILS**									
1500 Hot water heating, 1 row, 24" x 48"	Q-5	4	4	Ea.	1,425	178		1,603	1,850
9000 Minimum labor/equipment charge	1 Plum	1	8	Job		390		390	605

23 82 16.20 Duct Heaters

	Crew	Daily Output	Labor-Hours	Unit	Material	2009 Bare Costs Labor	2009 Bare Costs Equipment	Total	Total Incl O&P
0010 **DUCT HEATERS**, Electric, 480 V, 3 Ph.									
0020 Finned tubular insert, 500 ° F									
0100 8" wide x 6" high, 4.0 kW	Q-20	16	1.250	Ea.	705	54		759	860
0120 12" high, 8.0 kW		15	1.333		1,175	58		1,233	1,375
0140 18" high, 12.0 kW		14	1.429		1,650	62		1,712	1,900
0160 24" high, 16.0 kW		13	1.538		2,100	66.50		2,166.50	2,425
0180 30" high, 20.0 kW		12	1.667		2,575	72.50		2,647.50	2,950
0300 12" wide x 6" high, 6.7 kW		15	1.333		750	58		808	915
0320 12" high, 13.3 kW		14	1.429		1,200	62		1,262	1,425
0340 18" high, 20.0 kW	↓	13	1.538	↓	1,700	66.50		1,766.50	1,975
8000 To obtain BTU multiply kW by 3413									
9900 Minimum labor/equipment charge	1 Shee	4	2	Job		94.50		94.50	150

23 82 19 – Fan Coil Units

23 82 19.10 Fan Coil Air Conditioning

	Crew	Daily Output	Labor-Hours	Unit	Material	2009 Bare Costs Labor	2009 Bare Costs Equipment	Total	Total Incl O&P
0010 **FAN COIL AIR CONDITIONING** Cabinet mounted, filters, controls									
0020									
0100 Chilled water, 1/2 ton cooling	Q-5	8	2	Ea.	675	89		764	885
0120 1 ton cooling		6	2.667		775	118		893	1,050
0160 2.5 ton cooling		5	3.200		1,450	142		1,592	1,825
0180 3 ton cooling	↓	4	4	↓	1,650	178		1,828	2,100
0262 For hot water coil, add					40%	10%			

23 82 Convection Heating and Cooling Units

23 82 19 – Fan Coil Units

23 82 19.10 Fan Coil Air Conditioning	Crew	Daily Output	Labor-Hours	Unit	Material	2009 Bare Costs Labor	Equipment	Total	Total Incl O&P
0940 Direct expansion, for use w/air cooled condensing unit, 1.5 ton cooling	Q-5	5	3.200	Ea.	540	142		682	815
1000 5 ton cooling	"	3	5.333		1,025	237		1,262	1,500
1040 10 ton cooling	Q-6	2.60	9.231		2,750	425		3,175	3,675
1060 20 ton cooling	"	.70	34.286		4,850	1,575		6,425	7,775
1512 For condensing unit add see Div. 23 62									
3000 Chilled water, horizontal unit, housing, 2 pipe, controls									
3100 1/2 ton cooling	Q-5	8	2	Ea.	1,050	89		1,139	1,325
3110 1 ton cooling		6	2.667		1,275	118		1,393	1,600
3120 1.5 ton cooling		5.50	2.909		1,525	129		1,654	1,900
3130 2 ton cooling		5.25	3.048		1,850	135		1,985	2,225
3140 3 ton cooling		4	4		2,125	178		2,303	2,625
3150 3.5 ton cooling		3.80	4.211		2,350	187		2,537	2,875
3160 4 ton cooling		3.80	4.211		2,350	187		2,537	2,875
4000 With electric heat, 2 pipe									
4100 1/2 ton cooling	Q-5	8	2	Ea.	1,300	89		1,389	1,575
4105 3/4 ton cooling		7	2.286		1,425	102		1,527	1,725
4110 1 ton cooling		6	2.667		1,575	118		1,693	1,900
4120 1.5 ton cooling		5.50	2.909		1,750	129		1,879	2,125
4130 2 ton cooling		5.25	3.048		2,200	135		2,335	2,625
4135 2.5 ton cooling		4.80	3.333		3,050	148		3,198	3,575
4140 3 ton cooling		4	4		4,175	178		4,353	4,875
4150 3.5 ton cooling		3.80	4.211		5,075	187		5,262	5,875
4160 4 ton cooling		3.60	4.444		5,875	197		6,072	6,750
9000 Minimum labor/equipment charge		3.75	4.267	Job		190		190	294

23 82 29 – Radiators

23 82 29.10 Hydronic Heating

	Crew	Daily Output	Labor-Hours	Unit	Material	Labor	Equipment	Total	Total Incl O&P
0010 **HYDRONIC HEATING**, Terminal units, not incl. main supply pipe									
1000 Radiation									
3000 Radiators, cast iron									
3100 Free standing or wall hung, 6 tube, 25" high	Q-5	96	.167	Section	30	7.40		37.40	44.50
9000 Minimum labor/equipment charge	"	3.20	5	Job		222		222	345

23 82 36 – Finned-Tube Radiation Heaters

23 82 36.10 Finned Tube Radiation

	Crew	Daily Output	Labor-Hours	Unit	Material	Labor	Equipment	Total	Total Incl O&P
0010 **FINNED TUBE RADIATION**, Terminal units, not incl. main supply pipe									
1150 Fin tube, wall hung, 14" slope top cover, with damper									
1200 1-1/4" copper tube, 4-1/4" alum. fin	Q-5	38	.421	L.F.	39	18.70		57.70	71.50
1250 1-1/4" steel tube, 4-1/4" steel fin		36	.444		36.50	19.75		56.25	71
1255 2" steel tube, 4-1/4" steel fin		32	.500		39.50	22		61.50	78
1310 Baseboard, pkgd, 1/2" copper tube, alum. fin, 7" high		60	.267		7.20	11.85		19.05	26.50
1320 3/4" copper tube, alum. fin, 7" high		58	.276		7.65	12.25		19.90	27.50
1340 1" copper tube, alum. fin, 8-7/8" high		56	.286		24.50	12.70		37.20	46
1360 1-1/4" copper tube, alum. fin, 8-7/8" high		54	.296		25	13.15		38.15	48
1500 Note: fin tube may also require corners, caps, etc.									

23 82 39 – Unit Heaters

23 82 39.16 Propeller Unit Heaters

	Crew	Daily Output	Labor-Hours	Unit	Material	Labor	Equipment	Total	Total Incl O&P
0010 **PROPELLER UNIT HEATERS**									
3950 Unit heaters, propeller, 115 V 2 psi steam, 60° F entering air									
4000 Horizontal, 12 MBH	Q-5	12	1.333	Ea.	315	59		374	435
4060 43.9 MBH		8	2		475	89		564	665
4240 286.9 MBH		2	8		1,425	355		1,780	2,125
4250 326.0 MBH		1.90	8.421		1,825	375		2,200	2,575

23 82 Convection Heating and Cooling Units

23 82 39 – Unit Heaters

23 82 39.16 Propeller Unit Heaters

		Crew	Daily Output	Labor-Hours	Unit	Material	2009 Bare Costs Labor	Equipment	Total	Total Incl O&P
4260	364 MBH	Q-5	1.80	8.889	Ea.	2,000	395		2,395	2,775
4270	404 MBH	↓	1.60	10		2,500	445		2,945	3,450
4300	For vertical diffuser, add					174			174	191
4310	Vertical flow, 40 MBH	Q-5	11	1.455		465	64.50		529.50	615
4326	131.0 MBH	"	4	4		730	178		908	1,075
4354	420 MBH, (460 V)	Q-6	1.80	13.333		1,875	615		2,490	3,000
4358	500 MBH, (460 V)		1.71	14.035		2,475	645		3,120	3,725
4362	570 MBH, (460 V)		1.40	17.143		3,725	790		4,515	5,325
4366	620 MBH, (460 V)		1.30	18.462		4,275	850		5,125	6,025
4370	960 MBH, (460 V)	↓	1.10	21.818	↓	7,200	1,000		8,200	9,450

23 83 Radiant Heating Units

23 83 33 – Electric Radiant Heaters

23 83 33.10 Electric Heating

		Crew	Daily Output	Labor-Hours	Unit	Material	2009 Bare Costs Labor	Equipment	Total	Total Incl O&P
0010	**ELECTRIC HEATING**, not incl. conduit or feed wiring.									
1300	Baseboard heaters, 2' long, 375 watt	1 Elec	8	1	Ea.	40	47		87	117
1400	3' long, 500 watt		8	1		46	47		93	123
1600	4' long, 750 watt		6.70	1.194		54	56		110	146
1800	5' long, 935 watt		5.70	1.404		68	66		134	176
2000	6' long, 1125 watt		5	1.600		74	75		149	198
2400	8' long, 1500 watt		4	2		92	94		186	246
2600	9' long, 1680 watt		3.60	2.222		104	104		208	275
2800	10' long, 1875 watt	↓	3.30	2.424	↓	193	114		307	385
2950	Wall heaters with fan, 120 to 277 volt									
3190	1500 watt	1 Elec	4	2	Ea.	111	94		205	267
3230	2500 watt		3.50	2.286		252	107		359	445
3250	4000 watt		2.70	2.963		385	139		524	635
3600	Thermostats, integral		16	.500		20.50	23.50		44	58.50
3800	Line voltage, 1 pole		8	1		31	47		78	107
3810	2 pole		8	1	↓	33.50	47		80.50	110
9990	Minimum labor/equipment charge	↓	4	2	Job		94		94	145

23 84 Humidity Control Equipment

23 84 13 – Humidifiers

23 84 13.10 Humidifier Units

		Crew	Daily Output	Labor-Hours	Unit	Material	2009 Bare Costs Labor	Equipment	Total	Total Incl O&P
0010	**HUMIDIFIER UNITS**									
0520	Steam, room or duct, filter, regulators, auto. controls, 220 V									
0560	22 lb. per hour	Q-5	5	3.200	Ea.	2,525	142		2,667	3,000
0580	33 lb. per hour		4	4		2,575	178		2,753	3,125
0620	100 lb. per hour		3	5.333	↓	3,800	237		4,037	4,550
9000	Minimum labor/equipment charge	↓	3.50	4.571	Job		203		203	315

23 91 Prefabricated Equipment Supports

23 91 10 – Prefabricated Curbs, Pads and Stands

23 91 10.10 Prefabricated Pads and Stands	Crew	Daily Output	Labor-Hours	Unit	Material	2009 Bare Costs Labor	2009 Bare Costs Equipment	Total	Total Incl O&P
0010 **PREFABRICATED PADS AND STANDS**									
6000 Pad, fiberglass reinforced concrete with polystyrene foam core									
6050 Condenser, 2" thick, 20" x 38"	1 Shee	8	1	Ea.	22	47		69	99.50
6090 24" x 36"	"	12	.667		28	31.50		59.50	81
6280 36" x 36"	Q-9	8	2		40.50	85		125.50	180
6300 36" x 40"		7	2.286		46	97		143	205
6320 36" x 48"		7	2.286		54	97		151	214

Division Notes

	CREW	DAILY OUTPUT	LABOR-HOURS	UNIT	2009 BARE COSTS				TOTAL INCL O&P
					MAT.	LABOR	EQUIP.	TOTAL	

Estimating Tips

26 05 00 Common Work Results for Electrical

- Conduit should be taken off in three main categories—power distribution, branch power, and branch lighting—so the estimator can concentrate on systems and components, therefore making it easier to ensure all items have been accounted for.

- For cost modifications for elevated conduit installation, add the percentages to labor according to the height of installation, and only to the quantities exceeding the different height levels, not to the total conduit quantities.

- Remember that aluminum wiring of equal ampacity is larger in diameter than copper and may require larger conduit.

- If more than three wires at a time are being pulled, deduct percentages from the labor hours of that grouping of wires.

- When taking off grounding systems, identify separately the type and size of wire, and list each unique type of ground connection.

- The estimator should take the weights of materials into consideration when completing a takeoff. Topics to consider include: How will the materials be supported? What methods of support are available? How high will the support structure have to reach? Will the final support structure be able to withstand the total burden? Is the support material included or separate from the fixture, equipment, and material specified?

- Do not overlook the costs for equipment used in the installation. If scaffolding or highlifts are available in the field, contractors may use them in lieu of the proposed ladders and rolling staging.

26 20 00 Low-Voltage Electrical Transmission

- Supports and concrete pads may be shown on drawings for the larger equipment, or the support system may be only a piece of plywood for the back of a panelboard. In either case, it must be included in the costs.

26 40 00 Electrical and Cathodic Protection

- When taking off cathodic protections systems, identify the type and size of cable, and list each unique type of anode connection.

26 50 00 Lighting

- Fixtures should be taken off room by room, using the fixture schedule, specifications, and the ceiling plan. For large concentrations of lighting fixtures in the same area, deduct the percentages from labor hours.

Reference Numbers

Reference numbers are shown in shaded boxes at the beginning of some major classifications. These numbers refer to related items in the Reference Section. The reference information may be an estimating procedure, an alternate pricing method, or technical information.

Note: Not all subdivisions listed here necessarily appear in this publication.

Note: **Trade Service,** *in part, has been used as a reference source for some of the material prices used in Division 26.*

26 05 05 – Selective Electrical Demolition

26 05 05.10 Electrical Demolition		Crew	Daily Output	Labor-Hours	Unit	Material	2009 Bare Costs		Total	Total Incl O&P
							Labor	Equipment		
0010	**ELECTRICAL DEMOLITION** R260105-30									
0020	· Conduit to 15' high, including fittings & hangers									
0100	Rigid galvanized steel, 1/2" to 1" diameter R024119-10	1 Elec	242	.033	L.F.		1.55		1.55	2.39
0120	1-1/4" to 2"	"	200	.040			1.88		1.88	2.89
0140	2-1/2" to 3-1/2"	2 Elec	302	.053			2.49		2.49	3.83
0200	Electric metallic tubing (EMT), 1/2" to 1"	1 Elec	394	.020			.95		.95	1.47
0220	1-1/4" to 1-1/2"		326	.025			1.15		1.15	1.77
0240	2" to 3"		236	.034			1.59		1.59	2.45
0400	Wiremold raceway, including fittings & hangers									
0420	No. 3000	1 Elec	250	.032	L.F.		1.50		1.50	2.31
0440	No. 4000	"	217	.037	"		1.73		1.73	2.67
0500	Channels, steel, including fittings & hangers									
0520	3/4" x 1-1/2"	1 Elec	308	.026	L.F.		1.22		1.22	1.88
0540	1-1/2" x 1-1/2"	"	269	.030	"		1.40		1.40	2.15
0600	Copper bus duct, indoor, 3 phase									
0610	Including hangers & supports									
0620	225 amp	2 Elec	135	.119	L.F.		5.55		5.55	8.55
0640	400 amp		106	.151			7.10		7.10	10.90
0660	600 amp		86	.186			8.75		8.75	13.45
0680	1000 amp		60	.267			12.55		12.55	19.30
0700	1600 amp		40	.400			18.80		18.80	29
0720	3000 amp		10	1.600			75		75	116
0800	Plug-in switches, 600V 3 ph, incl. disconnecting									
0820	wire, conduit terminations, 30 amp	1 Elec	15.50	.516	Ea.		24.50		24.50	37.50
0840	60 amp		13.90	.576			27		27	41.50
0850	100 amp		10.40	.769			36		36	55.50
0860	200 amp		6.20	1.290			60.50		60.50	93.50
0940	1200 amp	2 Elec	2	8			375		375	580
0960	1600 amp	"	1.70	9.412			440		440	680
1010	Safety switches, 250 or 600V, incl. disconnection									
1050	of wire & conduit terminations									
1100	30 amp	1 Elec	12.30	.650	Ea.		30.50		30.50	47
1120	60 amp		8.80	.909			42.50		42.50	65.50
1140	100 amp		7.30	1.096			51.50		51.50	79
1160	200 amp		5	1.600			75		75	116
1210	Panel boards, incl. removal of all breakers,									
1220	conduit terminations & wire connections									
1230	3 wire, 120/240V, 100A, to 20 circuits	1 Elec	2.60	3.077	Ea.		145		145	222
1240	200 amps, to 42 circuits	2 Elec	2.60	6.154			289		289	445
1260	4 wire, 120/208V, 125A, to 20 circuits	1 Elec	2.40	3.333			157		157	241
1270	200 amps, to 42 circuits	2 Elec	2.40	6.667			315		315	480
1300	Transformer, dry type, 1 phase, incl. removal of									
1320	supports, wire & conduit terminations									
1340	1 kVA	1 Elec	7.70	1.039	Ea.		49		49	75
1360	5 kVA	"	4.70	1.702			80		80	123
1420	75 kVA	2 Elec	2.50	6.400			300		300	465
1440	3 phase to 600V, primary									
1460	3 kVA	1 Elec	3.85	2.078	Ea.		97.50		97.50	150
1480	15 kVA	2 Elec	4.20	3.810			179		179	275
1500	30 kVA	"	3.50	4.571			215		215	330
1530	112.5 kVA	R-3	2.90	6.897			320	44.50	364.50	545
1560	500 kVA		1.40	14.286			660	92.50	752.50	1,125
1570	750 kVA		1.10	18.182			840	117	957	1,425

26 05 05.10 Electrical Demolition	Crew	Daily Output	Labor-Hours	Unit	Material	2009 Bare Costs Labor	Equipment	Total	Total Incl O&P
1600 Pull boxes & cabinets, sheet metal, incl. removal									
1620 of supports and conduit terminations									
1640 6" x 6" x 4"	1 Elec	31.10	.257	Ea.		12.10		12.10	18.60
1660 12" x 12" x 4"		23.30	.343			16.15		16.15	25
1720 Junction boxes, 4" sq. & oct.		80	.100			4.70		4.70	7.25
1740 Handy box		107	.075			3.51		3.51	5.40
1760 Switch box		107	.075			3.51		3.51	5.40
1780 Receptacle & switch plates	▼	257	.031	▼		1.46		1.46	2.25
1800 Wire, THW-THWN-THHN, removed from									
1810 in place conduit, to 15' high									
1830 #14	1 Elec	65	.123	C.L.F.		5.80		5.80	8.90
1840 #12		55	.145			6.85		6.85	10.50
1850 #10	▼	45.50	.176			8.25		8.25	12.70
1880 #4	2 Elec	53	.302			14.20		14.20	22
1890 #3		50	.320			15.05		15.05	23
1910 1/0		33.20	.482			22.50		22.50	35
1920 2/0		29.20	.548			26		26	39.50
1930 3/0		25	.640			30		30	46.50
1980 400 kcmil		17	.941			44		44	68
1990 500 kcmil	▼	16.20	.988	▼		46.50		46.50	71.50
2000 Interior fluorescent fixtures, incl. supports									
2010 & whips, to 15' high									
2100 Recessed drop-in 2' x 2', 2 lamp	2 Elec	35	.457	Ea.		21.50		21.50	33
2120 2' x 4', 2 lamp		33	.485			23		23	35
2140 2' x 4', 4 lamp		30	.533			25		25	38.50
2160 4' x 4', 4 lamp	▼	20	.800	▼		37.50		37.50	58
2180 Surface mount, acrylic lens & hinged frame									
2200 1' x 4', 2 lamp	2 Elec	44	.364	Ea.		17.10		17.10	26.50
2220 2' x 2', 2 lamp		44	.364			17.10		17.10	26.50
2260 2' x 4', 4 lamp		33	.485			23		23	35
2280 4' x 4', 4 lamp	▼	23	.696	▼		32.50		32.50	50.50
2300 Strip fixtures, surface mount									
2320 4' long, 1 lamp	2 Elec	53	.302	Ea.		14.20		14.20	22
2340 4' long, 2 lamp		50	.320			15.05		15.05	23
2360 8' long, 1 lamp		42	.381			17.90		17.90	27.50
2380 8' long, 2 lamp	▼	40	.400	▼		18.80		18.80	29
2400 Pendant mount, industrial, incl. removal									
2410 of chain or rod hangers, to 15' high									
2420 4' long, 2 lamp	2 Elec	35	.457	Ea.		21.50		21.50	33
2440 8' long, 2 lamp	"	27	.593	"		28		28	43
2460 Interior incandescent, surface, ceiling									
2470 or wall mount, to 12' high									
2480 Metal cylinder type, 75 Watt	2 Elec	62	.258	Ea.		12.15		12.15	18.65
2500 150 Watt	"	62	.258	"		12.15		12.15	18.65
2520 Metal halide, high bay									
2540 400 Watt	2 Elec	15	1.067	Ea.		50		50	77
2560 1000 Watt		12	1.333			62.50		62.50	96.50
2580 150 Watt, low bay	▼	20	.800	▼		37.50		37.50	58
2600 Exterior fixtures, incandescent, wall mount									
2620 100 Watt	2 Elec	50	.320	Ea.		15.05		15.05	23
2640 Quartz, 500 Watt		33	.485			23		23	35
2660 1500 Watt	▼	27	.593	▼		28		28	43
2680 Wall pack, mercury vapor									

26 05 Common Work Results for Electrical

26 05 05 – Selective Electrical Demolition

26 05 05.10 Electrical Demolition	Crew	Daily Output	Labor-Hours	Unit	Material	2009 Bare Costs Labor	Equipment	Total	Total Incl O&P
2700	175 Watt	2 Elec	25	.640	Ea.		30		30	46.50
2720	250 Watt	"	25	.640	"		30		30	46.50
9000	Minimum labor/equipment charge	1 Elec	4	2	Job		94		94	145

26 05 19 – Low-Voltage Electrical Power Conductors and Cables

26 05 19.20 Armored Cable

		Crew	Daily Output	Labor-Hours	Unit	Material	Labor	Equipment	Total	Total Incl O&P
0010	**ARMORED CABLE**									
0050	600 volt, copper (BX), #14, 2 conductor, solid	1 Elec	2.40	3.333	C.L.F.	77.50	157		234.50	325
0100	3 conductor, solid		2.20	3.636		121	171		292	395
0150	#12, 2 conductor, solid		2.30	3.478		79	163		242	340
0200	3 conductor, solid		2	4		128	188		316	430
0250	#10, 2 conductor, solid		2	4		150	188		338	455
0300	3 conductor, solid		1.60	5		221	235		456	605
0350	#8, 3 conductor, solid		1.30	6.154		450	289		739	940
9010	600 volt, copper (MC) steel clad, #14, 2 wire		2.40	3.333		77.50	157		234.50	325
9020	3 wire		2.20	3.636		120	171		291	395
9040	#12, 2 wire		2.30	3.478		79	163		242	340
9050	3 wire		2	4		128	188		316	430
9070	#10, 2 wire		2	4		150	188		338	455
9080	3 wire		1.60	5		221	235		456	605
9100	#8, 2 wire, stranded		1.80	4.444		305	209		514	655
9110	3 wire, stranded		1.30	6.154		455	289		744	945
9900	Minimum labor/equipment charge		4	2	Job		94		94	145

26 05 19.50 Mineral Insulated Cable

		Crew	Daily Output	Labor-Hours	Unit	Material	Labor	Equipment	Total	Total Incl O&P
0010	**MINERAL INSULATED CABLE** 600 volt									
0100	1 conductor, #12	1 Elec	1.60	5	C.L.F.	310	235		545	700
1500	2 conductor, #12		1.40	5.714		640	269		909	1,125
1600	#10		1.20	6.667		775	315		1,090	1,325
1800	#8		1.10	7.273		965	340		1,305	1,575
2000	#6		1.05	7.619		1,225	360		1,585	1,900
2100	#4	2 Elec	2	8		1,625	375		2,000	2,350
2200	3 conductor, #12	1 Elec	1.20	6.667		805	315		1,120	1,375
2400	#10		1.10	7.273		930	340		1,270	1,550
2600	#8		1.05	7.619		1,125	360		1,485	1,775
2800	#6		1	8		1,450	375		1,825	2,175
3000	#4	2 Elec	1.80	8.889		1,775	420		2,195	2,625
3100	4 conductor, #12	1 Elec	1.20	6.667		855	315		1,170	1,425
3200	#10		1.10	7.273		1,025	340		1,365	1,650
3400	#8		1	8		1,325	375		1,700	2,025
3600	#6		.90	8.889		1,650	420		2,070	2,475
3620	7 conductor, #12		1.10	7.273		1,075	340		1,415	1,725
3640	#10		1	8		1,350	375		1,725	2,050
3800	Terminations, 600 volt, 1 conductor, #12		8	1	Ea.	13.55	47		60.55	87.50
5500	2 conductor, #12		6.70	1.194		13.55	56		69.55	101
5600	#10		6.40	1.250		20.50	59		79.50	113
5800	#8		6.20	1.290		20.50	60.50		81	116
6000	#6		5.70	1.404		20.50	66		86.50	124
6200	#4		5.30	1.509		45.50	71		116.50	159
6400	3 conductor, #12		5.70	1.404		20.50	66		86.50	124
6500	#10		5.50	1.455		20.50	68.50		89	128
6600	#8		5.20	1.538		20.50	72.50		93	134
6800	#6		4.80	1.667		20.50	78.50		99	144
7200	#4		4.60	1.739		45.50	81.50		127	176

26 05 19.50 Mineral Insulated Cable

	26 05 19.50 Mineral Insulated Cable	Crew	Daily Output	Labor-Hours	Unit	Material	2009 Bare Costs Labor	Equipment	Total	Total Incl O&P
7400	4 conductor, #12	1 Elec	4.60	1.739	Ea.	22.50	81.50		104	151
7500	#10		4.40	1.818		22.50	85.50		108	156
7600	#8		4.20	1.905		22.50	89.50		112	163
8400	#6		4	2		47.50	94		141.50	198
8500	7 conductor, #12		3.50	2.286		22.50	107		129.50	190
8600	#10		3	2.667		49	125		174	247
8800	Crimping tool, plier type					49.50			49.50	54.50
9000	Stripping tool					187			187	205
9200	Hand vise					55.50			55.50	61.50
9500	Minimum labor/equipment charge	1 Elec	4	2	Job		94		94	145

26 05 19.55 Non-Metallic Sheathed Cable

	26 05 19.55 Non-Metallic Sheathed Cable	Crew	Daily Output	Labor-Hours	Unit	Material	2009 Bare Costs Labor	Equipment	Total	Total Incl O&P
0010	**NON-METALLIC SHEATHED CABLE** 600 volt									
0100	Copper with ground wire, (Romex)									
0150	#14, 2 conductor	1 Elec	2.70	2.963	C.L.F.	32.50	139		171.50	250
0200	3 conductor		2.40	3.333		50	157		207	296
0250	#12, 2 conductor		2.50	3.200		50	150		200	286
0300	3 conductor		2.20	3.636		70	171		241	340
0350	#10, 2 conductor		2.20	3.636		78	171		249	350
0400	3 conductor		1.80	4.444		111	209		320	440
0450	#8, 3 conductor		1.50	5.333		185	251		436	590
0500	#6, 3 conductor		1.40	5.714		300	269		569	745
0550	SE type SER aluminum cable, 3 RHW and									
0600	1 bare neutral, 3 #8 & 1 #8	1 Elec	1.60	5	C.L.F.	161	235		396	535
0650	3 #6 & 1 #6	"	1.40	5.714		182	269		451	615
0700	3 #4 & 1 #6	2 Elec	2.40	6.667		204	315		519	705
0750	3 #2 & 1 #4		2.20	7.273		300	340		640	855
0800	3 #1/0 & 1 #2		2	8		455	375		830	1,075
0850	3 #2/0 & 1 #1		1.80	8.889		535	420		955	1,225
0900	3 #4/0 & 1 #2/0		1.60	10		765	470		1,235	1,575
6500	Service entrance cap for copper SEU									
6700	150 amp	1 Elec	10	.800	Ea.	22	37.50		59.50	82.50
6800	200 amp		8	1	"	33.50	47		80.50	110
9000	Minimum labor/equipment charge		4	2	Job		94		94	145

26 05 19.90 Wire

	26 05 19.90 Wire	Crew	Daily Output	Labor-Hours	Unit	Material	2009 Bare Costs Labor	Equipment	Total	Total Incl O&P
0010	**WIRE** R260533-22									
0020	600 volt type THW, copper, solid, #14	1 Elec	13	.615	C.L.F.	10.30	29		39.30	56
0030	#12		11	.727		15.90	34		49.90	70
0040	#10		10	.800		25	37.50		62.50	85.50
0051	Wire, 600 volt, stranded									
0140	#8	1 Elec	8	1	C.L.F.	43.50	47		90.50	121
0160	#6	"	6.50	1.231		67.50	58		125.50	164
0180	#4	2 Elec	10.60	1.509		106	71		177	226
0200	#3		10	1.600		134	75		209	264
0220	#2		9	1.778		168	83.50		251.50	315
0240	#1		8	2		213	94		307	380
0260	1/0		6.60	2.424		259	114		373	460
0280	2/0		5.80	2.759		325	130		455	560
0300	3/0		5	3.200		410	150		560	680
0350	4/0		4.40	3.636		515	171		686	830
0400	250 kcmil	3 Elec	6	4		610	188		798	960
0420	300 kcmil		5.70	4.211		725	198		923	1,100
0450	350 kcmil		5.40	4.444		850	209		1,059	1,250

26 05 19 – Low-Voltage Electrical Power Conductors and Cables

26 05 19.90 Wire		Crew	Daily Output	Labor-Hours	Unit	Material	2009 Bare Costs Labor	Equipment	Total	Total Incl O&P
0480	400 kcmil	3 Elec	5.10	4.706	C.L.F.	970	221		1,191	1,425
0490	500 kcmil		4.80	5		1,175	235		1,410	1,650
0510	750 kcmil		3.30	7.273		2,600	340		2,940	3,400
0540	Aluminum, stranded, #6	1 Elec	8	1		31.50	47		78.50	108
0560	#4	2 Elec	13	1.231		39.50	58		97.50	133
0580	#2		10.60	1.509		53.50	71		124.50	168
0600	#1		9	1.778		78.50	83.50		162	215
0620	1/0		8	2		94	94		188	248
0640	2/0		7.20	2.222		112	104		216	284
0680	3/0		6.60	2.424		138	114		252	325
0700	4/0		6.20	2.581		154	121		275	355
0720	250 kcmil	3 Elec	8.70	2.759		188	130		318	405
0740	300 kcmil		8.10	2.963		259	139		398	500
0760	350 kcmil		7.50	3.200		264	150		414	520
0780	400 kcmil		6.90	3.478		310	163		473	590
0800	500 kcmil		6	4		340	188		528	665
0850	600 kcmil		5.70	4.211		430	198		628	780
0880	700 kcmil		5.10	4.706		500	221		721	885
9000	Minimum labor/equipment charge	1 Elec	4	2	Job		94		94	145

26 05 26 – Grounding and Bonding for Electrical Systems

26 05 26.80 Grounding

		Crew	Daily Output	Labor-Hours	Unit	Material	2009 Bare Costs Labor	Equipment	Total	Total Incl O&P
0010	**GROUNDING**									
0030	Rod, copper clad, 8' long, 1/2" diameter	1 Elec	5.50	1.455	Ea.	14.80	68.50		83.30	121
0040	5/8" diameter		5.50	1.455		16.40	68.50		84.90	123
0050	3/4" diameter		5.30	1.509		27.50	71		98.50	140
0080	10' long, 1/2" diameter		4.80	1.667		17.35	78.50		95.85	140
0090	5/8" diameter		4.60	1.739		19.35	81.50		100.85	148
0100	3/4" diameter		4.40	1.818		33	85.50		118.50	168
0130	15' long, 3/4" diameter		4	2		92	94		186	246
0260	Wire ground bare armored, #8-1 conductor		2	4	C.L.F.	109	188		297	410
0280	#4-1 conductor		1.60	5		175	235		410	555
0390	Bare copper wire, stranded, #8		11	.727		42	34		76	98.50
0400	#6		10	.800		66	37.50		103.50	131
0600	#2	2 Elec	10	1.600		167	75		242	299
0800	3/0		6.60	2.424		380	114		494	590
1000	4/0		5.70	2.807		480	132		612	730
1200	250 kcmil	3 Elec	7.20	3.333		565	157		722	860
1800	Water pipe ground clamps, heavy duty									
2000	Bronze, 1/2" to 1" diameter	1 Elec	8	1	Ea.	26	47		73	101
2100	1-1/4" to 2" diameter		8	1		34	47		81	110
2200	2-1/2" to 3" diameter		6	1.333		57	62.50		119.50	160
2800	Brazed connections, #6 wire		12	.667		13.45	31.50		44.95	63
3000	#2 wire		10	.800		18	37.50		55.50	78
3100	3/0 wire		8	1		27	47		74	102
3200	4/0 wire		7	1.143		31	53.50		84.50	117
3400	250 kcmil wire		5	1.600		36	75		111	156
3600	500 kcmil wire		4	2		44.50	94		138.50	194
9000	Minimum labor/equipment charge		4	2	Job		94		94	145

26 05 33 – Raceway and Boxes for Electrical Systems

26 05 33.05 Conduit

0010	**CONDUIT** To 15' high, includes 2 terminations, 2 elbows,	R260533-22								
0020	11 beam clamps, and 11 couplings per 100 L.F.									

26 05 33 – Raceway and Boxes for Electrical Systems

26 05 33.05 Conduit		Crew	Daily Output	Labor-Hours	Unit	Material	2009 Bare Costs Labor	Equipment	Total	Total Incl O&P
1161	Field bends, 45° to 90°, 1/2" diameter	1 Elec	53	.151	Ea.		7.10		7.10	10.90
1162	3/4" diameter		47	.170			8		8	12.30
1163	1" diameter		44	.182			8.55		8.55	13.15
1164	1-1/4" diameter		23	.348			16.35		16.35	25
1165	1-1/2" diameter		21	.381			17.90		17.90	27.50
1166	2" diameter		16	.500			23.50		23.50	36
1991	Field bends, 45° to 90°, 1/2" diameter		44	.182			8.55		8.55	13.15
1992	3/4" diameter		40	.200			9.40		9.40	14.45
1993	1" diameter		36	.222			10.45		10.45	16.05
1994	1-1/4" diameter		19	.421			19.80		19.80	30.50
1995	1-1/2" diameter		18	.444			21		21	32
1996	2" diameter		13	.615			29		29	44.50
2500	Steel, intermediate conduit (IMC), 1/2" diameter		100	.080	L.F.	2.04	3.76		5.80	8.05
2530	3/4" diameter		90	.089		2.44	4.18		6.62	9.15
2550	1" diameter		70	.114		3.41	5.35		8.76	12
2570	1-1/4" diameter		65	.123		4.47	5.80		10.27	13.80
2600	1-1/2" diameter		60	.133		5.40	6.25		11.65	15.60
2630	2" diameter		50	.160		6.90	7.50		14.40	19.15
2650	2-1/2" diameter		40	.200		13.40	9.40		22.80	29
2670	3" diameter	2 Elec	60	.267		17.85	12.55		30.40	39
2700	3-1/2" diameter		54	.296		23	13.95		36.95	46.50
2730	4" diameter		50	.320		25	15.05		40.05	50.50
2731	Field bends, 45° to 90°, 1/2" diameter	1 Elec	44	.182	Ea.		8.55		8.55	13.15
2732	3/4" diameter		40	.200			9.40		9.40	14.45
2733	1" diameter		36	.222			10.45		10.45	16.05
2734	1-1/4" diameter		19	.421			19.80		19.80	30.50
2735	1-1/2" diameter		18	.444			21		21	32
2736	2" diameter		13	.615			29		29	44.50
5000	Electric metallic tubing (EMT), 1/2" diameter		170	.047	L.F.	.66	2.21		2.87	4.12
5020	3/4" diameter		130	.062		1.05	2.89		3.94	5.60
5040	1" diameter		115	.070		1.84	3.27		5.11	7.05
5060	1-1/4" diameter		100	.080		2.95	3.76		6.71	9.05
5080	1-1/2" diameter		90	.089		3.78	4.18		7.96	10.60
5100	2" diameter		80	.100		4.88	4.70		9.58	12.60
5120	2-1/2" diameter		60	.133		11.70	6.25		17.95	22.50
5140	3" diameter	2 Elec	100	.160		13.75	7.50		21.25	26.50
5160	3-1/2" diameter		90	.178		17.40	8.35		25.75	32
5180	4" diameter		80	.200		18.80	9.40		28.20	35
5200	Field bends, 45° to 90°, 1/2" diameter	1 Elec	89	.090	Ea.		4.22		4.22	6.50
5220	3/4" diameter		80	.100			4.70		4.70	7.25
5240	1" diameter		73	.110			5.15		5.15	7.90
5260	1-1/4" diameter		38	.211			9.90		9.90	15.20
5280	1-1/2" diameter		36	.222			10.45		10.45	16.05
5300	2" diameter		26	.308			14.45		14.45	22.50
5320	Offsets, 1/2" diameter		65	.123			5.80		5.80	8.90
5340	3/4" diameter		62	.129			6.05		6.05	9.35
5360	1" diameter		53	.151			7.10		7.10	10.90
5380	1-1/4" diameter		30	.267			12.55		12.55	19.30
5400	1-1/2" diameter		28	.286			13.45		13.45	20.50
7600	EMT, "T" fittings with covers, 1/2" diameter, set screw		16	.500		14.45	23.50		37.95	52
9990	Minimum labor/equipment charge		4	2	Job		94		94	145

26 05 Common Work Results for Electrical

26 05 33 – Raceway and Boxes for Electrical Systems

26 05 33.45 Wireway

26 05 33.45 Wireway		Crew	Daily Output	Labor-Hours	Unit	Material	2009 Bare Costs Labor	Equipment	Total	Total Incl O&P
0010	**WIREWAY** to 15' high									
0100	Screw cover, NEMA 1 w/ fittings and supports, 2-1/2" x 2-1/2"	1 Elec	45	.178	L.F.	11.45	8.35		19.80	25.50
0200	4" x 4"	"	40	.200		12.65	9.40		22.05	28.50
0400	6" x 6"	2 Elec	60	.267		19.70	12.55		32.25	41
0600	8" x 8"	"	40	.400		33	18.80		51.80	65.50
4475	Screw cover, NEMA 3R w/ fittings and supports, 4" x 4"	1 Elec	36	.222		27	10.45		37.45	45.50
4480	6" x 6"	2 Elec	55	.291		34	13.65		47.65	58.50
4485	8" x 8"		36	.444		55	21		76	92.50
4490	12" x 12"		18	.889		77	42		119	149

26 05 33.50 Outlet Boxes

26 05 33.50 Outlet Boxes		Crew	Daily Output	Labor-Hours	Unit	Material	2009 Bare Costs Labor	Equipment	Total	Total Incl O&P
0010	**OUTLET BOXES**									
0020	Pressed steel, octagon, 4"	1 Elec	20	.400	Ea.	3.09	18.80		21.89	32.50
0060	Covers, blank		64	.125		1.29	5.90		7.19	10.45
0100	Extension rings		40	.200		5.15	9.40		14.55	20
0150	Square, 4"		20	.400		2.38	18.80		21.18	31.50
0200	Extension rings		40	.200		5.20	9.40		14.60	20
0250	Covers, blank		64	.125		1.40	5.90		7.30	10.60
0300	Plaster rings		64	.125		2.84	5.90		8.74	12.15
0652	Switchbox		24	.333		4.97	15.65		20.62	29.50
1100	Concrete, floor, 1 gang		5.30	1.509		85	71		156	203
9000	Minimum labor/equipment charge		4	2	Job		94		94	145

26 05 33.60 Outlet Boxes, Plastic

26 05 33.60 Outlet Boxes, Plastic		Crew	Daily Output	Labor-Hours	Unit	Material	2009 Bare Costs Labor	Equipment	Total	Total Incl O&P
0010	**OUTLET BOXES, PLASTIC**									
0050	4" diameter, round with 2 mounting nails	1 Elec	25	.320	Ea.	2.96	15.05		18.01	26.50
0100	Bar hanger mounted		25	.320		5.35	15.05		20.40	29
0200	4", square with 2 mounting nails		25	.320		4.44	15.05		19.49	28
0300	Plaster ring		64	.125		1.84	5.90		7.74	11.05
0400	Switch box with 2 mounting nails, 1 gang		30	.267		2.01	12.55		14.56	21.50
0500	2 gang		25	.320		4.12	15.05		19.17	27.50
0600	3 gang		20	.400		6.50	18.80		25.30	36
0700	Old work box		30	.267		3.96	12.55		16.51	23.50
9000	Minimum labor/equipment charge		4	2	Job		94		94	145

26 05 33.65 Pull Boxes

26 05 33.65 Pull Boxes		Crew	Daily Output	Labor-Hours	Unit	Material	2009 Bare Costs Labor	Equipment	Total	Total Incl O&P
0010	**PULL BOXES**									
0100	Sheet metal, pull box, NEMA 1, type SC, 6" W x 6" H x 4" D	1 Elec	8	1	Ea.	12.55	47		59.55	86.50
0200	8" W x 8" H x 4" D		8	1		17.20	47		64.20	91.50
0300	10" W x 12" H x 6" D		5.30	1.509		30.50	71		101.50	143
0400	16" W x 20" H x 8" D		4	2		114	94		208	270
0500	20" W x 24" H x 8" D		3.20	2.500		134	118		252	330
0600	24" W x 36" H x 8" D		2.70	2.963		190	139		329	425
0650	Pull box, hinged , NEMA 1, 6" W x 6" H x 4" D		8	1		12.90	47		59.90	86.50
0800	12" W x 16" H x 6" D		4.70	1.702		42	80		122	170
1000	20" W x 20" H x 6" D		3.60	2.222		85	104		189	255
1200	20" W x 20" H x 8" D		3.20	2.500		170	118		288	370
1400	24" W x 36" H x 8" D		2.70	2.963		272	139		411	515
1600	24" W x 42" H x 8" D		2	4		410	188		598	745

26 05 33.95 Cutting and Drilling

26 05 33.95 Cutting and Drilling		Crew	Daily Output	Labor-Hours	Unit	Material	2009 Bare Costs Labor	Equipment	Total	Total Incl O&P
0010	**CUTTING AND DRILLING**									
0100	Hole drilling to 10' high, concrete wall									
0110	8" thick, 1/2" pipe size	R-31	12	.667	Ea.	4.71	31.50	4.41	40.62	58
0120	3/4" pipe size		12	.667		4.71	31.50	4.41	40.62	58

26 05 33.95 Cutting and Drilling	Crew	Daily Output	Labor-Hours	Unit	Material	2009 Bare Costs Labor	Equipment	Total	Total Incl O&P
0130	1" pipe size	R-31	9.50	.842	Ea.	9.50	39.50	5.55	54.55	77.50
0140	1-1/4" pipe size		9.50	.842		9.50	39.50	5.55	54.55	77.50
0150	1-1/2" pipe size		9.50	.842		9.50	39.50	5.55	54.55	77.50
0160	2" pipe size		4.40	1.818		10.15	85.50	12.05	107.70	155
0170	2-1/2" pipe size		4.40	1.818		10.15	85.50	12.05	107.70	155
0180	3" pipe size		4.40	1.818		10.15	85.50	12.05	107.70	155
0190	3-1/2" pipe size		3.30	2.424		11	114	16.05	141.05	205
0200	4" pipe size		3.30	2.424		11	114	16.05	141.05	205
0500	12" thick, 1/2" pipe size		9.40	.851		7.25	40	5.65	52.90	75.50
0520	3/4" pipe size		9.40	.851		7.25	40	5.65	52.90	75.50
0540	1" pipe size		7.30	1.096		13.75	51.50	7.25	72.50	102
0560	1-1/4" pipe size		7.30	1.096		13.75	51.50	7.25	72.50	102
0570	1-1/2" pipe size		7.30	1.096		13.75	51.50	7.25	72.50	102
0580	2" pipe size		3.60	2.222		15.70	104	14.70	134.40	194
0590	2-1/2" pipe size		3.60	2.222		15.70	104	14.70	134.40	194
0600	3" pipe size		3.60	2.222		15.70	104	14.70	134.40	194
0610	3-1/2" pipe size		2.80	2.857		18.25	134	18.90	171.15	248
0630	4" pipe size		2.50	3.200		18.25	150	21	189.25	275
0650	16" thick, 1/2" pipe size		7.60	1.053		9.75	49.50	6.95	66.20	94.50
0670	3/4" pipe size		7	1.143		9.75	53.50	7.55	70.80	102
0690	1" pipe size		6	1.333		18	62.50	8.85	89.35	126
0710	1-1/4" pipe size		5.50	1.455		18	68.50	9.65	96.15	135
0730	1-1/2" pipe size		5.50	1.455		18	68.50	9.65	96.15	135
0750	2" pipe size		3	2.667		21	125	17.65	163.65	236
0770	2-1/2" pipe size		2.70	2.963		21	139	19.60	179.60	259
0790	3" pipe size		2.50	3.200		21	150	21	192	278
0810	3-1/2" pipe size		2.30	3.478		25.50	163	23	211.50	305
0830	4" pipe size		2	4		25.50	188	26.50	240	345
0850	20" thick, 1/2" pipe size		6.40	1.250		12.25	59	8.30	79.55	113
0870	3/4" pipe size		6	1.333		12.25	62.50	8.85	83.60	120
0890	1" pipe size		5	1.600		22	75	10.60	107.60	152
0910	1-1/4" pipe size		4.80	1.667		22	78.50	11.05	111.55	158
0930	1-1/2" pipe size		4.60	1.739		22	81.50	11.50	115	163
0950	2" pipe size		2.70	2.963		26.50	139	19.60	185.10	265
0970	2-1/2" pipe size		2.40	3.333		26.50	157	22	205.50	295
0990	3" pipe size		2.20	3.636		26.50	171	24	221.50	320
1010	3-1/2" pipe size		2	4		32.50	188	26.50	247	355
1030	4" pipe size		1.70	4.706		32.50	221	31	284.50	410
1050	24" thick, 1/2" pipe size		5.50	1.455		14.80	68.50	9.65	92.95	132
1070	3/4" pipe size		5.10	1.569		14.80	73.50	10.40	98.70	141
1090	1" pipe size		4.30	1.860		26.50	87.50	12.30	126.30	178
1110	1-1/4" pipe size		4	2		26.50	94	13.25	133.75	189
1130	1-1/2" pipe size		4	2		26.50	94	13.25	133.75	189
1150	2" pipe size		2.40	3.333		32	157	22	211	300
1170	2-1/2" pipe size		2.20	3.636		32	171	24	227	325
1190	3" pipe size		2	4		32	188	26.50	246.50	355
1210	3-1/2" pipe size		1.80	4.444		40	209	29.50	278.50	395
1230	4" pipe size		1.50	5.333		40	251	35.50	326.50	470
1500	Brick wall, 8" thick, 1/2" pipe size		18	.444		4.71	21	2.94	28.65	40.50
1520	3/4" pipe size		18	.444		4.71	21	2.94	28.65	40.50
1540	1" pipe size		13.30	.602		9.50	28.50	3.98	41.98	58.50
1560	1-1/4" pipe size		13.30	.602		9.50	28.50	3.98	41.98	58.50
1580	1-1/2" pipe size		13.30	.602		9.50	28.50	3.98	41.98	58.50

26 05 33.95 Cutting and Drilling		Crew	Daily Output	Labor-Hours	Unit	Material	2009 Bare Costs			Total Incl O&P
							Labor	Equipment	Total	
1600	2" pipe size	R-31	5.70	1.404	Ea.	10.15	66	9.30	85.45	122
1620	2-1/2" pipe size		5.70	1.404		10.15	66	9.30	85.45	122
1640	3" pipe size		5.70	1.404		10.15	66	9.30	85.45	122
1660	3-1/2" pipe size		4.40	1.818		11	85.50	12.05	108.55	156
1680	4" pipe size		4	2		11	94	13.25	118.25	172
1700	12" thick, 1/2" pipe size		14.50	.552		7.25	26	3.65	36.90	52
1720	3/4" pipe size		14.50	.552		7.25	26	3.65	36.90	52
1740	1" pipe size		11	.727		13.75	34	4.81	52.56	73
1760	1-1/4" pipe size		11	.727		13.75	34	4.81	52.56	73
1780	1-1/2" pipe size		11	.727		13.75	34	4.81	52.56	73
1800	2" pipe size		5	1.600		15.70	75	10.60	101.30	145
1820	2-1/2" pipe size		5	1.600		15.70	75	10.60	101.30	145
1840	3" pipe size		5	1.600		15.70	75	10.60	101.30	145
1860	3-1/2" pipe size		3.80	2.105		18.25	99	13.95	131.20	187
1880	4" pipe size		3.30	2.424		18.25	114	16.05	148.30	213
1900	16" thick, 1/2" pipe size		12.30	.650		9.75	30.50	4.31	44.56	62.50
1920	3/4" pipe size		12.30	.650		9.75	30.50	4.31	44.56	62.50
1940	1" pipe size		9.30	.860		18	40.50	5.70	64.20	88
1960	1-1/4" pipe size		9.30	.860		18	40.50	5.70	64.20	88
1980	1-1/2" pipe size		9.30	.860		18	40.50	5.70	64.20	88
2000	2" pipe size		4.40	1.818		21	85.50	12.05	118.55	168
2010	2-1/2" pipe size		4.40	1.818		21	85.50	12.05	118.55	168
2030	3" pipe size		4.40	1.818		21	85.50	12.05	118.55	168
2050	3-1/2" pipe size		3.30	2.424		25.50	114	16.05	155.55	221
2070	4" pipe size		3	2.667		25.50	125	17.65	168.15	240
2090	20" thick, 1/2" pipe size		10.70	.748		12.25	35	4.95	52.20	73
2110	3/4" pipe size		10.70	.748		12.25	35	4.95	52.20	73
2130	1" pipe size		8	1		22	47	6.60	75.60	104
2150	1-1/4" pipe size		8	1		22	47	6.60	75.60	104
2170	1-1/2" pipe size		8	1		22	47	6.60	75.60	104
2190	2" pipe size		4	2		26.50	94	13.25	133.75	189
2210	2-1/2" pipe size		4	2		26.50	94	13.25	133.75	189
2230	3" pipe size		4	2		26.50	94	13.25	133.75	189
2250	3-1/2" pipe size		3	2.667		32.50	125	17.65	175.15	248
2270	4" pipe size		2.70	2.963		32.50	139	19.60	191.10	272
2290	24" thick, 1/2" pipe size		9.40	.851		14.80	40	5.65	60.45	84
2310	3/4" pipe size		9.40	.851		14.80	40	5.65	60.45	84
2330	1" pipe size		7.10	1.127		26.50	53	7.45	86.95	119
2350	1-1/4" pipe size		7.10	1.127		26.50	53	7.45	86.95	119
2370	1-1/2" pipe size		7.10	1.127		26.50	53	7.45	86.95	119
2390	2" pipe size		3.60	2.222		32	104	14.70	150.70	213
2410	2-1/2" pipe size		3.60	2.222		32	104	14.70	150.70	213
2430	3" pipe size		3.60	2.222		32	104	14.70	150.70	213
2450	3-1/2" pipe size		2.80	2.857		40	134	18.90	192.90	272
2470	4" pipe size	↓	2.50	3.200	↓	40	150	21	211	299
2480										
3000	Knockouts to 8' high, metal boxes & enclosures									
3020	With hole saw, 1/2" pipe size	1 Elec	53	.151	Ea.		7.10		7.10	10.90
3040	3/4" pipe size		47	.170			8		8	12.30
3050	1" pipe size		40	.200			9.40		9.40	14.45
3060	1-1/4" pipe size		36	.222			10.45		10.45	16.05
3070	1-1/2" pipe size		32	.250			11.75		11.75	18.10
3080	2" pipe size	↓	27	.296	↓		13.95		13.95	21.50

26 05 Common Work Results for Electrical

26 05 33 – Raceway and Boxes for Electrical Systems

26 05 33.95 Cutting and Drilling

		Crew	Daily Output	Labor-Hours	Unit	Material	2009 Bare Costs Labor	Equipment	Total	Total Incl O&P
3090	2-1/2" pipe size	1 Elec	20	.400	Ea.		18.80		18.80	29
4010	3" pipe size		16	.500			23.50		23.50	36
4030	3-1/2" pipe size		13	.615			29		29	44.50
4050	4" pipe size		11	.727			34		34	52.50
4070	With hand punch set, 1/2" pipe size		40	.200			9.40		9.40	14.45
4090	3/4" pipe size		32	.250			11.75		11.75	18.10
4110	1" pipe size		30	.267			12.55		12.55	19.30
4130	1-1/4" pipe size		28	.286			13.45		13.45	20.50
4150	1-1/2" pipe size		26	.308			14.45		14.45	22.50
4170	2" pipe size		20	.400			18.80		18.80	29
4190	2-1/2" pipe size		17	.471			22		22	34
4200	3" pipe size		15	.533			25		25	38.50
4220	3-1/2" pipe size		12	.667			31.50		31.50	48
4240	4" pipe size		10	.800			37.50		37.50	58
4260	With hydraulic punch, 1/2" pipe size		44	.182			8.55		8.55	13.15
4280	3/4" pipe size		38	.211			9.90		9.90	15.20
4300	1" pipe size		38	.211			9.90		9.90	15.20
4320	1-1/4" pipe size		38	.211			9.90		9.90	15.20
4340	1-1/2" pipe size		38	.211			9.90		9.90	15.20
4360	2" pipe size		32	.250			11.75		11.75	18.10
4380	2-1/2" pipe size		27	.296			13.95		13.95	21.50
4400	3" pipe size		23	.348			16.35		16.35	25
4420	3-1/2" pipe size		20	.400			18.80		18.80	29

26 05 39 – Underfloor Raceways for Electrical Systems

26 05 39.30 Conduit In Concrete Slab

		Crew	Daily Output	Labor-Hours	Unit	Material	2009 Bare Costs Labor	Equipment	Total	Total Incl O&P
0010	**CONDUIT IN CONCRETE SLAB** Including terminations,									
0020	fittings and supports									
3230	PVC, schedule 40, 1/2" diameter	1 Elec	270	.030	L.F.	.93	1.39		2.32	3.16
3250	3/4" diameter		230	.035		1.14	1.63		2.77	3.76
3270	1" diameter		200	.040		1.55	1.88		3.43	4.60
3300	1-1/4" diameter		170	.047		2.18	2.21		4.39	5.80
3330	1-1/2" diameter		140	.057		2.60	2.69		5.29	7
3350	2" diameter		120	.067		3.35	3.13		6.48	8.50
4350	Rigid galvanized steel, 1/2" diameter		200	.040		2.73	1.88		4.61	5.90
4400	3/4" diameter		170	.047		3.01	2.21		5.22	6.70
4450	1" diameter		130	.062		4.21	2.89		7.10	9.10
4500	1-1/4" diameter		110	.073		5.65	3.42		9.07	11.45
4600	1-1/2" diameter		100	.080		6.70	3.76		10.46	13.15
4800	2" diameter		90	.089		8.55	4.18		12.73	15.90
9000	Minimum labor/equipment charge		4	2	Job		94		94	145

26 05 39.40 Conduit In Trench

		Crew	Daily Output	Labor-Hours	Unit	Material	2009 Bare Costs Labor	Equipment	Total	Total Incl O&P
0010	**CONDUIT IN TRENCH** Includes terminations and fittings									
0020	Does not include excavation or backfill, see Div. 31 23 16.00									
0200	Rigid galvanized steel, 2" diameter	1 Elec	150	.053	L.F.	8.20	2.51		10.71	12.90
0400	2-1/2" diameter	"	100	.080		15.60	3.76		19.36	23
0600	3" diameter	2 Elec	160	.100		19.60	4.70		24.30	29
0800	3-1/2" diameter		140	.114		25.50	5.35		30.85	36.50
1000	4" diameter		100	.160		28.50	7.50		36	43
1200	5" diameter		80	.200		58.50	9.40		67.90	79
1400	6" diameter		60	.267		85.50	12.55		98.05	113
9000	Minimum labor/equipment charge	1 Elec	4	2	Job		94		94	145

26 05 80.10 Motor Connections		Crew	Daily Output	Labor-Hours	Unit	Material	2009 Bare Costs Labor	Equipment	Total	Total Incl O&P
0010	**MOTOR CONNECTIONS**									
0020	Flexible conduit and fittings, 115 volt, 1 phase, up to 1 HP motor	1 Elec	8	1	Ea.	9.10	47		56.10	82.50
9000	Minimum labor/equipment charge	"	4	2	Job		94		94	145

26 05 90.10 Residential Wiring		Crew	Daily Output	Labor-Hours	Unit	Material	2009 Bare Costs Labor	Equipment	Total	Total Incl O&P
0010	**RESIDENTIAL WIRING**									
0020	20' avg. runs and #14/2 wiring incl. unless otherwise noted									
1000	Service & panel, includes 24' SE-AL cable, service eye, meter,									
1010	Socket, panel board, main bkr., ground rod, 15 or 20 amp									
1020	1-pole circuit breakers, and misc. hardware									
1100	100 amp, with 10 branch breakers	1 Elec	1.19	6.723	Ea.	585	315		900	1,125
1110	With PVC conduit and wire		.92	8.696		660	410		1,070	1,350
1120	With RGS conduit and wire		.73	10.959		840	515		1,355	1,725
1150	150 amp, with 14 branch breakers		1.03	7.767		895	365		1,260	1,550
1170	With PVC conduit and wire		.82	9.756		1,050	460		1,510	1,850
1180	With RGS conduit and wire		.67	11.940		1,400	560		1,960	2,400
1200	200 amp, with 18 branch breakers	2 Elec	1.80	8.889		1,175	420		1,595	1,950
1220	With PVC conduit and wire		1.46	10.959		1,325	515		1,840	2,275
1230	With RGS conduit and wire		1.24	12.903		1,800	605		2,405	2,900
1800	Lightning surge suppressor for above services, add	1 Elec	32	.250		50.50	11.75		62.25	73.50
2000	Switch devices									
2100	Single pole, 15 amp, Ivory, with a 1-gang box, cover plate,									
2110	Type NM (Romex) cable	1 Elec	17.10	.468	Ea.	11.70	22		33.70	47
2120	Type MC (BX) cable		14.30	.559		28	26.50		54.50	71.50
2130	EMT & wire		5.71	1.401		34.50	66		100.50	139
2150	3-way, #14/3, type NM cable		14.55	.550		17.25	26		43.25	59
2170	Type MC cable		12.31	.650		39	30.50		69.50	90
2180	EMT & wire		5	1.600		38.50	75		113.50	159
2200	4-way, #14/3, type NM cable		14.55	.550		33.50	26		59.50	77
2220	Type MC cable		12.31	.650		55.50	30.50		86	108
2230	EMT & wire		5	1.600		55	75		130	177
2250	S.P., 20 amp, #12/2, type NM cable		13.33	.600		19.70	28		47.70	65
2270	Type MC cable		11.43	.700		33	33		66	87
2280	EMT & wire		4.85	1.649		44.50	77.50		122	168
2290	S.P. rotary dimmer, 600W, no wiring		17	.471		19.90	22		41.90	56
2300	S.P. rotary dimmer, 600W, type NM cable		14.55	.550		26.50	26		52.50	69
2320	Type MC cable		12.31	.650		43	30.50		73.50	94
2330	EMT & wire		5	1.600		51.50	75		126.50	173
2350	3-way rotary dimmer, type NM cable		13.33	.600		22.50	28		50.50	68
2370	Type MC cable		11.43	.700		39	33		72	93.50
2380	EMT & wire		4.85	1.649		47.50	77.50		125	171
2400	Interval timer wall switch, 20 amp, 1-30 min., #12/2									
2410	Type NM cable	1 Elec	14.55	.550	Ea.	53	26		79	98
2420	Type MC cable		12.31	.650		60	30.50		90.50	113
2430	EMT & wire		5	1.600		77.50	75		152.50	202
2500	Decorator style									
2510	S.P., 15 amp, type NM cable	1 Elec	17.10	.468	Ea.	16.15	22		38.15	52
2520	Type MC cable		14.30	.559		32.50	26.50		59	76.50
2530	EMT & wire		5.71	1.401		39	66		105	144
2550	3-way, #14/3, type NM cable		14.55	.550		21.50	26		47.50	64
2570	Type MC cable		12.31	.650		43.50	30.50		74	94.50
2580	EMT & wire		5	1.600		43	75		118	164

26 05 90 – Residential Applications

26 05 90.10 Residential Wiring	Crew	Daily Output	Labor-Hours	Unit	Material	2009 Bare Costs Labor	Equipment	Total	Total Incl O&P
2600	4-way, #14/3, type NM cable	1 Elec	14.55	.550	Ea.	38	26		64	82
2620	Type MC cable		12.31	.650		60	30.50		90.50	113
2630	EMT & wire		5	1.600		59.50	75		134.50	182
2650	S.P., 20 amp, #12/2, type NM cable		13.33	.600		24	28		52	70
2670	Type MC cable		11.43	.700		37.50	33		70.50	92
2680	EMT & wire		4.85	1.649		49	77.50		126.50	173
2700	S.P., slide dimmer, type NM cable		17.10	.468		30	22		52	67.50
2720	Type MC cable		14.30	.559		46.50	26.50		73	92
2730	EMT & wire		5.71	1.401		55	66		121	162
2770	Type MC cable		14.30	.559		43.50	26.50		70	88.50
2780	EMT & wire		5.71	1.401		52	66		118	159
2800	3-way touch dimmer, type NM cable		13.33	.600		47	28		75	95.50
2820	Type MC cable		11.43	.700		63.50	33		96.50	121
2830	EMT & wire	▼	4.85	1.649	▼	72	77.50		149.50	198
3100	S.P. switch/15 amp recpt., Ivory, 1-gang box, plate									
3110	Type NM cable	1 Elec	11.43	.700	Ea.	22.50	33		55.50	75.50
3120	Type MC cable		10	.800		39	37.50		76.50	101
3130	EMT & wire		4.40	1.818		47.50	85.50		133	183
3150	S.P. switch/pilot light, type NM cable		11.43	.700		23.50	33		56.50	76
3170	Type MC cable		10	.800		40	37.50		77.50	102
3180	EMT & wire		4.43	1.806		48	85		133	184
3190	2-S.P. switches, 2-#14/2, no wiring		14	.571		7.80	27		34.80	50
3200	2-S.P. switches, 2-#14/2, type NM cables		10	.800		28	37.50		65.50	89
3220	Type MC cable		8.89	.900		53.50	42.50		96	124
3230	EMT & wire		4.10	1.951		52.50	91.50		144	199
3250	3-way switch/15 amp recpt., #14/3, type NM cable		10	.800		31.50	37.50		69	93
3270	Type MC cable		8.89	.900		53.50	42.50		96	124
3280	EMT & wire		4.10	1.951		53	91.50		144.50	200
3300	2-3 way switches, 2-#14/3, type NM cables		8.89	.900		44.50	42.50		87	114
3320	Type MC cable		8	1		80	47		127	161
3330	EMT & wire		4	2		62	94		156	213
3350	S.P. switch/20 amp recpt., #12/2, type NM cable		10	.800		36	37.50		73.50	97.50
3370	Type MC cable		8.89	.900		43	42.50		85.50	113
3380	EMT & wire	▼	4.10	1.951	▼	60.50	91.50		152	208
3400	Decorator style									
3410	S.P. switch/15 amp recpt., type NM cable	1 Elec	11.43	.700	Ea.	27	33		60	80
3420	Type MC cable		10	.800		43.50	37.50		81	106
3430	EMT & wire		4.40	1.818		52	85.50		137.50	188
3450	S.P. switch/pilot light, type NM cable		11.43	.700		28	33		61	81
3470	Type MC cable		10	.800		44.50	37.50		82	107
3480	EMT & wire		4.40	1.818		52.50	85.50		138	189
3500	2-S.P. switches, 2-#14/2, type NM cables		10	.800		32.50	37.50		70	94
3520	Type MC cable		8.89	.900		58	42.50		100.50	129
3530	EMT & wire		4.10	1.951		57	91.50		148.50	204
3550	3-way/15 amp recpt., #14/3, type NM cable		10	.800		36	37.50		73.50	97.50
3580	EMT & wire		4.10	1.951		57.50	91.50		149	205
3650	2-3 way switches, 2-#14/3, type NM cables		8.89	.900		48.50	42.50		91	119
3670	Type MC cable		8	1		84.50	47		131.50	166
3680	EMT & wire		4	2		66.50	94		160.50	218
3700	S.P. switch/20 amp recpt., #12/2, type NM cable		10	.800		40.50	37.50		78	103
3720	Type MC cable		8.89	.900		47.50	42.50		90	118
3730	EMT & wire	▼	4.10	1.951	▼	65	91.50		156.50	213
4000	Receptacle devices									

26 05 90 – Residential Applications

26 05 90.10 Residential Wiring	Crew	Daily Output	Labor-Hours	Unit	Material	2009 Bare Costs Labor	Equipment	Total	Total Incl O&P
4010 Duplex outlet, 15 amp recpt., Ivory, 1-gang box, plate									
4015 Type NM cable	1 Elec	14.55	.550	Ea.	10.20	26		36.20	51
4020 Type MC cable		12.31	.650		26.50	30.50		57	76.50
4030 EMT & wire		5.33	1.501		33	70.50		103.50	146
4050 With #12/2, type NM cable		12.31	.650		13.65	30.50		44.15	62
4070 Type MC cable		10.67	.750		27	35		62	83.50
4080 EMT & wire		4.71	1.699		38.50	80		118.50	166
4100 20 amp recpt., #12/2, type NM cable		12.31	.650		22.50	30.50		53	71.50
4120 Type MC cable		10.67	.750		35.50	35		70.50	93
4130 EMT & wire		4.71	1.699		47	80		127	175
4140 For GFI see Div. 26 05 90.10 line 4300 below									
4150 Decorator style, 15 amp recpt., type NM cable	1 Elec	14.55	.550	Ea.	14.65	26		40.65	56
4170 Type MC cable		12.31	.650		31	30.50		61.50	81
4180 EMT & wire		5.33	1.501		37.50	70.50		108	150
4200 With #12/2, type NM cable		12.31	.650		18.10	30.50		48.60	67
4220 Type MC cable		10.67	.750		31.50	35		66.50	88.50
4230 EMT & wire		4.71	1.699		43	80		123	170
4250 20 amp recpt. #12/2, type NM cable		12.31	.650		26.50	30.50		57	76.50
4270 Type MC cable		10.67	.750		40	35		75	98
4280 EMT & wire		4.71	1.699		51.50	80		131.50	180
4300 GFI, 15 amp recpt., type NM cable		12.31	.650		41.50	30.50		72	92.50
4320 Type MC cable		10.67	.750		58	35		93	118
4330 EMT & wire		4.71	1.699		64.50	80		144.50	194
4350 GFI with #12/2, type NM cable		10.67	.750		45	35		80	104
4370 Type MC cable		9.20	.870		58.50	41		99.50	127
4380 EMT & wire		4.21	1.900		69.50	89.50		159	214
4400 20 amp recpt., #12/2 type NM cable		10.67	.750		47	35		82	106
4420 Type MC cable		9.20	.870		60	41		101	129
4430 EMT & wire		4.21	1.900		71.50	89.50		161	216
4500 Weather-proof cover for above receptacles, add		32	.250		5.25	11.75		17	24
4550 Air conditioner outlet, 20 amp-240 volt recpt.									
4560 30' of #12/2, 2 pole circuit breaker									
4570 Type NM cable	1 Elec	10	.800	Ea.	63	37.50		100.50	127
4580 Type MC cable		9	.889		79	42		121	152
4590 EMT & wire		4	2		87.50	94		181.50	241
4600 Decorator style, type NM cable		10	.800		67.50	37.50		105	132
4620 Type MC cable		9	.889		84	42		126	157
4630 EMT & wire		4	2		92	94		186	246
4650 Dryer outlet, 30 amp-240 volt recpt., 20' of #10/3									
4660 2 pole circuit breaker									
4670 Type NM cable	1 Elec	6.41	1.248	Ea.	75	58.50		133.50	173
4680 Type MC cable		5.71	1.401		83	66		149	193
4690 EMT & wire		3.48	2.299		89	108		197	264
4700 Range outlet, 50 amp-240 volt recpt., 30' of #8/3									
4710 Type NM cable	1 Elec	4.21	1.900	Ea.	110	89.50		199.50	257
4720 Type MC cable		4	2		195	94		289	360
4730 EMT & wire		2.96	2.703		121	127		248	330
4750 Central vacuum outlet, Type NM cable		6.40	1.250		69.50	59		128.50	167
4770 Type MC cable		5.71	1.401		94	66		160	204
4780 EMT & wire		3.48	2.299		96.50	108		204.50	272
4800 30 amp-110 volt locking recpt., #10/2 circ. bkr.									
4810 Type NM cable	1 Elec	6.20	1.290	Ea.	80.50	60.50		141	182
4820 Type MC cable		5.40	1.481		111	69.50		180.50	229

26 05 90.10 Residential Wiring	Crew	Daily Output	Labor-Hours	Unit	Material	2009 Bare Costs Labor	Equipment	Total	Total Incl O&P
4830	EMT & wire	1 Elec	3.20	2.500	Ea.	110	118		228	300
4900	Low voltage outlets									
4910	Telephone recpt., 20' of 4/C phone wire	1 Elec	26	.308	Ea.	10.30	14.45		24.75	34
4920	TV recpt., 20' of RG59U coax wire, F type connector	"	16	.500	"	18.35	23.50		41.85	56
4950	Door bell chime, transformer, 2 buttons, 60' of bellwire									
4970	Economy model	1 Elec	11.50	.696	Ea.	75	32.50		107.50	133
4980	Custom model		11.50	.696		116	32.50		148.50	179
4990	Luxury model, 3 buttons		9.50	.842		310	39.50		349.50	400
6000	Lighting outlets									
6050	Wire only (for fixture), type NM cable	1 Elec	32	.250	Ea.	8.45	11.75		20.20	27.50
6070	Type MC cable		24	.333		18.55	15.65		34.20	44.50
6080	EMT & wire		10	.800		23.50	37.50		61	83.50
6100	Box (4"), and wire (for fixture), type NM cable		25	.320		17.45	15.05		32.50	42
6120	Type MC cable		20	.400		27.50	18.80		46.30	59.50
6130	EMT & wire		11	.727		32.50	34		66.50	88
6200	Fixtures (use with lines 6050 or 6100 above)									
6210	Canopy style, economy grade	1 Elec	40	.200	Ea.	29	9.40		38.40	46.50
6220	Custom grade		40	.200		51	9.40		60.40	70.50
6250	Dining room chandelier, economy grade		19	.421		82.50	19.80		102.30	122
6270	Luxury grade		15	.533		540	25		565	635
6310	Kitchen fixture (fluorescent), economy grade		30	.267		61.50	12.55		74.05	87
6320	Custom grade		25	.320		168	15.05		183.05	208
6350	Outdoor, wall mounted, economy grade		30	.267		29.50	12.55		42.05	52
6360	Custom grade		30	.267		108	12.55		120.55	138
6370	Luxury grade		25	.320		244	15.05		259.05	291
6410	Outdoor PAR floodlights, 1 lamp, 150 watt		20	.400		32	18.80		50.80	64
6420	2 lamp, 150 watt each		20	.400		52	18.80		70.80	86
6425	Motion sensing, 2 lamp, 150 watt each		20	.400		70	18.80		88.80	106
6430	For infrared security sensor, add		32	.250		111	11.75		122.75	140
6450	Outdoor, quartz-halogen, 300 watt flood		20	.400		41	18.80		59.80	74
6600	Recessed downlight, round, pre-wired, 50 or 75 watt trim		30	.267		42	12.55		54.55	65.50
6610	With shower light trim		30	.267		52	12.55		64.55	76.50
6620	With wall washer trim		28	.286		62.50	13.45		75.95	89.50
6630	With eye-ball trim		28	.286		62.50	13.45		75.95	89.50
6640	For direct contact with insulation, add					2.10			2.10	2.31
6700	Porcelain lamp holder	1 Elec	40	.200		4.10	9.40		13.50	18.95
6710	With pull switch		40	.200		4.59	9.40		13.99	19.50
6750	Fluorescent strip, 1-20 watt tube, wrap around diffuser, 24"		24	.333		54	15.65		69.65	83.50
6760	1-40 watt tube, 48"		24	.333		66	15.65		81.65	96.50
6770	2-40 watt tubes, 48"		20	.400		80	18.80		98.80	117
6780	With residential ballast		20	.400		90	18.80		108.80	128
6800	Bathroom heat lamp, 1-250 watt		28	.286		46	13.45		59.45	71
6810	2-250 watt lamps		28	.286		71.50	13.45		84.95	99
6820	For timer switch, see Div. 26 05 90.10 line 2400									
6900	Outdoor post lamp, incl. post, fixture, 35' of #14/2									
6910	Type NMC cable	1 Elec	3.50	2.286	Ea.	203	107		310	390
6920	Photo-eye, add		27	.296		32.50	13.95		46.45	57.50
6950	Clock dial time switch, 24 hr., w/enclosure, type NM cable		11.43	.700		63	33		96	120
6970	Type MC cable		11	.727		79.50	34		113.50	140
6980	EMT & wire		4.85	1.649		85.50	77.50		163	213
7000	Alarm systems									
7050	Smoke detectors, box, #14/3, type NM cable	1 Elec	14.55	.550	Ea.	37.50	26		63.50	81
7070	Type MC cable		12.31	.650		53	30.50		83.50	105

26 05 90.10 Residential Wiring	Crew	Daily Output	Labor-Hours	Unit	Material	2009 Bare Costs Labor	Equipment	Total	Total Incl O&P
7080	EMT & wire	1 Elec	5	1.600	Ea.	52.50	75		127.50	174
7090	For relay output to security system, add	↓			↓	12.90			12.90	14.20
8000	Residential equipment									
8050	Disposal hook-up, incl. switch, outlet box, 3' of flex									
8060	20 amp-1 pole circ. bkr., and 25' of #12/2									
8070	Type NM cable	1 Elec	10	.800	Ea.	33.50	37.50		71	95
8080	Type MC cable	↓	8	1		48.50	47		95.50	126
8090	EMT & wire	↓	5	1.600	↓	61	75		136	184
8100	Trash compactor or dishwasher hook-up, incl. outlet box,									
8110	3' of flex, 15 amp-1 pole circ. bkr., and 25' of #14/2									
8120	Type NM cable	1 Elec	10	.800	Ea.	24.50	37.50		62	85
8130	Type MC cable	↓	8	1		43.50	47		90.50	121
8140	EMT & wire	↓	5	1.600	↓	52.50	75		127.50	174
8150	Hot water sink dispensor hook-up, use line 8100									
8200	Vent/exhaust fan hook-up, type NM cable	1 Elec	32	.250	Ea.	8.45	11.75		20.20	27.50
8220	Type MC cable		24	.333		18.55	15.65		34.20	44.50
8230	EMT & wire	↓	10	.800	↓	23.50	37.50		61	83.50
8250	Bathroom vent fan, 50 CFM (use with above hook-up)									
8260	Economy model	1 Elec	15	.533	Ea.	24.50	25		49.50	65.50
8270	Low noise model		15	.533		32	25		57	73.50
8280	Custom model	↓	12	.667	↓	117	31.50		148.50	177
8300	Bathroom or kitchen vent fan, 110 CFM									
8310	Economy model	1 Elec	15	.533	Ea.	66.50	25		91.50	112
8320	Low noise model	"	15	.533	"	86	25		111	133
8350	Paddle fan, variable speed (w/o lights)									
8360	Economy model (AC motor)	1 Elec	10	.800	Ea.	105	37.50		142.50	174
8370	Custom model (AC motor)		10	.800		182	37.50		219.50	258
8380	Luxury model (DC motor)		8	1		360	47		407	470
8390	Remote speed switch for above, add	↓	12	.667	↓	28.50	31.50		60	79.50
8500	Whole house exhaust fan, ceiling mount, 36", variable speed									
8510	Remote switch, incl. shutters, 20 amp-1 pole circ. bkr.									
8520	30' of #12/2, type NM cable	1 Elec	4	2	Ea.	985	94		1,079	1,225
8530	Type MC cable		3.50	2.286		1,000	107		1,107	1,275
8540	EMT & wire	↓	3	2.667	↓	1,025	125		1,150	1,325
8600	Whirlpool tub hook-up, incl. timer switch, outlet box									
8610	3' of flex, 20 amp-1 pole GFI circ. bkr.									
8620	30' of #12/2, type NM cable	1 Elec	5	1.600	Ea.	127	75		202	255
8630	Type MC cable		4.20	1.905		135	89.50		224.50	287
8640	EMT & wire	↓	3.40	2.353	↓	146	111		257	330
8650	Hot water heater hook-up, incl. 1-2 pole circ. bkr., box;									
8660	3' of flex, 20' of #10/2, type NM cable	1 Elec	5	1.600	Ea.	36	75		111	156
8670	Type MC cable		4.20	1.905		58	89.50		147.50	202
8680	EMT & wire	↓	3.40	2.353	↓	55.50	111		166.50	231
9000	Heating/air conditioning									
9050	Furnace/boiler hook-up, incl. firestat, local on-off switch									
9060	Emergency switch, and 40' of type NM cable	1 Elec	4	2	Ea.	57	94		151	208
9070	Type MC cable		3.50	2.286		82.50	107		189.50	256
9080	EMT & wire	↓	1.50	5.333		94	251		345	490
9100	Air conditioner hook-up, incl. local 60 amp disc. switch									
9110	3' sealtite, 40 amp, 2 pole circuit breaker									
9130	40' of #8/2, type NM cable	1 Elec	3.50	2.286	Ea.	229	107		336	415
9140	Type MC cable		3	2.667		360	125		485	590
9150	EMT & wire	↓	1.30	6.154	↓	268	289		557	740

26 05 Common Work Results for Electrical

26 05 90 – Residential Applications

26 05 90.10 Residential Wiring	Crew	Daily Output	Labor-Hours	Unit	Material	2009 Bare Costs Labor	Equipment	Total	Total Incl O&P
9200	Heat pump hook-up, 1-40 & 1-100 amp 2 pole circ. bkr.									
9210	Local disconnect switch, 3' sealtite									
9220	40' of #8/2 & 30' of #3/2									
9230	Type NM cable	1 Elec	1.30	6.154	Ea.	595	289		884	1,100
9240	Type MC cable		1.08	7.407		865	350		1,215	1,475
9250	EMT & wire	↓	.94	8.511	↓	675	400		1,075	1,350
9500	Thermostat hook-up, using low voltage wire									
9520	Heating only, 25' of #18-3	1 Elec	24	.333	Ea.	7.70	15.65		23.35	32.50
9530	Heating/cooling, 25' of #18-4	"	20	.400	"	9.15	18.80		27.95	39

26 22 Low-Voltage Transformers

26 22 13 – Low-Voltage Distribution Transformers

26 22 13.10 Transformer, Dry-Type

		Crew	Daily Output	Labor-Hours	Unit	Material	2009 Bare Costs Labor	Equipment	Total	Total Incl O&P
0010	**TRANSFORMER, DRY-TYPE**									
0050	Single phase, 240/480 volt primary, 120/240 volt secondary									
0100	1 kVA	1 Elec	2	4	Ea.	266	188		454	580
0300	2 kVA		1.60	5		400	235		635	800
0500	3 kVA		1.40	5.714		495	269		764	960
0700	5 kVA	↓	1.20	6.667		680	315		995	1,225
0900	7.5 kVA	2 Elec	2.20	7.273		945	340		1,285	1,575
1100	10 kVA		1.60	10		1,200	470		1,670	2,050
1300	15 kVA	↓	1.20	13.333	↓	1,650	625		2,275	2,800
2300	3 phase, 480 volt primary 120/208 volt secondary									
2310	Ventilated, 3 kVA	1 Elec	1	8	Ea.	875	375		1,250	1,550
2700	6 kVA		.80	10		1,200	470		1,670	2,050
2900	9 kVA	↓	.70	11.429		1,375	535		1,910	2,325
3100	15 kVA	2 Elec	1.10	14.545		1,825	685		2,510	3,050
3300	30 kVA		.90	17.778		2,125	835		2,960	3,625
3500	45 kVA	↓	.80	20	↓	2,550	940		3,490	4,275
9000	Minimum labor/equipment charge	1 Elec	1	8	Job		375		375	580

26 24 Switchboards and Panelboards

26 24 16 – Panelboards

26 24 16.20 Panelboard and Load Center Circuit Breakers

		Crew	Daily Output	Labor-Hours	Unit	Material	2009 Bare Costs Labor	Equipment	Total	Total Incl O&P
0010	**PANELBOARD AND LOAD CENTER CIRCUIT BREAKERS**									
0050	Bolt-on, 10,000 amp IC, 120 volt, 1 pole									
0100	15 to 50 amp	1 Elec	10	.800	Ea.	15.80	37.50		53.30	75.50
0200	60 amp		8	1		15.80	47		62.80	90
0300	70 amp	↓	8	1	↓	30	47		77	106
0350	240 volt, 2 pole									
0400	15 to 50 amp	1 Elec	8	1	Ea.	35	47		82	111
0500	60 amp		7.50	1.067		35	50		85	115
0600	80 to 100 amp		5	1.600		89.50	75		164.50	214
0700	3 pole, 15 to 60 amp		6.20	1.290		110	60.50		170.50	215
0800	70 amp		5	1.600		138	75		213	268
0900	80 to 100 amp		3.60	2.222		157	104		261	335
1000	22,000 amp I.C., 240 volt, 2 pole, 70 – 225 amp		2.70	2.963		675	139		814	960
1100	3 pole, 70 – 225 amp		2.30	3.478		750	163		913	1,075
1200	14,000 amp I.C., 277 volts, 1 pole, 15 – 30 amp	↓	8	1	↓	42	47		89	119

26 24 16 – Panelboards

26 24 16.20 Panelboard and Load Center Circuit Breakers	Crew	Daily Output	Labor-Hours	Unit	Material	2009 Bare Costs Labor	Equipment	Total	Total Incl O&P
1300	22,000 amp I.C., 480 volts, 2 pole, 70 – 225 amp	1 Elec	2.70	2.963	Ea.	675	139		814	960
1400	3 pole, 70 – 225 amp		2.30	3.478		835	163		998	1,175
2060	Plug-in tandem, 120/240 V, 2-15 A, 1 pole		11	.727		31.50	34		65.50	87.50
2070	1-15 A & 1-20 A		11	.727		31.50	34		65.50	87.50
2080	2-20 A		11	.727		31.50	34		65.50	87.50
2082	Arc fault circuit interrupter, 120/240 V, 1-15 A & 1-20 A, 1 pole		11	.727		75	34		109	135
9000	Minimum labor/equipment charge		3	2.667	Job		125		125	193

26 24 16.30 Panelboards Commercial Applications

		Crew	Daily Output	Labor-Hours	Unit	Material	Labor	Equipment	Total	Total Incl O&P
0010	**PANELBOARDS COMMERCIAL APPLICATIONS**									
0050	NQOD, w/20 amp 1 pole bolt-on circuit breakers									
0100	3 wire, 120/240 volts, 100 amp main lugs									
0150	10 circuits	1 Elec	1	8	Ea.	495	375		870	1,125
0200	14 circuits		.88	9.091		580	425		1,005	1,300
0250	18 circuits		.75	10.667		635	500		1,135	1,475
0300	20 circuits		.65	12.308		715	580		1,295	1,675
0600	4 wire, 120/208 volts, 100 amp main lugs, 12 circuits		1	8		560	375		935	1,200
0650	16 circuits		.75	10.667		645	500		1,145	1,475
0700	20 circuits		.65	12.308		750	580		1,330	1,725
0750	24 circuits		.60	13.333		815	625		1,440	1,850
0800	30 circuits		.53	15.094		940	710		1,650	2,125
0850	225 amp main lugs, 32 circuits	2 Elec	.90	17.778		1,050	835		1,885	2,425
0900	34 circuits		.84	19.048		1,075	895		1,970	2,575
0950	36 circuits		.80	20		1,100	940		2,040	2,675
1000	42 circuits		.68	23.529		1,250	1,100		2,350	3,075
1200	NEHB,w/20 amp, 1 pole bolt-on circuit breakers									
1250	4 wire, 277/480 volts, 100 amp main lugs, 12 circuits	1 Elec	.88	9.091	Ea.	1,075	425		1,500	1,825
1300	20 circuits	"	.60	13.333		1,575	625		2,200	2,700
1350	225 amp main lugs, 24 circuits	2 Elec	.90	17.778		1,825	835		2,660	3,275
1400	30 circuits		.80	20		2,200	940		3,140	3,850
1450	36 circuits		.72	22.222		2,550	1,050		3,600	4,400
1600	NQOD panel, w/20 amp, 1 pole, circuit breakers									
2000	4 wire, 120/208 volts with main circuit breaker									
2050	100 amp main, 24 circuits	1 Elec	.47	17.021	Ea.	1,025	800		1,825	2,350
2100	30 circuits	"	.40	20		1,150	940		2,090	2,725
2200	225 amp main, 32 circuits	2 Elec	.72	22.222		1,950	1,050		3,000	3,750
2250	42 circuits		.56	28.571		2,150	1,350		3,500	4,425
2300	400 amp main, 42 circuits		.48	33.333		2,900	1,575		4,475	5,575
2350	600 amp main, 42 circuits		.40	40		4,300	1,875		6,175	7,625
2400	NEHB, with 20 amp, 1 pole circuit breaker									
2450	4 wire, 277/480 volts with main circuit breaker									
2500	100 amp main, 24 circuits	1 Elec	.42	19.048	Ea.	2,100	895		2,995	3,700
2550	30 circuits	"	.38	21.053		2,475	990		3,465	4,250
2600	225 amp main, 30 circuits	2 Elec	.72	22.222		3,100	1,050		4,150	5,025
2650	42 circuits	"	.56	28.571		3,850	1,350		5,200	6,300
9000	Minimum labor/equipment charge	1 Elec	1	8	Job		375		375	580

26 27 Low-Voltage Distribution Equipment

26 27 13 – Electricity Metering

26 27 13.10 Meter Centers and Sockets

		Crew	Daily Output	Labor-Hours	Unit	Material	2009 Bare Costs Labor	Equipment	Total	Total Incl O&P
0010	**METER CENTERS AND SOCKETS**									
0100	Sockets, single position, 4 terminal, 100 amp	1 Elec	3.20	2.500	Ea.	43	118		161	229
0200	150 amp		2.30	3.478		50.50	163		213.50	305
0300	200 amp		1.90	4.211		72.50	198		270.50	385
0400	Transformer rated, 20 amp		3.20	2.500		111	118		229	305
0500	Double position, 4 terminal, 100 amp		2.80	2.857		177	134		311	400
0600	150 amp		2.10	3.810		208	179		387	505
0700	200 amp		1.70	4.706		430	221		651	810
9000	Minimum labor/equipment charge		3	2.667	Job		125		125	193

26 27 16 – Electrical Cabinets and Enclosures

26 27 16.10 Cabinets

		Crew	Daily Output	Labor-Hours	Unit	Material	2009 Bare Costs Labor	Equipment	Total	Total Incl O&P
0010	**CABINETS**									
7000	Cabinets, current transformer									
7050	Single door, 24" H x 24" W x 10" D	1 Elec	1.60	5	Ea.	206	235		441	585
7100	30" H x 24" W x 10" D		1.30	6.154		210	289		499	675
7150	36" H x 24" W x 10" D		1.10	7.273		215	340		555	760
7200	30" H x 30" W x 10" D		1	8		225	375		600	830
7250	36" H x 30" W x 10" D		.90	8.889		305	420		725	980
7300	36" H x 36" W x 10" D		.80	10		325	470		795	1,075
7500	Double door, 48" H x 36" W x 10" D		.60	13.333		625	625		1,250	1,650
7550	24" H x 24" W x 12" D		1	8		274	375		649	880
9990	Minimum labor/equipment charge		2	4	Job		188		188	289

26 27 23 – Indoor Service Poles

26 27 23.40 Surface Raceway

		Crew	Daily Output	Labor-Hours	Unit	Material	2009 Bare Costs Labor	Equipment	Total	Total Incl O&P
0010	**SURFACE RACEWAY**									
0090	Metal, straight section									
0100	No. 500	1 Elec	100	.080	L.F.	.98	3.76		4.74	6.90
0110	No. 700		100	.080		1.10	3.76		4.86	7
0400	No. 1500, small pancake		90	.089		1.99	4.18		6.17	8.65
0600	No. 2000, base & cover, blank		90	.089		2.03	4.18		6.21	8.70
0800	No. 3000, base & cover, blank		75	.107		3.91	5		8.91	12
1000	No. 4000, base & cover, blank		65	.123		6.40	5.80		12.20	15.95
1200	No. 6000, base & cover, blank		50	.160		10.60	7.50		18.10	23
2400	Fittings, elbows, No. 500		40	.200	Ea.	1.76	9.40		11.16	16.40
2800	Elbow cover, No. 2000		40	.200		3.35	9.40		12.75	18.15
2880	Tee, No. 500		42	.190		3.40	8.95		12.35	17.50
2900	No. 2000		27	.296		11.15	13.95		25.10	34
3000	Switch box, No. 500		16	.500		11.55	23.50		35.05	48.50
3400	Telephone outlet, No. 1500		16	.500		13.80	23.50		37.30	51
3600	Junction box, No. 1500		16	.500		9.40	23.50		32.90	46.50
3800	Plugmold wired sections, No. 2000									
4000	1 circuit, 6 outlets, 3 ft. long	1 Elec	8	1	Ea.	35.50	47		82.50	112
4100	2 circuits, 8 outlets, 6 ft. long	"	5.30	1.509	"	57	71		128	172
9300	Non-metallic, straight section									
9310	7/16" x 7/8", base & cover, blank	1 Elec	160	.050	L.F.	1.60	2.35		3.95	5.40
9320	Base & cover w/ adhesive		160	.050		1.75	2.35		4.10	5.55
9340	7/16" x 1-5/16", base & cover, blank		145	.055		1.75	2.59		4.34	5.90
9350	Base & cover w/ adhesive		145	.055		2.05	2.59		4.64	6.25
9370	11/16" x 2-1/4", base & cover, blank		130	.062		2.50	2.89		5.39	7.20
9380	Base & cover w/ adhesive		130	.062		2.85	2.89		5.74	7.60
9400	Fittings, elbows, 7/16" x 7/8"		50	.160	Ea.	1.75	7.50		9.25	13.50
9410	7/16" x 1-5/16"		45	.178		1.82	8.35		10.17	14.85

26 27 23 – Indoor Service Poles

26 27 23.40 Surface Raceway		Crew	Daily Output	Labor-Hours	Unit	Material	2009 Bare Costs Labor	Equipment	Total	Total Incl O&P
9420	11/16" x 2-1/4"	1 Elec	40	.200	Ea.	1.97	9.40		11.37	16.60
9430	Tees, 7/16" x 7/8"		35	.229		2.30	10.75		13.05	19.10
9440	7/16" x 1-5/16"		32	.250		2.37	11.75		14.12	20.50
9450	11/16" x 2-1/4"		30	.267		2.40	12.55		14.95	22
9460	Cover clip, 7/16" x 7/8"		80	.100		.46	4.70		5.16	7.75
9470	7/16" x 1-5/16"		72	.111		.41	5.20		5.61	8.50
9480	11/16" x 2-1/4"		64	.125		.70	5.90		6.60	9.80
9490	Blank end, 7/16" x 7/8"		50	.160		.66	7.50		8.16	12.30
9500	7/16" x 1-5/16"		45	.178		.74	8.35		9.09	13.65
9510	11/16" x 2-1/4"		40	.200		1.12	9.40		10.52	15.70
9520	Round fixture box 5.5" dia x 1"		25	.320		10.55	15.05		25.60	34.50
9530	Device box, 1 gang		30	.267		4.70	12.55		17.25	24.50
9540	2 gang		25	.320		6.95	15.05		22	30.50
9990	Minimum labor/equipment charge		5	1.600	Job		75		75	116

26 27 26 – Wiring Devices

26 27 26.20 Wiring Devices Elements

		Crew	Daily Output	Labor-Hours	Unit	Material	2009 Bare Costs Labor	Equipment	Total	Total Incl O&P
0010	**WIRING DEVICES ELEMENTS**									
0200	Toggle switch, quiet type, single pole, 15 amp	1 Elec	40	.200	Ea.	5.35	9.40		14.75	20.50
0500	20 amp		27	.296		7.40	13.95		21.35	29.50
0550	Rocker, 15 amp		40	.200		6	9.40		15.40	21
0560	20 amp		27	.296		13.70	13.95		27.65	36.50
0600	3 way, 15 amp		23	.348		7.65	16.35		24	33.50
0850	Rocker, 15 amp		23	.348		8.45	16.35		24.80	34.50
0860	20 amp		18	.444		19.85	21		40.85	54
0900	4 way, 15 amp		15	.533		25	25		50	66
1030	Rocker, 15 amp		15	.533		32.50	25		57.50	74.50
1040	20 amp		11	.727		49	34		83	106
1650	Dimmer switch, 120 volt, incandescent, 600 watt, 1 pole [G]		16	.500		13.10	23.50		36.60	50.50
2460	Receptacle, duplex, 120 volt, grounded, 15 amp		40	.200		1.35	9.40		10.75	15.95
2470	20 amp		27	.296		9.95	13.95		23.90	32.50
2480	Ground fault interrupting, 15 amp		27	.296		32.50	13.95		46.45	57.50
2482	20 amp		27	.296		34.50	13.95		48.45	59.50
2490	Dryer, 30 amp		15	.533		6.05	25		31.05	45
2500	Range, 50 amp		11	.727		13.10	34		47.10	67
2600	Wall plates, stainless steel, 1 gang		80	.100		2.30	4.70		7	9.80
2800	2 gang		53	.151		4.20	7.10		11.30	15.50
3200	Lampholder, keyless		26	.308		12.30	14.45		26.75	36
3400	Pullchain with receptacle		22	.364		13.35	17.10		30.45	41
9000	Minimum labor/equipment charge		4	2	Job		94		94	145

26 27 73 – Door Chimes

26 27 73.10 Doorbell System

		Crew	Daily Output	Labor-Hours	Unit	Material	2009 Bare Costs Labor	Equipment	Total	Total Incl O&P
0010	**DOORBELL SYSTEM**, incl. transformer, button & signal									
1000	Door chimes, 2 notes, minimum	1 Elec	16	.500	Ea.	29.50	23.50		53	68
1020	Maximum		12	.667		150	31.50		181.50	213
1100	Tube type, 3 tube system		12	.667		222	31.50		253.50	293
1180	4 tube system		10	.800		355	37.50		392.50	450
1900	For transformer & button, minimum add		5	1.600		16.15	75		91.15	134
1960	Maximum, add		4.50	1.778		48.50	83.50		132	183
3000	For push button only, minimum		24	.333		3.27	15.65		18.92	27.50
3100	Maximum		20	.400		25.50	18.80		44.30	57
9000	Minimum labor/equipment charge		4	2	Job		94		94	145

26 28 Low-Voltage Circuit Protective Devices

26 28 16 – Enclosed Switches and Circuit Breakers

26 28 16.10 Circuit Breakers

		Crew	Daily Output	Labor-Hours	Unit	Material	2009 Bare Costs Labor	Equipment	Total	Total Incl O&P
0010	**CIRCUIT BREAKERS** (in enclosure)									
0100	Enclosed (NEMA 1), 600 volt, 3 pole, 30 amp	1 Elec	3.20	2.500	Ea.	610	118		728	850
0200	60 amp		2.80	2.857		610	134		744	875
0400	100 amp		2.30	3.478		700	163		863	1,025
0600	225 amp		1.50	5.333		1,625	251		1,876	2,150
0700	400 amp	2 Elec	1.60	10		2,750	470		3,220	3,750
9000	Minimum labor/equipment charge	1 Elec	4	2	Job		94		94	145

26 28 16.20 Safety Switches

		Crew	Daily Output	Labor-Hours	Unit	Material	2009 Bare Costs Labor	Equipment	Total	Total Incl O&P
0010	**SAFETY SWITCHES**									
0100	General duty 240 volt, 3 pole NEMA 1, fusible, 30 amp	1 Elec	3.20	2.500	Ea.	113	118		231	305
0200	60 amp		2.30	3.478		192	163		355	460
0300	100 amp		1.90	4.211		330	198		528	665
0400	200 amp		1.30	6.154		710	289		999	1,225
0500	400 amp	2 Elec	1.80	8.889		1,775	420		2,195	2,625
9990	Minimum labor/equipment charge	1 Elec	3	2.667	Job		125		125	193

26 28 16.40 Time Switches

		Crew	Daily Output	Labor-Hours	Unit	Material	2009 Bare Costs Labor	Equipment	Total	Total Incl O&P
0010	**TIME SWITCHES**									
0100	Single pole, single throw, 24 hour dial	1 Elec	4	2	Ea.	107	94		201	263
0200	24 hour dial with reserve power		3.60	2.222		460	104		564	665
0300	Astronomic dial		3.60	2.222		184	104		288	365
0400	Astronomic dial with reserve power		3.30	2.424		595	114		709	830
0500	7 day calendar dial		3.30	2.424		146	114		260	335
0600	7 day calendar dial with reserve power		3.20	2.500		655	118		773	900
0700	Photo cell 2000 watt		8	1		19	47		66	93.50
1080	Load management device, 4 loads		2	4		1,100	188		1,288	1,500
1100	8 loads		1	8		1,800	375		2,175	2,550
9000	Minimum labor/equipment charge		3.50	2.286	Job		107		107	165

26 29 Low-Voltage Controllers

26 29 13 – Enclosed Controllers

26 29 13.40 Relays

		Crew	Daily Output	Labor-Hours	Unit	Material	2009 Bare Costs Labor	Equipment	Total	Total Incl O&P
0010	**RELAYS** Enclosed (NEMA 1)									
0100	2 pole, 12 amp	1 Elec	5	1.600	Ea.	92	75		167	217
0200	4 pole, 10 amp	"	4.50	1.778	"	122	83.50		205.50	264

26 29 23 – Variable-Frequency Motor Controllers

26 29 23.10 Variable Frequency Drives/Adj. Frequency Drives

			Crew	Daily Output	Labor-Hours	Unit	Material	2009 Bare Costs Labor	Equipment	Total	Total Incl O&P
0010	**VARIABLE FREQUENCY DRIVES/ADJ. FREQUENCY DRIVES**										
0100	Enclosed (NEMA 1), 460 volt, for 3 HP motor size	G	1 Elec	.80	10	Ea.	1,575	470		2,045	2,450
0110	5 HP motor size	G		.80	10		1,700	470		2,170	2,600
0120	7.5 HP motor size	G		.67	11.940		2,025	560		2,585	3,100
0130	10 HP motor size	G		.67	11.940		2,025	560		2,585	3,100
0140	15 HP motor size	G	2 Elec	.89	17.978		2,325	845		3,170	3,875
0150	20 HP motor size	G		.89	17.978		3,450	845		4,295	5,100
0160	25 HP motor size	G		.67	23.881		4,025	1,125		5,150	6,150
0170	30 HP motor size	G		.67	23.881		4,900	1,125		6,025	7,100
0180	40 HP motor size	G		.67	23.881		7,175	1,125		8,300	9,625
0190	50 HP motor size	G		.53	30.189		7,825	1,425		9,250	10,800
0200	60 HP motor size	G	R-3	.56	35.714		8,575	1,650	231	10,456	12,200
0210	75 HP motor size	G		.56	35.714		11,700	1,650	231	13,581	15,700
0220	100 HP motor size	G		.50	40		12,000	1,850	258	14,108	16,200

26 29 Low-Voltage Controllers

26 29 23 – Variable-Frequency Motor Controllers

26 29 23.10 Variable Frequency Drives/Adj. Frequency Drives

		Crew	Daily Output	Labor-Hours	Unit	Material	2009 Bare Costs Labor	Equipment	Total	Total Incl O&P
0230	125 HP motor size G	R-3	.50	40	Ea.	13,200	1,850	258	15,308	17,700
0240	150 HP motor size G		.50	40		16,900	1,850	258	19,008	21,700
0250	200 HP motor size G		.42	47.619		21,000	2,200	310	23,510	26,800
1100	Custom-engineered, 460 volt, for 3 HP motor size G	1 Elec	.56	14.286		2,600	670		3,270	3,875
1110	5 HP motor size G		.56	14.286		2,725	670		3,395	4,025
1120	7.5 HP motor size G		.47	17.021		3,275	800		4,075	4,825
1130	10 HP motor size G		.47	17.021		3,275	800		4,075	4,825
1140	15 HP motor size G	2 Elec	.62	25.806		3,575	1,225		4,800	5,825
1150	20 HP motor size G		.62	25.806		4,900	1,225		6,125	7,250
1160	25 HP motor size G		.47	34.043		5,325	1,600		6,925	8,300
1170	30 HP motor size G		.47	34.043		7,250	1,600		8,850	10,400
1180	40 HP motor size G		.47	34.043		8,075	1,600		9,675	11,400
1190	50 HP motor size G		.37	43.243		8,075	2,025		10,100	12,000
1200	60 HP motor size G	R-3	.39	51.282		12,100	2,375	330	14,805	17,300
1210	75 HP motor size G		.39	51.282		14,000	2,375	330	16,705	19,300
1220	100 HP motor size G		.35	57.143		14,500	2,650	370	17,520	20,500
1230	125 HP motor size G		.35	57.143		15,300	2,650	370	18,320	21,400
1240	150 HP motor size G		.35	57.143		17,200	2,650	370	20,220	23,400
1250	200 HP motor size G		.29	68.966		23,800	3,200	445	27,445	31,600
2000	For complex & special design systems to meet specific									
2010	requirements, obtain quote from vendor.									

26 32 Packaged Generator Assemblies

26 32 13 – Engine Generators

26 32 13.16 Gas-Engine-Driven Generator Sets

		Crew	Daily Output	Labor-Hours	Unit	Material	2009 Bare Costs Labor	Equipment	Total	Total Incl O&P
0010	**GAS-ENGINE-DRIVEN GENERATOR SETS**									
0020	Gas or gasoline operated, includes battery,									
0050	charger, muffler & transfer switch									
0200	3 phase 4 wire, 277/480 volt, 7.5 kW	R-3	.83	24.096	Ea.	7,700	1,125	156	8,981	10,400
0300	11.5 kW		.71	28.169		10,900	1,300	182	12,382	14,200
0400	20 kW		.63	31.746		12,900	1,475	205	14,580	16,600
0500	35 kW		.55	36.364		15,300	1,675	235	17,210	19,700

26 51 Interior Lighting

26 51 13 – Interior Lighting Fixtures, Lamps, and Ballasts

26 51 13.50 Interior Lighting Fixtures

		Crew	Daily Output	Labor-Hours	Unit	Material	2009 Bare Costs Labor	Equipment	Total	Total Incl O&P
0010	**INTERIOR LIGHTING FIXTURES** Including lamps, mounting									
0030	hardware and connections									
0100	Fluorescent, C.W. lamps, troffer, recess mounted in grid, RS									
0200	Acrylic lens, 1'W x 4'L, two 40 watt	1 Elec	5.70	1.404	Ea.	53	66		119	159
0300	2'W x 2'L, two U40 watt		5.70	1.404		56	66		122	163
0600	2'W x 4'L, four 40 watt		4.70	1.702		64	80		144	194
0910	Acrylic lens, 1'W x 4'L, two 32 watt G		5.70	1.404		66	66		132	174
0930	2'W x 2'L, two U32 watt G		5.70	1.404		80	66		146	189
0940	2'W x 4'L, two 32 watt G		5.30	1.509		72	71		143	189
0950	2'W x 4'L, three 32 watt G		5	1.600		76.50	75		151.50	200
0960	2'W x 4'L, four 32 watt G		4.70	1.702		78.50	80		158.50	210
1000	Surface mounted, RS									
1030	Acrylic lens with hinged & latched door frame									

26 51 13 – Interior Lighting Fixtures, Lamps, and Ballasts

26 51 13.50 Interior Lighting Fixtures		Crew	Daily Output	Labor-Hours	Unit	Material	2009 Bare Costs Labor	Equipment	Total	Total Incl O&P
1100	1'W x 4'L, two 40 watt	1 Elec	7	1.143	Ea.	74	53.50		127.50	164
1200	2'W x 2'L, two U40 watt		7	1.143		79.50	53.50		133	170
1500	2'W x 4'L, four 40 watt		5.30	1.509		94	71		165	212
2100	Strip fixture									
2130	Surface mounted									
2200	4' long, one 40 watt, RS	1 Elec	8.50	.941	Ea.	31	44		75	102
2300	4' long, two 40 watt, RS	"	8	1		33.50	47		80.50	110
2600	8' long, one 75 watt, SL	2 Elec	13.40	1.194		46.50	56		102.50	138
2700	8' long, two 75 watt, SL	"	12.40	1.290		56	60.50		116.50	155
3000	Strip, pendent mounted, industrial, white porcelain enamel									
3100	4' long, two 40 watt, RS	1 Elec	5.70	1.404	Ea.	51	66		117	157
3200	4' long, two 60 watt, HO	"	5	1.600		81	75		156	205
3300	8' long, two 75 watt, SL	2 Elec	8.80	1.818		96	85.50		181.50	237
4220	Metal halide, integral ballast, ceiling, recess mounted									
4230	prismatic glass lens, floating door									
4240	2'W x 2'L, 250 watt	1 Elec	3.20	2.500	Ea.	310	118		428	520
4250	2'W x 2'L, 400 watt	2 Elec	5.80	2.759		355	130		485	590
4260	Surface mounted, 2'W x 2'L, 250 watt	1 Elec	2.70	2.963		335	139		474	585
4270	400 watt	2 Elec	4.80	3.333		395	157		552	675
4280	High bay, aluminum reflector,									
4290	Single unit, 400 watt	2 Elec	4.60	3.478	Ea.	405	163		568	695
4300	Single unit, 1000 watt		4	4		580	188		768	930
4310	Twin unit, 400 watt		3.20	5		805	235		1,040	1,250
4320	Low bay, aluminum reflector, 250W DX lamp	1 Elec	3.20	2.500		355	118		473	570
4340	High pressure sodium integral ballast ceiling, recess mounted									
4350	prismatic glass lens, floating door									
4360	2'W x 2'L, 150 watt lamp	1 Elec	3.20	2.500	Ea.	380	118		498	595
4370	2'W x 2'L, 400 watt lamp	2 Elec	5.80	2.759		450	130		580	695
4380	Surface mounted, 2'W x 2'L, 150 watt lamp	1 Elec	2.70	2.963		460	139		599	725
4390	400 watt lamp	2 Elec	4.80	3.333		515	157		672	810
4400	High bay, aluminum reflector,									
4410	Single unit, 400 watt lamp	2 Elec	4.60	3.478	Ea.	370	163		533	660
4430	Single unit, 1000 watt lamp	"	4	4		535	188		723	880
4440	Low bay, aluminum reflector, 150 watt lamp	1 Elec	3.20	2.500		320	118		438	535
4450	Incandescent, high hat can, round alzak reflector, prewired									
4470	100 watt	1 Elec	8	1	Ea.	69	47		116	149
4480	150 watt		8	1		102	47		149	185
4500	300 watt		6.70	1.194		236	56		292	345
4600	Square glass lens with metal trim, prewired									
4630	100 watt	1 Elec	6.70	1.194	Ea.	52.50	56		108.50	145
4700	200 watt		6.70	1.194		95	56		151	192
6010	Vapor tight, incandescent, ceiling mounted, 200 watt		6.20	1.290		64.50	60.50		125	164
6100	Fluorescent, surface mounted, 2 lamps, 4'L, RS, 40 watt		3.20	2.500		113	118		231	305
6850	Vandalproof, surface mounted, fluorescent, two 40 watt		3.20	2.500		239	118		357	445
6860	Incandescent, one 150 watt		8	1		73	47		120	153
6900	Mirror light, fluorescent, RS, acrylic enclosure, two 40 watt		8	1		116	47		163	201
6910	One 40 watt		8	1		94	47		141	176
6920	One 20 watt		12	.667		74	31.50		105.50	130
7500	Ballast replacement, by weight of ballast, to 15' high									
7520	Indoor fluorescent, less than 2 lbs.	1 Elec	10	.800	Ea.	22	37.50		59.50	82.50
7540	Two 40W, watt reducer, 2 to 5 lbs.		9.40	.851		39.50	40		79.50	105
7560	Two F96 slimline, over 5 lbs.		8	1		73	47		120	153
7580	Vaportite ballast, less than 2 lbs.		9.40	.851		22	40		62	86

26 51 Interior Lighting

26 51 13 – Interior Lighting Fixtures, Lamps, and Ballasts

26 51 13.50 Interior Lighting Fixtures		Crew	Daily Output	Labor-Hours	Unit	Material	2009 Bare Costs Labor	Equipment	Total	Total Incl O&P
7600	2 lbs. to 5 lbs.	1 Elec	8.90	.899	Ea.	39.50	42.50		82	109
7620	Over 5 lbs.		7.60	1.053		73	49.50		122.50	157
7630	Electronic ballast for two tubes		8	1		35.50	47		82.50	112
7640	Dimmable ballast one lamp ⏁		8	1		81.50	47		128.50	162
7650	Dimmable ballast two-lamp ⏁		7.60	1.053		133	49.50		182.50	222
9000	Minimum labor/equipment charge		3	2.667	Job		125		125	193

26 51 13.70 Residential Fixtures

26 51 13.70 Residential Fixtures		Crew	Daily Output	Labor-Hours	Unit	Material	2009 Bare Costs Labor	Equipment	Total	Total Incl O&P
0010	**RESIDENTIAL FIXTURES**									
0400	Fluorescent, interior, surface, circline, 32 watt & 40 watt	1 Elec	20	.400	Ea.	91	18.80		109.80	129
0500	2' x 2', two U 40 watt		8	1		110	47		157	194
0700	Shallow under cabinet, two 20 watt		16	.500		47.50	23.50		71	88.50
0900	Wall mounted, 4'L, one 40 watt, with baffle		10	.800		135	37.50		172.50	207
2000	Incandescent, exterior lantern, wall mounted, 60 watt		16	.500		37	23.50		60.50	76.50
2100	Post light, 150W, with 7' post		4	2		117	94		211	274
2500	Lamp holder, weatherproof with 150W PAR		16	.500		23	23.50		46.50	61
2550	With reflector and guard		12	.667		58.50	31.50		90	113
2600	Interior pendent, globe with shade, 150 watt		20	.400		152	18.80		170.80	196
9000	Minimum labor/equipment charge		4	2	Job		94		94	145

26 52 Emergency Lighting

26 52 13 – Emergency Lighting Equipments

26 52 13.10 Emergency Lighting and Battery Units

26 52 13.10 Emergency Lighting and Battery Units		Crew	Daily Output	Labor-Hours	Unit	Material	2009 Bare Costs Labor	Equipment	Total	Total Incl O&P
0010	**EMERGENCY LIGHTING AND BATTERY UNITS**									
0300	Emergency light units, battery operated									
0350	Twin sealed beam light, 25 watt, 6 volt each									
0500	Lead battery operated	1 Elec	4	2	Ea.	129	94		223	287
0700	Nickel cadmium battery operated		4	2	"	605	94		699	810
9000	Minimum labor/equipment charge		4	2	Job		94		94	145

26 53 Exit Signs

26 53 13 – Exit Lighting

26 53 13.10 Exit Lighting Fixtures

26 53 13.10 Exit Lighting Fixtures		Crew	Daily Output	Labor-Hours	Unit	Material	2009 Bare Costs Labor	Equipment	Total	Total Incl O&P
0010	**EXIT LIGHTING FIXTURES**									
0080	Exit light ceiling or wall mount, incandescent, single face	1 Elec	8	1	Ea.	42.50	47		89.50	120
0100	Double face		6.70	1.194	"	49	56		105	141
9000	Minimum labor/equipment charge		4	2	Job		94		94	145

26 55 Special Purpose Lighting

26 55 59 – Display Lighting

26 55 59.10 Track Lighting

26 55 59.10 Track Lighting		Crew	Daily Output	Labor-Hours	Unit	Material	2009 Bare Costs Labor	Equipment	Total	Total Incl O&P
0010	**TRACK LIGHTING**									
0080	Track, 1 circuit, 4' section	1 Elec	6.70	1.194	Ea.	49.50	56		105.50	141
0100	8' section		5.30	1.509		81	71		152	199
0300	3 circuits, 4' section		6.70	1.194		68	56		124	162
0400	8' section		5.30	1.509		106	71		177	226
9000	Minimum labor/equipment charge		3	2.667	Job		125		125	193

26 56 Exterior Lighting

26 56 23 – Area Lighting

26 56 23.10 Exterior Fixtures

		Crew	Daily Output	Labor-Hours	Unit	Material	2009 Bare Costs Labor	2009 Bare Costs Equipment	Total	Total Incl O&P
0010	**EXTERIOR FIXTURES** With lamps									
0200	Wall mounted, incandescent, 100 watt	1 Elec	8	1	Ea.	31.50	47		78.50	107
0400	Quartz, 500 watt		5.30	1.509		54	71		125	169
1100	Wall pack, low pressure sodium, 35 watt		4	2		234	94		328	400
1150	55 watt		4	2		278	94		372	450
1160	High pressure sodium, 70 watt		4	2		220	94		314	385
1170	150 watt		4	2		251	94		345	420
1180	Metal Halide, 175 watt		4	2		255	94		349	425
1190	250 watt		4	2		290	94		384	465
1195	400 watt		4	2		335	94		429	510

26 56 33 – Walkway Lighting

26 56 33.10 Walkway Luminaire

		Crew	Daily Output	Labor-Hours	Unit	Material	2009 Bare Costs Labor	2009 Bare Costs Equipment	Total	Total Incl O&P
0010	**WALKWAY LUMINAIRE**									
9000	Minimum labor/equipment charge	1 Elec	3.75	2.133	Job		100		100	154

26 56 36 – Flood Lighting

26 56 36.20 Floodlights

		Crew	Daily Output	Labor-Hours	Unit	Material	2009 Bare Costs Labor	2009 Bare Costs Equipment	Total	Total Incl O&P
0010	**FLOODLIGHTS** with ballast and lamp,									
1400	pole mounted, pole not included									
2250	Low pressure sodium, 55 watt	1 Elec	2.70	2.963	Ea.	565	139		704	835
2270	90 watt	"	2	4	"	620	188		808	975

26 61 Lighting Systems and Accessories

26 61 23 – Lamps Applications

26 61 23.10 Lamps

		Crew	Daily Output	Labor-Hours	Unit	Material	2009 Bare Costs Labor	2009 Bare Costs Equipment	Total	Total Incl O&P
0010	**LAMPS**									
0080	Fluorescent, rapid start, cool white, 2' long, 20 watt	1 Elec	1	8	C	335	375		710	945
0100	4' long, 40 watt		.90	8.889		365	420		785	1,050
0170	4' long, 34 watt energy saver	[G]	.90	8.889		345	420		765	1,025
0176	2' long, T8, 17 W energy saver	[G]	1	8		490	375		865	1,125
0178	3' long, T8, 25 W energy saver	[G]	.90	8.889		490	420		910	1,175
0180	4' long, T8, 32 watt energy saver	[G]	.90	8.889		315	420		735	990
0560	Twin tube compact lamp	[G]	.90	8.889		575	420		995	1,275
0570	Double twin tube compact lamp	[G]	.80	10		1,550	470		2,020	2,450
0600	Mercury vapor, mogul base, deluxe white, 100 watt		.30	26.667		3,100	1,250		4,350	5,350
0800	400 watt		.30	26.667		3,550	1,250		4,800	5,825
1000	Metal halide, mogul base, 175 watt		.30	26.667		3,900	1,250		5,150	6,200
1100	250 watt		.30	26.667		4,400	1,250		5,650	6,775
1350	High pressure sodium, 70 watt		.30	26.667		4,525	1,250		5,775	6,900
1370	150 watt		.30	26.667		4,850	1,250		6,100	7,275
1500	Low pressure sodium, 35 watt		.30	26.667		9,650	1,250		10,900	12,500
1600	90 watt		.30	26.667		11,300	1,250		12,550	14,300
1800	Incandescent, interior, A21, 100 watt		1.60	5		180	235		415	560
1900	A21, 150 watt		1.60	5		176	235		411	555
2300	R30, 75 watt		1.30	6.154		595	289		884	1,100
2500	Exterior, PAR 38, 75 watt		1.30	6.154		1,350	289		1,639	1,925
2600	PAR 38, 150 watt		1.30	6.154		1,550	289		1,839	2,150
9000	Minimum labor/equipment charge		4	2	Job		94		94	145

Division Notes

	CREW	DAILY OUTPUT	LABOR-HOURS	UNIT	2009 BARE COSTS				TOTAL INCL O&P
					MAT.	LABOR	EQUIP.	TOTAL	

Estimating Tips

27 20 00 Data Communications
27 30 00 Voice Communications
27 40 00 Audio-Video Communications

When estimating material costs for special systems, it is always prudent to obtain manufacturers' quotations for equipment prices and special installation requirements which will affect the total costs.

Reference Numbers

Reference numbers are shown in shaded boxes at the beginning of some major classifications. These numbers refer to related items in the Reference Section. The reference information may be an estimating procedure, an alternate pricing method, or technical information.

Note: Not all subdivisions listed here necessarily appear in this publication.

Note: **Trade Service,** *in part, has been used as a reference source for some of the material prices used in Division 27.*

27 15 Communications Horizontal Cabling

27 15 10 – Special Communications Cabling

27 15 10.23 Sound and Video Cables and Fittings	Crew	Daily Output	Labor-Hours	Unit	Material	2009 Bare Costs Labor	Equipment	Total	Total Incl O&P
0010 **SOUND AND VIDEO CABLES & FITTINGS**									
1250　Nonshielded, #22-2 conductor	1 Elec	10	.800	C.L.F.	15.35	37.50		52.85	75

27 15 13 – Communications Copper Horizontal Cabling

27 15 13.13 Communication Cables

	Crew	Daily Output	Labor-Hours	Unit	Material	Labor	Equipment	Total	Total Incl O&P
0010 **COMMUNICATION CABLES**									
5000　High performance unshielded twisted pair (UTP)									
7000　　Category 5, #24, 4 pair solid, PVC jacket	1 Elec	7	1.143	C.L.F.	14.50	53.50		68	98.50
7100　　　4 pair solid, plenum		7	1.143		44.50	53.50		98	132
7200　　　4 pair stranded, PVC jacket		7	1.143		21.50	53.50		75	107
7210　　Category 5e, #24, 4 pair solid, PVC jacket		7	1.143		12.85	53.50		66.35	96.50
7212　　　4 pair solid, plenum		7	1.143		37	53.50		90.50	123
7214　　　4 pair stranded, PVC jacket		7	1.143		26	53.50		79.50	111
7240　　Category 6, #24, 4 pair solid, PVC jacket		7	1.143		22	53.50		75.50	107
7242　　　4 pair solid, plenum		7	1.143		62	53.50		115.50	151
7244　　　4 pair stranded, PVC jacket		7	1.143		25	53.50		78.50	110
7300　Connector, RJ-45, category 5		80	.100	Ea.	1.21	4.70		5.91	8.60
7302　　Shielded RJ-45, category 5		72	.111		3.45	5.20		8.65	11.85
7310　Jack, UTP RJ-45, category 3		72	.111		3.51	5.20		8.71	11.90
7312　　Category 5		65	.123		4.54	5.80		10.34	13.90
7314　　Category 5e		65	.123		4.54	5.80		10.34	13.90
7316　　Category 6		65	.123		4.54	5.80		10.34	13.90

27 41 Audio-Video Systems

27 41 19 – Portable Audio-Video Equipment

27 41 19.10 T.V. Systems

	Crew	Daily Output	Labor-Hours	Unit	Material	Labor	Equipment	Total	Total Incl O&P
0010 **T.V. SYSTEMS**, not including rough-in wires, cables & conduits									
5000　T.V. Antenna only, minimum	1 Elec	6	1.333	Ea.	44.50	62.50		107	146
5100　　Maximum	"	4	2	"	187	94		281	350

27 51 Distributed Audio-Video Communications Systems

27 51 16 – Public Address and Mass Notification Systems

27 51 16.10 Public Address System

	Crew	Daily Output	Labor-Hours	Unit	Material	Labor	Equipment	Total	Total Incl O&P
0010 **PUBLIC ADDRESS SYSTEM**									
0100　Conventional, office	1 Elec	5.33	1.501	Speaker	127	70.50		197.50	248
0200　　Industrial		2.70	2.963	"	244	139		383	485
9000　Minimum labor/equipment charge		3.50	2.286	Job		107		107	165

27 51 19 – Sound Masking Systems

27 51 19.10 Sound System

	Crew	Daily Output	Labor-Hours	Unit	Material	Labor	Equipment	Total	Total Incl O&P
0010 **SOUND SYSTEM**, not including rough-in wires, cables & conduits									
2000　Intercom, 25 station capacity, master station	2 Elec	2	8	Ea.	1,925	375		2,300	2,675
2020　　11 station capacity	"	4	4		900	188		1,088	1,275
3600　House telephone, talking station	1 Elec	1.60	5		440	235		675	840
3800　　Press to talk, release to listen	"	5.30	1.509		102	71		173	221
4000　　System-on button					61			61	67
4200　Door release	1 Elec	4	2		109	94		203	265
4400　Combination speaker and microphone		8	1		186	47		233	277
4600　Termination box		3.20	2.500		58.50	118		176.50	246
4800　Amplifier or power supply		5.30	1.509		670	71		741	850

27 51 19.10 Sound System	Crew	Daily Output	Labor-Hours	Unit	Material	2009 Bare Costs Labor	Equipment	Total	Total Incl O&P
5000	Vestibule door unit	1 Elec	16	.500	Name	124	23.50	.	147.50	172
5200	Strip cabinet		27	.296	Ea.	233	13.95		246.95	279
5400	Directory		16	.500	"	110	23.50		133.50	157
9000	Minimum labor/equipment charge		3.50	2.286	Job		107		107	165

Division Notes

	CREW	DAILY OUTPUT	LABOR-HOURS	UNIT	2009 BARE COSTS				TOTAL INCL O&P
					MAT.	LABOR	EQUIP.	TOTAL	

Estimating Tips

- When estimating material costs for electronic safety and security systems, it is always prudent to obtain manufacturers' quotations for equipment prices and special installation requirements that affect the total cost.

- Fire alarm systems consist of control panels, annunciator panels, battery with rack, charger, and fire alarm actuating and indicating devices. Some fire alarm systems include speakers, telephone lines, door closer controls, and other components. Be careful not to overlook the costs related to installation for these items.

Also be aware of costs for integrated automation instrumentation and terminal devices, control equipment, control wiring, and programming.

- Security equipment includes items such as CCTV, access control, and other detection and identification systems to perform alert and alarm functions. Be sure to consider the costs related to installation for this security equipment, such as for integrated automation instrumentation and terminal devices, control equipment, control wiring, and programming.

Reference Numbers

Reference numbers are shown in shaded boxes at the beginning of some major classifications. These numbers refer to related items in the Reference Section. The reference information may be an estimating procedure, an alternate pricing method, or technical information.

Note: Not all subdivisions listed here necessarily appear in this publication.

Division 28 - Electronic Safety and Security

28 13 Access Control

28 13 53 – Security Access Detection

28 13 53.13 Security Access Metal Detectors	Crew	Daily Output	Labor-Hours	Unit	Material	2009 Bare Costs Labor	Equipment	Total	Total Incl O&P
0010 **SECURITY ACCESS METAL DETECTORS**									
0240 Metal detector, hand-held, wand type, unit only				Ea.	81.50			81.50	90
0250 Metal detector, walk through portal type, single zone	1 Elec	2	4		2,750	188		2,938	3,325
0260 Multi-zone	"	2	4		3,500	188		3,688	4,150

28 13 53.16 Security Access X-Ray Equipment

	Crew	Daily Output	Labor-Hours	Unit	Material	2009 Bare Costs Labor	Equipment	Total	Total Incl O&P
0010 **SECURITY ACCESS X-RAY EQUIPMENT**									
0290 X-ray machine, desk top, for mail/small packages/letters	1 Elec	4	2	Ea.	3,000	94		3,094	3,450
0300 Conveyor type, incl monitor, minimum		2	4		14,000	188		14,188	15,700
0310 Maximum		2	4		25,000	188		25,188	27,800
0320 X-ray machine, large unit, for airports, incl monitor, min	2 Elec	1	16		35,000	750		35,750	39,700
0330 Maximum	"	.50	32		60,000	1,500		61,500	68,500

28 13 53.23 Security Access Explosive Detection Equipment

	Crew	Daily Output	Labor-Hours	Unit	Material	2009 Bare Costs Labor	Equipment	Total	Total Incl O&P
0010 **SECURITY ACCESS EXPLOSIVE DETECTION EQUIPMENT**									
0270 Explosives detector, walk through portal type	1 Elec	2	4	Ea.	3,500	188		3,688	4,150
0280 Hand-held, battery operated				"				25,500	28,050

28 16 Intrusion Detection

28 16 16 – Intrusion Detection Systems Infrastructure

28 16 16.50 Intrusion Detection

	Crew	Daily Output	Labor-Hours	Unit	Material	2009 Bare Costs Labor	Equipment	Total	Total Incl O&P
0010 **INTRUSION DETECTION**, not including wires & conduits									
0100 Burglar alarm, battery operated, mechanical trigger	1 Elec	4	2	Ea.	272	94		366	445
0200 Electrical trigger		4	2		325	94		419	500
0400 For outside key control, add		8	1		77	47		124	157
0600 For remote signaling circuitry, add		8	1		122	47		169	208
0800 Card reader, flush type, standard		2.70	2.963		910	139		1,049	1,225
1000 Multi-code		2.70	2.963		1,175	139		1,314	1,525
1200 Door switches, hinge switch		5.30	1.509		57.50	71		128.50	172
1400 Magnetic switch		5.30	1.509		67.50	71		138.50	184
1600 Exit control locks, horn alarm		4	2		340	94		434	520
1800 Flashing light alarm		4	2		380	94		474	565
2000 Indicating panels, 1 channel		2.70	2.963		360	139		499	610
2200 10 channel	2 Elec	3.20	5		1,225	235		1,460	1,700
2400 20 channel		2	8		2,400	375		2,775	3,225
2600 40 channel		1.14	14.035		4,375	660		5,035	5,825
2800 Ultrasonic motion detector, 12 volt	1 Elec	2.30	3.478		225	163		388	500
3000 Infrared photoelectric detector		2.30	3.478		186	163		349	455
3200 Passive infrared detector		2.30	3.478		278	163		441	555
3400 Glass break alarm switch		8	1		46.50	47		93.50	124
3420 Switchmats, 30" x 5'		5.30	1.509		83	71		154	201
3440 30" x 25'		4	2		199	94		293	365
3460 Police connect panel		4	2		239	94		333	410
3480 Telephone dialer		5.30	1.509		375	71		446	525
3500 Alarm bell		4	2		76	94		170	229
3520 Siren		4	2		143	94		237	300
3540 Microwave detector, 10' to 200'		2	4		655	188		843	1,000
3560 10' to 350'		2	4		1,900	188		2,088	2,400

28 23 Video Surveillance

28 23 23 – Video Surveillance Systems Infrastructure

28 23 23.50 Video Surveillance Equipments

28 23 23.50 Video Surveillance Equipments	Crew	Daily Output	Labor-Hours	Unit	Material	2009 Bare Costs Labor	Equipment	Total	Total Incl O&P
0010 **VIDEO SURVEILLANCE EQUIPMENTS**									
0200 Video cameras, wireless, hidden in exit signs, clocks, etc, incl receiver	1 Elec	3	2.667	Ea.	149	125		274	355
0210 Accessories for VCR, single camera		3	2.667		570	125		695	820
0220 For multiple cameras		3	2.667		1,450	125		1,575	1,800
0230 Video cameras, wireless, for under vehicle searching, complete		2	4		9,975	188		10,163	11,300

28 31 Fire Detection and Alarm

28 31 23 – Fire Detection and Alarm Annunciation Panels and Fire Stations

28 31 23.50 Alarm Panels and Devices

28 31 23.50 Alarm Panels and Devices	Crew	Daily Output	Labor-Hours	Unit	Material	2009 Bare Costs Labor	Equipment	Total	Total Incl O&P
0010 **ALARM PANELS AND DEVICES**, not including wires & conduits									
3594 Fire, alarm control panel									
3600 4 zone	2 Elec	2	8	Ea.	615	375		990	1,250
3800 8 zone		1	16		675	750		1,425	1,900
4000 12 zone		.67	23.988		2,300	1,125		3,425	4,250
4200 Battery and rack	1 Elec	4	2		805	94		899	1,025
4400 Automatic charger		8	1		495	47		542	620
4600 Signal bell		8	1		57.50	47		104.50	136
4800 Trouble buzzer or manual station		8	1		41	47		88	118
5600 Strobe and horn		5.30	1.509		105	71		176	225
5800 Fire alarm horn		6.70	1.194		40.50	56		96.50	131
6000 Door holder, electro-magnetic		4	2		86	94		180	240
6200 Combination holder and closer		3.20	2.500		475	118		593	705
6400 Code transmitter		4	2		765	94		859	985
6600 Drill switch		8	1		96	47		143	178
6800 Master box		2.70	2.963		3,425	139		3,564	4,000
7000 Break glass station		8	1		55.50	47		102.50	134
7800 Remote annunciator, 8 zone lamp		1.80	4.444		201	209		410	540
8000 12 zone lamp	2 Elec	2.60	6.154		345	289		634	825
8200 16 zone lamp	"	2.20	7.273		345	340		685	905
8400 Standpipe or sprinkler alarm, alarm device	1 Elec	8	1		139	47		186	225
8600 Actuating device	"	8	1		320	47		367	430

28 31 43 – Fire Detection Sensors

28 31 43.50 Fire and Heat Detectors

28 31 43.50 Fire and Heat Detectors	Crew	Daily Output	Labor-Hours	Unit	Material	2009 Bare Costs Labor	Equipment	Total	Total Incl O&P
0010 **FIRE & HEAT DETECTORS**									
5000 Detector, rate of rise	1 Elec	8	1	Ea.	39	47		86	115
5100 Fixed temperature	"	8	1	"	32.50	47		79.50	109

28 31 46 – Smoke Detection Sensors

28 31 46.50 Smoke Detectors

28 31 46.50 Smoke Detectors	Crew	Daily Output	Labor-Hours	Unit	Material	2009 Bare Costs Labor	Equipment	Total	Total Incl O&P
0010 **SMOKE DETECTORS**									
5200 Smoke detector, ceiling type	1 Elec	6.20	1.290	Ea.	87.50	60.50		148	190
5400 Duct type	"	3.20	2.500	"	277	118		395	485

Division Notes

	CREW	DAILY OUTPUT	LABOR-HOURS	UNIT	2009 BARE COSTS				TOTAL INCL O&P
					MAT.	LABOR	EQUIP.	TOTAL	

Estimating Tips

31 05 00 Common Work Results for Earthwork

- Estimating the actual cost of performing earthwork requires careful consideration of the variables involved. This includes items such as type of soil, whether water will be encountered, dewatering, whether banks need bracing, disposal of excavated earth, and length of haul to fill or spoil sites, etc. If the project has large quantities of cut or fill, consider raising or lowering the site to reduce costs, while paying close attention to the effect on site drainage and utilities.

- If the project has large quantities of fill, creating a borrow pit on the site can significantly lower the costs.

- It is very important to consider what time of year the project is scheduled for completion. Bad weather can create large cost overruns from dewatering, site repair, and lost productivity from cold weather.

Reference Numbers

Reference numbers are shown in shaded boxes at the beginning of some major classifications. These numbers refer to related items in the Reference Section. The reference information may be an estimating procedure, an alternate pricing method, or technical information.

Note: Not all subdivisions listed here necessarily appear in this publication.

Division 31 – Earthwork

31 05 Common Work Results for Earthwork

31 05 13 – Soils for Earthwork

31 05 13.10 Borrow

31 05 13.10 Borrow	Crew	Daily Output	Labor-Hours	Unit	Material	2009 Bare Costs Labor	Equipment	Total	Total Incl O&P
0010 **BORROW**									
0020 Spread, 200 H.P. dozer, no compaction, 2 mi. RT haul									
0200 Common borrow	B-15	600	.047	C.Y.	7.35	1.61	3.58	12.54	14.65
0700 Screened loam		600	.047		20.50	1.61	3.58	25.69	29.50
0800 Topsoil, weed free		600	.047		22.50	1.61	3.58	27.69	31.50
0900 For 5 mile haul, add	B-34B	200	.040			1.28	2.67	3.95	5

31 05 16 – Aggregates for Earthwork

31 05 16.10 Borrow

31 05 16.10 Borrow	Crew	Daily Output	Labor-Hours	Unit	Material	2009 Bare Costs Labor	Equipment	Total	Total Incl O&P
0010 **BORROW**									
0020 Spread, with 200 H.P. dozer, no compaction, 2 mi. RT haul									
0100 Bank run gravel	B-15	600	.047	C.Y.	27.50	1.61	3.58	32.69	36.50
0300 Crushed stone (1.40 tons per CY) , 1-1/2"		600	.047		43.50	1.61	3.58	48.69	54.50
0320 3/4"		600	.047		43.50	1.61	3.58	48.69	54.50
0340 1/2"		600	.047		37.50	1.61	3.58	42.69	48
0360 3/8"		600	.047		37	1.61	3.58	42.19	47
0400 Sand, washed, concrete		600	.047		34.50	1.61	3.58	39.69	44.50
0500 Dead or bank sand		600	.047		10.50	1.61	3.58	15.69	18.10
0600 Select structural fill		600	.047		12.60	1.61	3.58	17.79	20.50
0900 For 5 mile haul, add	B-34B	200	.040			1.28	2.67	3.95	5

31 06 Schedules for Earthwork

31 06 60 – Schedules for Special Foundations and Load Bearing Elements

31 06 60.14 Piling Special Costs

31 06 60.14 Piling Special Costs	Crew	Daily Output	Labor-Hours	Unit	Material	2009 Bare Costs Labor	Equipment	Total	Total Incl O&P
0010 **PILING SPECIAL COSTS**									
0011 Piling special costs, pile caps, see Div. 03 30 53.40									
0500 Cutoffs, concrete piles, plain	1 Pile	5.50	1.455	Ea.		56		56	94
0600 With steel thin shell, add		38	.211			8.10		8.10	13.55
0700 Steel pile or "H" piles		19	.421			16.20		16.20	27
0800 Wood piles		38	.211			8.10		8.10	13.55
1000 Testing, any type piles, test load is twice the design load									
1050 50 ton design load, 100 ton test				Ea.				17,000	18,700
1100 100 ton design load, 200 ton test								22,000	24,200
1200 200 ton design load, 400 ton test								30,000	33,000

31 06 60.15 Mobilization

31 06 60.15 Mobilization	Crew	Daily Output	Labor-Hours	Unit	Material	2009 Bare Costs Labor	Equipment	Total	Total Incl O&P
0010 **MOBILIZATION**									
0020 Set up & remove, air compressor, 600 C.F.M.	A-5	3.30	5.455	Ea.		172	14.75	186.75	300
0100 1200 C.F.M.	"	2.20	8.182			258	22	280	450
0200 Crane, with pile leads and pile hammer, 75 ton	B-19	.60	106			4,225	2,950	7,175	10,100
0300 150 ton	"	.36	177			7,025	4,900	11,925	16,900

31 23 Excavation and Fill

31 23 16 – Excavation

31 23 16.13 Excavating, Trench

		Crew	Daily Output	Labor-Hours	Unit	Material	2009 Bare Costs Labor	Equipment	Total	Total Incl O&P
0010	**EXCAVATING, TRENCH**									
0011	Or continuous footing									
0020	Common earth with no sheeting or dewatering included									
1400	By hand with pick and shovel 2' to 6' deep, light soil	1 Clab	8	1	B.C.Y.		31.50		31.50	52
1500	Heavy soil	"	4	2	"		63		63	104
1700	For tamping backfilled trenches, air tamp, add	A-1G	100	.080	E.C.Y.		2.53	.49	3.02	4.71
1900	Vibrating plate, add	B-18	180	.133	"		4.30	.22	4.52	7.35
2100	Trim sides and bottom for concrete pours, common earth		1500	.016	S.F.		.52	.03	.55	.88
2300	Hardpan		600	.040	"		1.29	.07	1.36	2.20
9000	Minimum labor/equipment charge	1 Clab	4	2	Job		63		63	104

31 23 16.14 Excavating, Utility Trench

		Crew	Daily Output	Labor-Hours	Unit	Material	2009 Bare Costs Labor	Equipment	Total	Total Incl O&P
0010	**EXCAVATING, UTILITY TRENCH**									
0011	Common earth									
0050	Trenching with chain trencher, 12 H.P., operator walking									
0100	4" wide trench, 12" deep	B-53	800	.010	L.F.		.39	.07	.46	.69
0150	18" deep		750	.011			.42	.08	.50	.75
0200	24" deep		700	.011			.45	.09	.54	.79
0300	6" wide trench, 12" deep		650	.012			.48	.09	.57	.86
0350	18" deep		600	.013			.52	.10	.62	.93
0400	24" deep		550	.015			.57	.11	.68	1.01
0450	36" deep		450	.018			.69	.13	.82	1.24
0600	8" wide trench, 12" deep		475	.017			.66	.13	.79	1.17
0650	18" deep		400	.020			.78	.15	.93	1.39
0700	24" deep		350	.023			.89	.17	1.06	1.59
0750	36" deep		300	.027			1.04	.20	1.24	1.86
0900	Minimum labor/equipment charge		2	4	Job		156	30	186	279
1000	Backfill by hand including compaction, add									
1050	4" wide trench, 12" deep	A-1G	800	.010	L.F.		.32	.06	.38	.59
1100	18" deep		530	.015			.48	.09	.57	.89
1150	24" deep		400	.020			.63	.12	.75	1.18
1300	6" wide trench, 12" deep		540	.015			.47	.09	.56	.87
1350	18" deep		405	.020			.62	.12	.74	1.16
1400	24" deep		270	.030			.94	.18	1.12	1.75
1450	36" deep		180	.044			1.40	.27	1.67	2.62
1600	8" wide trench, 12" deep		400	.020			.63	.12	.75	1.18
1650	18" deep		265	.030			.95	.19	1.14	1.78
1700	24" deep		200	.040			1.26	.25	1.51	2.36
1750	36" deep		135	.059			1.87	.37	2.24	3.49
2000	Chain trencher, 40 H.P. operator riding									
2050	6" wide trench and backfill, 12" deep	B-54	1200	.007	L.F.		.26	.25	.51	.69
2100	18" deep		1000	.008			.31	.30	.61	.82
2150	24" deep		975	.008			.32	.31	.63	.84
2200	36" deep		900	.009			.35	.33	.68	.92
2250	48" deep		750	.011			.42	.40	.82	1.10
2300	60" deep		650	.012			.48	.46	.94	1.27
2400	8" wide trench and backfill, 12" deep		1000	.008			.31	.30	.61	.82
2450	18" deep		950	.008			.33	.32	.65	.87
2500	24" deep		900	.009			.35	.33	.68	.92
2550	36" deep		800	.010			.39	.38	.77	1.02
2600	48" deep		650	.012			.48	.46	.94	1.27
2700	12" wide trench and backfill, 12" deep		975	.008			.32	.31	.63	.84
2750	18" deep		860	.009			.36	.35	.71	.95

31 23 16.14 Excavating, Utility Trench

		Crew	Daily Output	Labor-Hours	Unit	Material	2009 Bare Costs Labor	Equipment	Total	Total Incl O&P
2800	24" deep	B-54	800	.010	L.F.		.39	.38	.77	1.02
2850	36" deep		725	.011			.43	.42	.85	1.14
3000	16" wide trench and backfill, 12" deep		835	.010			.37	.36	.73	.99
3050	18" deep		750	.011			.42	.40	.82	1.10
3100	24" deep		700	.011			.45	.43	.88	1.17
3200	Compaction with vibratory plate, add								50%	50%
5100	Hand excavate and trim for pipe bells after trench excavation									
5200	8" pipe	1 Clab	155	.052	L.F.		1.63		1.63	2.69
5300	18" pipe	"	130	.062	"		1.94		1.94	3.21
9000	Minimum labor/equipment charge	A-1G	4	2	Job		63	12.35	75.35	118

31 23 16.16 Structural Excavation for Minor Structures

		Crew	Daily Output	Labor-Hours	Unit	Material	2009 Bare Costs Labor	Equipment	Total	Total Incl O&P
0010	**STRUCTURAL EXCAVATION FOR MINOR STRUCTURES**									
0015	Hand, pits to 6' deep, sandy soil	1 Clab	8	1	B.C.Y.		31.50		31.50	52
0100	Heavy soil or clay		4	2			63		63	104
0300	Pits 6' to 12' deep, sandy soil		5	1.600			50.50		50.50	83.50
0500	Heavy soil or clay		3	2.667			84.50		84.50	139
0700	Pits 12' to 18' deep, sandy soil		4	2			63		63	104
0900	Heavy soil or clay		2	4			126		126	209
1100	Hand loading trucks from stock pile, sandy soil		12	.667			21		21	35
1300	Heavy soil or clay		8	1			31.50		31.50	52
1500	For wet or muck hand excavation, add to above				%				50%	50%
6000	Machine excavation, for spread and mat footings, elevator pits,									
6001	and small building foundations									
6030	Common earth, hydraulic backhoe, 1/2 C.Y. bucket	B-12E	55	.291	B.C.Y.		10.80	7.15	17.95	25
6035	3/4 C.Y. bucket	B-12F	90	.178			6.60	6.65	13.25	17.90
6040	1 C.Y. bucket	B-12A	108	.148			5.50	6.20	11.70	15.60
6050	1-1/2 C.Y. bucket	B-12B	144	.111			4.12	6	10.12	13.20
6060	2 C.Y. bucket	B-12C	200	.080			2.97	5.90	8.87	11.25
6070	Sand and gravel, 3/4 C.Y. bucket	B-12F	100	.160			5.95	6	11.95	16.15
6080	1 C.Y. bucket	B-12A	120	.133			4.94	5.60	10.54	14.10
6090	1-1/2 C.Y. bucket	B-12B	160	.100			3.71	5.40	9.11	11.90
6100	2 C.Y. bucket	B-12C	220	.073			2.70	5.35	8.05	10.25
6110	Clay, till, or blasted rock, 3/4 C.Y. bucket	B-12F	80	.200			7.40	7.50	14.90	20
6120	1 C.Y. bucket	B-12A	95	.168			6.25	7.05	13.30	17.80
6130	1-1/2 C.Y. bucket	B-12B	130	.123			4.56	6.65	11.21	14.70
6140	2 C.Y. bucket	B-12C	175	.091			3.39	6.75	10.14	12.90
6230	Sandy clay & loam, hydraulic backhoe, 1/2 C.Y. bucket	B-12E	60	.267			9.90	6.60	16.50	23
6235	3/4 C.Y. bucket	B-12F	98	.163			6.05	6.10	12.15	16.40
6240	1 C.Y. bucket	B-12A	116	.138			5.10	5.75	10.85	14.55
6250	1-1/2 C.Y. bucket	B-12B	156	.103			3.80	5.55	9.35	12.20
9000	Minimum labor/equipment charge	1 Clab	4	2	Job		63		63	104

31 23 16.42 Excavating, Bulk Bank Measure

			Crew	Daily Output	Labor-Hours	Unit	Material	2009 Bare Costs Labor	Equipment	Total	Total Incl O&P
0010	**EXCAVATING, BULK BANK MEASURE**	R312316-40									
0011	Common earth piled										
0020	For loading onto trucks, add									15%	15%
0200	Excavator, hydraulic, crawler mtd., 1 C.Y. cap. = 100 C.Y./hr.	R312316-45	B-12A	800	.020	B.C.Y.		.74	.84	1.58	2.11
1200	Front end loader, track mtd., 1-1/2 C.Y. cap. = 70 C.Y./hr.		B-10N	560	.021			.82	.66	1.48	2.02
1500	Wheel mounted, 3/4 C.Y. cap. = 45 C.Y./hr.		B-10R	360	.033			1.27	.66	1.93	2.75
5000	Excavating, bulk bank measure, sandy clay & loam piled										
5020	For loading onto trucks, add									15%	15%
5100	Excavator, hydraulic, crawler mtd., 1 C.Y. cap. = 120 C.Y./hr.		B-12A	960	.017	B.C.Y.		.62	.70	1.32	1.76
9000	Minimum labor/equipment charge		B-10L	2	6	Job		229	200	429	585

31 23 Excavation and Fill

31 23 16 — Excavation

31 23 16.46 Excavating, Bulk, Dozer

		Crew	Daily Output	Labor-Hours	Unit	Material	2009 Bare Costs Labor	2009 Bare Costs Equipment	Total	Total Incl O&P
0010	**EXCAVATING, BULK, DOZER**									
0011	Open site									
2000	80 H.P., 50' haul, sand & gravel	B-10L	460	.026	B.C.Y.		.99	.87	1.86	2.54
2200	150' haul, sand & gravel		230	.052			1.99	1.74	3.73	5.10
2400	300' haul, sand & gravel	↓	120	.100			3.81	3.33	7.14	9.75
3000	105 H.P., 50' haul, sand & gravel	B-10W	700	.017			.65	.86	1.51	1.98
3200	150' haul, sand & gravel		310	.039			1.47	1.94	3.41	4.48
3300	300' haul, sand & gravel	↓	140	.086			3.27	4.29	7.56	9.90
4000	200 H.P., 50' haul, sand & gravel	B-10B	1400	.009			.33	.77	1.10	1.37
4200	150' haul, sand & gravel		595	.020			.77	1.82	2.59	3.23
4400	300' haul, sand & gravel	↓	310	.039			1.47	3.49	4.96	6.20
5040	Clay	B-10M	1025	.012			.45	1.39	1.84	2.24
5400	300' haul, sand & gravel	"	470	.026	↓		.97	3.03	4	4.88

31 23 23 — Fill

31 23 23.13 Backfill

		Crew	Daily Output	Labor-Hours	Unit	Material	2009 Bare Costs Labor	2009 Bare Costs Equipment	Total	Total Incl O&P
0010	**BACKFILL** R312323-30									
0015	By hand, no compaction, light soil	1 Clab	14	.571	L.C.Y.		18.05		18.05	30
0100	Heavy soil	↓	11	.727	"		23		23	38
0300	Compaction in 6" layers, hand tamp, add to above	↓	20.60	.388	E.C.Y.		12.25		12.25	20.50
0400	Roller compaction operator walking, add	B-10A	100	.120			4.57	1.45	6.02	8.90
0500	Air tamp, add	B-9D	190	.211			6.75	1.17	7.92	12.40
0600	Vibrating plate, add	A-1D	60	.133			4.21	.53	4.74	7.55
0800	Compaction in 12" layers, hand tamp, add to above	1 Clab	34	.235			7.45		7.45	12.25
1000	Air tamp, add	B-9	285	.140			4.49	.67	5.16	8.15
1100	Vibrating plate, add	A-1E	90	.089	↓		2.81	.44	3.25	5.10
1200	Trench, dozer, no compaction, 60 HP	B-10L	425	.028	L.C.Y.		1.08	.94	2.02	2.75
1300	Dozer backfilling, bulk, up to 300' haul, no compaction	B-10B	1200	.010	"		.38	.90	1.28	1.60
1400	Air tamped, add	B-11B	80	.200	E.C.Y.		7.05	3.10	10.15	14.75
1900	Dozer backfilling, trench, up to 300' haul, no compaction	B-10B	900	.013	L.C.Y.		.51	1.20	1.71	2.13
2000	Air tamped, add	B-11B	80	.200	E.C.Y.		7.05	3.10	10.15	14.75
2350	Spreading in 8" layers, small dozer	B-10B	1060	.011	L.C.Y.		.43	1.02	1.45	1.81
2450	Compacting with vibrating plate, 8" lifts	A-1D	73	.110	E.C.Y.		3.46	.44	3.90	6.20

31 23 23.14 Backfill, Structural

		Crew	Daily Output	Labor-Hours	Unit	Material	2009 Bare Costs Labor	2009 Bare Costs Equipment	Total	Total Incl O&P
0010	**BACKFILL, STRUCTURAL**									
0011	Dozer or F.E. loader									
0020	From existing stockpile, no compaction									
2000	80 H.P., 50' haul, sand & gravel	B-10L	1100	.011	L.C.Y.		.42	.36	.78	1.06
2010	Sandy clay & loam		1070	.011			.43	.37	.80	1.09
2020	Common earth		975	.012			.47	.41	.88	1.20
2040	Clay		850	.014			.54	.47	1.01	1.38
2400	300' haul, sand & gravel		370	.032			1.24	1.08	2.32	3.16
2410	Sandy clay & loam		360	.033			1.27	1.11	2.38	3.24
2420	Common earth		330	.036			1.39	1.21	2.60	3.54
2440	Clay	↓	290	.041			1.58	1.38	2.96	4.02
3000	105 H.P., 50' haul, sand & gravel	B-10W	1350	.009			.34	.44	.78	1.03
3010	Sandy clay & loam		1325	.009			.35	.45	.80	1.05
3020	Common earth		1225	.010			.37	.49	.86	1.14
3040	Clay		1100	.011			.42	.55	.97	1.26
3300	300' haul, sand & gravel		465	.026			.98	1.29	2.27	2.99
3310	Sandy clay & loam		455	.026			1	1.32	2.32	3.05
3320	Common earth	↓	415	.029	↓		1.10	1.45	2.55	3.35

433

31 23 23.14 Backfill, Structural	Crew	Daily Output	Labor-Hours	Unit	Material	2009 Bare Costs Labor	2009 Bare Costs Equipment	Total	Total Incl O&P
3340 Clay	B-10W	370	.032	L.C.Y.		1.24	1.62	2.86	3.76

31 23 23.16 Fill By Borrow and Utility Bedding

		Crew	Daily Output	Labor-Hours	Unit	Material	Labor	Equipment	Total	Total Incl O&P
0010	**FILL BY BORROW AND UTILITY BEDDING**									
0049	Utility bedding, for pipe & conduit, not incl. compaction									
0050	Crushed or screened bank run gravel	B-6	150	.160	L.C.Y.	31	5.45	1.96	38.41	45
0100	Crushed stone 3/4" to 1/2"		150	.160		43.50	5.45	1.96	50.91	59
0200	Sand, dead or bank	↓	150	.160	↓	10.50	5.45	1.96	17.91	22.50
0500	Compacting bedding in trench	A-1D	90	.089	E.C.Y.		2.81	.35	3.16	5.05
0600	If material source exceeds 2 miles, add for extra mileage.									

31 23 23.17 General Fill

		Crew	Daily Output	Labor-Hours	Unit	Material	Labor	Equipment	Total	Total Incl O&P
0010	**GENERAL FILL**									
0011	Spread dumped material, no compaction									
0020	By dozer, no compaction	B-10B	1000	.012	L.C.Y.		.46	1.08	1.54	1.92
0100	By hand	1 Clab	12	.667	"		21		21	35
9000	Minimum labor/equipment charge	"	4	2	Job		63		63	104

31 23 23.20 Hauling

		Crew	Daily Output	Labor-Hours	Unit	Material	Labor	Equipment	Total	Total Incl O&P
0010	**HAULING**									
0011	Excavated or borrow, loose cubic yards									
0012	no loading equipment, including hauling,waiting, loading/dumping									
0013	time per cycle (wait, load, travel, unload or dump & return)									
0014	8 CY truck, 15 MPH ave, cycle 0.5 miles, 10 min. wait/Ld./Uld.	B-34A	320	.025	L.C.Y.		.80	1.02	1.82	2.43
0016	cycle 1 mile		272	.029			.94	1.20	2.14	2.86
0018	cycle 2 miles		208	.038			1.23	1.57	2.80	3.74
0020	cycle 4 miles		144	.056			1.78	2.27	4.05	5.40
0022	cycle 6 miles		112	.071			2.28	2.91	5.19	6.95
0024	cycle 8 miles		88	.091			2.90	3.71	6.61	8.85
0026	20 MPH ave,cycle 0.5 mile		336	.024			.76	.97	1.73	2.32
0028	cycle 1 mile		296	.027			.86	1.10	1.96	2.63
0030	cycle 2 miles		240	.033			1.06	1.36	2.42	3.24
0032	cycle 4 miles		176	.045			1.45	1.85	3.30	4.42
0034	cycle 6 miles		136	.059			1.88	2.40	4.28	5.70
0036	cycle 8 miles		112	.071			2.28	2.91	5.19	6.95
0044	25 MPH ave, cycle 4 miles		192	.042			1.33	1.70	3.03	4.05
0046	cycle 6 miles		160	.050			1.60	2.04	3.64	4.86
0048	cycle 8 miles		128	.063			2	2.55	4.55	6.10
0050	30 MPH ave, cycle 4 miles		216	.037			1.18	1.51	2.69	3.60
0052	cycle 6 miles		176	.045			1.45	1.85	3.30	4.42
0054	cycle 8 miles		144	.056			1.78	2.27	4.05	5.40
0114	15 MPH ave, cycle 0.5 mile, 15 min. wait/Ld./Uld.		224	.036			1.14	1.46	2.60	3.47
0116	cycle 1 mile		200	.040			1.28	1.63	2.91	3.89
0118	cycle 2 miles		168	.048			1.52	1.94	3.46	4.63
0120	cycle 4 miles		120	.067			2.13	2.72	4.85	6.50
0122	cycle 6 miles		96	.083			2.66	3.40	6.06	8.10
0124	cycle 8 miles		80	.100			3.20	4.08	7.28	9.75
0126	20 MPH ave, cycle 0.5 mile		232	.034			1.10	1.41	2.51	3.36
0128	cycle 1 mile		208	.038			1.23	1.57	2.80	3.74
0130	cycle 2 miles		184	.043			1.39	1.77	3.16	4.23
0132	cycle 4 miles		144	.056			1.78	2.27	4.05	5.40
0134	cycle 6 miles		112	.071			2.28	2.91	5.19	6.95
0136	cycle 8 miles		96	.083			2.66	3.40	6.06	8.10
0144	25 MPH ave, cycle 4 miles		152	.053			1.68	2.15	3.83	5.10
0146	cycle 6 miles	↓	128	.063	↓		2	2.55	4.55	6.10

31 23 23.20 Hauling		Crew	Daily Output	Labor-Hours	Unit	Material	2009 Bare Costs		Total	Total Incl O&P
							Labor	Equipment		
0148	cycle 8 miles	B-34A	112	.071	L.C.Y.		2.28	2.91	5.19	6.95
0150	30 MPH ave, cycle 4 miles		168	.048			1.52	1.94	3.46	4.63
0152	cycle 6 miles		144	.056			1.78	2.27	4.05	5.40
0154	cycle 8 miles		120	.067			2.13	2.72	4.85	6.50
0214	15 MPH ave, cycle 0.5 mile, 20 min wait/Ld./Uld.		176	.045			1.45	1.85	3.30	4.42
0216	cycle 1 mile		160	.050			1.60	2.04	3.64	4.86
0218	cycle 2 miles		136	.059			1.88	2.40	4.28	5.70
0220	cycle 4 miles		104	.077			2.46	3.14	5.60	7.50
0222	cycle 6 miles		88	.091			2.90	3.71	6.61	8.85
0224	cycle 8 miles		72	.111			3.55	4.53	8.08	10.80
0226	20 MPH ave, cycle 0.5 mile		176	.045			1.45	1.85	3.30	4.42
0228	cycle 1 mile		168	.048			1.52	1.94	3.46	4.63
0230	cycle 2 miles		144	.056			1.78	2.27	4.05	5.40
0232	cycle 4 miles		120	.067			2.13	2.72	4.85	6.50
0234	cycle 6 miles		96	.083			2.66	3.40	6.06	8.10
0236	cycle 8 miles		88	.091			2.90	3.71	6.61	8.85
0244	25 MPH ave, cycle 4 miles		128	.063			2	2.55	4.55	6.10
0246	cycle 6 miles		112	.071			2.28	2.91	5.19	6.95
0248	cycle 8 miles		96	.083			2.66	3.40	6.06	8.10
0250	30 MPH ave, cycle 4 miles		136	.059			1.88	2.40	4.28	5.70
0252	cycle 6 miles		120	.067			2.13	2.72	4.85	6.50
0254	cycle 8 miles		104	.077			2.46	3.14	5.60	7.50
0314	15 MPH ave, cycle 0.5 mile, 25 min wait/Ld./Uld.		144	.056			1.78	2.27	4.05	5.40
0316	cycle 1 mile		128	.063			2	2.55	4.55	6.10
0318	cycle 2 miles		112	.071			2.28	2.91	5.19	6.95
0320	cycle 4 miles		96	.083			2.66	3.40	6.06	8.10
0322	cycle 6 miles		80	.100			3.20	4.08	7.28	9.75
0324	cycle 8 miles		64	.125			3.99	5.10	9.09	12.15
0326	20 MPH ave, cycle 0.5 mile		144	.056			1.78	2.27	4.05	5.40
0328	cycle 1 mile		136	.059			1.88	2.40	4.28	5.70
0330	cycle 2 miles		120	.067			2.13	2.72	4.85	6.50
0332	cycle 4 miles		104	.077			2.46	3.14	5.60	7.50
0334	cycle 6 miles		88	.091			2.90	3.71	6.61	8.85
0336	cycle 8 miles		80	.100			3.20	4.08	7.28	9.75
0344	25 MPH ave, cycle 4 miles		112	.071			2.28	2.91	5.19	6.95
0346	cycle 6 miles		96	.083			2.66	3.40	6.06	8.10
0348	cycle 8 miles		88	.091			2.90	3.71	6.61	8.85
0350	30 MPH ave, cycle 4 miles		112	.071			2.28	2.91	5.19	6.95
0352	cycle 6 miles		104	.077			2.46	3.14	5.60	7.50
0354	cycle 8 miles		96	.083			2.66	3.40	6.06	8.10
0414	15 MPH ave, cycle 0.5 mile, 30 min wait/Ld./Uld.		120	.067			2.13	2.72	4.85	6.50
0416	cycle 1 mile		112	.071			2.28	2.91	5.19	6.95
0418	cycle 2 miles		96	.083			2.66	3.40	6.06	8.10
0420	cycle 4 miles		80	.100			3.20	4.08	7.28	9.75
0422	cycle 6 miles		72	.111			3.55	4.53	8.08	10.80
0424	cycle 8 miles		64	.125			3.99	5.10	9.09	12.15
0426	20 MPH ave, cycle 0.5 mile		120	.067			2.13	2.72	4.85	6.50
0428	cycle 1 mile		112	.071			2.28	2.91	5.19	6.95
0430	cycle 2 miles		104	.077			2.46	3.14	5.60	7.50
0432	cycle 4 miles		88	.091			2.90	3.71	6.61	8.85
0434	cycle 6 miles		80	.100			3.20	4.08	7.28	9.75
0436	cycle 8 miles		72	.111			3.55	4.53	8.08	10.80
0444	25 MPH ave, cycle 4 miles		96	.083			2.66	3.40	6.06	8.10

31 23 23.20 Hauling		Crew	Daily Output	Labor-Hours	Unit	Material	2009 Bare Costs Labor	Equipment	Total	Total Incl O&P
0446	cycle 6 miles	B-34A	88	.091	L.C.Y.		2.90	3.71	6.61	8.85
0448	cycle 8 miles		80	.100			3.20	4.08	7.28	9.75
0450	30 MPH ave, cycle 4 miles		96	.083			2.66	3.40	6.06	8.10
0452	cycle 6 miles		88	.091			2.90	3.71	6.61	8.85
0454	cycle 8 miles		80	.100			3.20	4.08	7.28	9.75
0514	15 MPH ave, cycle 0.5 mile, 35 min wait/Ld./Uld.		104	.077			2.46	3.14	5.60	7.50
0516	cycle 1 mile		96	.083			2.66	3.40	6.06	8.10
0518	cycle 2 miles		88	.091			2.90	3.71	6.61	8.85
0520	cycle 4 miles		72	.111			3.55	4.53	8.08	10.80
0522	cycle 6 miles		64	.125			3.99	5.10	9.09	12.15
0524	cycle 8 miles		56	.143			4.56	5.85	10.41	13.90
0526	20 MPH ave, cycle 0.5 mile		104	.077			2.46	3.14	5.60	7.50
0528	cycle 1 mile		96	.083			2.66	3.40	6.06	8.10
0530	cycle 2 miles		96	.083			2.66	3.40	6.06	8.10
0532	cycle 4 miles		80	.100			3.20	4.08	7.28	9.75
0534	cycle 6 miles		72	.111			3.55	4.53	8.08	10.80
0536	cycle 8 miles		64	.125			3.99	5.10	9.09	12.15
0544	25 MPH ave, cycle 4 miles		88	.091			2.90	3.71	6.61	8.85
0546	cycle 6 miles		80	.100			3.20	4.08	7.28	9.75
0548	cycle 8 miles		72	.111			3.55	4.53	8.08	10.80
0550	30 MPH ave, cycle 4 miles		88	.091			2.90	3.71	6.61	8.85
0552	cycle 6 miles		80	.100			3.20	4.08	7.28	9.75
0554	cycle 8 miles		72	.111			3.55	4.53	8.08	10.80
1014	12 CY truck, cycle 0.5 mile, 15 MPH ave, 15 min. wait/Ld./Uld.	B-34B	336	.024			.76	1.59	2.35	3
1016	cycle 1 mile		300	.027			.85	1.78	2.63	3.35
1018	cycle 2 miles		252	.032			1.01	2.12	3.13	3.99
1020	cycle 4 miles		180	.044			1.42	2.96	4.38	5.60
1022	cycle 6 miles		144	.056			1.78	3.70	5.48	7
1024	cycle 8 miles		120	.067			2.13	4.44	6.57	8.40
1025	cycle 10 miles		96	.083			2.66	5.55	8.21	10.45
1026	20 MPH ave, cycle 0.5 mile		348	.023			.73	1.53	2.26	2.88
1028	cycle 1 mile		312	.026			.82	1.71	2.53	3.22
1030	cycle 2 miles		276	.029			.93	1.93	2.86	3.64
1032	cycle 4 miles		216	.037			1.18	2.47	3.65	4.65
1034	cycle 6 miles		168	.048			1.52	3.17	4.69	6
1036	cycle 8 miles		144	.056			1.78	3.70	5.48	7
1038	cycle 10 miles		120	.067			2.13	4.44	6.57	8.40
1040	25 MPH ave, cycle 4 miles		228	.035			1.12	2.34	3.46	4.41
1042	cycle 6 miles		192	.042			1.33	2.78	4.11	5.25
1044	cycle 8 miles		168	.048			1.52	3.17	4.69	6
1046	cycle 10 miles		144	.056			1.78	3.70	5.48	7
1050	30 MPH ave, cycle 4 miles		252	.032			1.01	2.12	3.13	3.99
1052	cycle 6 miles		216	.037			1.18	2.47	3.65	4.65
1054	cycle 8 miles		180	.044			1.42	2.96	4.38	5.60
1056	cycle 10 miles		156	.051			1.64	3.42	5.06	6.45
1060	35 MPH ave, cycle 4 miles		264	.030			.97	2.02	2.99	3.81
1062	cycle 6 miles		228	.035			1.12	2.34	3.46	4.41
1064	cycle 8 miles		204	.039			1.25	2.61	3.86	4.92
1066	cycle 10 miles		180	.044			1.42	2.96	4.38	5.60
1068	cycle 20 miles		120	.067			2.13	4.44	6.57	8.40
1069	cycle 30 miles		84	.095			3.04	6.35	9.39	12
1070	cycle 40 miles		72	.111			3.55	7.40	10.95	13.95
1072	40 MPH ave, cycle 6 miles		240	.033			1.06	2.22	3.28	4.18

31 23 23.20 Hauling		Crew	Daily Output	Labor-Hours	Unit	Material	2009 Bare Costs Labor	Equipment	Total	Total Incl O&P
1074	cycle 8 miles	B-34B	216	.037	L.C.Y.		1.18	2.47	3.65	4.65
1076	cycle 10 miles		192	.042			1.33	2.78	4.11	5.25
1078	cycle 20 miles		120	.067			2.13	4.44	6.57	8.40
1080	cycle 30 miles		96	.083			2.66	5.55	8.21	10.45
1082	cycle 40 miles		72	.111			3.55	7.40	10.95	13.95
1084	cycle 50 miles		60	.133			4.26	8.90	13.16	16.75
1094	45 MPH ave, cycle 8 miles		216	.037			1.18	2.47	3.65	4.65
1096	cycle 10 miles		204	.039			1.25	2.61	3.86	4.92
1098	cycle 20 miles		132	.061			1.94	4.04	5.98	7.60
1100	cycle 30 miles		108	.074			2.37	4.94	7.31	9.35
1102	cycle 40 miles		84	.095			3.04	6.35	9.39	12
1104	cycle 50 miles		72	.111			3.55	7.40	10.95	13.95
1106	50 MPH ave, cycle 10 miles		216	.037			1.18	2.47	3.65	4.65
1108	cycle 20 miles		144	.056			1.78	3.70	5.48	7
1110	cycle 30 miles		108	.074			2.37	4.94	7.31	9.35
1112	cycle 40 miles		84	.095			3.04	6.35	9.39	12
1114	cycle 50 miles		72	.111			3.55	7.40	10.95	13.95
1214	15 MPH ave, cycle 0.5 mile, 20 min. wait/Ld./Uld.		264	.030			.97	2.02	2.99	3.81
1216	cycle 1 mile		240	.033			1.06	2.22	3.28	4.18
1218	cycle 2 miles		204	.039			1.25	2.61	3.86	4.92
1220	cycle 4 miles		156	.051			1.64	3.42	5.06	6.45
1222	cycle 6 miles		132	.061			1.94	4.04	5.98	7.60
1224	cycle 8 miles		108	.074			2.37	4.94	7.31	9.35
1225	cycle 10 miles		96	.083			2.66	5.55	8.21	10.45
1226	20 MPH ave, cycle 0.5 mile		264	.030			.97	2.02	2.99	3.81
1228	cycle 1 mile		252	.032			1.01	2.12	3.13	3.99
1230	cycle 2 miles		216	.037			1.18	2.47	3.65	4.65
1232	cycle 4 miles		180	.044			1.42	2.96	4.38	5.60
1234	cycle 6 miles		144	.056			1.78	3.70	5.48	7
1236	cycle 8 miles		132	.061			1.94	4.04	5.98	7.60
1238	cycle 10 miles		108	.074			2.37	4.94	7.31	9.35
1240	25 MPH ave, cycle 4 miles		192	.042			1.33	2.78	4.11	5.25
1242	cycle 6 miles		168	.048			1.52	3.17	4.69	6
1244	cycle 8 miles		144	.056			1.78	3.70	5.48	7
1246	cycle 10 miles		132	.061			1.94	4.04	5.98	7.60
1250	30 MPH ave, cycle 4 miles		204	.039			1.25	2.61	3.86	4.92
1252	cycle 6 miles		180	.044			1.42	2.96	4.38	5.60
1254	cycle 8 miles		156	.051			1.64	3.42	5.06	6.45
1256	cycle 10 miles		144	.056			1.78	3.70	5.48	7
1260	35 MPH ave, cycle 4 miles		216	.037			1.18	2.47	3.65	4.65
1262	cycle 6 miles		192	.042			1.33	2.78	4.11	5.25
1264	cycle 8 miles		168	.048			1.52	3.17	4.69	6
1266	cycle 10 miles		156	.051			1.64	3.42	5.06	6.45
1268	cycle 20 miles		108	.074			2.37	4.94	7.31	9.35
1269	cycle 30 miles		72	.111			3.55	7.40	10.95	13.95
1270	cycle 40 miles		60	.133			4.26	8.90	13.16	16.75
1272	40 MPH ave, cycle 6 miles		192	.042			1.33	2.78	4.11	5.25
1274	cycle 8 miles		180	.044			1.42	2.96	4.38	5.60
1276	cycle 10 miles		156	.051			1.64	3.42	5.06	6.45
1278	cycle 20 miles		108	.074			2.37	4.94	7.31	9.35
1280	cycle 30 miles		84	.095			3.04	6.35	9.39	12
1282	cycle 40 miles		72	.111			3.55	7.40	10.95	13.95
1284	cycle 50 miles		60	.133			4.26	8.90	13.16	16.75

31 23 23.20 Hauling		Crew	Daily Output	Labor-Hours	Unit	Material	2009 Bare Costs		Total	Total Incl O&P
							Labor	Equipment		
1294	45 MPH ave, cycle 8 miles	B-34B	180	.044	L.C.Y.		1.42	2.96	4.38	5.60
1296	cycle 10 miles		168	.048			1.52	3.17	4.69	6
1298	cycle 20 miles		120	.067			2.13	4.44	6.57	8.40
1300	cycle 30 miles		96	.083			2.66	5.55	8.21	10.45
1302	cycle 40 miles		72	.111			3.55	7.40	10.95	13.95
1304	cycle 50 miles		60	.133			4.26	8.90	13.16	16.75
1306	50 MPH ave, cycle 10 miles		180	.044			1.42	2.96	4.38	5.60
1308	cycle 20 miles		132	.061			1.94	4.04	5.98	7.60
1310	cycle 30 miles		96	.083			2.66	5.55	8.21	10.45
1312	cycle 40 miles		84	.095			3.04	6.35	9.39	12
1314	cycle 50 miles		72	.111			3.55	7.40	10.95	13.95
1414	15 MPH ave, cycle 0.5 mile, 25 min. wait/Ld./Uld.		204	.039			1.25	2.61	3.86	4.92
1416	cycle 1 mile		192	.042			1.33	2.78	4.11	5.25
1418	cycle 2 miles		168	.048			1.52	3.17	4.69	6
1420	cycle 4 miles		132	.061			1.94	4.04	5.98	7.60
1422	cycle 6 miles		120	.067			2.13	4.44	6.57	8.40
1424	cycle 8 miles		96	.083			2.66	5.55	8.21	10.45
1425	cycle 10 miles		84	.095			3.04	6.35	9.39	12
1426	20 MPH ave, cycle 0.5 mile		216	.037			1.18	2.47	3.65	4.65
1428	cycle 1 mile		204	.039			1.25	2.61	3.86	4.92
1430	cycle 2 miles		180	.044			1.42	2.96	4.38	5.60
1432	cycle 4 miles		156	.051			1.64	3.42	5.06	6.45
1434	cycle 6 miles		132	.061			1.94	4.04	5.98	7.60
1436	cycle 8 miles		120	.067			2.13	4.44	6.57	8.40
1438	cycle 10 miles		96	.083			2.66	5.55	8.21	10.45
1440	25 MPH ave, cycle 4 miles		168	.048			1.52	3.17	4.69	6
1442	cycle 6 miles		144	.056			1.78	3.70	5.48	7
1444	cycle 8 miles		132	.061			1.94	4.04	5.98	7.60
1446	cycle 10 miles		108	.074			2.37	4.94	7.31	9.35
1450	30 MPH ave, cycle 4 miles		168	.048			1.52	3.17	4.69	6
1452	cycle 6 miles		156	.051			1.64	3.42	5.06	6.45
1454	cycle 8 miles		132	.061			1.94	4.04	5.98	7.60
1456	cycle 10 miles		120	.067			2.13	4.44	6.57	8.40
1460	35 MPH ave, cycle 4 miles		180	.044			1.42	2.96	4.38	5.60
1462	cycle 6 miles		156	.051			1.64	3.42	5.06	6.45
1464	cycle 8 miles		144	.056			1.78	3.70	5.48	7
1466	cycle 10 miles		132	.061			1.94	4.04	5.98	7.60
1468	cycle 20 miles		96	.083			2.66	5.55	8.21	10.45
1469	cycle 30 miles		72	.111			3.55	7.40	10.95	13.95
1470	cycle 40 miles		60	.133			4.26	8.90	13.16	16.75
1472	40 MPH ave, cycle 6 miles		168	.048			1.52	3.17	4.69	6
1474	cycle 8 miles		156	.051			1.64	3.42	5.06	6.45
1476	cycle 10 miles		144	.056			1.78	3.70	5.48	7
1478	cycle 20 miles		96	.083			2.66	5.55	8.21	10.45
1480	cycle 30 miles		84	.095			3.04	6.35	9.39	12
1482	cycle 40 miles		60	.133			4.26	8.90	13.16	16.75
1484	cycle 50 miles		60	.133			4.26	8.90	13.16	16.75
1494	45 MPH ave, cycle 8 miles		156	.051			1.64	3.42	5.06	6.45
1496	cycle 10 miles		144	.056			1.78	3.70	5.48	7
1498	cycle 20 miles		108	.074			2.37	4.94	7.31	9.35
1500	cycle 30 miles		84	.095			3.04	6.35	9.39	12
1502	cycle 40 miles		72	.111			3.55	7.40	10.95	13.95
1504	cycle 50 miles		60	.133			4.26	8.90	13.16	16.75

31 23 23 – Fill

31 23 23.20 Hauling		Crew	Daily Output	Labor-Hours	Unit	Material	2009 Bare Costs Labor	Equipment	Total	Total Incl O&P
1506	50 MPH ave, cycle 10 miles	B-34B	156	.051	L.C.Y.		1.64	3.42	5.06	6.45
1508	cycle 20 miles		120	.067			2.13	4.44	6.57	8.40
1510	cycle 30 miles		96	.083			2.66	5.55	8.21	10.45
1512	cycle 40 miles		72	.111			3.55	7.40	10.95	13.95
1514	cycle 50 miles		60	.133			4.26	8.90	13.16	16.75
1614	15 MPH, cycle 0.5 mile, 30 min. wait/Ld./Uld.		180	.044			1.42	2.96	4.38	5.60
1616	cycle 1 mile		168	.048			1.52	3.17	4.69	6
1618	cycle 2 miles		144	.056			1.78	3.70	5.48	7
1620	cycle 4 miles		120	.067			2.13	4.44	6.57	8.40
1622	cycle 6 miles		108	.074			2.37	4.94	7.31	9.35
1624	cycle 8 miles		84	.095			3.04	6.35	9.39	12
1625	cycle 10 miles		84	.095			3.04	6.35	9.39	12
1626	20 MPH ave, cycle 0.5 mile		180	.044			1.42	2.96	4.38	5.60
1628	cycle 1 mile		168	.048			1.52	3.17	4.69	6
1630	cycle 2 miles		156	.051			1.64	3.42	5.06	6.45
1632	cycle 4 miles		132	.061			1.94	4.04	5.98	7.60
1634	cycle 6 miles		120	.067			2.13	4.44	6.57	8.40
1636	cycle 8 miles		108	.074			2.37	4.94	7.31	9.35
1638	cycle 10 miles		96	.083			2.66	5.55	8.21	10.45
1640	25 MPH ave, cycle 4 miles		144	.056			1.78	3.70	5.48	7
1642	cycle 6 miles		132	.061			1.94	4.04	5.98	7.60
1644	cycle 8 miles		108	.074			2.37	4.94	7.31	9.35
1646	cycle 10 miles		108	.074			2.37	4.94	7.31	9.35
1650	30 MPH ave, cycle 4 miles		144	.056			1.78	3.70	5.48	7
1652	cycle 6 miles		132	.061			1.94	4.04	5.98	7.60
1654	cycle 8 miles		120	.067			2.13	4.44	6.57	8.40
1656	cycle 10 miles		108	.074			2.37	4.94	7.31	9.35
1660	35 MPH ave, cycle 4 miles		156	.051			1.64	3.42	5.06	6.45
1662	cycle 6 miles		144	.056			1.78	3.70	5.48	7
1664	cycle 8 miles		132	.061			1.94	4.04	5.98	7.60
1666	cycle 10 miles		120	.067			2.13	4.44	6.57	8.40
1668	cycle 20 miles		84	.095			3.04	6.35	9.39	12
1669	cycle 30 miles		72	.111			3.55	7.40	10.95	13.95
1670	cycle 40 miles		60	.133			4.26	8.90	13.16	16.75
1672	40 MPH, cycle 6 miles		144	.056			1.78	3.70	5.48	7
1674	cycle 8 miles		132	.061			1.94	4.04	5.98	7.60
1676	cycle 10 miles		120	.067			2.13	4.44	6.57	8.40
1678	cycle 20 miles		96	.083			2.66	5.55	8.21	10.45
1680	cycle 30 miles		72	.111			3.55	7.40	10.95	13.95
1682	cycle 40 miles		60	.133			4.26	8.90	13.16	16.75
1684	cycle 50 miles		48	.167			5.35	11.10	16.45	21
1694	45 MPH ave, cycle 8 miles		144	.056			1.78	3.70	5.48	7
1696	cycle 10 miles		132	.061			1.94	4.04	5.98	7.60
1698	cycle 20 miles		96	.083			2.66	5.55	8.21	10.45
1700	cycle 30 miles		84	.095			3.04	6.35	9.39	12
1702	cycle 40 miles		60	.133			4.26	8.90	13.16	16.75
1704	cycle 50 miles		60	.133			4.26	8.90	13.16	16.75
1706	50 MPH ave, cycle 10 miles		132	.061			1.94	4.04	5.98	7.60
1708	cycle 20 miles		108	.074			2.37	4.94	7.31	9.35
1710	cycle 30 miles		84	.095			3.04	6.35	9.39	12
1712	cycle 40 miles		72	.111			3.55	7.40	10.95	13.95
1714	cycle 50 miles		60	.133			4.26	8.90	13.16	16.75
2000	Hauling, 8 CY truck, small project cost per hour	B-34A	8	1	Hr.		32	41	73	97.50

31 23 Excavation and Fill

31 23 23 – Fill

31 23 23.20 Hauling		Crew	Daily Output	Labor-Hours	Unit	Material	2009 Bare Costs Labor	2009 Bare Costs Equipment	Total	Total Incl O&P
2100	12 CY Truck	B-34B	8	1	Hr.		32	66.50	98.50	126
2150	16.5 CY Truck	B-34C	8	1			32	74.50	106.50	134
2200	20 CY Truck	B-34D	8	1	↓		32	76	108	136
2300	Grading at dump, or embankment if required, by dozer	B-10B	1000	.012	L.C.Y.		.46	1.08	1.54	1.92
2310	Spotter at fill or cut, if required	1 Clab	8	1	Hr.		31.50		31.50	52
3014	16.5 CY truck, 15 min. wait/Ld./Uld., 15 MPH, cycle 0.5 mile	B-34C	462	.017	L.C.Y.		.55	1.29	1.84	2.33
3016	cycle 1 mile		413	.019			.62	1.44	2.06	2.59
3018	cycle 2 miles		347	.023			.74	1.71	2.45	3.09
3020	cycle 4 miles		248	.032			1.03	2.40	3.43	4.33
3022	cycle 6 miles		198	.040			1.29	3	4.29	5.40
3024	cycle 8 miles		165	.048			1.55	3.60	5.15	6.50
3025	cycle 10 miles		132	.061			1.94	4.50	6.44	8.10
3026	20 MPH ave, cycle 0.5 mile		479	.017			.53	1.24	1.77	2.23
3028	cycle 1 mile		429	.019			.60	1.39	1.99	2.50
3030	cycle 2 miles		380	.021			.67	1.56	2.23	2.82
3032	cycle 4 miles		281	.028			.91	2.12	3.03	3.82
3034	cycle 6 miles		231	.035			1.11	2.57	3.68	4.64
3036	cycle 8 miles		198	.040			1.29	3	4.29	5.40
3038	cycle 10 miles		165	.048			1.55	3.60	5.15	6.50
3040	25 MPH ave, cycle 4 miles		314	.025			.81	1.89	2.70	3.41
3042	cycle 6 miles		264	.030			.97	2.25	3.22	4.07
3044	cycle 8 miles		231	.035			1.11	2.57	3.68	4.64
3046	cycle 10 miles		198	.040			1.29	3	4.29	5.40
3050	30 MPH ave, cycle 4 miles		347	.023			.74	1.71	2.45	3.09
3052	cycle 6 miles		281	.028			.91	2.12	3.03	3.82
3054	cycle 8 miles		248	.032			1.03	2.40	3.43	4.33
3056	cycle 10 miles		215	.037			1.19	2.76	3.95	4.99
3060	35 MPH ave, cycle 4 miles		363	.022			.70	1.64	2.34	2.95
3062	cycle 6 miles		314	.025			.81	1.89	2.70	3.41
3064	cycle 8 miles		264	.030			.97	2.25	3.22	4.07
3066	cycle 10 miles		248	.032			1.03	2.40	3.43	4.33
3068	cycle 20 miles		149	.054			1.72	3.99	5.71	7.20
3070	cycle 30 miles		116	.069			2.20	5.10	7.30	9.25
3072	cycle 40 miles		83	.096			3.08	7.15	10.23	12.95
3074	40 MPH ave, cycle 6 miles		330	.024			.77	1.80	2.57	3.25
3076	cycle 8 miles		281	.028			.91	2.12	3.03	3.82
3078	cycle 10 miles		264	.030			.97	2.25	3.22	4.07
3080	cycle 20 miles		165	.048			1.55	3.60	5.15	6.50
3082	cycle 30 miles		132	.061			1.94	4.50	6.44	8.10
3084	cycle 40 miles		99	.081			2.58	6	8.58	10.85
3086	cycle 50 miles		83	.096			3.08	7.15	10.23	12.95
3094	45 MPH ave, cycle 8 miles		297	.027			.86	2	2.86	3.61
3096	cycle 10 miles		281	.028			.91	2.12	3.03	3.82
3098	cycle 20 miles		182	.044			1.40	3.27	4.67	5.90
3100	cycle 30 miles		132	.061			1.94	4.50	6.44	8.10
3102	cycle 40 miles		116	.069			2.20	5.10	7.30	9.25
3104	cycle 50 miles		99	.081			2.58	6	8.58	10.85
3106	50 MPH ave, cycle 10 miles		281	.028			.91	2.12	3.03	3.82
3108	cycle 20 miles		198	.040			1.29	3	4.29	5.40
3110	cycle 30 miles		149	.054			1.72	3.99	5.71	7.20
3112	cycle 40 miles		116	.069			2.20	5.10	7.30	9.25
3114	cycle 50 miles		99	.081			2.58	6	8.58	10.85
3214	20 min. wait/Ld./Uld., 15 MPH, cycle 0.5 mile		363	.022			.70	1.64	2.34	2.95

31 23 23.20 Hauling		Crew	Daily Output	Labor-Hours	Unit	Material	2009 Bare Costs Labor	Equipment	Total	Total Incl O&P
3216	cycle 1 mile	B-34C	330	.024	L.C.Y.		.77	1.80	2.57	3.25
3218	cycle 2 miles		281	.028			.91	2.12	3.03	3.82
3220	cycle 4 miles		215	.037			1.19	2.76	3.95	4.99
3222	cycle 6 miles		182	.044			1.40	3.27	4.67	5.90
3224	cycle 8 miles		149	.054			1.72	3.99	5.71	7.20
3225	cycle 10 miles		132	.061			1.94	4.50	6.44	8.10
3226	20 MPH ave, cycle 0.5 mile		363	.022			.70	1.64	2.34	2.95
3228	cycle 1 mile		347	.023			.74	1.71	2.45	3.09
3230	cycle 2 miles		297	.027			.86	2	2.86	3.61
3232	cycle 4 miles		248	.032			1.03	2.40	3.43	4.33
3234	cycle 6 miles		198	.040			1.29	3	4.29	5.40
3236	cycle 8 miles		182	.044			1.40	3.27	4.67	5.90
3238	cycle 10 miles		149	.054			1.72	3.99	5.71	7.20
3240	25 MPH ave, cycle 4 miles		264	.030			.97	2.25	3.22	4.07
3242	cycle 6 miles		231	.035			1.11	2.57	3.68	4.64
3244	cycle 8 miles		198	.040			1.29	3	4.29	5.40
3246	cycle 10 miles		182	.044			1.40	3.27	4.67	5.90
3250	30 MPH ave, cycle 4 miles		281	.028			.91	2.12	3.03	3.82
3252	cycle 6 miles		248	.032			1.03	2.40	3.43	4.33
3254	cycle 8 miles		215	.037			1.19	2.76	3.95	4.99
3256	cycle 10 miles		198	.040			1.29	3	4.29	5.40
3260	35 MPH ave, cycle 4 miles		297	.027			.86	2	2.86	3.61
3262	cycle 6 miles		264	.030			.97	2.25	3.22	4.07
3264	cycle 8 miles		231	.035			1.11	2.57	3.68	4.64
3266	cycle 10 miles		215	.037			1.19	2.76	3.95	4.99
3268	cycle 20 miles		149	.054			1.72	3.99	5.71	7.20
3270	cycle 30 miles		99	.081			2.58	6	8.58	10.85
3272	cycle 40 miles		83	.096			3.08	7.15	10.23	12.95
3274	40 MPH ave, cycle 6 miles		264	.030			.97	2.25	3.22	4.07
3276	cycle 8 miles		248	.032			1.03	2.40	3.43	4.33
3278	cycle 10 miles		215	.037			1.19	2.76	3.95	4.99
3280	cycle 20 miles		149	.054			1.72	3.99	5.71	7.20
3282	cycle 30 miles		116	.069			2.20	5.10	7.30	9.25
3284	cycle 40 miles		99	.081			2.58	6	8.58	10.85
3286	cycle 50 miles		83	.096			3.08	7.15	10.23	12.95
3294	45 MPH ave, cycle 8 miles		248	.032			1.03	2.40	3.43	4.33
3296	cycle 10 miles		231	.035			1.11	2.57	3.68	4.64
3298	cycle 20 miles		165	.048			1.55	3.60	5.15	6.50
3300	cycle 30 miles		132	.061			1.94	4.50	6.44	8.10
3302	cycle 40 miles		99	.081			2.58	6	8.58	10.85
3304	cycle 50 miles		83	.096			3.08	7.15	10.23	12.95
3306	50 MPH ave, cycle 10 miles		248	.032			1.03	2.40	3.43	4.33
3308	cycle 20 miles		182	.044			1.40	3.27	4.67	5.90
3310	cycle 30 miles		132	.061			1.94	4.50	6.44	8.10
3312	cycle 40 miles		116	.069			2.20	5.10	7.30	9.25
3314	cycle 50 miles		99	.081			2.58	6	8.58	10.85
3414	25 min. wait/Ld./Uld., 15 MPH, cycle 0.5 mile		281	.028			.91	2.12	3.03	3.82
3416	cycle 1 mile		264	.030			.97	2.25	3.22	4.07
3418	cycle 2 miles		231	.035			1.11	2.57	3.68	4.64
3420	cycle 4 miles		182	.044			1.40	3.27	4.67	5.90
3422	cycle 6 miles		165	.048			1.55	3.60	5.15	6.50
3424	cycle 8 miles		132	.061			1.94	4.50	6.44	8.10
3425	cycle 10 miles		116	.069			2.20	5.10	7.30	9.25

31 23 23.20 Hauling	Crew	Daily Output	Labor-Hours	Unit	Material	2009 Bare Costs Labor	Equipment	Total	Total Incl O&P
3426	20 MPH ave, cycle 0.5 mile	B-34C	297	.027	L.C.Y.		.86	2	2.86	3.61
3428	cycle 1 mile		281	.028			.91	2.12	3.03	3.82
3430	cycle 2 miles		248	.032			1.03	2.40	3.43	4.33
3432	cycle 4 miles		215	.037			1.19	2.76	3.95	4.99
3434	cycle 6 miles		182	.044			1.40	3.27	4.67	5.90
3436	cycle 8 miles		165	.048			1.55	3.60	5.15	6.50
3438	cycle 10 miles		132	.061			1.94	4.50	6.44	8.10
3440	25 MPH ave, cycle 4 miles		231	.035			1.11	2.57	3.68	4.64
3442	cycle 6 miles		198	.040			1.29	3	4.29	5.40
3444	cycle 8 miles		182	.044			1.40	3.27	4.67	5.90
3446	cycle 10 miles		165	.048			1.55	3.60	5.15	6.50
3450	30 MPH ave, cycle 4 miles		231	.035			1.11	2.57	3.68	4.64
3452	cycle 6 miles		215	.037			1.19	2.76	3.95	4.99
3454	cycle 8 miles		182	.044			1.40	3.27	4.67	5.90
3456	cycle 10 miles		165	.048			1.55	3.60	5.15	6.50
3460	35 MPH ave, cycle 4 miles		248	.032			1.03	2.40	3.43	4.33
3462	cycle 6 miles		215	.037			1.19	2.76	3.95	4.99
3464	cycle 8 miles		198	.040			1.29	3	4.29	5.40
3466	cycle 10 miles		182	.044			1.40	3.27	4.67	5.90
3468	cycle 20 miles		132	.061			1.94	4.50	6.44	8.10
3470	cycle 30 miles		99	.081			2.58	6	8.58	10.85
3472	cycle 40 miles		83	.096			3.08	7.15	10.23	12.95
3474	40 MPH, cycle 6 miles		231	.035			1.11	2.57	3.68	4.64
3476	cycle 8 miles		215	.037			1.19	2.76	3.95	4.99
3478	cycle 10 miles		198	.040			1.29	3	4.29	5.40
3480	cycle 20 miles		132	.061			1.94	4.50	6.44	8.10
3482	cycle 30 miles		116	.069			2.20	5.10	7.30	9.25
3484	cycle 40 miles		83	.096			3.08	7.15	10.23	12.95
3486	cycle 50 miles		83	.096			3.08	7.15	10.23	12.95
3494	45 MPH ave, cycle 8 miles		215	.037			1.19	2.76	3.95	4.99
3496	cycle 10 miles		198	.040			1.29	3	4.29	5.40
3498	cycle 20 miles		149	.054			1.72	3.99	5.71	7.20
3500	cycle 30 miles		116	.069			2.20	5.10	7.30	9.25
3502	cycle 40 miles		99	.081			2.58	6	8.58	10.85
3504	cycle 50 miles		83	.096			3.08	7.15	10.23	12.95
3506	50 MPH ave, cycle 10 miles		215	.037			1.19	2.76	3.95	4.99
3508	cycle 20 miles		165	.048			1.55	3.60	5.15	6.50
3510	cycle 30 miles		132	.061			1.94	4.50	6.44	8.10
3512	cycle 40 miles		99	.081			2.58	6	8.58	10.85
3514	cycle 50 miles		83	.096			3.08	7.15	10.23	12.95
3614	30 min. wait/Ld./Uld., 15 MPH, cycle 0.5 mile		248	.032			1.03	2.40	3.43	4.33
3616	cycle 1 mile		231	.035			1.11	2.57	3.68	4.64
3618	cycle 2 miles		198	.040			1.29	3	4.29	5.40
3620	cycle 4 miles		165	.048			1.55	3.60	5.15	6.50
3622	cycle 6 miles		149	.054			1.72	3.99	5.71	7.20
3624	cycle 8 miles		116	.069			2.20	5.10	7.30	9.25
3625	cycle 10 miles		116	.069			2.20	5.10	7.30	9.25
3626	20 MPH ave, cycle 0.5 mile		248	.032			1.03	2.40	3.43	4.33
3628	cycle 1 mile		231	.035			1.11	2.57	3.68	4.64
3630	cycle 2 miles		215	.037			1.19	2.76	3.95	4.99
3632	cycle 4 miles		182	.044			1.40	3.27	4.67	5.90
3634	cycle 6 miles		165	.048			1.55	3.60	5.15	6.50
3636	cycle 8 miles		149	.054			1.72	3.99	5.71	7.20

31 23 23.20 Hauling		Crew	Daily Output	Labor-Hours	Unit	Material	2009 Bare Costs		Total	Total Incl O&P
							Labor	Equipment		
3638	cycle 10 miles	B-34C	132	.061	L.C.Y.		1.94	4.50	6.44	8.10
3640	25 MPH ave, cycle 4 miles		198	.040			1.29	3	4.29	5.40
3642	cycle 6 miles		182	.044			1.40	3.27	4.67	5.90
3644	cycle 8 miles		149	.054			1.72	3.99	5.71	7.20
3646	cycle 10 miles		149	.054			1.72	3.99	5.71	7.20
3650	30 MPH ave, cycle 4 miles		198	.040			1.29	3	4.29	5.40
3652	cycle 6 miles		182	.044			1.40	3.27	4.67	5.90
3654	cycle 8 miles		165	.048			1.55	3.60	5.15	6.50
3656	cycle 10 miles		149	.054			1.72	3.99	5.71	7.20
3660	35 MPH ave, cycle 4 miles		215	.037			1.19	2.76	3.95	4.99
3662	cycle 6 miles		198	.040			1.29	3	4.29	5.40
3664	cycle 8 miles		182	.044			1.40	3.27	4.67	5.90
3666	cycle 10 miles		165	.048			1.55	3.60	5.15	6.50
3668	cycle 20 miles		116	.069			2.20	5.10	7.30	9.25
3670	cycle 30 miles		99	.081			2.58	6	8.58	10.85
3672	cycle 40 miles		83	.096			3.08	7.15	10.23	12.95
3674	40 MPH, cycle 6 miles		198	.040			1.29	3	4.29	5.40
3676	cycle 8 miles		182	.044			1.40	3.27	4.67	5.90
3678	cycle 10 miles		165	.048			1.55	3.60	5.15	6.50
3680	cycle 20 miles		132	.061			1.94	4.50	6.44	8.10
3682	cycle 30 miles		99	.081			2.58	6	8.58	10.85
3684	cycle 40 miles		83	.096			3.08	7.15	10.23	12.95
3686	cycle 50 miles		66	.121			3.87	9	12.87	16.25
3694	45 MPH ave, cycle 8 miles		198	.040			1.29	3	4.29	5.40
3696	cycle 10 miles		182	.044			1.40	3.27	4.67	5.90
3698	cycle 20 miles		132	.061			1.94	4.50	6.44	8.10
3700	cycle 30 miles		116	.069			2.20	5.10	7.30	9.25
3702	cycle 40 miles		83	.096			3.08	7.15	10.23	12.95
3704	cycle 50 miles		83	.096			3.08	7.15	10.23	12.95
3706	50 MPH ave, cycle 10 miles		182	.044			1.40	3.27	4.67	5.90
3708	cycle 20 miles		149	.054			1.72	3.99	5.71	7.20
3710	cycle 30 miles		116	.069			2.20	5.10	7.30	9.25
3712	cycle 40 miles		99	.081			2.58	6	8.58	10.85
3714	cycle 50 miles		83	.096			3.08	7.15	10.23	12.95
4014	20 CY truck, 15 min. wait/Ld./Uld., 15 MPH, cycle 0.5 mile	B-34D	560	.014			.46	1.09	1.55	1.95
4016	cycle 1 mile		500	.016			.51	1.22	1.73	2.18
4018	cycle 2 miles		420	.019			.61	1.45	2.06	2.60
4020	cycle 4 miles		300	.027			.85	2.03	2.88	3.63
4022	cycle 6 miles		240	.033			1.06	2.54	3.60	4.53
4024	cycle 8 miles		200	.040			1.28	3.05	4.33	5.45
4025	cycle 10 miles		160	.050			1.60	3.81	5.41	6.80
4026	20 MPH ave, cycle 0.5 mile		580	.014			.44	1.05	1.49	1.87
4028	cycle 1 mile		520	.015			.49	1.17	1.66	2.10
4030	cycle 2 miles		460	.017			.56	1.32	1.88	2.37
4032	cycle 4 miles		340	.024			.75	1.79	2.54	3.20
4034	cycle 6 miles		280	.029			.91	2.18	3.09	3.89
4036	cycle 8 miles		240	.033			1.06	2.54	3.60	4.53
4038	cycle 10 miles		200	.040			1.28	3.05	4.33	5.45
4040	25 MPH ave, cycle 4 miles		380	.021			.67	1.60	2.27	2.86
4042	cycle 6 miles		320	.025			.80	1.90	2.70	3.40
4044	cycle 8 miles		280	.029			.91	2.18	3.09	3.89
4046	cycle 10 miles		240	.033			1.06	2.54	3.60	4.53
4050	30 MPH ave, cycle 4 miles		420	.019			.61	1.45	2.06	2.60

31 23 23.20 Hauling	Crew	Daily Output	Labor-Hours	Unit	Material	2009 Bare Costs Labor	2009 Bare Costs Equipment	Total	Total Incl O&P
4052	cycle 6 miles	B-34D	340	.024	L.C.Y.		.75	1.79	2.54	3.20
4054	cycle 8 miles		300	.027			.85	2.03	2.88	3.63
4056	cycle 10 miles		260	.031			.98	2.34	3.32	4.19
4060	35 MPH ave, cycle 4 miles		440	.018			.58	1.38	1.96	2.47
4062	cycle 6 miles		380	.021			.67	1.60	2.27	2.86
4064	cycle 8 miles		320	.025			.80	1.90	2.70	3.40
4066	cycle 10 miles		300	.027			.85	2.03	2.88	3.63
4068	cycle 20 miles		180	.044			1.42	3.38	4.80	6.05
4070	cycle 30 miles		140	.057			1.83	4.35	6.18	7.75
4072	cycle 40 miles		100	.080			2.56	6.10	8.66	10.90
4074	40 MPH ave, cycle 6 miles		400	.020			.64	1.52	2.16	2.72
4076	cycle 8 miles		340	.024			.75	1.79	2.54	3.20
4078	cycle 10 miles		320	.025			.80	1.90	2.70	3.40
4080	cycle 20 miles		200	.040			1.28	3.05	4.33	5.45
4082	cycle 30 miles		160	.050			1.60	3.81	5.41	6.80
4084	cycle 40 miles		120	.067			2.13	5.10	7.23	9.10
4086	cycle 50 miles		100	.080			2.56	6.10	8.66	10.90
4094	45 MPH ave, cycle 8 miles		360	.022			.71	1.69	2.40	3.02
4096	cycle 10 miles		340	.024			.75	1.79	2.54	3.20
4098	cycle 20 miles		220	.036			1.16	2.77	3.93	4.94
4100	cycle 30 miles		160	.050			1.60	3.81	5.41	6.80
4102	cycle 40 miles		140	.057			1.83	4.35	6.18	7.75
4104	cycle 50 miles		120	.067			2.13	5.10	7.23	9.10
4106	50 MPH ave, cycle 10 miles		340	.024			.75	1.79	2.54	3.20
4108	cycle 20 miles		240	.033			1.06	2.54	3.60	4.53
4110	cycle 30 miles		180	.044			1.42	3.38	4.80	6.05
4112	cycle 40 miles		140	.057			1.83	4.35	6.18	7.75
4114	cycle 50 miles		120	.067			2.13	5.10	7.23	9.10
4214	20 min. wait/Ld./Uld., 15 MPH, cycle 0.5 mile		440	.018			.58	1.38	1.96	2.47
4216	cycle 1 mile		400	.020			.64	1.52	2.16	2.72
4218	cycle 2 miles		340	.024			.75	1.79	2.54	3.20
4220	cycle 4 miles		260	.031			.98	2.34	3.32	4.19
4222	cycle 6 miles		220	.036			1.16	2.77	3.93	4.94
4224	cycle 8 miles		180	.044			1.42	3.38	4.80	6.05
4225	cycle 10 miles		160	.050			1.60	3.81	5.41	6.80
4226	20 MPH ave, cycle 0.5 mile		440	.018			.58	1.38	1.96	2.47
4228	cycle 1 mile		420	.019			.61	1.45	2.06	2.60
4230	cycle 2 miles		360	.022			.71	1.69	2.40	3.02
4232	cycle 4 miles		300	.027			.85	2.03	2.88	3.63
4234	cycle 6 miles		240	.033			1.06	2.54	3.60	4.53
4236	cycle 8 miles		220	.036			1.16	2.77	3.93	4.94
4238	cycle 10 miles		180	.044			1.42	3.38	4.80	6.05
4240	25 MPH ave, cycle 4 miles		320	.025			.80	1.90	2.70	3.40
4242	cycle 6 miles		280	.029			.91	2.18	3.09	3.89
4244	cycle 8 miles		240	.033			1.06	2.54	3.60	4.53
4246	cycle 10 miles		220	.036			1.16	2.77	3.93	4.94
4250	30 MPH ave, cycle 4 miles		340	.024			.75	1.79	2.54	3.20
4252	cycle 6 miles		300	.027			.85	2.03	2.88	3.63
4254	cycle 8 miles		260	.031			.98	2.34	3.32	4.19
4256	cycle 10 miles		240	.033			1.06	2.54	3.60	4.53
4260	35 MPH ave, cycle 4 miles		360	.022			.71	1.69	2.40	3.02
4262	cycle 6 miles		320	.025			.80	1.90	2.70	3.40
4264	cycle 8 miles		280	.029			.91	2.18	3.09	3.89

31 23 23.20 Hauling		Crew	Daily Output	Labor-Hours	Unit	Material	2009 Bare Costs Labor	2009 Bare Costs Equipment	Total	Total Incl O&P
4266	cycle 10 miles	B-34D	260	.031	L.C.Y.		.98	2.34	3.32	4.19
4268	cycle 20 miles		180	.044			1.42	3.38	4.80	6.05
4270	cycle 30 miles		120	.067			2.13	5.10	7.23	9.10
4272	cycle 40 miles		100	.080			2.56	6.10	8.66	10.90
4274	40 MPH ave, cycle 6 miles		320	.025			.80	1.90	2.70	3.40
4276	cycle 8 miles		300	.027			.85	2.03	2.88	3.63
4278	cycle 10 miles		260	.031			.98	2.34	3.32	4.19
4280	cycle 20 miles		180	.044			1.42	3.38	4.80	6.05
4282	cycle 30 miles		140	.057			1.83	4.35	6.18	7.75
4284	cycle 40 miles		120	.067			2.13	5.10	7.23	9.10
4286	cycle 50 miles		100	.080			2.56	6.10	8.66	10.90
4294	45 MPH ave, cycle 8 miles		300	.027			.85	2.03	2.88	3.63
4296	cycle 10 miles		280	.029			.91	2.18	3.09	3.89
4298	cycle 20 miles		200	.040			1.28	3.05	4.33	5.45
4300	cycle 30 miles		160	.050			1.60	3.81	5.41	6.80
4302	cycle 40 miles		120	.067			2.13	5.10	7.23	9.10
4304	cycle 50 miles		100	.080			2.56	6.10	8.66	10.90
4306	50 MPH ave, cycle 10 miles		300	.027			.85	2.03	2.88	3.63
4308	cycle 20 miles		220	.036			1.16	2.77	3.93	4.94
4310	cycle 30 miles		180	.044			1.42	3.38	4.80	6.05
4312	cycle 40 miles		140	.057			1.83	4.35	6.18	7.75
4314	cycle 50 miles		120	.067			2.13	5.10	7.23	9.10
4414	25 min. wait/Ld./Uld., 15 MPH, cycle 0.5 mile		340	.024			.75	1.79	2.54	3.20
4416	cycle 1 mile		320	.025			.80	1.90	2.70	3.40
4418	cycle 2 miles		280	.029			.91	2.18	3.09	3.89
4420	cycle 4 miles		220	.036			1.16	2.77	3.93	4.94
4422	cycle 6 miles		200	.040			1.28	3.05	4.33	5.45
4424	cycle 8 miles		160	.050			1.60	3.81	5.41	6.80
4425	cycle 10 miles		140	.057			1.83	4.35	6.18	7.75
4426	20 MPH ave, cycle 0.5 mile		360	.022			.71	1.69	2.40	3.02
4428	cycle 1 mile		340	.024			.75	1.79	2.54	3.20
4430	cycle 2 miles		300	.027			.85	2.03	2.88	3.63
4432	cycle 4 miles		260	.031			.98	2.34	3.32	4.19
4434	cycle 6 miles		220	.036			1.16	2.77	3.93	4.94
4436	cycle 8 miles		200	.040			1.28	3.05	4.33	5.45
4438	cycle 10 miles		160	.050			1.60	3.81	5.41	6.80
4440	25 MPH ave, cycle 4 miles		280	.029			.91	2.18	3.09	3.89
4442	cycle 6 miles		240	.033			1.06	2.54	3.60	4.53
4444	cycle 8 miles		220	.036			1.16	2.77	3.93	4.94
4446	cycle 10 miles		200	.040			1.28	3.05	4.33	5.45
4450	30 MPH ave, cycle 4 miles		280	.029			.91	2.18	3.09	3.89
4452	cycle 6 miles		260	.031			.98	2.34	3.32	4.19
4454	cycle 8 miles		220	.036			1.16	2.77	3.93	4.94
4456	cycle 10 miles		200	.040			1.28	3.05	4.33	5.45
4460	35 MPH ave, cycle 4 miles		300	.027			.85	2.03	2.88	3.63
4462	cycle 6 miles		260	.031			.98	2.34	3.32	4.19
4464	cycle 8 miles		240	.033			1.06	2.54	3.60	4.53
4466	cycle 10 miles		220	.036			1.16	2.77	3.93	4.94
4468	cycle 20 miles		160	.050			1.60	3.81	5.41	6.80
4470	cycle 30 miles		120	.067			2.13	5.10	7.23	9.10
4472	cycle 40 miles		100	.080			2.56	6.10	8.66	10.90
4474	40 MPH, cycle 6 miles		280	.029			.91	2.18	3.09	3.89
4476	cycle 8 miles		260	.031			.98	2.34	3.32	4.19

31 23 23.20 Hauling		Crew	Daily Output	Labor-Hours	Unit	Material	2009 Bare Costs Labor	Equipment	Total	Total Incl O&P
4478	cycle 10 miles	B-34D	240	.033	L.C.Y.		1.06	2.54	3.60	4.53
4480	cycle 20 miles		160	.050			1.60	3.81	5.41	6.80
4482	cycle 30 miles		140	.057			1.83	4.35	6.18	7.75
4484	cycle 40 miles		100	.080			2.56	6.10	8.66	10.90
4486	cycle 50 miles		100	.080			2.56	6.10	8.66	10.90
4494	45 MPH ave, cycle 8 miles		260	.031			.98	2.34	3.32	4.19
4496	cycle 10 miles		240	.033			1.06	2.54	3.60	4.53
4498	cycle 20 miles		180	.044			1.42	3.38	4.80	6.05
4500	cycle 30 miles		140	.057			1.83	4.35	6.18	7.75
4502	cycle 40 miles		120	.067			2.13	5.10	7.23	9.10
4504	cycle 50 miles		100	.080			2.56	6.10	8.66	10.90
4506	50 MPH ave, cycle 10 miles		260	.031			.98	2.34	3.32	4.19
4508	cycle 20 miles		200	.040			1.28	3.05	4.33	5.45
4510	cycle 30 miles		160	.050			1.60	3.81	5.41	6.80
4512	cycle 40 miles		120	.067			2.13	5.10	7.23	9.10
4514	cycle 50 miles		100	.080			2.56	6.10	8.66	10.90
4614	30 min. wait/Ld./Uld., 15 MPH, cycle 0.5 mile		300	.027			.85	2.03	2.88	3.63
4616	cycle 1 mile		280	.029			.91	2.18	3.09	3.89
4618	cycle 2 miles		240	.033			1.06	2.54	3.60	4.53
4620	cycle 4 miles		200	.040			1.28	3.05	4.33	5.45
4622	cycle 6 miles		180	.044			1.42	3.38	4.80	6.05
4624	cycle 8 miles		140	.057			1.83	4.35	6.18	7.75
4625	cycle 10 miles		140	.057			1.83	4.35	6.18	7.75
4626	20 MPH ave, cycle 0.5 mile		300	.027			.85	2.03	2.88	3.63
4628	cycle 1 mile		280	.029			.91	2.18	3.09	3.89
4630	cycle 2 miles		260	.031			.98	2.34	3.32	4.19
4632	cycle 4 miles		220	.036			1.16	2.77	3.93	4.94
4634	cycle 6 miles		200	.040			1.28	3.05	4.33	5.45
4636	cycle 8 miles		180	.044			1.42	3.38	4.80	6.05
4638	cycle 10 miles		160	.050			1.60	3.81	5.41	6.80
4640	25 MPH ave, cycle 4 miles		240	.033			1.06	2.54	3.60	4.53
4642	cycle 6 miles		220	.036			1.16	2.77	3.93	4.94
4644	cycle 8 miles		180	.044			1.42	3.38	4.80	6.05
4646	cycle 10 miles		180	.044			1.42	3.38	4.80	6.05
4650	30 MPH ave, cycle 4 miles		240	.033			1.06	2.54	3.60	4.53
4652	cycle 6 miles		220	.036			1.16	2.77	3.93	4.94
4654	cycle 8 miles		200	.040			1.28	3.05	4.33	5.45
4656	cycle 10 miles		180	.044			1.42	3.38	4.80	6.05
4660	35 MPH ave, cycle 4 miles		260	.031			.98	2.34	3.32	4.19
4662	cycle 6 miles		240	.033			1.06	2.54	3.60	4.53
4664	cycle 8 miles		220	.036			1.16	2.77	3.93	4.94
4666	cycle 10 miles		200	.040			1.28	3.05	4.33	5.45
4668	cycle 20 miles		140	.057			1.83	4.35	6.18	7.75
4670	cycle 30 miles		120	.067			2.13	5.10	7.23	9.10
4672	cycle 40 miles		100	.080			2.56	6.10	8.66	10.90
4674	40 MPH, cycle 6 miles		240	.033			1.06	2.54	3.60	4.53
4676	cycle 8 miles		220	.036			1.16	2.77	3.93	4.94
4678	cycle 10 miles		200	.040			1.28	3.05	4.33	5.45
4680	cycle 20 miles		160	.050			1.60	3.81	5.41	6.80
4682	cycle 30 miles		120	.067			2.13	5.10	7.23	9.10
4684	cycle 40 miles		100	.080			2.56	6.10	8.66	10.90
4686	cycle 50 miles		80	.100			3.20	7.60	10.80	13.60
4694	45 MPH ave, cycle 8 miles		220	.036			1.16	2.77	3.93	4.94

31 23 23.20 Hauling		Crew	Daily Output	Labor-Hours	Unit	Material	2009 Bare Costs Labor	Equipment	Total	Total Incl O&P
4696	cycle 10 miles	B-34D	220	.036	L.C.Y.		1.16	2.77	3.93	4.94
4698	cycle 20 miles		160	.050			1.60	3.81	5.41	6.80
4700	cycle 30 miles		140	.057			1.83	4.35	6.18	7.75
4702	cycle 40 miles		100	.080			2.56	6.10	8.66	10.90
4704	cycle 50 miles		100	.080			2.56	6.10	8.66	10.90
4706	50 MPH ave, cycle 10 miles		220	.036			1.16	2.77	3.93	4.94
4708	cycle 20 miles		180	.044			1.42	3.38	4.80	6.05
4710	cycle 30 miles		140	.057			1.83	4.35	6.18	7.75
4712	cycle 40 miles		120	.067			2.13	5.10	7.23	9.10
4714	cycle 50 miles		100	.080			2.56	6.10	8.66	10.90
5000	22 CY off-road, 15 min. wait/Ld./Uld., 5 MPH, cycle 2000 ft	B-34F	528	.015			.48	2.17	2.65	3.17
5010	cycle 3000 ft		484	.017			.53	2.36	2.89	3.47
5020	cycle 4000 ft		440	.018			.58	2.60	3.18	3.81
5030	cycle 0.5 mile		506	.016			.51	2.26	2.77	3.32
5040	cycle 1 mile		374	.021			.68	3.06	3.74	4.48
5050	cycle 2 miles		264	.030			.97	4.33	5.30	6.35
5060	10 MPH, cycle 2000 ft		594	.013			.43	1.93	2.36	2.83
5070	cycle 3000 ft		572	.014			.45	2	2.45	2.93
5080	cycle 4000 ft		528	.015			.48	2.17	2.65	3.17
5090	cycle 0.5 mile		572	.014			.45	2	2.45	2.93
5100	cycle 1 mile		506	.016			.51	2.26	2.77	3.32
5110	cycle 2 miles		374	.021			.68	3.06	3.74	4.48
5120	cycle 4 miles		264	.030			.97	4.33	5.30	6.35
5130	15 MPH, cycle 2000 ft		638	.013			.40	1.79	2.19	2.63
5140	cycle 3000 ft		594	.013			.43	1.93	2.36	2.83
5150	cycle 4000 ft		572	.014			.45	2	2.45	2.93
5160	cycle 0.5 mile		616	.013			.42	1.86	2.28	2.72
5170	cycle 1 mile		550	.015			.46	2.08	2.54	3.05
5180	cycle 2 miles		462	.017			.55	2.48	3.03	3.63
5190	cycle 4 miles		330	.024			.77	3.47	4.24	5.10
5200	20 MPH, cycle 2 miles		506	.016			.51	2.26	2.77	3.32
5210	cycle 4 miles		374	.021			.68	3.06	3.74	4.48
5220	25 MPH, cycle 2 miles		528	.015			.48	2.17	2.65	3.17
5230	cycle 4 miles		418	.019			.61	2.74	3.35	4.01
5300	20 min. wait/Ld./Uld., 5 MPH, cycle 2000 ft		418	.019			.61	2.74	3.35	4.01
5310	cycle 3000 ft		396	.020			.65	2.89	3.54	4.24
5320	cycle 4000 ft		352	.023			.73	3.25	3.98	4.77
5330	cycle 0.5 mile		396	.020			.65	2.89	3.54	4.24
5340	cycle 1 mile		330	.024			.77	3.47	4.24	5.10
5350	cycle 2 miles		242	.033			1.06	4.73	5.79	6.95
5360	10 MPH, cycle 2000 ft		462	.017			.55	2.48	3.03	3.63
5370	cycle 3000 ft		440	.018			.58	2.60	3.18	3.81
5380	cycle 4000 ft		418	.019			.61	2.74	3.35	4.01
5390	cycle 0.5 mile		462	.017			.55	2.48	3.03	3.63
5400	cycle 1 mile		396	.020			.65	2.89	3.54	4.24
5410	cycle 2 miles		330	.024			.77	3.47	4.24	5.10
5420	cycle 4 miles		242	.033			1.06	4.73	5.79	6.95
5430	15 MPH, cycle 2000 ft		484	.017			.53	2.36	2.89	3.47
5440	cycle 3000 ft		462	.017			.55	2.48	3.03	3.63
5450	cycle 4000 ft		462	.017			.55	2.48	3.03	3.63
5460	cycle 0.5 mile		484	.017			.53	2.36	2.89	3.47
5470	cycle 1 mile		440	.018			.58	2.60	3.18	3.81
5480	cycle 2 miles		374	.021			.68	3.06	3.74	4.48

31 23 23.20 Hauling	Crew	Daily Output	Labor-Hours	Unit	Material	2009 Bare Costs Labor	Equipment	Total	Total Incl O&P
5490	cycle 4 miles	B-34F	286	.028	L.C.Y.		.89	4	4.89	5.85
5500	20 MPH, cycle 2 miles		396	.020			.65	2.89	3.54	4.24
5510	cycle 4 miles		330	.024			.77	3.47	4.24	5.10
5520	25 MPH, cycle 2 miles		418	.019			.61	2.74	3.35	4.01
5530	cycle 4 miles		352	.023			.73	3.25	3.98	4.77
5600	25 min. wait/Ld./Uld., 5 MPH, cycle 2000 ft		352	.023			.73	3.25	3.98	4.77
5610	cycle 3000 ft		330	.024			.77	3.47	4.24	5.10
5620	cycle 4000 ft		308	.026			.83	3.71	4.54	5.45
5630	cycle 0.5 mile		330	.024			.77	3.47	4.24	5.10
5640	cycle 1 mile		286	.028			.89	4	4.89	5.85
5650	cycle 2 miles		220	.036			1.16	5.20	6.36	7.60
5660	10 MPH, cycle 2000 ft		374	.021			.68	3.06	3.74	4.48
5670	cycle 3000 ft		374	.021			.68	3.06	3.74	4.48
5680	cycle 4000 ft		352	.023			.73	3.25	3.98	4.77
5690	cycle 0.5 mile		374	.021			.68	3.06	3.74	4.48
5700	cycle 1 mile		330	.024			.77	3.47	4.24	5.10
5710	cycle 2 miles		286	.028			.89	4	4.89	5.85
5720	cycle 4 miles		220	.036			1.16	5.20	6.36	7.60
5730	15 MPH, cycle 2000 ft		396	.020			.65	2.89	3.54	4.24
5740	cycle 3000 ft		374	.021			.68	3.06	3.74	4.48
5750	cycle 4000 ft		374	.021			.68	3.06	3.74	4.48
5760	cycle 0.5 mile		374	.021			.68	3.06	3.74	4.48
5770	cycle 1 mile		352	.023			.73	3.25	3.98	4.77
5780	cycle 2 miles		308	.026			.83	3.71	4.54	5.45
5790	cycle 4 miles		242	.033			1.06	4.73	5.79	6.95
5800	20 MPH, cycle 2 miles		330	.024			.77	3.47	4.24	5.10
5810	cycle 4 miles		286	.028			.89	4	4.89	5.85
5820	25 MPH, cycle 2 miles		352	.023			.73	3.25	3.98	4.77
5830	cycle 4 miles		308	.026			.83	3.71	4.54	5.45
6000	34 CY off-road, 15 min. wait/Ld./Uld., 5 MPH, cycle 2000 ft	B-34G	816	.010			.31	1.94	2.25	2.64
6010	cycle 3000 ft		748	.011			.34	2.12	2.46	2.89
6020	cycle 4000 ft		680	.012			.38	2.33	2.71	3.18
6030	cycle 0.5 mile		782	.010			.33	2.02	2.35	2.77
6040	cycle 1 mile		578	.014			.44	2.74	3.18	3.73
6050	cycle 2 miles		408	.020			.63	3.88	4.51	5.30
6060	10 MPH, cycle 2000 ft		918	.009			.28	1.72	2	2.35
6070	cycle 3000 ft		884	.009			.29	1.79	2.08	2.44
6080	cycle 4000 ft		816	.010			.31	1.94	2.25	2.64
6090	cycle 0.5 mile		884	.009			.29	1.79	2.08	2.44
6100	cycle 1 mile		782	.010			.33	2.02	2.35	2.77
6110	cycle 2 miles		578	.014			.44	2.74	3.18	3.73
6120	cycle 4 miles		408	.020			.63	3.88	4.51	5.30
6130	15 MPH, cycle 2000 ft		986	.008			.26	1.60	1.86	2.18
6140	cycle 3000 ft		918	.009			.28	1.72	2	2.35
6150	cycle 4000 ft		884	.009			.29	1.79	2.08	2.44
6160	cycle 0.5 mile		952	.008			.27	1.66	1.93	2.27
6170	cycle 1 mile		850	.009			.30	1.86	2.16	2.54
6180	cycle 2 miles		714	.011			.36	2.21	2.57	3.03
6190	cycle 4 miles		510	.016			.50	3.10	3.60	4.23
6200	20 MPH, cycle 2 miles		782	.010			.33	2.02	2.35	2.77
6210	cycle 4 miles		578	.014			.44	2.74	3.18	3.73
6220	25 MPH, cycle 2 miles		816	.010			.31	1.94	2.25	2.64
6230	cycle 4 miles		646	.012			.40	2.45	2.85	3.34

31 23 23 – Fill

31 23 23.20 Hauling	Crew	Daily Output	Labor-Hours	Unit	Material	2009 Bare Costs Labor	Equipment	Total	Total Incl O&P
6300	20 min. wait/Ld./Uld., 5 MPH, cycle 2000 ft	B-34G	646	.012	L.C.Y.		.40	2.45	2.85	3.34
6310	cycle 3000 ft		612	.013			.42	2.58	3	3.52
6320	cycle 4000 ft		544	.015			.47	2.91	3.38	3.97
6330	cycle 0.5 mile		612	.013			.42	2.58	3	3.52
6340	cycle 1 mile		510	.016			.50	3.10	3.60	4.23
6350	cycle 2 miles		374	.021			.68	4.23	4.91	5.75
6360	10 MPH, cycle 2000 ft		714	.011			.36	2.21	2.57	3.03
6370	cycle 3000 ft		680	.012			.38	2.33	2.71	3.18
6380	cycle 4000 ft		646	.012			.40	2.45	2.85	3.34
6390	cycle 0.5 mile		714	.011			.36	2.21	2.57	3.03
6400	cycle 1 mile		612	.013			.42	2.58	3	3.52
6410	cycle 2 miles		510	.016			.50	3.10	3.60	4.23
6420	cycle 4 miles		374	.021			.68	4.23	4.91	5.75
6430	15 MPH, cycle 2000 ft		748	.011			.34	2.12	2.46	2.89
6440	cycle 3000 ft		714	.011			.36	2.21	2.57	3.03
6450	cycle 4000 ft		714	.011			.36	2.21	2.57	3.03
6460	cycle 0.5 mile		748	.011			.34	2.12	2.46	2.89
6470	cycle 1 mile		680	.012			.38	2.33	2.71	3.18
6480	cycle 2 miles		578	.014			.44	2.74	3.18	3.73
6490	cycle 4 miles		442	.018			.58	3.58	4.16	4.89
6500	20 MPH, cycle 2 miles		612	.013			.42	2.58	3	3.52
6510	cycle 4 miles		510	.016			.50	3.10	3.60	4.23
6520	25 MPH, cycle 2 miles		646	.012			.40	2.45	2.85	3.34
6530	cycle 4 miles		544	.015			.47	2.91	3.38	3.97
6600	25 min. wait/Ld./Uld., 5 MPH, cycle 2000 ft		544	.015			.47	2.91	3.38	3.97
6610	cycle 3000 ft		510	.016			.50	3.10	3.60	4.23
6620	cycle 4000 ft		476	.017			.54	3.32	3.86	4.54
6630	cycle 0.5 mile		510	.016			.50	3.10	3.60	4.23
6640	cycle 1 mile		442	.018			.58	3.58	4.16	4.89
6650	cycle 2 miles		340	.024			.75	4.65	5.40	6.35
6660	10 MPH, cycle 2000 ft		578	.014			.44	2.74	3.18	3.73
6670	cycle 3000 ft		578	.014			.44	2.74	3.18	3.73
6680	cycle 4000 ft		544	.015			.47	2.91	3.38	3.97
6690	cycle 0.5 mile		578	.014			.44	2.74	3.18	3.73
6700	cycle 1 mile		510	.016			.50	3.10	3.60	4.23
6710	cycle 2 miles		442	.018			.58	3.58	4.16	4.89
6720	cycle 4 miles		340	.024			.75	4.65	5.40	6.35
6730	15 MPH, cycle 2000 ft		612	.013			.42	2.58	3	3.52
6740	cycle 3000 ft		578	.014			.44	2.74	3.18	3.73
6750	cycle 4000 ft		578	.014			.44	2.74	3.18	3.73
6760	cycle 0.5 mile		612	.013			.42	2.58	3	3.52
6770	cycle 1 mile		544	.015			.47	2.91	3.38	3.97
6780	cycle 2 miles		476	.017			.54	3.32	3.86	4.54
6790	cycle 4 miles		374	.021			.68	4.23	4.91	5.75
6800	20 MPH, cycle 2 miles		510	.016			.50	3.10	3.60	4.23
6810	cycle 4 miles		442	.018			.58	3.58	4.16	4.89
6820	25 MPH, cycle 2 miles		544	.015			.47	2.91	3.38	3.97
6830	cycle 4 miles		476	.017			.54	3.32	3.86	4.54
7000	42 CY off-road, 20 min. wait/Ld./Uld., 5 MPH, cycle 2000 ft	B-34H	798	.010			.32	2	2.32	2.73
7010	cycle 3000 ft		756	.011			.34	2.11	2.45	2.87
7020	cycle 4000 ft		672	.012			.38	2.37	2.75	3.23
7030	cycle 0.5 mile		756	.011			.34	2.11	2.45	2.87
7040	cycle 1 mile		630	.013			.41	2.53	2.94	3.44

31 23 23.20 Hauling	Crew	Daily Output	Labor-Hours	Unit	Material	2009 Bare Costs Labor	Equipment	Total	Total Incl O&P
7050	cycle 2 miles	B-34H	462	.017	L.C.Y.		.55	3.45	4	4.71
7060	10 MPH, cycle 2000 ft		882	.009			.29	1.81	2.10	2.46
7070	cycle 3000 ft		840	.010			.30	1.90	2.20	2.59
7080	cycle 4000 ft		798	.010			.32	2	2.32	2.73
7090	cycle 0.5 mile		882	.009			.29	1.81	2.10	2.46
7100	cycle 1 mile		798	.010			.32	2	2.32	2.73
7110	cycle 2 miles		630	.013			.41	2.53	2.94	3.44
7120	cycle 4 miles		462	.017			.55	3.45	4	4.71
7130	15 MPH, cycle 2000 ft		924	.009			.28	1.73	2.01	2.35
7140	cycle 3000 ft		882	.009			.29	1.81	2.10	2.46
7150	cycle 4000 ft		882	.009			.29	1.81	2.10	2.46
7160	cycle 0.5 mile		882	.009			.29	1.81	2.10	2.46
7170	cycle 1 mile		840	.010			.30	1.90	2.20	2.59
7180	cycle 2 miles		714	.011			.36	2.23	2.59	3.04
7190	cycle 4 miles		546	.015			.47	2.92	3.39	3.98
7200	20 MPH, cycle 2 miles		756	.011			.34	2.11	2.45	2.87
7210	cycle 4 miles		630	.013			.41	2.53	2.94	3.44
7220	25 MPH, cycle 2 miles		798	.010			.32	2	2.32	2.73
7230	cycle 4 miles		672	.012			.38	2.37	2.75	3.23
7300	25 min. wait/Ld./Uld., 5 MPH, cycle 2000 ft		672	.012			.38	2.37	2.75	3.23
7310	cycle 3000 ft		630	.013			.41	2.53	2.94	3.44
7320	cycle 4000 ft		588	.014			.43	2.71	3.14	3.69
7330	cycle 0.5 mile		630	.013			.41	2.53	2.94	3.44
7340	cycle 1 mile		546	.015			.47	2.92	3.39	3.98
7350	cycle 2 miles		378	.021			.68	4.22	4.90	5.75
7360	10 MPH, cycle 2000 ft		714	.011			.36	2.23	2.59	3.04
7370	cycle 3000 ft		714	.011			.36	2.23	2.59	3.04
7380	cycle 4000 ft		672	.012			.38	2.37	2.75	3.23
7390	cycle 0.5 mile		714	.011			.36	2.23	2.59	3.04
7400	cycle 1 mile		630	.013			.41	2.53	2.94	3.44
7410	cycle 2 miles		546	.015			.47	2.92	3.39	3.98
7420	cycle 4 miles		378	.021			.68	4.22	4.90	5.75
7430	15 MPH, cycle 2000 ft		756	.011			.34	2.11	2.45	2.87
7440	cycle 3000 ft		714	.011			.36	2.23	2.59	3.04
7450	cycle 4000 ft		714	.011			.36	2.23	2.59	3.04
7460	cycle 0.5 mile		714	.011			.36	2.23	2.59	3.04
7470	cycle 1 mile		672	.012			.38	2.37	2.75	3.23
7480	cycle 2 miles		588	.014			.43	2.71	3.14	3.69
7490	cycle 4 miles		462	.017			.55	3.45	4	4.71
7500	20 MPH, cycle 2 miles		630	.013			.41	2.53	2.94	3.44
7510	cycle 4 miles		546	.015			.47	2.92	3.39	3.98
7520	25 MPH, cycle 2 miles		672	.012			.38	2.37	2.75	3.23
7530	cycle 4 miles		588	.014			.43	2.71	3.14	3.69
8000	60 CY off-road, 20 min. wait/Ld./Uld., 5 MPH, cycle 2000 ft	B-34J	1140	.007			.22	1.80	2.02	2.35
8010	cycle 3000 ft		1080	.007			.24	1.90	2.14	2.48
8020	cycle 4000 ft		960	.008			.27	2.13	2.40	2.78
8030	cycle 0.5 mile		1080	.007			.24	1.90	2.14	2.48
8040	cycle 1 mile		900	.009			.28	2.27	2.55	2.97
8050	cycle 2 miles		660	.012			.39	3.10	3.49	4.04
8060	10 MPH, cycle 2000 ft		1260	.006			.20	1.62	1.82	2.12
8070	cycle 3000 ft		1200	.007			.21	1.71	1.92	2.23
8080	cycle 4000 ft		1140	.007			.22	1.80	2.02	2.35
8090	cycle 0.5 mile		1260	.006			.20	1.62	1.82	2.12

31 23 Excavation and Fill

31 23 23 – Fill

31 23 23.20 Hauling		Crew	Daily Output	Labor-Hours	Unit	Material	2009 Bare Costs Labor	2009 Bare Costs Equipment	Total	Total Incl O&P
8100	cycle 1 mile	B-34J	1080	.007	L.C.Y.		.24	1.90	2.14	2.48
8110	cycle 2 miles		900	.009			.28	2.27	2.55	2.97
8120	cycle 4 miles		660	.012			.39	3.10	3.49	4.04
8130	15 MPH, cycle 2000 ft		1320	.006			.19	1.55	1.74	2.03
8140	cycle 3000 ft		1260	.006			.20	1.62	1.82	2.12
8150	cycle 4000 ft		1260	.006			.20	1.62	1.82	2.12
8160	cycle 0.5 mile		1320	.006			.19	1.55	1.74	2.03
8170	cycle 1 mile		1200	.007			.21	1.71	1.92	2.23
8180	cycle 2 miles		1020	.008			.25	2.01	2.26	2.62
8190	cycle 4 miles		780	.010			.33	2.63	2.96	3.43
8200	20 MPH, cycle 2 miles		1080	.007			.24	1.90	2.14	2.48
8210	cycle 4 miles		900	.009			.28	2.27	2.55	2.97
8220	25 MPH, cycle 2 miles		1140	.007			.22	1.80	2.02	2.35
8230	cycle 4 miles		960	.008			.27	2.13	2.40	2.78
8300	25 min. wait/Ld./Uld., 5 MPH, cycle 2000 ft		960	.008			.27	2.13	2.40	2.78
8310	cycle 3000 ft		900	.009			.28	2.27	2.55	2.97
8320	cycle 4000 ft		840	.010			.30	2.44	2.74	3.18
8330	cycle 0.5 mile		900	.009			.28	2.27	2.55	2.97
8340	cycle 1 mile		780	.010			.33	2.63	2.96	3.43
8350	cycle 2 miles		600	.013			.43	3.41	3.84	4.45
8360	10 MPH, cycle 2000 ft		1020	.008			.25	2.01	2.26	2.62
8370	cycle 3000 ft		1020	.008			.25	2.01	2.26	2.62
8380	cycle 4000 ft		960	.008			.27	2.13	2.40	2.78
8390	cycle 0.5 mile		1020	.008			.25	2.01	2.26	2.62
8400	cycle 1 mile		900	.009			.28	2.27	2.55	2.97
8410	cycle 2 miles		780	.010			.33	2.63	2.96	3.43
8420	cycle 4 miles		600	.013			.43	3.41	3.84	4.45
8430	15 MPH, cycle 2000 ft		1080	.007			.24	1.90	2.14	2.48
8440	cycle 3000 ft		1020	.008			.25	2.01	2.26	2.62
8450	cycle 4000 ft		1020	.008			.25	2.01	2.26	2.62
8460	cycle 0.5 mile		1080	.007			.24	1.90	2.14	2.48
8470	cycle 1 mile		960	.008			.27	2.13	2.40	2.78
8480	cycle 2 miles		840	.010			.30	2.44	2.74	3.18
8490	cycle 4 miles		660	.012			.39	3.10	3.49	4.04
8500	20 MPH, cycle 2 miles		900	.009			.28	2.27	2.55	2.97
8510	cycle 4 miles		780	.010			.33	2.63	2.96	3.43
8520	25 MPH, cycle 2 miles		960	.008			.27	2.13	2.40	2.78
8530	cycle 4 miles		840	.010			.30	2.44	2.74	3.18

31 23 23.23 Compaction

31 23 23.23 Compaction		Crew	Daily Output	Labor-Hours	Unit	Material	2009 Bare Costs Labor	2009 Bare Costs Equipment	Total	Total Incl O&P
0010	**COMPACTION**									
5000	Riding, vibrating roller, 6" lifts, 2 passes	B-10Y	3000	.004	E.C.Y.		.15	.16	.31	.41
5020	3 passes		2300	.005			.20	.20	.40	.54
5040	4 passes		1900	.006			.24	.25	.49	.65
5060	12" lifts, 2 passes		5200	.002			.09	.09	.18	.24
5080	3 passes		3500	.003			.13	.13	.26	.36
5100	4 passes		2600	.005			.18	.18	.36	.48
5600	Sheepsfoot or wobbly wheel roller, 6" lifts, 2 passes	B-10G	2400	.005			.19	.48	.67	.82
5620	3 passes		1735	.007			.26	.66	.92	1.14
5640	4 passes		1300	.009			.35	.88	1.23	1.53
5680	12" lifts, 2 passes		5200	.002			.09	.22	.31	.38
5700	3 passes		3500	.003			.13	.33	.46	.57
5720	4 passes		2600	.005			.18	.44	.62	.76

31 23 Excavation and Fill

31 23 23 – Fill

31 23 23.23 Compaction

		Crew	Daily Output	Labor-Hours	Unit	Material	2009 Bare Costs Labor	Equipment	Total	Total Incl O&P
7000	Walk behind, vibrating plate 18" wide, 6" lifts, 2 passes	A-1D	200	.040	E.C.Y.		1.26	.16	1.42	2.26
7020	3 passes		185	.043			1.37	.17	1.54	2.45
7040	4 passes		140	.057			1.81	.23	2.04	3.23
7200	12" lifts, 2 passes	A-1E	560	.014			.45	.07	.52	.83
7220	3 passes		375	.021			.67	.10	.77	1.23
7240	4 passes		280	.029			.90	.14	1.04	1.64
7500	Vibrating roller 24" wide, 6" lifts, 2 passes	B-10A	420	.029			1.09	.35	1.44	2.12
7520	3 passes		280	.043			1.63	.52	2.15	3.17
7540	4 passes		210	.057			2.18	.69	2.87	4.23
7600	12" lifts, 2 passes		840	.014			.54	.17	.71	1.06
7620	3 passes		560	.021			.82	.26	1.08	1.59
7640	4 passes		420	.029			1.09	.35	1.44	2.12
8000	Rammer tamper, 6" to 11", 4" lifts, 2 passes	A-1F	130	.062			1.94	.33	2.27	3.58
8050	3 passes		97	.082			2.61	.45	3.06	4.79
8100	4 passes		65	.123			3.89	.67	4.56	7.15
8200	8" lifts, 2 passes		260	.031			.97	.17	1.14	1.78
8250	3 passes		195	.041			1.30	.22	1.52	2.38
8300	4 passes		130	.062			1.94	.33	2.27	3.58
8400	13" to 18", 4" lifts, 2 passes	A-1G	390	.021			.65	.13	.78	1.21
8450	3 passes		290	.028			.87	.17	1.04	1.63
8500	4 passes		195	.041			1.30	.25	1.55	2.42
8600	8" lifts, 2 passes		780	.010			.32	.06	.38	.61
8650	3 passes		585	.014			.43	.08	.51	.80
8700	4 passes		390	.021			.65	.13	.78	1.21
9000	Water, 3000 gal. truck, 3 mile haul	B-45	1888	.008		1.14	.31	.41	1.86	2.22
9010	6 mile haul		1444	.011		1.14	.41	.54	2.09	2.51
9020	12 mile haul		1000	.016		1.14	.59	.78	2.51	3.06
9030	6000 gal. wagon, 3 mile haul	B-59	2000	.004		1.14	.13	.21	1.48	1.70
9040	6 mile haul	"	1600	.005		1.14	.16	.27	1.57	1.81

31 23 23.24 Compaction, Structural

		Crew	Daily Output	Labor-Hours	Unit	Material	2009 Bare Costs Labor	Equipment	Total	Total Incl O&P
0010	**COMPACTION, STRUCTURAL** R312323-30									
0020	Steel wheel tandem roller, 5 tons	B-10E	8	1.500	Hr.		57	17.20	74.20	110
0050	Air tamp, 6" to 8" lifts, common fill	B-9	250	.160	E.C.Y.		5.10	.77	5.87	9.30
0060	Select fill	"	300	.133			4.27	.64	4.91	7.75
0600	Vibratory plate, 8" lifts, common fill	A-1D	200	.040			1.26	.16	1.42	2.26
0700	Select fill	"	216	.037			1.17	.15	1.32	2.09
9000	Minimum labor/equipment charge	1 Clab	4	2	Job		63		63	104

31 25 Erosion and Sedimentation Controls

31 25 13 – Erosion Controls

31 25 13.10 Synthetic Erosion Control

			Crew	Daily Output	Labor-Hours	Unit	Material	2009 Bare Costs Labor	Equipment	Total	Total Incl O&P
0010	**SYNTHETIC EROSION CONTROL**										
0020	Jute mesh, 100 SY per roll, 4' wide, stapled	G	B-80A	2400	.010	S.Y.	1.06	.32	.10	1.48	1.80
0100	Plastic netting, stapled, 2" x 1" mesh, 20 mil	G	B-1	2500	.010		.81	.31		1.12	1.40
0200	Polypropylene mesh, stapled, 6.5 oz./S.Y.	G		2500	.010		1.67	.31		1.98	2.35
0300	Tobacco netting, or jute mesh #2, stapled	G		2500	.010		.09	.31		.40	.61
1000	Silt fence, polypropylene, 3' high, ideal conditions	G	2 Clab	1600	.010	L.F.	.39	.32		.71	.95
1100	Adverse conditions	G	"	950	.017	"	.39	.53		.92	1.31
1200	Place and remove hay bales	G	A-2	3	8	Ton	171	251	65	487	675
1250	Hay bales, staked	G	"	2500	.010	L.F.	6.85	.30	.08	7.23	8.15

452

31 31 Soil Treatment

31 31 16 – Termite Control

31 31 16.13 Chemical Termite Control

		Crew	Daily Output	Labor-Hours	Unit	Material	2009 Bare Costs Labor	2009 Bare Costs Equipment	Total	Total Incl O&P
0010	**CHEMICAL TERMITE CONTROL**									
0020	Slab and walls, residential	1 Skwk	1200	.007	SF Flr.	.32	.27		.59	.79
0100	Commercial, minimum		2496	.003		.34	.13		.47	.58
0200	Maximum		1645	.005		.51	.20		.71	.88
0390	Minimum labor/equipment charge		4	2	Job		81.50		81.50	133
0400	Insecticides for termite control, minimum		14.20	.563	Gal.	13.05	23		36.05	52
0500	Maximum		11	.727	"	22.50	29.50		52	73

31 41 Shoring

31 41 13 – Timber Shoring

31 41 13.10 Building Shoring

		Crew	Daily Output	Labor-Hours	Unit	Material	2009 Bare Costs Labor	2009 Bare Costs Equipment	Total	Total Incl O&P
0010	**BUILDING SHORING**									
0020	Shoring, existing building, with timber, no salvage allowance	B-51	2.20	21.818	M.B.F.	695	695	88.50	1,478.50	2,000
1000	On cribbing with 35 ton screw jacks, per box and jack		3.60	13.333	Jack	63.50	425	54	542.50	830
1090	Minimum labor/equipment charge		2	24	Ea.		765	97	862	1,350
1100	Masonry openings in walls, see Div. 02 41 19.16									

31 41 16 – Sheet Piling

31 41 16.10 Sheet Piling Systems

		Crew	Daily Output	Labor-Hours	Unit	Material	2009 Bare Costs Labor	2009 Bare Costs Equipment	Total	Total Incl O&P
0010	**SHEET PILING SYSTEMS**									
0020	Sheet piling steel, not incl. wales, 22 psf, 15' excav., left in place	B-40	10.81	5.920	Ton	1,225	234	305	1,764	2,075
0100	Drive, extract & salvage R314116-40		6	10.667	"	505	420	550	1,475	1,850
1200	15' deep excavation, 22 psf, left in place		983	.065	S.F.	14.35	2.58	3.37	20.30	23.50
1300	Drive, extract & salvage		545	.117	"	5.65	4.64	6.10	16.39	20.50
2100	Rent steel sheet piling and wales, first month				Ton	263			263	289
2200	Per added month				"	26.50			26.50	29
3900	Wood, solid sheeting, incl. wales, braces and spacers,									
3910	drive, extract & salvage, 8' deep excavation	B-31	330	.121	S.F.	1.47	4.08	.55	6.10	8.95
4520	Left in place, 8' deep, 55 S.F./hr.		440	.091	"	2.65	3.06	.41	6.12	8.40
4990	Minimum labor/equipment charge		2	20	Job		675	90.50	765.50	1,200
5000	For treated lumber add cost of treatment to lumber									

31 43 Concrete Raising

31 43 13 – Pressure Grouting

31 43 13.13 Concrete Pressure Grouting

		Crew	Daily Output	Labor-Hours	Unit	Material	2009 Bare Costs Labor	2009 Bare Costs Equipment	Total	Total Incl O&P
0010	**CONCRETE PRESSURE GROUTING**									
0020	Grouting, pressure, cement & sand, 1:1 mix, minimum	B-61	124	.323	Bag	12.40	10.80	2.64	25.84	34
0100	Maximum		51	.784	"	12.40	26.50	6.40	45.30	63.50
0200	Cement and sand, 1:1 mix, minimum		250	.160	C.F.	25	5.35	1.31	31.66	37.50
0300	Maximum		100	.400		37	13.40	3.27	53.67	66.50
0400	Epoxy cement grout, minimum		137	.292		109	9.80	2.39	121.19	139
0500	Maximum		57	.702		109	23.50	5.75	138.25	165
0600	Structural epoxy grout					82			82	90
0700	Alternate pricing method: (Add for materials)									
0710	5 person crew and equipment	B-61	1	40	Day		1,350	325	1,675	2,525

31 46 Needle Beams

31 46 13 – Cantilever Needle Beams

31 46 13.10 Needle Beams

		Crew	Daily Output	Labor-Hours	Unit	Material	2009 Bare Costs Labor	Equipment	Total	Total Incl O&P
0010	**NEEDLE BEAMS**									
0011	Incl. wood shoring 10' x 10' opening									
0400	Block, concrete, 8" thick	B-9	7.10	5.634	Ea.	45.50	180	27	252.50	380
0420	12" thick		6.70	5.970		52.50	191	28.50	272	405
0800	Brick, 4" thick with 8" backup block		5.70	7.018		52.50	225	33.50	311	465
1000	Brick, solid, 8" thick		6.20	6.452		45.50	206	31	282.50	425
1040	12" thick		4.90	8.163		52.50	261	39	352.50	530
1080	16" thick		4.50	8.889		67	284	42.50	393.50	590
2000	Add for additional floors of shoring	B-1	6	4		45.50	129		174.50	263
9000	Minimum labor/equipment charge	"	2	12	Job		385		385	640

31 48 Underpinning

31 48 13 – Underpinning Piers

31 48 13.10 Underpinning Foundations

		Crew	Daily Output	Labor-Hours	Unit	Material	2009 Bare Costs Labor	Equipment	Total	Total Incl O&P
0010	**UNDERPINNING FOUNDATIONS**									
0011	Including excavation,									
0020	forming, reinforcing, concrete and equipment									
0100	5' to 16' below grade, 100 to 500 C.Y.	B-52	2.30	24.348	C.Y.	278	895	215	1,388	2,025
0200	Over 500 C.Y.		2.50	22.400		250	825	198	1,273	1,850
0400	16' to 25' below grade, 100 to 500 C.Y.		2	28		305	1,025	248	1,578	2,300
0500	Over 500 C.Y.		2.10	26.667		289	985	236	1,510	2,175
0700	26' to 40' below grade, 100 to 500 C.Y.		1.60	35		335	1,300	310	1,945	2,800
0800	Over 500 C.Y.		1.80	31.111		305	1,150	275	1,730	2,500
0900	For under 50 C.Y., add					10%	40%			

31 62 Driven Piles

31 62 13 – Concrete Piles

31 62 13.23 Prestressed Concrete Piles

		Crew	Daily Output	Labor-Hours	Unit	Material	2009 Bare Costs Labor	Equipment	Total	Total Incl O&P
0010	**PRESTRESSED CONCRETE PILES**, 200 piles									
0020	Unless specified otherwise, not incl. pile caps or mobilization									
3100	Precast, prestressed, 40' long, 10" thick, square	B-19	700	.091	V.L.F.	14.40	3.62	2.52	20.54	24.50
3200	12" thick, square		680	.094		18.15	3.72	2.59	24.46	29
3400	14" thick, square		600	.107		21.50	4.22	2.94	28.66	33.50
4000	18" thick, square	B-19A	520	.123		41	4.87	4.41	50.28	58

31 62 16 – Steel Piles

31 62 16.13 Sheet Steel Piles

		Crew	Daily Output	Labor-Hours	Unit	Material	2009 Bare Costs Labor	Equipment	Total	Total Incl O&P
0010	**SHEET STEEL PILES**									
0100	Step tapered, round, concrete filled									
0110	8" tip, 60 ton capacity, 30' depth	B-19	760	.084	V.L.F.	10.15	3.33	2.32	15.80	19.15
0120	60' depth		740	.086		11.45	3.42	2.38	17.25	21
0250	"H" Sections, 50' long, HP8 x 36		640	.100		17.40	3.96	2.76	24.12	28.50
1300	HP14 X 102	B-19A	510	.125		50	4.96	4.50	59.46	68.50

31 62 19 – Timber Piles

31 62 19.10 Wood Piles

		Crew	Daily Output	Labor-Hours	Unit	Material	2009 Bare Costs Labor	Equipment	Total	Total Incl O&P
0010	**WOOD PILES**									
0011	Friction or end bearing, not including									
0050	mobilization or demobilization									
0100	Untreated piles, up to 30' long, 12" butts, 8" points	B-19	625	.102	V.L.F.	7.25	4.05	2.82	14.12	17.65

31 62 Driven Piles

31 62 19 – Timber Piles

31 62 19.10 Wood Piles

		Crew	Daily Output	Labor-Hours	Unit	Material	2009 Bare Costs Labor	Equipment	Total	Total Incl O&P
0200	30' to 39' long, 12" butts, 8" points	B-19	700	.091	V.L.F.	7.25	3.62	2.52	13.39	16.60
0800	Treated piles, 12 lb. per C.F.,									
0810	friction or end bearing, ASTM class B									
1000	Up to 30' long, 12" butts, 8" points	B-19	625	.102	V.L.F.	10.70	4.05	2.82	17.57	21.50
1100	30' to 39' long, 12" butts, 8" points		700	.091		10.45	3.62	2.52	16.59	20
2700	Mobilization for 10,000 L.F. pile job, add		3300	.019			.77	.53	1.30	1.84
2800	25,000 L.F. pile job, add		8500	.008			.30	.21	.51	.72

31 62 23 – Composite Piles

31 62 23.13 Concrete Filled Steel Piles

		Crew	Daily Output	Labor-Hours	Unit	Material	2009 Bare Costs Labor	Equipment	Total	Total Incl O&P
0010	**CONCRETE FILLED STEEL PILES** no mobilization or demobilization									
2600	Pipe piles, 50' lg. 8" diam., 29 lb. per L.F., no concrete	B-19	500	.128	V.L.F.	18.80	5.05	3.53	27.38	32.50
2700	Concrete filled		460	.139		19.80	5.50	3.83	29.13	35
3500	14" diameter, 46 lb. per L.F., no concrete		430	.149		31	5.90	4.10	41	48
3600	Concrete filled		355	.180		32.50	7.15	4.97	44.62	53
4100	18" diameter, 59 lb. per L.F., no concrete		355	.180		45.50	7.15	4.97	57.62	67
4200	Concrete filled		310	.206		44	8.15	5.70	57.85	68

31 63 Bored Piles

31 63 26 – Drilled Caissons

31 63 26.13 Fixed End Cassion Piles

		Crew	Daily Output	Labor-Hours	Unit	Material	2009 Bare Costs Labor	Equipment	Total	Total Incl O&P
0010	**FIXED END CASSION PILES**									
0015	Including excavation, concrete, 50 lbs reinforcing									
0020	per C.Y., not incl. mobilization, boulder removal, disposal									
0100	Open style, machine drilled, to 50' deep, in stable ground, no R316326-60									
0110	casings or ground water, 18" diam., 0.065 C.Y./L.F.	B-43	200	.240	V.L.F.	9.05	8.30	12.65	30	37.50
0200	24" diameter, 0.116 C.Y./L.F.		190	.253	"	16.20	8.75	13.35	38.30	46.50
0210	4' bell diameter , add		20	2.400	Ea.	164	83	127	374	455
0500	48" diameter, 0.465 C.Y./L.F.		100	.480	V.L.F.	65	16.60	25.50	107.10	127
0510	9' bell diameter , add		2	24	Ea.	1,325	830	1,275	3,430	4,200
1200	Open style, machine drilled, to 50' deep, in wet ground, pulled									
1220	pulled casing and pumping									
1400	24" diameter, 0.116 C.Y./L.F.	B-48	125	.448	V.L.F.	16.20	15.80	23	55	68.50
1410	4' bell diameter , add	"	19.80	2.828	Ea.	164	99.50	144	407.50	500
1700	48" diameter, 0.465 C.Y./L.F.	B-49	55	1.600	V.L.F.	65	58.50	66.50	190	240
1710	9' bell diameter , add	"	3.30	26.667	Ea.	1,325	975	1,100	3,400	4,250
2300	Open style, machine drilled, to 50' deep, in soft rocks and									
2320	medium hard shales									
2500	24" diameter, 0.116 C.Y./L.F.	B-49	30	2.933	V.L.F.	16.20	108	122	246.20	325
2510	4' bell diameter, add		10.90	8.073	Ea.	164	296	335	795	1,025
2800	48" diameter, 0.465 C.Y./L.F.		10	8.800	V.L.F.	65	325	365	755	990
2810	9' bell diameter , add		1.10	80	Ea.	990	2,925	3,325	7,240	9,475
3600	For rock excavation, sockets, add, minimum		120	.733	C.F.		27	30.50	57.50	77
3650	Average		95	.926			34	38.50	72.50	97.50
3700	Maximum		48	1.833			67	76	143	193
3900	For 50' to 100' deep, add				V.L.F.				7%	7%
4000	For 100' to 150' deep, add								25%	25%
4100	For 150' to 200' deep, add								30%	30%
4200	For casings left in place, add				Lb.	.95			.95	1.05
4300	For other than 50 lb. reinf. per C.Y., add or deduct				"	1.01			1.01	1.11
4400	For steel "I" beam cores, add	B-49	8.30	10.602	Ton	2,400	390	440	3,230	3,775

31 63 Bored Piles

31 63 26 – Drilled Caissons

31 63 26.13 **Fixed End Cassion Piles**	Crew	Daily Output	Labor-Hours	Unit	Material	2009 Bare Costs Labor	Equipment	Total	Total Incl O&P
4500	Load and haul excess excavation, 2 miles	B-34B	178	.045	L.C.Y.		1.44	2.99	4.43	5.65
4600	For mobilization, 50 mile radius, rig to 36"	B-43	2	24	Ea.		830	1,275	2,105	2,750
4650	Rig to 84"	B-48	1.75	32			1,125	1,625	2,750	3,625
4700	For low headroom, add								50%	
4750	For difficult access, add								25%	

31 63 29 – Drilled Concrete Piers and Shafts

31 63 29.13 Uncased Drilled Concrete Piers

		Crew	Daily Output	Labor-Hours	Unit	Material	2009 Bare Costs Labor	Equipment	Total	Total Incl O&P
0010	**UNCASED DRILLED CONCRETE PIERS**									
0020	Unless specified otherwise, not incl. pile caps or mobilization									
0800	Cast in place friction pile, 50' long, fluted,									
0810	tapered steel, 4000 psi concrete, no reinforcing									
0900	12" diameter, 7 ga.	B-19	600	.107	V.L.F.	28	4.22	2.94	35.16	40.50
1200	18" diameter, 7 ga.	"	480	.133	"	41.50	5.25	3.67	50.42	58.50
1300	End bearing, fluted, constant diameter,									
1320	4000 psi concrete, no reinforcing									
1340	12" diameter, 7 ga.	B-19	600	.107	V.L.F.	29	4.22	2.94	36.16	42
1400	18" diameter, 7 ga.	"	480	.133	"	46.50	5.25	3.67	55.42	63.50

Estimating Tips

32 01 00 Operations and Maintenance of Exterior Improvements

• Recycling of asphalt pavement is becoming very popular and is an alternative to removal and replacement of asphalt pavement. It can be a good value engineering proposal if removed pavement can be recycled either at the site or another site that is reasonably close to the project site.

32 10 00 Bases, Ballasts, and Paving

• When estimating paving, keep in mind the project schedule. If an asphaltic paving project is in a colder climate and runs through to the spring, consider placing the base course in the autumn and then topping it in the spring, just prior to completion. This could save considerable costs in spring repair. Keep in mind that prices for asphalt and concrete are generally higher in the cold seasons.

32 90 00 Planting

• The timing of planting and guarantee specifications often dictate the costs for establishing tree and shrub growth and a stand of grass or ground cover. Establish the work performance schedule to coincide with the local planting season. Maintenance and growth guarantees can add from 20%–100% to the total landscaping cost. The cost to replace trees and shrubs can be as high as 5% of the total cost, depending on the planting zone, soil conditions, and time of year.

Reference Numbers

Reference numbers are shown in shaded boxes at the beginning of some major classifications. These numbers refer to related items in the Reference Section. The reference information may be an estimating procedure, an alternate pricing method, or technical information.

Note: Not all subdivisions listed here necessarily appear in this publication.

32 01 13 – Flexible Paving Surface Treatment

32 01 13.61 Slurry Seal (Latex Modified)	Crew	Daily Output	Labor-Hours	Unit	Material	2009 Bare Costs Labor	Equipment	Total	Total Incl O&P
0010 **SLURRY SEAL (LATEX MODIFIED)**									
3600 Waterproofing, membrane, tar and fabric, small area	B-63	233	.172	S.Y.	11.60	5.70	.64	17.94	22.50
3640 Large area		1435	.028		10.65	.92	.10	11.67	13.35
3680 Preformed rubberized asphalt, small area		100	.400		15.70	13.25	1.49	30.44	40.50
3720 Large area		367	.109		14.30	3.61	.41	18.32	22
3780 Rubberized asphalt (latex) seal	B-45	5000	.003		2.26	.12	.16	2.54	2.85
9000 Minimum labor/equipment charge	1 Clab	2	4	Job		126		126	209

32 01 13.62 Asphalt Surface Treatment	Crew	Daily Output	Labor-Hours	Unit	Material	Labor	Equipment	Total	Total Incl O&P
0010 **ASPHALT SURFACE TREATMENT**									
3000 Pavement overlay, polypropylene									
3040 6 oz. per S.Y., ideal conditions	B-63	10000	.004	S.Y.	1.82	.13	.01	1.96	2.24
3080 Adverse conditions		1000	.040		2.44	1.32	.15	3.91	5
3120 4 oz. per S.Y., ideal conditions		10000	.004		1.25	.13	.01	1.39	1.62
3160 Adverse conditions		1000	.040		1.61	1.32	.15	3.08	4.09
3200 Tack coat, emulsion, .05 gal per S.Y., 1000 S.Y	B-45	2500	.006		.46	.23	.31	1	1.23
3240 10,000 S.Y.		10000	.002		.37	.06	.08	.51	.58
3270 .10 gal per S.Y., 1000 S.Y.		2500	.006		.86	.23	.31	1.40	1.67
3275 10,000 S.Y.		10000	.002		.86	.06	.08	1	1.13
3280 .15 gal per S.Y., 1000 S.Y.		2500	.006		1.27	.23	.31	1.81	2.11
3320 10,000 S.Y.		10000	.002		1.01	.06	.08	1.15	1.29

32 01 13.64 Sand Seal	Crew	Daily Output	Labor-Hours	Unit	Material	Labor	Equipment	Total	Total Incl O&P
0010 **SAND SEAL**									
2080 Sand sealing, sharp sand, asphalt emulsion, small area	B-91	10000	.006	S.Y.	1.24	.24	.22	1.70	1.98
2120 Roadway or large area	"	18000	.004	"	1.07	.13	.12	1.32	1.51
3000 Sealing random cracks, min 1/2" wide, to 1-1/2", 1,000 L.F.	B-77	2800	.014	L.F.	.98	.46	.18	1.62	2.02
3040 10,000 L.F.		4000	.010	"	.75	.32	.12	1.19	1.50
3080 Alternate method, 1,000 L.F.		200	.200	Gal.	14.65	6.35	2.46	23.46	29.50
3120 10,000 L.F.		325	.123	"	11.80	3.92	1.51	17.23	21
3200 Multi-cracks (flooding), 1 coat, small area	B-92	460	.070	S.Y.	4.41	2.23	1.09	7.73	9.75
3240 Large area		2850	.011		3.92	.36	.18	4.46	5.10
3280 2 coat, small area		230	.139		14.50	4.47	2.19	21.16	25.50
3320 Large area		1425	.022		13.20	.72	.35	14.27	16.10
3360 Alternate method, small area		115	.278	Gal.	15.15	8.95	4.37	28.47	36
3400 Large area		715	.045	"	14.25	1.44	.70	16.39	18.80

32 01 13.66 Fog Seal	Crew	Daily Output	Labor-Hours	Unit	Material	Labor	Equipment	Total	Total Incl O&P
0010 **FOG SEAL**									
0012 Sealcoating, 2 coat coal tar pitch emulsion over 10,000 SY	B-45	5000	.003	S.Y.	1	.12	.16	1.28	1.46
0030 1000 to 10,000 S.Y.	"	3000	.005		1	.20	.26	1.46	1.70
0100 Under 1000 S.Y.	B-1	1050	.023		1	.74		1.74	2.32
0300 Petroleum resistant, over 10,000 S.Y.	B-45	5000	.003		1.24	.12	.16	1.52	1.72
0320 1000 to 10,000 S.Y.	"	3000	.005		1.24	.20	.26	1.70	1.96
0400 Under 1000 S.Y.	B-1	1050	.023		1.24	.74		1.98	2.58
0600 Non-skid pavement renewal, over 10,000 S.Y.	B-45	5000	.003		1.43	.12	.16	1.71	1.93
0620 1000 to 10,000 S.Y.	"	3000	.005		1.43	.20	.26	1.89	2.17
0700 Under 1000 S.Y.	B-1	1050	.023		1.43	.74		2.17	2.79
0800 Prepare and clean surface for above	A-2	8545	.003			.09	.02	.11	.18
1000 Hand seal asphalt curbing	B-1	4420	.005	L.F.	.73	.18		.91	1.09
1900 Asphalt surface treatment, single course, small area									
1901 0.30 gal/S.Y. asphalt material, 20#/S.Y. aggregate	B-91	5000	.013	S.Y.	1.70	.47	.44	2.61	3.11
1910 Roadway or large area		10000	.006		1.56	.24	.22	2.02	2.34
1950 Asphalt surface treatment, dbl. course for small area		3000	.021		3.15	.78	.73	4.66	5.55
1960 Roadway or large area		6000	.011		2.84	.39	.36	3.59	4.15

32 01 13 – Flexible Paving Surface Treatment

32 01 13.66 Fog Seal

		Crew	Daily Output	Labor-Hours	Unit	Material	2009 Bare Costs Labor	Equipment	Total	Total Incl O&P
1980	Asphalt surface treatment, single course, for shoulders	B-91	7500	.009	S.Y.	1.81	.31	.29	2.41	2.81

32 01 13.68 Slurry Seal

		Crew	Daily Output	Labor-Hours	Unit	Material	2009 Bare Costs Labor	Equipment	Total	Total Incl O&P
0010	**SLURRY SEAL**									
0100	Slurry seal, type I, 8 lbs agg./S.Y., 1 coat, small or irregular area	B-90	2800	.023	S.Y.	1.48	.77	.78	3.03	3.74
0150	Roadway or large area		10000	.006		1.48	.22	.22	1.92	2.22
0200	Type II, 12 lbs aggregate/S.Y., 2 coats, small or irregular area		2000	.032		2.96	1.08	1.09	5.13	6.20
0250	Roadway or large area		8000	.008		2.96	.27	.27	3.50	4
0300	Type III, 20 lbs aggregate/S.Y., 2 coats, small or irregular area		1800	.036		3.49	1.20	1.21	5.90	7.10
0350	Roadway or large area		6000	.011		3.49	.36	.36	4.21	4.83
0400	Slurry seal, thermoplastic coal-tar, type I, small or irregular area		2400	.027		3.08	.90	.91	4.89	5.85
0450	Roadway or large area		8000	.008		3.08	.27	.27	3.62	4.13
0500	Type II, small or irregular area		2400	.027		3.94	.90	.91	5.75	6.80
0550	Roadway or large area		7800	.008		3.94	.28	.28	4.50	5.10

32 01 16 – Flexible Paving Rehabilitation

32 01 16.71 Cold Milling Asphalt Paving

		Crew	Daily Output	Labor-Hours	Unit	Material	2009 Bare Costs Labor	Equipment	Total	Total Incl O&P
0010	**COLD MILLING ASPHALT PAVING**									
5200	Cold planing & cleaning, 1" to 3" asphalt pavmt., over 25,000 S.Y.	B-71	6000	.009	S.Y.		.34	1.07	1.41	1.72
5280	5,000 S.Y. to 10,000 S.Y.	"	4000	.014	"		.50	1.61	2.11	2.58
5300	Asphalt pavement removal from conc. base, no haul									
5320	Rip, load & sweep 1" to 3"	B-70	8000	.007	S.Y.		.25	.20	.45	.62
5330	3" to 6" deep	"	5000	.011			.40	.31	.71	.99
5340	Profile grooving, asphalt pavement load & sweep, 1" deep	B-71	12500	.004			.16	.51	.67	.83
5350	3" deep		9000	.006			.22	.71	.93	1.15
5360	6" deep		5000	.011			.40	1.29	1.69	2.07

32 01 16.73 In Place Cold Reused Asphalt Paving

		Crew	Daily Output	Labor-Hours	Unit	Material	2009 Bare Costs Labor	Equipment	Total	Total Incl O&P
0010	**IN PLACE COLD REUSED ASPHALT PAVING**									
5000	Reclamation, pulverizing and blending with existing base									
5040	Aggregate base, 4" thick pavement, over 15,000 S.Y.	B-73	2400	.027	S.Y.		1.01	2.04	3.05	3.87
5080	5,000 S.Y. to 15,000 S.Y.		2200	.029			1.10	2.23	3.33	4.21
5120	8" thick pavement, over 15,000 S.Y.		2200	.029			1.10	2.23	3.33	4.21
5160	5,000 S.Y. to 15,000 S.Y.		2000	.032			1.21	2.45	3.66	4.64

32 01 16.74 In Place Hot Reused Asphalt Paving

		Crew	Daily Output	Labor-Hours	Unit	Material	2009 Bare Costs Labor	Equipment	Total	Total Incl O&P
0010	**IN PLACE HOT REUSED ASPHALT PAVING**									
5500	Recycle asphalt pavement at site									
5520	Remove, rejuvenate and spread 4" deep [G]	B-72	2500	.026	S.Y.	3.85	.94	4.34	9.13	10.50
5521	6" deep [G]	"	2000	.032	"	5.65	1.18	5.45	12.28	14.05

32 06 Schedules for Exterior Improvements

32 06 10 – Schedules for Bases, Ballasts, and Paving

32 06 10.10 Sidewalks, Driveways and Patios

		Crew	Daily Output	Labor-Hours	Unit	Material	2009 Bare Costs Labor	Equipment	Total	Total Incl O&P
0010	**SIDEWALKS, DRIVEWAYS AND PATIOS** No base									
0020	Asphaltic concrete, 2" thick	B-37	720	.067	S.Y.	6.05	2.21	.19	8.45	10.50
0100	2-1/2" thick	"	660	.073	"	7.65	2.41	.21	10.27	12.60
0110	Bedding for brick or stone, mortar, 1" thick	D-1	300	.053	S.F.	.77	1.94		2.71	3.98
0120	2" thick	"	200	.080		1.92	2.91		4.83	6.80
0130	Sand, 2" thick	B-18	8000	.003		.25	.10		.35	.44
0140	4" thick	"	4000	.006		.49	.19	.01	.69	.87
0300	Concrete, 3000 psi, CIP, 6 x 6 - W1.4 x W1.4 mesh,									
0310	broomed finish, no base, 4" thick	B-24	600	.040	S.F.	1.68	1.46		3.14	4.22
0350	5" thick		545	.044		2.24	1.61		3.85	5.10

32 06 Schedules for Exterior Improvements

32 06 10 – Schedules for Bases, Ballasts, and Paving

32 06 10.10 Sidewalks, Driveways and Patios

		Crew	Daily Output	Labor-Hours	Unit	Material	2009 Bare Costs Labor	2009 Bare Costs Equipment	Total	Total Incl O&P
0400	6" thick	B-24	510	.047	S.F.	2.62	1.72		4.34	5.65
0450	For bank run gravel base, 4" thick, add	B-18	2500	.010		.64	.31	.02	.97	1.23
0520	8" thick, add	"	1600	.015		1.28	.48	.02	1.78	2.24
0550	Exposed aggregate finish, add to above, minimum	B-24	1875	.013		.15	.47		.62	.92
0600	Maximum	"	455	.053		.49	1.93		2.42	3.66
0950	Concrete tree grate, 5' square	B-6	25	.960	Ea.	420	32.50	11.75	464.25	530
0960	Cast iron tree grate with frame, 2 piece, round, 5' diameter		25	.960		1,175	32.50	11.75	1,219.25	1,350
0980	Square, 5' side		25	.960		1,175	32.50	11.75	1,219.25	1,350
1700	Redwood, prefabricated, 4' x 4' sections	2 Carp	316	.051	S.F.	8.25	2.02		10.27	12.45
1750	Redwood planks, 1" thick, on sleepers	"	240	.067	"	5.80	2.66		8.46	10.75
2250	Stone dust, 4" thick	B-62	900	.027	S.Y.	3.33	.91	.17	4.41	5.30
9000	Minimum labor/equipment charge	D-1	2	8	Job		291		291	470

32 06 10.20 Steps

		Crew	Daily Output	Labor-Hours	Unit	Material	2009 Bare Costs Labor	2009 Bare Costs Equipment	Total	Total Incl O&P
0010	**STEPS**									
0011	Incl. excav., borrow & concrete base as required									
0100	Brick steps	B-24	35	.686	LF Riser	11.40	25		36.40	53
0200	Railroad ties	2 Clab	25	.640		3.30	20		23.30	37
0300	Bluestone treads, 12" x 2" or 12" x 1-1/2"	B-24	30	.800		27	29.50		56.50	77
0490	Minimum labor/equipment charge	D-1	2	8	Job		291		291	470
0500	Concrete, cast in place, see Div. 03 30 53.40									
0600	Precast concrete, see Div. 03 41 23.50									

32 11 Base Courses

32 11 23 – Aggregate Base Courses

32 11 23.23 Base Course Drainage Layers

		Crew	Daily Output	Labor-Hours	Unit	Material	2009 Bare Costs Labor	2009 Bare Costs Equipment	Total	Total Incl O&P
0010	**BASE COURSE DRAINAGE LAYERS**									
0011	For roadways and large areas									
0050	Crushed 3/4" stone base, compacted, 3" deep	B-36C	5200	.008	S.Y.	4.23	.29	.63	5.15	5.80
0100	6" deep		5000	.008		8.45	.30	.66	9.41	10.50
0200	9" deep		4600	.009		12.70	.33	.71	13.74	15.30
0300	12" deep		4200	.010		16.95	.36	.78	18.09	20
0301	Crushed 1-1/2" stone base, compacted to 4" deep	B-36B	6000	.011		5.85	.39	.62	6.86	7.70
0302	6" deep		5400	.012		8.75	.44	.69	9.88	11.05
0303	8" deep		4500	.014		11.65	.52	.83	13	14.55
0304	12" deep		3800	.017		17.50	.62	.99	19.11	21.50
0310	Minimum labor/equipment charge	B-36	1	40	Job		1,425	1,350	2,775	3,800
0350	Bank run gravel, spread and compacted									
0370	6" deep	B-32	6000	.005	S.Y.	5.35	.21	.31	5.87	6.50
0390	9" deep		4900	.007		8	.25	.38	8.63	9.60
0400	12" deep		4200	.008		10.65	.30	.44	11.39	12.65
6900	For small and irregular areas, add						50%	50%		
6990	Minimum labor/equipment charge	1 Clab	3	2.667	Job		84.50		84.50	139
7000	Prepare and roll sub-base, small areas to 2500 S.Y.	B-32A	1500	.016	S.Y.		.61	.76	1.37	1.81
8000	Large areas over 2500 S.Y.	B-32	3700	.009	"		.34	.50	.84	1.08
9000	Minimum labor/equipment charge	1 Clab	4	2	Job		63		63	104

32 12 Flexible Paving

32 12 16 – Asphalt Paving

32 12 16.13 Plant-Mix Asphalt Paving

		Crew	Daily Output	Labor-Hours	Unit	Material	2009 Bare Costs Labor	Equipment	Total	Total Incl O&P
0010	**PLANT-MIX ASPHALT PAVING**									
0080	Binder course, 1-1/2" thick	B-25	7725	.011	S.Y.	4.66	.39	.32	5.37	6.15
0120	2" thick		6345	.014		6.20	.48	.39	7.07	8
0130	2-1/2" thick		5620	.016		7.75	.54	.44	8.73	9.85
0160	3" thick		4905	.018		9.30	.62	.51	10.43	11.80
0170	3-1/2" thick		4520	.019		10.85	.67	.55	12.07	13.65
0300	Wearing course, 1" thick	B-25B	10575	.009		3.08	.32	.26	3.66	4.18
0340	1-1/2" thick		7725	.012		5.15	.44	.35	5.94	6.80
0380	2" thick		6345	.015		6.95	.53	.43	7.91	9
0420	2-1/2" thick		5480	.018		8.55	.61	.49	9.65	10.95
0460	3" thick		4900	.020		10.20	.69	.55	11.44	12.95
0470	3-1/2" thick		4520	.021		12	.74	.60	13.34	15
0480	4" thick		4140	.023		13.70	.81	.65	15.16	17.10
0500	Open graded friction course	B-25C	5000	.010		3.29	.34	.43	4.06	4.64
0800	Alternate method of figuring paving costs									
0810	Binder course, 1-1/2" thick	B-25	630	.140	Ton	57	4.81	3.93	65.74	74.50
0811	2" thick		690	.128		57	4.39	3.59	64.98	73.50
0812	3" thick		800	.110		57	3.79	3.10	63.89	72
0813	4" thick		900	.098		57	3.37	2.75	63.12	71
0850	Wearing course, 1" thick	B-25B	575	.167		62	5.85	4.71	72.56	82.50
0851	1-1/2" thick		630	.152		62	5.35	4.30	71.65	81.50
0852	2" thick		690	.139		62	4.87	3.93	70.80	80
0853	2-1/2" thick		765	.125		62	4.39	3.54	69.93	79
0854	3" thick		800	.120		62	4.20	3.39	69.59	78.50
1000	Pavement replacement over trench, 2" thick	B-37	90	.533	S.Y.	6.40	17.70	1.53	25.63	37.50
1050	4" thick		70	.686		12.65	23	1.97	37.62	53
1080	6" thick		55	.873		20	29	2.50	51.50	72.50

32 12 16.14 Asphaltic Concrete Paving, Parking Lots and Driveways

		Crew	Daily Output	Labor-Hours	Unit	Material	2009 Bare Costs Labor	Equipment	Total	Total Incl O&P
0011	**ASPHALTIC CONCRETE PAVING, PARKING LOTS & DRIVEWAYS**									
0020	6" stone base, 2" binder course, 1" topping	B-25C	9000	.005	S.F.	2.07	.19	.24	2.50	2.84
0300	Binder course, 1-1/2" thick		35000	.001		.52	.05	.06	.63	.72
0400	2" thick		25000	.002		.67	.07	.09	.83	.95
0500	3" thick		15000	.003		1.04	.11	.14	1.29	1.48
0600	4" thick		10800	.004		1.36	.16	.20	1.72	1.96
0800	Sand finish course, 3/4" thick		41000	.001		.30	.04	.05	.39	.45
0900	1" thick		34000	.001		.36	.05	.06	.47	.55
1000	Fill pot holes, hot mix, 2" thick	B-16	4200	.008		.77	.25	.13	1.15	1.38
1100	4" thick		3500	.009		1.12	.29	.15	1.56	1.88
1120	6" thick		3100	.010		1.50	.33	.17	2	2.39
1140	Cold patch, 2" thick	B-51	3000	.016		.78	.51	.06	1.35	1.76
1160	4" thick		2700	.018		1.48	.57	.07	2.12	2.63
1180	6" thick		1900	.025		2.30	.80	.10	3.20	3.97

32 13 Rigid Paving

32 13 13 — Concrete Paving

32 13 13.23 Concrete Paving Surface Treatment

		Crew	Daily Output	Labor-Hours	Unit	Material	2009 Bare Costs Labor	Equipment	Total	Total Incl O&P
0010	**CONCRETE PAVING SURFACE TREATMENT**									
0015	Including joints, finishing and curing									
0020	Fixed form, 12' pass, unreinforced, 6" thick	B-26	3000	.029	S.Y.	21.50	1.04	1.02	23.56	27
0100	8" thick		2750	.032		32.50	1.13	1.11	34.74	39
0400	12" thick		1800	.049		45.50	1.73	1.69	48.92	54.50
0410	Conc pavement, w/jt,fnsh&curing,fix form,24' pass,unreinforced,6"T		6000	.015		20	.52	.51	21.03	23.50
0430	8" thick		5500	.016		29	.57	.55	30.12	33.50
0460	12" thick		3600	.024		42	.86	.85	43.71	48.50
0505	Minimum labor/equipment charge	1 Cefi	1	8	Job		305		305	480
0520	Welded wire fabric, sheets for rigid paving 2.33 lbs/SY	2 Rodm	389	.041	S.Y.	1.62	1.83		3.45	4.91
0530	Reinforcing steel for rigid paving 12 lbs/SY		666	.024		9.80	1.07		10.87	12.55
0540	Reinforcing steel for rigid paving 18 lbs/SY		444	.036		14.65	1.61		16.26	18.90
0620	Slip form, 12' pass, unreinforced, 6" thick	B-26A	5600	.016		21	.56	.57	22.13	24.50
0624	8" thick		5300	.017		31.50	.59	.60	32.69	36
0630	12" thick		3470	.025		44.50	.90	.92	46.32	51
0632	15" thick		2890	.030		57	1.08	1.10	59.18	65.50
0640	Slip form, 24' pass, unreinforced, 6" thick		11200	.008		20	.28	.29	20.57	23
0644	8" thick		10600	.008		28.50	.29	.30	29.09	32.50
0650	12" thick		6940	.013		41.50	.45	.46	42.41	46.50
0652	15" thick		5780	.015		52.50	.54	.55	53.59	59.50
0700	Finishing, broom finish small areas	2 Cefi	120	.133			5.10		5.10	8
1000	Curing, with sprayed membrane by hand	2 Clab	1500	.011		.47	.34		.81	1.08

32 14 Unit Paving

32 14 16 — Brick Unit Paving

32 14 16.10 Brick Paving

		Crew	Daily Output	Labor-Hours	Unit	Material	2009 Bare Costs Labor	Equipment	Total	Total Incl O&P
0010	**BRICK PAVING**									
0012	4" x 8" x 1-1/2", without joints (4.5 brick/S.F.)	D-1	110	.145	S.F.	2.78	5.30		8.08	11.60
0100	Grouted, 3/8" joint (3.9 brick/S.F.)		90	.178		3.38	6.45		9.83	14.15
0200	4" x 8" x 2-1/4", without joints (4.5 bricks/S.F.)		110	.145		3.74	5.30		9.04	12.65
0300	Grouted, 3/8" joint (3.9 brick/S.F.)		90	.178		3.45	6.45		9.90	14.25
0500	Bedding, asphalt, 3/4" thick	B-25	5130	.017		.57	.59	.48	1.64	2.12
0540	Course washed sand bed, 1" thick	B-18	5000	.005		.28	.15	.01	.44	.57
0580	Mortar, 1" thick	D-1	300	.053		.64	1.94		2.58	3.83
0620	2" thick		200	.080		1.28	2.91		4.19	6.10
1500	Brick on 1" thick sand bed laid flat, 4.5 per S.F.		100	.160		2.99	5.80		8.79	12.70
2000	Brick pavers, laid on edge, 7.2 per S.F.		70	.229		2.74	8.30		11.04	16.45
2500	For 4" thick concrete bed and joints, add		595	.027		1.19	.98		2.17	2.89
2800	For steam cleaning, add	A-1H	950	.008		.06	.27	.05	.38	.57
9000	Minimum labor/equipment charge	1 Bric	2	4	Job		162		162	262

32 14 23 — Asphalt Unit Paving

32 14 23.10 Asphalt Blocks

		Crew	Daily Output	Labor-Hours	Unit	Material	2009 Bare Costs Labor	Equipment	Total	Total Incl O&P
0010	**ASPHALT BLOCKS**									
0020	Rectangular, 6" x 12" x 1-1/4", w/bed & neopr. adhesive	D-1	135	.119	S.F.	7.65	4.31		11.96	15.40
0100	3" thick		130	.123		10.75	4.47		15.22	19.05
0300	Hexagonal tile, 8" wide, 1-1/4" thick		135	.119		7.65	4.31		11.96	15.40
0400	2" thick		130	.123		10.75	4.47		15.22	19.05
0500	Square, 8" x 8", 1-1/4" thick		135	.119		7.65	4.31		11.96	15.40
0600	2" thick		130	.123		10.75	4.47		15.22	19.05
9000	Minimum labor/equipment charge	1 Bric	2	4	Job		162		162	262

32 14 Unit Paving

32 14 40 – Stone Paving

32 14 40.10 Stone Pavers

32 14 40.10 Stone Pavers	Crew	Daily Output	Labor-Hours	Unit	Material	2009 Bare Costs Labor	Equipment	Total	Total Incl O&P
0010 **STONE PAVERS**									
1300 Slate, natural cleft, irregular, 3/4" thick	D-1	92	.174	S.F.	6.95	6.30		13.25	17.85
1350 Random rectangular, gauged, 1/2" thick		105	.152		15.05	5.55		20.60	25.50
1400 Random rectangular, butt joint, gauged, 1/4" thick		150	.107		16.20	3.88		20.08	24
1450 For sand rubbed finish, add					7.55			7.55	8.30
1550 Granite blocks, 3-1/2" x 3-1/2" x 3-1/2"	D-1	92	.174		10	6.30		16.30	21
1600 4" to 12" long, 3" to 5" wide, 3" to 5" thick	"	98	.163		8.35	5.95		14.30	18.75

32 16 Curbs and Gutters

32 16 13 – Concrete Curbs and Gutters

32 16 13.13 Cast-in-Place Concrete Curbs and Gutters

32 16 13.13 Cast-in-Place Concrete Curbs and Gutters	Crew	Daily Output	Labor-Hours	Unit	Material	2009 Bare Costs Labor	Equipment	Total	Total Incl O&P
0010 **CAST-IN-PLACE CONCRETE CURBS AND GUTTERS**									
0300 Concrete, wood forms, 6" x 18", straight	C-2A	500	.096	L.F.	4.84	3.71		8.55	11.35
0400 6" x 18", radius	"	200	.240		4.92	9.25		14.17	20.50
0415 Machine formed, 6" x 18", straight	B-69A	2000	.024		3.89	.83	.38	5.10	6.05
0416 6" x 18", radius	"	900	.053		4.06	1.85	.84	6.75	8.40

32 16 13.26 Precast Concrete Curbs

32 16 13.26 Precast Concrete Curbs	Crew	Daily Output	Labor-Hours	Unit	Material	2009 Bare Costs Labor	Equipment	Total	Total Incl O&P
0010 **PRECAST CONCRETE CURBS**									
0550 Precast, 6" x 18", straight	B-29	700	.080	L.F.	10.50	2.74	1.37	14.61	17.50
0600 6" x 18", radius	"	325	.172	"	12.10	5.90	2.95	20.95	26

32 16 19 – Asphalt Curbs

32 16 19.10 Bituminous Concrete Curbs

32 16 19.10 Bituminous Concrete Curbs	Crew	Daily Output	Labor-Hours	Unit	Material	2009 Bare Costs Labor	Equipment	Total	Total Incl O&P
0010 **BITUMINOUS CONCRETE CURBS**									
0012 Curbs, asphaltic, machine formed, 8" wide, 6" high, 40 L.F./ton	B-27	1000	.032	L.F.	1.24	1.03	.27	2.54	3.36
0100 8" wide, 8" high, 30 L.F. per ton		900	.036		1.43	1.14	.30	2.87	3.78
0150 Asphaltic berm, 12" W, 3"-6" H, 35 L.F./ton, before pavement		700	.046		1.59	1.47	.39	3.45	4.60
0200 12" W, 1-1/2" to 4" H, 60 L.F. per ton, laid with pavement	B-2	1050	.038		.97	1.22		2.19	3.07

32 16 40 – Stone Curbs

32 16 40.13 Cut Stone Curbs

32 16 40.13 Cut Stone Curbs	Crew	Daily Output	Labor-Hours	Unit	Material	2009 Bare Costs Labor	Equipment	Total	Total Incl O&P
0010 **CUT STONE CURBS**									
1000 Granite, split face, straight, 5" x 16"	D-13	500	.096	L.F.	12.05	3.68	1.20	16.93	20.50
1100 6" x 18"	"	450	.107		15.85	4.09	1.34	21.28	25.50
1300 Radius curbing, 6" x 18", over 10' radius	B-29	260	.215		19.40	7.35	3.69	30.44	37.50
1400 Corners, 2' radius		80	.700	Ea.	65	24	12	101	124
1600 Edging, 4-1/2" x 12", straight		300	.187	L.F.	6.05	6.40	3.20	15.65	20.50
1800 Curb inlets, (guttermouth) straight		41	1.366	Ea.	145	46.50	23.50	215	261
2000 Indian granite (belgian block)									
2100 Jumbo, 10-1/2" x 7-1/2" x 4", grey	D-1	150	.107	L.F.	2.21	3.88		6.09	8.70
2150 Pink		150	.107		2.88	3.88		6.76	9.40
2200 Regular, 9" x 4-1/2" x 4-1/2", grey		160	.100		2	3.63		5.63	8.10
2250 Pink		160	.100		2.76	3.63		6.39	8.95
2300 Cubes, 4" x 4" x 4", grey		175	.091		1.91	3.32		5.23	7.45
2350 Pink		175	.091		2	3.32		5.32	7.55
2400 6" x 6" x 6", pink		155	.103		4.96	3.75		8.71	11.50
2500 Alternate pricing method for indian granite									
2550 Jumbo, 10-1/2" x 7-1/2" x 4" (30 lb), grey				Ton	126			126	139
2600 Pink					167			167	184
2650 Regular, 9" x 4-1/2" x 4-1/2" (20 lb), grey					141			141	156
2700 Pink					193			193	212

32 16 Curbs and Gutters

32 16 40 – Stone Curbs

32 16 40.13 Cut Stone Curbs

32 16 40.13 Cut Stone Curbs		Crew	Daily Output	Labor-Hours	Unit	Material	2009 Bare Costs Labor	Equipment	Total	Total Incl O&P
2750	Cubes, 4" x 4" x 4" (5 lb), grey				Ton	231			231	255
2800	Pink					257			257	283
2850	6" x 6" x 6" (25 lb), pink					193			193	212
2900	For pallets, add					20			20	22

32 17 Paving Specialties

32 17 13 – Parking Bumpers

32 17 13.13 Metal Parking Bumpers

32 17 13.13 Metal Parking Bumpers		Crew	Daily Output	Labor-Hours	Unit	Material	2009 Bare Costs Labor	Equipment	Total	Total Incl O&P
0010	**METAL PARKING BUMPERS**									
0015	Bumper rails for garages, 12 Ga. rail, 6" wide, with steel									
0020	posts 12'-6" O.C., minimum	E-4	190	.168	L.F.	17.90	7.60	.71	26.21	34.50
0030	Average		165	.194		22.50	8.75	.81	32.06	41.50
0100	Maximum		140	.229		27	10.35	.96	38.31	49.50
1300	Pipe bollards, conc filled/paint, 8' L x 4' D hole, 6" diam.	B-6	20	1.200	Ea.	410	41	14.70	465.70	535
1400	8" diam.		15	1.600		620	54.50	19.60	694.10	790
1500	12" diam.		12	2		805	68	24.50	897.50	1,025
1592	Bollards, steel, 3' H, retractable, incl hydraulic controls, min		4	6		40,700	204	73.50	40,977.50	45,100
1594	Max		2	12		45,300	410	147	45,857	50,500
9000	Minimum labor/equipment charge	E-4	2	16	Job		725	67	792	1,425

32 17 13.16 Plastic Parking Bumpers

32 17 13.16 Plastic Parking Bumpers			Crew	Daily Output	Labor-Hours	Unit	Material	2009 Bare Costs Labor	Equipment	Total	Total Incl O&P
0010	**PLASTIC PARKING BUMPERS**										
1600	Bollards, recycled plastic, 5" x 5" x 5' L	G	B-6	20	1.200	Ea.	58.50	41	14.70	114.20	147
1610	Wheel stops, recycled plastic, yellow, 4" x 6" x 6' L	G		20	1.200		58.50	41	14.70	114.20	147
1620	Speed bumps, recycled plastic, yellow, 3"H x 10" W x 6' L	G		20	1.200		126	41	14.70	181.70	222
1630	3"H x 10" W x 9' L	G		20	1.200		174	41	14.70	229.70	274

32 17 13.19 Precast Concrete Parking Bumpers

32 17 13.19 Precast Concrete Parking Bumpers		Crew	Daily Output	Labor-Hours	Unit	Material	2009 Bare Costs Labor	Equipment	Total	Total Incl O&P
0010	**PRECAST CONCRETE PARKING BUMPERS**									
1000	Wheel stops, precast concrete incl. dowels, 6" x 10" x 6'-0"	B-2	120	.333	Ea.	58.50	10.65		69.15	82
1100	8" x 13" x 6'-0"	"	120	.333		68	10.65		78.65	92
1540	Bollards, precast concrete, free standing, round, 15" Dia x 36" H	B-6	14	1.714		179	58.50	21	258.50	315
1550	16" Dia x 30" H		14	1.714		179	58.50	21	258.50	315
1560	18" Dia x 34" H		14	1.714		365	58.50	21	444.50	525
1570	18" Dia x 72" H		14	1.714		445	58.50	21	524.50	610
1572	26" Dia x 18" H		12	2		445	68	24.50	537.50	630
1574	26" Dia x 30" H		12	2		660	68	24.50	752.50	865
1578	26" Dia x 50" H		12	2		895	68	24.50	987.50	1,125
1580	Square, 22" x 32" H		12	2		475	68	24.50	567.50	660
1585	22" x 41" H		12	2		840	68	24.50	932.50	1,075
1590	Hexagon, 26" x 41" H		10	2.400		875	82	29.50	986.50	1,125

32 17 13.26 Wood Parking Bumpers

32 17 13.26 Wood Parking Bumpers		Crew	Daily Output	Labor-Hours	Unit	Material	2009 Bare Costs Labor	Equipment	Total	Total Incl O&P
0010	**WOOD PARKING BUMPERS**									
0020	Parking barriers, timber w/saddles, treated type									
0100	4" x 4" for cars	B-2	520	.077	L.F.	2.98	2.46		5.44	7.35
0200	6" x 6" for trucks	"	520	.077	"	5.95	2.46		8.41	10.60

32 17 23 – Pavement Markings

32 17 23.13 Painted Pavement Markings

32 17 23.13 Painted Pavement Markings		Crew	Daily Output	Labor-Hours	Unit	Material	2009 Bare Costs Labor	Equipment	Total	Total Incl O&P
0010	**PAINTED PAVEMENT MARKINGS**									
0020	Acrylic waterborne, white or yellow, 4" wide	B-78	20000	.002	L.F.	.25	.08	.03	.36	.44
0200	6" wide		11000	.004		.19	.14	.05	.38	.49
0500	8" wide		10000	.005		.33	.15	.05	.53	.67

32 17 Paving Specialties

32 17 23 – Pavement Markings

32 17 23.13 Painted Pavement Markings

		Crew	Daily Output	Labor-Hours	Unit	Material	2009 Bare Costs Labor	Equipment	Total	Total Incl O&P
0600	12" wide	B-78	4000	.012	L.F.	.60	.38	.13	1.11	1.43
0620	Arrows or gore lines		2300	.021	S.F.	.68	.66	.22	1.56	2.09
0640	Temporary paint, white or yellow		15000	.003	L.F.	.20	.10	.03	.33	.43
0660	Removal	1 Clab	300	.027			.84		.84	1.39
0680	Temporary tape	2 Clab	1500	.011		1.74	.34		2.08	2.47
0710	Thermoplastic, white or yellow, 4" wide	B-79	15000	.003		.87	.09	.08	1.04	1.19
0730	6" wide		14000	.003		1.26	.09	.08	1.43	1.63
0740	8" wide		12000	.003		1.70	.11	.10	1.91	2.15
0750	12" wide		6000	.007		2.52	.21	.20	2.93	3.35
0760	Arrows		660	.061	S.F.	1.82	1.93	1.79	5.54	7.15
0770	Gore lines		2500	.016		1.22	.51	.47	2.20	2.70
0780	Letters		660	.061		1.51	1.93	1.79	5.23	6.80
1000	Airport painted markings									
1100	Painting, white or yellow, taxiway markings	B-79	4000	.010	S.F.	.25	.32	.29	.86	1.13
1200	Runway markings		3500	.011		.25	.36	.34	.95	1.25
1300	Pavement location or direction signs		2500	.016		.25	.51	.47	1.23	1.64
1350	Mobilization airport pavement painting		4	10	Ea.		320	295	615	850
1400	Paint markings or pavement signs removal daytime	B-78B	400	.045	S.F.		1.46	.80	2.26	3.27
1500	Removal nighttime		335	.054	"		1.74	.96	2.70	3.91
1600	Mobilization pavement paint removal		4	4.500	Ea.		146	80.50	226.50	330

32 17 23.14 Pavement Parking Markings

		Crew	Daily Output	Labor-Hours	Unit	Material	2009 Bare Costs Labor	Equipment	Total	Total Incl O&P
0010	**PAVEMENT PARKING MARKINGS**									
0790	Layout of pavement marking	A-2	25000	.001	L.F.		.03	.01	.04	.06
0800	Lines on pvmt, parking stall, paint, white, 4" wide	B-78	440	.109	Stall	6	3.47	1.15	10.62	13.55
1000	Street letters and numbers	"	1600	.030	S.F.	.63	.95	.32	1.90	2.61

32 31 Fences and Gates

32 31 13 – Chain Link Fences and Gates

32 31 13.20 Fence, Chain Link Industrial

		Crew	Daily Output	Labor-Hours	Unit	Material	2009 Bare Costs Labor	Equipment	Total	Total Incl O&P
0010	**FENCE, CHAIN LINK INDUSTRIAL**									
0011	Schedule 40, including concrete									
0020	3 strands barb wire, 2" post @ 10' O.C., set in concrete, 6' H									
0200	9 ga. wire, galv. steel, in concrete	B-80C	240	.100	L.F.	15.25	3.14	.84	19.23	23
0300	Aluminized steel		240	.100		19.55	3.14	.84	23.53	27.50
0500	6 ga. wire, galv. steel		240	.100		24	3.14	.84	27.98	32.50
0600	Aluminized steel		240	.100		27.50	3.14	.84	31.48	36
0800	6 ga. wire, 6' high but omit barbed wire, galv. steel		250	.096		23	3.01	.81	26.82	31.50
0900	Aluminized steel, in concrete		250	.096		32.50	3.01	.81	36.32	42
0920	8' H, 6 ga. wire, 2-1/2" line post, galv. steel, in concrete		180	.133		37	4.18	1.12	42.30	49
0940	Aluminized steel, in concrete		180	.133		45.50	4.18	1.12	50.80	58
1100	Add for corner posts, 3" diam., galv. steel, in concrete		40	.600	Ea.	108	18.85	5.05	131.90	156
1200	Aluminized steel, in concrete		40	.600		129	18.85	5.05	152.90	179
1300	Add for braces, galv. steel		80	.300		29	9.40	2.52	40.92	50.50
1350	Aluminized steel		80	.300		38.50	9.40	2.52	50.42	61
1400	Gate for 6' high fence, 1-5/8" frame, 3' wide, galv. steel		10	2.400		177	75.50	20	272.50	340
1500	Aluminized steel, in concrete		10	2.400		218	75.50	20	313.50	385
2000	5'-0" high fence, 9 ga., no barbed wire, 2" line post, in concrete									
2010	10' O.C., 1-5/8" top rail, in concrete									
2100	Galvanized steel, in concrete	B-80C	300	.080	L.F.	13	2.51	.67	16.18	19.15
2200	Aluminized steel, in concrete		300	.080	"	15.65	2.51	.67	18.83	22
2400	Gate, 4' wide, 5' high, 2" frame, galv. steel, in concrete		10	2.400	Ea.	191	75.50	20	286.50	355

32 31 Fences and Gates

32 31 13 – Chain Link Fences and Gates

32 31 13.20 Fence, Chain Link Industrial

		Crew	Daily Output	Labor-Hours	Unit	Material	2009 Bare Costs		Total	Total Incl O&P
							Labor	Equipment		
2500	Aluminized steel, in concrete	B-80C	10	2.400	Ea.	214	75.50	20	309.50	380
3100	Overhead slide gate, chain link, 6' high, to 18' wide, in concrete	↓	38	.632	L.F.	167	19.80	5.30	192.10	221
3110	Cantilever type, in concrete	B-80	48	.667		77.50	22.50	13.65	113.65	137
3120	8' high, in concrete		24	1.333		111	45	27.50	183.50	226
3130	10' high, in concrete		18	1.778	↓	131	60	36.50	227.50	282
9000	Minimum labor/equipment charge	↓	2	16	Job		540	325	865	1,250

32 31 13.25 Fence, Chain Link Residential

		Crew	Daily Output	Labor-Hours	Unit	Material	Labor	Equipment	Total	Total Incl O&P
0010	**FENCE, CHAIN LINK RESIDENTIAL**									
0011	Schedule 20, 11 gauge wire, 1-5/8" post									
3300	Residential, 11 ga. wire, 1-5/8" line post @ 10' O.C.									
3310	1-3/8" top rail									
3350	4' high, galvanized steel	B-80C	475	.051	L.F.	10.85	1.59	.43	12.87	15.05
3400	Aluminized		475	.051	"	10.95	1.59	.43	12.97	15.15
3600	Gate, 3' wide, 1-3/8" frame, galv. steel		10	2.400	Ea.	89	75.50	20	184.50	244
3700	Aluminized		10	2.400	"	100	75.50	20	195.50	257
3900	3' high, galvanized steel		620	.039	L.F.	6.55	1.21	.33	8.09	9.55
4000	Aluminized		620	.039	"	8.05	1.21	.33	9.59	11.20
4200	Gate, 3' wide, 1-3/8" frame, galv. steel		12	2	Ea.	73	63	16.80	152.80	202
4300	Aluminized	↓	12	2	"	82	63	16.80	161.80	212
9000	Minimum labor/equipment charge	B-1	2	12	Job		385		385	640

32 31 13.26 Tennis Court Fences and Gates

		Crew	Daily Output	Labor-Hours	Unit	Material	Labor	Equipment	Total	Total Incl O&P
0010	**TENNIS COURT FENCES AND GATES**									
3150	Tennis courts, 11 ga. wire, 1-3/4" mesh, 2-1/2" line posts									
3170	1-5/8" top rail, 3" corner and gate posts									
3190	10' high, galvanized steel	B-80	190	.168	L.F.	19.70	5.70	3.45	28.85	34.50
3210	Vinyl covered 9 ga. wire		190	.168	"	21.50	5.70	3.45	30.65	37
3240	Corner posts for above, 3" diameter, 10' high	↓	30	1.067	Ea.	207	36	22	265	310

32 31 13.40 Fence, Fabric and Accessories

		Crew	Daily Output	Labor-Hours	Unit	Material	Labor	Equipment	Total	Total Incl O&P
0010	**FENCE, FABRIC & ACCESSORIES**									
8000	Components only, fabric, galvanized, 4' high	B-80C	585	.041	L.F.	3.98	1.29	.35	5.62	6.90
8040	6' high		430	.056	"	5.25	1.75	.47	7.47	9.15
8200	Posts, line post, 4' high		34	.706	Ea.	14.80	22	5.95	42.75	59.50
8240	6' high		26	.923	"	15	29	7.75	51.75	72.50
8400	Top rails, 1-3/8" O.D.		1000	.024	L.F.	1.90	.75	.20	2.85	3.55
8440	1-5/8" O.D.	↓	1000	.024	"	2.15	.75	.20	3.10	3.83

32 31 19 – Decorative Metal Fences and Gates

32 31 19.10 Decorative Fence

		Crew	Daily Output	Labor-Hours	Unit	Material	Labor	Equipment	Total	Total Incl O&P
0010	**DECORATIVE FENCE**									
5300	Tubular picket, steel, 6' sections, 1-9/16" posts, 4' high	B-80C	300	.080	L.F.	31	2.51	.67	34.18	39.50
5400	2" posts, 5' high		240	.100		43	3.14	.84	46.98	53.50
5600	2" posts, 6' high		200	.120		49	3.77	1.01	53.78	61
5700	Staggered picket 1-9/16" posts, 4' high		300	.080		28	2.51	.67	31.18	36
5800	2" posts, 5' high		240	.100		46	3.14	.84	49.98	57
5900	2" posts, 6' high	↓	200	.120	↓	48	3.77	1.01	52.78	60
6200	Gates, 4' high, 3' wide	B-1	10	2.400	Ea.	271	77.50		348.50	425
6300	5' high, 3' wide		10	2.400		350	77.50		427.50	515
6400	6' high, 3' wide		10	2.400		360	77.50		437.50	530
6500	4' wide	↓	10	2.400	↓	420	77.50		497.50	595

32 31 23 – Plastic Fences and Gates

32 31 23.20 Fence, Recycled Plastic

0010	**FENCE, RECYCLED PLASTIC**									

32 31 Fences and Gates

32 31 23 – Plastic Fences and Gates

32 31 23.20 Fence, Recycled Plastic

			Crew	Daily Output	Labor-Hours	Unit	Material	2009 Bare Costs Labor	Equipment	Total	Total Incl O&P
9015	Fence rail, made from recycled plastic, various colors, 2 rail	G	B-1	150	.160	L.F.	4.70	5.15		9.85	13.65
9018	3 rail	G		150	.160		6.30	5.15		11.45	15.45
9020	4 rail	G		150	.160		7.60	5.15		12.75	16.85
9030	Fence pole, made from recycled plastic, various colors, 7'	G		96	.250	Ea.	18.10	8.05		26.15	33
9040	Stockade fence, made from recycled plastic, various colors, 4' high	G	B-80C	160	.150	L.F.	31.50	4.71	1.26	37.47	43.50
9050	6' high	G		160	.150	"	38.50	4.71	1.26	44.47	51.50
9060	6' pole	G		96	.250	Ea.	19.30	7.85	2.10	29.25	36
9070	9' pole	G		96	.250	"	29	7.85	2.10	38.95	47
9080	Picket fence, made from recycled plastic, various colors, 3' high	G		160	.150	L.F.	31.50	4.71	1.26	37.47	43.50
9090	4' high	G		160	.150	"	38.50	4.71	1.26	44.47	51.50
9100	3' high gate	G		8	3	Ea.	60.50	94	25	179.50	250
9110	4' high gate	G		8	3		70	94	25	189	260
9120	5' high pole	G		96	.250		10.85	7.85	2.10	20.80	27
9130	6' high pole	G		96	.250		13.25	7.85	2.10	23.20	30
9140	Pole cap only	G					3.65			3.65	4.02
9150	Keeper pins only	G					.25			.25	.28

32 31 26 – Wire Fences and Gates

32 31 26.10 Fences, Misc. Metal

			Crew	Daily Output	Labor-Hours	Unit	Material	2009 Bare Costs Labor	Equipment	Total	Total Incl O&P
0010	**FENCES, MISC. METAL**										
0012	Chicken wire, posts @ 4', 1" mesh, 4' high		B-80C	410	.059	L.F.	2.16	1.84	.49	4.49	5.95
0100	2" mesh, 6' high			350	.069		1.95	2.15	.58	4.68	6.30
0200	Galv. steel, 12 ga., 2" x 4" mesh, posts 5' O.C., 3' high			300	.080		3.12	2.51	.67	6.30	8.30
0300	5' high			300	.080		4.16	2.51	.67	7.34	9.45
0400	14 ga., 1" x 2" mesh, 3' high			300	.080		3.31	2.51	.67	6.49	8.50
0500	5' high			300	.080		4.58	2.51	.67	7.76	9.90
1000	Kennel fencing, 1-1/2" mesh, 6' long, 3'-6" wide, 6'-2" high		2 Clab	4	4	Ea.	520	126		646	780
1050	12' long			4	4		625	126		751	895
1200	Top covers, 1-1/2" mesh, 6' long			15	1.067		106	33.50		139.50	172
1250	12' long			12	1.333		169	42		211	256
4500	Security fence, prison grade, set in concrete, 12' high		B-80	25	1.280	L.F.	46	43.50	26	115.50	150
4600	16' high		"	20	1.600	"	55	54	32.50	141.50	185

32 31 29 – Wood Fences and Gates

32 31 29.20 Fence, Wood Rail

			Crew	Daily Output	Labor-Hours	Unit	Material	2009 Bare Costs Labor	Equipment	Total	Total Incl O&P
0010	**FENCE, WOOD RAIL**										
0012	Picket, No. 2 cedar, Gothic, 2 rail, 3' high		B-1	160	.150	L.F.	5.85	4.84		10.69	14.45
0050	Gate, 3'-6" wide		B-80C	9	2.667	Ea.	50.50	83.50	22.50	156.50	218
0600	Open rail, rustic, No. 1 cedar, 2 rail, 3' high			160	.150	L.F.	5.25	4.71	1.26	11.22	14.90
0650	Gate, 3' wide			9	2.667	Ea.	58.50	83.50	22.50	164.50	227
1200	Stockade, No. 2 cedar, treated wood rails, 6' high			160	.150	L.F.	7.30	4.71	1.26	13.27	17.20
1250	Gate, 3' wide			9	2.667	Ea.	60	83.50	22.50	166	229
3300	Board, shadow box, 1" x 6", treated pine, 6' high			160	.150	L.F.	10.50	4.71	1.26	16.47	20.50
3400	No. 1 cedar, 6' high			150	.160		20.50	5	1.35	26.85	32.50
3900	Basket weave, No. 1 cedar, 6' high			160	.150		20.50	4.71	1.26	26.47	31.50
3950	Gate, 3'-6" wide		B-1	8	3	Ea.	140	97		237	315
4000	Treated pine, 6' high			150	.160	L.F.	13.45	5.15		18.60	23.50
4200	Gate, 3'-6" wide			9	2.667	Ea.	65.50	86		151.50	214
8000	Posts only, 4' high			40	.600		15.20	19.35		34.55	48.50
8040	6' high			30	.800		21	26		47	65.50
9000	Minimum labor/equipment charge		1 Clab	2	4	Job		126		126	209

32 32 Retaining Walls

32 32 60 – Stone Retaining Walls

32 32 60.10 Retaining Walls, Stone

	Crew	Daily Output	Labor-Hours	Unit	Material	2009 Bare Costs Labor	2009 Bare Costs Equipment	Total	Total Incl O&P
0010 **RETAINING WALLS, STONE**									
0015 Including excavation, concrete footing and									
0020 stone 3' below grade. Price is exposed face area.									
0200 Decorative random stone, to 6' high, 1'-6" thick, dry set	D-1	35	.457	S.F.	47	16.60		63.60	79
0300 Mortar set		40	.400		47	14.55		61.55	75.50
0500 Cut stone, to 6' high, 1'-6" thick, dry set		35	.457		47	16.60		63.60	79
0600 Mortar set		40	.400		47	14.55		61.55	75.50
0800 Random stone, 6' to 10' high, 2' thick, dry set		45	.356		47	12.90		59.90	73
0900 Mortar set		50	.320		47	11.65		58.65	71
1100 Cut stone, 6' to 10' high, 2' thick, dry set		45	.356		47	12.90		59.90	73
1200 Mortar set		50	.320		47	11.65		58.65	71
9000 Minimum labor/equipment charge		2	8	Job		291		291	470

32 84 Planting Irrigation

32 84 23 – Underground Sprinklers

32 84 23.10 Sprinkler Irrigation System

	Crew	Daily Output	Labor-Hours	Unit	Material	2009 Bare Costs Labor	2009 Bare Costs Equipment	Total	Total Incl O&P
0010 **SPRINKLER IRRIGATION SYSTEM**									
0011 For lawns									
0100 Golf course with fully automatic system	C-17	.05	1600	9 holes	97,500	66,000		163,500	215,000
0200 24' diam. head at 15' O.C incl. piping, auto oper., minimum	B-20	70	.343	Head	22	12.10		34.10	44
0300 Maximum		40	.600		50.50	21		71.50	90.50
0500 60' diam. head at 40' O.C. incl. piping, auto oper., minimum		28	.857		66.50	30.50		97	123
0600 Maximum		23	1.043		186	37		223	265
0800 Residential system, custom, 1" supply		2000	.012	S.F.	.32	.42		.74	1.05
0900 1-1/2" supply		1800	.013	"	.39	.47		.86	1.20

32 91 Planting Preparation

32 91 13 – Soil Preparation

32 91 13.23 Structural Soil Mixing

	Crew	Daily Output	Labor-Hours	Unit	Material	2009 Bare Costs Labor	2009 Bare Costs Equipment	Total	Total Incl O&P
0010 **STRUCTURAL SOIL MIXING**									
0100 Rake topsoil, site material, harley rock rake, ideal	B-6	33	.727	M.S.F.		25	8.90	33.90	50
0200 Adverse	"	7	3.429			117	42	159	235
0300 Screened loam, york rake and finish, ideal	B-62	24	1			34	6.20	40.20	62.50
0400 Adverse	"	20	1.200			41	7.45	48.45	74.50
1000 Remove topsoil & stock pile on site, 75 HP dozer, 6" deep, 50' haul	B-10L	30	.400			15.25	13.30	28.55	39
1050 300' haul		6.10	1.967			75	65.50	140.50	192
1100 12" deep, 50' haul		15.50	.774			29.50	26	55.50	75.50
1150 300' haul		3.10	3.871			147	129	276	375
1200 200 HP dozer, 6" deep, 50' haul	B-10B	125	.096			3.66	8.65	12.31	15.35
1250 300' haul		30.70	.391			14.90	35.50	50.40	63
1300 12" deep, 50' haul		62	.194			7.35	17.45	24.80	31
1350 300' haul		15.40	.779			29.50	70.50	100	125
1400 Alternate method, 75 HP dozer, 50' haul	B-10L	860	.014	C.Y.		.53	.46	.99	1.36
1450 300' haul	"	114	.105			4.01	3.50	7.51	10.25
1500 200 HP dozer, 50' haul	B-10B	2660	.005			.17	.41	.58	.72
1600 300' haul	"	570	.021			.80	1.90	2.70	3.37
1800 Rolling topsoil, hand push roller	1 Clab	3200	.003	S.F.		.08		.08	.13
1850 Tractor drawn roller	B-66	10666	.001	"		.03	.02	.05	.07
2000 Root raking and loading, residential, no boulders	B-6	53.30	.450	M.S.F.		15.35	5.50	20.85	31

32 91 Planting Preparation

32 91 13 – Soil Preparation

32 91 13.23 Structural Soil Mixing

		Crew	Daily Output	Labor-Hours	Unit	Material	2009 Bare Costs Labor	Equipment	Total	Total Incl O&P
2100	With boulders	B-6	32	.750	M.S.F.		25.50	9.20	34.70	51.50
2200	Municipal, no boulders	↓	200	.120	↓		4.09	1.47	5.56	8.25
2300	With boulders	↓	120	.200			6.80	2.45	9.25	13.75
2400	Large commercial, no boulders	B-10B	400	.030			1.14	2.71	3.85	4.80
2500	With boulders	"	240	.050			1.91	4.51	6.42	8
3000	Scarify subsoil, residential, skid steer loader w/scarifiers, 50 HP	B-66	32	.250			9.75	7.25	17	23.50
3050	Municipal, skid steer loader w/scarifiers, 50 HP	"	120	.067			2.60	1.93	4.53	6.20
3100	Large commercial, 75 HP, dozer w/scarifier	B-10L	240	.050	↓		1.91	1.66	3.57	4.87
3500	Screen topsoil from stockpile, vibrating screen, wet material (organic)	B-10P	200	.060	C.Y.		2.29	4.95	7.24	9.10
3550	Dry material	"	300	.040			1.52	3.30	4.82	6.05
3600	Mixing with conditioners, manure and peat	B-10R	550	.022	↓		.83	.43	1.26	1.81
3650	Mobilization add for 2 days or less operation	B-34K	3	2.667	Job		85	261	346	430
3800	Spread conditioned topsoil, 6" deep, by hand	B-1	360	.067	S.Y.	5	2.15		7.15	9.05
3850	300 HP dozer	B-10M	27	.444	M.S.F.	540	16.95	52.50	609.45	680
4000	Spread soil conditioners, alum. sulfate, 1#/S.Y., hand push spreader	1 Clab	17500	.001	S.Y.	16.70	.01		16.71	18.35
4050	Tractor spreader	B-66	700	.011	M.S.F.	1,850	.45	.33	1,850.78	2,050
4100	Fertilizer, 0.2#/S.Y., push spreader	1 Clab	17500	.001	S.Y.	.08	.01		.09	.11
4150	Tractor spreader	B-66	700	.011	M.S.F.	8.90	.45	.33	9.68	10.85
4200	Ground limestone, 1#/S.Y., push spreader	1 Clab	17500	.001	S.Y.	.09	.01		.10	.12
4250	Tractor spreader	B-66	700	.011	M.S.F.	10	.45	.33	10.78	12.05
4300	Lusoil, 3#/S.Y., push spreader	1 Clab	17500	.001	S.Y.	.52	.01		.53	.59
4350	Tractor spreader	B-66	700	.011	M.S.F.	58	.45	.33	58.78	64.50
4400	Manure, 18#/S.Y., push spreader	1 Clab	2500	.003	S.Y.	3.05	.10		3.15	3.53
4450	Tractor spreader	B-66	280	.029	M.S.F.	340	1.12	.83	341.95	380
4500	Perlite, 1" deep, push spreader	1 Clab	17500	.001	S.Y.	9.35	.01		9.36	10.30
4550	Tractor spreader	B-66	700	.011	M.S.F.	1,050	.45	.33	1,050.78	1,150
4600	Vermiculite, push spreader	1 Clab	17500	.001	S.Y.	2.90	.01		2.91	3.21
4650	Tractor spreader	B-66	700	.011	M.S.F.	320	.45	.33	320.78	355
5000	Spread topsoil, skid steer loader and hand dress	B-62	270	.089	C.Y.	22.50	3.03	.55	26.08	30.50
5100	Articulated loader and hand dress	B-100	320	.038		22.50	1.43	2.40	26.33	30
5200	Articulated loader and 75HP dozer	B-10M	500	.024		22.50	.91	2.85	26.26	29.50
5300	Road grader and hand dress	B-11L	1000	.016	↓	22.50	.58	.55	23.63	26.50
6000	Tilling topsoil, 20 HP tractor, disk harrow, 2" deep	B-66	450	.018	M.S.F.		.69	.51	1.20	1.66
6050	4" deep	↓	360	.022			.87	.64	1.51	2.08
6100	6" deep	↓	270	.030	↓		1.16	.86	2.02	2.76
6150	26" rototiller, 2" deep	A-1J	1250	.006	S.Y.		.20	.05	.25	.38
6200	4" deep	↓	1000	.008			.25	.06	.31	.48
6250	6" deep	↓	750	.011	↓		.34	.08	.42	.64

32 91 13.26 Planting Beds

		Crew	Daily Output	Labor-Hours	Unit	Material	2009 Bare Costs Labor	Equipment	Total	Total Incl O&P
0010	**PLANTING BEDS**									
0100	Backfill planting pit, by hand, on site topsoil	2 Clab	18	.889	C.Y.		28		28	46.50
0200	Prepared planting mix, by hand	"	24	.667			21		21	35
0300	Skid steer loader, on site topsoil	B-62	340	.071			2.41	.44	2.85	4.38
0400	Prepared planting mix	"	410	.059			2	.36	2.36	3.63
1000	Excavate planting pit, by hand, sandy soil	2 Clab	16	1			31.50		31.50	52
1100	Heavy soil or clay	"	8	2			63		63	104
1200	1/2 C.Y. backhoe, sandy soil	B-11C	150	.107			3.89	1.96	5.85	8.40
1300	Heavy soil or clay	"	115	.139			5.10	2.55	7.65	10.95
2000	Mix planting soil, incl. loam, manure, peat, by hand	2 Clab	60	.267		39.50	8.45		47.95	57.50
2100	Skid steer loader	B-62	150	.160	↓	39.50	5.45	1	45.95	53.50
3000	Pile sod, skid steer loader	"	2800	.009	S.Y.		.29	.05	.34	.53
3100	By hand	2 Clab	400	.040	↓		1.26		1.26	2.09

32 91 Planting Preparation

32 91 13 – Soil Preparation

32 91 13.26 Planting Beds

	Crew	Daily Output	Labor-Hours	Unit	Material	2009 Bare Costs Labor	2009 Bare Costs Equipment	Total	Total Incl O&P
4000	Remove sod, F.E. loader	B-10S	2000	.006	S.Y.		.23	.16	.39	.54
4100	Sod cutter	B-12K	3200	.005			.19	.35	.54	.69
4200	By hand	2 Clab	240	.067	↓		2.11		2.11	3.48

32 91 19 – Landscape Grading

32 91 19.13 Topsoil Placement and Grading

	Crew	Daily Output	Labor-Hours	Unit	Material	Labor	Equipment	Total	Total Incl O&P
0010	**TOPSOIL PLACEMENT AND GRADING**									
0700	Furnish and place, truck dumped, screened, 4" deep	B-10S	1300	.009	S.Y.	2.59	.35	.25	3.19	3.69
0800	6" deep	↓	820	.015	"	3.32	.56	.40	4.28	4.98
0810	Minimum labor/equipment charge	↓	2	6	Ea.		229	164	393	545
0900	Fine grading and seeding, incl. lime, fertilizer & seed,									
1000	With equipment	B-14	1000	.048	S.Y.	.35	1.59	.29	2.23	3.30
2000	Minimum labor/equipment charge	1 Clab	4	2	Job		63		63	104

32 92 Turf and Grasses

32 92 19 – Seeding

32 92 19.13 Mechanical Seeding

		Crew	Daily Output	Labor-Hours	Unit	Material	Labor	Equipment	Total	Total Incl O&P
0010	**MECHANICAL SEEDING**	R329219-50									
0020	Mechanical seeding, 215 lb./acre	B-66	1.50	5.333	Acre	565	208	154	927	1,125
0100	44 lb./M.S.Y.	"	2500	.003	S.Y.	.17	.13	.09	.39	.49
0600	Limestone hand push spreader, 50 lbs. per M.S.F.	1 Clab	180	.044	M.S.F.	3.97	1.40		5.37	6.70
9000	Minimum labor/equipment charge	"	4	2	Job		63		63	104

32 92 23 – Sodding

32 92 23.10 Sodding Systems

		Crew	Daily Output	Labor-Hours	Unit	Material	Labor	Equipment	Total	Total Incl O&P
0010	**SODDING SYSTEMS**									
0020	Sodding, 1" deep, bluegrass sod, on level ground, over 8 MSF	B-63	22	1.818	M.S.F.	245	60	6.80	311.80	375
0200	4 M.S.F.		17	2.353		270	78	8.80	356.80	435
0300	1000 S.F.		13.50	2.963		295	98	11.05	404.05	495
0500	Sloped ground, over 8 M.S.F.		6	6.667		245	221	25	491	660
0600	4 M.S.F.		5	8		270	265	30	565	760
0700	1000 S.F.		4	10		295	330	37.50	662.50	905
1000	Bent grass sod, on level ground, over 6 M.S.F.		20	2		545	66	7.45	618.45	710
1100	3 M.S.F.		18	2.222		605	73.50	8.30	686.80	795
1200	Sodding 1000 S.F. or less		14	2.857		695	94.50	10.65	800.15	925
1500	Sloped ground, over 6 M.S.F.		15	2.667		545	88	9.95	642.95	750
1600	3 M.S.F.		13.50	2.963		605	98	11.05	714.05	835
1700	1000 S.F.	↓	12	3.333	↓	695	110	12.45	817.45	955

32 93 Plants

32 93 13 – Ground Covers

32 93 13.10 Ground Cover Plants

		Crew	Daily Output	Labor-Hours	Unit	Material	Labor	Equipment	Total	Total Incl O&P
0010	**GROUND COVER PLANTS**									
0012	Plants, pachysandra, in prepared beds	B-1	15	1.600	C	27	51.50		78.50	115
0200	Vinca minor, 1 yr, bare root, in prepared beds		12	2	"	40	64.50		104.50	151
0600	Stone chips, in 50 lb. bags, Georgia marble		520	.046	Bag	5.20	1.49		6.69	8.15
0700	Onyx gemstone		260	.092	↓	22.50	2.98		25.48	30
0800	Quartz		260	.092	↓	8.45	2.98		11.43	14.20
0900	Pea gravel, truckload lots	↓	28	.857	Ton	34	27.50		61.50	83

470

32 93 33.10 Shrubs and Trees

		Daily Output	Labor-Hours	Unit	Material	2009 Bare Costs Labor	Equipment	Total	Total Incl O&P
		Crew							
0010	**SHRUBS AND TREES**								
0011	Evergreen, in prepared beds, B & B								
0100	Arborvitae pyramidal, 4'-5'	B-17	30	1.067	Ea.	47	36	20.50	103.50	133
0150	Globe, 12"-15"	B-1	96	.250		21	8.05		29.05	36.50
0300	Cedar, blue, 8'-10'	B-17	18	1.778		233	59.50	34.50	327	390
0500	Hemlock, canadian, 2-1/2'-3'	B-1	36	.667		33.50	21.50		55	72.50
0550	Holly, Savannah, 8' - 10' H		9.68	2.479		475	80		555	655
0600	Juniper, andorra, 18"-24"		80	.300		16.75	9.70		26.45	34.50
0620	Wiltoni, 15"-18"		80	.300		18.10	9.70		27.80	36
0640	Skyrocket, 4-1/2'-5'	B-17	55	.582		55	19.50	11.30	85.80	104
0660	Blue pfitzer, 2'-2-1/2'	B-1	44	.545		25.50	17.60		43.10	57
0680	Ketleerie, 2-1/2'-3'		50	.480		33.50	15.50		49	62.50
0700	Pine, black, 2-1/2'-3'		50	.480		45.50	15.50		61	75.50
0720	Mugo, 18"-24"		60	.400		36	12.90		48.90	61
0740	White, 4'-5'	B-17	75	.427		56.50	14.30	8.25	79.05	95
0800	Spruce, blue, 18"-24"	B-1	60	.400		38	12.90		50.90	63.50
0840	Norway, 4'-5'	B-17	75	.427		65	14.30	8.25	87.55	104
0900	Yew, denisforma, 12"-15"	B-1	60	.400		25.50	12.90		38.40	50
1000	Capitata, 18"-24"		30	.800		24	26		50	68.50
1100	Hicksi, 2'-2-1/2'		30	.800		32	26		58	77.50

32 93 33.20 Shrubs

		Crew	Daily Output	Labor-Hours	Unit	Material	2009 Bare Costs Labor	Equipment	Total	Total Incl O&P
0010	**SHRUBS**									
0011	Broadleaf Evergreen, planted in prepared beds									
0100	Andromeda, 15"-18", container	B-1	96	.250	Ea.	26.50	8.05		34.55	42.50
0200	Azalea, 15" - 18", container		96	.250		32	8.05		40.05	48.50
0300	Barberry, 9"-12", container		130	.185		11.75	5.95		17.70	23
0400	Boxwood, 15"-18", B&B		96	.250		29	8.05		37.05	45
0500	Euonymus, emerald gaiety, 12" to 15", container		115	.209		21	6.75		27.75	34
0600	Holly, 15"-18", B & B		96	.250		18.60	8.05		26.65	34
0900	Mount laurel, 18" - 24", B & B		80	.300		56	9.70		65.70	77.50
1000	Paxistema, 9 – 12" high		130	.185		18.25	5.95		24.20	30
1100	Rhododendron, 18"-24", container		48	.500		31.50	16.15		47.65	61
1200	Rosemary, 1 gal container		600	.040		62.50	1.29		63.79	71
2000	Deciduous, planted in prepared beds, amelanchier, 2'-3', B & B		57	.421		90	13.60		103.60	122
2100	Azalea, 15"-18", B & B		96	.250		24.50	8.05		32.55	40.50
2300	Bayberry, 2'-3', B & B		57	.421		28	13.60		41.60	53
2600	Cotoneaster, 15"-18", B & B		80	.300		16.50	9.70		26.20	34
2800	Dogwood, 3'-4', B & B	B-17	40	.800		26	27	15.50	68.50	89
2900	Euonymus, alatus compacta, 15" to 18", container	B-1	80	.300		22	9.70		31.70	40
3200	Forsythia, 2'-3', container	"	60	.400		20	12.90		32.90	43.50
3300	Hibiscus, 3'-4', B & B	B-17	75	.427		15	14.30	8.25	37.55	49
3400	Honeysuckle, 3'-4', B & B	B-1	60	.400		23	12.90		35.90	47
3500	Hydrangea, 2'-3', B & B	"	57	.421		28.50	13.60		42.10	54
3600	Lilac, 3'-4', B & B	B-17	40	.800		27	27	15.50	69.50	90
3900	Privet, bare root, 18"-24"	B-1	80	.300		13.15	9.70		22.85	30.50
4100	Quince, 2'-3', B & B	"	57	.421		24	13.60		37.60	49
4200	Russian olive, 3'-4', B & B	B-17	75	.427		24	14.30	8.25	46.55	58.50
4400	Spirea, 3'-4', B & B	B-1	70	.343		17.15	11.05		28.20	37
4500	Viburnum, 3'-4', B & B	B-17	40	.800		29.50	27	15.50	72	93

32 93 Plants

32 93 43 – Trees

32 93 43.20 Trees

32 93 43.20 Trees		Crew	Daily Output	Labor-Hours	Unit	Material	2009 Bare Costs Labor	Equipment	Total	Total Incl O&P
0010	**TREES**									
0011	Deciduous, in prep. beds, balled & burlapped (B&B)									
0100	Ash, 2" caliper	G B-17	8	4	Ea.	115	134	77.50	326.50	430
0200	Beech, 5'-6'	G	50	.640		223	21.50	12.40	256.90	294
0300	Birch, 6'-8', 3 stems	G	20	1.600		123	53.50	31	207.50	257
0500	Crabapple, 6'-8'	G	20	1.600		160	53.50	31	244.50	298
0600	Dogwood, 4'-5'	G	40	.800		68.50	27	15.50	111	136
0700	Eastern redbud 4'-5'	G	40	.800		133	27	15.50	175.50	208
0800	Elm, 8'-10'	G	20	1.600		119	53.50	31	203.50	253
0900	Ginkgo, 6'-7'	G	24	1.333		163	44.50	26	233.50	280
1000	Hawthorn, 8'-10', 1" caliper	G	20	1.600		136	53.50	31	220.50	272
1100	Honeylocust, 10'-12', 1-1/2" caliper	G	10	3.200		149	107	62	318	405
1300	Larch, 8'	G	32	1		95	33.50	19.40	147.90	181
1400	Linden, 8'-10', 1" caliper	G	20	1.600		107	53.50	31	191.50	240
1500	Magnolia, 4'-5'	G	20	1.600		79.50	53.50	31	164	209
1600	Maple, red, 8'-10', 1-1/2" caliper	G	10	3.200		174	107	62	343	435
1700	Mountain ash, 8'-10', 1" caliper	G	16	2		181	67	39	287	350
1800	Oak, 2-1/2"-3" caliper	G	6	5.333		258	179	103	540	690
2100	Planetree, 9'-11', 1-1/4" caliper	G	10	3.200		105	107	62	274	360
2200	Plum, 6'-8', 1" caliper	G	20	1.600		91	53.50	31	175.50	222
2300	Poplar, 9'-11', 1-1/4" caliper	G	10	3.200		50	107	62	219	298
2500	Sumac, 2'-3'	G	75	.427		23	14.30	8.25	45.55	57.50
2700	Tulip, 5'-6'	G	40	.800		49	27	15.50	91.50	114
2800	Willow, 6'-8', 1" caliper	G	20	1.600		64	53.50	31	148.50	192
9000	Minimum labor/equipment charge	1 Clab	4	2	Job		63		63	104

32 94 Planting Accessories

32 94 13 – Landscape Edging

32 94 13.20 Edging

32 94 13.20 Edging		Crew	Daily Output	Labor-Hours	Unit	Material	2009 Bare Costs Labor	Equipment	Total	Total Incl O&P
0010	**EDGING**									
0050	Aluminum alloy, including stakes, 1/8" x 4", mill finish	B-1	390	.062	L.F.	3.10	1.99		5.09	6.70
0051	Black paint		390	.062		3.60	1.99		5.59	7.25
0052	Black anodized		390	.062		4.15	1.99		6.14	7.85
0100	Brick, set horizontally, 1-1/2 bricks per L.F.	D-1	370	.043		1.67	1.57		3.24	4.38
0150	Set vertically, 3 bricks per L.F.	"	135	.119		4.20	4.31		8.51	11.55
0200	Corrugated aluminum, roll, 4" wide	1 Carp	650	.012		.70	.49		1.19	1.58
0250	6" wide	"	550	.015		.88	.58		1.46	1.92
0600	Railroad ties, 6" x 8"	2 Carp	170	.094		3.10	3.76		6.86	9.60
0650	7" x 9"		136	.118		3.44	4.70		8.14	11.55
0750	Redwood 2" x 4"		330	.048		2.27	1.94		4.21	5.70
0800	Steel edge strips, incl. stakes, 1/4" x 5"	B-1	390	.062		4.25	1.99		6.24	7.95
0850	3/16" x 4"	"	390	.062		3.36	1.99		5.35	6.95
9000	Minimum labor/equipment charge	1 Carp	4	2	Job		80		80	132

32 94 50 – Tree Guying

32 94 50.10 Tree Guying Systems

32 94 50.10 Tree Guying Systems		Crew	Daily Output	Labor-Hours	Unit	Material	2009 Bare Costs Labor	Equipment	Total	Total Incl O&P
0010	**TREE GUYING SYSTEMS**									
0015	Tree guying Including stakes, guy wire and wrap									
0100	Less than 3" caliper, 2 stakes	2 Clab	35	.457	Ea.	17.50	14.45		31.95	43.50
0200	3" to 4" caliper, 3 stakes	"	21	.762	"	20	24		44	61.50
1000	Including arrowhead anchor, cable, turnbuckles and wrap									

32 94 Planting Accessories

32 94 50 – Tree Guying

32 94 50.10 Tree Guying Systems	Crew	Daily Output	Labor-Hours	Unit	Material	2009 Bare Costs Labor	Equipment	Total	Total Incl O&P
1100	Less than 3" caliper, 3" anchors	2 Clab	20	.800	Ea.	51	25.50		76.50	97.50
1200	3" to 6" caliper, 4" anchors		15	1.067		51	33.50		84.50	112
1300	6" caliper, 6" anchors		12	1.333		95	42		137	175
1400	8" caliper, 8" anchors		9	1.778		107	56		163	210

32 96 Transplanting

32 96 23 – Plant and Bulb Transplanting

32 96 23.23 Planting

		Crew	Daily Output	Labor-Hours	Unit	Material	2009 Bare Costs Labor	Equipment	Total	Total Incl O&P
0010	**PLANTING**									
0012	Moving shrubs on site, 12" ball	B-62	28	.857	Ea.	29	5.35		34.35	53.50
0100	24" ball	"	22	1.091	"	37	6.80		43.80	68

32 96 23.43 Moving Trees

		Crew	Daily Output	Labor-Hours	Unit	Material	2009 Bare Costs Labor	Equipment	Total	Total Incl O&P
0290	**MOVING TREES**, On site									
0300	Moving trees on site, 36" ball	B-6	3.75	6.400	Ea.	218	78.50		296.50	440
0400	60" ball	"	1	24	"	820	294		1,114	1,650

32 96 43 – Tree Transplanting

32 96 43.20 Tree Removal

		Crew	Daily Output	Labor-Hours	Unit	Material	2009 Bare Costs Labor	Equipment	Total	Total Incl O&P
0010	**TREE REMOVAL**									
0100	Dig & lace, shrubs, broadleaf evergreen, 18"-24" high	B-1	55	.436	Ea.		14.10		14.10	23
0200	2'-3'	"	35	.686			22		22	36.50
0300	3'-4'	B-6	30	.800			27.50	9.80	37.30	55
0400	4'-5'	"	20	1.200			41	14.70	55.70	82.50
1000	Deciduous, 12"-15"	B-1	110	.218			7.05		7.05	11.60
1100	18"-24"		65	.369			11.90		11.90	19.65
1200	2'-3'		55	.436			14.10		14.10	23
1300	3'-4'	B-6	50	.480			16.35	5.90	22.25	33
2000	Evergreen, 18"-24"	B-1	55	.436			14.10		14.10	23
2100	2'-0" to 2'-6"		50	.480			15.50		15.50	25.50
2200	2'-6" to 3'-0"		35	.686			22		22	36.50
2300	3'-0" to 3'-6"		20	1.200			38.50		38.50	64
3000	Trees, deciduous, small, 2'-3'		55	.436			14.10		14.10	23
3100	3'-4'	B-6	50	.480			16.35	5.90	22.25	33
3200	4'-5'		35	.686			23.50	8.40	31.90	47.50
3300	5'-6'		30	.800			27.50	9.80	37.30	55
4000	Shade, 5'-6'		50	.480			16.35	5.90	22.25	33
4100	6'-8'		35	.686			23.50	8.40	31.90	47.50
4200	8'-10'		25	.960			32.50	11.75	44.25	66
4300	2" caliper		12	2			68	24.50	92.50	138
5000	Evergreen, 4'-5'		35	.686			23.50	8.40	31.90	47.50
5100	5'-6'		25	.960			32.50	11.75	44.25	66
5200	6'-7'		19	1.263			43	15.45	58.45	87
5300	7'-8'		15	1.600			54.50	19.60	74.10	110
5400	8'-10'		11	2.182			74.50	26.50	101	151

Division Notes

	CREW	DAILY OUTPUT	LABOR-HOURS	UNIT	2009 BARE COSTS				TOTAL INCL O&P
					MAT.	LABOR	EQUIP.	TOTAL	

Estimating Tips

33 10 00 Water Utilities
33 30 00 Sanitary Sewerage Utilities
33 40 00 Storm Drainage Utilities

• Never assume that the water, sewer, and drainage lines will go in at the early stages of the project. Consider the site access needs before dividing the site in half with open trenches, loose pipe, and machinery obstructions. Always inspect the site to establish that the site drawings are complete. Check off all existing utilities on your drawings as you locate them. If you find any discrepancies, mark up the site plan for further research. Differing site conditions can be very costly if discovered later in the project.

• See also Section 33 01 00 for restoration of pipe where removal/replacement may be undesirable. Use of new types of piping materials can reduce the overall project cost. Owners/design engineers should consider the installing contractor as a valuable source of current information on utility products and local conditions that could lead to significant cost savings.

Reference Numbers

Reference numbers are shown in shaded boxes at the beginning of some major classifications. These numbers refer to related items in the Reference Section. The reference information may be an estimating procedure, an alternate pricing method, or technical information.

Note: Not all subdivisions listed here necessarily appear in this publication.

Note: **Trade Service,** *in part, has been used as a reference source for some of the material prices used in Division 33.*

33 01 30.72 Relining Sewers

		Crew	Daily Output	Labor-Hours	Unit	Material	2009 Bare Costs Labor	Equipment	Total	Total Incl O&P
0010	**RELINING SEWERS**									
0011	With cement inc bypass & cleaning									
0020	Less than 10,000 L.F., urban, 6" to 10"	C-17E	130	.615	L.F.	8.50	25.50	.66	34.66	51.50
0050	10" to 12"		125	.640		10.45	26.50	.68	37.63	55
0070	12" to 16"		115	.696		10.70	28.50	.74	39.94	59
0100	16" to 20"		95	.842		12.55	34.50	.90	47.95	71.50
0200	24" to 36"		90	.889		13.55	36.50	.95	51	75.50
0300	48" to 72"		80	1		21.50	41.50	1.07	64.07	91.50
0500	Rural, 6" to 10"		180	.444		8.50	18.35	.48	27.33	40
0550	10" to 12"		175	.457		10.45	18.85	.49	29.79	42.50
0570	12" to 16"		160	.500		10.90	20.50	.54	31.94	46
0600	16" to 20"		135	.593		11.50	24.50	.63	36.63	53.50
0700	24" to 36"		125	.640		13.75	26.50	.68	40.93	59
0800	48" to 72"		100	.800		21.50	33	.86	55.36	78
1000	Greater than 10,000 L.F., urban, 6" to 10"		160	.500		8.50	20.50	.54	29.54	43.50
1050	10" to 12"		155	.516		10.30	21.50	.55	32.35	46.50
1070	12" to 16"		140	.571		10.70	23.50	.61	34.81	51
1100	16" to 20"		120	.667		11.50	27.50	.71	39.71	58.50
1200	24" to 36"		115	.696		13.75	28.50	.74	42.99	62.50
1300	48" to 72"		95	.842		21.50	34.50	.90	56.90	81
1500	Rural, 6" to 10"		215	.372		8.50	15.35	.40	24.25	35
1550	10" to 12"		210	.381		10.45	15.70	.41	26.56	37.50
1570	12" to 16"		185	.432		10.70	17.85	.46	29.01	41.50
1600	16" to 20"		150	.533		11.50	22	.57	34.07	49.50
1700	24" to 36"		140	.571		13.75	23.50	.61	37.86	54.50
1800	48" to 72"		120	.667		21.50	27.50	.71	49.71	70

33 05 Common Work Results for Utilities

33 05 16 – Utility Structures

33 05 16.13 Utility Boxes

		Crew	Daily Output	Labor-Hours	Unit	Material	2009 Bare Costs Labor	Equipment	Total	Total Incl O&P
0010	**UTILITY BOXES** Precast concrete, 6" thick									
0050	5' x 10' x 6' high, I.D.	B-13	2	28	Ea.	2,225	955	395	3,575	4,425
0350	Hand hole, precast concrete, 1-1/2" thick									
0400	1'-0" x 2'-0" x 1'-9", I.D., light duty	B-1	4	6	Ea.	380	194		574	740
0450	4'-6" x 3'-2" x 2'-0", O.D., heavy duty	B-6	3	8	"	1,175	273	98	1,546	1,825

33 05 23 – Trenchless Utility Installation

33 05 23.19 Microtunneling

		Crew	Daily Output	Labor-Hours	Unit	Material	2009 Bare Costs Labor	Equipment	Total	Total Incl O&P
0010	**MICROTUNNELING**									
0011	Not including excavation, backfill, shoring,									
0020	or dewatering, average 50'/day, slurry method									
0100	24" to 48" outside diameter, minimum				L.F.				800	880
0110	Adverse conditions, add				%				50%	50%
1000	Rent microtunneling machine, average monthly lease				Month				90,000	99,000
1010	Operating technician				Day				560	620
1100	Mobilization and demobilization, minimum				Job				40,000	44,000
1110	Maximum				"				400,000	440,000

33 05 23.20 Horizontal Boring

		Crew	Daily Output	Labor-Hours	Unit	Material	2009 Bare Costs Labor	Equipment	Total	Total Incl O&P
0010	**HORIZONTAL BORING**									
0011	Casing only, 100' minimum,									
0020	not incl. jacking pits or dewatering									

33 05 Common Work Results for Utilities

33 05 23 – Trenchless Utility Installation

33 05 23.20 Horizontal Boring	Crew	Daily Output	Labor-Hours	Unit	Material	2009 Bare Costs Labor	Equipment	Total	Total Incl O&P
0100	Roadwork, 1/2" thick wall, 24" diameter casing	B-42	20	3.200	L.F.	119	114	68.50	301.50	395
0200	36" diameter		16	4		189	142	85.50	416.50	540
0300	48" diameter		15	4.267		278	152	91	521	655
0500	Railroad work, 24" diameter		15	4.267		119	152	91	362	485
0600	36" diameter		14	4.571		189	162	97.50	448.50	585
0700	48" diameter	↓	12	5.333		278	189	114	581	745
0900	For ledge, add									20%
1000	Small diameter boring, 3", sandy soil	B-82	900	.018		29	.63	.09	29.72	32.50
1040	Rocky soil	"	500	.032	↓	29	1.13	.16	30.29	33.50

33 05 26 – Utility Line Signs, Markers, and Flags

33 05 26.10 Utility Accessories

		Crew	Daily Output	Labor-Hours	Unit	Material	Labor	Equipment	Total	Total Incl O&P
0010	**UTILITY ACCESSORIES** R312316-40									
0400	Underground tape, detectable, reinforced, alum. foil core, 2"	1 Clab	150	.053	C.L.F.	2.10	1.69		3.79	5.10
0500	6"		140	.057	"	5.25	1.81		7.06	8.80
9000	Minimum labor/equipment charge	↓	4	2	Job		63		63	104

33 11 Water Utility Distribution Piping

33 11 13 – Public Water Utility Distribution Piping

33 11 13.15 Water Supply, Ductile Iron Pipe

		Crew	Daily Output	Labor-Hours	Unit	Material	Labor	Equipment	Total	Total Incl O&P
0010	**WATER SUPPLY, DUCTILE IRON PIPE** R331113-80									
0020	Not including excavation or backfill									
2000	Pipe, class 50 water piping, 18' lengths									
2020	Mechanical joint, 4" diameter	B-21A	200	.200	L.F.	15.35	7.80	3.01	26.16	32.50
3000	Tyton, push-on joint, 4" diameter		400	.100		11.50	3.91	1.51	16.92	20.50
3020	6" diameter	↓	333.33	.120	↓	14.70	4.69	1.81	21.20	25.50
8000	Fittings, mechanical joint									
8006	90° bend, 4" diameter	B-20A	16	2	Ea.	213	76.50		289.50	355
8020	6" diameter		12.80	2.500		292	95.50		387.50	470
8200	Wye or tee, 4" diameter		10.67	2.999		325	115		440	540
8220	6" diameter		8.53	3.751		435	143		578	705
8398	45° bends, 4" diameter		16	2		167	76.50		243.50	305
8450	Decreaser, 6" x 4" diameter		14.22	2.250		179	86		265	335
8460	8" x 6" diameter	↓	11.64	2.749	↓	229	105		334	420
8550	Piping, butterfly valves, cast iron									
8560	4" diameter	B-20	6	4	Ea.	860	141		1,001	1,175
9600	Steel sleeve and tap, 4" diameter		3	8		610	283		893	1,125
9620	6" diameter	↓	2	12	↓	650	425		1,075	1,400

33 12 Water Utility Distribution Equipment

33 12 13 – Water Service Connections

33 12 13.15 Tapping, Crosses and Sleeves

		Crew	Daily Output	Labor-Hours	Unit	Material	Labor	Equipment	Total	Total Incl O&P
0010	**TAPPING, CROSSES AND SLEEVES**									
4000	Drill and tap pressurized main (labor only)									
4100	6" main, 1" to 2" service	Q-1	3	5.333	Ea.		234		234	365
4150	8" main, 1" to 2" service	"	2.75	5.818	"		255		255	395
4500	Tap and insert gate valve									
4600	8" main, 4" branch	B-21	3.20	8.750	Ea.		320	40.50	360.50	565
4650	6" branch		2.70	10.370			375	48	423	670
4800	12" main, 6" branch	↓	2.35	11.915	↓		435	55	490	765

33 21 Water Supply Wells

33 21 13 – Public Water Supply Wells

33 21 13.10 Wells and Accessories

		Crew	Daily Output	Labor-Hours	Unit	Material	2009 Bare Costs Labor	2009 Bare Costs Equipment	Total	Total Incl O&P
0010	**WELLS & ACCESSORIES** R221113-50									
0011	Domestic									
0100	Drilled, 4" to 6" diameter	B-23	120	.333	L.F.		10.65	23	33.65	43
0200	8" diameter	"	95.20	.420	"		13.45	29	42.45	54
1400	Remove & reset pump, minimum	B-21	4	7	Ea.		255	32.50	287.50	450
1420	Maximum	"	2	14	"		510	64.50	574.50	900
1500	Pumps, installed in wells to 100' deep, 4" submersible									
1510	1/2 H.P.	Q-1	3.22	4.969	Ea.	460	218		678	845
1520	3/4 H.P.		2.66	6.015		565	264		829	1,025
1600	1 H.P.		2.29	6.987		615	305		920	1,150
1700	1-1/2 H.P.	Q-22	1.60	10		800	440	480	1,720	2,100
1800	2 H.P.		1.33	12.030		1,200	530	580	2,310	2,750
1900	3 H.P.		1.14	14.035		1,975	615	675	3,265	3,875
2000	5 H.P.		1.14	14.035		2,775	615	675	4,065	4,750
2050	Remove and install motor only, 4 H.P.		1.14	14.035		2,375	615	675	3,665	4,300
5000	Wells to 180 ft. deep, 4" submersible, 1 HP	B-21	1.10	25.455		2,050	925	117	3,092	3,875
5500	2 HP		1.10	25.455		2,925	925	117	3,967	4,850
6000	3 HP		1	28		3,925	1,025	129	5,079	6,100
7000	5 HP		.90	31.111		5,525	1,125	143	6,793	8,075
9000	Minimum labor/equipment charge		1.80	15.556	Job		565	71.50	636.50	1,000

33 31 Sanitary Utility Sewerage Piping

33 31 13 – Public Sanitary Utility Sewerage Piping

33 31 13.15 Sewage Collection, Concrete Pipe

		Crew	Daily Output	Labor-Hours	Unit	Material	2009 Bare Costs Labor	2009 Bare Costs Equipment	Total	Total Incl O&P
0010	**SEWAGE COLLECTION, CONCRETE PIPE**									
0020	See Div. 33 41 13.60 for sewage/drainage collection, concrete pipe									

33 31 13.25 Sewage Collection, Polyvinyl Chloride Pipe

		Crew	Daily Output	Labor-Hours	Unit	Material	2009 Bare Costs Labor	2009 Bare Costs Equipment	Total	Total Incl O&P
0010	**SEWAGE COLLECTION, POLYVINYL CHLORIDE PIPE**									
0020	Not including excavation or backfill									
2000	20' lengths, S.D.R. 35, B&S, 4" diameter	B-20	375	.064	L.F.	1.66	2.26		3.92	5.55
2040	6" diameter		350	.069		3.40	2.42		5.82	7.70
2080	13' lengths, S.D.R. 35, B&S, 8" diameter		335	.072		7.05	2.53		9.58	11.90
2120	10" diameter	B-21	330	.085		11.15	3.09	.39	14.63	17.75
4000	Piping, DWV PVC, no exc/bkfill, 10' L, Sch 40, 4" diameter	B-20	375	.064		3.84	2.26		6.10	7.95
4010	6" diameter		350	.069		7.65	2.42		10.07	12.40
4020	8" diameter		335	.072		13.30	2.53		15.83	18.80

33 36 Utility Septic Tanks

33 36 13 – Utility Septic Tank and Effluent Wet Wells

33 36 13.10 Septic Tanks

		Crew	Daily Output	Labor-Hours	Unit	Material	2009 Bare Costs Labor	2009 Bare Costs Equipment	Total	Total Incl O&P
0010	**SEPTIC TANKS**									
0011	Not including excavation or piping									
0015	Septic tanks, not incl exc or piping, precast, 1,000 gal	B-21	8	3.500	Ea.	730	127	16.15	873.15	1,025
0020	1,250 gallon		8	3.500		950	127	16.15	1,093.15	1,275
0060	1,500 gallon		7	4		1,125	146	18.45	1,289.45	1,500
0100	2,000 gallon		5	5.600		1,625	204	26	1,855	2,150
0140	2,500 gallon		5	5.600		2,300	204	26	2,530	2,875
0180	4,000 gallon		4	7		5,675	255	32.50	5,962.50	6,675
0220	5,000 gal., 4 piece	B-13	3	18.667		9,075	640	263	9,978	11,300

33 36 Utility Septic Tanks

33 36 13 – Utility Septic Tank and Effluent Wet Wells

33 36 13.10 Septic Tanks

33 36 13.10 Septic Tanks		Crew	Daily Output	Labor-Hours	Unit	Material	2009 Bare Costs Labor	Equipment	Total	Total Incl O&P
0300	15,000 gallon, 4 piece	B-13B	1.70	32.941	Ea.	20,500	1,125	660	22,285	25,200
0400	25,000 gallon, 4 piece		1.10	50.909		43,400	1,750	1,025	46,175	52,000
0500	40,000 gallon, 4 piece	↓	.80	70		51,000	2,400	1,400	54,800	61,500
0520	50,000 gallon, 5 piece	B-13C	.60	93.333		59,000	3,200	2,675	64,875	72,500
0540	75,000 gallon, cast in place	C-14C	.25	448		71,500	17,100	103	88,703	107,000
0560	100,000 gallon	"	.15	746		88,500	28,600	172	117,272	145,000
0600	High density polyethylene, 1,000 gallon	B-21	6	4.667		1,125	170	21.50	1,316.50	1,525
0700	1,500 gallon		4	7		1,450	255	32.50	1,737.50	2,050
0900	Galley, 4' x 4' x 4'	↓	16	1.750		270	63.50	8.05	341.55	410
1000	Distribution boxes, concrete, 7 outlets	2 Clab	16	1		75	31.50		106.50	135
1100	9 outlets	"	8	2		440	63		503	590
1150	Leaching field chambers, 13' x 3'-7" x 1'-4", standard	B-13	16	3.500		510	120	49.50	679.50	810
1200	Heavy duty, 8' x 4' x 1'-6"		14	4		475	137	56.50	668.50	810
1300	13' x 3'-9" x 1'-6"		12	4.667		1,000	160	66	1,226	1,425
1350	20' x 4' x 1'-6"	↓	5	11.200		1,050	385	158	1,593	1,950
1420	Leaching pit, 6', dia, 3' deep complete					1,025			1,025	1,125
1600	Leaching pit, 6'-6" diameter, 6' deep	B-21	5	5.600		1,325	204	26	1,555	1,825
1620	8' deep		4	7		1,550	255	32.50	1,837.50	2,175
1700	8' diameter, H-20 load, 6' deep		4	7		2,200	255	32.50	2,487.50	2,875
1720	8' deep	↓	3	9.333		2,600	340	43	2,983	3,450
2000	Velocity reducing pit, precast conc., 6' diameter, 3' deep	↓	4.70	5.957	↓	1,475	217	27.50	1,719.50	2,000
2200	Excavation for septic tank, 3/4 C.Y. backhoe	B-12F	145	.110	C.Y.		4.09	4.13	8.22	11.10
2400	4' trench for disposal field, 3/4 C.Y. backhoe	"	335	.048	L.F.		1.77	1.79	3.56	4.81
2600	Gravel fill, run of bank	B-6	150	.160	C.Y.	25	5.45	1.96	32.41	38.50
2800	Crushed stone, 3/4"	"	150	.160	"	29	5.45	1.96	36.41	42.50

33 41 Storm Utility Drainage Piping

33 41 13 – Public Storm Utility Drainage Piping

33 41 13.40 Piping, Storm Drainage, Corrugated Metal

		Crew	Daily Output	Labor-Hours	Unit	Material	2009 Bare Costs Labor	Equipment	Total	Total Incl O&P
0010	**PIPING, STORM DRAINAGE, CORRUGATED METAL** R221113-50									
0020	Not including excavation or backfill									
2000	Corrugated metal pipe, galvanized									
2020	Bituminous coated with paved invert, 20' lengths									
2040	8" diameter, 16 ga.	B-14	330	.145	L.F.	12.65	4.83	.89	18.37	23
2080	12" diameter, 16 ga.	"	210	.229	"	18.20	7.60	1.40	27.20	34

33 41 13.60 Sewage/Drainage Collection, Concrete Pipe

		Crew	Daily Output	Labor-Hours	Unit	Material	2009 Bare Costs Labor	Equipment	Total	Total Incl O&P
0010	**SEWAGE/DRAINAGE COLLECTION, CONCRETE PIPE**									
0020	Not including excavation or backfill									
1000	Non-reinforced pipe, extra strength, B&S or T&G joints									
1010	6" diameter	B-14	265.04	.181	L.F.	6.05	6	1.11	13.16	17.70
1020	8" diameter		224	.214		6.70	7.10	1.31	15.11	20.50
1040	12" diameter	↓	200	.240	↓	9.10	7.95	1.47	18.52	24.50
2000	Reinforced culvert, class 3, no gaskets									
2010	12" diameter	B-14	150	.320	L.F.	10.20	10.60	1.96	22.76	30.50
2020	15" diameter	"	150	.320	"	14.95	10.60	1.96	27.51	36

33 44 Storm Utility Water Drains

33 44 13 – Utility Area Drains

33 44 13.13 Catchbasins

		Crew	Daily Output	Labor-Hours	Unit	Material	2009 Bare Costs Labor	Equipment	Total	Total Incl O&P
0010	**CATCHBASINS**									
0011	Not including footing & excavation									
1600	Frames & covers, C.I., 24" square, 500 lb.	B-6	7.80	3.077	Ea.	298	105	37.50	440.50	540
1700	26" D shape, 600 lb.	"	7	3.429	"	510	117	42	669	800
3320	Frames and covers, existing, raised for paving, 2", including									
3340	row of brick, concrete collar, up to 12" wide frame	B-6	18	1.333	Ea.	43.50	45.50	16.30	105.30	139

33 44 13.15 Remove and Replace Catch Basin Cover

		Crew	Daily Output	Labor-Hours	Unit	Material	2009 Bare Costs Labor	Equipment	Total	Total Incl O&P
0010	**REMOVE AND REPLACE CATCH BASIN COVER**									
0023	Remove catch basin cover	1 Clab	80	.100	Ea.		3.16		3.16	5.20
0033	Replace catch basin cover	"	80	.100	"		3.16		3.16	5.20

33 49 Storm Drainage Structures

33 49 13 – Storm Drainage Manholes, Frames, and Covers

33 49 13.10 Storm Drainage Manholes, Frames and Covers

		Crew	Daily Output	Labor-Hours	Unit	Material	2009 Bare Costs Labor	Equipment	Total	Total Incl O&P
0010	**STORM DRAINAGE MANHOLES, FRAMES & COVERS**									
0020	Excludes footing, excavation, backfill (See line items for frame & cover)									
0050	Brick, 4' inside diameter, 4' deep	D-1	1	16	Ea.	400	580		980	1,375
0100	6' deep		.70	22.857		555	830		1,385	1,950
0150	8' deep		.50	32		710	1,175		1,885	2,650
0200	For depths over 8', add		4	4	V.L.F.	162	145		307	415
1110	Precast, 4' I.D., 4' deep	B-22	4.10	7.317	Ea.	880	269	47.50	1,196.50	1,450
1120	6' deep		3	10		1,100	370	64.50	1,534.50	1,900
1130	8' deep		2	15		1,325	550	97	1,972	2,450
1140	For depths over 8', add		16	1.875	V.L.F.	181	69	12.10	262.10	325

33 51 Natural-Gas Distribution

33 51 13 – Natural-Gas Piping

33 51 13.10 Piping, Gas Service and Distribution, P.E.

		Crew	Daily Output	Labor-Hours	Unit	Material	2009 Bare Costs Labor	Equipment	Total	Total Incl O&P
0010	**PIPING, GAS SERVICE AND DISTRIBUTION, POLYETHYLENE**									
0020	Not including excavation or backfill									
1000	60 psi coils, compression coupling @ 100', 1/2" diameter, SDR 11	B-20A	608	.053	L.F.	1.61	2.01		3.62	4.98
1040	1-1/4" diameter, SDR 11		544	.059		2.49	2.25		4.74	6.30
1100	2" diameter, SDR 11		488	.066		3.09	2.51		5.60	7.40
1160	3" diameter, SDR 11		408	.078		6.45	3		9.45	11.90
1500	60 PSI 40' joints with coupling, 3" diameter, SDR 11	B-21A	408	.098		6.45	3.83	1.48	11.76	14.80
1540	4" diameter, SDR 11		352	.114		14.75	4.44	1.71	20.90	25
1600	6" diameter, SDR 11		328	.122		46	4.77	1.84	52.61	60
1640	8" diameter, SDR 11		272	.147		62.50	5.75	2.21	70.46	80.50
9000	Minimum labor/equipment charge	B-20	2	12	Job		425		425	695

33 52 16.13 Gasoline Piping	Crew	Daily Output	Labor-Hours	Unit	Material	2009 Bare Costs Labor	Equipment	Total	Total Incl O&P
0010 **GASOLINE PIPING**									
0020 Primary containment pipe, fiberglass-reinforced									
0030 Plastic pipe 15' & 30' lengths									
0040 2" diameter	Q-6	425	.056	L.F.	4.94	2.60		7.54	9.50
0050 3" diameter		400	.060		6.45	2.76		9.21	11.40
0060 4" diameter		375	.064		8.35	2.95		11.30	13.70
0100 Fittings									
0110 Elbows, 90° & 45°, bell-ends, 2"	Q-6	24	1	Ea.	50	46		96	127
0120 3" diameter		22	1.091		52.50	50.50		103	136
0130 4" diameter		20	1.200		69.50	55.50		125	162
0200 Tees, bell ends, 2"		21	1.143		60.50	52.50		113	148
0210 3" diameter		18	1.333		61	61.50		122.50	163
0220 4" diameter		15	1.600		83.50	73.50		157	206
0230 Flanges bell ends, 2"		24	1		20	46		66	93.50
0240 3" diameter		22	1.091		25	50.50		75.50	106
0250 4" diameter		20	1.200		34.50	55.50		90	124
0260 Sleeve couplings, 2"		21	1.143		12.75	52.50		65.25	95.50
0270 3" diameter		18	1.333		18.05	61.50		79.55	115
0280 4" diameter		15	1.600		25	73.50		98.50	142
0290 Threaded adapters 2"		21	1.143		16.75	52.50		69.25	100
0300 3" diameter		18	1.333		29.50	61.50		91	128
0310 4" diameter		15	1.600		40	73.50		113.50	158
0320 Reducers, 2"		27	.889		22.50	41		63.50	88
0330 3" diameter		22	1.091		26	50.50		76.50	107
0340 4" diameter		20	1.200		33.50	55.50		89	122
1010 Gas station product line for secondary containment (double wall)									
1100 Fiberglass reinforced plastic pipe 25' lengths									
1120 Pipe, plain end, 3" diameter	Q-6	375	.064	L.F.	8.50	2.95		11.45	13.90
1130 4" diameter		350	.069		13.90	3.16		17.06	20
1140 5" diameter		325	.074		18.05	3.40		21.45	25
1150 6" diameter		300	.080		18.60	3.69		22.29	26
1200 Fittings									
1230 Elbows, 90° & 45°, 3" diameter	Q-6	18	1.333	Ea.	61	61.50		122.50	163
1240 4" diameter		16	1.500		106	69		175	224
1250 5" diameter		14	1.714		247	79		326	395
1260 6" diameter		12	2		250	92		342	420
1270 Tees, 3" diameter		15	1.600		90.50	73.50		164	214
1280 4" diameter		12	2		133	92		225	289
1290 5" diameter		9	2.667		269	123		392	485
1300 6" diameter		6	4		281	184		465	595
1310 Couplings, 3" diameter		18	1.333		43	61.50		104.50	143
1320 4" diameter		16	1.500		111	69		180	229
1330 5" diameter		14	1.714		230	79		309	375
1340 6" diameter		12	2		238	92		330	405
1350 Cross-over nipples, 3" diameter		18	1.333		9.80	61.50		71.30	106
1360 4" diameter		16	1.500		11.50	69		80.50	120
1370 5" diameter		14	1.714		17.05	79		96.05	141
1380 6" diameter		12	2		17.85	92		109.85	163
1400 Telescoping, reducers, concentric 4" x 3"		18	1.333		32.50	61.50		94	132
1410 5" x 4"		17	1.412		85	65		150	195
1420 6" x 5"		16	1.500		204	69		273	330

33 71 Electrical Utility Transmission and Distribution

33 71 19 – Electrical Underground Ducts and Manholes

33 71 19.17 Electric and Telephone Underground	Crew	Daily Output	Labor-Hours	Unit	Material	2009 Bare Costs Labor	Equipment	Total	Total Incl O&P
0010 **ELECTRIC AND TELEPHONE UNDERGROUND**									
0011 Not including excavation									
0200 backfill and cast in place concrete									
4200 Underground duct, banks ready for concrete fill, min. of 7.5"									
4400 between conduits, center to center									
4600 2 @ 2" diameter	2 Elec	240	.067	L.F.	1.74	3.13		4.87	6.75
4800 4 @ 2" diameter		120	.133		3.47	6.25		9.72	13.45
5600 4 @ 4" diameter		80	.200		7.20	9.40		16.60	22.50
6200 Rigid galvanized steel, 2 @ 2" diameter		180	.089		16.60	4.18		20.78	24.50
6400 4 @ 2" diameter		90	.178		33	8.35		41.35	49.50
7400 4 @ 4" diameter		34	.471		112	22		134	157
9990 Minimum labor/equipment charge	1 Elec	3.50	2.286	Job		107		107	165

Estimating Tips

34 11 00 Rail Tracks

This subdivision includes items that may involve either repair of existing, or construction of new, railroad tracks. Additional preparation work, such as the roadbed earthwork, would be found in Division 31. Additional new construction siding and turnouts are found in Subdivision 34 72. Maintenance of railroads is found under 34 01 23 Operation and Maintenance of Railways.

34 41 13 Traffic Signals

This subdivision includes traffic signal systems. Other traffic control devices such as traffic signs are found in Subdivision 10 14 53 Traffic Signage.

34 71 13 Vehicle Barriers

This subdivision includes security vehicle barriers, guide and guard rails, crash barriers, and delineators. The actual maintenance and construction of concrete and asphalt pavement is found in Division 32.

Reference Numbers

Reference numbers are shown in shaded boxes at the beginning of some major classifications. These numbers refer to related items in the Reference Section. The reference information may be an estimating procedure, an alternate pricing method, or technical information.

Note: Not all subdivisions listed here necessarily appear in this publication.

Division 34 - Transportation

34 71 13.26 Vehicle Guide Rails	Crew	Daily Output	Labor-Hours	Unit	Material	2009 Bare Costs Labor	Equipment	Total	Total Incl O&P
0010 **VEHICLE GUIDE RAILS**									
0012 Corrugated stl, galv. stl posts, 6'-3" O.C.	B-80	850	.038	L.F.	22.50	1.27	.77	24.54	27.50
0200 End sections, galvanized, flared		50	.640	Ea.	93	21.50	13.10	127.60	151
0300 Wrap around end		50	.640	"	139	21.50	13.10	173.60	202
0400 Timber guide rail, 4" x 8" with 6" x 8" wood posts, treated	↓	960	.033	L.F.	26	1.13	.68	27.81	31

Estimating Tips

Products such as conveyors, material handling cranes and hoists, as well as other items specified in this division, may require trained installers. The general contractor may not have any choice as to who will perform the installation or when it will be performed. Long lead times are often required for these products, making early decisions in purchasing and scheduling necessary.

The installation of this type of equipment may require the embedment of mounting hardware during construction of floors, structural walls, or interior walls/partitions. Electrical connections will require coordination with the electrical contractor.

Reference Numbers

Reference numbers are shown in shaded boxes at the beginning of some major classifications. These numbers refer to related items in the Reference Section. The reference information may be an estimating procedure, an alternate pricing method, or technical information.

Note: Not all subdivisions listed here necessarily appear in this publication.

41 22 13.10 Crane Rail		Crew	Daily Output	Labor-Hours	Unit	Material	2009 Bare Costs Labor	Equipment	Total	Total Incl O&P
0010	**CRANE RAIL**									
0020	Box beam bridge, no equipment included	E-4	3400	.009	Lb.	1.19	.43	.04	1.66	2.14
0210	Running track only, 104 lb per yard, 20' piece	"	160	.200	L.F.	21.50	9.05	.84	31.39	41.50

41 22 13.13 Bridge Cranes		Crew	Daily Output	Labor-Hours	Unit	Material	2009 Bare Costs Labor	Equipment	Total	Total Incl O&P
0010	**BRIDGE CRANES**									
0100	1 girder, 20' span, 3 ton	M-3	1	34	Ea.	23,800	1,550	175	25,525	28,800
0125	5 ton		1	34		26,000	1,550	175	27,725	31,200
0150	7.5 ton		1	34		31,000	1,550	175	32,725	36,700
0175	10 ton		.80	42.500		41,000	1,925	219	43,144	48,200
0200	15 ton		.80	42.500		52,500	1,925	219	54,644	61,000
0225	30' span, 3 ton		1	34		24,700	1,550	175	26,425	29,800
0250	5 ton		1	34		27,200	1,550	175	28,925	32,500
0275	7.5 ton		1	34		32,600	1,550	175	34,325	38,400
0300	10 ton		.80	42.500		42,400	1,925	219	44,544	49,800
0325	15 ton		.80	42.500		55,000	1,925	219	57,144	63,500
0350	2 girder, 40' span, 3 ton	M-4	.50	72		40,900	3,225	470	44,595	50,500
0375	5 ton		.50	72		42,800	3,225	470	46,495	52,500
0400	7.5 ton		.50	72		46,900	3,225	470	50,595	57,000
0425	10 ton		.40	90		55,000	4,050	585	59,635	67,500
0450	15 ton		.40	90		74,500	4,050	585	79,135	89,000
0475	25 ton		.30	120		88,000	5,400	780	94,180	106,500
0500	50' span, 3 ton		.50	72		46,500	3,225	470	50,195	56,500
0525	5 ton		.50	72		48,400	3,225	470	52,095	58,500
0550	7.5 ton		.50	72		52,000	3,225	470	55,695	62,500
0575	10 ton		.40	90		59,500	4,050	585	64,135	72,500
0600	15 ton		.40	90		78,000	4,050	585	82,635	93,000
0625	25 ton		.30	120		92,500	5,400	780	98,680	111,000

Assemblies Section

Table of Contents

How to Use the Assemblies Cost Tables

The following is a detailed explanation of a sample Assemblies Cost Table. Most Assembly Tables are separated into three parts: 1) an illustration of the system to be estimated; 2) the components and related costs of a typical system; and 3) the costs for similar systems with dimensional and/or size variations. For costs of the components that comprise these systems or "assemblies," refer to the Unit Price Section. Next to each bold number below is the item being described with the appropriate component of the sample entry following in parenthesis. In most cases, if the work is to be subcontracted, the general contractor will need to add an additional markup (RSMeans suggests using 10%) to the "Total" figures.

1

System/Line Numbers (B1010 263 1700)

Each Assemblies Cost Line has been assigned a unique identification number based on the UNIFORMAT II classification system.

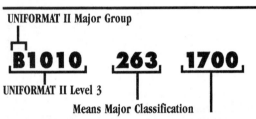

UNIFORMAT II Major Group

B1010 **263** **1700**

UNIFORMAT II Level 3

Means Major Classification

Means Individual Line Number

B10 Superstructure

B1010 Floor Construction

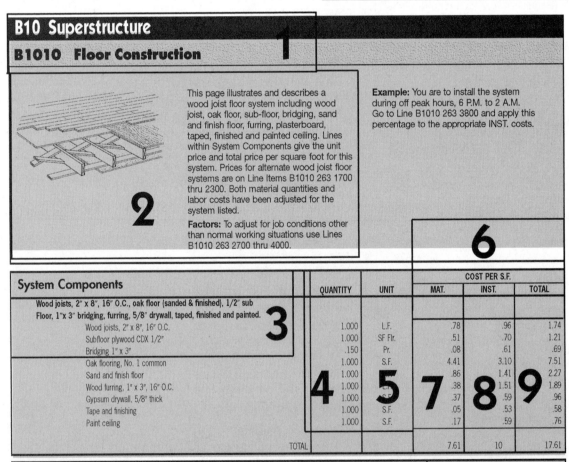

This page illustrates and describes a wood joist floor system including wood joist, oak floor, sub-floor, bridging, sand and finish floor, furring, plasterboard, taped, finished and painted ceiling. Lines within System Components give the unit price and total price per square foot for this system. Prices for alternate wood joist floor systems are on Line Items B1010 263 1700 thru 2300. Both material quantities and labor costs have been adjusted for the system listed.

Factors: To adjust for job conditions other than normal working situations use Lines B1010 263 2700 thru 4000.

Example: You are to install the system during off peak hours, 6 P.M. to 2 A.M. Go to Line B1010 263 3800 and apply this percentage to the appropriate INST. costs.

System Components	QUANTITY	UNIT	COST PER S.F. MAT.	COST PER S.F. INST.	COST PER S.F. TOTAL
Wood joists, 2" x 8", 16" O.C., oak floor (sanded & finished), 1/2" sub Floor, 1"x 3" bridging, furring, 5/8" drywall, taped, finished and painted.					
Wood joists, 2" x 8", 16" O.C.	1.000	L.F.	.78	.96	1.74
Subfloor plywood CDX 1/2"	1.000	SF Flr.	.51	.70	1.21
Bridging 1" x 3"	.150	Pr.	.08	.61	.69
Oak flooring, No. 1 common	1.000	S.F.	4.41	3.10	7.51
Sand and finish floor	1.000		.86	1.41	2.27
Wood furring, 1" x 3", 16" O.C.	1.000		.38	1.51	1.89
Gypsum drywall, 5/8" thick	1.000	S.F.	.37	.59	.96
Tape and finishing	1.000	S.F.	.05	.53	.58
Paint ceiling	1.000	S.F.	.17	.59	.76
TOTAL			7.61	10	17.61

B1010 263	Floor-Ceiling, Wood Joists	COST PER S.F. MAT.	COST PER S.F. INST.	COST PER S.F. TOTAL
1600	For alternate wood joist systems:			
1700	16" on center, 2" x 6" joists	7.40	9.90	17.30
1800	2" x 10"	7.90	10.20	18.10
1900	2" x 12"	8.35	10.25	18.60
2000	2" x 14"	9.20	10.40	19.60
2100	12" on center, 2" x 10" joists	8.20	10.50	18.70
2200	2" x 12"	8.75	10.55	19.30
2300	2" x 14"	9.75	10.75	20.50

2 Illustration

At the top of most assembly tables are an illustration, a brief description, and the design criteria used to develop the cost.

3 System Components

The components of a typical system are listed separately to show what has been included in the development of the total system price. The table below contains prices for other similar systems with dimensional and/or size variations.

4 Quantity

This is the number of line item units required for one system unit. For example, we assume that it will take .15 pair of 1″ × 3″ bridging on a square foot basis.

5 Unit of Measure for Each Item

The abbreviated designation indicates the unit of measure, as defined by industry standards, upon which the price of the component is based. For example, wood joists are priced by L.F. (linear foot) while plywood is priced by S.F. (square foot).

6 Unit of Measure for Each System (Cost per S.F.)

Costs shown in the three right-hand columns have been adjusted by the component quantity and unit of measure for the entire system. In this example, "Cost per S.F." is the unit of measure for this system, or assembly.

7 Materials (7.40)

This column contains the Materials Cost of each component. These cost figures are bare costs plus 10% for profit.

8 Installation (9.90)

Installation includes labor and equipment plus the installing contractor's overhead and profit. Equipment costs are the bare rental costs plus 10% for profit. The labor overhead and profit are defined on the inside back cover of this book.

9 Total (17.30)

The figure in this column is the sum of the material and installation costs.

Material Cost	+	Installation Cost	=	Total
$7.40	+	$9.90	=	$17.30

A1010 Standard Foundations

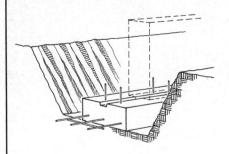

This page illustrates and describes a strip footing system including concrete, forms, reinforcing, keyway and dowels. Lines within System Components give the unit price and total price per linear foot for this system. Prices for alternate strip footing systems are on Line Items A1010 120 1500 thru 2500. Both material quantities and labor costs have been adjusted for the system listed.

Factors: To adjust for job conditions other than normal working situations use Lines A1010 120 2900 thru 4000.

Example: You are to install this footing, and due to a lack of accessibility, only hand tools can be used. Material handling is also a problem. Go to Lines A1010 120 3400 and 3600 and apply these percentages to the appropriate MAT. and INST. costs.

System Components	QUANTITY	UNIT	COST PER L.F. MAT.	COST PER L.F. INST.	COST PER L.F. TOTAL
Strip footing, 2'-0" wide x 1'-0" thick, 2000 psi concrete including forms					
Reinforcing, keyway, and dowels.					
Concrete, 2000 psi	.074	C.Y.	7.92		7.92
Placing concrete	.074	C.Y.		1.63	1.63
Forms, footing, 4 uses	2.000	S.F.	5.34	8.24	13.58
Reinforcing	3.170	Lb.	2.82	1.84	4.66
Keyway, 2" x 4", 4 uses	1.000	L.F.	.17	1	1.17
Dowels, #4 bars, 2' long, 24" O.C.	.500	Ea.	.63	1.27	1.90
TOTAL			16.88	13.98	30.86

A1010 120	Strip Footing	COST PER L.F. MAT.	COST PER L.F. INST.	COST PER L.F. TOTAL
1400	Above system with the following:			
1500	2'-0" wide x 1' thick, 3000 psi concrete	17.15	14	31.15
1600	4000 psi concrete	17.55	14	31.55
1800	For alternate footing systems:			
1900	2'-6" wide x 1' thick, 2000 psi concrete	19.65	14.90	34.55
2000	3000 psi concrete	20	14.90	34.90
2100	4000 psi concrete	20.50	14.90	35.40
2300	3'-0" wide x 1' thick, 2000 psi concrete	22	15.70	37.70
2400	3000 psi concrete	22.50	15.70	38.20
2500	4000 psi concrete	23	15.70	38.70
2700				
2800				
2900	Cut & patch to match existing construction, add, minimum	2%	3%	
3000	Maximum	5%	9%	
3100	Dust protection, add, minimum	1%	2%	
3200	Maximum	4%	11%	
3300	Equipment usage curtailment, add, minimum	1%	1%	
3400	Maximum	3%	10%	
3500	Material handling & storage limitation, add, minimum	1%	1%	
3600	Maximum	6%	7%	
3700	Protection of existing work, add, minimum	2%	2%	
3800	Maximum	5%	7%	
3900	Shift work requirements, add, minimum		5%	
4000	Maximum		30%	

A10 Foundations

A1010 Standard Foundations

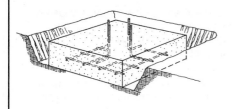

This page illustrates and describes a spread footing system including concrete, forms, reinforcing and anchor bolts. Lines within System Components give the unit price and total price on a cost each basis for this system. Prices for alternate spread footing systems are on Line Items A1010 230 1300 thru 2300. Both material quantities and labor costs have been adjusted for the system listed.

Factors: To adjust for job conditions other than normal working situations use Lines A1010 230 2900 thru 4000.

Example: You are to install the system in an existing occupied building. Access to the site and protection to the building are mandatory. Go to Lines A1010 230 3600 and 3800 and apply these percentages to the appropriate MAT. and INST. costs.

System Components	QUANTITY	UNIT	COST EACH MAT.	COST EACH INST.	COST EACH TOTAL
Interior column footing, 3' square 1' thick, 2000 psi concrete, Including forms, reinforcing, and anchor bolts.					
Concrete, 2000 psi	.330	C.Y.	35.31		35.31
Placing concrete	.330	C.Y.		15.85	15.85
Forms, footing, 4 uses	12.000	SFCA	9.24	57.96	67.20
Reinforcing	11.000	Lb.	9.79	6.38	16.17
Anchor bolts, 3/4" diameter	2.000	Ea.	1.82	5.24	7.06
TOTAL			56.16	85.43	141.59

A1010 230	Spread Footing	COST EACH MAT.	COST EACH INST.	COST EACH TOTAL
1200	Above system with the following:			
1300	3' square x 1' thick, 3000 psi concrete	57.50	85.50	143
1400	4000 psi concrete	59	85.50	144.50
1600	For alternate footing systems:			
1700	4' square x 1' thick, 2000 psi concrete	94.50	123	217.50
1800	3000 psi concrete	97	123	220
1900	4000 psi concrete	100	123	223
2100	5' square x 1'-3" thick, 2000 psi concrete	179	204	383
2200	3000 psi concrete	184	204	388
2300	4000 psi concrete	189	204	393
2600				
2700				
2900	Cut & patch to match existing construction, add, minimum	2%	3%	
3000	Maximum	5%	9%	
3100	Dust protection, add, minimum	1%	2%	
3200	Maximum	4%	11%	
3300	Equipment usage curtailment, add, minimum	1%	1%	
3400	Maximum	3%	10%	
3500	Material handling & storage limitation, add, minimum	1%	1%	
3600	Maximum	6%	7%	
3700	Protection of existing work, add, minimum	2%	2%	
3800	Maximum	5%	7%	
3900	Shift work requirements, add, minimum		5%	
4000	Maximum		30%	

A1010 Standard Foundations

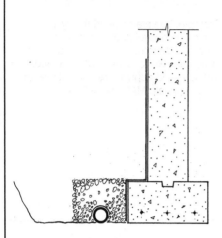

General: Footing drains can be placed either inside or outside of foundation walls depending upon the source of water to be intercepted. If the source of subsurface water is principally from grade or a subsurface stream above the bottom of the footing, outside drains should be used. For high water tables, use inside drains or both inside and outside.

The effectiveness of underdrains depends on good waterproofing. This must be carefully installed and protected during construction.

Costs below include the labor and materials for the pipe and 6″ only of gravel or crushed stone around pipe. Excavation and backfill are not included.

System Components	QUANTITY	UNIT	COST PER L.F.		
			MAT.	INST.	TOTAL
Foundation underdrain, outside, PVC 4″ diameter.					
PVC pipe 4″ diam. S.D.R. 35	1.000	L.F.	1.83	3.71	5.54
Pipe bedding, graded gravel 3/4″ to 1/2″	.070	C.Y.	3.36	.77	4.13
TOTAL			5.19	4.48	9.67

A1010 310	Foundation Underdrain	COST PER L.F.		
		MAT.	INST.	TOTAL
0900	For alternate drain systems:			
1000	Foundation underdrain, outside only, PVC, 4″ diameter	5.20	4.48	9.68
1100	6″ diameter	8.05	4.97	13.02
1400	Porous concrete, 6″ diameter	9.05	5.40	14.45
1450	8″ diameter	11.15	7	18.15
1500	12″ diameter	20	8.10	28.10
1600	Corrugated metal, 16 ga. asphalt coated, 6″ diameter	9.45	8.70	18.15
1650	8″ diameter	13.15	9.10	22.25
1700	10″ diameter	16.05	9.60	25.65
3000	Outside and inside, PVC, 4″ diameter	10.40	8.95	19.35
3100	6″ diameter	16.10	9.95	26.05
3400	Porous concrete, 6″ diameter	18.10	10.85	28.95
3450	8″ diameter	22.50	14.05	36.55
3500	12″ diameter	40	16.25	56.25
3600	Corrugated metal, 16 ga., asphalt coated, 6″ diameter	18.95	17.40	36.35
3650	8″ diameter	26.50	18.25	44.75
3700	10″ diameter	32	19.15	51.15
4700	Cut & patch to match existing construction, add, minimum	2%	3%	
4800	Maximum	5%	9%	
4900	Dust protection, add, minimum	1%	2%	
5000	Maximum	4%	11%	
5100	Equipment usage curtailment, add, minimum	1%	1%	
5200	Maximum	3%	10%	
5300	Material handling & storage limitation, add, minimum	1%	1%	
5400	Maximum	6%	7%	
5500	Protection of existing work, add, minimum	2%	2%	
5600	Maximum	5%	7%	
5700	Shift work requirements, add, minimum		5%	
5800	Maximum		30%	
5900	Temporary shoring and bracing, add, minimum	2%	5%	
6000	Maximum	5%	12%	

A10 Foundations

A1030 Slab on Grade

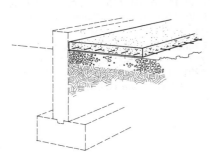

This page illustrates and describes a slab on grade system including slab, bank run gravel, bulkhead forms, placing concrete, welded wire fabric, vapor barrier, steel trowel finish and curing paper. Lines within System Components give the unit price and total price per square foot for this system. Prices for alternate slab on grade systems are on Line Items A1030 110 1500 thru 2600. Both material quantities and labor costs have been adjusted for the system listed.

Factors: To adjust for job conditions other than normal working situations use Lines A1030 110 2900 thru 4000.

Example: You are to install the system at a site where protection of the existing building is required. Go to Line A1030 110 3800 and apply these percentages to the appropriate MAT. and INST. costs.

System Components	QUANTITY	UNIT	COST PER S.F. MAT.	COST PER S.F. INST.	COST PER S.F. TOTAL
Ground slab, 4″ thick, 3000 psi concrete, 4″ granular base, vapor barrier					
Welded wire fabric, screed and steel trowel finish.					
Concrete, 4″ thick, 3000 psi concrete	.012	C.Y.	1.33		1.33
Bank run gravel, 4″ deep	.074	C.Y.	.36	.08	.44
Polyethylene vapor barrier, 10 mil.	.011	C.S.F.	.08	.16	.24
Bulkhead forms, expansion material	.100	L.F.	.04	.33	.37
Welded wire fabric, 6 x 6 - #10/10	.011	C.S.F.	.22	.38	.60
Place concrete	.012	C.Y.		.29	.29
Screed & steel trowel finish	1.000	S.F.		.83	.83
TOTAL			2.03	2.07	4.10

A1030 110	Interior Slab on Grade	COST PER S.F. MAT.	COST PER S.F. INST.	COST PER S.F. TOTAL
1400	Above system with the following:			
1500	4″ thick slab, 3000 psi concrete, 6″ deep bank run gravel	2.31	2.07	4.38
1600	12″ deep bank run gravel	2.96	2.09	5.05
1700				
1800				
1900				
2000	For alternate slab systems:			
2100	5″ thick slab, 3000 psi concrete, 6″ deep bank run gravel	2.65	2.14	4.79
2200	12″ deep bank run gravel	3.30	2.16	5.46
2300				
2400				
2500	6″ thick slab, 3000 psi concrete, 6″ deep bank run gravel	3.09	2.24	5.33
2600	12″ deep bank run gravel	3.74	2.26	6
2700				
2900	Cut & patch to match existing construction, add, minimum	2%	3%	
3000	Maximum	5%	9%	
3100	Dust protection, add, minimum	1%	2%	
3200	Maximum	4%	11%	
3300	Equipment usage curtailment, add, minimum	1%	1%	
3400	Maximum	3%	10%	
3500	Material handling & storage limitation, add, minimum	1%	1%	
3600	Maximum	6%	7%	
3700	Protection of existing work, add, minimum	2%	2%	
3800	Maximum	5%	7%	
3900	Shift work requirements, add, minimum		5%	
4000	Maximum		30%	

A20 Basement Construction

A2010 Basement Excavation

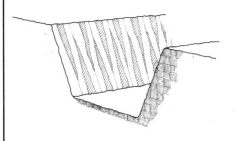

This page illustrates and describes continuous footing and trench excavation systems including a wheel mounted backhoe, operator, equipment rental, fuel, oil, mobilization, no hauling or backfill. Lines within System Components give the unit price and total price per linear foot for this system. Prices for alternate continuous footing and trench excavation systems are on Line Items A2010 130 1300 thru 2700. Both material quantities and labor costs have been adjusted for the system listed.

Factors: To adjust for job conditions other than normal working situations use Lines A2010 130 2900 thru 4000.

Example: You are to install the system and supply dust protection. Go to Line A2010 130 3000 and apply this percentage to the appropriate TOTAL costs.

System Components			COST PER L.F.		
	QUANTITY	UNIT	EQUIP.	LABOR	TOTAL
Continuous footing or trench excav. w/ 3/4 C.Y. wheel mntd backhoe					
Including operator, equipment rental, fuel, oil and mobilization. Trench					
Is 4' wide at bottom, 3' deep w/sides sloped 1 to 2 and 200' long. Costs					
Are based on a production rate of 240 C.Y./day. No hauling/backfill incl.					
Equipment operator	4.000	Hr.		260	260
3/4 C.Y. backhoe, wheel mounted	.500	Day	91.85		91.85
Operating expense (fuel, oil)	4.000	Hr.	69.76		69.76
Mobilization	1.000	Ea.	240	116	356
TOTAL			401.61	376	777.61
COST PER L.F.		L.F.	2.01	1.88	3.89

A2010 130	Excavation, Footings or Trench	COST PER L.F.		
		EQUIP.	LABOR	TOTAL
1200	For alternate trench sizes, 4' bottom with sloped sides:			
1300	2' deep, 50' long	6.90	7.50	14.40
1400	100' long	3.58	3.76	7.34
1500	300' long	1.33	1.26	2.59
1600	4' deep, 50' long	7.15	7.50	14.65
1700	100' long	3.84	3.76	7.60
1800	300' long	1.92	2.23	4.15
1900	6' deep, 50' long	7.55	7.50	15.05
2000	100' long	4.05	4.74	8.79
2100	300' long	2.95	3.86	6.81
2200	8' deep, 50' long	8.05	7.50	15.55
2300	100' long	5.65	6.35	12
2400	300' long	4.04	3.86	7.90
2500	10' deep, 50' long	9.65	10.10	19.75
2600	100' long	7.05	8.65	15.70
2700	300' long	5.30	7.65	12.95
2800				
2900	Dust protection, add, minimum		2%	2%
3000	Maximum		11%	11%
3100	Equipment usage curtailment, add, minimum		1%	1%
3200	Maximum		10%	10%
3300	Material handling & storage limitation, add, minimum		1%	1%
3400	Maximum		7%	7%
3500	Protection of existing work, add, minimum		2%	2%
3600	Maximum		7%	7%
3700	Shift work requirements, add, minimum		5%	5%
3800	Maximum		30%	30%
3900	Temporary shoring and bracing, add, minimum		5%	5%
4000	Maximum		12%	12%

A20 Basement Construction

A2010 Basement Excavation

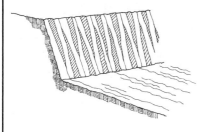

This page illustrates and describes foundation excavation systems including a backhoe-loader, operator, equipment rental, fuel, oil, mobilization, hauling material, no backfilling. Lines within System Components give the unit price and total price per cubic yard for this system. Prices for alternate foundation excavation systems are on Line Items A2010 140 1600 thru 2200. Both material quantities and labor costs have been adjusted for the system listed.

Factors: To adjust for job conditions other than normal working situations use Lines A2010 140 2900 thru 4000.

Example: You are to install the system with the use of temporary shoring and bracing. Go to Line A2010 140 3900 and apply these percentages to the appropriate TOTAL costs.

System Components	QUANTITY	UNIT	COST PER C.Y. EQUIP.	COST PER C.Y. LABOR	COST PER C.Y. TOTAL
Foundation excav w/ 3/4 C.Y. backhoe-loader, incl. operator, equip Rental, fuel, oil, and mobilization. Hauling of excavated material is Included. Prices based on one day production of 360 C.Y. in medium soil Without backfilling.					
Equipment operator	8.000	Hr.		520	520
Backhoe-loader, 3/4 C.Y.	1.000	Day	183.70		183.70
Operating expense (fuel, oil)	8.000	Hr.	139.52		139.52
Hauling, 12 C.Y. trucks, 1 mile round trip	360.000	L.C.Y.	838.80	597.60	1,436.40
Mobilization	1.000	Ea.	240	116	356
TOTAL			1,402.02	1,233.60	2,635.62
COST PER C.Y.		C.Y.	3.88	3.42	7.32

A2010 140	Excavation, Foundation	COST PER C.Y. EQUIP.	COST PER C.Y. LABOR	COST PER C.Y. TOTAL
1400	For alternate size excavations:			
1500				
1600	100 C.Y.	5.15	5.75	10.90
1700	200 C.Y.	4.43	3.70	8.13
1800	300 C.Y.	4.06	3.50	7.56
1900	400 C.Y.	3.82	3.38	7.20
2000	500 C.Y.	3.71	3.35	7.06
2100	600 C.Y.	3.61	3.29	6.90
2200	700 C.Y.	3.56	3.24	6.80
2300				
2400				
2500				
2900	Dust protection, add, minimum		2%	2%
3000	Maximum		11%	11%
3100	Equipment usage curtailment, add, minimum		1%	1%
3200	Maximum		10%	10%
3300	Material handling & storage limitation, add, minimum		1%	1%
3400	Maximum		7%	7%
3500	Protection of existing work, add, minimum		2%	2%
3600	Maximum		7%	7%
3700	Shift work requirements, add, minimum		5%	5%
3800	Maximum		30%	30%
3900	Temporary shoring and bracing, add, minimum		5%	5%
4000	Maximum		12%	12%

A2020 Basement Walls

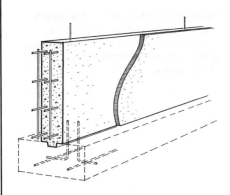

This page illustrates and describes a concrete wall system including concrete, placing concrete, forms, reinforcing, insulation, waterproofing and anchor bolts. Lines within System Components give the unit price and total price per linear foot for this system. Prices for alternate concrete wall systems are on Line Items A2020 120 1500 thru 2600. Both material quantities and labor costs have been adjusted for the system listed.

Factors: To adjust for job conditions other than normal working situations use Lines A2020 120 2900 thru 4000.

Example: You are to install this wall system where delivery of material is difficult. Go to Line A2020 120 3600 and apply these percentages to the appropriate MAT. and INST. costs.

System Components	QUANTITY	UNIT	COST PER L.F. MAT.	COST PER L.F. INST.	COST PER L.F. TOTAL
Cast in place concrete foundation wall, 8″ thick, 3′ high, 2500 psi					
Concrete including forms, reinforcing, waterproofing, and anchor bolts.					
Concrete, 2500 psi, 8″ thick, 3′ high	.070	C.Y.	7.56		7.56
Forms, wall, 4 uses	6.000	S.F.	6.30	36.60	42.90
Reinforcing	6.000	Lb.	5.34	2.40	7.74
Placing concrete	.070	C.Y.		3.10	3.10
Waterproofing	3.000	S.F.	1.08	2.94	4.02
Rigid insulaton, 1″ polystyrene	3.000	S.F.	1.02	2.34	3.36
Anchor bolts, 1/2″ diameter, 4′ O.C.	.250	Ea.	.35	.69	1.04
TOTAL			21.65	48.07	69.72

A2020 120	Concrete Wall	COST PER L.F. MAT.	COST PER L.F. INST.	COST PER L.F. TOTAL
1400	For alternate wall systems:			
1500	8″ thick, 2500 psi concrete, 4′ high	29.50	64.50	94
1600	6′ high	44	96	140
1700	8′ high	58.50	128	186.50
1800	3500 psi concrete, 4′ high	30	64.50	94.50
1900	6′ high	45	96	141
2000	8′ high	60	128	188
2100	12″ thick, 2500 psi concrete, 4′ high	38.50	68	106.50
2200	6′ high	55	100	155
2300	8′ high	73	134	207
2400	3500 psi concrete, 4′ high	39.50	68	107.50
2500	8′ high	75	134	209
2600	10′ high	93	166	259
2700				
2900	Cut & patch to match existing construction, add, minimum	2%	3%	
3000	Maximum	5%	9%	
3100	Dust protection, add, minimum	1%	2%	
3200	Maximum	4%	11%	
3300	Equipment usage curtailment, add, minimum	1%	1%	
3400	Maximum	3%	10%	
3500	Material handling & storage limitation, add, minimum	1%	1%	
3600	Maximum	6%	7%	
3700	Protection of existing work, add, minimum	2%	2%	
3800	Maximum	5%	7%	
3900	Shift work requirements, add, minimum		5%	
4000	Maximum		30%	
4100				

A2020 Basement Walls

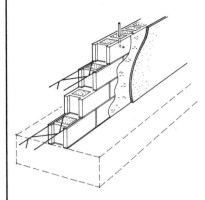

This page illustrates and describes a concrete block wall system including concrete block, masonry reinforcing, parging, waterproofing insulation and anchor bolts. Lines within System Components give the unit price and total price per linear foot for this system. Prices for alternate concrete block wall systems are on Line Items A2020 140 1300 thru 2600. Both material quantities and labor costs have been adjusted for the system listed.

Factors: To adjust for job conditions other than normal working situations use Lines A2020 140 2900 thru 4000.

Example: You are to install the system to match an existing foundation wall. Go to Line A2020 140 3000 and apply these percentages to the appropriate MAT. and INST. costs.

System Components	QUANTITY	UNIT	COST PER L.F. MAT.	INST.	TOTAL
Concrete block, 8" thick, masonry reinforcing, parged and					
Waterproofed, insulation and anchor bolts, wall 2'-8" high.					
Concrete block, 8" x 8" x 16"	2.670	S.F.	7.85	15.09	22.94
Masonry reinforcing	2.000	L.F.	.36	.34	.70
Parging	.193	S.Y.	.77	2.29	3.06
Waterproofing	2.670	S.F.	.96	2.62	3.58
Insulation, 1" rigid polystyrene	2.670	S.F.	.91	2.08	2.99
Anchor bolts, 1/2" diameter, 4' O.C.	.250	Ea.	.35	.69	1.04
TOTAL			11.20	23.11	34.31

A2020 140	Concrete Block Wall	COST PER L.F. MAT.	INST.	TOTAL
1200	For alternate wall systems:			
1300	8" thick block, 4' high	16.60	34	50.60
1400	6' high	24.50	51	75.50
1500	8' high	33	68	101
1600	Grouted solid, 4' high	21.50	46	67.50
1700	6' high	32	68.50	100.50
1800	8' high	42.50	91.50	134
2100	12" thick block, 4' high	22	49.50	71.50
2200	6' high	33	74	107
2300	8' high	44	98.50	142.50
2400	Grouted solid, 4' high	30	62	92
2500	6' high	44.50	92.50	137
2600	8' high	59.50	123	182.50
2700				
2900	Cut & patch to match existing construction, add, minimum	2%	3%	
3000	Maximum	5%	9%	
3100	Dust protection, add, minimum	1%	2%	
3200	Maximum	4%	11%	
3300	Equipment usage curtailment, add, minimum	1%	1%	
3400	Maximum	3%	10%	
3500	Material handling & storage limitation, add, minimum	1%	1%	
3600	Maximum	6%	7%	
3700	Protection of existing work, add, minimum	2%	2%	
3800	Maximum	5%	7%	
3900	Shift work requirements, add, minimum		5%	
4000	Maximum		30%	

This page illustrates and describes a reinforced concrete slab system including concrete, placing concrete, formwork, reinforcing, steel trowel finish, V.A. floor tile and acoustical spray ceiling finish. Lines within System Components give the unit price per square foot for this system. Prices for alternate reinforced concrete slab systems are on Line Items B1010 225 1500 thru 1800. Both material quantities and labor costs have been adjusted for the system listed.

Factors: To adjust for job conditions other than normal working situations use Lines B1010 225 2700 thru 4000.

Example: You are to install the system to match an existing floor system. Go to Line B1010 225 2800 and apply these percentages to the appropriate MAT. and INST. costs.

System Components	QUANTITY	UNIT	COST PER S.F. MAT.	COST PER S.F. INST.	COST PER S.F. TOTAL
Flat slab system, reinforced concrete with vinyl tile floor, sprayed Acoustical ceiling finish, not including columns.					
Concrete, 4000 psi, 6" thick	.020	C.Y.	2.32		2.32
Placing concrete	.020	C.Y.		.62	.62
Formwork, 4 use	1.000	S.F.	1.62	5.50	7.12
Edge form	.120	S.F.	.05	1.07	1.12
Reinforcing steel	3.000	Lb.	2.67	1.74	4.41
Steel trowel finish	1.000	S.F.		1.09	1.09
Vinyl asbestos floor tile	1.000	S.F.	4.62	.95	5.57
Acoustical spray ceiling finish	1.000	S.F.	.26	.15	.41
TOTAL			11.54	11.12	22.66

B1010 225	Floor - Ceiling, Concrete Slab		COST PER S.F. MAT.	COST PER S.F. INST.	COST PER S.F. TOTAL
1400	For alternate slab systems:				
1500	Concrete, 4000 psi, 7" thick		12.15	11.60	23.75
1600	8" thick		13.05	12.20	25.25
1700	9" thick		13.75	12.70	26.45
1800	10" thick		14.65	13.35	28
1900					
2000					
2100					
2200					
2300					
2400					
2500					
2700	Cut & patch to match existing construction, add, minimum		2%	3%	
2800	Maximum		5%	9%	
2900	Dust protection, add, minimum		1%	2%	
3000	Maximum		4%	11%	
3100	Equipment usage curtailment, add, minimum		1%	1%	
3200	Maximum		3%	10%	
3300	Material handling & storage limitation, add, minimum		1%	1%	
3400	Maximum		6%	7%	
3500	Protection of existing work, add, minimum		2%	2%	
3600	Maximum		5%	7%	
3700	Shift work requirements, add, minimum			5%	
3800	Maximum			30%	
3900	Temporary shoring and bracing, add, minimum		2%	5%	
4000	Maximum		5%	12%	

B1010 Floor Construction

This page illustrates and describes a hollow core prestressed concrete panel system including hollow core slab, grout, carpet, carpet padding, and sprayed ceiling. Lines within System Components give the unit price and total price per square foot for this system. Prices for alternate hollow core prestressed concrete panel systems are on Line Items B1010 232 1300 thru 1600. Both material and labor costs have been adjusted for the system listed.

Factors: To adjust for job conditions other than normal working situations use Lines B1010 232 2700 thru 3800.

Example: You are to install the system where dust control is a major concern. Go to Line B1010 232 2800 and apply these percentages to the appropriate MAT. and INST. costs.

System Components			COST PER S.F.		
	QUANTITY	UNIT	MAT.	INST.	TOTAL
Precast hollow core plank with carpeted floors, padding					
And sprayed textured ceiling.					
Hollow core plank, 4″ thick with grout topping	1.000	S.F.	6.25	3.23	9.48
Nylon carpet, 26 oz. medium traffic	.110	S.Y.	3.91	.70	4.61
Carpet padding, minimum quality	.110	S.Y.	.51	.35	.86
Sprayed texture ceiling	1.000	S.F.	.09	.56	.65
TOTAL			10.76	4.84	15.60

B1010 232	Floor - Ceiling, Conc. Panel	COST PER S.F.		
		MAT.	INST.	TOTAL
1200	For alternate floor systems:			
1300	Hollow core concrete plank, with grout, 6″ thick	11.85	4.37	16.22
1400	8″ thick	12.55	4.03	16.58
1500	10″ thick	13	3.76	16.76
1600	12″ thick	13.65	3.55	17.20
1700				
1800				
1900				
2000				
2100				
2200				
2300				
2400				
2500				
2700	Dust protection, add, minimum	1%	2%	
2800	Maximum	4%	11%	
2900	Equipment usage curtailment, add, minimum	1%	1%	
3000	Maximum	3%	10%	
3100	Material handling & storage limitation, add, minimum	1%	1%	
3200	Maximum	6%	7%	
3300	Protection of existing work, add, minimum	2%	2%	
3400	Maximum	5%	7%	
3500	Shift work requirements, add, minimum		5%	
3600	Maximum		30%	
3700	Temporary shoring and bracing, add, minimum	2%	5%	
3800	Maximum	5%	12%	

B1010 Floor Construction

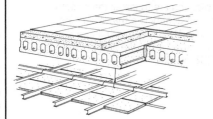

This page illustrates and describes a hollow core plank system including precast hollow core plank, concrete topping, V.A. tile, concealed suspension system and ceiling tile. Lines within System Components give the unit price and total price per square foot for this system. Prices for alternate hollow core plank systems are on Line Items B1010 233 1500 thru 1800. Both material quantities and labor costs have been adjusted for the system listed.

Factors: To adjust for job conditions other than normal working situations use Lines B1010 233 2900 thru 4000.

Example: You are to install the system where equipment usage is a problem. Go to Line B1010 233 3200 and apply these percentages to the appropriate MAT. and INST. costs.

System Components				COST PER S.F.		
	QUANTITY	UNIT		MAT.	INST.	TOTAL
Precast hollow core plank with 2″ topping, vinyl composition floor tile						
And suspended acoustical tile ceiling.						
Hollow core concrete plank, 4″ thick	1.000	S.F.		6.25	3.23	9.48
Concrete topping, 2″ thick	1.000	S.F.		1.26	4.95	6.21
Vinyl composition tile, .08″ thick	1.000	S.F.		1.43	.95	2.38
Concealed Z bar suspension system, 12″ module	1.000	S.F.		.72	1.01	1.73
Ceiling tile, mineral fiber, 3/4″ thick	1.000	S.F.		2.06	1.76	3.82
TOTAL				11.72	11.90	23.62

B1010 233	Floor - Ceiling, Conc. Plank		COST PER S.F.		
			MAT.	INST.	TOTAL
1400	For alternate floor systems:				
1500	Hollow core concrete plank, 6″ thick		12.80	11.40	24.20
1600	8″ thick		13.50	11.10	24.60
1700	10″ thick		13.95	10.80	24.75
1800	12″ thick		14.60	10.60	25.20
1900					
2000					
2100					
2200					
2300					
2400					
2500					
2600					
2700					
2900	Dust protection, add, minimum		1%	2%	
3000	Maximum		4%	11%	
3100	Equipment usage curtailment, add, minimum		1%	1%	
3200	Maximum		3%	10%	
3300	Material handling & storage limitation, add, minimum		1%	1%	
3400	Maximum		6%	7%	
3500	Protection of existing work, add, minimum		2%	2%	
3600	Maximum		5%	7%	
3700	Shift work requirements, add, minimum			5%	
3800	Maximum			30%	
3900	Temporary shoring and bracing, add, minimum		2%	5%	
4000	Maximum		5%	12%	

B1010 Floor Construction

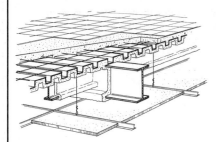

This page illustrates and describes a structural steel w/metal decking and concrete system including steel beams, steel decking, shear studs, concrete, placing concrete, edge form, steel trowel finish, curing, wire fabric, fireproofing, beams and decking, tile floor, and suspended ceiling. Lines within System Components give the unit price and total price per square foot for this system. Prices for alternate structural steel w/metal decking and concrete systems are on Line Items B1010 243 1900 thru 2400. Both material quantities and labor costs have been adjusted for the system listed.

Factors: To adjust for job conditions other than normal working situations use Lines B1010 243 2900 thru 4000.

Example: You are to install the system where material handling and storage are a problem. Go to Line B1010 243 3400 and apply these percentages to the appropriate MAT. and INST. costs.

System Components	QUANTITY	UNIT	COST PER S.F.		
			MAT.	INST.	TOTAL
Composite structural beams, 20 ga. 3″ deep steel decking, shear studs,					
3000 psi, concrete, placing concrete, edge forms, steel trowel finish,					
Welded wire fabric fireproofing, tile floor and suspended ceiling.					
Steel deck, 20 gage, 3″ deep, galvanized	1.000	S.F.	3.28	.94	4.22
Structural steel framing	.004	Ton	13.20	2.62	15.82
Shear studs, 3/4″	.250	Ea.	.13	.43	.56
Concrete, 3000 psi, 5″ thick	.015	C.Y.	1.67		1.67
Place concrete	.015	C.Y.		.47	.47
Edge form	.140	L.F.	.03	.56	.59
Steel trowel finish	1.000	S.F.		1.09	1.09
Welded wire fabric 6 x 6 - #10/10	.011	C.S.F.	.22	.38	.60
Fireproofing, sprayed	1.880	S.F.	1.09	1.81	2.90
Vinyl composition floor tile	1.000	S.F.	.96	.95	1.91
Suspended acoustical ceiling	1.000	S.F.	2.55	1.39	3.94
TOTAL			23.13	10.64	33.77

B1010 243	Floor - Ceiling, Struc. Steel	COST PER S.F.		
		MAT.	INST.	TOTAL
1800	For alternate floor systems:			
1900	Composite deck, galvanized, 3″ deep, 22 ga.	23	10.60	33.60
2000	18 ga.	24	10.70	34.70
2100	16 ga.	25.50	10.75	36.25
2200	Composite deck, galvanized, 2″ deep, 22 ga.	22.50	10.45	32.95
2300	20 ga.	23	10.50	33.50
2400	16 ga.	24.50	10.60	35.10
2500				
2900	Dust protection, add, minimum	1%	2%	
3000	Maximum	4%	11%	
3100	Equipment usage curtailment, add, minimum	1%	1%	
3200	Maximum	3%	10%	
3300	Material handling & storage limitation, add, minimum	1%	1%	
3400	Maximum	6%	7%	
3500	Protection of existing work, add, minimum	2%	2%	
3600	Maximum	5%	7%	
3700	Shift work requirements, add, minimum		5%	
3800	Maximum		30%	
3900	Temporary shoring and bracing, add, minimum	2%	5%	
4000	Maximum	5%	12%	

B10 Superstructure

B1010 Floor Construction

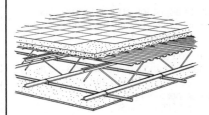

This page illustrates and describes an open web joist and steel slab-form system including open web steel joist, slab form, concrete, placing concrete, wire fabric, steel trowel finish, tile floor and plasterboard. Lines within System Components give the unit price and total price per square foot for this system. Prices for alternate open web joists and steel slab-form systems are on Line Items B1010 245 1900 thru 2200. Both material quantities and labor costs have been adjusted for the systems listed.

Factors: To adjust for job conditions other than normal working situations use Lines B1010 245 2900 thru 4000.

Example: You are to install the system in a congested commercial area and most work will be done at night. Go to Line B1010 245 3800 and apply this percentage to the appropriate INST. cost.

System Components	QUANTITY	UNIT	COST PER S.F. MAT.	COST PER S.F. INST.	COST PER S.F. TOTAL
Open web steel joists, 24″ O.C., slab form, 3000 psi concrete, welded					
Wire fabric, steel trowel finish, floor tile, 5/8″ drywall ceiling.					
Open web steel joists, 12″ deep, 5.2#/L.F., 24″ O.C.	2.630	Lb.	3.18	1.13	4.31
Slab form, 28 gage, 9/16″ deep, galvanized	1.000	S.F.	1.67	.71	2.38
Concrete, 3000 psi, 2-1/2″ thick	.008	C.Y.	.89		.89
Placing concrete	.008	C.Y.		.25	.25
Welded wire fabric 6 x 6 - #10/10	.011	C.S.F.	.22	.38	.60
Steel trowel finish	1.000	S.F.		1.09	1.09
Vinyl composition floor tile	1.000	S.F.	.96	.95	1.91
Ceiling furring, 3/4″ channels, 24″ O.C.	1.000	S.F.	.23	1.07	1.30
Gypsum drywall, 5/8″ thick, finished	1.000	S.F.	.42	1.38	1.80
Paint ceiling	1.000	S.F.	.17	.59	.76
TOTAL			7.74	7.55	15.29

B1010 245	Floor - Ceiling, Steel Joists	COST PER S.F. MAT.	COST PER S.F. INST.	COST PER S.F. TOTAL
1800	For alternate floor systems:			
1900	Open web joists, 16″ deep, 6.6#/L.F.	8.55	7.85	16.40
2000	20″ deep, 8.4# L.F.	9.65	8.20	17.85
2100	24″ deep, 11.5# L.F.	11.50	8.85	20.35
2200	26″ deep, 12.8# L.F.	12.30	9.20	21.50
2300				
2400				
2500				
2600				
2700				
2900	Dust protection, add, minimum	1%	2%	
3000	Maximum	4%	11%	
3100	Equipment usage curtailment, add, minimum	1%	1%	
3200	Maximum	3%	10%	
3300	Material handling & storage limitation, add, minimum	1%	1%	
3400	Maximum	6%	7%	
3500	Protection of existing work, add, minimum	2%	2%	
3600	Maximum	5%	7%	
3700	Shift work requirements, add, minimum		5%	
3800	Maximum		30%	
3900	Temporary shoring and bracing, add, minimum	2%	5%	
4000	Maximum	5%	12%	

B10 Superstructure

B1010 Floor Construction

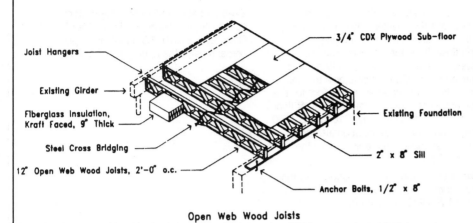

Joist Hangers

Existing Girder

Fiberglass Insulation, Kraft Faced, 9" Thick

Steel Cross Bridging

12" Open Web Wood Joists, 2'-0" o.c.

3/4" CDX Plywood Sub-floor

Existing Foundation

2" x 8" Sill

Anchor Bolts, 1/2" x 8"

Open Web Wood Joists

This page illustrates and describes floor framing systems including sills, joist hangers, joists, bridging, subfloor and insulation. Lines within systems components give the unit price and total price per square foot for this system. Prices for alternate joist types are on line items B1010 262 0100 through 0700. Both material quantities and labor costs have been adjusted for the system listed.

Factors: To adjust for job conditions other than normal working situations use lines B1010 262 2700 through 4000.

Example: You are to install the system where delivery of material is difficult. Go to line B1010 262 3400 and apply these percentages to the appropriate MAT. and INST. costs.

System Components	QUANTITY	UNIT	COST PER S.F.		
			MAT.	INST.	TOTAL
Open web wood joists, sill, butt to girder, 5/8" subfloor, insul.					
Anchor bolt, hook type, with nut and washer, 1/2" x 8" long	.250	Ea.	.02	.07	.09
Joist & beam hanger, 18 ga. galvanized	.750	Ea.	.10	.24	.34
Bridging, steel, compression, for 2" x 12" joists	.001	C.Pr.	.18	.26	.44
Framing, open web wood joists, 12" deep	.001	M.L.F.	3.35	1.20	4.55
Framing, sills, 2" x 8"	.001	M.B.F.	.59	1.58	2.17
Sub-floor, plywood, CDX, 3/4" thick	1.000	S.F.	.74	.78	1.52
Insulation, fiberglass, kraft face, 9" thick x 23" wide	1.000	S.F.	.73	.39	1.12
TOTAL			5.71	4.52	10.23

B1010 262	Open Web Wood Joists	COST PER S.F.		
		MAT.	INST.	TOTAL
0080	For alternate floor systems:			
0100	12" open web joists, sill, butt to girder, 5/8" subflr	5.70	4.52	10.22
0200	14" joists	5.95	4.60	10.55
0300	16" joists	5.95	4.67	10.62
0400	18" joists	6.20	4.75	10.95
0450	Wood struc. "I" joists, 24" O.C., to 24' span, 50 PSF LL, butt to girder	4.87	4.21	9.08
0500	55 PSF LL	5.10	4.27	9.37
0600	to 30' span, 45 PSF LL	5.85	4.14	9.99
0700	55 PSF LL	6.45	4.21	10.66
2700	Cut & patch to match existing construction, add, minimum	2%	3%	
2800	Maximum	5%	9%	
2900	Dust protection, add, minimum	1%	2%	
3000	Maximum	4%	11%	
3100	Equipment usage curtailment, add, minimum	1%	1%	
3200	Maximum	3%	10%	
3300	Material handling & storage limitation, add, minimum	1%	1%	
3400	Maximum	6%	7%	
3500	Protection of existing work, add, minimum	2%	2%	
3600	Maximum	5%	7%	
3700	Shift work requirements, add, minimum		5%	
3800	Maximum		30%	
3900	Temporary shoring and bracing, add, minimum	2%	5%	
4000	Maximum	5%	12%	

B10 Superstructure

B1010 Floor Construction

This page illustrates and describes a wood joist floor system including wood joist, oak floor, sub-floor, bridging, sand and finish floor, furring, plasterboard, taped, finished and painted ceiling. Lines within System Components give the unit price and total price per square foot for this system. Prices for alternate wood joist floor systems are on Line Items B1010 263 1700 thru 2300. Both material quantities and labor costs have been adjusted for the system listed.

Factors: To adjust for job conditions other than normal working situations use Lines B1010 263 2700 thru 4000.

Example: You are to install the system during off peak hours, 6 P.M. to 2 A.M. Go to Line B1010 263 3800 and apply this percentage to the appropriate INST. costs.

System Components	QUANTITY	UNIT	COST PER S.F.		
			MAT.	INST.	TOTAL
Wood joists, 2″ x 8″, 16″ O.C., oak floor (sanded & finished), 1/2″ sub Floor, 1″x 3″ bridging, furring, 5/8″ drywall, taped, finished and painted.					
Wood joists, 2″ x 8″, 16″ O.C.	1.000	L.F.	.78	.96	1.74
Subfloor plywood CDX 1/2″	1.000	SF Flr.	.51	.70	1.21
Bridging 1″ x 3″	.150	Pr.	.08	.61	.69
Oak flooring, No. 1 common	1.000	S.F.	4.41	3.10	7.51
Sand and finish floor	1.000	S.F.	.86	1.41	2.27
Wood furring, 1″ x 3″, 16″ O.C.	1.000	L.F.	.38	1.51	1.89
Gypsum drywall, 5/8″ thick	1.000	S.F.	.37	.59	.96
Tape and finishing	1.000	S.F.	.05	.53	.58
Paint ceiling	1.000	S.F.	.17	.59	.76
TOTAL			7.61	10	17.61

B1010 263	Floor-Ceiling, Wood Joists		COST PER S.F.		
			MAT.	INST.	TOTAL
1600	For alternate wood joist systems:				
1700	16″ on center, 2″ x 6″ joists		7.40	9.90	17.30
1800	2″ x 10″		7.90	10.20	18.10
1900	2″ x 12″		8.35	10.25	18.60
2000	2″ x 14″		9.20	10.40	19.60
2100	12″ on center, 2″ x 10″ joists		8.20	10.50	18.70
2200	2″ x 12″		8.75	10.55	19.30
2300	2″ x 14″		9.75	10.75	20.50
2400					
2500					
2700	Cut & patch to match existing construction, add, minimum		2%	3%	
2800	Maximum		5%	9%	
2900	Dust protection, add, minimum		1%	2%	
3000	Maximum		4%	11%	
3100	Equipment usage curtailment, add, minimum		1%	1%	
3200	Maximum		3%	10%	
3300	Material handling & storage limitation, add, minimum		1%	1%	
3400	Maximum		6%	7%	
3500	Protection of existing work, add, minimum		2%	2%	
3600	Maximum		5%	7%	
3700	Shift work requirements, add, minimum			5%	
3800	Maximum			30%	
3900	Temporary shoring and bracing, add, minimum		2%	5%	
4000	Maximum		5%	12%	

B1020 Roof Coverings

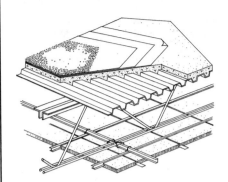

This page illustrates and describes a flat roof bar joist system including asphalt and gravel roof, lightweight concrete, metal decking, open web joists, and suspended acoustic ceiling. Lines within System Components give the unit price and total price per square foot for this system. Prices for alternate flat roof bar joist systems are on Line Items B1020 106 1300 thru 1600. Both material quantities and labor costs have been adjusted for the system listed.

Factors: To adjust for job conditions other than normal working situations use Lines B1020 106 2700 thru 4000.

Example: You are to install the system where material handling and storage present some problem. Go to Line B1020 106 3300 and apply these percentages to the appropriate MAT. and INST. costs.

System Components	QUANTITY	UNIT	COST PER S.F. MAT.	COST PER S.F. INST.	COST PER S.F. TOTAL
Three ply asphalt and gravel roof on lightweight concrete, metal decking on					
Web joist 4' O.C., with suspended acoustic ceiling.					
Open web joist, 10" deep, 4' O.C.	1.250	Lb.	1.51	.54	2.05
Metal decking, 1-1/2" deep, 22 Ga., galvanized	1.000	S.F.	2.83	.63	3.46
Gravel roofing, 3 ply asphalt	.010	C.S.F.	1.20	1.67	2.87
Perlite roof fill, 3" thick	1.000	S.F.	1.09	.52	1.61
Suspended ceiling, 3/4" mineral fiber on "Z" bar suspension	1.000	S.F.	2.54	3.52	6.06
TOTAL			9.17	6.88	16.05

B1020 106	Steel Joist Roof & Ceiling	COST PER S.F. MAT.	COST PER S.F. INST.	COST PER S.F. TOTAL
1200	For alternate roof systems:			
1300	Open web bar joist, 16" deep, 5#/L.F.	9.65	7.05	16.70
1400	18" deep, 6.6#/L.F.	10.10	7.20	17.30
1500	20" deep, 9.6#/L.F.	10.55	7.35	17.90
1600	24" deep, 11.52#/L.F.	11.15	7.60	18.75
1700				
1800				
1900				
2000				
2100				
2200				
2300				
2400				
2500				
2700	Cut & patch to match existing construction, add, minimum	2%	3%	
2800	Maximum	5%	9%	
2900	Dust protection, add, minimum	1%	2%	
3000	Maximum	4%	11%	
3100	Equipment usage curtailment, add, minimum	1%	1%	
3200	Maximum	3%	10%	
3300	Material handling & storage limitation, add, minimum	1%	1%	
3400	Maximum	6%	7%	
3500	Protection of existing work, add, minimum	2%	2%	
3600	Maximum	5%	7%	
3700	Shift work requirements, add, minimum		5%	
3800	Maximum		30%	
3900	Temporary shoring and bracing, add, minimum	2%	5%	
4000	Maximum	5%	12%	

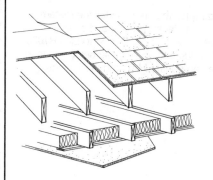

This page illustrates and describes a wood frame roof system including rafters, ceiling joists, sheathing, building paper, asphalt shingles, roof trim, furring, insulation, plaster and paint. Lines within System Components give the unit price and total price per square foot for this system. Prices for alternate wood frame roof systems are on Line Items B1020 202 1900 thru 2700. Both material quantities and labor costs have been adjusted for the system listed.

Factors: To adjust for job conditions other than normal working situations use Lines B1020 202 3300 thru 4000.

Example: You are to install the system while protecting existing work. Go to Line B1020 202 3800 and apply these percentages to the appropriate MAT. and INST. costs.

System Components	QUANTITY	UNIT	COST PER S.F. MAT.	COST PER S.F. INST.	COST PER S.F. TOTAL
Wood frame roof system, 4 in 12 pitch, including rafters, sheathing, Shingles, insulation, drywall, thin coat plaster, and painting.					
Rafters, 2" x 6", 16" O.C., 4 in 12 pitch	1.080	L.F.	.62	1.14	1.76
Ceiling joists, 2" x 6", 16 O.C.	1.000	L.F.	.57	.84	1.41
Sheathing, 1/2" CDX	1.080	S.F.	.55	.81	1.36
Building paper, 15# felt	.011	C.S.F.	.06	.16	.22
Asphalt shingles, 240#	.011	C.S.F.	.62	1.08	1.70
Roof trim	.100	L.F.	.16	.24	.40
Furring, 1" x 3", 16" O.C.	1.000	L.F.	.38	1.51	1.89
Fiberglass insulation, 6" batts	1.000	S.F.	.85	.39	1.24
Gypsum board, 1/2" thick	1.000	S.F.	.34	.59	.93
Thin coat plaster	1.000	S.F.	.09	.66	.75
Paint, roller, 2 coats	1.000	S.F.	.17	.59	.76
TOTAL			4.41	8.01	12.42

B1020 202	Wood Frame Roof & Ceiling	COST PER S.F. MAT.	COST PER S.F. INST.	COST PER S.F. TOTAL
1800	For alternate roof systems:			
1900	Rafters 16" O.C., 2" x 8"	4.63	8.10	12.73
2000	2" x 10"	4.97	8.70	13.67
2100	2" x 12"	5.45	8.85	14.30
2200	Rafters 24" O.C., 2" x 6"	4.11	7.50	11.61
2300	2" x 8"	4.27	7.55	11.82
2400	2" x 10"	4.52	8	12.52
2500	2" x 12"	4.87	8.15	13.02
2600	Roof pitch, 6 in 12, add	3%	10%	
2700	8 in 12, add	5%	12%	
2800				
2900				
3000				
3100				
3300	Cut & patch to match existing construction, add, minimum	2%	3%	
3400	Maximum	5%	9%	
3500	Material handling & storage limitation, add, minimum	1%	1%	
3600	Maximum	6%	7%	
3700	Protection of existing work, add, minimum	2%	2%	
3800	Maximum	5%	7%	
3900	Shift work requirements, add, minimum		5%	
4000	Maximum		30%	

B1020 Roof Construction

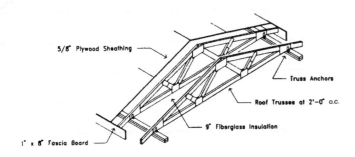

5/8" Plywood Sheathing

Truss Anchors

Roof Trusses at 2'-0" o.c.

9" Fiberglass Insulation

1" x 8" Fascia Board

Wood Fabricated Roof Truss System

This page illustrates and describes a sloped roof truss framing system including trusses, truss anchors, sheathing, fascia and insulation. Lines B1020 210 1500 through 1800 are for a flat roof framing system including flat roof trusses, truss anchors, sheathing, fascia and insulation. Lines within system components give the unit price and total price per square foot for this system. Prices for alternate sizes and systems are given in line items B1020 210 0100 through 1800. Both material quantities and labor costs have been adjusted for the system listed.

Factors: To adjust for job conditions other than normal working situations use lines B1020 210 2700 through 4000.

Example: You are to install the system where crane placement and movement will be difficult. Go to line B1020 210 3200 and apply these percentages to the appropriate MAT. and INST. costs.

System Components	QUANTITY	UNIT	COST PER S.F.		
			MAT.	INST.	TOTAL
Roof truss, 2' O.C., 4/12 slope, 12' span, 5/8 sheath., fascia, insul.					
Timber connectors, rafter anchors, galv., 1 1/2" x 5 1/4"	.084	Ea.	.04	.31	.35
Plywood sheathing, CDX, 5/8" thick	1.054	S.F.	.78	.85	1.63
Truss, 2' O.C., metal plate connectors, 12' span	.042	Ea.	2.16	1.64	3.80
Molding, fascia trim, 1" x 8"	.167	L.F.	.26	.39	.65
Insulation, fiberglass, kraft face, 9" thick x 23" wide	1.000	S.F.	.73	.39	1.12
TOTAL			3.97	3.58	7.55

B1020 210	Roof Truss, Wood	COST PER S.F.		
		MAT.	INST.	TOTAL
0090	For alternate roof systems:			
0100	Roof truss, 2' O.C., 4/12 p, 1' ovhg, 12' span, 5/8" sheath, fascia, insul.	3.97	3.58	7.55
0200	20' span	3.70	3.11	6.81
0300	24' span	3.62	2.90	6.52
0400	26' span	5.30	3.91	9.21
0500	28' span	3.19	2.84	6.03
0600	30' span	3.68	2.80	6.48
0700	32' span	3.62	2.75	6.37
0800	34' span	3.89	2.71	6.60
0890	60' span	5.30	3.09	8.39
0900	8/12 pitch, 20' span	3.92	3.34	7.26
1000	24' span	4.04	3.11	7.15
1100	26' span	3.89	3.04	6.93
1200	28' span	3.98	3.04	7.02
1300	32' span	4.11	2.98	7.09
1400	36' span	4.31	2.92	7.23
1500	Flat roof frame, fab. strl. joists, 2' O.C., 15'-24' span, 50 PSF LL	4.16	2.42	6.58
1600	55 PSF LL	4.37	2.48	6.85
1700	24'-30' span, 45 PSF LL	5.15	2.31	7.46
1800	55 PSF LL	5.70	2.38	8.08
2700	Cut & patch to match existing construction, add, minimum	2%	3%	
2800	Maximum	5%	9%	
2900	Dust protection, add, minimum	1%	2%	
3000	Maximum	4%	11%	
3100	Equipment usage curtailment, add, minimum	1%	1%	
3200	Maximum	3%	10%	

B10 Superstructure

B1020 Roof Construction

B1020 210	Roof Truss, Wood	COST PER S.F.		
		MAT.	INST.	TOTAL
3300	Material handling & storage limitation, add, minimum	1%	1%	
3400	Maximum	6%	7%	
3500	Protection of existing work, add, minimum	2%	2%	
3600	Maximum	5%	7%	
3700	Shift work requirements, add, minimum		5%	
3800	Maximum		30%	
3900	Temporary shoring and bracing, add, minimum	2%	5%	
4000	Maximum	5%	12%	

B2010 Exterior Walls

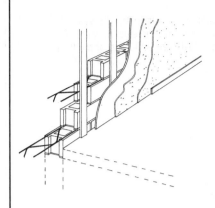

This page illustrates and describes a masonry concrete block system including concrete block wall, pointed, reinforcing, waterproofing, gypsum plaster on gypsum lath, and furring. Lines within System Components give the unit price and total price per square foot for this system. Prices for alternate masonry concrete block wall systems are on Line Items B2010 108 1500 thru 2200. Both material quantities and labor costs have been adjusted for the system listed.

Factors: To adjust for job conditions other than normal working situations use Lines B2010 108 2700 thru 4000.

Example: You are to install the system and match existing construction at several locations. Go to Line B2010 108 2800 and apply these percentages to the appropriate MAT. and INST. costs.

System Components	QUANTITY	UNIT	COST PER S.F.		
			MAT.	INST.	TOTAL
Concrete block, 8″ thick, reinforced every 2 courses, waterproofing gypsum Plaster over gypsum lath on 1″ x 3″ furring, interior painting & baseboard.					
Concrete block, 8″ x 8″ x 16″, reinforced	1.000	S.F.	4.32	5.80	10.12
Silicone waterproofing, 2 coats	1.000	S.F.	.80	.16	.96
Bituminous coating, 1/16″ thick	1.000	S.F.	.41	.98	1.39
Furring, 1″ x 3″, 16″ O.C.	1.000	L.F.	.38	2.03	2.41
Gypsum lath, 3/8″ thick	.110	S.Y.	.67	.58	1.25
Gypsum plaster, 2 coats	.110	S.Y.	.46	2.46	2.92
Painting, 2 coats	1.000	S.F.	.11	.46	.57
Baseboard wood, 9/16″ x 2-5/8″	.100	L.F.	.24	.22	.46
Paint baseboard, primer + 1 coat enamel	.100	L.F.	.01	.06	.07
TOTAL			7.40	12.75	20.15

B2010 108	Masonry Wall, Concrete Block	COST PER S.F.		
		MAT.	INST.	TOTAL
1400	For alternate exterior wall systems:			
1500	8″ thick block, fluted 2 sides	10.30	14.95	25.25
1600	Deep grooved	8.25	14.95	23.20
1700	Slump block	9.75	13	22.75
1800	Split rib	7.65	14.95	22.60
1900				
2000	12″ thick block, regular	6.75	15.25	22
2100	Slump block	15.65	14.20	29.85
2200	Split rib	8.50	17.20	25.70
2300				
2700	Cut & patch to match existing construction, add, minimum	2%	3%	
2800	Maximum	5%	9%	
2900	Dust protection, add, minimum	1%	2%	
3000	Maximum	4%	11%	
3100	Equipment usage curtailment, add, minimum	1%	1%	
3200	Maximum	3%	10%	
3300	Material handling & storage limitation, add, minimum	1%	1%	
3400	Maximum	6%	7%	
3500	Protection of existing work, add, minimum	2%	2%	
3600	Maximum	5%	7%	
3700	Shift work requirements, add, minimum		5%	
3800	Maximum		30%	
3900	Temporary shoring and bracing, add, minimum	2%	5%	
4000	Maximum	5%	12%	

B2010 Exterior Walls

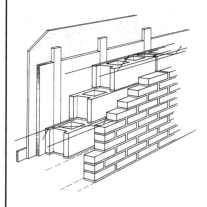

This page illustrates and describes a masonry wall, brick-stone system including brick, concrete block, durawall, insulation, plasterboard, taped and finished, furring, baseboard and painting interior. Lines within System Components give the unit price and total price per square foot for this system. Prices for alternate masonry wall, brick-stone systems are on Line Item B2010 129 1500 thru 2500. Both material quantities and labor costs have been adjusted for the system listed.

Factors: To adjust for job conditions other than normal working situations use Lines B2010 129 3100 thru 4200.

Example: You are to install the system without damaging the existing work. Go to Line B2010 129 3900 and apply these percentages to the appropriate MAT. and INST. costs.

System Components	QUANTITY	UNIT	COST PER S.F.		
			MAT.	INST.	TOTAL
Face brick, 4″thick, concrete block back-up, reinforce every second course, 3/4″insulation, furring, 1/2″drywall, taped, finish, and painted, baseboard					
Face brick, 4″ brick	1.000	S.F.	6.25	10.95	17.20
Concrete back-up block, reinforced 8″ thick	1.000	S.F.	2.60	6.10	8.70
3/4″ rigid polystyrene insulation	1.000	S.F.	.56	.66	1.22
Furring, 1″ x 3″, wood, 16″ O.C.	1.000	L.F.	.38	1.07	1.45
Drywall, 1/2″ thick	1.000	S.F.	.34	.53	.87
Taping & finishing	1.000	S.F.	.05	.53	.58
Painting, 2 coats	1.000	S.F.	.17	.59	.76
Baseboard, wood, 9/16″ x 2-5/8″	.100	L.F.	.24	.22	.46
Paint baseboard, primer + 1 coat enamel	.100	L.F.	.01	.06	.07
TOTAL			10.60	20.71	31.31

B2010 131	Masonry Wall, Brick - Stone	COST PER S.F.		
		MAT.	INST.	TOTAL
1400	For alternate exterior wall systems:			
1500	Face brick, Norman, 4″ x 2-2/3″ x 12″ (4.5 per S.F.)	10.60	17.25	27.85
1600	Roman, 4″ x 2″ x 12″ (6.0 per S.F.)	10.65	19.35	30
1700	Engineer, 4″ x 3-1/5″ x 8″ (5.63 per S.F.)	8.15	19	27.15
1800	S.C.R., 6″ x 2-2/3″ x 12″ (4.5 per S.F.)	10.10	17.50	27.60
1900	Jumbo, 6″ x 4″ x 12″ (3.0 per S.F.)	9.70	15.30	25
2000	Norwegian, 6″ x 3-1/5″ x 12″ (3.75 per S.F.)	8.90	16.15	25.05
2100				
2200				
2300	Stone, veneer, fieldstone, 6″ thick	10.50	17.60	28.10
2400	Marble, 2″ thick	48	31	79
2500	Limestone, 2″ thick	43	19.50	62.50
3100	Cut & patch to match existing construction, add, minimum	2%	3%	
3200	Maximum	5%	9%	
3300	Dust protection, add, minimum	1%	2%	
3400	Maximum	4%	11%	
3500	Equipment usage curtailment, add, minimum	1%	1%	
3600	Maximum	3%	10%	
3700	Material handling & storage limitation, add, minimum	1%	1%	
3800	Maximum	6%	7%	
3900	Protection of existing work, add, minimum	2%	2%	
4000	Maximum	5%	7%	
4100	Shift work requirements, add, minimum		5%	
4200	Maximum		30%	

B2010 Exterior Walls

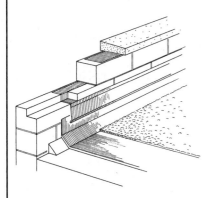

This page illustrates and describes a parapet wall system including wall, coping, flashing and cant strip. Lines within System Components give the unit price and cost per lineal foot for this system. Prices for alternate parapet wall systems are on Lines B2010 142 1300 thru 2400. Both material quantities and labor costs have been adjusted for the system listed.

Factors: To adjust for job conditions other than normal working situations, use Lines B2010 142 2900 thru 4000.

Example: You are to install the system without damaging the adjacent property. Go to Line B2010 142 3600 and apply these percentages to the appropriate MAT. and INST. costs.

System Components	QUANTITY	UNIT	COST PER L.F. MAT.	COST PER L.F. INST.	COST PER L.F. TOTAL
Concrete block parapet, incl. reinf., coping, flashing & cant strip, 2'high					
8" concrete block	2.000	S.F.	8.64	11.60	20.24
Masonry reinforcing	1.000	L.F.	.18	.17	.35
Coping, precast	1.000	L.F.	19.15	13.45	32.60
Roof cant	1.000	L.F.	2.45	1.51	3.96
Through wall flashing	1.000	L.F.	2.65	2.35	5
Cap flashing	1.000	L.F.	7.45	4.25	11.70
TOTAL			40.52	33.33	73.85

B2010 142	Parapet Wall	COST PER L.F. MAT.	COST PER L.F. INST.	COST PER L.F. TOTAL
1200	For alternate systems:			
1300	Concrete block			
1400	12" thick	44.50	43.50	88
1500	Split rib block, 8" thick	41	37.50	78.50
1600	12" thick	42.50	44.50	87
1700	Brick, common brick, 8" wall	45	57	102
1800	12" wall	51.50	72.50	124
1900	4" brick, 4" backup block	48	55	103
2000	8" backup block	49.50	56.50	106
2100	Stucco on masonry	41	37	78
2200	On wood frame	17.90	30	47.90
2300	Wood, T 1-11 siding	26.50	36.50	63
2400	Boards, 1" x 6" cedar	28	40.50	68.50
2500				
2600				
2700				
2900	Dust protection, add, minimum	1%	2%	
3000	Maximum	4%	11%	
3100	Equipment usage curtailment, add, minimum	1%	1%	
3200	Maximum	3%	10%	
3300	Material handling & storage limitation, add, minimum	1%	1%	
3400	Maximum	6%	7%	
3500	Protection of existing work, add, minimum	2%	2%	
3600	Maximum	5%	7%	
3700	Shift work requirements, add, minimum		5%	
3800	Maximum		30%	
3900	Temporary shoring and bracing, add, minimum	2%	5%	
4000	Maximum	5%	12%	

B2010 Exterior Walls

This page illustrates and describes a masonry cleaning and restoration system, including staging, cleaning, repointing. Lines within System Components give the unit price and cost per square foot for this system. Prices for alternate systems are on Lines B2010 143 1300 thru 2500. Both material quantities and labor costs have been adjusted for the system listed.

Factors: To adjust for conditions other than normal working situations, use Lines B2010 143 2900 thru 4200.

Example: You are to clean a wall and be concerned about dust control. Go to line B2010 143 3100 and apply these percentages to the appropriate MAT. and INST. costs.

System Components	QUANTITY	UNIT	COST PER S.F. MAT.	COST PER S.F. INST.	COST PER S.F. TOTAL
Repoint existing building, brick, common bond, high pressure cleaning, Water only, soft old mortar.					
Scaffolding, steel tubular, bldg ext wall face, labor only, 1 to 5 stories	.010	C.S.F.		1.98	1.98
Cleaning, high pressure water only	1.000	S.F.		1.70	1.70
Repoint, brick common bond	1.000	S.F.	.58	5.45	6.03
TOTAL			.58	9.13	9.71

B2010 143	Masonry Restoration - Cleaning	COST PER S.F. MAT.	COST PER S.F. INST.	COST PER S.F. TOTAL
1200	For alternate masonry surfaces:			
1300	Brick, common bond	.58	9.15	9.73
1400	Flemish bond	.61	9.50	10.11
1500	English bond	.61	10.10	10.71
1700	Stone, 2' x 2' blocks	.77	6.95	7.72
1800	2' x 4' blocks	.58	6.15	6.73
2000	Add to above prices for alternate cleaning systems:			
2100	Chemical brush and wash	.04	.59	.63
2200	High pressure chemical and water	.04	.58	.62
2300	Sandblasting, wet system	.58	1.58	2.16
2400	Dry system	.54	.93	1.47
2500	Steam cleaning		.47	.47
2600				
2900	Cut & patch to match existing construction, add, minimum	2%	3%	
3000	Maximum	5%	9%	
3100	Dust protection, add, minimum	1%	2%	
3200	Maximum	4%	11%	
3300	Equipment usage curtailment, add, minimum	1%	1%	
3400	Maximum	3%	10%	
3500	Material handling & storage limitation, add, minimum	1%	1%	
3600	Maximum	6%	7%	
3800				
3900	Protection of existing work, add, minimum	2%	2%	
4000	Maximum	5%	7%	
4100	Shift work requirements, add, minimum		5%	
4200	Maximum		30%	

B2010 Exterior Walls

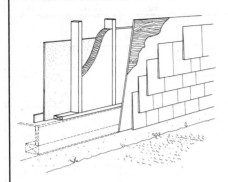

This page illustrates and describes a wood frame exterior wall system including wood studs, sheathing, felt, insulation, plasterboard, taped and finished, baseboard and painted interior. Lines within System Components give the unit price and total price per square foot for this system. Prices for alternate wood frame exterior wall systems are on Line Items B2010 150 1700 thru 2700. Both material quantities and labor costs have been adjusted for the system listed.

Factors: To adjust for job conditions other than normal working situations use Lines B2010 150 3100 thru 4200.

Example: You are to install the system with need for complete temporary bracing. Go to Line B2010 150 4200 and apply these percentages to the appropriate MAT. and INST. costs.

System Components	QUANTITY	UNIT	COST PER S.F. MAT.	COST PER S.F. INST.	COST PER S.F. TOTAL
Wood stud wall, cedar shingle siding, building paper, plywood sheathing,					
Insulation, 5/8" drywall, taped, finished and painted, baseboard.					
2" x 4" wood studs, 16" O.C.	.100	L.F.	.37	1.06	1.43
1/2" CDX sheathing	1.000	S.F.	.51	.75	1.26
18" No. 1 red cedar shingles, 7-1/2" exposure	.008	C.S.F.	2.13	1.87	4
15# felt paper	.010	C.S.F.	.05	.14	.19
3-1/2" fiberglass insulation	1.000	S.F.	.63	.39	1.02
5/8" drywall	1.000	S.F.	.37	.53	.90
Drywall finishing adder	1.000	S.F.	.05	.53	.58
Baseboard trim, stock pine, 9/16" x 3-1/2", painted	.100	L.F.	.24	.22	.46
Paint baseboard, primer + 1 coat enamel	.100	L.F.	.01	.06	.07
Paint, 2 coats, interior	1.000	S.F.	.17	.59	.76
TOTAL			4.53	6.14	10.67

B2010 150	Wood Frame Exterior Wall	COST PER S.F. MAT.	COST PER S.F. INST.	COST PER S.F. TOTAL
1600	For alternate exterior wall systems:			
1700	Aluminum siding, horizontal clapboard	4.13	6.30	10.43
1800	Cedar bevel siding, 1/2" x 6", vertical , painted	7.50	6.05	13.55
1900	Redwood siding 1" x 4" to 1" x 6" vertical, T & G	5.80	6.65	12.45
2000	Board and batten	6.75	5.55	12.30
2100	Ship lap siding	4.97	5.85	10.82
2200	Plywood, grooved (T1-11) fir	3.64	5.85	9.49
2300	Redwood	5.10	5.85	10.95
2400	Southern yellow pine	3.73	5.85	9.58
2500	Masonry on stud wall, stucco, wire and plaster	3.21	9.20	12.41
2600	Stone veneer	8.55	12.10	20.65
2700	Brick veneer, brick $275 per M	8.65	15.20	23.85
3100	Cut & patch to match existing construction, add, minimum	2%	3%	
3200	Maximum	5%	9%	
3300	Dust protection, add, minimum	1%	2%	
3400	Maximum	4%	11%	
3500	Material handling & storage limitation, add, minimum	1%	1%	
3600	Maximum	6%	7%	
3700	Protection of existing work, add, minimum	2%	2%	
3800	Maximum	5%	7%	
3900	Shift work requirements, add, minimum		5%	
4000	Maximum		30%	
4100	Temporary shoring and bracing, add, minimum	2%	5%	
4200	Maximum	5%	12%	

B2010 Exterior Walls

B2010 190	Selective Price Sheet	COST PER S.F.		
		MAT.	INST.	TOTAL
0100	Exterior surface, masonry, concrete block, standard 4" thick	1.58	5.60	7.18
0200	6" thick	2.24	6	8.24
0300	8" thick	2.36	6.40	8.76
0400	12" thick	3.69	8.30	11.99
0500	Split rib, 4" thick	3.47	6.95	10.42
0600	8" thick	4.55	8	12.55
0700	Brick running bond, standard size, 6.75/S.F.	6.25	10.95	17.20
0800	Buff, 6.75/S.F.	6.60	10.95	17.55
0900	Stucco, on frame	.82	4.97	5.79
1000	On masonry	.29	3.89	4.18
1100	Metal, aluminum, horizontal, plain	1.73	2.05	3.78
1200	Insulated	1.69	2.05	3.74
1300	Vertical, plain	1.68	2.05	3.73
1400	Insulated	1.75	2.14	3.89
1500	Wood, beveled siding, "A" grade cedar, 1/2" x 6"	5.10	1.79	6.89
1600	1/2" x 8"	5.45	1.60	7.05
1700	Shingles, 16" #1 red, 7-1/2" exposure	2.43	2.57	5
1800	18" perfections, 7-1/2" exposure	2.96	2.15	5.11
1900	Handsplit, 10" exposure	2.78	2.11	4.89
2000	White cedar, 7-1/2" exposure	1.40	2.64	4.04
2100	Vertical, board & batten, redwood	3.41	2.40	5.81
2200	White pine	2.42	1.60	4.02
2300	T. & G. boards, redwood, 1" x 4"	3.41	2.40	5.81
2400	1' x 8"	3.56	1.60	5.16
2500				
2600	Interior surface, drywall, taped & finished, standard, 1/2"	.39	1.09	1.48
2700	5/8" thick	.42	1.09	1.51
2800	Fire resistant, 1/2" thick	.40	1.09	1.49
2900	5/8" thick	.44	1.09	1.53
3000	Moisture resistant, 1/2" thick	.57	1.09	1.66
3100	5/8" thick	.56	1.09	1.65
3200	Core board, 1" thick	.97	2.20	3.17
3300	Plaster, gypsum, 2 coats	.46	2.48	2.94
3400	3 coats	.66	3.01	3.67
3500	Perlite or vermiculite, 2 coats	.50	2.83	3.33
3600	3 coats	.77	3.54	4.31
3700	Gypsum lath, standard, 3/8" thick	.67	.59	1.26
3800	1/2" thick	.72	.62	1.34
3900	Fire resistant, 3/8" thick	.35	.71	1.06
4000	1/2" thick	.40	.77	1.17
4100	Metal lath, diamond, 2.5 lb.	.41	.59	1
4200	Rib, 3.4 lb.	.68	.71	1.39
4300	Framing metal studs including top and bottom			
4400	Runners, walls 10' high			
4500	24" O.C., non load bearing 20 gauge, 2-1/2" wide	.48	.88	1.36
4600	3-5/8" wide	.52	.89	1.41
4700	4" wide	.61	.89	1.50
4800	6" wide	.72	.91	1.63
4900	Load bearing 18 gauge, 2-1/2" wide	.87	.99	1.86
5000	3-5/8" wide	1.03	1.01	2.04
5100	4" wide	1.16	1.28	2.44
5200	6" wide	1.38	1.05	2.43
5300	16" O.C., non load bearing 20 gauge, 2-1/2" wide	.65	1.35	2
5400	3-5/8" wide	.70	1.37	2.07
5500	4" wide	.83	1.39	2.22
5600	6" wide	.98	1.41	2.39
5700	Load bearing 18 gauge, 2-1/2" wide	1.18	1.37	2.55
5800	3-5/8" wide	1.40	1.39	2.79

B2010 190	Selective Price Sheet	COST PER S.F.		
		MAT.	INST.	TOTAL
5900	4" wide	1.47	1.43	2.90
6000	6" wide	1.87	1.45	3.32
6100	Framing wood studs incl. double top plate and			
6200	Single bottom plate, walls 10' high			
6300	24" O.C., 2" x 4"	.28	.85	1.13
6400	2" x 6"	.51	.92	1.43
6500	16" O.C., 2" x 4"	.37	1.06	1.43
6600	2" x 6"	.66	1.17	1.83
6700	Sheathing, boards, 1" x 6"	1.53	1.62	3.15
6800	1" x 8"	1.47	1.38	2.85
6900	Plywood, 3/8" thick	.49	.88	1.37
7000	1/2" thick	.51	.94	1.45
7100	5/8" thick	.74	1.01	1.75
7200	3/4" thick	.91	1.08	1.99
7300	Wood fiber, 5/8" thick	.78	.88	1.66
7400	Gypsum weatherproof, 1/2" thick	.67	1.01	1.68
7500	Insulation, fiberglass batts, 3-1/2" thick, R13	.46	.75	1.21
7600	6" thick, R19	.57	.88	1.45
7700	Poured 4" thick, fiberglass wool, R4/inch	.51	2.64	3.15
7800	Mineral wool, R3/inch	.43	2.64	3.07
7900	Polystyrene, R4/inch	3.40	2.64	6.04
8000	Perlite or vermiculite, R2.7/inch	1.93	2.64	4.57
8100	Rigid, fiberglass, R4.3/inch, 1" thick	.54	.53	1.07
8200	R8.7/inch, 2" thick	1.27	.59	1.86

B2020 Exterior Windows

This page illustrates and describes an aluminum window system including double hung aluminum window, exterior and interior trim, hardware and insulating glass. Lines within System Components give the unit price and total price on a cost each basis for this system. Prices for alternate aluminum window systems are on Line Items B2020 110 1100 thru 2400. Both material quantities and labor costs have been adjusted for the system listed.

Factors: To adjust for job conditions other than normal working situations use Lines B2020 110 3100 thru 4000.

Example: You are to install the system and cut and patch to match existing construction. Go to Line B2020 110 3200 and apply these percentages to the appropriate MAT. and INST. costs.

System Components	QUANTITY	UNIT	COST EACH MAT.	COST EACH INST.	TOTAL
Double hung aluminum window, 2'-4" x 2'-6" exterior and interior trim, Hardware glazed with insulating glass.					
Double hung aluminum sash, 2'-4" x 2'-6"	1.000	Ea.	233.20	38.48	271.68
Exterior and interior trim	1.000	Set	35.50	40.50	76
Hardware	1.000	Set	3.74	22	25.74
Insulating glass	5.830	S.F.	72.88	29.03	101.91
TOTAL			345.32	130.01	475.33

B2020 110	Windows - Aluminum	COST EACH MAT.	COST EACH INST.	TOTAL
1000	For alternate window systems:			
1100	Aluminum, double hung, 2'-8" x 4'-6"	675	214	889
1200	3'-4" x 5'-6"	1,025	320	1,345
1300	Casement, 3'-6" x 2'-4"	465	157	622
1400	4'-6" x 2'-4"	590	196	786
1500	5'-6" x 2'-4"	740	258	998
1600	Projected window, 2'-1" x 3'-0"	340	135	475
1700	3'-5" x 3'-0"	545	194	739
1800	4'-0" x 4'-0"	850	295	1,145
1900	Horizontal sliding 3'-0" x 3'-0"	355	167	522
2000	3'-6" x 4'-0"	535	237	772
2100	5'-0" x 6'-0"	1,125	455	1,580
2200	Picture window, 3'-8" x 3'-1"	385	193	578
2300	4'-4" x 4'-5"	635	296	931
2400	5'-8" x 4'-9"	900	420	1,320
2500				
2600				
2700				
2800				
2900				
3100	Cut & patch to match existing construction, add, minimum	2%	3%	
3200	Maximum	5%	9%	
3300	Dust protection, add, minimum	1%	2%	
3400	Maximum	4%	11%	
3500	Material handling & storage limitation, add, minimum	1%	1%	
3600	Maximum	6%	7%	
3700	Protection of existing work, add, minimum	2%	2%	
3800	Maximum	5%	7%	
3900	Shift work requirements, add, minimum		5%	
4000	Maximum		30%	

B20 Exterior Enclosure

B2020 Exterior Windows

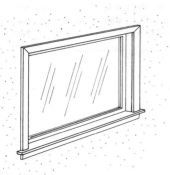

This page illustrates and describes a wood window system including wood picture window, exterior and interior trim, hardware and insulating glass. Lines within System Components give the unit price and total price on a cost each basis for this system. Prices for alternate wood window systems are on Line Items B2020 112 1300 thru 1800. Both material quantities and labor costs have been adjusted for the system listed.

Factors: To adjust for job conditions other than normal working situations use Lines B2020 112 3100 thru 4000.

Example: You are to install the above system where dust control is vital. Go to Line B2020 112 3400 and apply these percentages to the appropriate MAT. and INST. costs.

System Components	QUANTITY	UNIT	COST EACH		
			MAT.	INST.	TOTAL
Wood picture window 4'-0" x 4'-6", exterior and interior trim, hardware,					
Glazed with insulating glass.					
4'-0" x 4'-6" wood picture window with insulating glass	1.000	Ea.	550	96	646
Exterior and interior trim	1.000	Set	68.50	88	156.50
Hardware	1.000	Set	3.74	22	25.74
TOTAL			622.24	206	828.24

B2020 112	Windows - Wood	COST EACH		
		MAT.	INST.	TOTAL
1200	For alternate window systems:			
1300	Picture window 5'-0" x 4'-0"	705	206	911
1400	6'-0" x 4'-6"	745	216	961
1500	Bow, bay window, 8'-0" x 5'-0", standard	2,525	216	2,741
1600	Deluxe	1,850	216	2,066
1700	Bow, bay window, 12'-0" x 6'-0" standard	3,225	286	3,511
1800	Deluxe	2,600	286	2,886
1900	Double hung, 3'-0" x 4'-0"	350	134	484
2000	4'-0" x 4'-6"	445	176	621
2100	Casement 2'-0" x 3'-0"	287	116	403
2200	2 leaf, 4'-0" x 4'-0"	625	192	817
2300	3 leaf, 6'-0" x 6'-0"	1,625	310	1,935
2400	Awning, 2'-10" x 1'-10"	325	116	441
2500	3'-6" x 2'-4"	410	134	544
2600	4'-0" x 3'-0"	595	176	771
2700	Horizontal sliding 3'-0" x 2'-0"	330	116	446
2800	4'-0" x 3'-6"	380	134	514
2900	6'-0" x 5'-0"	575	176	751
3100	Cut & patch to match existing construction, add, minimum	2%	3%	
3200	Maximum	5%	9%	
3300	Dust protection, add, minimum	1%	2%	
3400	Maximum	4%	11%	
3500	Material handling & storage limitation, add, minimum	1%	1%	
3600	Maximum	6%	7%	
3700	Protection of existing work, add, minimum	2%	2%	
3800	Maximum	5%	7%	
3900	Shift work requirements, add, minimum		5%	
4000	Maximum		30%	

B20 Exterior Enclosure

B2020 Exterior Windows

This page illustrates and describes storm window and door systems based on a cost each price. Prices for alternate storm window and door systems are on Line Items B2020 116 0400 thru 2400. Both material quantities and labor costs have been adjusted for the system listed.

Factors: To adjust for job conditions other than normal working situations use Lines B2020 116 3100 thru 4000 and apply these percentages to the appropriate MAT and INST. costs.

Example: You are to install the system and protect all existing construction. Go to Line B2020 116 3800 and apply these percentages to the appropriate MAT. and INST. costs.

B2020 116	Storm Windows & Doors	COST EACH		
		MAT.	INST.	TOTAL
0350	Storm Window and door systems, single glazed			
0400	Window, custom aluminum anodized, 2'-0" x 3'-5"	90	35	125
0500	2'-6" x 5'-0"	119	37.50	156.50
0600	4'-0" x 6'-0"	249	42	291
0700	White painted aluminum, 2'-0" x 3'-5"	106	35	141
0800	2'-6" x 5'-0"	169	37.50	206.50
0900	4'-0" x 6'-0"	300	42	342
1000	Average quality aluminum, anodized, 2'-0" x 3'-5"	90	35	125
1100	2'-6" x 5'-0"	112	37.50	149.50
1200	4'-0" x 6'-0"	132	42	174
1300	White painted aluminum, 2'-0" x 3'-5"	89	35	124
1400	2'-6" x 5'-0"	97	37.50	134.50
1500	4'-0" x 6'-0"	106	42	148
1600	Mill finish, 2'-0" x 3'-5"	80.50	35	115.50
1700	2'-6" x 5'-0"	90	37.50	127.50
1800	4'-0" x 6'-0"	101	42	143
1900				
2000	Door, aluminum anodized 3'-0" x 6'-8"	171	75.50	246.50
2100	White painted aluminum, 3'-0" x 6'-8"	335	75.50	410.50
2200	Mill finish, 3'-0" x 6'-8"	271	75.50	346.50
2300				
2400	Wood, storm and screen, painted	310	117	427
2500				
2600				
2700				
2800				
3100	Cut & patch to match existing construction, add, minimum	2%	3%	
3200	Maximum	5%	9%	
3300	Dust protection, add, minimum	1%	2%	
3400	Maximum	4%	11%	
3500	Material handling & storage limitation, add, minimum	1%	1%	
3600	Maximum	6%	7%	
3700	Protection of existing work, add, minimum	2%	2%	
3800	Maximum	5%	7%	
3900	Shift work requirements, add, minimum		5%	
4000	Maximum		30%	

B2020 Exterior Windows

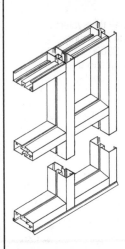

This page illustrates and describes a window wall system including aluminum tube framing, caulking, and glass. Lines within System Components give the unit price and total price per square foot for this system. Prices for alternate window wall systems are on Line Items B2020 124 1400 thru 2600. Both material quantities and labor costs have been adjusted for the system listed.

Factors: To adjust for job conditions other than normal working situations use Lines B2020 124 3100 thru 4000.

Example: You are to install the system with need for complete dust protection. Go to Line B2020 124 3400 and apply these percentages to the appropriate MAT. and INST. costs.

System Components	QUANTITY	UNIT	COST PER S.F. MAT.	INST.	TOTAL
Window wall, including aluminum header, sill, mullions, caulking And glass.					
Header, mill finish, 1-3/4" x 4-1/2" deep	.167	L.F.	3.09	2	5.09
Sill, mill finish, 1-3/4" x 4-1/2" deep	.167	L.F.	3	1.95	4.95
Vertical mullion, 1-3/4" x 4-1/2" deep, 6' O.C.	.191	L.F.	7.52	4.89	12.41
Caulking	.381	L.F.	.07	.80	.87
Glass, 1/4" plate	1.000	S.F.	5.80	8.30	14.10
TOTAL			19.48	17.94	37.42

B2020 124	Aluminum Frame, Window Wall	COST PER S.F. MAT.	INST.	TOTAL
1200	For alternate systems:			
1300				
1400	Mill finish, 2" x 4-1/2" deep, insulating glass	21.50	21	42.50
1500	Thermo-break, 2-1/4" x 4-1/2" deep, insulating glass	22.50	21	43.50
1600	Bronze finish, 1-3/4" x 4-1/2" deep, 1/4" plate glass	22	19.50	41.50
1700	2" x 4-1/2" deep, insulating glass	23.50	22.50	46
1800	Thermo-break, 2-1/4" x 4-1/2" deep, insulating glass	25	23	48
1900				
2000	Black finish, 1-3/4" x 4-1/2" deep, 1/4" plate glass	23	20.50	43.50
2100	2" x 4-1/2" deep, insulating glass	24.50	23.50	48
2200	Thermo-break, 2-1/4" x 4-1/2" deep, insulating glass	26	23.50	49.50
2300				
2400	Stainless steel, 1-3/4" x 4-1/2" deep, 1/4" plate glass	29.50	24.50	54
2500	2" x 4-1/2" deep, insulating glass	30.50	28	58.50
2600	Thermo-break, 2-1/4" x 4-1/2" deep, insulating glass	32.50	28.50	61
2700				
3100	Cut & patch to match existing construction, add, minimum	2%	3%	
3200	Maximum	5%	9%	
3300	Dust protection, add, minimum	1%	2%	
3400	Maximum	4%	11%	
3500	Material handling, & storage limitation, add, minimum	1%	1%	
3600	Maximum	6%	7%	
3700	Protection of existing work, add, minimum	2%	2%	
3800	Maximum	5%	7%	
3900	Shift work requirements, add, minimum		5%	
4000	Maximum		30%	

B2030 Exterior Doors

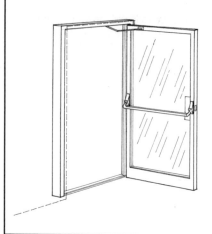

This page illustrates and describes a commercial metal door system, including a single aluminum and glass door, narrow stiles, jamb, hardware weatherstripping, panic hardware and closer. Lines within System Components give the unit price and total price on a cost each basis for this system. Prices for alternate commercial metal door systems are on Line Items B2030 125 1300 thru 2500. Both material quantities and labor costs have been adjusted for the system listed.

Factors: To adjust for job conditions other than normal working situations, use Lines B2030 125 3100 thru 4000.

Example: You are to install the system and cut and patch to match existing construction. Go to Line B2030 125 3200 and apply these percentages to the appropriate MAT. and INST. costs.

System Components			COST EACH		
	QUANTITY	UNIT	MAT.	INST.	TOTAL
Single aluminum and glass door, 3'-0"x7'-0", with narrow stiles, ext. jamb. **Weatherstripping, 1/2" tempered insul. glass, panic hardware, and closer.**					
Aluminum door, 3'-0" x 7'-0" x 1-3/4", narrow stiles	1.000	Ea.	740	660	1,400
Tempered insulating glass, 1/2" thick	20.000	S.F.	490	362	852
Panic hardware	1.000	Set	525	88	613
Automatic closer	1.000	Ea.	107	81	188
TOTAL			1,862	1,191	3,053

B2030 125	Doors, Metal - Commercial	COST EACH		
		MAT.	INST.	TOTAL
1200	For alternate door systems:			
1300	Single aluminum and glass with transom, 3'-0" x 10'-0"	2,625	1,400	4,025
1400	Anodized aluminum and glass, 3'-0" x 7'-0"	2,125	1,425	3,550
1500	With transom, 3'-0" x 10'-0"	3,075	1,675	4,750
1600	Steel, deluxe, hollow metal 3'-0" x 7'-0"	1,350	455	1,805
1700	With transom 3'-0" x 10'-0"	1,800	515	2,315
1800	Fire door, "A" label, 3'-0" x 7'-0"	1,475	455	1,930
1900	Double, aluminum and glass, 6'-0" x 7'-0"	2,850	2,050	4,900
2000	With transom, 6'-0" x 10'-0"	3,525	2,500	6,025
2100	Anodized aluminum and glass 6'-0" x 7'-0"	3,250	2,400	5,650
2200	With transom, 6'-0" x 10'-0"	4,000	2,925	6,925
2300	Steel, deluxe, hollow metal, 6'-0" x 7'-0"	2,125	790	2,915
2400	With transom, 6'-0" x 10'-0"	3,025	915	3,940
2500	Fire door, "A" label, 6'-0" x 7'-0"	2,250	785	3,035
2800				
2900				
3100	Cut & patch to match existing construction, add, minimum	2%	3%	
3200	Maximum	5%	9%	
3300	Dust protection, add, minimum	1%	2%	
3400	Maximum	4%	11%	
3500	Material handling & storage limitation, add, minimum	1%	1%	
3600	Maximum	6%	7%	
3700	Protection of existing work, add, minimum	2%	2%	
3800	Maximum	5%	7%	
3900	Shift work requirements, add, minimum		5%	
4000	Maximum		30%	

B2030 Exterior Doors

This page illustrates and describes sliding door systems including a sliding door, frame, interior and exterior trim with exterior staining. Lines within System Components give the unit price and total price on a cost each basis for this system. Prices for alternate sliding door systems are on Line Items B2030 150 1100 thru 2400. Both material quantities and labor costs have been adjusted for the system listed.

Factors: To adjust for job conditions other than normal working situations use Lines B2030 150 2700 thru 4000.

Example: You are to install the system with temporary shoring and bracing. Go to Line B2030 150 3900 and apply these percentages to the appropriate MAT. and INST. costs.

System Components	QUANTITY	UNIT	COST EACH		
			MAT.	INST.	TOTAL
Sliding wood door, 6'-0" x 6'-8", with wood frame, interior and exterior Trim and exterior staining.					
Sliding wood door, standard, 6'-0" x 6'-8", insulated glass	1.000	Ea.	935	264	1,199
Interior & exterior trim	1.000	Set	29.20	44	73.20
Stain door	1.000	Ea.	1.98	26.50	28.48
Stain trim	1.000	Ea.	2.40	11.20	13.60
TOTAL			968.58	345.70	1,314.28

B2030 150	Doors, Sliding - Patio	COST EACH		
		MAT.	INST.	TOTAL
1000	For alternate sliding door systems:			
1100	Wood, standard, 8'-0" x 6'-8", insulated glass	1,050	435	1,485
1200	12'-0" x 6'-8"	2,400	530	2,930
1300	Vinyl coated, 6'-0" x 6'-8"	1,350	345	1,695
1400	8'-0" x 6'-8"	1,600	435	2,035
1500	12'-0" x 6'-8"	3,225	530	3,755
1700	Aluminum, standard, 6'-0" x 6'-8", insulated glass	995	370	1,365
1800	8'-0" x 6'-8"	1,575	470	2,045
1900	12'-0" x 6'-8"	1,675	560	2,235
2000	Anodized, 6'-0" x 6'-8"	1,550	370	1,920
2100	8'-0" x 6'-8"	1,825	470	2,295
2200	12'-0" x 6'-8"	3,050	560	3,610
2300				
2400	Deduct for single glazing	86.50		86.50
2500				
2700	Cut & patch to match existing construction, add, minimum	2%	3%	
2800	Maximum	5%	9%	
2900	Dust protection, add, minimum	1%	2%	
3000	Maximum	4%	11%	
3100	Equipment usage curtailment, add, minimum	1%	1%	
3200	Maximum	3%	10%	
3300	Material handling & storage limitation, add, minimum	1%	1%	
3400	Maximum	6%	7%	
3500	Protection of existing, work, add, minimum	2%	2%	
3600	Maximum	5%	7%	
3700	Shift work requirements, add, minimum		5%	
3800	Maximum		30%	
3900	Temporary shoring and bracing, add, minimum	2%	5%	
4000	Maximum	5%	12%	

B2030 Exterior Doors

This page illustrates and describes residential door systems including a door, frame, trim, hardware, weatherstripping, stained and finished. Lines within System Components give the unit price and total price on a cost each basis for this system. Prices for alternate residential door systems are on Line Items B2030 240 1500 thru 2200. Both material quantities and labor costs have been adjusted for the system listed.

Factors: To adjust for job conditions other than normal working situations use Lines B2030 240 3100 thru 4000.

Example: You are to install the system with a material handling and storage limitation. Go to Line B2030 240 3600 and apply these percentages to the appropriate MAT. and INST. costs.

System Components	QUANTITY	UNIT	COST EACH MAT.	COST EACH INST.	COST EACH TOTAL
Single solid wood colonial door, 3' x 6'-8" with wood frame and trim, Stained and finished, including hardware and weatherstripping.					
Solid wood colonial door, fir 3' x 6'-8" x 1-3/4", hinges	1.000	Ea.	475	70.50	545.50
Hinges, ball bearing	1.000	Ea.	34		34
Exterior frame, with trim	1.000	Set	158.95	47.77	206.72
Interior trim	1.000	Set	25	89.50	114.50
Lockset	1.000	Set	40	37.50	77.50
Sill, oak 8" deep	3.500	Ea.	73.50	36.93	110.43
Weatherstripping	1.000	Set	23.50	69.50	93
Stained and finished	1.000	Ea.	4.23	56.50	60.73
TOTAL			834.18	408.20	1,242.38

B2030 240	Doors, Residential - Exterior	COST EACH MAT.	COST EACH INST.	COST EACH TOTAL
1400	For alternate exterior door system:			
1500	Single doors			
1600	Hollow metal exterior door, plain	665	385	1,050
1700	Solid core wood door, plain	485	395	880
1800				
1900	Double doors, 6' x 6'-8"			
2000	Solid colonial double doors, fir	1,400	535	1,935
2100	Hollow metal exterior doors, plain	1,075	505	1,580
2200	Hollow core wood doors, plain	705	520	1,225
2300				
2800				
2900				
3100	Cut & patch to match existing construction, add, minimum	2%	3%	
3200	Maximum	5%	9%	
3300	Dust protection, add, minimum	1%	2%	
3400	Maximum	4%	11%	
3500	Material handling & storage limitation, add, minimum	1%	1%	
3600	Maximum	6%	7%	
3700	Protection of existing work, add, minimum	2%	2%	
3800	Maximum	5%	7%	
3900	Shift work requirements, add, minimum		5%	
4000	Maximum		30%	

B2030 Exterior Doors

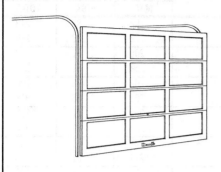

This page illustrates and describes overhead door systems including an overhead door, track, hardware, trim, and electric door opener. Lines within System Components give the unit price and total price on a cost each basis for this system. Prices for alternate overhead door systems are on Line Items B2030 410 1300 thru 2300. Both material quantities and labor costs have been adjusted for the system listed.

Factors: To adjust for job conditions other than normal working situations use Lines B2030 410 3100 thru 4000.

Example: You are to install the system and match existing construction. Go to Line B2030 410 3200 and apply these percentages to the appropriate MAT. and INST. costs.

System Components	QUANTITY	UNIT	COST EACH		
			MAT.	INST.	TOTAL
Wood overhead door, commercial sectional, including track, door, hardware And trim, electrically operated.					
Commercial, heavy duty wood door, 8' x 8' x 1-3/4" thick	1.000	Ea.	830	530	1,360
Frame, 2 x 8, pressure treated	1.000	Set	27.60	66.72	94.32
Wood trim	1.000	Set	58.56	52.80	111.36
Painting, two coats	1.000	Ea.	18.72	180	198.72
Electric trolley operator	1.000	Ea.	1,025	264	1,289
TOTAL			1,959.88	1,093.52	3,053.40

B2030 410	Doors, Overhead	COST EACH		
		MAT.	INST.	TOTAL
1200	For alternate sectional overhead door systems:			
1300	Commercial, wood, 1-3/4" thick, 12' x 12'	3,000	1,550	4,550
1400	14' x 14'	4,025	1,800	5,825
1500	Fiberglass & aluminum 12' x 12'	4,000	1,575	5,575
1600	20' x 20'	8,625	3,975	12,600
1700	Residential, wood, 9' x 7'	865	415	1,280
1800	16' x 7'	1,675	620	2,295
1900	Hardboard faced, 9' x 7'	685	415	1,100
2000	16' x 7'	1,350	620	1,970
2100	Fiberglass & aluminum, 9' x 7'	875	485	1,360
2200	16' x 7'	1,825	620	2,445
2300	For residential electric opener, add	650	66	716
2400				
2500				
2600				
2700				
2800				
2900				
3100	Cut & patch to match existing construction, add, minumum	2%	3%	
3200	Maximum	5%	9%	
3300	Dust protection, add, minimum	1%	2%	
3400	Maximum	4%	11%	
3500	Material handling & storage limitation, add, minimum	1%	1%	
3600	Maximum	6%	7%	
3700	Protection of existing work, add, minimum	2%	2%	
3800	Maximum	5%	7%	
3900	Shift work requirements, add, minimum		5%	
4000	Maximum		30%	

B20 Exterior Enclosure

B2030 Exterior Doors

B2030 810	Selective Price Sheet	COST EACH		
		MAT.	INST.	TOTAL
0100	Door closer, rack and pinion	162	81	243
0200	Backcheck and adjustable power	172	88	260
0300	Regular, hinge face mount, all sizes, regular arm	206	81	287
0400	Hold open arm	259	81	340
0500	Top jamb mount, all sizes, regular arm	202	88	290
0600	Hold open arm	223	88	311
0700	Stop face mount, all sizes, regular arm	202	81	283
0800	Hold open arm	223	81	304
0900	Fusible link, hinge face mount, all sizes, regular arm	221	81	302
1000	Hold open arm	274	81	355
1100	Top jamb mount, all sizes, regular arm	217	88	305
1200	Hold open arm	238	88	326
1300	Stop face mount, all sizes, regular arm	217	81	298
1400	Hold open arm	238	81	319
1500				
1600				
1700	Door stops			
1800				
1900	Holder & bumper, floor or wall	37	16.50	53.50
2000	Wall bumper	11.30	16.50	27.80
2100	Floor bumper	6.45	16.50	22.95
2200	Plunger type, door mounted	29.50	16.50	46
2300	Hinges, full mortise, material only, per pair			
2400	Low frequency, 4-1/2" x 4-1/2", steel base, USP	13.25		13.25
2500	Brass base, US10	46		46
2600	Stainless steel base, US32	79.50		79.50
2700	Average frequency, 4-1/2" x 4-1/2", steel base, USP	27		27
2800	Brass base, US10	55		55
2900	Stainless steel base, US32	82		82
3000	High frequency, 4-1/2" x 4-1/2", steel base, USP	64.50		64.50
3100	Brass base, US10	86.50		86.50
3200	Stainless steel base, US32	137		137
3300	Kick plate			
3400				
3500	6" high, for 3'-0" door, aluminum	34	35	69
3600	Bronze	48	35	83
3700	Panic device			
3800				
3900	For rim locks, single door, exit	525	88	613
4000	Outside key and pull	570	106	676
4100	Bar and vertical rod, exit only	765	106	871
4200	Outside key and pull	855	132	987
4300	Lockset			
4400				
4500	Heavy duty, cylindrical, passage doors	152	44	196
4600	Classroom	340	66	406
4700	Bedroom, bathroom, and inner office doors	190	44	234
4800	Apartment, office, and corridor doors	259	53	312
4900	Standard duty, cylindrical, exit doors	92	53	145
5100	Passage doors	67	44	111
5200	Public restroom, classroom, & office doors	132	66	198
5300	Deadlock, mortise, heavy duty	165	58.50	223.50
5400	Double cylinder	178	58.50	236.50
5500	Entrance lock, cylinder, deadlocking latch	140	58.50	198.50
5600	Deadbolt	169	66	235
5700	Commercial, mortise, wrought knob, keyed, minimum	208	66	274
5800	Maximum	385	75.50	460.50
5900	Cast knob, keyed, minimum	275	58.50	333.50

B2030 Exterior Doors

B2030 810	Selective Price Sheet	COST EACH		
		MAT.	INST.	TOTAL
6000	Maximum	540	58.50	598.50
6100	Push-pull			
6200				
6300	Aluminum	9.50	44	53.50
6400	Bronze	22.50	44	66.50
6500	Door pull, designer style, minimum	74.50	44	118.50
6600	Maximum	400	66	466
6700	Threshold			
6800				
6900	3'-0" long door saddles, aluminum, minimum	4.42	11	15.42
7000	Maximum	42	44	86
7100	Bronze, minimum	42.50	8.80	51.30
7200	Maximum	72	44	116
7300	Rubber, 1/2" thick, 5-1/2" wide	41.50	26.50	68
7400	2-3/4" wide	17.60	26.50	44.10
7500	Weatherstripping, per set			
7600				
7700	Doors, wood frame, interlocking for 3' x 7' door, zinc	17.05	176	193.05
7800	Bronze	26.50	176	202.50
7900	Wood frame, spring type for 3' x 7' door, bronze	23.50	69.50	93
8000	Metal frame, spring type for 3' x 7' door, bronze	39	176	215
8100	For stainless steel, spring type, add	133%		
8200				
8300	Metal frame, extruded sections, 3' x 7' door, aluminum	49.50	264	313.50
8400	Bronze	125	264	389

B3010 Roof Coverings

B3010 160	Selective Price Sheet	COST PER S.F.		
		MAT.	INST.	TOTAL
0100	Roofing, built-up, asphalt roll roof, 3 ply organic/mineral surface	.75	1.36	2.11
0200	3 plies glass fiber felt type iv, 1 ply mineral surface	1.14	1.47	2.61
0300	Cold applied, 3 ply		.49	.49
0400	Coal tar pitch, 4 ply tarred felt	1.72	1.75	3.47
0500	Mopped, 3 ply glass fiber	1.40	1.94	3.34
0600	4 ply organic felt	1.72	1.75	3.47
0700	Elastomeric, hypalon, neoprene unreinforced	2.83	3.25	6.08
0800	Polyester reinforced	2.87	3.83	6.70
0900	Neoprene, 5 coats 60 mils	6.30	11.35	17.65
1000	Over 10,000 S.F.	5.90	5.85	11.75
1100	PVC, traffic deck sprayed	1.85	5.85	7.70
1200	With neoprene	1.95	2.37	4.32
1300	Shingles, fiber cement, strip, 14" x 30", 325#/sq.	3.70	2.22	5.92
1500	Shake, 9.35" x 16" 500#/sq.	3.35	2.22	5.57
1600				
1700	Asphalt, strip, 210-235#/sq.	.55	.89	1.44
1800	235-240#/sq.	.56	.98	1.54
1900	Class A laminated	.68	1.09	1.77
2000	Class C laminated	.74	1.22	1.96
2100	Slate, buckingham, 3/16" thick	5.25	2.78	8.03
2200	Black, 1/4" thick	5.25	2.78	8.03
2300	Wood, shingles, 16" no. 1, 5" exp.	3.65	2.11	5.76
2400	Red cedar, 18" perfections	3.60	1.92	5.52
2500	Shakes, 24", 10" exposure	2.78	2.11	4.89
2600	18", 8-1/2" exposure	2.63	2.64	5.27
2700	Insulation, ceiling batts, fiberglass, 3-1/2" thick, R13	.37	.39	.76
2800	6" thick, R19	.48	.39	.87
2900	9" thick, R30	.73	.46	1.19
3000	12" thick, R38	1.05	.46	1.51
3100	Mineral, 3-1/2" thick, R13	.42	.33	.75
3200	Fiber, 6" thick, R19	.53	.33	.86
3300	Roof deck, fiberboard, 1" thick, R2.78	.46	.61	1.07
3400	Mineral, 2" thick, R5.26	.94	.61	1.55
3500	Perlite boards, 3/4" thick, R2.08	.41	.61	1.02
3600	2" thick, R5.26	.88	.70	1.58
3700	Polystyrene extruded, R5.26, 1" thick,	.54	.33	.87
3800	2" thick R10	.68	.39	1.07
3900	40 PSI compressive strength, 1" thick R5	.56	.33	.89
4000	Tapered for drainage	.78	.35	1.13
4100	Foamglass, 1 1/2" thick R4.55	1.49	.61	2.10
4200	3" thick R9.00	3.03	.70	3.73
4300	Ceiling, plaster, gypsum, 2 coats	.46	2.83	3.29
4400	3 coats	.66	3.36	4.02
4500	Perlite or vermiculite, 2 coats	.50	3.31	3.81
4600	3 coats	.77	4.19	4.96
4700	Gypsum lath, plain 3/8" thick	.67	.59	1.26
4800	1/2" thick	.72	.62	1.34
4900	Firestop, 3/8" thick	.35	.71	1.06
5000	1/2" thick	.40	.77	1.17
5100	Metal lath, rib, 2.75 lb.	.44	.67	1.11
5200	3.40 lb.	.68	.71	1.39
5300	Diamond, 2.50 lb.	.41	.67	1.08
5400	3.40 lb.	.56	.83	1.39
5500	Drywall, taped and finished standard, 1/2" thick	.39	1.38	1.77
5600	5/8" thick	.42	1.38	1.80
5700	Fire resistant, 1/2" thick	.40	1.38	1.78
5800	5/8" thick	.44	1.38	1.82
5900	Water resist., 1/2" thick	.57	1.38	1.95

B3010 Roof Coverings

B3010 160	Selective Price Sheet	COST PER S.F.		
		MAT.	INST.	TOTAL
6000	5/8" thick	.56	1.38	1.94
6100	Finish, instead of taping			
6200	For thin coat plaster, add	.09	.66	.75
6300	Finish textured spray, add	.09	.56	.65
6400				
6500	Tile, stapled glued, mineral fiber plastic coated, 5/8" thick	2.04	1.76	3.80
6600	3/4" thick	2.06	1.76	3.82
6700	Wood fiber, 1/2" thick	1.01	1.32	2.33
6800	3/4" thick	1.29	1.32	2.61
6900	Suspended, fiberglass film faced, 5/8" thick	.80	.84	1.64
7000	3" thick	1.68	1.17	2.85
7100	Mineral fiber 5/8" thick, standard face	.86	.78	1.64
7200	Aluminum faced	2.05	.88	2.93
7300	Wood fiber reveal edge, 1" thick			
7400	3" thick			
7500	Ceiling suspension systems, for tile, "T" bar, class "A", 2' x 4' grid	.69	.66	1.35
7600	2' x 2' grid	.86	.81	1.67
7700	Concealed "Z" bar, 12" module	.72	1.01	1.73
7800				
7900	Plaster/drywall, 3/4" channels, steel furring, 16" O.C.	.34	1.55	1.89
8000	24" O.C.	.23	1.07	1.30
8100	1-1/2" channels, 16" O.C.	.46	1.73	2.19
8200	24" O.C.	.30	1.15	1.45
8300	Ceiling framing, 2" x 4" studs, 16" O.C.	.24	.88	1.12
8400	24" O.C.	.16	.59	.75

B3020 Roof Openings

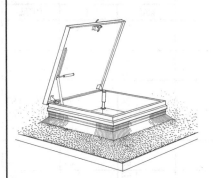

This page illustrates and describes a roof hatch system. Lines within System Components give the unit price and total cost each for this system. Prices for alternate systems are on Lines B3020 230 1500 thru 2400. Both material quantities and labor costs have been adjusted for the system listed.

Factors: To adjust for job conditions other than normal working situations use Lines B3020 230 2900 thru 4000.

Example: You are to install the system and cut and patch to match existing construction. Use line B3020 230 3000 and apply these percentages to the appropriate MAT. and INST. costs.

System Components			QUANTITY	UNIT	COST EACH		
					MAT.	INST.	TOTAL
Roof hatch, 2'-6" x 3'-0", aluminum, curb and cover included.							
Through steel construction.							
	Roof hatch, 2'-6" x 3'-0", aluminum		1.000	Ea.	1,075	203	1,278
	Cutout decking		11.000	L.F.		27.28	27.28
	Frame opening		44.000	L.F.	43.56	359.04	402.60
	Flashing, 16 oz. copper		16.000	S.F.	119.20	68	187.20
	Cant strip, 4 x 4, foamglass		12.000	L.F.	29.40	18.12	47.52
		TOTAL			1,267.16	675.44	1,942.60

B3020 230	Roof Hatches, Skylights		COST EACH		
			MAT.	INST.	TOTAL
1400	For alternate systems:				
1500	Roof hatch, aluminum, curb and cover, 2'-6" x 4'-6"		1,175	825	2,000
1600	2'-6" x 8'-0"		2,175	1,200	3,375
1700	Skylight, plexiglass dome, with curb mounting, 30" x 32"		610	645	1,255
1800	30" x 45"		740	800	1,540
1900	40" x 45"		885	1,075	1,960
2000	Smoke hatch, unlabeled, 2'-6" x 3'-0"		1,525	730	2,255
2100	2'-6" x 4'-6"		1,400	880	2,280
2200	2'-6" x 8'-0"		2,350	1,225	3,575
2300					
2400	For steel ladder, add per vertical linear foot		45	33	78
2500					
2600					
2700					
2900	Cut & patch to match existing construction, add, minimum		2%	3%	
3000	Maximum		5%	9%	
3100	Dust protection, add, minimum		1%	2%	
3200	Maximum		4%	11%	
3300	Equipment usage curtailment, add, minimum		1%	1%	
3400	Maximum		3%	10%	
3500	Material handling & storage limitation, add, minimum		1%	1%	
3600	Maximum		6%	7%	
3700	Protection of existing work, add, minimum		2%	2%	
3800	Maximum		5%	7%	
3900	Shift work requirements, add, minimum			5%	
4000	Maximum			30%	

B3020 Roof Openings

B3020 240	Selective Price Sheet	COST PER L.F.		
		MAT.	INST.	TOTAL
0100	Downspouts per L.F., aluminum, enameled .024" thick, 2" x 3"	2.02	3.33	5.35
0200	3" x 4"	2.81	4.29	7.10
0300	Round .025" thick, 3" diam.	1.56	3.16	4.72
0400	4" diam.	2.63	4.29	6.92
0500	Copper, round 16 oz. stock, 2" diam.	9.90	3.16	13.06
0600	3" diam.	9.45	3.16	12.61
0700	4" diam.	9.90	4.14	14.04
0800	5" diam.	16.25	4.62	20.87
0900	Rectangular, 2" x 3"	7.70	3.16	10.86
1000	3" x 4"	10	4.14	14.14
1100	Lead coated copper, round, 2" diam.	12.60	3.16	15.76
1200	3" diam.	13.85	3.16	17.01
1300	4" diam.	16.20	4.14	20.34
1400	5" diam.	25.50	4.62	30.12
1500	Rectangular, 2" x 3"	10.75	3.16	13.91
1600	3" x 4"	11	4.14	15.14
1700	Steel galvanized, round 28 gauge, 3" diam.	1.91	3.16	5.07
1800	4" diam.	2.73	4.14	6.87
1900	5" diam.	4.05	4.62	8.67
2000	6" diam.	5.30	5.70	11
2100	Rectangular, 2" x 3"	1.91	3.16	5.07
2200	3" x 4"	2.76	4.14	6.90
2300	Elbows, aluminum, round, 3" diam.	2.62	6	8.62
2400	4" diam.	7.15	6	13.15
2500	Rectangular, 2" x 3"	1.29	6	7.29
2600	3" x 4"	4.39	6	10.39
2700	Copper, round 16 oz., 2" diam.	14.75	6	20.75
2800	3" diam.	8.45	6	14.45
2900	4" diam.	14.10	6	20.10
3100	Rectangular, 2" x 3"	7.80	6	13.80
3200	3" x 4"	13.20	6	19.20
3300	Drip edge per L.F., aluminum, 5" wide	.46	1.32	1.78
3400	8" wide	.62	1.32	1.94
3500	28" wide	4.18	5.30	9.48
3600				
3700	Steel galvanized, 5" wide	.59	1.32	1.91
3800	8" wide	.86	1.32	2.18
3900				
4000				
4100				
4200				
4300	Flashing 12" wide per S.F., aluminum, mill finish, .013" thick	.74	3.37	4.11
4400	.019" thick	1.12	3.37	4.49
4500	.040" thick	2.43	3.37	5.80
4600	.050" thick	2.77	3.37	6.14
4700	Copper, mill finish, 16 oz.	7.45	4.25	11.70
4800	20 oz.	8.70	4.45	13.15
4900	24 oz.	10.60	4.66	15.26
5000	32 oz.	14.10	4.89	18.99
5100	Lead, 2.5 lb./S.F., 12" wide	4.55	3.62	8.17
5200	Over 12" wide	4.55	3.62	8.17
5300	Lead-coated copper, fabric backed, 2 oz.	2.66	1.48	4.14
5400	5 oz.	3.06	1.48	4.54
5500	Mastic backed, 2 oz.	2.08	1.48	3.56
5600	5 oz.	2.57	1.48	4.05
5700	Paper backed, 2 oz.	1.79	1.48	3.27
5800	3 oz.	2.11	1.48	3.59
5900	Polyvinyl chloride, black, .010" thick	.29	1.72	2.01

B3020 Roof Openings

B3020 240	Selective Price Sheet	COST PER L.F.		
		MAT.	INST.	TOTAL
6000	.020″ thick	.40	1.72	2.12
6100	.030″ thick	.52	1.72	2.24
6200	.056″ thick	1.22	1.72	2.94
6300	Steel, galvanized, 20 gauge	1.31	3.76	5.07
6400	30 gauge	.55	3.06	3.61
6500	Stainless, 32 gauge, .010″ thick	4.17	3.16	7.33
6600	28 gauge, .015″ thick	5.15	3.16	8.31
6700	26 gauge, .018″ thick	6.25	3.16	9.41
6800	24 gauge, .025″ thick	8.15	3.16	11.31
6900	Gutters per L.F., aluminum, 5″ box, .027″ thick	2.55	5	7.55
7000	.032″ thick	3.60	5	8.60
7100	Copper, half round, 4″ wide	6.90	5	11.90
7200	6″ wide	9.80	5.20	15
7300	Steel, 26 gauge galvanized, 5″ wide	1.75	5	6.75
7400	6″ wide	2.53	5	7.53
7500	Wood, treated hem-fir, 3″ x 4″	7.70	5.30	13
7600	4″ x 5″	8.80	5.30	14.10
7700	Reglet per L.F., aluminum, .025″ thick	1.40	2.35	3.75
7800	Copper, 10 oz.	2.65	2.35	5
7900	Steel, galvanized, 24 gauge	.95	2.35	3.30
8000	Stainless, .020″ thick	3.19	2.35	5.54
8100	Counter flashing 12″ wide per L.F., aluminum, .032″ thick	1.65	4	5.65
8200	Copper, 10 oz.	5	4	9
8300	Steel, galvanized, 24 gauge	.94	4	4.94
8400	Stainless, .020″ thick	5.50	4	9.50

C1010 Partitions

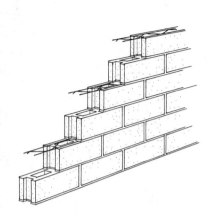

This page illustrates and describes a concrete block wall system including concrete block, horizontal reinforcing alternate courses, mortar, and tooled joints both sides. Lines within System Components give the unit price and total price per square foot for this system. Prices for alternate concrete block wall systems are on Line Items C1010 106 0900 thru 2100. Both material quantities and labor costs have been adjusted for the system listed.

Factors: To adjust for job conditions other than normal working situations use Lines C1010 106 2900 thru 4000.

Example: You are to install the system and protect all existing work. Go to Line C1010 106 3600 and apply these percentages to the appropriate MAT. and INST. costs.

System Components	QUANTITY	UNIT	COST PER S.F. MAT.	COST PER S.F. INST.	COST PER S.F. TOTAL
Concrete block partition, including horizontal reinforcing every second Course, mortar, tooled joints both sides.					
8" x 16" concrete block, normal weight, 4" thick	1.000	S.F.	1.58	5.60	7.18
Horizontal reinforcing every second course	.750	L.F.	.14	.13	.27
TOTAL			1.72	5.73	7.45

C1010 106	Partitions, Concrete Block	MAT.	INST.	TOTAL
0800	For alternate block partition systems:			
0900	8" x 16" concrete block, normal weight, 6" thick	2.38	6.15	8.53
1000	8" thick	2.50	6.55	9.05
1100	10" thick	3.38	6.85	10.23
1200	12" thick	3.83	8.45	12.28
1300	8" x 16" concrete block, lightweight, 4" thick	2.03	5.60	7.63
1400	6" thick	2.70	6	8.70
1500	8" thick	3.28	6.40	9.68
1600	10" thick	4.24	6.65	10.89
1700	12" thick	4.82	8.20	13.02
1800	8" x 16" glazed concrete block, 4" thick	9.50	7.10	16.60
1900	8" thick	10.30	7.90	18.20
2000	Structural facing tile, 6T series, glazed 2 sides, 4" thick	15.75	12.50	28.25
2100	6" thick	21.50	13.15	34.65
2200				
2300				
2400				
2500				
2600				
2900	Cut & patch to match existing construction, add, minimum	2%	3%	
3000	Maximum	5%	9%	
3100	Dust protection, add, minimum	1%	2%	
3200	Maximum	4%	11%	
3300	Material handling & storage limitation, add, minimum	1%	1%	
3400	Maximum	6%	7%	
3500	Protection of existing work, add, minimum	2%	2%	
3600	Maximum	5%	7%	
3700	Shift work requirements, add, minimum		5%	
3800	Maximum		30%	
3900	Temporary shoring and bracing, add, minimum	2%	5%	
4000	Maximum	5%	12%	

C1010 Partitions

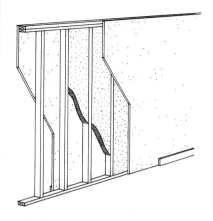

This page illustrates and describes a wood stud partition system including wood studs with plates, gypsum plasterboard – taped and finished, insulation, baseboard and painting. Lines within System Components give the unit price and total price per square foot for this system. Prices for alternate wood stud partition systems are on Line Items C1010 132 1300 thru 2700. Both material quantities and labor costs have been adjusted for the system listed.

Factors: To adjust for job conditions other than normal working situations use Lines C1010 132 2900 thru 4000.

Example: You are to install the system where material handling and storage present a serious problem. Go to Line C1010 132 3400 and apply these percentages to the appropriate MAT. and INST. costs.

System Components	QUANTITY	UNIT	COST PER S.F. MAT.	COST PER S.F. INST.	COST PER S.F. TOTAL
Wood stud wall, 2"x4", 16" O.C., dbl. top plate, sngl bot. plate, 5/8" dwl.					
Taped, finished and painted on 2 faces, insulation, baseboard, wall 8' high.					
Wood studs, 2" x 4", 16" O.C., 8' high	1.000	S.F.	.40	1.32	1.72
Gypsum drywall, 5/8" thick	2.000	S.F.	.74	1.06	1.80
Taping and finishing	2.000	S.F.	.10	1.06	1.16
Insulation, 3-1/2" fiberglass batts	1.000	S.F.	.63	.39	1.02
Baseboard	.200	L.F.	.61	.55	1.16
Paint baseboard, primer + 2 coats	.200	L.F.	.04	.33	.37
Painting, roller, 2 coats	2.000	S.F.	.34	1.18	1.52
TOTAL			2.86	5.89	8.75

C1010 132	Partitions, Wood Stud		MAT.	INST.	TOTAL
1200	For alternate wood stud systems:				
1300	2" x 3" studs, 8' high, 16" O.C.		2.83	5.80	8.63
1400	24" O.C.		2.75	5.55	8.30
1500	10' high, 16" O.C.		2.81	5.55	8.36
1600	24" O.C.		2.73	5.35	8.08
1700	2" x 4" studs, 8' high, 24" O.C.		2.77	5.65	8.42
1800	10' high, 16" O.C.		2.83	5.65	8.48
1900	24" O.C.		2.74	5.40	8.14
2000	12' high, 16" O.C.		2.80	5.65	8.45
2100	24" O.C.		2.69	5.40	8.09
2200	2" x 6" studs, 8' high, 16" O.C.		3.17	6.05	9.22
2300	24" O.C.		3.01	5.70	8.71
2400	10' high, 16" O.C.		3.12	5.75	8.87
2500	24" O.C.		2.97	5.50	8.47
2600	12' high, 16" O.C.		3.06	5.75	8.81
2700	24" O.C.		2.91	5.50	8.41
2900	Cut & patch to match existing construction, add, minimum		2%	3%	
3000	Maximum		5%	9%	
3100	Dust protection, add, minimum		1%	2%	
3200	Maximum		4%	11%	
3300	Material handling & storage limitation, add, minimum		1%	1%	
3400	Maximum		6%	7%	
3500	Protection of existing work, add, minimum		2%	2%	
3600	Maximum		5%	7%	
3700	Shift work requirements, add, minimum			5%	
3800	Maximum			30%	
3900	Temporary shoring and bracing, add, minimum		2%	5%	
4000	Maximum		5%	12%	

C10 Interior Construction

C1010 Partitions

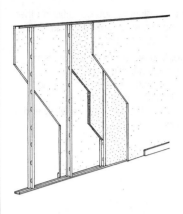

This page illustrates and describes a non-load bearing metal stud partition system including metal studs with runners, gypsum plasterboard, taped and finished, insulation, baseboard and painting. Lines within System Components give the unit price and total price per square foot for this system. Prices for alternate non-load bearing metal stud partition systems are on Line Items C1010 134 1300 thru 2300. Both material quantities and labor costs have been adjusted for the system listed.

Factors: To adjust for job conditions other than normal working situations use Lines C1010 134 2900 thru 4000.

Example: You are to install the system and cut and patch to match existing construction. Go to Line C1010 134 3000 and apply these percentages to the appropriate MAT. and INST. costs.

System Components	QUANTITY	UNIT	COST PER S.F. MAT.	COST PER S.F. INST.	COST PER S.F. TOTAL
Non-load bearing metal studs, including top & bottom runners, 5/8" drywall, Taped, finished and painted 2 faces, insulation, painted baseboard.					
Metal studs, 25 ga., 3-5/8" wide, 24" O.C.	1.000	S.F.	.36	.71	1.07
Gypsum drywall, 5/8" thick	2.000	S.F.	.74	1.06	1.80
Taping & finishing	2.000	S.F.	.10	1.06	1.16
Insulation, 3-1/2" fiberglass batts	1.000	S.F.	.63	.39	1.02
Baseboard	.200	L.F.	.49	.44	.93
Paint baseboard, primer + 2 coats	.200	L.F.	.03	.26	.29
Painting, roller work, 2 coats	2.000	S.F.	.34	1.18	1.52
TOTAL			2.69	5.10	7.79

C1010 134	Partitions, Metal Stud, NLB	COST PER S.F. MAT.	COST PER S.F. INST.	COST PER S.F. TOTAL
1200	For alternate metal stud systems:			
1300	Non-load bearing, 25 ga., 24" O.C., 2-1/2" wide	2.65	5.10	7.75
1400	6" wide	2.88	5.10	7.98
1500	16" O.C., 2-1/2" wide	2.73	5.25	7.98
1600	3-5/8" wide	2.78	5.30	8.08
1700	6" wide	3.02	5.30	8.32
1800	20 ga., 24" O.C., 2-1/2" wide	2.81	5.25	8.06
1900	3-5/8" wide	2.94	5.30	8.24
2000	6" wide	3.05	5.30	8.35
2100	16" O.C., 2-1/2" wide	2.93	5.50	8.43
2200	3-5/8" wide	3.09	5.50	8.59
2300	6" wide	3.23	5.55	8.78
2400				
2500				
2600				
2700				
2900	Cut & patch to match existing construction, add, minimum	2%	3%	
3000	Maximum	5%	9%	
3100	Dust protection, add, minimum	1%	2%	
3200	Maximum	4%	11%	
3300	Material handling & storage limitation, add, minimum	1%	1%	
3400	Maximum	6%	7%	
3500	Protection of existing work, add, minimum	2%	2%	
3600	Maximum	5%	7%	
3700	Shift work requirements, add, minimum		5%	
3800	Maximum		30%	
3900	Temporary shoring and bracing, add, minimum	2%	5%	
4000	Maximum	5%	12%	

C1010 Partitions

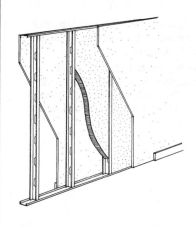

This page illustrates and describes a drywall system including gypsum plasterboard, taped and finished, metal studs with runners, insulation, baseboard and painting. Lines within System Components give the unit price and total price per square foot for this system. Prices for alternate drywall systems are on Line Items C1010 136 1300 thru 1900. Both material quantities and labor costs have been adjusted for the system listed.

Factors: To adjust for job conditions other than normal working situations use Lines C1010 136 2900 thru 4000.

Example: You are to install the system and control dust in the work area. Go to Line C1010 136 3100 and apply these percentages to the appropriate MAT. and INST. costs.

System Components	QUANTITY	UNIT	COST PER S.F. MAT.	COST PER S.F. INST.	COST PER S.F. TOTAL
Gypsum drywall, taped, finished and painted 2 faces, galvanized metal studs					
Including top & bottom runners, insulation, painted baseboard, wall 10'high.					
Gypsum drywall, 5/8" thick, standard	2.000	S.F.	.74	1.06	1.80
Taping and finishing	2.000	S.F.	.10	1.06	1.16
Metal studs, 20 ga., 3-5/8" wide, 24" O.C.	1.000	S.F.	.52	.89	1.41
Insulation, 3-1/2" fiberglass batts	1.000	S.F.	.63	.39	1.02
Baseboard	.200	L.F.	.49	.44	.93
Paint baseboard, primer + 2 coats	.200	L.F.	.02	.11	.13
Painting, roller 2 coats	2.000	S.F.	.26	1.12	1.38
TOTAL			2.76	5.07	7.83

C1010 136	Partitions, Drywall	COST PER S.F. MAT.	COST PER S.F. INST.	COST PER S.F. TOTAL
1200	For alternate drywall systems:			
1300	Gypsum drywall, 5/8" thick, fire resistant	2.66	4.89	7.55
1400	Water resistant	2.90	4.89	7.79
1500	1/2" thick, standard	2.54	4.89	7.43
1600	Fire resistant	2.56	4.89	7.45
1700	Water resistant	2.92	4.89	7.81
1800	3/8" thick, vinyl faced, standard	3.68	6.15	9.83
1900	5/8" thick, vinyl faced, fire resistant	4.10	6.15	10.25
2000				
2100				
2200				
2300				
2400				
2500				
2600				
2700				
2900	Cut & patch to match existing construction, add, minimum	2%	3%	
3000	Maximum	5%	9%	
3100	Dust protection, add, minimum	1%	2%	
3200	Maximum	4%	11%	
3300	Material handling & storage limitation, add, minimum	1%	1%	
3400	Maximum	6%	7%	
3500	Protection of existing work, add, minimum	2%	2%	
3600	Maximum	5%	7%	
3700	Shift work requirements, add, minimum		5%	
3800	Maximum		30%	
3900	Temporary shoring and bracing, add, minimum	2%	5%	
4000	Maximum	5%	12%	

C1010 Partitions

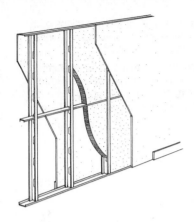

This page illustrates and describes a load bearing metal stud wall system including metal studs, sheetrock–taped and finished, insulation, baseboard and painting. Lines within System Components give the unit price and total price per square foot for this system. Prices for alternate load bearing metal stud wall systems are on Line Items C1010 138 1500 thru 2500. Both material quantities and labor costs have been adjusted for the system listed.

Factors: To adjust for job conditions other than normal working situations use Lines C1010 138 2900 thru 4000.

Example: You are to install the system using temporary shoring and bracing. Go to Line C1010 138 3900 and apply these percentages to the appropriate MAT. and INST. costs.

System Components	QUANTITY	UNIT	COST PER S.F.		
			MAT.	INST.	TOTAL
Load bearing, 18 ga., 3-5/8″, galvanized metal studs, 24″ O.C.,including					
Top and bottom runners, 1/2″ drywall, taped, finished and painted 2					
Faces, 3″ insulation, and painted baseboard, wall 10′ high.					
Metal studs, 24″ O.C., 18 ga., 3-5/8″ wide, galvanized	1.000	S.F.	1.03	1.01	2.04
Gypsum drywall 1/2″ thick	2.000	S.F.	.68	1.06	1.74
Taping and finishing	2.000	S.F.	.10	1.06	1.16
Insulation, 3-1/2″ fiberglass batts	1.000	S.F.	.63	.39	1.02
Baseboard	.200	L.F.	.49	.44	.93
Paint baseboard, primer + 2 coats	.200	L.F.	.03	.26	.29
Painting, roller 2 coats	2.000	S.F.	.34	1.18	1.52
TOTAL			3.30	5.40	8.70

C1010 138	Partitions, Metal Stud, LB	COST PER S.F.		
		MAT.	INST.	TOTAL
1400	For alternate metal stud systems:			
1500	Load bearing, 18 ga., 24″ O.C., 2-1/2″ wide	3.14	5.40	8.54
1600	6″ wide	3.65	5.45	9.10
1700	16″ O.C. 2-1/2″ wide	3.35	5.60	8.95
1800	3-5/8″ wide	3.56	5.65	9.21
1900	6″ wide	4	5.70	9.70
2000	16 ga., 24″ O.C., 2-1/2″ wide	3.27	5.50	8.77
2100	3-5/8″ wide	3.47	5.55	9.02
2200	6″ wide	3.87	5.60	9.47
2300	16″ O.C., 2-1/2″ wide	3.52	5.80	9.32
2400	3-5/8″ wide	3.77	5.80	9.57
2500	6″ wide	4.27	5.90	10.17
2600				
2700				
2900	Cut & patch to match existing construction, add, minimum	2%	3%	
3000	Maximum	5%	9%	
3100	Dust protection, add, minumum	1%	2%	
3200	Maximum	4%	11%	
3300	Material handling & storage limitation, add, minimum	1%	1%	
3400	Maximum	6%	7%	
3500	Protection of existing work, add, minimum	2%	2%	
3600	Maximum	5%	7%	
3700	Shift work requirements, add, minimum		5%	
3800	Maximum		30%	
3900	Temporary shoring and bracing, add, minimum	2%	5%	
4000	Maximum	5%	12%	

C10 Interior Construction

C1010 Partitions

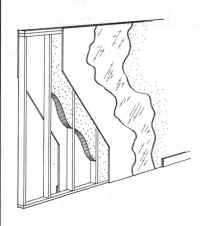

This page illustrates and describes a plaster and lath system including gypsum plaster, gypsum lath, wood studs with plates, insulation, baseboard and painting. Lines within System Components give the unit price and price per square foot for this system. Prices for alternate plaster and lath systems are on Line Items C1010 148 1500 thru 2500. Both material quantities and labor costs have been adjusted for the system listed.

Factors: To adjust for job conditions other than normal working situations use Lines C1010 148 2900 thru 4000.

Example: You are to install the system during evening hours only. Go to Line C1010 148 3800 and apply this percentage to the appropriate INST. costs.

System Components	QUANTITY	UNIT	MAT.	INST.	TOTAL
Gypsum plaster, 2 coats, over 3/8″ lath, 2 faces, 2″ x 4″ wood					
Stud partition, 24″ O.C. incl. double top plate, single bottom plate, 3-1/2″					
Insulation, baseboard and painting.					
Gypsum plaster, 2 coats	.220	S.Y.	.91	4.91	5.82
Lath, gypsum, 3/8″ thick	.220	S.Y.	1.33	1.17	2.50
Wood studs, 2″ x 4″, 24″ O.C.	1.000	S.F.	.28	.85	1.13
Insulation, 3-1/2″ fiberglass batts	1.000	S.F.	.63	.39	1.02
Baseboard, 9/16″ x 3-1/2″	.200	L.F.	.49	.44	.93
Paint baseboard, primer + 2 coats	.200	L.F.	.02	.11	.13
Paint, 2 coats	2.000	S.F.	.34	1.18	1.52
TOTAL			4	9.05	13.05

C1010 148	Partitions, Plaster & Lath	MAT.	INST.	TOTAL
1400	For alternate plaster systems:			
1500	Gypsum plaster, 3 coats	4.40	10.10	14.50
1600	Perlite plaster, 2 coats	4.07	9.75	13.82
1700	3 coats	4.62	11.15	15.77
1800				
1900				
2000	For alternate lath systems:			
2100	Gypsum lath, 1/2″ thick	4.16	9.85	14.01
2200	Foil back, 3/8″ thick	3.59	9.90	13.49
2300	1/2″ thick	3.66	10	13.66
2400	Metal lath, 2.5 Lb. diamond	3.54	9.75	13.29
2500	3.4 Lb. diamond	3.85	9.90	13.75
2600				
2700				
2900	Cut & patch to match existing construction, add, minimum	2%	3%	
3000	Maximum	5%	9%	
3100	Dust protection, add, minimum	1%	2%	
3200	Maximum	4%	11%	
3300	Material handling & storage limitation, add, minimum	1%	1%	
3400	Maximum	6%	7%	
3500	Protection of existing work, add, minimum	2%	2%	
3600	Maximum	5%	7%	
3700	Shift work requirements, add, minimum		5%	
3800	Maximum		30%	
3900	Temporary shoring and bracing, add, minimum	2%	5%	
4000	Maximum	5%	12%	

C10 Interior Construction

C1010 Partitions

C1010 170	Selective Price Sheet	COST PER S.F.		
		MAT.	INST.	TOTAL
0100	Studs			
0200				
0300	24" O.C. metal, 10' high wall, including			
0400	Top and bottom runners			
0500	Non load bearing, galvanized 25 Ga., 1-5/8" wide	.25	.69	.94
0600	2-1/2" wide	.32	.70	1.02
0700	3-5/8" wide	.36	.71	1.07
0800	4" wide	.39	.71	1.10
0900	6" wide	.55	.73	1.28
1000				
1100	Galvanized 20 Ga., 2-1/2" wide	.48	.88	1.36
1200	3-5/8" wide	.52	.89	1.41
1300	4" wide	.61	.89	1.50
1400	6" wide	.72	.91	1.63
1500	Load bearing, painted 18 Ga., 2-1/2" wide	.87	.99	1.86
1600	3-5/8" wide	1.03	1.01	2.04
1700	4" wide	.93	1.03	1.96
1800	6" wide	1.38	1.05	2.43
1900	Galvanized 18 Ga., 2-1/2" wide	.87	.99	1.86
2000	3-5/8" wide	1.03	1.01	2.04
2100	4" wide	.93	1.03	1.96
2200	6" wide	1.38	1.05	2.43
2300	Galvanized 16 Ga., 2-1/2" wide	1	1.13	2.13
2400	3-5/8" wide	1.20	1.15	2.35
2500	4" wide	1.27	1.17	2.44
2600	6" wide	1.60	1.20	2.80
2700				
2800				
2900	24" O.C. wood, 10' high wall, including			
3000	Double top plate and shoe			
3100	2" x 4"	.33	1	1.33
3200	2" x 6"	.59	1.07	1.66
3300				
3400				
3500	Furring, 24" O.C. 10' high wall, metal, 3/4" channels	.23	1.07	1.30
3600	1-1/2" channels	.30	1.15	1.45
3700	Wood, on wood, 1" x 2" strips	.14	.48	.62
3800	1" x 3" strips	.19	.48	.67
3900	On masonry, 1" x 2" strips	.14	.54	.68
4000	1" x 3" strips	.19	.54	.73
4100	On concrete, 1" x 2" strips	.14	1.02	1.16
4200	1" x 3" strips	.19	1.02	1.21
4300	Studs			
4400				
4500	16" O.C., metal, 10' high wall, including			
4600	Top and bottom runners			
4700	Non load bearing, galvanized 25 Ga., 1-5/8" wide	.34	1.07	1.41
4800	2-1/2" wide	.43	1.08	1.51
4900	3-5/8" wide	.50	1.10	1.60
5000	4" wide	.53	1.11	1.64
5100	6" wide	.75	1.12	1.87
5200				
5300	Galvanized 20 Ga., 2-1/2" wide	.65	1.35	2
5400	3-5/8" wide	.70	1.37	2.07
5500	4" wide	.83	1.39	2.22
5600	6" wide	.98	1.41	2.39

C1010 Partitions

C1010 170	Selective Price Sheet	COST PER S.F.		
		MAT.	INST.	TOTAL
5700	Load bearing, painted 18 Ga., 2-1/2" wide	1.18	1.37	2.55
5800	3-5/8" wide	1.40	1.39	2.79
5900	4" wide	1.47	1.43	2.90
6000	6" wide	1.87	1.45	3.32
6100	Galvanized 18 Ga., 2-1/2" wide	1.18	1.37	2.55
6200	3-5/8" wide	1.40	1.39	2.79
6300	4" wide	1.47	1.43	2.90
6400	6" wide	1.87	1.45	3.32
6500	Galvanized 16 Ga., 2-1/2" wide	1.38	1.55	2.93
6600	3-5/8" wide	1.66	1.60	3.26
6700	4" wide	1.75	1.63	3.38
6800	6" wide	2.20	1.65	3.85
6900				
7000				
7100	16" O.C., wood, 10' high wall, including			
7200	Double top plate and shoe			
7300	2" x 4"	.40	1.15	1.55
7400	2" x 6"	.58	1.03	1.61
7500				
7600				
7700	Furring, 16" O.C. 10' high wall, metal, 3/4" channels	.34	1.55	1.89
7800	1-1/2" channels	.46	1.73	2.19
7900	Wood, on wood, 1" x 2" strips	.21	.72	.93
8000	1" x 3" strips	.29	.72	1.01
8100	On masonry, 1" x 2" strips	.21	.80	1.01
8200	1" x 3" strips	.29	.80	1.09
8300	On concrete, 1" x 2" strips	.21	1.52	1.73
8400	1" x 3" strips	.29	1.52	1.81

C10 Interior Construction

C1010 Partitions

C1010 180	Selective Price Sheet, Drywall & Plaster	COST PER S.F.		
		MAT.	INST.	TOTAL
0100	Lath, gypsum perforated			
0200				
0300	Lath, gypsum, perforated, regular, 3/8" thick	.67	.59	1.26
0400	1/2" thick	.72	.71	1.43
0500	Fire resistant, 3/8" thick	.35	.71	1.06
0600	1/2" thick	.40	.77	1.17
0700	Foil back, 3/8" thick	.43	.67	1.10
0800	1/2" thick	.46	.71	1.17
0900				
1000	Metal lath			
1100	Metal lath, diamond, painted, 2.5 lb.	.41	.59	1
1200	3.4 lb.	.56	.67	1.23
1300	Rib painted, 2.75 lb	.44	.67	1.11
1400	3.40 lb	.68	.71	1.39
1500				
1600				
1700	Plaster, gypsum, 2 coats	.46	2.48	2.94
1800	3 coats	.66	3.01	3.67
1900	Perlite/vermiculite, 2 coats	.50	2.83	3.33
2000	3 coats	.77	3.54	4.31
2100	Bondcrete, 1 coat	.44	1.32	1.76
2200				
2500				
2600				
2700	Drywall, standard, 3/8" thick, no finish included	.34	.53	.87
2800	1/2" thick, no finish included	.34	.53	.87
2900	Taped and finished	.39	1.09	1.48
3000	5/8" thick, no finish included	.37	.53	.90
3100	Taped and finished	.42	1.09	1.51
3200	Fire resistant, 1/2" thick, no finish included	.35	.53	.88
3300	Taped and finished	.40	1.09	1.49
3400	5/8" thick, no finish included	.40	.53	.93
3500	Taped and finished	.44	1.09	1.53
3600	Water resistant, 1/2" thick, no finish included	.53	.53	1.06
3700	Taped and finished	.57	1.09	1.66
3800	5/8" thick, no finish included	.52	.53	1.05
3900	Taped and finished	.56	1.09	1.65
4000	Finish, instead of taping			
4100	For thin coat plaster, add	.09	.66	.75
4200	Finish, textured spray, add	.09	.56	.65

C1010 Partitions

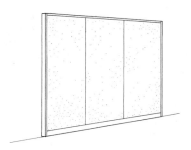

This page illustrates and describes a movable office partition system including demountable office partitions of various styles and sizes based on a cost per square foot basis. Prices for alternate movable office partition systems are on Line Items C1010 210 0300 thru 3600.

C1010 210	Partitions, Movable Office	COST PER S.F.		
		MAT.	INST.	TOTAL
0250	Office partition, demountable, no deduction for door opening, add for doors			
0300	Air wall, cork finish, semi acoustic, 1-5/8" thick, minimum	42.50	3.25	45.75
0400	Maximum	40	5.55	45.55
0500	Acoustic, 2" thick, minimum	34	3.46	37.46
0600	Maximum	59	4.69	63.69
0700				
0800				
0900	In-plant modular office system, w/prehung steel door			
1000	3" thick honeycomb core panels			
1100	12' x 12', 2 wall	6.90	.44	7.34
1200	4 wall	14.50	1.32	15.82
1300				
1400				
1500	Gypsum, demountable, 3" to 3-3/4" thick x 9' high, vinyl clad	6.95	2.44	9.39
1600	Fabric clad	17.20	2.66	19.86
1700	1.75 system, vinyl clad hardboard, paper honeycomb core panel			
1800	1-3/4" to 2-1/2" thick x 9' high	11.65	2.44	14.09
1900	Unitized gypsum panel, 2" to 2-1/2" thick x 9' high, vinyl clad	15	2.44	17.44
2000	Fabric clad	24.50	2.66	27.16
2100				
2200				
2300	Unitized mineral fiber panel system, 2-1/4" thick x 9' high			
2400	Vinyl clad mineral fiber	14.85	2.44	17.29
2500	Fabric clad mineral fiber	22	2.66	24.66
2600				
2700	Movable steel walls, modular system			
2800	Unitized panels, 48" wide x 9' high			
2900	Baked enamel, pre-finished	16.85	1.95	18.80
3000	Fabric clad	24.50	2.09	26.59
3100	For acoustical partitions, add, minimum	2.44		2.44
3200	Maximum	11.40		11.40
3300				
3400				
3500	Note: For door prices, see Div. 08			
3600	For door hardware prices, see Div. 08			
3700				
3800				
3900				
4000				

C10 Interior Construction

C1020 Interior Doors

This page illustrates and describes flush interior door systems including hollow core door, jamb, header and trim with hardware. Lines within System Components give the unit price and total price on a cost each basis for this system. Prices for alternate flush interior door systems are on Line items C1020 106 1100 thru 2400. Both material quantities and labor costs have been adjusted for the system listed.

Factors: To adjust for job conditions other than normal working situations use Lines C1020 106 2700 thru 3800.

Example: You are to install the system in an area where dust protection is vital. Go to Line C1020 106 3000 and apply these percentages to the appropriate MAT. and INST. costs.

System Components	QUANTITY	UNIT	COST EACH		
			MAT.	INST.	TOTAL
Single hollow core door, include jamb, header, trim and hardware, painted.					
Hollow core Lauan, 1-3/8" thick, 2'-0" x 6'-8", painted	1.000	Ea.	30.50	58.50	89
Paint door and frame, 1 coat	1.000	Ea.	1.98	26.50	28.48
Wood jamb, 4-9/16" deep	1.000	Set	113.60	44.96	158.56
Trim, casing	1.000	Set	46.72	70.40	117.12
Hardware, hinges	1.000	Set	13.25		13.25
Hardware, lockset	1.000	Set	18.70	33	51.70
TOTAL			224.75	233.36	458.11

C1020 106	Doors, Interior Flush, Wood	COST EACH		
		MAT.	INST.	TOTAL
1000	For alternate door systems:			
1100	Lauan (Mahogany) hollow core, 1-3/8" x 2'-6" x 6'-8"	234	237	471
1200	2'-8" x 6'-8"	239	240	479
1300	3'-0" x 6'-8"	245	248	493
1500	Birch, hollow core, 1-3/8" x 2'-0" x 6'-8"	239	233	472
1600	2'-6" x 6'-8"	253	237	490
1700	2'-8" x 6'-8"	257	240	497
1800	3'-0" x 6'-8"	263	248	511
1900	Solid core, pre-hung, 1-3/8" x 2'-6" x 6'-8"	385	240	625
2000	2'-8" x 6'-8"	420	243	663
2100	3'-0" x 6'-8"	435	252	687
2200				
2400	For metal frame instead of wood, add	50%	20%	
2600				
2700	Cut & patch to match existing construction, add, minimum	2%	3%	
2800	Maximum	5%	9%	
2900	Dust protection, add, minimum	1%	2%	
3000	Maximum	4%	11%	
3100	Equipment usage curtailment, add, minimum	1%	1%	
3200	Maximum	3%	10%	
3300	Material handling & storage limitation, add, minimum	1%	1%	
3400	Maximum	6%	7%	
3500	Protection of existing work, add, minimum	2%	2%	
3600	Maximum	5%	7%	
3700	Shift work requirements, add, minimum		5%	
3800	Maximum		30%	
3900				

C1020 Interior Doors

This page illustrates and describes interior, solid and louvered door systems including a pine panel door, wood jambs, header, and trim with hardware. Lines within System Components give the unit price and total price on a cost each basis for this system. Prices for alternate interior, solid and louvered systems are on Line Items C1020 108 1200 thru 2400. Both material quantities and labor costs have been adjusted for the system listed.

Factors: To adjust for job conditions other than normal working situations use Lines C1020 108 2900 thru 4000.

Example: You are to install the system during night hours only. Go to Line C1020 108 4000 and apply these percentages to the appropriate INST. costs.

System Components	QUANTITY	UNIT	COST EACH		
			MAT.	INST.	TOTAL
Single interior door, including jamb, header, trim and hardware, painted.					
Solid pine panel door, 1-3/8" thick, 2'-0" x 6'-8"	1.000	Ea.	140	58.50	198.50
Paint door and frame, 1 coat	1.000	Ea.	1.98	26.50	28.48
Wooden jamb, 4-5/8" deep	1.000	Set	113.60	44.96	158.56
Trim, casing	1.000	Set	46.72	70.40	117.12
Hardware, hinges	1.000	Set	13.25		13.25
Hardware, lockset	1.000	Set	18.70	33	51.70
TOTAL			334.25	233.36	567.61

C1020 108	Doors, Interior Solid & Louvered	COST EACH		
		MAT.	INST.	TOTAL
1000				
1100	For alternate door systems:			
1200	Solid pine, painted raised panel, 1-3/8" x 2'-6" x 6'-8"	370	237	607
1300	2'-8" x 6'-8"	380	240	620
1400	3'-0" x 6'-8"	390	248	638
1500				
1600	Louvered pine, painted 1'-6" x 6'-8"	305	230	535
1700	2'-0" x 6'-8"	340	237	577
1800	2'-6" x 6'-8"	355	240	595
1900	3'-0" x 6'-8"	380	248	628
2200	For prehung door, deduct	5%	30%	
2300				
2400	For metal frame instead of wood, add	50%	20%	
2500				
2900	Cut & patch to match existing construction, add, minimum	2%	3%	
3000	Maximum	5%	9%	
3100	Dust protection, add, minimum	1%	2%	
3200	Maximum	4%	11%	
3300	Equipment usage curtailment, add, minimum	1%	1%	
3400	Maximum	3%	10%	
3500	Material handling & storage limitation, add, minimum	1%	1%	
3600	Maximum	6%	7%	
3700	Protection of existing work, add, minimum	2%	2%	
3800	Maximum	5%	7%	
3900	Shift work requirements, add, minimum		5%	
4000	Maximum		30%	

C1020 Interior Doors

This page illustrates and describes interior metal door systems including a metal door, metal frame and hardware. Lines within System Components give the unit price and total price on a cost each for this system. Prices for alternate interior metal door systems are on Line Items C1020 110 1100 thru 2000. Both material quantities and labor costs have been adjusted for the system listed.

Factors: To adjust for job conditions other than normal working situations use Lines C1020 110 2900 thru 4000.

Example: You are to install the system while protecting existing construction. Go to Line C1020 110 3700 and apply these percentages to the appropriate MAT. and INST. costs.

System Components	QUANTITY	UNIT	COST EACH		
			MAT.	INST.	TOTAL
Single metal door, including frame and hardware.					
Hollow metal door, 18 ga., 1-3/4" x 7'-0" x 2'-6" wide	1.000	Ea.	370	70.50	440.50
Metal frame, 6-1/4" deep	1.000	Set	153	66	219
Paint door and frame, 1 coat	1.000	Ea.	1.98	26.50	28.48
Hardware, hinges	1.000	Set	81		81
Hardware, passage lockset	1.000	Set	18.70	33	51.70
TOTAL			624.68	196	820.68

C1020 110	Doors, Interior Flush, Metal	COST EACH		
		MAT.	INST.	TOTAL
1000	For alternate systems:			
1100	Hollow metal doors, 1-3/8" thick, 2'-8" x 6'-8"	625	196	821
1200	3'-0" x 7'-0"	615	206	821
1300				
1400	Interior fire door, 1-3/8" thick, 2'-6" x 6'-8"	770	196	966
1500	2'-8" x 6'-8"	770	196	966
1600	3'-0" x 7'-0"	770	206	976
1700				
1800	Add to fire doors:			
1900	Baked enamel finish	30%	15%	
2000	Galvanizing	15%		
2200				
2300				
2400				
2900	Cut & patch to match existing construction, add, minimum	2%	3%	
3000	Maximum	5%	9%	
3100	Dust protection, add, minimum	1%	2%	
3200	Maximum	4%	11%	
3300	Equipment usage curtailment, add, minimum	1%	1%	
3400	Maximum	3%	10%	
3500	Material handling & storage limitation, add, minimum	1%	1%	
3600	Maximum	6%	7%	
3700	Protection of existing work, add, minimum	2%	2%	
3800	Maximum	5%	7%	
3900	Shift work requirements, add, minimum		5%	
4000	Maximum		30%	

C1020 Interior Doors

This page illustrates and describes an interior closet door system including an interior closet door, painted, with trim and hardware. Prices for alternate interior closet door systems are on Line Items C1020 112 0500 thru 2200. Both material quantities and labor costs have been adjusted for the system listed.

Factors: To adjust for job conditions other than normal working situations use Lines C1020 112 2900 thru 4000.

Example: You are to install the system and match the existing construction. Go to Line C1020 112 2900 and apply these percentages to the appropriate MAT. and INST. costs.

C1020 112	Doors, Closet	COST PER SET		
		MAT.	INST.	TOTAL
0350	Interior closet door painted, including frame, trim and hardware, prehung.			
0400	Bi-fold doors			
0500	Pine paneled, 3'-0" x 6'-8"	410	193	603
0600	6'-0" x 6'-8"	595	259	854
0700	Birch, hollow core, 3'-0" x 6'-8"	202	209	411
0800	6'-0" x 6'-8"	298	280	578
0900	Lauan, hollow core, 3'-0" x 6'-8"	190	193	383
1000	6'-0" x 6'-8"	274	259	533
1100	Louvered pine, 3'-0" x 6'-8"	310	193	503
1200	6'-0" x 6'-8"	425	259	684
1300				
1400	Sliding, bi-passing closet doors			
1500	Pine paneled, 4'-0" x 6'-8"	680	205	885
1600	6'-0" x 6'-8"	805	259	1,064
1700	Birch, hollow core, 4'-0" x 6'-8"	370	205	575
1800	6'-0" x 6'-8"	435	259	694
1900	Lauan, hollow core, 4'-0" x 6'-8"	320	205	525
2000	6'-0" x 6'-8"	370	259	629
2100	Louvered pine, 4'-0" x 6'-8"	595	205	800
2200	6'-0" x 6'-8"	720	259	979
2300				
2400				
2500				
2600				
2700				
2800				
2900	Cut & patch to match existing construction, add, minimum	2%	3%	
3000	Maximum	5%	9%	
3100	Dust protection, add, minimum	1%	2%	
3200	Maximum	4%	11%	
3300	Equipment usage curtailment, add, minimum	1%	1%	
3400	Maximum	3%	10%	
3500	Material handling & storage limitation, add, minimum	1%	1%	
3600	Maximum	6%	7%	
3700	Protection of existing work, add, minimum	2%	2%	
3800	Maximum	5%	7%	
3900	Shift work requirements, add, minimum		5%	
4000	Maximum		30%	

C2010   Stair Construction

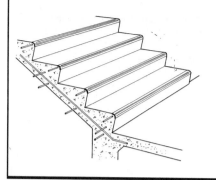

This page illustrates and describes a stair system based on a cost per flight price. Prices for various stair systems are on Line Items C2010 130 0700 thru 3200. Both material quantities and labor costs have been adjusted for the system listed.

Factors: To adjust for job conditions other than normal working situations use Lines C2010 130 3500 thru 4200.

Example: You are to install the system during evenings only. Go to Line C2010 130 4200 and apply this percentage to the appropriate MAT. and INST. costs.

System Components	QUANTITY	UNIT	COST PER FLIGHT		
			MAT.	INST.	TOTAL
Below are various stair systems based on cost per flight of stairs, no side Walls. Stairs are 4'-0" wide, railings are included unless otherwise noted.					

C2010 130	Stairs	COST PER FLIGHT		
		MAT.	INST.	TOTAL
0700	Concrete, cast in place, no nosings, no railings, 12 risers	330	1,825	2,155
0800	24 risers	660	3,625	4,285
0900	Add for 1 intermediate landing	184	505	689
1000	Concrete, cast in place, with nosings, no railings, 12 risers	1,075	2,150	3,225
1100	24 risers	2,150	4,250	6,400
1200	Add for 1 intermediate landing	246	530	776
1300	Steel, grating tread, safety nosing, 12 risers	5,400	1,050	6,450
1400	24 risers	10,800	2,125	12,925
1500	Add for intermediate landing	1,325	283	1,608
1600	Steel, cement fill pan tread, 12 risers	8,175	1,050	9,225
1700	24 risers	16,400	2,125	18,525
1800	Add for intermediate landing	1,325	283	1,608
1900	Spiral, industrial, 4' - 6" diameter, 12 risers	7,975	755	8,730
2000	24 risers	16,000	1,500	17,500
2100	Wood, box stairs, oak treads, 12 risers	2,225	595	2,820
2200	24 risers	4,450	1,200	5,650
2300	Add for 1 intermediate landing	152	135	287
2400	Wood, basement stairs, no risers, 12 steps	725	246	971
2500	24 steps	1,450	490	1,940
2600	Add for 1 intermediate landing	38.50	26	64.50
2700	Wood, open, rough sawn cedar, 12 steps	2,525	480	3,005
2800	24 steps	5,025	965	5,990
2900	Add for 1 intermediate landing	23.50	26.50	50
3000	Wood, residential, oak treads, 12 risers	1,900	2,650	4,550
3100	24 risers	3,800	5,275	9,075
3200	Add for 1 intermediate landing	133	122	255
3500	Dust protection, add, minimum	1%	2%	
3600	Maximum	4%	11%	
3700	Material handling & storage limitation, add, minimum	1%	1%	
3800	Maximum	6%	7%	
3900	Protection of existing work, add, minimum	2%	2%	
4000	Maximum	5%	7%	
4100	Shift work requirements, add, minimum		5%	
4200	Maximum		30%	

C30 Interior Finishes

C3010 Wall Finishes

C3010 210	Selective Price Sheet	COST PER S.F.		
		MAT.	INST.	TOTAL
0100	Painting, on plaster or drywall, brushwork, primer and 1 ct.	.13	.66	.79
0200	Primer and 2 ct.	.19	1.05	1.24
0300	Rollerwork, primer and 1 ct.	.13	.56	.69
0400	Primer and 2 ct.	.18	.89	1.07
0500	Woodwork incl. puttying, brushwork, primer and 1 ct.	.11	1	1.11
0600	Primer and 2 ct.	.17	1.32	1.49
0700	Wood trim to 6" wide, enamel, primer and 1 ct.	.12	.56	.68
0800	Primer and 2 ct.	.18	.71	.89
0900	Cabinets and casework, enamel, primer and 1 ct.	.12	1.13	1.25
1000	Primer and 2 ct.	.18	1.38	1.56
1100	On masonry or concrete, latex, brushwork, primer and 1 ct.	.23	.94	1.17
1200	Primer and 2 ct.	.30	1.34	1.64
1300	For block filler, add	.15	1.15	1.30
1400				
1500	Varnish, wood trim, sealer 1 ct., sanding, puttying, quality work	.14	2.03	2.17
1600	Medium work	.11	1.58	1.69
1700	Without sanding	.17	.24	.41
1800				
1900	Wall coverings, wall paper, at low price per double roll, avg workmanship	.39	.71	1.10
2000	At medium price per double roll, average workmanship	.69	.85	1.54
2100	At high price per double roll, quality workmanship	2.59	1.04	3.63
2200				
2300	Grass cloths with lining paper, minimum	.81	1.13	1.94
2400	Maximum	2.61	1.30	3.91
2500	Vinyl, fabric backed, light weight	.72	.71	1.43
2600	Medium weight	.85	.94	1.79
2700	Heavy weight	1.75	1.04	2.79
2800				
2900	Cork tiles, 12" x 12", 3/16" thick	4.68	1.89	6.57
3000	5/16" thick	3.98	1.93	5.91
3100	Granular surface, 12" x 36", 1/2" thick	1.32	1.18	2.50
3200	1" thick	1.71	1.22	2.93
3300	Aluminum foil	1.07	1.65	2.72
3400				
3500	Tile, ceramic, adhesive set, 4-1/4" x 4-1/4"	2.74	4.49	7.23
3600	6" x 6"	3.68	4.26	7.94
3700	Decorated, 4-1/4" x 4-1/4", minimum	4.03	3.16	7.19
3800	Maximum	46	4.74	50.74
3900	For epoxy grout, add	.43	1.07	1.50
4000	Pregrouted sheets	5.20	3.55	8.75
4100	Glass mosaics, 3/4" tile on 12" sheets, minimum	19.05	11.70	30.75
4200	Color group 8	62.50	13.35	75.85
4300	Metal, tile pattern, 4' x 4' sheet, 24 ga., nailed			
4400	Stainless steel	27	2.06	29.06
4500	Aluminized steel	14.45	2.06	16.51
4600				
4700	Brick, interior veneer, 4" face brick, running bond, minimum	6.15	11.20	17.35
4800	Maximum	6.15	11.20	17.35
4900	Simulated, urethane pieces, set in mastic	7.35	3.52	10.87
5000	Fiberglass panels	4.26	2.64	6.90
5100	Wall coating, on drywall, thin coat, plain	.09	.66	.75
5200	Stipple	.09	.66	.75
5300	Textured spray	.09	.56	.65
5400				
5500	Paneling not incl. furring or trim, hardboard, tempered, 1/8" thick	.42	2.11	2.53
5600	1/4" thick	.58	2.11	2.69
5700	Plastic faced, 1/8" thick	.70	2.11	2.81
5800	1/4" thick	.94	2.11	3.05

C3010 Wall Finishes

C3010 210	Selective Price Sheet	COST PER S.F.		
		MAT.	INST.	TOTAL
5900	Woodgrained, 1/4" thick, minimum	.63	2.11	2.74
6000	Maximum	1.31	2.48	3.79
6100	Plywood, 4' x 8' shts. 1/4" thick, prefin., birch faced, min.	.98	2.11	3.09
6200	Maximum	2.17	3.01	5.18
6300	Walnut, minimum	3.03	2.11	5.14
6400	Maximum	5.75	2.64	8.39
6500	Mahogany, African	2.76	2.64	5.40
6600	Philippine	1.19	2.11	3.30
6700	Chestnut	5.25	2.81	8.06
6800	Pecan	2.27	2.64	4.91
6900	Rosewood	5.05	3.30	8.35
7000	Teak	3.54	2.64	6.18
7100	Aromatic cedar, plywood	2.29	2.64	4.93
7200	Particle board	1.11	2.64	3.75
7300	Wood board, 3/4" thick, knotty pine	1.58	3.52	5.10
7400	Rough sawn cedar	2.04	3.52	5.56
7500	Redwood, clear	4.73	3.52	8.25
7600	Aromatic cedar	3.65	3.84	7.49

C3020 Floor Finishes

C3020 430	Selective Price Sheet	MAT.	INST.	TOTAL
0100	Flooring, carpet, acrylic, 26 oz. light traffic	3.16	.70	3.86
0200	35 oz. heavy traffic	5.90	.70	6.60
0300	Nylon anti-static, 15 oz. light traffic	1.99	.93	2.92
0400	22 oz. medium traffic	3.89	.70	4.59
0500	26 oz. heavy traffic	5.10	.75	5.85
0600	28 oz. heavy traffic	5.60	.75	6.35
0700	Tile, foamed back, needle punch	4.08	.84	4.92
0800	Tufted loop	2.70	.84	3.54
0900	Wool, 36 oz. medium traffic	13.30	.75	14.05
1000	42 oz. heavy traffic	13.65	.75	14.40
1100	Composition, epoxy, with colored chips, minimum	2.74	3.92	6.66
1200	Maximum	3.32	5.40	8.72
1300	Trowelled, minimum	3.53	4.73	8.26
1400	Maximum	5.15	5.50	10.65
1500	Terrazzo, 1/4" thick, chemical resistant, minimum	5.80	5.70	11.50
1600	Maximum	8.50	7.55	16.05
1700	Resilient, asphalt tile, 1/8" thick	1.24	1.19	2.43
1800	Conductive flooring, rubber, 1/8" thick	5.50	1.51	7.01
1900	Cork tile 1/8" thick, standard finish	6.35	1.51	7.86
2000	Urethane finish	7.55	1.51	9.06
2100	PVC sheet goods for gyms, 1/4" thick	8.25	5.95	14.20
2200	3/8" thick	9.35	7.95	17.30
2300	Vinyl composition 12" x 12" tile, plain, 1/16" thick	.96	.95	1.91
2400	1/8" thick	2.35	.95	3.30
2500	Vinyl tile, 12" x 12" x 1/8" thick, minimum	4.62	.95	5.57
2600	Maximum	13.60	.95	14.55
2700	Vinyl sheet goods, backed, .093" thick	3.85	2.07	5.92
2900	Slate, random rectangular, 1/4" thick	17.85	6.25	24.10
3000	1/2" thick	16.60	8.95	25.55
3100	Natural cleft, irregular, 3/4" thick	7.65	10.20	17.85
3200	For sand rubbed finish, add	8.30		8.30
3300	Terrazzo, cast in place, bonded 1-3/4" thick, gray cement	3.16	15.15	18.31
3400	White cement	3.60	15.15	18.75
3500	Not bonded 3" thick, gray cement	3.95	16.25	20.20
3600	White cement	4.31	16.25	20.56
3700	Precast, 12" x 12" x 1" thick	19.45	15.65	35.10
3800	1-1/4" thick	21.50	15.65	37.15
3900	16" x 16" x 1-1/4" thick	23.50	18.80	42.30
4000	1-1/2" thick	21.50	21	42.50
4100	Marble travertine, standard, 12" x 12" x 3/4" thick	13.40	14.20	27.60
4200 / 4300	Tile, ceramic, natural clay, thin set	4.78	4.66	9.44
4400	Porcelain, thin set	5.95	4.49	10.44
4500	Specialty, decorator finish	11	4.66	15.66
4600 / 4700	Quarry, red, mud set, 4" x 4" x 1/2" thick	8.25	7.10	15.35
4800	6" x 6" x 1/2" thick	7.75	6.10	13.85
4900	Brown, imported, 6" x 6" x 3/4" thick	9.25	7.10	16.35
5000	8" x 8" x 1" thick	10	7.75	17.75
5100	Slate, Vermont, thin set, 6" x 6" x 1/4" thick	7.45	4.74	12.19
5200 / 5300	Wood, maple strip, 25/32" x 2-1/4", finished, select	5.95	3.34	9.29
5400	2nd and better	3.82	3.34	7.16
5500	Oak, 25/32" x 2-1/4" finished, clear	3.67	3.34	7.01
5600	No. 1 common	4.58	3.34	7.92
5700	Parquet, standard, 5/16", finished, minimum	5	3.54	8.54
5800	Maximum	5.60	5.55	11.15
5900	Custom, finished, minimum	17	5.30	22.30

C3020 Floor Finishes

C3020 430	Selective Price Sheet	COST PER S.F.		
		MAT.	INST.	TOTAL
6000	Maximum	22.50	10.55	33.05
6100	Prefinished, oak, 2-1/4" wide	7.05	3.10	10.15
6200	Ranch plank	8.70	3.64	12.34
6300	Sleepers on concrete, treated, 24" O.C., 1" x 2"	.33	.35	.68
6400	1" x 3"	.48	.45	.93
6500	2" x 4"	.43	.62	1.05
6600	2" x 6"	.66	.71	1.37
6700	Refinish old floors, minimum	.86	1.04	1.90
6800	Maximum	1.28	3.21	4.49
6900	Subflooring, plywood, CDX, 1/2" thick	.51	.70	1.21
7000	5/8" thick	.74	.78	1.52
7100	3/4" thick	.91	.84	1.75
7200				
7300	1" x 10" boards, S4S, laid regular	1.70	.96	2.66
7400	Laid diagonal	1.70	1.17	2.87
7500	1" x 8" boards, S4S, laid regular	1.47	1.06	2.53
7600	Laid diagonal	1.47	1.24	2.71
7700	Underlayment, plywood, underlayment grade, 3/8" thick	1.13	.70	1.83
7800	1/2" thick	1.10	.73	1.83
7900	5/8" thick	1.42	.75	2.17
8000	3/4" thick	1.46	.81	2.27
8100	Particle board, 3/8" thick	.45	.70	1.15
8200	1/2" thick	.50	.73	1.23
8300	5/8" thick	.59	.75	1.34
8400	3/4" thick	.72	.81	1.53

C3030 Ceiling Finishes

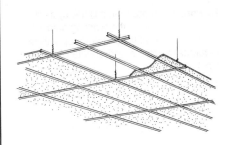

This page illustrates suspended acoustical board systems including acoustic ceiling board, hangers, and T bar suspension. Lines within System Components give the unit price and total price per square foot for this system. Prices for alternate suspended acoustical board systems are on Line Items C3030 210 1100 thru 2400. Both material quantities and labor costs have been adjusted for the system listed.

Factors: To adjust for job conditions other than normal working situations use Lines C3030 210 2900 thru 4000.

Example: You are to install the system and protect existing construction. Go to Line C3030 210 3800 and apply these percentages to the appropriate MAT. and INST. costs.

System Components	QUANTITY	UNIT	COST PER S.F.		
			MAT.	INST.	TOTAL
Suspended acoustical ceiling board installed on exposed grid system.					
Fiberglass boards, film faced, 2' x 4', 5/8" thick	1.000	S.F.	.80	.84	1.64
Hangers, #12 wire	1.000	S.F.	.53	.51	1.04
T bar suspension system, 2' x 4' grid	1.000	S.F.	.69	.66	1.35
TOTAL			2.02	2.01	4.03

C3030 210	Ceiling, Suspended Acoustical	COST PER S.F.		
		MAT.	INST.	TOTAL
1000	For alternate suspended ceiling systems:			
1100	2' x 4' grid, mineral fiber board, aluminum faced, 5/8" thick	2.61	1.98	4.59
1200	Standard faced	2.08	1.95	4.03
1300	Plastic faced	2.57	2.49	5.06
1400	Fiberglass, film faced, 3" thick, R11	2.90	2.34	5.24
1500	Grass cloth faced, 3/4" thick	4.56	2.23	6.79
1600	1" thick	3.60	2.26	5.86
1700	1-1/2" thick, nubby face	4.15	2.28	6.43
2200				
2300				
2400	Add for 2' x 2' grid system	.17	.16	.33
2500				
2600				
2700				
2900	Cut & patch to match existing construction, add, minimum	2%	3%	
3000	Maximum	5%	9%	
3100	Dust protection, add, minimum	1%	2%	
3200	Maximum	4%	11%	
3300	Equipment usage curtailment, add, minimum	1%	1%	
3400	Maximum	3%	10%	
3500	Material handling & storage limitation, add, minimum	1%	1%	
3600	Maximum	6%	7%	
3700	Protection of existing work, add, minimum	2%	2%	
3800	Maximum	5%	7%	
3900	Shift work requirements, add, minimum		5%	
4000	Maximum		30%	

C3030 Ceiling Finishes

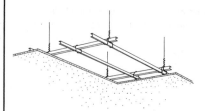

This page illustrates and describes suspended gypsum board systems including gypsum board, metal furring, taping, finished and painted. Lines within System Components give the unit price and total price per square foot for this system. Prices for alternate suspended gypsum board systems are on Line Items C3030 220 1500 thru 1700. Both material quantities and labor costs have been adjusted for the system listed.

Factors: To adjust for job conditions other than normal working situations use Lines C3030 220 2900 thru 4000.

Example: You are to install the system and control dust in the working area. Go to Line C3030 220 3200 and apply these percentages to the appropriate MAT. and INST. costs.

System Components	QUANTITY	UNIT	COST PER S.F. MAT.	COST PER S.F. INST.	COST PER S.F. TOTAL
Suspended ceiling gypsum board, 4' x 8' x 5/8" thick,					
On metal furring, taped, finished and painted.					
Gypsum drywall, 4' x 8', 5/8" thick, screwed	1.000	S.F.	.37	.59	.96
Main runners, 1-1/2" C.R.C., 4' O.C.	.500	S.F.	.15	.58	.73
25 ga., channels, 2' O.C.	1.000	S.F.	.23	1.07	1.30
Taped and finished	1.000	S.F.	.05	.53	.58
Paint, 2 coats, roller work	1.000	S.F.	.17	.59	.76
TOTAL			.97	3.36	4.33

C3030 220	Ceilings, Suspended Gypsum Board		MAT.	INST.	TOTAL
1400	For alternate drywall ceiling systems:				
1500	Thin coat plaster, 2 coats paint		.84	2.90	3.74
1600	Spray-on sand finish, no paint		.84	2.80	3.64
1700	12" x 12" x 3/4" acoustical wood fiber tile		2.04	3.56	5.60
1800					
1900					
2000					
2100					
2200					
2300					
2400					
2500					
2600					
2700					
2900	Cut & patch to match existing construction, add, minimum		2%	3%	
3000	Maximum		5%	9%	
3100	Dust protection, add, minimum		1%	2%	
3200	Maximum		4%	11%	
3300	Equipment usage curtailment, add, minimum		1%	1%	
3400	Maximum		3%	10%	
3500	Material handling & storage limitation, add, minimum		1%	1%	
3600	Maximum		6%	7%	
3700	Protection of existing work, add, minimum		2%	2%	
3800	Maximum		5%	7%	
3900	Shift work requirements, add, minimum			5%	
4000	Maximum			30%	

C3030 Ceiling Finishes

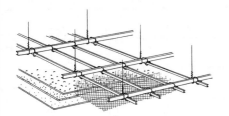

This page illustrates and describes suspended plaster and lath systems including gypsum plaster, lath, furring and runners with ceiling painted. Lines within System Components give the unit price and total price per square foot for this system. Prices for alternate suspended plaster and lath systems are on Line Items C3030 230 1300 thru 2300. Both material quantities and labor costs have been adjusted for the system listed.

Factors: To adjust for job conditions other than normal working situations use Lines C3030 230 2900 thru 4000.

Example: You are to install the system to match existing construction. Go to Line C3030 230 2900 and apply these percentages to the appropriate MAT. and INST. costs.

System Components	QUANTITY	UNIT	COST PER S.F. MAT.	COST PER S.F. INST.	COST PER S.F. TOTAL
Gypsum plaster, 3 coats, on 3.4# rib lath, on 3/4″ C.R.C. furring on 1-1/2″ main runners, ceiling painted.					
Gypsum plaster, 3 coats	.110	S.Y.	.65	3.34	3.99
3.4# rib lath	.110	S.Y.	.68	.70	1.38
Main runners, 1-1/2″ C.R.C. 24″ O.C.	.333	S.F.	.30	1.15	1.45
Furring 3/4″ C.R.C. 16″ O.C.	1.000	S.F.	.34	1.55	1.89
Painting, 2 coats, roller work	1.000	S.F.	.17	.59	.76
TOTAL			2.14	7.33	9.47

C3030 230	Ceiling, Suspended Plaster	COST PER S.F. MAT.	COST PER S.F. INST.	COST PER S.F. TOTAL
1200	For alternate plaster ceiling systems:			
1300	Gypsum plaster, 3 coats, on 2.5# diamond lath	1.86	7.30	9.16
1400	On 3/8″ gypsum lath	2.13	7.30	9.43
1500	2 coats, on 3.4# rib lath	2.17	7.35	9.52
1600	On 2.5# diamond lath	1.67	6.75	8.42
1700	On 3/8″ gypsum lath	1.94	6.75	8.69
1800	Perlite plaster, 3 coats, on 3.4# rib lath	2.25	8.15	10.40
1900	On 2.5# diamond lath	1.97	8.10	10.07
2000	On 3/8″ gypsum lath	2.24	8.10	10.34
2100	2 coats, on 3.4# rib lath	1.98	7.25	9.23
2200	On 2.5# diamond lath	1.70	7.25	8.95
2300	On 3/8″ gypsum lath	1.97	7.25	9.22
2400				
2500				
2600				
2700				
2900	Cut & patch to match existing construction, add, minimum	2%	3%	
3000	Maximum	5%	9%	
3100	Dust protection, add, minimum	1%	2%	
3200	Maximum	4%	11%	
3300	Equipment usage curtailment, add, minimum	1%	1%	
3400	Maximum	3%	10%	
3500	Material handling & storage limitation, add, minimum	1%	1%	
3600	Maximum	6%	7%	
3700	Protection of existing work, add, minimum	2%	2%	
3800	Maximum	5%	7%	
3900	Shift work requirements, add, minimum		5%	
4000	Maximum		30%	

C30 Interior Finishes

C3030 Ceiling Finishes

C3030 250	Selective Price Sheet	COST PER S.F.		
		MAT.	INST.	TOTAL
0100	Ceiling, plaster, gypsum, 2 coats	.46	2.83	3.29
0200	3 coats	.66	3.36	4.02
0300	Perlite or vermiculite, 2 coats	.50	3.31	3.81
0400	3 coats	.77	4.19	4.96
0500	Gypsum lath, plain, 3/8" thick	.67	.59	1.26
0600	1/2" thick	.72	.62	1.34
0700	Firestop, 3/8" thick	.35	.71	1.06
0800	1/2" thick	.40	.77	1.17
0900	Metal lath, rib, 2.75 lb.	.44	.67	1.11
1000	3.40 lb.	.68	.71	1.39
1100	Diamond, 2.50 lb.	.41	.67	1.08
1200	3.40 lb.	.56	.83	1.39
1500	Drywall, standard, 1/2" thick, no finish included	.34	.59	.93
1600	Taped and finished	.39	1.38	1.77
1700	5/8" thick, no finish included	.37	.59	.96
1800	Taped and finished	.42	1.38	1.80
1900	Fire resistant, 1/2" thick, no finish included	.35	.59	.94
2000	Taped and finished	.40	1.38	1.78
2100	5/8" thick, no finish included	.40	.59	.99
2200	Taped and finished	.44	1.38	1.82
2300	Water resistant, 1/2" thick, no finish included	.53	.59	1.12
2400	Taped and finished	.57	1.38	1.95
2500	5/8" thick, no finish included	.52	.59	1.11
2600	Taped and finished	.56	1.38	1.94
2700	Finish, instead of taping			
2800	For thin coat plaster, add	.09	.66	.75
2900	Finish, textured spray, add	.09	.56	.65
3000				
3100				
3200				
3300	Tile, stapled or glued, mineral fiber plastic coated, 5/8" thick	2.04	1.76	3.80
3400	3/4" thick	2.06	1.76	3.82
3500	Wood fiber, 1/2" thick	1.01	1.32	2.33
3600	3/4" thick	1.29	1.32	2.61
3700	Suspended, fiberglass, film faced, 5/8" thick	.80	.84	1.64
3800	3" thick	1.68	1.17	2.85
3900	Mineral fiber, 5/8" thick, standard	.86	.78	1.64
4000	Aluminum	2.05	.88	2.93
4100	Wood fiber, reveal edge, 1" thick			
4200	3" thick			
4300	Framing, metal furring, 3/4" channels, 12" O.C.	.38	2.14	2.52
4400	16" O.C.	.34	1.55	1.89
4500	24" O.C.	.23	1.07	1.30
4600				
4700	1-1/2" channels, 12" O.C.	.51	2.36	2.87
4800	16" O.C.	.46	1.73	2.19
4900	24" O.C.	.30	1.15	1.45
5000				
5100				
5200				
5300	Ceiling suspension systems, for tile,			
5400	Concealed "Z" bar, 12" module	.72	1.01	1.73
5500	Class A, "T" bar 2'-0" x 4'-0" grid	.69	.66	1.35
5600	2'-0" x 2'-0" grid	.86	.81	1.67
5700	Carrier channels for lighting fixtures, add	.30	1.15	1.45
5800				
5900				
6000				

C30 Interior Finishes

C3030 Ceiling Finishes

C3030 250	Selective Price Sheet	COST PER S.F.		
		MAT.	INST.	TOTAL
6100	Wood, furring 1" x 3", on wood, 12" O.C.	.38	1.51	1.89
6200	16" O.C..	.29	1.13	1.42
6300	24" O.C.	.19	.76	.95
6400				
6500	On concrete, 12" O.C.	.38	2.51	2.89
6600	16" O.C.	.29	1.88	2.17
6700	24" O.C.	.19	1.26	1.45
6800				
6900	Joists, 2" x 4", 12" O.C.	.41	1.08	1.49
7000	16" O.C.	.35	.91	1.26
7100	24" O.C.	.28	.74	1.02
7200	32" O.C.	.21	.56	.77
7300	2" x 6", 12" O.C.	.74	1.10	1.84
7400	16" O.C.	.62	.91	1.53
7500	24" O.C.	.49	.72	1.21
7600	32" O.C.	.36	.54	.90

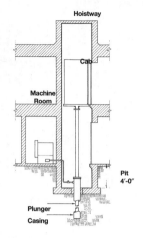

This page illustrates and describes oil hydraulic elevator systems. Prices for various oil hydraulic elevator systems are on Line Items D1010 130 0400 thru 2000. Both material quantities and labor costs have been adjusted for the system listed.

Factors: To adjust for job conditions other than normal working situations use Lines D1010 130 3100 thru 4000.

Example: You are to install the system with limited equipment usage. Go to Line D1010 130 3400 and apply this percentage to the appropriate TOTAL costs.

D1010 130	Oil Hydraulic Elevators	COST EACH		
		MAT.	**INST.**	**TOTAL**
0310	Oil-hydraulic elevator systems			
0320	Including piston and piston shaft			
0330				
0400	1500 Lb. passenger, 2 floors	43,800	13,900	57,700
0500	3 floors	58,000	21,500	79,500
0600	4 floors	97,000	30,200	127,200
0700	5 floors	111,000	37,600	148,600
0800	6 floors	124,000	44,900	168,900
0900				
1000	2500 Lb. passenger, 2 floors	47,400	13,900	61,300
1100	3 floors	62,000	21,500	83,500
1200	4 floors	96,500	30,200	126,700
1300	5 floors	110,500	37,600	148,100
1400	6 floors	127,500	44,900	172,400
1500				
1600	4000 Lb. passenger, 2 floors	52,000	13,900	65,900
1700	3 floors	70,000	21,500	91,500
1800	4 floors	105,500	30,200	135,700
1900	5 floors	119,500	37,600	157,100
2000	6 floors	132,500	44,900	177,400
2100				
2200				
2300				
2400				
2500				
2600				
2700				
2800				
2900				
3000				
3100	Dust protection, add, minimum	1%	2%	
3200	Maximum	4%	11%	
3300	Equipment usage curtailment, add, minimum	1%	1%	
3400	Maximum	3%	10%	
3500	Material handling & storage limitation, add, minimum	1%	1%	
3600	Maximum	6%	7%	
3700	Protection of existing work, add, minimum	2%	2%	
3800	Maximum	5%	7%	
3900	Shift work requirements, add, minimum		5%	
4000	Maximum		30%	

D2010 Plumbing Fixtures

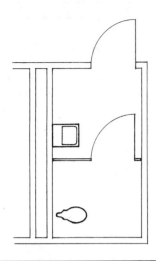

This page illustrates and describes a women's public restroom system including a water closet, lavatory, accessories, and service piping. Lines within System Components give the unit price and total price on a cost each basis for this system. Prices for alternate women's public restroom systems are on Line Items D2010 956 1400 thru 1700. Both material quantities and labor costs have been adjusted for the system listed.

Factors: To adjust for job conditions other than normal working situations use Lines D2010 956 2900 thru 4000.

Example: You are to install the system and protect surrounding area from dust. Go to Line D2010 956 3100 and apply these percentages to the MAT. and INST. costs.

System Components	QUANTITY	UNIT	COST EACH		
			MAT.	INST.	TOTAL
Public women's restroom incl. water closet, lavatory, accessories and					
Necessary service piping to install this system in one wall.					
Water closet, wall mounted, one piece	1.000	Ea.	315	188	503
Rough-in waste and vent for water closet	1.000	Set	745	425	1,170
Lavatory, 20″ x 18″ P.E. cast iron with accessories	1.000	Ea.	325	136	461
Rough-in waste and vent for lavatory	1.000	Set	400	655	1,055
Toilet partition, painted metal between walls, floor mounted	1.000	Ea.	485	176	661
For handicap unit, add	1.000	Ea.	365		365
Grab bar, 36″ long	1.000	Ea.	79	53	132
Mirror, 18″ x 24″, with stainless steel shelf	1.000	Ea.	249	26.50	275.50
Napkin/tampon dispenser, recessed	1.000	Ea.	905	35	940
Soap Dispenser, chrome, surface mounted, liquid	1.000	Ea.	52	26.50	78.50
Toilet tissue dispenser, surface mounted, stainless steel	1.000	Ea.	21.50	17.60	39.10
Towel dispenser, surface mounted, stainless steel	1.000	Ea.	46	33	79
TOTAL			3,987.50	1,771.60	5,759.10

D2010 956		Plumbing - Public Restroom	COST EACH		
			MAT.	INST.	TOTAL
1200					
1300	For alternate size restrooms:				
1400	Two water closets, two lavatories		6,800	3,375	10,175
1500					
1600	For each additional water closet over 2, add		1,675	780	2,455
1700	For each additional lavatory over 2, add		1,025	845	1,870
1800					
1900					
2400	NOTE: PLUMBING APPROXIMATIONS				
2500	WATER CONTROL: water meter, backflow preventer,				
2600	Shock absorbers, vacuum breakers, mixer....10 to 15% of fixtures				
2700	PIPE AND FITTINGS: .... 30 to 60% of fixtures				
2800					
2900	Cut & patch to match existing construction, add, minimum		2%	3%	
3000	Maximum		5%	9%	
3100	Dust protection, add, minimum		1%	2%	
3200	Maximum		4%	11%	
3300	Equipment usage curtailment, add, minimum		1%	1%	
3400	Maximum		3%	10%	
3500	Material handling & storage limitation, add, minimum		1%	1%	

D20 Plumbing

D2010 Plumbing Fixtures

D2010 956	Plumbing - Public Restroom	COST EACH		
		MAT.	INST.	TOTAL
3600	Maximum	6%	7%	
3700	Protection of existing work, add, minimum	2%	2%	
3800	Maximum	5%	7%	
3900	Shift work requirements, add, minimum		5%	
4000	Maximum		30%	

D20 Plumbing

D2010 Plumbing Fixtures

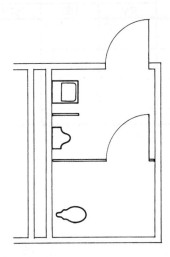

This page illustrates and describes a men's public restroom system including a water closet, urinal, lavatory, accessories and service piping. Lines within System Components give the unit price and total price on a cost each basis for this system. Prices for alternate men's public restroom systems are on Line Items D2010 957 1800 thru 2200. Both material quantities and labor costs have been adjusted for the system listed.

Factors: To adjust for job conditions other than normal working situations use Lines D2010 957 2900 thru 4000.

Example: You are to install the system and match existing construction. Go to Line D2010 957 3000 and apply these percentages to the appropriate MAT. and INST. costs.

System Components	QUANTITY	UNIT	COST EACH		
			MAT.	INST.	TOTAL
Public men's restroom incl. water closet, urinal, lavatory, accessories,					
And necessary service piping to install this system in one wall.					
Water closet, wall mounted, one piece	1.000	Ea.	315	188	503
Rough-in waste & vent for water closet	1.000	Set	745	425	1,170
Urinal, wall hung	1.000	Ea.	273	365	638
Rough-in waste & vent for urinal	1.000	Set	264	385	649
Lavatory, 20" x 18", P.E. cast iron with accessories	1.000	Ea.	325	136	461
Rough-in waste & vent for lavatory	1.000	Set	400	655	1,055
Partition, painted mtl., between walls, floor mntd.	1.000	Ea.	485	176	661
For handicap unit, add	1.000	Ea.	365		365
Grab bars	1.000	Ea.	79	53	132
Urinal screen, painted metal, wall mounted	1.000	Ea.	320	106	426
Mirror, 18" x 24" with stainless steel shelf	1.000	Ea.	249	26.50	275.50
Soap dispenser, surface mounted, liquid	1.000	Ea.	52	26.50	78.50
Toilet tissue dispenser, surface mounted, stainless steel	1.000	Ea.	21.50	17.60	39.10
Towel dispenser, surface mounted, stainless steel	1.000	Ea.	46	33	79
TOTAL			3,939.50	2,592.60	6,532.10

D2010 957	Plumbing - Public Restroom	COST EACH		
		MAT.	INST.	TOTAL
1600				
1700	For alternate size restrooms:			
1800	Two water closets, two urinals, two lavatories	7,600	5,050	12,650
1900				
2000	For each additional water closet over 2, add	1,675	780	2,455
2100	For each additional urinal over 2, add	855	855	1,710
2200	For each additional lavatory over 2, add	1,025	845	1,870
2300				
2400	NOTE: PLUMBING APPROXIMATIONS			
2500	WATER CONTROL: water meter, backflow preventer,			
2600	Shock absorbers, vacuum breakers, mixer....10 to 15% of fixtures			
2700	PIPE AND FITTINGS: .... 30 to 60% of fixtures			
2800				
2900	Cut & patch to match existing construction, add, minimum	2%	3%	
3000	Maximum	5%	9%	
3100	Dust protection, add, minimum	1%	2%	
3200	Maximum	4%	11%	
3300	Equipment usage curtailment, add, minimum	1%	1%	

D20 Plumbing

D2010 Plumbing Fixtures

D2010 957	Plumbing - Public Restroom	COST EACH		
		MAT.	INST.	TOTAL
3400	Maximum	3%	10%	
3500	Material handling, & storage limitation, add, minimum	1%	1%	
3600	Maximum	6%	7%	
3700	Protection of existing work, add, minimum	2%	2%	
3800	Maximum	5%	7%	
3900	Shift work requirements, add, minimum		5%	
4000	Maximum		30%	

D20 Plumbing

D2010 Plumbing Fixtures

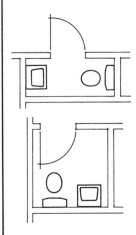

This page illustrates and describes a two fixture lavatory system including a water closet, lavatory, accessories and all service piping. Lines within System Components give the unit price and total price on a cost each basis for this system. Prices for an alternate two fixture lavatory system are on Line Item D2010 958 1900. Both material quantities and labor costs have been adjusted for the system listed.

Factors: To adjust for job conditions other than normal working situations use Lines D2010 958 2900 thru 4000.

Example: You are to install the system while controlling dust in the work area. Go to Line D2010 958 3200 and apply these percentages to the appropriate MAT. and INST. costs.

System Components	QUANTITY	UNIT	COST EACH MAT.	COST EACH INST.	COST EACH TOTAL
Two fixture bathroom incl. water closet, lavatory, accessories and					
Necessary service piping to install this system in 2 walls.					
Water closet, floor mounted, 2 piece, close coupled	1.000	Ea.	203	205	408
Rough in waste & vent for water closet	1.000	Set	300	355	655
Lavatory, 20" x 18", P.E. cast iron with accessories	1.000	Ea.	325	136	461
Rough in waste & vent for lavatory	1.000	Set	400	655	1,055
Additional service piping					
1/2" copper pipe with sweat solder joints	10.000	L.F.	42.10	74.50	116.60
2" black schedule 40 steel pipe with threaded couplings	12.000	L.F.	101.40	204	305.40
4" cast iron soil pipe with lead and oakum joints	7.000	L.F.	88.55	138.60	227.15
Accessories					
Toilet tissue dispenser, chrome, single roll	1.000	Ea.	21.50	17.60	39.10
18" long stainless steel towel bar	1.000	Ea.	43.50	23	66.50
Medicine cabinet with mirror, 20" x 16", unlighted	1.000	Ea.	89.50	37.50	127
TOTAL			1,614.55	1,846.20	3,460.75

D2010 958	Plumbing - Two Fixture Bathroom	COST EACH MAT.	COST EACH INST.	COST EACH TOTAL
1800 1900	Above system installed in 1 wall with all necessary service piping	1,500	1,600	3,100
2000 2400	NOTE: PLUMBING APPROXIMATIONS			
2500 2600	WATER CONTROL: water meter, backflow preventer, Shock absorbers, vacuum breakers, mixer....10 to 15% of fixtures			
2700 2800	PIPE AND FITTINGS: .... 30 to 60% of fixtures			
2900 3000	Cut & patch to match existing construction, add, minimum Maximum	2% 5%	3% 9%	
3100 3200	Dust protection, add, minimum Maximum	1% 4%	2% 11%	
3300 3400	Equipment usage curtailment, add, minimum Maximum	1% 3%	1% 10%	
3500 3600	Material handling & storage limitation, add, minimum Maximum	1% 6%	1% 7%	
3700 3800	Protection of existing work, add, minimum Maximum	2% 5%	2% 7%	
3900 4000	Shift work requirements, add, minimum Maximum		5% 30%	

D2010 Plumbing Fixtures

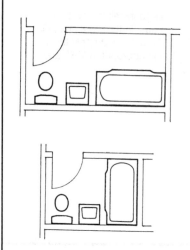

This page illustrates and describes a three fixture bathroom system including a water closet, tub, lavatory, accessories and service piping. Lines within System Components give the unit price and total price on a cost each basis for this system. Prices for an alternate three fixture bathroom system are on Line Item D2010 959 1700. Both material quantities and labor costs have been adjusted for the system listed.

Factors: To adjust for job conditions other than normal working situations use Lines D2010 959 2900 thru 4000.

Example: You are to install the system and protect all existing work. Go to Line D2010 959 3800 and apply these percentages to the appropriate MAT. and INST. costs.

System Components

	QUANTITY	UNIT	COST EACH MAT.	COST EACH INST.	COST EACH TOTAL
Three fixture bathroom incl. water closet, bathtub, lavatory, accessories, And necessary service piping to install this system in 1 wall.					
Water closet, floor mounted, 2 piece, close coupled	1.000	Ea.	203	205	408
Rough-in waste & vent for water closet	1.000	Set	279	330.15	609.15
Bathtub, P.E. cast iron 5' long with accessories	1.000	Ea.	890	247	1,137
Rough-in waste & vent for bathtub	1.000	Set	313.50	498.75	812.25
Lavatory, 20" x 18" P.E. cast iron with accessories	1.000	Ea.	325	136	461
Rough-in waste & vent for lavatory	1.000	Set	400	655	1,055
Accessories					
Toilet tissue dispenser, chrome, single roll	1.000	Ea.	21.50	17.60	39.10
18" long stainless steel towel bar	2.000	Ea.	87	46	133
Medicine cabinet with mirror, 20" x 16", unlighted	1.000	Ea.	89.50	37.50	127
TOTAL			2,608.50	2,173	4,781.50

D2010 959	Plumbing - Three Fixture Bathroom	COST EACH MAT.	COST EACH INST.	COST EACH TOTAL
1600				
1700	Above system installed in one wall with all necessary service piping	2,575	2,125	4,700
2400	NOTE: PLUMBING APPROXIMATIONS			
2500	WATER CONTROL: water meter, backflow preventer,			
2600	Shock absorbers, vacuum breakers, mixer....10 to 15% of fixtures			
2700	PIPE AND FITTINGS: .... 30 to 60% of fixtures			
2800				
2900	Cut & patch to match existing construction, add, minimum	2%	3%	
3000	Maximum	5%	9%	
3100	Dust protection, add, minimum	1%	2%	
3200	Maximum	4%	11%	
3300	Equipment usage curtailment, add, minimum	1%	1%	
3400	Maximum	3%	10%	
3500	Material handling & storage limitation, add, minimum	1%	1%	
3600	Maximum	6%	7%	
3700	Protection of existing work, add, minimum	2%	2%	
3800	Maximum	5%	7%	
3900	Shift work requirements, add, minimum		5%	
4000	Maximum		30%	

D20 Plumbing

D2010 Plumbing Fixtures

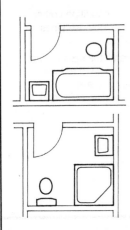

This page illustrates and describes a three fixture bathroom system including a water closet, tub, lavatory, accessories, and service piping. Lines within System Components give the unit price and total price on a cost each basis for this system. Prices for an alternate three fixture bathroom system are on Line Item D2010 960 2000. Both material quantities and labor costs have been adjusted for the system listed.

Factors: To adjust for job conditions other than normal working situations use Lines D2010 960 2900 thru 4000.

Example: You are to install the system and protect the surrounding area from dust. Go to Line D2010 960 3100 and apply these percentages to the appropriate MAT. and INST. costs.

System Components	QUANTITY	UNIT	COST EACH MAT.	COST EACH INST.	COST EACH TOTAL
Three fixture bathroom incl. water closet, bathtub, lavatory, accessories,					
And necessary service piping to install this system in 2 walls.					
Water closet, floor mounted, 2 piece, close coupled	1.000	Ea.	203	205	408
Rough-in waste & vent for water closet	1.000	Set	300	355	655
Bathtub, P.E. cast iron, 5' long with accessories	1.000	Ea.	890	247	1,137
Rough-in waste & vent for bathtub	1.000	Set	330	525	855
Lavatory, 20" x 18" P.E. cast iron with accessories	1.000	Ea.	325	136	461
Rough-in waste & vent for lavatory	1.000	Set	400	655	1,055
Additional service piping					
1-1/4" copper DWV type tubing, sweat solder joints	6.000	L.F.	72	60.60	132.60
2" black schedule 40 steel pipe with threaded couplings	12.000	L.F.	101.40	204	305.40
Accessories					
Toilet tissue dispenser, chrome, single roll	1.000	Ea.	21.50	17.60	39.10
18" long stainless steel towel bar	2.000	Ea.	87	46	133
Medicine cabinet with mirror, 20" x 16", unlighted	1.000	Ea.	89.50	37.50	127
TOTAL			2,819.40	2,488.70	5,308.10

D2010 960	Plumbing - Three Fixture Bathroom	COST EACH MAT.	COST EACH INST.	COST EACH TOTAL
2000	Above system with corner contour tub, P.E. cast iron	3,900	2,500	6,400
2300				
2400	NOTE: PLUMBING APPROXIMATIONS			
2500	WATER CONTROL: water meter, backflow preventer,			
2600	Shock absorbers, vacuum breakers, mixer....10 to 15% of fixtures			
2700	PIPE AND FITTINGS: .... 30 to 60% of fixtures			
2800				
2900	Cut & patch to match existing construction, add, minimum	2%	3%	
3000	Maximum	5%	9%	
3100	Dust protection, add, minimum	1%	2%	
3200	Maximum	4%	11%	
3300	Equipment usage curtailment, add, minimum	1%	1%	
3400	Maximum	3%	10%	
3500	Material handling & storage limitation, add, minimum	1%	1%	
3600	Maximum	6%	7%	
3700	Protection of existing work, add, minimum	2%	2%	
3800	Maximum	5%	7%	
3900	Shift work requirements, add, minimum		5%	
4000	Maximum		30%	

D20 Plumbing

D2010 Plumbing Fixtures

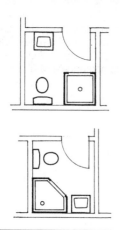

This page illustrates and describes a three fixture bathroom system including a water closet, shower, lavatory, accessories, and service piping, Lines within System Components give the unit price and total price on a cost each basis for this system. Prices for an alternate three fixture bathroom system are on Line Item D2010 961 2200. Both material quantities and labor costs have been adjusted for the system listed.

Factors: To adjust for job conditions other than normal working situations use Lines D2010 961 2900 thru 4000.

Example: You are to install the system and protect existing construction. Go to Line D2010 961 3700 and apply these percentages to the appropriate MAT. and INST. costs.

System Components	QUANTITY	UNIT	COST EACH		
			MAT.	INST.	TOTAL
Three fixture bathroom incl. water closet, shower, lavatory, accessories, And necessary service piping to install this system in 2 walls.					
Water closet, floor mounted, 2 piece, close coupled	1.000	Ea.	203	205	408
Rough-in waste & vent for water closet	1.000	Set	300	355	655
32" shower, enameled stall, molded stone receptor	1.000	Ea.	400	218	618
Rough-in waste & vent for shower	1.000	Set	435	530	965
Lavatory, 20" x 18" P.E. cast iron with accessories	1.000	Ea.	325	136	461
Rough-in waste & vent for lavatory	1.000	Set	400	655	1,055
Additional service piping					
1-1/4" Copper DWV type tubing, sweat solder joints	4.000	L.F.	48	40.40	88.40
2" black schedule 40 steel pipe with threaded couplings	6.000	L.F.	50.70	102	152.70
4" cast iron soil with lead and oakum joints	7.000	L.F.	88.55	138.60	227.15
Accessories					
Toilet tissue dispenser, chrome, single roll	1.000	Ea.	21.50	17.60	39.10
18" long stainless steel towel bar	2.000	Ea.	87	46	133
Medicine cabinet with mirror, 20" x 16", unlighted	1.000	Ea.	89.50	37.50	127
TOTAL			**2,448.25**	**2,481.10**	**4,929.35**

D2010 961	Plumbing - Three Fixture Bathroom	COST EACH		
		MAT.	INST.	TOTAL
2100	Above system installed with 36" corner angle shower, enameled steel,			
2200	Molded stone receptor	3,450	2,500	5,950
2300				
2400	NOTE: PLUMBING APPROXIMATIONS			
2500	WATER CONTROL: water meter, backflow preventer,			
2600	Shock absorbers, vacuum breakers, mixer....10 to 15% of fixtures			
2700	PIPE AND FITTINGS: .... 30 to 60% of fixtures			
2800				
2900	Cut & patch to match existing construction, add, minimum	2%	3%	
3000	Maximum	5%	9%	
3100	Dust protection, add, minimum	1%	2%	
3200	Maximum	4%	11%	
3300	Equipment usage curtailment, add, minimum	1%	1%	
3400	Maximum	3%	10%	
3500	Material handling & storage limitation, add, minimum	1%	1%	
3600	Maximum	6%	7%	
3700	Protection of existing work, add, minimum	2%	2%	
3800	Maximum	5%	7%	
3900	Shift work requirements, add, minimum		5%	
4000	Maximum		30%	

D2010 Plumbing Fixtures

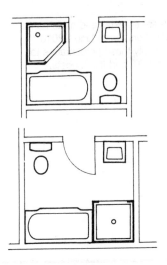

This page illustrates and describes a four fixture bathroom system including a water closet, shower, bathtub, lavatory, accessories, and service piping. Lines within System Components give the unit price and total price on a cost each basis for this system. Prices for an alternate four fixture bathroom system are on Line Item D2010 962 1900. Both material quantities and labor costs have been adjusted for the system listed.

Factors: To adjust for job conditions other than normal working situations use Lines D2010 962 2900 thru 4000.

Example: You are to install the system during weekends and evenings. Go to Line D2010 962 4000 and apply these percentages to the appropriate MAT. and INST. costs.

System Components	QUANTITY	UNIT	COST EACH MAT.	COST EACH INST.	COST EACH TOTAL
Four fixture bathroom incl. water closet, shower, bathtub, lavatory					
Accessories and necessary service piping to install this system in 2 walls.					
Water closet, floor mounted, 2 piece, close coupled	1.000	Ea.	203	205	408
Rough-in waste & vent for water closet	1.000	Set	300	355	655
32" shower, enameled steel stall, molded stone receptor	1.000	Ea.	400	218	618
Rough-in waste & vent for shower	1.000	Set	435	530	965
Bathtub, P.E. cast iron, 5' long with accessories	1.000	Ea.	890	247	1,137
Rough-in waste & vent for bathtub	1.000	Set	330	525	855
Lavatory, 20" x 18" P.E. cast iron with accessories	1.000	Ea.	325	136	461
Rough-in waste & vent for lavatory	1.000	Set	400	655	1,055
Accessories					
Toilet tissue dispenser, chrome, single roll	1.000	Ea.	21.50	17.60	39.10
18" long stainless steel towel bar	2.000	Ea.	87	46	133
Medicine cabinet with mirror, 20" x 16" unlighted	1.000	Ea.	89.50	37.50	127
TOTAL			3,481	2,972.10	6,453.10

D2010 962	Plumbing - Four Fixture Bathroom	COST EACH MAT.	COST EACH INST.	COST EACH TOTAL
1900	Above system with 36" corner angle shower, plumbing in 3 walls	4,625	3,250	7,875
2000				
2300				
2400	NOTE: PLUMBING APPROXIMATIONS			
2500	WATER CONTROL: water meter, backflow preventer,			
2600	Shock absorbers, vacuum breakers, mixer....10 to 15% of fixtures			
2700	PIPE AND FITTINGS: .... 30 to 60% of fixtures			
2800				
2900	Cut & patch to match existing construction, add, minimum	2%	3%	
3000	Maximum	5%	9%	
3100	Dust protection, add, minimum	1%	2%	
3200	Maximum	4%	11%	
3300	Equipment usage curtailment, add, minimum	1%	1%	
3400	Maximum	3%	10%	
3500	Material handling & storage limitation, add, minimum	1%	1%	
3600	Maximum	6%	7%	
3700	Protection of existing work, add, minimum	2%	2%	
3800	Maximum	5%	7%	
3900	Shift work requirements, add, minimum		5%	
4000	Maximum		30%	

D2010 Plumbing Fixtures

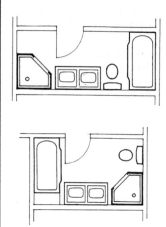

This page illustrates and describes a five fixture bathroom system including a water closet, shower, bathtub, lavatories, accessories, and service piping. Lines within System Components give the unit price and total price on a cost each basis for this system. Prices for an alternate five fixture bathroom system are on Line Item D2010 963 1900. Both material quantities and labor costs have been adjusted for the system listed.

Factors: To adjust for job conditions other than normal working situations use Lines D2010 963 2900 thru 4000.

Example: You are to install and match any existing construction. Go to Line D2010 963 2900 and apply these percentages to the appropriate MAT. and INST. costs.

System Components	QUANTITY	UNIT	COST EACH MAT.	COST EACH INST.	COST EACH TOTAL
Five fixture bathroom incl. water closet, shower, bathtub, 2 lavatories, Accessories and necessary service piping to install this system in 1 wall.					
Water closet, floor mounted, 2 piece, close coupled	1.000	Ea.	203	205	408
Rough-in waste & vent for water closet	1.000	Set	300	355	655
36" corner angle shower, enameled steel stall, molded stone receptor	1.000	Ea.	1,400	227	1,627
Rough-in waste & vent for shower	1.000	Set	435	530	965
Bathtub, P.E. cast iron, 5' long with accessories	1.000	Ea.	890	247	1,137
Rough-in waste & vent for bathtub	1.000	Set	330	525	855
Countertop, plastic laminate, with backsplash	2.000	Ea.	140	70.40	210.40
Vanity base, 2 doors, 24" wide	2.000	Ea.	243	96	339
Lavatories, 20" x 18" cabinet mntd, PECI	2.000	Ea.	508	340	848
Rough-in waste & vent for lavatories	1.600	Set	393.60	760	1,153.60
Accessories					
Toilet tissue dispenser, chrome, single roll	1.000	Ea.	21.50	17.60	39.10
18" long stainless steel towel bars	2.000	Ea.	87	46	133
Medicine cabinet with mirror, 20" x 16", unlighted	2.000	Ea.	179	75	254
TOTAL			5,130.10	3,494	8,624.10

D2010 963	Plumbing - Five Fixture Bathroom	COST EACH MAT.	COST EACH INST.	COST EACH TOTAL
1900 2000	Above system installed in 2 walls with all necessary service piping	5,200	3,625	8,825
2100 2200				
2300 2400	NOTE: PLUMBING APPROXIMATIONS			
2500	WATER CONTROL: water meter, backflow preventer,			
2600	Shock absorbers, vacuum breakers, mixer....10 to 15% of fixtures			
2700 2800	PIPE AND FITTINGS: .... 30 to 60% of fixtures			
2900	Cut & patch to match existing construction, add, minimum	2%	3%	
3000	Maximum	5%	9%	
3100	Dust protection, add, minimum	1%	2%	
3200	Maximum	4%	11%	
3300	Equipment usage curtailment, add, minimum	1%	1%	
3400	Maximum	3%	10%	
3500	Material handling & storage limitation, add, minimum	1%	1%	
3600	Maximum	6%	7%	

D2010 Plumbing Fixtures

D2010 963	Plumbing - Five Fixture Bathroom	COST EACH		
		MAT.	INST.	TOTAL
3700	Protection of existing work, add, minimum	2%	2%	
3800	Maximum	5%	7%	
3900	Shift work requirements, add, minimum		5%	
4000	Maximum		30%	

D3010 Energy Supply

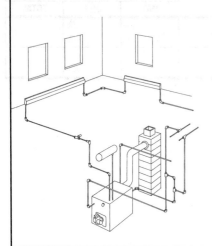

This page illustrates and describes an oil fired hot water baseboard system including an oil fired boiler, fin tube radiation and all fittings and piping. Lines within System Components give the unit price and total price per square foot for this system. Prices for alternate oil fired hot water baseboard systems are on Line Items D3010 540 1700 thru 2100. Both material quantities and labor costs have been adjusted for the system listed.

Factors: To adjust for job conditions other than normal working situations use Lines D3010 540 3000 thru 4000.

Example: You are to install the system while protecting all existing work. Go to Line D3010 540 3700 and apply these percentages to the appropriate MAT. and INST. costs.

System Components	QUANTITY	UNIT	COST PER S.F.		
			MAT.	INST.	TOTAL
Oil fired hot water baseboard system including boiler, fin tube radiation					
And all necessary fittings and piping.					
Area to 800 S.F.					
Boiler, cast iron, w/oil piping, 97 MBH	1.000	Ea.	2,312.50	1,343.75	3,656.25
Copper piping	130.000	L.F.	1,787.50	1,358.50	3,146
Fin tube radiation	68.000	L.F.	571.20	1,292	1,863.20
Circulator	1.000	Ea.	490	181	671
Oil tank	1.000	Ea.	420	220	640
Expansion tank, ASME	1.000	Ea.	520	78.50	598.50
TOTAL			6,101.20	4,473.75	10,574.95
COST PER S.F.		S.F.	7.63	5.59	13.22

D3010 540	Heating - Oil Fired Hot Water	COST PER S.F.		
		MAT.	INST.	TOTAL
1600	For alternate hot water systems:			
1700	Cast iron boiler, area to 1000 S.F.	6.55	4.82	11.37
1800	To 1200 S.F.	5.70	4.37	10.07
1900	To 1600 S.F.	6.75	4.36	11.11
2000	To 2000 S.F.	5.70	3.87	9.57
2100	To 3000 S.F.	4.28	3.27	7.55
2200				
2300				
2700				
2900	Cut & patch to mach existing construction, add, minimum	2%	3%	
3000	Maximum	5%	9%	
3100	Dust protection, add, minimum	1%	2%	
3200	Maximum	4%	11%	
3300	Equipment usage curtailment, add, minimum	1%	1%	
3400	Maximum	3%	10%	
3500	Material handling & storage limitation, add, minimum	1%	1%	
3600	Maximum	6%	7%	
3700	Protection of existing work, add, minimum	2%	2%	
3800	Maximum	5%	7%	
3900	Shift work requirements, add, minimum		5%	
4000	Maximum		30%	

D3010 Energy Supply

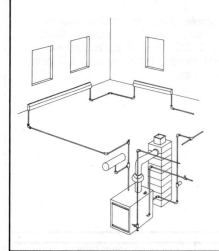

This page illustrates and describes a gas fired hot water baseboard system including a gas fired boiler, fin tube radiation, fittings and piping. Lines within System Components give the unit price and total price per square foot for this system. Prices for alternate gas fired hot water baseboard systems are on Line Items D3010 550 1500 thru 1900. Both material quantities and labor costs have been adjusted for the system listed.

Factors: To adjust for job conditions other than normal working situations use Lines D3010 550 2900 thru 4000.

Example: You are to install the system with minimal equipment usage. Go to Line D3010 550 3400 and apply these percentages to the appropriate MAT. and INST. costs.

System Components		QUANTITY	UNIT	COST PER S.F. MAT.	COST PER S.F. INST.	COST PER S.F. TOTAL
Gas fired hot water baseboard system including boiler, fin tube radiation,						
All necessary fittings and piping.						
Area to 800 S.F.						
Cast iron boiler, insulating jacket, gas piping, 80 MBH		1.000	Ea.	2,343.75	2,000	4,343.75
Copper piping		130.000	L.F.	1,787.50	1,358.50	3,146
Fin tube radiation		68.000	L.F.	571.20	1,292	1,863.20
Circulator, flange connection		1.000	Ea.	490	181	671
Expansion tank, ASME		1.000	Ea.	520	78.50	598.50
TOTAL				5,712.45	4,910	10,622.45
COST PER S.F.			S.F.	7.14	6.14	13.28

D3010 550	Heating - Gas Fired Hot Water	COST PER S.F. MAT.	COST PER S.F. INST.	COST PER S.F. TOTAL
1400	For alternate hot water systems:			
1500	Cast iron boiler, area to 1000 S.F.	6.05	5.40	11.45
1600	To 1200 S.F.	5.30	4.86	10.16
1700	To 1600 S.F.	6.45	4.72	11.17
1800	To 2000 S.F.	5.60	4.22	9.82
1900	To 3000 S.F.	4.43	3.70	8.13
2000				
2100				
2900	Cut & patch to match existing construction, add, minimum	2%	3%	
3000	Maximum	5%	9%	
3100	Dust protection, add, minimum	1%	2%	
3200	Maximum	4%	11%	
3300	Equipment usage curtailment, add, minimum	1%	1%	
3400	Maximum	3%	10%	
3500	Material handling & storage limitation, add, minimum	1%	1%	
3600	Maximum	6%	7%	
3700	Protection of existing work, add, minimum	2%	2%	
3800	Maximum	5%	7%	
3900	Shift work requirements, add, minimum		5%	
4000	Maximum		30%	

D30 HVAC

D3010 Energy Supply

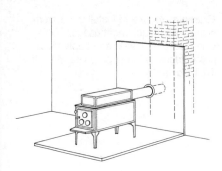

This page illustrates and describes a wood burning stove system including a free standing stove, preformed hearth, masonry chimney, and necessary piping and fittings. Lines within System Components give the unit price and total price on a cost each basis for this system. Prices for alternate wood burning stove systems are on Line Items D3010 991 1600 thru 2000. Both material quantities and labor costs have been adjusted for the system listed.

Factors: To adjust for job conditions other than normal working situations use Lines D3010 991 2900 thru 4000.

Example: You are to install the system and protect existing construction. Go to Line D3010 991 3500 and apply these percentages to the appropriate MAT., and INST. costs.

System Components	QUANTITY	UNIT	COST EACH		
			MAT.	INST.	TOTAL
Cast iron, free standing,wood burning stove with preformed hearth, masonry Chimney, and all necessary piping and fittings to install in chimney.					
Cast iron wood burning stove, stove pipe	1.000	Ea.	1,250	810	2,060
Wall panel or hearth, non-combustible	1.000	Ea.	570	149.15	719.15
16" x 16" brick chimney, 8" x 8" flue	20.000	V.L.F.	575	1,287.50	1,862.50
Foundation	.500	C.Y.	91	110.45	201.45
TOTAL			2,486	2,357.10	4,843.10

D3010 991	Wood Burning Stoves	COST EACH		
		MAT.	INST.	TOTAL
1400	For alternate wood burning systems:			
1500				
1600	Installed in existing fireplace	1,550	890	2,440
1700				
1800	Installed with insulated metal chimney			
1900	System including ceiling package,			
2000	Metal chimney to 10 L.F.	2,700	1,250	3,950
2100				
2200				
2300				
2400				
2500				
2600				
2700				
2900	Dust protection, add, minimum	1%	2%	
3000	Maximum	4%	11%	
3100	Equipment usage curtailment, add, minimum	1%	1%	
3200	Maximum	3%	10%	
3300	Material handling & storage limitation, add, minimum	1%	1%	
3400	Maximum	6%	7%	
3500	Protection of existing work, add, minimum	2%	2%	
3600	Maximum	5%	7%	
3700	Shift work requirements, add, minimum		5%	
3800	Maximum		30%	
3900	Temporary shoring and bracing, add, minimum	2%	5%	
4000	Maximum	5%	12%	

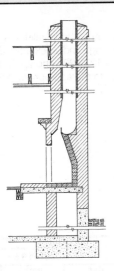

This page illustrates and describes masonry fireplace systems including a brick fireplace, footing, foundation, hearth, firebox, chimney and flue. Lines within System Components give the unit price and total price on a cost each basis for this system. Prices for alternate masonry fireplace systems are on Line Items D3010 992 1300 thru 1500. Both material quantities and labor costs have been adjusted for the system listed.

Factors: To adjust for job conditions other than normal working situations use Lines D3010 992 2700 thru 4000.

Example: You are to install the system with some temporary shoring and bracing. Go to Line D3010 992 3900 and apply these percentages to the appropriate MAT. and INST. costs.

System Components	QUANTITY	UNIT	COST EACH		
			MAT.	INST.	TOTAL
Brick masonry fireplace, including footing, foundation, hearth, firebox,					
Chimney and flue, chimney 12' above firebox.					
Footing, 4' x 7' x 12" thick, 3000 psi concrete	1.040	C.Y.	180.96	192.49	373.45
Foundation, 12" concrete block	180.000	S.F.	774	1,692	2,466
Fireplace, brick faced, 6'-0" wide x 5'-0" high	1.000	Ea.	590	2,350	2,940
Hearth	1.000	Ea.	217	470	687
Chimney, 20" x 20", one 12" x 12" flue, 12 V.L.F.	12.000	V.L.F.	504	822	1,326
Mantle, wood	7.000	L.F.	46.20	102.55	148.75
TOTAL			2,312.16	5,629.04	7,941.20

D3010 992	Masonry Fireplace	COST EACH		
		MAT.	INST.	TOTAL
1200	Above system with the following:	2,575	6,050	8,625
1300	Chimney, 20" x 20", one 12" x 12" flue, 18 V.L.F.			
1400	24 V.L.F.	2,825	6,450	9,275
1500	Fieldstone face instead of brick	2,775	5,625	8,400
1600				
1700				
1800				
1900				
2000				
2100				
2200				
2300				
2700	Cut & patch to match existing construction, add, minimum	2%	3%	
2800	Maximum	5%	9%	
2900	Dust protection, add, minimum	1%	2%	
3000	Maximum	4%	11%	
3100	Equipment usage curtailment, add, minimum	1%	1%	
3200	Maximum	3%	10%	
3300	Material handling & storage limitation, add, minimum	1%	1%	
3400	Maximum	6%	7%	
3500	Protection of existing work, add, minimum	2%	2%	
3600	Maximum	5%	7%	
3700	Shift work requirements, add, minimum		5%	
3800	Maximum		30%	
3900	Temporary shoring and bracing, add, minimum	2%	5%	
4000	Maximum	5%	12%	

D3020 Heat Generating Systems

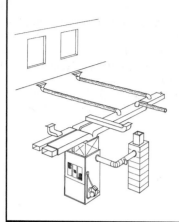

This page illustrates and describes an oil fired forced air system including an oil fired furnace, ductwork, registers and hookups. Lines within System Components give the unit price and total price per square foot for this system. Prices for alternate oil fired forced air systems are on Line Items D3020 122 1700 thru 2800. Both material quantities and labor costs have been adjusted for the system listed.

Factors: To adjust for job conditions other than normal working situations use Lines D3020 122 3100 thru 4200.

Example: You are to install the system during evenings and weekends. Go to Line D3020 122 4200 and apply this percentage to the appropriate INST. cost.

System Components			COST PER S.F.		
	QUANTITY	UNIT	MAT.	INST.	TOTAL
Oil fired hot air heating system including furnace, ductwork, registers And all necessary hookups.					
Area to 800 S.F., heat only					
Furnace, oil, atomizing gun type burner, w/oil piping	1.000	Ea.	2,531.25	375	2,906.25
Oil tank, steel, 275 gallon	1.000	Ea.	420	220	640
Duct, galvanized, steel	312.000	Lb.	383.76	2,230.80	2,614.56
Insulation, blanket type, duct work	270.000	S.F.	202.50	791.10	993.60
Flexible duct, 6" diameter, insulated	100.000	L.F.	254	415	669
Registers, baseboard, gravity, 12" x 6"	8.000	Ea.	133.60	208	341.60
Return, damper, 36" x 18"	1.000	Ea.	139	60	199
TOTAL			4,064.11	4,299.90	8,364.01
COST PER S.F.		S.F.	5.08	5.37	10.45

D3020 122	Heating-Cooling, Oil, Forced Air	COST PER S.F.		
		MAT.	INST.	TOTAL
1600	For alternate heating systems:			
1700	Oil fired, area to 1000 S.F.	4.09	4.36	8.45
1800	To 1200 S.F.	3.56	3.84	7.40
1900	To 1600 S.F.	2.87	3.20	6.07
2000	To 2000 S.F.	2.80	4.13	6.93
2100	To 3000 S.F.	2.20	3.28	5.48
2200	For combined heating and cooling systems:			
2300	Oil fired, heating and cooling, area to 800 S.F.	7.45	6.40	13.85
2400	To 1000 S.F.	6.10	5.20	11.30
2500	To 1200 S.F.	5.25	4.55	9.80
2600	To 1600 S.F.	4.29	3.76	8.05
2700	To 2000 S.F.	3.94	4.64	8.58
2800	To 3000 S.F.	2.97	3.64	6.61
3100	Cut & patch to match existing construction, add, minimum	2%	3%	
3200	Maximum	5%	9%	
3300	Dust protection, add, minimum	1%	2%	
3400	Maximum	4%	11%	
3500	Equipment usage curtailment, add, minimum	1%	1%	
3600	Maximum	3%	10%	
3700	Material handling & storage limitation, add, minimum	1%	1%	
3800	Maximum	6%	7%	
3900	Protection of existing work, add, minimum	2%	2%	
4000	Maximum	5%	7%	
4100	Shift work requirements, add, minimum		5%	
4200	Maximum		30%	

D3020 Heat Generating Systems

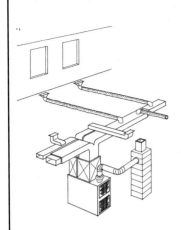

This page illustrates and describes a gas fired forced air system including a gas fired furnace, ductwork, registers and hookups. Lines within System Components give the unit price and total price per square foot for this system. Prices for alternate gas fired forced air systems are on Line Items D3020 124 1500 thru 2600. Both material quantities and labor costs have been adjusted for the system listed.

Factors: To adjust for job conditions other than normal working situations use Lines D3020 124 2900 thru 4000.

Example: You are to install the system with material handling and storage limitations. Go to Line D3020 124 3500 and apply these percentages to the appropriate MAT. and INST. costs.

System Components	QUANTITY	UNIT	COST PER S.F.		
			MAT.	INST.	TOTAL
Gas fired hot air heating system including furnace, ductwork, registers And all necessary hookups.					
Area to 800 S.F., heat only					
Furnace, gas, AGA certified, direct drive, w/gas piping, 44 MBH	1.000	Ea.	793.75	337.50	1,131.25
Duct, galvanized steel	312.000	Lb.	383.76	2,230.80	2,614.56
Insulation, blanket type, ductwork	270.000	S.F.	202.50	791.10	993.60
Flexible duct, 6″ diameter, insulated	100.000	L.F.	254	415	669
Registers, baseboard, gravity, 12″ x 6″	8.000	Ea.	133.60	208	341.60
Return, damper, 36″ x 18″	1.000	Ea.	139	60	199
TOTAL			1,906.61	4,042.40	5,949.01
COST PER S.F.		S.F.	2.38	5.05	7.43

D3020 124	Heating-Cooling, Gas, Forced Air	COST PER S.F.		
		MAT.	INST.	TOTAL
1400	For alternate heating systems:			
1500	Gas fired, area to 1000 S.F.	1.98	4.11	6.09
1600	To 1200 S.F.	1.79	3.65	5.44
1700	To 1600 S.F.	1.74	3.41	5.15
1800	To 2000 S.F.	1.76	4.02	5.78
1900	To 3000 S.F.	1.37	3.20	4.57
2000	For combined heating and cooling systems:			
2100	Gas fired, heating and cooling, area to 800 S.F.	4.76	6.05	10.81
2200	To 1000 S.F.	4	4.97	8.97
2300	To 1200 S.F.	3.48	4.36	7.84
2400	To 1600 S.F.	3.16	3.97	7.13
2500	To 2000 S.F.	2.90	4.53	7.43
2600	To 3000 S.F.	2.14	3.56	5.70
2900	Cut & patch to match existing construction, add, minimum	2%	3%	
3000	Maximum	5%	9%	
3100	Dust protection, add, minimum	1%	2%	
3200	Maximum	4%	11%	
3300	Equipment usage curtailment, add, minimum	1%	1%	
3400	Maximum	3%	10%	
3500	Material handling & storage limitation, add, minimum	1%	1%	
3600	Maximum	6%	7%	
3700	Protection of existing work, add, minimum	2%	2%	
3800	Maximum	5%	7%	
3900	Shift work requirements, add, minimum		5%	
4000	Maximum		30%	

D3020 Heat Generating Systems

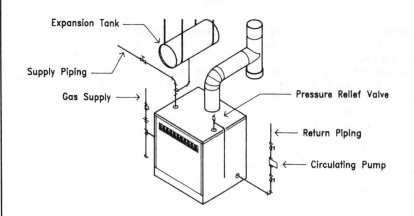

Expansion Tank

Supply Piping

Gas Supply

Pressure Relief Valve

Return Piping

Circulating Pump

Cast Iron Boiler, Hot Water, Gas Fired

This page illustrates and describes boilers including expansion tank, circulating pump and all service piping. Lines within Systems Components give the material and installation price on a cost each basis for the components. Prices for alternate boiler systems are on Line Items D3020 130 1010 thru D3020 138 1040. Material quantities and labor costs have been adjusted for the system listed.

Factors: To adjust for job conditions other than normal working situations use Line D3020 138 2700 thru 4000.

System Components	QUANTITY	UNIT	COST EACH MAT.	COST EACH INST.	COST EACH TOTAL
Cast iron boiler, including piping, chimney and accessories.					
Boilers, gas fired, std controls, CI, insulated, HW, gross output 100 MBH	1.000	Ea.	2,150	1,725	3,875
Pipe, black steel, Sch 40, threaded, W/coupling & hangers, 10' OC, 3/4" dia	20.000	L.F.	65.25	222.75	288
Pipe, black steel, Sch 40, threaded, W/coupling & hangers, 10' OC, 1" dia.	20.000	L.F.	91.59	245.10	336.69
Elbow, 90°, black, straight, 3/4" dia.	9.000	Ea.	33.39	387	420.39
Elbow, 90°, black, straight, 1" dia.	6.000	Ea.	28.20	279	307.20
Tee, black, straight, 3/4" dia.	2.000	Ea.	8.60	134	142.60
Tee, black, straight, 1" dia.	2.000	Ea.	14.70	151	165.70
Tee, black, reducing, 1" dia.	2.000	Ea.	23.60	151	174.60
Pipe cap, black, 3/4" dia.	1.000	Ea.	3.07	18.90	21.97
Union, black with brass seat, 3/4" dia.	2.000	Ea.	22.70	93	115.70
Union, black with brass seat, 1" dia.	2.000	Ea.	29.70	101	130.70
Pipe nipples, black, 3/4" dia.	5.000	Ea.	7.25	24.75	32
Pipe nipples, black, 1" dia.	3.000	Ea.	6.39	17.10	23.49
Valves, bronze, gate, N.R.S., threaded, class 150, 3/4" size	2.000	Ea.	124	60	184
Valves, bronze, gate, N.R.S., threaded, class 150, 1" size	2.000	Ea.	157	64	221
Gas cock, brass, 3/4" size	1.000	Ea.	15.05	27.50	42.55
Thermometer, stem type, 9" case, 8" stem, 3/4" NPT	2.000	Ea.	240	44	284
Tank, steel, liquid expansion, ASME, painted, 15 gallon capacity	1.000	Ea.	485	65	550
Pump, circulating, bronze, flange connection, 3/4" to 1-1/2" size, 1/8 HP	1.000	Ea.	825	181	1,006
Vent chimney, all fuel, pressure tight, double wall, SS, 6" dia.	20.000	L.F.	1,040	360	1,400
Vent chimney, elbow, 90° fixed, 6" dia.	2.000	Ea.	586	72	658
Vent chimney, Tee, 6" dia.	2.000	Ea.	386	90	476
Vent chimney, ventilated roof thimble, 6" dia.	1.000	Ea.	247	41.50	288.50
Vent chimney, adjustable roof flashing, 6" dia.	1.000	Ea.	69	36	105
Vent chimney, stack cap, 6" diameter	1.000	Ea.	218	23.50	241.50
Insulation, fiberglass pipe covering, 1" wall, 1" IPS	20.000	L.F.	22.80	93.40	116.20
TOTAL			6,899.29	4,707.50	11,606.79

D3020 130	Boiler, Cast Iron, Hot Water, Gas		COST EACH MAT.	COST EACH INST.	COST EACH TOTAL
1000	For alternate boilers:				
1010	Boiler, cast iron, gas, hot water, 100 MBH		6,900	4,700	11,600
1020	200 MBH		8,000	5,375	13,375
1030	320 MBH		8,650	5,875	14,525
1040	440 MBH		11,700	7,550	19,250
1050	544 MBH		19,300	11,900	31,200
1060	765 MBH		24,000	12,700	36,700
1070	1088 MBH		26,700	13,600	40,300

D3020 Heat Generating Systems

D3020 130	Boiler, Cast Iron, Hot Water, Gas	COST EACH		
		MAT.	INST.	TOTAL
1080	1530 MBH	31,300	17,900	49,200
1090	2312 MBH	38,000	21,100	59,100

D3020 134	Boiler, Cast Iron, Steam, Gas	COST EACH		
		MAT.	INST.	TOTAL
1010	Boiler, cast iron, gas, steam, 100 MBH	5,875	4,575	10,450
1020	200 MBH	11,900	8,600	20,500
1030	320 MBH	13,700	9,625	23,325
1040	544 MBH	20,600	15,100	35,700
1050	765 MBH	23,600	15,700	39,300
1060	1275 MBH	28,200	17,300	45,500
1070	2675 MBH	39,100	25,100	64,200

D3020 136	Boiler, Cast Iron, Hot Water, Gas/Oil	COST EACH		
		MAT.	INST.	TOTAL
1010	Boiler, cast iron, gas & oil, hot water, 584 MBH	25,400	13,100	38,500
1020	876 MBH	29,500	14,100	43,600
1030	1168 MBH	31,500	15,900	47,400
1040	1460 MBH	35,600	19,100	54,700
1050	2044 MBH	42,200	19,800	62,000
1060	2628 MBH	53,000	24,000	77,000

D3020 138	Boiler, Cast Iron, Steam, Gas/Oil	COST EACH		
		MAT.	INST.	TOTAL
1010	Boiler, cast iron, gas & oil, steam, 810 MBH	27,900	17,100	45,000
1020	1360 MBH	31,300	18,200	49,500
1030	2040 MBH	40,900	23,300	64,200
1040	2700 MBH	44,800	26,000	70,800
2700	Cut & patch to match existing construction, add, minimum	2%	3%	
2800	Maximum	5%	9%	
2900	Dust protection, add, minimum	1%	2%	
3000	Maximum	4%	11%	
3100	Equipment usage curtailment, add, minimum	1%	1%	
3200	Maximum	3%	10%	
3300	Material handling & storage limitation, add, minimum	1%	1%	
3400	Maximum	6%	7%	
3500	Protection of existing work, add, minimum	2%	2%	
3600	Maximum	5%	7%	
3700	Shift work requirements, add, minimum		5%	
3800	Maximum		30%	
3900	Temporary shoring and bracing, add, minimum	2%	5%	
4000	Maximum	5%	12%	

D3020 Heat Generating Systems

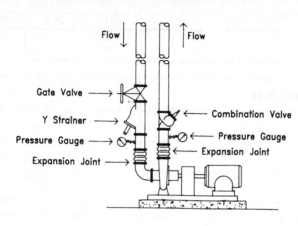

Base Mounted End—Suction Pump

This page illustrates and describes end suction base mounted circulating pumps, including strainer and piping. Lines within system components give the material and installation price on a cost each basis for the components. Prices for alternate circulating pumps are on Line items D3020 330 1010 thru 1050. Material quantities and labor costs have been adjusting for the system listed.

Factors: To adjust for job conditions other than normal working situations uses Lines D3020 330 2700 thru 4000.

System Components	QUANTITY	UNIT	COST EACH		
			MAT.	INST.	TOTAL
Circulator pump, end suction, including pipe,					
Valves, fittings and guages.					
Pump, circulating, CI, base mounted, 2-1/2" size, 3 HP, to 150 GPM	1.000	Ea.	4,075	605	4,680
Pipe, black steel, Sch. 40, on yoke & roll hangers, 10' O.C., 2-1/2" dia.	12.000	L.F.	129.60	291.60	421.20
Elbow, 90°, weld joint, steel, 2-1/2" pipe size	1.000	Ea.	18.65	143.65	162.30
Flange, weld neck, 150 LB, 2-1/2" pipe size	8.000	Ea.	204	574.64	778.64
Valve, iron body, gate, 125 lb., N.R.S., flanged, 2-1/2" size	1.000	Ea.	560	218	778
Strainer, Y type, iron body, flanged, 125 lb., 2-1/2" pipe size	1.000	Ea.	141	220	361
Multipurpose valve, CI body, 2" size	1.000	Ea.	415	76.50	491.50
Expansion joint, flanged spool, 6" F to F, 2-1/2" dia.	2.000	Ea.	572	178	750
T-O-L, weld joint, socket, 1/4" pipe size, nozzle	2.000	Ea.	13.40	100.32	113.72
T-O-L, weld joint, socket, 1/2" pipe size, nozzle	2.000	Ea.	13.40	104.58	117.98
Control gauges, pressure or vacuum, 3-1/2" diameter dial	2.000	Ea.	79	38.30	117.30
Pressure/temperature relief plug, 316 SS, 3/4" OD, 7-1/2" insertion	2.000	Ea.	150	38.30	188.30
Insulation, fiberglass pipe covering, 1-1/2" wall, 2-1/2" IPS	12.000	L.F.	31.92	68.40	100.32
Pump control system	1.000	Ea.	1,200	560	1,760
Pump balancing	1.000	Ea.		269	269
TOTAL			7,602.97	3,486.29	11,089.26

D3020 330		Circulating Pump Systems, End Suction	COST EACH		
			MAT.	INST.	TOTAL
1000	For alternate pump sizes:				
1010	Pump, base mtd with motor, end-suction, 2-1/2" size, 3 HP, to 150 GPM		7,600	3,500	11,100
1020	3" size, 5 HP, to 225 GPM		8,300	4,000	12,300
1030	4" size, 7-1/2 HP, to 350 GPM		9,475	4,825	14,300
1040	5" size, 15 HP, to 1000 GPM		12,300	7,275	19,575
1050	6" size, 25 HP, to 1550 GPM		17,100	8,850	25,950
2700	Cut & patch to match existing construction, add, minimum		2%	3%	
2800	Maximum		5%	9%	
2900	Dust protection, add, minimum		1%	2%	
3000	Maximum		4%	11%	
3100	Equipment usage curtailment, add, minimum		1%	1%	
3200	Maximum		3%	10%	
3300	Material handling & storage limitation, add, minimum		1%	1%	
3400	Maximum		6%	7%	
3500	Protection of existing work, add, minimum		2%	2%	
3600	Maximum		5%	7%	

D3020 Heat Generating Systems

D3020 330	Circulating Pump Systems, End Suction	COST EACH		
		MAT.	INST.	TOTAL
3700	Shift work requirements, add, minimum		5%	
3800	Maximum		30%	
3900	Temporary shoring and bracing, add, minimum	2%	5%	
4000	Maximum	5%	12%	

D3040 Distribution Systems

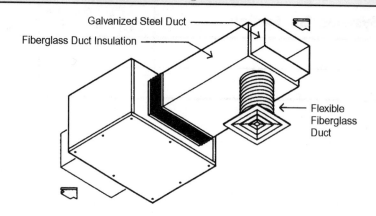

Galvanized Steel Duct

Fiberglass Duct Insulation

Flexible Fiberglass Duct

This page illustrates and describes fan coil air conditioners including duct work, duct installation, piping and diffusers. Lines within Systems Components give the material and installation price on a cost each basis for the components. Prices for alternate fan coil A/C unit systems are on Line Items D3040 118 1010 thru D3040 126 1120. Material quantities and labor costs have been adjusted for the system listed.

Factors: To adjust for job conditions other than normal working situations use Lines D3040 126 2700 thru 4000.

Horizontal Fan Coil Air Conditioning System

System Components	QUANTITY	UNIT	COST EACH		
			MAT.	INST.	TOTAL
Fan coil A/C system, horizontal, with housing, controls, 2 pipe, 1/2 ton.					
Fan coil A/C, horizontal housing, filters, chilled water, 1/2 ton cooling	1.000	Ea.	1,175	138	1,313
Pipe, black steel, Sch 40, threaded, W/cplgs & hangers, 10' OC, 3/4" dia.	20.000	L.F.	62.35	212.85	275.20
Elbow, 90°, black, straight, 3/4" dia.	6.000	Ea.	22.26	258	280.26
Tee, black, straight, 3/4" dia.	2.000	Ea.	8.60	134	142.60
Union, black with brass seat, 3/4" dia.	2.000	Ea.	22.70	93	115.70
Pipe nipples, 3/4" diam.	3.000	Ea.	4.35	14.85	19.20
Valves, bronze, gate, N.R.S., threaded, class 150, 3/4" size	2.000	Ea.	124	60	184
Circuit setter, balance valve, bronze body, threaded, 3/4" pipe size	1.000	Ea.	68.50	30.50	99
Insulation, fiberglass pipe covering, 1" wall, 3/4" IPS	20.000	L.F.	21.40	89.40	110.80
Ductwork, 12" x 8" fabricated, galvanized steel, 12 LF	55.000	Lb.	67.65	393.25	460.90
Insulation, ductwork, blanket type, fiberglass, 1" thk, 1-1/2 LB density	40.000	S.F.	30	117.20	147.20
Diffusers, aluminum, OB damper, ceiling, perf, 24"x24" panel size, 6"x6"	2.000	Ea.	123	75	198
Ductwork, flexible, fiberglass fabric, insulated, 1"thk, PE jacket, 6" dia.	16.000	L.F.	40.64	66.40	107.04
Round volume control damper 6" dia.	2.000	Ea.	51	55	106
Fan coil unit balancing	1.000	Ea.		78	78
Re-heat coil balancing	1.000	Ea.		113	113
Diffuser/register, high, balancing	2.000	Ea.		216	216
TOTAL			1,821.45	2,144.45	3,965.90

D3040 118	Fan Coil A/C Unit, Two Pipe	COST EACH		
		MAT.	INST.	TOTAL
1000	For alternate A/C units:			
1010	Fan coil A/C system, cabinet mounted, controls, 2 pipe, 1/2 Ton	1,075	1,150	2,225
1020	1 Ton	1,300	1,275	2,575
1030	1-1/2 Ton	1,400	1,300	2,700
1040	2 Ton	2,075	1,575	3,650
1050	3 Ton	2,700	1,650	4,350

D3040 120	Fan Coil A/C Unit, Two Pipe, Electric Heat	COST EACH		
		MAT.	INST.	TOTAL
1010	Fan coil A/C system, cabinet mntd, elect. ht, controls, 2 pipe, 1/2 Ton	1,575	1,150	2,725
1020	1 Ton	1,925	1,275	3,200
1030	1-1/2 Ton	2,275	1,300	3,575
1040	2 Ton	3,300	1,600	4,900
1050	3 Ton	5,700	1,650	7,350

D30 HVAC

D3040 Distribution Systems

D3040 122	Fan Coil A/C Unit, Four Pipe	COST EACH		
		MAT.	INST.	TOTAL
1010	Fan coil A/C system, cabinet mounted, controls, 4 pipe, 1/2 Ton	1,725	2,225	3,950
1020	1 Ton	1,975	2,350	4,325
1030	1-1/2 Ton	2,125	2,375	4,500
1040	2 Ton	2,625	2,475	5,100
1050	3 Ton	3,600	2,650	6,250

D3040 124	Fan Coil A/C, Horizontal, Duct Mount, 2 Pipe	COST EACH		
		MAT.	INST.	TOTAL
1010	Fan coil A/C system, horizontal w/housing, controls, 2 pipe, 1/2 Ton	1,825	2,150	3,975
1020	1 Ton	2,375	3,125	5,500
1030	1-1/2 Ton	2,950	4,050	7,000
1040	2 Ton	3,800	4,875	8,675
1050	3 Ton	4,575	6,475	11,050
1060	3-1/2 Ton	4,800	6,700	11,500
1070	4 Ton	4,975	7,600	12,575
1080	5 Ton	5,925	9,675	15,600
1090	6 Ton	6,050	10,400	16,450
1100	7 Ton	6,225	10,800	17,025
1110	8 Ton	6,225	11,000	17,225
1120	10 Ton	6,925	11,400	18,325

D3040 126	Fan Coil A/C, Horiz., Duct Mount, 2 Pipe, Elec. Ht.	COST EACH		
		MAT.	INST.	TOTAL
1010	Fan coil A/C system, horiz. hsng, elect. ht, ctrls, 2 pipe, 1/2 Ton	2,075	2,150	4,225
1020	1 Ton	2,675	3,125	5,800
1030	1-1/2 Ton	3,175	4,050	7,225
1040	2 Ton	4,200	4,875	9,075
1050	3 Ton	6,825	6,675	13,500
1060	3-1/2 Ton	7,800	6,700	14,500
1070	4 Ton	8,850	7,625	16,475
1080	5 Ton	10,500	9,675	20,175
1090	6 Ton	10,600	10,400	21,000
1100	7 Ton	10,600	10,300	20,900
1110	8 Ton	11,600	10,500	22,100
1120	10 Ton	12,400	10,800	23,200
2700	Cut & patch to match existing construction, add, minimum	2%	3%	
2800	Maximum	5%	9%	
2900	Dust protection, add, minimum	1%	2%	
3000	Maximum	4%	11%	
3100	Equipment usage curtailment, add, minimum	1%	1%	
3200	Maximum	3%	10%	
3300	Material handling & storage limitation, add, minimum	1%	1%	
3400	Maximum	6%	7%	
3500	Protection of existing work, add, minimum	2%	2%	
3600	Maximum	5%	7%	
3700	Shift work requirements, add, minimum		5%	
3800	Maximum		30%	
3900	Temporary shoring and bracing, add, minimum	2%	5%	
4000	Maximum	5%	12%	

D30 HVAC

D3040 Distribution Systems

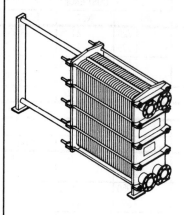

Plate Heat Exchanger

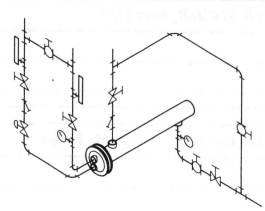

Shell and Tube Heat Exchanger

This page illustrates and describes plate and shell and tube type heat exchangers including pressure gauge and exchanger control system. Lines within Systems Components give the material and installation price on a cost each basis for the components. Prices for alternate heat exchanger systems are on Line Items D3040 610 1010 thru D3040 620 1040. Material quantities and labor costs have been adjusted for the system listed.

Factors To adjust for job conditions other then normal working situations use Lines D3040 620 2700 thru 4000.

System Components	QUANTITY	UNIT	COST EACH		
			MAT.	INST.	TOTAL
Shell and tube type heat exchanger					
with related valves and piping.					
Heat exchanger, 4 pass, 3/4" O.D. copper tubes, by steam at 10 psi, 40 GPM	1.000	Ea.	4,000	276	4,276
Pipe, black steel, Sch 40, threaded, W/couplings & hangers, 10' OC, 2" dia.	40.000	L.F.	426.73	858.50	1,285.23
Elbow, 90°, straight, 2" dia.	16.000	Ea.	293.60	968	1,261.60
Reducer, black steel, concentric, 2" dia.	2.000	Ea.	39.60	104	143.60
Valves, bronze, gate, rising stem, threaded, class 150, 2" size	6.000	Ea.	1,038	330	1,368
Valves, bronze, globe, class 150, rising stem, threaded, 2" size	2.000	Ea.	900	110	1,010
Strainers, Y type, bronze body, screwed, 150 lb., 2" pipe size	2.000	Ea.	145	94	239
Valve, electric motor actuated, brass, 2 way, screwed, 1-1/2" pipe size	1.000	Ea.	435	47	482
Union, black with brass seat, 2" dia.	8.000	Ea.	248	512	760
Tee, black, straight, 2" dia.	9.000	Ea.	193.50	891	1,084.50
Tee, black, reducing run and outlet, 2" dia.	4.000	Ea.	100	396	496
Pipe nipple, black, 2" dia.	21.000	Ea.	4.23	8.50	12.73
Thermometers, stem type, 9" case, 8" stem, 3/4" NPT	2.000	Ea.	240	44	284
Gauges, pressure or vacuum, 3-1/2" diameter dial	2.000	Ea.	79	38.30	117.30
Insulation, fiberglass pipe covering, 1" wall, 2" IPS	40.000	L.F.	57.60	206	263.60
Heat exchanger control system	1.000	Ea.	2,500	1,950	4,450
Coil balancing	1.000	Ea.		113	113
TOTAL			10,700.26	6,946.30	17,646.56

D3040 610	Heat Exchanger, Plate Type	COST EACH		
		MAT.	INST.	TOTAL
1000	For alternate heat exchangers:			
1010	Plate heat exchanger, 400 GPM	46,700	14,100	60,800
1020	800 GPM	75,000	17,800	92,800
1030	1200 GPM	113,000	24,100	137,100
1040	1800 GPM	152,000	30,100	182,100

D3040 620	Heat Exchanger, Shell & Tube	COST EACH		
		MAT.	INST.	TOTAL
1010	Shell & tube heat exchanger, 40 GPM	10,700	6,950	17,650
1020	96 GPM	19,700	12,400	32,100
1030	240 GPM	35,400	16,700	52,100
1040	600 GPM	77,500	23,500	101,000
2700	Cut & patch to match existing construction, add, minimum	2%	3%	
2800	Maximum	5%	9%	
2900	Dust protection, add, minimum	1%	2%	
3000	Maximum	4%	11%	

D30 HVAC

D3040 Distribution Systems

D3040 620	Heat Exchanger, Shell & Tube	COST EACH		
		MAT.	INST.	TOTAL
3100	Equipment usage curtailment, add, minimum	1%	1%	
3200	Maximum	3%	10%	
3300	Material handling & storage limitation, add, minimum	1%	1%	
3400	Maximum	6%	7%	
3500	Protection of existing work, add, minimum	2%	2%	
3600	Maximum	5%	7%	
3700	Shift work requirements, add, minimum		5%	
3800	Maximum		30%	
3900	Temporary shoring and bracing, add, minimum	2%	5%	
4000	Maximum	5%	12%	

D3050 Terminal & Package Units

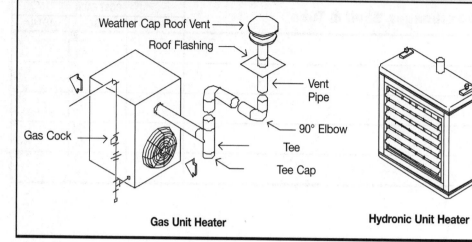

Weather Cap Roof Vent

Roof Flashing

Vent Pipe

90° Elbow

Tee

Tee Cap

Gas Cock

Gas Unit Heater

Hydronic Unit Heater

This page illustrates and describes unit heaters including piping and vents. Lines within Systems Components give the material and installation price on a cost each basis for this system. Prices for alternate unit heater systems are on Line items D3050 120 1010 thru D3050 140 1040. Material quantities and labor costs have been adjusted for the system listed.

Factors: To adjust for job conditions other than normal working situations use Lines D3050 140 2700 thru 4000.

System Components	QUANTITY	UNIT	COST EACH		
			MAT.	INST.	TOTAL
Unit heaters complete with piping, thermostats and chimney as required.					
Space heater, propeller fan, 20 MBH output	1.000	Ea.	1,075	130	1,205
Pipe, black steel, Sch 40, threaded, W/coupling & hangers, 10' OC, 3/4" dia.	20.000	L.F.	60.90	207.90	268.80
Elbow, 90°, black steel, straight, 1/2" dia.	3.000	Ea.	9.69	121.50	131.19
Elbow, 90°, black steel, straight, 3/4" dia.	3.000	Ea.	11.13	129	140.13
Tee, black steel, reducing, 3/4" dia.	1.000	Ea.	7.40	67	74.40
Union, black with brass seat, 3/4" dia.	1.000	Ea.	11.35	46.50	57.85
Pipe nipples, black, 1/2" dia.	2.000	Ea.	4.92	19.20	24.12
Pipe nipples, black, 3/4" dia.	2.000	Ea.	2.90	9.90	12.80
Pipe cap, black, 3/4" dia.	1.000	Ea.	3.07	18.90	21.97
Gas cock, brass, 3/4" size	1.000	Ea.	15.05	27.50	42.55
Thermostat, 1 set back, electric, timed	1.000	Ea.	104	75	179
Wiring, thermostat hook-up	1.000	Ea.	8.45	24	32.45
Vent chimney, prefab metal, U.L. listed, gas, double wall, galv. st, 4" dia.	12.000	L.F.	88.20	190.80	279
Vent chimney, gas, double wall, galv. steel, elbow 90°, 4" dia.	2.000	Ea.	59	64	123
Vent chimney, gas, double wall, galv. steel, Tee, 4" dia.	1.000	Ea.	37.50	41.50	79
Vent chimney, gas, double wall, galv. steel, T cap, 4" dia.	1.000	Ea.	2.89	25.50	28.39
Vent chimney, gas, double wall, galv. steel, roof flashing, 4" dia.	1.000	Ea.	31	32	63
Vent chimney, gas, double wall, galv. steel, top, 4" dia.	1.000	Ea.	17.80	24.50	42.30
TOTAL			1,550.25	1,254.70	2,804.95

D3050 120		Unit Heaters, Gas	COST EACH		
			MAT.	INST.	TOTAL
1000	For alternate unit heaters:				
1010	Space heater, suspended, gas fired, propeller fan, 20 MBH	1,550	1,250	2,800
1020	60 MBH	1,825	1,300	3,125
1030	100 MBH	2,050	1,375	3,425
1040	160 MBH	2,350	1,525	3,875
1050	200 MBH	2,775	1,675	4,450
1060	280 MBH	3,475	1,800	5,275
1070	320 MBH	4,550	2,025	6,575

D30 HVAC

D3050 Terminal & Package Units

D3050 130	Unit Heaters, Hydronic	COST EACH		
		MAT.	INST.	TOTAL
1010	Space heater, suspended, horiz. mount, HW, prop. fan, 20 MBH	1,700	1,350	3,050
1020	60 MBH	2,150	1,500	3,650
1030	100 MBH	2,575	1,650	4,225
1040	150 MBH	2,975	1,825	4,800
1050	200 MBH	3,425	2,050	5,475
1060	300 MBH	4,075	2,400	6,475

D3050 140	Cabinet Unit Heaters, Hydronic	COST EACH		
		MAT.	INST.	TOTAL
1010	Unit heater, cabinet type, horizontal blower, hot water, 20 MBH	1,975	1,300	3,275
1020	60 MBH	2,775	1,450	4,225
1030	100 MBH	3,050	1,600	4,650
1040	120 MBH	3,125	1,650	4,775
2700	Cut & patch to match existing construction, add, minimum	2%	3%	
2800	Maximum	5%	9%	
2900	Dust protection, add, minimum	1%	2%	
3000	Maximum	4%	11%	
3100	Equipment usage curtailment, add, minimum	1%	1%	
3200	Maximum	3%	10%	
3300	Material handling & storage limitation, add, minimum	1%	1%	
3400	Maximum	6%	7%	
3500	Protection of existing work, add, minimum	2%	2%	
3600	Maximum	5%	7%	
3700	Shift work requirements, add, minimum		5%	
3800	Maximum		30%	
3900	Temporary shoring and bracing, add, minimum	2%	5%	
4000	Maximum	5%	12%	

D3050 Terminal & Package Units

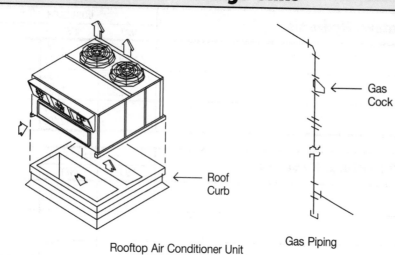

Gas Cock

Roof Curb

Rooftop Air Conditioner Unit

Gas Piping

This page illustrates and describes packaged air conditioners including rooftop DX unit, pipes and fittings. Lines within Systems Components give the material and installation price on a cost each basis for the components. Prices for alternate A/C systems are on Line Items D3050 175 1010 thru 1060. Material quantities and labor costs have been adjusted for the system listed.

Factors: To adjust for job conditions other than normal working situations use Lines D3050 175 2700 thru 4000.

System Components	QUANTITY	UNIT	COST EACH		
			MAT.	INST.	TOTAL
Packaged rooftop DX unit, gas, with related piping and valves.					
Roof top A/C, curb, economizer, sgl zone, elec cool, gas ht, 5 ton, 112 MBH	1.000	Ea.	4,575	1,975	6,550
Pipe, black steel, Sch 40, threaded, W/cplgs & hangers, 10' OC, 1" dia.	20.000	L.F.	89.46	239.40	328.86
Elbow, 90°, black, straight, 3/4" dia.	3.000	Ea.	11.13	129	140.13
Elbow, 90°, black, straight, 1" dia.	3.000	Ea.	14.10	139.50	153.60
Tee, black, reducing, 1" dia.	1.000	Ea.	11.80	75.50	87.30
Union, black with brass seat, 1" dia.	1.000	Ea.	14.85	50.50	65.35
Pipe nipple, black, 3/4" dia.	2.000	Ea.	5.80	19.80	25.60
Pipe nipple, black, 1" dia.	2.000	Ea.	4.26	11.40	15.66
Cap, black, 1" dia.	1.000	Ea.	3.71	20	23.71
Gas cock, brass, 1" size	1.000	Ea.	19.70	32	51.70
Control system, pneumatic, Rooftop A/C unit	1.000	Ea.	2,575	2,150	4,725
Rooftop unit heat/cool balancing	1.000	Ea.		420	420
TOTAL			7,324.81	5,262.10	12,586.91

D3050 175	Rooftop Air Conditioner, Const. Volume	COST EACH		
		MAT.	INST.	TOTAL
1000	For alternate A/C systems:			
1010	A/C, Rooftop, DX cool, gas heat, curb, economizer, fltrs, 5 Ton	7,325	5,250	12,575
1020	7-1/2 Ton	9,650	5,475	15,125
1030	12-1/2 Ton	13,100	6,025	19,125
1040	18 Ton	19,700	6,575	26,275
1050	25 Ton	28,800	7,500	36,300
1060	40 Ton	39,100	10,000	49,100
1010	A/C, Rooftop, DX cool, gas heat, curb, ecmizr, fltrs, VAV, 12-1/2 Ton	15,400	6,425	21,825
1020	18 Ton	22,200	7,075	29,275
1030	25 Ton	32,600	8,125	40,725
1040	40 Ton	47,000	11,000	58,000
1050	60 Ton	67,500	15,000	82,500
1060	80 Ton	99,500	18,700	118,200
2700	Cut & patch to match existing construction, add, minimum	2%	3%	
2800	Maximum	5%	9%	
2900	Dust protection, add, minimum	1%	2%	
3000	Maximum	4%	11%	
3100	Equipment usage curtailment, add, minimum	1%	1%	
3200	Maximum	3%	10%	
3300	Material handling & storage limitation, add, minimum	1%	1%	
3400	Maximum	6%	7%	
3500	Protection of existing work, add, minimum	2%	2%	

D3050 Terminal & Package Units

D3050 175	Rooftop Air Conditioner, Const. Volume	COST EACH		
		MAT.	INST.	TOTAL
3600	Maximum	5%	7%	
3700	Shift work requirements, add, minimum		5%	
3800	Maximum		30%	
3900	Temporary shoring and bracing, add, minimum	2%	5%	
4000	Maximum	5%	12%	

D3050 Terminal & Package Units

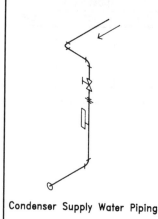

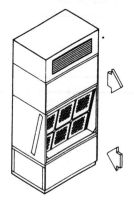

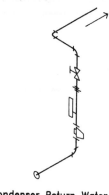

This page illustrates and describes packaged electrical air conditioning units including multizone control and related piping. Lines within Systems Components give the material and installation price on a cost each basis for the components. Prices for alternate packaged electrical A/C unit systems are on Line Items D3050 210 1010 thru D3050 225 1060. Material quantities and labor costs have been adjusted for the system listed.

Factors: To adjust for job conditions other than normal working situations use Lines D3050 225 2700 thru 4000.

Condenser Supply Water Piping

Condenser Return Water Piping

Self—Contained Air Conditioner Unit, Water Cooled

System Components

	QUANTITY	UNIT	COST EACH		
			MAT.	INST.	TOTAL
Packaged water cooled electric air conditioning unit with related piping and valves.					
Self-contained, water cooled, elect. heat, not inc tower, 5 ton, const vol	1.000	Ea.	7,975	1,425	9,400
Pipe, black steel, Sch 40, threaded, W/cplg & hangers, 10' OC, 1-1/4" dia.	20.000	L.F.	116.10	263.38	379.48
Elbow, 90°, black, straight, 1-1/4" dia.	6.000	Ea.	46.50	297	343.50
Tee, black, straight, 1-1/4" dia.	2.000	Ea.	23.80	156	179.80
Tee, black, reducing, 1-1/4" dia.	2.000	Ea.	40	156	196
Thermometers, stem type, 9" case, 8" stem, 3/4" NPT	2.000	Ea.	240	44	284
Union, black with brass seat, 1-1/4" dia.	2.000	Ea.	43	104	147
Pipe nipple, black, 1-1/4" dia.	3.000	Ea.	8.10	18.38	26.48
Valves, bronze, gate, N.R.S., threaded, class 150, 1-1/4" size	2.000	Ea.	210	81	291
Circuit setter, bal valve, bronze body, threaded, 1-1/4" pipe size	1.000	Ea.	131	41	172
Insulation, fiberglass pipe covering, 1" wall, 1-1/4" IPS	20.000	L.F.	24.60	97.80	122.40
Control system, pneumatic, A/C Unit with heat	1.000	Ea.	2,575	2,150	4,725
Re-heat coil balancing	1.000	Ea.		113	113
Rooftop unit heat/cool balancing	1.000	Ea.		420	420
TOTAL			11,433.10	5,366.56	16,799.66

D3050 210	AC Unit, Package, Elec. Ht., Water Cooled		COST EACH		
			MAT.	INST.	TOTAL
1000	For alternate A/C systems:				
1010	A/C, Self contained, single pkg., water cooled, elect. heat, 5 Ton		11,400	5,375	16,775
1020	10 Ton		18,000	6,600	24,600
1030	20 Ton		24,500	8,400	32,900
1040	30 Ton		38,400	9,175	47,575
1050	40 Ton		57,000	10,200	67,200
1060	50 Ton		73,000	12,100	85,100

D3050 215	AC Unit, Package, Elec. Ht., Water Cooled, VAV		COST EACH		
			MAT.	INST.	TOTAL
1010	A/C, Self contained, single pkg., water cooled, elect. ht, VAV, 10 Ton		20,200	6,600	26,800
1020	20 Ton		28,300	8,400	36,700
1030	30 Ton		43,800	9,175	52,975
1040	40 Ton		65,000	10,200	75,200
1050	50 Ton		81,500	12,100	93,600
1060	60 Ton		88,000	14,700	102,700

D30 HVAC

D3050 Terminal & Package Units

D3050 220 — AC Unit, Package, Hot Water Coil, Water Cooled

		COST EACH		
		MAT.	INST.	TOTAL
1010	A/C, Self contn'd, single pkg., water cool, H/W ht, const. vol, 5 Ton	9,600	7,450	17,050
1020	10 Ton	14,600	8,875	23,475
1030	20 Ton	37,700	11,100	48,800
1040	30 Ton	49,500	14,000	63,500
1050	40 Ton	61,000	17,100	78,100
1060	50 Ton	75,000	21,000	96,000

D3050 225 — AC Unit, Package, HW Coil, Water Cooled, VAV

		COST EACH		
		MAT.	INST.	TOTAL
1010	A/C, Self contn'd, single pkg., water cool, H/W ht, VAV, 10 Ton	16,800	8,350	25,150
1020	20 Ton	24,200	10,800	35,000
1030	30 Ton	36,600	13,500	50,100
1040	40 Ton	59,000	14,700	73,700
1050	50 Ton	72,000	17,300	89,300
1060	60 Ton	80,000	18,800	98,800
2700	Cut & patch to match existing construction, add, minimum	2%	3%	
2800	Maximum	5%	9%	
2900	Dust protection, add, minimum	1%	2%	
3000	Maximum	4%	11%	
3100	Equipment usage curtailment, add, minimum	1%	1%	
3200	Maximum	3%	10%	
3300	Material handling & storage limitation, add, minimum	1%	1%	
3400	Maximum	6%	7%	
3500	Protection of existing work, add, minimum	2%	2%	
3600	Maximum	5%	7%	
3700	Shift work requirements, add, minimum		5%	
3800	Maximum		30%	
3900	Temporary shoring and bracing, add, minimum	2%	5%	
4000	Maximum	5%	12%	

D3050 Terminal & Package Units

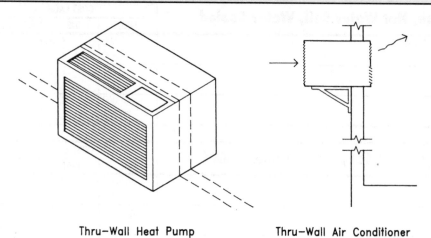

Thru—Wall Heat Pump   Thru—Wall Air Conditioner

This page illustrates and describes thru-wall air conditioning and heat pump units. Lines within Systems Components give the material and installation price on a cost each basis for the components. Prices for alternate thru-wall unit systems are on Line Items D3050 255 1010 thru 1050. Material quantities and labor costs have been adjusted for the system listed.

Factors: To adjust for job conditions other than normal working conditions use Lines D3050 255 2700 thru 4000.

System Components	QUANTITY	UNIT	COST EACH		
			MAT.	INST.	TOTAL
Electric thru-wall air conditioning unit.					
A/C unit, thru-wall, electric heat., cabinet, louver, 1/2 ton	1.000	Ea.	690	184	874
TOTAL			690	184	874

D3050 255	Thru-Wall A/C Unit	COST EACH		
		MAT.	INST.	TOTAL
1000	For alternate A/C units:			
1010	A/C Unit, thru-the-wall, supplemental elect. heat, cabinet, louver, 1/2 Ton	690	184	874
1020	3/4 Ton	790	220	1,010
1030	1 Ton	860	276	1,136
1040	1-1/2 Ton	1,750	460	2,210
1050	2 Ton	1,900	580	2,480
1010	Heat pump, thru-the-wall, cabinet, louver, 1/2 Ton	1,850	138	1,988
1020	3/4 Ton	2,025	184	2,209
1030	1 Ton	2,225	276	2,501
1040	Supplemental. elect. heat, 1-1/2 Ton	2,650	710	3,360
1050	2 Ton	2,825	735	3,560
2700	Cut & patch to match existing construction, add, minimum	2%	3%	
2800	Maximum	5%	9%	
2900	Dust protection, add, minimum	1%	2%	
3000	Maximum	4%	11%	
3100	Equipment usage curtailment, add, minimum	1%	1%	
3200	Maximum	3%	10%	
3300	Material handling & storage limitation, add, minimum	1%	1%	
3400	Maximum	6%	7%	
3500	Protection of existing work, add, minimum	2%	2%	
3600	Maximum	5%	7%	
3700	Shift work requirements, add, minimum		5%	
3800	Maximum		30%	
3900	Temporary shoring and bracing, add, minimum	2%	5%	
4000	Maximum	5%	12%	

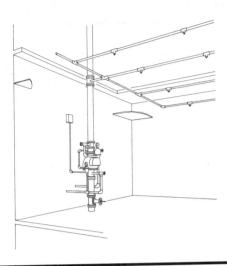

This page illustrates and describes a dry type fire sprinkler system. Lines within System Components give the unit cost of a system for a 2,000 square foot building. Lines D4010 320 1700 thru 2400 give the square foot costs for alternate systems. Both material quantities and labor costs have been adjusted for the system listed.

Factors: To adjust for conditions other than normal working conditions, use Lines D4010 320 2900 thru 4000.

System Components	QUANTITY	UNIT	COST EACH MAT.	COST EACH INST.	COST EACH TOTAL
Pre-action fire sprinkler system, ordinary hazard, open area to 2000 S.F.					
On one floor.					
Air compressor	1.000	Ea.	765	460	1,225
4″ OS & Y valve	1.000	Ea.	560	365	925
Pipe riser 4″ diameter	10.000	L.F.	199	311.60	510.60
Sprinkler head supply piping, 1″	163.000	L.F.	319.50	855	1,174.50
Sprinkler head supply piping, 1-1/4″	163.000	L.F.	297	673.75	970.75
Sprinkler head supply piping, 2-1/2″	163.000	L.F.	503.50	836	1,339.50
Pipe fittings, 1″	8.000	Ea.	58.80	604	662.80
Pipe fittings, 1-1/4″	8.000	Ea.	95.20	624	719.20
Pipe fittings, 2-1/2″	4.000	Ea.	218	484	702
Pipe fittings, 4″	2.000	Ea.	388	544	932
Detectors	2.000	Ea.	1,150	75	1,225
Sprinkler heads	16.000	Ea.	129.60	600	729.60
Fire department connection	1.000	Ea.	470	215	685
TOTAL			5,153.60	6,647.35	11,800.95

D4010 320	Fire Sprinkler Systems, Dry	COST PER S.F. MAT.	COST PER S.F. INST.	COST PER S.F. TOTAL
1700	Ordinary hazard, one floor, area to 2000 S.F./floor	2.58	3.33	5.91
1800	For each additional floor, add per floor	1.10	2.77	3.87
1900	Area to 3200 S.F./ floor	2.05	3.29	5.34
2000	For each additional floor, add per floor	1.12	2.95	4.07
2100	Area to 5000 S.F./ floor	2.37	3.33	5.70
2200	For each additional floor, add per floor	1.56	3.10	4.66
2300	Area to 8000 S.F./ floor	1.81	2.84	4.65
2400	For each additional floor, add per floor	1.22	2.68	3.90
2500				
2600				
2700				
2800				
2900	Cut & patch to match existing construction, add, minimum	2%	3%	
3000	Maximum	5%	9%	
3100	Dust protection, add, minimum	1%	2%	
3200	Maximum	4%	11%	
3300	Equipment usage curtailment, add, minimum	1%	1%	
3400	Maximum	3%	10%	

D4010 Sprinklers

D4010 320	Fire Sprinkler Systems, Dry	COST PER S.F.		
		MAT.	INST.	TOTAL
3500	Material handling & storage limitation, add, minimum	1%	1%	
3600	Maximum	6%	7%	
3700	Protection of existing work, add, minimum	2%	2%	
3800	Maximum	5%	7%	
3900	Shift work requirements, add, minimum		5%	
4000	Maximum		30%	

D4010 Sprinklers

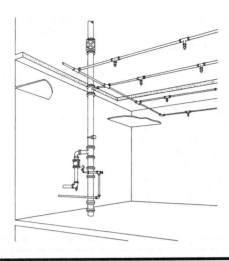

This page illustrates and describes a wet type fire sprinkler system. Lines within System Components give the unit cost of a system for a 2,000 square foot building. Lines D4010 420 1900 thru 2600 give the square foot costs for alternate systems. Both material quantities and labor costs have been adjusted for the system listed.

Factors: To adjust for conditions other than normal working conditions, use Lines D4010 420 2900 thru 4000.

System Components	QUANTITY	UNIT	COST EACH		
			MAT.	INST.	TOTAL
Wet pipe fire sprinkler system, ordinary hazard, open area to 2000 S.F.					
On one floor.					
4" OS & Y valve	1.000	Ea.	560	365	925
Wet pipe alarm valve	1.000	Ea.	1,400	540	1,940
Water motor alarm	1.000	Ea.	335	149	484
3" check valve	1.000	Ea.	280	360	640
Pipe riser, 4" diameter	10.000	L.F.	199	311.60	510.60
Water gauges and trim	1.000	Set	2,225	1,075	3,300
Electric fire horn	1.000	Ea.	44.50	86.50	131
Sprinkler head supply piping, 1" diameter	168.000	L.F.	319.50	855	1,174.50
Sprinkler head supply piping, 1-1/2" diameter	168.000	L.F.	349.25	748	1,097.25
Sprinkler head supply piping, 2-1/2" diameter	168.000	L.F.	503.50	836	1,339.50
Pipe fittings, 1"	8.000	Ea.	58.80	604	662.80
Pipe fittings, 1-1/2"	8.000	Ea.	118.80	672	790.80
Pipe fittings, 2-1/2"	4.000	Ea.	218	484	702
Pipe fittings, 4"	2.000	Ea.	388	544	932
Sprinkler heads	16.000	Ea.	129.60	600	729.60
Fire department connection	1.000	Ea.	470	215	685
TOTAL			7,598.95	8,445.10	16,044.05

D4010 420	Fire Sprinkler Systems, Wet	COST PER S.F.		
		MAT.	INST.	TOTAL
1900	Ordinary hazard, one floor, area to 2000 S.F./floor	3.79	4.22	8.01
2000	For each additional floor, add per floor	1.13	2.83	3.96
2100	Area to 3200 S.F./floor	2.79	3.83	6.62
2200	For each additional floor, add per floor	1.12	2.95	4.07
2300	Area to 5000 S.F./ floor	2.63	3.66	6.29
2400	For each additional floor, add per floor	1.56	3.10	4.66
2500	Area to 8000 S.F./ floor	1.90	3.04	4.94
2600	For each additional floor, add per floor	1.31	2.84	4.15
2700				
2900	Cut & patch to match existing construction, add minimum	2%	3%	
3000	Maximum	5%	9%	
3100	Dust protection, add, minimum	1%	2%	
3200	Maximum	4%	11%	
3300	Equipment usage curtailment, add, minimum	1%	1%	
3400	Maximum	3%	10%	
3500	Material handling & storage limitation, add, minimum	1%	1%	
3600	Maximum	6%	7%	

D5010 Electrical Service/Distribution

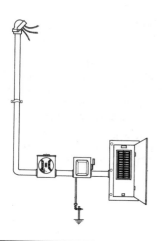

This page illustrates and describes commercial service systems including a meter socket, service head and cable, entrance switch, steel conduit, copper wire, panel board, ground rod, wire, and conduit. Lines within System Components give the unit price and total price on a cost each basis for this system. Prices for alternate commercial service systems are on Line Items D5010 210 1500 thru 1700. Both material quantities and labor costs have been adjusted for the system listed.

Factors: To adjust for job conditions other than normal working situations use Lines D5010 210 2900 thru 4000.

Example: You are to install the system with maximum equipment usage curtailment. Go to Line D5010 210 3400 and apply these percentages to the appropriate MAT. and INST. costs.

System Components	QUANTITY	UNIT	COST EACH		
			MAT.	INST.	TOTAL
Commercial electric service including service breakers, metering 120/208 Volt, 3 phase, 4 wire, feeder, and panel board.					
100 Amp Service					
Meter socket	1.000	Ea.	47.50	181	228.50
Service head	.200	C.L.F.	66	116	182
Service entrance switch	1.000	Ea.	360	305	665
Rigid steel conduit	20.000	L.F.	98.40	178	276.40
600 volt copper wire #3	1.000	C.L.F.	148	116	264
Panel board, 24 circuits, 20 Amp breakers	1.000	Ea.	895	965	1,860
Ground rod	1.000	Ea.	16.25	105	121.25
TOTAL			1,631.15	1,966	3,597.15

D5010 210	Commercial Service - 3 Phase	COST EACH		
		MAT.	INST.	TOTAL
1400	For alternate size services:			
1500	120/208 Volt, 3 phase, 4 wire service, 60 Amp	1,100	1,525	2,625
1600	200 Amp	3,400	3,450	6,850
1700	400 Amp	7,800	4,325	12,125
1800				
1900				
2000				
2100				
2200				
2300				
2400				
2500				
2900	Cut & patch to match existing construction, add, minimum	2%	3%	
3000	Maximum	5%	9%	
3100	Dust protection, add, minimum	1%	2%	
3200	Maximum	4%	11%	
3300	Equipment usage curtailment, add, minimum	1%	1%	
3400	Maximum	3%	10%	
3500	Material handling & storage limitation, add, minimum	1%	1%	
3600	Maximum	6%	7%	
3700	Protection of existing work, add, minimum	2%	2%	
3800	Maximum	5%	7%	
3900	Shift work requirements, add, minimum		5%	
4000	Maximum		30%	

D5010 Electrical Service/Distribution

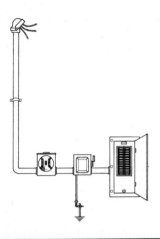

This page illustrates and describes a residential, single phase system including a weather cap, service entrance cable, meter socket, entrance switch, ground rod, ground cable, EMT, and panelboard. Lines with System Components give the unit price and total price on a cost each basis for this system. Prices for an alternate residential, single phase system are also given. Both material quantities and labor costs have been adjusted for the system listed.

Factors: To adjust for job conditions other than normal working situations use Lines D5010 220 2900 thru 4000.

Example: You are to install the system with a minimum equipment usage curtailment. Go to Line D5010 220 3300 and apply these percentages to the appropriate MAT. and INST. costs.

System Components	QUANTITY	UNIT	MAT.	INST.	TOTAL
			COST EACH		
100 Amp Service, single phase					
Weathercap	1.000	Ea.	13.40	48	61.40
Service entrance cable	.200	C.L.F.	66	116	182
Meter socket	1.000	Ea.	47.50	181	228.50
Entrance disconnect switch	1.000	Ea.	360	305	665
Ground rod, with clamp	1.000	Ea.	19.10	121	140.10
Ground cable	.100	C.L.F.	19.30	36	55.30
Panelboard, 12 circuit	1.000	Ea.	270	480	750
TOTAL			795.30	1,287	2,082.30
200 Amp Service , single phase					
Weathercap	1.000	Ea.	37	72.50	109.50
Service entrance cable	.200	C.L.F.	168	165	333
Meter socket	1.000	Ea.	80	305	385
Entrance disconnect switch	1.000	Ea.	780	445	1,225
Ground rod, with clamp	1.000	Ea.	36.50	131	167.50
Ground cable	.100	C.L.F.	36	19.90	55.90
3/4" EMT	10.000	L.F.	11.60	44.50	56.10
Panelboard, 24 circuit	1.000	Ea.	615	750	1,365
TOTAL			1,764.10	1,932.90	3,697

D5010 220	Residential Service - Single Phase	MAT.	INST.	TOTAL
		COST EACH		
2800				
2810	100 Amp Service, single phase	795	1,275	2,070
2820	200 Amp Service, single phase	1,775	1,925	3,700
2900	Cut & patch to match existing construction, add, minimum	2%	3%	
3000	Maximum	5%	9%	
3100	Dust protection, add, minimum	1%	2%	
3200	Maximum	4%	11%	
3300	Equipment usage curtailment, add, minimum	1%	1%	
3400	Maximum	3%	10%	
3500	Material handling & storage limitation, add, minimum	1%	1%	
3600	Maximum	6%	7%	
3700	Protection of existing work, add, minimum	2%	2%	
3800	Maximum	5%	7%	
3900	Shift work requirements, add, minimum		5%	
4000	Maximum		30%	

D5020 Lighting and Branch Wiring

D5020 180	Selective Price Sheet	COST EACH		
		MAT.	INST.	TOTAL
0100	Using non-metallic sheathed, cable, air conditioning receptacle	24.50	58	82.50
0200	Disposal wiring	19.75	64.50	84.25
0300	Dryer circuit	52	105	157
0400	Duplex receptacle	24.50	44.50	69
0500	Fire alarm or smoke detector	80.50	58	138.50
0600	Furnace circuit & switch	27	96.50	123.50
0700	Ground fault receptacle	56	72.50	128.50
0800	Heater circuit	30	72.50	102.50
0900	Lighting wiring	30.50	36	66.50
1000	Range circuit	100	145	245
1100	Switches single pole	21.50	36	57.50
1200	3-way	28	48	76
1300	Water heater circuit	35	116	151
1400	Weatherproof receptacle	156	96.50	252.50
1500	Using BX cable, air conditioning receptacle	39	69.50	108.50
1600	Disposal wiring	33.50	77	110.50
1700	Dryer circuit	61	126	187
1800	Duplex receptacle	39	53.50	92.50
1900	Fire alarm or smoke detector	80.50	58	138.50
2000	Furnace circuit & switch	45	116	161
2100	Ground fault receptacle	70	87.50	157.50
2200	Heater circuit	37.50	87.50	125
2300	Lighting wiring	38.50	43.50	82
2400	Range circuit	164	175	339
2500	Switches, single pole	36.50	43.50	80
2600	3-way	40.50	58	98.50
2700	Water heater circuit	59	138	197
2800	Weatherproof receptacle	164	116	280
2900	Using EMT conduit, air conditioning receptacle	52	86.50	138.50
3000	Disposal wiring	48.50	96.50	145
3100	Dryer circuit	67.50	156	223.50
3200	Duplex receptacle	52	66.50	118.50
3300	Fire alarm or smoke detector	99.50	86.50	186
3400	Furnace circuit & switch	52.50	145	197.50
3500	Ground fault receptacle	91	107	198
3600	Heater circuit	49.50	107	156.50
3700	Lighting wiring	47.50	54	101.50
3800	Range circuit	109	214	323
3900	Switches, single pole	49	54	103
4000	3-way	44	72.50	116.50
4100	Water heater circuit	56	170	226
4200	Weatherproof receptacle	176	145	321
4300	Using aluminum conduit, air conditioning receptacle	59.50	116	175.50
4400	Disposal wiring	56.50	129	185.50
4500	Dryer circuit	89	207	296
4600	Duplex receptacle	56.50	89	145.50
4700	Fire alarm or smoke detector	116	116	232
4800	Furnace circuit & switch	66	193	259
4900	Ground fault receptacle	100	145	245
5000	Heater circuit	60	145	205
5100	Lighting wiring	67.50	72.50	140
5200	Range circuit	137	289	426
5300	Switches, single pole	66.50	72.50	139
5400	3-way	61.50	96.50	158
5500	Water heater circuit	75	231	306
5600	Weatherproof receptacle	204	193	397
5700	Using galvanized steel conduit	61.50	123	184.50
5800	Disposal wiring	50.50	138	188.50

D5020 Lighting and Branch Wiring

D5020 180	Selective Price Sheet	COST EACH		
		MAT.	INST.	TOTAL
5900	Dryer circuit	93.50	222	315.50
6000	Duplex receptacle	61.50	95	156.50
6100	Fire alarm or smoke detector	111	123	234
6200	Furnace circuit & switch	61	207	268
6300	Ground fault receptacle	99.50	152	251.50
6400	Heater circuit	59.50	152	211.50
6500	Lighting wiring	67.50	77	144.50
6600	Range circuit	136	305	441
6700	Switches, single pole	66	77	143
6800	3-way	62.50	99.50	162
6900	Water heater circuit	70.50	241	311.50
7000	Weatherproof receptacle	194	207	401
7100				
7200				
7300				
7400				
7500				
7600				
7700				
7800				
7900				
8000				
8100				
8200				
8300				
8400				

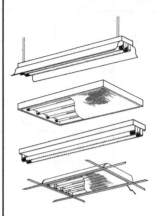

This page illustrates and describes fluorescent lighting systems including a fixture, lamp, outlet box and wiring. Lines within System Components give the unit price and total price on a cost each basis for this system. Prices for alternate fluorescent lighting systems are on Line Items D5020 248 1300 thru 1500. Both material quantities and labor costs have been adjusted for the system listed.

Factors: To adjust for job conditions other than normal working situations use Lines D5020 248 2900 thru 4000.

Example: You are to install the system during evening hours. Go to Line D5020 248 3900 and apply this percentage to the appropriate INST. cost.

System Components

Fluorescent lighting, including fixture, lamp, outlet box and wiring.	QUANTITY	UNIT	COST EACH MAT.	COST EACH INST.	COST EACH TOTAL
Recessed lighting fixture, on suspended system	1.000	Ea.	70.50	123	193.50
Outlet box	1.000	Ea.	2.62	32	34.62
#12 wire	.660	C.L.F.	11.55	34.65	46.20
Conduit, EMT, 1/2" conduit	20.000	L.F.	14.40	68	82.40
TOTAL			99.07	257.65	356.72

D5020 248	Lighting, Fluorescent	COST EACH MAT.	COST EACH INST.	COST EACH TOTAL
1200	For alternate lighting fixtures:			
1300	Surface mounted, 2' x 4', acrylic prismatic diffuser	133	271	404
1400	Strip fixture, 8' long, two 8' lamps	91	255	346
1500	Pendant mounted, industrial, 8' long, with reflectors	136	293	429
1600				
1700				
1800				
1900				
2000				
2100				
2200				
2300				
2900	Cut & patch to match existing construction, add, minimum	2%	3%	
3000	Maximum	5%	9%	
3100	Dust protection, add, minimum	1%	2%	
3200	Maximum	4%	11%	
3300	Equipment usage curtailment, add, minimum	1%	1%	
3400	Maximum	3%	10%	
3500	Material handling & storage limitation, add, minimum	1%	1%	
3600	Maximum	6%	7%	
3700	Protection of existing work, add, minimum	2%	2%	
3800	Maximum	5%	7%	
3900	Shift work requirements, add, minimum		5%	
4000	Maximum		30%	

D5020 Lighting and Branch Wiring

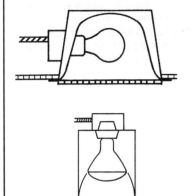

This page illustrates and describes incandescent lighting systems including a fixture, lamp, outlet box, conduit and wiring. Lines within System Components give the unit price and total price on a cost each basis for this system. Prices for an alternate incandescent lighting system are also given. Both material quantities and labor costs have been adjusted for the system listed.

Factors: To adjust for conditions other than normal working situations use Lines D5020 250 2900 thru 4000.

Example: You are to install the system and cut and match existing construction. Go to Line D5020 250 3000 and apply this percentage to the appropriate INST. costs.

System Components	QUANTITY	UNIT	COST EACH MAT.	COST EACH INST.	COST EACH TOTAL
Incandescent light fixture, including lamp, outlet box, conduit and wiring.					
Recessed wide reflector with flat glass lens	1.000	Ea.	105	86.50	191.50
Outlet box	1.000	Ea.	2.62	32	34.62
Armored cable, 3 wire	.200	C.L.F.	26.60	52.60	79.20
TOTAL			134.22	171.10	305.32
Recessed flood light fixture, including lamp, outlet box, conduit & wire.					
Recessed, R-40 flood lamp with refractor	1.000	Ea.	112	72.50	184.50
150 watt R-40 flood lamp	.010	Ea.	6.65	4.45	11.10
Outlet box	1.000	Ea.	2.62	32	34.62
Outlet box cover	1.000	Ea.	1.54	9.05	10.59
Romex, 12-2 with ground	.200	C.L.F.	11	52.60	63.60
Conduit, 1/2" EMT	20.000	L.F.	14.40	68	82.40
TOTAL			148.21	238.60	386.81

D5020 250	Lighting, Incandescent	COST EACH MAT.	COST EACH INST.	COST EACH TOTAL
2810	Incandescent light fixture, including lamp, outlet box, conduit and wiring.	134	171	305
2820	Recessed flood light fixture, including lamp, outlet box, conduit & wire	148	239	387
2900	Cut & patch to match existing construction, add, minimum	2%	3%	
3000	Maximum	5%	9%	
3100	Dust protection, add, minimum	1%	2%	
3200	Maximum	4%	11%	
3300	Equipment usage curtailment, add, minimum	1%	1%	
3400	Maximum	3%	10%	
3500	Material handling & storage limitation, add, minimum	1%	1%	
3600	Maximum	6%	7%	
3700	Protection of existing work, add, minimum	2%	2%	
3800	Maximum	5%	7%	
3900	Shift work requirements, add, minimum		5%	
4000	Maximum		30%	

D5020 Lighting and Branch Wiring

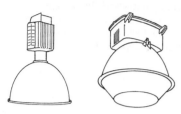

This page illustrates and describes high intensity lighting systems including a lamp, EMT conduit, EMT "T" fitting with cover, and wire. Lines within System Components give the unit price and total price on a cost each basis for this system. Prices for alternate high intensity lighting systems are on Line Items D5020 252 1200 thru 1800. Both material quantities and labor costs have been adjusted for the system listed.

Factors: To adjust for job conditions other than normal working situations use Lines D5020 252 2900 thru 4000.

Example: You are to install the system and protect existing construction. Go to Line D5020 252 3700 and apply these percentages to the appropriate MAT. and INST. costs.

System Components

System Components	QUANTITY	UNIT	COST EACH MAT.	COST EACH INST.	COST EACH TOTAL
High intensity lighting system consisting of 400 watt metal halide fixture					
And lamp with 1/2″ EMT conduit and fittings using #12 wire, high bay.					
Single unit, 400 watt	1.000	Ea.	445	251	696
Electric metallic tubing (EMT), 1/2″ diameter	30.000	Ea.	21.60	102	123.60
1/2″ EMT "T" fitting with cover	1.000	Ea.	15.90	36	51.90
Wire, 600 volt, type THWN-THHN, copper, solid, #12	.600	Ea.	10.50	31.50	42
TOTAL			493	420.50	913.50

D5020 252	Lighting, High Intensity	COST EACH MAT.	COST EACH INST.	COST EACH TOTAL
1000	For alternate high intensity systems:			
1200	High bay: 400 watt, high pressure sodium and lamp	460	420	880
1400	1000 watt, metal halide	690	460	1,150
1500	High pressure sodium	640	460	1,100
1700	Low bay: 250 watt, metal halide	440	350	790
1800	150 watt, high pressure sodium	405	350	755
1900				
2000				
2100				
2200				
2300				
2400				
2500				
2600				
2700				
2900	Cut & patch to match existing construction, add, minimum	2%	3%	
3000	Maximum	5%	9%	
3100	Dust protection, add, minimum	1%	2%	
3200	Maximum	4%	11%	
3300	Equipment usage curtailment, add, minimum	1%	1%	
3400	Maximum	3%	10%	
3500	Material handling & storage limitation, add, minimum	1%	1%	
3600	Maximum	6%	7%	
3700	Protection of existing work, add, minimum	2%	2%	
3800	Maximum	5%	7%	
3900	Shift work requirements, add, minimum		5%	
4000	Maximum		30%	

D5090 Other Electrical Systems

This page illustrates and describes baseboard heat systems including a thermostat, outlet box, breaker, and feed. Lines within System Components give the unit price and total price on a cost each basis for this system. Prices for alternate baseboard heat systems are on Line Items D5090 430 1500 thru 1800. Both material quantities and labor costs have been adjusted for the system listed.

Factors: To adjust for job conditions other than normal working situations use Lines D5090 430 2900 thru 4000.

Example: You are to install the system during evenings and weekends only. Go to Line D5090 430 4000 and apply this percentage to the appropriate INST. cost.

System Components	QUANTITY	UNIT	COST EACH		
			MAT.	INST.	TOTAL
Baseboard heat including thermostat, outlet box, breaker and feed.	1.000	Ea.	59.50	86.50	146
Electric baseboard heater, 4' long	1.000	Ea.	22.50	36	58.50
Thermostat, integral	.400	C.L.F.	30.80	115.60	146.40
Romex, 12-3 with ground	1.000	Ea.	38	72.50	110.50
Panel board breaker, 20 Amp	1.000	Ea.	2.62	32	34.62
Outlet box					
TOTAL			153.42	342.60	496.02

D5090 530	Heat, Baseboard	COST EACH		
		MAT.	INST.	TOTAL
1400	For alternate baseboard heating systems:			
1500	Electric baseboard, 2' long	138	330	468
1600	6' long	175	370	545
1700	8' long	195	400	595
1800	10' long	305	430	735
1900				
2000				
2100				
2200				
2300				
2400				
2500				
2600				
2700				
2900	Cut & patch to match existing construction, add, minimum	2%	3%	
3000	Maximum	5%	9%	
3100	Dust protection, add, minimum	1%	2%	
3200	Maximum	4%	11%	
3300	Equipment usage curtailment, add, minimum	1%	1%	
3400	Maximum	3%	10%	
3500	Material handling & storage limitation, add, minimum	1%	1%	
3600	Maximum	6%	7%	
3700	Protection of existing work, add, minimum	2%	2%	
3800	Maximum	5%	7%	
3900	Shift work requirements, add, minimum		5%	
4000	Maximum		30%	

E1090 Other Equipment

This page illustrates and describes kitchen systems including top and bottom cabinets, custom laminated plastic top, single bowl sink, and appliances. Lines within System Components give the unit price and total price on a cost each basis for this system. Prices for alternate kitchen systems are on Line Items E1090 310 1500 and 1600. Both material quantities and labor costs have been adjusted for the system listed.

Factors: To adjust for job conditions other than normal working situations use Lines E1090 310 2900 thru 4000.

Example: You are to install the system and protect the work area from dust. Go to Line E1090 310 3200 and apply these percentages to the appropriate MAT. and INST. costs.

System Components	QUANTITY	UNIT	COST EACH		
			MAT.	INST.	TOTAL
Kitchen cabinets including wall and base cabinets, custom laminated Plastic top, sink & appliances, no plumbing or electrical rough-in included.					
Prefinished wood cabinets, average quality, wall and base	20.000	L.F.	2,860	700	3,560
Custom laminated plastic counter top	20.000	L.F.	191	352	543
Stainless steel sink, 22" x 25"	1.000	Ea.	625	194	819
Faucet, top mount	1.000	Ea.	60.50	60.50	121
Dishwasher, built-in	1.000	Ea.	227	296	523
Compactor, built-in	1.000	Ea.	530	106	636
Range hood, vented, 30" wide	1.000	Ea.	58	223	281
TOTAL			4,551.50	1,931.50	6,483

E1090 310	Kitchens	COST EACH		
		MAT.	INST.	TOTAL
1400	For alternate kitchen systems:			
1500	Prefinished wood cabinets, high quality			
1600	Custom cabinets, built in place, high quality	8,875	2,200	11,075
1700		11,500	2,550	14,050
1800				
1900				
2000				
2100				
2200	NOTE: No plumbing or electric rough-ins are included in the above			
2300	Prices, for plumbing see Division 22, for electric see Division 26.			
2400				
2500				
2600				
2700				
2900	Cut & patch to match existing construction, add, minimum	2%	3%	
3000	Maximum	5%	9%	
3100	Dust protection, add, minimum	1%	2%	
3200	Maximum	4%	11%	
3300	Equipment usage curtailment, add, minimum	1%	1%	
3400	Maximum	3%	10%	
3500	Material handling & storage limitation, add, minimum	1%	1%	
3600	Maximum	6%	7%	
3700	Protection of existing work, add, minimum	2%	2%	
3800	Maximum	5%	7%	
3900	Shift work requirements, add, minimum		5%	
4000	Maximum		30%	

E1090 Other Equipment

E1090 315	Selective Price Sheet	MAT.	INST.	TOTAL
0100	Cabinets standard wood, base, one drawer one door, 12" wide	250	42.50	292.50
0200	15" wide	299	44	343
0300	18" wide	325	45.50	370.50
0400	21" wide	340	46.50	386.50
0500	24" wide	400	47.50	447.50
0600				
0700	Two drawers two doors, 27" wide	410	48	458
0800	30" wide	410	49.50	459.50
0900	33" wide	415	50.50	465.50
1000	36" wide	430	52	482
1100	42" wide	460	53.50	513.50
1200	48" wide	495	56	551
1300	Drawer base (4 drawers), 12" wide	365	42.50	407.50
1400	15" wide	400	44	444
1500	18" wide	420	45.50	465.50
1600	24" wide	450	47.50	497.50
1700	Sink or range base, 30" wide	330	49.50	379.50
1800	33" wide	355	50.50	405.50
1900	36" wide	370	52	422
2000	42" wide	400	53.50	453.50
2100	Corner base, 36" wide	690	58.50	748.50
2200	Lazy susan with revolving door	595	64	659
2300	Cabinets standard wood, wall two doors, 12" high, 30" wide	217	42.50	259.50
2400	36" wide	251	44	295
2500	15" high, 30" high	218	44	262
2600	36" wide	267	46.50	313.50
2700	24" high, 30" wide	340	45.50	385.50
2800	36" wide	380	46.50	426.50
2900	30" high, 30" wide	350	54.50	404.50
3000	36" wide	395	56	451
3100	42" wide	425	57	482
3200	48" wide	440	57.50	497.50
3300	One door, 30" high, 12" wide	211	48	259
3400	15" wide	231	49.50	280.50
3500	18" wide	258	50.50	308.50
3600	24" wide	298	52	350
3700	Corner, 30" high, 24" wide	230	58.50	288.50
3800	36" wide	320	64	384
3900	Broom, 84" high, 24" deep, 18" wide	565	106	671
4000	Oven, 84" high, 24" deep, 27" wide	860	132	992
4100	Valance board, 4' long	54	10.65	64.65
4200	6" long	81	15.95	96.95
4300	Counter tops, laminated plastic, stock 25" wide w/backsplash, min.	9.55	17.60	27.15
4400	Maximum	19.80	21	40.80
4500	Custom, 7/8" thick, no splash	22	17.60	39.60
4600	Cove splash	28.50	17.60	46.10
4700	1-1/4" thick, no splash	32.50	18.85	51.35
4800	Square splash	31	18.85	49.85
4900	Post formed	11.25	17.60	28.85
5000				
5100	Maple laminated 1-1/2" thick, no splash	65.50	18.85	84.35
5200	Square splash	77.50	18.85	96.35
5300				
5400				
5500	Appliances, range, free standing, minimum	350	83.50	433.50
5600	Maximum	1,950	209	2,159
5700	Built-in, minimum	760	96.50	856.50
5800	Maximum	1,525	530	2,055

E1090 Other Equipment

E1090 315	Selective Price Sheet	COST EACH		
		MAT.	INST.	TOTAL
5900	Counter top range 4 burner, maximum	268	96.50	364.50
6000	Maximum	645	193	838
6100	Compactor, built-in, minimum	530	106	636
6200	Maximum	565	176	741
6300	Dishwasher, built-in, minimum	227	296	523
6400	Maximum	340	590	930
6500	Garbage disposer, sink-pipe, minimum	73.50	118	191.50
6600	Maximum	211	118	329
6700	Range hood, 30" wide, 2 speed, minimum	58	223	281
6800	Maximum	895	370	1,265
6900	Refrigerator, no frost, 12 cu. ft.	490	83.50	573.50
7000	20 cu. ft.	575	139	714
7100	Plumb. not incl. rough-ins, sinks porc. C.I., single bowl, 21" x 24"	299	194	493
7200	21" x 30"	305	194	499
7300	Double bowl, 20" x 32"	410	227	637
7400				
7500	Stainless steel, single bowl, 19" x 18"	560	194	754
7600	22" x 25"	625	194	819
7700				
7800				
7900				
8000				
8100				
8200				
8300				
8400				

G1030 Site Clearing

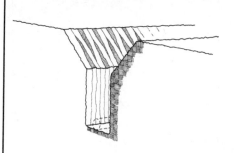

This page illustrates and describes utility trench excavation with backfill systems including a trench, backfill, excavated material, utility pipe or ductwork not included. Lines within System Components give the unit price and total price on a linear foot basis for this system. Prices for alternate utility trench excavation with backfill systems are on Line Items G1030 810 1100 thru 2500. Both material quantities and labor costs have been adjusted for the system listed.

Factors: To adjust for job conditions other than normal working situations use Lines G1030 810 3100 thru 4000.

Example: You are to install the system and protect existing construction. Go to Line G1030 810 3600 and apply these percentages to the appropriate TOTAL cost.

System Components	QUANTITY	UNIT	COST PER L.F.		
			EQUIP.	LABOR	TOTAL
Cont. utility trench excav. with 3/8 C.Y. wheel mtd. backhoe, Backfilling with excav. mat. and comp. in 12″ lifts. Trench is 2′ wide by 4′ dp. Cost of utility piping or ductwork is not incl. Cost based on an Excavation production rate of 150 C.Y. daily no hauling included.					
Machine excavate trench, 2′ wide by 4′ deep	.296	C.Y.	.64	1.85	2.49
Trench, dozer, no compaction, 60 HP	.296	C.Y.	.30	.51	.81
Compact in 12″ lifts, vibrating plate	.296	C.Y.	.07	2.10	2.17
TOTAL			1.01	4.46	5.47

G1030 810	Excavation, Utility Trench	COST PER L.F.		
		EQUIP.	LABOR	TOTAL
1000	Alternate size trenches:			
1100	2′ wide x 2′ deep	.51	2.23	2.74
1200	3′ deep	.76	3.35	4.11
1300	With sloping sides, 2′ wide x 5′ deep	2.85	12.55	15.40
1400	6′ deep	3.80	16.75	20.50
1500	7′ deep	4.88	21.50	26.50
1600	8′ deep	6.10	27	33
1700	9′ deep	7.40	32.50	40
1800	10′ deep	8.85	39	48
2000	For hauling excavated material up to 2 miles & backfilling w/ gravel			
2100	Gravel, 2′ wide by 2′ deep, add	5.05	3.07	8.12
2200	4′ deep, add	10.05	6.15	16.20
2300	6′ deep, add	38	23	61
2400	8′ deep, add	60.50	37	97.50
2500	10′ deep, add	88	54	142
2600				
2700				
2800	For shallow, hand excavated trenches, no backfill, to 6′ dp, light soil		52	52
2900	Heavy soil		104	104
3000				
3100	Equipment usage curtailment, add, minimum	1%	1%	
3200	Maximum	3%	10%	
3300	Material handling & storage limitation, add, minimum	1%	1%	
3400	Maximum	6%	7%	
3500	Protection of existing work, add, minimum	2%	2%	
3600	Maximum	5%	7%	
3700	Shift work requirements, add, minimum		5%	
3800	Maximum		30%	
3900	Temporary shoring and bracing, add, minimum	2%	5%	
4000	Maximum	5%	12%	

Reed Construction Data
The leader in construction information and BIM solutions

Reed Construction Data, Inc. is a leading provider of construction information and building information modeling (BIM) solutions. The company's portfolio of information products and services is designed specifically to help construction industry professionals advance their business with timely and accurate project, product, and cost data. Reed Construction Data is a division of Reed Business Information, a member of the Reed Elsevier PLC group of companies.

Cost Information

RSMeans, the undisputed market leader in construction costs, provides current cost and estimating information through its innovative MeansCostworks.com® web-based solution. In addition, RSMeans publishes annual cost books, estimating software, and a rich library of reference books. RSMeans also conducts a series of professional seminars and provides construction cost consulting for owners, manufacturers, designers, and contractors to sharpen personal skills and maximize the effective use of cost estimating and management tools.

Project Data

Reed Construction Data assembles one of the largest databases of public and private project data for use by contractors, distributors, and building product manufacturers in the U.S. and Canadian markets. In addition, Reed Construction Data is the North American construction community's premier resource for project leads and bid documents. Reed Bulletin and Reed CONNECT™ provide project leads and project data through all stages of construction for many of the country's largest public sector, commercial, industrial, and multi-family residential projects.

Research and Analytics

Reed Construction Data's forecasting tools cover most aspects of the construction business in the U.S. and Canada. With a vast network of resources, Reed Construction Data is uniquely qualified to give you the information you need to keep your business profitable.

SmartBIM Solutions

Reed Construction Data has emerged as the leader in the field of building information modeling (BIM) with products and services that have helped to advance the evolution of BIM. Through an in-depth SmartBIM Object Creation Program, BPMs can rely on Reed to create high-quality, real world objects embedded with superior cost data (RSMeans). In addition, Reed has made it easy for architects to manage these manufacturer-specific objects as well as generic objects in Revit with the SmartBIM Library.

SmartBuilding Index

The leading industry source for product research, product documentation, BIM objects, design ideas, and source locations. Search, select, and specify available building products with our online directories of manufacturer profiles, MANU-SPEC, SPEC-DATA, guide specs, manufacturer catalogs, building codes, historical project data, and BIM objects.

SmartBuilding Studio

A vibrant online resource library which brings together a comprehensive catalog of commercial interior finishes and products under a single standard for high-definition imagery, product data, and searchable attributes.

Associated Construction Publications (ACP)

Reed Construction Data's regional construction magazines cover the nation through a network of 14 regional magazines. Serving the construction market for more than 100 years, our magazines are a trusted source of news and information in the local and national construction communities.

For more information, please visit our website at www.reedconstructiondata.com

The leader in construction information and BIM solutions.

G2010 Roadways

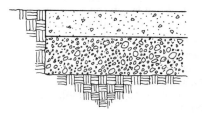

This page illustrates and describes driveway systems including concrete slab, gravel base, broom finish, compaction, and joists. Lines within System Components give the unit price and total price on a cost each basis for this system. Prices for alternate driveway systems are on Line Items G2010 240 1300 thru 2700. Both material quantities and labor costs have been adjusted for the system listed.

Factors: To adjust for job conditions other than normal working situations use Lines G2010 240 2900 thru 4000.

Example: You are to install the system and match existing construction. Go to Line G2010 240 3000 and apply these percentages to the appropriate MAT. and INST. costs.

System Components	QUANTITY	UNIT	COST EACH MAT.	COST EACH INST.	COST EACH TOTAL
Complete driveway, 10′ x 30′, 4″ concrete slab, 3000 psi, on 6″ compacted Gravel base, concrete broom finished and cured with 10′ x 10′ joints.					
Grade and compact subgrade	33.000	S.Y.		59.73	59.73
Place and compact 6″ crushed stone base	33.000	S.Y.	306.90	39.60	346.50
Place and remove edgeforms (4 use)	80.000	L.F.	33.60	266.40	300
Concrete for slab, 3000 psi	4.000	C.Y.	444		444
Place concrete, 4″ thick slab, direct chute	4.000	C.Y.		96.08	96.08
Screed, float and broom finish slab	4.000	C.Y.		222	222
Spray on membrane curing compound	3.000	S.F.	17.25	26.40	43.65
Saw cut 3″ deep joints	60.000	L.F.	32.40	82.20	114.60
TOTAL			834.15	792.41	1,626.56

G2010 240	Driveways	COST EACH MAT.	COST EACH INST.	COST EACH TOTAL
1200	For alternate driveway systems:			
1300	Concrete: 10′ x 30′ with 6″ concrete on 6″ crushed stone	1,050	840	1,890
1400	20′ x 30′ with 4″ concrete	1,700	1,500	3,200
1500	With 6″ concrete	2,150	1,600	3,750
1600	10′ wide, for each additional 10′ length over 30′, 4″ thick, add	272	254	526
1700	6″ thick, add	355	272	627
1800	20′ wide, for each additional 10′ length over 30′, 4″ thick, add	550	480	1,030
1900	6″ thick, add	690	510	1,200
2000	Asphalt: 10′ x 30′ with 2″ binder, 1″ topping on 6″ cr. st., sealed	670	570	1,240
2100	3″ binder, 1″ topping	785	585	1,370
2200	20′ x 30′ with 2″ binder, 1″ topping	1,350	780	2,130
2300	3″ binding, 1″ topping	1,575	805	2,380
2400	10′ wide for each add'l. 10′ length over 30′, 2″ b + 1″ t, add	223	71	294
2500	3″ binder, 1″ topping	261	75	336
2600	20′ wide for each add'l. 10′ length over 30′, 2″ b + 1″ t, add	445	143	588
2700	3″ binder, 1″ topping	520	151	671
2800				
2900	Cut & patch to match existing construction, add, minimum	2%	3%	
3000	Maximum	5%	9%	
3100	Dust protection, add, minimum	1%	2%	
3200	Maximum	4%	11%	
3300	Equipment usage curtailment, add, minimum	1%	1%	
3400	Maximum	3%	10%	
3500	Material handling & storage limitation, add, minimum	1%	1%	
3600	Maximum	6%	7%	
3700	Protection of existing work, add, minimum	2%	2%	
3800	Maximum	5%	7%	
3900	Shift work requirements, add, minimum		5%	
4000	Maximum		30%	

G2020 Parking Lots

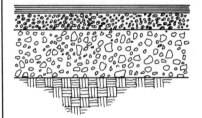

This page illustrates and describes asphalt parking lot systems including asphalt binder, topping, crushed stone base, painted parking stripes and concrete parking blocks. Lines within System Components give the unit price and total price per square yard for this system. Prices for alternate asphalt parking lot systems are on Line Items G2020 230 1500 thru 2500. Both material quantities and labor costs have been adjusted for the system listed.

Factors: To adjust for job conditions other than normal working situations use Lines G2020 230 2900 thru 4000.

Example: You are to install the system and match existing construction. Go to Line G2020 230 2900 and apply these percentages to the appropriate MAT. and INST. costs.

System Components	QUANTITY	UNIT	COST PER S.Y. MAT.	COST PER S.Y. INST.	COST PER S.Y. TOTAL
Parking lot consisting of 2″ asphalt binder and 1″ topping on 6″					
Crushed stone base with painted parking stripes and concrete parking blocks.					
Fine grade and compact subgrade	1.000	S.Y.		1.81	1.81
6″ crushed stone base, stone	.320	Ton	9.30	1.20	10.50
2″ asphalt binder	1.000	S.Y.	6.80	1.21	8.01
1″ asphalt topping	1.000	S.Y.	3.38	.80	4.18
Paint parking stripes	.500	L.F.	.14	.09	.23
6″ x 10″ x 6′ precast concrete parking blocks	.020	Ea.	1.29	.35	1.64
Mobilization of equipment	.005	Ea.		1.78	1.78
TOTAL			20.91	7.24	28.15

G2020 230	Parking Lots, Asphalt	COST PER S.Y. MAT.	COST PER S.Y. INST.	COST PER S.Y. TOTAL
1400	For alternate parking lot systems:			
1500	Above system on 9″ crushed stone	25.50	7.35	32.85
1600	12″ crushed stone	30.50	7.50	38
1700	On bank run gravel, 6″ deep	17.45	6.70	24.15
1800	9″ deep	20.50	6.85	27.35
1900	12″ deep	23.50	7	30.50
2000	3″ binder plus 1″ topping on 6″ crushed stone	24.50	7.60	32.10
2100	9″ deep crushed stone	29	7.70	36.70
2200	12″ deep crushed stone	33.50	7.85	41.35
2300	On bank run gravel, 6″ deep	21	7.05	28.05
2400	9″ deep	24	7.20	31.20
2500	12″ deep	27	7.35	34.35
2600				
2700				
2900	Cut & patch to match existing construction, add, minimum	2%	3%	
3000	Maximum	5%	9%	
3100	Dust protection, add, minimum	1%	2%	
3200	Maximum	4%	11%	
3300	Equipment usage curtailment, add, minimum	1%	1%	
3400	Maximum	3%	10%	
3500	Material handling & storage limitation, add, minimum	1%	1%	
3600	Maximum	6%	7%	
3700	Protection of existing work, add, minimum	2%	2%	
3800	Maximum	5%	7%	
3900	Shift work requirements, add, minimum		5%	
4000	Maximum		30%	

G20 Site Improvements

G2020 Parking Lots

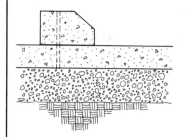

This page illustrates and describes concrete parking lot systems including a concrete slab, unreinforced, compacted gravel base, broom finished, cured, joints, painted parking stripes, and parking blocks. Lines within System Components give the unit price and total price per square yard for this system. Prices for alternate concrete parking lot systems are on Line Items G2020 240 1500 thru 2500. All materials have been adjusted according to the system listed.

Factors: To adjust for job conditions other than normal working situations use Lines G2020 240 2900 thru 4000.

Example: You are to install the system and protect existing construction. Go to Line G2020 240 3800 and apply these percentages to the appropriate MAT. and INST. costs.

System Components	QUANTITY	UNIT	COST PER S.Y. MAT.	COST PER S.Y. INST.	COST PER S.Y. TOTAL
Parking lot, 4″ slab, 3000 psi concrete on 6″ compacted gravel base					
With 12′ x 12′ joints, painted parking strips, conc. parking blocks.					
Grade and compact subgrade	1.000	S.Y.		1.81	1.81
6″ crushed stone base, stone	.320	Ton	9.30	1.20	10.50
Place and remove edge forms, 4 use	.250	L.F.	.11	.83	.94
Concrete for slab, 3000 psi	1.000	S.F.	12.21		12.21
Place concrete, 4″ thick slab, direct chute	1.000	S.F.		2.65	2.65
Spray on membrane curing compound	1.000	S.F.	.52	.56	1.08
Saw cut 3″ deep joints	2.250	L.F.	1.22	3.08	4.30
Paint parking stripes	.500	L.F.	.14	.09	.23
6″ x 10″ x 6′ precast concrete parking blocks	.020	Ea.	1.29	.35	1.64
TOTAL			24.79	10.57	35.36

G2020 240	Parking Lots, Concrete	COST PER S.Y. MAT.	COST PER S.Y. INST.	COST PER S.Y. TOTAL
1400	For alternate parking lot systems:			
1500	Above system on 3″ crushed stone	25	10.55	35.55
1600	9″ crushed stone	29.50	10.65	40.15
1700	On bank run gravel, 6″ deep	21.50	10	31.50
1800	9″ deep	24.50	10.20	34.70
1900	On compacted subgrade only	15.50	9.40	24.90
2000	6″ concrete slab, 3000 psi on 3″ crushed stone	30.50	11.80	42.30
2100	6″ crushed stone	30.50	11.80	42.30
2200	9″ crushed stone	35.50	11.95	47.45
2300	On bank run gravel, 6″ deep	27	11.30	38.30
2400	9″ deep	30	11.45	41.45
2500	On compacted subgrade only	30.50	11.80	42.30
2600				
2700				
2900	Cut & patch to match existing construction, add, minimum	2%	3%	
3000	Maximum	5%	9%	
3100	Dust protection, add, minimum	1%	2%	
3200	Maximum	4%	11%	
3300	Equipment usage curtailment, add, minimum	1%	1%	
3400	Maximum	3%	10%	
3500	Material handling & storage limitation, add, minimum	1%	1%	
3600	Maximum	6%	7%	
3700	Protection of existing work, add, minimum	2%	2%	
3800	Maximum	5%	7%	
3900	Shift work requirements, add, minimum		5%	
4000	Maximum		30%	

G2030 Pedestrian Paving

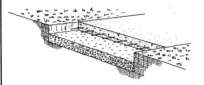

This page illustrates and describes sidewalk systems including concrete, welded wire and broom finish. Lines within System Components give unit price and total price per square foot for this system. Prices for alternate sidewalk systems are on Line Items G2030 105 1700 thru 1900. Both material quantities and labor costs have been adjusted for the system listed.

Factors: To adjust for job conditions other than normal working situations use Lines G2030 105 2900 thru 4000.

Example: You are to install the system and match existing construction. Go to Line G2030 105 2900 and apply these percentages to the appropriate MAT. and INST. costs.

System Components	QUANTITY	UNIT	COST PER S.F. MAT.	COST PER S.F. INST.	COST PER S.F. TOTAL
4″ thick concrete sidewalk with welded wire fabric					
3000 psi air entrained concrete, broom finish.		S.F.			
Gravel fill, 4″ deep	.012	C.Y.	.36	.08	.44
Compact fill	.012	C.Y.		.02	.02
Hand grade	1.000	S.F.		1.89	1.89
Edge form	.250	L.F.	.11	.83	.94
Welded wire fabric	.011	S.F.	.22	.38	.60
Concrete, 3000 psi air entrained	.012	C.Y.	1.33		1.33
Place concrete	.012	C.Y.		.29	.29
Broom finish	1.000	S.F.		.74	.74
TOTAL			2.02	4.23	6.25

G2030 105	Sidewalks	COST PER S.F. MAT.	COST PER S.F. INST.	COST PER S.F. TOTAL
1600	For alternate sidewalk systems:			
1700	Asphalt (bituminous), 2″ thick	1.09	2.41	3.50
1800	Brick, on sand, bed, 4.5 brick per S.F.	3.28	9.40	12.68
1900	Flagstone, slate, 1″ thick, rectangular	9.90	10.20	20.10
2000				
2100				
2200				
2300				
2400				
2500				
2600				
2700				
2900	Cut & patch to match existing construction, add, minimum	2%	3%	
3000	Maximum	5%	9%	
3100	Dust protection, add, minimum	1%	2%	
3200	Maximum	4%	11%	
3300	Equipment usage curtailment, add, minimum	1%	1%	
3400	Maximum	3%	10%	
3500	Material handling & storage limitation, add, minimum	1%	1%	
3600	Maximum	6%	7%	
3700	Protection of existing work, add, minimum	2%	2%	
3800	Maximum	5%	7%	
3900	Shift work requirements, add, minimum		5%	
4000	Maximum		30%	

G2050 Landscaping

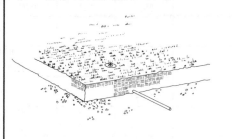

This page describes landscaping—lawn establishment systems including loam, lime, fertilizer, side and top mulching. Lines within system components give the unit price and total price per square yard for this system. Prices for alternate landscaping—lawn establishment systems are on Line Items G2050 420 1100 and 1200. Both material quantities and labor costs have been adjusted for the system listed.

Factors: To adjust for job conditions other than normal working situations use Lines G2050 420 2900 thru 4000.

Example: You are to install the system and provide dust protection. Go to Line G2050 420 3100 and apply these percentages to the appropriate MAT. and INST. costs.

System Components				COST PER S.Y.		
	QUANTITY	UNIT		MAT.	INST.	TOTAL
Establishing lawns with loam, lime, fertilizer, seed and top mulching						
On rough graded areas.						
Furnish and place loam 4" deep	.110	C.Y.		2.85	.84	3.69
Fine grade, lime, fertilize and seed	1.000	S.Y.		.38	2.92	3.30
Tobacco netting, or jute mesh #2, stapled	1.000	S.Y.		.10	.51	.61
Rolling with hand roller	1.000	S.Y.			.89	.89
TOTAL				3.33	5.16	8.49

G2050 420	Landscaping - Lawn Establishment	COST PER S.Y.		
		MAT.	INST.	TOTAL
1000	For alternate lawn systems:			
1100	Above system with jute mesh in place of hay mulch	4.40	5.30	9.70
1200	Above system with sod in place of seed	5.40	1.83	7.23
1300				
1400				
1500				
1600				
1700				
1800				
1900				
2000				
2100				
2200				
2300				
2400				
2500				
2600				
2700				
2900	Cut & patch to match existing construction, add, minimum	2%	3%	
3000	Maximum	5%	9%	
3100	Dust protection, add, minimum	1%	2%	
3300	Equipment usage curtailment, add, minimum	1%	1%	
3400	Maximum	3%	10%	
3500	Material handling & storage limitation, add, minimum	1%	1%	
3600	Maximum	6%	7%	
3700	Protection of existing work, add, minimum	2%	2%	
3800	Maximum	5%	7%	
3900	Shift work requirements, add, minimum		5%	
4000	Maximum		30%	

G20 Site Improvements

G2050 Landscaping

G2050 510	Selective Price Sheet	COST PER UNIT		
		MAT.	INST.	TOTAL
0100	Shrubs and trees, Evergreen, in prepared beds			
0200				
0300	Arborvitae pyramidal, 4'-5'	52	80.50	132.50
0400	Globe, 12"-15"	23	13.30	36.30
0500	Cedar, blue, 8'-10'	256	135	391
0600	Hemlock, Canadian, 2-1/2'-3'	37	35.50	72.50
0700	Juniper, andora, 18"-24"	18.45	16	34.45
0800	Wiltoni, 15"-18"	19.90	16	35.90
0900	Skyrocket, 4-1/2'-5'	60.50	44	104.50
1000	Blue pfitzer, 2'-2-1/2'	28	29	57
1100	Ketleerie, 2-1/2'-3'	37	25.50	62.50
1200	Pine, black, 2-1/2'-3'	50	25.50	75.50
1300	Mugo, 18"-24"	39.50	21.50	61
1400	White, 4'-5'	62.50	32.50	95
1500	Spruce, blue, 18"-24"	42	21.50	63.50
1600	Norway, 4'-5'	71.50	32.50	104
1700	Yew, denisforma, 12"-15"	28.50	21.50	50
1800	Capitata, 18"-24"	26	42.50	68.50
1900	Hicksi, 2'-2-1/2'	35	42.50	77.50
2000				
2100	Trees, Deciduous, in prepared beds			
2200				
2300	Beech, 5'-6'	245	48.50	293.50
2400	Dogwood, 4'-5'	75.50	60.50	136
2500	Elm, 8'-10'	131	122	253
2600	Magnolia, 4'-5'	87.50	122	209.50
2700	Maple, red, 8'-10', 1-1/2" caliper	191	243	434
2800	Oak, 2-1/2"-3" caliper	284	405	689
2900	Willow, 6'-8', 1" caliper	70	122	192
3000				
3100	Shrubs, Broadleaf Evergreen, in prepared beds			
3200				
3300	Andromeda, 15"-18", container	29	13.30	42.30
3400	Azalea, 15"-18", container	35	13.30	48.30
3500	Barberry, 9"-12", container	12.95	9.85	22.80
3600	Boxwood, 15"-18", B & B	31.50	13.30	44.80
3700	Euonymus, emerald gaiety, 12"-15", container	23	11.10	34.10
3800	Holly, 15"-18", B & B	20.50	13.30	33.80
3900	Mount laurel, 18"-24", B & B	61.50	16	77.50
4000	Privet, 18"-24", B & B	20	9.85	29.85
4100	Rhodendron, 18"-24", container	34.50	26.50	61
4200	Rosemary, 1 gal. container	69	2.13	71.13
4300	Deciduous, amalanchier, 2'-3', B & B	99	22.50	121.50
4400	Azalea, 15"-18", B & B	27	13.30	40.30
4500	Bayberry, 2'-3', B & B	30.50	22.50	53
4600	Cotoneaster, 15"-18", B & B	18.15	16	34.15
4700	Dogwood, 3'-4', B & B	28.50	60.50	89
4800	Euonymus, alatus compacta, 15"-18", container	24	16	40
4900	Forsythia, 2'-3', container	22	21.50	43.50
5000	Honeysuckle, 3'-4', B & B	25.50	21.50	47
5100	Hydrangea, 2'-3', B & B	31.50	22.50	54
5200	Lilac, 3'-4', B & B	29.50	60.50	90
5300	Quince, 2'-3', B & B	26.50	22.50	49
5400				
5500	Ground Cover			
5600				
5700	Plants, Pachysandra, prepared beds, per hundred	29.50	85	114.50
5800	Vinca minor or English ivy, per hundred	44	107	151

G2050 Landscaping

G2050 510	Selective Price Sheet	COST PER UNIT		
		MAT.	INST.	TOTAL
5900	Plant bed prep. 18" dp., mach., per square foot	2.41	.37	2.78
6000	By hand, per square foot	2.41	2.69	5.10
6100	Stone chips, Georgia marble, 50# bags, per bag	5.70	2.46	8.16
6200	Onyx gemstone, per bag	25	4.92	29.92
6300	Quartz, per bag	9.30	4.92	14.22
6400	Pea gravel, truck load lots, per cubic yard	37.50	45.50	83
6500	Mulch, polyethylene mulch, per square yard	.21	.42	.63
6600	Wood chips, 2" deep, per square yard	2.07	1.90	3.97
6700	Peat moss, 1" deep, per square yard	2.04	.46	2.50
6800	Erosion control per square yard, Jute mesh stapled	1.17	.63	1.80
6900	Plastic netting, stapled	.89	.51	1.40
7000	Polypropylene mesh, stapled	1.84	.51	2.35
7100	Tobacco netting, stapled	.10	.51	.61
7200	Lawns per square yard, seeding incl. fine grade, limestone, fert. and seed	.38	2.92	3.30
7300	Sodding, incl. fine grade, level ground	2.43	.95	3.38
7500	Edging, per linear foot, Redwood, untreated, 1" x 4"	1.25	2.11	3.36
7600	2" x 4"	2.49	3.20	5.69
7700	Stl. edge strips, 1/4" x 5" inc. stakes	5.25	3.20	8.45
7800	3/16" x 4"	4.13	3.20	7.33
7900	Brick edging, set on edge	3.67	6.95	10.62
8000	Set flat	1.84	2.54	4.38
8100				
8200				
8300				
8400				

G30 Site Mechanical Utilities

G3020 Sanitary Sewer

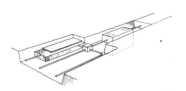

This page illustrates and describes drainage and utilities – septic system including a tank, distribution box, excavation, piping, crushed stone and backfill. Lines within System Components give the unit price and total price on a cost each basis for this system. Prices for alternate drainage and utilities – septic systems are on Line Items G3020 610 1700 thru 2000. Both material quantities and labor costs have been adjusted for the system listed.

Factors: To adjust for job conditions other than normal working situations use Lines G3020 610 3100 thru 4000.

Example: You are to install the system with a material handling limitation. Go to Line G3020 610 3500 and apply these percentages to the appropriate MAT. and INST. costs.

System Components			COST EACH		
	QUANTITY	UNIT	MAT.	INST.	TOTAL
Septic system including tank, distribution box, excavation, piping, crushed Stone and backfill for a 1000 S.F. leaching field.					
Precast concrete septic tank, 1000 gal. capacity	1.000	Ea.	800	225.75	1,025.75
Concrete distribution box	1.000	Ea.	82.50	52	134.50
Sch. 40 sewer pipe, PVC, 4″ diameter	25.000	L.F.	311.10	630.70	941.80
Sch. 40 sewer pipe fittings, PVC, 4″ diameter	8.000	Ea.	126.40	708	834.40
Excavation for septic tank	120.000	C.Y.		199.62	199.62
Excavation for disposal field	120.000	C.Y.		485.81	485.81
Crushed stone backfill	76.000	C.Y.	2,090	836.76	2,926.76
Backfill with excavated material	26.000	C.Y.		41.60	41.60
Building paper	6.000	C.S.F.	18.84	85.50	104.34
TOTAL			3,428.84	3,265.74	6,694.58

G3020 610		Septic Systems	COST EACH		
			MAT.	INST.	TOTAL
1600	For alternate septic systems:				
1700	1000 gal. tank with 2000 S.F. field		5,125	4,725	9,850
1800	With leaching pits		3,275	1,050	4,325
1900	2000 gal. tank with 2000 S.F. field		6,125	4,875	11,000
2000	With leaching pits		4,350	1,575	5,925
2100					
2200					
2300					
2400					
2500					
2600					
2700					
2800					
2900					
3100	Dust protection, add, minimum		1%	2%	
3200	Maximum		4%	11%	
3300	Equipment usage curtailment, add, minimum		1%	1%	
3400	Maximum		3%	10%	
3500	Material handling & storage limitation, add, minimum		1%	1%	
3600	Maximum		6%	7%	
3700	Protection of existing work, add, minimum		2%	2%	
3800	Maximum		5%	7%	
3900	Shift work requirements, add, minimum			5%	
4000	Maximum			30%	

G3060 Fuel Distribution

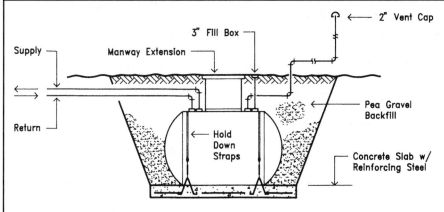

This page illustrates and describes fiberglass tanks including hold down slab, reinforcing steel and peastone gravel. Lines within Systems Components give the material and installation price on a cost each basis for the components. Prices for alternate fiberglass tank systems are on Line Items G3060 305 310 thru G3060 310 1100. Material quantities and labor costs have been adjusted for the system listed.

Factors: To adjust for job conditions other than normal working situations use Lines G3060 310 2700 thru 4000.

Fiberglass Underground Fuel Storage Tank, Double Wall

System Components	QUANTITY	UNIT	COST EACH MAT.	COST EACH INST.	COST EACH TOTAL
Double wall underground fiberglass storage tanks including, excavation,		Ea.			
hold down slab, backfill, manway extension and related piping.					
Excavating trench, no sheeting or dewatering, 6'-10 ' D, 3/4 CY hyd backhoe	18.000	Ea.		128.88	128.88
Corrosion resistance wrap & coat, small diam pipe, 1" diam, add	120.000	Ea.	198		198
Fiberglass-reinforced elbows 90 & 45 degree, 2 inch	7.000	Ea.	385	500.50	885.50
Fiberglass-reinforced threaded adapters, 2 inch	2.000	Ea.	36.80	163	199.80
Remote tank gauging system, 30', 5" pointer travel	1.000	Ea.	4,850	305	5,155
Corrosion resistance wrap & coat, small diam pipe. 4 " diam, add	30.000	Ea.	66		66
Stone back-fill, hand spread, pea gravel	9.000	Ea.	675	409.50	1,084.50
Reinforcing in hold down pad, #3 to #7	120.000	Ea.	102	69.60	171.60
Tank leak detector system, 8 channel, external monitoring	1.000	Ea.	6,600		6,600
Tubbing copper, type L, 3/4" diam	120.000	Ea.	792	954	1,746
Elbow 90 degree ,wrought cu x cu 3/4 " diam	4.000	Ea.	12.16	128	140.16
Pipe, black steel welded, sch 40, 3" diam	30.000	Ea.	513	807.90	1,320.90
Elbow 90 degree, steel welded butt joint, 3 " diam	3.000	Ea.	67.50	494.25	561.75
Fiberglass-reinforced plastic pipe, 2 inch	156.000	Ea.	850.20	628.68	1,478.88
Fiberglass-reinforced plastic pipe, 3 inch	36.000	Ea.	336.60	164.52	501.12
Fiberglass-reinforced elbows 90 & 45 degree, 3 inch	3.000	Ea.	201	286.50	487.50
Tank leak detection system, probes, well monitoring, liquid phase detection	2.000	Ea.	1,670		1,670
Double wall fiberglass annular space	1.000	Ea.	275		275
Flange, weld neck, 150 lb, 3 " pipe size	1.000	Ea.	27.50	82.38	109.88
Vent protector / breather, 2" dia.	1.000	Ea.	23	19.15	42.15
Corrosion resistance wrap & coat, small diam pipe, 2" diam, add	36.000	Ea.	65.52		65.52
Concrete hold down pad, 6' x 5' x 8" thick	30.000	Ea.	88.20	41.70	129.90
600 gallon capacity	1.000	Ea.	7,300	455	7,755
Foot valve, single poppet, 3/4" dia	1.000	Ea.	62.50	34	96.50
Fuel fill box, locking inner cover, 3" dia.	1.000	Ea.	130	122	252
Fuel oil specialties, valve,ball chk, globe type fusible, 3/4" dia	2.000	Ea.	134	61	195
Tank, for hold-downs 500-4000 gal, add	1.000	Ea.	460	146	606
TOTAL			25,920.98	6,001.56	31,922.54

G3060 310	Fiberglass Fuel Tank, Double Wall		COST EACH MAT.	COST EACH INST.	COST EACH TOTAL
1000	For alternate storage tanks:				
1010	Storage tank, fuel, underground, double wall fiberglass, 550 Gal.		25,900	6,000	31,900
1020	2500 Gal.		36,100	8,375	44,475
1030	4000 Gal.		43,200	10,400	53,600
1040	6000 Gal.		44,600	11,700	56,300
1050	8000 Gal.		54,000	14,100	68,100

G3060 Fuel Distribution

G3060 310	Fiberglass Fuel Tank, Double Wall	COST EACH		
		MAT.	INST.	TOTAL
1060	10,000 Gal.	60,500	16,300	76,800
1070	15,000 Gal.	78,000	18,800	96,800
1080	20,000 Gal.	96,000	22,700	118,700
1090	25,000 Gal.	117,500	26,500	144,000
1100	30,000 Gal.	133,000	28,000	161,000
2700	Cut & patch to match existing construction, add, minimum	2%	3%	
2800	Maximum	5%	9%	
2900	Dust protection, add, minimum	1%	2%	
3000	Maximum	4%	11%	
3100	Equipment usage curtailment, add, minimum	1%	1%	
3200	Maximum	3%	10%	
3300	Material handling & storage limitation, add, minimum	1%	1%	
3400	Maximum	6%	7%	
3500	Protection of existing work, add, minimum	2%	2%	
3600	Maximum	5%	7%	
3700	Shift work requirements, add, minimum		5%	
3800	Maximum		30%	
3900	Temporary shoring and bracing, add, minimum	2%	5%	
4000	Maximum	5%	12%	

G3060 Fuel Distribution

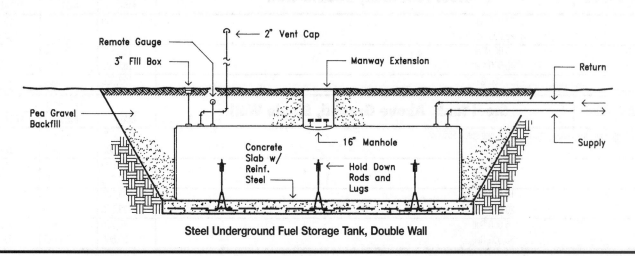

Steel Underground Fuel Storage Tank, Double Wall

System Components	QUANTITY	UNIT	COST EACH		
			MAT.	INST.	TOTAL
Double wall underground steel storage tank including, excavation, hold down slab, backfill, manway extension and related piping.		Ea.			
Excavating trench, no sheeting or dewatering, 6'-10' D, 3/4 CY hyd backhoe	18.000	Ea.		128.88	128.88
Concrete hold down pad, 6' x 5' x 8" thick	30.000	Ea.	88.20	41.70	129.90
Tanks, for hold-downs, 500-2000 gal, add	1.000	Ea.	166	146	312
Reinforcing in hold down pad, #3 to #7	120.000	Ea.	102	69.60	171.60
Double wall steel tank annular space	1.000	Ea.	275		275
Tank, for manways, add	1.000	Ea.	1,650		1,650
Foot valve, single poppet, 3/4" dia	1.000	Ea.	62.50	34	96.50
Fuel fill box, locking inner cover, 3" dia.	1.000	Ea.	130	122	252
Stone back-fill, hand spread, pea gravel	9.000	Ea.	675	409.50	1,084.50
Pipe, black steel welded, sch 40, 3" diam	30.000	Ea.	513	807.90	1,320.90
Elbow 90 degree, steel welded butt joint, 3" diam	3.000	Ea.	67.50	494.25	561.75
Fiberglass-reinforced elbows 90 & 45 degree, 2 inch	7.000	Ea.	385	500.50	885.50
Fiberglass-reinforced threaded adapters, 2 inch	2.000	Ea.	36.80	163	199.80
Fiberglass-reinforced plastic pipe, 2 inch	156.000	Ea.	850.20	628.68	1,478.88
Fiberglass-reinforced plastic pipe, 3 inch	36.000	Ea.	336.60	164.52	501.12
Fiberglass-reinforced elbows 90 & 45 degree, 3 inch	3.000	Ea.	201	286.50	487.50
Flange, weld neck, 150 lb, 3" pipe size	1.000	Ea.	27.50	82.38	109.88
Vent protector / breather, 2" dia.	1.000	Ea.	23	19.15	42.15
500 gallon capacity	1.000	Ea.	4,300	460	4,760
Remote tank gauging system, 30', 5" pointer travel	1.000	Ea.	4,850	305	5,155
Tank leak detector system, 8 channel, external monitoring	1.000	Ea.	6,600		6,600
Tubing copper, type L, 3/4" diam	120.000	Ea.	792	954	1,746
Elbow 90 degree, wrought cu x cu 3/4" diam	8.000	Ea.	24.32	256	280.32
Tank leak detection system, probes, well monitoring, liquid phase detection	2.000	Ea.	1,670		1,670
Fuel oil specialties, valve, ball chk, globe type fusible, 3/4" dia	2.000	Ea.	134	61	195
TOTAL			23,959.62	6,134.56	30,094.18

G3060 320		Steel Fuel Tank, Double Wall	COST EACH		
			MAT.	INST.	TOTAL
1000	For alternate storage tanks:				
1010	Storage tank, fuel, underground, double wall steel, 500 Gal.		24,000	6,125	30,125
1020	2000 Gal.		30,700	8,075	38,775
1030	4000 Gal.		40,100	10,500	50,600
1040	6000 Gal.		45,900	11,600	57,500
1050	8000 Gal.		53,000	14,300	67,300
1060	10,000 Gal.		60,500	16,500	77,000
1070	15,000 Gal.		71,000	19,300	90,300

G30 Site Mechanical Utilities

G3060 Fuel Distribution

G3060 320	Steel Fuel Tank, Double Wall	COST EACH		
		MAT.	INST.	TOTAL
1080	20,000 Gal.	90,500	23,000	113,500
1090	25,000 Gal.	118,000	26,400	144,400
1100	30,000 Gal.	134,500	28,200	162,700
1110	40,000 Gal.	165,500	35,500	201,000
1120	50,000 Gal.	198,000	42,600	240,600

G3060 325	Steel Tank, Above Ground, Single Wall	COST EACH		
		MAT.	INST.	TOTAL
1010	Storage tank, fuel, above ground, single wall steel, 550 Gal.	11,400	4,275	15,675
1020	2000 Gal.	14,300	4,375	18,675
1030	5000 Gal.	17,600	6,725	24,325
1040	10,000 Gal.	30,900	7,175	38,075
1050	15,000 Gal.	33,700	7,375	41,075
1060	20,000 Gal.	39,500	7,600	47,100
1070	25,000 Gal.	45,700	7,800	53,500
1080	30,000 Gal.	50,500	8,125	58,625

G3060 330	Steel Tank, Above Ground, Double Wall	COST EACH		
		MAT.	INST.	TOTAL
1010	Storage tank, fuel, above ground, double wall steel, 500 Gal.	18,300	4,325	22,625
1020	2000 Gal.	24,800	4,425	29,225
1030	4000 Gal.	32,200	4,525	36,725
1040	6000 Gal.	36,800	6,975	43,775
1050	8000 Gal.	42,500	7,175	49,675
1060	10,000 Gal.	45,500	7,300	52,800
1070	15,000 Gal.	59,500	7,550	67,050
1080	20,000 Gal.	65,500	7,800	73,300
1090	25,000 Gal.	75,500	8,025	83,525
1100	30,000 Gal.	81,500	8,325	89,825
2700	Cut & patch to match existing construction, add, minimum	2%	3%	
2800	Maximum	5%	9%	
2900	Dust protection, add, minimum	1%	2%	
3000	Maximum	4%	11%	
3100	Equipment usage curtailment, add, minimum	1%	1%	
3200	Maximum	3%	10%	
3300	Material handling & storage limitation, add, minimum	1%	1%	
3400	Maximum	6%	7%	
3500	Protection of existing work, add, minimum	2%	2%	
3600	Maximum	5%	7%	
3700	Shift work requirements, add, minimum		5%	
3800	Maximum		30%	
3900	Temporary shoring and bracing, add, minimum	2%	5%	
4000	Maximum	5%	12%	

G4020 Site Lighting

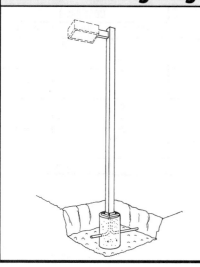

This page illustrates and describes light poles for parking or walkway area lighting. Included are aluminum or steel light poles, single or multiple fixture bracket arms, excavation, concrete footing, backfill and compaction. Lines within system components give the unit price and total price on a cost per each basis. Prices for alternate systems are shown on lines D4020 0200 through 1440. Both material quantities and labor costs have been adjusted for the system listed.

Factors: To adjust for job conditions other than normal working situations use lines D4020 2700 through 3600.

Example: You are to install the system where you must be careful not to damage existing walks or parking areas. Go to line D4020 3300 and apply these percentages to the appropriate MAT. and INST. costs.

System Components	QUANTITY	UNIT	COST EACH		
			MAT.	INST.	TOTAL
Light poles, aluminum, 20' high, 1 arm bracket.					
Aluminum light pole, 20', no concrete base	1.000	Ea.	950	544	1,494
Bracket arm for Aluminum light pole	1.000	Ea.	125	72.50	197.50
Excavation by hand, pits to 6' deep, heavy soil or clay	2.368	C.Y.		246.27	246.27
Footing, concrete incl forms, reinforcing, spread, under 1 C.Y.	.465	C.Y.	80.91	86.07	166.98
Backfill by hand	1.903	C.Y.		72.31	72.31
Compaction vibrating plate	1.903	C.Y.		9.74	9.74
TOTAL			1,155.91	1,030.89	2,186.80

G4020 210	Light Pole (Installed)	COST EACH		
		MAT.	INST.	TOTAL
0195	For alternate light pole systems:			
0200	Light pole, aluminum, 20' high, 1 arm bracket	1,150	1,025	2,175
0240	2 arm brackets	1,275	1,025	2,300
0280	3 arm brackets	1,400	1,075	2,475
0320	4 arm brackets	1,525	1,075	2,600
0360	30' high, 1 arm bracket	2,075	1,300	3,375
0400	2 arm brackets	2,200	1,300	3,500
0440	3 arm brackets	2,325	1,325	3,650
0480	4 arm brackets	2,450	1,325	3,775
0680	40' high, 1 arm bracket	2,550	1,750	4,300
0720	2 arm brackets	2,675	1,750	4,425
0760	3 arm brackets	2,800	1,775	4,575
0800	4 arm brackets	2,925	1,775	4,700
0840	Steel, 20' high, 1 arm bracket	1,400	1,075	2,475
0880	2 arm brackets	1,500	1,075	2,575
0920	3 arm brackets	1,525	1,125	2,650
0960	4 arm brackets	1,650	1,125	2,775
1000	30' high, 1 arm bracket	1,600	1,400	3,000
1040	2 arm brackets	1,725	1,400	3,125
1080	3 arm brackets	1,725	1,425	3,150
1120	4 arm brackets	1,850	1,425	3,275
1320	40' high, 1 arm bracket	2,100	1,875	3,975
1360	2 arm brackets	2,200	1,875	4,075
1400	3 arm brackets	2,225	1,900	4,125
1440	4 arm brackets	2,350	1,900	4,250
2700	Cut & patch to match existing construction, add, minimum	2%	3%	

G40 Site Electrical Utilities

G4020 Site Lighting

G4020 210	Light Pole (Installed)	COST EACH		
		MAT.	INST.	TOTAL
2800	Maximum	5%	9%	
2900	Equipment usage curtailment, add, minimum	1%	1%	
3000	Maximum	3%	10%	
3100	Material handling & storage limitation, add, minimum	1%	1%	
3200	Maximum	6%	7%	
3300	Protection of existing work, add, minimum	2%	2%	
3400	Maximum	5%	7%	
3500	Shift work requirements, add, minimum		5%	
3600	Maximum		30%	

Reference Section

All the reference information is in one section, making it easy to find what you need to know . . . and easy to use the book on a daily basis. This section is visually identified by a vertical gray bar on the page edges.

In this Reference Section, we've included Equipment Rental Costs, a listing of rental and operating costs; Crew Listings, a full listing of all crews and equipment, and their costs; Historical Cost Indexes for cost comparisons over time; City Cost Indexes and Location Factors for adjusting costs to the region you are in; Reference Tables, where you will find explanations, estimating information and procedures, or technical data; Change Orders, information on pricing changes to contract documents; Square Foot Costs that allow you to make a rough estimate for the overall cost of a project; and an explanation of all the Abbreviations in the book.

Table of Contents

Estimating Tips

- This section contains the average costs to rent and operate hundreds of pieces of construction equipment. This is useful information when estimating the time and material requirements of any particular operation in order to establish a unit or total cost. Equipment costs include not only rental, but also operating costs for equipment under normal use.

Rental Costs

- Equipment rental rates are obtained from industry sources throughout North America–contractors, suppliers, dealers, manufacturers, and distributors.
- Rental rates vary throughout the country, with larger cities generally having lower rates. Lease plans for new equipment are available for periods in excess of six months, with a percentage of payments applying toward purchase.
- Monthly rental rates vary from 2% to 5% of the purchase price of the equipment depending on the anticipated life of the equipment and its wearing parts.
- Weekly rental rates are about 1/3 the monthly rates, and daily rental rates are about 1/3 the weekly rate.

Operating Costs

- The operating costs include parts and labor for routine servicing, such as repair and replacement of pumps, filters and worn lines. Normal operating expendables, such as fuel, lubricants, tires and electricity (where applicable), are also included.
- Extraordinary operating expendables with highly variable wear patterns, such as diamond bits and blades, are excluded. These costs can be found as material costs in the Unit Price section.
- The hourly operating costs listed do not include the operator's wages.

Equipment Cost/Day

- Any power equipment required by a crew is shown in the Crew Listings with a daily cost.
- The daily cost of equipment needed by a crew is based on dividing the weekly rental rate by 5 (number of working days in the week), and then adding the hourly operating cost times 8 (the number of hours in a day). This "Equipment Cost/Day" is shown in the far right column of the Equipment Rental pages.
- If equipment is needed for only one or two days, it is best to develop your own cost by including components for daily rent and hourly operating cost. This is important when the listed Crew for a task does not contain the equipment needed, such as a crane for lifting mechanical heating/cooling equipment up onto a roof.

- If the quantity of work is less than the crew's Daily Output shown for a Unit Price line item that includes a bare unit equipment cost, it is recommended to estimate one day's rental cost and operating cost for equipment shown in the Crew Listing for that line item.

Mobilization/ Demobilization

- The cost to move construction equipment from an equipment yard or rental company to the jobsite and back again is not included in equipment rental costs listed in the Reference section, nor in the bare equipment cost of any Unit Price line item, nor in any equipment costs shown in the Crew listings.
- Mobilization (to the site) and demobilization (from the site) costs can be found in the Unit Price section.
- If a piece of equipment is already at the jobsite, it is not appropriate to utilize mobil./ demob. costs again in an estimate.

01 54 33 | Equipment Rental

			UNIT	HOURLY OPER. COST	RENT PER DAY	RENT PER WEEK	RENT PER MONTH	EQUIPMENT COST/DAY	
10	0010	**CONCRETE EQUIPMENT RENTAL** without operators	R015433 -10						10
	0200	Bucket, concrete lightweight, 1/2 C.Y.	Ea.	.65	20	60	180	17.20	
	0300	1 C.Y.		.70	24.50	73	219	20.20	
	0400	1-1/2 C.Y.		.90	33	99	297	27	
	0500	2 C.Y.		1.00	40	120	360	32	
	0580	8 C.Y.		5.50	262	785	2,350	201	
	0600	Cart, concrete, self propelled, operator walking, 10 C.F.		2.65	56.50	170	510	55.20	
	0700	Operator riding, 18 C.F.		4.30	86.50	260	780	86.40	
	0800	Conveyer for concrete, portable, gas, 16" wide, 26' long		10.00	123	370	1,100	154	
	0900	46' long		10.40	150	450	1,350	173.20	
	1000	56' long		10.50	158	475	1,425	179	
	1100	Core drill, electric, 2-1/2 H.P., 1" to 8" bit diameter		1.72	65.50	196	590	52.95	
	1150	11 HP, 8" to 18" cores		5.15	115	345	1,025	110.20	
	1200	Finisher, concrete floor, gas, riding trowel, 96" wide		9.95	145	435	1,300	166.60	
	1300	Gas, walk-behind, 3 blade, 36" trowel		1.65	19.65	59	177	25	
	1400	4 blade, 48" trowel		3.20	28	84	252	42.40	
	1500	Float, hand-operated (Bull float) 48" wide		.08	13.65	41	123	8.85	
	1570	Curb builder, 14 H.P., gas, single screw		11.85	243	730	2,200	240.80	
	1590	Double screw		12.55	285	855	2,575	271.40	
	1600	Grinder, concrete and terrazzo, electric, floor		2.45	103	310	930	81.60	
	1700	Wall grinder		1.23	51.50	155	465	40.85	
	1800	Mixer, powered, mortar and concrete, gas, 6 C.F., 18 H.P.		6.95	118	355	1,075	126.60	
	1900	10 C.F., 25 H.P.		8.65	143	430	1,300	155.20	
	2000	16 C.F.		8.95	167	500	1,500	171.60	
	2100	Concrete, stationary, tilt drum, 2 C.Y.		6.10	232	695	2,075	187.80	
	2120	Pump, concrete, truck mounted 4" line 80' boom		22.85	925	2,775	8,325	737.80	
	2140	5" line, 110' boom		29.75	1,225	3,695	11,100	977	
	2160	Mud jack, 50 C.F. per hr.		6.10	130	390	1,175	126.80	
	2180	225 C.F. per hr.		7.95	147	440	1,325	151.60	
	2190	Shotcrete pump rig, 12 CY/hr		12.60	228	685	2,050	237.80	
	2600	Saw, concrete, manual, gas, 18 H.P.		5.15	36.50	110	330	63.20	
	2650	Self-propelled, gas, 30 H.P.		10.25	98.50	295	885	141	
	2700	Vibrators, concrete, electric, 60 cycle, 2 H.P.		.46	8.35	25	75	8.70	
	2800	3 H.P.		.64	10.35	31	93	11.30	
	2900	Gas engine, 5 H.P.		1.55	14.35	43	129	21	
	3000	8 H.P.		2.10	15.35	46	138	26	
	3050	Vibrating screed, gas engine, 8 H.P.		3.32	73	219	655	70.35	
	3120	Concrete transit mixer, 6 x 4, 250 H.P., 8 C.Y., rear discharge		47.15	575	1,720	5,150	721.20	
	3200	Front discharge		55.50	710	2,125	6,375	869	
	3300	6 x 6, 285 H.P., 12 C.Y., rear discharge		54.60	665	2,000	6,000	836.80	
	3400	Front discharge	▼	57.05	715	2,150	6,450	886.40	
20	0010	**EARTHWORK EQUIPMENT RENTAL** without operators	R015433 -10						20
	0040	Aggregate spreader, push type 8' to 12' wide	Ea.	2.45	25.50	77	231	35	
	0045	Tailgate type, 8' wide		2.35	33.50	100	300	38.80	
	0055	Earth auger, truck-mounted, for fence & sign posts, utility poles		13.50	505	1,520	4,550	412	
	0060	For borings and monitoring wells		38.20	670	2,005	6,025	706.60	
	0070	Portable, trailer mounted		2.65	26	78	234	36.80	
	0075	Truck-mounted, for caissons, water wells		85.35	3,075	9,250	27,800	2,533	
	0080	Horizontal boring machine, 12" to 36" diameter, 45 H.P.		20.30	202	605	1,825	283.40	
	0090	12" to 48" diameter, 65 H.P.		28.65	360	1,075	3,225	444.20	
	0095	Auger, for fence posts, gas engine, hand held		.50	5.65	17	51	7.40	
	0100	Excavator, diesel hydraulic, crawler mounted, 1/2 C.Y. cap.		20.95	380	1,135	3,400	394.60	
	0120	5/8 C.Y. capacity		25.40	525	1,570	4,700	517.20	
	0140	3/4 C.Y. capacity		29.70	600	1,805	5,425	598.60	
	0150	1 C.Y. capacity		36.90	625	1,870	5,600	669.20	
	0200	1-1/2 C.Y. capacity		45.10	840	2,525	7,575	865.80	
	0300	2 C.Y. capacity	▼	61.75	1,150	3,440	10,300	1,182	

01 54 33 | Equipment Rental

		UNIT	HOURLY OPER. COST	RENT PER DAY	RENT PER WEEK	RENT PER MONTH	EQUIPMENT COST/DAY	
20	0320	2-1/2 C.Y. capacity	Ea.	83.30	1,600	4,765	14,300	1,619	20
	0325	3-1/2 C.Y. capacity		111.30	2,125	6,360	19,100	2,162	
	0330	4-1/2 C.Y. capacity		134.65	2,600	7,790	23,400	2,635	
	0335	6 C.Y. capacity		168.45	2,875	8,660	26,000	3,080	
	0340	7 C.Y. capacity		173.20	3,125	9,350	28,100	3,256	
	0342	Excavator attachments, bucket thumbs		2.70	215	645	1,925	150.60	
	0345	Grapples		2.45	188	565	1,700	132.60	
	0347	Hydraulic hammer for boom mounting, 5000 ft-lb		12.10	405	1,220	3,650	340.80	
	0349	11,000 ft-lb		20.40	730	2,190	6,575	601.20	
	0350	Gradall type, truck mounted, 3 ton @ 15' radius, 5/8 C.Y.		49.60	935	2,810	8,425	958.80	
	0370	1 C.Y. capacity		55.10	1,150	3,445	10,300	1,130	
	0400	Backhoe-loader, 40 to 45 H.P., 5/8 C.Y. capacity		11.90	227	680	2,050	231.20	
	0450	45 H.P. to 60 H.P., 3/4 C.Y. capacity		15.85	278	835	2,500	293.80	
	0460	80 H.P., 1-1/4 C.Y. capacity		19.00	320	965	2,900	345	
	0470	112 H.P., 1-1/2 C.Y. capacity		25.45	465	1,400	4,200	483.60	
	0482	Backhoe-loader attachment, compactor, 20,000 lb		4.70	127	380	1,150	113.60	
	0485	Hydraulic hammer, 750 ft-lbs		2.80	90	270	810	76.40	
	0486	Hydraulic hammer, 1200 ft-lbs		5.25	185	555	1,675	153	
	0500	Brush chipper, gas engine, 6" cutter head, 35 H.P.		8.95	108	325	975	136.60	
	0550	12" cutter head, 130 H.P.		14.95	177	530	1,600	225.60	
	0600	15" cutter head, 165 H.P.		22.45	217	650	1,950	309.60	
	0750	Bucket, clamshell, general purpose, 3/8 C.Y.		1.10	36.50	110	330	30.80	
	0800	1/2 C.Y.		1.20	45	135	405	36.60	
	0850	3/4 C.Y.		1.35	55	165	495	43.80	
	0900	1 C.Y.		1.40	58.50	175	525	46.20	
	0950	1-1/2 C.Y.		2.25	80	240	720	66	
	1000	2 C.Y.		2.40	88.50	265	795	72.20	
	1010	Bucket, dragline, medium duty, 1/2 C.Y.		.65	23.50	70	210	19.20	
	1020	3/4 C.Y.		.65	24.50	74	222	20	
	1030	1 C.Y.		.70	26.50	80	240	21.60	
	1040	1-1/2 C.Y.		1.10	40	120	360	32.80	
	1050	2 C.Y.		1.15	45	135	405	36.20	
	1070	3 C.Y.		1.65	55	165	495	46.20	
	1200	Compactor, manually guided 2-drum vibratory smooth roller, 7.5 H.P.		5.65	167	500	1,500	145.20	
	1250	Rammer/tamper, gas, 8"		2.30	41.50	125	375	43.40	
	1260	15"		2.55	48.50	145	435	49.40	
	1300	Vibratory plate, gas, 18" plate, 3000 lb. blow		2.25	23	69	207	31.80	
	1350	21" plate, 5000 lb. blow		2.80	28	84	252	39.20	
	1370	Curb builder/extruder, 14 H.P., gas, single screw		11.85	243	730	2,200	240.80	
	1390	Double screw		12.55	285	855	2,575	271.40	
	1500	Disc harrow attachment, for tractor		.39	65.50	197	590	42.50	
	1810	Feller buncher, shearing & accumulating trees, 100 H.P.		28.55	555	1,665	5,000	561.40	
	1860	Grader, self-propelled, 25,000 lb.		24.85	450	1,345	4,025	467.80	
	1910	30,000 lb.		29.65	520	1,565	4,700	550.20	
	1920	40,000 lb.		45.50	995	2,985	8,950	961	
	1930	55,000 lb.		59.10	1,400	4,215	12,600	1,316	
	1950	Hammer, pavement breaker, self-propelled, diesel, 1000 to 1250 lb		23.55	340	1,025	3,075	393.40	
	2000	1300 to 1500 lb.		35.33	685	2,050	6,150	692.65	
	2050	Pile driving hammer, steam or air, 4150 ft.-lb. @ 225 BPM		9.35	470	1,405	4,225	355.80	
	2100	8750 ft.-lb. @ 145 BPM		11.35	650	1,955	5,875	481.80	
	2150	15,000 ft.-lb. @ 60 BPM		12.95	790	2,375	7,125	578.60	
	2200	24,450 ft.-lb. @ 111 BPM		13.95	885	2,655	7,975	642.60	
	2250	Leads, 60' high for pile driving hammers up to 20,000 ft.-lb.		2.85	78	234	700	69.60	
	2300	90' high for hammers over 20,000 ft.-lb.		4.35	140	420	1,250	118.80	
	2350	Diesel type hammer, 22,400 ft.-lb.		25.75	640	1,920	5,750	590	
	2400	41,300 ft.-lb.		32.70	695	2,080	6,250	677.60	
	2450	141,000 ft.-lb.		49.25	1,200	3,570	10,700	1,108	
	2500	Vib. elec. hammer/extractor, 200 KW diesel generator, 34 H.P.		41.05	665	2,000	6,000	728.40	

01 54 33 | Equipment Rental

		UNIT	HOURLY OPER. COST	RENT PER DAY	RENT PER WEEK	RENT PER MONTH	EQUIPMENT COST/DAY	
20	2550	80 H.P.	Ea.	73.75	980	2,940	8,825	1,178	20
	2600	150 H.P.		139.15	1,900	5,720	17,200	2,257	
	2700	Extractor, steam or air, 700 ft.-lb.		17.15	510	1,530	4,600	443.20	
	2750	1000 ft.-lb.		19.55	625	1,875	5,625	531.40	
	2800	Log chipper, up to 22" diam, 600 H.P.		40.05	445	1,340	4,025	588.40	
	2850	Logger, for skidding & stacking logs, 150 H.P.		44.95	895	2,690	8,075	897.60	
	2860	Mulcher, diesel powered, trailer mounted		17.60	218	655	1,975	271.80	
	2900	Rake, spring tooth, with tractor		11.34	278	835	2,500	257.70	
	3000	Roller, vibratory, tandem, smooth drum, 20 H.P.		7.35	132	395	1,175	137.80	
	3050	35 H.P.		9.50	262	785	2,350	233	
	3100	Towed type vibratory compactor, smooth drum, 50 H.P.		22.75	380	1,145	3,425	411	
	3150	Sheepsfoot, 50 H.P.		25.05	440	1,325	3,975	465.40	
	3170	Landfill compactor, 220 HP		69.10	1,450	4,315	12,900	1,416	
	3200	Pneumatic tire roller, 80 H.P.		13.10	340	1,015	3,050	307.80	
	3250	120 H.P.		19.90	600	1,805	5,425	520.20	
	3300	Sheepsfoot vibratory roller, 240 H.P.		56.65	1,150	3,445	10,300	1,142	
	3320	340 H.P.		75.60	1,425	4,240	12,700	1,453	
	3350	Smooth drum vibratory roller, 75 H.P.		19.70	520	1,560	4,675	469.60	
	3400	125 H.P.		25.70	650	1,945	5,825	594.60	
	3410	Rotary mower, brush, 60", with tractor		15.30	252	755	2,275	273.40	
	3420	Rototiller, walk-behind, gas, 5 H.P.		2.43	62.50	188	565	57.05	
	3422	8 H.P.		3.89	100	300	900	91.10	
	3440	Scrapers, towed type, 7 CY capacity		4.90	105	315	945	102.20	
	3450	10 C.Y. capacity		6.20	173	520	1,550	153.60	
	3500	15 C.Y. capacity		6.85	190	570	1,700	168.80	
	3525	Self-propelled, single engine, 14 C.Y. capacity		83.90	1,400	4,190	12,600	1,509	
	3550	Dual engine, 21 C.Y. capacity		119.25	1,625	4,850	14,600	1,924	
	3600	31 C.Y. capacity		165.60	2,600	7,765	23,300	2,878	
	3640	44 C.Y. capacity		195.70	3,150	9,430	28,300	3,452	
	3650	Elevating type, single engine, 11 CY capacity		56.70	1,075	3,205	9,625	1,095	
	3700	22 C.Y. capacity		105.60	2,125	6,410	19,200	2,127	
	3710	Screening plant 110 H.P. w/ 5' x 10' screen		30.75	410	1,230	3,700	492	
	3720	5' x 16' screen		32.90	515	1,540	4,625	571.20	
	3850	Shovel, crawler-mounted, front-loading, 7 C.Y. capacity		171.90	2,850	8,575	25,700	3,090	
	3855	12 C.Y. capacity		244.40	3,650	10,965	32,900	4,148	
	3860	Shovel/backhoe bucket, 1/2 C.Y.		2.10	60	180	540	52.80	
	3870	3/4 C.Y.		2.15	68.50	205	615	58.20	
	3880	1 C.Y.		2.25	76.50	230	690	64	
	3890	1-1/2 C.Y.		2.40	91.50	275	825	74.20	
	3910	3 C.Y.		2.70	127	380	1,150	97.60	
	3950	Stump chipper, 18" deep, 30 H.P.		8.99	215	646	1,950	201.10	
	4110	Tractor, crawler, with bulldozer, torque converter, diesel 80 H.P.		21.90	375	1,120	3,350	399.20	
	4150	105 H.P.		31.70	580	1,735	5,200	600.60	
	4200	140 H.P.		36.50	650	1,950	5,850	682	
	4260	200 H.P.		55.85	1,050	3,175	9,525	1,082	
	4310	300 H.P.		71.45	1,425	4,255	12,800	1,423	
	4360	410 H.P.		95.30	1,800	5,435	16,300	1,849	
	4370	500 H.P.		127.20	2,400	7,230	21,700	2,464	
	4380	700 H.P.		204.15	4,325	12,995	39,000	4,232	
	4400	Loader, crawler, torque conv., diesel, 1-1/2 C.Y., 80 H.P.		20.00	345	1,035	3,100	367	
	4450	1-1/2 to 1-3/4 C.Y., 95 H.P.		23.75	455	1,365	4,100	463	
	4510	1-3/4 to 2-1/4 C.Y., 130 H.P.		35.05	810	2,435	7,300	767.40	
	4530	2-1/2 to 3-1/4 C.Y., 190 H.P.		47.75	1,025	3,040	9,125	990	
	4560	3-1/2 to 5 C.Y., 275 H.P.		68.30	1,500	4,520	13,600	1,450	
	4610	Front end loader, 4WD, articulated frame, 1 to 1-1/4 C.Y., 70 H.P.		14.25	208	625	1,875	239	
	4620	1-1/2 to 1-3/4 C.Y., 95 H.P.		18.90	295	885	2,650	328.20	
	4650	1-3/4 to 2 C.Y., 130 H.P.		21.60	345	1,035	3,100	379.80	
	4710	2-1/2 to 3-1/2 C.Y., 145 H.P.		22.85	355	1,065	3,200	395.80	

01 54 33 | Equipment Rental

		UNIT	HOURLY OPER. COST	RENT PER DAY	RENT PER WEEK	RENT PER MONTH	EQUIPMENT COST/DAY	
20								**20**
4730	3 to 4-1/2 C.Y., 185 H.P.	Ea.	29.75	505	1,515	4,550	541	
4760	5-1/4 to 5-3/4 C.Y., 270 H.P.		48.05	785	2,350	7,050	854.40	
4810	7 to 9 C.Y., 475 H.P.		81.85	1,400	4,210	12,600	1,497	
4870	9 - 11 C.Y., 620 H.P.		112.20	2,200	6,635	19,900	2,225	
4880	Skid steer loader, wheeled, 10 C.F., 30 H.P. gas		8.15	140	420	1,250	149.20	
4890	1 C.Y., 78 H.P., diesel		14.95	233	700	2,100	259.60	
4892	Skid-steer attachment, auger		.54	89.50	268	805	57.90	
4893	Backhoe		.67	112	336	1,000	72.55	
4894	Broom		.70	117	352	1,050	76	
4895	Forks		.24	39.50	118	355	25.50	
4896	Grapple		.57	95.50	287	860	61.95	
4897	Concrete hammer		1.06	176	528	1,575	114.10	
4898	Tree spade		.80	133	398	1,200	86	
4899	Trencher		.77	128	385	1,150	83.15	
4900	Trencher, chain, boom type, gas, operator walking, 12 H.P.		3.95	46.50	140	420	59.60	
4910	Operator riding, 40 H.P.		15.50	295	885	2,650	301	
5000	Wheel type, diesel, 4' deep, 12" wide		65.30	810	2,435	7,300	1,009	
5100	6' deep, 20" wide		77.15	1,900	5,680	17,000	1,753	
5150	Chain type, diesel, 5' deep, 8" wide		28.90	580	1,745	5,225	580.20	
5200	Diesel, 8' deep, 16" wide		74.70	1,975	5,895	17,700	1,777	
5210	Tree spade, self-propelled		16.78	267	800	2,400	294.25	
5250	Truck, dump, 2-axle, 12 ton, 8 CY payload, 220 H.P.		24.55	217	650	1,950	326.40	
5300	Three axle dump, 16 ton, 12 CY payload, 400 H.P.		43.25	310	935	2,800	533	
5350	Dump trailer only, rear dump, 16-1/2 C.Y.		4.70	130	390	1,175	115.60	
5400	20 C.Y.		5.15	148	445	1,325	130.20	
5450	Flatbed, single axle, 1-1/2 ton rating		19.30	66.50	200	600	194.40	
5500	3 ton rating		23.20	95	285	855	242.60	
5550	Off highway rear dump, 25 ton capacity		59.95	1,325	4,000	12,000	1,280	
5600	35 ton capacity		55.80	1,175	3,490	10,500	1,144	
5610	50 ton capacity		76.40	1,625	4,855	14,600	1,582	
5620	65 ton capacity		78.80	1,600	4,820	14,500	1,594	
5630	100 ton capacity		101.35	2,050	6,180	18,500	2,047	
6000	Vibratory plow, 25 H.P., walking		6.65	61.50	185	555	90.20	
40								**40**
0010	**GENERAL EQUIPMENT RENTAL** without operators	R015433 -10 Ea.	2.70	56.50	170	510	55.60	
0150	Aerial lift, scissor type, to 15' high, 1000 lb. cap., electric		3.15	80	240	720	73.20	
0160	To 25' high, 2000 lb. capacity		17.10	320	960	2,875	328.80	
0170	Telescoping boom to 40' high, 500 lb. capacity, gas		18.10	365	1,100	3,300	364.80	
0180	To 45' high, 500 lb. capacity		20.45	505	1,510	4,525	465.60	
0190	To 60' high, 600 lb. capacity		.47	12.65	38	114	11.35	
0195	Air compressor, portable, 6.5 CFM, electric		.96	19	57	171	19.10	
0196	Gasoline		11.70	48.50	145	435	122.60	
0200	Towed type, gas engine, 60 C.F.M.		13.60	50	150	450	138.80	
0300	160 C.F.M.		12.80	100	300	900	162.40	
0400	Diesel engine, rotary screw, 250 C.F.M.		17.15	123	370	1,100	211.20	
0500	365 C.F.M.		21.75	153	460	1,375	266	
0550	450 C.F.M.		38.05	217	650	1,950	434.40	
0600	600 C.F.M.		38.20	223	670	2,000	439.60	
0700	750 C.F.M.		3%	5%	5%	5%		
0800	For silenced models, small sizes, add to rent		5%	7%	7%	7%		
0900	Large sizes, add to rent		.40	8.35	25	75	8.20	
0930	Air tools, breaker, pavement, 60 lb.	Ea.	.40	8.35	25	75	8.20	
0940	80 lb.		.50	16	48	144	13.60	
0950	Drills, hand (jackhammer) 65 lb.		49.20	760	2,285	6,850	850.60	
0960	Track or wagon, swing boom, 4" drifter		60.00	825	2,470	7,400	974	
0970	5" drifter		82.25	1,250	3,780	11,300	1,414	
0975	Track mounted quarry drill, 6" diameter drill		.99	19	57	171	19.30	
0980	Dust control per drill							

01 54 33 | Equipment Rental

		UNIT	HOURLY OPER. COST	RENT PER DAY	RENT PER WEEK	RENT PER MONTH	EQUIPMENT COST/DAY
40								**40**
0990	Hammer, chipping, 12 lb.	Ea.	.45	22.50	67	201	17
1000	Hose, air with couplings, 50' long, 3/4" diameter		.03	4.33	13	39	2.85
1100	1" diameter		.04	6.35	19	57	4.10
1200	1-1/2" diameter		.06	9.65	29	87	6.30
1300	2" diameter		.13	21.50	65	195	14.05
1400	2-1/2" diameter		.12	19.65	59	177	12.75
1410	3" diameter		.15	25	75	225	16.20
1450	Drill, steel, 7/8" x 2'		.05	8.35	25	75	5.40
1460	7/8" x 6'		.06	9.35	28	84	6.10
1520	Moil points		.02	3.67	11	33	2.35
1525	Pneumatic nailer w/accessories		.45	30	90	270	21.60
1530	Sheeting driver for 60 lb. breaker		.04	6	18	54	3.90
1540	For 90 lb. breaker		.12	8	24	72	5.75
1550	Spade, 25 lb.		.40	6	18	54	6.80
1560	Tamper, single, 35 lb.		.50	33.50	100	300	24
1570	Triple, 140 lb.		.75	50	150	450	36
1580	Wrenches, impact, air powered, up to 3/4" bolt		.30	8	24	72	7.20
1590	Up to 1-1/4" bolt		.45	21	63	189	16.20
1600	Barricades, barrels, reflectorized, 1 to 50 barrels		.03	4.60	13.80	41.50	3
1610	100 to 200 barrels		.02	3.53	10.60	32	2.30
1620	Barrels with flashers, 1 to 50 barrels		.03	5.25	15.80	47.50	3.40
1630	100 to 200 barrels		.03	4.20	12.60	38	2.75
1640	Barrels with steady burn type C lights		.04	7	21	63	4.50
1650	Illuminated board, trailer mounted, with generator		.70	120	360	1,075	77.60
1670	Portable barricade, stock, with flashers, 1 to 6 units		.03	5.25	15.80	47.50	3.40
1680	25 to 50 units		.03	4.90	14.70	44	3.20
1690	Butt fusion machine, electric		20.85	55	165	495	199.80
1695	Electro fusion machine		15.35	40	120	360	146.80
1700	Carts, brick, hand powered, 1000 lb. capacity		.38	63	189	565	40.85
1800	Gas engine, 1500 lb., 7-1/2' lift		4.73	119	357	1,075	109.25
1822	Dehumidifier, medium, 6 lb/hr, 150 CFM		.85	51.50	155	465	37.80
1824	Large, 18 lb/hr, 600 CFM		1.64	100	300	900	73.10
1830	Distributor, asphalt, trailer mtd, 2000 gal., 38 H.P. diesel		9.15	340	1,020	3,050	277.20
1840	3000 gal., 38 H.P. diesel		10.40	370	1,105	3,325	304.20
1850	Drill, rotary hammer, electric, 1-1/2" diameter		.70	22	66	198	18.80
1860	Carbide bit for above		.05	9	27	81	5.80
1865	Rotary, crawler, 250 H.P.		122.45	2,075	6,240	18,700	2,228
1870	Emulsion sprayer, 65 gal., 5 H.P. gas engine		3.20	99	297	890	85
1880	200 gal., 5 H.P. engine		6.85	165	495	1,475	153.80
1930	Floodlight, mercury vapor, or quartz, on tripod, 1000 watt		.34	13	39	117	10.50
1940	2000 watt		.59	24.50	73	219	19.30
1950	Floodlights, trailer mounted with generator, 1 - 300 watt light		3.00	70	210	630	66
1960	2 - 1000 watt lights		4.15	107	320	960	97.20
2000	4 - 300 watt lights		3.65	86.50	260	780	81.20
2005	Foam spray rig, incl. box trailer, compressor, generator, proportioner		24.90	480	1,445	4,325	488.20
2020	Forklift, straight mast, 12' lift, 5000 lb., 2 wheel drive, gas		22.00	198	595	1,775	295
2040	21' lift, 5000 lb., 4 wheel drive, diesel		16.65	247	740	2,225	281.20
2050	For rough terrain, 42' lift, 35' reach, 9000 lb., 110 HP		22.50	445	1,330	4,000	446
2060	For plant, 4 T. capacity, 80 H.P., 2 wheel drive, gas		13.55	105	315	945	171.40
2080	10 T. capacity, 120 H.P., 2 wheel drive, diesel		18.65	183	550	1,650	259.20
2100	Generator, electric, gas engine, 1.5 KW to 3 KW		3.10	10.35	31	93	31
2200	5 KW		3.95	14.35	43	129	40.20
2300	10 KW		7.45	33.50	100	300	79.60
2400	25 KW		9.10	65	195	585	111.80
2500	Diesel engine, 20 KW		9.25	73.50	220	660	118
2600	50 KW		16.60	98.50	295	885	191.80
2700	100 KW		31.90	120	360	1,075	327.20
2800	250 KW		63.05	232	695	2,075	643.40

01 54 33 | Equipment Rental

		UNIT	HOURLY OPER. COST	RENT PER DAY	RENT PER WEEK	RENT PER MONTH	EQUIPMENT COST/DAY	
2850	Hammer, hydraulic, for mounting on boom, to 500 ft.-lb.	Ea.	2.15	68.50	205	615	58.20	40
2860	1000 ft.-lb.		3.65	112	335	1,000	96.20	
2900	Heaters, space, oil or electric, 50 MBH		2.50	6.65	20	60	24	
3000	100 MBH		4.52	9	27	81	41.55	
3100	300 MBH		14.52	36.50	110	330	138.15	
3150	500 MBH		29.33	45	135	405	261.65	
3200	Hose, water, suction with coupling, 20' long, 2" diameter		.02	3	9	27	1.95	
3210	3" diameter		.03	4.67	14	42	3.05	
3220	4" diameter		.03	5.35	16	48	3.45	
3230	6" diameter		.11	18.35	55	165	11.90	
3240	8" diameter		.20	33.50	100	300	21.60	
3250	Discharge hose with coupling, 50' long, 2" diameter		.01	1.67	5	15	1.10	
3260	3" diameter		.02	2.67	8	24	1.75	
3270	4" diameter		.02	3.67	11	33	2.35	
3280	6" diameter		.06	9.65	29	87	6.30	
3290	8" diameter		.20	33.50	100	300	21.60	
3295	Insulation blower		.22	6	18	54	5.35	
3300	Ladders, extension type, 16' to 36' long		.14	23	69	207	14.90	
3400	40' to 60' long		.20	33.50	100	300	21.60	
3405	Lance for cutting concrete		2.97	104	313	940	86.35	
3407	Lawn mower, rotary, 22", 5HP		2.48	65	195	585	58.85	
3408	48" self propelled		4.48	128	385	1,150	112.85	
3410	Level, laser type, for pipe and sewer leveling		1.46	97.50	292	875	70.10	
3430	Electronic		.75	50	150	450	36	
3440	Laser type, rotating beam for grade control		1.17	77.50	233	700	55.95	
3460	Builders level with tripod and rod		.08	14	42	126	9.05	
3500	Light towers, towable, with diesel generator, 2000 watt		3.65	86.50	260	780	81.20	
3600	4000 watt		4.15	107	320	960	97.20	
3700	Mixer, powered, plaster and mortar, 6 C.F., 7 H.P.		2.10	19.65	59	177	28.60	
3800	10 C.F., 9 H.P.		2.25	31.50	95	285	37	
3850	Nailer, pneumatic		.45	30	90	270	21.60	
3900	Paint sprayers complete, 8 CFM		.67	44.50	134	400	32.15	
4000	17 CFM		1.17	77.50	233	700	55.95	
4020	Pavers, bituminous, rubber tires, 8' wide, 50 H.P., diesel		35.70	1,050	3,155	9,475	916.60	
4030	10' wide, 150 H.P.		75.65	1,700	5,095	15,300	1,624	
4050	Crawler, 8' wide, 100 H.P., diesel		75.65	1,775	5,355	16,100	1,676	
4060	10' wide, 150 H.P.		84.80	2,175	6,505	19,500	1,979	
4070	Concrete paver, 12' to 24' wide, 250 H.P.		81.00	1,575	4,755	14,300	1,599	
4080	Placer-spreader-trimmer, 24' wide, 300 H.P.		116.90	2,600	7,825	23,500	2,500	
4100	Pump, centrifugal gas pump, 1-1/2" diam., 65 GPM		3.35	45	135	405	53.80	
4200	2" diameter, 130 GPM		4.55	50	150	450	66.40	
4300	3" diameter, 250 GPM		4.80	51.50	155	465	69.40	
4400	6" diameter, 1500 GPM		25.55	175	525	1,575	309.40	
4500	Submersible electric pump, 1-1/4" diameter, 55 GPM		.39	15.65	47	141	12.50	
4600	1-1/2" diameter, 83 GPM		.42	18	54	162	14.15	
4700	2" diameter, 120 GPM		1.20	22.50	67	201	23	
4800	3" diameter, 300 GPM		2.05	41.50	125	375	41.40	
4900	4" diameter, 560 GPM		8.90	163	490	1,475	169.20	
5000	6" diameter, 1590 GPM		13.05	222	665	2,000	237.40	
5100	Diaphragm pump, gas, single, 1-1/2" diameter		1.24	45.50	136	410	37.10	
5200	2" diameter		3.65	56.50	170	510	63.20	
5300	3" diameter		3.65	56.50	170	510	63.20	
5400	Double, 4" diameter		4.90	78.50	235	705	86.20	
5450	Pressure Washer 5 Ga.Pm,3000 PSI		4.00	35	105	315	53	
5500	Trash pump, self-priming, gas, 2" diameter		3.85	20	60	180	42.80	
5600	Diesel, 4" diameter		10.25	58.50	175	525	117	
5650	Diesel, 6" diameter		34.00	130	390	1,175	350	
5655	Grout Pump		13.80	93.50	280	840	166.40	

01 54 33 | Equipment Rental

		UNIT	HOURLY OPER. COST	RENT PER DAY	RENT PER WEEK	RENT PER MONTH	EQUIPMENT COST/DAY	
40	5700	Salamanders, L.P. gas fired, 100,000 BTU	Ea.	4.56	15.35	46	138	45.70	**40**
	5705	50,000 BTU		3.40	8.65	26	78	32.40	
	5720	Sandblaster, portable, open top, 3 C.F. capacity		.55	27	81	243	20.60	
	5730	6 C.F. capacity		.85	40	120	360	30.80	
	5740	Accessories for above		.12	19.65	59	177	12.75	
	5750	Sander, floor		.81	18.65	56	168	17.70	
	5760	Edger		.75	25.50	77	231	21.40	
	5800	Saw, chain, gas engine, 18" long		1.90	17.65	53	159	25.80	
	5900	Hydraulic powered, 36" long		.60	55	165	495	37.80	
	5950	60" long		.65	56.50	170	510	39.20	
	6000	Masonry, table mounted, 14" diameter, 5 H.P.		1.25	53	159	475	41.80	
	6050	Portable cut-off, 8 H.P.		2.05	28	84	252	33.20	
	6100	Circular, hand held, electric, 7-1/4" diameter		.19	4	12	36	3.90	
	6200	12" diameter		.26	7.35	22	66	6.50	
	6250	Wall saw, w/hydraulic power, 10 H.P		7.50	60	180	540	96	
	6275	Shot blaster, walk behind, 20" wide		4.60	300	905	2,725	217.80	
	6280	Sidewalk broom, walk-behind		2.31	57	171	515	52.70	
	6300	Steam cleaner, 100 gallons per hour		2.90	46.50	140	420	51.20	
	6310	200 gallons per hour		4.15	55	165	495	66.20	
	6340	Tar Kettle/Pot, 400 gallon		6.15	80	240	720	97.20	
	6350	Torch, cutting, acetylene-oxygen, 150' hose		.50	21.50	65	195	17	
	6360	Hourly operating cost includes tips and gas		9.45				75.60	
	6410	Toilet, portable chemical		.11	19	57	171	12.30	
	6420	Recycle flush type		.14	23	69	207	14.90	
	6430	Toilet, fresh water flush, garden hose,		.16	26.50	79	237	17.10	
	6440	Hoisted, non-flush, for high rise		.14	22.50	68	204	14.70	
	6450	Toilet, trailers, minimum		.24	39.50	118	355	25.50	
	6460	Maximum		.72	119	358	1,075	77.35	
	6465	Tractor, farm with attachment		14.40	262	785	2,350	272.20	
	6500	Trailers, platform, flush deck, 2 axle, 25 ton capacity		4.75	107	320	960	102	
	6600	40 ton capacity		6.20	150	450	1,350	139.60	
	6700	3 axle, 50 ton capacity		6.70	165	495	1,475	152.60	
	6800	75 ton capacity		8.35	218	655	1,975	197.80	
	6810	Trailer mounted cable reel for H.V. line work		4.84	231	692	2,075	177.10	
	6820	Trailer mounted cable tensioning rig		9.59	455	1,370	4,100	350.70	
	6830	Cable pulling rig		69.62	2,575	7,710	23,100	2,099	
	6900	Water tank, engine driven discharge, 5000 gallons		6.25	143	430	1,300	136	
	6925	10,000 gallons		8.50	202	605	1,825	189	
	6950	Water truck, off highway, 6000 gallons		66.95	775	2,320	6,950	999.60	
	7010	Tram car for H.V. line work, powered, 2 conductor		6.48	125	375	1,125	126.85	
	7020	Transit (builder's level) with tripod		.08	14	42	126	9.05	
	7030	Trench box, 3000 lbs. 6'x8'		.56	93	279	835	60.30	
	7040	7200 lbs. 6'x20'		1.05	175	525	1,575	113.40	
	7050	8000 lbs., 8' x 16'		.95	158	475	1,425	102.60	
	7060	9500 lbs., 8'x20'		1.16	194	581	1,750	125.50	
	7065	11,000 lbs., 8'x24'		1.27	212	637	1,900	137.55	
	7070	12,000 lbs., 10' x 20'		1.71	285	855	2,575	184.70	
	7100	Truck, pickup, 3/4 ton, 2 wheel drive		10.35	56.50	170	510	116.80	
	7200	4 wheel drive		10.65	71.50	215	645	128.20	
	7250	Crew carrier, 9 passenger		14.70	86.50	260	780	169.60	
	7290	Flat bed truck, 20,000 G.V.W.		15.55	122	365	1,100	197.40	
	7300	Tractor, 4 x 2, 220 H.P.		21.80	190	570	1,700	288.40	
	7410	330 H.P.		32.15	262	785	2,350	414.20	
	7500	6 x 4, 380 H.P.		36.85	305	920	2,750	478.80	
	7600	450 H.P.		44.90	370	1,110	3,325	581.20	
	7620	Vacuum truck, hazardous material, 2500 gallon		10.80	310	925	2,775	271.40	
	7625	5,000 gallon		16.90	435	1,300	3,900	395.20	
	7640	Tractor, with A frame, boom and winch, 225 H.P.		24.35	267	800	2,400	354.80	

01 54 33 | Equipment Rental

		UNIT	HOURLY OPER. COST	RENT PER DAY	RENT PER WEEK	RENT PER MONTH	EQUIPMENT COST/DAY	
40	7650	Vacuum, H.E.P.A., 16 gal., wet/dry	Ea.	.82	18	54	162	17.35	**40**
	7655	55 gal, wet/dry		.83	27	81	243	22.85	
	7660	Water tank, portable		.17	28.50	85.50	257	18.45	
	7690	Sewer/catch basin vacuum, 14 CY, 1500 Gallon		21.45	650	1,950	5,850	561.60	
	7700	Welder, electric, 200 amp		3.63	17.65	53	159	39.65	
	7800	300 amp		5.36	21.50	64	192	55.70	
	7900	Gas engine, 200 amp		12.85	25.50	77	231	118.20	
	8000	300 amp		14.70	27.50	83	249	134.20	
	8100	Wheelbarrow, any size		.07	11.65	35	105	7.55	
	8200	Wrecking ball, 4000 lb.		2.10	73.50	220	660	60.80	
50	0010	**HIGHWAY EQUIPMENT RENTAL** without operators R015433 -10							**50**
	0050	Asphalt batch plant, portable drum mixer, 100 ton/hr.	Ea.	63.70	1,450	4,330	13,000	1,376	
	0060	200 ton/hr.		71.10	1,525	4,565	13,700	1,482	
	0070	300 ton/hr.		82.55	1,800	5,395	16,200	1,739	
	0100	Backhoe attachment, long stick, up to 185 HP, 10.5' long		.32	21.50	64	192	15.35	
	0140	Up to 250 HP, 12' long		.35	23	69	207	16.60	
	0180	Over 250 HP, 15' long		.47	31	93	279	22.35	
	0200	Special dipper arm, up to 100 HP, 32' long		.96	63.50	191	575	45.90	
	0240	Over 100 HP, 33' long		1.19	79.50	238	715	57.10	
	0280	Catch basin/sewer cleaning truck, 3 ton, 9 CY, 1000 Gal		34.00	405	1,210	3,625	514	
	0300	Concrete batch plant, portable, electric, 200 CY/Hr		17.20	510	1,530	4,600	443.60	
	0520	Grader/dozer attachment, ripper/scarifier, rear mounted, up to 135 HP		3.10	65	195	585	63.80	
	0540	Up to 180 HP		3.70	83.50	250	750	79.60	
	0580	Up to 250 HP		4.05	95	285	855	89.40	
	0700	Pvmt. removal bucket, for hyd. excavator, up to 90 HP		1.65	48.50	145	435	42.20	
	0740	Up to 200 HP		1.85	70	210	630	56.80	
	0780	Over 200 HP		1.95	83.50	250	750	65.60	
	0900	Aggregate spreader, self-propelled, 187 HP		54.90	695	2,080	6,250	855.20	
	1000	Chemical spreader, 3 C.Y.		2.75	43.50	130	390	48	
	1900	Hammermill, traveling, 250 HP		92.70	1,850	5,540	16,600	1,850	
	2000	Horizontal borer, 3" diam, 13 HP gas driven		5.45	56.50	170	510	77.60	
	2150	Horizontal directional drill, 20,000 lb. thrust, 78 H.P. diesel		26.15	700	2,095	6,275	628.20	
	2160	30,000 lb. thrust, 115 H.P. diesel		32.45	1,050	3,185	9,550	896.60	
	2170	50,000 lb. thrust, 170 H.P. diesel		45.85	1,350	4,080	12,200	1,183	
	2190	Mud trailer for HDD, 1500 gallon, 175 H.P., gas		24.40	147	440	1,325	283.20	
	2200	Hydromulchers, gas power, 3000 gal., for truck mounting		17.00	258	775	2,325	291	
	2400	Joint & crack cleaner, walk behind, 25 HP		3.10	50	150	450	54.80	
	2500	Filler, trailer mounted, 400 gal., 20 HP		7.90	218	655	1,975	194.20	
	3000	Paint striper, self propelled, double line, 30 HP		6.25	162	485	1,450	147	
	3200	Post drivers, 6" I-Beam frame, for truck mounting		13.40	435	1,305	3,925	368.20	
	3400	Road sweeper, self propelled, 8' wide, 90 HP		31.95	575	1,730	5,200	601.60	
	3450	Road sweeper, vacuum assisted, 4 CY, 220 Gal		56.60	630	1,895	5,675	831.80	
	4000	Road mixer, self-propelled, 130 HP		39.70	770	2,315	6,950	780.60	
	4100	310 HP		71.00	2,225	6,640	19,900	1,896	
	4220	Cold mix paver, incl pug mill and bitumen tank, 165 HP		86.60	2,175	6,515	19,500	1,996	
	4250	Paver, asphalt, wheel or crawler, 130 H.P., diesel		83.50	2,125	6,340	19,000	1,936	
	4300	Paver, road widener, gas 1' to 6', 67 HP		40.10	825	2,480	7,450	816.80	
	4400	Diesel, 2' to 14', 88 HP		52.30	1,050	3,175	9,525	1,053	
	4600	Slipform pavers, curb and gutter, 2 track, 75 HP		36.80	765	2,300	6,900	754.40	
	4700	4 track, 165 HP		45.85	815	2,445	7,325	855.80	
	4800	Median barrier, 215 HP		46.50	845	2,535	7,600	879	
	4901	Trailer, low bed, 75 ton capacity		9.00	218	655	1,975	203	
	5000	Road planer, walk behind, 10" cutting width, 10 HP		2.90	30.50	91	273	41.40	
	5100	Self propelled, 12" cutting width, 64 HP		8.25	127	380	1,150	142	
	5120	Traffic line remover, metal ball blaster, truck mounted, 115 HP		46.45	760	2,285	6,850	828.60	
	5140	Grinder, truck mounted, 115 HP		49.55	805	2,420	7,250	880.40	
	5160	Walk-behind, 11 HP		3.25	46.50	140	420	54	
	5200	Pavement profiler, 4' to 6' wide, 450 HP		210.55	3,400	10,220	30,700	3,728	

01 54 33 | Equipment Rental

		UNIT	HOURLY OPER. COST	RENT PER DAY	RENT PER WEEK	RENT PER MONTH	EQUIPMENT COST/DAY	
50	5300	8' to 10' wide, 750 HP	Ea.	333.90	4,725	14,160	42,500	5,503	**50**
	5400	Roadway plate, steel, 1"x8'x20'		.07	11.35	34	102	7.35	
	5600	Stabilizer, self-propelled, 150 HP		39.45	610	1,830	5,500	681.60	
	5700	310 HP		66.45	1,350	4,060	12,200	1,344	
	5800	Striper, thermal, truck mounted 120 gal. paint, 150 H.P.		49.90	505	1,520	4,550	703.20	
	6000	Tar kettle, 330 gal., trailer mounted		5.77	61.50	185	555	83.15	
	7000	Tunnel locomotive, diesel, 8 to 12 ton		26.70	585	1,750	5,250	563.60	
	7005	Electric, 10 ton		23.25	665	2,000	6,000	586	
	7010	Muck cars, 1/2 C.Y. capacity		1.75	23	69	207	27.80	
	7020	1 C.Y. capacity		1.95	31.50	94	282	34.40	
	7030	2 C.Y. capacity		2.10	36.50	110	330	38.80	
	7040	Side dump, 2 C.Y. capacity		2.30	45	135	405	45.40	
	7050	3 C.Y. capacity		3.10	51.50	155	465	55.80	
	7060	5 C.Y. capacity		4.40	65	195	585	74.20	
	7100	Ventilating blower for tunnel, 7-1/2 H.P.		1.35	51.50	155	465	41.80	
	7110	10 H.P.		1.57	53.50	160	480	44.55	
	7120	20 H.P.		2.58	69.50	208	625	62.25	
	7140	40 H.P.		4.56	98.50	295	885	95.50	
	7160	60 H.P.		6.90	152	455	1,375	146.20	
	7175	75 H.P.		8.81	202	607	1,825	191.90	
	7180	200 H.P.		19.95	305	910	2,725	341.60	
	7800	Windrow loader, elevating		47.65	1,350	4,080	12,200	1,197	
60	0010	**LIFTING AND HOISTING EQUIPMENT RENTAL** without operators							**60**
	0120	Aerial lift truck, 2 person, to 80' [R015433-10]	Ea.	25.55	725	2,180	6,550	640.40	
	0140	Boom work platform, 40' snorkel		15.40	260	780	2,350	279.20	
	0150	Crane, flatbed mntd, 3 ton cap. [R015433-15]		13.50	195	585	1,750	225	
	0200	Crane, climbing, 106' jib, 6000 lb. capacity, 410 FPM [R312316-45]		37.40	1,475	4,400	13,200	1,179	
	0300	101' jib, 10,250 lb. capacity, 270 FPM		43.30	1,850	5,580	16,700	1,462	
	0500	Tower, static, 130' high, 106' jib, 6200 lb. capacity at 400 FPM		40.85	1,700	5,090	15,300	1,345	
	0600	Crawler mounted, lattice boom, 1/2 C.Y., 15 tons at 12' radius		30.66	615	1,850	5,550	615.30	
	0700	3/4 C.Y., 20 tons at 12' radius		40.88	770	2,310	6,925	789.05	
	0800	1 C.Y., 25 tons at 12' radius		54.50	1,025	3,080	9,250	1,052	
	0900	1-1/2 C.Y., 40 tons at 12' radius		54.50	1,025	3,095	9,275	1,055	
	1000	2 C.Y., 50 tons at 12' radius		54.20	1,125	3,380	10,100	1,110	
	1100	3 C.Y., 75 tons at 12' radius		72.95	1,525	4,560	13,700	1,496	
	1200	100 ton capacity, 60' boom		71.55	1,725	5,195	15,600	1,611	
	1300	165 ton capacity, 60' boom		89.60	1,925	5,750	17,300	1,867	
	1400	200 ton capacity, 70' boom		111.30	2,350	7,085	21,300	2,307	
	1500	350 ton capacity, 80' boom		158.35	3,775	11,335	34,000	3,534	
	1600	Truck mounted, lattice boom, 6 x 4, 20 tons at 10' radius		44.77	1,200	3,610	10,800	1,080	
	1700	25 tons at 10' radius		48.31	1,300	3,930	11,800	1,172	
	1800	30 tons at 10' radius		52.66	1,400	4,180	12,500	1,257	
	1900	40 tons at 12' radius		56.59	1,450	4,370	13,100	1,327	
	2000	60 tons at 15' radius		64.84	1,550	4,620	13,900	1,443	
	2050	82 tons at 15' radius		73.58	1,650	4,940	14,800	1,577	
	2100	90 tons at 15' radius		83.16	1,800	5,380	16,100	1,741	
	2200	115 tons at 15' radius		94.14	2,000	6,020	18,100	1,957	
	2300	150 tons at 18' radius		77.90	2,100	6,335	19,000	1,890	
	2350	165 tons at 18' radius		112.04	2,250	6,720	20,200	2,240	
	2400	Truck mounted, hydraulic, 12 ton capacity		49.10	625	1,880	5,650	768.80	
	2500	25 ton capacity		49.30	660	1,975	5,925	789.40	
	2550	33 ton capacity		50.35	695	2,080	6,250	818.80	
	2560	40 ton capacity		48.50	670	2,015	6,050	791	
	2600	55 ton capacity		72.00	915	2,745	8,225	1,125	
	2700	80 ton capacity		85.50	1,025	3,060	9,175	1,296	
	2720	100 ton capacity		104.90	1,550	4,685	14,100	1,776	
	2740	120 ton capacity		83.95	2,000	5,975	17,900	1,867	
	2760	150 ton capacity		89.85	2,175	6,490	19,500	2,017	

01 54 33 | Equipment Rental

		UNIT	HOURLY OPER. COST	RENT PER DAY	RENT PER WEEK	RENT PER MONTH	EQUIPMENT COST/DAY	
60	2800	Self-propelled, 4 x 4, with telescoping boom, 5 ton	Ea.	14.65	235	705	2,125	258.20	60
	2900	12-1/2 ton capacity		34.60	540	1,625	4,875	601.80	
	3000	15 ton capacity		32.30	575	1,725	5,175	603.40	
	3050	20 ton capacity		37.95	660	1,985	5,950	700.60	
	3100	25 ton capacity		38.85	675	2,030	6,100	716.80	
	3150	40 ton capacity		50.15	895	2,680	8,050	937.20	
	3200	Derricks, guy, 20 ton capacity, 60' boom, 75' mast		33.55	360	1,078	3,225	484	
	3300	100' boom, 115' mast		51.95	615	1,850	5,550	785.60	
	3400	Stiffleg, 20 ton capacity, 70' boom, 37' mast		35.80	465	1,400	4,200	566.40	
	3500	100' boom, 47' mast		54.68	745	2,240	6,725	885.45	
	3550	Helicopter, small, lift to 1250 lbs. maximum, w/pilot		105.67	2,900	8,710	26,100	2,587	
	3600	Hoists, chain type, overhead, manual, 3/4 ton		.10	1	3	9	1.40	
	3900	10 ton		.70	9.65	29	87	11.40	
	4000	Hoist and tower, 5000 lb. cap., portable electric, 40' high		4.65	207	621	1,875	161.40	
	4100	For each added 10' section, add		.10	16.35	49	147	10.60	
	4200	Hoist and single tubular tower, 5000 lb. electric, 100' high		6.26	289	867	2,600	223.50	
	4300	For each added 6'-6" section, add		.17	27.50	83	249	17.95	
	4400	Hoist and double tubular tower, 5000 lb., 100' high		6.70	320	955	2,875	244.60	
	4500	For each added 6'-6" section, add		.19	31	93	279	20.10	
	4550	Hoist and tower, mast type, 6000 lb., 100' high		7.26	330	990	2,975	256.10	
	4570	For each added 10' section, add		.12	19.65	59	177	12.75	
	4600	Hoist and tower, personnel, electric, 2000 lb., 100' @ 125 FPM		14.74	880	2,640	7,925	645.90	
	4700	3000 lb., 100' @ 200 FPM		16.88	995	2,990	8,975	733.05	
	4800	3000 lb., 150' @ 300 FPM		18.63	1,125	3,340	10,000	817.05	
	4900	4000 lb., 100' @ 300 FPM		19.36	1,125	3,410	10,200	836.90	
	5000	6000 lb., 100' @ 275 FPM	▼	20.98	1,200	3,580	10,700	883.85	
	5100	For added heights up to 500', add	L.F.	.01	1.67	5	15	1.10	
	5200	Jacks, hydraulic, 20 ton	Ea.	.05	1.67	5	15	1.40	
	5500	100 ton		.35	10.65	32	96	9.20	
	6100	Jacks, hydraulic, climbing w/ 50' jackrods, control console, 30 ton cap		1.79	119	357	1,075	85.70	
	6150	For each added 10' jackrod section, add		.05	3.33	10	30	2.40	
	6300	50 ton capacity		2.87	191	574	1,725	137.75	
	6350	For each added 10' jackrod section, add		.06	4	12	36	2.90	
	6500	125 ton capacity		7.55	505	1,510	4,525	362.40	
	6550	For each added 10' jackrod section, add		.52	34.50	103	310	24.75	
	6600	Cable jack, 10 ton capacity with 200' cable		1.50	99.50	299	895	71.80	
	6650	For each added 50' of cable, add	▼	.17	11.35	34	102	8.15	
70	0010	**WELLPOINT EQUIPMENT RENTAL** without operators [R015433 -10]							70
	0020	Based on 2 months rental							
	0100	Combination jetting & wellpoint pump, 60 H.P. diesel	Ea.	21.79	295	884	2,650	351.10	
	0200	High pressure gas jet pump, 200 H.P., 300 psi	"	57.29	252	756	2,275	609.50	
	0300	Discharge pipe, 8" diameter	L.F.	.01	.48	1.44	4.32	.35	
	0350	12" diameter		.01	.71	2.12	6.35	.50	
	0400	Header pipe, flows up to 150 G.P.M., 4" diameter		.01	.43	1.29	3.87	.35	
	0500	400 G.P.M., 6" diameter		.01	.51	1.54	4.62	.40	
	0600	800 G.P.M., 8" diameter		.01	.71	2.12	6.35	.50	
	0700	1500 G.P.M., 10" diameter		.01	.74	2.22	6.65	.50	
	0800	2500 G.P.M., 12" diameter		.02	1.40	4.21	12.65	1	
	0900	4500 G.P.M., 16" diameter		.03	1.80	5.39	16.15	1.30	
	0950	For quick coupling aluminum and plastic pipe, add	▼	.03	1.86	5.58	16.75	1.35	
	1100	Wellpoint, 25' long, with fittings & riser pipe, 1-1/2" or 2" diameter	Ea.	.06	3.71	11.13	33.50	2.70	
	1200	Wellpoint pump, diesel powered, 4" diameter, 20 H.P.		8.77	170	510	1,525	172.15	
	1300	6" diameter, 30 H.P.		12.22	211	632	1,900	224.15	
	1400	8" suction, 40 H.P.		16.47	289	867	2,600	305.15	
	1500	10" suction, 75 H.P.		26.59	340	1,013	3,050	415.30	
	1600	12" suction, 100 H.P.		37.34	540	1,620	4,850	622.70	
	1700	12" suction, 175 H.P.	▼	57.89	590	1,770	5,300	817.10	

01 54 33 | Equipment Rental

		UNIT	HOURLY OPER. COST	RENT PER DAY	RENT PER WEEK	RENT PER MONTH	EQUIPMENT COST/DAY
80	0010 **MARINE EQUIPMENT RENTAL** without operators							80
	0200 Barge, 400 Ton, 30' wide x 90' long R015433 -10	Ea.	14.80	1,050	3,140	9,425	746.40
	0240 800 Ton, 45' wide x 90' long		17.95	1,275	3,800	11,400	903.60
	2000 Tugboat, diesel, 100 HP		27.75	195	585	1,750	339
	2040 250 HP		56.05	360	1,075	3,225	663.40
	2080 380 HP		115.60	1,075	3,195	9,575	1,564

Crews

Crew No.	Bare Costs		Incl. Subs O & P		Cost Per Labor-Hour	

Crew A-1	Hr.	Daily	Hr.	Daily	Bare Costs	Incl. O&P
1 Building Laborer	$31.60	$252.80	$52.15	$417.20	$31.60	$52.15
1 Concrete saw, gas manual		63.20		69.52	7.90	8.69
8 L.H., Daily Totals		$316.00		$486.72	$39.50	$60.84

Crew A-1A	Hr.	Daily	Hr.	Daily	Bare Costs	Incl. O&P
1 Skilled Worker	$40.85	$326.80	$66.50	$532.00	$40.85	$66.50
1 Shot Blaster, 20"		217.80		239.58	27.23	29.95
8 L.H., Daily Totals		$544.60		$771.58	$68.08	$96.45

Crew A-1B	Hr.	Daily	Hr.	Daily	Bare Costs	Incl. O&P
1 Building Laborer	$31.60	$252.80	$52.15	$417.20	$31.60	$52.15
1 Concrete Saw		141.00		155.10	17.63	19.39
8 L.H., Daily Totals		$393.80		$572.30	$49.23	$71.54

Crew A-1C	Hr.	Daily	Hr.	Daily	Bare Costs	Incl. O&P
1 Building Laborer	$31.60	$252.80	$52.15	$417.20	$31.60	$52.15
1 Chain saw, gas, 18"		25.80		28.38	3.23	3.55
8 L.H., Daily Totals		$278.60		$445.58	$34.83	$55.70

Crew A-1D	Hr.	Daily	Hr.	Daily	Bare Costs	Incl. O&P
1 Building Laborer	$31.60	$252.80	$52.15	$417.20	$31.60	$52.15
1 Vibrating plate, gas, 18"		31.80		34.98	3.98	4.37
8 L.H., Daily Totals		$284.60		$452.18	$35.58	$56.52

Crew A-1E	Hr.	Daily	Hr.	Daily	Bare Costs	Incl. O&P
1 Building Laborer	$31.60	$252.80	$52.15	$417.20	$31.60	$52.15
1 Vibratory Plate, Gas, 21"		39.20		43.12	4.90	5.39
8 L.H., Daily Totals		$292.00		$460.32	$36.50	$57.54

Crew A-1F	Hr.	Daily	Hr.	Daily	Bare Costs	Incl. O&P
1 Building Laborer	$31.60	$252.80	$52.15	$417.20	$31.60	$52.15
1 Rammer/tamper, gas, 8"		43.40		47.74	5.42	5.97
8 L.H., Daily Totals		$296.20		$464.94	$37.02	$58.12

Crew A-1G	Hr.	Daily	Hr.	Daily	Bare Costs	Incl. O&P
1 Building Laborer	$31.60	$252.80	$52.15	$417.20	$31.60	$52.15
1 Rammer/tamper, gas, 15"		49.40		54.34	6.17	6.79
8 L.H., Daily Totals		$302.20		$471.54	$37.77	$58.94

Crew A-1H	Hr.	Daily	Hr.	Daily	Bare Costs	Incl. O&P
1 Building Laborer	$31.60	$252.80	$52.15	$417.20	$31.60	$52.15
1 Exterior Steam Cleaner		51.20		56.32	6.40	7.04
8 L.H., Daily Totals		$304.00		$473.52	$38.00	$59.19

Crew A-1J	Hr.	Daily	Hr.	Daily	Bare Costs	Incl. O&P
1 Building Laborer	$31.60	$252.80	$52.15	$417.20	$31.60	$52.15
1 Cultivator, Walk-Behind, 5 H.P.		57.05		62.76	7.13	7.84
8 L.H., Daily Totals		$309.85		$479.95	$38.73	$59.99

Crew A-1K	Hr.	Daily	Hr.	Daily	Bare Costs	Incl. O&P
1 Building Laborer	$31.60	$252.80	$52.15	$417.20	$31.60	$52.15
1 Cultivator, Walk-Behind, 8 H.P.		91.10		100.21	11.39	12.53
8 L.H., Daily Totals		$343.90		$517.41	$42.99	$64.68

Crew A-1M	Hr.	Daily	Hr.	Daily	Bare Costs	Incl. O&P
1 Building Laborer	$31.60	$252.80	$52.15	$417.20	$31.60	$52.15
1 Snow Blower, Walk-Behind		52.70		57.97	6.59	7.25
8 L.H., Daily Totals		$305.50		$475.17	$38.19	$59.40

Crew A-2	Hr.	Daily	Hr.	Daily	Bare Costs	Incl. O&P
2 Laborers	$31.60	$505.60	$52.15	$834.40	$31.38	$51.68
1 Truck Driver (light)	30.95	247.60	50.75	406.00		
1 Flatbed Truck, Gas, 1.5 Ton		194.40		213.84	8.10	8.91
24 L.H., Daily Totals		$947.60		$1454.24	$39.48	$60.59

Crew A-2A	Hr.	Daily	Hr.	Daily	Bare Costs	Incl. O&P
2 Laborers	$31.60	$505.60	$52.15	$834.40	$31.38	$51.68
1 Truck Driver (light)	30.95	247.60	50.75	406.00		
1 Flatbed Truck, Gas, 1.5 Ton		194.40		213.84		
1 Concrete Saw		141.00		155.10	13.98	15.37
24 L.H., Daily Totals		$1088.60		$1609.34	$45.36	$67.06

Crew A-2B	Hr.	Daily	Hr.	Daily	Bare Costs	Incl. O&P
1 Truck Driver (light)	$30.95	$247.60	$50.75	$406.00	$30.95	$50.75
1 Flatbed Truck, Gas, 1.5 Ton		194.40		213.84	24.30	26.73
8 L.H., Daily Totals		$442.00		$619.84	$55.25	$77.48

Crew A-3A	Hr.	Daily	Hr.	Daily	Bare Costs	Incl. O&P
1 Truck Driver (light)	$30.95	$247.60	$50.75	$406.00	$30.95	$50.75
1 Pickup truck, 4 x 4, 3/4 ton		128.20		141.02	16.02	17.63
8 L.H., Daily Totals		$375.80		$547.02	$46.98	$68.38

Crew A-3B	Hr.	Daily	Hr.	Daily	Bare Costs	Incl. O&P
1 Equip. Oper. (medium)	$41.35	$330.80	$65.05	$520.40	$36.65	$58.70
1 Truck Driver (heavy)	31.95	255.60	52.35	418.80		
1 Dump Truck, 12 C.Y., 400 H.P.		533.00		586.30		
1 F.E. Loader, W.M., 2.5 C.Y.		395.80		435.38	58.05	63.85
16 L.H., Daily Totals		$1515.20		$1960.88	$94.70	$122.56

Crew A-3C	Hr.	Daily	Hr.	Daily	Bare Costs	Incl. O&P
1 Equip. Oper. (light)	$39.05	$312.40	$61.45	$491.60	$39.05	$61.45
1 Loader, Skid Steer, 78 H.P.		259.60		285.56	32.45	35.70
8 L.H., Daily Totals		$572.00		$777.16	$71.50	$97.14

Crew A-3D	Hr.	Daily	Hr.	Daily	Bare Costs	Incl. O&P
1 Truck Driver, Light	$30.95	$247.60	$50.75	$406.00	$30.95	$50.75
1 Pickup truck, 4 x 4, 3/4 ton		128.20		141.02		
1 Flatbed Trailer, 25 Ton		102.00		112.20	28.77	31.65
8 L.H., Daily Totals		$477.80		$659.22	$59.73	$82.40

Crew A-3E	Hr.	Daily	Hr.	Daily	Bare Costs	Incl. O&P
1 Equip. Oper. (crane)	$42.55	$340.40	$66.95	$535.60	$37.25	$59.65
1 Truck Driver (heavy)	31.95	255.60	52.35	418.80		
1 Pickup truck, 4 x 4, 3/4 ton		128.20		141.02	8.01	8.81
16 L.H., Daily Totals		$724.20		$1095.42	$45.26	$68.46

Crew A-3F	Hr.	Daily	Hr.	Daily	Bare Costs	Incl. O&P
1 Equip. Oper. (crane)	$42.55	$340.40	$66.95	$535.60	$37.25	$59.65
1 Truck Driver (heavy)	31.95	255.60	52.35	418.80		
1 Pickup truck, 4 x 4, 3/4 ton		128.20		141.02		
1 Truck Tractor, 6x4, 380 H.P.		478.80		526.68		
1 Lowbed Trailer, 75 Ton		203.00		223.30	50.63	55.69
16 L.H., Daily Totals		$1406.00		$1845.40	$87.88	$115.34

Crew No.	Bare Costs Hr.	Bare Costs Daily	Incl. Subs O & P Hr.	Incl. Subs O & P Daily	Cost Per Labor-Hour Bare Costs	Cost Per Labor-Hour Incl. O&P
Crew A-3G	Hr.	Daily	Hr.	Daily	Bare Costs	Incl. O&P
1 Equip. Oper. (crane)	$42.55	$340.40	$66.95	$535.60	$37.25	$59.65
1 Truck Driver (heavy)	31.95	255.60	52.35	418.80		
1 Pickup truck, 4 x 4, 3/4 ton		128.20		141.02		
1 Truck Tractor, 6x4, 450 H.P.		581.20		639.32		
1 Lowbed Trailer, 75 Ton		203.00		223.30	57.02	62.73
16 L.H., Daily Totals		$1508.40		$1958.04	$94.28	$122.38
Crew A-3H	Hr.	Daily	Hr.	Daily	Bare Costs	Incl. O&P
1 Equip. Oper. (crane)	$42.55	$340.40	$66.95	$535.60	$42.55	$66.95
1 Hyd. crane, 12 Ton (daily)		1018.00		1119.80	127.25	139.97
8 L.H., Daily Totals		$1358.40		$1655.40	$169.80	$206.93
Crew A-3I	Hr.	Daily	Hr.	Daily	Bare Costs	Incl. O&P
1 Equip. Oper. (crane)	$42.55	$340.40	$66.95	$535.60	$42.55	$66.95
1 Hyd. crane, 25 Ton (daily)		1054.00		1159.40	131.75	144.93
8 L.H., Daily Totals		$1394.40		$1695.00	$174.30	$211.88
Crew A-3J	Hr.	Daily	Hr.	Daily	Bare Costs	Incl. O&P
1 Equip. Oper. (crane)	$42.55	$340.40	$66.95	$535.60	$42.55	$66.95
1 Hyd. crane, 40 Ton (daily)		1058.00		1163.80	132.25	145.47
8 L.H., Daily Totals		$1398.40		$1699.40	$174.80	$212.43
Crew A-3K	Hr.	Daily	Hr.	Daily	Bare Costs	Incl. O&P
1 Equip. Oper. (crane)	$42.55	$340.40	$66.95	$535.60	$39.67	$62.42
1 Equip. Oper. Oiler	36.80	294.40	57.90	463.20		
1 Hyd. crane, 55 Ton (daily)		1491.00		1640.10		
1 P/U Truck, 3/4 Ton (daily)		137.80		151.58	101.80	111.98
16 L.H., Daily Totals		$2263.60		$2790.48	$141.47	$174.41
Crew A-3L	Hr.	Daily	Hr.	Daily	Bare Costs	Incl. O&P
1 Equip. Oper. (crane)	$42.55	$340.40	$66.95	$535.60	$39.67	$62.42
1 Equip. Oper. Oiler	36.80	294.40	57.90	463.20		
1 Hyd. crane, 80 Ton (daily)		1704.00		1874.40		
1 P/U Truck, 3/4 Ton (daily)		137.80		151.58	115.11	126.62
16 L.H., Daily Totals		$2476.60		$3024.78	$154.79	$189.05
Crew A-3M	Hr.	Daily	Hr.	Daily	Bare Costs	Incl. O&P
1 Equip. Oper. (crane)	$42.55	$340.40	$66.95	$535.60	$39.67	$62.42
1 Equip. Oper. Oiler	36.80	294.40	57.90	463.20		
1 Hyd. crane, 100 Ton (daily)		2399.00		2638.90		
1 P/U Truck, 3/4 Ton (daily)		137.80		151.58	158.55	174.41
16 L.H., Daily Totals		$3171.60		$3789.28	$198.22	$236.83
Crew A-3N	Hr.	Daily	Hr.	Daily	Bare Costs	Incl. O&P
1 Equip. Oper. (crane)	$42.55	$340.40	$66.95	$535.60	$42.55	$66.95
1 Tower crane (monthly)		1022.00		1124.20	127.75	140.53
8 L.H., Daily Totals		$1362.40		$1659.80	$170.30	$207.47
Crew A-3P	Hr.	Daily	Hr.	Daily	Bare Costs	Incl. O&P
1 Equip. Oper., Light	$39.05	$312.40	$61.45	$491.60	$39.05	$61.45
1 A.T. Forklift, 42' lift		446.00		490.60	55.75	61.33
8 L.H., Daily Totals		$758.40		$982.20	$94.80	$122.78
Crew A-4	Hr.	Daily	Hr.	Daily	Bare Costs	Incl. O&P
2 Carpenters	$39.95	$639.20	$65.95	$1055.20	$38.37	$62.72
1 Painter, Ordinary	35.20	281.60	56.25	450.00		
24 L.H., Daily Totals		$920.80		$1505.20	$38.37	$62.72

Crew No.	Bare Costs Hr.	Bare Costs Daily	Incl. Subs O & P Hr.	Incl. Subs O & P Daily	Cost Per Labor-Hour Bare Costs	Cost Per Labor-Hour Incl. O&P
Crew A-5	Hr.	Daily	Hr.	Daily	Bare Costs	Incl. O&P
2 Laborers	$31.60	$505.60	$52.15	$834.40	$31.53	$51.99
.25 Truck Driver (light)	30.95	61.90	50.75	101.50		
.25 Flatbed Truck, Gas, 1.5 Ton		48.60		53.46	2.70	2.97
18 L.H., Daily Totals		$616.10		$989.36	$34.23	$54.96
Crew A-6	Hr.	Daily	Hr.	Daily	Bare Costs	Incl. O&P
1 Instrument Man	$40.85	$326.80	$66.50	$532.00	$39.73	$64.35
1 Rodman/Chainman	38.60	308.80	62.20	497.60		
1 Laser Transit/Level		70.10		77.11	4.38	4.82
16 L.H., Daily Totals		$705.70		$1106.71	$44.11	$69.17
Crew A-7	Hr.	Daily	Hr.	Daily	Bare Costs	Incl. O&P
1 Chief Of Party	$50.20	$401.60	$79.65	$637.20	$43.22	$69.45
1 Instrument Man	40.85	326.80	66.50	532.00		
1 Rodman/Chainman	38.60	308.80	62.20	497.60		
1 Laser Transit/Level		70.10		77.11	2.92	3.21
24 L.H., Daily Totals		$1107.30		$1743.91	$46.14	$72.66
Crew A-8	Hr.	Daily	Hr.	Daily	Bare Costs	Incl. O&P
1 Chief of Party	$50.20	$401.60	$79.65	$637.20	$42.06	$67.64
1 Instrument Man	40.85	326.80	66.50	532.00		
2 Rodmen/Chainmen	38.60	617.60	62.20	995.20		
1 Laser Transit/Level		70.10		77.11	2.19	2.41
32 L.H., Daily Totals		$1416.10		$2241.51	$44.25	$70.05
Crew A-9	Hr.	Daily	Hr.	Daily	Bare Costs	Incl. O&P
1 Asbestos Foreman	$44.60	$356.80	$72.10	$576.80	$44.16	$71.40
7 Asbestos Workers	44.10	2469.60	71.30	3992.80		
64 L.H., Daily Totals		$2826.40		$4569.60	$44.16	$71.40
Crew A-10A	Hr.	Daily	Hr.	Daily	Bare Costs	Incl. O&P
1 Asbestos Foreman	$44.60	$356.80	$72.10	$576.80	$44.27	$71.57
2 Asbestos Workers	44.10	705.60	71.30	1140.80		
24 L.H., Daily Totals		$1062.40		$1717.60	$44.27	$71.57
Crew A-10B	Hr.	Daily	Hr.	Daily	Bare Costs	Incl. O&P
1 Asbestos Foreman	$44.60	$356.80	$72.10	$576.80	$44.23	$71.50
3 Asbestos Workers	44.10	1058.40	71.30	1711.20		
32 L.H., Daily Totals		$1415.20		$2288.00	$44.23	$71.50
Crew A-10C	Hr.	Daily	Hr.	Daily	Bare Costs	Incl. O&P
3 Asbestos Workers	$44.10	$1058.40	$71.30	$1711.20	$44.10	$71.30
1 Flatbed Truck, Gas, 1.5 Ton		194.40		213.84	8.10	8.91
24 L.H., Daily Totals		$1252.80		$1925.04	$52.20	$80.21
Crew A-10D	Hr.	Daily	Hr.	Daily	Bare Costs	Incl. O&P
2 Asbestos Workers	$44.10	$705.60	$71.30	$1140.80	$41.89	$66.86
1 Equip. Oper. (crane)	42.55	340.40	66.95	535.60		
1 Equip. Oper. Oiler	36.80	294.40	57.90	463.20		
1 Hydraulic Crane, 33 Ton		818.80		900.68	25.59	28.15
32 L.H., Daily Totals		$2159.20		$3040.28	$67.47	$95.01
Crew A-11	Hr.	Daily	Hr.	Daily	Bare Costs	Incl. O&P
1 Asbestos Foreman	$44.60	$356.80	$72.10	$576.80	$44.16	$71.40
7 Asbestos Workers	44.10	2469.60	71.30	3992.80		
2 Chipping Hammers, 12 Lb., Elec.		34.00		37.40	.53	.58
64 L.H., Daily Totals		$2860.40		$4607.00	$44.69	$71.98

Crew No.	Bare Costs		Incl. Subs O & P		Cost Per Labor-Hour	
Crew A-12	Hr.	Daily	Hr.	Daily	Bare Costs	Incl. O&P
1 Asbestos Foreman	$44.60	$356.80	$72.10	$576.80	$44.16	$71.40
7 Asbestos Workers	44.10	2469.60	71.30	3992.80		
1 Trk-mtd vac, 14 CY, 1500 Gal.		561.60		617.76		
1 Flatbed Truck, 20,000 GVW		197.40		217.14	11.86	13.05
64 L.H., Daily Totals		$3585.40		$5404.50	$56.02	$84.45

Crew No.	Bare Costs		Incl. Subs O & P		Cost Per Labor-Hour	
Crew A-13	Hr.	Daily	Hr.	Daily	Bare Costs	Incl. O&P
1 Equip. Oper. (light)	$39.05	$312.40	$61.45	$491.60	$39.05	$61.45
1 Trk-mtd vac, 14 CY, 1500 Gal.		561.60		617.76		
1 Flatbed Truck, 20,000 GVW		197.40		217.14	94.88	104.36
8 L.H., Daily Totals		$1071.40		$1326.50	$133.93	$165.81

Crew No.	Bare Costs		Incl. Subs O & P		Cost Per Labor-Hour	
Crew B-1	Hr.	Daily	Hr.	Daily	Bare Costs	Incl. O&P
1 Labor Foreman (outside)	$33.60	$268.80	$55.45	$443.60	$32.27	$53.25
2 Laborers	31.60	505.60	52.15	834.40		
24 L.H., Daily Totals		$774.40		$1278.00	$32.27	$53.25

Crew No.	Bare Costs		Incl. Subs O & P		Cost Per Labor-Hour	
Crew B-1A	Hr.	Daily	Hr.	Daily	Bare Costs	Incl. O&P
1 Laborer Foreman	$33.60	$268.80	$55.45	$443.60	$32.27	$53.25
2 Laborers	31.60	505.60	52.15	834.40		
2 Cutting Torches		34.00		37.40		
2 Gases		151.20		166.32	7.72	8.49
24 L.H., Daily Totals		$959.60		$1481.72	$39.98	$61.74

Crew No.	Bare Costs		Incl. Subs O & P		Cost Per Labor-Hour	
Crew B-1B	Hr.	Daily	Hr.	Daily	Bare Costs	Incl. O&P
1 Laborer Foreman	$33.60	$268.80	$55.45	$443.60	$34.84	$56.67
2 Laborers	31.60	505.60	52.15	834.40		
1 Equip. Oper. (crane)	42.55	340.40	66.95	535.60		
2 Cutting Torches		34.00		37.40		
2 Gases		151.20		166.32		
1 Hyd. Crane, 12 Ton		768.80		845.68	29.81	32.79
32 L.H., Daily Totals		$2068.80		$2863.00	$64.65	$89.47

Crew No.	Bare Costs		Incl. Subs O & P		Cost Per Labor-Hour	
Crew B-2	Hr.	Daily	Hr.	Daily	Bare Costs	Incl. O&P
1 Labor Foreman (outside)	$33.60	$268.80	$55.45	$443.60	$32.00	$52.81
4 Laborers	31.60	1011.20	52.15	1668.80		
40 L.H., Daily Totals		$1280.00		$2112.40	$32.00	$52.81

Crew No.	Bare Costs		Incl. Subs O & P		Cost Per Labor-Hour	
Crew B-3	Hr.	Daily	Hr.	Daily	Bare Costs	Incl. O&P
1 Labor Foreman (outside)	$33.60	$268.80	$55.45	$443.60	$33.67	$54.92
2 Laborers	31.60	505.60	52.15	834.40		
1 Equip. Oper. (med.)	41.35	330.80	65.05	520.40		
2 Truck Drivers (heavy)	31.95	511.20	52.35	837.60		
1 Crawler Loader, 3 C.Y.		990.00		1089.00		
2 Dump Trucks 12 C.Y., 400 H.P.		1066.00		1172.60	42.83	47.12
48 L.H., Daily Totals		$3672.40		$4897.60	$76.51	$102.03

Crew No.	Bare Costs		Incl. Subs O & P		Cost Per Labor-Hour	
Crew B-3A	Hr.	Daily	Hr.	Daily	Bare Costs	Incl. O&P
4 Laborers	$31.60	$1011.20	$52.15	$1668.80	$33.55	$54.73
1 Equip. Oper. (med.)	41.35	330.80	65.05	520.40		
1 Hyd. Excavator, 1.5 C.Y.		865.80		952.38	21.65	23.81
40 L.H., Daily Totals		$2207.80		$3141.58	$55.20	$78.54

Crew No.	Bare Costs		Incl. Subs O & P		Cost Per Labor-Hour	
Crew B-3B	Hr.	Daily	Hr.	Daily	Bare Costs	Incl. O&P
2 Laborers	$31.60	$505.60	$52.15	$834.40	$34.13	$55.42
1 Equip. Oper. (med.)	41.35	330.80	65.05	520.40		
1 Truck Driver (heavy)	31.95	255.60	52.35	418.80		
1 Backhoe Loader, 80 H.P.		345.00		379.50		
1 Dump Truck, 12 C.Y., 400 H.P.		533.00		586.30	27.44	30.18
32 L.H., Daily Totals		$1970.00		$2739.40	$61.56	$85.61

Crew No.	Bare Costs		Incl. Subs O & P		Cost Per Labor-Hour	
Crew B-3C	Hr.	Daily	Hr.	Daily	Bare Costs	Incl. O&P
3 Laborers	$31.60	$758.40	$52.15	$1251.60	$34.04	$55.38
1 Equip. Oper. (med.)	41.35	330.80	65.05	520.40		
1 Crawler Loader, 4 C.Y.		1450.00		1595.00	45.31	49.84
32 L.H., Daily Totals		$2539.20		$3367.00	$79.35	$105.22

Crew No.	Bare Costs		Incl. Subs O & P		Cost Per Labor-Hour	
Crew B-4	Hr.	Daily	Hr.	Daily	Bare Costs	Incl. O&P
1 Labor Foreman (outside)	$33.60	$268.80	$55.45	$443.60	$31.99	$52.73
4 Laborers	31.60	1011.20	52.15	1668.80		
1 Truck Driver (heavy)	31.95	255.60	52.35	418.80		
1 Truck Tractor, 220 H.P.		288.40		317.24		
1 Flatbed Trailer, 40 Ton		139.60		153.56	8.92	9.81
48 L.H., Daily Totals		$1963.60		$3002.00	$40.91	$62.54

Crew No.	Bare Costs		Incl. Subs O & P		Cost Per Labor-Hour	
Crew B-5	Hr.	Daily	Hr.	Daily	Bare Costs	Incl. O&P
1 Labor Foreman (outside)	$33.60	$268.80	$55.45	$443.60	$34.67	$56.31
4 Laborers	31.60	1011.20	52.15	1668.80		
2 Equip. Oper. (med.)	41.35	661.60	65.05	1040.80		
1 Air Compressor, 250 cfm		162.40		178.64		
2 Breakers, Pavement, 60 lb.		16.40		18.04		
2 -50' Air Hoses, 1.5"		12.60		13.86		
1 Crawler Loader, 3 C.Y.		990.00		1089.00	21.10	23.21
56 L.H., Daily Totals		$3123.00		$4452.74	$55.77	$79.51

Crew No.	Bare Costs		Incl. Subs O & P		Cost Per Labor-Hour	
Crew B-6	Hr.	Daily	Hr.	Daily	Bare Costs	Incl. O&P
2 Laborers	$31.60	$505.60	$52.15	$834.40	$34.08	$55.25
1 Equip. Oper. (light)	39.05	312.40	61.45	491.60		
1 Backhoe Loader, 48 H.P.		293.80		323.18	12.24	13.47
24 L.H., Daily Totals		$1111.80		$1649.18	$46.33	$68.72

Crew No.	Bare Costs		Incl. Subs O & P		Cost Per Labor-Hour	
Crew B-6B	Hr.	Daily	Hr.	Daily	Bare Costs	Incl. O&P
2 Labor Foremen (out)	$33.60	$537.60	$55.45	$887.20	$32.27	$53.25
4 Laborers	31.60	1011.20	52.15	1668.80		
1 S.P. Crane, 4x4, 5 Ton		258.20		284.02		
1 Flatbed Truck, Gas, 1.5 Ton		194.40		213.84		
1 Butt Fusion Machine		199.80		219.78	13.59	14.95
48 L.H., Daily Totals		$2201.20		$3273.64	$45.86	$68.20

Crew No.	Bare Costs		Incl. Subs O & P		Cost Per Labor-Hour	
Crew B-7	Hr.	Daily	Hr.	Daily	Bare Costs	Incl. O&P
1 Labor Foreman (outside)	$33.60	$268.80	$55.45	$443.60	$33.56	$54.85
4 Laborers	31.60	1011.20	52.15	1668.80		
1 Equip. Oper. (med.)	41.35	330.80	65.05	520.40		
1 Brush Chipper, 12", 130 H.P.		225.60		248.16		
1 Crawler Loader, 3 C.Y.		990.00		1089.00		
2 Chainsaws, Gas, 36" Long		75.60		83.16	26.90	29.59
48 L.H., Daily Totals		$2902.00		$4053.12	$60.46	$84.44

Crew No.	Bare Costs		Incl. Subs O & P		Cost Per Labor-Hour	
Crew B-7A	Hr.	Daily	Hr.	Daily	Bare Costs	Incl. O&P
2 Laborers	$31.60	$505.60	$52.15	$834.40	$34.08	$55.25
1 Equip. Oper. (light)	39.05	312.40	61.45	491.60		
1 Rake w/Tractor		257.70		283.47		
2 Chain Saws, gas, 18"		51.60		56.76	12.89	14.18
24 L.H., Daily Totals		$1127.30		$1666.23	$46.97	$69.43

Crew B-8

	Bare Costs Hr.	Bare Costs Daily	Incl. Subs O&P Hr.	Incl. Subs O&P Daily	Cost Per Labor-Hour Bare Costs	Cost Per Labor-Hour Incl. O&P
1 Labor Foreman (outside)	$33.60	$268.80	$55.45	$443.60	$35.02	$56.56
2 Laborers	31.60	505.60	52.15	834.40		
2 Equip. Oper. (med.)	41.35	661.60	65.05	1040.80		
1 Equip. Oper. Oiler	36.80	294.40	57.90	463.20		
2 Truck Drivers (heavy)	31.95	511.20	52.35	837.60		
1 Hyd. Crane, 25 Ton		789.40		868.34		
1 Crawler Loader, 3 C.Y.		990.00		1089.00		
2 Dump Trucks, 12 C.Y., 400 H.P.		1066.00		1172.60	44.46	48.91
64 L.H., Daily Totals		$5087.00		$6749.54	$79.48	$105.46

Crew B-9

	Bare Costs Hr.	Bare Costs Daily	Incl. Subs O&P Hr.	Incl. Subs O&P Daily	Cost Per Labor-Hour Bare Costs	Cost Per Labor-Hour Incl. O&P
1 Labor Foreman (outside)	$33.60	$268.80	$55.45	$443.60	$32.00	$52.81
4 Laborers	31.60	1011.20	52.15	1668.80		
1 Air Compressor, 250 cfm		162.40		178.64		
2 Breakers, Pavement, 60 lb.		16.40		18.04		
2 -50' Air Hoses, 1.5"		12.60		13.86	4.79	5.26
40 L.H., Daily Totals		$1471.40		$2322.94	$36.78	$58.07

Crew B-9A

	Bare Costs Hr.	Bare Costs Daily	Incl. Subs O&P Hr.	Incl. Subs O&P Daily	Cost Per Labor-Hour Bare Costs	Cost Per Labor-Hour Incl. O&P
2 Laborers	$31.60	$505.60	$52.15	$834.40	$31.72	$52.22
1 Truck Driver (heavy)	31.95	255.60	52.35	418.80		
1 Water Tanker, 5000 Gal.		136.00		149.60		
1 Truck Tractor, 220 H.P.		288.40		317.24		
2 -50' Discharge Hoses, 3"		3.50		3.85	17.83	19.61
24 L.H., Daily Totals		$1189.10		$1723.89	$49.55	$71.83

Crew B-9B

	Bare Costs Hr.	Bare Costs Daily	Incl. Subs O&P Hr.	Incl. Subs O&P Daily	Cost Per Labor-Hour Bare Costs	Cost Per Labor-Hour Incl. O&P
2 Laborers	$31.60	$505.60	$52.15	$834.40	$31.72	$52.22
1 Truck Driver (heavy)	31.95	255.60	52.35	418.80		
2 -50' Discharge Hoses, 3"		3.50		3.85		
1 Water Tanker, 5000 Gal.		136.00		149.60		
1 Truck Tractor, 220 H.P.		288.40		317.24		
1 Pressure Washer		53.00		58.30	20.04	22.04
24 L.H., Daily Totals		$1242.10		$1782.19	$51.75	$74.26

Crew B-9D

	Bare Costs Hr.	Bare Costs Daily	Incl. Subs O&P Hr.	Incl. Subs O&P Daily	Cost Per Labor-Hour Bare Costs	Cost Per Labor-Hour Incl. O&P
1 Labor Foreman (Outside)	$33.60	$268.80	$55.45	$443.60	$32.00	$52.81
4 Common Laborers	31.60	1011.20	52.15	1668.80		
1 Air Compressor, 250 cfm		162.40		178.64		
2 -50' Air Hoses, 1.5"		12.60		13.86		
2 Air Powered Tampers		48.00		52.80	5.58	6.13
40 L.H., Daily Totals		$1503.00		$2357.70	$37.58	$58.94

Crew B-10

	Bare Costs Hr.	Bare Costs Daily	Incl. Subs O&P Hr.	Incl. Subs O&P Daily	Cost Per Labor-Hour Bare Costs	Cost Per Labor-Hour Incl. O&P
1 Equip. Oper. (med.)	$41.35	$330.80	$65.05	$520.40	$38.10	$60.75
.5 Laborer	31.60	126.40	52.15	208.60		
12 L.H., Daily Totals		$457.20		$729.00	$38.10	$60.75

Crew B-10A

	Bare Costs Hr.	Bare Costs Daily	Incl. Subs O&P Hr.	Incl. Subs O&P Daily	Cost Per Labor-Hour Bare Costs	Cost Per Labor-Hour Incl. O&P
1 Equip. Oper. (med.)	$41.35	$330.80	$65.05	$520.40	$38.10	$60.75
.5 Laborer	31.60	126.40	52.15	208.60		
1 Roller, 2-Drum, W.B., 7.5 H.P.		145.20		159.72	12.10	13.31
12 L.H., Daily Totals		$602.40		$888.72	$50.20	$74.06

Crew B-10B

	Bare Costs Hr.	Bare Costs Daily	Incl. Subs O&P Hr.	Incl. Subs O&P Daily	Cost Per Labor-Hour Bare Costs	Cost Per Labor-Hour Incl. O&P
1 Equip. Oper. (med.)	$41.35	$330.80	$65.05	$520.40	$38.10	$60.75
.5 Laborer	31.60	126.40	52.15	208.60		
1 Dozer, 200 H.P.		1082.00		1190.20	90.17	99.18
12 L.H., Daily Totals		$1539.20		$1919.20	$128.27	$159.93

Crew B-10C

	Bare Costs Hr.	Bare Costs Daily	Incl. Subs O&P Hr.	Incl. Subs O&P Daily	Cost Per Labor-Hour Bare Costs	Cost Per Labor-Hour Incl. O&P
1 Equip. Oper. (med.)	$41.35	$330.80	$65.05	$520.40	$38.10	$60.75
.5 Laborer	31.60	126.40	52.15	208.60		
1 Dozer, 200 H.P.		1082.00		1190.20		
1 Vibratory Roller, Towed, 23 Ton		411.00		452.10	124.42	136.86
12 L.H., Daily Totals		$1950.20		$2371.30	$162.52	$197.61

Crew B-10D

	Bare Costs Hr.	Bare Costs Daily	Incl. Subs O&P Hr.	Incl. Subs O&P Daily	Cost Per Labor-Hour Bare Costs	Cost Per Labor-Hour Incl. O&P
1 Equip. Oper. (med.)	$41.35	$330.80	$65.05	$520.40	$38.10	$60.75
.5 Laborer	31.60	126.40	52.15	208.60		
1 Dozer, 200 H.P.		1082.00		1190.20		
1 Sheepsft. Roller, Towed		465.40		511.94	128.95	141.85
12 L.H., Daily Totals		$2004.60		$2431.14	$167.05	$202.60

Crew B-10E

	Bare Costs Hr.	Bare Costs Daily	Incl. Subs O&P Hr.	Incl. Subs O&P Daily	Cost Per Labor-Hour Bare Costs	Cost Per Labor-Hour Incl. O&P
1 Equip. Oper. (med.)	$41.35	$330.80	$65.05	$520.40	$38.10	$60.75
.5 Laborer	31.60	126.40	52.15	208.60		
1 Tandem Roller, 5 Ton		137.80		151.58	11.48	12.63
12 L.H., Daily Totals		$595.00		$880.58	$49.58	$73.38

Crew B-10F

	Bare Costs Hr.	Bare Costs Daily	Incl. Subs O&P Hr.	Incl. Subs O&P Daily	Cost Per Labor-Hour Bare Costs	Cost Per Labor-Hour Incl. O&P
1 Equip. Oper. (med.)	$41.35	$330.80	$65.05	$520.40	$38.10	$60.75
.5 Laborer	31.60	126.40	52.15	208.60		
1 Tandem Roller, 10 Ton		233.00		256.30	19.42	21.36
12 L.H., Daily Totals		$690.20		$985.30	$57.52	$82.11

Crew B-10G

	Bare Costs Hr.	Bare Costs Daily	Incl. Subs O&P Hr.	Incl. Subs O&P Daily	Cost Per Labor-Hour Bare Costs	Cost Per Labor-Hour Incl. O&P
1 Equip. Oper. (med.)	$41.35	$330.80	$65.05	$520.40	$38.10	$60.75
.5 Laborer	31.60	126.40	52.15	208.60		
1 Sheepsft. Roll., 240 H.P.		1142.00		1256.20	95.17	104.68
12 L.H., Daily Totals		$1599.20		$1985.20	$133.27	$165.43

Crew B-10H

	Bare Costs Hr.	Bare Costs Daily	Incl. Subs O&P Hr.	Incl. Subs O&P Daily	Cost Per Labor-Hour Bare Costs	Cost Per Labor-Hour Incl. O&P
1 Equip. Oper. (med.)	$41.35	$330.80	$65.05	$520.40	$38.10	$60.75
.5 Laborer	31.60	126.40	52.15	208.60		
1 Diaphragm Water Pump, 2"		63.20		69.52		
1 -20' Suction Hose, 2"		1.95		2.15		
2 -50' Discharge Hoses, 2"		2.20		2.42	5.61	6.17
12 L.H., Daily Totals		$524.55		$803.09	$43.71	$66.92

Crew B-10I

	Bare Costs Hr.	Bare Costs Daily	Incl. Subs O&P Hr.	Incl. Subs O&P Daily	Cost Per Labor-Hour Bare Costs	Cost Per Labor-Hour Incl. O&P
1 Equip. Oper. (med.)	$41.35	$330.80	$65.05	$520.40	$38.10	$60.75
.5 Laborer	31.60	126.40	52.15	208.60		
1 Diaphragm Water Pump, 4"		86.20		94.82		
1 -20' Suction Hose, 4"		3.45		3.79		
2 -50' Discharge Hoses, 4"		4.70		5.17	7.86	8.65
12 L.H., Daily Totals		$551.55		$832.78	$45.96	$69.40

Crew B-10J

	Bare Costs Hr.	Bare Costs Daily	Incl. Subs O&P Hr.	Incl. Subs O&P Daily	Cost Per Labor-Hour Bare Costs	Cost Per Labor-Hour Incl. O&P
1 Equip. Oper. (med.)	$41.35	$330.80	$65.05	$520.40	$38.10	$60.75
.5 Laborer	31.60	126.40	52.15	208.60		
1 Centrifugal Water Pump, 3"		69.40		76.34		
1 -20' Suction Hose, 3"		3.05		3.36		
2 -50' Discharge Hoses, 3"		3.50		3.85	6.33	6.96
12 L.H., Daily Totals		$533.15		$812.54	$44.43	$67.71

Crew No.	Bare Costs		Incl. Subs O & P		Cost Per Labor-Hour	
Crew B-10K	**Hr.**	**Daily**	**Hr.**	**Daily**	**Bare Costs**	**Incl. O&P**
1 Equip. Oper. (med.)	$41.35	$330.80	$65.05	$520.40	$38.10	$60.75
.5 Laborer	31.60	126.40	52.15	208.60		
1 Centr. Water Pump, 6"		309.40		340.34		
1 -20' Suction Hose, 6"		11.90		13.09		
2 -50' Discharge Hoses, 6"		12.60		13.86	27.82	30.61
12 L.H., Daily Totals		$791.10		$1096.29	$65.92	$91.36
Crew B-10L	**Hr.**	**Daily**	**Hr.**	**Daily**	**Bare Costs**	**Incl. O&P**
1 Equip. Oper. (med.)	$41.35	$330.80	$65.05	$520.40	$38.10	$60.75
.5 Laborer	31.60	126.40	52.15	208.60		
1 Dozer, 80 H.P.		399.20		439.12	33.27	36.59
12 L.H., Daily Totals		$856.40		$1168.12	$71.37	$97.34
Crew B-10M	**Hr.**	**Daily**	**Hr.**	**Daily**	**Bare Costs**	**Incl. O&P**
1 Equip. Oper. (med.)	$41.35	$330.80	$65.05	$520.40	$38.10	$60.75
.5 Laborer	31.60	126.40	52.15	208.60		
1 Dozer, 300 H.P.		1423.00		1565.30	118.58	130.44
12 L.H., Daily Totals		$1880.20		$2294.30	$156.68	$191.19
Crew B-10N	**Hr.**	**Daily**	**Hr.**	**Daily**	**Bare Costs**	**Incl. O&P**
1 Equip. Oper. (med.)	$41.35	$330.80	$65.05	$520.40	$38.10	$60.75
.5 Laborer	31.60	126.40	52.15	208.60		
1 F.E. Loader, T.M., 1.5 C.Y		367.00		403.70	30.58	33.64
12 L.H., Daily Totals		$824.20		$1132.70	$68.68	$94.39
Crew B-10O	**Hr.**	**Daily**	**Hr.**	**Daily**	**Bare Costs**	**Incl. O&P**
1 Equip. Oper. (med.)	$41.35	$330.80	$65.05	$520.40	$38.10	$60.75
.5 Laborer	31.60	126.40	52.15	208.60		
1 F.E. Loader, T.M., 2.25 C.Y.		767.40		844.14	63.95	70.34
12 L.H., Daily Totals		$1224.60		$1573.14	$102.05	$131.10
Crew B-10P	**Hr.**	**Daily**	**Hr.**	**Daily**	**Bare Costs**	**Incl. O&P**
1 Equip. Oper. (med.)	$41.35	$330.80	$65.05	$520.40	$38.10	$60.75
.5 Laborer	31.60	126.40	52.15	208.60		
1 Crawler Loader, 3 C.Y.		990.00		1089.00	82.50	90.75
12 L.H., Daily Totals		$1447.20		$1818.00	$120.60	$151.50
Crew B-10Q	**Hr.**	**Daily**	**Hr.**	**Daily**	**Bare Costs**	**Incl. O&P**
1 Equip. Oper. (med.)	$41.35	$330.80	$65.05	$520.40	$38.10	$60.75
.5 Laborer	31.60	126.40	52.15	208.60		
1 Crawler Loader, 4 C.Y.		1450.00		1595.00	120.83	132.92
12 L.H., Daily Totals		$1907.20		$2324.00	$158.93	$193.67
Crew B-10R	**Hr.**	**Daily**	**Hr.**	**Daily**	**Bare Costs**	**Incl. O&P**
1 Equip. Oper. (med.)	$41.35	$330.80	$65.05	$520.40	$38.10	$60.75
.5 Laborer	31.60	126.40	52.15	208.60		
1 F.E. Loader, W.M., 1 C.Y.		239.00		262.90	19.92	21.91
12 L.H., Daily Totals		$696.20		$991.90	$58.02	$82.66
Crew B-10S	**Hr.**	**Daily**	**Hr.**	**Daily**	**Bare Costs**	**Incl. O&P**
1 Equip. Oper. (med.)	$41.35	$330.80	$65.05	$520.40	$38.10	$60.75
.5 Laborer	31.60	126.40	52.15	208.60		
1 F.E. Loader, W.M., 1.5 C.Y.		328.20		361.02	27.35	30.09
12 L.H., Daily Totals		$785.40		$1090.02	$65.45	$90.83
Crew B-10T	**Hr.**	**Daily**	**Hr.**	**Daily**	**Bare Costs**	**Incl. O&P**
1 Equip. Oper. (med.)	$41.35	$330.80	$65.05	$520.40	$38.10	$60.75
.5 Laborer	31.60	126.40	52.15	208.60		
1 F.E. Loader, W.M.,2.5 C.Y.		395.80		435.38	32.98	36.28
12 L.H., Daily Totals		$853.00		$1164.38	$71.08	$97.03

Crew No.	Bare Costs		Incl. Subs O & P		Cost Per Labor-Hour	
Crew B-10U	**Hr.**	**Daily**	**Hr.**	**Daily**	**Bare Costs**	**Incl. O&P**
1 Equip. Oper. (med.)	$41.35	$330.80	$65.05	$520.40	$38.10	$60.75
.5 Laborer	31.60	126.40	52.15	208.60		
1 F.E. Loader, W.M., 5.5 C.Y.		854.40		939.84	71.20	78.32
12 L.H., Daily Totals		$1311.60		$1668.84	$109.30	$139.07
Crew B-10V	**Hr.**	**Daily**	**Hr.**	**Daily**	**Bare Costs**	**Incl. O&P**
1 Equip. Oper. (med.)	$41.35	$330.80	$65.05	$520.40	$38.10	$60.75
.5 Laborer	31.60	126.40	52.15	208.60		
1 Dozer, 700 H.P.		4232.00		4655.20	352.67	387.93
12 L.H., Daily Totals		$4689.20		$5384.20	$390.77	$448.68
Crew B-10W	**Hr.**	**Daily**	**Hr.**	**Daily**	**Bare Costs**	**Incl. O&P**
1 Equip. Oper. (med.)	$41.35	$330.80	$65.05	$520.40	$38.10	$60.75
.5 Laborer	31.60	126.40	52.15	208.60		
1 Dozer, 105 H.P.		600.60		660.66	50.05	55.06
12 L.H., Daily Totals		$1057.80		$1389.66	$88.15	$115.81
Crew B-10X	**Hr.**	**Daily**	**Hr.**	**Daily**	**Bare Costs**	**Incl. O&P**
1 Equip. Oper. (med.)	$41.35	$330.80	$65.05	$520.40	$38.10	$60.75
.5 Laborer	31.60	126.40	52.15	208.60		
1 Dozer, 410 H.P.		1849.00		2033.90	154.08	169.49
12 L.H., Daily Totals		$2306.20		$2762.90	$192.18	$230.24
Crew B-10Y	**Hr.**	**Daily**	**Hr.**	**Daily**	**Bare Costs**	**Incl. O&P**
1 Equip. Oper. (med.)	$41.35	$330.80	$65.05	$520.40	$38.10	$60.75
.5 Laborer	31.60	126.40	52.15	208.60		
1 Vibr. Roller, Towed, 12 Ton		469.60		516.56	39.13	43.05
12 L.H., Daily Totals		$926.80		$1245.56	$77.23	$103.80
Crew B-11A	**Hr.**	**Daily**	**Hr.**	**Daily**	**Bare Costs**	**Incl. O&P**
1 Equipment Oper. (med.)	$41.35	$330.80	$65.05	$520.40	$36.48	$58.60
1 Laborer	31.60	252.80	52.15	417.20		
1 Dozer, 200 H.P.		1082.00		1190.20	67.63	74.39
16 L.H., Daily Totals		$1665.60		$2127.80	$104.10	$132.99
Crew B-11B	**Hr.**	**Daily**	**Hr.**	**Daily**	**Bare Costs**	**Incl. O&P**
1 Equipment Oper. (light)	$39.05	$312.40	$61.45	$491.60	$35.33	$56.80
1 Laborer	31.60	252.80	52.15	417.20		
1 Air Powered Tamper		24.00		26.40		
1 Air Compressor, 365 cfm		211.20		232.32		
2 -50' Air Hoses, 1.5"		12.60		13.86	15.49	17.04
16 L.H., Daily Totals		$813.00		$1181.38	$50.81	$73.84
Crew B-11C	**Hr.**	**Daily**	**Hr.**	**Daily**	**Bare Costs**	**Incl. O&P**
1 Equipment Oper. (med.)	$41.35	$330.80	$65.05	$520.40	$36.48	$58.60
1 Laborer	31.60	252.80	52.15	417.20		
1 Backhoe Loader, 48 H.P.		293.80		323.18	18.36	20.20
16 L.H., Daily Totals		$877.40		$1260.78	$54.84	$78.80
Crew B-11K	**Hr.**	**Daily**	**Hr.**	**Daily**	**Bare Costs**	**Incl. O&P**
1 Equipment Oper. (med.)	$41.35	$330.80	$65.05	$520.40	$36.48	$58.60
1 Laborer	31.60	252.80	52.15	417.20		
1 Trencher, Chain Type, 8' D		1777.00		1954.70	111.06	122.17
16 L.H., Daily Totals		$2360.60		$2892.30	$147.54	$180.77
Crew B-11L	**Hr.**	**Daily**	**Hr.**	**Daily**	**Bare Costs**	**Incl. O&P**
1 Equipment Oper. (med.)	$41.35	$330.80	$65.05	$520.40	$36.48	$58.60
1 Laborer	31.60	252.80	52.15	417.20		
1 Grader, 30,000 Lbs.		550.20		605.22	34.39	37.83
16 L.H., Daily Totals		$1133.80		$1542.82	$70.86	$96.43

Crews

Crew B-11M

Crew No.	Bare Costs Hr.	Daily	Incl. Subs O & P Hr.	Daily	Cost Per Labor-Hour Bare Costs	Incl. O&P
1 Equipment Oper. (med.)	$41.35	$330.80	$65.05	$520.40	$36.48	$58.60
1 Laborer	31.60	252.80	52.15	417.20		
1 Backhoe Loader, 80 H.P.		345.00		379.50	21.56	23.72
16 L.H., Daily Totals		$928.60		$1317.10	$58.04	$82.32

Crew B-11S

Crew No.	Bare Costs Hr.	Daily	Incl. Subs O & P Hr.	Daily	Cost Per Labor-Hour Bare Costs	Incl. O&P
1 Equipment Operator (med.)	$41.35	$330.80	$65.05	$520.40	$38.10	$60.75
.5 Laborer	31.60	126.40	52.15	208.60		
1 Dozer, 300 H.P.		1423.00		1565.30		
1 Ripper, beam & 1 shank		79.60		87.56	125.22	137.74
12 L.H., Daily Totals		$1959.80		$2381.86	$163.32	$198.49

Crew B-11T

Crew No.	Bare Costs Hr.	Daily	Incl. Subs O & P Hr.	Daily	Cost Per Labor-Hour Bare Costs	Incl. O&P
1 Equipment Operator (med.)	$41.35	$330.80	$65.05	$520.40	$38.10	$60.75
.5 Laborer	31.60	126.40	52.15	208.60		
1 Dozer, 410 H.P.		1849.00		2033.90		
1 Ripper, beam & 2 shanks		89.40		98.34	161.53	177.69
12 L.H., Daily Totals		$2395.60		$2861.24	$199.63	$238.44

Crew B-11W

Crew No.	Bare Costs Hr.	Daily	Incl. Subs O & P Hr.	Daily	Cost Per Labor-Hour Bare Costs	Incl. O&P
1 Equipment Operator (med.)	$41.35	$330.80	$65.05	$520.40	$32.70	$53.39
1 Common Laborer	31.60	252.80	52.15	417.20		
10 Truck Drivers (hvy.)	31.95	2556.00	52.35	4188.00		
1 Dozer, 200 H.P.		1082.00		1190.20		
1 Vibratory Roller, Towed, 23 Ton		411.00		452.10		
10 Dump Trucks, 8 C.Y., 220 H.P.		3264.00		3590.40	49.55	54.51
96 L.H., Daily Totals		$7896.60		$10358.30	$82.26	$107.90

Crew B-11Y

Crew No.	Bare Costs Hr.	Daily	Incl. Subs O & P Hr.	Daily	Cost Per Labor-Hour Bare Costs	Incl. O&P
1 Labor Foreman (Outside)	$33.60	$268.80	$55.45	$443.60	$35.07	$56.82
5 Common Laborers	31.60	1264.00	52.15	2086.00		
3 Equipment Operators (med.)	41.35	992.40	65.05	1561.20		
1 Dozer, 80 H.P.		399.20		439.12		
2 Roller, 2-Drum, W.B., 7.5 H.P.		290.40		319.44		
4 Vibratory Plates, Gas, 21"		156.80		172.48	11.76	12.93
72 L.H., Daily Totals		$3371.60		$5021.84	$46.83	$69.75

Crew B-12A

Crew No.	Bare Costs Hr.	Daily	Incl. Subs O & P Hr.	Daily	Cost Per Labor-Hour Bare Costs	Incl. O&P
1 Equip. Oper. (crane)	$42.55	$340.40	$66.95	$535.60	$37.08	$59.55
1 Laborer	31.60	252.80	52.15	417.20		
1 Hyd. Excavator, 1 C.Y.		669.20		736.12	41.83	46.01
16 L.H., Daily Totals		$1262.40		$1688.92	$78.90	$105.56

Crew B-12B

Crew No.	Bare Costs Hr.	Daily	Incl. Subs O & P Hr.	Daily	Cost Per Labor-Hour Bare Costs	Incl. O&P
1 Equip. Oper. (crane)	$42.55	$340.40	$66.95	$535.60	$37.08	$59.55
1 Laborer	31.60	252.80	52.15	417.20		
1 Hyd. Excavator, 1.5 C.Y.		865.80		952.38	54.11	59.52
16 L.H., Daily Totals		$1459.00		$1905.18	$91.19	$119.07

Crew B-12C

Crew No.	Bare Costs Hr.	Daily	Incl. Subs O & P Hr.	Daily	Cost Per Labor-Hour Bare Costs	Incl. O&P
1 Equip. Oper. (crane)	$42.55	$340.40	$66.95	$535.60	$37.08	$59.55
1 Laborer	31.60	252.80	52.15	417.20		
1 Hyd. Excavator, 2 C.Y.		1182.00		1300.20	73.88	81.26
16 L.H., Daily Totals		$1775.20		$2253.00	$110.95	$140.81

Crew B-12D

Crew No.	Bare Costs Hr.	Daily	Incl. Subs O & P Hr.	Daily	Cost Per Labor-Hour Bare Costs	Incl. O&P
1 Equip. Oper. (crane)	$42.55	$340.40	$66.95	$535.60	$37.08	$59.55
1 Laborer	31.60	252.80	52.15	417.20		
1 Hyd. Excavator, 3.5 C.Y.		2162.00		2378.20	135.13	148.64
16 L.H., Daily Totals		$2755.20		$3331.00	$172.20	$208.19

Crew B-12E

Crew No.	Bare Costs Hr.	Daily	Incl. Subs O & P Hr.	Daily	Cost Per Labor-Hour Bare Costs	Incl. O&P
1 Equip. Oper. (crane)	$42.55	$340.40	$66.95	$535.60	$37.08	$59.55
1 Laborer	31.60	252.80	52.15	417.20		
1 Hyd. Excavator, .5 C.Y.		394.60		434.06	24.66	27.13
16 L.H., Daily Totals		$987.80		$1386.86	$61.74	$86.68

Crew B-12F

Crew No.	Bare Costs Hr.	Daily	Incl. Subs O & P Hr.	Daily	Cost Per Labor-Hour Bare Costs	Incl. O&P
1 Equip. Oper. (crane)	$42.55	$340.40	$66.95	$535.60	$37.08	$59.55
1 Laborer	31.60	252.80	52.15	417.20		
1 Hyd. Excavator, .75 C.Y.		598.60		658.46	37.41	41.15
16 L.H., Daily Totals		$1191.80		$1611.26	$74.49	$100.70

Crew B-12G

Crew No.	Bare Costs Hr.	Daily	Incl. Subs O & P Hr.	Daily	Cost Per Labor-Hour Bare Costs	Incl. O&P
1 Equip. Oper. (crane)	$42.55	$340.40	$66.95	$535.60	$37.08	$59.55
1 Laborer	31.60	252.80	52.15	417.20		
1 Crawler Crane, 15 Ton		615.30		676.83		
1 Clamshell Bucket, .5 C.Y.		36.60		40.26	40.74	44.82
16 L.H., Daily Totals		$1245.10		$1669.89	$77.82	$104.37

Crew B-12H

Crew No.	Bare Costs Hr.	Daily	Incl. Subs O & P Hr.	Daily	Cost Per Labor-Hour Bare Costs	Incl. O&P
1 Equip. Oper. (crane)	$42.55	$340.40	$66.95	$535.60	$37.08	$59.55
1 Laborer	31.60	252.80	52.15	417.20		
1 Crawler Crane, 25 Ton		1052.00		1157.20		
1 Clamshell Bucket, 1 C.Y.		46.20		50.82	68.64	75.50
16 L.H., Daily Totals		$1691.40		$2160.82	$105.71	$135.05

Crew B-12I

Crew No.	Bare Costs Hr.	Daily	Incl. Subs O & P Hr.	Daily	Cost Per Labor-Hour Bare Costs	Incl. O&P
1 Equip. Oper. (crane)	$42.55	$340.40	$66.95	$535.60	$37.08	$59.55
1 Laborer	31.60	252.80	52.15	417.20		
1 Crawler Crane, 20 Ton		789.05		867.96		
1 Dragline Bucket, .75 C.Y.		20.00		22.00	50.57	55.62
16 L.H., Daily Totals		$1402.25		$1842.76	$87.64	$115.17

Crew B-12J

Crew No.	Bare Costs Hr.	Daily	Incl. Subs O & P Hr.	Daily	Cost Per Labor-Hour Bare Costs	Incl. O&P
1 Equip. Oper. (crane)	$42.55	$340.40	$66.95	$535.60	$37.08	$59.55
1 Laborer	31.60	252.80	52.15	417.20		
1 Gradall, 5/8 C.Y.		958.80		1054.68	59.92	65.92
16 L.H., Daily Totals		$1552.00		$2007.48	$97.00	$125.47

Crew B-12K

Crew No.	Bare Costs Hr.	Daily	Incl. Subs O & P Hr.	Daily	Cost Per Labor-Hour Bare Costs	Incl. O&P
1 Equip. Oper. (crane)	$42.55	$340.40	$66.95	$535.60	$37.08	$59.55
1 Laborer	31.60	252.80	52.15	417.20		
1 Gradall, 3 Ton, 1 C.Y.		1130.00		1243.00	70.63	77.69
16 L.H., Daily Totals		$1723.20		$2195.80	$107.70	$137.24

Crew B-12L

Crew No.	Bare Costs Hr.	Daily	Incl. Subs O & P Hr.	Daily	Cost Per Labor-Hour Bare Costs	Incl. O&P
1 Equip. Oper. (crane)	$42.55	$340.40	$66.95	$535.60	$37.08	$59.55
1 Laborer	31.60	252.80	52.15	417.20		
1 Crawler Crane, 15 Ton		615.30		676.83		
1 F.E. Attachment, .5 C.Y.		52.80		58.08	41.76	45.93
16 L.H., Daily Totals		$1261.30		$1687.71	$78.83	$105.48

Crew B-12M

Crew No.	Bare Costs Hr.	Daily	Incl. Subs O & P Hr.	Daily	Cost Per Labor-Hour Bare Costs	Incl. O&P
1 Equip. Oper. (crane)	$42.55	$340.40	$66.95	$535.60	$37.08	$59.55
1 Laborer	31.60	252.80	52.15	417.20		
1 Crawler Crane, 20 Ton		789.05		867.96		
1 F.E. Attachment, .75 C.Y.		58.20		64.02	52.95	58.25
16 L.H., Daily Totals		$1440.45		$1884.78	$90.03	$117.80

Crew B-12N

	Bare Costs Hr.	Bare Costs Daily	Incl. Subs O & P Hr.	Incl. Subs O & P Daily	Cost Per Labor-Hour Bare Costs	Cost Per Labor-Hour Incl. O&P
1 Equip. Oper. (crane)	$42.55	$340.40	$66.95	$535.60	$37.08	$59.55
1 Laborer	31.60	252.80	52.15	417.20		
1 Crawler Crane, 25 Ton		1052.00		1157.20		
1 F.E. Attachment, 1 C.Y.		64.00		70.40	69.75	76.72
16 L.H., Daily Totals		$1709.20		$2180.40	$106.83	$136.28

Crew B-12O

	Bare Costs Hr.	Bare Costs Daily	Incl. Subs O & P Hr.	Incl. Subs O & P Daily	Cost Per Labor-Hour Bare Costs	Cost Per Labor-Hour Incl. O&P
1 Equip. Oper. (crane)	$42.55	$340.40	$66.95	$535.60	$37.08	$59.55
1 Laborer	31.60	252.80	52.15	417.20		
1 Crawler Crane, 40 Ton		1055.00		1160.50		
1 F.E. Attachment, 1.5 C.Y.		74.20		81.62	70.58	77.63
16 L.H., Daily Totals		$1722.40		$2194.92	$107.65	$137.18

Crew B-12P

	Bare Costs Hr.	Bare Costs Daily	Incl. Subs O & P Hr.	Incl. Subs O & P Daily	Cost Per Labor-Hour Bare Costs	Cost Per Labor-Hour Incl. O&P
1 Equip. Oper. (crane)	$42.55	$340.40	$66.95	$535.60	$37.08	$59.55
1 Laborer	31.60	252.80	52.15	417.20		
1 Crawler Crane, 40 Ton		1055.00		1160.50		
1 Dragline Bucket, 1.5 C.Y.		32.80		36.08	67.99	74.79
16 L.H., Daily Totals		$1681.00		$2149.38	$105.06	$134.34

Crew B-12Q

	Bare Costs Hr.	Bare Costs Daily	Incl. Subs O & P Hr.	Incl. Subs O & P Daily	Cost Per Labor-Hour Bare Costs	Cost Per Labor-Hour Incl. O&P
1 Equip. Oper. (crane)	$42.55	$340.40	$66.95	$535.60	$37.08	$59.55
1 Laborer	31.60	252.80	52.15	417.20		
1 Hyd. Excavator, 5/8 C.Y.		517.20		568.92	32.33	35.56
16 L.H., Daily Totals		$1110.40		$1521.72	$69.40	$95.11

Crew B-12S

	Bare Costs Hr.	Bare Costs Daily	Incl. Subs O & P Hr.	Incl. Subs O & P Daily	Cost Per Labor-Hour Bare Costs	Cost Per Labor-Hour Incl. O&P
1 Equip. Oper. (crane)	$42.55	$340.40	$66.95	$535.60	$37.08	$59.55
1 Laborer	31.60	252.80	52.15	417.20		
1 Hyd. Excavator, 2.5 C.Y.		1619.00		1780.90	101.19	111.31
16 L.H., Daily Totals		$2212.20		$2733.70	$138.26	$170.86

Crew B-12T

	Bare Costs Hr.	Bare Costs Daily	Incl. Subs O & P Hr.	Incl. Subs O & P Daily	Cost Per Labor-Hour Bare Costs	Cost Per Labor-Hour Incl. O&P
1 Equip. Oper. (crane)	$42.55	$340.40	$66.95	$535.60	$37.08	$59.55
1 Laborer	31.60	252.80	52.15	417.20		
1 Crawler Crane, 75 Ton		1496.00		1645.60		
1 F.E. Attachment, 3 C.Y.		97.60		107.36	99.60	109.56
16 L.H., Daily Totals		$2186.80		$2705.76	$136.68	$169.11

Crew B-12V

	Bare Costs Hr.	Bare Costs Daily	Incl. Subs O & P Hr.	Incl. Subs O & P Daily	Cost Per Labor-Hour Bare Costs	Cost Per Labor-Hour Incl. O&P
1 Equip. Oper. (crane)	$42.55	$340.40	$66.95	$535.60	$37.08	$59.55
1 Laborer	31.60	252.80	52.15	417.20		
1 Crawler Crane, 75 Ton		1496.00		1645.60		
1 Dragline Bucket, 3 C.Y.		46.20		50.82	96.39	106.03
16 L.H., Daily Totals		$2135.40		$2649.22	$133.46	$165.58

Crew B-13

	Bare Costs Hr.	Bare Costs Daily	Incl. Subs O & P Hr.	Incl. Subs O & P Daily	Cost Per Labor-Hour Bare Costs	Cost Per Labor-Hour Incl. O&P
1 Labor Foreman (outside)	$33.60	$268.80	$55.45	$443.60	$34.19	$55.56
4 Laborers	31.60	1011.20	52.15	1668.80		
1 Equip. Oper. (crane)	42.55	340.40	66.95	535.60		
1 Equip. Oper. Oiler	36.80	294.40	57.90	463.20		
1 Hyd. Crane, 25 Ton		789.40		868.34	14.10	15.51
56 L.H., Daily Totals		$2704.20		$3979.54	$48.29	$71.06

Crew B-13A

	Bare Costs Hr.	Bare Costs Daily	Incl. Subs O & P Hr.	Incl. Subs O & P Daily	Cost Per Labor-Hour Bare Costs	Cost Per Labor-Hour Incl. O&P
1 Foreman	$33.60	$268.80	$55.45	$443.60	$34.77	$56.36
2 Laborers	31.60	505.60	52.15	834.40		
2 Equipment Operators	41.35	661.60	65.05	1040.80		
2 Truck Drivers (heavy)	31.95	511.20	52.35	837.60		
1 Crawler Crane, 75 Ton		1496.00		1645.60		
1 Crawler Loader, 4 C.Y.		1450.00		1595.00		
2 Dump Trucks, 8 C.Y., 220 H.P.		652.80		718.08	64.26	70.69
56 L.H., Daily Totals		$5546.00		$7115.08	$99.04	$127.06

Crew B-13B

	Bare Costs Hr.	Bare Costs Daily	Incl. Subs O & P Hr.	Incl. Subs O & P Daily	Cost Per Labor-Hour Bare Costs	Cost Per Labor-Hour Incl. O&P
1 Labor Foreman (outside)	$33.60	$268.80	$55.45	$443.60	$34.19	$55.56
4 Laborers	31.60	1011.20	52.15	1668.80		
1 Equip. Oper. (crane)	42.55	340.40	66.95	535.60		
1 Equip. Oper. Oiler	36.80	294.40	57.90	463.20		
1 Hyd. Crane, 55 Ton		1125.00		1237.50	20.09	22.10
56 L.H., Daily Totals		$3039.80		$4348.70	$54.28	$77.66

Crew B-13C

	Bare Costs Hr.	Bare Costs Daily	Incl. Subs O & P Hr.	Incl. Subs O & P Daily	Cost Per Labor-Hour Bare Costs	Cost Per Labor-Hour Incl. O&P
1 Labor Foreman (outside)	$33.60	$268.80	$55.45	$443.60	$34.19	$55.56
4 Laborers	31.60	1011.20	52.15	1668.80		
1 Equip. Oper. (crane)	42.55	340.40	66.95	535.60		
1 Equip. Oper. Oiler	36.80	294.40	57.90	463.20		
1 Crawler Crane, 100 Ton		1611.00		1772.10	28.77	31.64
56 L.H., Daily Totals		$3525.80		$4883.30	$62.96	$87.20

Crew B-13D

	Bare Costs Hr.	Bare Costs Daily	Incl. Subs O & P Hr.	Incl. Subs O & P Daily	Cost Per Labor-Hour Bare Costs	Cost Per Labor-Hour Incl. O&P
1 Laborer	$31.60	$252.80	$52.15	$417.20	$37.08	$59.55
1 Equip. Oper. (crane)	42.55	340.40	66.95	535.60		
1 Hyd. Excavator, 1 C.Y.		669.20		736.12		
1 Trench Box		113.40		124.74	48.91	53.80
16 L.H., Daily Totals		$1375.80		$1813.66	$85.99	$113.35

Crew B-13E

	Bare Costs Hr.	Bare Costs Daily	Incl. Subs O & P Hr.	Incl. Subs O & P Daily	Cost Per Labor-Hour Bare Costs	Cost Per Labor-Hour Incl. O&P
1 Laborer	$31.60	$252.80	$52.15	$417.20	$37.08	$59.55
1 Equip. Oper. (crane)	42.55	340.40	66.95	535.60		
1 Hyd. Excavator, 1.5 C.Y.		865.80		952.38		
1 Trench Box		113.40		124.74	61.20	67.32
16 L.H., Daily Totals		$1572.40		$2029.92	$98.28	$126.87

Crew B-13F

	Bare Costs Hr.	Bare Costs Daily	Incl. Subs O & P Hr.	Incl. Subs O & P Daily	Cost Per Labor-Hour Bare Costs	Cost Per Labor-Hour Incl. O&P
1 Laborer	$31.60	$252.80	$52.15	$417.20	$37.08	$59.55
1 Equip. Oper. (crane)	42.55	340.40	66.95	535.60		
1 Hyd. Excavator, 3.5 C.Y.		2162.00		2378.20		
1 Trench Box		113.40		124.74	142.21	156.43
16 L.H., Daily Totals		$2868.60		$3455.74	$179.29	$215.98

Crew B-13G

	Bare Costs Hr.	Bare Costs Daily	Incl. Subs O & P Hr.	Incl. Subs O & P Daily	Cost Per Labor-Hour Bare Costs	Cost Per Labor-Hour Incl. O&P
1 Laborer	$31.60	$252.80	$52.15	$417.20	$37.08	$59.55
1 Equip. Oper. (crane)	42.55	340.40	66.95	535.60		
1 Hyd. Excavator, .75 C.Y.		598.60		658.46		
1 Trench Box		113.40		124.74	44.50	48.95
16 L.H., Daily Totals		$1305.20		$1736.00	$81.58	$108.50

Crew B-13H

	Bare Costs Hr.	Bare Costs Daily	Incl. Subs O & P Hr.	Incl. Subs O & P Daily	Cost Per Labor-Hour Bare Costs	Cost Per Labor-Hour Incl. O&P
1 Laborer	$31.60	$252.80	$52.15	$417.20	$37.08	$59.55
1 Equip. Oper. (crane)	42.55	340.40	66.95	535.60		
1 Gradall, 5/8 C.Y.		958.80		1054.68		
1 Trench Box		113.40		124.74	67.01	73.71
16 L.H., Daily Totals		$1665.40		$2132.22	$104.09	$133.26

Crew B-13I

Crew No.	Bare Costs Hr.	Daily	Incl. Subs O & P Hr.	Daily	Cost Per Labor-Hour Bare Costs	Incl. O&P
1 Laborer	$31.60	$252.80	$52.15	$417.20	$37.08	$59.55
1 Equip. Oper. (crane)	42.55	340.40	66.95	535.60		
1 Gradall, 3 Ton, 1 C.Y.		1130.00		1243.00		
1 Trench Box		113.40		124.74	77.71	85.48
16 L.H., Daily Totals		$1836.60		$2320.54	$114.79	$145.03

Crew B-13J

Crew No.	Hr.	Daily	Hr.	Daily	Bare Costs	Incl. O&P
1 Laborer	$31.60	$252.80	$52.15	$417.20	$37.08	$59.55
1 Equip. Oper. (crane)	42.55	340.40	66.95	535.60		
1 Hyd. Excavator, 2.5 C.Y.		1619.00		1780.90		
1 Trench Box		113.40		124.74	108.28	119.10
16 L.H., Daily Totals		$2325.60		$2858.44	$145.35	$178.65

Crew B-14

Crew No.	Hr.	Daily	Hr.	Daily	Bare Costs	Incl. O&P
1 Labor Foreman (outside)	$33.60	$268.80	$55.45	$443.60	$33.17	$54.25
4 Laborers	31.60	1011.20	52.15	1668.80		
1 Equip. Oper. (light)	39.05	312.40	61.45	491.60		
1 Backhoe Loader, 48 H.P.		293.80		323.18	6.12	6.73
48 L.H., Daily Totals		$1886.20		$2927.18	$39.30	$60.98

Crew B-14A

Crew No.	Hr.	Daily	Hr.	Daily	Bare Costs	Incl. O&P
1 Equip. Oper. (crane)	$42.55	$340.40	$66.95	$535.60	$38.90	$62.02
.5 Laborer	31.60	126.40	52.15	208.60		
1 Hyd. Excavator, 4.5 C.Y.		2635.00		2898.50	219.58	241.54
12 L.H., Daily Totals		$3101.80		$3642.70	$258.48	$303.56

Crew B-14B

Crew No.	Hr.	Daily	Hr.	Daily	Bare Costs	Incl. O&P
1 Equip. Oper. (crane)	$42.55	$340.40	$66.95	$535.60	$38.90	$62.02
.5 Laborer	31.60	126.40	52.15	208.60		
1 Hyd. Excavator, 6 C.Y.		3080.00		3388.00	256.67	282.33
12 L.H., Daily Totals		$3546.80		$4132.20	$295.57	$344.35

Crew B-14C

Crew No.	Hr.	Daily	Hr.	Daily	Bare Costs	Incl. O&P
1 Equip. Oper. (crane)	$42.55	$340.40	$66.95	$535.60	$38.90	$62.02
.5 Laborer	31.60	126.40	52.15	208.60		
1 Hyd. Excavator, 7 C.Y.		3256.00		3581.60	271.33	298.47
12 L.H., Daily Totals		$3722.80		$4325.80	$310.23	$360.48

Crew B-14F

Crew No.	Hr.	Daily	Hr.	Daily	Bare Costs	Incl. O&P
1 Equip. Oper. (crane)	$42.55	$340.40	$66.95	$535.60	$38.90	$62.02
.5 Laborer	31.60	126.40	52.15	208.60		
1 Hyd. Shovel, 7 C.Y.		3090.00		3399.00	257.50	283.25
12 L.H., Daily Totals		$3556.80		$4143.20	$296.40	$345.27

Crew B-14G

Crew No.	Hr.	Daily	Hr.	Daily	Bare Costs	Incl. O&P
1 Equip. Oper. (crane)	$42.55	$340.40	$66.95	$535.60	$38.90	$62.02
.5 Laborer	31.60	126.40	52.15	208.60		
1 Hyd. Shovel, 12 C.Y.		4148.00		4562.80	345.67	380.23
12 L.H., Daily Totals		$4614.80		$5307.00	$384.57	$442.25

Crew B-14J

Crew No.	Hr.	Daily	Hr.	Daily	Bare Costs	Incl. O&P
1 Equip. Oper. (med.)	$41.35	$330.80	$65.05	$520.40	$38.10	$60.75
.5 Laborer	31.60	126.40	52.15	208.60		
1 F.E. Loader, 8 C.Y.		1497.00		1646.70	124.75	137.22
12 L.H., Daily Totals		$1954.20		$2375.70	$162.85	$197.97

Crew B-14K

Crew No.	Hr.	Daily	Hr.	Daily	Bare Costs	Incl. O&P
1 Equip. Oper. (med.)	$41.35	$330.80	$65.05	$520.40	$38.10	$60.75
.5 Laborer	31.60	126.40	52.15	208.60		
1 F.E. Loader, 10 C.Y.		2225.00		2447.50	185.42	203.96
12 L.H., Daily Totals		$2682.20		$3176.50	$223.52	$264.71

Crew B-15

Crew No.	Hr.	Daily	Hr.	Daily	Bare Costs	Incl. O&P
1 Equipment Oper. (med)	$41.35	$330.80	$65.05	$520.40	$34.59	$55.95
.5 Laborer	31.60	126.40	52.15	208.60		
2 Truck Drivers (heavy)	31.95	511.20	52.35	837.60		
2 Dump Trucks, 12 C.Y., 400 H.P.		1066.00		1172.60		
1 Dozer, 200 H.P.		1082.00		1190.20	76.71	84.39
28 L.H., Daily Totals		$3116.40		$3929.40	$111.30	$140.34

Crew B-16

Crew No.	Hr.	Daily	Hr.	Daily	Bare Costs	Incl. O&P
1 Labor Foreman (outside)	$33.60	$268.80	$55.45	$443.60	$32.19	$53.02
2 Laborers	31.60	505.60	52.15	834.40		
1 Truck Driver (heavy)	31.95	255.60	52.35	418.80		
1 Dump Truck, 12 C.Y., 400 H.P.		533.00		586.30	16.66	18.32
32 L.H., Daily Totals		$1563.00		$2283.10	$48.84	$71.35

Crew B-17

Crew No.	Hr.	Daily	Hr.	Daily	Bare Costs	Incl. O&P
2 Laborers	$31.60	$505.60	$52.15	$834.40	$33.55	$54.52
1 Equip. Oper. (light)	39.05	312.40	61.45	491.60		
1 Truck Driver (heavy)	31.95	255.60	52.35	418.80		
1 Backhoe Loader, 48 H.P.		293.80		323.18		
1 Dump Truck, 8 C.Y., 220 H.P.		326.40		359.04	19.38	21.32
32 L.H., Daily Totals		$1693.80		$2427.02	$52.93	$75.84

Crew B-17A

Crew No.	Hr.	Daily	Hr.	Daily	Bare Costs	Incl. O&P
2 Laborer Foremen	$33.60	$537.60	$55.45	$887.20	$34.05	$56.01
6 Laborers	31.60	1516.80	52.15	2503.20		
1 Skilled Worker Foreman	42.85	342.80	69.75	558.00		
1 Skilled Worker	40.85	326.80	66.50	532.00		
80 L.H., Daily Totals		$2724.00		$4480.40	$34.05	$56.01

Crew B-18

Crew No.	Hr.	Daily	Hr.	Daily	Bare Costs	Incl. O&P
1 Labor Foreman (outside)	$33.60	$268.80	$55.45	$443.60	$32.27	$53.25
2 Laborers	31.60	505.60	52.15	834.40		
1 Vibratory Plate, Gas, 21"		39.20		43.12	1.63	1.80
24 L.H., Daily Totals		$813.60		$1321.12	$33.90	$55.05

Crew B-19

Crew No.	Hr.	Daily	Hr.	Daily	Bare Costs	Incl. O&P
1 Pile Driver Foreman	$40.50	$324.00	$67.80	$542.40	$39.55	$64.67
4 Pile Drivers	38.50	1232.00	64.45	2062.40		
2 Equip. Oper. (crane)	42.55	680.80	66.95	1071.20		
1 Equip. Oper. Oiler	36.80	294.40	57.90	463.20		
1 Crawler Crane, 40 Ton		1055.00		1160.50		
1 Lead, 90' high		118.80		130.68		
1 Hammer, Diesel, 22k Ft-Lb		590.00		649.00	27.56	30.32
64 L.H., Daily Totals		$4295.00		$6079.38	$67.11	$94.99

Crew B-19A

Crew No.	Hr.	Daily	Hr.	Daily	Bare Costs	Incl. O&P
1 Pile Driver Foreman	$40.50	$324.00	$67.80	$542.40	$39.55	$64.67
4 Pile Drivers	38.50	1232.00	64.45	2062.40		
2 Equip. Oper. (crane)	42.55	680.80	66.95	1071.20		
1 Equip. Oper. Oiler	36.80	294.40	57.90	463.20		
1 Crawler Crane, 75 Ton		1496.00		1645.60		
1 Lead, 90' high		118.80		130.68		
1 Hammer, Diesel, 41k Ft-Lb		677.60		745.36	35.82	39.40
64 L.H., Daily Totals		$4823.60		$6660.84	$75.37	$104.08

Crews

Crew No.	Bare Costs		Incl. Subs O & P		Cost Per Labor-Hour	

Crew B-20

	Hr.	Daily	Hr.	Daily	Bare Costs	Incl. O&P
1 Labor Foreman (out)	$33.60	$268.80	$55.45	$443.60	$35.35	$58.03
1 Skilled Worker	40.85	326.80	66.50	532.00		
1 Laborer	31.60	252.80	52.15	417.20		
24 L.H., Daily Totals		$848.40		$1392.80	$35.35	$58.03

Crew B-20A

	Hr.	Daily	Hr.	Daily	Bare Costs	Incl. O&P
1 Labor Foreman	$33.60	$268.80	$55.45	$443.60	$38.24	$60.92
1 Laborer	31.60	252.80	52.15	417.20		
1 Plumber	48.75	390.00	75.60	604.80		
1 Plumber Apprentice	39.00	312.00	60.50	484.00		
32 L.H., Daily Totals		$1223.60		$1949.60	$38.24	$60.92

Crew B-21

	Hr.	Daily	Hr.	Daily	Bare Costs	Incl. O&P
1 Labor Foreman (out)	$33.60	$268.80	$55.45	$443.60	$36.38	$59.31
1 Skilled Worker	40.85	326.80	66.50	532.00		
1 Laborer	31.60	252.80	52.15	417.20		
.5 Equip. Oper. (crane)	42.55	170.20	66.95	267.80		
.5 S.P. Crane, 4x4, 5 Ton		129.10		142.01	4.61	5.07
28 L.H., Daily Totals		$1147.70		$1802.61	$40.99	$64.38

Crew B-21A

	Hr.	Daily	Hr.	Daily	Bare Costs	Incl. O&P
1 Labor Foreman	$33.60	$268.80	$55.45	$443.60	$39.10	$62.13
1 Laborer	31.60	252.80	52.15	417.20		
1 Plumber	48.75	390.00	75.60	604.80		
1 Plumber Apprentice	39.00	312.00	60.50	484.00		
1 Equip. Oper. (crane)	42.55	340.40	66.95	535.60		
1 S.P. Crane, 4x4, 12 Ton		601.80		661.98	15.05	16.55
40 L.H., Daily Totals		$2165.80		$3147.18	$54.15	$78.68

Crew B-21B

	Hr.	Daily	Hr.	Daily	Bare Costs	Incl. O&P
1 Laborer Foreman	$33.60	$268.80	$55.45	$443.60	$34.19	$55.77
3 Laborers	31.60	758.40	52.15	1251.60		
1 Equip. Oper. (crane)	42.55	340.40	66.95	535.60		
1 Hyd. Crane, 12 Ton		768.80		845.68	19.22	21.14
40 L.H., Daily Totals		$2136.40		$3076.48	$53.41	$76.91

Crew B-21C

	Hr.	Daily	Hr.	Daily	Bare Costs	Incl. O&P
1 Laborer Foreman	$33.60	$268.80	$55.45	$443.60	$34.19	$55.56
4 Laborers	31.60	1011.20	52.15	1668.80		
1 Equip. Oper. (crane)	42.55	340.40	66.95	535.60		
1 Equip. Oper. Oiler	36.80	294.40	57.90	463.20		
2 Cutting Torches		34.00		37.40		
2 Gases		151.20		166.32		
1 Lattice Boom Crane, 90 Ton		1741.00		1915.10	34.40	37.84
56 L.H., Daily Totals		$3841.00		$5230.02	$68.59	$93.39

Crew B-22

	Hr.	Daily	Hr.	Daily	Bare Costs	Incl. O&P
1 Labor Foreman (out)	$33.60	$268.80	$55.45	$443.60	$36.79	$59.82
1 Skilled Worker	40.85	326.80	66.50	532.00		
1 Laborer	31.60	252.80	52.15	417.20		
.75 Equip. Oper. (crane)	42.55	255.30	66.95	401.70		
.75 S.P. Crane, 4x4, 5 Ton		193.65		213.01	6.46	7.10
30 L.H., Daily Totals		$1297.35		$2007.52	$43.24	$66.92

Crew B-22A

	Hr.	Daily	Hr.	Daily	Bare Costs	Incl. O&P
1 Labor Foreman (out)	$33.60	$268.80	$55.45	$443.60	$35.70	$58.20
1 Skilled Worker	40.85	326.80	66.50	532.00		
2 Laborers	31.60	505.60	52.15	834.40		
.75 Equipment Oper. (crane)	42.55	255.30	66.95	401.70		
.75 S.P. Crane, 4x4, 5 Ton		193.65		213.01		
1 Generator, 5 kW		40.20		44.22		
1 Butt Fusion Machine		199.80		219.78	11.41	12.55
38 L.H., Daily Totals		$1790.15		$2688.72	$47.11	$70.76

Crew B-22B

	Hr.	Daily	Hr.	Daily	Bare Costs	Incl. O&P
1 Skilled Worker	$40.85	$326.80	$66.50	$532.00	$36.23	$59.33
1 Laborer	31.60	252.80	52.15	417.20		
1 Electro Fusion Machine		146.80		161.48	9.18	10.09
16 L.H., Daily Totals		$726.40		$1110.68	$45.40	$69.42

Crew B-23

	Hr.	Daily	Hr.	Daily	Bare Costs	Incl. O&P
1 Labor Foreman (outside)	$33.60	$268.80	$55.45	$443.60	$32.00	$52.81
4 Laborers	31.60	1011.20	52.15	1668.80		
1 Drill Rig, Truck-Mounted		2533.00		2786.30		
1 Flatbed Truck, Gas, 3 Ton		242.60		266.86	69.39	76.33
40 L.H., Daily Totals		$4055.60		$5165.56	$101.39	$129.14

Crew B-23A

	Hr.	Daily	Hr.	Daily	Bare Costs	Incl. O&P
1 Labor Foreman (outside)	$33.60	$268.80	$55.45	$443.60	$35.52	$57.55
1 Laborer	31.60	252.80	52.15	417.20		
1 Equip. Operator (medium)	41.35	330.80	65.05	520.40		
1 Drill Rig, Truck-Mounted		2533.00		2786.30		
1 Pickup Truck, 3/4 Ton		116.80		128.48	110.41	121.45
24 L.H., Daily Totals		$3502.20		$4295.98	$145.93	$179.00

Crew B-23B

	Hr.	Daily	Hr.	Daily	Bare Costs	Incl. O&P
1 Labor Foreman (outside)	$33.60	$268.80	$55.45	$443.60	$35.52	$57.55
1 Laborer	31.60	252.80	52.15	417.20		
1 Equip. Operator (medium)	41.35	330.80	65.05	520.40		
1 Drill Rig, Truck-Mounted		2533.00		2786.30		
1 Pickup Truck, 3/4 Ton		116.80		128.48		
1 Centr. Water Pump, 6"		309.40		340.34	123.30	135.63
24 L.H., Daily Totals		$3811.60		$4636.32	$158.82	$193.18

Crew B-24

	Hr.	Daily	Hr.	Daily	Bare Costs	Incl. O&P
1 Cement Finisher	$38.30	$306.40	$59.90	$479.20	$36.62	$59.33
1 Laborer	31.60	252.80	52.15	417.20		
1 Carpenter	39.95	319.60	65.95	527.60		
24 L.H., Daily Totals		$878.80		$1424.00	$36.62	$59.33

Crew B-25

	Hr.	Daily	Hr.	Daily	Bare Costs	Incl. O&P
1 Labor Foreman	$33.60	$268.80	$55.45	$443.60	$34.44	$55.97
7 Laborers	31.60	1769.60	52.15	2920.40		
3 Equip. Oper. (med.)	41.35	992.40	65.05	1561.20		
1 Asphalt Paver, 130 H.P.		1936.00		2129.60		
1 Tandem Roller, 10 Ton		233.00		256.30		
1 Roller, Pneum. Whl, 12 Ton		307.80		338.58	28.15	30.96
88 L.H., Daily Totals		$5507.60		$7649.68	$62.59	$86.93

Crew No.	Bare Costs		Incl. Subs O & P		Cost Per Labor-Hour	
Crew B-25B	Hr.	Daily	Hr.	Daily	Bare Costs	Incl. O&P
1 Labor Foreman	$33.60	$268.80	$55.45	$443.60	$35.02	$56.73
7 Laborers	31.60	1769.60	52.15	2920.40		
4 Equip. Oper. (medium)	41.35	1323.20	65.05	2081.60		
1 Asphalt Paver, 130 H.P.		1936.00		2129.60		
2 Tandem Rollers, 10 Ton		466.00		512.60		
1 Roller, Pneum. Whl, 12 Ton		307.80		338.58	28.23	31.05
96 L.H., Daily Totals		$6071.40		$8426.38	$63.24	$87.77
Crew B-25C	Hr.	Daily	Hr.	Daily	Bare Costs	Incl. O&P
1 Labor Foreman	$33.60	$268.80	$55.45	$443.60	$35.18	$57.00
3 Laborers	31.60	758.40	52.15	1251.60		
2 Equip. Oper. (medium)	41.35	661.60	65.05	1040.80		
1 Asphalt Paver, 130 H.P.		1936.00		2129.60		
1 Tandem Roller, 10 Ton		233.00		256.30	45.19	49.71
48 L.H., Daily Totals		$3857.80		$5121.90	$80.37	$106.71
Crew B-26	Hr.	Daily	Hr.	Daily	Bare Costs	Incl. O&P
1 Labor Foreman (outside)	$33.60	$268.80	$55.45	$443.60	$35.34	$57.66
6 Laborers	31.60	1516.80	52.15	2503.20		
2 Equip. Oper. (med.)	41.35	661.60	65.05	1040.80		
1 Rodman (reinf.)	44.55	356.40	75.95	607.60		
1 Cement Finisher	38.30	306.40	59.90	479.20		
1 Grader, 30,000 Lbs.		550.20		605.22		
1 Paving Mach. & Equip.		2500.00		2750.00	34.66	38.13
88 L.H., Daily Totals		$6160.20		$8429.62	$70.00	$95.79
Crew B-26A	Hr.	Daily	Hr.	Daily	Bare Costs	Incl. O&P
1 Labor Foreman (outside)	$33.60	$268.80	$55.45	$443.60	$35.34	$57.66
6 Laborers	31.60	1516.80	52.15	2503.20		
2 Equip. Oper. (med.)	41.35	661.60	65.05	1040.80		
1 Rodman (reinf.)	44.55	356.40	75.95	607.60		
1 Cement Finisher	38.30	306.40	59.90	479.20		
1 Grader, 30,000 Lbs.		550.20		605.22		
1 Paving Mach. & Equip.		2500.00		2750.00		
1 Concrete Saw		141.00		155.10	36.26	39.89
88 L.H., Daily Totals		$6301.20		$8584.72	$71.60	$97.55
Crew B-26B	Hr.	Daily	Hr.	Daily	Bare Costs	Incl. O&P
1 Labor Foreman (outside)	$33.60	$268.80	$55.45	$443.60	$35.84	$58.28
6 Laborers	31.60	1516.80	52.15	2503.20		
3 Equip. Oper. (med.)	41.35	992.40	65.05	1561.20		
1 Rodman (reinf.)	44.55	356.40	75.95	607.60		
1 Cement Finisher	38.30	306.40	59.90	479.20		
1 Grader, 30,000 Lbs.		550.20		605.22		
1 Paving Mach. & Equip.		2500.00		2750.00		
1 Concrete Pump, 110' Boom		977.00		1074.70	41.95	46.15
96 L.H., Daily Totals		$7468.00		$10024.72	$77.79	$104.42
Crew B-27	Hr.	Daily	Hr.	Daily	Bare Costs	Incl. O&P
1 Labor Foreman (outside)	$33.60	$268.80	$55.45	$443.60	$32.10	$52.98
3 Laborers	31.60	758.40	52.15	1251.60		
1 Berm Machine		271.40		298.54	8.48	9.33
32 L.H., Daily Totals		$1298.60		$1993.74	$40.58	$62.30
Crew B-28	Hr.	Daily	Hr.	Daily	Bare Costs	Incl. O&P
2 Carpenters	$39.95	$639.20	$65.95	$1055.20	$37.17	$61.35
1 Laborer	31.60	252.80	52.15	417.20		
24 L.H., Daily Totals		$892.00		$1472.40	$37.17	$61.35

Crew No.	Bare Costs		Incl. Subs O & P		Cost Per Labor-Hour	
Crew B-29	Hr.	Daily	Hr.	Daily	Bare Costs	Incl. O&P
1 Labor Foreman (outside)	$33.60	$268.80	$55.45	$443.60	$34.19	$55.56
4 Laborers	31.60	1011.20	52.15	1668.80		
1 Equip. Oper. (crane)	42.55	340.40	66.95	535.60		
1 Equip. Oper. Oiler	36.80	294.40	57.90	463.20		
1 Gradall, 5/8 C.Y.		958.80		1054.68	17.12	18.83
56 L.H., Daily Totals		$2873.60		$4165.88	$51.31	$74.39
Crew B-30	Hr.	Daily	Hr.	Daily	Bare Costs	Incl. O&P
1 Equip. Oper. (med.)	$41.35	$330.80	$65.05	$520.40	$35.08	$56.58
2 Truck Drivers (heavy)	31.95	511.20	52.35	837.60		
1 Hyd. Excavator, 1.5 C.Y.		865.80		952.38		
2 Dump Trucks, 12 C.Y., 400 H.P.		1066.00		1172.60	80.49	88.54
24 L.H., Daily Totals		$2773.80		$3482.98	$115.58	$145.12
Crew B-31	Hr.	Daily	Hr.	Daily	Bare Costs	Incl. O&P
1 Labor Foreman (outside)	$33.60	$268.80	$55.45	$443.60	$33.67	$55.57
3 Laborers	31.60	758.40	52.15	1251.60		
1 Carpenter	39.95	319.60	65.95	527.60		
1 Air Compressor, 250 cfm		162.40		178.64		
1 Sheeting Driver		5.75		6.33		
2 -50' Air Hoses, 1.5"		12.60		13.86	4.52	4.97
40 L.H., Daily Totals		$1527.55		$2421.63	$38.19	$60.54
Crew B-32	Hr.	Daily	Hr.	Daily	Bare Costs	Incl. O&P
1 Laborer	$31.60	$252.80	$52.15	$417.20	$38.91	$61.83
3 Equip. Oper. (med.)	41.35	992.40	65.05	1561.20		
1 Grader, 30,000 Lbs.		550.20		605.22		
1 Tandem Roller, 10 Ton		233.00		256.30		
1 Dozer, 200 H.P.		1082.00		1190.20	58.29	64.12
32 L.H., Daily Totals		$3110.40		$4030.12	$97.20	$125.94
Crew B-32A	Hr.	Daily	Hr.	Daily	Bare Costs	Incl. O&P
1 Laborer	$31.60	$252.80	$52.15	$417.20	$38.10	$60.75
2 Equip. Oper. (medium)	41.35	661.60	65.05	1040.80		
1 Grader, 30,000 Lbs.		550.20		605.22		
1 Roller, Vibratory, 25 Ton		594.60		654.06	47.70	52.47
24 L.H., Daily Totals		$2059.20		$2717.28	$85.80	$113.22
Crew B-32B	Hr.	Daily	Hr.	Daily	Bare Costs	Incl. O&P
1 Laborer	$31.60	$252.80	$52.15	$417.20	$38.10	$60.75
2 Equip. Oper. (medium)	41.35	661.60	65.05	1040.80		
1 Dozer, 200 H.P.		1082.00		1190.20		
1 Roller, Vibratory, 25 Ton		594.60		654.06	69.86	76.84
24 L.H., Daily Totals		$2591.00		$3302.26	$107.96	$137.59
Crew B-32C	Hr.	Daily	Hr.	Daily	Bare Costs	Incl. O&P
1 Labor Foreman	$33.60	$268.80	$55.45	$443.60	$36.81	$59.15
2 Laborers	31.60	505.60	52.15	834.40		
3 Equip. Oper. (medium)	41.35	992.40	65.05	1561.20		
1 Grader, 30,000 Lbs.		550.20		605.22		
1 Tandem Roller, 10 Ton		233.00		256.30		
1 Dozer, 200 H.P.		1082.00		1190.20	38.86	42.74
48 L.H., Daily Totals		$3632.00		$4890.92	$75.67	$101.89
Crew B-33A	Hr.	Daily	Hr.	Daily	Bare Costs	Incl. O&P
1 Equip. Oper. (med.)	$41.35	$330.80	$65.05	$520.40	$38.56	$61.36
.5 Laborer	31.60	126.40	52.15	208.60		
.25 Equip. Oper. (med.)	41.35	82.70	65.05	130.10		
1 Scraper, Towed, 7 C.Y.		102.20		112.42		
1.25 Dozers, 300 H.P.		1778.75		1956.63	134.35	147.79
14 L.H., Daily Totals		$2420.85		$2928.15	$172.92	$209.15

Crew B-33B

Crew No.	Bare Costs Hr.	Daily	Incl. Subs O&P Hr.	Daily	Cost Per Labor-Hour Bare Costs	Incl. O&P
1 Equip. Oper. (med.)	$41.35	$330.80	$65.05	$520.40	$38.56	$61.36
.5 Laborer	31.60	126.40	52.15	208.60		
.25 Equip. Oper. (med.)	41.35	82.70	65.05	130.10		
1 Scraper, Towed, 10 C.Y.		153.60		168.96		
1.25 Dozers, 300 H.P.		1778.75		1956.63	138.03	151.83
14 L.H., Daily Totals		$2472.25		$2984.68	$176.59	$213.19

Crew B-33C

Crew No.	Bare Costs Hr.	Daily	Incl. Subs O&P Hr.	Daily	Cost Per Labor-Hour Bare Costs	Incl. O&P
1 Equip. Oper. (med.)	$41.35	$330.80	$65.05	$520.40	$38.56	$61.36
.5 Laborer	31.60	126.40	52.15	208.60		
.25 Equip. Oper. (med.)	41.35	82.70	65.05	130.10		
1 Scraper, Towed, 15 C.Y.		168.80		185.68		
1.25 Dozers, 300 H.P.		1778.75		1956.63	139.11	153.02
14 L.H., Daily Totals		$2487.45		$3001.41	$177.68	$214.39

Crew B-33D

Crew No.	Bare Costs Hr.	Daily	Incl. Subs O&P Hr.	Daily	Cost Per Labor-Hour Bare Costs	Incl. O&P
1 Equip. Oper. (med.)	$41.35	$330.80	$65.05	$520.40	$38.56	$61.36
.5 Laborer	31.60	126.40	52.15	208.60		
.25 Equip. Oper. (med.)	41.35	82.70	65.05	130.10		
1 S.P. Scraper, 14 C.Y.		1509.00		1659.90		
.25 Dozer, 300 H.P.		355.75		391.32	133.20	146.52
14 L.H., Daily Totals		$2404.65		$2910.32	$171.76	$207.88

Crew B-33E

Crew No.	Bare Costs Hr.	Daily	Incl. Subs O&P Hr.	Daily	Cost Per Labor-Hour Bare Costs	Incl. O&P
1 Equip. Oper. (med.)	$41.35	$330.80	$65.05	$520.40	$38.56	$61.36
.5 Laborer	31.60	126.40	52.15	208.60		
.25 Equip. Oper. (med.)	41.35	82.70	65.05	130.10		
1 S.P. Scraper, 21 C.Y.		1924.00		2116.40		
.25 Dozer, 300 H.P.		355.75		391.32	162.84	179.12
14 L.H., Daily Totals		$2819.65		$3366.82	$201.40	$240.49

Crew B-33F

Crew No.	Bare Costs Hr.	Daily	Incl. Subs O&P Hr.	Daily	Cost Per Labor-Hour Bare Costs	Incl. O&P
1 Equip. Oper. (med.)	$41.35	$330.80	$65.05	$520.40	$38.56	$61.36
.5 Laborer	31.60	126.40	52.15	208.60		
.25 Equip. Oper. (med.)	41.35	82.70	65.05	130.10		
1 Elev. Scraper, 11 C.Y.		1095.00		1204.50		
.25 Dozer, 300 H.P.		355.75		391.32	103.63	113.99
14 L.H., Daily Totals		$1990.65		$2454.93	$142.19	$175.35

Crew B-33G

Crew No.	Bare Costs Hr.	Daily	Incl. Subs O&P Hr.	Daily	Cost Per Labor-Hour Bare Costs	Incl. O&P
1 Equip. Oper. (med.)	$41.35	$330.80	$65.05	$520.40	$38.56	$61.36
.5 Laborer	31.60	126.40	52.15	208.60		
.25 Equip. Oper. (med.)	41.35	82.70	65.05	130.10		
1 Elev. Scraper, 22 C.Y.		2127.00		2339.70		
.25 Dozer, 300 H.P.		355.75		391.32	177.34	195.07
14 L.H., Daily Totals		$3022.65		$3590.13	$215.90	$256.44

Crew B-33K

Crew No.	Bare Costs Hr.	Daily	Incl. Subs O&P Hr.	Daily	Cost Per Labor-Hour Bare Costs	Incl. O&P
1 Equipment Operator (med.)	$41.35	$330.80	$65.05	$520.40	$38.56	$61.36
.25 Equipment Operator (med.)	41.35	82.70	65.05	130.10		
.5 Laborer	31.60	126.40	52.15	208.60		
1 S.P. Scraper, 31 C.Y.		2878.00		3165.80		
.25 Dozer, 410 H.P.		462.25		508.48	238.59	262.45
14 L.H., Daily Totals		$3880.15		$4533.38	$277.15	$323.81

Crew B-34A

Crew No.	Bare Costs Hr.	Daily	Incl. Subs O&P Hr.	Daily	Cost Per Labor-Hour Bare Costs	Incl. O&P
1 Truck Driver (heavy)	$31.95	$255.60	$52.35	$418.80	$31.95	$52.35
1 Dump Truck, 8 C.Y., 220 H.P.		326.40		359.04	40.80	44.88
8 L.H., Daily Totals		$582.00		$777.84	$72.75	$97.23

Crew B-34B

Crew No.	Bare Costs Hr.	Daily	Incl. Subs O&P Hr.	Daily	Cost Per Labor-Hour Bare Costs	Incl. O&P
1 Truck Driver (heavy)	$31.95	$255.60	$52.35	$418.80	$31.95	$52.35
1 Dump Truck, 12 C.Y., 400 H.P.		533.00		586.30	66.63	73.29
8 L.H., Daily Totals		$788.60		$1005.10	$98.58	$125.64

Crew B-34C

Crew No.	Bare Costs Hr.	Daily	Incl. Subs O&P Hr.	Daily	Cost Per Labor-Hour Bare Costs	Incl. O&P
1 Truck Driver (heavy)	$31.95	$255.60	$52.35	$418.80	$31.95	$52.35
1 Truck Tractor, 6x4, 380 H.P.		478.80		526.68		
1 Dump Trailer, 16.5 C.Y.		115.60		127.16	74.30	81.73
8 L.H., Daily Totals		$850.00		$1072.64	$106.25	$134.08

Crew B-34D

Crew No.	Bare Costs Hr.	Daily	Incl. Subs O&P Hr.	Daily	Cost Per Labor-Hour Bare Costs	Incl. O&P
1 Truck Driver (heavy)	$31.95	$255.60	$52.35	$418.80	$31.95	$52.35
1 Truck Tractor, 6x4, 380 H.P.		478.80		526.68		
1 Dump Trailer, 20 C.Y.		130.20		143.22	76.13	83.74
8 L.H., Daily Totals		$864.60		$1088.70	$108.08	$136.09

Crew B-34E

Crew No.	Bare Costs Hr.	Daily	Incl. Subs O&P Hr.	Daily	Cost Per Labor-Hour Bare Costs	Incl. O&P
1 Truck Driver (heavy)	$31.95	$255.60	$52.35	$418.80	$31.95	$52.35
1 Dump Truck, Off Hwy., 25 Ton		1280.00		1408.00	160.00	176.00
8 L.H., Daily Totals		$1535.60		$1826.80	$191.95	$228.35

Crew B-34F

Crew No.	Bare Costs Hr.	Daily	Incl. Subs O&P Hr.	Daily	Cost Per Labor-Hour Bare Costs	Incl. O&P
1 Truck Driver (heavy)	$31.95	$255.60	$52.35	$418.80	$31.95	$52.35
1 Dump Truck, Off Hwy., 35 Ton		1144.00		1258.40	143.00	157.30
8 L.H., Daily Totals		$1399.60		$1677.20	$174.95	$209.65

Crew B-34G

Crew No.	Bare Costs Hr.	Daily	Incl. Subs O&P Hr.	Daily	Cost Per Labor-Hour Bare Costs	Incl. O&P
1 Truck Driver (heavy)	$31.95	$255.60	$52.35	$418.80	$31.95	$52.35
1 Dump Truck, Off Hwy., 50 Ton		1582.00		1740.20	197.75	217.53
8 L.H., Daily Totals		$1837.60		$2159.00	$229.70	$269.88

Crew B-34H

Crew No.	Bare Costs Hr.	Daily	Incl. Subs O&P Hr.	Daily	Cost Per Labor-Hour Bare Costs	Incl. O&P
1 Truck Driver (heavy)	$31.95	$255.60	$52.35	$418.80	$31.95	$52.35
1 Dump Truck, Off Hwy., 65 Ton		1594.00		1753.40	199.25	219.18
8 L.H., Daily Totals		$1849.60		$2172.20	$231.20	$271.52

Crew B-34J

Crew No.	Bare Costs Hr.	Daily	Incl. Subs O&P Hr.	Daily	Cost Per Labor-Hour Bare Costs	Incl. O&P
1 Truck Driver (heavy)	$31.95	$255.60	$52.35	$418.80	$31.95	$52.35
1 Dump Truck, Off Hwy., 100 Ton		2047.00		2251.70	255.88	281.46
8 L.H., Daily Totals		$2302.60		$2670.50	$287.82	$333.81

Crew B-34K

Crew No.	Bare Costs Hr.	Daily	Incl. Subs O&P Hr.	Daily	Cost Per Labor-Hour Bare Costs	Incl. O&P
1 Truck Driver (heavy)	$31.95	$255.60	$52.35	$418.80	$31.95	$52.35
1 Truck Tractor, 6x4, 450 H.P.		581.20		639.32		
1 Lowbed Trailer, 75 Ton		203.00		223.30	98.03	107.83
8 L.H., Daily Totals		$1039.80		$1281.42	$129.97	$160.18

Crew B-34L

Crew No.	Bare Costs Hr.	Daily	Incl. Subs O&P Hr.	Daily	Cost Per Labor-Hour Bare Costs	Incl. O&P
1 Equip. Oper. (light)	$39.05	$312.40	$61.45	$491.60	$39.05	$61.45
1 Flatbed Truck, Gas, 1.5 Ton		194.40		213.84	24.30	26.73
8 L.H., Daily Totals		$506.80		$705.44	$63.35	$88.18

Crew B-34N

Crew No.	Bare Costs Hr.	Daily	Incl. Subs O&P Hr.	Daily	Cost Per Labor-Hour Bare Costs	Incl. O&P
1 Truck Driver (heavy)	$31.95	$255.60	$52.35	$418.80	$31.95	$52.35
1 Dump Truck, 8 C.Y., 220 H.P.		326.40		359.04		
1 Flatbed Trailer, 40 Ton		139.60		153.56	58.25	64.08
8 L.H., Daily Totals		$721.60		$931.40	$90.20	$116.43

Crew No.	Bare Costs		Incl. Subs O & P		Cost Per Labor-Hour	
Crew B-34P	Hr.	Daily	Hr.	Daily	Bare Costs	Incl. O&P
1 Pipe Fitter	$49.35	$394.80	$76.55	$612.40	$40.55	$64.12
1 Truck Driver (light)	30.95	247.60	50.75	406.00		
1 Equip. Oper. (medium)	41.35	330.80	65.05	520.40		
1 Flatbed Truck, Gas, 3 Ton		242.60		266.86		
1 Backhoe Loader, 48 H.P.		293.80		323.18	22.35	24.59
24 L.H., Daily Totals		$1509.60		$2128.84	$62.90	$88.70
Crew B-34Q	Hr.	Daily	Hr.	Daily	Bare Costs	Incl. O&P
1 Pipe Fitter	$49.35	$394.80	$76.55	$612.40	$40.95	$64.75
1 Truck Driver (light)	30.95	247.60	50.75	406.00		
1 Eqip. Oper. (crane)	42.55	340.40	66.95	535.60		
1 Flatbed Trailer, 25 Ton		102.00		112.20		
1 Dump Truck, 8 C.Y., 220 H.P.		326.40		359.04		
1 Hyd. Crane, 25 Ton		789.40		868.34	50.74	55.82
24 L.H., Daily Totals		$2200.60		$2893.58	$91.69	$120.57
Crew B-34R	Hr.	Daily	Hr.	Daily	Bare Costs	Incl. O&P
1 Pipe Fitter	$49.35	$394.80	$76.55	$612.40	$40.95	$64.75
1 Truck Driver (light)	30.95	247.60	50.75	406.00		
1 Eqip. Oper. (crane)	42.55	340.40	66.95	535.60		
1 Flatbed Trailer, 25 Ton		102.00		112.20		
1 Dump Truck, 8 C.Y., 220 H.P.		326.40		359.04		
1 Hyd. Crane, 25 Ton		789.40		868.34		
1 Hyd. Excavator, 1 C.Y.		669.20		736.12	78.63	86.49
24 L.H., Daily Totals		$2869.80		$3629.70	$119.58	$151.24
Crew B-34S	Hr.	Daily	Hr.	Daily	Bare Costs	Incl. O&P
2 Pipe Fitters	$49.35	$789.60	$76.55	$1224.80	$43.30	$68.10
1 Truck Driver (heavy)	31.95	255.60	52.35	418.80		
1 Eqip. Oper. (crane)	42.55	340.40	66.95	535.60		
1 Flatbed Trailer, 40 Ton		139.60		153.56		
1 Truck Tractor, 6x4, 380 H.P.		478.80		526.68		
1 Hyd. Crane, 80 Ton		1296.00		1425.60		
1 Hyd. Excavator, 2 C.Y.		1182.00		1300.20	96.76	106.44
32 L.H., Daily Totals		$4482.00		$5585.24	$140.06	$174.54
Crew B-34T	Hr.	Daily	Hr.	Daily	Bare Costs	Incl. O&P
2 Pipe Fitters	$49.35	$789.60	$76.55	$1224.80	$43.30	$68.10
1 Truck Driver (heavy)	31.95	255.60	52.35	418.80		
1 Eqip. Oper. (crane)	42.55	340.40	66.95	535.60		
1 Flatbed Trailer, 40 Ton		139.60		153.56		
1 Truck Tractor, 6x4, 380 H.P.		478.80		526.68		
1 Hyd. Crane, 80 Ton		1296.00		1425.60	59.83	65.81
32 L.H., Daily Totals		$3300.00		$4285.04	$103.13	$133.91
Crew B-35	Hr.	Daily	Hr.	Daily	Bare Costs	Incl. O&P
1 Laborer Foreman (out)	$33.60	$268.80	$55.45	$443.60	$39.02	$62.42
1 Skilled Worker	40.85	326.80	66.50	532.00		
1 Welder (plumber)	48.75	390.00	75.60	604.80		
1 Laborer	31.60	252.80	52.15	417.20		
1 Equip. Oper. (crane)	42.55	340.40	66.95	535.60		
1 Equip. Oper. Oiler	36.80	294.40	57.90	463.20		
1 Welder, electric, 300 amp		55.70		61.27		
1 Hyd. Excavator, .75 C.Y.		598.60		658.46	13.63	14.99
48 L.H., Daily Totals		$2527.50		$3716.13	$52.66	$77.42

Crew No.	Bare Costs		Incl. Subs O & P		Cost Per Labor-Hour	
Crew B-35A	Hr.	Daily	Hr.	Daily	Bare Costs	Incl. O&P
1 Laborer Foreman (out)	$33.60	$268.80	$55.45	$443.60	$37.96	$60.96
2 Laborers	31.60	505.60	52.15	834.40		
1 Skilled Worker	40.85	326.80	66.50	532.00		
1 Welder (plumber)	48.75	390.00	75.60	604.80		
1 Equip. Oper. (crane)	42.55	340.40	66.95	535.60		
1 Equip. Oper. Oiler	36.80	294.40	57.90	463.20		
1 Welder, gas engine, 300 amp		134.20		147.62		
1 Crawler Crane, 75 Ton		1496.00		1645.60	29.11	32.02
56 L.H., Daily Totals		$3756.20		$5206.82	$67.08	$92.98
Crew B-36	Hr.	Daily	Hr.	Daily	Bare Costs	Incl. O&P
1 Labor Foreman (outside)	$33.60	$268.80	$55.45	$443.60	$35.90	$57.97
2 Laborers	31.60	505.60	52.15	834.40		
2 Equip. Oper. (med.)	41.35	661.60	65.05	1040.80		
1 Dozer, 200 H.P.		1082.00		1190.20		
1 Aggregate Spreader		35.00		38.50		
1 Tandem Roller, 10 Ton		233.00		256.30	33.75	37.13
40 L.H., Daily Totals		$2786.00		$3803.80	$69.65	$95.09
Crew B-36A	Hr.	Daily	Hr.	Daily	Bare Costs	Incl. O&P
1 Labor Foreman (outside)	$33.60	$268.80	$55.45	$443.60	$37.46	$59.99
2 Laborers	31.60	505.60	52.15	834.40		
4 Equip. Oper. (med.)	41.35	1323.20	65.05	2081.60		
1 Dozer, 200 H.P.		1082.00		1190.20		
1 Aggregate Spreader		35.00		38.50		
1 Tandem Roller, 10 Ton		233.00		256.30		
1 Roller, Pneum. Whl, 12 Ton		307.80		338.58	29.60	32.56
56 L.H., Daily Totals		$3755.40		$5183.18	$67.06	$92.56
Crew B-36B	Hr.	Daily	Hr.	Daily	Bare Costs	Incl. O&P
1 Labor Foreman (outside)	$33.60	$268.80	$55.45	$443.60	$36.77	$59.04
2 Laborers	31.60	505.60	52.15	834.40		
4 Equip. Oper. (medium)	41.35	1323.20	65.05	2081.60		
1 Truck Driver, Heavy	31.95	255.60	52.35	418.80		
1 Grader, 30,000 Lbs.		550.20		605.22		
1 F.E. Loader, crl, 1.5 C.Y.		463.00		509.30		
1 Dozer, 300 H.P.		1423.00		1565.30		
1 Roller, Vibratory, 25 Ton		594.60		654.06		
1 Truck Tractor, 6x4, 450 H.P.		581.20		639.32		
1 Water Tanker, 5000 Gal.		136.00		149.60	58.56	64.42
64 L.H., Daily Totals		$6101.20		$7901.20	$95.33	$123.46
Crew B-36C	Hr.	Daily	Hr.	Daily	Bare Costs	Incl. O&P
1 Labor Foreman (outside)	$33.60	$268.80	$55.45	$443.60	$37.92	$60.59
3 Equip. Oper. (medium)	41.35	992.40	65.05	1561.20		
1 Truck Driver, Heavy	31.95	255.60	52.35	418.80		
1 Grader, 30,000 Lbs.		550.20		605.22		
1 Dozer, 300 H.P.		1423.00		1565.30		
1 Roller, Vibratory, 25 Ton		594.60		654.06		
1 Truck Tractor, 6x4, 450 H.P.		581.20		639.32		
1 Water Tanker, 5000 Gal.		136.00		149.60	82.13	90.34
40 L.H., Daily Totals		$4801.80		$6037.10	$120.05	$150.93
Crew B-37	Hr.	Daily	Hr.	Daily	Bare Costs	Incl. O&P
1 Labor Foreman (outside)	$33.60	$268.80	$55.45	$443.60	$33.17	$54.25
4 Laborers	31.60	1011.20	52.15	1668.80		
1 Equip. Oper. (light)	39.05	312.40	61.45	491.60		
1 Tandem Roller, 5 Ton		137.80		151.58	2.87	3.16
48 L.H., Daily Totals		$1730.20		$2755.58	$36.05	$57.41

Crew B-38

Crew No.	Bare Costs Hr.	Bare Costs Daily	Incl. Subs O&P Hr.	Incl. Subs O&P Daily	Cost Per Labor-Hour Bare Costs	Cost Per Labor-Hour Incl. O&P
1 Labor Foreman (outside)	$33.60	$268.80	$55.45	$443.60	$35.44	$57.25
2 Laborers	31.60	505.60	52.15	834.40		
1 Equip. Oper. (light)	39.05	312.40	61.45	491.60		
1 Equip. Oper. (medium)	41.35	330.80	65.05	520.40		
1 Backhoe Loader, 48 H.P.		293.80		323.18		
1 Hyd.Hammer, (1200 lb.)		153.00		168.30		
1 F.E. Loader, W.M., 4 C.Y.		541.00		595.10		
1 Pavt. Rem. Bucket		56.80		62.48	26.11	28.73
40 L.H., Daily Totals		$2462.20		$3439.06	$61.56	$85.98

Crew B-39

Crew No.	Bare Costs Hr.	Bare Costs Daily	Incl. Subs O&P Hr.	Incl. Subs O&P Daily	Bare Costs	Incl. O&P
1 Labor Foreman (outside)	$33.60	$268.80	$55.45	$443.60	$33.17	$54.25
4 Laborers	31.60	1011.20	52.15	1668.80		
1 Equip. Oper. (light)	39.05	312.40	61.45	491.60		
1 Air Compressor, 250 cfm		162.40		178.64		
2 Breakers, Pavement, 60 lb.		16.40		18.04		
2 -50' Air Hoses, 1.5"		12.60		13.86	3.99	4.39
48 L.H., Daily Totals		$1783.80		$2814.54	$37.16	$58.64

Crew B-40

Crew No.	Bare Costs Hr.	Bare Costs Daily	Incl. Subs O&P Hr.	Incl. Subs O&P Daily	Bare Costs	Incl. O&P
1 Pile Driver Foreman (out)	$40.50	$324.00	$67.80	$542.40	$39.55	$64.67
4 Pile Drivers	38.50	1232.00	64.45	2062.40		
2 Equip. Oper. (crane)	42.55	680.80	66.95	1071.20		
1 Equip. Oper. Oiler	36.80	294.40	57.90	463.20		
1 Crawler Crane, 40 Ton		1055.00		1160.50		
1 Vibratory Hammer & Gen.		2257.00		2482.70	51.75	56.92
64 L.H., Daily Totals		$5843.20		$7782.40	$91.30	$121.60

Crew B-40B

Crew No.	Bare Costs Hr.	Bare Costs Daily	Incl. Subs O&P Hr.	Incl. Subs O&P Daily	Bare Costs	Incl. O&P
1 Laborer Foreman	$33.60	$268.80	$55.45	$443.60	$34.63	$56.13
3 Laborers	31.60	758.40	52.15	1251.60		
1 Equip. Oper. (crane)	42.55	340.40	66.95	535.60		
1 Equip. Oper. Oiler	36.80	294.40	57.90	463.20		
1 Lattice Boom Crane, 40 Ton		1327.00		1459.70	27.65	30.41
48 L.H., Daily Totals		$2989.00		$4153.70	$62.27	$86.54

Crew B-41

Crew No.	Bare Costs Hr.	Bare Costs Daily	Incl. Subs O&P Hr.	Incl. Subs O&P Daily	Bare Costs	Incl. O&P
1 Labor Foreman (outside)	$33.60	$268.80	$55.45	$443.60	$32.70	$53.68
4 Laborers	31.60	1011.20	52.15	1668.80		
.25 Equip. Oper. (crane)	42.55	85.10	66.95	133.90		
.25 Equip. Oper. Oiler	36.80	73.60	57.90	115.80		
.25 Crawler Crane, 40 Ton		263.75		290.13	5.99	6.59
44 L.H., Daily Totals		$1702.45		$2652.22	$38.69	$60.28

Crew B-42

Crew No.	Bare Costs Hr.	Bare Costs Daily	Incl. Subs O&P Hr.	Incl. Subs O&P Daily	Bare Costs	Incl. O&P
1 Labor Foreman (outside)	$33.60	$268.80	$55.45	$443.60	$35.51	$58.96
4 Laborers	31.60	1011.20	52.15	1668.80		
1 Equip. Oper. (crane)	42.55	340.40	66.95	535.60		
1 Equip. Oper. Oiler	36.80	294.40	57.90	463.20		
1 Welder	44.70	357.60	82.80	662.40		
1 Hyd. Crane, 25 Ton		789.40		868.34		
1 Welder, gas engine, 300 amp		134.20		147.62		
1 Horz. Boring Csg. Mch.		444.20		488.62	21.37	23.51
64 L.H., Daily Totals		$3640.20		$5278.18	$56.88	$82.47

Crew B-43

Crew No.	Bare Costs Hr.	Bare Costs Daily	Incl. Subs O&P Hr.	Incl. Subs O&P Daily	Bare Costs	Incl. O&P
1 Labor Foreman (outside)	$33.60	$268.80	$55.45	$443.60	$34.63	$56.13
3 Laborers	31.60	758.40	52.15	1251.60		
1 Equip. Oper. (crane)	42.55	340.40	66.95	535.60		
1 Equip. Oper. Oiler	36.80	294.40	57.90	463.20		
1 Drill Rig, Truck-Mounted		2533.00		2786.30	52.77	58.05
48 L.H., Daily Totals		$4195.00		$5480.30	$87.40	$114.17

Crew B-44

Crew No.	Bare Costs Hr.	Bare Costs Daily	Incl. Subs O&P Hr.	Incl. Subs O&P Daily	Bare Costs	Incl. O&P
1 Pile Driver Foreman	$40.50	$324.00	$67.80	$542.40	$38.90	$63.96
4 Pile Drivers	38.50	1232.00	64.45	2062.40		
2 Equip. Oper. (crane)	42.55	680.80	66.95	1071.20		
1 Laborer	31.60	252.80	52.15	417.20		
1 Crawler Crane, 40 Ton		1055.00		1160.50		
1 Lead, 60' high		69.60		76.56		
1 Hammer, diesel, 15K Ft-Lbs.		578.60		636.46	26.61	29.27
64 L.H., Daily Totals		$4192.80		$5966.72	$65.51	$93.23

Crew B-45

Crew No.	Bare Costs Hr.	Bare Costs Daily	Incl. Subs O&P Hr.	Incl. Subs O&P Daily	Bare Costs	Incl. O&P
1 Equip. Oper. (med.)	$41.35	$330.80	$65.05	$520.40	$36.65	$58.70
1 Truck Driver (heavy)	31.95	255.60	52.35	418.80		
1 Dist. Tanker, 3000 Gallon		304.20		334.62		
1 Truck Tractor, 6x4, 380 H.P.		478.80		526.68	48.94	53.83
16 L.H., Daily Totals		$1369.40		$1800.50	$85.59	$112.53

Crew B-46

Crew No.	Bare Costs Hr.	Bare Costs Daily	Incl. Subs O&P Hr.	Incl. Subs O&P Daily	Bare Costs	Incl. O&P
1 Pile Driver Foreman	$40.50	$324.00	$67.80	$542.40	$35.38	$58.86
2 Pile Drivers	38.50	616.00	64.45	1031.20		
3 Laborers	31.60	758.40	52.15	1251.60		
1 Chainsaw, Gas, 36" Long		37.80		41.58	.79	.87
48 L.H., Daily Totals		$1736.20		$2866.78	$36.17	$59.72

Crew B-47

Crew No.	Bare Costs Hr.	Bare Costs Daily	Incl. Subs O&P Hr.	Incl. Subs O&P Daily	Bare Costs	Incl. O&P
1 Blast Foreman	$33.60	$268.80	$55.45	$443.60	$34.75	$56.35
1 Driller	31.60	252.80	52.15	417.20		
1 Equip. Oper. (light)	39.05	312.40	61.45	491.60		
1 Air Track Drill, 4"		850.60		935.66		
1 Air Compressor, 600 cfm		434.40		477.84		
2 -50' Air Hoses, 3"		32.40		35.64	54.89	60.38
24 L.H., Daily Totals		$2151.40		$2801.54	$89.64	$116.73

Crew B-47A

Crew No.	Bare Costs Hr.	Bare Costs Daily	Incl. Subs O&P Hr.	Incl. Subs O&P Daily	Bare Costs	Incl. O&P
1 Drilling Foreman	$33.60	$268.80	$55.45	$443.60	$37.65	$60.10
1 Equip. Oper. (heavy)	42.55	340.40	66.95	535.60		
1 Oiler	36.80	294.40	57.90	463.20		
1 Air Track Drill, 5"		974.00		1071.40	40.58	44.64
24 L.H., Daily Totals		$1877.60		$2513.80	$78.23	$104.74

Crew B-47C

Crew No.	Bare Costs Hr.	Bare Costs Daily	Incl. Subs O&P Hr.	Incl. Subs O&P Daily	Bare Costs	Incl. O&P
1 Laborer	$31.60	$252.80	$52.15	$417.20	$35.33	$56.80
1 Equip. Oper. (light)	39.05	312.40	61.45	491.60		
1 Air Compressor, 750 cfm		439.60		483.56		
2 -50' Air Hoses, 3"		32.40		35.64		
1 Air Track Drill, 4"		850.60		935.66	82.66	90.93
16 L.H., Daily Totals		$1887.80		$2363.66	$117.99	$147.73

Crew B-47E

Crew No.	Bare Costs Hr.	Bare Costs Daily	Incl. Subs O&P Hr.	Incl. Subs O&P Daily	Bare Costs	Incl. O&P
1 Laborer Foreman	$33.60	$268.80	$55.45	$443.60	$32.10	$52.98
3 Laborers	31.60	758.40	52.15	1251.60		
1 Flatbed Truck, Gas, 3 Ton		242.60		266.86	7.58	8.34
32 L.H., Daily Totals		$1269.80		$1962.06	$39.68	$61.31

Crews

Crew B-47G	Hr.	Daily	Hr.	Daily	Bare Costs	Incl. O&P
1 Laborer Foreman	$33.60	$268.80	$55.45	$443.60	$33.96	$55.30
2 Laborers	31.60	505.60	52.15	834.40		
1 Equip. Oper. (light)	39.05	312.40	61.45	491.60		
1 Air Track Drill, 4"		850.60		935.66		
1 Air Compressor, 600 cfm		434.40		477.84		
2 -50' Air Hoses, 3"		32.40		35.64		
1 Grout Pump		166.40		183.04	46.37	51.01
32 L.H., Daily Totals		$2570.60		$3401.78	$80.33	$106.31

Crew B-48	Hr.	Daily	Hr.	Daily	Bare Costs	Incl. O&P
1 Labor Foreman (outside)	$33.60	$268.80	$55.45	$443.60	$35.26	$56.89
3 Laborers	31.60	758.40	52.15	1251.60		
1 Equip. Oper. (crane)	42.55	340.40	66.95	535.60		
1 Equip. Oper. Oiler	36.80	294.40	57.90	463.20		
1 Equip. Oper. (light)	39.05	312.40	61.45	491.60		
1 Centr. Water Pump, 6"		309.40		340.34		
1 -20' Suction Hose, 6"		11.90		13.09		
1 -50' Discharge Hose, 6"		6.30		6.93		
1 Drill Rig, Truck-Mounted		2533.00		2786.30	51.08	56.19
56 L.H., Daily Totals		$4835.00		$6332.26	$86.34	$113.08

Crew B-49	Hr.	Daily	Hr.	Daily	Bare Costs	Incl. O&P
1 Labor Foreman (outside)	$33.60	$268.80	$55.45	$443.60	$36.65	$59.27
3 Laborers	31.60	758.40	52.15	1251.60		
2 Equip. Oper. (crane)	42.55	680.80	66.95	1071.20		
2 Equip. Oper. Oilers	36.80	588.80	57.90	926.40		
1 Equip. Oper. (light)	39.05	312.40	61.45	491.60		
2 Pile Drivers	38.50	616.00	64.45	1031.20		
1 Hyd. Crane, 25 Ton		789.40		868.34		
1 Centr. Water Pump, 6"		309.40		340.34		
1 -20' Suction Hose, 6"		11.90		13.09		
1 -50' Discharge Hose, 6"		6.30		6.93		
1 Drill Rig, Truck-Mounted		2533.00		2786.30	41.48	45.63
88 L.H., Daily Totals		$6875.20		$9230.60	$78.13	$104.89

Crew B-50	Hr.	Daily	Hr.	Daily	Bare Costs	Incl. O&P
2 Pile Driver Foremen	$40.50	$648.00	$67.80	$1084.80	$37.76	$62.18
6 Pile Drivers	38.50	1848.00	64.45	3093.60		
2 Equip. Oper. (crane)	42.55	680.80	66.95	1071.20		
1 Equip. Oper. Oiler	36.80	294.40	57.90	463.20		
3 Laborers	31.60	758.40	52.15	1251.60		
1 Crawler Crane, 40 Ton		1055.00		1160.50		
1 Lead, 60' high		69.60		76.56		
1 Hammer, diesel, 15K Ft.-Lbs.		578.60		636.46		
1 Air Compressor, 600 cfm		434.40		477.84		
2 -50' Air Hoses, 3"		32.40		35.64		
1 Chainsaw, Gas, 36" Long		37.80		41.58	19.71	21.68
112 L.H., Daily Totals		$6437.40		$9392.98	$57.48	$83.87

Crew B-51	Hr.	Daily	Hr.	Daily	Bare Costs	Incl. O&P
1 Labor Foreman (outside)	$33.60	$268.80	$55.45	$443.60	$31.82	$52.47
4 Laborers	31.60	1011.20	52.15	1668.80		
1 Truck Driver (light)	30.95	247.60	50.75	406.00		
1 Flatbed Truck, Gas, 1.5 Ton		194.40		213.84	4.05	4.46
48 L.H., Daily Totals		$1722.00		$2732.24	$35.88	$56.92

Crew B-52	Hr.	Daily	Hr.	Daily	Bare Costs	Incl. O&P
1 Carpenter Foreman	$41.95	$335.60	$69.25	$554.00	$36.85	$60.29
1 Carpenter	39.95	319.60	65.95	527.60		
3 Laborers	31.60	758.40	52.15	1251.60		
1 Cement Finisher	38.30	306.40	59.90	479.20		
.5 Rodman (reinf.)	44.55	178.20	75.95	303.80		
.5 Equip. Oper. (med.)	41.35	165.40	65.05	260.20		
.5 Crawler Loader, 3 C.Y.		495.00		544.50	8.84	9.72
56 L.H., Daily Totals		$2558.60		$3920.90	$45.69	$70.02

Crew B-53	Hr.	Daily	Hr.	Daily	Bare Costs	Incl. O&P
1 Equip. Oper. (light)	$39.05	$312.40	$61.45	$491.60	$39.05	$61.45
1 Trencher, Chain, 12 H.P.		59.60		65.56	7.45	8.20
8 L.H., Daily Totals		$372.00		$557.16	$46.50	$69.64

Crew B-54	Hr.	Daily	Hr.	Daily	Bare Costs	Incl. O&P
1 Equip. Oper. (light)	$39.05	$312.40	$61.45	$491.60	$39.05	$61.45
1 Trencher, Chain, 40 H.P.		301.00		331.10	37.63	41.39
8 L.H., Daily Totals		$613.40		$822.70	$76.67	$102.84

Crew B-54A	Hr.	Daily	Hr.	Daily	Bare Costs	Incl. O&P
.17 Labor Foreman (outside)	$33.60	$45.70	$55.45	$75.41	$40.22	$63.66
1 Equipment Operator (med.)	41.35	330.80	65.05	520.40		
1 Wheel Trencher, 67 H.P.		1009.00		1109.90	107.80	118.58
9.36 L.H., Daily Totals		$1385.50		$1705.71	$148.02	$182.23

Crew B-54B	Hr.	Daily	Hr.	Daily	Bare Costs	Incl. O&P
.25 Labor Foreman (outside)	$33.60	$67.20	$55.45	$110.90	$39.80	$63.13
1 Equipment Operator (med.)	41.35	330.80	65.05	520.40		
1 Wheel Trencher, 150 H.P.		1753.00		1928.30	175.30	192.83
10 L.H., Daily Totals		$2151.00		$2559.60	$215.10	$255.96

Crew B-55	Hr.	Daily	Hr.	Daily	Bare Costs	Incl. O&P
2 Laborers	$31.60	$505.60	$52.15	$834.40	$31.38	$51.68
1 Truck Driver (light)	30.95	247.60	50.75	406.00		
1 Truck-mounted earth auger		706.60		777.26		
1 Flatbed Truck, Gas, 3 Ton		242.60		266.86	39.55	43.51
24 L.H., Daily Totals		$1702.40		$2284.52	$70.93	$95.19

Crew B-56	Hr.	Daily	Hr.	Daily	Bare Costs	Incl. O&P
1 Laborer	$31.60	$252.80	$52.15	$417.20	$35.33	$56.80
1 Equip. Oper. (light)	39.05	312.40	61.45	491.60		
1 Air Track Drill, 4"		850.60		935.66		
1 Air Compressor, 600 cfm		434.40		477.84		
1 -50' Air Hose, 3"		16.20		17.82	81.33	89.46
16 L.H., Daily Totals		$1866.40		$2340.12	$116.65	$146.26

Crew B-57	Hr.	Daily	Hr.	Daily	Bare Costs	Incl. O&P
1 Labor Foreman (outside)	$33.60	$268.80	$55.45	$443.60	$35.87	$57.67
2 Laborers	31.60	505.60	52.15	834.40		
1 Equip. Oper. (crane)	42.55	340.40	66.95	535.60		
1 Equip. Oper. (light)	39.05	312.40	61.45	491.60		
1 Equip. Oper. Oiler	36.80	294.40	57.90	463.20		
1 Crawler Crane, 25 Ton		1052.00		1157.20		
1 Clamshell Bucket, 1 C.Y.		46.20		50.82		
1 Centr. Water Pump, 6"		309.40		340.34		
1 -20' Suction Hose, 6"		11.90		13.09		
20 -50' Discharge Hoses, 6"		126.00		138.60	32.20	35.42
48 L.H., Daily Totals		$3267.10		$4468.45	$68.06	$93.09

Crew B-58	Hr.	Daily	Hr.	Daily	Bare Costs	Incl. O&P
2 Laborers	$31.60	$505.60	$52.15	$834.40	$34.08	$55.25
1 Equip. Oper. (light)	39.05	312.40	61.45	491.60		
1 Backhoe Loader, 48 H.P.		293.80		323.18		
1 Small Helicopter, w/pilot		2587.00		2845.70	120.03	132.04
24 L.H., Daily Totals		$3698.80		$4494.88	$154.12	$187.29

Crew B-59	Hr.	Daily	Hr.	Daily	Bare Costs	Incl. O&P
1 Truck Driver (heavy)	$31.95	$255.60	$52.35	$418.80	$31.95	$52.35
1 Truck Tractor, 220 H.P.		288.40		317.24		
1 Water Tanker, 5000 Gal.		136.00		149.60	53.05	58.35
8 L.H., Daily Totals		$680.00		$885.64	$85.00	$110.71

Crew B-59A	Hr.	Daily	Hr.	Daily	Bare Costs	Incl. O&P
2 Laborers	$31.60	$505.60	$52.15	$834.40	$31.72	$52.22
1 Truck Driver (heavy)	31.95	255.60	52.35	418.80		
1 Water Tanker, 5000 Gal.		136.00		149.60		
1 Truck Tractor, 220 H.P.		288.40		317.24	17.68	19.45
24 L.H., Daily Totals		$1185.60		$1720.04	$49.40	$71.67

Crew B-60	Hr.	Daily	Hr.	Daily	Bare Costs	Incl. O&P
1 Labor Foreman (outside)	$33.60	$268.80	$55.45	$443.60	$36.32	$58.21
2 Laborers	31.60	505.60	52.15	834.40		
1 Equip. Oper. (crane)	42.55	340.40	66.95	535.60		
2 Equip. Oper. (light)	39.05	624.80	61.45	983.20		
1 Equip. Oper. Oiler	36.80	294.40	57.90	463.20		
1 Crawler Crane, 40 Ton		1055.00		1160.50		
1 Lead, 60' high		69.60		76.56		
1 Hammer, diesel, 15K Ft.-Lbs.		578.60		636.46		
1 Backhoe Loader, 48 H.P.		293.80		323.18	35.66	39.23
56 L.H., Daily Totals		$4031.00		$5456.70	$71.98	$97.44

Crew B-61	Hr.	Daily	Hr.	Daily	Bare Costs	Incl. O&P
1 Labor Foreman (outside)	$33.60	$268.80	$55.45	$443.60	$33.49	$54.67
3 Laborers	31.60	758.40	52.15	1251.60		
1 Equip. Oper. (light)	39.05	312.40	61.45	491.60		
1 Cement Mixer, 2 C.Y.		187.80		206.58		
1 Air Compressor, 160 cfm		138.80		152.68	8.16	8.98
40 L.H., Daily Totals		$1666.20		$2546.06	$41.66	$63.65

Crew B-62	Hr.	Daily	Hr.	Daily	Bare Costs	Incl. O&P
2 Laborers	$31.60	$505.60	$52.15	$834.40	$34.08	$55.25
1 Equip. Oper. (light)	39.05	312.40	61.45	491.60		
1 Loader, Skid Steer, 30 H.P., gas		149.20		164.12	6.22	6.84
24 L.H., Daily Totals		$967.20		$1490.12	$40.30	$62.09

Crew B-63	Hr.	Daily	Hr.	Daily	Bare Costs	Incl. O&P
4 Laborers	$31.60	$1011.20	$52.15	$1668.80	$33.09	$54.01
1 Equip. Oper. (light)	39.05	312.40	61.45	491.60		
1 Loader, Skid Steer, 30 H.P., gas		149.20		164.12	3.73	4.10
40 L.H., Daily Totals		$1472.80		$2324.52	$36.82	$58.11

Crew B-64	Hr.	Daily	Hr.	Daily	Bare Costs	Incl. O&P
1 Laborer	$31.60	$252.80	$52.15	$417.20	$31.27	$51.45
1 Truck Driver (light)	30.95	247.60	50.75	406.00		
1 Power Mulcher (small)		136.60		150.26		
1 Flatbed Truck, Gas, 1.5 Ton		194.40		213.84	20.69	22.76
16 L.H., Daily Totals		$831.40		$1187.30	$51.96	$74.21

Crew B-65	Hr.	Daily	Hr.	Daily	Bare Costs	Incl. O&P
1 Laborer	$31.60	$252.80	$52.15	$417.20	$31.27	$51.45
1 Truck Driver (light)	30.95	247.60	50.75	406.00		
1 Power Mulcher (large)		271.80		298.98		
1 Flatbed Truck, Gas, 1.5 Ton		194.40		213.84	29.14	32.05
16 L.H., Daily Totals		$966.60		$1336.02	$60.41	$83.50

Crew B-66	Hr.	Daily	Hr.	Daily	Bare Costs	Incl. O&P
1 Equip. Oper. (light)	$39.05	$312.40	$61.45	$491.60	$39.05	$61.45
1 Loader-Backhoe		231.20		254.32	28.90	31.79
8 L.H., Daily Totals		$543.60		$745.92	$67.95	$93.24

Crew B-67	Hr.	Daily	Hr.	Daily	Bare Costs	Incl. O&P
1 Millwright	$41.60	$332.80	$65.25	$522.00	$40.33	$63.35
1 Equip. Oper. (light)	39.05	312.40	61.45	491.60		
1 Forklift, R/T, 4,000 Lb.		281.20		309.32	17.57	19.33
16 L.H., Daily Totals		$926.40		$1322.92	$57.90	$82.68

Crew B-68	Hr.	Daily	Hr.	Daily	Bare Costs	Incl. O&P
2 Millwrights	$41.60	$665.60	$65.25	$1044.00	$40.75	$63.98
1 Equip. Oper. (light)	39.05	312.40	61.45	491.60		
1 Forklift, R/T, 4,000 Lb.		281.20		309.32	11.72	12.89
24 L.H., Daily Totals		$1259.20		$1844.92	$52.47	$76.87

Crew B-69	Hr.	Daily	Hr.	Daily	Bare Costs	Incl. O&P
1 Labor Foreman (outside)	$33.60	$268.80	$55.45	$443.60	$34.63	$56.13
3 Laborers	31.60	758.40	52.15	1251.60		
1 Equip Oper. (crane)	42.55	340.40	66.95	535.60		
1 Equip Oper. Oiler	36.80	294.40	57.90	463.20		
1 Hyd. Crane, 80 Ton		1296.00		1425.60	27.00	29.70
48 L.H., Daily Totals		$2958.00		$4119.60	$61.63	$85.83

Crew B-69A	Hr.	Daily	Hr.	Daily	Bare Costs	Incl. O&P
1 Labor Foreman	$33.60	$268.80	$55.45	$443.60	$34.67	$56.14
3 Laborers	31.60	758.40	52.15	1251.60		
1 Equip. Oper. (medium)	41.35	330.80	65.05	520.40		
1 Concrete Finisher	38.30	306.40	59.90	479.20		
1 Curb/Gutter Paver, 2-Track		754.40		829.84	15.72	17.29
48 L.H., Daily Totals		$2418.80		$3524.64	$50.39	$73.43

Crew B-69B	Hr.	Daily	Hr.	Daily	Bare Costs	Incl. O&P
1 Labor Foreman	$33.60	$268.80	$55.45	$443.60	$34.67	$56.14
3 Laborers	31.60	758.40	52.15	1251.60		
1 Equip. Oper. (medium)	41.35	330.80	65.05	520.40		
1 Cement Finisher	38.30	306.40	59.90	479.20		
1 Curb/Gutter Paver, 4-Track		855.80		941.38	17.83	19.61
48 L.H., Daily Totals		$2520.20		$3636.18	$52.50	$75.75

Crew B-70	Hr.	Daily	Hr.	Daily	Bare Costs	Incl. O&P
1 Labor Foreman (outside)	$33.60	$268.80	$55.45	$443.60	$36.06	$58.15
3 Laborers	31.60	758.40	52.15	1251.60		
3 Equip. Oper. (med.)	41.35	992.40	65.05	1561.20		
1 Grader, 30,000 Lbs.		550.20		605.22		
1 Ripper, beam & 1 shank		79.60		87.56		
1 Road Sweeper, S.P., 8' wide		601.60		661.76		
1 F.E. Loader, W.M., 1.5 C.Y.		328.20		361.02	27.85	30.64
56 L.H., Daily Totals		$3579.20		$4971.96	$63.91	$88.78

Crew No.	Bare Costs		Incl. Subs O & P		Cost Per Labor-Hour	
Crew B-71	Hr.	Daily	Hr.	Daily	Bare Costs	Incl. O&P
1 Labor Foreman (outside)	$33.60	$268.80	$55.45	$443.60	$36.06	$58.15
3 Laborers	31.60	758.40	52.15	1251.60		
3 Equip. Oper. (med.)	41.35	992.40	65.05	1561.20		
1 Pvmt. Profiler, 750 H.P.		5503.00		6053.30		
1 Road Sweeper, S.P., 8' wide		601.60		661.76		
1 F.E. Loader, W.M., 1.5 C.Y.		328.20		361.02	114.87	126.36
56 L.H., Daily Totals		$8452.40		$10332.48	$150.94	$184.51
Crew B-72	Hr.	Daily	Hr.	Daily	Bare Costs	Incl. O&P
1 Labor Foreman (outside)	$33.60	$268.80	$55.45	$443.60	$36.73	$59.01
3 Laborers	31.60	758.40	52.15	1251.60		
4 Equip. Oper. (med.)	41.35	1323.20	65.05	2081.60		
1 Pvmt. Profiler, 750 H.P.		5503.00		6053.30		
1 Hammermill, 250 H.P.		1850.00		2035.00		
1 Windrow Loader		1197.00		1316.70		
1 Mix Paver 165 H.P.		1996.00		2195.60		
1 Roller, Pneum. Whl, 12 Ton		307.80		338.58	169.59	186.55
64 L.H., Daily Totals		$13204.20		$15715.98	$206.32	$245.56
Crew B-73	Hr.	Daily	Hr.	Daily	Bare Costs	Incl. O&P
1 Labor Foreman (outside)	$33.60	$268.80	$55.45	$443.60	$37.94	$60.63
2 Laborers	31.60	505.60	52.15	834.40		
5 Equip. Oper. (med.)	41.35	1654.00	65.05	2602.00		
1 Road Mixer, 310 H.P.		1896.00		2085.60		
1 Tandem Roller, 10 Ton		233.00		256.30		
1 Hammermill, 250 H.P.		1850.00		2035.00		
1 Grader, 30,000 Lbs.		550.20		605.22		
.5 F.E. Loader, W.M., 1.5 C.Y.		164.10		180.51		
.5 Truck Tractor, 220 H.P.		144.20		158.62		
.5 Water Tanker, 5000 Gal.		68.00		74.80	76.65	84.31
64 L.H., Daily Totals		$7333.90		$9276.05	$114.59	$144.94
Crew B-74	Hr.	Daily	Hr.	Daily	Bare Costs	Incl. O&P
1 Labor Foreman (outside)	$33.60	$268.80	$55.45	$443.60	$36.81	$59.06
1 Laborer	31.60	252.80	52.15	417.20		
4 Equip. Oper. (med.)	41.35	1323.20	65.05	2081.60		
2 Truck Drivers (heavy)	31.95	511.20	52.35	837.60		
1 Grader, 30,000 Lbs.		550.20		605.22		
1 Ripper, beam & 1 shank		79.60		87.56		
2 Stabilizers, 310 H.P.		2688.00		2956.80		
1 Flatbed Truck, Gas, 3 Ton		242.60		266.86		
1 Chem. Spreader, Towed		48.00		52.80		
1 Roller, Vibratory, 25 Ton		594.60		654.06		
1 Water Tanker, 5000 Gal.		136.00		149.60		
1 Truck Tractor, 220 H.P.		288.40		317.24	72.30	79.53
64 L.H., Daily Totals		$6983.40		$8870.14	$109.12	$138.60
Crew B-75	Hr.	Daily	Hr.	Daily	Bare Costs	Incl. O&P
1 Labor Foreman (outside)	$33.60	$268.80	$55.45	$443.60	$37.51	$60.02
1 Laborer	31.60	252.80	52.15	417.20		
4 Equip. Oper. (med.)	41.35	1323.20	65.05	2081.60		
1 Truck Driver (heavy)	31.95	255.60	52.35	418.80		
1 Grader, 30,000 Lbs.		550.20		605.22		
1 Ripper, beam & 1 shank		79.60		87.56		
2 Stabilizers, 310 H.P.		2688.00		2956.80		
1 Dist. Tanker, 3000 Gallon		304.20		334.62		
1 Truck Tractor, 6x4, 380 H.P.		478.80		526.68		
1 Roller, Vibratory, 25 Ton		594.60		654.06	83.85	92.23
56 L.H., Daily Totals		$6795.80		$8526.14	$121.35	$152.25

Crew No.	Bare Costs		Incl. Subs O & P		Cost Per Labor-Hour	
Crew B-76	Hr.	Daily	Hr.	Daily	Bare Costs	Incl. O&P
1 Dock Builder Foreman	$40.50	$324.00	$67.80	$542.40	$39.43	$64.65
5 Dock Builders	38.50	1540.00	64.45	2578.00		
2 Equip. Oper. (crane)	42.55	680.80	66.95	1071.20		
1 Equip. Oper. Oiler	36.80	294.40	57.90	463.20		
1 Crawler Crane, 50 Ton		1110.00		1221.00		
1 Barge, 400 Ton		746.40		821.04		
1 Hammer, diesel, 15K Ft.-Lbs.		578.60		636.46		
1 Lead, 60' high		69.60		76.56		
1 Air Compressor, 600 cfm		434.40		477.84		
2 -50' Air Hoses, 3"		32.40		35.64	41.27	45.40
72 L.H., Daily Totals		$5810.60		$7923.34	$80.70	$110.05
Crew B-76A	Hr.	Daily	Hr.	Daily	Bare Costs	Incl. O&P
1 Laborer Foreman	$33.60	$268.80	$55.45	$443.60	$33.87	$55.13
5 Laborers	31.60	1264.00	52.15	2086.00		
1 Equip. Oper. (crane)	42.55	340.40	66.95	535.60		
1 Equip. Oper. Oiler	36.80	294.40	57.90	463.20		
1 Crawler Crane, 50 Ton		1110.00		1221.00		
1 Barge, 400 Ton		746.40		821.04	29.01	31.91
64 L.H., Daily Totals		$4024.00		$5570.44	$62.88	$87.04
Crew B-77	Hr.	Daily	Hr.	Daily	Bare Costs	Incl. O&P
1 Labor Foreman	$33.60	$268.80	$55.45	$443.60	$31.87	$52.53
3 Laborers	31.60	758.40	52.15	1251.60		
1 Truck Driver (light)	30.95	247.60	50.75	406.00		
1 Crack Cleaner, 25 H.P.		54.80		60.28		
1 Crack Filler, Trailer Mtd.		194.20		213.62		
1 Flatbed Truck, Gas, 3 Ton		242.60		266.86	12.29	13.52
40 L.H., Daily Totals		$1766.40		$2641.96	$44.16	$66.05
Crew B-78	Hr.	Daily	Hr.	Daily	Bare Costs	Incl. O&P
1 Labor Foreman	$33.60	$268.80	$55.45	$443.60	$31.82	$52.47
4 Laborers	31.60	1011.20	52.15	1668.80		
1 Truck Driver (light)	30.95	247.60	50.75	406.00		
1 Paint Striper, S.P.		147.00		161.70		
1 Flatbed Truck, Gas, 3 Ton		242.60		266.86		
1 Pickup Truck, 3/4 Ton		116.80		128.48	10.55	11.61
48 L.H., Daily Totals		$2034.00		$3075.44	$42.38	$64.07
Crew B-78A	Hr.	Daily	Hr.	Daily	Bare Costs	Incl. O&P
1 Equip. Oper. (light)	$39.05	$312.40	$61.45	$491.60	$39.05	$61.45
1 Line Rem. (metal balls) 115 H.P.		828.60		911.46	103.58	113.93
8 L.H., Daily Totals		$1141.00		$1403.06	$142.63	$175.38
Crew B-78B	Hr.	Daily	Hr.	Daily	Bare Costs	Incl. O&P
2 Laborers	$31.60	$505.60	$52.15	$834.40	$32.43	$53.18
.25 Equip. Oper. (light)	39.05	78.10	61.45	122.90		
1 Pickup Truck, 3/4 Ton		116.80		128.48		
1 Line Rem., 11 H.P., walk behind		54.00		59.40		
.25 Road Sweeper, S.P., 8' wide		150.40		165.44	17.84	19.63
18 L.H., Daily Totals		$904.90		$1310.62	$50.27	$72.81
Crew B-79	Hr.	Daily	Hr.	Daily	Bare Costs	Incl. O&P
1 Labor Foreman	$33.60	$268.80	$55.45	$443.60	$31.87	$52.53
3 Laborers	31.60	758.40	52.15	1251.60		
1 Truck Driver (light)	30.95	247.60	50.75	406.00		
1 Thermo. Striper, T.M.		703.20		773.52		
1 Flatbed Truck, Gas, 3 Ton		242.60		266.86		
2 Pickup Trucks, 3/4 Ton		233.60		256.96	29.48	32.43
40 L.H., Daily Totals		$2454.20		$3398.54	$61.35	$84.96

Crew No.	Bare Costs Hr.	Bare Costs Daily	Incl. Subs O&P Hr.	Incl. Subs O&P Daily	Cost Per Labor-Hour Bare Costs	Cost Per Labor-Hour Incl. O&P
Crew B-79A						
1.5 Equip. Oper. (light)	$39.05	$468.60	$61.45	$737.40	$39.05	$61.45
.5 Line Remov. (grinder) 115 H.P.		440.20		484.22		
1 Line Rem. (metal balls) 115 H.P.		828.60		911.46	105.73	116.31
12 L.H., Daily Totals		$1737.40		$2133.08	$144.78	$177.76
Crew B-80						
1 Labor Foreman	$33.60	$268.80	$55.45	$443.60	$33.80	$54.95
1 Laborer	31.60	252.80	52.15	417.20		
1 Truck Driver (light)	30.95	247.60	50.75	406.00		
1 Equip. Oper. (light)	39.05	312.40	61.45	491.60		
1 Flatbed Truck, Gas, 3 Ton		242.60		266.86		
1 Earth Auger, Truck-Mtd.		412.00		453.20	20.46	22.50
32 L.H., Daily Totals		$1736.20		$2478.46	$54.26	$77.45
Crew B-80A						
3 Laborers	$31.60	$758.40	$52.15	$1251.60	$31.60	$52.15
1 Flatbed Truck, Gas, 3 Ton		242.60		266.86	10.11	11.12
24 L.H., Daily Totals		$1001.00		$1518.46	$41.71	$63.27
Crew B-80B						
3 Laborers	$31.60	$758.40	$52.15	$1251.60	$33.46	$54.48
1 Equip. Oper. (light)	39.05	312.40	61.45	491.60		
1 Crane, Flatbed Mounted, 3 Ton		225.00		247.50	7.03	7.73
32 L.H., Daily Totals		$1295.80		$1990.70	$40.49	$62.21
Crew B-80C						
2 Laborers	$31.60	$505.60	$52.15	$834.40	$31.38	$51.68
1 Truck Driver (light)	30.95	247.60	50.75	406.00		
1 Flatbed Truck, Gas, 1.5 Ton		194.40		213.84		
1 Manual fence post auger, gas		7.40		8.14	8.41	9.25
24 L.H., Daily Totals		$955.00		$1462.38	$39.79	$60.93
Crew B-81						
1 Laborer	$31.60	$252.80	$52.15	$417.20	$34.97	$56.52
1 Equip. Oper. (med.)	41.35	330.80	65.05	520.40		
1 Truck Driver (heavy)	31.95	255.60	52.35	418.80		
1 Hydromulcher, T.M.		291.00		320.10		
1 Truck Tractor, 220 H.P.		288.40		317.24	24.14	26.56
24 L.H., Daily Totals		$1418.60		$1993.74	$59.11	$83.07
Crew B-82						
1 Laborer	$31.60	$252.80	$52.15	$417.20	$35.33	$56.80
1 Equip. Oper. (light)	39.05	312.40	61.45	491.60		
1 Horiz. Borer, 6 H.P.		77.60		85.36	4.85	5.34
16 L.H., Daily Totals		$642.80		$994.16	$40.17	$62.13
Crew B-82A						
1 Laborer	$31.60	$252.80	$52.15	$417.20	$35.33	$56.80
1 Equip. Oper. (light)	39.05	312.40	61.45	491.60		
1 Flatbed Truck, Gas, 3 Ton		242.60		266.86		
1 Flatbed Trailer, 25 Ton		102.00		112.20		
1 Horiz. Dir. Drill, 20k lb. thrust		628.20		691.02	60.80	66.88
16 L.H., Daily Totals		$1538.00		$1978.88	$96.13	$123.68
Crew B-82B						
2 Laborers	$31.60	$505.60	$52.15	$834.40	$34.08	$55.25
1 Equip. Oper. (light)	39.05	312.40	61.45	491.60		
1 Flatbed Truck, Gas, 3 Ton		242.60		266.86		
1 Flatbed Trailer, 25 Ton		102.00		112.20		
1 Horiz. Dir. Drill, 30k lb. thrust		896.60		986.26	51.72	56.89
24 L.H., Daily Totals		$2059.20		$2691.32	$85.80	$112.14
Crew B-82C						
2 Laborers	$31.60	$505.60	$52.15	$834.40	$34.08	$55.25
1 Equip. Oper. (light)	39.05	312.40	61.45	491.60		
1 Flatbed Truck, Gas, 3 Ton		242.60		266.86		
1 Flatbed Trailer, 25 Ton		102.00		112.20		
1 Horiz. Dir. Drill, 50k lb. thrust		1183.00		1301.30	63.65	70.02
24 L.H., Daily Totals		$2345.60		$3006.36	$97.73	$125.27
Crew B-82D						
1 Equip. Oper. (light)	$39.05	$312.40	$61.45	$491.60	$39.05	$61.45
1 Mud Trailer for HDD, 1500 gallon		283.20		311.52	35.40	38.94
8 L.H., Daily Totals		$595.60		$803.12	$74.45	$100.39
Crew B-83						
1 Tugboat Captain	$41.35	$330.80	$65.05	$520.40	$36.48	$58.60
1 Tugboat Hand	31.60	252.80	52.15	417.20		
1 Tugboat, 250 H.P.		663.40		729.74	41.46	45.61
16 L.H., Daily Totals		$1247.00		$1667.34	$77.94	$104.21
Crew B-84						
1 Equip. Oper. (med.)	$41.35	$330.80	$65.05	$520.40	$41.35	$65.05
1 Rotary Mower/Tractor		273.40		300.74	34.17	37.59
8 L.H., Daily Totals		$604.20		$821.14	$75.53	$102.64
Crew B-85						
3 Laborers	$31.60	$758.40	$52.15	$1251.60	$33.62	$54.77
1 Equip. Oper. (med.)	41.35	330.80	65.05	520.40		
1 Truck Driver (heavy)	31.95	255.60	52.35	418.80		
1 Aerial Lift Truck, 80'		640.40		704.44		
1 Brush Chipper, 12", 130 H.P.		225.60		248.16		
1 Pruning Saw, Rotary		6.50		7.15	21.81	23.99
40 L.H., Daily Totals		$2217.30		$3150.55	$55.43	$78.76
Crew B-86						
1 Equip. Oper. (med.)	$41.35	$330.80	$65.05	$520.40	$41.35	$65.05
1 Stump Chipper, S.P.		201.10		221.21	25.14	27.65
8 L.H., Daily Totals		$531.90		$741.61	$66.49	$92.70
Crew B-86A						
1 Equip. Oper. (medium)	$41.35	$330.80	$65.05	$520.40	$41.35	$65.05
1 Grader, 30,000 Lbs.		550.20		605.22	68.78	75.65
8 L.H., Daily Totals		$881.00		$1125.62	$110.13	$140.70
Crew B-86B						
1 Equip. Oper. (medium)	$41.35	$330.80	$65.05	$520.40	$41.35	$65.05
1 Dozer, 200 H.P.		1082.00		1190.20	135.25	148.78
8 L.H., Daily Totals		$1412.80		$1710.60	$176.60	$213.82

Crew No.	Bare Costs		Incl. Subs O & P		Cost Per Labor-Hour	

Crew B-87

	Hr.	Daily	Hr.	Daily	Bare Costs	Incl. O&P
1 Laborer	$31.60	$252.80	$52.15	$417.20	$39.40	$62.47
4 Equip. Oper. (med.)	41.35	1323.20	65.05	2081.60		
2 Feller Bunchers, 100 H.P.		1122.80		1235.08		
1 Log Chipper, 22" Tree		588.40		647.24		
1 Dozer, 105 H.P.		600.60		660.66		
1 Chainsaw, Gas, 36" Long		37.80		41.58	58.74	64.61
40 L.H., Daily Totals		$3925.60		$5083.36	$98.14	$127.08

Crew B-88

	Hr.	Daily	Hr.	Daily	Bare Costs	Incl. O&P
1 Laborer	$31.60	$252.80	$52.15	$417.20	$39.96	$63.21
6 Equip. Oper. (med.)	41.35	1984.80	65.05	3122.40		
2 Feller Bunchers, 100 H.P.		1122.80		1235.08		
1 Log Chipper, 22" Tree		588.40		647.24		
2 Log Skidders, 50 H.P.		1795.20		1974.72		
1 Dozer, 105 H.P.		600.60		660.66		
1 Chainsaw, Gas, 36" Long		37.80		41.58	74.01	81.42
56 L.H., Daily Totals		$6382.40		$8098.88	$113.97	$144.62

Crew B-89

	Hr.	Daily	Hr.	Daily	Bare Costs	Incl. O&P
1 Equip. Oper. (light)	$39.05	$312.40	$61.45	$491.60	$35.00	$56.10
1 Truck Driver (light)	30.95	247.60	50.75	406.00		
1 Flatbed Truck, Gas, 3 Ton		242.60		266.86		
1 Concrete Saw		141.00		155.10		
1 Water Tank, 65 Gal.		18.45		20.30	25.13	27.64
16 L.H., Daily Totals		$962.05		$1339.86	$60.13	$83.74

Crew B-89A

	Hr.	Daily	Hr.	Daily	Bare Costs	Incl. O&P
1 Skilled Worker	$40.85	$326.80	$66.50	$532.00	$36.23	$59.33
1 Laborer	31.60	252.80	52.15	417.20		
1 Core Drill (large)		110.20		121.22	6.89	7.58
16 L.H., Daily Totals		$689.80		$1070.42	$43.11	$66.90

Crew B-89B

	Hr.	Daily	Hr.	Daily	Bare Costs	Incl. O&P
1 Equip. Oper. (light)	$39.05	$312.40	$61.45	$491.60	$35.00	$56.10
1 Truck Driver, Light	30.95	247.60	50.75	406.00		
1 Wall Saw, Hydraulic, 10 H.P.		96.00		105.60		
1 Generator, Diesel, 100 kW		327.20		359.92		
1 Water Tank, 65 Gal.		18.45		20.30		
1 Flatbed Truck, Gas, 3 Ton		242.60		266.86	42.77	47.04
16 L.H., Daily Totals		$1244.25		$1650.28	$77.77	$103.14

Crew B-90

	Hr.	Daily	Hr.	Daily	Bare Costs	Incl. O&P
1 Labor Foreman (outside)	$33.60	$268.80	$55.45	$443.60	$33.80	$54.94
3 Laborers	31.60	758.40	52.15	1251.60		
2 Equip. Oper. (light)	39.05	624.80	61.45	983.20		
2 Truck Drivers (heavy)	31.95	511.20	52.35	837.60		
1 Road Mixer, 310 H.P.		1896.00		2085.60		
1 Dist. Truck, 2000 Gal.		277.20		304.92	33.96	37.35
64 L.H., Daily Totals		$4336.40		$5906.52	$67.76	$92.29

Crew B-90A

	Hr.	Daily	Hr.	Daily	Bare Costs	Incl. O&P
1 Labor Foreman	$33.60	$268.80	$55.45	$443.60	$37.46	$59.99
2 Laborers	31.60	505.60	52.15	834.40		
4 Equip. Oper. (medium)	41.35	1323.20	65.05	2081.60		
2 Graders, 30,000 Lbs.		1100.40		1210.44		
1 Tandem Roller, 10 Ton		233.00		256.30		
1 Roller, Pneum. Whl, 12 Ton		307.80		338.58	29.31	32.24
56 L.H., Daily Totals		$3738.80		$5164.92	$66.76	$92.23

Crew B-90B

	Hr.	Daily	Hr.	Daily	Bare Costs	Incl. O&P
1 Labor Foreman	$33.60	$268.80	$55.45	$443.60	$36.81	$59.15
2 Laborers	31.60	505.60	52.15	834.40		
3 Equip. Oper. (medium)	41.35	992.40	65.05	1561.20		
1 Tandem Roller, 10 Ton		233.00		256.30		
1 Roller, Pneum. Whl, 12 Ton		307.80		338.58		
1 Road Mixer, 310 H.P.		1896.00		2085.60	50.77	55.84
48 L.H., Daily Totals		$4203.60		$5519.68	$87.58	$114.99

Crew B-91

	Hr.	Daily	Hr.	Daily	Bare Costs	Incl. O&P
1 Labor Foreman (outside)	$33.60	$268.80	$55.45	$443.60	$36.77	$59.04
2 Laborers	31.60	505.60	52.15	834.40		
4 Equip. Oper. (med.)	41.35	1323.20	65.05	2081.60		
1 Truck Driver (heavy)	31.95	255.60	52.35	418.80		
1 Dist. Tanker, 3000 Gallon		304.20		334.62		
1 Truck Tractor, 6x4, 380 H.P.		478.80		526.68		
1 Aggreg. Spreader, S.P.		855.20		940.72		
1 Roller, Pneum. Whl, 12 Ton		307.80		338.58		
1 Tandem Roller, 10 Ton		233.00		256.30	34.05	37.45
64 L.H., Daily Totals		$4532.20		$6175.30	$70.82	$96.49

Crew B-92

	Hr.	Daily	Hr.	Daily	Bare Costs	Incl. O&P
1 Labor Foreman (outside)	$33.60	$268.80	$55.45	$443.60	$32.10	$52.98
3 Laborers	31.60	758.40	52.15	1251.60		
1 Crack Cleaner, 25 H.P.		54.80		60.28		
1 Air Compressor, 60 cfm		122.60		134.86		
1 Tar Kettle, T.M.		83.15		91.47		
1 Flatbed Truck, Gas, 3 Ton		242.60		266.86	15.72	17.30
32 L.H., Daily Totals		$1530.35		$2248.67	$47.82	$70.27

Crew B-93

	Hr.	Daily	Hr.	Daily	Bare Costs	Incl. O&P
1 Equip. Oper. (med.)	$41.35	$330.80	$65.05	$520.40	$41.35	$65.05
1 Feller Buncher, 100 H.P.		561.40		617.54	70.17	77.19
8 L.H., Daily Totals		$892.20		$1137.94	$111.53	$142.24

Crew B-94A

	Hr.	Daily	Hr.	Daily	Bare Costs	Incl. O&P
1 Laborer	$31.60	$252.80	$52.15	$417.20	$31.60	$52.15
1 Diaphragm Water Pump, 2"		63.20		69.52		
1 -20' Suction Hose, 2"		1.95		2.15		
2 -50' Discharge Hoses, 2"		2.20		2.42	8.42	9.26
8 L.H., Daily Totals		$320.15		$491.29	$40.02	$61.41

Crew B-94B

	Hr.	Daily	Hr.	Daily	Bare Costs	Incl. O&P
1 Laborer	$31.60	$252.80	$52.15	$417.20	$31.60	$52.15
1 Diaphragm Water Pump, 4"		86.20		94.82		
1 -20' Suction Hose, 4"		3.45		3.79		
2 -50' Discharge Hoses, 4"		4.70		5.17	11.79	12.97
8 L.H., Daily Totals		$347.15		$520.99	$43.39	$65.12

Crew B-94C

	Hr.	Daily	Hr.	Daily	Bare Costs	Incl. O&P
1 Laborer	$31.60	$252.80	$52.15	$417.20	$31.60	$52.15
1 Centrifugal Water Pump, 3"		69.40		76.34		
1 -20' Suction Hose, 3"		3.05		3.36		
2 -50' Discharge Hoses, 3"		3.50		3.85	9.49	10.44
8 L.H., Daily Totals		$328.75		$500.75	$41.09	$62.59

Crew B-94D

	Hr.	Daily	Hr.	Daily	Bare Costs	Incl. O&P
1 Laborer	$31.60	$252.80	$52.15	$417.20	$31.60	$52.15
1 Centr. Water Pump, 6"		309.40		340.34		
1 -20' Suction Hose, 6"		11.90		13.09		
2 -50' Discharge Hoses, 6"		12.60		13.86	41.74	45.91
8 L.H., Daily Totals		$586.70		$784.49	$73.34	$98.06

Crew No.	Bare Costs		Incl. Subs O & P		Cost Per Labor-Hour	
Crew C-1	Hr.	Daily	Hr.	Daily	Bare Costs	Incl. O&P
3 Carpenters	$39.95	$958.80	$65.95	$1582.80	$37.86	$62.50
1 Laborer	31.60	252.80	52.15	417.20		
32 L.H., Daily Totals		$1211.60		$2000.00	$37.86	$62.50
Crew C-2	Hr.	Daily	Hr.	Daily	Bare Costs	Incl. O&P
1 Carpenter Foreman (out)	$41.95	$335.60	$69.25	$554.00	$38.89	$64.20
4 Carpenters	39.95	1278.40	65.95	2110.40		
1 Laborer	31.60	252.80	52.15	417.20		
48 L.H., Daily Totals		$1866.80		$3081.60	$38.89	$64.20
Crew C-2A	Hr.	Daily	Hr.	Daily	Bare Costs	Incl. O&P
1 Carpenter Foreman (out)	$41.95	$335.60	$69.25	$554.00	$38.62	$63.19
3 Carpenters	39.95	958.80	65.95	1582.80		
1 Cement Finisher	38.30	306.40	59.90	479.20		
1 Laborer	31.60	252.80	52.15	417.20		
48 L.H., Daily Totals		$1853.60		$3033.20	$38.62	$63.19
Crew C-3	Hr.	Daily	Hr.	Daily	Bare Costs	Incl. O&P
1 Rodman Foreman	$46.55	$372.40	$79.35	$634.80	$40.88	$68.61
4 Rodmen (reinf.)	44.55	1425.60	75.95	2430.40		
1 Equip. Oper. (light)	39.05	312.40	61.45	491.60		
2 Laborers	31.60	505.60	52.15	834.40		
3 Stressing Equipment		27.60		30.36		
.5 Grouting Equipment		75.80		83.38	1.62	1.78
64 L.H., Daily Totals		$2719.40		$4504.94	$42.49	$70.39
Crew C-4	Hr.	Daily	Hr.	Daily	Bare Costs	Incl. O&P
1 Rodman Foreman	$46.55	$372.40	$79.35	$634.80	$45.05	$76.80
3 Rodmen (reinf.)	44.55	1069.20	75.95	1822.80		
3 Stressing Equipment		27.60		30.36	.86	.95
32 L.H., Daily Totals		$1469.20		$2487.96	$45.91	$77.75
Crew C-5	Hr.	Daily	Hr.	Daily	Bare Costs	Incl. O&P
1 Rodman Foreman	$46.55	$372.40	$79.35	$634.80	$43.44	$72.57
4 Rodmen (reinf.)	44.55	1425.60	75.95	2430.40		
1 Equip. Oper. (crane)	42.55	340.40	66.95	535.60		
1 Equip. Oper. Oiler	36.80	294.40	57.90	463.20		
1 Hyd. Crane, 25 Ton		789.40		868.34	14.10	15.51
56 L.H., Daily Totals		$3222.20		$4932.34	$57.54	$88.08
Crew C-6	Hr.	Daily	Hr.	Daily	Bare Costs	Incl. O&P
1 Labor Foreman (outside)	$33.60	$268.80	$55.45	$443.60	$33.05	$53.99
4 Laborers	31.60	1011.20	52.15	1668.80		
1 Cement Finisher	38.30	306.40	59.90	479.20		
2 Gas Engine Vibrators		52.00		57.20	1.08	1.19
48 L.H., Daily Totals		$1638.40		$2648.80	$34.13	$55.18
Crew C-7	Hr.	Daily	Hr.	Daily	Bare Costs	Incl. O&P
1 Labor Foreman (outside)	$33.60	$268.80	$55.45	$443.60	$34.23	$55.45
5 Laborers	31.60	1264.00	52.15	2086.00		
1 Cement Finisher	38.30	306.40	59.90	479.20		
1 Equip. Oper. (med.)	41.35	330.80	65.05	520.40		
1 Equip. Oper. (oiler)	36.80	294.40	57.90	463.20		
2 Gas Engine Vibrators		52.00		57.20		
1 Concrete Bucket, 1 C.Y.		20.20		22.22		
1 Hyd. Crane, 55 Ton		1125.00		1237.50	16.63	18.29
72 L.H., Daily Totals		$3661.60		$5309.32	$50.86	$73.74
Crew C-8	Hr.	Daily	Hr.	Daily	Bare Costs	Incl. O&P
1 Labor Foreman (outside)	$33.60	$268.80	$55.45	$443.60	$35.19	$56.68
3 Laborers	31.60	758.40	52.15	1251.60		
2 Cement Finishers	38.30	612.80	59.90	958.40		
1 Equip. Oper. (med.)	41.35	330.80	65.05	520.40		
1 Concrete Pump (small)		737.80		811.58	13.18	14.49
56 L.H., Daily Totals		$2708.60		$3985.58	$48.37	$71.17
Crew C-8A	Hr.	Daily	Hr.	Daily	Bare Costs	Incl. O&P
1 Labor Foreman (outside)	$33.60	$268.80	$55.45	$443.60	$34.17	$55.28
3 Laborers	31.60	758.40	52.15	1251.60		
2 Cement Finishers	38.30	612.80	59.90	958.40		
48 L.H., Daily Totals		$1640.00		$2653.60	$34.17	$55.28
Crew C-8B	Hr.	Daily	Hr.	Daily	Bare Costs	Incl. O&P
1 Labor Foreman (outside)	$33.60	$268.80	$55.45	$443.60	$33.95	$55.39
3 Laborers	31.60	758.40	52.15	1251.60		
1 Equip. Oper. (med.)	41.35	330.80	65.05	520.40		
1 Vibrating Power Screed		70.35		77.39		
1 Roller, Vibratory, 25 Ton		594.60		654.06		
1 Dozer, 200 H.P.		1082.00		1190.20	43.67	48.04
40 L.H., Daily Totals		$3104.95		$4137.24	$77.62	$103.43
Crew C-8C	Hr.	Daily	Hr.	Daily	Bare Costs	Incl. O&P
1 Labor Foreman (outside)	$33.60	$268.80	$55.45	$443.60	$34.67	$56.14
3 Laborers	31.60	758.40	52.15	1251.60		
1 Cement Finisher	38.30	306.40	59.90	479.20		
1 Equip. Oper. (med.)	41.35	330.80	65.05	520.40		
1 Shotcrete Rig, 12 C.Y./hr		237.80		261.58	4.95	5.45
48 L.H., Daily Totals		$1902.20		$2956.38	$39.63	$61.59
Crew C-8D	Hr.	Daily	Hr.	Daily	Bare Costs	Incl. O&P
1 Labor Foreman (outside)	$33.60	$268.80	$55.45	$443.60	$35.64	$57.24
1 Laborer	31.60	252.80	52.15	417.20		
1 Cement Finisher	38.30	306.40	59.90	479.20		
1 Equipment Oper. (light)	39.05	312.40	61.45	491.60		
1 Air Compressor, 250 cfm		162.40		178.64		
2 -50' Air Hoses, 1"		8.20		9.02	5.33	5.86
32 L.H., Daily Totals		$1311.00		$2019.26	$40.97	$63.10
Crew C-8E	Hr.	Daily	Hr.	Daily	Bare Costs	Incl. O&P
1 Labor Foreman (outside)	$33.60	$268.80	$55.45	$443.60	$35.64	$57.24
1 Laborer	31.60	252.80	52.15	417.20		
1 Cement Finisher	38.30	306.40	59.90	479.20		
1 Equipment Oper. (light)	39.05	312.40	61.45	491.60		
1 Air Compressor, 250 cfm		162.40		178.64		
2 -50' Air Hoses, 1"		8.20		9.02		
1 Concrete Pump (small)		737.80		811.58	28.39	31.23
32 L.H., Daily Totals		$2048.80		$2830.84	$64.03	$88.46
Crew C-10	Hr.	Daily	Hr.	Daily	Bare Costs	Incl. O&P
1 Laborer	$31.60	$252.80	$52.15	$417.20	$36.07	$57.32
2 Cement Finishers	38.30	612.80	59.90	958.40		
24 L.H., Daily Totals		$865.60		$1375.60	$36.07	$57.32
Crew C-10B	Hr.	Daily	Hr.	Daily	Bare Costs	Incl. O&P
3 Laborers	$31.60	$758.40	$52.15	$1251.60	$34.28	$55.25
2 Cement Finishers	38.30	612.80	59.90	958.40		
1 Concrete Mixer, 10 C.F.		155.20		170.72		
2 Trowels, 48" Walk-Behind		84.80		93.28	6.00	6.60
40 L.H., Daily Totals		$1611.20		$2474.00	$40.28	$61.85

Crew No.	Bare Costs		Incl. Subs O & P		Cost Per Labor-Hour	
Crew C-10C	**Hr.**	**Daily**	**Hr.**	**Daily**	**Bare Costs**	**Incl. O&P**
1 Laborer	$31.60	$252.80	$52.15	$417.20	$36.07	$57.32
2 Cement Finishers	38.30	612.80	59.90	958.40		
1 Trowel, 48" Walk-Behind		42.40		46.64	1.77	1.94
24 L.H., Daily Totals		$908.00		$1422.24	$37.83	$59.26
Crew C-10D	**Hr.**	**Daily**	**Hr.**	**Daily**	**Bare Costs**	**Incl. O&P**
1 Laborer	$31.60	$252.80	$52.15	$417.20	$36.07	$57.32
2 Cement Finishers	38.30	612.80	59.90	958.40		
1 Vibrating Power Screed		70.35		77.39		
1 Trowel, 48" Walk-Behind		42.40		46.64	4.70	5.17
24 L.H., Daily Totals		$978.35		$1499.63	$40.76	$62.48
Crew C-10E	**Hr.**	**Daily**	**Hr.**	**Daily**	**Bare Costs**	**Incl. O&P**
1 Laborer	$31.60	$252.80	$52.15	$417.20	$36.07	$57.32
2 Cement Finishers	38.30	612.80	59.90	958.40		
1 Vibrating Power Screed		70.35		77.39		
1 Cement Trowel, 96" Ride-On		166.60		183.26	9.87	10.86
24 L.H., Daily Totals		$1102.55		$1636.24	$45.94	$68.18
Crew C-11	**Hr.**	**Daily**	**Hr.**	**Daily**	**Bare Costs**	**Incl. O&P**
1 Struc. Steel Foreman	$46.70	$373.60	$86.50	$692.00	$43.81	$78.68
6 Struc. Steel Workers	44.70	2145.60	82.80	3974.40		
1 Equip. Oper. (crane)	42.55	340.40	66.95	535.60		
1 Equip. Oper. Oiler	36.80	294.40	57.90	463.20		
1 Lattice Boom Crane, 150 Ton		1890.00		2079.00	26.25	28.88
72 L.H., Daily Totals		$5044.00		$7744.20	$70.06	$107.56
Crew C-12	**Hr.**	**Daily**	**Hr.**	**Daily**	**Bare Costs**	**Incl. O&P**
1 Carpenter Foreman (out)	$41.95	$335.60	$69.25	$554.00	$39.33	$64.37
3 Carpenters	39.95	958.80	65.95	1582.80		
1 Laborer	31.60	252.80	52.15	417.20		
1 Equip. Oper. (crane)	42.55	340.40	66.95	535.60		
1 Hyd. Crane, 12 Ton		768.80		845.68	16.02	17.62
48 L.H., Daily Totals		$2656.40		$3935.28	$55.34	$81.98
Crew C-13	**Hr.**	**Daily**	**Hr.**	**Daily**	**Bare Costs**	**Incl. O&P**
1 Struc. Steel Worker	$44.70	$357.60	$82.80	$662.40	$43.12	$77.18
1 Welder	44.70	357.60	82.80	662.40		
1 Carpenter	39.95	319.60	65.95	527.60		
1 Welder, gas engine, 300 amp		134.20		147.62	5.59	6.15
24 L.H., Daily Totals		$1169.00		$2000.02	$48.71	$83.33
Crew C-14	**Hr.**	**Daily**	**Hr.**	**Daily**	**Bare Costs**	**Incl. O&P**
1 Carpenter Foreman (out)	$41.95	$335.60	$69.25	$554.00	$39.01	$64.22
5 Carpenters	39.95	1598.00	65.95	2638.00		
4 Laborers	31.60	1011.20	52.15	1668.80		
4 Rodmen (reinf.)	44.55	1425.60	75.95	2430.40		
2 Cement Finishers	38.30	612.80	59.90	958.40		
1 Equip. Oper. (crane)	42.55	340.40	66.95	535.60		
1 Equip. Oper. Oiler	36.80	294.40	57.90	463.20		
1 Hyd. Crane, 80 Ton		1296.00		1425.60	9.00	9.90
144 L.H., Daily Totals		$6914.00		$10674.00	$48.01	$74.13

Crew No.	Bare Costs		Incl. Subs O & P		Cost Per Labor-Hour	
Crew C-14A	**Hr.**	**Daily**	**Hr.**	**Daily**	**Bare Costs**	**Incl. O&P**
1 Carpenter Foreman (out)	$41.95	$335.60	$69.25	$554.00	$40.09	$66.30
16 Carpenters	39.95	5113.60	65.95	8441.60		
4 Rodmen (reinf.)	44.55	1425.60	75.95	2430.40		
2 Laborers	31.60	505.60	52.15	834.40		
1 Cement Finisher	38.30	306.40	59.90	479.20		
1 Equip. Oper. (med.)	41.35	330.80	65.05	520.40		
1 Gas Engine Vibrator		26.00		28.60		
1 Concrete Pump (small)		737.80		811.58	3.82	4.20
200 L.H., Daily Totals		$8781.40		$14100.18	$43.91	$70.50
Crew C-14B	**Hr.**	**Daily**	**Hr.**	**Daily**	**Bare Costs**	**Incl. O&P**
1 Carpenter Foreman (out)	$41.95	$335.60	$69.25	$554.00	$40.02	$66.05
16 Carpenters	39.95	5113.60	65.95	8441.60		
4 Rodmen (reinf.)	44.55	1425.60	75.95	2430.40		
2 Laborers	31.60	505.60	52.15	834.40		
2 Cement Finishers	38.30	612.80	59.90	958.40		
1 Equip. Oper. (med.)	41.35	330.80	65.05	520.40		
1 Gas Engine Vibrator		26.00		28.60		
1 Concrete Pump (small)		737.80		811.58	3.67	4.04
208 L.H., Daily Totals		$9087.80		$14579.38	$43.69	$70.09
Crew C-14C	**Hr.**	**Daily**	**Hr.**	**Daily**	**Bare Costs**	**Incl. O&P**
1 Carpenter Foreman (out)	$41.95	$335.60	$69.25	$554.00	$38.25	$63.24
6 Carpenters	39.95	1917.60	65.95	3165.60		
2 Rodmen (reinf.)	44.55	712.80	75.95	1215.20		
4 Laborers	31.60	1011.20	52.15	1668.80		
1 Cement Finisher	38.30	306.40	59.90	479.20		
1 Gas Engine Vibrator		26.00		28.60	.23	.26
112 L.H., Daily Totals		$4309.60		$7111.40	$38.48	$63.49
Crew C-14D	**Hr.**	**Daily**	**Hr.**	**Daily**	**Bare Costs**	**Incl. O&P**
1 Carpenter Foreman (out)	$41.95	$335.60	$69.25	$554.00	$39.72	$65.50
18 Carpenters	39.95	5752.80	65.95	9496.80		
2 Rodmen (reinf.)	44.55	712.80	75.95	1215.20		
2 Laborers	31.60	505.60	52.15	834.40		
1 Cement Finisher	38.30	306.40	59.90	479.20		
1 Equip. Oper. (med.)	41.35	330.80	65.05	520.40		
1 Gas Engine Vibrator		26.00		28.60		
1 Concrete Pump (small)		737.80		811.58	3.82	4.20
200 L.H., Daily Totals		$8707.80		$13940.18	$43.54	$69.70
Crew C-14E	**Hr.**	**Daily**	**Hr.**	**Daily**	**Bare Costs**	**Incl. O&P**
1 Carpenter Foreman (out)	$41.95	$335.60	$69.25	$554.00	$39.38	$65.57
2 Carpenters	39.95	639.20	65.95	1055.20		
4 Rodmen (reinf.)	44.55	1425.60	75.95	2430.40		
3 Laborers	31.60	758.40	52.15	1251.60		
1 Cement Finisher	38.30	306.40	59.90	479.20		
1 Gas Engine Vibrator		26.00		28.60	.30	.33
88 L.H., Daily Totals		$3491.20		$5799.00	$39.67	$65.90
Crew C-14F	**Hr.**	**Daily**	**Hr.**	**Daily**	**Bare Costs**	**Incl. O&P**
1 Laborer Foreman (out)	$33.60	$268.80	$55.45	$443.60	$36.29	$57.68
2 Laborers	31.60	505.60	52.15	834.40		
6 Cement Finishers	38.30	1838.40	59.90	2875.20		
1 Gas Engine Vibrator		26.00		28.60	.36	.40
72 L.H., Daily Totals		$2638.80		$4181.80	$36.65	$58.08

Crew No.	Bare Costs		Incl. Subs O & P		Cost Per Labor-Hour	
Crew C-14G	Hr.	Daily	Hr.	Daily	Bare Costs	Incl. O&P
1 Laborer Foreman (out)	$33.60	$268.80	$55.45	$443.60	$35.71	$57.05
2 Laborers	31.60	505.60	52.15	834.40		
4 Cement Finishers	38.30	1225.60	59.90	1916.80		
1 Gas Engine Vibrator		26.00		28.60	.46	.51
56 L.H., Daily Totals		$2026.00		$3223.40	$36.18	$57.56
Crew C-14H	Hr.	Daily	Hr.	Daily	Bare Costs	Incl. O&P
1 Carpenter Foreman (out)	$41.95	$335.60	$69.25	$554.00	$39.38	$64.86
2 Carpenters	39.95	639.20	65.95	1055.20		
1 Rodman (reinf.)	44.55	356.40	75.95	607.60		
1 Laborer	31.60	252.80	52.15	417.20		
1 Cement Finisher	38.30	306.40	59.90	479.20		
1 Gas Engine Vibrator		26.00		28.60	.54	.60
48 L.H., Daily Totals		$1916.40		$3141.80	$39.92	$65.45
Crew C-14L	Hr.	Daily	Hr.	Daily	Bare Costs	Incl. O&P
1 Carpenter Foreman (out)	$41.95	$335.60	$69.25	$554.00	$37.20	$61.12
6 Carpenters	39.95	1917.60	65.95	3165.60		
4 Laborers	31.60	1011.20	52.15	1668.80		
1 Cement Finisher	38.30	306.40	59.90	479.20		
1 Gas Engine Vibrator		26.00		28.60	.27	.30
96 L.H., Daily Totals		$3596.80		$5896.20	$37.47	$61.42
Crew C-15	Hr.	Daily	Hr.	Daily	Bare Costs	Incl. O&P
1 Carpenter Foreman (out)	$41.95	$335.60	$69.25	$554.00	$37.53	$61.48
2 Carpenters	39.95	639.20	65.95	1055.20		
3 Laborers	31.60	758.40	52.15	1251.60		
2 Cement Finishers	38.30	612.80	59.90	958.40		
1 Rodman (reinf.)	44.55	356.40	75.95	607.60		
72 L.H., Daily Totals		$2702.40		$4426.80	$37.53	$61.48
Crew C-16	Hr.	Daily	Hr.	Daily	Bare Costs	Incl. O&P
1 Labor Foreman (outside)	$33.60	$268.80	$55.45	$443.60	$37.27	$60.96
3 Laborers	31.60	758.40	52.15	1251.60		
2 Cement Finishers	38.30	612.80	59.90	958.40		
1 Equip. Oper. (med.)	41.35	330.80	65.05	520.40		
2 Rodmen (reinf.)	44.55	712.80	75.95	1215.20		
1 Concrete Pump (small)		737.80		811.58	10.25	11.27
72 L.H., Daily Totals		$3421.40		$5200.78	$47.52	$72.23
Crew C-17	Hr.	Daily	Hr.	Daily	Bare Costs	Incl. O&P
2 Skilled Worker Foremen	$42.85	$685.60	$69.75	$1116.00	$41.25	$67.15
8 Skilled Workers	40.85	2614.40	66.50	4256.00		
80 L.H., Daily Totals		$3300.00		$5372.00	$41.25	$67.15
Crew C-17A	Hr.	Daily	Hr.	Daily	Bare Costs	Incl. O&P
2 Skilled Worker Foremen	$42.85	$685.60	$69.75	$1116.00	$41.27	$67.15
8 Skilled Workers	40.85	2614.40	66.50	4256.00		
.125 Equip. Oper. (crane)	42.55	42.55	66.95	66.95		
.125 Hyd. Crane, 80 Ton		162.00		178.20	2.00	2.20
81 L.H., Daily Totals		$3504.55		$5617.15	$43.27	$69.35
Crew C-17B	Hr.	Daily	Hr.	Daily	Bare Costs	Incl. O&P
2 Skilled Worker Foremen	$42.85	$685.60	$69.75	$1116.00	$41.28	$67.15
8 Skilled Workers	40.85	2614.40	66.50	4256.00		
.25 Equip. Oper. (crane)	42.55	85.10	66.95	133.90		
.25 Hyd. Crane, 80 Ton		324.00		356.40		
.25 Trowel, 48" Walk-Behind		10.60		11.66	4.08	4.49
82 L.H., Daily Totals		$3719.70		$5873.96	$45.36	$71.63

Crew No.	Bare Costs		Incl. Subs O & P		Cost Per Labor-Hour	
Crew C-17C	Hr.	Daily	Hr.	Daily	Bare Costs	Incl. O&P
2 Skilled Worker Foremen	$42.85	$685.60	$69.75	$1116.00	$41.30	$67.14
8 Skilled Workers	40.85	2614.40	66.50	4256.00		
.375 Equip. Oper. (crane)	42.55	127.65	66.95	200.85		
.375 Hyd. Crane, 80 Ton		486.00		534.60	5.86	6.44
83 L.H., Daily Totals		$3913.65		$6107.45	$47.15	$73.58
Crew C-17D	Hr.	Daily	Hr.	Daily	Bare Costs	Incl. O&P
2 Skilled Worker Foremen	$42.85	$685.60	$69.75	$1116.00	$41.31	$67.14
8 Skilled Workers	40.85	2614.40	66.50	4256.00		
.5 Equip. Oper. (crane)	42.55	170.20	66.95	267.80		
.5 Hyd. Crane, 80 Ton		648.00		712.80	7.71	8.49
84 L.H., Daily Totals		$4118.20		$6352.60	$49.03	$75.63
Crew C-17E	Hr.	Daily	Hr.	Daily	Bare Costs	Incl. O&P
2 Skilled Worker Foremen	$42.85	$685.60	$69.75	$1116.00	$41.25	$67.15
8 Skilled Workers	40.85	2614.40	66.50	4256.00		
1 Hyd. Jack with Rods		85.70		94.27	1.07	1.18
80 L.H., Daily Totals		$3385.70		$5466.27	$42.32	$68.33
Crew C-18	Hr.	Daily	Hr.	Daily	Bare Costs	Incl. O&P
.125 Labor Foreman (out)	$33.60	$33.60	$55.45	$55.45	$31.82	$52.52
1 Laborer	31.60	252.80	52.15	417.20		
1 Concrete Cart, 10 C.F.		55.20		60.72	6.13	6.75
9 L.H., Daily Totals		$341.60		$533.37	$37.96	$59.26
Crew C-19	Hr.	Daily	Hr.	Daily	Bare Costs	Incl. O&P
.125 Labor Foreman (out)	$33.60	$33.60	$55.45	$55.45	$31.82	$52.52
1 Laborer	31.60	252.80	52.15	417.20		
1 Concrete Cart, 18 C.F.		86.40		95.04	9.60	10.56
9 L.H., Daily Totals		$372.80		$567.69	$41.42	$63.08
Crew C-20	Hr.	Daily	Hr.	Daily	Bare Costs	Incl. O&P
1 Labor Foreman (outside)	$33.60	$268.80	$55.45	$443.60	$33.91	$55.14
5 Laborers	31.60	1264.00	52.15	2086.00		
1 Cement Finisher	38.30	306.40	59.90	479.20		
1 Equip. Oper. (med.)	41.35	330.80	65.05	520.40		
2 Gas Engine Vibrators		52.00		57.20		
1 Concrete Pump (small)		737.80		811.58	12.34	13.57
64 L.H., Daily Totals		$2959.80		$4397.98	$46.25	$68.72
Crew C-21	Hr.	Daily	Hr.	Daily	Bare Costs	Incl. O&P
1 Labor Foreman (outside)	$33.60	$268.80	$55.45	$443.60	$33.91	$55.14
5 Laborers	31.60	1264.00	52.15	2086.00		
1 Cement Finisher	38.30	306.40	59.90	479.20		
1 Equip. Oper. (med.)	41.35	330.80	65.05	520.40		
2 Gas Engine Vibrators		52.00		57.20		
1 Concrete Conveyer		179.00		196.90	3.61	3.97
64 L.H., Daily Totals		$2401.00		$3783.30	$37.52	$59.11
Crew C-22	Hr.	Daily	Hr.	Daily	Bare Costs	Incl. O&P
1 Rodman Foreman	$46.55	$372.40	$79.35	$634.80	$44.70	$75.95
4 Rodmen (reinf.)	44.55	1425.60	75.95	2430.40		
.125 Equip. Oper. (crane)	42.55	42.55	66.95	66.95		
.125 Equip. Oper. Oiler	36.80	36.80	57.90	57.90		
.125 Hyd. Crane, 25 Ton		98.67		108.54	2.35	2.58
42 L.H., Daily Totals		$1976.03		$3298.59	$47.05	$78.54

Crews

Crew No.	Bare Costs Hr.	Daily	Incl. Subs O & P Hr.	Daily	Bare Costs	Incl. O&P
Crew C-23	Hr.	Daily	Hr.	Daily	Bare Costs	Incl. O&P
2 Skilled Worker Foremen	$42.85	$685.60	$69.75	$1116.00	$41.02	$66.33
6 Skilled Workers	40.85	1960.80	66.50	3192.00		
1 Equip. Oper. (crane)	42.55	340.40	66.95	535.60		
1 Equip. Oper. Oiler	36.80	294.40	57.90	463.20		
1 Lattice Boom Crane, 90 Ton		1741.00		1915.10	21.76	23.94
80 L.H., Daily Totals		$5022.20		$7221.90	$62.78	$90.27
Crew C-24	Hr.	Daily	Hr.	Daily	Bare Costs	Incl. O&P
2 Skilled Worker Foremen	$42.85	$685.60	$69.75	$1116.00	$41.02	$66.33
6 Skilled Workers	40.85	1960.80	66.50	3192.00		
1 Equip. Oper. (crane)	42.55	340.40	66.95	535.60		
1 Equip. Oper. Oiler	36.80	294.40	57.90	463.20		
1 Lattice Boom Crane, 150 Ton		1890.00		2079.00	23.63	25.99
80 L.H., Daily Totals		$5171.20		$7385.80	$64.64	$92.32
Crew C-25	Hr.	Daily	Hr.	Daily	Bare Costs	Incl. O&P
2 Rodmen (reinf.)	$44.55	$712.80	$75.95	$1215.20	$34.95	$60.60
2 Rodmen Helpers	25.35	405.60	45.25	724.00		
32 L.H., Daily Totals		$1118.40		$1939.20	$34.95	$60.60
Crew C-27	Hr.	Daily	Hr.	Daily	Bare Costs	Incl. O&P
2 Cement Finishers	$38.30	$612.80	$59.90	$958.40	$38.30	$59.90
1 Concrete Saw		141.00		155.10	8.81	9.69
16 L.H., Daily Totals		$753.80		$1113.50	$47.11	$69.59
Crew C-28	Hr.	Daily	Hr.	Daily	Bare Costs	Incl. O&P
1 Cement Finisher	$38.30	$306.40	$59.90	$479.20	$38.30	$59.90
1 Portable Air Compressor, Gas		19.10		21.01	2.39	2.63
8 L.H., Daily Totals		$325.50		$500.21	$40.69	$62.53
Crew D-1	Hr.	Daily	Hr.	Daily	Bare Costs	Incl. O&P
1 Bricklayer	$40.50	$324.00	$65.50	$524.00	$36.33	$58.75
1 Bricklayer Helper	32.15	257.20	52.00	416.00		
16 L.H., Daily Totals		$581.20		$940.00	$36.33	$58.75
Crew D-2	Hr.	Daily	Hr.	Daily	Bare Costs	Incl. O&P
3 Bricklayers	$40.50	$972.00	$65.50	$1572.00	$37.41	$60.63
2 Bricklayer Helpers	32.15	514.40	52.00	832.00		
.5 Carpenter	39.95	159.80	65.95	263.80		
44 L.H., Daily Totals		$1646.20		$2667.80	$37.41	$60.63
Crew D-3	Hr.	Daily	Hr.	Daily	Bare Costs	Incl. O&P
3 Bricklayers	$40.50	$972.00	$65.50	$1572.00	$37.29	$60.38
2 Bricklayer Helpers	32.15	514.40	52.00	832.00		
.25 Carpenter	39.95	79.90	65.95	131.90		
42 L.H., Daily Totals		$1566.30		$2535.90	$37.29	$60.38
Crew D-4	Hr.	Daily	Hr.	Daily	Bare Costs	Incl. O&P
1 Bricklayer	$40.50	$324.00	$65.50	$524.00	$35.96	$57.74
2 Bricklayer Helpers	32.15	514.40	52.00	832.00		
1 Equip. Oper. (light)	39.05	312.40	61.45	491.60		
1 Grout Pump, 50 C.F./hr.		126.80		139.48	3.96	4.36
32 L.H., Daily Totals		$1277.60		$1987.08	$39.92	$62.10
Crew D-5	Hr.	Daily	Hr.	Daily	Bare Costs	Incl. O&P
1 Bricklayer	$40.50	$324.00	$65.50	$524.00	$40.50	$65.50
8 L.H., Daily Totals		$324.00		$524.00	$40.50	$65.50

Crew No.	Bare Costs Hr.	Daily	Incl. Subs O & P Hr.	Daily	Bare Costs	Incl. O&P
Crew D-6	Hr.	Daily	Hr.	Daily	Bare Costs	Incl. O&P
3 Bricklayers	$40.50	$972.00	$65.50	$1572.00	$36.47	$59.04
3 Bricklayer Helpers	32.15	771.60	52.00	1248.00		
.25 Carpenter	39.95	79.90	65.95	131.90		
50 L.H., Daily Totals		$1823.50		$2951.90	$36.47	$59.04
Crew D-7	Hr.	Daily	Hr.	Daily	Bare Costs	Incl. O&P
1 Tile Layer	$38.10	$304.80	$59.60	$476.80	$34.08	$53.30
1 Tile Layer Helper	30.05	240.40	47.00	376.00		
16 L.H., Daily Totals		$545.20		$852.80	$34.08	$53.30
Crew D-8	Hr.	Daily	Hr.	Daily	Bare Costs	Incl. O&P
3 Bricklayers	$40.50	$972.00	$65.50	$1572.00	$37.16	$60.10
2 Bricklayer Helpers	32.15	514.40	52.00	832.00		
40 L.H., Daily Totals		$1486.40		$2404.00	$37.16	$60.10
Crew D-9	Hr.	Daily	Hr.	Daily	Bare Costs	Incl. O&P
3 Bricklayers	$40.50	$972.00	$65.50	$1572.00	$36.33	$58.75
3 Bricklayer Helpers	32.15	771.60	52.00	1248.00		
48 L.H., Daily Totals		$1743.60		$2820.00	$36.33	$58.75
Crew D-10	Hr.	Daily	Hr.	Daily	Bare Costs	Incl. O&P
1 Bricklayer Foreman	$42.50	$340.00	$68.70	$549.60	$39.42	$63.29
1 Bricklayer	40.50	324.00	65.50	524.00		
1 Bricklayer Helper	32.15	257.20	52.00	416.00		
1 Equip. Oper. (crane)	42.55	340.40	66.95	535.60		
1 S.P. Crane, 4x4, 12 Ton		601.80		661.98	18.81	20.69
32 L.H., Daily Totals		$1863.40		$2687.18	$58.23	$83.97
Crew D-11	Hr.	Daily	Hr.	Daily	Bare Costs	Incl. O&P
1 Bricklayer Foreman	$42.50	$340.00	$68.70	$549.60	$38.38	$62.07
1 Bricklayer	40.50	324.00	65.50	524.00		
1 Bricklayer Helper	32.15	257.20	52.00	416.00		
24 L.H., Daily Totals		$921.20		$1489.60	$38.38	$62.07
Crew D-12	Hr.	Daily	Hr.	Daily	Bare Costs	Incl. O&P
1 Bricklayer Foreman	$42.50	$340.00	$68.70	$549.60	$36.83	$59.55
1 Bricklayer	40.50	324.00	65.50	524.00		
2 Bricklayer Helpers	32.15	514.40	52.00	832.00		
32 L.H., Daily Totals		$1178.40		$1905.60	$36.83	$59.55
Crew D-13	Hr.	Daily	Hr.	Daily	Bare Costs	Incl. O&P
1 Bricklayer Foreman	$42.50	$340.00	$68.70	$549.60	$38.30	$61.85
1 Bricklayer	40.50	324.00	65.50	524.00		
2 Bricklayer Helpers	32.15	514.40	52.00	832.00		
1 Carpenter	39.95	319.60	65.95	527.60		
1 Equip. Oper. (crane)	42.55	340.40	66.95	535.60		
1 S.P. Crane, 4x4, 12 Ton		601.80		661.98	12.54	13.79
48 L.H., Daily Totals		$2440.20		$3630.78	$50.84	$75.64
Crew E-1	Hr.	Daily	Hr.	Daily	Bare Costs	Incl. O&P
1 Welder Foreman	$46.70	$373.60	$86.50	$692.00	$43.48	$76.92
1 Welder	44.70	357.60	82.80	662.40		
1 Equip. Oper. (light)	39.05	312.40	61.45	491.60		
1 Welder, gas engine, 300 amp		134.20		147.62	5.59	6.15
24 L.H., Daily Totals		$1177.80		$1993.62	$49.08	$83.07

Crew No.	Bare Costs		Incl. Subs O & P		Cost Per Labor-Hour	
Crew E-2	Hr.	Daily	Hr.	Daily	Bare Costs	Incl. O&P
1 Struc. Steel Foreman	$46.70	$373.60	$86.50	$692.00	$43.55	$77.51
4 Struc. Steel Workers	44.70	1430.40	82.80	2649.60		
1 Equip. Oper. (crane)	42.55	340.40	66.95	535.60		
1 Equip. Oper. Oiler	36.80	294.40	57.90	463.20		
1 Lattice Boom Crane, 90 Ton		1741.00		1915.10	31.09	34.20
56 L.H., Daily Totals		$4179.80		$6255.50	$74.64	$111.71
Crew E-3	Hr.	Daily	Hr.	Daily	Bare Costs	Incl. O&P
1 Struc. Steel Foreman	$46.70	$373.60	$86.50	$692.00	$45.37	$84.03
1 Struc. Steel Worker	44.70	357.60	82.80	662.40		
1 Welder	44.70	357.60	82.80	662.40		
1 Welder, gas engine, 300 amp		134.20		147.62	5.59	6.15
24 L.H., Daily Totals		$1223.00		$2164.42	$50.96	$90.18
Crew E-4	Hr.	Daily	Hr.	Daily	Bare Costs	Incl. O&P
1 Struc. Steel Foreman	$46.70	$373.60	$86.50	$692.00	$45.20	$83.72
3 Struc. Steel Workers	44.70	1072.80	82.80	1987.20		
1 Welder, gas engine, 300 amp		134.20		147.62	4.19	4.61
32 L.H., Daily Totals		$1580.60		$2826.82	$49.39	$88.34
Crew E-5	Hr.	Daily	Hr.	Daily	Bare Costs	Incl. O&P
2 Struc. Steel Foremen	$46.70	$747.20	$86.50	$1384.00	$44.09	$79.47
5 Struc. Steel Workers	44.70	1788.00	82.80	3312.00		
1 Equip. Oper. (crane)	42.55	340.40	66.95	535.60		
1 Welder	44.70	357.60	82.80	662.40		
1 Equip. Oper. Oiler	36.80	294.40	57.90	463.20		
1 Lattice Boom Crane, 90 Ton		1741.00		1915.10		
1 Welder, gas engine, 300 amp		134.20		147.62	23.44	25.78
80 L.H., Daily Totals		$5402.80		$8419.92	$67.53	$105.25
Crew E-6	Hr.	Daily	Hr.	Daily	Bare Costs	Incl. O&P
3 Struc. Steel Foremen	$46.70	$1120.80	$86.50	$2076.00	$44.09	$79.61
9 Struc. Steel Workers	44.70	3218.40	82.80	5961.60		
1 Equip. Oper. (crane)	42.55	340.40	66.95	535.60		
1 Welder	44.70	357.60	82.80	662.40		
1 Equip. Oper. Oiler	36.80	294.40	57.90	463.20		
1 Equip. Oper. (light)	39.05	312.40	61.45	491.60		
1 Lattice Boom Crane, 90 Ton		1741.00		1915.10		
1 Welder, gas engine, 300 amp		134.20		147.62		
1 Air Compressor, 160 cfm		138.80		152.68		
2 Impact Wrenches		32.40		35.64	15.99	17.59
128 L.H., Daily Totals		$7690.40		$12441.44	$60.08	$97.20
Crew E-7	Hr.	Daily	Hr.	Daily	Bare Costs	Incl. O&P
1 Struc. Steel Foreman	$46.70	$373.60	$86.50	$692.00	$44.09	$79.47
4 Struc. Steel Workers	44.70	1430.40	82.80	2649.60		
1 Equip. Oper. (crane)	42.55	340.40	66.95	535.60		
1 Equip. Oper. Oiler	36.80	294.40	57.90	463.20		
1 Welder Foreman	46.70	373.60	86.50	692.00		
2 Welders	44.70	715.20	82.80	1324.80		
1 Lattice Boom Crane, 90 Ton		1741.00		1915.10		
2 Welders, gas engine, 300 amp		268.40		295.24	25.12	27.63
80 L.H., Daily Totals		$5537.00		$8567.54	$69.21	$107.09

Crew No.	Bare Costs		Incl. Subs O & P		Cost Per Labor-Hour	
Crew E-8	Hr.	Daily	Hr.	Daily	Bare Costs	Incl. O&P
1 Struc. Steel Foreman	$46.70	$373.60	$86.50	$692.00	$43.80	$78.59
4 Struc. Steel Workers	44.70	1430.40	82.80	2649.60		
1 Welder Foreman	46.70	373.60	86.50	692.00		
4 Welders	44.70	1430.40	82.80	2649.60		
1 Equip. Oper. (crane)	42.55	340.40	66.95	535.60		
1 Equip. Oper. Oiler	36.80	294.40	57.90	463.20		
1 Equip. Oper. (light)	39.05	312.40	61.45	491.60		
1 Lattice Boom Crane, 90 Ton		1741.00		1915.10		
4 Welders, gas engine, 300 amp		536.80		590.48	21.90	24.09
104 L.H., Daily Totals		$6833.00		$10679.18	$65.70	$102.68
Crew E-9	Hr.	Daily	Hr.	Daily	Bare Costs	Incl. O&P
2 Struc. Steel Foremen	$46.70	$747.20	$86.50	$1384.00	$44.09	$79.61
5 Struc. Steel Workers	44.70	1788.00	82.80	3312.00		
1 Welder Foreman	46.70	373.60	86.50	692.00		
5 Welders	44.70	1788.00	82.80	3312.00		
1 Equip. Oper. (crane)	42.55	340.40	66.95	535.60		
1 Equip. Oper. Oiler	36.80	294.40	57.90	463.20		
1 Equip. Oper. (light)	39.05	312.40	61.45	491.60		
1 Lattice Boom Crane, 90 Ton		1741.00		1915.10		
5 Welders, gas engine, 300 amp		671.00		738.10	18.84	20.73
128 L.H., Daily Totals		$8056.00		$12843.60	$62.94	$100.34
Crew E-10	Hr.	Daily	Hr.	Daily	Bare Costs	Incl. O&P
1 Welder Foreman	$46.70	$373.60	$86.50	$692.00	$45.70	$84.65
1 Welder	44.70	357.60	82.80	662.40		
1 Welder, gas engine, 300 amp		134.20		147.62		
1 Flatbed Truck, Gas, 3 Ton		242.60		266.86	23.55	25.91
16 L.H., Daily Totals		$1108.00		$1768.88	$69.25	$110.56
Crew E-11	Hr.	Daily	Hr.	Daily	Bare Costs	Incl. O&P
2 Painters, Struc. Steel	$36.00	$576.00	$68.95	$1103.20	$35.66	$62.88
1 Building Laborer	31.60	252.80	52.15	417.20		
1 Equip. Oper. (light)	39.05	312.40	61.45	491.60		
1 Air Compressor, 250 cfm		162.40		178.64		
1 Sandblaster, portable, 3 C.F.		20.60		22.66		
1 Set Sand Blasting Accessories		12.75		14.03	6.12	6.73
32 L.H., Daily Totals		$1336.95		$2227.32	$41.78	$69.60
Crew E-12	Hr.	Daily	Hr.	Daily	Bare Costs	Incl. O&P
1 Welder Foreman	$46.70	$373.60	$86.50	$692.00	$42.88	$73.97
1 Equip. Oper. (light)	39.05	312.40	61.45	491.60		
1 Welder, gas engine, 300 amp		134.20		147.62	8.39	9.23
16 L.H., Daily Totals		$820.20		$1331.22	$51.26	$83.20
Crew E-13	Hr.	Daily	Hr.	Daily	Bare Costs	Incl. O&P
1 Welder Foreman	$46.70	$373.60	$86.50	$692.00	$44.15	$78.15
.5 Equip. Oper. (light)	39.05	156.20	61.45	245.80		
1 Welder, gas engine, 300 amp		134.20		147.62	11.18	12.30
12 L.H., Daily Totals		$664.00		$1085.42	$55.33	$90.45
Crew E-14	Hr.	Daily	Hr.	Daily	Bare Costs	Incl. O&P
1 Welder Foreman	$46.70	$373.60	$86.50	$692.00	$46.70	$86.50
1 Welder, gas engine, 300 amp		134.20		147.62	16.77	18.45
8 L.H., Daily Totals		$507.80		$839.62	$63.48	$104.95
Crew E-16	Hr.	Daily	Hr.	Daily	Bare Costs	Incl. O&P
1 Welder Foreman	$46.70	$373.60	$86.50	$692.00	$45.70	$84.65
1 Welder	44.70	357.60	82.80	662.40		
1 Welder, gas engine, 300 amp		134.20		147.62	8.39	9.23
16 L.H., Daily Totals		$865.40		$1502.02	$54.09	$93.88

Crew No.	Bare Costs		Incl. Subs O & P		Cost Per Labor-Hour	

Crew E-17

	Hr.	Daily	Hr.	Daily	Bare Costs	Incl. O&P
1 Structural Steel Foreman	$46.70	$373.60	$86.50	$692.00	$45.70	$84.65
1 Structural Steel Worker	44.70	357.60	82.80	662.40		
16 L.H., Daily Totals		$731.20		$1354.40	$45.70	$84.65

Crew E-18

	Hr.	Daily	Hr.	Daily	Bare Costs	Incl. O&P
1 Structural Steel Foreman	$46.70	$373.60	$86.50	$692.00	$44.43	$79.99
3 Structural Steel Workers	44.70	1072.80	82.80	1987.20		
1 Equipment Operator (med.)	41.35	330.80	65.05	520.40		
1 Lattice Boom Crane, 20 Ton		1080.00		1188.00	27.00	29.70
40 L.H., Daily Totals		$2857.20		$4387.60	$71.43	$109.69

Crew E-19

	Hr.	Daily	Hr.	Daily	Bare Costs	Incl. O&P
1 Structural Steel Worker	$44.70	$357.60	$82.80	$662.40	$43.48	$76.92
1 Structural Steel Foreman	46.70	373.60	86.50	692.00		
1 Equip. Oper. (light)	39.05	312.40	61.45	491.60		
1 Lattice Boom Crane, 20 Ton		1080.00		1188.00	45.00	49.50
24 L.H., Daily Totals		$2123.60		$3034.00	$88.48	$126.42

Crew E-20

	Hr.	Daily	Hr.	Daily	Bare Costs	Incl. O&P
1 Structural Steel Foreman	$46.70	$373.60	$86.50	$692.00	$43.69	$78.17
5 Structural Steel Workers	44.70	1788.00	82.80	3312.00		
1 Equip. Oper. (crane)	42.55	340.40	66.95	535.60		
1 Oiler	36.80	294.40	57.90	463.20		
1 Lattice Boom Crane, 40 Ton		1327.00		1459.70	20.73	22.81
64 L.H., Daily Totals		$4123.40		$6462.50	$64.43	$100.98

Crew E-22

	Hr.	Daily	Hr.	Daily	Bare Costs	Incl. O&P
1 Skilled Worker Foreman	$42.85	$342.80	$69.75	$558.00	$41.52	$67.58
2 Skilled Workers	40.85	653.60	66.50	1064.00		
24 L.H., Daily Totals		$996.40		$1622.00	$41.52	$67.58

Crew E-24

	Hr.	Daily	Hr.	Daily	Bare Costs	Incl. O&P
3 Structural Steel Workers	$44.70	$1072.80	$82.80	$1987.20	$43.86	$78.36
1 Equipment Operator (medium)	41.35	330.80	65.05	520.40		
1 Hyd. Crane, 25 Ton		789.40		868.34	24.67	27.14
32 L.H., Daily Totals		$2193.00		$3375.94	$68.53	$105.50

Crew E-25

	Hr.	Daily	Hr.	Daily	Bare Costs	Incl. O&P
1 Welder Foreman	$46.70	$373.60	$86.50	$692.00	$46.70	$86.50
1 Cutting Torch		17.00		18.70		
1 Gases		75.60		83.16	11.57	12.73
8 L.H., Daily Totals		$466.20		$793.86	$58.27	$99.23

Crew F-3

	Hr.	Daily	Hr.	Daily	Bare Costs	Incl. O&P
4 Carpenters	$39.95	$1278.40	$65.95	$2110.40	$40.47	$66.15
1 Equip. Oper. (crane)	42.55	340.40	66.95	535.60		
1 Hyd. Crane, 12 Ton		768.80		845.68	19.22	21.14
40 L.H., Daily Totals		$2387.60		$3491.68	$59.69	$87.29

Crew F-4

	Hr.	Daily	Hr.	Daily	Bare Costs	Incl. O&P
4 Carpenters	$39.95	$1278.40	$65.95	$2110.40	$39.86	$64.78
1 Equip. Oper. (crane)	42.55	340.40	66.95	535.60		
1 Equip. Oper. Oiler	36.80	294.40	57.90	463.20		
1 Hyd. Crane, 55 Ton		1125.00		1237.50	23.44	25.78
48 L.H., Daily Totals		$3038.20		$4346.70	$63.30	$90.56

Crew F-5

	Hr.	Daily	Hr.	Daily	Bare Costs	Incl. O&P
1 Carpenter Foreman	$41.95	$335.60	$69.25	$554.00	$40.45	$66.78
3 Carpenters	39.95	958.80	65.95	1582.80		
32 L.H., Daily Totals		$1294.40		$2136.80	$40.45	$66.78

Crew F-6

	Hr.	Daily	Hr.	Daily	Bare Costs	Incl. O&P
2 Carpenters	$39.95	$639.20	$65.95	$1055.20	$37.13	$60.63
2 Building Laborers	31.60	505.60	52.15	834.40		
1 Equip. Oper. (crane)	42.55	340.40	66.95	535.60		
1 Hyd. Crane, 12 Ton		768.80		845.68	19.22	21.14
40 L.H., Daily Totals		$2254.00		$3270.88	$56.35	$81.77

Crew F-7

	Hr.	Daily	Hr.	Daily	Bare Costs	Incl. O&P
2 Carpenters	$39.95	$639.20	$65.95	$1055.20	$35.77	$59.05
2 Building Laborers	31.60	505.60	52.15	834.40		
32 L.H., Daily Totals		$1144.80		$1889.60	$35.77	$59.05

Crew G-1

	Hr.	Daily	Hr.	Daily	Bare Costs	Incl. O&P
1 Roofer Foreman	$36.25	$290.00	$64.70	$517.60	$31.99	$57.11
4 Roofers, Composition	34.25	1096.00	61.15	1956.80		
2 Roofer Helpers	25.35	405.60	45.25	724.00		
1 Application Equipment		173.20		190.52		
1 Tar Kettle/Pot		97.20		106.92		
1 Crew Truck		169.60		186.56	7.86	8.64
56 L.H., Daily Totals		$2231.60		$3682.40	$39.85	$65.76

Crew G-2

	Hr.	Daily	Hr.	Daily	Bare Costs	Incl. O&P
1 Plasterer	$36.15	$289.20	$58.15	$465.20	$33.35	$54.08
1 Plasterer Helper	32.30	258.40	51.95	415.60		
1 Building Laborer	31.60	252.80	52.15	417.20		
1 Grout Pump, 50 C.F./hr.		126.80		139.48	5.28	5.81
24 L.H., Daily Totals		$927.20		$1437.48	$38.63	$59.90

Crew G-2A

	Hr.	Daily	Hr.	Daily	Bare Costs	Incl. O&P
1 Roofer, composition	$34.25	$274.00	$61.15	$489.20	$30.40	$52.85
1 Roofer Helper	25.35	202.80	45.25	362.00		
1 Building Laborer	31.60	252.80	52.15	417.20		
1 Foam spray rig, trailer-mtd.		488.20		537.02		
1 Pickup Truck, 3/4 Ton		116.80		128.48	25.21	27.73
24 L.H., Daily Totals		$1334.60		$1933.90	$55.61	$80.58

Crew G-3

	Hr.	Daily	Hr.	Daily	Bare Costs	Incl. O&P
2 Sheet Metal Workers	$47.20	$755.20	$75.00	$1200.00	$39.40	$63.58
2 Building Laborers	31.60	505.60	52.15	834.40		
32 L.H., Daily Totals		$1260.80		$2034.40	$39.40	$63.58

Crew G-4

	Hr.	Daily	Hr.	Daily	Bare Costs	Incl. O&P
1 Labor Foreman (outside)	$33.60	$268.80	$55.45	$443.60	$32.27	$53.25
2 Building Laborers	31.60	505.60	52.15	834.40		
1 Flatbed Truck, Gas, 1.5 Ton		194.40		213.84		
1 Air Compressor, 160 cfm		138.80		152.68	13.88	15.27
24 L.H., Daily Totals		$1107.60		$1644.52	$46.15	$68.52

Crew G-5

	Hr.	Daily	Hr.	Daily	Bare Costs	Incl. O&P
1 Roofer Foreman	$36.25	$290.00	$64.70	$517.60	$31.09	$55.50
2 Roofers, Composition	34.25	548.00	61.15	978.40		
2 Roofer Helpers	25.35	405.60	45.25	724.00		
1 Application Equipment		173.20		190.52	4.33	4.76
40 L.H., Daily Totals		$1416.80		$2410.52	$35.42	$60.26

Crew G-6A

	Hr.	Daily	Hr.	Daily	Bare Costs	Incl. O&P
2 Roofers Composition	$34.25	$548.00	$61.15	$978.40	$34.25	$61.15
1 Small Compressor, Electric		11.35		12.48		
2 Pneumatic Nailers		43.20		47.52	3.41	3.75
16 L.H., Daily Totals		$602.55		$1038.41	$37.66	$64.90

Crew No.	Bare Costs		Incl. Subs O & P		Cost Per Labor-Hour	
Crew G-7	Hr.	Daily	Hr.	Daily	Bare Costs	Incl. O&P
1 Carpenter	$39.95	$319.60	$65.95	$527.60	$39.95	$65.95
1 Small Compressor, Electric		11.35		12.48		
1 Pneumatic Nailer		21.60		23.76	4.12	4.53
8 L.H., Daily Totals		$352.55		$563.85	$44.07	$70.48

Crew No.	Bare Costs		Incl. Subs O & P		Cost Per Labor-Hour	
Crew H-1	Hr.	Daily	Hr.	Daily	Bare Costs	Incl. O&P
2 Glaziers	$38.60	$617.60	$62.20	$995.20	$41.65	$72.50
2 Struc. Steel Workers	44.70	715.20	82.80	1324.80		
32 L.H., Daily Totals		$1332.80		$2320.00	$41.65	$72.50

Crew H-2	Hr.	Daily	Hr.	Daily	Bare Costs	Incl. O&P
2 Glaziers	$38.60	$617.60	$62.20	$995.20	$36.27	$58.85
1 Building Laborer	31.60	252.80	52.15	417.20		
24 L.H., Daily Totals		$870.40		$1412.40	$36.27	$58.85

Crew H-3	Hr.	Daily	Hr.	Daily	Bare Costs	Incl. O&P
1 Glazier	$38.60	$308.80	$62.20	$497.60	$34.45	$56.02
1 Helper	30.30	242.40	49.85	398.80		
16 L.H., Daily Totals		$551.20		$896.40	$34.45	$56.02

Crew J-1	Hr.	Daily	Hr.	Daily	Bare Costs	Incl. O&P
3 Plasterers	$36.15	$867.60	$58.15	$1395.60	$34.61	$55.67
2 Plasterer Helpers	32.30	516.80	51.95	831.20		
1 Mixing Machine, 6 C.F.		126.60		139.26	3.17	3.48
40 L.H., Daily Totals		$1511.00		$2366.06	$37.77	$59.15

Crew J-2	Hr.	Daily	Hr.	Daily	Bare Costs	Incl. O&P
3 Plasterers	$36.15	$867.60	$58.15	$1395.60	$34.77	$55.75
2 Plasterer Helpers	32.30	516.80	51.95	831.20		
1 Lather	35.55	284.40	56.15	449.20		
1 Mixing Machine, 6 C.F.		126.60		139.26	2.64	2.90
48 L.H., Daily Totals		$1795.40		$2815.26	$37.40	$58.65

Crew J-3	Hr.	Daily	Hr.	Daily	Bare Costs	Incl. O&P
1 Terrazzo Worker	$37.70	$301.60	$58.95	$471.60	$34.27	$53.60
1 Terrazzo Helper	30.85	246.80	48.25	386.00		
1 Terrazzo Grinder, Electric		81.60		89.76		
1 Terrazzo Mixer		171.60		188.76	15.82	17.41
16 L.H., Daily Totals		$801.60		$1136.12	$50.10	$71.01

Crew K-1	Hr.	Daily	Hr.	Daily	Bare Costs	Incl. O&P
1 Carpenter	$39.95	$319.60	$65.95	$527.60	$35.45	$58.35
1 Truck Driver (light)	30.95	247.60	50.75	406.00		
1 Flatbed Truck, Gas, 3 Ton		242.60		266.86	15.16	16.68
16 L.H., Daily Totals		$809.80		$1200.46	$50.61	$75.03

Crew K-2	Hr.	Daily	Hr.	Daily	Bare Costs	Incl. O&P
1 Struc. Steel Foreman	$46.70	$373.60	$86.50	$692.00	$40.78	$73.35
1 Struc. Steel Worker	44.70	357.60	82.80	662.40		
1 Truck Driver (light)	30.95	247.60	50.75	406.00		
1 Flatbed Truck, Gas, 3 Ton		242.60		266.86	10.11	11.12
24 L.H., Daily Totals		$1221.40		$2027.26	$50.89	$84.47

Crew L-1	Hr.	Daily	Hr.	Daily	Bare Costs	Incl. O&P
1 Electrician	$47.00	$376.00	$72.30	$578.40	$47.88	$73.95
1 Plumber	48.75	390.00	75.60	604.80		
16 L.H., Daily Totals		$766.00		$1183.20	$47.88	$73.95

Crew No.	Bare Costs		Incl. Subs O & P		Cost Per Labor-Hour	
Crew L-2	Hr.	Daily	Hr.	Daily	Bare Costs	Incl. O&P
1 Carpenter	$39.95	$319.60	$65.95	$527.60	$35.13	$57.90
1 Carpenter Helper	30.30	242.40	49.85	398.80		
16 L.H., Daily Totals		$562.00		$926.40	$35.13	$57.90

Crew L-3	Hr.	Daily	Hr.	Daily	Bare Costs	Incl. O&P
1 Carpenter	$39.95	$319.60	$65.95	$527.60	$43.52	$69.80
.5 Electrician	47.00	188.00	72.30	289.20		
.5 Sheet Metal Worker	47.20	188.80	75.00	300.00		
16 L.H., Daily Totals		$696.40		$1116.80	$43.52	$69.80

Crew L-3A	Hr.	Daily	Hr.	Daily	Bare Costs	Incl. O&P
1 Carpenter Foreman (outside)	$41.95	$335.60	$69.25	$554.00	$43.70	$71.17
.5 Sheet Metal Worker	47.20	188.80	75.00	300.00		
12 L.H., Daily Totals		$524.40		$854.00	$43.70	$71.17

Crew L-4	Hr.	Daily	Hr.	Daily	Bare Costs	Incl. O&P
2 Skilled Workers	$40.85	$653.60	$66.50	$1064.00	$37.33	$60.95
1 Helper	30.30	242.40	49.85	398.80		
24 L.H., Daily Totals		$896.00		$1462.80	$37.33	$60.95

Crew L-5	Hr.	Daily	Hr.	Daily	Bare Costs	Incl. O&P
1 Struc. Steel Foreman	$46.70	$373.60	$86.50	$692.00	$44.68	$81.06
5 Struc. Steel Workers	44.70	1788.00	82.80	3312.00		
1 Equip. Oper. (crane)	42.55	340.40	66.95	535.60		
1 Hyd. Crane, 25 Ton		789.40		868.34	14.10	15.51
56 L.H., Daily Totals		$3291.40		$5407.94	$58.77	$96.57

Crew L-5A	Hr.	Daily	Hr.	Daily	Bare Costs	Incl. O&P
1 Structural Steel Foreman	$46.70	$373.60	$86.50	$692.00	$44.66	$79.76
2 Structural Steel Workers	44.70	715.20	82.80	1324.80		
1 Equip. Oper. (crane)	42.55	340.40	66.95	535.60		
1 S.P. Crane, 4x4, 25 Ton		716.80		788.48	22.40	24.64
32 L.H., Daily Totals		$2146.00		$3340.88	$67.06	$104.40

Crew L-5B	Hr.	Daily	Hr.	Daily	Bare Costs	Incl. O&P
1 Structural Steel Foreman	$46.70	$373.60	$86.50	$692.00	$45.35	$74.96
2 Structural Steel Workers	44.70	715.20	82.80	1324.80		
2 Electricians	47.00	752.00	72.30	1156.80		
2 Steamfitters/Pipefitters	49.35	789.60	76.55	1224.80		
1 Equip. Oper. (crane)	42.55	340.40	66.95	535.60		
1 Equip. Oper. Oiler	36.80	294.40	57.90	463.20		
1 Hyd. Crane, 80 Ton		1296.00		1425.60	18.00	19.80
72 L.H., Daily Totals		$4561.20		$6822.80	$63.35	$94.76

Crew L-6	Hr.	Daily	Hr.	Daily	Bare Costs	Incl. O&P
1 Plumber	$48.75	$390.00	$75.60	$604.80	$48.17	$74.50
.5 Electrician	47.00	188.00	72.30	289.20		
12 L.H., Daily Totals		$578.00		$894.00	$48.17	$74.50

Crew L-7	Hr.	Daily	Hr.	Daily	Bare Costs	Incl. O&P
2 Carpenters	$39.95	$639.20	$65.95	$1055.20	$38.57	$62.91
1 Building Laborer	31.60	252.80	52.15	417.20		
.5 Electrician	47.00	188.00	72.30	289.20		
28 L.H., Daily Totals		$1080.00		$1761.60	$38.57	$62.91

Crew L-8	Hr.	Daily	Hr.	Daily	Bare Costs	Incl. O&P
2 Carpenters	$39.95	$639.20	$65.95	$1055.20	$41.71	$67.88
.5 Plumber	48.75	195.00	75.60	302.40		
20 L.H., Daily Totals		$834.20		$1357.60	$41.71	$67.88

Crew No.	Bare Costs		Incl. Subs O & P		Cost Per Labor-Hour	
Crew L-9	Hr.	Daily	Hr.	Daily	Bare Costs	Incl. O&P
1 Labor Foreman (inside)	$32.10	$256.80	$53.00	$424.00	$36.33	$61.39
2 Building Laborers	31.60	505.60	52.15	834.40		
1 Struc. Steel Worker	44.70	357.60	82.80	662.40		
.5 Electrician	47.00	188.00	72.30	289.20		
36 L.H., Daily Totals		$1308.00		$2210.00	$36.33	$61.39
Crew L-10	Hr.	Daily	Hr.	Daily	Bare Costs	Incl. O&P
1 Structural Steel Foreman	$46.70	$373.60	$86.50	$692.00	$44.65	$78.75
1 Structural Steel Worker	44.70	357.60	82.80	662.40		
1 Equip. Oper. (crane)	42.55	340.40	66.95	535.60		
1 Hyd. Crane, 12 Ton		768.80		845.68	32.03	35.24
24 L.H., Daily Totals		$1840.40		$2735.68	$76.68	$113.99
Crew L-11	Hr.	Daily	Hr.	Daily	Bare Costs	Incl. O&P
2 Wreckers	$31.60	$505.60	$57.30	$916.80	$36.20	$60.75
1 Equip. Oper. (crane)	42.55	340.40	66.95	535.60		
1 Equip. Oper. (light)	39.05	312.40	61.45	491.60		
1 Hyd. Excavator, 2.5 C.Y.		1619.00		1780.90		
1 Loader, Skid Steer, 78 H.P.		259.60		285.56	58.71	64.58
32 L.H., Daily Totals		$3037.00		$4010.46	$94.91	$125.33
Crew M-1	Hr.	Daily	Hr.	Daily	Bare Costs	Incl. O&P
3 Elevator Constructors	$56.60	$1358.40	$87.10	$2090.40	$53.77	$82.75
1 Elevator Apprentice	45.30	362.40	69.70	557.60		
5 Hand Tools		57.00		62.70	1.78	1.96
32 L.H., Daily Totals		$1777.80		$2710.70	$55.56	$84.71
Crew M-3	Hr.	Daily	Hr.	Daily	Bare Costs	Incl. O&P
1 Electrician Foreman (out)	$49.00	$392.00	$75.35	$602.80	$45.37	$70.72
1 Common Laborer	31.60	252.80	52.15	417.20		
.25 Equipment Operator, Medium	41.35	82.70	65.05	130.10		
1 Elevator Constructor	56.60	452.80	87.10	696.80		
1 Elevator Apprentice	45.30	362.40	69.70	557.60		
.25 S.P. Crane, 4x4, 20 Ton		175.15		192.66	5.15	5.67
34 L.H., Daily Totals		$1717.85		$2597.17	$50.52	$76.39
Crew M-4	Hr.	Daily	Hr.	Daily	Bare Costs	Incl. O&P
1 Electrician Foreman (out)	$49.00	$392.00	$75.35	$602.80	$44.96	$70.11
1 Common Laborer	31.60	252.80	52.15	417.20		
.25 Equipment Operator, Crane	42.55	85.10	66.95	133.90		
.25 Equipment Operator, Oiler	36.80	73.60	57.90	115.80		
1 Elevator Constructor	56.60	452.80	87.10	696.80		
1 Elevator Apprentice	45.30	362.40	69.70	557.60		
.25 S.P. Crane, 4x4, 40 Ton		234.30		257.73	6.51	7.16
36 L.H., Daily Totals		$1853.00		$2781.83	$51.47	$77.27
Crew Q-1	Hr.	Daily	Hr.	Daily	Bare Costs	Incl. O&P
1 Plumber	$48.75	$390.00	$75.60	$604.80	$43.88	$68.05
1 Plumber Apprentice	39.00	312.00	60.50	484.00		
16 L.H., Daily Totals		$702.00		$1088.80	$43.88	$68.05
Crew Q-1A	Hr.	Daily	Hr.	Daily	Bare Costs	Incl. O&P
.25 Plumber Foreman (out)	$50.75	$101.50	$78.70	$157.40	$49.15	$76.22
1 Plumber	48.75	390.00	75.60	604.80		
10 L.H., Daily Totals		$491.50		$762.20	$49.15	$76.22

Crew No.	Bare Costs		Incl. Subs O & P		Cost Per Labor-Hour	
Crew Q-1C	Hr.	Daily	Hr.	Daily	Bare Costs	Incl. O&P
1 Plumber	$48.75	$390.00	$75.60	$604.80	$43.03	$67.05
1 Plumber Apprentice	39.00	312.00	60.50	484.00		
1 Equip. Oper. (medium)	41.35	330.80	65.05	520.40		
1 Trencher, Chain Type, 8' D		1777.00		1954.70	74.04	81.45
24 L.H., Daily Totals		$2809.80		$3563.90	$117.08	$148.50
Crew Q-2	Hr.	Daily	Hr.	Daily	Bare Costs	Incl. O&P
2 Plumbers	$48.75	$780.00	$75.60	$1209.60	$45.50	$70.57
1 Plumber Apprentice	39.00	312.00	60.50	484.00	45.50	70.57
24 L.H., Daily Totals		$1092.00		$1693.60	$45.50	$70.57
Crew Q-3	Hr.	Daily	Hr.	Daily	Bare Costs	Incl. O&P
1 Plumber Foreman (inside)	$49.25	$394.00	$76.40	$611.20	$46.44	$72.03
2 Plumbers	48.75	780.00	75.60	1209.60		
1 Plumber Apprentice	39.00	312.00	60.50	484.00		
32 L.H., Daily Totals		$1486.00		$2304.80	$46.44	$72.03
Crew Q-4	Hr.	Daily	Hr.	Daily	Bare Costs	Incl. O&P
1 Plumber Foreman (inside)	$49.25	$394.00	$76.40	$611.20	$46.44	$72.03
1 Plumber	48.75	390.00	75.60	604.80		
1 Welder (plumber)	48.75	390.00	75.60	604.80		
1 Plumber Apprentice	39.00	312.00	60.50	484.00		
1 Welder, electric, 300 amp		55.70		61.27	1.74	1.91
32 L.H., Daily Totals		$1541.70		$2366.07	$48.18	$73.94
Crew Q-5	Hr.	Daily	Hr.	Daily	Bare Costs	Incl. O&P
1 Steamfitter	$49.35	$394.80	$76.55	$612.40	$44.42	$68.90
1 Steamfitter Apprentice	39.50	316.00	61.25	490.00		
16 L.H., Daily Totals		$710.80		$1102.40	$44.42	$68.90
Crew Q-6	Hr.	Daily	Hr.	Daily	Bare Costs	Incl. O&P
2 Steamfitters	$49.35	$789.60	$76.55	$1224.80	$46.07	$71.45
1 Steamfitter Apprentice	39.50	316.00	61.25	490.00	46.07	71.45
24 L.H., Daily Totals		$1105.60		$1714.80	$46.07	$71.45
Crew Q-7	Hr.	Daily	Hr.	Daily	Bare Costs	Incl. O&P
1 Steamfitter Foreman (inside)	$49.85	$398.80	$77.30	$618.40	$47.01	$72.91
2 Steamfitters	49.35	789.60	76.55	1224.80		
1 Steamfitter Apprentice	39.50	316.00	61.25	490.00		
32 L.H., Daily Totals		$1504.40		$2333.20	$47.01	$72.91
Crew Q-8	Hr.	Daily	Hr.	Daily	Bare Costs	Incl. O&P
1 Steamfitter Foreman (inside)	$49.85	$398.80	$77.30	$618.40	$47.01	$72.91
1 Steamfitter	49.35	394.80	76.55	612.40		
1 Welder (steamfitter)	49.35	394.80	76.55	612.40		
1 Steamfitter Apprentice	39.50	316.00	61.25	490.00		
1 Welder, electric, 300 amp		55.70		61.27	1.74	1.91
32 L.H., Daily Totals		$1560.10		$2394.47	$48.75	$74.83
Crew Q-9	Hr.	Daily	Hr.	Daily	Bare Costs	Incl. O&P
1 Sheet Metal Worker	$47.20	$377.60	$75.00	$600.00	$42.48	$67.50
1 Sheet Metal Apprentice	37.75	302.00	60.00	480.00		
16 L.H., Daily Totals		$679.60		$1080.00	$42.48	$67.50
Crew Q-10	Hr.	Daily	Hr.	Daily	Bare Costs	Incl. O&P
2 Sheet Metal Workers	$47.20	$755.20	$75.00	$1200.00	$44.05	$70.00
1 Sheet Metal Apprentice	37.75	302.00	60.00	480.00		
24 L.H., Daily Totals		$1057.20		$1680.00	$44.05	$70.00

Crews

Crew No.	Bare Costs Hr.	Daily	Incl. Subs O & P Hr.	Daily	Cost Per Labor-Hour Bare Costs	Incl. O&P
Crew Q-11	Hr.	Daily	Hr.	Daily	Bare Costs	Incl. O&P
1 Sheet Metal Foreman (inside)	$47.70	$381.60	$75.80	$606.40	$44.96	$71.45
2 Sheet Metal Workers	47.20	755.20	75.00	1200.00		
1 Sheet Metal Apprentice	37.75	302.00	60.00	480.00		
32 L.H., Daily Totals		$1438.80		$2286.40	$44.96	$71.45
Crew Q-12	Hr.	Daily	Hr.	Daily	Bare Costs	Incl. O&P
1 Sprinkler Installer	$48.15	$385.20	$74.70	$597.60	$43.33	$67.20
1 Sprinkler Apprentice	38.50	308.00	59.70	477.60		
16 L.H., Daily Totals		$693.20		$1075.20	$43.33	$67.20
Crew Q-13	Hr.	Daily	Hr.	Daily	Bare Costs	Incl. O&P
1 Sprinkler Foreman (inside)	$48.65	$389.20	$75.45	$603.60	$45.86	$71.14
2 Sprinkler Installers	48.15	770.40	74.70	1195.20		
1 Sprinkler Apprentice	38.50	308.00	59.70	477.60		
32 L.H., Daily Totals		$1467.60		$2276.40	$45.86	$71.14
Crew Q-14	Hr.	Daily	Hr.	Daily	Bare Costs	Incl. O&P
1 Asbestos Worker	$44.10	$352.80	$71.30	$570.40	$39.70	$64.20
1 Asbestos Apprentice	35.30	282.40	57.10	456.80		
16 L.H., Daily Totals		$635.20		$1027.20	$39.70	$64.20
Crew Q-15	Hr.	Daily	Hr.	Daily	Bare Costs	Incl. O&P
1 Plumber	$48.75	$390.00	$75.60	$604.80	$43.88	$68.05
1 Plumber Apprentice	39.00	312.00	60.50	484.00		
1 Welder, electric, 300 amp		55.70		61.27	3.48	3.83
16 L.H., Daily Totals		$757.70		$1150.07	$47.36	$71.88
Crew Q-16	Hr.	Daily	Hr.	Daily	Bare Costs	Incl. O&P
2 Plumbers	$48.75	$780.00	$75.60	$1209.60	$45.50	$70.57
1 Plumber Apprentice	39.00	312.00	60.50	484.00		
1 Welder, electric, 300 amp		55.70		61.27	2.32	2.55
24 L.H., Daily Totals		$1147.70		$1754.87	$47.82	$73.12
Crew Q-17	Hr.	Daily	Hr.	Daily	Bare Costs	Incl. O&P
1 Steamfitter	$49.35	$394.80	$76.55	$612.40	$44.42	$68.90
1 Steamfitter Apprentice	39.50	316.00	61.25	490.00		
1 Welder, electric, 300 amp		55.70		61.27	3.48	3.83
16 L.H., Daily Totals		$766.50		$1163.67	$47.91	$72.73
Crew Q-17A	Hr.	Daily	Hr.	Daily	Bare Costs	Incl. O&P
1 Steamfitter	$49.35	$394.80	$76.55	$612.40	$43.80	$68.25
1 Steamfitter Apprentice	39.50	316.00	61.25	490.00		
1 Equip. Oper. (crane)	42.55	340.40	66.95	535.60		
1 Hyd. Crane, 12 Ton		768.80		845.68		
1 Welder, electric, 300 amp		55.70		61.27	34.35	37.79
24 L.H., Daily Totals		$1875.70		$2544.95	$78.15	$106.04
Crew Q-18	Hr.	Daily	Hr.	Daily	Bare Costs	Incl. O&P
2 Steamfitters	$49.35	$789.60	$76.55	$1224.80	$46.07	$71.45
1 Steamfitter Apprentice	39.50	316.00	61.25	490.00		
1 Welder, electric, 300 amp		55.70		61.27	2.32	2.55
24 L.H., Daily Totals		$1161.30		$1776.07	$48.39	$74.00
Crew Q-19	Hr.	Daily	Hr.	Daily	Bare Costs	Incl. O&P
1 Steamfitter	$49.35	$394.80	$76.55	$612.40	$45.28	$70.03
1 Steamfitter Apprentice	39.50	316.00	61.25	490.00		
1 Electrician	47.00	376.00	72.30	578.40		
24 L.H., Daily Totals		$1086.80		$1680.80	$45.28	$70.03

Crew No.	Bare Costs Hr.	Daily	Incl. Subs O & P Hr.	Daily	Cost Per Labor-Hour Bare Costs	Incl. O&P
Crew Q-20	Hr.	Daily	Hr.	Daily	Bare Costs	Incl. O&P
1 Sheet Metal Worker	$47.20	$377.60	$75.00	$600.00	$43.38	$68.46
1 Sheet Metal Apprentice	37.75	302.00	60.00	480.00		
.5 Electrician	47.00	188.00	72.30	289.20		
20 L.H., Daily Totals		$867.60		$1369.20	$43.38	$68.46
Crew Q-21	Hr.	Daily	Hr.	Daily	Bare Costs	Incl. O&P
2 Steamfitters	$49.35	$789.60	$76.55	$1224.80	$46.30	$71.66
1 Steamfitter Apprentice	39.50	316.00	61.25	490.00		
1 Electrician	47.00	376.00	72.30	578.40		
32 L.H., Daily Totals		$1481.60		$2293.20	$46.30	$71.66
Crew Q-22	Hr.	Daily	Hr.	Daily	Bare Costs	Incl. O&P
1 Plumber	$48.75	$390.00	$75.60	$604.80	$43.88	$68.05
1 Plumber Apprentice	39.00	312.00	60.50	484.00		
1 Hyd. Crane, 12 Ton		768.80		845.68	48.05	52.85
16 L.H., Daily Totals		$1470.80		$1934.48	$91.92	$120.91
Crew Q-22A	Hr.	Daily	Hr.	Daily	Bare Costs	Incl. O&P
1 Plumber	$48.75	$390.00	$75.60	$604.80	$40.48	$63.80
1 Plumber Apprentice	39.00	312.00	60.50	484.00		
1 Laborer	31.60	252.80	52.15	417.20		
1 Equip. Oper. (crane)	42.55	340.40	66.95	535.60		
1 Hyd. Crane, 12 Ton		768.80		845.68	24.02	26.43
32 L.H., Daily Totals		$2064.00		$2887.28	$64.50	$90.23
Crew Q-23	Hr.	Daily	Hr.	Daily	Bare Costs	Incl. O&P
1 Plumber Foreman	$50.75	$406.00	$78.70	$629.60	$46.95	$73.12
1 Plumber	48.75	390.00	75.60	604.80		
1 Equip. Oper. (medium)	41.35	330.80	65.05	520.40		
1 Lattice Boom Crane, 20 Ton		1080.00		1188.00	45.00	49.50
24 L.H., Daily Totals		$2206.80		$2942.80	$91.95	$122.62
Crew R-1	Hr.	Daily	Hr.	Daily	Bare Costs	Incl. O&P
1 Electrician Foreman	$47.50	$380.00	$73.05	$584.40	$41.52	$64.94
3 Electricians	47.00	1128.00	72.30	1735.20		
2 Helpers	30.30	484.80	49.85	797.60		
48 L.H., Daily Totals		$1992.80		$3117.20	$41.52	$64.94
Crew R-1A	Hr.	Daily	Hr.	Daily	Bare Costs	Incl. O&P
1 Electrician	$47.00	$376.00	$72.30	$578.40	$38.65	$61.08
1 Helper	30.30	242.40	49.85	398.80		
16 L.H., Daily Totals		$618.40		$977.20	$38.65	$61.08
Crew R-2	Hr.	Daily	Hr.	Daily	Bare Costs	Incl. O&P
1 Electrician Foreman	$47.50	$380.00	$73.05	$584.40	$41.66	$65.23
3 Electricians	47.00	1128.00	72.30	1735.20		
2 Helpers	30.30	484.80	49.85	797.60		
1 Equip. Oper. (crane)	42.55	340.40	66.95	535.60		
1 S.P. Crane, 4x4, 5 Ton		258.20		284.02	4.61	5.07
56 L.H., Daily Totals		$2591.40		$3936.82	$46.27	$70.30
Crew R-3	Hr.	Daily	Hr.	Daily	Bare Costs	Incl. O&P
1 Electrician Foreman	$47.50	$380.00	$73.05	$584.40	$46.31	$71.53
1 Electrician	47.00	376.00	72.30	578.40		
.5 Equip. Oper. (crane)	42.55	170.20	66.95	267.80		
.5 S.P. Crane, 4x4, 5 Ton		129.10		142.01	6.46	7.10
20 L.H., Daily Totals		$1055.30		$1572.61	$52.77	$78.63

Crews

Crew R-4	Hr.	Daily	Hr.	Daily	Bare Costs	Incl. O&P
1 Struc. Steel Foreman	$46.70	$373.60	$86.50	$692.00	$45.56	$81.44
3 Struc. Steel Workers	44.70	1072.80	82.80	1987.20		
1 Electrician	47.00	376.00	72.30	578.40		
1 Welder, gas engine, 300 amp		134.20		147.62	3.36	3.69
40 L.H., Daily Totals		$1956.60		$3405.22	$48.91	$85.13

Crew R-5	Hr.	Daily	Hr.	Daily	Bare Costs	Incl. O&P
1 Electrician Foreman	$47.50	$380.00	$73.05	$584.40	$40.97	$64.20
4 Electrician Linemen	47.00	1504.00	72.30	2313.60		
2 Electrician Operators	47.00	752.00	72.30	1156.80		
4 Electrician Groundmen	30.30	969.60	49.85	1595.20		
1 Crew Truck		169.60		186.56		
1 Flatbed Truck, 20,000 GVW		197.40		217.14		
1 Pickup Truck, 3/4 Ton		116.80		128.48		
.2 Hyd. Crane, 55 Ton		225.00		247.50		
.2 Hyd. Crane, 12 Ton		153.76		169.14		
.2 Earth Auger, Truck-Mtd.		82.40		90.64		
1 Tractor w/Winch		354.80		390.28	14.77	16.25
88 L.H., Daily Totals		$4905.36		$7079.74	$55.74	$80.45

Crew R-6	Hr.	Daily	Hr.	Daily	Bare Costs	Incl. O&P
1 Electrician Foreman	$47.50	$380.00	$73.05	$584.40	$40.97	$64.20
4 Electrician Linemen	47.00	1504.00	72.30	2313.60		
2 Electrician Operators	47.00	752.00	72.30	1156.80		
4 Electrician Groundmen	30.30	969.60	49.85	1595.20		
1 Crew Truck		169.60		186.56		
1 Flatbed Truck, 20,000 GVW		197.40		217.14		
1 Pickup Truck, 3/4 Ton		116.80		128.48		
.2 Hyd. Crane, 55 Ton		225.00		247.50		
.2 Hyd. Crane, 12 Ton		153.76		169.14		
.2 Earth Auger, Truck-Mtd.		82.40		90.64		
1 Tractor w/Winch		354.80		390.28		
3 Cable Trailers		531.30		584.43		
.5 Tensioning Rig		175.35		192.88		
.5 Cable Pulling Rig		1049.50		1154.45	34.73	38.20
88 L.H., Daily Totals		$6661.51		$9011.50	$75.70	$102.40

Crew R-7	Hr.	Daily	Hr.	Daily	Bare Costs	Incl. O&P
1 Electrician Foreman	$47.50	$380.00	$73.05	$584.40	$33.17	$53.72
5 Electrician Groundmen	30.30	1212.00	49.85	1994.00		
1 Crew Truck		169.60		186.56	3.53	3.89
48 L.H., Daily Totals		$1761.60		$2764.96	$36.70	$57.60

Crew R-8	Hr.	Daily	Hr.	Daily	Bare Costs	Incl. O&P
1 Electrician Foreman	$47.50	$380.00	$73.05	$584.40	$41.52	$64.94
3 Electrician Linemen	47.00	1128.00	72.30	1735.20		
2 Electrician Groundmen	30.30	484.80	49.85	797.60		
1 Pickup Truck, 3/4 Ton		116.80		128.48		
1 Crew Truck		169.60		186.56	5.97	6.56
48 L.H., Daily Totals		$2279.20		$3432.24	$47.48	$71.50

Crew R-9	Hr.	Daily	Hr.	Daily	Bare Costs	Incl. O&P
1 Electrician Foreman	$47.50	$380.00	$73.05	$584.40	$38.71	$61.17
1 Electrician Lineman	47.00	376.00	72.30	578.40		
2 Electrician Operators	47.00	752.00	72.30	1156.80		
4 Electrician Groundmen	30.30	969.60	49.85	1595.20		
1 Pickup Truck, 3/4 Ton		116.80		128.48		
1 Crew Truck		169.60		186.56	4.47	4.92
64 L.H., Daily Totals		$2764.00		$4229.84	$43.19	$66.09

Crew R-10	Hr.	Daily	Hr.	Daily	Bare Costs	Incl. O&P
1 Electrician Foreman	$47.50	$380.00	$73.05	$584.40	$44.30	$68.68
4 Electrician Linemen	47.00	1504.00	72.30	2313.60		
1 Electrician Groundman	30.30	242.40	49.85	398.80		
1 Crew Truck		169.60		186.56		
3 Tram Cars		380.55		418.61	11.46	12.61
48 L.H., Daily Totals		$2676.55		$3901.97	$55.76	$81.29

Crew R-11	Hr.	Daily	Hr.	Daily	Bare Costs	Incl. O&P
1 Electrician Foreman	$47.50	$380.00	$73.05	$584.40	$44.24	$68.76
4 Electricians	47.00	1504.00	72.30	2313.60		
1 Equip. Oper. (crane)	42.55	340.40	66.95	535.60		
1 Common Laborer	31.60	252.80	52.15	417.20		
1 Crew Truck		169.60		186.56		
1 Hyd. Crane, 12 Ton		768.80		845.68	16.76	18.43
56 L.H., Daily Totals		$3415.60		$4883.04	$60.99	$87.20

Crew R-12	Hr.	Daily	Hr.	Daily	Bare Costs	Incl. O&P
1 Carpenter Foreman	$40.45	$323.60	$66.80	$534.40	$37.52	$62.46
4 Carpenters	39.95	1278.40	65.95	2110.40		
4 Common Laborers	31.60	1011.20	52.15	1668.80		
1 Equip. Oper. (med.)	41.35	330.80	65.05	520.40		
1 Steel Worker	44.70	357.60	82.80	662.40		
1 Dozer, 200 H.P.		1082.00		1190.20		
1 Pickup Truck, 3/4 Ton		116.80		128.48	13.62	14.98
88 L.H., Daily Totals		$4500.40		$6815.08	$51.14	$77.44

Crew R-13	Hr.	Daily	Hr.	Daily	Bare Costs	Incl. O&P
1 Electrician Foreman	$47.50	$380.00	$73.05	$584.40	$44.94	$69.45
3 Electricians	47.00	1128.00	72.30	1735.20		
.25 Equip. Oper. (crane)	42.55	85.10	66.95	133.90		
1 Equipment Oiler	36.80	294.40	57.90	463.20		
.25 Hydraulic Crane, 33 Ton		204.70		225.17	4.87	5.36
42 L.H., Daily Totals		$2092.20		$3141.87	$49.81	$74.81

Crew R-15	Hr.	Daily	Hr.	Daily	Bare Costs	Incl. O&P
1 Electrician Foreman	$47.50	$380.00	$73.05	$584.40	$45.76	$70.62
4 Electricians	47.00	1504.00	72.30	2313.60		
1 Equipment Oper. (Light)	39.05	312.40	61.45	491.60		
1 Aerial Lift Truck		328.80		361.68	6.85	7.54
48 L.H., Daily Totals		$2525.20		$3751.28	$52.61	$78.15

Crew R-18	Hr.	Daily	Hr.	Daily	Bare Costs	Incl. O&P
.25 Electrician Foreman	$47.50	$95.00	$73.05	$146.10	$36.76	$58.54
1 Electrician	47.00	376.00	72.30	578.40		
2 Helpers	30.30	484.80	49.85	797.60		
26 L.H., Daily Totals		$955.80		$1522.10	$36.76	$58.54

Crew R-19	Hr.	Daily	Hr.	Daily	Bare Costs	Incl. O&P
.5 Electrician Foreman	$47.50	$190.00	$73.05	$292.20	$47.10	$72.45
2 Electricians	47.00	752.00	72.30	1156.80		
20 L.H., Daily Totals		$942.00		$1449.00	$47.10	$72.45

Crew R-21	Hr.	Daily	Hr.	Daily	Bare Costs	Incl. O&P
1 Electrician Foreman	$47.50	$380.00	$73.05	$584.40	$46.98	$72.31
3 Electricians	47.00	1128.00	72.30	1735.20		
.1 Equip. Oper. (med.)	41.35	33.08	65.05	52.04		
.1 S.P. Crane, 4x4, 25 Ton		71.68		78.85	2.19	2.40
32.8 L.H., Daily Totals		$1612.76		$2450.49	$49.17	$74.71

Crews

Crew No.	Bare Costs		Incl. Subs O & P		Cost Per Labor-Hour	
Crew R-22	**Hr.**	**Daily**	**Hr.**	**Daily**	**Bare Costs**	**Incl. O&P**
.66 Electrician Foreman	$47.50	$250.80	$73.05	$385.70	$39.90	$62.77
2 Helpers	30.30	484.80	49.85	797.60		
2 Electricians	47.00	752.00	72.30	1156.80		
37.28 L.H., Daily Totals		$1487.60		$2340.10	$39.90	$62.77
Crew R-30	**Hr.**	**Daily**	**Hr.**	**Daily**	**Bare Costs**	**Incl. O&P**
.25 Electrician Foreman (out)	$49.00	$98.00	$75.35	$150.70	$37.68	$60.13
1 Electrician	47.00	376.00	72.30	578.40		
2 Laborers, (Semi-Skilled)	31.60	505.60	52.15	834.40		
26 L.H., Daily Totals		$979.60		$1563.50	$37.68	$60.13
Crew R-31	**Hr.**	**Daily**	**Hr.**	**Daily**	**Bare Costs**	**Incl. O&P**
1 Electrician	$47.00	$376.00	$72.30	$578.40	$47.00	$72.30
1 Core Drill, Electric, 2.5 H.P.		52.95		58.24	6.62	7.28
8 L.H., Daily Totals		$428.95		$636.64	$53.62	$79.58

Historical Cost Indexes

The table below lists both the RSMeans Historical Cost Index based on Jan. 1, 1993 = 100 as well as the computed value of an index based on Jan. 1, 2009 costs. Since the Jan. 1, 2009 figure is estimated, space is left to write in the actual index figures as they become available through either the quarterly "RSMeans Construction Cost Indexes" or as printed in the "Engineering News-Record." To compute the actual index based on Jan. 1, 2009 = 100, divide the Historical Cost Index for a particular year by the actual Jan. 1, 2009 Construction Cost Index. Space has been left to advance the index figures as the year progresses.

Year	Historical Cost Index Jan. 1, 1993 = 100		Current Index Based on Jan. 1, 2009 = 100		Year	Historical Cost Index Jan. 1, 1993 = 100	Current Index Based on Jan. 1, 2009 = 100		Year	Historical Cost Index Jan. 1, 1993 = 100	Current Index Based on Jan. 1, 2009 = 100	
	Est.	Actual	Est.	Actual		Actual	Est.	Actual		Actual	Est.	Actual
Oct 2009					July 1994	104.4	60.3		July 1976	46.9	27.1	
July 2009					1993	101.7	54.7		1975	44.8	24.1	
April 2009					1992	99.4	53.5		1974	41.4	22.3	
Jan 2009	185.9		100.0	100.0	1991	96.8	52.1		1973	37.7	20.3	
July 2008		180.4	97.0		1990	94.3	50.7		1972	34.8	18.7	
2007		169.4	91.1		1989	92.1	49.6		1971	32.1	17.3	
2006		162.0	87.1		1988	89.9	48.3		1970	28.7	15.4	
2005		151.6	81.5		1987	87.7	47.2		1969	26.9	14.5	
2004		143.7	77.3		1986	84.2	45.3		1968	24.9	13.4	
2003		132.0	71.0		1985	82.6	44.4		1967	23.5	12.6	
2002		128.7	69.2		1984	82.0	44.1		1966	22.7	12.2	
2001		125.1	67.3		1983	80.2	43.1		1965	21.7	11.7	
2000		120.9	65.0		1982	76.1	41.0		1964	21.2	11.4	
1999		117.6	63.3		1981	70.0	37.6		1963	20.7	11.1	
1998		115.1	61.9		1980	62.9	33.8		1962	20.2	10.9	
1997		112.8	60.7		1979	57.8	31.1		1961	19.8	10.7	
1996		110.2	59.3		1978	53.5	28.8		1960	19.7	10.6	
1995		107.6	57.9		1977	49.5	26.6		1959	19.3	10.4	

Adjustments to Costs

The Historical Cost Index can be used to convert National Average building costs at a particular time to the approximate building costs for some other time.

Example:

Estimate and compare construction costs for different years in the same city.

To estimate the National Average construction cost of a building in 1970, knowing that it cost $900,000 in 2009:

INDEX in 1970 = 28.7

INDEX in 2009 = 185.9

Note: The City Cost Indexes for Canada can be used to convert U.S. National averages to local costs in Canadian dollars.

Time Adjustment using the Historical Cost Indexes:

$$\frac{\text{Index for Year A}}{\text{Index for Year B}} \times \text{Cost in Year B} = \text{Cost in Year A}$$

$$\frac{\text{INDEX 1970}}{\text{INDEX 2009}} \times \text{Cost 2009} = \text{Cost 1970}$$

$$\frac{28.7}{185.9} \times \$900{,}000 = .154 \times \$900{,}000 = \$138{,}600$$

The construction cost of the building in 1970 is $138,600.

How to Use the City Cost Indexes

What you should know before you begin

RSMeans City Cost Indexes (CCI) are an extremely useful tool to use when you want to compare costs from city to city and region to region.

This publication contains average construction cost indexes for 316 U.S. and Canadian cities covering over 930 three-digit zip code locations.

Keep in mind that a City Cost Index number is a percentage ratio of a specific city's cost to the national average cost of the same item at a stated time period.

In other words, these index figures represent relative construction factors (or, if you prefer, multipliers) for Material and Installation costs, as well as the weighted average for Total In Place costs for each CSI MasterFormat division. Installation costs include both labor and equipment rental costs. When estimating equipment rental rates only, for a specific location, use 01543 CONTRACTOR EQUIPMENT index.

The 30 City Average Index is the average of 30 major U.S. cities and serves as a National Average.

Index figures for both material and installation are based on the 30 major city average of 100 and represent the cost relationship as of July 1, 2008. The index for each division is computed from representative material and labor quantities for that division. The weighted average for each city is a weighted total of the components listed above it, but does not include relative productivity between trades or cities.

As changes occur in local material prices, labor rates, and equipment rental rates, (including fuel costs) the impact of these changes should be accurately measured by the change in the City Cost Index for each particular city (as compared to the 30 City Average).

Therefore, if you know (or have estimated) building costs in one city today, you can easily convert those costs to expected building costs in another city.

In addition, by using the Historical Cost Index, you can easily convert National Average building costs at a particular time to the approximate building costs for some other time. The City Cost Indexes can then be applied to calculate the costs for a particular city.

Quick Calculations

Location Adjustment Using the City Cost Indexes:

$$\frac{\text{Index for City A}}{\text{Index for City B}} \times \text{Cost in City B} = \text{Cost in City A}$$

Time Adjustment for the National Average Using the Historical Cost Index:

$$\frac{\text{Index for Year A}}{\text{Index for Year B}} \times \text{Cost in Year B} = \text{Cost in Year A}$$

Adjustment from the National Average:

$$\frac{\text{Index for City A}}{100} \times \text{National Average Cost} = \text{Cost in City A}$$

Since each of the other RSMeans publications contains many different items, any *one* item multiplied by the particular city index may give incorrect results. However, the larger the number of items compiled, the closer the results should be to actual costs for that particular city.

The City Cost Indexes for Canadian cities are calculated using Canadian material and equipment prices and labor rates, in Canadian dollars. Therefore, indexes for Canadian cities can be used to convert U.S. National Average prices to local costs in Canadian dollars.

How to use this section

1. Compare costs from city to city.

In using the RSMeans Indexes, remember that an index number is not a fixed number but a ratio: It's a percentage ratio of a building component's cost at any stated time to the National Average cost of that same component at the same time period. Put in the form of an equation:

$$\frac{\text{Specific City Cost}}{\text{National Average Cost}} \times 100 = \text{City Index Number}$$

Therefore, when making cost comparisons between cities, do not subtract one city's index number from the index number of another city and read the result as a percentage difference. Instead, divide one city's index number by that of the other city. The resulting number may then be used as a multiplier to calculate cost differences from city to city.

The formula used to find cost differences between cities for the purpose of comparison is as follows:

$$\frac{\text{City A Index}}{\text{City B Index}} \times \text{City B Cost (Known)} = \text{City A Cost (Unknown)}$$

In addition, you can use RSMeans CCI to calculate and compare costs division by division between cities using the same basic formula. (Just be sure that you're comparing similar divisions.)

2. Compare a specific city's construction costs with the National Average.

When you're studying construction location feasibility, it's advisable to compare a prospective project's cost index with an index of the National Average cost.

For example, divide the weighted average index of construction costs of a specific city by that of the 30 City Average, which = 100.

$$\frac{\text{City Index}}{100} = \% \text{ of National Average}$$

As a result, you get a ratio that indicates the relative cost of construction in that city in comparison with the National Average.

3. Convert U.S. National Average to actual costs in Canadian City.

$$\frac{\text{Index for Canadian City}}{100} \times \text{National Average Cost} = \text{Cost in Canadian City in \$ CAN}$$

4. **Adjust construction cost data based on a National Average.**

When you use a source of construction cost data which is based on a National Average (such as RSMeans cost data publications), it is necessary to adjust those costs to a specific location.

$$\frac{\text{City Index}}{100} \quad \text{x} \quad \frac{\text{``Book'' Cost Based on}}{\text{National Average Costs}} \quad = \quad \frac{\text{City Cost}}{\text{(Unknown)}}$$

5. **When applying the City Cost Indexes to demolition projects, use the appropriate division installation index.** For example, for removal of existing doors and windows, use Division 8 (Openings) index.

What you might like to know about how we developed the Indexes

The information presented in the CCI is organized according to the Construction Specifications Institute (CSI) MasterFormat 2004.

To create a reliable index, RSMeans researched the building type most often constructed in the United States and Canada. Because it was concluded that no one type of building completely represented the building construction industry, nine different types of buildings were combined to create a composite model.

The exact material, labor and equipment quantities are based on detailed analysis of these nine building types, then each quantity is weighted in proportion to expected usage. These various material items, labor hours, and equipment rental rates are thus combined to form a composite building representing as closely as possible the actual usage of materials, labor and equipment used in the North American Building Construction Industry.

The following structures were chosen to make up that composite model:

1. Factory, 1 story
2. Office, 2–4 story
3. Store, Retail
4. Town Hall, 2–3 story
5. High School, 2–3 story
6. Hospital, 4–8 story
7. Garage, Parking
8. Apartment, 1–3 story
9. Hotel/Motel, 2–3 story

For the purposes of ensuring the timeliness of the data, the components of the index for the composite model have been streamlined. They currently consist of:

- specific quantities of 66 commonly used construction materials;
- specific labor-hours for 21 building construction trades; and
- specific days of equipment rental for 6 types of construction equipment (normally used to install the 66 material items by the 21 trades.) Fuel costs and routine maintenance costs are included in the equipment costs.

A sophisticated computer program handles the updating of all costs for each city on a quarterly basis. Material and equipment price quotations are gathered quarterly from 316 cities in the United States and Canada. These prices and the latest negotiated labor wage rates for 21 different building trades are used to compile the quarterly update of the City Cost Index.

The 30 major U.S. cities used to calculate the National Average are:

Atlanta, GA	Memphis, TN
Baltimore, MD	Milwaukee, WI
Boston, MA	Minneapolis, MN
Buffalo, NY	Nashville, TN
Chicago, IL	New Orleans, LA
Cincinnati, OH	New York, NY
Cleveland, OH	Philadelphia, PA
Columbus, OH	Phoenix, AZ
Dallas, TX	Pittsburgh, PA
Denver, CO	St. Louis, MO
Detroit, MI	San Antonio, TX
Houston, TX	San Diego, CA
Indianapolis, IN	San Francisco, CA
Kansas City, MO	Seattle, WA
Los Angeles, CA	Washington, DC

What the CCI does not indicate

The weighted average for each city is a total of the divisional components weighted to reflect typical usage, but it does not include the productivity variations between trades or cities.

In addition, the CCI does not take into consideration factors such as the following:

- managerial efficiency
- competitive conditions
- automation
- restrictive union practices
- unique local requirements
- regional variations due to specific building codes

City Cost Indexes

DIVISION		UNITED STATES ANYTOWN MAT.	INST.	TOTAL	BIRMINGHAM MAT.	INST.	TOTAL	HUNTSVILLE MAT.	INST.	TOTAL	MOBILE MAT.	INST.	TOTAL	MONTGOMERY MAT.	INST.	TOTAL	TUSCALOOSA MAT.	INST.	TOTAL
015433	CONTRACTOR EQUIPMENT		100.0	100.0		100.5	100.5		100.4	100.4		98.9	98.9		98.9	98.9		100.4	100.4
0241, 31 - 34	SITE & INFRASTRUCTURE, DEMOLITION	100.0	100.0	100.0	86.5	91.9	90.2	82.2	91.4	88.5	95.7	87.7	90.2	93.7	87.8	89.6	82.7	90.8	88.2
0310	Concrete Forming & Accessories	100.0	100.0	100.0	91.0	72.9	75.3	94.9	67.1	70.9	95.0	59.2	64.1	92.7	49.9	55.7	94.8	53.8	59.4
0320	Concrete Reinforcing	100.0	100.0	100.0	89.7	80.3	85.8	89.7	76.3	84.2	92.5	52.8	76.3	92.5	78.9	87.0	89.7	79.3	85.4
0330	Cast-in-Place Concrete	100.0	100.0	100.0	99.2	71.4	88.6	89.7	68.5	81.6	94.7	64.3	83.1	97.0	52.0	79.8	93.4	54.5	78.6
03	CONCRETE	100.0	100.0	100.0	92.8	74.8	84.6	88.6	70.6	80.4	91.5	61.3	77.8	92.5	57.8	76.7	90.4	60.4	76.8
04	MASONRY	100.0	100.0	100.0	87.3	81.1	83.6	86.8	66.0	74.3	87.5	60.7	71.4	86.0	41.3	59.2	85.9	52.9	66.1
05	METALS	100.0	100.0	100.0	99.9	91.5	97.4	101.3	90.4	98.1	100.1	80.0	94.2	99.9	88.7	96.6	100.5	89.2	97.2
06	WOOD, PLASTICS & COMPOSITES	100.0	100.0	100.0	91.7	70.1	79.9	94.6	67.4	79.7	94.8	56.1	73.7	91.3	49.6	68.4	94.6	52.0	71.2
07	THERMAL & MOISTURE PROTECTION	100.0	100.0	100.0	94.8	85.8	91.3	92.4	78.2	86.9	92.4	77.4	86.6	91.8	68.2	82.6	92.7	72.1	84.7
08	OPENINGS	100.0	100.0	100.0	96.8	73.7	91.1	97.1	64.9	89.1	97.1	56.5	87.0	97.1	57.2	87.2	97.1	63.8	88.8
0920	Plaster & Gypsum Board	100.0	100.0	100.0	103.6	69.6	81.5	98.3	66.7	77.8	100.7	58.1	73.0	103.6	48.3	67.7	100.7	50.8	68.3
0950, 0980	Ceilings & Acoustic Treatment	100.0	100.0	100.0	97.1	69.6	80.0	100.4	66.7	79.6	100.4	58.1	74.2	97.1	48.3	66.9	100.4	50.8	69.7
0960	Flooring	100.0	100.0	100.0	103.5	67.6	93.8	103.5	60.6	91.9	113.4	62.8	99.7	111.6	33.2	90.4	103.5	42.1	86.9
0970, 0990	Wall Finishes & Painting/Coating	100.0	100.0	100.0	100.1	63.9	78.4	100.1	60.5	76.4	104.2	62.4	79.1	100.1	61.3	76.8	100.1	52.8	71.7
09	FINISHES	100.0	100.0	100.0	97.7	70.6	83.4	97.4	64.8	80.2	102.5	59.9	79.9	101.0	46.9	72.4	97.7	50.8	72.9
COVERS	DIVS. 10 - 14, 25, 28, 41, 43, 44	100.0	100.0	100.0	100.0	87.4	97.5	100.0	83.7	96.8	100.0	81.4	96.3	100.0	78.7	95.8	100.0	80.9	96.2
21, 22, 23	FIRE SUPPRESSION, PLUMBING & HVAC	100.0	100.0	100.0	99.9	67.3	86.9	99.8	62.1	84.8	99.8	63.7	85.4	99.9	40.6	76.3	99.9	40.4	76.2
26, 27, 3370	ELECTRICAL, COMMUNICATIONS & UTIL.	100.0	100.0	100.0	108.6	60.1	85.0	100.5	67.3	84.3	100.5	58.4	80.0	103.4	71.0	87.6	100.6	60.1	80.5
MF2004	WEIGHTED AVERAGE	100.0	100.0	100.0	98.3	74.9	88.5	97.0	71.1	86.1	97.9	65.9	84.5	98.0	59.2	81.7	97.0	60.5	81.7

UNITED STATES columns are ANYTOWN; BIRMINGHAM, HUNTSVILLE, MOBILE, MONTGOMERY, TUSCALOOSA are ALABAMA.

DIVISION		ANCHORAGE MAT.	INST.	TOTAL	FAIRBANKS MAT.	INST.	TOTAL	JUNEAU MAT.	INST.	TOTAL	FLAGSTAFF MAT.	INST.	TOTAL	MESA/TEMPE MAT.	INST.	TOTAL	PHOENIX MAT.	INST.	TOTAL
015433	CONTRACTOR EQUIPMENT		113.0	113.0		113.0	113.0		113.0	113.0		96.2	96.2		99.2	99.2		99.8	99.8
0241, 31 - 34	SITE & INFRASTRUCTURE, DEMOLITION	140.1	126.1	130.5	126.3	126.2	126.3	139.2	126.1	130.2	81.1	102.1	95.4	86.2	104.8	98.9	86.7	106.0	99.9
0310	Concrete Forming & Accessories	131.5	114.1	116.5	137.2	116.5	119.3	133.2	114.0	116.6	103.2	65.4	70.6	103.8	60.0	66.0	104.8	69.2	74.0
0320	Concrete Reinforcing	136.8	103.8	123.3	139.4	103.8	124.9	104.9	103.7	104.4	99.9	79.4	91.5	98.8	79.4	90.9	97.4	79.7	90.2
0330	Cast-in-Place Concrete	168.8	111.3	146.8	148.0	112.6	134.5	169.7	111.2	147.3	96.0	77.4	88.9	104.8	65.6	89.8	104.0	77.9	94.6
03	CONCRETE	138.1	110.7	125.6	125.3	112.1	119.3	132.2	110.6	122.4	121.0	72.3	98.9	108.5	65.9	89.1	108.1	74.3	92.7
04	MASONRY	207.3	116.1	152.7	216.8	117.7	157.5	198.8	116.1	149.3	97.8	61.8	76.3	110.7	47.6	72.9	97.8	64.9	78.1
05	METALS	124.0	100.9	117.3	124.1	101.1	117.4	124.1	100.8	117.3	95.3	74.9	89.3	93.5	75.4	88.2	95.1	76.6	89.7
06	WOOD, PLASTICS & COMPOSITES	114.3	113.6	113.9	118.5	115.7	117.0	114.3	113.6	113.9	105.5	65.2	83.4	96.2	65.3	79.3	97.1	70.0	82.2
07	THERMAL & MOISTURE PROTECTION	181.2	111.1	153.9	178.9	113.9	153.6	179.5	111.1	152.9	98.9	67.0	86.5	107.4	58.4	88.3	107.3	68.1	92.1
08	OPENINGS	125.0	109.8	121.3	122.2	111.0	119.4	122.2	109.8	119.1	100.8	68.9	92.9	96.3	66.7	89.0	97.3	71.5	90.9
0920	Plaster & Gypsum Board	139.3	113.8	122.7	156.6	116.0	130.2	139.3	113.8	122.7	85.7	64.0	71.6	84.9	64.0	71.3	87.3	68.8	75.2
0950, 0980	Ceilings & Acoustic Treatment	125.5	113.8	118.3	131.4	116.0	121.8	125.5	113.8	118.3	112.0	64.0	82.3	101.2	64.0	78.2	108.7	68.8	84.0
0960	Flooring	158.9	120.7	148.6	158.9	120.7	148.6	158.9	120.7	148.6	98.3	44.5	83.7	102.9	55.5	90.1	103.2	58.0	91.0
0970, 0990	Wall Finishes & Painting/Coating	155.7	112.9	130.0	155.7	121.0	134.8	155.7	112.9	130.0	90.4	55.1	69.2	94.0	55.1	70.7	94.0	57.2	71.9
09	FINISHES	148.5	115.6	131.1	150.3	118.2	133.3	147.3	115.6	130.5	101.0	59.8	79.2	100.2	58.1	78.0	102.4	65.3	82.8
COVERS	DIVS. 10 - 14, 25, 28, 41, 43, 44	100.0	110.2	102.0	100.0	111.0	102.2	100.0	110.2	102.0	100.0	81.5	96.3	100.0	76.7	95.4	100.0	82.4	96.5
21, 22, 23	FIRE SUPPRESSION, PLUMBING & HVAC	100.5	103.0	101.5	100.4	109.8	104.2	100.5	94.4	98.1	100.2	76.8	90.9	100.1	66.6	86.8	100.1	76.9	90.9
26, 27, 3370	ELECTRICAL, COMMUNICATIONS & UTIL.	140.0	109.5	125.1	148.2	109.5	129.4	141.5	109.5	125.9	101.6	61.3	81.9	92.5	61.3	77.3	100.3	64.5	82.9
MF2004	WEIGHTED AVERAGE	128.3	110.3	120.8	127.6	112.6	121.3	127.0	108.5	119.2	101.4	71.6	88.9	99.2	66.7	85.6	99.9	74.0	89.0

ANCHORAGE, FAIRBANKS, JUNEAU are ALASKA; FLAGSTAFF, MESA/TEMPE, PHOENIX are ARIZONA.

DIVISION		PRESCOTT MAT.	INST.	TOTAL	TUCSON MAT.	INST.	TOTAL	FORT SMITH MAT.	INST.	TOTAL	JONESBORO MAT.	INST.	TOTAL	LITTLE ROCK MAT.	INST.	TOTAL	PINE BLUFF MAT.	INST.	TOTAL
015433	CONTRACTOR EQUIPMENT		96.2	96.2		99.2	99.2		92.0	92.0		105.6	105.6		92.0	92.0		92.0	92.0
0241, 31 - 34	SITE & INFRASTRUCTURE, DEMOLITION	69.8	100.8	91.0	82.0	105.6	98.1	78.5	90.7	86.9	101.1	96.1	97.7	87.0	90.8	89.6	80.6	90.7	87.5
0310	Concrete Forming & Accessories	99.4	49.6	56.4	104.1	68.8	73.6	99.9	40.9	48.9	88.5	55.8	60.3	92.2	68.5	71.7	81.9	68.2	70.1
0320	Concrete Reinforcing	99.9	74.3	89.4	81.4	79.4	80.6	99.6	69.8	87.4	95.3	71.7	85.7	99.7	66.5	86.2	99.7	66.4	86.1
0330	Cast-in-Place Concrete	95.9	59.1	81.8	107.9	77.8	96.4	89.9	68.1	81.5	85.6	62.4	76.7	90.3	68.3	81.9	82.4	68.2	77.0
03	CONCRETE	106.7	57.9	84.5	106.3	74.0	91.6	89.0	56.8	74.4	85.5	62.5	75.2	88.8	68.4	79.5	86.9	68.2	78.4
04	MASONRY	98.1	50.9	69.8	95.8	61.8	75.4	97.4	57.0	73.2	93.6	48.6	66.6	94.5	57.0	72.0	118.5	57.0	81.6
05	METALS	95.3	69.2	87.7	94.3	75.5	88.8	102.8	72.3	93.9	96.8	83.8	93.0	98.6	71.7	90.7	101.3	71.4	92.6
06	WOOD, PLASTICS & COMPOSITES	101.1	47.9	71.9	96.4	70.0	81.9	103.9	36.4	66.9	90.3	56.5	71.8	98.0	73.2	84.4	83.1	73.2	77.6
07	THERMAL & MOISTURE PROTECTION	97.6	55.0	81.0	108.7	64.4	91.5	99.2	52.7	81.1	102.5	56.3	84.5	98.5	56.6	82.2	98.0	56.6	81.9
08	OPENINGS	100.8	53.1	89.0	92.6	71.5	87.4	97.2	44.6	84.2	99.1	60.4	89.5	97.2	65.1	89.3	92.5	65.1	85.7
0920	Plaster & Gypsum Board	83.1	46.0	59.0	90.4	68.8	76.3	86.9	34.7	53.0	98.2	55.1	70.2	83.1	72.7	76.4	80.6	72.7	75.5
0950, 0980	Ceilings & Acoustic Treatment	110.3	46.0	70.6	102.1	68.8	81.5	92.8	34.7	56.9	91.6	55.1	69.1	90.3	72.7	79.4	88.6	72.7	78.8
0960	Flooring	96.7	44.3	82.5	93.7	44.5	80.4	112.0	68.2	100.1	74.9	59.9	70.8	113.1	68.2	101.0	101.8	68.2	92.7
0970, 0990	Wall Finishes & Painting/Coating	90.4	42.4	61.6	94.0	55.1	70.7	98.5	61.9	76.5	87.4	55.6	68.3	98.5	63.5	77.5	98.5	63.5	77.5
09	FINISHES	98.6	47.0	71.3	98.3	62.6	79.4	95.6	47.1	70.0	88.2	56.3	71.4	95.0	68.9	81.2	90.9	68.9	79.3
COVERS	DIVS. 10 - 14, 25, 28, 41, 43, 44	100.0	77.8	95.6	100.0	82.4	96.5	100.0	69.7	94.0	100.0	56.6	91.4	100.0	74.0	94.9	100.0	74.0	94.9
21, 22, 23	FIRE SUPPRESSION, PLUMBING & HVAC	100.2	71.2	88.7	100.1	68.8	87.7	100.1	48.1	79.5	100.3	48.1	79.6	100.1	66.4	86.7	100.1	50.8	80.5
26, 27, 3370	ELECTRICAL, COMMUNICATIONS & UTIL.	101.3	61.2	81.8	94.8	58.8	77.3	95.8	75.3	85.8	103.8	52.2	78.7	100.5	76.6	88.9	95.1	76.6	86.1
MF2004	WEIGHTED AVERAGE	99.2	63.7	84.3	98.0	70.6	86.5	97.4	60.1	81.8	96.9	60.2	81.5	97.1	69.9	85.7	96.9	66.6	84.2

PRESCOTT, TUCSON are ARIZONA; FORT SMITH, JONESBORO, LITTLE ROCK, PINE BLUFF are ARKANSAS.

City Cost Indexes

DIVISION		ARKANSAS — TEXARKANA			CALIFORNIA — ANAHEIM			BAKERSFIELD			FRESNO			LOS ANGELES			OAKLAND		
		MAT.	INST.	TOTAL	MAT.	INST.	TOTAL	MAT.	INST.	TOTAL	MAT.	INST.	TOTAL	MAT.	INST.	TOTAL	MAT.	INST.	TOTAL
015433	CONTRACTOR EQUIPMENT		90.7	90.7		102.2	102.2		100.7	100.7		100.7	100.7		100.3	100.3		104.6	104.6
0241, 31 - 34	SITE & INFRASTRUCTURE, DEMOLITION	95.9	87.8	90.4	101.8	109.4	107.0	108.0	107.8	107.9	107.6	107.2	107.4	98.5	110.1	106.4	141.8	105.6	117.1
0310	Concrete Forming & Accessories	88.6	46.1	51.9	106.2	126.7	123.9	102.1	126.1	122.8	102.5	126.4	123.2	104.2	126.7	123.6	108.4	142.5	137.8
0320	Concrete Reinforcing	99.2	69.7	87.2	93.2	118.2	103.4	105.0	118.3	110.4	88.9	118.5	101.0	104.4	118.6	110.2	98.2	119.2	106.8
0330	Cast-in-Place Concrete	89.9	49.0	74.2	101.2	123.5	109.7	108.3	122.4	113.7	106.7	117.0	110.6	108.8	122.0	113.9	132.9	120.7	128.2
03	CONCRETE	86.1	52.5	70.8	104.4	122.9	112.8	107.6	122.3	114.3	105.9	120.5	112.5	108.7	122.4	115.0	122.1	129.1	125.3
04	MASONRY	97.4	33.2	59.0	88.8	115.3	104.7	109.8	117.4	114.3	112.8	118.2	116.1	95.5	121.8	111.2	156.6	127.6	139.2
05	METALS	93.7	69.9	86.8	106.4	103.8	105.6	101.3	103.1	101.8	106.5	102.7	105.4	104.0	103.2	103.8	98.8	106.9	101.2
06	WOOD, PLASTICS & COMPOSITES	91.9	49.2	68.4	94.5	126.5	112.1	87.6	126.6	109.0	99.6	127.0	114.6	86.8	126.3	108.5	106.4	146.2	128.2
07	THERMAL & MOISTURE PROTECTION	98.9	45.9	78.3	102.5	116.7	108.1	100.5	109.8	104.1	96.4	112.5	102.7	104.2	118.8	109.9	107.3	125.6	114.4
08	OPENINGS	97.7	52.0	86.4	104.5	121.5	108.7	102.0	119.4	106.3	103.8	118.1	107.3	98.3	121.4	104.0	104.2	133.2	111.4
0920	Plaster & Gypsum Board	84.8	47.9	60.8	105.6	127.3	119.7	100.8	127.3	118.0	101.3	127.8	118.5	104.4	127.3	119.3	108.3	147.3	133.6
0950, 0980	Ceilings & Acoustic Treatment	94.5	47.9	65.7	114.8	127.3	122.6	116.4	127.3	123.2	119.5	127.8	124.6	126.1	127.3	126.9	120.5	147.3	137.0
0960	Flooring	104.4	57.2	91.6	114.2	113.6	114.1	114.2	87.9	107.1	121.1	133.7	124.5	102.3	113.6	105.3	117.6	116.2	117.2
0970, 0990	Wall Finishes & Painting/Coating	98.5	32.6	59.0	104.7	105.8	105.3	106.2	101.6	103.5	124.4	100.4	110.0	97.5	111.7	106.0	108.1	144.5	129.9
09	FINISHES	94.2	46.6	69.0	109.0	122.4	116.1	108.5	117.3	114.1	114.9	126.0	120.7	109.6	122.9	116.6	115.6	139.4	128.2
COVERS	DIVS. 10 - 14, 25, 28, 41, 43, 44	100.0	41.1	88.3	100.0	114.8	102.9	100.0	115.2	103.0	100.0	126.3	105.2	100.0	114.3	102.8	100.0	130.1	106.0
21, 22, 23	FIRE SUPPRESSION, PLUMBING & HVAC	100.1	35.6	74.5	100.1	115.5	106.3	100.1	99.9	100.0	100.3	111.6	104.8	100.3	115.5	106.3	100.3	143.0	117.3
26, 27, 3370	ELECTRICAL, COMMUNICATIONS & UTIL.	97.0	42.0	70.2	91.5	105.2	98.2	104.2	97.6	101.0	90.7	94.5	92.5	100.4	114.5	107.3	104.9	130.0	117.1
MF2004	WEIGHTED AVERAGE	95.9	48.8	76.1	101.5	114.7	107.0	103.3	109.4	105.8	103.2	112.6	107.2	102.2	116.7	108.3	108.8	129.5	117.5

DIVISION		CALIFORNIA — OXNARD			REDDING			RIVERSIDE			SACRAMENTO			SAN DIEGO			SAN FRANCISCO		
		MAT.	INST.	TOTAL	MAT.	INST.	TOTAL	MAT.	INST.	TOTAL	MAT.	INST.	TOTAL	MAT.	INST.	TOTAL	MAT.	INST.	TOTAL
015433	CONTRACTOR EQUIPMENT		100.0	100.0		100.6	100.6		101.4	101.4		102.9	102.9		99.6	99.6		108.1	108.1
0241, 31 - 34	SITE & INFRASTRUCTURE, DEMOLITION	108.0	106.3	106.9	112.2	106.2	108.1	100.1	107.8	105.4	111.5	110.7	110.9	105.6	104.1	104.6	144.3	111.2	121.7
0310	Concrete Forming & Accessories	105.7	126.6	123.8	105.4	126.0	123.2	106.4	126.5	123.7	106.9	127.0	124.2	108.1	111.0	110.6	108.2	144.2	139.3
0320	Concrete Reinforcing	105.0	117.9	110.3	101.9	118.0	108.5	104.1	118.0	109.8	90.8	118.6	102.1	99.3	118.2	107.0	111.8	119.9	115.1
0330	Cast-in-Place Concrete	107.5	122.7	113.3	121.8	116.7	119.8	105.0	123.4	112.0	112.3	118.0	114.5	107.0	105.8	106.6	136.3	122.7	131.1
03	CONCRETE	107.5	122.6	114.3	116.4	120.1	118.1	106.2	122.8	113.7	110.9	121.1	115.6	108.5	109.9	109.1	126.4	130.8	128.4
04	MASONRY	114.7	109.9	111.8	117.9	110.1	113.3	87.8	110.8	101.6	125.0	118.3	121.0	98.7	111.5	106.4	157.2	136.0	144.5
05	METALS	101.1	103.1	101.7	105.9	101.6	104.7	106.6	103.3	105.6	94.5	103.4	97.1	103.5	103.1	103.4	104.9	110.5	106.5
06	WOOD, PLASTICS & COMPOSITES	94.3	126.6	112.0	94.5	127.0	112.3	94.5	126.5	112.1	100.3	127.2	115.1	95.3	107.6	102.0	106.4	146.4	128.3
07	THERMAL & MOISTURE PROTECTION	105.8	112.5	108.4	106.2	108.9	107.3	102.8	111.8	106.3	112.8	114.5	113.4	108.4	105.6	107.3	109.9	138.3	120.9
08	OPENINGS	100.9	121.6	106.0	103.5	118.9	107.3	103.0	121.5	107.6	117.4	120.2	118.1	99.9	109.8	102.3	108.6	135.8	115.3
0920	Plaster & Gypsum Board	105.6	127.3	119.7	104.0	127.8	119.4	105.0	127.3	119.5	103.4	127.8	119.2	117.3	107.6	111.0	110.7	147.3	134.5
0950, 0980	Ceilings & Acoustic Treatment	121.4	127.3	125.1	126.4	127.8	127.2	118.9	127.3	124.1	118.6	127.8	124.3	107.6	107.6	107.4	127.8	147.3	139.8
0960	Flooring	112.7	113.6	113.0	113.5	109.6	112.4	118.4	113.6	117.1	117.0	114.3	116.3	105.4	107.6	106.0	117.6	126.2	119.9
0970, 0990	Wall Finishes & Painting/Coating	105.5	96.5	100.1	105.5	104.7	105.0	102.7	105.8	104.6	105.2	114.3	110.7	103.8	111.7	108.5	108.1	156.9	137.4
09	FINISHES	111.8	121.4	116.9	113.4	122.1	118.0	110.7	122.4	116.9	112.2	124.3	118.6	107.6	110.1	108.9	117.8	142.9	131.1
COVERS	DIVS. 10 - 14, 25, 28, 41, 43, 44	100.0	115.0	103.0	100.0	126.3	105.2	100.0	114.7	102.9	100.0	126.9	105.3	100.0	111.7	102.3	100.0	130.7	106.1
21, 22, 23	FIRE SUPPRESSION, PLUMBING & HVAC	100.2	115.6	106.3	100.2	103.5	101.5	100.1	115.5	106.2	100.1	112.9	105.2	100.3	113.8	105.6	100.3	164.9	126.0
26, 27, 3370	ELECTRICAL, COMMUNICATIONS & UTIL.	96.4	99.5	97.9	99.4	98.1	98.8	91.6	96.3	93.9	99.6	103.7	101.6	97.7	104.1	100.8	104.9	162.2	132.8
MF2004	WEIGHTED AVERAGE	102.9	112.8	107.0	105.8	109.8	107.5	101.6	112.7	106.3	104.9	114.6	108.9	102.4	108.8	105.1	111.2	141.3	123.8

DIVISION		CALIFORNIA — SAN JOSE			SANTA BARBARA			STOCKTON			VALLEJO			COLORADO — COLORADO SPRINGS			DENVER		
		MAT.	INST.	TOTAL	MAT.	INST.	TOTAL	MAT.	INST.	TOTAL	MAT.	INST.	TOTAL	MAT.	INST.	TOTAL	MAT.	INST.	TOTAL
015433	CONTRACTOR EQUIPMENT		102.4	102.4		100.7	100.7		100.6	100.6		103.0	103.0		94.5	94.5		98.3	98.3
0241, 31 - 34	SITE & INFRASTRUCTURE, DEMOLITION	144.1	103.4	116.3	107.9	107.8	107.9	105.7	107.1	106.7	110.6	110.1	110.2	89.9	94.0	92.7	89.0	102.2	98.1
0310	Concrete Forming & Accessories	107.9	143.1	138.3	106.5	126.4	123.7	106.0	126.3	123.6	107.3	141.4	136.8	93.6	80.7	82.5	99.9	82.1	84.5
0320	Concrete Reinforcing	92.9	119.3	103.7	105.0	118.2	110.4	105.6	118.1	110.7	97.8	118.8	106.4	106.9	79.6	95.8	106.9	79.7	95.8
0330	Cast-in-Place Concrete	127.8	120.7	125.1	107.1	122.6	113.0	107.0	117.2	110.9	122.3	119.1	121.1	101.8	82.9	94.5	96.3	84.2	91.7
03	CONCRETE	114.6	129.5	121.4	107.3	122.4	114.2	107.4	120.5	113.4	117.0	127.9	122.0	108.0	81.5	96.0	102.5	82.6	93.5
04	MASONRY	148.2	128.9	136.6	111.1	114.7	113.2	114.3	110.2	111.9	88.9	121.4	108.3	101.7	75.9	86.1	103.3	79.0	88.7
05	METALS	100.5	109.6	103.2	101.5	103.2	102.0	102.9	102.9	102.9	97.7	104.4	99.6	101.7	86.0	97.1	104.2	86.1	98.9
06	WOOD, PLASTICS & COMPOSITES	101.9	145.9	126.1	94.3	126.6	112.0	96.1	127.0	113.1	98.0	145.9	124.3	94.4	84.4	88.9	102.1	84.2	92.3
07	THERMAL & MOISTURE PROTECTION	102.1	135.8	115.2	101.9	109.6	104.9	105.8	110.1	107.4	110.0	122.1	114.7	100.0	80.2	92.3	98.9	76.6	90.2
08	OPENINGS	93.9	135.5	104.2	102.0	121.6	106.8	101.2	120.1	105.9	117.0	135.5	121.6	99.5	85.2	96.0	99.4	85.1	95.9
0920	Plaster & Gypsum Board	100.9	147.3	131.0	105.6	127.3	119.7	106.0	127.8	120.1	108.2	147.3	133.6	82.4	83.9	83.4	95.8	83.9	88.1
0950, 0980	Ceilings & Acoustic Treatment	110.5	147.3	133.3	121.4	127.3	125.1	121.4	127.8	125.3	128.0	147.3	139.9	107.2	83.9	92.8	106.3	83.9	92.4
0960	Flooring	106.2	126.2	111.6	114.2	87.9	107.1	114.2	105.4	111.8	125.9	126.2	126.0	104.0	76.5	96.6	109.2	88.8	103.7
0970, 0990	Wall Finishes & Painting/Coating	106.2	146.2	130.2	105.5	96.5	100.1	105.5	105.1	105.3	105.1	128.2	119.0	105.7	46.6	70.2	106.0	76.5	88.3
09	FINISHES	108.6	141.4	126.0	112.5	117.0	114.8	112.3	121.4	117.1	115.6	139.1	128.0	103.1	76.7	87.1	103.1	83.0	92.4
COVERS	DIVS. 10 - 14, 25, 28, 41, 43, 44	100.0	129.5	105.8	100.0	115.0	103.0	100.0	126.3	105.2	100.0	128.9	105.7	100.0	87.8	97.6	100.0	88.6	97.7
21, 22, 23	FIRE SUPPRESSION, PLUMBING & HVAC	100.2	145.6	118.3	100.2	115.5	106.3	100.2	103.7	101.6	100.2	117.5	107.1	100.1	75.1	90.2	100.0	86.7	94.7
26, 27, 3370	ELECTRICAL, COMMUNICATIONS & UTIL.	103.1	141.0	121.6	88.4	100.3	94.2	99.1	111.7	105.2	95.6	116.5	105.8	104.9	83.3	94.3	106.7	85.4	96.3
MF2004	WEIGHTED AVERAGE	106.0	132.5	117.1	101.9	112.9	106.5	103.5	112.0	107.0	104.1	121.6	111.4	101.5	81.0	92.9	101.8	85.5	95.0

COLORADO / CONNECTICUT

DIVISION		FORT COLLINS MAT.	INST.	TOTAL	GRAND JUNCTION MAT.	INST.	TOTAL	GREELEY MAT.	INST.	TOTAL	PUEBLO MAT.	INST.	TOTAL	BRIDGEPORT MAT.	INST.	TOTAL	BRISTOL MAT.	INST.	TOTAL
015433	CONTRACTOR EQUIPMENT		96.0	96.0		98.9	98.9		96.0	96.0		95.8	95.8		100.4	100.4		100.4	100.4
0241, 31 - 34	SITE & INFRASTRUCTURE, DEMOLITION	99.8	96.8	97.8	119.3	99.5	105.7	87.3	95.5	92.9	111.1	93.0	98.7	98.6	104.2	102.5	97.8	104.2	102.2
0310	Concrete Forming & Accessories	100.5	74.6	78.2	110.1	75.0	79.8	98.5	47.2	54.2	107.5	81.0	84.6	96.4	121.9	118.4	96.4	121.6	118.2
0320	Concrete Reinforcing	107.9	77.7	95.6	107.5	77.6	95.3	107.8	77.2	95.3	103.6	79.6	93.8	100.4	127.1	111.3	100.4	127.1	111.3
0330	Cast-in-Place Concrete	111.9	78.1	98.9	118.5	79.0	103.4	93.5	56.1	79.2	106.2	84.0	97.7	100.0	120.2	107.7	93.7	120.1	103.8
03	CONCRETE	114.9	76.8	97.6	119.9	77.2	100.5	100.2	57.0	80.6	109.6	82.1	97.1	107.4	122.0	114.0	104.5	121.8	112.4
04	MASONRY	112.1	52.7	76.5	139.9	54.8	88.9	106.6	35.0	63.7	103.5	74.2	86.0	106.6	132.0	121.8	99.2	132.0	118.8
05	METALS	99.8	81.5	94.5	98.1	80.9	93.1	99.8	80.1	94.0	99.7	86.8	96.0	94.9	123.3	103.2	94.9	123.1	103.1
06	WOOD, PLASTICS & COMPOSITES	101.9	78.6	89.1	103.9	78.7	90.1	99.4	46.9	70.6	101.2	84.7	92.2	101.4	119.6	111.4	101.4	119.6	111.4
07	THERMAL & MOISTURE PROTECTION	99.7	64.4	86.0	103.6	61.1	87.1	99.1	54.1	81.6	103.2	79.6	94.0	94.5	127.7	107.4	94.6	124.1	106.1
08	OPENINGS	95.3	81.8	92.0	99.3	81.9	95.0	95.3	64.6	87.7	93.8	85.4	91.7	104.9	129.4	110.9	104.9	127.4	110.4
0920	Plaster & Gypsum Board	94.3	78.0	83.7	105.1	78.0	87.5	93.1	45.2	62.0	80.6	83.9	82.7	104.8	119.5	114.4	104.8	119.5	114.4
0950, 0980	Ceilings & Acoustic Treatment	100.5	78.0	86.6	106.2	78.0	88.8	100.5	45.2	66.3	117.8	83.9	96.8	99.1	119.5	111.8	99.1	119.5	111.8
0960	Flooring	109.0	61.4	96.2	119.4	61.4	103.7	107.8	61.4	95.2	115.6	88.8	108.5	96.8	125.6	104.6	96.8	111.9	100.9
0970, 0990	Wall Finishes & Painting/Coating	106.0	46.2	70.1	108.9	76.5	89.5	106.0	28.8	59.7	108.9	43.2	69.5	91.7	116.7	106.7	91.7	116.7	106.7
09	FINISHES	102.0	69.4	84.8	110.7	73.4	91.0	100.8	46.5	72.1	108.6	78.8	92.9	100.5	121.9	111.8	100.5	119.5	110.5
COVERS	DIVS. 10 - 14, 25, 28, 41, 43, 44	100.0	85.9	97.2	100.0	87.0	97.4	100.0	78.8	95.8	100.0	88.6	97.8	100.0	111.7	102.3	100.0	111.7	102.3
21, 22, 23	FIRE SUPPRESSION, PLUMBING & HVAC	100.0	79.1	91.7	99.9	65.2	86.1	100.0	74.1	89.7	99.9	68.0	87.2	100.2	114.4	105.9	100.2	114.4	105.8
26, 27, 3370	ELECTRICAL, COMMUNICATIONS & UTIL.	101.2	85.3	93.5	94.7	61.7	78.7	101.2	85.3	93.5	95.7	77.6	86.9	97.0	109.9	103.3	97.0	109.6	103.2
MF2004	WEIGHTED AVERAGE	102.2	77.3	91.7	104.9	72.0	91.1	99.7	67.0	86.0	101.3	79.0	91.9	100.3	118.6	108.0	99.6	118.0	107.3

CONNECTICUT

DIVISION		HARTFORD MAT.	INST.	TOTAL	NEW BRITAIN MAT.	INST.	TOTAL	NEW HAVEN MAT.	INST.	TOTAL	NORWALK MAT.	INST.	TOTAL	STAMFORD MAT.	INST.	TOTAL	WATERBURY MAT.	INST.	TOTAL
015433	CONTRACTOR EQUIPMENT		100.4	100.4		100.4	100.4		100.6	100.6		100.4	100.4		100.4	100.4		100.4	100.4
0241, 31 - 34	SITE & INFRASTRUCTURE, DEMOLITION	95.1	104.2	101.3	98.0	104.2	102.2	97.9	104.6	102.5	98.4	104.2	102.4	99.0	104.3	102.6	98.2	104.2	102.3
0310	Concrete Forming & Accessories	94.5	121.6	117.9	96.6	121.6	118.2	96.1	121.6	118.1	96.4	122.1	118.6	96.4	122.4	118.8	96.4	121.8	118.3
0320	Concrete Reinforcing	100.4	127.1	111.3	100.4	127.1	111.3	100.4	127.1	111.3	100.4	127.2	111.3	100.4	127.3	111.3	100.4	127.1	111.3
0330	Cast-in-Place Concrete	96.1	120.1	105.3	95.2	120.1	104.7	96.9	110.0	102.3	98.4	128.8	110.0	100.0	128.9	111.1	100.0	120.1	107.7
03	CONCRETE	105.5	121.8	112.9	105.2	121.8	112.7	120.8	118.7	119.8	106.7	125.1	115.0	107.4	125.3	115.5	107.4	121.9	114.0
04	MASONRY	99.4	132.0	118.9	101.2	132.0	119.6	99.4	132.0	118.9	98.8	133.7	119.7	99.7	133.7	120.0	99.7	132.0	119.0
05	METALS	99.5	123.1	106.4	91.6	123.1	100.8	91.8	123.0	100.9	94.9	123.6	103.3	94.9	123.9	103.4	94.9	123.2	103.2
06	WOOD, PLASTICS & COMPOSITES	99.9	119.6	110.7	101.4	119.6	111.4	101.4	119.6	111.4	101.4	119.6	111.4	101.4	119.6	111.4	101.4	119.6	111.4
07	THERMAL & MOISTURE PROTECTION	95.1	124.1	106.4	94.6	124.5	106.2	94.7	123.0	105.7	94.7	129.5	108.2	94.6	129.5	108.2	94.6	124.5	106.2
08	OPENINGS	105.6	127.4	111.0	104.9	127.4	110.4	104.9	129.4	110.9	104.9	129.4	110.9	104.9	129.4	110.9	104.9	129.4	110.9
0920	Plaster & Gypsum Board	106.1	119.5	114.8	104.8	119.5	114.4	104.8	119.5	114.4	104.8	119.5	114.4	104.8	119.5	114.4	104.8	119.5	114.4
0950, 0980	Ceilings & Acoustic Treatment	99.1	119.5	111.8	99.1	119.5	111.8	99.1	119.5	111.8	99.1	119.5	111.8	99.1	119.5	111.8	99.1	119.5	111.8
0960	Flooring	96.8	111.9	100.9	96.8	111.9	100.9	96.8	125.6	104.6	96.8	120.6	103.2	96.8	120.6	103.2	96.8	125.6	104.6
0970, 0990	Wall Finishes & Painting/Coating	91.7	116.7	106.7	91.7	116.7	106.7	91.7	116.7	106.7	91.7	116.7	106.7	91.7	116.7	106.7	91.7	116.7	106.7
09	FINISHES	100.2	119.5	110.4	100.5	119.5	110.5	100.5	121.9	111.8	100.5	121.0	111.3	100.6	121.0	111.4	100.4	121.9	111.7
COVERS	DIVS. 10 - 14, 25, 28, 41, 43, 44	100.0	111.7	102.3	100.0	111.7	102.3	100.0	111.7	102.3	100.0	111.7	102.3	100.0	111.9	102.3	100.0	111.7	102.3
21, 22, 23	FIRE SUPPRESSION, PLUMBING & HVAC	100.1	114.4	105.8	100.2	114.4	105.8	100.2	114.4	105.8	100.2	114.5	105.9	100.2	114.5	105.9	100.2	114.4	105.8
26, 27, 3370	ELECTRICAL, COMMUNICATIONS & UTIL.	96.9	110.5	103.5	97.1	109.6	103.2	97.0	109.6	103.1	97.0	108.3	102.5	97.0	154.5	125.0	96.6	109.9	103.1
MF2004	WEIGHTED AVERAGE	100.5	118.1	107.9	99.2	118.0	107.1	101.0	118.0	108.2	99.9	119.0	107.9	100.0	125.4	110.7	99.9	118.5	107.7

D.C. / DELAWARE / FLORIDA

DIVISION		WASHINGTON MAT.	INST.	TOTAL	WILMINGTON MAT.	INST.	TOTAL	DAYTONA BEACH MAT.	INST.	TOTAL	FORT LAUDERDALE MAT.	INST.	TOTAL	JACKSONVILLE MAT.	INST.	TOTAL	MELBOURNE MAT.	INST.	TOTAL
015433	CONTRACTOR EQUIPMENT		103.6	103.6		114.7	114.7		98.9	98.9		93.2	93.2		98.9	98.9		98.9	98.9
0241, 31 - 34	SITE & INFRASTRUCTURE, DEMOLITION	110.3	91.8	97.7	91.3	109.0	103.4	116.1	89.5	97.9	101.4	78.7	85.9	116.2	88.7	97.4	124.6	88.7	100.0
0310	Concrete Forming & Accessories	101.3	81.9	84.5	99.5	102.8	102.4	95.8	77.8	80.2	94.0	68.1	71.6	95.5	59.1	64.1	92.0	79.7	81.4
0320	Concrete Reinforcing	113.5	91.1	104.3	99.5	100.1	99.7	92.5	82.2	88.3	92.5	80.4	87.6	92.5	51.9	75.9	93.5	82.2	88.9
0330	Cast-in-Place Concrete	127.6	88.5	112.7	98.3	89.1	94.8	91.9	77.2	86.3	96.4	74.4	88.0	92.8	66.0	82.5	110.8	79.5	98.8
03	CONCRETE	121.1	87.3	105.8	102.8	98.5	100.8	90.4	79.4	85.4	92.4	74.0	84.0	90.8	61.8	77.6	101.5	81.1	92.2
04	MASONRY	101.7	81.9	89.8	107.6	93.8	99.3	86.3	72.3	77.9	86.6	68.2	75.6	86.2	58.6	69.7	83.7	76.3	79.2
05	METALS	98.0	109.3	101.3	99.6	114.7	104.0	101.7	94.5	99.6	101.6	94.4	99.5	100.2	81.2	94.7	110.7	94.9	106.1
06	WOOD, PLASTICS & COMPOSITES	100.9	81.2	90.1	99.8	104.2	102.2	95.7	81.0	87.6	91.6	65.6	77.4	95.7	57.9	74.9	91.4	81.0	85.7
07	THERMAL & MOISTURE PROTECTION	105.3	84.7	97.3	99.8	102.9	101.0	95.1	79.8	89.1	95.1	82.3	90.1	95.3	64.7	83.4	95.5	85.9	91.7
08	OPENINGS	107.0	92.7	103.4	95.2	108.6	98.5	99.7	74.9	93.6	97.1	66.1	89.4	99.7	52.1	88.0	98.9	80.9	94.5
0920	Plaster & Gypsum Board	111.0	80.6	91.2	111.0	104.3	106.6	100.7	80.8	87.8	100.3	64.9	77.3	100.7	56.9	72.2	97.0	80.8	86.5
0950, 0980	Ceilings & Acoustic Treatment	107.4	80.6	90.8	94.7	104.3	100.6	100.4	80.8	88.3	100.4	64.9	78.4	100.4	56.9	73.5	96.2	80.8	86.7
0960	Flooring	127.7	100.2	120.3	82.1	98.2	86.5	117.2	79.9	107.1	117.2	65.4	103.2	117.2	55.9	100.6	114.0	79.9	104.7
0970, 0990	Wall Finishes & Painting/Coating	125.2	87.4	102.5	95.8	98.0	97.1	115.3	76.7	92.1	110.4	54.2	76.7	115.3	49.6	75.9	115.3	105.7	109.5
09	FINISHES	111.0	85.3	97.5	100.5	101.4	101.0	106.3	78.4	91.6	103.9	65.1	83.4	106.3	57.2	80.4	104.5	82.7	93.0
COVERS	DIVS. 10 - 14, 25, 28, 41, 43, 44	100.0	98.9	99.8	100.0	105.5	101.1	100.0	83.5	96.7	100.0	86.3	97.3	100.0	79.0	95.8	100.0	85.0	97.0
21, 22, 23	FIRE SUPPRESSION, PLUMBING & HVAC	100.1	93.8	97.6	100.2	112.4	105.1	99.9	69.7	87.8	99.9	67.2	86.9	99.9	48.1	79.3	99.9	71.9	88.7
26, 27, 3370	ELECTRICAL, COMMUNICATIONS & UTIL.	103.4	101.5	102.5	105.3	108.1	106.6	99.2	62.9	81.6	99.2	75.5	87.7	98.9	68.2	84.0	100.5	72.9	87.1
MF2004	WEIGHTED AVERAGE	104.6	92.8	99.7	100.6	105.9	102.8	99.1	76.5	89.6	98.5	73.6	88.0	98.8	63.1	83.8	102.0	80.0	92.7

FLORIDA

DIVISION		MIAMI MAT.	INST.	TOTAL	ORLANDO MAT.	INST.	TOTAL	PANAMA CITY MAT.	INST.	TOTAL	PENSACOLA MAT.	INST.	TOTAL	ST. PETERSBURG MAT.	INST.	TOTAL	TALLAHASSEE MAT.	INST.	TOTAL
015433	CONTRACTOR EQUIPMENT		93.2	93.2		98.9	98.9		98.9	98.9		98.9	98.9		98.9	98.9		98.9	98.9
0241, 31 - 34	SITE & INFRASTRUCTURE, DEMOLITION	104.8	79.0	87.2	118.7	88.3	98.0	131.2	86.4	100.6	130.8	87.9	101.5	117.0	87.8	97.0	114.3	87.6	96.0
0310	Concrete Forming & Accessories	98.8	68.8	72.9	95.5	79.4	81.6	94.8	43.6	50.6	92.8	55.9	60.9	93.6	49.9	55.9	97.3	45.9	52.9
0320	Concrete Reinforcing	92.5	80.6	87.6	92.5	79.0	87.0	96.6	48.8	77.1	99.0	48.7	78.5	95.8	55.4	79.3	92.5	51.6	75.8
0330	Cast-in-Place Concrete	104.6	75.8	93.6	121.1	76.9	104.2	97.5	46.3	77.9	120.5	62.6	98.4	104.0	61.8	87.8	102.3	54.3	83.9
03	CONCRETE	96.5	74.8	86.6	105.3	79.5	93.5	98.7	47.1	75.2	109.1	58.6	86.1	96.6	56.9	78.5	95.3	51.9	75.6
04	MASONRY	85.9	72.0	77.6	90.3	72.3	79.5	90.5	44.5	62.9	109.4	53.0	75.7	130.7	48.6	81.5	94.7	43.9	63.9
05	METALS	107.0	94.5	103.4	107.8	92.8	103.4	101.5	69.5	92.1	102.6	80.0	96.0	105.0	81.7	98.2	94.7	80.7	90.6
06	WOOD, PLASTICS & COMPOSITES	97.6	65.6	80.1	95.4	83.9	89.1	94.6	43.7	66.7	93.0	57.6	73.5	93.1	49.4	69.1	93.5	44.4	66.6
07	THERMAL & MOISTURE PROTECTION	103.2	79.5	94.0	96.2	80.1	89.9	95.5	47.4	76.8	95.5	59.2	81.3	95.0	55.6	79.7	100.5	74.5	90.4
08	OPENINGS	97.1	66.4	89.5	99.7	73.0	93.1	97.4	36.6	82.3	97.3	53.8	86.6	98.3	50.1	86.4	100.2	45.6	86.7
0920	Plaster & Gypsum Board	96.3	64.9	75.9	109.7	83.8	92.9	98.8	42.2	62.0	100.0	56.5	71.8	98.0	48.0	65.6	114.3	42.9	67.9
0950, 0980	Ceilings & Acoustic Treatment	97.1	64.9	77.2	100.4	83.8	90.2	94.6	42.2	62.2	96.2	56.5	71.7	94.6	48.0	65.8	100.4	42.9	64.8
0960	Flooring	124.4	68.9	109.4	117.4	79.9	107.2	116.6	46.9	97.8	110.9	59.4	97.0	115.6	60.3	100.7	117.2	50.3	99.1
0970, 0990	Wall Finishes & Painting/Coating	110.4	54.2	76.7	115.3	58.7	81.4	115.3	41.0	70.7	115.3	49.0	75.5	115.3	43.3	72.1	115.3	43.8	72.4
09	FINISHES	105.0	66.2	84.5	107.6	78.1	92.0	105.9	44.0	73.2	104.4	55.7	78.7	104.1	50.7	75.9	107.8	45.5	74.9
COVERS	DIVS. 10 - 14, 25, 28, 41, 43, 44	100.0	86.9	97.4	100.0	83.7	96.8	100.0	46.9	89.5	100.0	48.0	89.7	100.0	56.8	91.4	100.0	71.7	94.4
21, 22, 23	FIRE SUPPRESSION, PLUMBING & HVAC	99.9	72.1	88.8	99.8	64.3	85.7	99.9	37.8	75.2	99.9	57.1	82.8	99.9	55.6	82.3	99.8	42.2	76.9
26, 27, 3370	ELECTRICAL, COMMUNICATIONS & UTIL.	107.1	76.3	92.1	107.0	40.4	74.6	97.9	41.8	70.6	101.9	58.2	80.6	99.5	47.0	73.9	106.8	36.7	72.6
MF2004	WEIGHTED AVERAGE	101.2	75.4	90.3	103.1	72.0	90.0	100.3	48.6	78.6	102.9	61.3	85.4	102.3	58.1	83.7	99.6	53.0	80.1

FLORIDA / GEORGIA

DIVISION		TAMPA MAT.	INST.	TOTAL	ALBANY MAT.	INST.	TOTAL	ATLANTA MAT.	INST.	TOTAL	AUGUSTA MAT.	INST.	TOTAL	COLUMBUS MAT.	INST.	TOTAL	MACON MAT.	INST.	TOTAL
015433	CONTRACTOR EQUIPMENT		98.9	98.9		94.3	94.3		93.4	93.4		92.7	92.7		94.3	94.3		102.3	102.3
0241, 31 - 34	SITE & INFRASTRUCTURE, DEMOLITION	117.3	88.5	97.6	101.7	81.2	87.6	104.1	94.1	97.3	100.4	90.9	93.9	101.6	81.9	88.1	102.1	94.0	96.6
0310	Concrete Forming & Accessories	96.7	82.8	84.7	94.9	49.5	55.7	97.6	76.1	79.0	95.1	55.1	60.6	94.8	62.4	66.8	94.2	59.4	64.1
0320	Concrete Reinforcing	92.5	86.2	89.9	92.1	89.8	91.2	98.4	91.7	95.6	99.3	69.3	87.0	92.5	90.9	91.9	93.7	90.2	92.2
0330	Cast-in-Place Concrete	101.6	67.4	88.5	98.1	51.4	80.3	112.0	74.5	97.7	105.7	53.6	85.8	97.8	61.1	83.7	96.4	53.5	80.0
03	CONCRETE	94.9	79.0	87.7	93.1	59.2	77.7	101.6	78.6	91.1	97.3	57.9	79.4	93.0	68.6	81.9	92.6	64.5	79.8
04	MASONRY	87.5	73.0	78.8	89.4	45.4	63.1	87.7	73.7	79.3	87.9	45.1	62.3	88.4	64.0	73.8	101.5	43.2	66.6
05	METALS	103.9	95.5	101.5	101.0	89.0	97.5	93.6	83.5	90.7	92.4	71.5	86.3	100.5	92.8	98.2	96.1	92.0	94.9
06	WOOD, PLASTICS & COMPOSITES	97.0	86.4	91.2	94.6	44.5	67.1	97.3	77.4	86.4	94.5	55.6	73.1	94.6	61.0	76.2	103.4	59.9	79.5
07	THERMAL & MOISTURE PROTECTION	95.3	90.3	93.3	92.3	60.5	80.0	94.3	78.2	88.0	93.9	56.1	79.2	92.5	66.4	82.3	91.1	66.2	81.4
08	OPENINGS	99.7	77.8	94.3	97.1	51.0	85.7	99.1	75.4	93.2	93.3	54.8	83.8	97.1	62.1	88.4	95.2	60.5	86.5
0920	Plaster & Gypsum Board	100.7	86.4	91.4	98.0	43.0	62.3	118.8	77.0	91.7	117.7	54.3	76.6	100.5	60.1	74.3	108.5	58.9	76.3
0950, 0980	Ceilings & Acoustic Treatment	100.4	86.4	91.8	99.6	43.0	64.6	109.8	77.0	89.5	110.7	54.3	75.8	99.6	60.1	75.2	95.7	58.9	73.0
0960	Flooring	117.2	60.3	101.8	117.2	34.6	94.9	89.4	71.8	84.6	88.3	44.1	76.3	117.2	56.5	100.8	91.2	40.7	77.5
0970, 0990	Wall Finishes & Painting/Coating	115.3	61.6	83.1	110.4	46.5	72.1	102.8	86.2	92.9	102.8	42.0	66.3	110.4	54.9	77.1	112.7	51.8	76.1
09	FINISHES	106.4	76.7	90.7	103.3	44.8	72.4	100.7	76.2	87.7	100.2	51.2	74.3	103.6	59.7	80.4	92.4	54.1	72.2
COVERS	DIVS. 10 - 14, 25, 28, 41, 43, 44	100.0	84.7	97.0	100.0	81.4	96.3	100.0	85.7	97.2	100.0	59.5	92.0	100.0	83.3	96.7	100.0	81.1	96.3
21, 22, 23	FIRE SUPPRESSION, PLUMBING & HVAC	99.9	89.3	95.7	99.8	65.6	86.2	100.0	78.0	91.2	100.0	73.6	89.5	99.9	55.9	82.4	99.9	65.7	86.3
26, 27, 3370	ELECTRICAL, COMMUNICATIONS & UTIL.	99.1	47.0	73.8	94.5	58.3	76.9	98.7	76.7	88.0	99.6	68.7	84.5	94.6	61.4	78.4	93.1	63.8	78.8
MF2004	WEIGHTED AVERAGE	100.1	78.7	91.1	97.9	62.0	82.9	98.2	79.2	90.2	96.9	64.3	83.2	97.8	66.8	83.9	96.5	66.6	83.9

GEORGIA / HAWAII / IDAHO

DIVISION		SAVANNAH MAT.	INST.	TOTAL	VALDOSTA MAT.	INST.	TOTAL	HONOLULU MAT.	INST.	TOTAL	STATES & POSS., GUAM MAT.	INST.	TOTAL	BOISE MAT.	INST.	TOTAL	LEWISTON MAT.	INST.	TOTAL
015433	CONTRACTOR EQUIPMENT		95.0	95.0		94.3	94.3		101.7	101.7		166.8	166.8		101.1	101.1		97.4	97.4
0241, 31 - 34	SITE & INFRASTRUCTURE, DEMOLITION	104.1	81.4	88.6	112.2	81.1	90.9	137.0	107.0	116.5	175.5	104.4	126.9	77.2	101.1	93.6	84.6	100.5	95.4
0310	Concrete Forming & Accessories	95.0	54.7	60.2	84.2	47.4	52.4	116.5	134.3	131.9	110.8	65.8	71.9	101.1	75.5	79.0	119.6	63.6	71.2
0320	Concrete Reinforcing	93.4	69.5	83.6	94.4	48.1	75.5	105.0	112.4	108.1	185.6	35.3	124.3	102.2	72.7	90.2	109.4	95.6	103.8
0330	Cast-in-Place Concrete	114.0	52.5	90.5	96.0	54.8	80.3	174.6	120.7	154.0	177.7	117.2	154.6	96.8	81.3	90.9	108.8	90.1	101.7
03	CONCRETE	100.7	58.2	81.4	96.9	51.9	76.4	139.4	124.4	132.6	156.7	78.9	121.3	105.8	77.2	92.8	116.7	79.3	99.7
04	MASONRY	88.4	55.8	68.9	94.2	50.6	68.1	126.5	125.4	125.9	193.1	50.3	107.6	129.1	64.4	90.4	131.3	85.4	103.8
05	METALS	96.5	82.9	92.5	100.1	75.6	92.9	137.8	105.6	128.4	147.9	85.3	129.6	100.9	76.3	93.7	94.3	87.3	92.2
06	WOOD, PLASTICS & COMPOSITES	109.0	52.4	77.9	81.9	43.4	60.8	109.8	139.0	125.8	109.7	67.2	86.4	93.5	76.2	84.0	103.4	55.7	77.2
07	THERMAL & MOISTURE PROTECTION	95.1	57.8	80.6	92.7	63.3	81.2	115.3	122.7	118.2	123.5	73.1	103.9	95.4	71.0	85.9	146.2	78.1	119.7
08	OPENINGS	98.2	50.6	86.4	92.6	40.8	79.8	106.1	129.4	111.9	105.6	56.9	93.6	92.6	71.5	87.4	113.1	62.9	100.7
0920	Plaster & Gypsum Board	97.1	51.2	67.3	92.8	41.9	59.7	136.6	140.2	138.9	209.0	53.9	108.2	81.2	75.2	77.3	136.2	54.1	82.8
0950, 0980	Ceilings & Acoustic Treatment	100.4	51.2	70.0	94.6	41.9	62.0	119.7	140.2	132.3	244.1	53.9	126.5	112.2	75.2	89.3	143.8	54.1	88.3
0960	Flooring	118.0	52.0	100.1	109.3	43.4	91.5	164.8	125.5	154.2	164.5	52.0	134.1	100.3	71.3	92.5	142.5	89.7	128.2
0970, 0990	Wall Finishes & Painting/Coating	112.3	52.6	76.5	110.4	37.0	66.4	104.0	139.2	125.1	105.5	41.3	67.0	98.1	45.4	66.5	119.6	67.5	88.3
09	FINISHES	103.9	53.6	77.3	99.9	44.9	70.8	133.9	134.7	134.3	200.4	61.4	126.9	100.8	72.2	85.7	168.1	67.5	114.9
COVERS	DIVS. 10 - 14, 25, 28, 41, 43, 44	100.0	61.0	92.3	100.0	59.8	92.0	100.0	116.4	103.2	100.0	91.3	98.3	100.0	71.5	94.4	100.0	74.7	95.0
21, 22, 23	FIRE SUPPRESSION, PLUMBING & HVAC	99.9	59.6	83.9	99.9	40.8	76.4	100.2	107.3	103.0	102.8	44.9	79.8	99.9	69.2	87.7	101.0	83.4	94.0
26, 27, 3370	ELECTRICAL, COMMUNICATIONS & UTIL.	97.4	60.5	79.4	92.7	36.0	65.1	111.7	121.3	116.4	159.7	48.1	105.4	97.4	77.1	87.5	84.8	78.0	81.5
MF2004	WEIGHTED AVERAGE	98.8	62.0	83.3	97.8	51.0	78.2	118.7	118.7	118.7	138.1	64.3	107.1	100.6	74.8	89.8	109.0	80.4	97.0

IDAHO / ILLINOIS

DIVISION		POCATELLO MAT.	POCATELLO INST.	POCATELLO TOTAL	TWIN FALLS MAT.	TWIN FALLS INST.	TWIN FALLS TOTAL	CHICAGO MAT.	CHICAGO INST.	CHICAGO TOTAL	DECATUR MAT.	DECATUR INST.	DECATUR TOTAL	EAST ST. LOUIS MAT.	EAST ST. LOUIS INST.	EAST ST. LOUIS TOTAL	JOLIET MAT.	JOLIET INST.	JOLIET TOTAL
015433	CONTRACTOR EQUIPMENT		101.1	101.1		101.1	101.1		94.1	94.1		101.3	101.3		108.1	108.1		92.0	92.0
0241, 31 - 34	SITE & INFRASTRUCTURE, DEMOLITION	78.5	101.1	94.0	84.9	99.1	94.6	98.1	95.3	96.2	88.3	95.1	93.0	104.6	98.0	100.1	97.9	93.6	94.9
0310	Concrete Forming & Accessories	101.2	75.2	78.8	102.0	32.3	41.8	100.5	153.2	146.1	98.1	108.8	107.4	93.2	110.3	107.9	102.5	134.2	129.9
0320	Concrete Reinforcing	102.6	73.0	90.5	104.4	71.2	90.8	106.3	157.6	127.2	99.0	96.9	98.1	101.0	105.9	103.0	106.3	146.9	122.9
0330	Cast-in-Place Concrete	99.6	81.1	92.5	102.1	43.4	79.7	105.9	145.2	120.9	99.4	108.5	102.9	93.0	107.7	98.6	105.8	138.0	118.1
03	CONCRETE	105.2	77.0	92.4	112.9	44.8	81.9	104.5	149.8	125.1	97.3	106.6	101.5	89.1	109.2	98.2	104.6	136.9	119.3
04	MASONRY	126.4	70.6	93.0	129.2	33.8	72.1	94.4	152.2	129.0	72.1	109.4	94.5	75.6	113.5	98.3	97.2	139.5	122.5
05	METALS	108.9	76.4	99.4	108.9	73.0	98.4	90.1	130.7	102.0	95.0	104.6	97.8	92.7	117.4	99.9	88.2	124.1	98.7
06	WOOD, PLASTICS & COMPOSITES	93.5	76.2	84.0	94.1	31.3	59.6	105.2	152.3	131.1	102.9	108.2	105.8	100.2	109.1	105.1	107.0	131.5	120.5
07	THERMAL & MOISTURE PROTECTION	95.5	66.3	84.1	96.2	41.7	75.0	100.7	142.7	117.1	99.2	102.3	100.4	93.6	108.4	99.3	100.6	134.6	113.8
08	OPENINGS	93.3	65.7	86.5	96.6	38.8	82.3	101.6	155.3	114.8	96.9	105.3	99.0	86.6	113.9	93.3	99.4	141.2	109.8
0920	Plaster & Gypsum Board	78.7	75.2	76.4	78.4	28.6	46.1	92.0	154.1	132.3	98.5	108.4	104.9	95.1	109.4	104.4	89.4	132.6	117.5
0950, 0980	Ceilings & Acoustic Treatment	117.8	75.2	91.5	111.2	28.6	60.1	101.5	154.1	134.0	93.4	108.4	102.7	87.6	109.4	101.1	101.5	132.6	120.8
0960	Flooring	103.6	71.3	94.8	104.9	48.3	89.6	92.0	149.5	107.5	100.2	111.2	103.2	109.6	118.3	111.9	91.6	138.4	104.2
0970, 0990	Wall Finishes & Painting/Coating	98.0	47.2	67.5	98.1	27.8	55.9	86.5	148.0	123.4	93.0	106.6	101.2	100.6	103.3	102.2	84.6	135.3	115.0
09	FINISHES	102.8	72.4	86.7	102.2	33.8	66.0	93.8	152.7	124.9	97.1	109.9	103.8	96.2	110.9	104.0	93.2	135.2	115.4
COVERS	DIVS. 10 - 14, 25, 28, 41, 43, 44	100.0	71.5	94.4	100.0	41.3	88.4	100.0	126.8	105.3	100.0	95.4	99.1	100.0	96.9	99.4	100.0	121.3	104.2
21, 22, 23	FIRE SUPPRESSION, PLUMBING & HVAC	99.9	69.2	87.7	99.9	57.6	83.1	99.9	133.8	113.4	99.9	101.2	100.4	99.9	96.5	98.5	100.0	125.8	110.3
26, 27, 3370	ELECTRICAL, COMMUNICATIONS & UTIL.	93.8	72.8	83.6	87.9	63.6	76.1	97.5	135.8	116.1	96.5	91.2	93.9	93.5	100.0	96.7	96.3	118.0	106.9
MF2004	WEIGHTED AVERAGE	101.6	74.5	90.2	102.5	54.0	82.1	97.9	138.3	114.9	96.2	102.5	98.8	93.9	105.2	98.8	97.4	126.9	109.8

ILLINOIS / INDIANA

DIVISION		PEORIA MAT.	PEORIA INST.	PEORIA TOTAL	ROCKFORD MAT.	ROCKFORD INST.	ROCKFORD TOTAL	SPRINGFIELD MAT.	SPRINGFIELD INST.	SPRINGFIELD TOTAL	ANDERSON MAT.	ANDERSON INST.	ANDERSON TOTAL	BLOOMINGTON MAT.	BLOOMINGTON INST.	BLOOMINGTON TOTAL	EVANSVILLE MAT.	EVANSVILLE INST.	EVANSVILLE TOTAL
015433	CONTRACTOR EQUIPMENT		100.5	100.5		100.5	100.5		101.3	101.3		96.9	96.9		87.5	87.5		113.5	113.5
0241, 31 - 34	SITE & INFRASTRUCTURE, DEMOLITION	97.2	94.2	95.2	96.3	95.6	95.8	95.5	95.2	95.3	83.4	94.7	91.1	74.8	94.8	88.5	79.4	121.3	108.0
0310	Concrete Forming & Accessories	96.5	109.8	108.0	100.6	119.9	117.3	98.7	108.7	107.4	94.9	78.7	80.9	99.9	76.1	79.4	93.9	81.8	83.4
0320	Concrete Reinforcing	99.0	107.5	102.4	93.5	127.0	107.2	99.0	100.1	99.4	95.1	76.5	87.5	86.1	76.5	82.2	94.4	78.7	88.0
0330	Cast-in-Place Concrete	97.1	103.1	99.4	99.4	116.6	106.0	88.4	103.8	94.3	95.0	79.6	89.1	95.9	76.5	88.5	91.7	89.6	90.9
03	CONCRETE	96.1	107.2	101.1	96.3	120.0	107.1	92.1	105.6	98.2	90.2	79.3	85.2	97.9	76.3	88.1	98.6	84.2	92.1
04	MASONRY	112.5	109.2	110.5	87.6	110.3	101.2	76.8	110.3	96.8	85.8	80.4	82.6	86.5	74.2	79.1	82.8	84.6	83.9
05	METALS	95.0	109.6	99.2	95.0	119.6	102.2	97.4	106.0	99.9	100.2	87.8	96.6	94.3	76.1	89.0	87.0	86.1	86.8
06	WOOD, PLASTICS & COMPOSITES	102.9	106.1	104.7	102.9	117.7	111.1	104.0	107.4	105.8	106.6	78.4	91.1	115.7	75.4	93.6	95.2	80.5	87.1
07	THERMAL & MOISTURE PROTECTION	99.3	107.5	102.5	102.0	116.7	107.7	99.0	105.8	101.6	96.2	78.8	89.4	92.8	79.0	87.4	96.8	85.3	92.3
08	OPENINGS	97.1	107.5	99.6	97.1	119.8	102.7	98.1	105.7	100.0	95.6	78.1	91.3	103.1	76.5	96.5	95.5	79.1	91.5
0920	Plaster & Gypsum Board	98.5	106.2	103.5	98.5	118.3	111.4	94.9	107.6	103.1	95.7	78.1	84.3	88.8	75.2	80.0	84.7	79.1	81.1
0950, 0980	Ceilings & Acoustic Treatment	93.4	106.2	101.4	93.4	118.3	108.8	91.8	107.6	101.6	91.1	78.1	83.1	79.9	75.2	77.0	85.3	79.1	81.5
0960	Flooring	100.2	110.2	102.9	100.2	113.6	103.8	100.2	102.7	100.9	92.9	87.0	91.3	104.8	84.1	99.2	98.3	82.6	94.0
0970, 0990	Wall Finishes & Painting/Coating	93.0	112.8	104.9	93.0	122.6	110.8	93.0	106.2	100.9	95.1	69.4	79.7	91.6	85.5	87.9	97.7	82.0	88.3
09	FINISHES	97.2	110.3	104.1	97.2	119.2	108.8	96.0	107.8	102.3	91.4	79.3	85.0	93.7	78.6	85.7	92.4	81.9	86.9
COVERS	DIVS. 10 - 14, 25, 28, 41, 43, 44	100.0	102.0	100.4	100.0	102.4	100.5	100.0	95.6	99.1	100.0	89.9	98.0	100.0	80.3	96.1	100.0	85.6	97.1
21, 22, 23	FIRE SUPPRESSION, PLUMBING & HVAC	99.9	105.9	102.3	100.0	111.9	104.7	99.8	102.1	100.7	99.9	74.0	89.6	99.7	78.3	91.2	99.9	80.6	92.2
26, 27, 3370	ELECTRICAL, COMMUNICATIONS & UTIL.	95.6	95.1	95.4	96.0	122.8	109.0	101.8	96.9	99.4	83.2	93.0	88.0	98.6	78.8	89.0	93.8	87.3	90.6
MF2004	WEIGHTED AVERAGE	98.2	104.9	101.0	97.1	115.0	104.7	97.0	103.4	99.7	94.8	82.5	89.6	96.9	78.8	89.3	94.4	86.5	91.1

INDIANA

DIVISION		FORT WAYNE MAT.	FORT WAYNE INST.	FORT WAYNE TOTAL	GARY MAT.	GARY INST.	GARY TOTAL	INDIANAPOLIS MAT.	INDIANAPOLIS INST.	INDIANAPOLIS TOTAL	MUNCIE MAT.	MUNCIE INST.	MUNCIE TOTAL	SOUTH BEND MAT.	SOUTH BEND INST.	SOUTH BEND TOTAL	TERRE HAUTE MAT.	TERRE HAUTE INST.	TERRE HAUTE TOTAL
015433	CONTRACTOR EQUIPMENT		96.9	96.9		96.9	96.9		92.0	92.0		96.5	96.5		105.8	105.8		113.5	113.5
0241, 31 - 34	SITE & INFRASTRUCTURE, DEMOLITION	84.5	94.5	91.3	83.9	97.6	93.2	83.0	98.0	93.2	74.9	95.1	88.7	86.0	95.0	92.1	81.1	121.4	108.7
0310	Concrete Forming & Accessories	93.6	74.1	76.7	95.0	106.7	105.1	95.5	83.7	85.3	92.2	78.4	80.3	96.5	77.7	80.3	95.0	80.8	82.7
0320	Concrete Reinforcing	95.1	75.8	87.2	95.1	91.8	93.8	94.5	80.9	88.9	95.3	76.5	87.6	95.1	72.9	86.1	94.4	83.8	90.0
0330	Cast-in-Place Concrete	100.9	78.3	92.3	99.2	107.2	102.3	91.4	85.9	89.3	100.8	80.8	93.2	96.2	82.3	90.9	88.8	87.6	88.3
03	CONCRETE	92.9	76.6	85.5	92.2	104.0	97.6	91.7	83.8	88.1	97.5	79.5	89.3	87.4	79.9	84.0	101.0	84.0	93.3
04	MASONRY	89.4	77.8	82.5	89.2	104.4	97.4	93.5	83.6	87.6	88.5	80.4	83.6	86.0	79.1	81.8	89.4	81.2	84.5
05	METALS	100.2	86.9	96.3	100.2	98.8	99.8	102.2	80.7	95.9	95.9	87.5	93.5	100.2	98.0	99.5	87.7	88.6	88.0
06	WOOD, PLASTICS & COMPOSITES	106.3	73.1	88.1	104.5	107.0	105.9	103.5	83.3	92.4	109.8	78.2	92.5	103.7	77.1	89.1	97.3	80.2	87.9
07	THERMAL & MOISTURE PROTECTION	96.0	81.7	90.4	95.2	104.4	98.8	94.4	83.9	90.3	95.2	78.9	88.9	91.2	83.8	88.3	96.8	80.9	90.7
08	OPENINGS	95.6	72.7	90.0	95.6	104.7	97.9	103.6	83.4	98.6	97.9	78.0	93.0	88.9	76.2	85.8	96.1	80.9	92.3
0920	Plaster & Gypsum Board	94.9	72.6	80.4	90.5	107.8	101.7	94.0	83.0	86.8	84.5	78.1	80.4	88.6	76.7	80.9	84.7	78.8	80.9
0950, 0980	Ceilings & Acoustic Treatment	91.1	72.6	79.7	91.1	107.8	101.4	95.3	83.0	87.7	80.8	78.1	79.1	87.0	76.7	80.6	85.3	78.8	81.3
0960	Flooring	92.9	76.2	88.4	92.9	112.5	98.2	92.7	89.6	91.9	97.7	87.0	94.8	91.3	83.4	89.2	98.3	87.3	95.3
0970, 0990	Wall Finishes & Painting/Coating	95.1	76.1	83.7	95.1	104.9	100.9	95.1	90.8	92.5	91.6	69.4	78.3	94.6	81.4	86.7	97.7	84.1	89.6
09	FINISHES	91.2	74.6	82.5	90.7	108.0	99.8	92.5	85.6	88.9	90.9	79.1	84.7	89.0	79.1	83.8	92.4	82.3	87.1
COVERS	DIVS. 10 - 14, 25, 28, 41, 43, 44	100.0	79.4	95.9	100.0	86.2	97.3	100.0	91.7	98.4	100.0	89.3	97.9	100.0	80.2	96.1	100.0	92.8	98.6
21, 22, 23	FIRE SUPPRESSION, PLUMBING & HVAC	99.9	73.7	89.5	99.9	100.5	100.2	99.9	85.7	94.2	99.7	74.0	89.4	99.8	80.4	92.1	99.9	82.7	93.1
26, 27, 3370	ELECTRICAL, COMMUNICATIONS & UTIL.	84.0	77.4	80.8	96.0	97.4	96.7	98.3	93.0	95.7	89.1	79.7	84.5	103.7	82.7	93.5	92.1	83.7	88.0
MF2004	WEIGHTED AVERAGE	95.3	78.4	88.2	96.3	101.5	98.5	98.0	86.7	93.2	95.5	80.6	89.3	95.6	83.1	90.4	95.0	86.6	91.5

City Cost Indexes

IOWA

	DIVISION	CEDAR RAPIDS MAT.	INST.	TOTAL	COUNCIL BLUFFS MAT.	INST.	TOTAL	DAVENPORT MAT.	INST.	TOTAL	DES MOINES MAT.	INST.	TOTAL	DUBUQUE MAT.	INST.	TOTAL	SIOUX CITY MAT.	INST.	TOTAL
015433	CONTRACTOR EQUIPMENT		96.8	96.8		96.1	96.1		99.9	99.9		101.2	101.2		95.9	95.9		99.9	99.9
0241, 31 - 34	SITE & INFRASTRUCTURE, DEMOLITION	94.5	96.5	95.9	101.6	92.4	95.3	92.6	99.8	97.5	85.1	99.7	95.1	92.5	93.6	93.3	103.8	95.9	98.4
0310	Concrete Forming & Accessories	100.0	77.3	80.4	82.1	54.4	58.2	99.7	87.9	89.5	100.7	72.6	76.4	83.5	68.8	70.8	100.0	62.1	67.3
0320	Concrete Reinforcing	102.3	80.6	93.4	104.3	75.4	92.5	102.3	92.5	98.3	102.3	77.1	92.0	100.9	80.4	92.6	102.3	60.8	85.4
0330	Cast-in-Place Concrete	105.3	79.0	95.3	109.5	77.0	97.1	101.4	91.7	97.7	97.2	92.5	95.4	103.1	89.1	97.8	104.4	54.0	85.2
03	CONCRETE	100.6	79.3	90.9	103.0	67.6	86.9	98.8	90.7	95.1	95.8	81.3	89.2	97.6	79.0	89.1	100.0	60.3	81.9
04	MASONRY	102.5	75.8	86.5	102.9	76.2	86.9	98.6	84.9	90.4	94.1	69.4	79.3	103.4	67.3	81.8	96.2	53.0	70.3
05	METALS	87.7	91.0	88.7	93.1	88.5	91.8	87.7	99.1	91.3	87.7	90.6	88.7	86.5	90.6	87.7	87.7	81.0	85.8
06	WOOD, PLASTICS & COMPOSITES	108.4	76.4	90.8	87.1	47.8	65.5	108.4	86.6	96.4	108.2	72.1	88.4	88.9	68.4	77.7	108.4	62.4	83.1
07	THERMAL & MOISTURE PROTECTION	101.7	77.6	92.3	101.1	64.4	86.9	101.2	86.7	95.6	102.4	74.4	91.6	101.4	68.4	88.6	101.2	52.2	82.2
08	OPENINGS	99.0	79.8	94.2	98.0	63.1	89.4	99.0	89.5	96.6	99.0	77.0	93.6	98.0	77.4	92.9	99.0	59.9	89.3
0920	Plaster & Gypsum Board	98.5	75.8	83.8	89.9	46.2	61.5	98.5	86.2	90.5	88.9	71.2	77.4	90.3	67.6	75.6	98.5	61.1	74.2
0950, 0980	Ceilings & Acoustic Treatment	113.2	75.8	90.1	109.1	46.2	70.2	113.2	86.2	96.5	109.1	71.2	85.6	109.1	67.6	83.4	113.2	61.1	81.0
0960	Flooring	132.2	50.0	110.0	105.8	40.8	88.2	116.8	93.1	110.4	116.5	37.3	95.1	121.3	34.8	97.9	117.3	52.7	99.9
0970, 0990	Wall Finishes & Painting/Coating	105.6	73.3	86.2	98.4	70.2	81.5	102.0	88.5	93.9	102.0	76.4	86.6	104.6	73.3	85.8	103.0	61.0	77.8
09	FINISHES	114.3	71.0	91.4	103.7	51.5	76.1	109.5	88.8	98.5	106.0	65.6	84.6	108.7	62.0	84.0	111.2	60.2	84.2
COVERS	DIVS. 10 - 14, 25, 28, 41, 43, 44	100.0	84.7	97.0	100.0	83.5	96.7	100.0	88.0	97.6	100.0	83.3	96.7	100.0	81.5	96.3	100.0	80.5	96.1
21, 22, 23	FIRE SUPPRESSION, PLUMBING & HVAC	100.2	81.3	92.7	100.2	77.8	91.3	100.2	88.0	95.4	100.1	70.5	88.4	100.2	73.0	89.4	100.2	72.4	89.1
26, 27, 3370	ELECTRICAL, COMMUNICATIONS & UTIL.	96.5	78.6	87.7	101.6	82.4	92.3	93.9	87.3	90.7	109.1	72.8	91.4	100.2	75.0	87.9	96.5	71.5	84.3
MF2004	WEIGHTED AVERAGE	98.8	80.8	91.2	99.6	74.6	89.1	97.7	90.2	94.5	98.2	76.7	89.2	97.9	75.8	88.6	98.4	68.8	86.0

IOWA / KANSAS

	DIVISION	WATERLOO MAT.	INST.	TOTAL	DODGE CITY MAT.	INST.	TOTAL	KANSAS CITY MAT.	INST.	TOTAL	SALINA MAT.	INST.	TOTAL	TOPEKA MAT.	INST.	TOTAL	WICHITA MAT.	INST.	TOTAL
015433	CONTRACTOR EQUIPMENT		99.9	99.9		101.9	101.9		96.4	96.4		101.9	101.9		100.5	100.5		101.9	101.9
0241, 31 - 34	SITE & INFRASTRUCTURE, DEMOLITION	93.1	95.0	94.4	111.7	90.9	97.5	91.9	85.4	87.5	101.4	91.3	94.5	95.7	89.1	91.2	94.2	91.1	92.0
0310	Concrete Forming & Accessories	100.4	43.7	51.5	95.3	61.1	65.8	100.0	99.0	99.1	91.8	49.4	55.2	99.0	46.2	53.4	98.7	51.0	57.4
0320	Concrete Reinforcing	102.3	80.2	93.3	107.2	54.8	85.8	101.8	88.5	96.4	106.5	70.6	91.9	99.0	95.2	97.4	99.0	78.0	90.4
0330	Cast-in-Place Concrete	105.3	45.9	82.6	115.4	53.9	91.9	90.5	96.2	92.7	100.1	46.8	79.7	91.5	51.2	76.1	87.3	55.0	75.0
03	CONCRETE	100.6	53.0	79.0	113.8	58.8	88.8	95.7	96.3	96.0	101.7	54.2	80.1	93.6	58.8	77.8	91.6	58.9	76.8
04	MASONRY	97.6	55.6	72.5	104.7	51.3	72.8	105.5	94.8	99.1	119.4	45.0	74.9	99.6	58.9	75.2	93.5	58.3	72.4
05	METALS	87.7	89.0	88.1	92.4	79.3	88.6	97.2	96.9	97.1	92.3	85.6	90.3	97.4	96.0	97.0	97.4	87.2	94.4
06	WOOD, PLASTICS & COMPOSITES	109.0	43.4	73.0	96.3	67.8	80.6	102.4	100.4	101.3	92.8	52.0	70.4	98.2	43.7	68.3	99.8	49.8	72.4
07	THERMAL & MOISTURE PROTECTION	101.0	48.4	80.5	99.9	57.5	83.4	98.5	97.3	98.0	99.5	53.1	81.4	100.4	64.6	86.5	99.0	59.4	83.6
08	OPENINGS	94.7	56.2	85.2	96.8	57.9	87.2	96.8	92.5	95.0	96.7	51.9	85.6	97.1	59.6	87.8	97.1	57.1	87.2
0920	Plaster & Gypsum Board	98.5	41.4	61.4	95.6	66.5	76.7	92.1	100.3	97.4	94.9	50.2	65.8	93.8	41.5	59.8	90.9	47.9	62.9
0950, 0980	Ceilings & Acoustic Treatment	113.2	41.4	68.8	86.8	66.5	74.2	87.6	100.3	95.5	86.8	50.2	64.1	87.6	41.5	59.1	85.9	47.9	62.4
0960	Flooring	118.7	52.9	100.9	99.5	43.9	84.5	89.2	97.8	91.5	97.6	32.1	79.9	101.1	40.8	84.8	100.0	70.6	92.0
0970, 0990	Wall Finishes & Painting/Coating	103.0	32.0	60.4	93.0	44.7	64.0	100.6	77.8	86.9	93.0	44.7	64.0	93.0	60.4	73.4	93.0	55.1	70.3
09	FINISHES	110.1	43.3	74.8	96.1	57.0	75.4	94.0	95.9	95.0	94.7	45.1	68.5	95.3	44.8	68.6	94.2	54.6	73.3
COVERS	DIVS. 10 - 14, 25, 28, 41, 43, 44	100.0	74.6	95.0	100.0	46.0	89.3	100.0	74.5	94.9	100.0	56.1	91.3	100.0	59.3	91.9	100.0	58.5	91.8
21, 22, 23	FIRE SUPPRESSION, PLUMBING & HVAC	100.2	39.9	76.2	99.9	59.4	83.8	99.9	90.2	96.0	99.9	59.3	83.8	99.9	63.9	85.6	99.8	57.5	83.0
26, 27, 3370	ELECTRICAL, COMMUNICATIONS & UTIL.	96.5	64.9	81.1	96.7	71.8	84.5	101.8	98.7	100.3	96.4	71.8	84.4	108.4	72.2	90.7	103.2	71.8	87.9
MF2004	WEIGHTED AVERAGE	97.8	58.4	81.2	99.9	64.0	84.8	98.3	93.7	96.4	98.7	61.6	83.1	98.9	66.1	85.1	97.6	64.9	83.9

KENTUCKY / LOUISIANA

	DIVISION	BOWLING GREEN MAT.	INST.	TOTAL	LEXINGTON MAT.	INST.	TOTAL	LOUISVILLE MAT.	INST.	TOTAL	OWENSBORO MAT.	INST.	TOTAL	ALEXANDRIA MAT.	INST.	TOTAL	BATON ROUGE MAT.	INST.	TOTAL
015433	CONTRACTOR EQUIPMENT		94.2	94.2		101.0	101.0		94.2	94.2		113.5	113.5		90.7	90.7		89.6	89.6
0241, 31 - 34	SITE & INFRASTRUCTURE, DEMOLITION	67.9	96.2	87.2	75.3	99.3	91.7	65.8	95.9	86.4	79.4	121.8	108.4	103.3	88.4	93.1	117.8	86.5	96.4
0310	Concrete Forming & Accessories	85.2	83.5	83.7	97.8	75.8	78.8	93.5	82.0	83.6	90.6	75.0	77.1	84.2	39.7	45.8	100.0	63.7	68.7
0320	Concrete Reinforcing	85.5	88.2	86.6	94.4	93.4	94.0	94.4	93.8	94.1	85.6	91.5	88.0	101.1	63.1	85.6	100.9	63.3	85.6
0330	Cast-in-Place Concrete	83.7	106.4	92.4	90.4	84.0	87.9	94.3	78.8	88.4	86.3	103.0	92.7	93.9	42.6	74.3	91.5	62.6	80.4
03	CONCRETE	90.5	92.4	91.4	93.1	82.3	88.2	94.7	83.3	89.5	98.8	88.2	93.9	90.9	46.6	70.7	93.7	63.9	80.1
04	MASONRY	89.1	76.0	81.3	86.6	68.3	75.6	87.8	80.4	83.4	86.1	74.3	79.0	115.2	49.6	75.9	106.9	54.3	75.4
05	METALS	92.0	87.2	90.6	93.4	90.4	92.5	100.8	89.6	97.5	83.9	90.6	85.8	92.4	75.4	87.4	105.1	73.7	95.9
06	WOOD, PLASTICS & COMPOSITES	91.0	82.5	86.4	101.2	75.1	86.9	98.6	82.5	89.8	91.4	71.3	80.4	86.6	38.8	60.3	104.1	69.2	84.9
07	THERMAL & MOISTURE PROTECTION	85.3	86.6	85.8	97.1	89.0	94.0	91.1	80.8	87.1	96.8	78.5	89.7	99.3	49.9	80.1	101.3	59.7	85.1
08	OPENINGS	95.1	80.4	91.5	96.1	80.0	92.1	96.1	85.4	93.4	93.3	78.1	89.6	99.3	45.9	86.1	102.6	62.3	92.6
0920	Plaster & Gypsum Board	80.3	82.2	81.5	87.3	73.5	78.4	83.8	82.2	82.8	80.3	69.6	73.4	81.9	37.1	52.8	99.4	68.4	79.3
0950, 0980	Ceilings & Acoustic Treatment	85.3	82.2	83.4	89.5	73.5	79.6	86.1	82.2	83.7	77.0	69.6	72.4	92.0	37.1	58.0	98.6	68.4	80.0
0960	Flooring	94.1	85.0	91.7	99.0	62.2	89.1	97.5	71.4	90.5	96.6	85.0	93.5	102.0	64.0	91.7	107.2	68.7	96.8
0970, 0990	Wall Finishes & Painting/Coating	97.7	72.4	82.5	97.7	72.2	82.4	97.7	79.8	87.0	97.7	92.7	94.7	98.5	36.6	61.4	106.6	42.0	67.8
09	FINISHES	90.7	82.5	86.3	94.1	72.1	82.5	91.9	79.6	85.4	89.4	78.0	83.4	93.0	43.1	66.6	102.0	63.1	81.4
COVERS	DIVS. 10 - 14, 25, 28, 41, 43, 44	100.0	69.7	94.0	100.0	93.4	98.7	100.0	92.9	98.6	100.0	89.3	97.9	100.0	59.3	91.9	100.0	77.9	95.6
21, 22, 23	FIRE SUPPRESSION, PLUMBING & HVAC	99.9	85.3	94.1	99.9	68.9	87.6	99.9	83.4	93.4	99.9	73.0	89.2	100.1	55.4	82.3	99.9	54.4	81.8
26, 27, 3370	ELECTRICAL, COMMUNICATIONS & UTIL.	92.8	79.4	86.2	93.2	76.6	85.1	104.3	79.4	92.2	92.2	80.4	86.5	93.4	55.6	75.0	107.6	62.9	85.8
MF2004	WEIGHTED AVERAGE	93.6	84.6	89.8	95.1	78.5	88.1	97.1	83.9	91.6	93.4	83.2	89.1	97.0	56.0	79.8	102.2	63.8	86.1

LOUISIANA / MAINE

DIVISION		LAKE CHARLES MAT.	INST.	TOTAL	MONROE MAT.	INST.	TOTAL	NEW ORLEANS MAT.	INST.	TOTAL	SHREVEPORT MAT.	INST.	TOTAL	AUGUSTA MAT.	INST.	TOTAL	BANGOR MAT.	INST.	TOTAL
015433	CONTRACTOR EQUIPMENT		89.6	89.6		90.7	90.7		91.0	91.0		90.7	90.7		100.4	100.4		100.4	100.4
0241, 31 - 34	SITE & INFRASTRUCTURE, DEMOLITION	119.6	86.2	96.8	103.3	87.9	92.8	122.9	89.8	100.2	105.0	87.8	93.2	81.6	100.5	94.5	80.8	98.1	92.6
0310	Concrete Forming & Accessories	103.4	54.1	60.8	83.6	40.4	46.3	104.8	66.4	71.7	99.2	42.6	50.3	95.0	88.6	89.5	90.4	83.8	84.7
0320	Concrete Reinforcing	100.9	60.4	84.4	100.0	63.1	84.9	100.9	64.9	86.2	99.6	63.2	84.8	82.0	99.0	88.9	81.7	99.2	88.8
0330	Cast-in-Place Concrete	93.4	62.6	81.6	93.9	60.7	81.2	92.5	71.9	84.6	93.3	48.9	76.3	83.5	58.8	74.0	77.5	59.5	70.6
03	CONCRETE	94.8	59.1	78.6	90.6	53.0	73.5	94.4	68.7	82.7	90.6	49.9	72.1	100.6	79.8	91.1	97.0	77.8	88.3
04	MASONRY	105.0	51.8	73.1	109.9	50.2	74.2	106.3	59.7	78.4	101.0	51.9	71.6	97.6	45.5	66.4	111.9	52.0	76.0
05	METALS	99.3	72.2	91.4	92.4	74.3	87.1	113.3	76.7	102.6	88.6	74.2	84.4	82.7	83.3	82.9	82.4	79.5	81.6
06	WOOD, PLASTICS & COMPOSITES	108.8	56.3	80.0	85.9	39.1	60.2	105.0	69.2	85.3	103.1	42.5	69.8	99.9	97.7	98.7	94.8	90.4	92.4
07	THERMAL & MOISTURE PROTECTION	102.9	57.1	85.1	99.3	53.5	81.5	103.7	62.8	87.8	98.6	52.4	80.7	94.4	54.1	78.7	94.8	54.4	79.1
08	OPENINGS	102.6	55.8	91.0	99.3	50.2	87.2	103.4	68.1	94.7	97.2	48.0	85.0	101.7	77.4	95.7	101.6	74.9	95.0
0920	Plaster & Gypsum Board	105.2	55.1	72.6	81.5	37.5	52.9	104.0	68.4	80.9	86.3	40.9	56.8	106.9	96.8	100.4	100.4	89.3	93.2
0950, 0980	Ceilings & Acoustic Treatment	97.8	55.1	71.4	92.0	37.5	58.3	96.9	68.4	79.3	90.3	40.9	59.8	90.0	96.8	94.2	90.8	89.3	89.9
0960	Flooring	107.4	64.0	95.6	101.5	64.0	91.4	107.7	68.7	97.2	111.4	68.7	99.9	96.8	37.7	80.8	94.6	47.0	81.7
0970, 0990	Wall Finishes & Painting/Coating	106.6	38.7	65.8	98.5	43.7	65.6	108.1	63.4	81.3	98.5	36.6	61.4	91.7	31.7	55.7	91.7	29.2	54.2
09	FINISHES	102.5	54.2	77.0	92.8	44.2	67.1	102.9	66.8	83.8	96.3	46.2	69.8	97.3	74.7	85.3	96.0	72.5	83.6
COVERS	DIVS. 10 - 14, 25, 28, 41, 43, 44	100.0	76.4	95.3	100.0	54.5	91.0	100.0	79.8	96.0	100.0	69.0	93.9	100.0	58.2	91.7	100.0	70.4	94.1
21, 22, 23	FIRE SUPPRESSION, PLUMBING & HVAC	100.0	55.9	82.5	100.1	49.1	79.8	100.0	66.1	86.5	100.0	55.3	82.3	100.0	69.1	87.7	100.2	67.9	87.3
26, 27, 3370	ELECTRICAL, COMMUNICATIONS & UTIL.	95.3	61.1	78.7	95.2	56.7	76.5	96.3	68.6	82.8	102.6	68.7	86.1	98.3	80.1	89.5	95.3	80.0	87.9
MF2004	WEIGHTED AVERAGE	100.1	61.1	83.7	96.9	56.0	79.7	102.9	69.6	88.9	96.8	59.2	81.0	96.1	74.3	86.9	95.8	73.7	86.5

MAINE / MARYLAND / MASSACHUSETTS

DIVISION		LEWISTON MAT.	INST.	TOTAL	PORTLAND MAT.	INST.	TOTAL	BALTIMORE MAT.	INST.	TOTAL	HAGERSTOWN MAT.	INST.	TOTAL	BOSTON MAT.	INST.	TOTAL	BROCKTON MAT.	INST.	TOTAL
015433	CONTRACTOR EQUIPMENT		100.4	100.4		100.4	100.4		104.8	104.8		101.9	101.9		105.7	105.7		102.3	102.3
0241, 31 - 34	SITE & INFRASTRUCTURE, DEMOLITION	78.2	98.1	91.8	79.6	98.1	92.2	96.0	97.2	96.8	85.7	94.6	91.8	85.3	106.8	100.0	82.9	103.8	97.2
0310	Concrete Forming & Accessories	95.3	83.8	85.3	94.4	83.8	85.2	101.7	77.9	81.1	91.6	79.4	81.1	99.3	137.7	132.5	99.5	122.8	119.6
0320	Concrete Reinforcing	100.4	99.2	99.9	100.4	99.2	99.9	100.9	86.7	95.1	89.2	73.9	83.0	99.3	146.1	118.4	100.4	145.7	118.9
0330	Cast-in-Place Concrete	79.1	59.5	71.6	100.6	59.5	84.9	95.2	79.8	89.3	81.0	60.1	73.0	100.0	144.7	117.1	95.2	142.8	113.4
03	CONCRETE	97.6	77.8	88.6	107.6	77.8	94.0	101.3	81.3	92.2	87.0	72.8	80.5	110.3	140.6	124.1	107.9	133.2	119.4
04	MASONRY	96.5	52.0	69.8	96.8	52.0	70.0	96.9	71.5	81.7	100.6	73.3	84.3	114.9	155.6	139.3	111.4	146.4	132.4
05	METALS	85.3	79.5	83.6	86.5	79.5	84.5	97.3	97.9	97.5	93.8	91.7	93.2	96.8	126.2	105.4	93.8	123.2	102.4
06	WOOD, PLASTICS & COMPOSITES	100.6	90.4	95.0	99.8	90.4	94.6	97.9	79.9	88.0	86.7	79.8	82.9	104.1	136.6	121.9	102.9	120.1	112.3
07	THERMAL & MOISTURE PROTECTION	94.6	54.4	79.0	97.2	54.4	80.5	97.6	80.2	90.8	96.9	68.5	85.8	94.8	148.7	115.8	94.9	142.3	113.3
08	OPENINGS	104.9	74.9	97.5	104.9	74.9	97.5	93.7	85.8	91.8	90.9	78.9	88.0	100.9	141.0	110.8	98.9	129.0	106.3
0920	Plaster & Gypsum Board	105.5	89.3	95.0	111.1	89.3	96.9	104.6	79.4	88.2	99.8	79.4	86.5	108.0	137.2	127.0	103.1	120.0	114.1
0950, 0980	Ceilings & Acoustic Treatment	99.1	89.3	93.0	96.6	89.3	92.1	95.4	79.4	85.5	96.3	79.4	85.9	98.2	137.2	122.3	100.1	120.0	112.4
0960	Flooring	97.5	47.0	83.9	96.8	47.0	83.3	89.9	78.2	86.7	85.4	72.9	82.0	97.3	167.4	116.2	97.8	167.4	116.6
0970, 0990	Wall Finishes & Painting/Coating	91.7	29.2	54.2	91.7	29.2	54.2	94.6	81.0	86.4	94.6	35.7	59.3	94.4	158.2	132.7	93.9	130.3	115.7
09	FINISHES	99.1	72.5	85.1	99.0	72.5	85.0	96.0	77.8	86.4	93.4	73.7	83.0	99.1	145.7	123.7	99.0	131.6	116.3
COVERS	DIVS. 10 - 14, 25, 28, 41, 43, 44	100.0	70.4	94.1	100.0	70.4	94.1	100.0	86.0	97.2	100.0	88.7	97.8	100.0	121.9	104.3	100.0	118.1	103.6
21, 22, 23	FIRE SUPPRESSION, PLUMBING & HVAC	100.2	67.9	87.3	100.0	67.9	87.2	99.9	88.5	95.4	99.9	87.9	95.1	100.2	131.5	112.6	100.2	108.5	103.5
26, 27, 3370	ELECTRICAL, COMMUNICATIONS & UTIL.	97.2	80.0	88.8	103.7	80.0	92.2	98.2	91.7	95.0	95.4	81.4	88.6	96.3	137.1	116.2	95.4	99.5	97.4
MF2004	WEIGHTED AVERAGE	96.4	73.7	86.9	98.5	73.7	88.1	98.2	86.0	93.1	94.9	81.6	89.3	100.5	135.9	115.4	99.2	120.7	108.2

MASSACHUSETTS

DIVISION		FALL RIVER MAT.	INST.	TOTAL	HYANNIS MAT.	INST.	TOTAL	LAWRENCE MAT.	INST.	TOTAL	LOWELL MAT.	INST.	TOTAL	NEW BEDFORD MAT.	INST.	TOTAL	PITTSFIELD MAT.	INST.	TOTAL
015433	CONTRACTOR EQUIPMENT		103.2	103.2		102.3	102.3		102.3	102.3		100.4	100.4		103.2	103.2		100.4	100.4
0241, 31 - 34	SITE & INFRASTRUCTURE, DEMOLITION	82.0	103.7	96.8	79.8	103.8	96.2	83.6	103.8	97.4	82.5	103.8	97.0	80.5	103.7	96.4	83.6	101.8	96.0
0310	Concrete Forming & Accessories	99.5	122.5	119.4	91.6	122.2	118.0	99.3	122.8	119.6	96.5	122.9	119.3	99.5	122.5	119.4	96.5	103.9	102.9
0320	Concrete Reinforcing	100.4	122.5	109.4	80.5	122.4	97.6	99.5	131.4	112.5	100.4	131.6	113.1	100.4	122.5	109.4	83.3	111.1	94.7
0330	Cast-in-Place Concrete	92.1	144.9	112.3	86.9	144.3	108.8	96.0	140.7	113.1	87.3	140.8	107.7	81.1	144.9	105.5	95.2	117.3	103.7
03	CONCRETE	106.4	129.5	116.9	97.5	129.2	111.9	108.0	129.9	118.0	99.1	129.8	113.1	101.3	129.5	114.1	99.4	109.4	103.9
04	MASONRY	111.2	148.3	133.4	110.2	148.4	133.0	110.8	143.8	130.6	97.1	146.5	126.7	110.3	148.3	133.0	97.7	114.9	108.0
05	METALS	93.8	114.5	99.9	90.5	114.1	97.4	91.4	117.7	99.1	91.4	115.6	98.5	93.8	114.5	99.9	91.2	102.0	94.4
06	WOOD, PLASTICS & COMPOSITES	102.9	120.3	112.5	93.7	120.0	108.2	102.9	120.1	112.3	102.3	120.1	112.1	102.9	120.3	112.5	102.3	104.7	103.6
07	THERMAL & MOISTURE PROTECTION	94.9	137.4	111.4	94.5	138.2	111.5	94.9	141.3	112.9	94.7	141.9	113.0	94.9	137.4	111.4	94.7	111.7	101.3
08	OPENINGS	98.9	117.9	103.6	95.0	117.7	100.6	98.9	122.4	104.7	104.9	122.4	109.2	98.9	117.9	103.6	104.9	106.7	105.3
0920	Plaster & Gypsum Board	103.1	120.0	114.1	96.1	120.0	111.6	105.5	120.0	115.0	105.5	120.0	115.0	103.1	120.0	114.1	105.5	104.1	104.6
0950, 0980	Ceilings & Acoustic Treatment	100.1	120.0	112.4	91.7	120.0	109.2	99.1	120.0	112.1	99.1	120.0	112.1	100.1	120.0	112.4	99.1	104.1	102.2
0960	Flooring	97.6	167.4	116.5	94.1	167.4	113.9	96.8	167.4	115.9	96.8	167.4	115.9	97.6	167.4	116.5	97.0	131.4	106.3
0970, 0990	Wall Finishes & Painting/Coating	93.9	130.3	115.7	93.9	127.1	113.8	91.8	130.3	114.9	91.7	127.1	113.0	93.9	130.3	115.7	91.7	104.5	99.4
09	FINISHES	99.0	131.8	116.4	94.8	131.3	114.1	98.7	131.6	116.1	98.7	131.3	115.9	98.9	131.8	116.3	98.7	109.4	104.4
COVERS	DIVS. 10 - 14, 25, 28, 41, 43, 44	100.0	118.8	103.7	100.0	118.1	103.6	100.0	118.2	103.6	100.0	118.2	103.6	100.0	118.8	103.7	100.0	94.7	98.9
21, 22, 23	FIRE SUPPRESSION, PLUMBING & HVAC	100.2	108.3	103.4	100.2	108.2	103.4	100.2	112.9	105.3	100.2	122.9	109.2	100.2	108.3	103.4	100.2	87.5	95.1
26, 27, 3370	ELECTRICAL, COMMUNICATIONS & UTIL.	95.2	99.4	97.3	93.5	99.4	96.4	96.7	125.4	110.7	97.1	125.4	110.9	96.3	99.4	97.8	97.1	90.8	94.1
MF2004	WEIGHTED AVERAGE	99.0	118.9	107.3	96.2	118.7	105.7	98.9	123.6	109.3	97.7	125.7	109.5	98.4	118.9	107.0	97.8	101.0	99.1

City Cost Indexes

Table 1

015433 DIVISION		MASSACHUSETTS						MICHIGAN										
		SPRINGFIELD			WORCESTER			ANN ARBOR			DEARBORN			DETROIT			FLINT		
		MAT.	INST.	TOTAL	MAT.	INST.	TOTAL	MAT.	INST.	TOTAL	MAT.	INST.	TOTAL	MAT.	INST.	TOTAL	MAT.	INST.	TOTAL
015433	CONTRACTOR EQUIPMENT		100.4	100.4		100.4	100.4		111.3	111.3		111.3	111.3		98.5	98.5		111.3	111.3
0241, 31 - 34	SITE & INFRASTRUCTURE, DEMOLITION	83.0	102.1	96.0	82.9	103.7	97.1	77.1	96.5	90.4	76.9	96.7	90.4	90.9	98.3	95.9	67.3	95.7	86.7
0310	Concrete Forming & Accessories	96.7	105.4	104.2	97.0	126.0	122.0	98.2	110.1	108.5	98.1	119.5	116.6	99.1	119.6	116.8	101.2	95.0	95.8
0320	Concrete Reinforcing	100.4	109.0	103.9	100.4	141.6	117.2	96.1	128.4	109.3	96.1	128.7	109.4	95.4	128.7	109.0	96.1	127.9	109.1
0330	Cast-in-Place Concrete	90.8	119.2	101.7	90.3	142.6	110.3	85.0	114.5	96.3	83.1	118.0	96.4	91.3	118.0	101.5	85.6	100.4	91.3
03	CONCRETE	100.8	110.3	105.1	100.5	133.6	115.6	89.6	115.3	101.3	88.7	120.7	103.3	92.5	119.5	104.7	90.1	103.6	96.2
04	MASONRY	97.4	118.2	109.9	96.9	146.5	126.6	92.7	113.2	105.0	92.5	119.5	108.7	91.8	119.5	108.4	92.7	101.3	97.8
05	METALS	93.8	101.3	96.0	93.8	118.6	101.1	86.1	121.4	96.4	86.1	122.4	96.7	89.9	104.6	94.2	86.1	119.0	95.7
06	WOOD, PLASTICS & COMPOSITES	102.3	104.7	103.6	102.8	125.4	115.2	96.7	109.5	103.7	96.7	119.9	109.4	97.2	119.9	109.7	99.9	93.6	96.4
07	THERMAL & MOISTURE PROTECTION	94.7	113.1	101.8	94.7	134.8	110.3	102.5	111.9	106.2	101.1	121.3	109.0	99.3	121.3	107.8	100.2	96.9	99.0
08	OPENINGS	104.9	106.1	105.2	104.9	130.8	111.3	94.5	109.6	98.2	94.5	115.3	99.6	96.2	117.4	101.4	94.5	97.0	95.1
0920	Plaster & Gypsum Board	105.5	104.1	104.6	105.5	125.5	118.5	85.4	108.7	100.5	85.4	119.6	107.6	85.4	119.6	107.6	86.9	92.2	90.3
0950, 0980	Ceilings & Acoustic Treatment	99.1	104.1	102.2	99.1	125.5	115.4	88.7	108.7	101.1	88.7	119.6	107.8	89.7	119.6	108.2	88.7	92.2	90.9
0960	Flooring	96.7	131.4	106.1	96.8	163.3	114.8	92.7	120.3	100.2	92.4	126.2	101.5	92.4	126.2	101.6	92.5	82.5	89.8
0970, 0990	Wall Finishes & Painting/Coating	93.3	104.5	100.0	91.7	130.3	114.9	92.9	102.4	98.6	92.9	108.3	102.1	94.4	108.3	102.7	92.9	88.8	90.4
09	FINISHES	98.8	110.3	104.8	98.7	133.9	117.3	90.8	111.2	101.6	90.7	120.1	106.2	92.0	120.1	106.8	90.2	91.4	90.8
COVERS	DIVS. 10 - 14, 25, 28, 41, 43, 44	100.0	95.9	99.2	100.0	101.5	100.3	100.0	106.7	101.3	100.0	109.0	101.8	100.0	109.0	101.8	100.0	94.3	98.9
21, 22, 23	FIRE SUPPRESSION, PLUMBING & HVAC	100.2	101.9	100.9	100.2	105.4	102.3	99.9	99.5	99.7	99.9	110.7	104.2	99.9	112.8	105.0	99.9	101.4	100.5
26, 27, 3370	ELECTRICAL, COMMUNICATIONS & UTIL.	97.2	90.8	94.1	97.2	95.7	96.5	95.2	76.9	86.3	95.2	116.0	105.3	96.5	115.9	106.0	95.2	99.3	97.2
MF2004	WEIGHTED AVERAGE	98.4	104.5	101.0	98.3	118.9	107.0	93.7	104.5	98.2	93.5	115.5	102.7	95.3	114.2	103.2	93.4	100.8	96.5

Table 2

DIVISION		MICHIGAN															MINNESOTA		
		GRAND RAPIDS			KALAMAZOO			LANSING			MUSKEGON			SAGINAW			DULUTH		
		MAT.	INST.	TOTAL	MAT.	INST.	TOTAL	MAT.	INST.	TOTAL	MAT.	INST.	TOTAL	MAT.	INST.	TOTAL	MAT.	INST.	TOTAL
015433	CONTRACTOR EQUIPMENT		101.4	101.4		101.4	101.4		111.3	111.3		101.4	101.4		111.3	111.3		101.1	101.1
0241, 31 - 34	SITE & INFRASTRUCTURE, DEMOLITION	81.3	81.7	81.6	80.0	82.4	81.7	85.4	95.2	92.1	77.9	82.3	80.9	70.4	94.8	87.1	93.3	102.7	99.7
0310	Concrete Forming & Accessories	98.1	72.0	75.5	97.7	89.5	90.7	101.8	95.1	96.0	98.3	81.1	83.4	98.3	86.8	88.4	97.1	114.3	112.0
0320	Concrete Reinforcing	94.4	83.7	90.0	94.4	85.7	90.8	96.1	127.6	108.9	95.0	85.0	90.9	96.1	127.4	108.9	102.0	101.7	101.8
0330	Cast-in-Place Concrete	91.4	90.4	91.0	92.1	104.6	96.9	87.4	99.4	92.3	90.0	99.0	93.4	84.0	93.6	87.7	107.7	104.9	106.6
03	CONCRETE	93.6	80.4	87.6	96.0	93.5	94.9	90.9	103.3	96.5	91.6	87.7	89.8	89.2	97.6	93.0	100.8	109.3	104.7
04	MASONRY	88.8	47.0	63.8	91.9	86.7	88.8	86.5	96.5	92.5	90.5	73.9	80.6	94.0	89.5	91.3	102.0	116.9	110.9
05	METALS	93.8	80.7	90.0	92.3	85.1	90.2	84.7	118.2	94.5	90.1	83.7	88.2	86.2	117.5	95.3	89.3	119.0	98.0
06	WOOD, PLASTICS & COMPOSITES	94.9	73.7	83.2	97.3	89.9	93.2	99.3	94.1	96.4	93.7	79.6	86.0	92.6	86.2	89.1	111.8	114.3	113.2
07	THERMAL & MOISTURE PROTECTION	94.4	56.5	79.7	92.7	87.5	90.7	101.4	93.8	98.4	91.9	76.2	85.8	101.0	88.2	96.0	102.3	115.3	107.3
08	OPENINGS	94.1	66.1	87.2	90.7	83.4	88.9	94.5	97.3	95.2	88.9	76.2	86.5	92.4	93.0	92.6	94.7	116.1	100.0
0920	Plaster & Gypsum Board	84.3	68.4	74.0	85.8	85.2	85.4	83.8	92.8	89.6	72.6	74.6	73.9	85.4	84.6	84.9	94.5	115.3	108.0
0950, 0980	Ceilings & Acoustic Treatment	87.8	68.4	75.8	89.5	85.2	86.8	87.1	92.8	90.6	91.1	74.6	80.9	88.7	84.6	86.2	96.3	115.3	108.1
0960	Flooring	98.1	38.0	81.8	97.9	74.2	91.5	102.6	87.9	98.6	97.1	70.0	89.8	92.7	79.9	89.3	107.5	123.6	111.9
0970, 0990	Wall Finishes & Painting/Coating	97.7	37.1	61.4	97.7	75.3	84.3	105.2	91.7	97.1	97.0	53.8	71.1	92.9	83.8	87.4	91.3	100.9	97.1
09	FINISHES	93.1	61.8	76.5	93.4	85.2	89.1	94.8	93.3	94.0	91.2	75.6	83.0	90.5	84.8	87.5	101.7	115.2	108.9
COVERS	DIVS. 10 - 14, 25, 28, 41, 43, 44	100.0	93.8	98.8	100.0	98.9	99.8	100.0	94.5	98.9	100.0	97.5	99.5	100.0	91.6	98.3	100.0	95.1	99.0
21, 22, 23	FIRE SUPPRESSION, PLUMBING & HVAC	99.9	56.3	82.6	99.9	82.3	92.9	99.8	99.8	99.8	99.8	84.2	93.6	99.9	84.2	93.6	99.8	102.4	100.9
26, 27, 3370	ELECTRICAL, COMMUNICATIONS & UTIL.	103.2	53.6	79.1	92.6	81.8	87.3	101.2	77.2	89.5	93.2	79.1	86.3	92.9	72.8	83.1	102.5	96.9	99.7
MF2004	WEIGHTED AVERAGE	96.1	65.2	83.1	94.8	85.6	91.0	94.4	96.9	95.5	93.5	81.4	88.4	93.0	89.9	91.7	98.2	108.2	102.4

Table 3

DIVISION		MINNESOTA												MISSISSIPPI					
		MINNEAPOLIS			ROCHESTER			SAINT PAUL			ST. CLOUD			BILOXI			GREENVILLE		
		MAT.	INST.	TOTAL	MAT.	INST.	TOTAL	MAT.	INST.	TOTAL	MAT.	INST.	TOTAL	MAT.	INST.	TOTAL	MAT.	INST.	TOTAL
015433	CONTRACTOR EQUIPMENT		104.1	104.1		101.1	101.1		101.1	101.1		100.8	100.8		99.5	99.5		99.5	99.5
0241, 31 - 34	SITE & INFRASTRUCTURE, DEMOLITION	91.9	107.7	102.7	90.9	101.7	98.3	95.7	103.5	101.0	83.7	105.3	98.5	103.7	89.6	94.1	107.6	89.7	95.4
0310	Concrete Forming & Accessories	99.4	135.9	131.0	99.8	104.9	104.2	93.5	125.6	121.2	85.1	123.3	118.1	93.6	61.8	66.1	81.0	74.8	75.6
0320	Concrete Reinforcing	102.2	122.7	110.6	102.0	122.2	110.3	98.4	122.6	108.3	103.3	122.2	111.0	92.5	63.2	80.6	100.3	70.6	88.1
0330	Cast-in-Place Concrete	111.9	121.2	115.4	109.7	101.3	106.5	113.9	120.4	116.4	95.6	118.8	104.5	106.8	60.3	89.0	107.2	63.7	90.6
03	CONCRETE	104.4	128.4	115.3	101.9	107.8	104.6	106.4	123.6	114.2	91.9	121.0	105.5	97.1	63.2	81.7	100.6	71.4	87.3
04	MASONRY	100.9	131.8	119.4	100.1	115.7	109.5	109.8	131.8	123.0	105.0	124.3	116.5	88.7	50.1	65.6	129.1	56.0	85.3
05	METALS	91.8	131.8	103.5	89.2	129.7	101.0	88.8	131.3	101.2	90.7	129.1	101.9	97.6	87.9	94.7	95.7	91.2	94.4
06	WOOD, PLASTICS & COMPOSITES	115.2	135.9	126.6	115.2	102.8	108.4	107.8	122.3	115.7	94.4	120.8	108.9	94.9	65.7	78.8	78.9	79.7	79.4
07	THERMAL & MOISTURE PROTECTION	102.5	133.8	114.7	102.6	102.3	102.5	103.6	131.8	114.6	101.1	113.7	106.0	92.6	62.2	80.7	92.7	68.8	83.4
08	OPENINGS	98.0	142.0	108.9	94.7	124.0	102.0	92.3	134.5	102.8	91.2	133.8	101.8	97.1	66.2	89.4	96.4	71.8	90.3
0920	Plaster & Gypsum Board	98.9	137.5	124.0	98.5	103.4	101.7	91.8	123.6	112.4	83.9	122.2	108.8	104.0	64.9	78.6	92.8	79.5	84.2
0950, 0980	Ceilings & Acoustic Treatment	99.7	137.5	123.1	98.0	103.4	101.4	93.0	123.6	111.9	69.9	122.2	102.2	100.4	64.9	78.5	94.6	79.5	85.2
0960	Flooring	105.3	126.6	111.0	107.7	92.5	103.6	101.2	126.6	108.0	105.0	126.6	110.9	117.2	59.2	101.5	108.0	59.2	94.8
0970, 0990	Wall Finishes & Painting/Coating	98.9	128.0	116.4	94.2	101.5	98.6	98.9	119.8	111.4	103.2	128.0	118.1	110.4	49.9	74.1	110.4	75.7	89.6
09	FINISHES	102.7	134.3	119.4	102.6	102.8	102.7	99.4	125.3	113.1	93.2	124.8	109.9	104.3	63.8	83.1	99.1	72.9	85.3
COVERS	DIVS. 10 - 14, 25, 28, 41, 43, 44	100.0	106.0	101.2	100.0	96.9	99.4	100.0	104.0	100.8	100.0	98.8	99.8	100.0	69.6	94.0	100.0	73.5	94.8
21, 22, 23	FIRE SUPPRESSION, PLUMBING & HVAC	99.8	117.1	106.7	99.8	98.0	99.1	99.9	112.5	104.8	99.5	115.8	106.0	99.9	50.0	80.0	99.9	57.4	83.0
26, 27, 3370	ELECTRICAL, COMMUNICATIONS & UTIL.	101.9	115.3	108.5	100.6	85.7	93.4	100.1	102.4	101.2	100.8	102.4	101.6	99.9	55.5	78.3	99.2	73.2	86.5
MF2004	WEIGHTED AVERAGE	99.3	124.2	109.8	98.1	104.7	100.9	98.6	118.7	107.0	95.8	117.5	104.9	98.5	62.6	83.4	100.0	70.9	87.8

City Cost Indexes

MISSISSIPPI / MISSOURI

	DIVISION	JACKSON			MERIDIAN			CAPE GIRARDEAU			COLUMBIA			JOPLIN			KANSAS CITY		
		MAT.	INST.	TOTAL	MAT.	INST.	TOTAL	MAT.	INST.	TOTAL	MAT.	INST.	TOTAL	MAT.	INST.	TOTAL	MAT.	INST.	TOTAL
015433	CONTRACTOR EQUIPMENT		99.5	99.5		99.5	99.5		107.1	107.1		108.1	108.1		105.9	105.9		100.9	100.9
0241, 31 - 34	SITE & INFRASTRUCTURE, DEMOLITION	98.9	89.7	92.6	98.7	90.0	92.7	93.4	93.9	93.7	102.8	95.7	97.9	104.2	98.8	100.5	96.7	93.5	94.5
0310	Concrete Forming & Accessories	93.0	67.4	70.9	79.8	74.1	74.9	86.6	74.2	75.9	88.2	66.7	69.6	104.3	69.2	74.0	103.3	107.7	107.1
0320	Concrete Reinforcing	92.5	61.7	79.9	99.0	61.7	83.8	105.1	83.1	96.1	102.9	113.6	107.3	100.8	70.9	88.6	95.8	114.7	103.5
0330	Cast-in-Place Concrete	99.2	63.7	85.6	101.1	66.3	87.8	94.2	83.8	90.2	89.0	85.3	87.6	100.3	69.9	88.7	93.2	107.9	98.8
03	CONCRETE	93.5	66.5	81.3	93.6	70.4	83.1	93.6	80.9	87.8	86.6	83.8	85.3	97.9	71.0	85.6	93.9	109.3	100.9
04	MASONRY	89.8	56.0	69.6	88.4	60.6	71.8	119.3	77.3	94.1	131.0	84.4	103.1	93.1	58.8	72.6	98.8	107.6	104.1
05	METALS	97.5	87.6	94.6	95.6	87.8	93.3	94.2	104.9	97.4	90.0	118.4	98.3	94.9	87.6	92.8	102.4	113.4	105.6
06	WOOD, PLASTICS & COMPOSITES	93.7	69.8	80.6	77.8	76.3	77.0	88.5	72.1	79.5	94.8	59.6	75.5	104.2	68.2	84.4	103.5	107.6	105.8
07	THERMAL & MOISTURE PROTECTION	94.4	67.7	84.0	92.3	70.4	83.8	102.0	78.1	92.7	93.7	81.5	88.9	96.6	64.8	84.2	96.3	106.2	100.1
08	OPENINGS	97.4	63.9	89.2	96.4	73.9	90.8	98.8	72.5	92.3	91.4	86.9	90.3	91.5	70.4	86.3	97.5	110.3	100.7
0920	Plaster & Gypsum Board	96.8	69.2	78.9	92.8	75.9	81.8	95.2	71.0	79.5	93.6	58.1	70.5	97.0	66.8	77.4	93.7	107.7	102.8
0950, 0980	Ceilings & Acoustic Treatment	97.1	69.2	79.8	94.6	75.9	83.0	86.1	71.0	76.8	87.6	58.1	69.3	85.0	66.8	73.8	90.8	107.7	101.3
0960	Flooring	117.6	59.2	101.8	106.9	59.2	94.0	86.9	67.2	81.5	106.5	88.5	101.7	122.7	45.2	101.7	98.9	99.9	99.2
0970, 0990	Wall Finishes & Painting/Coating	110.4	75.7	89.6	110.4	87.5	96.7	98.3	77.8	86.0	100.6	75.5	85.5	94.6	40.3	62.0	99.5	120.1	111.9
09	FINISHES	102.7	67.1	83.9	98.1	73.4	85.0	90.8	72.3	81.1	94.9	68.1	80.7	103.1	61.8	81.3	98.3	107.5	103.2
COVERS	DIVS. 10 - 14, 25, 28, 41, 43, 44	100.0	72.3	94.5	100.0	74.7	95.0	100.0	60.5	92.2	100.0	94.1	98.8	100.0	69.5	94.0	100.0	99.4	99.9
21, 22, 23	FIRE SUPPRESSION, PLUMBING & HVAC	99.9	63.4	85.4	99.9	65.9	86.4	99.9	89.9	95.9	99.9	91.3	96.5	100.1	52.6	81.2	100.0	104.1	101.6
26, 27, 3370	ELECTRICAL, COMMUNICATIONS & UTIL.	104.6	73.2	89.3	98.9	58.9	79.4	100.9	108.3	104.5	96.0	85.3	90.8	90.9	68.8	80.1	101.2	105.2	103.1
MF2004	WEIGHTED AVERAGE	98.5	69.9	86.5	96.8	70.9	85.9	98.2	87.4	93.7	96.3	88.1	92.9	97.1	68.0	84.9	99.2	106.1	102.1

MISSOURI / MONTANA

	DIVISION	SPRINGFIELD			ST. JOSEPH			ST. LOUIS			BILLINGS			BUTTE			GREAT FALLS		
		MAT.	INST.	TOTAL	MAT.	INST.	TOTAL	MAT.	INST.	TOTAL	MAT.	INST.	TOTAL	MAT.	INST.	TOTAL	MAT.	INST.	TOTAL
015433	CONTRACTOR EQUIPMENT		101.2	101.2		99.5	99.5		108.0	108.0		99.1	99.1		98.7	98.7		98.7	98.7
0241, 31 - 34	SITE & INFRASTRUCTURE, DEMOLITION	95.9	91.6	92.9	98.2	88.6	91.7	93.2	97.5	96.1	91.0	97.7	95.6	99.9	96.7	97.7	103.6	97.5	99.4
0310	Concrete Forming & Accessories	101.2	71.7	75.7	103.1	79.4	82.6	98.8	105.3	104.4	96.9	65.3	69.6	85.4	69.7	71.8	96.8	66.7	70.8
0320	Concrete Reinforcing	99.0	113.3	104.8	94.6	104.7	98.7	96.1	109.4	101.5	102.3	70.6	89.4	111.0	70.7	94.6	102.3	70.5	89.3
0330	Cast-in-Place Concrete	96.6	62.9	83.8	93.2	103.3	97.0	94.2	112.3	101.1	108.7	69.2	93.6	119.7	70.9	101.0	126.5	68.4	104.3
03	CONCRETE	96.1	77.4	87.6	93.6	93.4	93.5	92.5	109.3	100.2	102.3	68.8	87.1	106.2	71.3	90.4	110.3	69.1	91.6
04	MASONRY	88.7	78.1	82.4	98.4	90.3	93.6	99.5	110.3	106.0	120.5	71.1	90.9	116.3	74.4	91.2	120.3	73.2	92.1
05	METALS	93.8	105.3	97.2	98.6	107.2	101.1	98.9	119.5	104.9	106.1	86.5	100.3	98.9	86.8	95.4	102.8	86.5	98.1
06	WOOD, PLASTICS & COMPOSITES	103.2	72.6	86.4	104.2	74.6	88.0	102.4	102.5	102.5	102.6	65.0	82.0	90.4	70.0	79.2	103.8	65.0	82.5
07	THERMAL & MOISTURE PROTECTION	98.2	73.3	88.5	96.6	92.3	94.9	101.9	107.8	104.2	101.3	69.3	88.9	101.2	73.8	90.5	101.7	68.2	88.7
08	OPENINGS	97.1	81.0	93.1	96.2	89.9	94.7	97.7	109.7	100.7	97.7	62.6	89.0	95.6	65.8	88.2	99.0	63.1	90.1
0920	Plaster & Gypsum Board	96.9	71.5	80.4	98.4	73.5	82.2	103.2	102.6	102.8	93.6	64.1	74.4	91.0	69.3	76.9	98.5	64.1	76.1
0950, 0980	Ceilings & Acoustic Treatment	87.6	71.5	77.7	90.0	73.5	79.8	91.1	102.6	98.2	106.7	64.1	80.3	111.6	69.3	85.4	113.2	64.1	82.8
0960	Flooring	109.2	45.2	91.9	102.0	97.0	100.7	94.0	102.4	96.3	110.1	61.4	96.9	106.9	83.2	100.5	115.5	83.2	106.7
0970, 0990	Wall Finishes & Painting/Coating	95.0	57.7	72.6	95.1	85.4	89.3	98.3	109.2	104.8	100.5	53.8	72.5	98.4	58.9	74.7	98.4	55.9	72.9
09	FINISHES	98.4	64.9	80.7	99.4	81.8	90.1	95.3	104.3	100.1	105.0	62.9	82.8	104.5	71.1	86.9	108.8	68.2	87.3
COVERS	DIVS. 10 - 14, 25, 28, 41, 43, 44	100.0	89.1	97.8	100.0	92.8	98.6	100.0	103.9	100.8	100.0	73.6	94.8	100.0	74.7	95.0	100.0	74.9	95.0
21, 22, 23	FIRE SUPPRESSION, PLUMBING & HVAC	99.9	68.3	87.4	100.1	90.1	96.1	99.9	108.7	103.4	100.1	76.1	90.6	100.2	69.9	88.2	100.2	74.3	89.9
26, 27, 3370	ELECTRICAL, COMMUNICATIONS & UTIL.	99.7	73.1	86.7	99.0	84.0	91.7	103.6	111.7	107.6	95.1	75.5	85.6	102.4	72.6	87.9	95.2	72.5	84.1
MF2004	WEIGHTED AVERAGE	97.3	77.6	89.1	98.3	90.2	94.9	98.6	108.8	102.9	101.9	74.6	90.4	101.6	75.0	90.4	103.1	74.8	91.2

MONTANA / NEBRASKA

	DIVISION	HELENA			MISSOULA			GRAND ISLAND			LINCOLN			NORTH PLATTE			OMAHA		
		MAT.	INST.	TOTAL	MAT.	INST.	TOTAL	MAT.	INST.	TOTAL	MAT.	INST.	TOTAL	MAT.	INST.	TOTAL	MAT.	INST.	TOTAL
015433	CONTRACTOR EQUIPMENT		98.7	98.7		98.7	98.7		100.5	100.5		100.5	100.5		100.5	100.5		91.9	91.9
0241, 31 - 34	SITE & INFRASTRUCTURE, DEMOLITION	100.7	96.6	97.9	81.3	96.6	91.8	99.7	90.7	93.6	89.3	90.7	90.3	100.6	90.2	93.5	80.1	90.7	87.3
0310	Concrete Forming & Accessories	97.1	66.1	70.3	88.2	61.0	64.7	97.9	82.8	84.8	102.5	65.5	70.5	97.8	81.6	83.8	94.4	72.4	75.4
0320	Concrete Reinforcing	105.8	70.5	91.4	112.9	80.2	99.6	107.8	73.6	93.8	99.0	73.0	88.4	109.2	74.3	95.0	103.6	73.2	91.2
0330	Cast-in-Place Concrete	106.7	67.0	91.5	88.1	66.0	79.7	116.1	74.6	100.2	92.6	77.3	86.7	116.1	71.2	98.9	99.3	79.2	91.6
03	CONCRETE	101.8	68.4	86.6	86.0	67.5	77.6	108.2	78.8	94.8	94.3	72.1	84.2	108.5	77.2	94.3	97.1	75.3	87.2
04	MASONRY	113.0	71.6	88.2	139.5	69.1	97.3	107.5	84.8	93.9	96.6	74.6	83.4	92.2	97.1	95.2	101.8	80.2	88.9
05	METALS	102.0	86.4	97.5	95.9	90.8	94.4	90.1	87.7	89.4	95.0	86.8	92.6	90.3	88.4	89.8	94.3	78.7	89.8
06	WOOD, PLASTICS & COMPOSITES	103.9	65.0	82.6	94.0	60.6	75.6	99.8	84.6	91.4	102.7	61.5	80.1	99.6	84.6	91.4	93.7	71.3	81.4
07	THERMAL & MOISTURE PROTECTION	101.4	73.7	90.6	100.5	70.8	89.0	99.0	80.5	91.8	98.5	75.3	89.5	99.0	87.9	94.6	94.4	82.9	89.9
08	OPENINGS	98.5	61.8	89.4	95.6	63.1	87.6	90.7	77.9	87.5	96.2	61.0	87.5	90.0	77.2	86.8	99.3	71.2	92.4
0920	Plaster & Gypsum Board	91.2	64.1	73.6	91.7	59.4	70.8	94.9	83.9	87.8	91.4	60.0	71.0	95.1	83.9	87.8	93.7	70.7	78.8
0950, 0980	Ceilings & Acoustic Treatment	108.2	64.1	80.9	111.6	59.4	79.3	86.8	83.9	85.0	89.3	60.0	71.2	87.6	83.9	85.3	102.1	70.7	82.7
0960	Flooring	115.5	66.3	102.2	109.0	83.2	102.0	98.5	120.2	104.3	100.2	93.7	98.4	98.4	112.5	102.2	128.1	54.5	108.2
0970, 0990	Wall Finishes & Painting/Coating	98.4	49.7	69.2	98.4	54.7	72.2	93.0	77.7	83.8	93.0	89.5	90.9	93.0	73.3	81.2	158.4	71.7	106.4
09	FINISHES	106.4	64.2	84.1	104.2	63.8	82.8	95.2	88.6	91.7	95.5	72.6	83.4	95.4	86.1	90.5	116.0	68.3	90.8
COVERS	DIVS. 10 - 14, 25, 28, 41, 43, 44	100.0	86.3	97.3	100.0	83.7	96.8	100.0	88.2	97.7	100.0	85.5	97.1	100.0	82.2	96.5	100.0	85.3	97.1
21, 22, 23	FIRE SUPPRESSION, PLUMBING & HVAC	100.1	70.3	88.3	100.2	66.6	86.8	99.9	80.1	92.1	99.8	80.1	92.0	99.9	78.2	91.3	99.7	80.0	91.8
26, 27, 3370	ELECTRICAL, COMMUNICATIONS & UTIL.	101.0	72.5	87.1	100.3	76.7	88.8	89.9	74.0	82.1	103.8	74.0	89.3	91.8	89.9	90.9	103.6	83.2	93.7
MF2004	WEIGHTED AVERAGE	101.8	73.5	89.9	99.0	73.2	88.2	97.3	82.4	91.0	97.6	77.1	89.0	96.8	85.0	91.8	99.5	78.8	90.8

NEVADA / NEW HAMPSHIRE

DIVISION		CARSON CITY			LAS VEGAS			RENO			MANCHESTER			NASHUA			PORTSMOUTH		
		MAT.	INST.	TOTAL	MAT.	INST.	TOTAL	MAT.	INST.	TOTAL	MAT.	INST.	TOTAL	MAT.	INST.	TOTAL	MAT.	INST.	TOTAL
015433	CONTRACTOR EQUIPMENT		101.1	101.1		101.1	101.1		101.1	101.1		100.4	100.4		100.4	100.4		100.4	100.4
0241, 31 - 34	SITE & INFRASTRUCTURE, DEMOLITION	68.7	103.0	92.2	63.5	106.4	92.9	61.8	103.0	90.0	82.1	98.4	93.2	84.7	98.4	94.1	79.1	99.7	93.2
0310	Concrete Forming & Accessories	101.7	92.8	94.0	103.7	110.5	109.6	98.7	92.9	93.7	94.5	81.3	83.1	96.7	81.3	83.4	86.0	83.7	84.0
0320	Concrete Reinforcing	107.5	113.2	109.8	99.6	125.4	110.2	102.1	124.4	111.2	100.4	93.2	97.4	100.4	93.2	97.4	80.3	93.3	85.6
0330	Cast-in-Place Concrete	109.5	91.2	102.5	106.6	115.0	109.8	116.4	91.2	106.8	101.0	117.1	107.2	88.2	117.1	99.3	83.6	120.9	97.9
03	CONCRETE	110.8	96.1	104.1	108.0	114.4	110.9	112.8	98.2	106.1	107.8	96.0	102.4	101.9	96.0	99.2	92.7	98.5	95.4
04	MASONRY	123.0	78.0	96.0	117.2	103.7	109.1	123.8	78.0	96.4	98.0	87.9	91.9	97.8	87.9	91.8	93.3	93.1	93.2
05	METALS	94.9	101.3	96.7	101.1	110.0	103.7	95.3	105.6	98.3	93.8	90.6	92.9	93.8	90.6	92.8	90.4	93.6	91.4
06	WOOD, PLASTICS & COMPOSITES	93.3	93.6	93.5	93.3	108.8	101.8	89.4	93.6	91.7	98.9	80.4	88.7	102.3	80.4	90.3	89.6	80.4	84.6
07	THERMAL & MOISTURE PROTECTION	99.4	87.1	94.6	115.4	102.4	110.4	100.8	87.1	95.5	94.2	92.3	93.4	94.8	92.3	93.8	94.5	113.5	101.9
08	OPENINGS	93.3	106.4	96.5	94.8	117.7	100.5	93.6	106.6	96.8	104.9	77.3	98.1	104.9	77.3	98.1	105.5	72.2	97.3
0920	Plaster & Gypsum Board	78.5	93.2	88.1	84.5	109.0	100.4	78.7	93.2	88.2	108.7	78.9	89.4	105.5	78.9	88.3	97.0	78.9	85.3
0950, 0980	Ceilings & Acoustic Treatment	107.8	93.2	98.8	121.2	109.0	113.6	117.8	93.2	102.6	96.6	78.9	85.7	99.1	78.9	86.6	90.8	78.9	83.5
0960	Flooring	104.9	57.3	92.0	98.8	98.1	98.6	104.5	57.3	91.8	97.9	100.4	98.5	96.8	100.4	97.8	91.8	100.4	94.1
0970, 0990	Wall Finishes & Painting/Coating	99.2	87.7	92.3	100.4	118.4	111.2	98.1	87.7	91.9	91.7	96.7	94.7	91.7	96.7	94.7	91.7	96.7	94.7
09	FINISHES	101.0	85.1	92.6	102.4	108.7	105.7	102.4	85.1	93.2	98.7	86.0	92.0	99.1	86.0	92.2	94.1	87.2	90.5
COVERS	DIVS. 10 - 14, 25, 28, 41, 43, 44	100.0	109.7	101.9	100.0	104.5	100.9	100.0	109.7	101.9	100.0	69.9	94.0	100.0	69.9	94.0	100.0	71.7	94.4
21, 22, 23	FIRE SUPPRESSION, PLUMBING & HVAC	99.9	87.3	94.9	100.0	110.2	104.1	99.9	87.5	95.0	100.0	85.0	94.1	100.2	85.0	94.2	100.2	87.7	95.2
26, 27, 3370	ELECTRICAL, COMMUNICATIONS & UTIL.	98.8	99.4	99.1	99.0	128.0	113.1	94.5	99.4	96.9	96.7	83.4	90.2	97.0	83.4	90.4	95.2	83.4	89.5
MF2004	WEIGHTED AVERAGE	99.9	93.1	97.1	101.1	112.0	105.7	99.8	93.9	97.3	99.1	87.7	94.3	98.6	87.7	94.0	95.9	90.1	93.5

NEW JERSEY

DIVISION		CAMDEN			ELIZABETH			JERSEY CITY			NEWARK			PATERSON			TRENTON		
		MAT.	INST.	TOTAL	MAT.	INST.	TOTAL	MAT.	INST.	TOTAL	MAT.	INST.	TOTAL	MAT.	INST.	TOTAL	MAT.	INST.	TOTAL
015433	CONTRACTOR EQUIPMENT		98.5	98.5		100.4	100.4		98.5	98.5		100.4	100.4		100.4	100.4		98.1	98.1
0241, 31 - 34	SITE & INFRASTRUCTURE, DEMOLITION	87.9	103.7	98.7	100.8	104.6	103.4	87.9	104.4	99.2	103.4	104.6	104.2	99.1	104.5	102.8	87.6	104.3	99.0
0310	Concrete Forming & Accessories	96.7	120.5	117.2	104.7	121.1	118.9	96.7	121.2	117.9	93.3	121.2	117.4	95.5	121.1	117.6	93.9	120.8	117.1
0320	Concrete Reinforcing	100.4	114.2	106.0	77.3	119.5	94.5	100.4	119.5	108.2	100.4	119.5	108.2	100.4	119.5	108.2	100.4	113.2	105.6
0330	Cast-in-Place Concrete	76.7	123.8	94.7	84.5	125.1	100.0	76.7	125.2	95.2	98.1	125.1	108.4	97.8	125.1	108.2	87.3	125.0	101.7
03	CONCRETE	96.5	119.4	106.9	97.8	121.4	108.5	96.5	121.2	107.8	106.3	121.4	113.2	106.3	121.3	113.2	101.4	119.8	109.7
04	MASONRY	91.2	120.3	108.6	110.7	120.3	116.4	88.6	120.3	107.6	99.9	120.3	112.1	95.4	120.3	110.3	87.2	120.3	107.0
05	METALS	93.6	102.4	96.2	90.2	108.8	95.6	93.7	106.5	97.4	93.7	108.9	98.1	88.9	108.8	94.7	88.9	102.3	92.8
06	WOOD, PLASTICS & COMPOSITES	102.3	120.8	112.5	116.2	120.8	118.8	102.3	120.8	112.5	100.9	120.8	111.8	104.1	120.8	113.3	99.3	120.8	111.1
07	THERMAL & MOISTURE PROTECTION	94.4	120.6	104.6	95.1	125.8	107.0	94.4	125.8	106.6	94.6	125.8	106.7	95.0	120.7	105.0	90.6	120.7	102.3
08	OPENINGS	104.9	117.4	108.0	105.9	119.9	109.4	104.9	119.9	108.6	110.9	119.9	113.1	110.9	119.9	113.1	105.4	117.1	108.3
0920	Plaster & Gypsum Board	105.5	120.8	115.5	109.3	120.8	116.8	105.5	120.8	115.5	108.7	120.8	116.6	105.5	120.8	115.5	108.7	120.8	116.6
0950, 0980	Ceilings & Acoustic Treatment	99.1	120.8	112.5	90.8	120.8	109.4	99.1	120.8	112.5	96.6	120.8	111.6	99.1	120.8	112.5	96.6	120.8	111.6
0960	Flooring	96.8	136.7	107.6	101.1	151.8	114.8	96.8	151.8	111.7	97.0	151.8	111.8	96.8	151.8	111.7	97.2	151.8	111.9
0970, 0990	Wall Finishes & Painting/Coating	91.7	124.2	111.2	91.6	126.9	112.8	91.7	126.9	112.8	91.6	126.9	112.8	91.6	126.9	112.8	91.7	124.2	111.2
09	FINISHES	99.5	124.6	112.8	100.1	127.5	114.6	99.5	127.5	114.3	99.6	127.5	114.4	99.6	127.5	114.4	99.3	127.2	114.0
COVERS	DIVS. 10 - 14, 25, 28, 41, 43, 44	100.0	110.2	102.0	100.0	118.0	103.6	100.0	118.0	103.6	100.0	118.0	103.6	100.0	118.0	103.6	100.0	110.1	102.0
21, 22, 23	FIRE SUPPRESSION, PLUMBING & HVAC	100.2	113.9	105.6	100.2	120.6	108.3	100.2	122.8	109.2	100.2	120.6	108.3	100.2	122.8	109.2	100.3	121.8	108.8
26, 27, 3370	ELECTRICAL, COMMUNICATIONS & UTIL.	97.5	134.0	115.3	95.0	137.5	115.6	98.8	141.1	119.4	99.0	141.1	119.5	98.8	137.5	117.6	101.4	139.2	119.8
MF2004	WEIGHTED AVERAGE	97.8	117.8	106.2	98.6	121.5	108.2	97.8	122.2	108.0	100.5	122.0	109.5	99.3	121.8	108.8	97.6	120.6	107.3

NEW MEXICO / NEW YORK

DIVISION		ALBUQUERQUE			FARMINGTON			LAS CRUCES			ROSWELL			SANTA FE			ALBANY		
		MAT.	INST.	TOTAL	MAT.	INST.	TOTAL	MAT.	INST.	TOTAL	MAT.	INST.	TOTAL	MAT.	INST.	TOTAL	MAT.	INST.	TOTAL
015433	CONTRACTOR EQUIPMENT		112.1	112.1		112.1	112.1		90.9	90.9		112.1	112.1		112.1	112.1		112.9	112.9
0241, 31 - 34	SITE & INFRASTRUCTURE, DEMOLITION	78.2	108.0	98.6	84.5	108.0	100.6	89.2	91.1	90.5	89.0	108.0	102.0	84.0	108.0	100.4	73.4	105.3	95.2
0310	Concrete Forming & Accessories	99.8	69.1	73.3	99.8	69.1	73.3	97.0	67.4	71.4	99.8	68.8	73.1	98.9	69.1	73.2	96.1	87.4	88.6
0320	Concrete Reinforcing	102.8	70.9	89.8	112.3	70.9	95.4	106.4	50.6	83.7	111.4	50.9	86.7	110.2	70.9	94.2	99.1	94.3	97.1
0330	Cast-in-Place Concrete	108.9	77.6	96.9	109.9	77.6	97.6	95.6	69.2	85.5	101.3	77.5	92.2	125.5	77.6	107.2	83.4	101.9	90.5
03	CONCRETE	109.5	73.5	93.1	113.8	73.5	95.5	91.2	65.7	79.6	114.0	69.6	93.8	118.7	73.5	98.1	99.0	94.8	97.1
04	MASONRY	113.9	65.3	84.8	120.3	65.3	87.3	107.2	62.7	80.6	121.7	65.3	87.9	116.6	65.3	85.9	91.8	99.1	96.2
05	METALS	101.4	89.8	98.0	99.1	89.8	96.4	99.0	73.7	91.6	99.1	80.8	93.7	99.1	89.8	96.4	94.7	105.7	97.9
06	WOOD, PLASTICS & COMPOSITES	93.4	69.5	80.3	93.5	69.5	80.3	84.5	68.3	75.6	93.5	69.5	80.3	92.8	69.5	80.0	95.1	84.5	89.3
07	THERMAL & MOISTURE PROTECTION	100.3	74.8	90.4	100.6	74.8	90.6	86.5	69.1	79.7	100.9	74.8	90.7	100.2	74.8	90.3	89.5	91.8	90.4
08	OPENINGS	93.4	71.8	88.1	96.2	71.8	90.1	86.0	64.9	80.8	92.2	65.6	85.6	92.4	71.8	87.3	96.0	82.8	92.7
0920	Plaster & Gypsum Board	83.2	68.0	73.3	76.1	68.0	70.8	78.3	68.0	71.6	76.1	68.0	70.8	84.5	68.0	73.8	107.9	83.7	92.2
0950, 0980	Ceilings & Acoustic Treatment	112.4	68.0	84.9	107.8	68.0	83.2	101.3	68.0	80.7	107.8	68.0	83.2	107.8	68.0	83.2	92.4	83.7	87.0
0960	Flooring	102.1	69.0	93.2	104.5	69.0	94.9	138.2	65.2	118.4	104.5	69.0	94.9	104.5	69.0	94.9	83.8	86.7	84.6
0970, 0990	Wall Finishes & Painting/Coating	102.6	54.9	74.0	98.1	54.9	72.2	92.8	54.9	69.3	98.1	54.9	72.2	98.1	54.9	72.2	82.4	79.3	80.5
09	FINISHES	101.7	67.5	83.6	100.4	67.5	83.0	115.2	65.8	89.0	100.8	67.5	83.2	101.6	67.5	83.6	94.3	85.8	89.8
COVERS	DIVS. 10 - 14, 25, 28, 41, 43, 44	100.0	75.1	95.1	100.0	75.1	95.1	100.0	71.8	94.4	100.0	75.1	95.1	100.0	75.1	95.1	100.0	93.2	98.7
21, 22, 23	FIRE SUPPRESSION, PLUMBING & HVAC	100.0	72.3	89.0	99.9	72.3	88.9	100.2	71.7	88.9	99.9	72.0	88.8	99.9	72.3	88.9	100.0	93.0	97.3
26, 27, 3370	ELECTRICAL, COMMUNICATIONS & UTIL.	88.6	73.8	81.4	86.8	73.8	80.5	88.0	57.1	72.9	87.5	73.8	80.9	101.6	73.8	88.1	106.9	93.9	100.6
MF2004	WEIGHTED AVERAGE	99.8	76.0	89.8	100.3	76.0	90.1	97.0	68.5	85.0	100.3	74.3	89.3	102.0	76.0	91.1	97.4	94.7	96.3

NEW YORK

DIVISION		BINGHAMTON			BUFFALO			HICKSVILLE			NEW YORK			RIVERHEAD			ROCHESTER		
		MAT.	INST.	TOTAL	MAT.	INST.	TOTAL	MAT.	INST.	TOTAL	MAT.	INST.	TOTAL	MAT.	INST.	TOTAL	MAT.	INST.	TOTAL
015433	CONTRACTOR EQUIPMENT		117.7	117.7		98.1	98.1		109.6	109.6		110.2	110.2		109.6	109.6		114.6	114.6
0241, 31 - 34	SITE & INFRASTRUCTURE, DEMOLITION	95.6	96.0	95.9	95.5	99.3	98.1	114.6	121.4	119.3	130.7	121.2	124.2	115.6	121.2	119.6	72.0	105.3	94.8
0310	Concrete Forming & Accessories	101.0	79.0	82.0	97.1	113.7	111.4	90.2	144.0	136.7	107.0	184.8	174.2	94.9	144.0	137.3	98.0	92.4	93.2
0320	Concrete Reinforcing	98.1	94.4	96.6	98.4	102.3	100.0	99.4	184.3	134.0	106.7	190.7	140.9	101.2	184.3	135.2	99.2	89.2	95.1
0330	Cast-in-Place Concrete	102.9	90.2	98.1	110.7	116.6	112.9	96.3	156.1	119.1	108.8	168.8	131.7	97.9	156.1	120.1	99.5	99.1	99.4
03	CONCRETE	99.7	87.7	94.2	107.4	111.8	109.4	103.5	154.3	126.6	112.0	177.7	141.9	104.3	154.3	127.1	111.1	95.2	103.9
04	MASONRY	107.0	81.0	91.5	109.3	118.4	114.8	110.9	159.6	140.0	107.8	171.5	145.9	116.4	159.6	142.3	105.3	95.8	99.6
05	METALS	91.3	114.7	98.1	94.1	95.5	94.5	101.1	138.4	112.0	108.4	140.9	117.9	101.5	138.4	112.3	90.8	105.9	95.2
06	WOOD, PLASTICS & COMPOSITES	105.3	79.7	91.3	98.4	113.4	106.6	87.5	140.7	116.7	110.5	188.2	153.2	92.9	140.7	119.2	95.4	92.8	94.0
07	THERMAL & MOISTURE PROTECTION	103.1	80.5	94.3	99.1	109.0	102.9	106.4	150.4	123.5	109.3	167.8	132.1	107.0	150.4	123.9	94.9	95.6	95.2
08	OPENINGS	91.0	78.8	88.0	95.2	102.2	96.9	88.8	150.6	104.1	98.0	177.2	117.5	88.8	150.6	104.1	98.8	87.9	96.1
0920	Plaster & Gypsum Board	119.9	78.5	93.0	99.0	113.6	108.5	107.3	142.2	129.9	128.3	191.1	169.1	109.4	142.2	130.7	98.2	92.5	94.5
0950, 0980	Ceilings & Acoustic Treatment	96.6	78.5	85.4	98.5	113.6	107.8	79.1	142.2	118.1	109.1	191.1	159.8	81.6	142.2	119.1	97.8	92.5	94.5
0960	Flooring	94.4	81.9	91.0	89.1	117.2	96.7	94.2	160.0	112.0	93.3	169.6	113.9	95.5	160.0	112.9	79.3	101.1	85.2
0970, 0990	Wall Finishes & Painting/Coating	86.7	85.2	85.8	86.1	112.6	102.0	111.0	154.5	137.1	90.3	156.7	130.2	111.0	154.5	137.1	84.7	96.0	91.5
09	FINISHES	97.3	79.8	88.1	92.5	115.0	104.4	104.7	147.1	127.1	109.2	180.1	146.7	105.9	147.1	127.7	92.1	94.6	93.5
COVERS	DIVS. 10 - 14, 25, 28, 41, 43, 44	100.0	93.3	98.7	100.0	107.5	101.5	100.0	128.9	105.7	100.0	139.6	107.8	100.0	128.9	105.7	100.0	88.4	97.7
21, 22, 23	FIRE SUPPRESSION, PLUMBING & HVAC	100.4	79.0	91.9	99.9	97.1	98.8	99.8	146.9	118.5	100.2	164.6	125.8	99.9	146.9	118.6	99.9	89.6	95.8
26, 27, 3370	ELECTRICAL, COMMUNICATIONS & UTIL.	102.2	86.8	94.7	99.8	98.5	99.1	102.2	149.3	125.1	109.4	173.1	140.4	103.8	149.3	126.0	101.0	86.2	93.8
MF2004	WEIGHTED AVERAGE	98.1	86.9	93.4	99.0	104.8	101.5	101.2	146.4	120.1	106.1	164.6	130.7	102.0	146.4	120.6	98.4	94.1	96.6

NEW YORK

DIVISION		SCHENECTADY			SYRACUSE			UTICA			WATERTOWN			WHITE PLAINS			YONKERS		
		MAT.	INST.	TOTAL	MAT.	INST.	TOTAL	MAT.	INST.	TOTAL	MAT.	INST.	TOTAL	MAT.	INST.	TOTAL	MAT.	INST.	TOTAL
015433	CONTRACTOR EQUIPMENT		112.9	112.9		112.9	112.9		112.9	112.9		112.9	112.9		109.7	109.7		109.7	109.7
0241, 31 - 34	SITE & INFRASTRUCTURE, DEMOLITION	73.4	105.4	95.3	94.4	104.9	101.6	71.6	102.8	92.9	80.1	105.4	97.4	120.5	114.7	116.5	130.0	114.5	119.4
0310	Concrete Forming & Accessories	100.7	88.1	89.8	99.9	88.9	90.4	101.2	85.6	87.8	86.2	92.4	91.6	106.4	128.8	125.7	106.7	128.8	125.8
0320	Concrete Reinforcing	97.7	94.3	96.3	99.1	94.7	97.3	99.1	91.4	96.0	99.7	94.6	97.6	99.8	183.6	134.0	103.8	183.6	136.3
0330	Cast-in-Place Concrete	94.5	102.9	97.7	95.5	96.9	96.0	87.4	90.1	88.4	101.8	101.0	101.5	95.9	133.8	110.4	107.3	133.8	117.4
03	CONCRETE	104.1	95.3	100.1	103.0	93.7	98.8	101.1	89.3	95.7	113.8	96.6	106.0	100.8	139.6	118.4	110.8	139.6	123.9
04	MASONRY	91.2	100.7	96.9	98.0	96.0	96.8	89.8	85.7	87.4	91.0	98.5	95.5	101.8	133.1	120.6	107.3	133.1	122.7
05	METALS	94.8	105.7	98.0	94.6	104.8	97.6	92.9	103.5	96.0	92.9	104.8	96.4	95.9	132.6	106.6	104.7	132.6	112.8
06	WOOD, PLASTICS & COMPOSITES	101.7	84.5	92.3	101.7	87.2	93.7	101.7	87.7	94.0	84.0	91.2	87.9	111.6	129.2	121.3	111.4	129.2	121.2
07	THERMAL & MOISTURE PROTECTION	90.1	92.6	91.1	100.2	94.7	98.0	89.9	90.2	90.0	90.2	96.4	92.6	109.7	141.0	121.9	110.0	141.0	122.1
08	OPENINGS	96.0	82.8	92.7	94.0	82.1	91.0	96.0	83.9	93.0	96.0	88.3	94.1	92.5	142.9	105.9	96.0	143.5	107.7
0920	Plaster & Gypsum Board	115.0	83.7	94.7	115.0	86.4	96.4	115.0	87.0	96.8	107.9	90.6	96.7	121.9	129.9	127.1	127.9	129.9	129.2
0950, 0980	Ceilings & Acoustic Treatment	96.6	83.7	88.6	96.6	86.4	90.3	96.6	87.0	90.7	96.6	90.6	92.9	83.3	129.9	112.1	107.5	129.9	121.3
0960	Flooring	83.4	86.7	84.3	85.1	88.4	86.0	83.4	93.3	86.1	76.1	93.3	80.8	89.7	164.6	110.0	89.3	164.6	109.6
0970, 0990	Wall Finishes & Painting/Coating	82.4	79.3	80.5	88.3	91.0	89.9	82.4	80.1	81.0	82.4	89.0	86.4	88.2	154.5	128.0	88.2	154.5	128.0
09	FINISHES	95.9	86.2	90.8	97.5	88.7	92.9	96.0	86.6	91.0	93.5	92.3	92.8	100.3	137.7	120.1	107.4	137.7	123.4
COVERS	DIVS. 10 - 14, 25, 28, 41, 43, 44	100.0	93.9	98.8	100.0	97.5	99.5	100.0	90.0	98.0	100.0	94.9	99.0	100.0	122.9	104.5	100.0	122.9	104.5
21, 22, 23	FIRE SUPPRESSION, PLUMBING & HVAC	100.2	93.9	97.7	100.2	86.9	94.9	100.2	81.7	92.9	100.2	76.0	90.6	100.4	127.3	111.1	100.4	127.3	111.1
26, 27, 3370	ELECTRICAL, COMMUNICATIONS & UTIL.	102.2	93.9	98.2	102.2	91.0	96.8	99.7	90.4	95.2	102.2	90.9	96.7	99.0	137.5	117.8	106.8	137.5	121.7
MF2004	WEIGHTED AVERAGE	97.7	95.3	96.7	98.7	93.1	96.4	96.7	89.5	93.6	96.7	92.2	95.8	99.8	132.8	113.6	104.7	132.8	116.5

NORTH CAROLINA

DIVISION		ASHEVILLE			CHARLOTTE			DURHAM			FAYETTEVILLE			GREENSBORO			RALEIGH		
		MAT.	INST.	TOTAL	MAT.	INST.	TOTAL	MAT.	INST.	TOTAL	MAT.	INST.	TOTAL	MAT.	INST.	TOTAL	MAT.	INST.	TOTAL
015433	CONTRACTOR EQUIPMENT		98.0	98.0		98.0	98.0		102.1	102.1		102.1	102.1		102.1	102.1		102.1	102.1
0241, 31 - 34	SITE & INFRASTRUCTURE, DEMOLITION	106.5	78.0	87.0	110.4	78.0	88.3	106.6	85.3	92.0	105.4	85.5	91.8	106.5	85.5	92.1	110.6	85.5	93.5
0310	Concrete Forming & Accessories	96.6	41.4	48.9	101.2	43.4	51.3	99.0	41.9	49.7	96.1	42.7	49.9	98.8	42.6	50.2	99.3	43.5	51.1
0320	Concrete Reinforcing	101.1	45.2	78.3	101.6	42.2	77.4	102.7	52.8	82.3	104.9	53.0	83.7	101.6	52.7	81.6	101.6	53.0	81.7
0330	Cast-in-Place Concrete	103.2	49.5	82.7	116.0	51.6	91.4	104.9	49.7	83.8	108.2	52.9	87.1	104.1	51.1	83.8	118.9	56.5	95.1
03	CONCRETE	106.2	47.1	79.3	111.9	48.1	82.9	106.8	48.8	80.4	108.3	50.2	81.9	106.4	49.5	80.4	113.1	51.8	85.2
04	MASONRY	85.3	41.9	59.3	92.1	45.9	64.4	90.9	37.7	59.0	89.0	42.9	61.4	89.0	38.4	58.7	91.3	43.9	62.9
05	METALS	88.3	77.6	85.2	92.5	76.7	87.9	102.1	80.7	95.8	106.4	80.7	98.9	95.7	80.3	91.2	93.3	80.7	89.6
06	WOOD, PLASTICS & COMPOSITES	96.4	40.8	65.9	101.7	43.0	69.5	98.9	42.3	67.8	95.2	42.3	66.1	98.6	42.2	67.7	98.6	43.0	68.1
07	THERMAL & MOISTURE PROTECTION	102.3	45.2	80.1	99.8	46.8	79.2	103.1	47.3	81.4	102.2	45.8	80.2	102.8	44.8	80.3	102.0	46.1	80.2
08	OPENINGS	94.2	40.5	80.9	98.3	42.1	84.4	98.3	44.8	85.1	94.3	44.8	82.0	98.3	44.7	85.0	95.1	44.9	82.6
0920	Plaster & Gypsum Board	106.6	38.4	62.3	108.1	40.7	64.4	110.6	40.0	64.8	110.0	40.0	64.5	112.3	39.9	65.3	108.1	40.7	64.3
0950, 0980	Ceilings & Acoustic Treatment	92.6	38.4	59.1	95.1	40.7	61.5	97.6	40.0	62.0	93.4	40.0	60.4	97.6	39.9	61.9	95.1	40.7	61.5
0960	Flooring	107.3	46.7	90.9	110.9	46.9	93.6	111.5	46.7	94.0	107.4	46.7	91.0	111.5	43.2	93.0	111.5	46.7	94.0
0970, 0990	Wall Finishes & Painting/Coating	114.4	36.8	67.8	114.4	42.0	71.0	114.4	38.6	68.9	114.4	36.8	67.8	114.4	35.8	67.1	114.4	40.6	70.1
09	FINISHES	101.8	41.6	70.0	103.5	43.7	71.9	104.8	42.1	71.6	102.5	42.5	70.7	104.0	41.7	71.6	104.0	43.3	72.0
COVERS	DIVS. 10 - 14, 25, 28, 41, 43, 44	100.0	75.0	95.1	100.0	75.6	95.2	100.0	70.2	94.1	100.0	71.1	94.3	100.0	75.3	95.1	100.0	71.5	94.4
21, 22, 23	FIRE SUPPRESSION, PLUMBING & HVAC	100.2	41.5	76.9	99.9	41.7	76.8	100.3	40.2	76.4	100.1	41.4	76.8	100.2	41.5	76.8	100.0	39.6	76.0
26, 27, 3370	ELECTRICAL, COMMUNICATIONS & UTIL.	97.7	40.2	69.7	100.3	52.3	76.9	98.6	50.9	75.4	97.2	46.2	72.4	97.5	40.4	69.7	104.9	40.1	73.3
MF2004	WEIGHTED AVERAGE	97.6	49.6	77.5	100.2	52.2	80.0	101.1	51.7	80.4	101.1	52.1	80.5	99.8	50.7	79.2	100.7	51.3	80.0

		NORTH CAROLINA						NORTH DAKOTA										
	DIVISION	WILMINGTON			WINSTON-SALEM			BISMARCK			FARGO			GRAND FORKS			MINOT		
		MAT.	INST.	TOTAL	MAT.	INST.	TOTAL	MAT.	INST.	TOTAL	MAT.	INST.	TOTAL	MAT.	INST.	TOTAL	MAT.	INST.	TOTAL
015433	CONTRACTOR EQUIPMENT		98.0	98.0		102.1	102.1		98.7	98.7		98.7	98.7		98.7	98.7		98.7	98.7
0241, 31 - 34	SITE & INFRASTRUCTURE, DEMOLITION	107.6	78.0	87.3	106.8	85.4	92.2	96.1	97.3	96.9	95.6	97.3	96.8	104.5	94.4	97.6	101.8	97.3	98.7
0310	Concrete Forming & Accessories	98.0	41.9	49.5	100.0	41.7	49.6	94.7	45.5	52.2	95.6	45.8	52.6	91.7	39.5	46.6	87.6	60.6	64.3
0320	Concrete Reinforcing	101.9	51.1	81.2	101.6	41.9	77.2	111.0	69.7	94.1	102.3	68.9	88.7	109.3	70.1	93.3	113.0	70.1	95.5
0330	Cast-in-Place Concrete	102.8	50.8	82.9	106.7	48.5	84.4	101.0	53.0	82.7	97.2	54.8	81.0	104.9	50.4	84.1	104.9	51.9	84.6
03	CONCRETE	106.2	48.8	80.1	107.5	46.2	79.6	100.1	54.1	79.2	105.6	54.8	82.5	106.5	50.4	81.0	105.5	60.4	85.0
04	MASONRY	75.1	39.7	53.9	89.2	39.5	59.5	100.2	58.0	74.9	101.6	43.4	66.8	102.7	65.2	80.2	103.0	65.2	80.4
05	METALS	88.0	79.8	85.6	93.3	76.2	88.3	92.7	80.8	89.2	94.9	79.7	90.4	92.7	77.0	88.1	92.9	81.6	89.6
06	WOOD, PLASTICS & COMPOSITES	98.1	41.6	67.1	98.6	41.5	67.3	87.0	40.7	61.6	86.9	41.3	61.8	83.7	37.5	58.3	79.2	60.6	69.1
07	THERMAL & MOISTURE PROTECTION	102.3	45.2	80.1	102.8	44.4	80.1	101.6	50.6	81.8	100.2	48.3	80.0	102.5	52.8	83.2	102.3	54.4	83.7
08	OPENINGS	94.3	43.3	81.7	98.3	41.4	84.2	99.1	46.4	86.1	99.1	46.8	86.1	99.1	40.9	84.7	99.3	57.4	88.9
0920	Plaster & Gypsum Board	107.5	39.3	63.2	112.3	39.2	64.8	99.2	38.8	60.0	101.6	39.4	61.2	106.4	35.5	60.4	104.9	59.7	75.5
0950, 0980	Ceilings & Acoustic Treatment	93.4	39.3	59.9	97.6	39.2	61.5	139.6	38.8	77.3	139.6	39.4	77.7	142.1	35.5	76.2	142.1	59.7	91.1
0960	Flooring	108.2	46.7	91.6	111.5	46.7	93.6	115.9	67.4	102.8	115.6	40.2	95.2	114.2	40.2	94.2	111.5	80.8	103.2
0970, 0990	Wall Finishes & Painting/Coating	114.4	35.7	67.2	114.4	37.5	68.3	98.4	33.8	59.7	98.4	69.7	81.2	98.4	25.8	54.9	98.4	28.3	56.4
09	FINISHES	102.4	41.8	70.4	105.1	42.0	71.7	115.4	46.9	79.2	115.7	45.8	78.7	117.1	37.4	75.0	115.9	60.8	86.8
COVERS	DIVS. 10 - 14, 25, 28, 41, 43, 44	100.0	70.8	94.2	100.0	74.9	95.0	100.0	80.0	96.0	100.0	80.1	96.1	100.0	40.9	88.3	100.0	82.4	96.5
21, 22, 23	FIRE SUPPRESSION, PLUMBING & HVAC	100.2	41.2	76.7	100.2	40.1	76.3	100.3	60.9	84.6	100.3	64.6	86.1	100.4	39.8	76.3	100.4	60.4	84.5
26, 27, 3370	ELECTRICAL, COMMUNICATIONS & UTIL.	98.0	38.8	69.2	97.5	44.0	71.5	103.0	66.2	85.1	104.9	66.3	86.1	97.4	62.5	80.4	100.6	67.4	84.4
MF2004	WEIGHTED AVERAGE	97.2	49.6	77.2	99.5	50.0	78.7	100.1	62.9	84.5	101.4	62.0	84.9	100.8	55.1	81.6	100.9	67.3	86.8

		OHIO																	
	DIVISION	AKRON			CANTON			CINCINNATI			CLEVELAND			COLUMBUS			DAYTON		
		MAT.	INST.	TOTAL	MAT.	INST.	TOTAL	MAT.	INST.	TOTAL	MAT.	INST.	TOTAL	MAT.	INST.	TOTAL	MAT.	INST.	TOTAL
015433	CONTRACTOR EQUIPMENT		95.4	95.4		95.4	95.4		100.2	100.2		96.1	96.1		95.3	95.3		95.1	95.1
0241, 31 - 34	SITE & INFRASTRUCTURE, DEMOLITION	89.0	102.8	98.4	89.1	102.8	98.5	73.9	105.2	95.3	89.0	104.5	99.6	88.5	100.1	96.4	73.0	104.9	94.8
0310	Concrete Forming & Accessories	101.0	96.3	96.9	101.0	85.4	87.5	96.2	85.4	86.8	101.1	103.3	103.0	98.3	83.6	85.6	96.1	80.6	82.7
0320	Concrete Reinforcing	98.0	93.3	96.1	98.0	77.1	89.5	89.7	84.8	87.7	98.6	93.8	96.6	92.8	86.2	90.1	89.7	80.5	85.9
0330	Cast-in-Place Concrete	93.2	100.7	96.1	94.1	98.3	95.7	81.6	87.3	83.8	91.4	109.9	98.5	89.5	93.0	90.8	75.9	90.4	81.5
03	CONCRETE	95.0	96.4	95.6	95.4	87.8	92.0	87.8	86.1	87.0	94.3	102.9	98.2	92.6	87.1	90.1	85.2	83.8	84.6
04	MASONRY	90.0	94.7	92.8	90.7	87.0	88.4	75.7	92.4	85.7	94.4	107.2	102.1	93.1	93.2	93.1	75.2	88.1	82.9
05	METALS	92.1	82.3	89.2	92.1	75.5	87.2	97.5	88.1	94.8	93.5	85.7	91.3	91.9	83.4	89.4	96.8	79.1	91.6
06	WOOD, PLASTICS & COMPOSITES	96.9	96.8	96.9	97.2	85.5	90.8	93.8	82.8	87.8	96.0	100.8	98.6	98.1	81.2	88.8	95.0	78.1	85.7
07	THERMAL & MOISTURE PROTECTION	104.4	97.5	101.7	105.3	93.7	100.8	98.8	93.5	96.8	103.2	110.2	106.0	100.2	93.8	97.7	104.3	89.3	98.5
08	OPENINGS	107.6	96.0	104.7	101.7	75.2	95.1	97.8	82.9	94.1	97.8	98.2	97.9	100.2	81.4	95.6	98.0	78.3	93.1
0920	Plaster & Gypsum Board	90.7	96.4	94.4	91.5	84.6	87.0	92.6	82.5	86.0	89.9	100.5	96.8	88.0	80.5	83.1	92.6	77.6	82.9
0950, 0980	Ceilings & Acoustic Treatment	92.3	96.4	94.8	92.3	84.6	87.5	99.2	82.5	88.9	90.6	100.5	96.8	91.1	80.5	84.5	100.2	77.6	86.2
0960	Flooring	102.6	88.2	98.7	102.8	80.9	96.9	106.6	93.6	103.1	102.4	103.3	102.6	93.6	91.5	93.0	109.3	86.2	103.0
0970, 0990	Wall Finishes & Painting/Coating	105.7	108.0	107.1	105.7	81.4	91.1	106.9	88.9	96.1	105.7	112.3	109.7	97.3	94.2	95.4	106.9	85.5	94.1
09	FINISHES	99.9	96.3	98.0	100.1	84.2	91.7	99.4	86.8	92.7	99.5	104.2	102.0	93.6	85.6	89.4	100.4	81.6	90.5
COVERS	DIVS. 10 - 14, 25, 28, 41, 43, 44	100.0	88.8	97.8	100.0	68.1	93.7	100.0	91.9	98.4	100.0	103.5	100.7	100.0	93.2	98.6	100.0	90.5	98.1
21, 22, 23	FIRE SUPPRESSION, PLUMBING & HVAC	99.9	96.5	98.2	99.9	81.4	92.6	100.0	87.0	94.9	99.9	105.1	102.1	99.7	89.8	95.8	101.0	85.2	94.7
26, 27, 3370	ELECTRICAL, COMMUNICATIONS & UTIL.	97.5	87.7	92.7	96.8	88.2	92.6	94.7	80.0	87.5	97.3	104.2	100.7	98.1	85.1	91.8	93.1	83.3	88.3
MF2004	WEIGHTED AVERAGE	97.8	93.7	96.1	97.3	85.0	92.1	95.3	88.1	92.3	97.2	102.6	99.5	96.3	88.6	93.0	95.2	85.5	91.1

		OHIO												OKLAHOMA					
	DIVISION	LORAIN			SPRINGFIELD			TOLEDO			YOUNGSTOWN			ENID			LAWTON		
		MAT.	INST.	TOTAL	MAT.	INST.	TOTAL	MAT.	INST.	TOTAL	MAT.	INST.	TOTAL	MAT.	INST.	TOTAL	MAT.	INST.	TOTAL
015433	CONTRACTOR EQUIPMENT		95.4	95.4		95.1	95.1		97.4	97.4		95.4	95.4		81.5	81.5		82.1	82.1
0241, 31 - 34	SITE & INFRASTRUCTURE, DEMOLITION	88.4	102.3	97.9	73.3	103.2	93.8	87.7	99.5	95.8	88.9	103.4	98.8	109.5	91.1	96.9	104.6	92.2	96.1
0310	Concrete Forming & Accessories	101.1	95.1	95.9	96.1	82.2	84.1	98.3	102.8	102.2	101.1	89.9	91.4	96.3	36.5	44.6	99.8	49.3	56.1
0320	Concrete Reinforcing	98.0	93.7	96.2	89.7	85.8	88.1	92.8	93.1	93.0	98.0	88.2	94.0	99.3	77.2	90.3	99.6	77.2	90.4
0330	Cast-in-Place Concrete	88.7	105.6	95.2	78.0	89.9	82.5	89.5	106.1	95.8	92.3	102.3	96.1	92.6	49.2	76.0	89.6	49.2	74.2
03	CONCRETE	92.9	97.7	95.1	86.1	85.3	85.8	92.0	101.5	96.7	94.6	93.2	94.0	92.1	49.2	72.6	88.9	54.8	73.4
04	MASONRY	86.7	100.9	95.2	75.4	87.3	82.5	102.1	100.3	101.0	90.2	93.6	92.3	103.5	55.6	74.8	97.4	55.6	72.3
05	METALS	92.6	83.9	90.1	96.8	80.8	92.1	91.7	89.1	91.0	92.1	80.6	88.8	95.9	70.0	88.3	99.5	70.1	90.9
06	WOOD, PLASTICS & COMPOSITES	96.9	93.9	95.2	96.3	80.9	87.9	98.1	103.6	101.1	96.9	88.6	92.4	100.9	33.3	63.8	103.9	50.5	74.6
07	THERMAL & MOISTURE PROTECTION	105.2	104.2	104.8	104.2	89.3	98.4	101.8	107.2	103.9	105.5	94.8	101.3	99.5	57.9	83.3	99.3	59.7	83.9
08	OPENINGS	101.7	94.4	99.9	95.9	81.1	92.2	98.1	98.5	98.2	101.7	89.7	98.7	95.5	46.5	83.4	97.2	55.8	87.0
0920	Plaster & Gypsum Board	90.7	93.3	92.4	92.6	80.5	84.7	88.0	103.7	98.2	90.7	87.9	88.9	84.5	31.6	50.2	86.9	49.4	62.6
0950, 0980	Ceilings & Acoustic Treatment	92.3	93.3	92.9	100.2	80.5	88.0	91.1	103.7	98.9	92.3	87.9	89.5	83.6	31.6	51.5	92.8	49.4	66.0
0960	Flooring	102.8	108.5	104.3	109.3	86.2	103.0	92.8	98.8	94.4	102.8	93.1	100.2	109.3	46.1	92.2	112.0	46.1	94.2
0970, 0990	Wall Finishes & Painting/Coating	105.7	112.3	109.7	106.9	85.5	94.1	97.3	104.5	101.6	105.7	93.8	98.5	98.5	58.9	74.7	98.5	58.9	74.7
09	FINISHES	100.0	99.4	99.7	100.4	83.0	91.2	93.4	102.6	98.2	100.1	90.7	95.1	94.4	38.6	64.9	97.2	48.7	71.5
COVERS	DIVS. 10 - 14, 25, 28, 41, 43, 44	100.0	100.2	100.0	100.0	90.5	98.1	100.0	91.3	98.3	100.0	87.7	97.6	100.0	65.0	93.1	100.0	67.0	93.5
21, 22, 23	FIRE SUPPRESSION, PLUMBING & HVAC	99.9	84.1	93.6	101.0	84.8	94.6	99.7	98.4	99.2	99.9	86.1	94.5	100.1	63.6	85.6	100.1	63.6	85.6
26, 27, 3370	ELECTRICAL, COMMUNICATIONS & UTIL.	97.0	91.6	94.4	93.1	82.2	87.8	98.0	105.0	101.4	97.0	86.0	91.6	95.5	66.6	81.4	97.1	66.6	82.3
MF2004	WEIGHTED AVERAGE	96.9	93.6	95.5	95.1	85.7	91.2	96.5	99.7	97.9	97.2	89.7	94.1	97.4	59.7	81.6	97.8	62.5	83.0

685

City Cost Indexes

OKLAHOMA / OREGON

DIVISION		MUSKOGEE MAT.	INST.	TOTAL	OKLAHOMA CITY MAT.	INST.	TOTAL	TULSA MAT.	INST.	TOTAL	EUGENE MAT.	INST.	TOTAL	MEDFORD MAT.	INST.	TOTAL	PORTLAND MAT.	INST.	TOTAL
015433	CONTRACTOR EQUIPMENT		90.7	90.7		82.3	82.3		90.7	90.7		100.7	100.7		100.7	100.7		100.7	100.7
0241, 31 - 34	SITE & INFRASTRUCTURE, DEMOLITION	95.0	87.8	90.1	104.0	92.7	96.3	101.6	89.3	93.2	106.9	104.9	105.5	116.1	104.9	108.4	109.7	104.9	106.5
0310	Concrete Forming & Accessories	101.0	32.8	42.0	98.9	42.7	50.4	100.8	42.8	50.7	105.7	100.0	100.8	101.4	100.0	100.2	107.1	100.4	101.3
0320	Concrete Reinforcing	99.3	34.3	72.8	99.6	77.2	90.4	99.6	77.1	90.4	102.3	95.5	99.5	99.8	95.5	98.0	103.1	95.8	100.1
0330	Cast-in-Place Concrete	83.3	38.8	66.3	92.3	52.0	76.9	90.8	48.1	74.5	104.9	102.8	104.1	108.3	102.8	106.2	107.8	103.0	106.0
03	CONCRETE	84.8	36.7	62.9	90.1	52.9	73.2	89.5	52.4	72.6	105.7	100.0	103.1	111.8	100.0	106.4	107.3	100.2	104.1
04	MASONRY	113.5	48.9	74.8	97.1	57.5	73.4	96.8	57.5	73.3	114.5	98.1	104.7	111.6	98.1	103.5	115.5	103.0	108.0
05	METALS	95.8	60.3	85.4	101.5	70.0	92.3	98.9	81.5	93.8	91.3	95.9	92.6	90.9	95.9	92.3	92.3	96.6	93.6
06	WOOD, PLASTICS & COMPOSITES	106.0	33.0	65.9	103.1	40.7	68.8	104.5	41.9	70.2	94.2	100.0	97.3	88.4	100.0	94.7	95.5	100.0	97.9
07	THERMAL & MOISTURE PROTECTION	99.1	43.7	77.6	96.8	59.5	82.3	99.1	56.8	82.7	105.4	91.5	100.0	106.0	93.5	101.1	105.2	99.9	103.1
08	OPENINGS	95.5	32.0	79.8	97.2	50.5	85.7	97.2	50.8	85.7	99.2	101.3	99.7	101.8	101.3	101.7	96.8	101.3	97.9
0920	Plaster & Gypsum Board	88.4	31.1	51.2	83.1	39.3	54.6	86.9	40.4	56.7	102.3	99.7	100.7	100.9	99.7	100.1	107.8	99.7	102.6
0950, 0980	Ceilings & Acoustic Treatment	92.8	31.1	54.7	90.3	39.3	58.7	92.8	40.4	60.4	103.3	99.7	101.1	118.9	99.7	107.0	107.0	99.7	102.5
0960	Flooring	113.2	37.2	92.6	112.0	46.1	94.2	111.8	48.0	94.6	112.4	99.0	108.8	110.0	99.0	107.0	109.9	99.0	106.9
0970, 0990	Wall Finishes & Painting/Coating	98.5	31.6	58.3	98.5	58.9	74.7	98.5	46.6	67.3	110.1	67.7	84.7	110.1	62.1	81.3	109.6	72.7	87.5
09	FINISHES	97.0	33.7	63.5	95.9	43.4	68.2	96.6	43.1	68.3	107.4	96.5	101.6	110.8	95.9	102.9	108.3	97.1	102.3
COVERS	DIVS. 10 - 14, 25, 28, 41, 43, 44	100.0	64.2	92.9	100.0	66.6	93.4	100.0	66.9	93.5	100.0	101.3	100.3	100.0	101.3	100.3	100.0	101.4	100.3
21, 22, 23	FIRE SUPPRESSION, PLUMBING & HVAC	100.1	25.8	70.6	100.0	64.6	86.0	100.1	53.1	81.4	100.2	99.5	99.9	100.2	99.5	99.9	100.1	104.3	101.8
26, 27, 3370	ELECTRICAL, COMMUNICATIONS & UTIL.	95.2	30.7	63.8	103.0	66.6	85.3	97.1	40.1	69.4	99.3	95.1	97.2	102.8	85.0	94.1	99.6	99.8	99.7
MF2004	WEIGHTED AVERAGE	96.9	41.6	73.7	98.7	61.7	83.1	97.6	56.3	80.3	100.7	98.4	99.7	102.3	97.0	100.1	101.1	101.0	101.0

OREGON / PENNSYLVANIA

DIVISION		SALEM MAT.	INST.	TOTAL	ALLENTOWN MAT.	INST.	TOTAL	ALTOONA MAT.	INST.	TOTAL	ERIE MAT.	INST.	TOTAL	HARRISBURG MAT.	INST.	TOTAL	PHILADELPHIA MAT.	INST.	TOTAL
015433	CONTRACTOR EQUIPMENT		100.7	100.7		112.9	112.9		112.9	112.9		112.9	112.9		112.1	112.1		96.8	96.8
0241, 31 - 34	SITE & INFRASTRUCTURE, DEMOLITION	103.5	104.9	104.5	93.2	104.4	100.9	98.1	104.1	102.2	94.4	104.8	101.5	83.6	102.8	96.7	103.0	97.5	99.3
0310	Concrete Forming & Accessories	104.5	100.2	100.7	99.3	112.3	110.5	84.5	80.5	81.0	98.8	88.7	90.0	92.5	84.3	85.4	100.8	135.8	131.0
0320	Concrete Reinforcing	103.2	95.8	99.2	99.1	103.5	102.9	96.1	92.7	94.7	98.1	93.8	96.4	99.1	100.9	99.8	101.3	140.6	117.3
0330	Cast-in-Place Concrete	94.7	102.9	97.9	86.5	100.6	91.9	96.4	84.4	91.8	94.8	86.3	91.6	96.1	93.8	95.2	109.1	127.5	116.1
03	CONCRETE	101.1	100.1	100.7	97.8	107.3	102.1	93.1	85.7	89.7	92.1	90.2	91.2	100.1	92.2	96.5	108.2	133.1	119.6
04	MASONRY	119.7	103.2	109.8	94.9	99.5	97.7	97.0	74.6	83.6	87.2	90.9	89.4	95.0	87.8	90.7	96.6	134.8	119.5
05	METALS	91.7	96.4	93.1	95.2	119.3	102.2	89.1	112.0	95.8	89.3	112.6	96.1	97.0	117.6	103.0	102.3	127.7	109.7
06	WOOD, PLASTICS & COMPOSITES	91.6	100.0	96.2	101.1	116.0	109.3	80.0	82.7	81.5	97.3	87.8	92.1	94.9	83.0	88.4	100.0	137.2	120.5
07	THERMAL & MOISTURE PROTECTION	102.8	95.6	100.0	100.2	114.4	105.7	99.4	88.0	95.0	99.6	94.5	97.6	105.4	105.4	105.4	99.5	133.9	112.9
08	OPENINGS	98.7	101.3	99.3	94.0	110.9	98.2	84.8	88.3	88.3	88.5	89.5	88.7	94.0	90.8	93.2	97.7	141.5	108.5
0920	Plaster & Gypsum Board	95.4	99.7	98.2	112.6	116.3	115.0	106.3	81.8	90.4	112.6	87.1	96.0	109.7	82.1	91.8	103.8	138.5	126.3
0950, 0980	Ceilings & Acoustic Treatment	104.7	99.7	101.6	87.4	116.3	105.3	91.6	81.8	85.5	87.4	87.1	87.2	89.9	82.1	85.1	96.4	138.5	122.4
0960	Flooring	112.6	99.0	108.9	85.1	94.7	87.7	79.2	51.1	71.6	86.5	76.1	83.7	85.7	90.0	86.8	84.3	136.6	98.5
0970, 0990	Wall Finishes & Painting/Coating	110.1	67.7	84.7	88.3	81.1	84.0	84.2	105.7	97.1	94.5	89.4	91.4	88.3	87.7	88.0	97.3	144.5	125.6
09	FINISHES	106.5	96.5	101.2	95.0	105.5	100.6	93.7	76.8	84.8	96.4	86.0	90.9	94.1	85.2	89.4	100.4	137.3	119.9
COVERS	DIVS. 10 - 14, 25, 28, 41, 43, 44	100.0	101.3	100.3	100.0	103.4	100.7	100.0	94.0	98.8	100.0	98.7	99.8	100.0	88.9	97.8	100.0	120.3	104.0
21, 22, 23	FIRE SUPPRESSION, PLUMBING & HVAC	100.0	99.6	99.9	100.2	109.1	103.8	99.8	84.8	93.8	99.8	86.4	94.5	100.1	91.6	96.7	100.0	127.6	111.0
26, 27, 3370	ELECTRICAL, COMMUNICATIONS & UTIL.	107.7	95.1	101.6	101.3	96.3	98.7	90.6	98.0	94.2	92.4	85.6	89.1	100.0	84.4	92.4	98.7	137.5	117.6
MF2004	WEIGHTED AVERAGE	101.0	99.1	100.2	97.7	106.5	101.4	94.3	89.4	92.2	94.2	92.0	93.3	97.9	93.1	95.9	101.0	130.0	113.2

PENNSYLVANIA / PUERTO RICO / RHODE ISLAND

DIVISION		PITTSBURGH MAT.	INST.	TOTAL	READING MAT.	INST.	TOTAL	SCRANTON MAT.	INST.	TOTAL	YORK MAT.	INST.	TOTAL	SAN JUAN MAT.	INST.	TOTAL	PROVIDENCE MAT.	INST.	TOTAL
015433	CONTRACTOR EQUIPMENT		111.7	111.7		114.5	114.5		112.9	112.9		112.1	112.1		85.9	85.9		102.0	102.0
0241, 31 - 34	SITE & INFRASTRUCTURE, DEMOLITION	108.6	105.1	106.2	100.1	107.9	105.4	93.7	104.3	100.9	82.4	102.8	96.4	118.7	82.9	94.2	81.6	103.3	96.4
0310	Concrete Forming & Accessories	99.7	93.0	93.9	100.2	84.7	86.8	99.4	82.7	85.0	82.3	84.5	84.2	90.3	18.4	28.2	98.4	112.2	110.3
0320	Concrete Reinforcing	96.1	104.8	99.7	99.5	97.5	98.7	99.1	104.7	101.4	98.1	100.9	99.3	180.4	10.8	111.2	100.4	113.9	105.9
0330	Cast-in-Place Concrete	96.2	91.5	94.4	79.7	96.6	86.1	90.4	88.9	89.8	85.2	94.1	88.6	100.0	31.3	73.7	94.2	117.2	103.0
03	CONCRETE	97.7	95.9	96.9	94.1	92.8	93.5	99.6	90.5	95.5	98.1	92.5	95.5	114.5	22.7	72.8	107.3	113.8	110.3
04	MASONRY	86.6	93.2	90.5	97.8	91.1	93.8	95.2	94.8	95.0	95.1	88.4	91.1	88.1	15.1	44.4	107.4	123.6	117.1
05	METALS	93.9	118.6	101.1	98.8	116.2	103.9	97.1	119.1	103.5	94.7	117.6	101.4	107.3	37.5	86.9	93.8	108.6	98.1
06	WOOD, PLASTICS & COMPOSITES	96.0	93.2	94.5	100.2	82.0	90.2	101.1	78.9	88.9	86.7	83.0	84.7	95.1	18.6	53.1	102.0	110.1	106.4
07	THERMAL & MOISTURE PROTECTION	102.6	96.6	100.3	99.1	107.7	102.5	100.1	98.2	99.4	99.3	105.7	101.8	131.3	21.7	88.6	93.0	114.7	101.4
08	OPENINGS	91.8	102.6	94.5	95.2	90.6	94.1	94.0	89.9	93.0	90.8	90.8	90.8	152.2	15.3	118.4	99.5	112.2	102.6
0920	Plaster & Gypsum Board	99.1	92.8	95.0	107.4	81.2	90.4	115.0	77.8	90.9	108.9	82.1	91.5	119.4	15.5	51.9	103.3	109.7	107.5
0950, 0980	Ceilings & Acoustic Treatment	85.5	92.8	90.0	86.3	81.2	83.2	96.6	77.8	85.0	89.1	82.1	84.8	219.3	15.5	93.2	95.9	109.7	104.4
0960	Flooring	97.3	99.5	97.9	82.5	91.3	84.9	85.1	107.4	91.1	81.1	90.0	83.5	199.5	14.7	149.6	97.2	126.8	105.2
0970, 0990	Wall Finishes & Painting/Coating	97.0	105.7	102.2	95.8	99.0	97.7	88.3	98.0	94.1	88.3	87.7	88.0	198.7	17.4	102.9	93.9	120.7	110.0
09	FINISHES	96.3	94.8	95.5	98.0	86.1	91.7	97.5	88.0	92.5	92.5	85.3	88.7	198.7	17.4	102.9	100.8	115.9	107.5
COVERS	DIVS. 10 - 14, 25, 28, 41, 43, 44	100.0	98.5	99.7	100.0	96.7	99.4	100.0	97.5	99.5	100.0	89.1	97.8	100.0	19.5	84.1	100.0	98.2	99.6
21, 22, 23	FIRE SUPPRESSION, PLUMBING & HVAC	99.9	93.8	97.5	100.2	105.6	102.3	100.2	88.8	95.7	100.2	91.9	96.9	104.1	13.7	68.1	100.0	107.2	102.9
26, 27, 3370	ELECTRICAL, COMMUNICATIONS & UTIL.	96.7	98.0	97.3	100.9	88.0	94.6	101.3	95.8	98.7	94.9	84.4	89.8	122.6	13.8	69.6	95.9	94.5	95.2
MF2004	WEIGHTED AVERAGE	96.9	98.7	97.6	98.5	97.6	98.1	98.5	95.2	97.1	96.1	93.3	94.9	120.1	23.9	79.7	98.8	109.1	103.2

SOUTH CAROLINA / SOUTH DAKOTA

DIVISION		CHARLESTON			COLUMBIA			FLORENCE			GREENVILLE			SPARTANBURG			ABERDEEN		
		MAT.	INST.	TOTAL	MAT.	INST.	TOTAL	MAT.	INST.	TOTAL	MAT.	INST.	TOTAL	MAT.	INST.	TOTAL	MAT.	INST.	TOTAL
015433	CONTRACTOR EQUIPMENT		101.7	101.7		101.7	101.7		101.7	101.7		101.7	101.7		101.7	101.7		98.7	98.7
0241, 31 - 34	SITE & INFRASTRUCTURE, DEMOLITION	101.6	85.5	90.5	102.8	85.4	90.9	113.3	85.4	94.2	107.6	85.0	92.2	107.4	85.0	92.1	90.7	95.3	93.8
0310	Concrete Forming & Accessories	98.5	45.4	52.6	100.5	44.3	52.0	86.2	44.5	50.2	98.2	44.3	51.7	101.5	44.3	52.1	94.9	43.0	50.0
0320	Concrete Reinforcing	101.6	64.9	86.6	101.6	64.7	86.5	101.2	64.8	86.4	101.1	51.7	81.0	101.1	51.7	81.0	109.5	44.4	82.9
0330	Cast-in-Place Concrete	95.5	54.1	79.7	102.3	54.7	84.1	81.4	53.9	70.9	81.4	53.8	70.8	81.4	53.8	70.8	100.0	49.9	80.8
03	CONCRETE	102.2	53.8	80.2	105.4	53.6	81.8	102.5	53.4	80.1	100.8	50.9	78.1	101.0	50.9	78.2	99.9	47.2	75.9
04	MASONRY	96.8	42.6	64.4	91.5	40.3	60.9	79.6	42.6	57.4	77.6	42.6	56.6	79.6	42.6	57.4	104.3	60.4	78.0
05	METALS	88.9	82.3	87.0	88.9	81.5	86.7	87.9	81.8	86.1	87.9	77.3	84.8	87.9	77.3	84.8	99.1	70.6	90.8
06	WOOD, PLASTICS & COMPOSITES	98.6	44.3	68.8	100.9	43.8	69.5	83.8	43.8	61.8	98.4	43.8	68.4	102.5	43.8	70.2	101.5	42.2	68.9
07	THERMAL & MOISTURE PROTECTION	102.5	46.4	80.6	101.6	45.0	79.6	102.7	46.3	80.8	102.7	46.3	80.8	102.7	46.3	80.8	101.1	53.0	82.4
08	OPENINGS	98.3	48.3	85.9	98.3	48.1	85.9	94.3	48.1	82.8	94.2	45.0	82.0	94.2	45.0	82.0	94.8	43.0	82.0
0920	Plaster & Gypsum Board	114.8	42.1	67.5	108.1	41.5	64.9	102.7	41.5	63.0	108.1	41.5	64.8	109.4	41.5	65.5	100.3	40.4	61.4
0950, 0980	Ceilings & Acoustic Treatment	97.6	42.1	63.2	95.1	41.5	62.0	93.4	41.5	61.3	92.6	41.5	61.0	92.6	41.5	61.0	112.0	40.4	67.7
0960	Flooring	111.5	65.3	99.0	107.9	48.9	92.0	101.4	48.9	87.2	108.7	64.1	96.6	110.4	64.1	97.9	116.2	56.4	100.1
0970, 0990	Wall Finishes & Painting/Coating	114.4	45.0	72.8	111.9	45.0	71.7	114.4	45.0	72.8	114.4	45.0	72.8	114.4	45.0	72.8	98.4	42.7	65.0
09	FINISHES	105.8	48.6	75.6	102.8	45.0	72.3	101.3	45.5	71.8	103.6	48.1	74.3	104.4	48.1	74.6	108.6	45.0	75.0
COVERS	DIVS. 10 - 14, 25, 28, 41, 43, 44	100.0	68.9	93.8	100.0	68.6	93.8	100.0	68.6	93.8	100.0	68.6	93.8	100.0	68.6	93.8	100.0	51.0	90.3
21, 22, 23	FIRE SUPPRESSION, PLUMBING & HVAC	100.2	44.8	78.1	99.9	39.6	76.0	100.2	39.7	76.1	100.2	38.0	75.5	100.2	38.0	75.5	100.1	47.9	79.4
26, 27, 3370	ELECTRICAL, COMMUNICATIONS & UTIL.	97.2	103.6	100.3	101.3	53.0	77.8	95.2	50.4	73.4	97.3	54.5	76.4	97.3	54.5	76.4	99.7	59.7	79.3
MF2004	WEIGHTED AVERAGE	98.4	62.2	83.2	98.6	53.3	79.6	96.6	53.3	78.4	96.7	52.9	78.3	96.9	52.9	78.4	99.9	56.3	81.6

SOUTH DAKOTA / TENNESSEE

DIVISION		PIERRE			RAPID CITY			SIOUX FALLS			CHATTANOOGA			JACKSON			JOHNSON CITY		
		MAT.	INST.	TOTAL	MAT.	INST.	TOTAL	MAT.	INST.	TOTAL	MAT.	INST.	TOTAL	MAT.	INST.	TOTAL	MAT.	INST.	TOTAL
015433	CONTRACTOR EQUIPMENT		98.7	98.7		98.7	98.7		99.4	99.4		103.7	103.7		103.8	103.8		98.8	98.8
0241, 31 - 34	SITE & INFRASTRUCTURE, DEMOLITION	89.8	95.3	93.6	88.9	95.4	93.3	90.5	96.6	94.7	102.0	95.5	97.6	100.7	93.9	96.1	111.5	86.6	94.5
0310	Concrete Forming & Accessories	93.5	43.9	50.7	101.2	41.7	49.8	94.3	46.7	53.2	94.6	42.0	49.2	88.1	34.0	41.4	82.1	42.7	48.1
0320	Concrete Reinforcing	109.0	55.9	87.3	102.3	55.7	83.3	102.3	55.9	83.4	84.7	59.6	74.5	87.3	37.9	67.2	85.3	58.6	74.4
0330	Cast-in-Place Concrete	92.0	48.0	75.2	96.4	48.6	78.1	91.9	49.0	75.5	97.3	46.6	78.0	102.7	40.9	79.1	78.3	50.5	67.7
03	CONCRETE	95.4	49.1	74.3	96.5	48.3	74.6	93.0	50.7	73.8	90.6	49.1	71.7	109.4	29.9	61.8	108.7	38.3	66.5
04	MASONRY	99.8	60.0	76.0	101.1	55.9	74.0	97.2	58.1	73.8	97.7	38.4	62.2	86.0	33.9	57.4	74.0	43.3	57.1
05	METALS	99.1	75.7	92.3	100.8	75.7	93.4	100.9	76.2	93.7	103.4	83.8	97.7	98.1	72.8	90.7	100.4	83.4	95.4
06	WOOD, PLASTICS & COMPOSITES	99.3	43.0	68.4	105.2	39.1	68.9	100.2	46.0	70.5	99.3	42.3	68.0	86.0	33.9	57.4	74.0	43.3	57.1
07	THERMAL & MOISTURE PROTECTION	100.9	51.3	81.6	101.5	52.9	82.6	102.6	54.7	84.0	95.4	54.6	79.6	92.0	36.3	70.3	90.2	54.7	76.4
08	OPENINGS	98.0	47.0	85.4	99.0	44.9	85.6	99.2	48.6	86.7	99.4	48.4	86.8	100.5	35.4	84.4	95.3	49.3	83.9
0920	Plaster & Gypsum Board	94.3	41.3	59.9	100.2	37.2	59.3	99.8	44.4	63.8	82.6	40.8	55.4	90.6	32.0	52.5	96.2	41.8	60.8
0950, 0980	Ceilings & Acoustic Treatment	107.0	41.3	66.4	116.2	37.2	67.3	109.5	44.4	69.2	97.1	40.8	62.3	95.1	32.0	56.1	93.5	41.8	61.5
0960	Flooring	115.5	40.2	95.1	115.5	71.0	103.5	115.5	76.7	105.0	99.5	44.3	84.6	91.7	21.5	72.7	93.8	45.6	80.7
0970, 0990	Wall Finishes & Painting/Coating	98.4	51.2	70.1	98.4	51.2	70.1	98.4	51.2	70.1	108.8	41.3	68.3	100.4	27.9	56.9	95.6	43.3	68.0
09	FINISHES	106.2	43.3	73.0	109.2	47.4	76.5	107.7	52.4	78.5	93.8	41.9	66.4	94.0	31.1	60.7	95.5	43.3	68.0
COVERS	DIVS. 10 - 14, 25, 28, 41, 43, 44	100.0	75.8	95.2	100.0	75.7	95.2	100.0	76.2	95.3	100.0	46.6	89.4	100.0	50.4	90.2	100.0	53.0	90.7
21, 22, 23	FIRE SUPPRESSION, PLUMBING & HVAC	100.0	47.5	79.2	100.1	48.0	79.4	100.0	46.2	78.6	100.0	39.7	76.0	99.9	48.8	79.6	99.7	52.2	80.8
26, 27, 3370	ELECTRICAL, COMMUNICATIONS & UTIL.	100.9	51.8	77.0	94.6	51.8	73.8	102.9	69.1	86.5	106.5	69.3	88.4	104.1	45.8	75.7	95.0	53.9	75.0
MF2004	WEIGHTED AVERAGE	99.5	56.5	81.4	99.7	56.5	81.5	99.9	60.1	83.2	99.4	55.0	80.8	99.0	47.9	77.5	98.4	55.4	80.3

TENNESSEE / TEXAS

DIVISION		KNOXVILLE			MEMPHIS			NASHVILLE			ABILENE			AMARILLO			AUSTIN		
		MAT.	INST.	TOTAL	MAT.	INST.	TOTAL	MAT.	INST.	TOTAL	MAT.	INST.	TOTAL	MAT.	INST.	TOTAL	MAT.	INST.	TOTAL
015433	CONTRACTOR EQUIPMENT		98.8	98.8		101.9	101.9		104.9	104.9		90.7	90.7		90.7	90.7		90.0	90.0
0241, 31 - 34	SITE & INFRASTRUCTURE, DEMOLITION	88.0	86.6	87.1	90.4	91.5	91.1	96.9	98.6	98.1	102.0	88.4	92.7	103.2	89.3	93.7	91.9	54.1	59.3
0310	Concrete Forming & Accessories	93.7	42.7	49.6	94.8	60.5	65.2	94.0	64.0	68.1	97.9	39.2	47.2	96.6	48.0	55.9	97.4	46.2	53.4
0320	Concrete Reinforcing	84.7	58.6	74.1	91.0	65.6	80.6	91.0	64.5	80.2	96.6	53.3	78.9	93.3	47.5	75.8	91.5	48.0	74.9
0330	Cast-in-Place Concrete	91.4	46.8	74.4	86.6	66.5	78.9	90.5	67.4	81.7	94.2	42.3	74.4	93.3	47.5	75.8	84.5	51.4	69.5
03	CONCRETE	88.1	49.2	70.4	87.1	65.3	77.2	89.3	66.9	79.1	90.3	44.4	69.4	90.0	49.6	71.6	84.5	51.4	69.5
04	MASONRY	75.1	38.3	53.0	87.8	68.0	76.0	83.4	62.6	71.0	100.4	50.3	70.4	100.1	45.8	67.6	94.0	48.8	66.9
05	METALS	104.0	83.6	98.1	105.1	91.0	101.0	105.1	89.4	100.5	102.8	70.5	93.4	102.8	68.4	92.8	98.1	65.3	88.5
06	WOOD, PLASTICS & COMPOSITES	87.6	43.3	63.3	91.7	61.1	74.9	97.1	64.9	79.4	100.6	38.3	66.4	99.0	51.7	73.0	96.5	58.1	75.4
07	THERMAL & MOISTURE PROTECTION	88.4	54.2	75.1	90.1	67.5	81.3	93.4	64.6	82.2	99.2	46.2	78.6	99.4	45.1	78.3	101.2	54.7	89.7
08	OPENINGS	92.2	49.3	81.6	98.4	63.3	89.7	97.6	65.6	89.7	93.2	42.3	80.6	93.2	46.2	81.6	89.9	57.1	85.4
0920	Plaster & Gypsum Board	102.7	41.8	63.1	91.4	60.1	71.1	98.7	64.0	76.2	86.9	36.6	54.3	82.7	50.5	61.8	89.9	57.1	68.6
0950, 0980	Ceilings & Acoustic Treatment	94.3	41.8	61.8	98.7	60.1	74.8	96.1	64.0	76.3	92.8	36.6	58.1	90.3	50.5	65.7	86.3	57.1	68.3
0960	Flooring	99.2	45.6	84.7	94.2	40.5	79.6	101.6	71.6	93.5	112.0	64.7	99.2	111.8	57.3	97.1	94.6	43.9	80.9
0970, 0990	Wall Finishes & Painting/Coating	106.1	48.4	71.5	101.3	57.0	74.7	108.4	66.4	83.2	97.1	49.2	68.4	97.1	34.1	59.3	92.7	38.3	60.0
09	FINISHES	90.7	43.3	65.6	90.8	55.8	72.3	99.3	65.7	81.5	96.6	44.5	69.0	95.7	49.1	71.1	89.5	50.6	68.9
COVERS	DIVS. 10 - 14, 25, 28, 41, 43, 44	100.0	53.0	90.7	100.0	79.0	95.9	100.0	79.9	96.0	100.0	73.3	94.7	100.0	66.5	93.4	100.0	67.0	93.5
21, 22, 23	FIRE SUPPRESSION, PLUMBING & HVAC	99.7	52.2	80.8	99.9	66.6	86.6	99.9	78.9	91.5	100.1	39.9	76.2	100.0	48.3	79.5	100.0	55.5	82.3
26, 27, 3370	ELECTRICAL, COMMUNICATIONS & UTIL.	100.8	58.3	80.1	103.2	68.7	86.4	103.0	65.4	84.7	97.3	43.8	71.2	101.9	68.6	85.7	104.2	67.6	86.4
MF2004	WEIGHTED AVERAGE	95.8	55.8	79.0	97.5	70.1	86.0	98.5	73.6	88.0	98.2	50.8	78.3	98.6	56.8	81.0	96.5	59.1	80.8

City Cost Indexes

		TEXAS																	
	DIVISION	BEAUMONT			CORPUS CHRISTI			DALLAS			EL PASO			FORT WORTH			HOUSTON		
		MAT.	INST.	TOTAL	MAT.	INST.	TOTAL	MAT.	INST.	TOTAL	MAT.	INST.	TOTAL	MAT.	INST.	TOTAL	MAT.	INST.	TOTAL
015433	CONTRACTOR EQUIPMENT		92.0	92.0		96.8	96.8		99.4	99.4		90.7	90.7		90.7	90.7		100.1	100.1
0241, 31 - 34	SITE & INFRASTRUCTURE, DEMOLITION	98.8	89.0	92.1	122.7	82.5	95.2	119.7	88.1	98.1	100.2	87.7	91.7	104.7	89.0	93.9	123.0	85.7	97.5
0310	Concrete Forming & Accessories	104.7	48.2	55.9	98.0	35.7	44.1	94.8	59.3	64.2	97.3	43.2	50.6	93.7	58.5	63.3	94.1	65.9	69.8
0320	Concrete Reinforcing	96.4	39.4	73.2	96.7	44.8	75.5	96.7	52.9	78.8	96.6	43.5	75.0	96.6	52.7	78.7	98.0	58.2	81.8
0330	Cast-in-Place Concrete	94.3	52.5	78.3	98.9	45.0	78.3	92.3	55.4	78.2	86.4	41.0	69.1	98.2	51.5	80.3	97.2	68.9	86.4
03	CONCRETE	92.7	49.3	72.9	90.0	42.8	68.5	88.7	58.4	75.0	86.7	43.7	67.1	92.0	55.9	75.6	92.2	67.2	80.8
04	MASONRY	101.3	56.1	74.2	83.8	48.5	62.7	102.3	58.6	76.1	98.0	46.2	67.0	94.7	58.5	73.0	96.1	60.8	75.1
05	METALS	102.3	65.8	91.6	97.6	76.1	91.3	102.5	82.2	96.6	102.6	63.5	91.1	99.5	70.4	91.0	106.9	87.5	101.2
06	WOOD, PLASTICS & COMPOSITES	111.4	48.3	76.8	115.1	34.9	71.0	98.7	60.2	77.6	100.2	45.8	70.3	101.1	60.1	78.6	96.8	66.7	80.3
07	THERMAL & MOISTURE PROTECTION	104.5	55.7	85.5	95.3	44.3	75.5	93.2	63.4	81.6	95.7	52.4	78.8	100.4	52.1	81.6	96.8	67.4	85.4
08	OPENINGS	94.9	44.3	82.4	108.6	36.1	90.7	103.4	54.0	91.2	93.2	41.3	80.3	87.1	53.9	78.9	104.5	63.6	94.4
0920	Plaster & Gypsum Board	96.3	47.1	64.3	93.9	32.9	54.3	90.5	59.2	70.1	85.2	44.4	58.7	86.2	59.2	68.6	94.8	65.9	76.0
0950, 0980	Ceilings & Acoustic Treatment	104.4	47.1	69.0	87.9	32.9	53.9	94.2	59.2	72.5	90.3	44.4	61.9	92.8	59.2	72.0	105.8	65.9	81.1
0960	Flooring	111.1	70.9	100.2	107.1	43.4	89.9	100.6	53.2	87.8	112.0	62.4	98.6	145.6	42.6	117.7	98.5	58.7	87.7
0970, 0990	Wall Finishes & Painting/Coating	93.9	46.0	65.1	105.5	53.5	74.3	103.5	53.1	73.2	97.1	32.7	58.5	98.5	48.6	68.6	101.4	61.2	77.3
09	FINISHES	95.0	51.8	72.2	96.3	38.2	65.6	98.4	57.5	76.7	95.7	46.0	69.4	107.1	54.7	79.4	99.9	64.0	80.9
COVERS	DIVS. 10 - 14, 25, 28, 41, 43, 44	100.0	80.4	96.1	100.0	77.8	95.6	100.0	79.2	95.9	100.0	64.9	93.0	100.0	78.9	95.8	100.0	86.0	97.2
21, 22, 23	FIRE SUPPRESSION, PLUMBING & HVAC	100.0	59.1	83.7	99.9	40.6	76.3	99.9	63.8	85.5	100.0	34.9	74.1	100.0	53.6	81.6	99.9	71.9	88.8
26, 27, 3370	ELECTRICAL, COMMUNICATIONS & UTIL.	95.0	65.0	80.4	97.6	50.1	74.5	97.3	69.2	83.6	98.6	51.9	75.9	98.1	60.3	79.7	97.2	68.5	83.2
MF2004	WEIGHTED AVERAGE	98.4	60.1	82.3	98.4	50.4	78.3	99.4	66.2	85.4	97.5	49.8	77.5	98.0	60.7	82.4	100.7	71.2	88.3

		TEXAS																	
	DIVISION	LAREDO			LUBBOCK			ODESSA			SAN ANTONIO			WACO			WICHITA FALLS		
		MAT.	INST.	TOTAL	MAT.	INST.	TOTAL	MAT.	INST.	TOTAL	MAT.	INST.	TOTAL	MAT.	INST.	TOTAL	MAT.	INST.	TOTAL
015433	CONTRACTOR EQUIPMENT		90.0	90.0		98.4	98.4		90.7	90.7		91.8	91.8		90.7	90.7		90.7	90.7
0241, 31 - 34	SITE & INFRASTRUCTURE, DEMOLITION	90.5	88.6	89.2	130.8	84.3	99.0	102.2	88.8	93.0	90.2	92.3	91.6	100.8	88.9	92.7	101.6	88.5	92.6
0310	Concrete Forming & Accessories	92.6	35.9	43.6	96.9	38.6	46.5	97.9	36.7	45.0	92.7	55.6	60.6	98.5	38.8	46.9	98.5	39.6	47.6
0320	Concrete Reinforcing	97.4	44.9	75.9	97.7	53.1	79.5	96.6	53.0	78.8	104.4	49.3	81.9	96.6	46.1	76.0	96.6	47.3	76.5
0330	Cast-in-Place Concrete	76.2	59.8	69.9	94.3	47.7	76.5	94.2	44.1	75.0	74.7	67.5	71.9	85.0	52.9	72.7	90.8	47.3	74.2
03	CONCRETE	81.2	47.3	65.8	89.5	46.7	70.0	90.3	43.8	69.2	82.0	59.5	71.7	86.1	46.5	68.1	88.8	45.3	69.0
04	MASONRY	91.7	49.6	66.5	99.7	44.4	66.6	100.4	44.7	67.0	91.5	60.6	73.0	97.0	55.3	72.0	97.5	56.3	72.8
05	METALS	99.1	64.5	89.0	106.5	82.3	99.4	102.1	69.5	92.6	99.7	69.4	90.9	102.7	67.4	92.4	102.6	71.3	93.5
06	WOOD, PLASTICS & COMPOSITES	95.8	34.7	62.3	100.4	38.4	66.4	100.6	36.5	65.4	95.8	54.0	72.9	106.9	34.0	66.9	106.9	38.3	69.2
07	THERMAL & MOISTURE PROTECTION	91.4	49.4	75.1	89.9	46.8	73.1	99.2	42.5	77.2	91.4	64.9	81.1	99.7	47.2	79.3	99.7	51.0	80.8
08	OPENINGS	101.9	36.6	85.7	104.2	40.7	88.5	93.2	39.3	79.9	103.9	53.6	91.4	87.1	35.0	74.2	87.1	42.2	76.0
0920	Plaster & Gypsum Board	90.9	32.9	53.3	87.4	36.6	54.4	86.9	34.7	53.0	90.9	52.9	66.2	86.9	32.1	51.3	86.9	36.6	54.2
0950, 0980	Ceilings & Acoustic Treatment	87.9	32.9	53.9	94.5	36.6	58.7	92.8	34.7	56.9	87.9	52.9	66.3	92.8	32.1	55.3	92.8	36.6	58.0
0960	Flooring	91.6	43.4	78.6	104.0	36.1	85.6	112.0	35.7	91.4	91.6	66.5	84.8	145.6	33.1	115.2	146.7	75.0	127.3
0970, 0990	Wall Finishes & Painting/Coating	92.7	50.0	67.1	108.4	30.6	61.7	97.1	30.6	57.2	92.7	50.0	67.1	98.5	31.6	58.3	102.1	50.0	70.8
09	FINISHES	89.2	37.7	62.0	98.8	36.3	65.7	96.6	35.1	64.1	89.2	55.9	71.6	106.9	35.5	69.1	107.5	46.6	75.3
COVERS	DIVS. 10 - 14, 25, 28, 41, 43, 44	100.0	71.7	94.4	100.0	73.1	94.7	100.0	64.1	92.9	100.0	78.0	95.7	100.0	75.8	95.2	100.0	64.9	93.0
21, 22, 23	FIRE SUPPRESSION, PLUMBING & HVAC	99.9	36.6	74.7	99.6	45.9	78.2	100.1	33.4	73.6	99.9	67.0	86.8	100.1	54.1	81.9	100.1	45.6	78.5
26, 27, 3370	ELECTRICAL, COMMUNICATIONS & UTIL.	99.3	64.9	82.6	95.8	41.2	69.2	97.4	38.3	68.7	99.8	64.9	82.8	97.4	71.9	85.0	99.7	60.7	80.7
MF2004	WEIGHTED AVERAGE	95.8	51.7	77.3	100.1	51.1	79.5	98.1	46.3	76.3	96.2	65.4	83.3	97.8	56.8	80.6	98.5	55.3	80.3

		UTAH												VERMONT					
	DIVISION	LOGAN			OGDEN			PROVO			SALT LAKE CITY			BURLINGTON			RUTLAND		
		MAT.	INST.	TOTAL	MAT.	INST.	TOTAL	MAT.	INST.	TOTAL	MAT.	INST.	TOTAL	MAT.	INST.	TOTAL	MAT.	INST.	TOTAL
015433	CONTRACTOR EQUIPMENT		101.4	101.4		101.4	101.4		100.7	100.7		101.4	101.4		100.4	100.4		100.4	100.4
0241, 31 - 34	SITE & INFRASTRUCTURE, DEMOLITION	88.6	101.7	97.5	77.5	101.7	94.0	85.2	100.4	95.6	77.3	101.7	93.9	76.5	96.8	90.3	76.6	96.8	90.4
0310	Concrete Forming & Accessories	105.2	55.4	62.2	105.3	55.4	62.2	106.5	55.5	62.5	104.7	55.5	62.2	93.5	55.8	60.9	97.0	56.3	61.8
0320	Concrete Reinforcing	102.6	74.0	90.9	102.2	74.0	90.7	110.9	74.0	95.9	104.6	74.0	92.1	100.4	81.1	92.5	100.4	81.2	92.5
0330	Cast-in-Place Concrete	92.3	71.4	84.3	93.7	71.4	85.2	92.4	71.4	84.4	102.4	71.4	90.6	94.2	107.0	99.1	88.1	107.2	95.4
03	CONCRETE	112.9	65.1	91.2	102.6	65.1	85.6	113.1	65.2	91.3	122.1	65.2	96.2	104.5	78.9	92.9	101.9	79.2	91.6
04	MASONRY	113.2	57.9	80.1	106.9	57.9	77.6	118.6	57.9	82.3	120.8	57.9	83.2	107.1	53.8	75.2	85.9	53.8	66.7
05	METALS	100.8	75.6	93.5	101.3	75.6	93.8	99.0	75.7	92.2	106.0	75.7	97.1	95.4	79.7	90.8	93.8	80.3	89.8
06	WOOD, PLASTICS & COMPOSITES	85.2	52.3	67.1	85.2	52.3	67.1	86.6	52.3	67.7	86.9	52.3	67.9	97.3	55.9	74.6	102.8	55.9	77.1
07	THERMAL & MOISTURE PROTECTION	99.5	63.9	85.7	98.3	63.9	84.9	101.3	63.9	86.8	101.8	63.9	87.1	94.1	54.6	78.7	94.1	63.0	82.0
08	OPENINGS	88.2	54.8	79.9	88.2	54.8	79.9	92.5	54.8	83.1	90.1	54.8	81.3	104.9	55.5	92.6	104.9	55.5	92.6
0920	Plaster & Gypsum Board	76.3	50.4	59.5	76.3	50.4	59.5	76.7	50.4	59.6	78.7	50.4	60.3	105.1	53.6	71.6	104.0	53.6	71.2
0950, 0980	Ceilings & Acoustic Treatment	108.7	50.4	72.6	108.7	50.4	72.6	108.7	50.4	72.6	103.1	50.4	70.5	92.5	53.6	68.4	93.3	53.6	68.7
0960	Flooring	104.5	53.5	90.7	102.2	53.5	89.0	105.3	53.5	91.3	103.3	53.5	89.9	97.0	61.8	87.5	96.8	61.8	87.4
0970, 0990	Wall Finishes & Painting/Coating	98.1	43.1	65.1	98.1	43.1	65.1	98.1	60.7	75.7	100.7	60.7	76.7	91.7	35.7	58.1	91.7	35.7	58.1
09	FINISHES	101.5	52.9	75.8	99.7	52.9	75.0	102.2	54.9	77.2	99.7	54.9	76.0	96.4	54.0	74.0	96.3	54.0	74.0
COVERS	DIVS. 10 - 14, 25, 28, 41, 43, 44	100.0	64.0	92.9	100.0	64.0	92.9	100.0	64.0	92.9	100.0	64.0	92.9	100.0	81.9	96.4	100.0	81.9	96.4
21, 22, 23	FIRE SUPPRESSION, PLUMBING & HVAC	99.9	69.5	87.8	99.9	69.5	87.8	99.9	69.5	87.8	100.0	69.5	87.9	100.0	64.9	86.1	100.2	65.0	86.2
26, 27, 3370	ELECTRICAL, COMMUNICATIONS & UTIL.	94.5	66.8	81.0	94.9	66.8	81.2	95.1	69.7	82.8	98.2	69.7	84.3	99.2	62.8	81.5	96.9	62.8	80.3
MF2004	WEIGHTED AVERAGE	100.3	67.3	86.4	98.4	67.3	85.3	100.8	67.8	87.0	102.9	67.9	88.2	99.3	67.7	86.1	97.5	68.1	85.2

VIRGINIA

DIVISION		ALEXANDRIA			ARLINGTON			NEWPORT NEWS			NORFOLK			PORTSMOUTH			RICHMOND		
		MAT.	INST.	TOTAL	MAT.	INST.	TOTAL	MAT.	INST.	TOTAL	MAT.	INST.	TOTAL	MAT.	INST.	TOTAL	MAT.	INST.	TOTAL
015433	CONTRACTOR EQUIPMENT		102.9	102.9		102.1	102.1		105.8	105.8		106.2	106.2		105.8	105.8		105.8	105.8
0241, 31 - 34	SITE & INFRASTRUCTURE, DEMOLITION	116.5	89.1	97.8	128.2	86.7	99.8	110.7	87.7	95.0	111.7	88.4	95.8	109.2	87.0	94.0	111.5	88.1	95.5
0310	Concrete Forming & Accessories	94.1	76.1	78.6	93.4	74.1	76.7	98.7	72.5	76.1	100.4	72.6	76.4	88.3	59.5	63.4	98.2	65.5	70.0
0320	Concrete Reinforcing	90.2	82.2	86.9	101.8	78.7	92.4	101.6	71.3	89.2	101.6	71.4	89.3	101.2	71.3	89.0	101.6	71.8	89.4
0330	Cast-in-Place Concrete	106.8	82.9	97.7	103.9	80.3	94.9	104.3	65.6	89.7	115.0	65.6	96.1	103.6	65.0	88.8	106.2	60.2	88.6
03	CONCRETE	108.2	80.8	95.7	113.7	77.9	97.4	106.4	71.2	90.4	111.4	71.2	93.1	105.3	65.1	87.0	107.1	66.3	88.5
04	MASONRY	91.6	72.3	80.1	104.3	69.2	83.3	97.7	58.4	74.2	103.1	58.4	76.3	102.5	58.0	75.9	96.0	60.9	75.0
05	METALS	97.2	96.7	97.0	95.9	86.6	93.2	97.2	91.7	95.6	96.2	91.7	94.9	96.2	89.8	94.3	99.3	92.4	97.3
06	WOOD, PLASTICS & COMPOSITES	97.7	75.7	85.6	94.4	75.7	84.2	98.6	77.4	87.0	100.5	77.4	87.8	86.4	60.1	72.0	97.4	68.4	81.5
07	THERMAL & MOISTURE PROTECTION	101.5	80.6	93.4	103.2	72.5	91.3	102.5	61.3	86.4	99.9	61.3	84.8	102.5	58.7	85.5	101.5	61.8	86.0
08	OPENINGS	98.3	78.3	93.3	96.2	72.3	90.3	98.3	69.5	91.2	98.3	69.5	91.2	98.4	60.1	88.9	98.3	65.7	90.2
0920	Plaster & Gypsum Board	112.3	74.7	87.9	109.0	74.7	86.7	112.3	75.6	88.5	108.8	75.6	87.2	105.3	57.7	74.4	117.0	66.4	84.1
0950, 0980	Ceilings & Acoustic Treatment	97.6	74.7	83.4	93.4	74.7	81.8	97.6	75.6	84.0	97.6	75.6	84.0	97.6	57.7	72.9	100.9	66.4	79.5
0960	Flooring	111.5	88.9	105.4	109.5	61.7	96.6	111.5	56.7	96.7	110.9	56.7	96.3	102.5	71.7	94.2	110.9	79.8	102.5
0970, 0990	Wall Finishes & Painting/Coating	127.6	83.3	101.0	127.6	83.3	101.0	114.4	48.3	74.8	114.4	64.6	84.5	114.4	64.6	84.5	114.4	67.6	86.3
09	FINISHES	105.9	78.9	91.6	105.1	73.4	88.4	105.1	67.6	85.3	104.4	64.6	85.9	101.5	61.9	80.6	106.2	68.7	86.4
COVERS	DIVS. 10 - 14, 25, 28, 41, 43, 44	100.0	89.2	97.9	100.0	80.8	96.2	100.0	80.8	96.2	100.0	80.7	96.2	100.0	71.9	94.4	100.0	79.0	95.8
21, 22, 23	FIRE SUPPRESSION, PLUMBING & HVAC	100.2	88.5	95.5	100.2	82.4	93.1	100.2	66.9	86.9	100.0	65.9	86.4	100.2	65.9	86.5	100.0	61.7	84.8
26, 27, 3370	ELECTRICAL, COMMUNICATIONS & UTIL.	97.6	98.2	97.9	95.1	93.4	94.3	97.2	63.8	80.9	105.5	60.4	83.5	95.4	60.4	78.4	104.2	69.8	87.5
MF2004	WEIGHTED AVERAGE	100.7	86.0	94.5	101.5	80.8	92.8	100.5	70.9	88.1	101.9	70.5	88.7	99.8	67.4	86.2	101.7	70.1	88.4

VIRGINIA / WASHINGTON

DIVISION		ROANOKE			EVERETT			RICHLAND			SEATTLE			SPOKANE			TACOMA		
		MAT.	INST.	TOTAL	MAT.	INST.	TOTAL	MAT.	INST.	TOTAL	MAT.	INST.	TOTAL	MAT.	INST.	TOTAL	MAT.	INST.	TOTAL
015433	CONTRACTOR EQUIPMENT		102.1	102.1		101.3	101.3		95.9	95.9		101.8	101.8		95.9	95.9		101.3	101.3
0241, 31 - 34	SITE & INFRASTRUCTURE, DEMOLITION	108.2	86.2	93.2	98.0	110.8	106.7	104.1	96.0	98.5	102.3	110.1	107.6	103.4	96.0	98.3	101.4	111.1	108.0
0310	Concrete Forming & Accessories	98.5	69.0	73.0	111.8	100.6	102.1	117.8	80.6	85.7	103.4	102.7	102.8	123.3	80.6	86.4	103.4	102.5	102.6
0320	Concrete Reinforcing	101.6	65.7	86.9	113.0	93.3	105.0	99.5	86.1	94.0	111.7	93.3	104.2	106.2	86.2	94.5	111.7	93.3	104.2
0330	Cast-in-Place Concrete	117.6	65.2	97.6	101.0	100.9	101.0	111.2	86.8	101.9	106.1	108.7	107.1	115.6	86.8	104.6	103.9	108.6	105.7
03	CONCRETE	112.5	68.5	92.4	98.0	98.8	98.3	103.1	83.8	94.3	100.7	102.5	101.5	105.6	83.8	95.7	99.7	102.3	100.9
04	MASONRY	98.2	61.2	76.1	138.6	96.3	113.3	115.6	81.3	95.1	133.2	100.8	113.8	116.3	83.7	96.8	133.0	100.8	113.7
05	METALS	97.0	88.0	94.3	109.5	88.6	103.4	91.2	83.8	89.0	111.2	90.3	105.1	93.7	83.9	90.8	111.2	88.8	104.7
06	WOOD, PLASTICS & COMPOSITES	98.6	73.3	84.7	103.9	102.1	102.9	96.8	79.9	87.5	94.8	102.1	98.8	105.7	79.9	91.5	93.7	102.1	98.3
07	THERMAL & MOISTURE PROTECTION	102.4	60.0	85.9	105.1	93.7	100.6	151.3	80.0	123.6	105.0	97.3	102.0	148.0	80.9	121.9	104.7	96.4	101.5
08	OPENINGS	98.3	65.2	90.1	103.7	97.9	102.3	111.6	74.8	102.5	106.0	97.9	104.0	112.2	74.8	103.0	104.4	97.9	102.8
0920	Plaster & Gypsum Board	112.3	72.1	86.2	112.7	102.1	105.8	140.1	78.9	100.4	107.3	102.1	103.9	137.2	78.9	99.4	109.7	102.1	104.8
0950, 0980	Ceilings & Acoustic Treatment	97.6	72.1	81.8	105.1	102.1	103.3	111.4	78.9	91.3	108.2	102.1	104.4	106.7	78.9	89.5	108.4	102.1	104.5
0960	Flooring	111.5	43.9	93.2	122.4	102.7	117.1	107.6	38.8	89.0	114.7	102.9	111.5	108.0	70.2	97.8	115.3	102.9	111.9
0970, 0990	Wall Finishes & Painting/Coating	114.4	47.8	74.4	109.3	81.6	92.7	109.0	67.3	84.0	109.3	89.6	97.5	108.9	67.3	83.9	109.3	89.6	97.5
09	FINISHES	105.0	62.9	82.7	110.2	99.4	104.5	122.6	70.8	95.2	108.1	101.4	104.5	121.3	77.2	98.0	108.5	101.4	104.8
COVERS	DIVS. 10 - 14, 25, 28, 41, 43, 44	100.0	71.9	94.4	100.0	98.2	99.6	100.0	76.1	95.3	100.0	101.4	100.3	100.0	76.0	95.3	100.0	101.4	100.3
21, 22, 23	FIRE SUPPRESSION, PLUMBING & HVAC	100.2	53.2	81.5	100.2	96.4	98.7	100.6	96.4	98.9	100.2	107.0	102.9	100.5	82.1	93.2	100.2	98.8	99.7
26, 27, 3370	ELECTRICAL, COMMUNICATIONS & UTIL.	97.2	44.4	71.5	103.3	90.3	97.0	97.0	90.8	93.9	103.1	99.4	101.3	95.3	76.3	86.1	103.1	97.1	100.2
MF2004	WEIGHTED AVERAGE	101.2	63.7	85.4	105.0	96.7	101.5	104.1	85.7	96.4	105.4	101.7	103.9	104.5	81.8	95.0	105.1	99.6	102.8

WASHINGTON / WEST VIRGINIA

DIVISION		VANCOUVER			YAKIMA			CHARLESTON			HUNTINGTON			PARKERSBURG			WHEELING		
		MAT.	INST.	TOTAL	MAT.	INST.	TOTAL	MAT.	INST.	TOTAL	MAT.	INST.	TOTAL	MAT.	INST.	TOTAL	MAT.	INST.	TOTAL
015433	CONTRACTOR EQUIPMENT		98.2	98.2		101.3	101.3		102.1	102.1		102.1	102.1		102.1	102.1		102.1	102.1
0241, 31 - 34	SITE & INFRASTRUCTURE, DEMOLITION	113.5	100.9	104.9	104.4	109.4	107.8	103.9	89.4	94.0	107.3	90.3	95.7	112.4	89.4	96.7	113.1	88.9	96.6
0310	Concrete Forming & Accessories	104.4	91.6	93.4	103.7	94.9	96.1	104.3	84.5	87.2	99.7	89.4	90.8	90.5	83.3	84.3	92.2	83.6	84.8
0320	Concrete Reinforcing	112.6	92.9	104.6	112.2	85.7	101.4	101.6	81.1	93.2	101.6	86.3	95.4	100.2	88.1	95.2	99.5	84.0	93.1
0330	Cast-in-Place Concrete	116.1	98.0	109.2	111.1	82.4	100.1	99.4	99.8	99.5	110.4	104.1	108.0	103.1	95.7	100.3	103.1	96.4	100.6
03	CONCRETE	109.1	94.0	102.2	104.2	88.5	97.1	104.3	90.0	97.8	109.2	94.6	102.6	109.2	89.4	100.2	109.2	89.0	100.0
04	MASONRY	133.4	93.0	109.2	125.2	63.4	88.2	94.7	89.4	91.0	97.6	86.6	91.0	81.7	83.9	83.1	105.3	81.5	91.0
05	METALS	108.4	89.5	102.9	109.4	84.2	102.0	97.2	96.7	97.1	97.3	98.2	97.5	95.8	98.7	96.6	95.9	97.4	96.4
06	WOOD, PLASTICS & COMPOSITES	86.4	91.6	89.3	94.0	102.1	98.4	104.8	82.8	92.7	98.6	87.8	92.7	88.3	81.2	84.4	90.1	82.8	86.1
07	THERMAL & MOISTURE PROTECTION	105.3	86.7	98.1	104.8	76.6	93.9	99.7	87.7	95.0	102.7	88.3	97.1	102.5	84.7	95.6	102.9	83.8	95.5
08	OPENINGS	102.2	91.4	99.6	103.8	84.5	99.1	99.5	77.5	94.1	98.3	81.5	94.1	98.9	78.3	93.8	99.8	82.1	95.4
0920	Plaster & Gypsum Board	107.8	91.5	97.2	109.1	102.1	104.6	107.9	82.0	91.1	111.0	87.1	95.5	105.2	80.4	89.1	105.6	82.0	90.3
0950, 0980	Ceilings & Acoustic Treatment	103.5	91.5	96.1	103.4	102.1	102.6	94.3	82.0	86.7	92.6	87.1	89.2	91.8	80.4	84.7	91.8	82.0	85.7
0960	Flooring	121.2	71.4	107.7	116.5	58.3	100.8	111.5	105.0	109.7	111.3	93.6	106.5	105.7	95.6	103.0	106.8	94.5	103.5
0970, 0990	Wall Finishes & Painting/Coating	115.7	64.1	84.7	109.3	67.3	84.1	114.4	87.1	98.0	114.4	87.1	98.0	114.4	87.2	98.1	114.4	91.3	100.5
09	FINISHES	107.0	84.7	95.2	107.8	86.0	96.3	103.5	88.5	95.6	103.5	89.9	96.3	101.5	85.7	93.1	101.9	86.2	93.6
COVERS	DIVS. 10 - 14, 25, 28, 41, 43, 44	100.0	72.3	94.5	100.0	95.7	99.2	100.0	83.9	96.8	100.0	85.7	97.2	100.0	83.8	96.8	100.0	93.5	98.7
21, 22, 23	FIRE SUPPRESSION, PLUMBING & HVAC	100.3	100.9	100.6	100.2	93.0	97.4	100.0	83.0	93.3	100.2	84.9	94.1	100.1	87.3	95.0	100.2	90.6	96.4
26, 27, 3370	ELECTRICAL, COMMUNICATIONS & UTIL.	109.7	97.6	103.8	106.1	90.8	98.6	102.5	88.0	95.4	97.2	95.8	96.5	97.7	94.3	96.0	94.8	92.8	93.8
MF2004	WEIGHTED AVERAGE	106.5	93.8	101.1	105.3	88.0	98.0	100.4	87.8	95.1	100.6	90.3	96.3	99.6	88.7	95.0	100.6	89.2	95.8

City Cost Indexes

WISCONSIN

DIVISION		EAU CLAIRE MAT.	INST.	TOTAL	GREEN BAY MAT.	INST.	TOTAL	KENOSHA MAT.	INST.	TOTAL	LA CROSSE MAT.	INST.	TOTAL	MADISON MAT.	INST.	TOTAL	MILWAUKEE MAT.	INST.	TOTAL
015433	CONTRACTOR EQUIPMENT		101.0	101.0		98.7	98.7		98.3	98.3		101.0	101.0		100.4	100.4		89.4	89.4
0241, 31 - 34	SITE & INFRASTRUCTURE, DEMOLITION	90.8	104.0	99.8	94.1	99.5	97.8	97.0	103.1	101.1	84.4	104.1	97.9	91.0	106.5	101.6	91.9	97.4	95.7
0310	Concrete Forming & Accessories	97.0	94.4	94.7	105.0	93.6	95.2	104.1	104.4	104.4	84.4	94.9	93.5	95.9	95.7	95.7	99.6	113.8	111.8
0320	Concrete Reinforcing	100.8	100.0	100.5	98.9	85.6	93.5	100.4	102.3	101.1	100.5	88.2	95.5	100.6	88.6	95.7	100.6	102.7	101.5
0330	Cast-in-Place Concrete	97.8	94.4	96.5	101.2	96.2	99.3	108.2	100.4	105.2	87.9	94.5	90.4	97.3	98.2	97.6	97.1	108.5	101.5
03	CONCRETE	97.7	95.8	96.8	98.3	93.4	96.1	102.5	102.6	102.5	89.7	93.8	91.6	96.9	95.4	96.2	97.2	109.0	102.6
04	MASONRY	89.5	95.1	92.8	119.3	94.2	104.3	100.4	107.7	104.8	88.7	98.0	94.2	101.2	104.0	102.9	103.2	117.5	111.8
05	METALS	91.9	101.6	94.8	94.1	95.7	94.5	99.3	102.6	100.3	91.8	96.5	93.2	101.1	95.7	99.5	101.8	96.2	100.2
06	WOOD, PLASTICS & COMPOSITES	113.5	94.8	103.3	115.8	94.8	104.3	112.2	103.9	107.6	98.3	94.8	96.4	107.0	94.7	100.3	112.3	113.6	113.0
07	THERMAL & MOISTURE PROTECTION	100.1	85.2	94.3	102.0	83.5	94.8	99.8	97.3	98.8	99.6	85.2	94.0	95.5	92.9	94.5	97.8	112.1	103.4
08	OPENINGS	100.7	93.5	98.9	98.8	91.8	97.1	98.4	105.9	100.3	100.7	82.9	96.3	103.3	95.2	101.3	105.6	111.3	107.0
0920	Plaster & Gypsum Board	100.4	95.0	96.9	92.7	95.0	94.2	85.0	104.4	97.6	94.8	95.0	94.9	92.1	95.0	94.0	96.5	114.3	108.1
0950, 0980	Ceilings & Acoustic Treatment	99.9	95.0	96.9	92.5	95.0	94.0	82.6	104.4	96.1	98.3	95.0	96.2	85.3	95.0	91.3	86.8	114.3	103.8
0960	Flooring	97.0	106.7	99.6	116.7	106.7	114.0	122.8	112.2	119.9	89.5	108.7	94.7	98.8	105.0	100.5	107.4	120.5	110.9
0970, 0990	Wall Finishes & Painting/Coating	89.2	91.8	90.8	101.0	74.5	85.1	102.8	101.8	102.2	89.2	71.5	78.6	93.6	93.5	93.5	95.6	113.1	106.1
09	FINISHES	99.4	96.8	98.0	103.8	94.6	99.0	101.3	106.3	103.9	95.6	95.6	95.6	94.8	97.6	96.3	98.8	115.5	107.5
COVERS	DIVS. 10 - 14, 25, 28, 41, 43, 44	100.0	87.3	97.5	100.0	86.6	97.3	100.0	102.3	100.5	100.0	88.3	97.7	100.0	87.9	-97.6	100.0	105.1	101.0
21, 22, 23	FIRE SUPPRESSION, PLUMBING & HVAC	100.1	80.1	92.2	100.4	83.8	93.8	100.1	88.5	95.5	100.1	81.3	92.7	99.7	89.3	95.6	99.9	101.4	100.5
26, 27, 3370	ELECTRICAL, COMMUNICATIONS & UTIL.	101.9	87.7	94.9	96.6	83.4	90.2	98.6	92.9	95.8	102.1	87.7	95.1	101.3	88.9	95.3	99.6	104.3	101.9
MF2004	WEIGHTED AVERAGE	97.9	92.0	95.4	99.7	90.4	95.8	100.0	99.3	99.7	96.3	91.3	94.2	99.5	95.0	97.6	100.3	106.4	102.9

WISCONSIN / WYOMING / CANADA

DIVISION		RACINE (WI) MAT.	INST.	TOTAL	CASPER MAT.	INST.	TOTAL	CHEYENNE MAT.	INST.	TOTAL	ROCK SPRINGS MAT.	INST.	TOTAL	CALGARY, ALBERTA MAT.	INST.	TOTAL	EDMONTON, ALBERTA MAT.	INST.	TOTAL
015433	CONTRACTOR EQUIPMENT		100.4	100.4		101.1	101.1		101.1	101.1		101.1	101.1		106.4	106.4		106.4	106.4
0241, 31 - 34	SITE & INFRASTRUCTURE, DEMOLITION	91.0	107.3	102.1	92.6	100.1	97.7	85.8	100.2	95.6	82.2	98.9	93.6	124.3	104.8	111.0	154.6	104.8	120.6
0310	Concrete Forming & Accessories	97.4	104.6	103.6	103.9	45.0	53.0	105.5	61.3	67.3	101.8	40.4	48.7	127.0	90.2	95.2	128.4	90.2	95.4
0320	Concrete Reinforcing	100.6	102.0	101.3	109.3	46.0	83.5	103.0	46.9	80.1	111.4	46.7	85.0	147.9	63.2	113.4	147.9	63.2	113.4
0330	Cast-in-Place Concrete	97.5	100.2	98.5	102.7	76.1	92.5	98.0	76.3	89.7	99.0	56.8	82.9	186.7	101.0	153.9	208.9	101.0	167.6
03	CONCRETE	97.1	102.6	99.6	108.1	57.1	84.9	105.3	64.5	86.7	106.7	48.4	80.2	157.0	89.4	126.3	167.5	89.4	131.9
04	MASONRY	103.3	107.7	105.9	107.3	39.3	66.6	104.0	54.2	74.2	162.7	48.6	94.4	213.5	91.4	140.4	202.0	91.4	135.8
05	METALS	100.2	102.8	100.9	98.3	64.7	88.5	99.5	66.3	89.8	95.5	64.1	86.3	151.1	88.1	132.7	151.6	88.0	133.0
06	WOOD, PLASTICS & COMPOSITES	110.3	103.9	106.8	97.7	42.4	67.3	99.3	63.7	79.8	96.1	39.1	64.8	114.3	89.5	100.7	110.9	89.5	99.2
07	THERMAL & MOISTURE PROTECTION	99.5	98.6	99.1	97.6	51.9	79.8	99.4	58.1	83.3	100.7	49.9	80.9	123.0	87.9	109.3	123.7	87.9	109.8
08	OPENINGS	103.3	105.9	104.0	92.7	41.9	80.2	94.8	53.5	84.6	99.1	40.1	84.5	92.0	79.9	89.0	92.0	79.9	89.0
0920	Plaster & Gypsum Board	95.6	104.4	101.3	86.7	40.2	56.5	82.1	62.2	69.2	84.1	36.7	53.3	162.0	88.6	114.3	158.2	88.6	113.0
0950, 0980	Ceilings & Acoustic Treatment	83.4	104.4	96.4	112.1	40.2	67.6	106.0	62.2	78.9	108.8	36.7	64.2	144.8	88.6	110.0	148.1	88.6	111.3
0960	Flooring	104.5	112.2	106.6	103.5	37.0	85.5	108.0	59.4	94.9	107.0	52.3	92.2	134.4	88.8	122.1	135.6	88.8	122.9
0970, 0990	Wall Finishes & Painting/Coating	93.5	101.8	98.5	98.0	49.5	68.9	101.2	49.5	70.2	97.3	31.7	57.9	109.6	100.7	104.3	109.7	92.2	99.2
09	FINISHES	96.7	106.3	101.8	104.7	42.9	72.1	104.8	60.0	81.1	102.3	40.4	69.6	132.7	91.6	111.0	135.2	90.7	111.6
COVERS	DIVS. 10 - 14, 25, 28, 41, 43, 44	100.0	102.3	100.5	100.0	79.4	95.9	100.0	82.0	96.4	100.0	65.3	93.1	140.0	102.3	132.5	140.0	102.3	132.5
21, 22, 23	FIRE SUPPRESSION, PLUMBING & HVAC	99.9	93.7	97.5	100.0	60.7	84.4	99.9	59.9	84.0	99.9	57.3	83.0	100.1	87.6	95.1	100.1	87.6	95.2
26, 27, 3370	ELECTRICAL, COMMUNICATIONS & UTIL.	98.0	96.2	97.1	101.2	59.3	80.8	97.6	72.7	85.5	93.3	60.4	77.3	125.2	84.7	105.5	118.5	84.7	102.1
MF2004	WEIGHTED AVERAGE	99.5	101.3	100.2	100.6	58.4	82.9	100.0	65.9	85.7	102.0	56.5	82.9	130.7	89.9	113.6	131.8	89.8	114.1

CANADA

DIVISION		HALIFAX, NOVA SCOTIA MAT.	INST.	TOTAL	HAMILTON, ONTARIO MAT.	INST.	TOTAL	KITCHENER, ONTARIO MAT.	INST.	TOTAL	LAVAL, QUEBEC MAT.	INST.	TOTAL	LONDON, ONTARIO MAT.	INST.	TOTAL	MONTREAL, QUEBEC MAT.	INST.	TOTAL
015433	CONTRACTOR EQUIPMENT		102.9	102.9		107.3	107.3		103.8	103.8		104.0	104.0		104.1	104.1		105.6	105.6
0241, 31 - 34	SITE & INFRASTRUCTURE, DEMOLITION	99.3	99.2	99.2	114.9	111.1	112.3	101.3	104.7	103.6	94.4	101.1	99.0	114.5	104.9	108.0	101.1	101.0	101.0
0310	Concrete Forming & Accessories	96.4	76.4	79.1	129.6	95.4	100.1	117.4	87.8	91.8	129.2	86.1	91.9	129.7	89.0	94.5	135.9	93.7	99.5
0320	Concrete Reinforcing	154.5	61.2	116.4	161.2	90.3	132.3	110.2	90.2	102.0	156.6	79.7	125.2	130.4	88.9	113.5	146.2	93.9	124.9
0330	Cast-in-Place Concrete	171.3	75.7	134.7	151.9	101.4	132.6	140.8	82.6	118.6	134.5	95.5	119.6	149.3	99.4	130.2	174.7	101.9	146.9
03	CONCRETE	150.7	74.0	115.8	143.6	96.6	122.2	121.6	86.7	105.7	134.5	88.5	113.6	136.2	92.8	116.5	151.6	96.7	126.6
04	MASONRY	173.5	82.9	117.9	181.0	101.5	133.4	161.9	97.8	123.5	160.9	85.4	115.7	182.4	99.1	132.5	167.3	93.7	123.2
05	METALS	120.4	81.4	109.0	127.9	94.2	118.1	116.1	93.8	109.6	105.5	89.3	100.8	125.3	92.1	115.6	123.5	94.8	115.4
06	WOOD, PLASTICS & COMPOSITES	86.1	75.5	80.3	115.5	94.6	104.0	109.3	86.1	96.6	127.9	85.7	104.8	115.5	87.0	99.9	127.9	93.9	109.3
07	THERMAL & MOISTURE PROTECTION	106.0	76.2	94.4	120.1	96.5	110.9	110.0	93.2	103.5	105.1	88.3	98.5	120.9	93.2	110.1	117.1	96.9	109.3
08	OPENINGS	82.8	68.9	79.4	92.0	91.4	91.9	83.3	85.2	83.7	92.0	74.7	87.7	93.2	85.4	91.2	92.0	82.1	89.6
0920	Plaster & Gypsum Board	163.1	74.4	105.5	195.8	94.3	129.9	152.4	85.5	109.0	145.7	84.9	106.2	199.5	86.4	126.1	152.5	93.2	114.0
0950, 0980	Ceilings & Acoustic Treatment	105.4	74.4	86.2	120.6	94.3	104.3	105.4	85.5	93.1	95.4	84.9	88.9	138.5	86.4	106.3	102.0	93.2	96.5
0960	Flooring	110.1	64.5	97.8	134.4	94.8	123.7	127.8	94.8	118.9	132.0	95.4	122.1	134.6	94.8	123.9	133.9	95.4	123.5
0970, 0990	Wall Finishes & Painting/Coating	109.6	74.4	88.5	109.6	100.5	104.2	109.6	89.7	97.7	109.6	88.6	97.0	109.6	97.0	102.0	109.6	104.2	106.4
09	FINISHES	114.2	74.2	93.1	130.8	95.8	112.3	119.2	86.9	102.6	116.0	88.2	101.3	135.6	90.6	111.8	119.2	95.8	106.9
COVERS	DIVS. 10 - 14, 25, 28, 41, 43, 44	140.0	73.8	126.9	140.0	106.6	133.4	140.0	104.6	133.0	140.0	91.6	130.4	140.0	105.2	133.1	140.0	94.1	130.9
21, 22, 23	FIRE SUPPRESSION, PLUMBING & HVAC	99.2	76.1	90.0	100.0	91.9	96.8	99.2	88.9	95.1	99.2	92.7	96.6	100.0	89.0	95.6	100.1	94.4	97.8
26, 27, 3370	ELECTRICAL, COMMUNICATIONS & UTIL.	128.4	74.6	102.2	128.8	96.1	112.9	124.1	93.4	109.1	123.5	74.4	99.6	123.3	93.4	108.8	129.8	86.1	108.5
MF2004	WEIGHTED AVERAGE	119.0	78.0	101.8	123.4	96.9	112.3	114.5	92.2	105.1	114.5	87.6	103.2	122.0	93.3	110.0	121.7	93.8	110.0

		CANADA																
	DIVISION	OSHAWA, ONTARIO			OTTAWA, ONTARIO			QUEBEC, QUEBEC			REGINA, SASKATCHEWAN			SASKATOON, SASKATCHEWAN			ST CATHARINES, ONTARIO		
		MAT.	INST.	TOTAL	MAT.	INST.	TOTAL	MAT.	INST.	TOTAL	MAT.	INST.	TOTAL	MAT.	INST.	TOTAL	MAT.	INST.	TOTAL
015433	CONTRACTOR EQUIPMENT		103.8	103.8		103.8	103.8		106.0	106.0		101.2	101.2		101.2	101.2		102.2	102.2
0241, 31 - 34	SITE & INFRASTRUCTURE, DEMOLITION	114.1	104.0	107.2	111.7	104.6	106.9	99.7	101.1	100.7	113.0	96.8	101.9	107.4	97.0	100.3	102.1	101.9	102.0
0310	Concrete Forming & Accessories	123.3	87.7	92.6	130.2	91.7	97.0	137.1	94.0	99.8	106.0	59.5	65.8	106.0	59.4	65.7	115.3	92.6	95.7
0320	Concrete Reinforcing	174.5	85.3	138.1	160.1	88.8	131.0	140.1	94.0	121.3	125.7	62.6	100.0	118.8	62.6	95.9	111.1	90.2	102.6
0330	Cast-in-Place Concrete	162.9	87.0	133.9	153.0	98.5	132.2	148.3	102.3	130.7	156.3	70.3	123.4	142.0	70.3	114.6	134.4	101.2	121.7
03	CONCRETE	151.0	87.4	122.1	143.9	93.7	121.1	138.1	96.9	119.4	131.1	64.8	100.9	123.1	64.7	96.5	118.6	95.3	108.0
04	MASONRY	165.4	93.8	122.5	179.7	97.2	130.3	174.3	93.7	126.0	166.4	63.6	104.8	165.9	63.6	104.6	161.4	101.5	125.5
05	METALS	107.1	92.7	102.9	126.2	93.7	116.7	125.2	95.1	116.4	105.5	76.4	97.0	105.5	76.3	97.0	106.4	93.7	102.7
06	WOOD, PLASTICS & COMPOSITES	116.4	85.9	99.6	115.2	90.9	101.9	128.6	94.0	109.6	95.8	57.8	74.9	94.2	57.8	74.2	106.9	91.5	98.4
07	THERMAL & MOISTURE PROTECTION	110.9	86.4	101.4	123.4	92.7	111.5	118.6	97.1	110.2	106.0	63.5	89.5	104.9	62.5	88.4	110.0	93.4	103.5
08	OPENINGS	90.8	85.8	89.6	92.0	88.4	91.1	92.0	89.6	91.4	87.0	55.6	79.2	86.0	55.6	78.5	82.7	88.5	84.2
0920	Plaster & Gypsum Board	155.3	85.3	108.9	234.7	90.5	141.1	191.0	93.2	127.4	165.1	56.2	94.3	145.5	56.2	87.5	137.2	91.1	107.2
0950, 0980	Ceilings & Acoustic Treatment	99.6	85.3	90.7	131.4	90.5	106.1	93.6	93.2	93.3	119.5	56.2	80.3	119.5	56.2	80.3	99.6	91.1	94.3
0960	Flooring	132.0	97.2	122.6	134.4	93.4	123.4	134.4	95.4	123.9	119.8	60.7	103.8	119.8	60.7	103.8	126.0	94.8	117.6
0970, 0990	Wall Finishes & Painting/Coating	109.6	104.9	106.8	109.6	91.4	98.7	110.2	104.2	106.6	109.6	64.5	82.6	109.6	55.0	76.9	109.6	100.5	104.2
09	FINISHES	119.0	91.1	104.2	138.2	92.0	113.8	123.1	95.9	108.7	122.0	59.8	89.1	119.1	58.7	87.2	114.0	94.0	103.4
COVERS	DIVS. 10 - 14, 25, 28, 41, 43, 44	140.0	104.7	133.0	140.0	103.1	132.7	140.0	94.3	130.9	140.0	69.7	126.1	140.0	69.7	126.1	140.0	81.2	128.4
21, 22, 23	FIRE SUPPRESSION, PLUMBING & HVAC	99.2	103.2	100.8	100.2	89.4	95.9	99.9	94.4	97.7	99.4	77.6	90.8	99.3	77.6	90.7	99.2	90.6	95.8
26, 27, 3370	ELECTRICAL, COMMUNICATIONS & UTIL.	125.3	92.7	109.4	119.7	94.4	107.4	121.7	86.1	104.4	126.7	64.4	96.4	127.0	64.4	96.5	126.3	94.5	110.8
MF2004	WEIGHTED AVERAGE	117.9	94.7	108.2	122.7	93.9	110.6	120.1	94.2	109.2	115.1	69.9	96.1	113.6	69.8	95.2	112.3	94.2	104.7

		CANADA																	
	DIVISION	ST JOHNS, NEWFOUNDLAND			THUNDER BAY, ONTARIO			TORONTO, ONTARIO			VANCOUVER, BRITISH COLUMBIA			WINDSOR, ONTARIO			WINNIPEG, MANITOBA		
		MAT.	INST.	TOTAL	MAT.	INST.	TOTAL	MAT.	INST.	TOTAL	MAT.	INST.	TOTAL	MAT.	INST.	TOTAL	MAT.	INST.	TOTAL
015433	CONTRACTOR EQUIPMENT		104.8	104.8		102.2	102.2		104.1	104.1		112.7	112.7		102.2	102.2		106.4	106.4
0241, 31 - 34	SITE & INFRASTRUCTURE, DEMOLITION	115.7	99.5	104.7	107.6	101.8	103.7	139.0	105.5	116.1	121.8	107.4	111.9	97.2	101.5	100.1	119.3	100.9	106.7
0310	Concrete Forming & Accessories	103.8	64.3	69.7	123.3	93.0	97.1	131.1	100.7	104.8	124.8	78.0	84.4	123.3	90.0	94.5	130.3	65.5	74.3
0320	Concrete Reinforcing	163.8	57.7	120.5	99.3	89.6	95.4	157.0	91.0	130.1	156.7	72.3	122.3	108.9	88.9	100.7	147.9	56.0	110.4
0330	Cast-in-Place Concrete	172.4	77.6	136.2	148.0	99.8	129.6	147.4	109.5	132.9	163.0	92.2	135.9	137.7	100.8	123.6	183.5	72.3	141.0
03	CONCRETE	164.6	68.7	121.0	126.3	94.8	112.0	140.8	101.8	123.1	153.2	82.7	121.1	120.2	93.7	108.2	155.5	67.1	115.3
04	MASONRY	164.6	65.1	105.1	162.1	101.3	125.7	193.0	106.7	141.4	177.4	81.1	119.8	161.5	99.9	124.6	180.0	62.1	109.4
05	METALS	108.2	79.1	99.7	106.2	92.7	102.3	127.7	94.9	118.2	165.9	89.2	143.5	106.3	93.4	102.5	159.9	78.6	136.1
06	WOOD, PLASTICS & COMPOSITES	95.2	64.3	78.3	116.4	91.6	102.8	116.4	99.4	107.1	114.3	75.1	92.8	116.4	88.3	101.0	114.2	66.0	87.7
07	THERMAL & MOISTURE PROTECTION	109.3	63.4	91.5	110.2	94.8	104.2	121.0	101.2	113.3	132.4	82.1	112.8	110.0	93.6	103.6	114.1	67.3	95.9
08	OPENINGS	98.2	60.2	88.8	81.7	87.9	83.3	90.8	96.0	92.1	93.8	75.8	89.3	81.5	86.6	82.7	92.0	60.7	84.3
0920	Plaster & Gypsum Board	168.5	62.7	99.8	166.5	91.2	117.6	177.3	99.3	126.6	154.9	73.5	102.0	156.7	87.8	111.9	157.0	64.1	96.7
0950, 0980	Ceilings & Acoustic Treatment	104.6	62.7	78.7	95.4	91.2	92.8	132.6	99.3	112.0	141.0	73.5	99.3	95.4	87.8	90.7	121.9	64.1	86.2
0960	Flooring	114.8	54.9	98.6	132.0	54.6	111.1	135.6	100.5	126.1	134.8	92.0	123.2	132.0	95.5	122.1	133.3	67.3	115.5
0970, 0990	Wall Finishes & Painting/Coating	109.6	64.5	82.6	109.6	93.2	99.8	111.5	104.9	107.5	109.6	88.7	97.1	109.6	92.8	99.5	109.7	48.1	72.7
09	FINISHES	117.4	62.5	88.3	119.2	86.7	102.0	133.2	101.2	116.3	130.7	81.2	104.5	117.5	91.3	103.6	125.4	64.2	93.1
COVERS	DIVS. 10 - 14, 25, 28, 41, 43, 44	140.0	71.2	126.4	140.0	81.6	128.4	140.0	108.5	133.8	140.0	99.9	132.1	140.0	80.6	128.2	140.0	71.7	126.5
21, 22, 23	FIRE SUPPRESSION, PLUMBING & HVAC	99.6	64.2	85.5	99.2	90.7	95.8	99.9	95.9	98.3	99.9	73.3	89.3	99.2	90.2	95.6	99.9	65.6	86.3
26, 27, 3370	ELECTRICAL, COMMUNICATIONS & UTIL.	122.9	65.8	95.1	124.1	92.7	108.8	122.0	96.8	109.8	124.7	75.0	100.5	130.2	94.4	112.8	127.9	67.0	98.3
MF2004	WEIGHTED AVERAGE	119.9	69.2	98.6	113.5	92.9	104.9	123.6	99.8	113.6	131.2	82.0	110.5	113.0	93.2	104.7	129.6	69.6	104.4

Location Factors

Costs shown in RSMeans cost data publications are based on national averages for materials and installation. To adjust these costs to a specific location, simply multiply the base cost by the factor and divide by 100 for that city. The data is arranged alphabetically by state and postal zip code numbers. For a city not listed, use the factor for a nearby city with similar economic characteristics.

STATE/ZIP	CITY	MAT.	INST.	TOTAL
ALABAMA				
350-352	Birmingham	98.3	74.9	88.5
354	Tuscaloosa	97.0	60.5	81.7
355	Jasper	97.2	54.0	79.0
356	Decatur	97.0	59.9	81.4
357-358	Huntsville	97.0	71.1	86.1
359	Gadsden	96.9	59.4	81.2
360-361	Montgomery	98.0	59.2	81.7
362	Anniston	96.3	52.9	78.1
363	Dothan	96.9	52.5	78.3
364	Evergreen	96.4	57.3	80.0
365-366	Mobile	97.9	65.9	84.5
367	Selma	96.6	54.4	78.9
368	Phenix City	97.3	59.4	81.4
369	Butler	96.8	55.7	79.5
ALASKA				
995-996	Anchorage	128.3	110.3	120.8
997	Fairbanks	127.6	112.6	121.3
998	Juneau	127.0	108.5	119.2
999	Ketchikan	139.6	108.1	126.4
ARIZONA				
850,853	Phoenix	99.9	74.0	89.0
852	Mesa/Tempe	99.2	66.7	85.6
855	Globe	99.5	63.1	84.2
856-857	Tucson	98.0	70.6	86.5
859	Show Low	99.5	64.2	84.7
860	Flagstaff	101.4	71.6	88.9
863	Prescott	99.2	63.7	84.3
864	Kingman	97.5	69.6	85.8
865	Chambers	97.5	64.4	83.6
ARKANSAS				
716	Pine Bluff	96.9	66.6	84.2
717	Camden	94.6	44.6	73.6
718	Texarkana	95.9	48.8	76.1
719	Hot Springs	93.8	47.9	74.5
720-722	Little Rock	97.1	69.9	85.7
723	West Memphis	96.0	59.6	80.8
724	Jonesboro	96.9	60.2	81.5
725	Batesville	94.7	54.3	77.7
726	Harrison	95.9	55.0	78.8
727	Fayetteville	93.4	54.7	77.1
728	Russellville	94.5	56.7	78.6
729	Fort Smith	97.4	60.1	81.8
CALIFORNIA				
900-902	Los Angeles	102.2	116.7	108.3
903-905	Inglewood	97.6	107.9	101.9
906-908	Long Beach	99.3	108.0	102.9
910-912	Pasadena	98.0	107.9	102.2
913-916	Van Nuys	101.2	107.9	104.0
917-918	Alhambra	100.2	107.9	103.4
919-921	San Diego	102.4	108.8	105.1
922	Palm Springs	99.2	106.8	102.4
923-924	San Bernardino	96.9	107.2	101.2
925	Riverside	101.6	112.7	106.3
926-927	Santa Ana	98.9	107.5	102.5
928	Anaheim	101.5	114.7	107.0
930	Oxnard	102.9	112.8	107.0
931	Santa Barbara	101.9	112.9	106.5
932-933	Bakersfield	103.3	109.4	105.8
934	San Luis Obispo	102.5	105.3	103.7
935	Mojave	99.6	104.3	101.6
936-938	Fresno	103.2	112.6	107.2
939	Salinas	103.3	116.4	108.8
940-941	San Francisco	111.2	141.3	123.8
942,956-958	Sacramento	104.9	114.6	108.9
943	Palo Alto	103.5	123.7	112.0
944	San Mateo	106.4	129.3	116.0
945	Vallejo	104.1	121.6	111.4
946	Oakland	108.8	129.5	117.5
947	Berkeley	108.3	123.8	114.8
948	Richmond	107.4	123.9	114.3
949	San Rafael	108.7	124.5	115.3
950	Santa Cruz	108.4	116.6	111.8

STATE/ZIP	CITY	MAT.	INST.	TOTAL
CALIFORNIA (CONT'D)				
951	San Jose	106.0	132.5	117.1
952	Stockton	103.5	112.0	107.0
953	Modesto	103.4	111.8	106.9
954	Santa Rosa	103.3	126.1	112.9
955	Eureka	104.7	109.6	106.7
959	Marysville	104.0	111.2	107.0
960	Redding	105.8	109.8	107.5
961	Susanville	104.6	110.0	106.8
COLORADO				
800-802	Denver	101.8	85.5	95.0
803	Boulder	98.6	82.0	91.6
804	Golden	100.7	80.8	92.4
805	Fort Collins	102.2	77.3	91.7
806	Greeley	99.7	67.0	86.0
807	Fort Morgan	99.2	80.6	91.4
808-809	Colorado Springs	101.5	81.0	92.9
810	Pueblo	101.3	79.0	91.9
811	Alamosa	102.4	75.3	91.0
812	Salida	102.2	76.2	91.3
813	Durango	102.8	76.0	91.6
814	Montrose	101.4	74.4	90.1
815	Grand Junction	104.9	72.0	91.1
816	Glenwood Springs	102.4	77.7	92.1
CONNECTICUT				
060	New Britain	99.2	118.0	107.1
061	Hartford	100.5	118.1	107.9
062	Willimantic	99.8	117.5	107.2
063	New London	95.8	117.1	104.8
064	Meriden	97.9	118.5	106.6
065	New Haven	101.0	118.0	108.2
066	Bridgeport	100.3	118.6	108.0
067	Waterbury	99.9	118.5	107.7
068	Norwalk	99.9	119.0	107.9
069	Stamford	100.0	125.4	110.7
D.C.				
200-205	Washington	104.6	92.8	99.7
DELAWARE				
197	Newark	100.5	105.8	102.7
198	Wilmington	100.6	105.9	102.8
199	Dover	101.2	105.8	103.1
FLORIDA				
320,322	Jacksonville	98.8	63.1	83.8
321	Daytona Beach	99.1	76.5	89.6
323	Tallahassee	99.6	53.0	80.1
324	Panama City	100.3	48.6	78.6
325	Pensacola	102.9	61.3	85.4
326,344	Gainesville	100.5	66.0	86.0
327-328,347	Orlando	103.1	72.0	90.0
329	Melbourne	102.0	80.0	92.7
330-332,340	Miami	101.2	75.4	90.3
333	Fort Lauderdale	98.5	73.6	88.0
334,349	West Palm Beach	97.2	69.8	85.7
335-336,346	Tampa	100.1	78.7	91.1
337	St. Petersburg	102.3	58.1	83.7
338	Lakeland	99.2	77.8	90.2
339,341	Fort Myers	98.6	70.1	86.6
342	Sarasota	100.4	71.1	88.1
GEORGIA				
300-303,399	Atlanta	98.2	79.2	90.2
304	Statesboro	97.9	48.1	77.0
305	Gainesville	96.5	63.2	82.5
306	Athens	95.9	66.9	83.7
307	Dalton	98.0	52.3	78.8
308-309	Augusta	96.9	64.3	83.2
310-312	Macon	96.5	66.6	83.9
313-314	Savannah	98.8	62.0	83.3
315	Waycross	97.7	57.2	80.7
316	Valdosta	97.8	51.0	78.2
317,398	Albany	97.9	62.0	82.9
318-319	Columbus	97.8	66.8	84.8

Location Factors

STATE/ZIP	CITY	MAT.	INST.	TOTAL
HAWAII				
967	Hilo	114.2	118.7	116.1
968	Honolulu	118.7	118.7	118.7
STATES & POSS.				
969	Guam	138.1	64.3	107.1
IDAHO				
832	Pocatello	101.6	74.5	90.2
833	Twin Falls	102.5	54.0	82.1
834	Idaho Falls	99.9	59.8	83.0
835	Lewiston	109.0	80.4	97.0
836-837	Boise	100.6	74.8	89.8
838	Coeur d'Alene	108.3	77.4	95.3
ILLINOIS				
600-603	North Suburban	97.4	123.7	108.5
604	Joliet	97.4	126.9	109.8
605	South Suburban	97.4	124.5	108.8
606-608	Chicago	97.9	138.3	114.9
609	Kankakee	93.6	107.0	99.2
610-611	Rockford	97.1	115.0	104.7
612	Rock Island	94.7	99.5	96.7
613	La Salle	95.9	106.3	100.3
614	Galesburg	95.7	99.0	97.1
615-616	Peoria	98.2	104.9	101.0
617	Bloomington	95.1	106.4	99.8
618-619	Champaign	98.4	105.0	101.2
620-622	East St. Louis	93.9	105.7	98.8
623	Quincy	94.9	94.8	94.9
624	Effingham	94.3	95.9	95.0
625	Decatur	96.2	102.5	98.8
626-627	Springfield	97.0	103.4	99.7
628	Centralia	92.4	100.5	95.8
629	Carbondale	92.1	94.1	92.9
INDIANA				
460	Anderson	94.8	82.5	89.6
461-462	Indianapolis	98.0	86.7	93.2
463-464	Gary	96.3	101.5	98.5
465-466	South Bend	95.6	83.1	90.4
467-468	Fort Wayne	95.3	78.4	88.2
469	Kokomo	92.5	80.8	87.6
470	Lawrenceburg	91.1	76.9	85.1
471	New Albany	92.4	74.4	84.9
472	Columbus	94.7	79.5	88.3
473	Muncie	95.5	80.6	89.3
474	Bloomington	96.9	78.8	89.3
475	Washington	92.7	82.4	88.4
476-477	Evansville	94.4	86.5	91.1
478	Terre Haute	95.0	86.6	91.5
479	Lafayette	94.3	82.1	89.2
IOWA				
500-503,509	Des Moines	98.2	76.7	89.2
504	Mason City	95.7	61.4	81.3
505	Fort Dodge	96.0	57.1	79.7
506-507	Waterloo	97.8	58.4	81.2
508	Creston	96.2	61.2	81.5
510-511	Sioux City	98.4	68.8	86.0
512	Sibley	96.9	47.6	76.2
513	Spencer	98.5	46.3	76.6
514	Carroll	95.6	51.4	77.0
515	Council Bluffs	99.6	74.6	89.1
516	Shenandoah	96.3	49.3	76.6
520	Dubuque	97.9	75.8	88.6
521	Decorah	96.5	48.8	76.5
522-524	Cedar Rapids	98.8	80.8	91.2
525	Ottumwa	96.6	69.3	85.1
526	Burlington	95.8	70.7	85.2
527-528	Davenport	97.7	90.2	94.5
KANSAS				
660-662	Kansas City	98.3	93.7	96.4
664-666	Topeka	98.9	66.1	85.1
667	Fort Scott	96.6	71.4	86.0
668	Emporia	96.7	59.7	81.2
669	Belleville	98.4	61.0	82.7
670-672	Wichita	97.6	64.9	83.9
673	Independence	98.1	63.3	83.5
674	Salina	98.7	61.6	83.1
675	Hutchinson	93.8	61.2	80.1
676	Hays	97.6	62.5	82.9
677	Colby	98.4	62.3	83.3

STATE/ZIP	CITY	MAT.	INST.	TOTAL
KANSAS (CONT'D)				
678	Dodge City	99.9	64.0	84.8
679	Liberal	97.5	62.2	82.7
KENTUCKY				
400-402	Louisville	97.1	83.9	91.6
403-405	Lexington	95.1	78.5	88.1
406	Frankfort	96.3	78.9	89.0
407-409	Corbin	92.3	65.8	81.2
410	Covington	92.8	99.3	95.5
411-412	Ashland	91.7	96.7	93.8
413-414	Campton	93.1	66.2	81.8
415-416	Pikeville	94.0	81.7	88.8
417-418	Hazard	92.4	58.7	78.3
420	Paducah	90.9	87.2	89.4
421-422	Bowling Green	93.6	84.6	89.8
423	Owensboro	93.4	83.2	89.1
424	Henderson	90.7	86.8	89.0
425-426	Somerset	90.6	70.3	82.1
427	Elizabethtown	90.1	83.6	87.3
LOUISIANA				
700-701	New Orleans	102.9	69.6	88.9
703	Thibodaux	100.4	63.7	85.0
704	Hammond	97.6	58.4	81.2
705	Lafayette	100.0	58.5	82.6
706	Lake Charles	100.1	61.1	83.7
707-708	Baton Rouge	102.2	63.8	86.1
710-711	Shreveport	96.8	59.2	81.0
712	Monroe	96.9	56.0	79.7
713-714	Alexandria	97.0	56.0	79.8
MAINE				
039	Kittery	93.0	74.4	85.2
040-041	Portland	98.5	73.7	88.1
042	Lewiston	96.4	73.7	86.9
043	Augusta	96.1	74.3	86.9
044	Bangor	95.8	73.7	86.5
045	Bath	94.2	74.4	85.9
046	Machias	93.7	73.9	85.4
047	Houlton	93.9	74.4	85.7
048	Rockland	92.9	74.5	85.2
049	Waterville	94.2	74.4	85.9
MARYLAND				
206	Waldorf	100.5	69.5	87.5
207-208	College Park	100.4	79.4	91.6
209	Silver Spring	99.7	75.5	89.5
210-212	Baltimore	98.2	86.0	93.1
214	Annapolis	98.7	80.2	90.9
215	Cumberland	94.0	80.4	88.3
216	Easton	95.5	42.8	73.4
217	Hagerstown	94.9	81.6	89.3
218	Salisbury	95.8	50.3	76.7
219	Elkton	93.1	62.9	80.4
MASSACHUSETTS				
010-011	Springfield	98.4	104.5	101.0
012	Pittsfield	97.8	101.0	99.1
013	Greenfield	95.7	102.3	98.4
014	Fitchburg	94.4	117.6	104.1
015-016	Worcester	98.3	118.9	107.0
017	Framingham	93.9	126.1	107.4
018	Lowell	97.7	125.7	109.5
019	Lawrence	98.9	123.6	109.3
020-022, 024	Boston	100.5	135.9	115.4
023	Brockton	99.2	120.7	108.2
025	Buzzards Bay	93.3	118.2	103.8
026	Hyannis	96.2	118.7	105.7
027	New Bedford	98.4	118.9	107.0
MICHIGAN				
480,483	Royal Oak	91.4	103.9	96.6
481	Ann Arbor	93.7	104.5	98.2
482	Detroit	95.3	114.2	103.2
484-485	Flint	93.4	100.8	96.5
486	Saginaw	93.0	89.9	91.7
487	Bay City	93.1	90.0	91.8
488-489	Lansing	94.4	96.9	95.5
490	Battle Creek	94.5	88.2	91.9
491	Kalamazoo	94.8	85.6	91.0
492	Jackson	92.7	91.6	92.3
493,495	Grand Rapids	96.1	65.2	83.1
494	Muskegon	93.5	81.4	88.4

Location Factors

STATE/ZIP	CITY	MAT.	INST.	TOTAL
MICHIGAN (CONT'D)				
496	Traverse City	92.3	69.1	82.5
497	Gaylord	93.3	72.1	84.4
498-499	Iron Mountain	95.3	83.2	90.2
MINNESOTA				
550-551	Saint Paul	98.6	118.7	107.0
553-555	Minneapolis	99.3	124.2	109.8
556-558	Duluth	98.2	108.2	102.4
MINNESOTA (CONT'd)				
559	Rochester	98.1	104.7	100.9
560	Mankato	96.0	103.2	99.0
561	Windom	94.8	77.6	87.6
562	Willmar	94.3	82.8	89.5
563	St. Cloud	95.8	117.5	104.9
564	Brainerd	95.9	99.5	97.4
565	Detroit Lakes	97.7	94.6	96.4
566	Bemidji	97.1	95.9	96.6
567	Thief River Falls	96.7	91.7	94.6
MISSISSIPPI				
386	Clarksdale	96.6	60.4	81.4
387	Greenville	100.0	70.9	87.8
388	Tupelo	97.9	63.1	83.2
389	Greenwood	97.8	59.9	81.9
390-392	Jackson	98.5	69.9	86.5
393	Meridian	96.8	70.9	85.9
394	Laurel	97.8	63.7	83.5
395	Biloxi	98.5	62.6	83.4
396	Mccomb	96.3	59.3	80.8
397	Columbus	97.7	61.0	82.3
MISSOURI				
630-631	St. Louis	98.6	108.8	102.9
633	Bowling Green	97.1	90.3	94.2
634	Hannibal	95.9	80.1	89.3
635	Kirksville	98.7	73.3	88.0
636	Flat River	98.1	91.1	95.1
637	Cape Girardeau	98.2	87.4	93.7
638	Sikeston	96.1	77.2	88.2
639	Poplar Bluff	95.6	77.7	88.1
640-641	Kansas City	99.2	106.1	102.1
644-645	St. Joseph	98.3	90.2	94.9
646	Chillicothe	95.1	68.5	83.9
647	Harrisonville	94.7	98.4	96.2
648	Joplin	97.1	68.0	84.9
650-651	Jefferson City	95.4	86.0	91.5
652	Columbia	96.3	88.1	92.9
653	Sedalia	95.4	81.7	89.6
654-655	Rolla	94.1	73.0	85.2
656-658	Springfield	97.3	77.6	89.1
MONTANA				
590-591	Billings	101.9	74.6	90.4
592	Wolf Point	101.1	71.5	88.7
593	Miles City	98.9	72.8	88.0
594	Great Falls	103.1	74.8	91.2
595	Havre	100.0	72.8	88.6
596	Helena	101.8	73.5	89.9
597	Butte	101.6	75.0	90.4
598	Missoula	99.0	73.2	88.2
599	Kalispell	97.9	72.0	87.0
NEBRASKA				
680-681	Omaha	99.5	78.8	90.8
683-685	Lincoln	97.6	77.1	89.0
686	Columbus	95.5	76.4	87.5
687	Norfolk	97.2	80.4	90.1
688	Grand Island	97.3	82.4	91.0
689	Hastings	96.5	84.8	91.6
690	Mccook	96.4	75.1	87.5
691	North Platte	96.8	85.0	91.8
692	Valentine	98.7	73.6	88.2
693	Alliance	98.6	71.6	87.3
NEVADA				
889-891	Las Vegas	101.1	112.0	105.7
893	Ely	99.5	72.0	88.0
894-895	Reno	99.8	93.9	97.3
897	Carson City	99.9	93.1	97.1
898	Elko	98.2	77.6	89.6
NEW HAMPSHIRE				
030	Nashua	98.6	87.7	94.0
031	Manchester	99.1	87.7	94.3

STATE/ZIP	CITY	MAT.	INST.	TOTAL
NEW HAMPSHIRE (CONT'D)				
032-033	Concord	97.0	83.8	91.5
034	Keene	95.1	53.4	77.6
035	Littleton	95.0	60.6	80.5
036	Charleston	94.6	50.8	76.2
037	Claremont	93.7	50.8	75.7
038	Portsmouth	95.9	90.1	93.5
NEW JERSEY				
070-071	Newark	100.5	122.0	109.5
072	Elizabeth	98.6	121.5	108.2
073	Jersey City	97.8	122.2	108.0
074-075	Paterson	99.3	121.8	108.8
076	Hackensack	97.3	122.3	107.8
077	Long Branch	97.1	121.2	107.2
078	Dover	97.6	122.0	107.8
079	Summit	97.6	121.5	107.6
080,083	Vineland	95.7	117.6	104.9
081	Camden	97.8	117.8	106.2
082,084	Atlantic City	96.3	117.6	105.2
085-086	Trenton	97.6	120.6	107.3
087	Point Pleasant	97.7	120.8	107.4
088-089	New Brunswick	98.1	121.4	107.9
NEW MEXICO				
870-872	Albuquerque	99.8	76.0	89.8
873	Gallup	99.9	76.0	89.9
874	Farmington	100.3	76.0	90.1
875	Santa Fe	102.0	76.0	91.1
877	Las Vegas	98.4	75.8	88.9
878	Socorro	97.9	76.0	88.7
879	Truth/Consequences	98.1	70.9	86.7
880	Las Cruces	97.0	68.5	85.0
881	Clovis	98.5	74.2	88.3
882	Roswell	100.3	74.3	89.3
883	Carrizozo	100.7	76.0	90.3
884	Tucumcari	99.3	74.2	88.8
NEW YORK				
100-102	New York	106.1	164.6	130.7
103	Staten Island	101.9	162.0	127.1
104	Bronx	100.0	162.0	126.0
105	Mount Vernon	99.8	132.8	113.6
106	White Plains	99.8	132.8	113.6
107	Yonkers	104.7	132.8	116.5
108	New Rochelle	100.1	132.8	113.8
109	Suffern	99.9	121.8	109.1
110	Queens	101.3	161.9	126.7
111	Long Island City	102.9	161.9	127.6
112	Brooklyn	103.1	161.9	127.8
113	Flushing	103.0	161.9	127.7
114	Jamaica	101.2	161.9	126.7
115,117,118	Hicksville	101.2	146.4	120.1
116	Far Rockaway	103.1	161.9	127.8
119	Riverhead	102.0	146.4	120.6
120-122	Albany	97.4	94.7	96.3
123	Schenectady	97.7	95.3	96.7
124	Kingston	100.8	112.5	105.7
125-126	Poughkeepsie	100.0	127.3	111.5
127	Monticello	99.3	115.7	106.2
128	Glens Falls	92.2	91.0	91.7
129	Plattsburgh	96.7	85.9	92.2
130-132	Syracuse	98.7	93.1	96.4
133-135	Utica	96.7	89.5	93.6
136	Watertown	98.4	92.2	95.8
137-139	Binghamton	98.1	86.9	93.4
140-142	Buffalo	99.0	104.8	101.5
143	Niagara Falls	96.7	101.5	98.7
144-146	Rochester	98.4	94.1	96.6
147	Jamestown	95.6	82.5	90.1
148-149	Elmira	95.5	84.4	90.8
NORTH CAROLINA				
270,272-274	Greensboro	99.8	50.7	79.2
271	Winston-Salem	99.5	50.0	78.7
275-276	Raleigh	100.7	51.3	80.0
277	Durham	101.1	51.7	80.4
278	Rocky Mount	96.9	41.6	73.7
279	Elizabeth City	97.7	43.0	74.7
280	Gastonia	98.5	49.6	78.0
281-282	Charlotte	100.2	52.2	80.0
283	Fayetteville	101.1	52.1	80.5
284	Wilmington	97.2	49.6	77.2
285	Kinston	95.2	43.3	73.4

Location Factors

STATE/ZIP	CITY	MAT.	INST.	TOTAL
NORTH CAROLINA (CONT'D)				
286	Hickory	95.5	45.9	74.7
287-288	Asheville	97.6	49.6	77.5
289	Murphy	96.3	35.7	70.9
NORTH DAKOTA				
580-581	Fargo	101.4	62.0	84.9
582	Grand Forks	100.8	55.1	81.6
583	Devils Lake	100.0	57.3	82.0
584	Jamestown	100.0	49.1	78.7
585	Bismarck	100.1	62.9	84.5
586	Dickinson	100.9	60.1	83.7
587	Minot	100.9	67.3	86.8
588	Williston	99.4	60.1	82.9
OHIO				
430-432	Columbus	96.3	88.6	93.0
433	Marion	92.6	83.5	88.8
434-436	Toledo	96.5	99.7	97.9
437-438	Zanesville	93.1	82.5	88.7
439	Steubenville	94.1	91.6	93.0
440	Lorain	96.9	93.6	95.5
441	Cleveland	97.2	102.6	99.5
442-443	Akron	97.8	93.7	96.1
444-445	Youngstown	97.2	89.7	94.1
446-447	Canton	97.3	85.0	92.1
448-449	Mansfield	94.4	88.3	91.9
450	Hamilton	95.0	85.8	91.1
451-452	Cincinnati	95.3	88.1	92.3
453-454	Dayton	95.2	85.5	91.1
455	Springfield	95.1	85.7	91.2
456	Chillicothe	93.8	91.4	92.8
457	Athens	96.7	75.9	88.0
458	Lima	97.0	85.3	92.1
OKLAHOMA				
730-731	Oklahoma City	98.7	61.7	83.1
734	Ardmore	95.2	61.1	80.9
735	Lawton	97.8	62.5	83.0
736	Clinton	96.6	59.7	81.1
737	Enid	97.4	59.7	81.6
738	Woodward	95.4	59.8	80.4
739	Guymon	96.4	32.0	69.4
740-741	Tulsa	97.6	56.3	80.3
743	Miami	94.1	64.5	81.7
744	Muskogee	96.9	41.6	73.7
745	Mcalester	93.8	52.4	76.5
746	Ponca City	94.4	60.4	80.1
747	Durant	94.4	60.4	80.2
748	Shawnee	95.9	57.8	79.9
749	Poteau	93.5	62.8	80.6
OREGON				
970-972	Portland	101.1	101.0	101.0
973	Salem	101.0	99.1	100.2
974	Eugene	100.7	98.4	99.7
975	Medford	102.3	97.0	100.1
976	Klamath Falls	102.2	96.7	99.9
977	Bend	101.0	98.5	100.0
978	Pendleton	96.1	99.0	97.3
979	Vale	93.8	90.3	92.3
PENNSYLVANIA				
150-152	Pittsburgh	96.9	98.7	97.6
153	Washington	93.8	98.4	95.7
154	Uniontown	94.1	95.5	94.7
155	Bedford	95.1	89.3	92.6
156	Greensburg	95.0	96.8	95.8
157	Indiana	93.9	95.6	94.6
158	Dubois	95.5	93.2	94.5
159	Johnstown	95.1	93.2	94.3
160	Butler	91.9	96.3	93.8
161	New Castle	91.9	94.7	93.1
162	Kittanning	92.4	98.6	95.0
163	Oil City	91.9	92.9	92.3
164-165	Erie	94.2	92.0	93.3
166	Altoona	94.3	89.4	92.2
167	Bradford	95.4	90.7	93.4
168	State College	94.9	90.1	92.9
169	Wellsboro	96.0	90.7	93.8
170-171	Harrisburg	97.9	93.1	95.9
172	Chambersburg	95.5	88.9	92.7
173-174	York	96.1	93.3	94.9
175-176	Lancaster	94.2	87.9	91.6

STATE/ZIP	CITY	MAT.	INST.	TOTAL
PENNSYLVANIA (CONT'D)				
177	Williamsport	92.9	81.1	87.9
178	Sunbury	95.0	92.2	93.8
179	Pottsville	94.1	92.3	93.4
180	Lehigh Valley	95.5	111.4	102.2
181	Allentown	97.7	106.5	101.4
182	Hazleton	94.9	92.4	93.9
183	Stroudsburg	94.8	98.8	96.5
184-185	Scranton	98.5	95.2	97.1
186-187	Wilkes-Barre	94.7	92.7	93.8
188	Montrose	94.4	93.7	94.1
189	Doylestown	94.4	118.6	104.5
190-191	Philadelphia	101.0	130.0	113.2
193	Westchester	97.2	119.9	106.8
194	Norristown	96.2	127.4	109.3
195-196	Reading	98.5	97.6	98.1
PUERTO RICO				
009	San Juan	120.1	23.9	79.7
RHODE ISLAND				
028	Newport	97.7	109.1	102.5
029	Providence	98.8	109.1	103.2
SOUTH CAROLINA				
290-292	Columbia	98.6	53.3	79.6
293	Spartanburg	96.9	52.9	78.4
294	Charleston	98.4	62.2	83.2
295	Florence	96.6	53.3	78.4
296	Greenville	96.7	52.9	78.3
297	Rock Hill	96.1	50.9	77.1
298	Aiken	97.1	70.0	85.7
299	Beaufort	97.9	46.2	76.2
SOUTH DAKOTA				
570-571	Sioux Falls	99.9	60.1	83.2
572	Watertown	98.1	55.3	80.1
573	Mitchell	97.1	55.2	79.5
574	Aberdeen	99.9	56.3	81.6
575	Pierre	99.5	56.5	81.4
576	Mobridge	97.7	55.0	79.8
577	Rapid City	99.7	56.5	81.5
TENNESSEE				
370-372	Nashville	98.5	73.6	88.0
373-374	Chattanooga	99.4	55.0	80.8
375,380-381	Memphis	97.5	70.1	86.0
376	Johnson City	98.4	55.4	80.3
377-379	Knoxville	95.8	55.8	79.0
382	McKenzie	97.1	55.2	79.5
383	Jackson	99.0	47.9	77.5
384	Columbia	95.6	57.0	79.4
385	Cookeville	97.0	58.1	80.6
TEXAS				
750	McKinney	98.7	51.5	78.8
751	Waxahachie	98.6	54.2	80.0
752-753	Dallas	99.4	66.2	85.4
754	Greenville	98.7	38.4	73.4
755	Texarkana	97.6	50.8	77.9
756	Longview	98.0	39.9	73.6
757	Tyler	98.8	54.2	80.1
758	Palestine	95.0	40.3	72.0
759	Lufkin	95.7	43.9	74.0
760-761	Fort Worth	98.0	60.7	82.4
762	Denton	97.6	48.1	76.8
763	Wichita Falls	98.5	55.3	80.3
764	Eastland	96.9	39.4	72.8
765	Temple	95.6	48.7	75.9
766-767	Waco	97.8	56.8	80.6
768	Brownwood	98.0	37.2	72.5
769	San Angelo	97.6	45.3	75.7
770-772	Houston	100.7	71.2	88.3
773	Huntsville	98.8	37.9	73.2
774	Wharton	100.2	42.2	75.8
775	Galveston	98.2	69.1	86.0
776-777	Beaumont	98.4	60.1	82.3
778	Bryan	95.4	62.6	81.6
779	Victoria	100.1	44.3	76.7
780	Laredo	95.8	51.7	77.3
781-782	San Antonio	96.2	65.4	83.3
783-784	Corpus Christi	98.4	50.4	78.3
785	McAllen	98.0	45.1	75.8
786-787	Austin	96.5	59.1	80.8

Location Factors

STATE/ZIP	CITY	MAT.	INST.	TOTAL
TEXAS (CONT'D)				
788	Del Rio	97.6	31.9	70.1
789	Giddings	94.9	39.5	71.7
790-791	Amarillo	98.6	56.8	81.0
792	Childress	97.6	49.6	77.4
793-794	Lubbock	100.1	51.1	79.5
795-796	Abilene	98.2	50.8	78.3
797	Midland	100.0	47.4	77.9
798-799,885	El Paso	97.5	49.8	77.5
UTAH				
840-841	Salt Lake City	102.9	67.9	88.2
842,844	Ogden	98.4	67.3	85.3
843	Logan	100.3	67.3	86.4
845	Price	100.7	46.5	78.0
846-847	Provo	100.8	67.8	87.0
VERMONT				
050	White River Jct.	96.3	56.6	79.7
051	Bellows Falls	94.8	64.0	81.9
052	Bennington	95.2	66.8	83.2
053	Brattleboro	95.6	69.0	84.4
054	Burlington	99.3	67.7	86.1
056	Montpelier	95.9	68.1	84.2
057	Rutland	97.5	68.1	85.2
058	St. Johnsbury	96.3	57.1	79.9
059	Guildhall	95.1	56.9	79.0
VIRGINIA				
220-221	Fairfax	100.2	82.2	92.6
222	Arlington	101.5	80.8	92.8
223	Alexandria	100.7	86.0	94.5
224-225	Fredericksburg	98.8	73.8	88.3
226	Winchester	99.4	66.5	85.6
227	Culpeper	99.3	72.0	87.9
228	Harrisonburg	99.6	68.2	86.4
229	Charlottesville	100.0	66.6	85.9
230-232	Richmond	101.7	70.1	88.4
233-235	Norfolk	101.9	70.5	88.7
236	Newport News	100.5	70.9	88.1
237	Portsmouth	99.8	67.4	86.2
238	Petersburg	99.9	70.1	87.4
239	Farmville	99.0	56.9	81.3
240-241	Roanoke	101.2	63.7	85.4
242	Bristol	98.7	57.4	81.4
243	Pulaski	98.4	54.7	80.1
244	Staunton	99.3	62.8	84.0
245	Lynchburg	99.4	68.1	86.2
246	Grundy	98.8	54.1	80.0
WASHINGTON				
980-981,987	Seattle	105.4	101.7	103.9
982	Everett	105.0	96.7	101.5
983-984	Tacoma	105.1	99.6	102.8
985	Olympia	103.1	99.5	101.6
986	Vancouver	106.5	93.8	101.1
988	Wenatchee	104.8	81.4	95.0
989	Yakima	105.3	88.0	98.0
990-992	Spokane	104.5	81.8	95.0
993	Richland	104.1	85.7	96.4
994	Clarkston	102.9	82.2	94.2
WEST VIRGINIA				
247-248	Bluefield	97.5	77.3	89.0
249	Lewisburg	99.3	82.0	92.0
250-253	Charleston	100.4	87.8	95.1
254	Martinsburg	98.9	76.5	89.5
255-257	Huntington	100.6	90.3	96.3
258-259	Beckley	97.3	86.0	92.6
260	Wheeling	100.6	89.2	95.8
261	Parkersburg	99.6	88.7	95.0
262	Buckhannon	98.9	88.9	94.7
263-264	Clarksburg	99.3	87.8	94.5
265	Morgantown	99.4	88.5	94.8
266	Gassaway	98.7	88.8	94.5
267	Romney	98.8	82.3	91.9
268	Petersburg	98.6	84.6	92.7
WISCONSIN				
530,532	Milwaukee	100.3	106.4	102.9
531	Kenosha	100.0	99.3	99.7
534	Racine	99.5	101.3	100.2
535	Beloit	99.3	94.6	97.3
537	Madison	99.5	95.0	97.6

STATE/ZIP	CITY	MAT.	INST.	TOTAL
WISCONSIN (CONT'D)				
538	Lancaster	96.9	90.3	94.1
539	Portage	95.6	94.3	95.0
540	New Richmond	95.5	93.6	94.7
541-543	Green Bay	99.7	90.4	95.8
544	Wausau	94.9	88.5	92.2
545	Rhinelander	98.0	88.7	94.1
546	La Crosse	96.3	91.3	94.2
547	Eau Claire	97.9	92.0	95.4
548	Superior	95.4	95.7	95.5
549	Oshkosh	95.5	89.9	93.2
WYOMING				
820	Cheyenne	100.0	65.9	85.7
821	Yellowstone Nat'l Park	97.5	58.6	81.2
822	Wheatland	98.6	58.6	81.8
823	Rawlins	100.2	58.4	82.6
824	Worland	98.1	56.5	80.6
825	Riverton	99.2	56.5	81.3
826	Casper	100.6	58.4	82.9
827	Newcastle	97.9	58.4	81.3
828	Sheridan	100.7	60.5	83.8
829-831	Rock Springs	102.0	56.5	82.9
CANADIAN FACTORS (reflect Canadian currency)				
ALBERTA				
	Calgary	130.7	89.9	113.6
	Edmonton	131.8	89.8	114.1
	Fort McMurray	127.7	92.6	112.9
	Lethbridge	121.9	92.0	109.3
	Lloydminster	116.4	88.8	104.8
	Medicine Hat	116.4	88.1	104.5
	Red Deer	116.9	88.1	104.8
BRITISH COLUMBIA				
	Kamloops	117.2	90.8	106.1
	Prince George	118.3	90.8	106.7
	Vancouver	131.2	82.0	110.5
	Victoria	118.3	78.8	101.7
MANITOBA				
	Brandon	116.4	76.1	99.5
	Portage la Prairie	116.5	74.7	98.9
	Winnipeg	129.6	69.6	104.4
NEW BRUNSWICK				
	Bathurst	114.7	67.8	95.0
	Dalhousie	114.6	67.8	95.0
	Fredericton	117.1	72.4	98.3
	Moncton	115.0	68.7	95.5
	Newcastle	114.7	67.8	95.0
	St. John	117.3	72.4	98.4
NEWFOUNDLAND				
	Corner Brook	119.7	68.2	98.0
	St. Johns	119.9	69.2	98.6
NORTHWEST TERRITORIES				
	Yellowknife	121.4	85.7	106.4
NOVA SCOTIA				
	Bridgewater	116.1	75.0	98.9
	Dartmouth	117.5	75.0	99.7
	Halifax	119.0	78.0	101.8
	New Glasgow	115.5	75.0	98.5
	Sydney	113.1	75.0	97.1
	Truro	115.5	75.0	98.5
	Yarmouth	115.4	75.0	98.5
ONTARIO				
	Barrie	119.2	93.0	108.2
	Brantford	118.4	96.9	109.4
	Cornwall	118.5	93.3	107.9
	Hamilton	123.4	96.9	112.3
	Kingston	119.4	93.4	108.5
	Kitchener	114.5	92.2	105.1
	London	122.0	93.3	110.0
	North Bay	118.4	91.2	107.0
	Oshawa	117.9	94.7	108.2
	Ottawa	122.7	93.9	110.6
	Owen Sound	119.4	91.2	107.6
	Peterborough	118.4	93.2	107.8
	Sarnia	117.9	97.5	109.3
ONTARIO (CONT'D)				

STATE/ZIP	CITY	MAT.	INST.	TOTAL
	Sault Ste Marie	112.9	92.6	104.4
	St. Catharines	112.3	94.2	104.7
	Sudbury	112.7	92.9	104.4
	Thunder Bay	113.5	92.9	104.9
	Timmins	118.6	91.2	107.1
	Toronto	123.6	99.8	113.6
	Windsor	113.0	93.2	104.7
PRINCE EDWARD ISLAND				
	Charlottetown	117.8	63.3	94.9
	Summerside	117.2	63.3	94.6
QUEBEC				
	Cap-de-la-Madeleine	115.3	87.8	103.8
	Charlesbourg	115.3	87.8	103.8
	Chicoutimi	114.0	93.5	105.4
	Gatineau	114.9	87.6	103.4
	Granby	115.1	87.6	103.5
	Hull	115.0	87.6	103.5
	Joliette	115.5	87.8	103.9
	Laval	114.5	87.6	103.2
	Montreal	121.7	93.8	110.0
	Quebec	120.1	94.2	109.2
	Rimouski	114.4	93.5	105.6
	Rouyn-Noranda	114.8	87.6	103.4
	Saint Hyacinthe	114.2	87.6	103.0
	Sherbrooke	115.2	87.6	103.6
	Sorel	115.4	87.8	103.8
	St. Jerome	114.8	87.6	103.4
	Trois Rivieres	115.4	87.8	103.8
SASKATCHEWAN				
	Moose Jaw	113.5	69.9	95.2
	Prince Albert	112.5	68.1	93.9
	Regina	115.1	69.9	96.1
	Saskatoon	113.6	69.8	95.2
YUKON				
	Whitehorse	113.2	68.9	94.6

697

R011105-05   Tips for Accurate Estimating

1. Use pre-printed or columnar forms for orderly sequence of dimensions and locations and for recording telephone quotations.

2. Use only the front side of each paper or form except for certain pre-printed summary forms.

3. Be consistent in listing dimensions: For example, length x width x height. This helps in rechecking to ensure that, the total length of partitions is appropriate for the building area.

4. Use printed (rather than measured) dimensions where given.

5. Add up multiple printed dimensions for a single entry where possible.

6. Measure all other dimensions carefully.

7. Use each set of dimensions to calculate multiple related quantities.

8. Convert foot and inch measurements to decimal feet when listing. Memorize decimal equivalents to .01 parts of a foot (1/8″ equals approximately .01′).

9. Do not "round off" quantities until the final summary.

10. Mark drawings with different colors as items are taken off.

11. Keep similar items together, different items separate.

12. Identify location and drawing numbers to aid in future checking for completeness.

13. Measure or list everything on the drawings or mentioned in the specifications.

14. It may be necessary to list items not called for to make the job complete.

15. Be alert for: Notes on plans such as N.T.S. (not to scale); changes in scale throughout the drawings; reduced size drawings; discrepancies between the specifications and the drawings.

16. Develop a consistent pattern of performing an estimate. For example:
   a. Start the quantity takeoff at the lower floor and move to the next higher floor.
   b. Proceed from the main section of the building to the wings.
   c. Proceed from south to north or vice versa, clockwise or counterclockwise.
   d. Take off floor plan quantities first, elevations next, then detail drawings.

17. List all gross dimensions that can be either used again for different quantities, or used as a rough check of other quantities for verification (exterior perimeter, gross floor area, individual floor areas, etc.).

18. Utilize design symmetry or repetition (repetitive floors, repetitive wings, symmetrical design around a center line, similar room layouts, etc.). Note: Extreme caution is needed here so as not to omit or duplicate an area.

19. Do not convert units until the final total is obtained. For instance, when estimating concrete work, keep all units to the nearest cubic foot, then summarize and convert to cubic yards.

20. When figuring alternatives, it is best to total all items involved in the basic system, then total all items involved in the alternates. Therefore you work with positive numbers in all cases. When adds and deducts are used, it is often confusing whether to add or subtract a portion of an item; especially on a complicated or involved alternate.

R011110-10   Architectural Fees

Tabulated below are typical percentage fees by project size, for good professional architectural service. Fees may vary from those listed depending upon degree of design difficulty and economic conditions in any particular area.

Rates can be interpolated horizontally and vertically. Various portions of the same project requiring different rates should be adjusted proportionately. For alterations, add 50% to the fee for the first $500,000 of project cost and add 25% to the fee for project cost over $500,000.

Architectural fees tabulated below include Structural, Mechanical and Electrical Engineering Fees. They do not include the fees for special consultants such as kitchen planning, security, acoustical, interior design, etc.

Civil Engineering fees are included in the Architectural fee for project sites requiring minimal design such as city sites. However, separate Civil Engineering fees must be added when utility connections require design, drainage calculations are needed, stepped foundations are required, or provisions are required to protect adjacent wetlands.

Building Types	Total Project Size in Thousands of Dollars						
	100	250	500	1,000	5,000	10,000	50,000
Factories, garages, warehouses, repetitive housing	9.0%	8.0%	7.0%	6.2%	5.3%	4.9%	4.5%
Apartments, banks, schools, libraries, offices, municipal buildings	12.2	12.3	9.2	8.0	7.0	6.6	6.2
Churches, hospitals, homes, laboratories, museums, research	15.0	13.6	12.7	11.9	9.5	8.8	8.0
Memorials, monumental work, decorative furnishings	—	16.0	14.5	13.1	10.0	9.0	8.3

Reference Tables

R011110-30 Engineering Fees

Typical **Structural Engineering Fees** based on type of construction and total project size. These fees are included in Architectural Fees.

Type of Construction	Total Project Size (in thousands of dollars)			
	$500	$500-$1,000	$1,000-$5,000	Over $5000
Industrial buildings, factories & warehouses	Technical payroll times 2.0 to 2.5	1.60%	1.25%	1.00%
Hotels, apartments, offices, dormitories, hospitals, public buildings, food stores		2.00%	1.70%	1.20%
Museums, banks, churches and cathedrals		2.00%	1.75%	1.25%
Thin shells, prestressed concrete, earthquake resistive		2.00%	1.75%	1.50%
Parking ramps, auditoriums, stadiums, convention halls, hangars & boiler houses		2.50%	2.00%	1.75%
Special buildings, major alterations, underpinning & future expansion		Add to above 0.5%	Add to above 0.5%	Add to above 0.5%

For complex reinforced concrete or unusually complicated structures, add 20% to 50%.

Typical **Mechanical and Electrical Engineering Fees** are based on the size of the subcontract. The fee structure for both are shown below. These fees are included in Architectural Fees.

Type of Construction	Subcontract Size							
	$25,000	$50,000	$100,000	$225,000	$350,000	$500,000	$750,000	$1,000,000
Simple structures	6.4%	5.7%	4.8%	4.5%	4.4%	4.3%	4.2%	4.1%
Intermediate structures	8.0	7.3	6.5	5.6	5.1	5.0	4.9	4.8
Complex structures	10.1	9.0	9.0	8.0	7.5	7.5	7.0	7.0

For renovations, add 15% to 25% to applicable fee.

R012153-10 Repair and Remodeling

Cost figures are based on new construction utilizing the most cost-effective combination of labor, equipment and material with the work scheduled in proper sequence to allow the various trades to accomplish their work in an efficient manner.

The costs for repair and remodeling work must be modified due to the following factors that may be present in any given repair and remodeling project.

1. Equipment usage curtailment due to the physical limitations of the project, with only hand-operated equipment being used.

2. Increased requirement for shoring and bracing to hold up the building while structural changes are being made and to allow for temporary storage of construction materials on above-grade floors.

3. Material handling becomes more costly due to having to move within the confines of an enclosed building. For multi-story construction, low capacity elevators and stairwells may be the only access to the upper floors.

4. Large amount of cutting and patching and attempting to match the existing construction is required. It is often more economical to remove entire walls rather than create many new door and window openings. This sort of trade-off has to be carefully analyzed.

5. Cost of protection of completed work is increased since the usual sequence of construction usually cannot be accomplished.

6. Economies of scale usually associated with new construction may not be present. If small quantities of components must be custom fabricated due to job requirements, unit costs will naturally increase. Also, if only small work areas are available at a given time, job scheduling between trades becomes difficult and subcontractor quotations may reflect the excessive start-up and shut-down phases of the job.

7. Work may have to be done on other than normal shifts and may have to be done around an existing production facility which has to stay in production during the course of the repair and remodeling.

8. Dust and noise protection of adjoining non-construction areas can involve substantial special protection and alter usual construction methods.

9. Job may be delayed due to unexpected conditions discovered during demolition or removal. These delays ultimately increase construction costs.

10. Piping and ductwork runs may not be as simple as for new construction. Wiring may have to be snaked through walls and floors.

11. Matching "existing construction" may be impossible because materials may no longer be manufactured. Substitutions may be expensive.

12. Weather protection of existing structure requires additional temporary structures to protect building at openings.

13. On small projects, because of local conditions, it may be necessary to pay a tradesman for a minimum of four hours for a task that is completed in one hour.

All of the above areas can contribute to increased costs for a repair and remodeling project. Each of the above factors should be considered in the planning, bidding and construction stage in order to minimize the increased costs associated with repair and remodeling jobs.

R012909-80   Sales Tax by State

State sales tax on materials is tabulated below (5 states have no sales tax). Many states allow local jurisdictions, such as a county or city, to levy additional sales tax.

Some projects may be sales tax exempt, particularly those constructed with public funds.

State	Tax (%)	State	Tax (%)	State	Tax (%)	State	Tax (%)
Alabama	4	Illinois	6.25	Montana	0	Rhode Island	7
Alaska	0	Indiana	6	Nebraska	5.5	South Carolina	6
Arizona	5.6	Iowa	5	Nevada	6.5	South Dakota	4
Arkansas	6	Kansas	5.3	New Hampshire	0	Tennessee	7
California	7.25	Kentucky	6	New Jersey	7	Texas	6.25
Colorado	2.9	Louisiana	4	New Mexico	5	Utah	4.65
Connecticut	6	Maine	5	New York	4	Vermont	6
Delaware	0	Maryland	6	North Carolina	4.25	Virginia	5
District of Columbia	5.75	Massachusetts	5	North Dakota	5	Washington	6.5
Florida	6	Michigan	6	Ohio	5.5	West Virginia	6
Georgia	4	Minnesota	6.5	Oklahoma	4.5	Wisconsin	5
Hawaii	4	Mississippi	7	Oregon	0	Wyoming	4
Idaho	6	Missouri	4.225	Pennsylvania	6	Average	4.91%

Sales Tax by Province (Canada)

GST - a value-added tax, which the government imposes on most goods and services provided in or imported into Canada. PST - a retail sales tax, which five of the provinces impose on the price of most goods and some services. QST - a value-added tax, similar to the federal GST, which Quebec imposes. HST - Three provinces have combined their retail sales tax with the federal GST into one harmonized tax.

Province	PST (%)	QST (%)	GST(%)	HST(%)
Alberta	0	0	6	0
British Columbia	7	0	6	0
Manitoba	7	0	6	0
New Brunswick	0	0	0	13
Newfoundland	0	0	0	13
Northwest Territories	0	0	6	0
Nova Scotia	0	0	0	13
Ontario	8	0	6	0
Prince Edward Island	10	0	6	0
Quebec	0	7.5	6	0
Saskatchewan	5	0	6	0
Yukon	0	0	6	0

R012909-80   Unemployment Taxes and Social Security Taxes

State Unemployment Tax rates vary not only from state to state, but also with the experience rating of the contractor. The Federal Unemployment Tax rate is 6.2% of the first $7,000 of wages. This is reduced by a credit of up to 5.4% for timely payment to the state. The minimum Federal Unemployment Tax is 0.8% after all credits.

Social Security (FICA) for 2009 is estimated at time of publication to be 7.65% of wages up to $102,000.

R013113-40 Builder's Risk Insurance

Builder's Risk Insurance is insurance on a building during construction. Premiums are paid by the owner or the contractor. Blasting, collapse and underground insurance would raise total insurance costs above those listed. Floater policy for materials delivered to the job runs .75 to $1.25 per $100 value. Contractor equipment insurance runs .50 to $1.50 per $100 value. Insurance for miscellaneous tools to $1,500 value runs from $3.00 to $7.50 per $100 value.

Tabulated below are New England Builder's Risk insurance rates in dollars per $100 value for $1,000 deductible. For $25,000 deductible, rates can be reduced 13% to 34%. On contracts over $1,000,000, rates may be lower than those tabulated. Policies are written annually for the total completed value in place. For "all risk" insurance (excluding flood, earthquake and certain other perils) add .025 to total rates below.

Coverage	Frame Construction (Class 1)			Brick Construction (Class 4)			Fire Resistive (Class 6)		
	Range		Average	Range		Average	Range		Average
Fire Insurance	$.350 to	$.850	$.600	$.158 to	$.189	$.174	$.052 to	$.080	$.070
Extended Coverage	.115 to	.200	.158	.080 to	.105	.101	.081 to	.105	.100
Vandalism	.012 to	.016	.014	.008 to	.011	.011	.008 to	.011	.010
Total Annual Rate	$.477 to	$1.066	$.772	$.246 to	$.305	$.286	$.141 to	$.196	$.180

R013113-50 General Contractor's Overhead

There are two distinct types of overhead on a construction project: Project Overhead and Main Office Overhead. Project Overhead includes those costs at a construction site not directly associated with the installation of construction materials. Examples of Project Overhead costs include the following:

1. Superintendent
2. Construction office and storage trailers
3. Temporary sanitary facilities
4. Temporary utilities
5. Security fencing
6. Photographs
7. Clean up
8. Performance and payment bonds

The above Project Overhead items are also referred to as General Requirements and therefore are estimated in Division 1. Division 1 is the first division listed in the CSI MasterFormat but it is usually the last division estimated. The sum of the costs in Divisions 1 through 49 is referred to as the sum of the direct costs.

All construction projects also include indirect costs. The primary components of indirect costs are the contractor's Main Office Overhead and profit. The amount of the Main Office Overhead expense varies depending on the the following:

1. Owner's compensation
2. Project managers and estimator's wages
3. Clerical support wages
4. Office rent and utilities
5. Corporate legal and accounting costs
6. Advertising
7. Automobile expenses
8. Association dues
9. Travel and entertainment expenses

These costs are usually calculated as a percentage of annual sales volume. This percentage can range from 35% for a small contractor doing less than $500,000 to 5% for a large contractor with sales in excess of $100 million.

R013113-60 Workers' Compensation Insurance Rates by Trade

The table below tabulates the national averages for Workers' Compensation insurance rates by trade and type of building. The average "Insurance Rate" is multiplied by the "% of Building Cost" for each trade. This produces the "Workers' Compensation Cost" by % of total labor cost, to be added for each trade by building type to determine the weighted average Workers' Compensation rate for the building types analyzed.

Trade	Insurance Rate (% Labor Cost)			% of Building Cost			Workers' Compensation		
	Range		Average	Office Bldgs.	Schools & Apts.	Mfg.	Office Bldgs.	Schools & Apts.	Mfg.
Excavation, Grading, etc.	4.2 % to	17.5%	10.0%	4.8%	4.9%	4.5%	0.48%	0.49%	0.45%
Piles & Foundations	5.9 to	40.3	20.1	7.1	5.2	8.7	1.43	1.05	1.75
Concrete	5.1 to	26.8	14.6	5.0	14.8	3.7	0.73	2.16	0.54
Masonry	4.8 to	43.3	14.4	6.9	7.5	1.9	0.99	1.08	0.27
Structural Steel	5.9 to	104.1	37.9	10.7	3.9	17.6	4.06	1.48	6.67
Miscellaneous & Ornamental Metals	3.4 to	22.8	10.7	2.8	4.0	3.6	0.30	0.43	0.39
Carpentry & Millwork	5.9 to	53.2	17.8	3.7	4.0	0.5	0.66	0.71	0.09
Metal or Composition Siding	5.9 to	38	16.6	2.3	0.3	4.3	0.38	0.05	0.71
Roofing	5.9 to	77.1	31.2	2.3	2.6	3.1	0.72	0.81	0.97
Doors & Hardware	4.9 to	32	11.6	0.9	1.4	0.4	0.10	0.16	0.05
Sash & Glazing	5.9 to	32.4	13.9	3.5	4.0	1.0	0.49	0.56	0.14
Lath & Plaster	3.3 to	43.9	13.6	3.3	6.9	0.8	0.45	0.94	0.11
Tile, Marble & Floors	3.1 to	17.9	9.1	2.6	3.0	0.5	0.24	0.27	0.05
Acoustical Ceilings	2.6 to	45.6	10.6	2.4	0.2	0.3	0.25	0.02	0.03
Painting	4.7 to	29.6	12.5	1.5	1.6	1.6	0.19	0.20	0.20
Interior Partitions	5.9 to	53.2	17.8	3.9	4.3	4.4	0.69	0.77	0.78
Miscellaneous Items	2.1 to	168.2	16.0	5.2	3.7	9.7	0.83	0.59	1.55
Elevators	2.8 to	11.8	6.6	2.1	1.1	2.2	0.14	0.07	0.15
Sprinklers	2.5 to	14.3	7.8	0.5	—	2.0	0.04	—	0.16
Plumbing	2.9 to	12.4	7.8	4.9	7.2	5.2	0.38	0.56	0.41
Heat., Vent., Air Conditioning	4.3 to	23	11.6	13.5	11.0	12.9	1.57	1.28	1.50
Electrical	2.8 to	11.5	6.5	10.1	8.4	11.1	0.66	0.55	0.72
Total	2.1 % to	168.2%	—	100.0%	100.0%	100.0%	15.78%	14.23%	17.69%
	Overall Weighted Average		15.90%						

Workers' Compensation Insurance Rates by States

The table below lists the weighted average Workers' Compensation base rate for each state with a factor comparing this with the national average of 15.5%.

State	Weighted Average	Factor	State	Weighted Average	Factor	State	Weighted Average	Factor
Alabama	24.2%	156	Kentucky	18.4%	119	North Dakota	13.7%	88
Alaska	21.7	140	Louisiana	28.3	183	Ohio	14.8	95
Arizona	9.5	61	Maine	15.4	99	Oklahoma	14.6	94
Arkansas	13.2	85	Maryland	16.7	108	Oregon	13.5	87
California	19.6	126	Massachusetts	12.5	81	Pennsylvania	13.8	89
Colorado	13.2	85	Michigan	17.3	112	Rhode Island	21.2	137
Connecticut	21.0	135	Minnesota	25.7	166	South Carolina	18.3	118
Delaware	15.1	97	Mississippi	19.1	123	South Dakota	19.1	123
District of Columbia	13.7	88	Missouri	17.2	111	Tennessee	15.1	97
Florida	13.8	89	Montana	16.6	107	Texas	12.2	79
Georgia	26.2	169	Nebraska	23.0	148	Utah	12.3	79
Hawaii	17.0	110	Nevada	11.6	75	Vermont	18.0	116
Idaho	10.8	70	New Hampshire	20.0	129	Virginia	11.7	75
Illinois	21.6	139	New Jersey	13.4	86	Washington	9.8	63
Indiana	5.9	38	New Mexico	18.5	119	West Virginia	9.2	59
Iowa	11.5	74	New York	12.6	81	Wisconsin	15.0	97
Kansas	8.9	57	North Carolina	18.9	122	Wyoming	6.5	42
			Weighted Average for U.S. is	15.9% of payroll = 100%				

Rates in the following table are the base or manual costs per $100 of payroll for Workers' Compensation in each state. Rates are usually applied to straight time wages only and not to premium time wages and bonuses.

The weighted average skilled worker rate for 35 trades is 15.5%. For bidding purposes, apply the full value of Workers' Compensation directly to total labor costs, or if labor is 38%, materials 42% and overhead and profit 20% of total cost, carry 38/80 x 15.5% =7.4% of cost (before overhead and profit) into overhead. Rates vary not only from state to state but also with the experience rating of the contractor.

Rates are the most current available at the time of publication.

R013113-060 Workers' Compensation Insurance Rates by Trade and State (cont.)

State	Carpentry – 3 stories or less 5651	Carpentry – interior cab. work 5437	Carpentry – general 5403	Concrete Work – NOC 5213	Concrete Work – flat (flr., sdwk.) 5221	Electrical Wiring – inside 5190	Excavation – earth NOC 6217	Excavation – rock 6217	Glaziers 5462	Insulation Work 5479	Lathing 5443	Masonry 5022	Painting & Decorating 5474	Pile Driving 6003	Plastering 5480	Plumbing 5183	Roofing 5551	Sheet Metal Work (HVAC) 5538	Steel Erection – door & sash 5102	Steel Erection – inter., ornam. 5102	Steel Erection – structure 5040	Steel Erection – NOC 5057	Tile Work – (interior ceramic) 5348	Waterproofing 9014	Wrecking 5701
AL	37.98	22.82	32.46	16.20	11.27	10.68	11.16	11.16	21.26	16.28	11.93	27.00	25.81	32.65	18.85	12.27	58.14	22.96	11.01	11.01	55.68	25.37	13.31	6.56	55.68
AK	14.73	12.27	17.44	11.07	9.71	8.72	12.14	12.14	32.36	27.78	9.74	43.27	21.95	32.40	43.85	9.97	35.12	8.61	9.39	9.39	43.01	33.00	6.76	6.61	43.01
AZ	11.60	6.45	15.40	8.56	4.45	4.59	4.90	4.90	6.93	12.48	6.61	6.51	6.33	11.45	5.78	5.24	13.38	7.38	10.66	10.66	21.10	15.34	3.17	2.59	21.10
AR	14.82	7.45	16.26	12.17	6.48	5.02	7.76	7.76	9.82	16.49	5.93	9.93	11.50	16.48	16.10	5.15	23.09	13.09	6.77	6.77	32.38	25.50	6.08	3.74	32.38
CA	31.95	31.95	31.95	11.89	11.89	7.98	13.79	13.79	15.74	12.50	11.24	16.21	17.08	15.27	20.53	12.03	44.43	15.60	12.46	12.46	17.17	20.88	8.54	17.08	20.88
CO	13.57	8.92	12.33	11.57	7.87	5.28	10.34	10.34	9.75	16.84	6.02	14.01	9.02	14.31	9.58	7.43	26.09	11.61	9.31	9.31	32.62	17.58	7.06	5.03	17.58
CT	15.67	14.66	28.76	24.79	10.09	8.01	12.91	12.91	22.98	15.94	21.94	22.24	17.61	22.88	14.48	11.16	41.95	15.61	18.57	18.57	42.72	23.45	11.58	5.41	42.72
DE	15.17	15.17	11.46	11.51	9.49	5.89	9.32	9.32	13.11	11.46	13.11	12.77	26.77	18.75	13.11	7.87	26.71	9.78	12.28	12.28	26.77	12.28	9.29	12.77	26.77
DC	11.01	12.78	10.19	9.66	9.14	6.01	12.54	12.54	17.02	8.18	9.40	11.97	7.51	12.11	12.48	10.38	19.21	8.93	10.61	10.61	37.21	16.49	17.86	4.00	37.21
FL	13.05	10.47	14.39	15.31	6.97	6.64	8.30	8.30	9.91	9.24	5.98	11.32	9.73	38.24	18.70	6.75	22.19	12.68	8.85	8.85	27.89	14.56	6.55	5.17	27.89
GA	35.68	20.54	25.67	16.85	13.26	10.82	17.38	17.38	18.77	23.64	15.09	21.06	27.77	27.36	22.83	11.4	54.71	20.76	19.93	19.93	64.78	42.21	12.42	8.50	64.78
HI	17.38	11.62	28.20	13.74	12.27	7.10	7.73	7.73	20.55	21.19	10.94	17.87	11.33	20.15	15.61	6.06	33.24	8.02	11.11	11.11	31.71	21.70	9.78	11.54	31.71
ID	9.16	6.04	10.35	11.25	6.13	3.70	5.72	5.72	8.70	7.67	8.89	7.59	8.45	12.13	13.37	5.36	30.27	8.89	5.72	5.72	29.16	10.28	9.79	4.75	29.16
IL	24.93	14.46	20.78	26.59	12.10	9.62	10.06	10.06	19.18	17.48	15.59	19.27	11.62	34.85	15.88	11.07	31.06	14.96	16.05	16.05	72.39	25.69	14.02	4.84	72.39
IN	9.65	4.92	7.29	5.10	3.22	2.84	4.87	4.87	6.28	7.44	2.62	4.84	4.72	6.44	3.72	2.91	10.88	4.45	3.43	3.43	12.12	7.04	3.09	2.47	12.12
IA	10.43	10.02	10.34	12.12	7.00	4.96	6.27	6.27	9.42	7.06	5.58	9.61	6.65	9.78	7.33	6.46	22.58	7.52	6.15	6.15	30.21	35.21	7.42	3.87	35.21
KS	10.05	7.85	9.89	7.19	6.42	3.37	4.41	4.41	9.22	7.45	4.68	6.91	7.57	11.31	6.25	4.85	15.06	7.14	4.82	4.82	26.19	13.21	6.35	3.46	13.21
KY	18.72	14.07	25.30	13.39	8.00	5.43	9.15	9.15	17.37	17.59	11.24	9.12	13.18	27.72	14.41	7.06	38.25	19.58	12.32	12.32	54.03	22.56	13.67	5.74	56.16
LA	22.02	24.93	53.17	26.43	15.61	9.88	17.49	17.49	20.14	21.43	24.51	28.33	29.58	31.25	22.37	8.64	77.12	21.20	18.98	18.98	51.76	24.21	13.77	13.78	66.41
ME	13.79	10.40	28.55	20.94	8.74	6.31	9.05	9.05	18.14	11.82	7.90	14.76	12.73	16.70	11.79	10.56	22.87	11.61	9.73	9.73	31.67	23.11	7.00	7.12	31.67
MD	15.05	11.06	14.11	15.67	6.67	9.24	9.20	9.20	20.03	15.02	8.83	11.43	7.48	20.09	15.02	6.11	34.74	13.54	9.51	9.51	59.03	28.60	7.60	5.36	28.60
MA	6.80	5.60	11.46	19.51	6.57	3.20	4.19	4.19	8.96	10.06	6.35	10.81	4.79	14.68	5.08	3.98	32.80	5.15	7.61	7.61	45.71	36.69	6.21	2.09	25.88
MI	18.40	11.39	20.64	19.45	8.95	4.72	10.20	10.20	12.41	14.75	13.59	15.54	13.64	40.27	15.18	6.84	30.67	10.00	10.16	10.16	40.27	22.21	11.11	4.87	40.27
MN	19.53	19.80	41.48	14.65	15.73	7.53	14.95	14.95	15.68	11.90	22.05	19.18	16.95	24.63	22.05	10.88	70.18	14.23	10.58	10.58	104.13	33.30	14.55	6.68	36.03
MS	21.07	11.11	22.29	15.49	9.26	6.28	11.91	11.91	13.07	14.85	7.91	14.09	13.14	35.34	31.72	9.80	44.11	16.91	14.23	14.23	36.11	25.03	9.84	4.18	36.11
MO	29.99	11.16	13.98	18.91	10.39	7.64	9.17	9.17	11.13	15.17	7.93	15.01	12.53	19.12	17.02	10.05	29.98	13.19	11.31	11.31	34.56	35.84	11.18	6.51	34.56
MT	17.12	11.15	22.57	12.37	12.79	6.39	15.54	15.54	10.67	32.03	11.06	13.75	9.61	25.18	10.02	8.75	37.61	11.53	8.70	8.70	27.45	15.73	7.53	7.46	15.73
NE	24.65	17.47	21.80	25.07	15.13	10.38	16.90	16.90	23.70	31.50	12.20	22.22	20.07	22.50	21.67	12.35	35.22	19.42	16.05	16.05	45.40	35.27	11.10	6.38	47.05
NV	15.25	7.14	11.58	10.76	8.23	7.17	9.77	9.77	10.19	8.12	5.27	7.50	7.97	13.33	7.30	5.98	14.08	18.10	8.67	8.67	25.10	21.54	5.39	5.55	21.54
NH	25.92	13.36	18.90	26.79	14.17	6.94	14.52	14.52	13.07	21.36	9.63	24.34	7.77	18.99	11.71	10.13	49.23	13.59	15.27	15.27	52.24	18.83	11.37	6.09	52.24
NJ	14.34	8.94	14.34	16.01	9.99	4.76	8.35	8.35	8.55	12.65	11.64	13.09	11.92	16.03	11.64	6.17	37.77	6.49	9.46	9.46	23.72	13.59	6.99	5.72	22.31
NM	18.91	7.15	20.11	17.36	10.05	7.51	11.20	11.20	21.26	13.72	8.39	15.34	11.64	21.86	13.44	8.13	36.37	13.47	13.69	13.69	55.78	38.81	6.59	6.29	55.78
NY	12.72	6.45	12.67	16.02	11.64	6.03	8.29	8.29	10.97	7.98	12.10	16.02	9.89	12.77	9.03	6.88	27.98	10.46	8.84	8.84	23.45	12.72	6.66	6.04	8.73
NC	18.76	14.56	15.36	18.15	6.88	11.50	11.70	11.70	15.21	16.00	15.28	12.51	12.90	17.50	15.68	9.16	29.70	15.90	10.44	10.44	79.66	26.21	9.28	5.55	79.66
ND	11.17	11.17	11.17	6.42	6.42	3.63	6.64	6.64	11.17	11.17	7.51	7.71	6.61	22.78	7.51	5.22	23.53	5.22	22.78	22.78	22.78	22.78	11.17	23.53	10.98
OH	11.04	8.52	11.68	12.33	9.99	6.14	8.70	8.70	7.86	16.94	45.64	12.71	12.37	19.06	3.28	6.99	31.63	9.71	7.59	7.59	29.23	16.59	9.60	6.42	16.59
OK	13.42	9.17	11.76	11.76	6.34	5.73	10.83	10.83	16.07	19.47	8.45	10.02	8.47	20.25	12.19	7.09	21.13	9.05	13.76	13.76	39.72	23.97	6.50	5.71	39.72
OR	15.43	8.10	15.69	14.22	8.97	5.14	9.33	9.33	15.22	11.80	7.60	14.07	10.63	16.17	12.73	5.50	26.30	9.41	7.93	7.93	26.94	18.78	11.04	4.54	26.94
PA	13.18	13.18	11.38	14.21	10.46	5.95	8.27	8.27	11.47	11.38	11.47	12.22	13.31	15.78	11.47	7.33	27.42	7.84	14.30	14.30	21.31	14.30	8.02	12.22	21.37
RI	19.53	11.65	18.07	18.23	16.24	4.43	10.38	10.38	12.85	22.78	11.97	25.11	24.13	37.66	17.25	8.52	33.92	10.29	14.07	14.07	59.49	37.50	14.36	7.70	59.49
SC	22.54	15.41	20.91	14.80	7.92	10.92	13.16	13.16	16.12	14.69	11.32	12.88	16.52	19.48	16.02	11.47	45.24	12.95	12.99	12.99	32.57	26.39	9.62	6.68	47.05
SD	27.77	8.13	29.58	25.56	9.55	5.52	11.01	11.01	12.16	12.05	7.70	11.31	9.53	21.32	12.44	11.75	20.73	12.48	8.98	8.98	80.86	40.98	8.74	4.98	80.86
TN	15.68	13.11	16.06	13.89	7.97	7.35	13.68	13.68	13.09	11.80	7.30	15.60	12.63	24.89	16.02	7.66	22.25	12.77	7.17	7.17	29.18	23.47	10.15	4.65	29.18
TX	11.25	8.89	11.25	9.00	7.26	6.15	8.54	8.54	9.65	12.87	6.36	10.91	8.98	22.43	8.98	6.75	19.59	12.84	9.09	9.09	29.68	13.37	7.07	6.69	9.96
UT	15.74	8.88	12.97	9.81	8.47	4.57	11.59	11.59	14.70	12.49	6.35	11.25	8.37	12.47	7.26	6.18	24.99	12.64	7.16	7.16	25.67	14.67	6.80	4.27	18.37
VT	15.70	10.75	17.37	17.27	7.21	6.98	12.59	12.59	19.75	17.78	8.33	15.82	11.62	16.74	10.78	10.44	36.43	11.59	12.51	12.51	44.95	42.80	11.24	7.28	44.95
VA	11.05	8.02	9.16	11.05	5.31	5.52	7.57	7.57	10.25	9.90	14.34	9.62	8.99	10.75	10.22	5.83	22.52	7.65	8.37	8.37	33.52	19.66	4.74	3.04	33.52
WA	8.48	8.48	8.48	7.32	7.32	2.81	6.69	6.69	12.27	9.77	8.43	9.09	10.93	16.12	9.95	4.17	16.36	4.27	7.60	7.60	7.60	7.60	8.45	16.36	7.60
WV	10.65	7.30	10.23	8.24	4.41	4.70	6.22	6.22	7.89	7.25	6.11	7.98	7.90	11.90	7.93	4.90	18.05	5.74	5.58	5.58	24.17	13.20	5.65	2.66	17.34
WI	9.99	11.82	14.45	11.15	9.69	6.02	8.19	8.19	10.44	13.58	6.25	15.31	13.03	16.47	9.96	6.24	37.36	7.21	9.36	9.36	28.23	44.93	13.80	4.86	28.23
WY	5.87	5.87	5.87	5.87	5.87	5.87	5.87	5.87	5.87	5.87	5.87	5.87	5.87	5.87	5.87	5.87	5.87	5.87	5.87	5.87	5.87	5.87	5.87	5.87	5.87
AVG.	16.63	11.62	17.80	14.58	9.14	6.46	10.01	10.01	13.89	14.44	10.63	14.37	12.49	20.09	13.60	7.84	31.18	11.57	10.74	10.74	37.94	23.15	9.13	6.69	34.13

R013113-060 Workers' Compensation (cont.) (Canada in Canadian dollars)

Province		Alberta	British Columbia	Manitoba	Ontario	New Brunswick	Newfndld. & Labrador	Northwest Territories	Nova Scotia	Prince Edward Island	Quebec	Saskat-chewan	Yukon
Carpentry—3 stories or less	Rate	7.32	4.74	4.36	4.35	3.79	10.08	3.84	7.82	5.96	15.05	6.23	8.89
	Code	42143	721028	40102	723	4226	4226	4-41	4226	401	80110	B1317	202
Carpentry—interior cab. work	Rate	2.18	4.90	4.36	4.35	4.40	4.69	3.84	4.52	3.71	15.05	3.51	8.89
	Code	42133	721021	40102	723	4279	4270	4-41	4274	402	80110	B11-27	202
CARPENTRY—general	Rate	7.32	4.74	4.36	4.35	3.79	4.69	3.84	7.82	5.96	15.05	6.23	8.89
	Code	42143	721028	40102	723	4226	4299	4-41	4226	401	80110	B1317	202
CONCRETE WORK—NOC	Rate	4.41	4.81	7.46	16.02	3.79	10.08	3.84	4.52	5.96	17.99	6.24	4.67
	Code	42104	721010	40110	748	4224	4224	4-41	4224	401	80100	B13-14	203
CONCRETE WORK—flat (flr. sidewalk)	Rate	4.41	4.81	7.46	16.02	3.79	10.08	3.84	4.52	5.96	17.99	6.24	4.67
	Code	42104	721010	40110	748	4224	4224	4-41	4224	401	80100	B13-14	203
ELECTRICAL Wiring—inside	Rate	2.13	1.77	2.51	3.79	2.17	3.01	3.52	2.31	3.71	6.12	3.51	4.67
	Code	42124	721019	40203	704	4261	4261	4-46	4261	402	80170	B11-05	206
EXCAVATION—earth NOC	Rate	2.51	3.56	3.76	4.55	2.62	4.37	3.64	3.34	3.82	7.13	3.93	4.67
	Code	40604	721031	40706	711	4214	4214	4-43	4214	404	80030	R11-06	207
EXCAVATION—rock	Rate	2.51	3.56	3.76	4.55	2.62	4.37	3.64	3.34	3.82	7.13	3.93	4.67
	Code	40604	721031	40706	711	4214	4214	4-43	4214	404	80030	R11-06	207
GLAZIERS	Rate	3.16	3.62	4.36	8.90	4.55	6.81	3.84	5.21	3.71	17.08	6.23	4.67
	Code	42121	715020	40109	751	4233	4233	4-41	4233	402	80150	B13-04	212
INSULATION WORK	Rate	2.57	8.42	4.36	8.90	4.55	6.81	3.84	5.21	5.96	15.05	5.98	8.89
	Code	42184	721029	40102	751	4234	4234	4-41	4234	401	80110	B12-07	202
LATHING	Rate	5.59	6.78	4.36	4.35	4.40	4.69	3.84	4.52	3.71	15.05	6.23	8.89
	Code	42135	721033	40102	723	4273	4279	4-41	4271	402	80110	B13-16	202
MASONRY	Rate	4.41	6.78	4.36	11.15	4.55	6.81	3.84	5.21	5.96	17.99	6.23	8.89
	Code	42102	721037	40102	741	4231	4231	4-41	4231	401	80100	B13-18	202
PAINTING & DECORATING	Rate	4.35	4.94	3.18	6.75	4.40	4.69	3.84	4.52	3.71	15.05	5.98	8.89
	Code	42111	721041	40105	719	4275	4275	4-41	4275	402	80110	B12-01	202
PILE DRIVING	Rate	4.41	5.18	3.76	6.34	3.79	9.78	3.64	4.64	5.96	7.13	6.23	8.89
	Code	42159	722004	40706	732	4221	4221	4-43	4221	401	80030	B13-10	202
PLASTERING	Rate	5.59	6.78	5.08	6.75	4.40	4.69	3.84	4.52	3.71	15.05	5.98	8.89
	Code	42135	721042	40108	719	4271	4271	4-41	4271	402	80110	B12-21	202
PLUMBING	Rate	2.13	3.11	3.05	4.02	2.32	3.37	3.52	2.43	3.71	6.93	3.51	4.67
	Code	42122	721043	40204	707	4241	4241	4-46	4241	402	80160	B11-01	214
ROOFING	Rate	8.12	8.68	7.02	12.98	4.55	10.08	3.84	10.57	5.96	23.01	6.23	8.89
	Code	42118	721036	40403	728	4236	4236	4-41	4236	401	80130	B13-20	202
SHEET METAL WORK (HVAC)	Rate	2.13	3.11	7.02	4.02	2.32	3.37	3.52	2.43	3.71	6.93	3.51	4.67
	Code	42117	721043	40402	707	4244	4244	4-46	4244	402	80160	B11-07	208
STEEL ERECTION—door & sash	Rate	2.57	14.07	11.32	16.02	3.79	10.08	3.84	5.21	5.96	25.32	6.23	8.89
	Code	42106	722005	40502	748	4227	4227	4-41	4227	401	80080	B13-22	202
STEEL ERECTION—inter., ornam.	Rate	2.57	14.07	11.32	16.02	3.79	10.08	3.84	5.21	5.96	25.32	6.23	8.89
	Code	42106	722005	40502	748	4227	4227	4-41	4227	401	80080	B13-22	202
STEEL ERECTION—structure	Rate	2.57	14.07	11.32	16.02	3.79	10.08	3.84	5.21	5.96	25.32	6.23	8.89
	Code	42106	722005	40502	748	4227	4227	4-41	4227	401	80080	B13-22	202
STEEL ERECTION—NOC	Rate	2.57	14.07	11.32	16.02	3.79	10.08	3.84	5.21	5.96	25.32	6.23	8.89
	Code	42106	722005	40502	748	4227	4227	4-41	4227	401	80080	B13-22	202
TILE WORK—inter. (ceramic)	Rate	3.63	5.62	1.97	6.75	4.40	10.08	3.84	4.52	3.71	15.05	6.23	8.89
	Code	42113	721054	40103	719	4276	4276	4-41	4276	402	80110	B13-01	202
WATERPROOFING	Rate	4.35	4.93	4.36	4.35	4.55	4.69	3.84	5.21	3.71	23.01	5.98	8.89
	Code	42139	721016	40102	723	4239	4299	4-41	4239	402	80130	B12-17	202
WRECKING	Rate	2.51	4.59	6.59	16.02	2.62	4.39	3.64	3.34	5.96	15.05	6.23	8.89
	Code	40604	721005	40106	748	4211	4211	4-43	4211	401	80110	B13-09	202

R013113-80 Performance Bond

This table shows the cost of a Performance Bond for a construction job scheduled to be completed in 12 months. Add 1% of the premium cost per month for jobs requiring more than 12 months to complete. The rates are "standard" rates offered to contractors that the bonding company considers financially sound and capable of doing the work. Preferred rates

are offered by some bonding companies based upon financial strength of the contractor. Actual rates vary from contractor to contractor and from bonding company to bonding company. Contractors should prequalify through a bonding agency before submitting a bid on a contract that requires a bond.

Contract Amount	Building Construction Class B Projects			Highways & Bridges					
				Class A New Construction			Class A-1 Highway Resurfacing		
First $ 100,000 bid	$25.00 per M			$15.00 per M			$9.40 per M		
Next 400,000 bid	$ 2,500	plus $15.00	per M	$ 1,500	plus $10.00	per M	$ 940	plus $7.20	per M
Next 2,000,000 bid	8,500	plus 10.00	per M	5,500	plus 7.00	per M	3,820	plus 5.00	per M
Next 2,500,000 bid	28,500	plus 7.50	per M	19,500	plus 5.50	per M	15,820	plus 4.50	per M
Next 2,500,000 bid	47,250	plus 7.00	per M	33,250	plus 5.00	per M	28,320	plus 4.50	per M
Over 7,500,000 bid	64,750	plus 6.00	per M	45,750	plus 4.50	per M	39,570	plus 4.00	per M

R015423-10 Steel Tubular Scaffolding

On new construction, tubular scaffolding is efficient up to 60' high or five stories. Above this it is usually better to use a hung scaffolding if construction permits. Swing scaffolding operations may interfere with tenants. In this case, the tubular is more practical at all heights.

In repairing or cleaning the front of an existing building the cost of tubular scaffolding per S.F. of building front increases as the height increases above the first tier. The first tier cost is relatively high due to leveling and alignment.

The minimum efficient crew for erecting and dismantling is three workers. They can set up and remove 18 frame sections per day up to 5 stories high. For 6 to 12 stories high, a crew of four is most efficient. Use two or more on top and two on the bottom for handing up or hoisting. They can

also set up and remove 18 frame sections per day. At 7' horizontal spacing, this will run about 800 S.F. per day of erecting and dismantling. Time for placing and removing planks must be added to the above. A crew of three can place and remove 72 planks per day up to 5 stories. For over 5 stories, a crew of four can place and remove 80 planks per day.

The table below shows the number of pieces required to erect tubular steel scaffolding for 1000 S.F. of building frontage. This area is made up of a scaffolding system that is 12 frames (11 bays) long by 2 frames high.

For jobs under twenty-five frames, add 50% to rental cost. Rental rates will be lower for jobs over three months duration. Large quantities for long periods can reduce rental rates by 20%.

Description of Component	Number of Pieces for 1000 S.F. of Building Front	Unit
5' Wide Standard Frame, 6'-4" High	24	Ea.
Leveling Jack & Plate	24	
Cross Brace	44	
Side Arm Bracket, 21"	12	
Guardrail Post	12	
Guardrail, 7' section	22	
Stairway Section	2	
Stairway Starter Bar	1	
Stairway Inside Handrail	2	
Stairway Outside Handrail	2	
Walk-Thru Frame Guardrail	2	

Scaffolding is often used as falsework over 15' high during construction of cast-in-place concrete beams and slabs. Two foot wide scaffolding is generally used for heavy beam construction. The span between frames depends upon the load to be carried with a maximum span of 5'.

Heavy duty shoring frames with a capacity of 10,000#/leg can be spaced up to 10' O.C. depending upon form support design and loading.

Scaffolding used as horizontal shoring requires less than half the material required with conventional shoring.

On new construction, erection is done by carpenters.

Rolling towers supporting horizontal shores can reduce labor and speed the job. For maintenance work, catwalks with spans up to 70' can be supported by the rolling towers.

R015423-20   Pump Staging

Pump staging is generally not available for rent. The table below shows the number of pieces required to erect pump staging for 2400 S.F. of building frontage. This area is made up of a pump jack system that is 3 poles (2 bays) wide by 2 poles high.

Item	Number of Pieces for 2400 S.F. of Building Front	Unit
Aluminum pole section, 24′ long	6	Ea.
Aluminum splice joint, 6′ long	3	
Aluminum foldable brace	3	
Aluminum pump jack	3	
Aluminum support for workbench/back safety rail	3	
Aluminum scaffold plank/workbench, 14″ wide x 24′ long	4	
Safety net, 22′ long	2	
Aluminum plank end safety rail	2	

The cost in place for this 2400 S.F. will depend on how many uses are realized during the life of the equipment.

R015433-10   Contractor Equipment

Rental Rates shown elsewhere in the book pertain to late model high quality machines in excellent working condition, rented from equipment dealers. Rental rates from contractors may be substantially lower than the rental rates from equipment dealers depending upon economic conditions; for older, less productive machines, reduce rates by a maximum of 15%. Any overtime must be added to the base rates. For shift work, rates are lower. Usual rule of thumb is 150% of one shift rate for two shifts; 200% for three shifts.

For periods of less than one week, operated equipment is usually more economical to rent than renting bare equipment and hiring an operator.

Costs to move equipment to a job site (mobilization) or from a job site (demobilization) are not included in rental rates, nor in any Equipment costs on any Unit Price line items or crew listings. These costs can be found elsewhere. If a piece of equipment is already at a job site, it is not appropriate to utilize mob/demob costs in an estimate again.

Rental rates vary throughout the country with larger cities generally having lower rates. Lease plans for new equipment are available for periods in excess of six months with a percentage of payments applying toward purchase.

Monthly rental rates vary from 2% to 5% of the cost of the equipment depending on the anticipated life of the equipment and its wearing parts. Weekly rates are about 1/3 the monthly rates and daily rental rates about 1/3 the weekly rate.

The hourly operating costs for each piece of equipment include costs to the user such as fuel, oil, lubrication, normal expendables for the equipment, and a percentage of mechanic's wages chargeable to maintenance. The hourly operating costs listed do not include the operator's wages.

The daily cost for equipment used in the standard crews is figured by dividing the weekly rate by five, then adding eight times the hourly operating cost to give the total daily equipment cost, not including the operator. This figure is in the right hand column of the Equipment listings under Equipment Cost/Day.

Pile Driving rates shown for pile hammer and extractor do not include leads, crane, boiler or compressor. Vibratory pile driving requires an added field specialist during set-up and pile driving operation for the electric model. The hydraulic model requires a field specialist for set-up only. Up to 125 reuses of sheet piling are possible using vibratory drivers. For normal conditions, crane capacity for hammer type and size are as follows.

Crane Capacity	Hammer Type and Size		
	Air or Steam	Diesel	Vibratory
25 ton	to 8,750 ft.-lb.		70 H.P.
40 ton	15,000 ft.-lb.	to 32,000 ft.-lb.	170 H.P.
60 ton	25,000 ft.-lb.		300 H.P.
100 ton		112,000 ft.-lb.	

Cranes should be specified for the job by size, building and site characteristics, availability, performance characteristics, and duration of time required.

Backhoes & Shovels rent for about the same as equivalent size cranes but maintenance and operating expense is higher. Crane operators rate must be adjusted for high boom heights. Average adjustments: for 150′ boom add 2% per hour; over 185′, add 4% per hour; over 210′, add 6% per hour; over 250′, add 8% per hour and over 295′, add 12% per hour.

Tower Cranes of the climbing or static type have jibs from 50′ to 200′ and capacities at maximum reach range from 4,000 to 14,000 pounds. Lifting capacities increase up to maximum load as the hook radius decreases.

Typical rental rates, based on purchase price are about 2% to 3% per month.

Erection and dismantling runs between 500 and 2000 labor hours. Climbing operation takes 10 labor hours per 20′ climb. Crane dead time is about 5 hours per 40′ climb. If crane is bolted to side of the building add cost of ties and extra mast sections. Climbing cranes have from 80′ to 180′ of mast while static cranes have 80′ to 800′ of mast.

Truck Cranes can be converted to tower cranes by using tower attachments. Mast heights over 400′ have been used.

A single 100′ high material **Hoist and Tower** can be erected and dismantled in about 400 labor hours; a double 100′ high hoist and tower in about 600 labor hours. Erection times for additional heights are 3 and 4 labor hours per vertical foot respectively up to 150′, and 4 to 5 labor hours per vertical foot over 150′ high. A 40′ high portable Buck hoist takes about 160 labor hours to erect and dismantle. Additional heights take 2 labor hours per vertical foot to 80′ and 3 labor hours per vertical foot for the next 100′. Most material hoists do not meet local code requirements for carrying personnel.

A 150′ high **Personnel Hoist** requires about 500 to 800 labor hours to erect and dismantle. Budget erection time at 5 labor hours per vertical foot for all trades. Local code requirements or labor scarcity requiring overtime can add up to 50% to any of the above erection costs.

Earthmoving Equipment: The selection of earthmoving equipment depends upon the type and quantity of material, moisture content, haul distance, haul road, time available, and equipment available. Short haul cut and fill operations may require dozers only, while another operation may require excavators, a fleet of trucks, and spreading and compaction equipment. Stockpiled material and granular material are easily excavated with front end loaders. Scrapers are most economically used with hauls between 300′ and 1-1/2 miles if adequate haul roads can be maintained. Shovels are often used for blasted rock and any material where a vertical face of 8′ or more can be excavated. Special conditions may dictate the use of draglines, clamshells, or backhoes. Spreading and compaction equipment must be matched to the soil characteristics, the compaction required and the rate the fill is being supplied.

R024119-10 Demolition Defined

Whole Building Demolition - Demolition of the whole building with no concern for any particular building element, component, or material type being demolished. This type of demolition is accomplished with large pieces of construction equipment that break up the structure, load it into trucks and haul it to a disposal site, but disposal or dump fees are not included. Demolition of below-grade foundation elements, such as footings, foundation walls, grade beams, slabs on grade, etc., is not included. Certain mechanical equipment containing flammable liquids or ozone-depleting refrigerants, electric lighting elements, communication equipment components, and other building elements may contain hazardous waste, and must be removed, either selectively or carefully, as hazardous waste before the building can be demolished.

Foundation Demolition - Demolition of below-grade foundation footings, foundation walls, grade beams, and slabs on grade. This type of demolition is accomplished by hand or pneumatic hand tools, and does not include saw cutting, or handling, loading, hauling, or disposal of the debris.

Gutting - Removal of building interior finishes and electrical/mechanical systems down to the load-bearing and sub-floor elements of the rough building frame, with no concern for any particular building element, component, or material type being demolished. This type of demolition is accomplished by hand or pneumatic hand tools, and includes loading into trucks, but not hauling, disposal or dump fees, scaffolding, or shoring. Certain mechanical equipment containing flammable liquids or ozone-depleting refrigerants, electric lighting elements, communication equipment components, and other building elements may contain hazardous waste, and must be removed, either selectively or carefully, as hazardous waste, before the building is gutted.

Selective Demolition - Demolition of a selected building element, component, or finish, with some concern for surrounding or adjacent elements, components, or finishes (see the first Subdivision (s) at the beginning of appropriate Divisions). This type of demolition is accomplished by hand or pneumatic hand tools, and does not include handling, loading, storing, hauling, or disposal of the debris, scaffolding, or shoring. "Gutting" methods may be used in order to save time, but damage that is caused to surrounding or adjacent elements, components, or finishes may have to be repaired at a later time.

Careful Removal - Removal of a piece of service equipment, building element or component, or material type, with great concern for both the removed item and surrounding or adjacent elements, components or finishes. The purpose of careful removal may be to protect the removed item for later re-use, preserve a higher salvage value of the removed item, or replace an item while taking care to protect surrounding or adjacent elements, components, connections, or finishes from cosmetic and/or structural damage. An approximation of the time required to perform this type of removal is 1/3 to 1/2 the time it would take to install a new item of like kind (see Reference Number R220105-10). This type of removal is accomplished by hand or pneumatic hand tools, and does not include loading, hauling, or storing the removed item, scaffolding, shoring, or lifting equipment.

Cutout Demolition - Demolition of a small quantity of floor, wall, roof, or other assembly, with concern for the appearance and structural integrity of the surrounding materials. This type of demolition is accomplished by hand or pneumatic hand tools, and does not include saw cutting, handling, loading, hauling, or disposal of debris, scaffolding, or shoring.

Rubbish Handling - Work activities that involve handling, loading or hauling of debris. Generally, the cost of rubbish handling must be added to the cost of all types of demolition, with the exception of whole building demolition.

Minor Site Demolition - Demolition of site elements outside the footprint of a building. This type of demolition is accomplished by hand or pneumatic hand tools, or with larger pieces of construction equipment, and may include loading a removed item onto a truck (check the Crew for equipment used). It does not include saw cutting, hauling or disposal of debris, and, sometimes, handling or loading.

R024119-20 Dumpsters

Dumpster rental costs on construction sites are presented in two ways.

The cost per week rental includes the delivery of the dumpster; its pulling or emptying once per week, and its final removal. The assumption is made that the dumpster contractor could choose to empty a dumpster by simply bringing in an empty unit and removing the full one. These costs also include the disposal of the materials in the dumpster.

The Alternate Pricing can be used when actual planned conditions are not approximated by the weekly numbers. For example, these lines can be used when a dumpster is needed for 4 weeks and will need to be emptied 2 or 3 times per week. Conversely the Alternate Pricing lines can be used when a dumpster will be rented for several weeks or months but needs to be emptied only a few times over this period.

R026510-20 Underground Storage Tank Removal

Underground Storage Tank Removal can be divided into two categories: Non-Leaking and Leaking. Prior to removing an underground storage tank, tests should be made, with the proper authorities present, to determine whether a tank has been leaking or the surrounding soil has been contaminated.

To safely remove Liquid Underground Storage Tanks:
1. Excavate to the top of the tank.
2. Disconnect all piping.
3. Open all tank vents and access ports.
4. Remove all liquids and/or sludge.
5. Purge the tank with an inert gas.
6. Provide access to the inside of the tank and clean out the interior using proper personal protective equipment (PPE).
7. Excavate soil surrounding the tank using proper PPE for on-site personnel.
8. Pull and properly dispose of the tank.
9. Clean up the site of all contaminated material.
10. Install new tanks or close the excavation.

R028213-20 Asbestos Removal Process

Asbestos removal is accomplished by a specialty contractor who understands the federal and state regulations regarding the handling and disposal of the material. The process of asbestos removal is divided into many individual steps. An accurate estimate can be calculated only after all the steps have been priced.

The steps are generally as follows:

1. Obtain an asbestos abatement plan from an industrial hygienist.
2. Monitor the air quality in and around the removal area and along the path of travel between the removal area and transport area. This establishes the background contamination.
3. Construct a two part decontamination chamber at entrance to removal area.
4. Install a HEPA filter to create a negative pressure in the removal area.
5. Install wall, floor and ceiling protection as required by the plan, usually 2 layers of fireproof 6 mil polyethylene.

6. Industrial hygienist visually inspects work area to verify compliance with plan.
7. Provide temporary supports for conduit and piping affected by the removal process.
8. Proceed with asbestos removal and bagging process. Monitor air quality as described in Step #2. Discontinue operations when contaminate levels exceed applicable standards.
9. Document the legal disposal of materials in accordance with EPA standards.
10. Thoroughly clean removal area including all ledges, crevices and surfaces.
11. Post abatement inspection by industrial hygienist to verify plan compliance.
12. Provide a certificate from a licensed industrial hygienist attesting that contaminate levels are within acceptable standards before returning area to regular use.

R028319-60 Lead Paint Remediation Methods

Lead paint remediation can be accomplished by the following methods.
1. Abrasive blast
2. Chemical stripping
3. Power tool cleaning with vacuum collection system
4. Encapsulation
5. Remove and replace
6. Enclosure

Each of these methods has strengths and weakness depending on the specific circumstances of the project. The following is an overview of each method.

1. **Abrasive blasting** is usually accomplished with sand or recyclable metallic blast. Before work can begin, the area must be contained to ensure the blast material with lead does not escape to the atmosphere. The use of vacuum blast greatly reduces the containment requirements. Lead abatement equipment that may be associated with this work includes a negative air machine. In addition, it is necessary to have an industrial hygienist monitor the project on a continual basis. When the work is complete, the spent blast sand with lead must be disposed of as a hazardous material. If metallic shot was used, the lead is separated from the shot and disposed of as hazardous material. Worker protection includes disposable clothing and respiratory protection.

2. **Chemical stripping** requires strong chemicals be applied to the surface to remove the lead paint. Before the work can begin, the area under/adjacent to the work area must be covered to catch the chemical and removed lead. After the chemical is applied to the painted surface it is usually covered with paper. The chemical is left in place for the specified period, then the paper with lead paint is pulled or scraped off. The process may require several chemical applications. The paper with chemicals and lead paint adhered to it, plus the containment and loose scrapings collected by a HEPA (High Efficiency Particulate Air Filter) vac, must be disposed of as a hazardous material. The chemical stripping process usually requires a neutralizing agent and several wash downs after the paint is removed. Worker protection includes a neoprene or other compatible protective clothing and respiratory protection with face shield. An industrial hygienist is required intermittently during the process.

3. **Power tool cleaning** is accomplished using shrouded needle blasting guns. The shrouding with different end configurations is held up against the surface to be cleaned. The area is blasted with hardened needles and the shroud captures the lead with a HEPA vac and deposits it in a holding tank. An industrial hygienist monitors the project, protective clothing and a respirator is required until air samples prove otherwise. When the work is complete the lead must be disposed of as a hazardous material.

4. **Encapsulation** is a method that leaves the well bonded lead paint in place after the peeling paint has been removed. Before the work can begin, the area under/adjacent to the work must be covered to catch the scrapings. The scraped surface is then washed with a detergent and rinsed. The prepared surface is covered with approximately 10 mils of paint. A reinforcing fabric can also be embedded in the paint covering. The scraped paint and containment must be disposed of as a hazardous material. Workers must wear protective clothing and respirators.

5. **Remove and replace** is an effective way to remove lead paint from windows, gypsum walls and concrete masonry surfaces. The painted materials are removed and new materials are installed. Workers should wear a respirator and tyvek suit. The demolished materials must be disposed of as hazardous waste if it fails the TCLP (Toxicity Characteristic Leachate Process) test.

6. **Enclosure** is the process that permanently seals lead painted materials in place. This process has many applications such as covering lead painted drywall with new drywall, covering exterior construction with tyvek paper then residing, or covering lead painted structural members with aluminum or plastic. The seams on all enclosing materials must be securely sealed. An industrial hygienist monitors the project, and protective clothing and a respirator is required until air samples prove otherwise.

All the processes require clearance monitoring and wipe testing as required by the hygienist.

R031113-10   Wall Form Materials

Aluminum Forms

Approximate weight is 3 lbs. per S.F.C.A. Standard widths are available from 4" to 36" with 36" most common. Standard lengths of 2', 4', 6' to 8' are available. Forms are lightweight and fewer ties are needed with the wider widths. The form face is either smooth or textured.

Metal Framed Plywood Forms

Manufacturers claim over 75 reuses of plywood and over 300 reuses of steel frames. Many specials such as corners, fillers, pilasters, etc. are available. Monthly rental is generally about 15% of purchase price for first month and 9% per month thereafter with 90% of rental applied to purchase for the first month and decreasing percentages thereafter. Aluminum framed forms cost 25% to 30% more than steel framed.

After the first month, extra days may be prorated from the monthly charge. Rental rates do not include ties, accessories, cleaning, loss of hardware or freight in and out. Approximate weight is 5 lbs. per S.F. for steel; 3 lbs. per S.F. for aluminum.

Forms can be rented with option to buy.

Plywood Forms, Job Fabricated

There are two types of plywood used for concrete forms.

1. Exterior plyform which is completely waterproof. This is face oiled to facilitate stripping. Ten reuses can be expected with this type with 25 reuses possible.
2. An overlaid type consists of a resin fiber fused to exterior plyform. No oiling is required except to facilitate cleaning. This is available in both high density (HDO) and medium density overlaid (MDO). Using HDO, 50 reuses can be expected with 200 possible.

Plyform is available in 5/8" and 3/4" thickness. High density overlaid is available in 3/8", 1/2", 5/8" and 3/4" thickness.

5/8" thick is sufficient for most building forms, while 3/4" is best on heavy construction.

Plywood Forms, Modular, Prefabricated

There are many plywood forming systems without frames. Most of these are manufactured from 1-1/8" (HDO) plywood and have some hardware attached. These are used principally for foundation walls 8' or less high. With care and maintenance, 100 reuses can be attained with decreasing quality of surface finish.

Steel Forms

Approximate weight is 6-1/2 lbs. per S.F.C.A. including accessories. Standard widths are available from 2" to 24", with 24" most common. Standard lengths are from 2' to 8', with 4' the most common. Forms are easily ganged into modular units.

Forms are usually leased for 15% of the purchase price per month prorated daily over 30 days.

Rental may be applied to sale price, and usually rental forms are bought. With careful handling and cleaning 200 to 400 reuses are possible.

Straight wall gang forms up to 12' x 20' or 8' x 30' can be fabricated. These crane handled forms usually lease for approx. 9% per month.

Individual job analysis is available from the manufacturer at no charge.

R031113-40   Forms for Reinforced Concrete

Design Economy

Avoid many sizes in proportioning beams and columns.

From story to story avoid changing column dimensions. Gain strength by adding steel or using a richer mix. If a change in size of column is necessary, vary one dimension only to minimize form alterations. Keep beams and columns the same width.

From floor to floor in a multi-story building vary beam depth, not width, as that will leave slab panel form unchanged. It is cheaper to vary the strength of a beam from floor to floor by means of steel area than by 2" changes in either width or depth.

Cost Factors

Material includes the cost of lumber, cost of rent for metal pans or forms if used, nails, form ties, form oil, bolts and accessories.

Labor includes the cost of carpenters to make up, erect, remove and repair, plus common labor to clean and move. Having carpenters remove forms minimizes repairs.

Improper alignment and condition of forms will increase finishing cost. When forms are heavily oiled, concrete surfaces must be neutralized before finishing. Special curing compounds will cause spillages to spall off in first frost. Gang forming methods will reduce costs on large projects.

Materials Used

Boards are seldom used unless their architectural finish is required. Generally, steel, fiberglass and plywood are used for contact surfaces. Labor on plywood is 10% less than with boards. The plywood is backed up with

2 x 4's at 12" to 32" O.C. Walers are generally 2 - 2 x 4's. Column forms are held together with steel yokes or bands. Shoring is with adjustable shoring or scaffolding for high ceilings.

Reuse

Floor and column forms can be reused four or possibly five times without excessive repair. Remember to allow for 10% waste on each reuse.

When modular sized wall forms are made, up to twenty uses can be expected with exterior plyform.

When forms are reused, the cost to erect, strip, clean and move will not be affected. 10% replacement of lumber should be included and about one hour of carpenter time for repairs on each reuse per 100 S.F.

The reuse cost for certain accessory items normally rented on a monthly basis will be lower than the cost for the first use.

After fifth use, new material required plus time needed for repair prevent form cost from dropping further and it may go up. Much depends on care in stripping, the number of special bays, changes in beam or column sizes and other factors.

Costs for multiple use of formwork may be developed as follows:

2 Uses	3 Uses	4 Uses
$\dfrac{\text{(1st Use + Reuse)}}{2}$ = avg. cost/2 uses	$\dfrac{\text{(1st Use + 2 Reuse)}}{3}$ = avg. cost/3 uses	$\dfrac{\text{(1st use + 3 Reuse)}}{4}$ = avg. cost/4 uses

R031113-60   Formwork Labor-Hours

Item	Unit	Hours Required			Total Hours	Multiple Use		
		Fabricate	Erect & Strip	Clean & Move	1 Use	2 Use	3 Use	4 Use
Beam and Girder, interior beams, 12" wide	100 S.F.	6.4	8.3	1.3	16.0	13.3	12.4	12.0
Hung from steel beams		5.8	7.7	1.3	14.8	12.4	11.6	11.2
Beam sides only, 36" high		5.8	7.2	1.3	14.3	11.9	11.1	10.7
Beam bottoms only, 24" wide		6.6	13.0	1.3	20.9	18.1	17.2	16.7
Box out for openings		9.9	10.0	1.1	21.0	16.6	15.1	14.3
Buttress forms, to 8' high		6.0	6.5	1.2	13.7	11.2	10.4	10.0
Centering, steel, 3/4" rib lath			1.0		1.0			
3/8" rib lath or slab form	▼		0.9		0.9			
Chamfer strip or keyway	100 L.F.		1.5		1.5	1.5	1.5	1.5
Columns, fiber tube 8" diameter			20.6		20.6			
12"			21.3		21.3			
16"			22.9		22.9			
20"			23.7		23.7			
24"			24.6		24.6			
30"	▼		25.6		25.6			
Columns, round steel, 12" diameter			22.0		22.0	22.0	22.0	22.0
16"			25.6		25.6	25.6	25.6	25.6
20"			30.5		30.5	30.5	30.5	30.5
24"	▼		37.7		37.7	37.7	37.7	37.7
Columns, plywood 8" x 8"	100 S.F.	7.0	11.0	1.2	19.2	16.2	15.2	14.7
12" x 12"		6.0	10.5	1.2	17.7	15.2	14.4	14.0
16" x 16"		5.9	10.0	1.2	17.1	14.7	13.8	13.4
24" x 24"		5.8	9.8	1.2	16.8	14.4	13.6	13.2
Columns, steel framed plywood 8" x 8"			10.0	1.0	11.0	11.0	11.0	11.0
12" x 12"			9.3	1.0	10.3	10.3	10.3	10.3
16" x 16"			8.5	1.0	9.5	9.5	9.5	9.5
24" x 24"			7.8	1.0	8.8	8.8	8.8	8.8
Drop head forms, plywood		9.0	12.5	1.5	23.0	19.0	17.7	17.0
Coping forms		8.5	15.0	1.5	25.0	21.3	20.0	19.4
Culvert, box			14.5	4.3	18.8	18.8	18.8	18.8
Curb forms, 6" to 12" high, on grade		5.0	8.5	1.2	14.7	12.7	12.1	11.7
On elevated slabs	▼	6.0	10.8	1.2	18.0	15.5	14.7	14.3
Edge forms to 6" high, on grade	100 L.F.	2.0	3.5	0.6	6.1	5.6	5.4	5.3
7" to 12" high	100 S.F.	2.5	5.0	1.0	8.5	7.8	7.5	7.4
Equipment foundations		10.0	18.0	2.0	30.0	25.5	24.0	23.3
Flat slabs, including drops		3.5	6.0	1.2	10.7	9.5	9.0	8.8
Hung from steel		3.0	5.5	1.2	9.7	8.7	8.4	8.2
Closed deck for domes		3.0	5.8	1.2	10.0	9.0	8.7	8.5
Open deck for pans		2.2	5.3	1.0	8.5	7.9	7.7	7.6
Footings, continuous, 12" high		3.5	3.5	1.5	8.5	7.3	6.8	6.6
Spread, 12" high		4.7	4.2	1.6	10.5	8.7	8.0	7.7
Pile caps, square or rectangular		4.5	5.0	1.5	11.0	9.3	8.7	8.4
Grade beams, 24" deep		2.5	5.3	1.2	9.0	8.3	8.0	7.9
Lintel or Sill forms		8.0	17.0	2.0	27.0	23.5	22.3	21.8
Spandrel beams, 12" wide		9.0	11.2	1.3	21.5	17.5	16.2	15.5
Stairs			25.0	4.0	29.0	29.0	29.0	29.0
Trench forms in floor		4.5	14.0	1.5	20.0	18.3	17.7	17.4
Walls, Plywood, at grade, to 8' high		5.0	6.5	1.5	13.0	11.0	9.7	9.5
8' to 16'		7.5	8.0	1.5	17.0	13.8	12.7	12.1
16' to 20'		9.0	10.0	1.5	20.5	16.5	15.2	14.5
Foundation walls, to 8' high		4.5	6.5	1.0	12.0	10.3	9.7	9.4
8' to 16' high		5.5	7.5	1.0	14.0	11.8	11.0	10.6
Retaining wall to 12' high, battered		6.0	8.5	1.5	16.0	13.5	12.7	12.3
Radial walls to 12' high, smooth		8.0	9.5	2.0	19.5	16.0	14.8	14.3
2' chords		7.0	8.0	1.5	16.5	13.5	12.5	12.0
Prefabricated modular, to 8' high		—	4.3	1.0	5.3	5.3	5.3	5.3
Steel, to 8' high		—	6.8	1.2	8.0	8.0	8.0	8.0
8' to 16' high		—	9.1	1.5	10.6	10.3	10.2	10.2
Steel framed plywood to 8' high		—	6.8	1.2	8.0	7.5	7.3	7.2
8' to 16' high	▼	—	9.3	1.2	10.5	9.5	9.2	9.0

R032110-10   Reinforcing Steel Weights and Measures

Bar Designation No.**	Nominal Weight Lb./Ft.	U.S. Customary Units Nominal Dimensions*			SI Units Nominal Dimensions*			
		Diameter in.	Cross Sectional Area, in.²	Perimeter in.	Nominal Weight kg/m	Diameter mm	Cross Sectional Area, cm²	Perimeter mm
3	.376	.375	.11	1.178	.560	9.52	.71	29.9
4	.668	.500	.20	1.571	.994	12.70	1.29	39.9
5	1.043	.625	.31	1.963	1.552	15.88	2.00	49.9
6	1.502	.750	.44	2.356	2.235	19.05	2.84	59.8
7	2.044	.875	.60	2.749	3.042	22.22	3.87	69.8
8	2.670	1.000	.79	3.142	3.973	25.40	5.10	79.8
9	3.400	1.128	1.00	3.544	5.059	28.65	6.45	90.0
10	4.303	1.270	1.27	3.990	6.403	32.26	8.19	101.4
11	5.313	1.410	1.56	4.430	7.906	35.81	10.06	112.5
14	7.650	1.693	2.25	5.320	11.384	43.00	14.52	135.1
18	13.600	2.257	4.00	7.090	20.238	57.33	25.81	180.1

* The nominal dimensions of a deformed bar are equivalent to those of a plain round bar having the same weight per foot as the deformed bar.
** Bar numbers are based on the number of eighths of an inch included in the nominal diameter of the bars.

R032110-80   Shop-Fabricated Reinforcing Steel

The material prices for reinforcing, shown in the unit cost sections of the book, are for 50 tons or more of shop-fabricated reinforcing steel and include:
1. Mill base price of reinforcing steel
2. Mill grade/size/length extras
3. Mill delivery to the fabrication shop
4. Shop storage and handling
5. Shop drafting/detailing
6. Shop shearing and bending
7. Shop listing
8. Shop delivery to the job site

Both material and installation costs can be considerably higher for small jobs consisting primarily of smaller bars, while material costs may be slightly lower for larger jobs.

R032205-30   Common Stock Styles of Welded Wire Fabric

This table provides some of the basic specifications, sizes, and weights of welded wire fabric used for reinforcing concrete.

	New Designation Spacing — Cross Sectional Area (in.) — (Sq. in. 100)		Old Designation Spacing — Wire Gauge (in.) — (AS & W)		Steel Area per Foot				Approximate Weight per 100 S.F.	
					Longitudinal		Transverse			
					in.	cm	in.	cm	lbs	kg
Rolls	6 x 6 — W1.4 x W1.4		6 x 6 — 10 x 10		.028	.071	.028	.071	21	9.53
	6 x 6 — W2.0 x W2.0		6 x 6 — 8 x 8	1	.040	.102	.040	.102	29	13.15
	6 x 6 — W2.9 x W2.9		6 x 6 — 6 x 6		.058	.147	.058	.147	42	19.05
	6 x 6 — W4.0 x W4.0		6 x 6 — 4 x 4		.080	.203	.080	.203	58	26.91
	4 x 4 — W1.4 x W1.4		4 x 4 — 10 x 10		.042	.107	.042	.107	31	14.06
	4 x 4 — W2.0 x W2.0		4 x 4 — 8 x 8	1	.060	.152	.060	.152	43	19.50
	4 x 4 — W2.9 x W2.9		4 x 4 — 6 x 6		.087	.227	.087	.227	62	28.12
	4 x 4 — W4.0 x W4.0		4 x 4 — 4 x 4		.120	.305	.120	.305	85	38.56
Sheets	6 x 6 — W2.9 x W2.9		6 x 6 — 6 x 6		.058	.147	.058	.147	42	19.05
	6 x 6 — W4.0 x W4.0		6 x 6 — 4 x 4		.080	.203	.080	.203	58	26.31
	6 x 6 — W5.5 x W5.5		6 x 6 — 2 x 2	2	.110	.279	.110	.279	80	36.29
	4 x 4 — W1.4 x W1.4		4 x 4 — 4 x 4		.120	.305	.120	.305	85	38.56

NOTES: 1. Exact W—number size for 8 gauge is W2.1
        2. Exact W—number size for 2 gauge is W5.4

Reference Tables

R033105-10   Proportionate Quantities

The tables below show both quantities per S.F. of floor areas as well as form and reinforcing quantities per C.Y. Unusual structural requirements would increase the ratios below. High strength reinforcing would reduce the steel weights. Figures are for 3000 psi concrete and 60,000 psi reinforcing unless specified otherwise.

Type of Construction	Live Load	Span	Per S.F. of Floor Area				Per C.Y. of Concrete		
			Concrete	Forms	Reinf.	Pans	Forms	Reinf.	Pans
Flat Plate	50 psf	15 Ft.	.46 C.F.	1.06 S.F.	1.71 lb.		62 S.F.	101 lb.	
		20	.63	1.02	2.40		44	104	
		25	.79	1.02	3.03		35	104	
	100	15	.46	1.04	2.14		61	126	
		20	.71	1.02	2.72		39	104	
		25	.83	1.01	3.47		33	113	
Flat Plate (waffle construction) 20″ domes	50	20	.43	1.00	2.10	.84 S.F.	63	135	53 S.F.
		25	.52	1.00	2.90	.89	52	150	46
		30	.64	1.00	3.70	.87	42	155	37
	100	20	.51	1.00	2.30	.84	53	125	45
		25	.64	1.00	3.20	.83	42	135	35
		30	.76	1.00	4.40	.81	36	160	29
Waffle Construction 30″ domes	50	25	.69	1.06	1.83	.68	42	72	40
		30	.74	1.06	2.39	.69	39	87	39
		35	.86	1.05	2.71	.69	33	85	39
		40	.78	1.00	4.80	.68	35	165	40
Flat Slab (two way with drop panels)	50	20	.62	1.03	2.34		45	102	
		25	.77	1.03	2.99		36	105	
		30	.95	1.03	4.09		29	116	
	100	20	.64	1.03	2.83		43	119	
		25	.79	1.03	3.88		35	133	
		30	.96	1.03	4.66		29	131	
	200	20	.73	1.03	3.03		38	112	
		25	.86	1.03	4.23		32	133	
		30	1.06	1.03	5.30		26	135	
One Way Joists 20″ Pans	50	15	.36	1.04	1.40	.93	78	105	70
		20	.42	1.05	1.80	.94	67	120	60
		25	.47	1.05	2.60	.94	60	150	54
	100	15	.38	1.07	1.90	.93	77	140	66
		20	.44	1.08	2.40	.94	67	150	58
		25	.52	1.07	3.50	.94	55	185	49
One Way Joists 8″ x 16″ filler blocks	50	15	.34	1.06	1.80	.81 Ea.	84	145	64 Ea.
		20	.40	1.08	2.20	.82	73	145	55
		25	.46	1.07	3.20	.83	63	190	49
	100	15	.39	1.07	1.90	.81	74	130	56
		20	.46	1.09	2.80	.82	64	160	48
		25	.53	1.10	3.60	.83	56	190	42
One Way Beam & Slab	50	15	.42	1.30	1.73		84	111	
		20	.51	1.28	2.61		68	138	
		25	.64	1.25	2.78		53	117	
	100	15	.42	1.30	1.90		84	122	
		20	.54	1.35	2.69		68	154	
		25	.69	1.37	3.93		54	145	
	200	15	.44	1.31	2.24		80	137	
		20	.58	1.40	3.30		65	163	
		25	.69	1.42	4.89		53	183	
Two Way Beam & Slab	100	15	.47	1.20	2.26		69	130	
		20	.63	1.29	3.06		55	131	
		25	.83	1.33	3.79		43	123	
	200	15	.49	1.25	2.70		41	149	
		20	.66	1.32	4.04		54	165	
		25	.88	1.32	6.08		41	187	

R033105-10 Proportionate Quantities (cont.)

Item	Size	Forms		Reinforcing	Minimum	Maximum
		4000 psi Concrete and 60,000 psi Reinforcing—Form and Reinforcing Quantities per C.Y.				
	10" x 10"	130 S.F.C.A.		#5 to #11	220 lbs.	875 lbs.
	12" x 12"	108		#6 to #14	200	955
	14" x 14"	92		#7 to #14	190	900
	16" x 16"	81		#6 to #14	187	1082
	18" x 18"	72		#6 to #14	170	906
	20" x 20"	65		#7 to #18	150	1080
Columns	22" x 22"	59		#8 to #18	153	902
(square tied)	24" x 24"	54		#8 to #18	164	884
	26" x 26"	50		#9 to #18	169	994
	28" x 28"	46		#9 to #18	147	864
	30" x 30"	43		#10 to #18	146	983
	32" x 32"	40		#10 to #18	175	866
	34" x 34"	38		#10 to #18	157	772
	36" x 36"	36		#10 to #18	175	852
	38" x 38"	34		#10 to #18	158	765
	40" x 40"	32		#10 to #18	143	692

Item	Size	Form	Spiral	Reinforcing	Minimum	Maximum
	12" diameter	34.5 L.F.	190 lbs.	#4 to #11	165 lbs.	1505 lb.
		34.5	190	#14 & #18	—	1100
	14"	25	170	#4 to #11	150	970
		25	170	#14 & #18	800	1000
	16"	19	160	#4 to #11	160	950
		19	160	#14 & #18	605	1080
	18"	15	150	#4 to #11	160	915
		15	150	#14 & #18	480	1075
	20"	12	130	#4 to #11	155	865
		12	130	#14 & #18	385	1020
	22"	10	125	#4 to #11	165	775
		10	125	#14 & #18	320	995
	24"	9	120	#4 to #11	195	800
		9	120	#14 & #18	290	1150
Columns	26"	7.3	100	#4 to #11	200	729
(spirally reinforced)		7.3	100	#14 & #18	235	1035
	28"	6.3	95	#4 to #11	175	700
		6.3	95	#14 & #18	200	1075
	30"	5.5	90	#4 to #11	180	670
		5.5	90	#14 & #18	175	1015
	32"	4.8	85	#4 to #11	185	615
		4.8	85	#14 & #18	155	955
	34"	4.3	80	#4 to #11	180	600
		4.3	80	#14 & #18	170	855
	36"	3.8	75	#4 to #11	165	570
		3.8	75	#14 & #18	155	865
	40"	3.0	70	#4 to #11	165	500
		3.0	70	#14 & #18	145	765

R033105-10　Proportionate Quantities (cont.)

3000 psi Concrete and 60,000 psi Reinforcing—Form and Reinforcing Quantities per C.Y.						
Item	Type	Loading	Height	C.Y./L.F.	Forms/C.Y.	Reinf./C.Y.
Retaining Walls	Cantilever	Level Backfill	4 Ft.	0.2 C.Y.	49 S.F.	35 lbs.
			8	0.5	42	45
			12	0.8	35	70
			16	1.1	32	85
			20	1.6	28	105
		Highway Surcharge	4	0.3	41	35
			8	0.5	36	55
			12	0.8	33	90
			16	1.2	30	120
			20	1.7	27	155
		Railroad Surcharge	4	0.4	28	45
			8	0.8	25	65
			12	1.3	22	90
			16	1.9	20	100
			20	2.6	18	120
	Gravity, with Vertical Face	Level Backfill	4	0.4	37	None
			7	0.6	27	
			10	1.2	20	
		Sloping Backfill	4	0.3	31	
			7	0.8	21	↓
			10	1.6	15	

		Live Load in Kips per Linear Foot							
	Span	Under 1 Kip		2 to 3 Kips		4 to 5 Kips		6 to 7 Kips	
		Forms	Reinf.	Forms	Reinf.	Forms	Reinf.	Forms	Reinf.
Beams	10 Ft.	—	—	90 S.F.	170 #	85 S.F.	175 #	75 S.F.	185 #
	16	130 S.F.	165 #	85	180	75	180	65	225
	20	110	170	75	185	62	200	51	200
	26	90	170	65	215	62	215	—	—
	30	85	175	60	200	—	—	—	—

Item	Size	Type	Forms per C.Y.	Reinforcing per C.Y.
Spread Footings	Under 1 C.Y.	1,000 psf soil	24 S.F.	44 lbs.
		5,000	24	42
		10,000	24	52
	1 C.Y. to 5 C.Y.	1,000	14	49
		5,000	14	50
		10,000	14	50
	Over 5 C.Y.	1,000	9	54
		5,000	9	52
		10,000	9	56
Pile Caps (30 Ton Concrete Piles)	Under 5 C.Y.	shallow caps	20	65
		medium	20	50
		deep	20	40
	5 C.Y. to 10 C.Y.	shallow	14	55
		medium	15	45
		deep	15	40
	10 C.Y. to 20 C.Y.	shallow	11	60
		medium	11	45
		deep	12	35
	Over 20 C.Y.	shallow	9	60
		medium	9	45
		deep	10	40

R033105-10 Proportionate Quantities (cont.)

Item	Size	Pile Spacing	50 T Pile	100 T Pile	50 T Pile	100 T Pile
				3000 psi Concrete and 60,000 psi Reinforcing — Form and Reinforcing Quantities per C.Y.		
Pile Caps (Steel H Piles)	Under 5 C.Y.	24" O.C.	24 S.F.	24 S.F.	75 lbs.	90 lbs.
		30"	25	25	80	100
		36"	24	24	80	110
	5 C.Y. to 10 C.Y.	24"	15	15	80	110
		30"	15	15	85	110
		36"	15	15	75	90
	Over 10 C.Y.	24"	13	13	85	90
		30"	11	11	85	95
		36"	10	10	85	90

Item	Height	8" Thick		10" Thick		12" Thick		15" Thick	
		Forms	Reinf.	Forms	Reinf.	Forms	Reinf.	Forms	Reinf.
Basement Walls	7 Ft.	81 S.F.	44 lbs.	65 S.F.	45 lbs.	54 S.F.	44 lbs.	41 S.F.	43 lbs.
	8		44		45		44		43
	9		46		45		44		43
	10		57		45		44		43
	12		83		50		52		43
	14		116		65		64		51
	16				86		90		65
	18						106		70

R033105-20 Materials for One C.Y. of Concrete

This is an approximate method of figuring quantities of cement, sand and coarse aggregate for a field mix with waste allowance included.

With crushed gravel as coarse aggregate, to determine barrels of cement required, divide 10 by total mix; that is, for 1:2:4 mix, 10 divided by 7 = 1-3/7 barrels.

If the coarse aggregate is crushed stone, use 10-1/2 instead of 10 as given for gravel.

To determine tons of sand required, multiply barrels of cement by parts of sand and then by 0.2; that is, for the 1:2:4 mix, as above, 1-3/7 x 2 x .2 = .57 tons.

Tons of crushed gravel are in the same ratio to tons of sand as parts in the mix, or 4/2 x .57 = 1.14 tons.

1 bag cement = 94#	1 C.Y. sand or crushed gravel = 2700#	1 C.Y. crushed stone = 2575#
4 bags = 1 barrel	1 ton sand or crushed gravel = 20 C.F.	1 ton crushed stone = 21 C.F.

Average carload of cement is 692 bags; of sand or gravel is 56 tons.

Do not stack stored cement over 10 bags high.

R033105-70 Placing Ready-Mixed Concrete

For ground pours allow for 5% waste when figuring quantities.

Prices in the front of the book assume normal deliveries. If deliveries are made before 8 A.M. or after 5 P.M. or on Saturday afternoons add 30%. Negotiated discounts for large volumes are not included in prices in front of book.

For the lower floors without truck access, concrete may be wheeled in rubber-tired buggies, conveyer handled, crane handled or pumped. Pumping is economical if there is top steel. Conveyers are more efficient for thick slabs.

At higher floors the rubber-tired buggies may be hoisted by a hoisting tower and wheeled to location. Placement by a conveyer is limited to three floors and is best for high-volume pours. Pumped concrete is best when building has no crane access. Concrete may be pumped directly as high as thirty-six stories using special pumping techniques. Normal maximum height is about fifteen stories.

Best pumping aggregate is screened and graded bank gravel rather than crushed stone.

Pumping downward is more difficult than pumping upward. Horizontal distance from pump to pour may increase preparation time prior to pour. Placing by cranes, either mobile, climbing or tower types, continues as the most efficient method for high-rise concrete buildings.

R033105-85   Lift Slabs

The cost advantage of the lift slab method is due to placing all concrete, reinforcing steel, inserts and electrical conduit at ground level and in reduction of formwork. Minimum economical project size is about 30,000 S.F. Slabs may be tilted for parking garage ramps.

It is now used in all types of buildings and has gone up to 22 stories high in apartment buildings. Current trend is to use post-tensioned flat plate slabs with spans from 22' to 35'. Cylindrical void forms are used when deep slabs are required. One pound of prestressing steel is about equal to seven pounds of conventional reinforcing.

To be considered cured for stressing and lifting, a slab must have attained 75% of design strength. Seven days are usually sufficient with four to five days possible if high early strength cement is used. Slabs can be stacked using two coats of a non-bonding agent to insure that slabs do not stick to each other. Lifting is done by companies specializing in this work. Lift rate is 5' to 15' per hour with an average of 10' per hour. Total areas up to 33,000 S.F. have been lifted at one time. 24 to 36 jacking columns are common. Most economical bay sizes are 24' to 28' with four to fourteen stories most efficient. Continuous design reduces reinforcing steel cost. Use of post-tensioned slabs allows larger bay sizes.

Concrete   **R0341 Precast Structural Concrete**

R034105-30   Prestressed Precast Concrete Structural Units

Type	Location	Depth	Span in Ft.		Live Load Lb. per S.F.
Double Tee	Floor	28" to 34"	60 to 80		50 to 80
	Roof	12" to 24"	30 to 50		40
	Wall	Width 8'	Up to 55' high		Wind
Multiple Tee	Roof	8" to 12"	15 to 40		40
	Floor	8" to 12"	15 to 30		100
Plank	Roof		Roof	Floor	40 for Roof
		4"	13	12	
		6"	22	18	
or	or	8"	26	25	
		10"	33	29	100 for Floor
	Floor	12"	42	32	
Single Tee	Roof	28"	40		
		32"	80		
		36"	100		40
		48"	120		
AASHO Girder	Bridges	Type 4	100		
		5	110		Highway
		6	125		
Box Beam	Bridges	15" 27" 33"	40 to 100		Highway

The majority of precast projects today utilize double tees rather than single tees because of speed and ease of installation. As a result casting beds at manufacturing plants are normally formed for double tees. Single tee projects will therefore require an initial set up charge to be spread over the individual single tee costs.

For floors, a 2" to 3" topping is field cast over the shapes. For roofs, insulating concrete or rigid insulation is placed over the shapes.

Member lengths up to 40' are standard haul, 40' to 60' require special permits and lengths over 60' must be escorted. Over width and/or over length can add up to 100% on hauling costs.

Large heavy members may require two cranes for lifting which would increase erection costs by about 45%. An eight man crew can install 12 to 20 double tees, or 45 to 70 quad tees or planks per day.

Grouting of connections must also be included.

Several system buildings utilizing precast members are available. Heights can go up to 22 stories for apartment buildings. Optimum design ratio is 3 S.F. of surface to 1 S.F. of floor area.

R035216-10   Lightweight Concrete

Lightweight aggregate concrete is usually purchased ready mixed, but it can also be field mixed.

Vermiculite or Perlite comes in bags of 4 C.F. under various trade names. Weight is about 8 lbs. per C.F. For insulating roof fill use 1:6 mix. For structural deck use 1:4 mix over gypsum boards, steeltex, steel centering, etc., supported by closely spaced joists or bulb trees. For structural slabs use 1:3:2 vermiculite sand concrete over steeltex, metal lath, steel centering, etc., on joists spaced 2'-0" O.C. for maximum L.L. of 80 P.S.F. Use same mix for slab base fill over steel flooring or regular reinforced concrete slab when tile, terrazzo or other finish is to be laid over.

For slabs on grade use 1:3:2 mix when tile, etc., finish is to be laid over. If radiant heating units are installed use a 1:6 mix for a base. After coils are in place, cover with a regular granolithic finish (mix 1:3:2) to a minimum depth of 1-1/2" over top of units.

Reinforce all slabs with 6 x 6 or 10 x 10 welded wire mesh.

R042110-20   Common and Face Brick

Common building brick manufactured according to ASTM C62 and facing brick manufactured according to ASTM C216 are the two standard bricks available for general building use.

Building brick is made in three grades; SW, where high resistance to damage caused by cyclic freezing is required; MW, where moderate resistance to cyclic freezing is needed; and NW, where little resistance to cyclic freezing is needed. Facing brick is made in only the two grades SW and MW. Additionally, facing brick is available in three types; FBS, for general use; FBX, for general use where a higher degree of precision and lower permissible variation in size than FBS is needed; and FBA, for general use to produce characteristic architectural effects resulting from non-uniformity in size and texture of the units.

In figuring the material cost of brickwork, an allowance of 25% mortar waste and 3% brick breakage was included. If bricks are delivered palletized with 280 to 300 per pallet, or packaged, allow only 1-1/2% for breakage. Packaged or palletized delivery is practical when a job is big enough to have a crane or other equipment available to handle a package of brick. This is so on all industrial work but not always true on small commercial buildings.

The use of buff and gray face is increasing, and there is a continuing trend to the Norman, Roman, Jumbo and SCR brick.

Common red clay brick for backup is not used that often. Concrete block is the most usual backup material with occasional use of sand lime or cement brick. Building brick is commonly used in solid walls for strength and as a fire stop.

Brick panels built on the ground and then crane erected to the upper floors have proven to be economical. This allows the work to be done under cover and without scaffolding.

R042110-50   Brick, Block & Mortar Quantities

Running Bond							For Other Bonds Standard Size Add to S.F. Quantities in Table to Left		
Number of Brick per S.F. of Wall - Single Wythe with 3/8" Joints					C.F. of Mortar per M Bricks, Waste Included				
Type Brick	Nominal Size (incl. mortar) L H W		Modular Coursing	Number of Brick per S.F.	3/8" Joint	1/2" Joint	Bond Type	Description	Factor
Standard	8 x 2-2/3 x 4		3C=8"	6.75	10.3	12.9	Common	full header every fifth course	+20%
Economy	8 x 4 x 4		1C=4"	4.50	11.4	14.6		full header every sixth course	+16.7%
Engineer	8 x 3-1/5 x 4		5C=16"	5.63	10.6	13.6	English	full header every second course	+50%
Fire	9 x 2-1/2 x 4-1/2		2C=5"	6.40	550 # Fireclay	—	Flemish	alternate headers every course	+33.3%
Jumbo	12 x 4 x 6 or 8		1C=4"	3.00	23.8	30.8		every sixth course	+5.6%
Norman	12 x 2-2/3 x 4		3C=8"	4.50	14.0	17.9	Header = W x H exposed		+100%
Norwegian	12 x 3-1/5 x 4		5C=16"	3.75	14.6	18.6	Rowlock = H x W exposed		+100%
Roman	12 x 2 x 4		2C=4"	6.00	13.4	17.0	Rowlock stretcher = L x W exposed		+33.3%
SCR	12 x 2-2/3 x 6		3C=8"	4.50	21.8	28.0	Soldier = H x L exposed		—
Utility	12 x 4 x 4		1C=4"	3.00	15.4	19.6	Sailor = W x L exposed		-33.3%

Concrete Blocks Nominal Size		Approximate Weight per S.F.		Blocks per 100 S.F.	Mortar per M block, waste included	
		Standard	Lightweight		Partitions	Back up
2"	x 8" x 16"	20 PSF	15 PSF	113	27 C.F.	36 C.F.
4"		30	20		41	51
6"		42	30		56	66
8"		55	38		72	82
10"		70	47		87	97
12"		85	55		102	112

R050521-20 Welded Structural Steel

Usual weight reductions with welded design run 10% to 20% compared with bolted or riveted connections. This amounts to about the same total cost compared with bolted structures since field welding is more expensive than bolts. For normal spans of 18′ to 24′ figure 6 to 7 connections per ton.

Trusses — For welded trusses add 4% to weight of main members for connections. Up to 15% less steel can be expected in a welded truss compared to one that is shop bolted. Cost of erection is the same whether shop bolted or welded.

General — Typical electrodes for structural steel welding are E6010, E6011, E60T and E70T. Typical buildings vary between 2# to 8# of weld rod per

ton of steel. Buildings utilizing continuous design require about three times as much welding as conventional welded structures. In estimating field erection by welding, it is best to use the average linear feet of weld per ton to arrive at the welding cost per ton. The type, size and position of the weld will have a direct bearing on the cost per linear foot. A typical field welder will deposit 1.8# to 2# of weld rod per hour manually. Using semiautomatic methods can increase production by as much as 50% to 75%.

R051223-10 Structural Steel

The bare material prices for structural steel, shown in the unit cost sections of the book, are for 100 tons of shop-fabricated structural steel and include:

1. Mill base price of structural steel
2. Mill scrap/grade/size/length extras
3. Mill delivery to a metals service center (warehouse)
4. Service center storage and handling
5. Service center delivery to a fabrication shop
6. Shop storage and handling
7. Shop drafting/detailing
8. Shop fabrication
9. Shop coat of primer paint
10. Shop listing
11. Shop delivery to the job site

In unit cost sections of the book that contain items for field fabrication of steel components, the bare material cost of steel includes:

1. Mill base price of structural steel
2. Mill scrap/grade/size/length extras
3. Mill delivery to a metals service center (warehouse)
4. Service center storage and handling
5. Service center delivery to the job site

R051223-15 Structural Steel Estimating for Repair and Remodeling Projects

The correct approach to estimating structural steel is dependent upon the amount of steel required for the particular project. If the project is a sizable addition to a building, the data can be used directly from the cost data book. This is not the case however if the project requires a small amount of steel, for instance to reinforce existing roof or floor structural systems. To better understand this, please refer to the unit price line for a W16x31 beam with bolted connections. Assume your project requires the reinforcement of the structural members of the roof system of a 3 story building required for the installation a new piece of HVAC equipment. The project will need 4 pieces of W16x31 that are 30′ long. After pricing this 120 L.F. job using the unit prices for a W16x31, an analysis will reveal that the price is wholly inadequate.

The first problem is apparent if you examine the amount of steel that can be installed per day. The unit price line indicates 900 linear feet per day can be installed by a 5 man crew with an 90 ton crane. This productivity is correct for new construction but certainly is not for repair and remodeling work. Installation of new structural steel considers that each member is installed from the foundation to the roof of the structure in a planned and systematic manner with a crane having unrestricted access to all parts of the project. Additionally each connection is planned and detailed with full field access for fit-up and final bolting. The erection is planned and progresses such that interferences and conflicts with other structural members are minimized, if not completely eliminated. All of these assumptions are clearly not the case with a repair and remodeling job, and a significant decrease in the stated productivity will be observed.

A crane will certainly be needed to lift the members into the general area of the project but in most cases will not be able to place the beams into

their final position. An opening in the existing roof may not be large enough to permit the beams to pass through and it may be necessary to bring them into the building through existing windows or doors. Moving the beams to the actual area where they will be installed may involve hand labor and the use of dollies. Finally, hoists and/or jacks may be needed for final positioning.

The connection of new members to existing can often be accomplished by field bolting with accurate field measurements and good planning but in many cases access to both sides of existing members is not possible and field welding becomes the only alternative. In addition to the cost of the actual welding, protection of existing finishes, systems and structure and fire protection must be considered.

New beams can never be installed tight to the existing decks or floors which they must support and the use of shims and tack welding becomes necessary. Additionally, further planning and cost is involved in assuring that existing loads are minimized during the installation and shimming process.

It is apparent that installation of structural steel as part of a repair and remodeling project involves more than simply installing the members and estimating the cost in the same manner as new construction. The best procedure for estimating the total cost is adequate planning and coordination of each process and activity that will be needed. Unit costs for the materials, labor and equipment can then be attached to each needed activity and a final, complete price can be determined.

R053100-10 Decking Descriptions

General - All Deck Products

Steel deck is made by cold forming structural grade sheet steel into a repeating pattern of parallel ribs. The strength and stiffness of the panels are the result of the ribs and the material properties of the steel. Deck lengths can be varied to suit job conditions, but because of shipping considerations, are usually less than 40 feet. Standard deck width varies with the product used but full sheets are usually 12″, 18″, 24″, 30″, or 36″. Deck is typically furnished in a standard width with the ends cut square. Any cutting for width, such as at openings or for angular fit, is done at the job site.

Deck is typically attached to the building frame with arc puddle welds, self-drilling screws, or powder or pneumatically driven pins. Sheet to sheet fastening is done with screws, button punching (crimping), or welds.

Composite Floor Deck

After installation and adequate fastening, floor deck serves several purposes. It (a) acts as a working platform, (b) stabilizes the frame, (c) serves as a concrete form for the slab, and (d) reinforces the slab to carry the design loads applied during the life of the building. Composite decks are distinguished by the presence of shear connector devices as part of the deck. These devices are designed to mechanically lock the concrete and deck together so that the concrete and the deck work together to carry subsequent floor loads. These shear connector devices can be rolled-in embossments, lugs, holes, or wires welded to the panels. The deck profile can also be used to interlock concrete and steel.

Composite deck finishes are either galvanized (zinc coated) or phosphatized/painted. Galvanized deck has a zinc coating on both the top and bottom surfaces. The phosphatized/painted deck has a bare (phosphatized) top surface that will come into contact with the concrete. This bare top surface can be expected to develop rust before the concrete is placed. The bottom side of the deck has a primer coat of paint.

Composite floor deck is normally installed so the panel ends do not overlap on the supporting beams. Shear lugs or panel profile shape often prevent a tight metal to metal fit if the panel ends overlap; the air gap caused by overlapping will prevent proper fusion with the structural steel supports when the panel end laps are shear stud welded.

Adequate end bearing of the deck must be obtained as shown on the drawings. If bearing is actually less in the field than shown on the drawings, further investigation is required.

Roof Deck

Roof deck is not designed to act compositely with other materials. Roof deck acts alone in transferring horizontal and vertical loads into the building frame. Roof deck rib openings are usually narrower than floor deck rib openings. This provides adequate support of rigid thermal insulation board.

Roof deck is typically installed to endlap approximately 2″ over supports. However, it can be butted (or lapped more than 2″) to solve field fit problems. Since designers frequently use the installed deck system as part of the horizontal bracing system (the deck as a diaphragm), any fastening substitution or change should be approved by the designer. Continuous perimeter support of the deck is necessary to limit edge deflection in the finished roof and may be required for diaphragm shear transfer.

Standard roof deck finishes are galvanized or primer painted. The standard factory applied paint for roof deck is a primer paint and is not intended to weather for extended periods of time. Field painting or touching up of abrasions and deterioration of the primer coat or other protective finishes is the responsibility of the contractor.

Cellular Deck

Cellular deck is made by attaching a bottom steel sheet to a roof deck or composite floor deck panel. Cellular deck can be used in the same manner as floor deck. Electrical, telephone, and data wires are easily run through the chase created between the deck panel and the bottom sheet.

When used as part of the electrical distribution system, the cellular deck must be installed so that the ribs line up and create a smooth cell transition at abutting ends. The joint that occurs at butting cell ends must be taped or otherwise sealed to prevent wet concrete from seeping into the cell. Cell interiors must be free of welding burrs, or other sharp intrusions, to prevent damage to wires.

When used as a roof deck, the bottom flat plate is usually left exposed to view. Care must be maintained during erection to keep good alignment and prevent damage.

Cellular deck is sometimes used with the flat plate on the top side to provide a flat working surface. Installation of the deck for this purpose requires special methods for attachment to the frame because the flat plate, now on the top, can prevent direct access to the deck material that is bearing on the structural steel. It may be advisable to treat the flat top surface to prevent slipping.

Cellular deck is always furnished galvanized or painted over galvanized.

Form Deck

Form deck can be any floor or roof deck product used as a concrete form. Connections to the frame are by the same methods used to anchor floor and roof deck. Welding washers are recommended when welding deck that is less than 20 gauge thickness.

Form deck is furnished galvanized, prime painted, or uncoated. Galvanized deck must be used for those roof deck systems where form deck is used to carry a lightweight insulating concrete fill.

Wood, Plastics & Comp.   R0611 Wood Framing

R061110-30   Lumber Product Material Prices

The price of forest products fluctuates widely from location to location and from season to season depending upon economic conditions. The bare material prices in the unit cost sections of the book show the National Average material prices in effect Jan. 1 of this book year. It must be noted that lumber prices in general may change significantly during the year.

Availability of certain items depends upon geographic location and must be checked prior to firm-price bidding.

Wood, Plastics & Comp.   R0616 Sheathing

R061636-20   Plywood

There are two types of plywood used in construction: interior, which is moisture-resistant but not waterproofed, and exterior, which is waterproofed.

The grade of the exterior surface of the plywood sheets is designated by the first letter: A, for smooth surface with patches allowed; B, for solid surface with patches and plugs allowed; C, which may be surface plugged or may have knot holes up to 1″ wide; and D, which is used only for interior type plywood and may have knot holes up to 2-1/2″ wide. "Structural Grade" is specifically designed for engineered applications such as box beams. All CC & DD grades have roof and floor spans marked on them.

Underlayment-grade plywood runs from 1/4″ to 1-1/4″ thick. Thicknesses 5/8″ and over have optional tongue and groove joints which eliminate the need for blocking the edges. Underlayment 19/32″ and over may be referred to as Sturd-i-Floor.

The price of plywood can fluctuate widely due to geographic and economic conditions.

Typical uses for various plywood grades are as follows:

AA-AD Interior — cupboards, shelving, paneling, furniture

BB Plyform — concrete form plywood

CDX — wall and roof sheathing

Structural — box beams, girders, stressed skin panels

AA-AC Exterior — fences, signs, siding, soffits, etc.

Underlayment — base for resilient floor coverings

Overlaid HDO — high density for concrete forms & highway signs

Overlaid MDO — medium density for painting, siding, soffits & signs

303 Siding — exterior siding, textured, striated, embossed, etc.

Thermal & Moist. Protec.   R0751 Built-Up Bituminous Roofing

R075113-20   Built-Up Roofing

Asphalt is available in kegs of 100 lbs. each; coal tar pitch in 560 lb. kegs. Prepared roofing felts are available in a wide range of sizes, weights and characteristics. However, the most commonly used are #15 (432 S.F. per roll, 13 lbs. per square) and #30 (216 S.F. per roll, 27 lbs. per square).

Inter-ply bitumen varies from 24 lbs. per sq. (asphalt) to 30 lbs. per sq. (coal tar) per ply, MF4@ 25%. Flood coat bitumen also varies from 60 lbs. per sq. (asphalt) to 75 lbs. per sq. (coal tar), MF4@ 25%. Expendable equipment (mops, brooms, screeds, etc.) runs about 16% of the bitumen cost. For new, inexperienced crews this factor may be much higher.

Rigid insulation board is typically applied in two layers. The first is mechanically attached to nailable decks or spot or solid mopped to non-nailable decks; the second layer is then spot or solid mopped to the first layer. Membrane application follows the insulation, except in protected membrane roofs, where the membrane goes down first and the insulation on top, followed with ballast (stone or concrete pavers). Insulation and related labor costs are NOT included in prices for built-up roofing.

Thermal & Moist. Protec. **R0752 Modified Bituminous Membrane Roofing**

Reference Tables

R075213-30   Modified Bitumen Roofing

The cost of modified bitumen roofing is highly dependent on the type of installation that is planned. Installation is based on the type of modifier used in the bitumen. The two most popular modifiers are atactic polypropylene (APP) and styrene butadiene styrene (SBS). The modifiers are added to heated bitumen during the manufacturing process to change its characteristics. A polyethylene, polyester or fiberglass reinforcing sheet is then sandwiched between layers of this bitumen. When completed, the result is a pre-assembled, built-up roof that has increased elasticity and weatherablility. Some manufacturers include a surfacing material such as ceramic or mineral granules, metal particles or sand.

The preferred method of adhering SBS-modified bitumen roofing to the substrate is with hot-mopped asphalt (much the same as built-up roofing). This installation method requires a tar kettle/pot to heat the asphalt, as well as the labor, tools and equipment necessary to distribute and spread the hot asphalt.

The alternative method for applying APP and SBS modified bitumen is as follows. A skilled installer uses a torch to melt a small pool of bitumen off the membrane. This pool must form across the entire roll for proper adhesion. The installer must unroll the roofing at a pace slow enough to melt the bitumen, but fast enough to prevent damage to the rest of the membrane.

Modified bitumen roofing provides the advantages of both built-up and single-ply roofing. Labor costs are reduced over those of built-up roofing because only a single ply is necessary. The elasticity of single-ply roofing is attained with the reinforcing sheet and polymer modifiers. Modifieds have some self-healing characteristics and because of their multi-layer construction, they offer the reliability and safety of built-up roofing.

R078413-30   Firestopping

Firestopping is the sealing of structural, mechanical, electrical and other penetrations through fire-rated assemblies. The basic components of firestop systems are safing insulation and firestop sealant on both sides of wall penetrations and the top side of floor penetrations.

Pipe penetrations are assumed to be through concrete, grout, or joint compound and can be sleeved or unsleeved. Costs for the penetrations and sleeves are not included. An annular space of 1″ is assumed. Escutcheons are not included.

Metallic pipe is assumed to be copper, aluminum, cast iron or similar metallic material. Insulated metallic pipe is assumed to be covered with a thermal insulating jacket of varying thickness and materials.

Non-metallic pipe is assumed to be PVC, CPVC, FR Polypropylene or similar plastic piping material. Intumescent firestop sealant or wrap strips are included. Collars on both sides of wall penetrations and a sheet metal plate on the underside of floor penetrations are included.

Ductwork is assumed to be sheet metal, stainless steel or similar metallic material. Duct penetrations are assumed to be through concrete, grout or joint compound. Costs for penetrations and sleeves are not included. An annular space of 1/2″ is assumed.

Multi-trade openings include costs for sheet metal forms, firestop mortar, wrap strips, collars and sealants as necessary.

Structural penetrations joints are assumed to be 1/2″ or less. CMU walls are assumed to be within 1-1/2″ of metal deck. Drywall walls are assumed to be tight to the underside of metal decking.

Metal panel, glass or curtain wall systems include a spandrel area of 5′ filled with mineral wool foil-faced insulation. Fasteners and stiffeners are included.

Openings   R0813 Metal Doors

R081313-20   Steel Door Selection Guide

Standard steel doors are classified into four levels, as recommended by the Steel Door Institute in the chart below. Each of the four levels offers a range of construction models and designs, to meet architectural requirements for preference and appearance, including full flush, seamless, and stile & rail. Recommended minimum gauge requirements are also included.

For complete standard steel door construction specifications and available sizes, refer to the Steel Door Institute Technical Data Series, ANSI A250.8-98 (SDI-100), and ANSI A250.4-94 Test Procedure and Acceptance Criteria for Physical Endurance of Steel Door and Hardware Reinforcements.

Level		Model	Construction	For Full Flush or Seamless		
				Min. Gauge	Thickness (in)	Thickness (mm)
I	Standard Duty	1	Full Flush	20	0.032	0.8
		2	Seamless			
II	Heavy Duty	1	Full Flush	18	0.042	1.0
		2	Seamless			
III	Extra Heavy Duty	1	Full Flush	16	0.053	1.3
		2	Seamless			
		3	*Stile & Rail			
IV	Maximum Duty	1	Full Flush	14	0.067	1.6
		2	Seamless			

*Stiles & rails are 16 gauge; flush panels, when specified, are 18 gauge

Openings   R0851 Metal Windows

R085123-10   Steel Sash

Ironworker crew will erect 25 S.F. or 1.3 sash unit per hour, whichever is less.

Mechanic will point 30 L.F. per hour.

Painter will paint 90 S.F. per coat per hour.

Glazier production depends on light size.

Allow 1 lb. special steel sash putty per 16″ x 20″ light.

Openings   R0853 Plastic Windows

R085313-20   Replacement Windows

Replacement windows are typically measured per United Inch.

United Inches are calculated by rounding the width and height of the window opening up to the nearest inch, then adding the two figures.

The labor cost for replacement windows includes removal of sash, existing sash balance or weights, parting bead where necessary and installation of new window.

Debris hauling and dump fees are not included.

Openings   R0871 Door Hardware

R087120-10   Hinges

All closer equipped doors should have ball bearing hinges. Lead lined or extremely heavy doors require special strength hinges. Usually 1-1/2 pair of hinges are used per door up to 7'-6" high openings. Table below shows typical hinge requirements.

Use Frequency	Type Hinge Required	Type of Opening	Type of Structure
High	Heavy weight	Entrances	Banks, Office buildings, Schools, Stores & Theaters
	ball bearing	Toilet Rooms	Office buildings and Schools
Average	Standard	Entrances	Dwellings
	weight	Corridors	Office buildings and Schools
	ball bearing	Toilet Rooms	Stores
Low	Plain bearing	Interior	Dwellings

Door Thickness	Weight of Doors in Pounds per Square Foot				
	White Pine	Oak	Hollow Core	Solid Core	Hollow Metal
1-3/8"	3psf	6psf	1-1/2psf	3-1/2 — 4psf	6-1/2psf
1-3/4"	3-1/2	7	2-1/2	4-1/2 — 5-1/4	6-1/2
2-1/4"	4-1/2	9	—	5-1/2 — 6-3/4	6-1/2

Openings   R0881 Glass Glazing

R088110-10   Glazing Productivity

Some glass sizes are estimated by the "united inch" (height + width). The table below shows the number of lights glazed in an eight-hour period by the crew size indicated, for glass up to 1/4" thick. Square or nearly square lights are more economical on a S.F. basis. Long slender lights will have a high S.F. installation cost. For insulated glass reduce production by 33%. For 1/2" float glass reduce production by 50%. Production time for glazing with two glaziers per day averages: 1/4" float glass 120 S.F.; 1/2" float glass 55 S.F.; 1/2" insulated glass 95 S.F.; 3/4" insulated glass 75 S.F.

Glazing Method	United Inches per Light							
	40"	60"	80"	100"	135"	165"	200"	240"
Number of Men in Crew	1	1	1	1	2	3	3	4
Industrial sash, putty	60	45	24	15	18	—	—	—
With stops, putty bed	50	36	21	12	16	8	4	3
Wood stops, rubber	40	27	15	9	11	6	3	2
Metal stops, rubber	30	24	14	9	9	6	3	2
Structural glass	10	7	4	3	—	—	—	—
Corrugated glass	12	9	7	4	4	4	3	—
Storefronts	16	15	13	11	7	6	4	4
Skylights, putty glass	60	36	21	12	16	—	—	—
Thiokol set	15	15	11	9	9	6	3	2
Vinyl set, snap on	18	18	13	12	12	7	5	4
Maximum area per light	2.8 S.F.	6.3 S.F.	11.1 S.F.	17.4 S.F.	31.6 S.F.	47 S.F.	69 S.F.	100 S.F.

R092000-50 Lath, Plaster and Gypsum Board

Gypsum board lath is available in 3/8" thick x 16" wide x 4' long sheets as a base material for multi-layer plaster applications. It is also available as a base for either multi-layer or veneer plaster applications in 1/2" and 5/8" thick–4' wide x 8', 10' or 12' long sheets. Fasteners are screws or blued ring shank nails for wood framing and screws for metal framing.

Metal lath is available in diamond mesh pattern with flat or self-furring profiles. Paper backing is available for applications where excessive plaster waste needs to be avoided. A slotted mesh ribbed lath should be used in areas where the span between structural supports is greater than normal. Most metal lath comes in 27" x 96" sheets. Diamond mesh weighs 1.75, 2.5 or 3.4 pounds per square yard, slotted mesh lath weighs 2.75 or 3.4 pounds per square yard. Metal lath can be nailed, screwed or tied in place.

Many **accessories** are available. Corner beads, flat reinforcing strips, casing beads, control and expansion joints, furring brackets and channels are some examples. Note that accessories are not included in plaster or stucco line items.

Plaster is defined as a material or combination of materials that when mixed with a suitable amount of water, forms a plastic mass or paste. When applied to a surface, the paste adheres to it and subsequently hardens, preserving in a rigid state the form or texture imposed during the period of elasticity.

Gypsum plaster is made from ground calcined gypsum. It is mixed with aggregates and water for use as a base coat plaster.

Vermiculite plaster is a fire-retardant plaster covering used on steel beams, concrete slabs and other heavy construction materials. Vermiculite is a group name for certain clay minerals, hydrous silicates or aluminum, magnesium and iron that have been expanded by heat.

Perlite plaster is a plaster using perlite as an aggregate instead of sand. Perlite is a volcanic glass that has been expanded by heat.

Gauging plaster is a mix of gypsum plaster and lime putty that when applied produces a quick drying finish coat.

Veneer plaster is a one or two component gypsum plaster used as a thin finish coat over special gypsum board.

Keenes cement is a white cementitious material manufactured from gypsum that has been burned at a high temperature and ground to a fine powder. Alum is added to accelerate the set. The resulting plaster is hard and strong and accepts and maintains a high polish, hence it is used as a finishing plaster.

Stucco is a Portland cement based plaster used primarily as an exterior finish.

Plaster is used on both interior and exterior surfaces. Generally it is applied in multiple-coat systems. A three-coat system uses the terms scratch, brown and finish to identify each coat. A two-coat system uses base and finish to describe each coat. Each type of plaster and application system has attributes that are chosen by the designer to best fit the intended use.

Gypsum Plaster	2 Coat, 5/8" Thick		3 Coat, 3/4" Thick		
	Base	Finish	Scratch	Brown	Finish
Quantities for 100 S.Y.	1:3 Mix	2:1 Mix	1:2 Mix	1:3 Mix	2:1 Mix
Gypsum plaster	1,300 lb.		1,350 lb.	650 lb.	
Sand	1.75 C.Y.		1.85 C.Y.	1.35 C.Y.	
Finish hydrated lime		340 lb.			340 lb.
Gauging plaster		170 lb.			170 lb.

Vermiculite or Perlite Plaster	2 Coat, 5/8" Thick		3 Coat, 3/4" Thick		
Quantities for 100 S.Y.	Base	Finish	Scratch	Brown	Finish
Gypsum plaster	1,250 lb.		1,450 lb.	800 lb.	
Vermiculite or perlite	7.8 bags		8.0 bags	3.3 bags	
Finish hydrated lime		340 lb.			340 lb.
Gauging plaster		170 lb.			170 lb.

Stucco–Three-Coat System	On Wood	On
Quantities for 100 S.Y.	Frame	Masonry
Portland cement	29 bags	21 bags
Sand	2.6 C.Y.	2.0 C.Y.
Hydrated lime	180 lb.	120 lb.

R092910-10   Levels of Gypsum Drywall Finish

In the past, contract documents often used phrases such as "industry standard" and "workmanlike finish" to specify the expected quality of gypsum board wall and ceiling installations. The vagueness of these descriptions led to unacceptable work and disputes.

In order to resolve this problem, four major trade associations concerned with the manufacture, erection, finish and decoration of gypsum board wall and ceiling systems have developed an industry-wide *Recommended Levels of Gypsum Board Finish*.

The finish of gypsum board walls and ceilings for specific final decoration is dependent on a number of factors. A primary consideration is the location of the surface and the degree of decorative treatment desired. Painted and unpainted surfaces in warehouses and other areas where appearance is normally not critical may simply require the taping of wallboard joints and 'spotting' of fastener heads. Blemish-free, smooth, monolithic surfaces often intended for painted and decorated walls and ceilings in habitated structures, ranging from single-family dwellings through monumental buildings, require additional finishing prior to the application of the final decoration.

Other factors to be considered in determining the level of finish of the gypsum board surface are (1) the type of angle of surface illumination (both natural and artificial lighting), and (2) the paint and method of application or the type and finish of wallcovering specified as the final decoration. Critical lighting conditions, gloss paints, and thin wallcoverings require a higher level of gypsum board finish than do heavily textured surfaces which are subsequently painted or surfaces which are to be decorated with heavy grade wallcoverings.

The following descriptions were developed jointly by the Association of the Wall and Ceiling Industries-International (AWCI), Ceiling & Interior Systems Construction Association (CISCA), Gypsum Association (GA), and Painting and Decorating Contractors of America (PDCA) as a guide.

Level 0: No taping, finishing, or accessories required. This level of finish may be useful in temporary construction or whenever the final decoration has not been determined.

Level 1: All joints and interior angles shall have tape set in joint compound. Surface shall be free of excess joint compound. Tool marks and ridges are acceptable. Frequently specified in plenum areas above ceilings, in attics, in areas where the assembly would generally be concealed or in building service corridors, and other areas not normally open to public view.

Level 2: All joints and interior angles shall have tape embedded in joint compound and wiped with a joint knife leaving a thin coating of joint compound over all joints and interior angles. Fastener heads and accessories shall be covered with a coat of joint compound. Surface shall be free of excess joint compound. Tool marks and ridges are acceptable. Joint compound applied over the body of the tape at the time of tape embedment shall be considered a separate coat of joint compound and shall satisfy the conditions of this level. Specified where water-resistant gypsum backing board is used as a substrate for tile; may be specified in garages, warehouse storage, or other similar areas where surface appearance is not of primary concern.

Level 3: All joints and interior angles shall have tape embedded in joint compound and one additional coat of joint compound applied over all joints and interior angles. Fastener heads and accessories shall be covered with two separate coats of joint compound. All joint compound shall be smooth and free of tool marks and ridges. Typically specified in appearance areas which are to receive heavy- or medium-texture (spray or hand applied) finishes before final painting, or where heavy-grade wallcoverings are to be applied as the final decoration. This level of finish is not recommended where smooth painted surfaces or light to medium wallcoverings are specified.

Level 4: All joints and interior angles shall have tape embedded in joint compound and two separate coats of joint compound applied over all flat joints and one separate coat of joint compound applied over interior angles. Fastener heads and accessories shall be covered with three separate coats of joint compound. All joint compound shall be smooth and free of tool marks and ridges. This level should be specified where flat paints, light textures, or wallcoverings are to be applied. In critical lighting areas, flat paints applied over light textures tend to reduce joint photographing. Gloss, semi-gloss, and enamel paints are not recommended over this level of finish. The weight, texture, and sheen level of wallcoverings applied over this level of finish should be carefully evaluated. Joints and fasteners must be adequately concealed if the wallcovering material is lightweight, contains limited pattern, has a gloss finish, or any combination of these finishes is present. Unbacked vinyl wallcoverings are not recommended over this level of finish.

Level 5: All joints and interior angles shall have tape embedded in joint compound and two separate coats of joint compound applied over all flat joints and one separate coat of joint compound applied over interior angles. Fastener heads and accessories shall be covered with three separate coats of joint compound. A thin skim coat of joint compound or a material manufactured especially for this purpose, shall be applied to the entire surface. The surface shall be smooth and free of tool marks and ridges. This level of finish is highly recommended where gloss, semi-gloss, enamel, or nontextured flat paints are specified or where severe lighting conditions occur. This highest quality finish is the most effective method to provide a uniform surface and minimize the possibility of joint photographing and of fasteners showing through the final decoration.

R142000-10   Freight Elevators

Capacities run from 2,000 lbs. to over 100,000 lbs. with 3,000 lbs. to 10,000 lbs. most common. Travel speeds are generally lower and control less intricate than on passenger elevators. Costs in the unit price sections are for hydraulic and geared elevators.

R142000-20   Elevator Selective Costs   See R142000-40 for cost development.

A. Base Unit	Passenger		Freight		Hospital	
	Hydraulic	**Electric**	**Hydraulic**	**Electric**	**Hydraulic**	**Electric**
Capacity	1,500 lb.	2,000 lb.	2,000 lb.	4,000 lb.	4,000 lb.	4,000 lb.
Speed	100 F.P.M.	200 F.P.M.	100 F.P.M.	200 F.P.M.	100 F.P.M.	200 F.P.M.
#Stops/Travel Ft.	2/12	4/40	2/20	4/40	2/20	4/40
Push Button Oper.	Yes	Yes	Yes	Yes	Yes	Yes
Telephone Box & Wire	"	"	"	"	"	"
Emergency Lighting	"	"	No	No	"	"
Cab	Plastic Lam. Walls	Plastic Lam. Walls	Painted Steel	Painted Steel	Plastic Lam. Walls	Plastic Lam. Walls
Cove Lighting	Yes	Yes	No	No	Yes	Yes
Floor	V.C.T.	V.C.T.	Wood w/Safety Treads	Wood w/Safety Treads	V.C.T.	V.C.T.
Doors, & Speedside Slide	Yes	Yes	Yes	Yes	Yes	Yes
Gates, Manual	No	No	No	No	No	No
Signals, Lighted Buttons	Car and Hall	Car and Hall	Car and Hall	Car and Hall	Car and Hall	Car and Hall
O.H. Geared Machine	N.A.	Yes	N.A.	Yes	N.A.	Yes
Variable Voltage Contr.	"	"	N.A.	"	"	"
Emergency Alarm	Yes	"	Yes	"	Yes	"
Class "A" Loading	N.A.	N.A.	"	"	N.A.	N.A.

R142000-30   Passenger Elevators

Electric elevators are used generally but hydraulic elevators can be used for lifts up to 70' and where large capacities are required. Hydraulic speeds are limited to 200 F.P.M. but cars are self leveling at the stops. On low rises, hydraulic installation runs about 15% less than standard electric types but on higher rises this installation cost advantage is reduced. Maintenance of hydraulic elevators is about the same as electric type but underground portion is not included in the maintenance contract.

In electric elevators there are several control systems available, the choice of which will be based upon elevator use, size, speed and cost criteria. The two types of drives are geared for low speeds and gearless for 450 F.P.M. and over.

The tables on the preceding pages illustrate typical installed costs of the various types of elevators available.

R142000-40   Elevator Cost Development

To price a new car or truck from the factory, you must start with the manufacturer's basic model, then add or exchange optional equipment and features. The same is true for pricing elevators.

Requirement: One-passenger elevator, five-story hydraulic, 2,500 lb. capacity, 12' floor to floor, speed 150 F.P.M., emergency power switching and maintenance contract.

Example:

Description	Adjustment
A. Base Elevator: Hydraulic Passenger, 1500 lb. Capacity, 100 fpm, 2 Stops, Standard Finish	1 Ea.
B. Capacity Adjustment (2,500 lb.)	1 Ea.
C. Excess Travel Adjustment: 48' Total Travel (4 x 12') minus 12' Base Unit Travel =	36 V.L.F.
D. Stops Adjustment: 5 Total Stops minus 2 Stops (Base Unit) =	3 Stops
E. Speed Adjustment (150 F.P.M.)	1 Ea.
F. Options:	
1. Intercom Service	1 Ea.
2. Emergency Power Switching, Automatic	1 Ea.
3. Stainless Steel Entrance Doors	5 Ea.
4. Maintenance Contract (12 Months)	1 Ea.
5. Position Indicator for main floor level (none indicated in Base Unit)	1 Ea.

R220105-10   Demolition (Selective vs. Removal for Replacement)

Demolition can be divided into two basic categories.

One type of demolition involves the removal of material with no concern for its replacement. The labor-hours to estimate this work are found under "Selective Demolition" in the Fire Protection, Plumbing and HVAC Divisions. It is selective in that individual items or all the material installed as a system or trade grouping such as plumbing or heating systems are removed. This may be accomplished by the easiest way possible, such as sawing, torch cutting, or sledge hammer as well as simple unbolting.

The second type of demolition is the removal of some item for repair or replacement. This removal may involve careful draining, opening of unions,

disconnecting and tagging of electrical connections, capping of pipes/ducts to prevent entry of debris or leakage of the material contained as well as transport of the item away from its in-place location to a truck/dumpster. An approximation of the time required to accomplish this type of demolition is to use half of the time indicated as necessary to install a new unit. For example; installation of a new pump might be listed as requiring 6 labor-hours so if we had to estimate the removal of the old pump we would allow an additional 3 hours for a total of 9 hours. That is, the complete replacement of a defective pump with a new pump would be estimated to take 9 labor-hours.

R221113-50   Pipe Material Considerations

1. Malleable fittings should be used for gas service.
2. Malleable fittings are used where there are stresses/strains due to expansion and vibration.
3. Cast fittings may be broken as an aid to disassembling of heating lines frozen by long use, temperature and minerals.
4. Cast iron pipe is extensively used for underground and submerged service.
5. Type M (light wall) copper tubing is available in hard temper only and is used for nonpressure and less severe applications than K and L.
6. Type L (medium wall) copper tubing, available hard or soft for interior service.
7. Type K (heavy wall) copper tubing, available in hard or soft temper for use where conditions are severe. For underground and interior service.
8. Hard drawn tubing requires fewer hangers or supports but should not be bent. Silver brazed fittings are recommended, however soft solder is normally used.
9. Type DMV (very light wall) copper tubing designed for drainage, waste and vent plus other non-critical pressure services.

Domestic/Imported Pipe and Fittings Cost

The prices shown in this publication for steel/cast iron pipe and steel, cast iron, malleable iron fittings are based on domestic production sold at the normal trade discounts. The above listed items of foreign manufacture may be available at prices of 1/3 to 1/2 those shown. Some imported items after minor machining or finishing operations are being sold as domestic to further complicate the system.

Caution: Most pipe prices in this book also include a coupling and pipe hangers which for the larger sizes can add significantly to the per foot cost and should be taken into account when comparing "book cost" with quoted supplier's cost.

R224000-40   Plumbing Fixture Installation Time

Item	Rough-In	Set	Total Hours	Item	Rough-In	Set	Total Hours
Bathtub	5	5	10	Shower head only	2	1	3
Bathtub and shower, cast iron	6	6	12	Shower drain	3	1	4
Fire hose reel and cabinet	4	2	6	Shower stall, slate		15	15
Floor drain to 4 inch diameter	3	1	4	Slop sink	5	3	8
Grease trap, single, cast iron	5	3	8	Test 6 fixtures			14
Kitchen gas range		4	4	Urinal, wall	6	2	8
Kitchen sink, single	4	4	8	Urinal, pedestal or floor	6	4	10
Kitchen sink, double	6	6	12	Water closet and tank	4	3	7
Laundry tubs	4	2	6	Water closet and tank, wall hung	5	3	8
Lavatory wall hung	5	3	8	Water heater, 45 gals. gas, automatic	5	2	7
Lavatory pedestal	5	3	8	Water heaters, 65 gals. gas, automatic	5	2	7
Shower and stall	6	4	10	Water heaters, electric, plumbing only	4	2	6

Fixture prices in front of book are based on the cost per fixture set in place. The rough-in cost, which must be added for each fixture, includes carrier, if required, some supply, waste and vent pipe connecting fittings and stops. The lengths of rough-in pipe are nominal runs which would connect to the larger runs and stacks. The supply runs and DWV runs and stacks must be accounted for in separate entries. In the eastern half of the United States it is common for the plumber to carry these to a point 5′ outside the building.

R233100-20   Ductwork

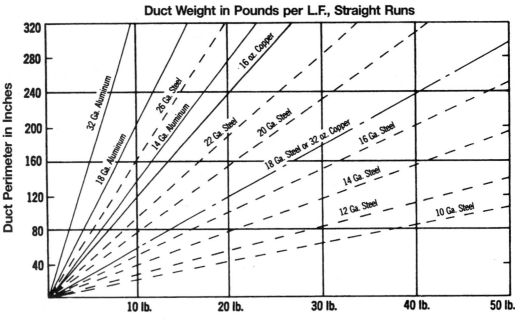

Add to the above for fittings; 90° elbow is 3 L.F.; 45° elbow is 2.5 L.F.; offset is 4 L.F.; transition offset is 6 L.F.; square-to-round transition is 4 L.F.; 90° reducing elbow is 5 L.F. For bracing and waste, add 20% to aluminum and copper, 15% to steel.

Electrical — R2601 Operation & Maint. of Elect. Systems

R260105-30 Electrical Demolition (Removal for Replacement)

The purpose of this reference number is to provide a guide to users for electrical "removal for replacement" by applying the rule of thumb: 1/3 of new installation time (typical range from 20% to 50%) for removal. Remember to use reasonable judgment when applying the suggested percentage factor. For example:

Contractors have been requested to remove an existing fluorescent lighting fixture and replace with a new fixture utilizing energy saver lamps and electronic ballast:

In order to fully understand the extent of the project, contractors should visit the job site and estimate the time to perform the renovation work in accordance with applicable national, state and local regulation and codes.

The contractor may need to add extra labor hours to his estimate if he discovers unknown concealed conditions such as: contaminated asbestos ceiling, broken acoustical ceiling tile and need to repair, patch and touch-up paint all the damaged or disturbed areas, tasks normally assigned to general contractors. In addition, the owner could request that the contractors salvage the materials removed and turn over the materials to the owner or dispose of the materials to a reclamation station. The normal removal item is 0.5 labor hour for a lighting fixture and 1.5 labor-hours for new installation time. Revise the estimate times from 2 labor-hours work up to a minimum 4 labor-hours work for just fluorescent lighting fixture.

For removal of large concentrations of lighting fixtures in the same area, apply an "economy of scale" to reduce estimating labor hours.

Electrical — R2605 Common Work Results for Electrical

R260533-22 Conductors in Conduit

Table below lists maximum number of conductors for various sized conduit using THW, TW or THWN insulations.

Copper Wire Size	1/2" TW	1/2" THW	1/2" THWN	3/4" TW	3/4" THW	3/4" THWN	1" TW	1" THW	1" THWN	1-1/4" TW	1-1/4" THW	1-1/4" THWN	1-1/2" TW	1-1/2" THW	1-1/2" THWN	2" TW	2" THW	2" THWN	2-1/2" TW	2-1/2" THW	2-1/2" THWN	3" THW	3" THWN	3-1/2" THW	3-1/2" THWN	4" THW	4" THWN
#14	9	6	13	15	10	24	25	16	39	44	29	69	60	40	94	99	65	154	142	93		143		192			
#12	7	4	10	12	8	18	19	13	29	35	24	51	47	32	70	78	53	114	111	76	164	117		157		163	
#10	5	4	6	9	6	11	15	11	18	26	19	32	36	26	44	60	43	73	85	61	104	95	160	127		163	
#8	2	1	3	4	3	5	7	5	9	12	10	16	17	13	22	28	22	36	40	32	51	49	79	66	106	85	136
#6		1	1		2	4		4	6		7	11		10	15		16	26		23	37	36	57	48	76	62	98
#4		1	1		1	2		3	4		5	7		7	9		12	16		17	22	27	35	36	47	47	60
#3		1	1		1	1		2	3		4	6		6	8		10	13		15	19	23	29	31	39	40	51
#2		1	1		1	1		2	3		4	5		5	7		9	11		13	16	20	25	27	33	34	43
#1					1	1		1	1		3	3		4	5		6	8		9	12	14	18	19	25	25	32
1/0					1	1		1	1		2	3		3	4		5	7		8	10	12	15	16	21	21	27
2/0					1	1		1	1		1	2		3	3		5	6		7	8	10	13	14	17	18	22
3/0					1	1		1	1		1	1		2	3		4	5		6	7	9	11	12	14	15	18
4/0						1		1	1		1	1		1	2		3	4		5	6	7	9	10	12	13	15
250 kcmil								1	1		1	1		1	1		2	3		4	4	6	7	8	10	10	12
300								1	1		1	1		1	1		2	3		3	4	5	6	7	8	9	11
350									1		1	1		1	1		1	2		3	3	4	5	6	7	8	9
400											1	1		1	1		1	1		2	3	4	5	5	6	7	8
500											1	1		1	1		1	1		1	2	3	4	4	5	6	7
600												1		1	1		1	1		1	1	3	3	4	4	5	5
700														1	1		1	1		1	1	2	3	3	4	4	5
750														1	1		1	1		1	1	2	2	3	3	4	4

R312316-40   Excavating

The selection of equipment used for structural excavation and bulk excavation or for grading is determined by the following factors.

1. Quantity of material
2. Type of material
3. Depth or height of cut
4. Length of haul
5. Condition of haul road
6. Accessibility of site
7. Moisture content and dewatering requirements
8. Availability of excavating and hauling equipment

Some additional costs must be allowed for hand trimming the sides and bottom of concrete pours and other excavation below the general excavation.

Number of B.C.Y. per truck = 1.5 C.Y. bucket x 8 passes = 12 loose C.Y.

$$= 12 \text{ x } \frac{100}{118} = 10.2 \text{ B.C.Y. per truck}$$

Truck Haul Cycle:

Load truck, 8 passes	=	4 minutes
Haul distance, 1 mile	=	9 minutes
Dump time	=	2 minutes
Return, 1 mile	=	7 minutes
Spot under machine	=	1 minute
		23 minute cycle

Add the mobilization and demobilization costs to the total excavation costs. When equipment is rented for more than three days, there is often no mobilization charge by the equipment dealer. On larger jobs outside of urban areas, scrapers can move earth economically provided a dump site or fill area and adequate haul roads are available. Excavation within sheeting bracing or cofferdam bracing is usually done with a clamshell and production

When planning excavation and fill, the following should also be considered.

1. Swell factor
2. Compaction factor
3. Moisture content
4. Density requirements

A typical example for scheduling and estimating the cost of excavation of a 15′ deep basement on a dry site when the material must be hauled off the site is outlined below.

Assumptions:

1. Swell factor, 18%
2. No mobilization or demobilization
3. Allowance included for idle time and moving on job
4. No dewatering, sheeting, or bracing
5. No truck spotter or hand trimming

Fleet Haul Production per day in B.C.Y.

$$4 \text{ trucks x } \frac{50 \text{ min. hour}}{23 \text{ min. haul cycle}} \text{ x 8 hrs. x 10.2 B.C.Y.}$$

$$= 4 \text{ x } 2.2 \text{ x } 8 \text{ x } 10.2 = 718 \text{ B.C.Y./day}$$

is low, since the clamshell may have to be guided by hand between the bracing. When excavating or filling an area enclosed with a wellpoint system, add 10% to 15% to the cost to allow for restricted access. When estimating earth excavation quantities for structures, allow work space outside the building footprint for construction of the foundation and a slope of 1:1 unless sheeting is used.

R312316-45 Excavating Equipment

The table below lists THEORETICAL hourly production in C.Y./hr. bank measure for some typical excavation equipment. Figures assume 50 minute hours, 83% job efficiency, 100% operator efficiency, 90° swing and properly sized hauling units, which must be modified for adverse digging and loading conditions. Actual production costs in the front of the book average about 50% of the theoretical values listed here.

Equipment	Soil Type	B.C.Y. Weight	% Swell	1 C.Y.	1-1/2 C.Y.	2 C.Y.	2-1/2 C.Y.	3 C.Y.	3-1/2 C.Y.	4 C.Y.
Hydraulic Excavator	Moist loam, sandy clay	3400 lb.	40%	165	195	200	275	330	385	440
"Backhoe"	Sand and gravel	3100	18	140	170	225	240	285	330	380
15' Deep Cut	Common earth	2800	30	150	180	230	250	300	350	400
	Clay, hard, dense	3000	33	120	140	190	200	240	260	320
Power Shovel Optimum Cut (Ft.)	Moist loam, sandy clay	3400	40	170 (6.0)	245 (7.0)	295 (7.8)	335 (8.4)	385 (8.8)	435 (9.1)	475 (9.4)
	Sand and gravel	3100	18	165 (6.0)	225 (7.0)	275 (7.8)	325 (8.4)	375 (8.8)	420 (9.1)	460 (9.4)
	Common earth	2800	30	145 (7.8)	200 (9.2)	250 (10.2)	295 (11.2)	335 (12.1)	375 (13.0)	425 (13.8)
	Clay, hard, dense	3000	33	120 (9.0)	175 (10.7)	220 (12.2)	255 (13.3)	300 (14.2)	335 (15.1)	375 (16.0)
Drag Line Optimum Cut (Ft.)	Moist loam, sandy clay	3400	40	130 (6.6)	180 (7.4)	220 (8.0)	250 (8.5)	290 (9.0)	325 (9.5)	385 (10.0)
	Sand and gravel	3100	18	130 (6.6)	175 (7.4)	210 (8.0)	245 (8.5)	280 (9.0)	315 (9.5)	375 (10.0)
	Common earth	2800	30	110 (8.0)	160 (9.0)	190 (9.9)	220 (10.5)	250 (11.0)	280 (11.5)	310 (12.0)
	Clay, hard, dense	3000	33	90 (9.3)	130 (10.7)	160 (11.8)	190 (12.3)	225 (12.8)	250 (13.3)	280 (12.0)

				Wheel Loaders				Track Loaders		
				3 C.Y.	4 C.Y.	6 C.Y.	8 C.Y.	2-1/4 C.Y.	3 C.Y.	4 C.Y.
Loading Tractors	Moist loam, sandy clay	3400	40	260	340	510	690	135	180	250
	Sand and gravel	3100	18	245	320	480	650	130	170	235
	Common earth	2800	30	230	300	460	620	120	155	220
	Clay, hard, dense	3000	33	200	270	415	560	110	145	200
	Rock, well-blasted	4000	50	180	245	380	520	100	130	180

R312323-30 Compacting Backfill

Compaction of fill in embankments, around structures, in trenches, and under slabs is important to control settlement. Factors affecting compaction are:

1. Soil gradation
2. Moisture content
3. Equipment used
4. Depth of fill per lift
5. Density required

Production Rate:

$$\frac{1.75' \text{ plate width x 50 F.P.M. x 50 min./hr. x .67' lift}}{27 \text{ C.F. per C.Y.}} = 108.5 \text{ C.Y./hr.}$$

Production Rate for 4 Passes:

$$\frac{108.5 \text{ C.Y.}}{4 \text{ passes}} = 27.125 \text{ C.Y./hr. x 8 hrs.} = 217 \text{ C.Y./day}$$

Example:

Compact granular fill around a building foundation using a 21″ wide x 24″ vibratory plate in 8″ lifts. Operator moves at 50 F.P.M. working a 50 minute hour to develop 95% Modified Proctor Density with 4 passes.

731

Earthwork   R3141 Shoring

R314116-40   Wood Sheet Piling

Wood sheet piling may be used for depths to 20' where there is no ground water. If moderate ground water is encountered Tongue & Groove

sheeting will help to keep it out. When considerable ground water is present, steel sheeting must be used.

For estimating purposes on trench excavation, sizes are as follows:

Depth	Sheeting	Wales	Braces	B.F. per S.F.
To 8'	3 x 12's	6 x 8's, 2 line	6 x 8's, @ 10'	4.0 @ 8'
8' x 12'	3 x 12's	10 x 10's, 2 line	10 x 10's, @ 9'	5.0 average
12' to 20'	3 x 12's	12 x 12's, 3 line	12 x 12's, @ 8'	7.0 average

Sheeting to be toed in at least 2' depending upon soil conditions. A five person crew with an air compressor and sheeting driver can drive and brace 440 SF/day at 8' deep, 360 SF/day at 12' deep, and 320 SF/day at 16' deep.

For normal soils, piling can be pulled in 1/3 the time to install. Pulling difficulty increases with the time in the ground. Production can be increased by high pressure jetting.

Earthwork   R3163 Drilled Caissons

R316326-60   Caissons

The three principal types of cassions are:

(1) Belled Caissons, which except for shallow depths and poor soil conditions, are generally recommended. They provide more bearing than shaft area. Because of its conical shape, no horizontal reinforcement of the bell is required.

(2) Straight Shaft Caissons are used where relatively light loads are to be supported by caissons that rest on high value bearing strata. While the shaft is larger in diameter than for belled types this is more than offset by the saving in time and labor.

(3) Keyed Caissons are used when extremely heavy loads are to be carried. A keyed or socketed caisson transfers its load into rock by a combination of end-bearing and shear reinforcing of the shaft. The most economical shaft often consists of a steel casing, a steel wide flange core and concrete. Allowable compressive stresses of .225 f'c for concrete, 16,000 psi for the wide flange core, and 9,000 psi for the steel casing are commonly used. The usual range of shaft diameter is 18" to 84". The number of sizes specified for any one project should be limited due to the problems of casing and auger storage. When hand work is to be performed, shaft diameters should not be less than 32". When inspection of borings is required a minimum shaft diameter of 30" is recommended. Concrete caissons are intended to be poured against earth excavation so permanent forms which add to cost should not be used if the excavation is clean and the earth sufficiently impervious to prevent excessive loss of concrete.

Soil Conditions for Belling		
Good	**Requires Handwork**	**Not Recommended**
Clay	Hard Shale	Silt
Sandy Clay	Limestone	Sand
Silty Clay	Sandstone	Gravel
Clayey Silt	Weathered Mica	Igneous Rock
Hard-pan		
Soft Shale		
Decomposed Rock		

Exterior Improvements   R3292 Turf & Grasses

R329219-50   Seeding

The type of grass is determined by light, shade and moisture content of soil plus intended use. Fertilizer should be disked 4" before seeding. For steep slopes disk five tons of mulch and lay two tons of hay or straw on surface per acre after seeding. Surface mulch can be staked, lightly disked or tar emulsion sprayed. Material for mulch can be wood chips, peat moss, partially

rotted hay or straw, wood fibers and sprayed emulsions. Hemp seed blankets with fertilizer are also available. For spring seeding, watering is necessary. Late fall seeding may have to be reseeded in the spring. Hydraulic seeding, power mulching, and aerial seeding can be used on large areas.

R331113-80   Piping Designations

There are several systems currently in use to describe pipe and fittings. The following paragraphs will help to identify and clarify classifications of piping systems used for water distribution.

Piping may be classified by schedule. Piping schedules include 5S, 10S, 10, 20, 30, Standard, 40, 60, Extra Strong, 80, 100, 120, 140, 160 and Double Extra Strong. These schedules are dependent upon the pipe wall thickness. The wall thickness of a particular schedule may vary with pipe size.

Ductile iron pipe for water distribution is classified by Pressure Classes such as Class 150, 200, 250, 300 and 350. These classes are actually the rated water working pressure of the pipe in pounds per square inch (psi). The pipe in these pressure classes is designed to withstand the rated water working pressure plus a surge allowance of 100 psi.

The American Water Works Association (AWWA) provides standards for various types of **plastic pipe.** C-900 is the specification for polyvinyl chloride (PVC) piping used for water distribution in sizes ranging from 4″ through 12″. C-901 is the specification for polyethylene (PE) pressure pipe, tubing and fittings used for water distribution in sizes ranging from 1/2″ through 3″. C-905 is the specification for PVC piping sizes 14″ and greater.

PVC pressure-rated pipe is identified using the standard dimensional ratio (SDR) method. This method is defined by the American Society for Testing and Materials (ASTM) Standard D 2241. This pipe is available in SDR numbers 64, 41, 32.5, 26, 21, 17, and 13.5. Pipe with an SDR of 64 will have the thinnest wall while pipe with an SDR of 13.5 will have the thickest wall. When the pressure rating (PR) of a pipe is given in psi, it is based on a line supplying water at 73 degrees F.

The National Sanitation Foundation (NSF) seal of approval is applied to products that can be used with potable water. These products have been tested to ANSI/NSF Standard 14.

Valves and strainers are classified by American National Standards Institute (ANSI) Classes. These Classes are 125, 150, 200, 250, 300, 400, 600, 900, 1500 and 2500. Within each class there is an operating pressure range dependent upon temperature. Design parameters should be compared to the appropriate material dependent, pressure-temperature rating chart for accurate valve selection.

Change Orders

Change Order Considerations

A Change Order is a written document, usually prepared by the design professional, and signed by the owner, the architect/ engineer and the contractor. A change order states the agreement of the parties to: an addition, deletion, or revision in the work; an adjustment in the contract sum, if any; or an adjustment in the contract time, if any. Change orders, or "extras" in the construction process occur after execution of the construction contract and impact architects/ engineers, contractors and owners.

Change orders that are properly recognized and managed can ensure orderly, professional and profitable progress for all who are involved in the project. There are many causes for change orders and change order requests. In all cases, change orders or change order requests should be addressed promptly and in a precise and prescribed manner. The following paragraphs include information regarding change order pricing and procedures.

The Causes of Change Orders

Reasons for issuing change orders include:

- Unforeseen field conditions that require a change in the work
- Correction of design discrepancies, errors or omissions in the contract documents
- Owner-requested changes, either by design criteria, scope of work, or project objectives
- Completion date changes for reasons unrelated to the construction process
- Changes in building code interpretations, or other public authority requirements that require a change in the work
- Changes in availability of existing or new materials and products

Procedures

Properly written contract documents must include the correct change order procedures for all parties—owners, design professionals and contractors—to follow in order to avoid costly delays and litigation.

Being "in the right" is not always a sufficient or acceptable defense. The contract provisions requiring notification and documentation must be adhered to within a defined or reasonable time frame.

The appropriate method of handling change orders is by a written proposal and acceptance by all parties involved. Prior to starting work on a project, all parties should identify their authorized agents who may sign and accept change orders, as well as any limits placed on their authority.

Time may be a critical factor when the need for a change arises. For such cases, the contractor might be directed to proceed on a "time and materials" basis, rather than wait for all paperwork to be processed—a delay that could impede progress. In this situation, the contractor must still follow the prescribed change order procedures, including but not limited to, notification and documentation.

All forms used for change orders should be dated and signed by the proper authority. Lack of documentation can be very costly, especially if legal judgments are to be made and if certain field personnel are no longer available. For time and material change orders, the contractor should keep accurate daily records of all labor and material allocated to the change. Forms that can be used to document change order work are available in *Means Forms for Building Construction Professionals.*

Owners or awarding authorities who do considerable and continual building construction (such as the federal government) realize the inevitability of change orders for numerous reasons, both predictable and unpredictable. As a result, the federal government, the American Institute of Architects (AIA), the Engineers Joint Contract Documents Committee (EJCDC) and other contractor, legal and technical organizations have developed standards and procedures to be followed by all parties to achieve contract continuance and timely completion, while being financially fair to all concerned.

In addition to the change order standards put forth by industry associations, there are also many books available on the subject.

Pricing Change Orders

When pricing change orders, regardless of their cause, the most significant factor is when the change occurs. The need for a change may be perceived in the field or requested by the architect/engineer *before* any of the actual installation has begun, or may evolve or appear *during* construction when the item of work in question is partially installed. In the latter cases, the original sequence of construction is disrupted, along with all contiguous and supporting systems. Change orders cause the greatest impact when they occur *after* the installation has been completed and must be uncovered, or even replaced. Post-completion changes may be caused by necessary design changes, product failure, or changes in the owner's requirements that are not discovered until the building or the systems begin to function.

Specified procedures of notification and record keeping must be adhered to and enforced regardless of the stage of construction: *before, during,* or *after* installation. Some bidding documents anticipate change orders by requiring that unit prices including overhead and profit percentages—for additional as well as deductible changes—be listed. Generally these unit prices do not fully take into account the ripple effect, or impact on other trades, and should be used for general guidance only.

When pricing change orders, it is important to classify the time frame in which the change occurs. There are two basic time frames for change orders: *pre-installation change orders,* which occur before the start of construction, and *post-installation change orders,* which involve reworking after the original installation. Change orders that occur between these stages may be priced according to the extent of work completed using a combination of techniques developed for pricing *pre-* and *post-installation* changes.

The following factors are the basis for a check list to use when preparing a change order estimate.

Factors To Consider When Pricing Change Orders

As an estimator begins to prepare a change order, the following questions should be reviewed to determine their impact on the final price.

General

- Is the change order work *pre-installation* or *post-installation*?

  Change order work costs vary according to how much of the installation has been completed. Once workers have the project scoped in their mind, even though they have not started, it can be difficult to refocus.

Consequently they may spend more than the normal amount of time understanding the change. Also, modifications to work in place such as trimming or refitting usually take more time than was initially estimated. The greater the amount of work in place, the more reluctant workers are to change it. Psychologically they may resent the change and as a result the rework takes longer than normal. Post-installation change order estimates must include demolition of existing work as required to accomplish the change. If the work is performed at a later time, additional obstacles such as building finishes may be present which must be protected. Regardless of whether the change

occurs pre-installation or post-installation, attempt to isolate the identifiable factors and price them separately. For example, add shipping costs that may be required pre-installation or any demolition required post-installation. Then analyze the potential impact on productivity of psychological and/or learning curve factors and adjust the output rates accordingly. One approach is to break down the typical workday into segments and quantify the impact on each segment. The following chart may be useful as a guide:

Activities (Productivity) Expressed as Percentages of a Workday			
Task	Means Mechanical Cost Data (for New Construction)	Pre-Installation Change Orders	Post-Installation Change Orders
1. Study plans	3%	6%	6%
2. Material procurement	3%	3%	3%
3. Receiving and storing	3%	3%	3%
4. Mobilization	5%	5%	5%
5. Site movement	5%	5%	8%
6. Layout and marking	8%	10%	12%
7. Actual installation	64%	59%	54%
8. Clean-up	3%	3%	3%
9. Breaks—non-productive	6%	6%	6%
Total	100%	100%	100%

Change Order Installation Efficiency

The labor-hours expressed (for new construction) are based on average installation time, using an efficiency level of approximately 60-65%. For change order situations, adjustments to this efficiency level should reflect the daily labor-hour allocation for that particular occurrence.

If any of the specific percentages expressed in the above chart do not apply to a particular project situation, then those percentage points should be reallocated to the appropriate task(s). Example: Using data for new construction, assume there is no new material being utilized. The percentages for Tasks 2 and 3 would therefore be reallocated to other tasks. If the time required for Tasks 2 and 3 can now be applied to installation, we can add the time allocated for *Material Procurement* and *Receiving and Storing* to the *Actual Installation* time for new construction, thereby increasing the Actual Installation percentage.

This chart shows that, due to reduced productivity, labor costs will be higher than those for new construction by 5% to 15% for pre-installation change orders and by 15% to 25% for post-installation change orders. Each job and change order is unique and must be examined individually. Many factors, covered elsewhere in this section, can each have a significant impact on productivity and change order costs. All such factors should be considered in every case.

- Will the change substantially delay the original completion date?

  A significant change in the project may cause the original completion date to be extended. The extended schedule may subject the contractor to new wage rates dictated by relevant labor contracts. Project supervision and other project overhead must also be extended beyond the original completion date. The schedule extension may also put installation into a new weather season. For example, underground piping scheduled for October installation was delayed until January. As a result, frost penetrated the trench area, thereby changing the degree of difficulty of the task. Changes and delays may have a ripple effect throughout the project. This effect must be analyzed and negotiated with the owner.

- What is the net effect of a deduct change order?

  In most cases, change orders resulting in a deduction or credit reflect only bare costs. The contractor may retain the overhead and profit based on the original bid.

Materials

- Will you have to pay more or less for the new material, required by the change order, than you paid for the original purchase?

  The same material prices or discounts will usually apply to materials purchased for change orders as new construction. In some instances, however, the contractor may forfeit the advantages of competitive pricing for change orders. Consider the following example:

  A contractor purchased over $20,000 worth of fan coil units for an installation, and obtained the maximum discount. Some time later it was determined the project required an additional matching unit. The contractor has to purchase this unit from the original supplier to ensure a match. The supplier at this time may not discount the unit because of the small quantity, and the fact that he is no longer in a competitive situation. The impact of quantity on purchase can add between 0% and 25% to material prices and/or subcontractor quotes.

- If materials have been ordered or delivered to the job site, will they be subject to a cancellation charge or restocking fee?

  Check with the supplier to determine if ordered materials are subject to a cancellation charge. Delivered materials not used as result of a change order may be subject to a restocking fee if returned to the supplier. Common restocking charges run between 20% and 40%. Also, delivery charges to return the goods to the supplier must be added.

Labor

- How efficient is the existing crew at the actual installation?

  Is the same crew that performed the initial work going to do the change order? Possibly the change consists of the installation of a unit identical to one already installed; therefore the change should take less time. Be sure to consider this potential productivity increase and modify the productivity rates accordingly.

- If the crew size is increased, what impact will that have on supervision requirements?

  Under most bargaining agreements or management practices, there is a point at which a working foreman is replaced by a nonworking foreman. This replacement increases project overhead by adding a nonproductive worker. If additional workers are added to accelerate the project or to perform changes while maintaining the schedule, be sure to add additional supervision time if warranted. Calculate the hours involved and the additional cost directly if possible.

- What are the other impacts of increased crew size?

  The larger the crew, the greater the potential for productivity to decrease. Some of the factors that cause this productivity loss are: overcrowding (producing restrictive conditions in the working space), and possibly a shortage of any special tools and equipment required. Such factors affect not only the crew working on the elements directly involved in the change order, but other crews whose movement may also be hampered.

As the crew increases, check its basic composition for changes by the addition or deletion of apprentices or nonworking foreman and quantify the potential effects of equipment shortages or other logistical factors.

- As new crews, unfamiliar with the project, are brought onto the site, how long will it take them to become oriented to the project requirements?

  The orientation time for a new crew to become 100% effective varies with the site and type of project. Orientation is easiest at a new construction site, and most difficult at existing, very restrictive renovation sites. The type of work also affects orientation time. When all elements of the work are exposed, such as concrete or masonry work, orientation is decreased. When the work is concealed or less visible, such as existing electrical systems, orientation takes longer. Usually orientation can be accomplished in one day or less. Costs for added orientation should be itemized and added to the total estimated cost.

- How much actual production can be gained by working overtime?

  Short term overtime can be used effectively to accomplish more work in a day. However, as overtime is scheduled to run beyond several weeks, studies have shown marked decreases in output. The following chart shows the effect of long term overtime on worker efficiency. If the anticipated change requires extended overtime to keep the job on schedule, these factors can be used as a guide to predict the impact on time and cost. Add project overhead, particularly supervision, that may also be incurred.

Days per Week	Hours per Day	Production Efficiency					Payroll Cost Factors	
		1 Week	2 Weeks	3 Weeks	4 Weeks	Average 4 Weeks	@ 1-1/2 Times	@ 2 Times
5	8	100%	100%	100%	100%	100%	100%	100%
	9	100	100	95	90	96.25	105.6	111.1
	10	100	95	90	85	91.25	110.0	120.0
	11	95	90	75	65	81.25	113.6	127.3
	12	90	85	70	60	76.25	116.7	133.3
6	8	100	100	95	90	96.25	108.3	116.7
	9	100	95	90	85	92.50	113.0	125.9
	10	95	90	85	80	87.50	116.7	133.3
	11	95	85	70	65	78.75	119.7	139.4
	12	90	80	65	60	73.75	122.2	144.4
7	8	100	95	85	75	88.75	114.3	128.6
	9	95	90	80	70	83.75	118.3	136.5
	10	90	85	75	65	78.75	121.4	142.9
	11	85	80	65	60	72.50	124.0	148.1
	12	85	75	60	55	68.75	126.2	152.4

Effects of Overtime

Caution: Under many labor agreements, Sundays and holidays are paid at a higher premium than the normal overtime rate.

The use of long-term overtime is counterproductive on almost any construction job; that is, the longer the period of overtime, the lower the actual production rate. Numerous studies have been conducted, and while they have resulted in slightly different numbers, all reach the same conclusion. The figure above tabulates the effects of overtime work on efficiency.

As illustrated, there can be a difference between the *actual* payroll cost per hour and the *effective* cost per hour for overtime work. This is due to the reduced production efficiency with the increase in weekly hours beyond 40. This difference between actual and effective cost results from overtime work over a prolonged period. Short-term overtime work does not result in as great a reduction in efficiency, and in such cases, effective cost may not vary significantly from the actual payroll cost. As the total hours per week are increased on a regular basis, more time is lost because of fatigue, lowered morale, and an increased accident rate.

As an example, assume a project where workers are working 6 days a week, 10 hours per day. From the figure above (based on productivity studies), the average effective productive hours over a four-week period are:

$$0.875 \times 60 = 52.5$$

Depending upon the locale and day of week, overtime hours may be paid at time and a half or double time. For time and a half, the overall (average) *actual* payroll cost (including regular and overtime hours) is determined as follows:

$$\frac{40 \text{ reg. hrs.} + (20 \text{ overtime hrs.} \times 1.5)}{60 \text{ hrs.}} = 1.167$$

Based on 60 hours, the payroll cost per hour will be 116.7% of the normal rate at 40 hours per week. However, because the effective production (efficiency) for 60 hours is reduced to the equivalent of 52.5 hours, the effective cost of overtime is calculated as follows:

For time and a half:

$$\frac{40 \text{ reg. hrs.} + (20 \text{ overtime hrs.} \times 1.5)}{52.5 \text{ hrs.}} = 1.33$$

Installed cost will be 133% of the normal rate (for labor).

Thus, when figuring overtime, the actual cost per unit of work will be higher than the apparent overtime payroll dollar increase, due to the reduced productivity of the longer workweek. These efficiency calculations are true only for those cost factors determined by hours worked. Costs that are applied weekly or monthly, such as equipment rentals, will not be similarly affected.

Equipment

• What equipment is required to complete the change order?

Change orders may require extending the rental period of equipment already on the job site, or the addition of special equipment brought in to accomplish the change work. In either case, the additional rental charges and operator labor charges must be added.

Summary

The preceding considerations and others you deem appropriate should be analyzed and applied to a change order estimate. The impact of each should be quantified and listed on the estimate to form an audit trail.

Change orders that are properly identified, documented, and managed help to ensure the orderly, professional and profitable progress of the work. They also minimize potential claims or disputes at the end of the project.

Estimating Tips

- The cost figures in this Square Foot Cost section were derived from approximately 11,200 projects contained in the RSMeans database of completed construction projects. They include the contractor's overhead and profit, but do not generally include architectural fees or land costs. The figures have been adjusted to January of the current year. New projects are added to our files each year, and outdated projects are discarded. For this reason, certain costs may not show a uniform annual progression. In no case are all subdivisions of a project listed.

- These projects were located throughout the U.S. and reflect a tremendous variation in square foot (S.F.) and cubic foot (C.F.) costs. This is due to differences, not only in labor and material costs, but also in individual owners' requirements. For instance, a bank in a large city would have different features than one in a rural area. This is true of all the different types of buildings analyzed. Therefore, caution should be exercised when using these Square Foot costs. For example, for court houses, costs in the database are local court house costs and will not apply to the larger, more elaborate federal court houses. As a general rule, the projects in the 1/4 column do not include any site work or equipment, while the projects in the 3/4 column may include both equipment and site work. The median figures do not generally include site work.

- None of the figures "go with" any others. All individual cost items were computed and tabulated separately. Thus, the sum of the median figures for Plumbing, HVAC and Electrical will not normally total up to the total Mechanical and Electrical costs arrived at by separate analysis and tabulation of the projects.

- Each building was analyzed as to total and component costs and percentages. The figures were arranged in ascending order with the results tabulated as shown. The 1/4 column shows that 25% of the projects had lower costs and 75% had higher. The 3/4 column shows that 75% of the projects had lower costs and 25% had higher. The median column shows that 50% of the projects had lower costs and 50% had higher.

- There are two times when square foot costs are useful. The first is in the conceptual stage when no details are available. Then square foot costs make a useful starting point. The second is after the bids are in and the costs can be worked back into their appropriate units for information purposes. As soon as details become available in the project design, the square foot approach should be discontinued and the project priced as to its particular components. When more precision is required, or for estimating the replacement cost of specific buildings, the current edition of *RSMeans Square Foot Costs* should be used.

- In using the figures in this section, it is recommended that the median column be used for preliminary figures if no additional information is available. The median figures, when multiplied by the total city construction cost index figures (see City Cost Indexes) and then multiplied by the project size modifier at the end of this section, should present a fairly accurate base figure, which would then have to be adjusted in view of the estimator's experience, local economic conditions, code requirements, and the owner's particular requirements. There is no need to factor the percentage figures, as these should remain constant from city to city. All tabulations mentioning air conditioning had at least partial air conditioning.

- The editors of this book would greatly appreciate receiving cost figures on one or more of your recent projects, which would then be included in the averages for next year. All cost figures received will be kept confidential, except that they will be averaged with other similar projects to arrive at Square Foot cost figures for next year's book. See the last page of the book for details and the discount available for submitting one or more of your projects.

50 17 00 | S.F. Costs		UNIT	UNIT COSTS			% OF TOTAL		
			1/4	MEDIAN	3/4	1/4	MEDIAN	3/4
01 0010	**APARTMENTS Low Rise (1 to 3 story)**	S.F.	67	84.50	112				**01**
0020	Total project cost	C.F.	6	8	9.85			
0100	Site work	S.F.	5.75	7.85	13.75	6.05%	10.55%	14.05%
0500	Masonry		1.32	3.07	5.30	1.54%	3.67%	6.35%
1500	Finishes		7.10	9.75	12.10	9.05%	10.75%	12.85%
1800	Equipment		2.19	3.32	4.94	2.73%	4.03%	5.95%
2720	Plumbing		5.20	6.70	8.55	6.65%	8.95%	10.05%
2770	Heating, ventilating, air conditioning		3.33	4.10	6.05	4.20%	5.60%	7.60%
2900	Electrical		3.88	5.15	6.90	5.20%	6.65%	8.40%
3100	Total: Mechanical & Electrical	↓	13.45	17.10	21.50	15.90%	18.05%	23%
9000	Per apartment unit, total cost	Apt.	62,500	95,500	140,500			
9500	Total: Mechanical & Electrical	"	11,800	18,600	24,300			
02 0010	**APARTMENTS Mid Rise (4 to 7 story)**	S.F.	89	107	133				**02**
0020	Total project costs	C.F.	6.95	9.60	13.10			
0100	Site work	S.F.	3.56	7.05	12.65	5.25%	6.70%	9.15%
0500	Masonry		5.90	8.15	11.15	5.10%	7.25%	10.50%
1500	Finishes		11.20	14.60	18.40	10.55%	13.45%	17.70%
1800	Equipment		2.58	3.90	5.05	2.54%	3.48%	4.31%
2500	Conveying equipment		1.91	2.44	2.96	1.94%	2.27%	2.69%
2720	Plumbing		5.20	8.35	8.85	5.70%	7.20%	8.95%
2900	Electrical		5.85	7.95	9.65	6.65%	7.20%	8.95%
3100	Total: Mechanical & Electrical	↓	18.75	23.50	28.50	18.50%	21%	23%
9000	Per apartment unit, total cost	Apt.	100,500	118,500	196,500			
9500	Total: Mechanical & Electrical	"	19,000	22,000	27,800			
03 0010	**APARTMENTS High Rise (8 to 24 story)**	S.F.	101	116	139				**03**
0020	Total project costs	C.F.	9.80	11.40	14.55			
0100	Site work	S.F.	3.66	5.90	8.30	2.58%	4.84%	6.15%
0500	Masonry		5.85	10.60	13.20	4.74%	9.65%	11.05%
1500	Finishes		11.20	14	16.50	9.75%	11.80%	13.70%
1800	Equipment		3.24	3.99	5.30	2.78%	3.49%	4.35%
2500	Conveying equipment		2.29	3.48	4.73	2.23%	2.78%	3.37%
2720	Plumbing		7.40	8.75	12.25	6.80%	7.20%	10.45%
2900	Electrical		6.95	8.75	11.85	6.45%	7.65%	8.80%
3100	Total: Mechanical & Electrical	↓	21	26.50	32	17.95%	22.50%	24.50%
9000	Per apartment unit, total cost	Apt.	105,000	115,500	160,000			
9500	Total: Mechanical & Electrical	"	22,700	25,900	27,400			
04 0010	**AUDITORIUMS**	S.F.	104	141	203				**04**
0020	Total project costs	C.F.	6.55	9.15	13.10			
2720	Plumbing	S.F.	6.35	9.20	11.15	5.85%	7.20%	8.70%
2900	Electrical		8.10	11.70	15.75	6.80%	8.95%	11.30%
3100	Total: Mechanical & Electrical	↓	23	46	56	24.50%	30.50%	31.50%
05 0010	**AUTOMOTIVE SALES**	S.F.	77.50	105	130				**05**
0020	Total project costs	C.F.	5.10	6.15	7.95			
2720	Plumbing	S.F.	3.54	6.15	6.70	2.89%	6.05%	6.50%
2770	Heating, ventilating, air conditioning		5.45	8.35	9	4.61%	10%	10.35%
2900	Electrical		6.25	9.80	13.30	7.25%	9.80%	12.15%
3100	Total: Mechanical & Electrical	↓	19.60	28	33.50	19.15%	20.50%	22%
06 0010	**BANKS**	S.F.	151	189	240				**06**
0020	Total project costs	C.F.	10.85	14.75	19.50			
0100	Site work	S.F.	17.35	26.50	38	7.75%	12.45%	17%
0500	Masonry		7.90	15.15	26.50	3.36%	6.95%	10.35%
1500	Finishes		13.45	19.50	24	5.80%	8.45%	11.25%
1800	Equipment		5.70	12.60	26	1.34%	5.95%	10.65%
2720	Plumbing		4.76	6.80	9.95	2.82%	3.90%	4.93%
2770	Heating, ventilating, air conditioning		9.05	12.10	16.10	4.86%	7.15%	8.50%
2900	Electrical		14.35	19.20	25	8.20%	10.20%	12.15%
3100	Total: Mechanical & Electrical	↓	33.50	45.50	53	16.25%	19.40%	23%
3500	See also Divisions 11020 & 11030 (MF2004 11 16 00 & 11 17 00)							

		50 17 00 | S.F. Costs		UNIT COSTS			% OF TOTAL			
			UNIT	1/4	MEDIAN	3/4	1/4	MEDIAN	3/4	
13	0010	**CHURCHES**	S.F.	102	130	168				13
	0020	Total project costs	C.F.	6.35	8	10.55				
	1800	Equipment	S.F.	1.22	2.93	6.25	.95%	2.11%	4.50%	
	2720	Plumbing		3.99	5.55	8.20	3.51%	4.96%	6.25%	
	2770	Heating, ventilating, air conditioning		9.30	12.15	17.20	7.50%	10%	12%	
	2900	Electrical		8.60	11.80	15.90	7.35%	8.80%	11%	
	3100	Total: Mechanical & Electrical	↓	26.50	34.50	46	18.30%	22%	24.50%	
	3500	See also Division 11040 (MF2004 11 91 00)								
15	0010	**CLUBS, COUNTRY**	S.F.	110	132	167				15
	0020	Total project costs	C.F.	8.85	10.80	14.90				
	2720	Plumbing	S.F.	6.65	9.85	22.50	5.60%	7.90%	10%	
	2900	Electrical		8.65	11.85	15.45	7%	8.95%	11%	
	3100	Total: Mechanical & Electrical	↓	46	57.50	60.50	19%	26.50%	29.50%	
17	0010	**CLUBS, SOCIAL Fraternal**	S.F.	87.50	126	169				17
	0020	Total project costs	C.F.	5.50	8.30	9.90				
	2720	Plumbing	S.F.	5.50	6.85	10.40	5.60%	6.90%	8.55%	
	2770	Heating, ventilating, air conditioning		7.95	9.60	12.35	8.20%	9.25%	14.40%	
	2900	Electrical		6.60	10.85	12.40	6.50%	9.50%	10.55%	
	3100	Total: Mechanical & Electrical	↓	19.45	37	47	21%	23%	23.50%	
18	0010	**CLUBS, Y.M.C.A.**	S.F.	110	143	179				18
	0020	Total project costs	C.F.	5.05	8.50	12.65				
	2720	Plumbing	S.F.	6.95	13.85	15.55	5.65%	7.60%	10.85%	
	2900	Electrical		8.80	11	15.15	6.05%	7.60%	9.25%	
	3100	Total: Mechanical & Electrical	↓	33.50	37.50	51.50	18.40%	21.50%	28.50%	
19	0010	**COLLEGES Classrooms & Administration**	S.F.	111	152	201				19
	0020	Total project costs	C.F.	8.10	11.80	18.20				
	0500	Masonry	S.F.	8.15	15.35	18.85	5.65%	8.25%	10.50%	
	2720	Plumbing		5.65	11.60	20.50	5.10%	6.60%	8.95%	
	2900	Electrical		9.25	14.05	19.20	7.70%	9.85%	12%	
	3100	Total: Mechanical & Electrical	↓	37	51.50	61.50	24%	28%	31.50%	
21	0010	**COLLEGES Science, Engineering, Laboratories**	S.F.	207	242	280				21
	0020	Total project costs	C.F.	11.85	17.30	19.65				
	1800	Equipment	S.F.	11.50	26	28.50	2%	6.45%	12.65%	
	2900	Electrical		17.05	23.50	37	7.10%	9.40%	12.10%	
	3100	Total: Mechanical & Electrical	↓	63.50	75	116	28.50%	31.50%	41%	
	3500	See also Division 11600 (MF2004 11 53 00)								
23	0010	**COLLEGES Student Unions**	S.F.	132	179	216				23
	0020	Total project costs	C.F.	7.35	9.65	11.90				
	3100	Total: Mechanical & Electrical	S.F.	49.50	53.50	63.50	23.50%	26%	29%	
25	0010	**COMMUNITY CENTERS**	S.F.	109	134	181				25
	0020	Total project costs	C.F.	7.10	10.15	13.15				
	1800	Equipment	S.F.	2.64	4.46	7.10	1.48%	3.01%	5.45%	
	2720	Plumbing		5.15	9	12.30	4.85%	7%	8.95%	
	2770	Heating, ventilating, air conditioning		8.25	12.10	17.20	6.80%	10.35%	12.90%	
	2900	Electrical		8.90	11.90	17.05	7.30%	9%	10.45%	
	3100	Total: Mechanical & Electrical	↓	31	38	54	20%	25%	31%	
28	0010	**COURT HOUSES**	S.F.	157	180	233				28
	0020	Total project costs	C.F.	12.05	14.40	18.15				
	2720	Plumbing	S.F.	7.45	10.45	15	5.95%	7.45%	8.20%	
	2900	Electrical		16.70	18.50	27	8.90%	10.45%	11.55%	
	3100	Total: Mechanical & Electrical	↓	42.50	58	64	22.50%	27.50%	30.50%	
30	0010	**DEPARTMENT STORES**	S.F.	58	78.50	99				30
	0020	Total project costs	C.F.	3.11	4.03	5.50				
	2720	Plumbing	S.F.	1.81	2.28	3.47	1.82%	4.21%	5.90%	
	2770	Heating, ventilating, air conditioning	↓	5.30	8.15	12.25	8.20%	9.10%	14.80%	

		50 17 00 | S.F. Costs		UNIT COSTS			% OF TOTAL			
			UNIT	1/4	MEDIAN	3/4	1/4	MEDIAN	3/4	
30	2900	Electrical	S.F.	6.65	9.15	10.80	9.05%	12.15%	14.95%	30
	3100	Total: Mechanical & Electrical	↓	11.75	15	26.50	13.20%	21.50%	50%	
31	0010	**DORMITORIES Low Rise (1 to 3 story)**	S.F.	109	143	185				31
	0020	Total project costs	C.F.	6.20	10.05	15.05				
	2720	Plumbing	S.F.	6.60	8.85	11.20	8.05%	9%	9.65%	
	2770	Heating, ventilating, air conditioning		7	8.40	11.15	4.61%	8.05%	10%	
	2900	Electrical		7.15	10.95	14.85	6.25%	8.65%	9.50%	
	3100	Total: Mechanical & Electrical	↓	36.50	40.50	63.50	21.50%	24%	27%	
	9000	Per bed, total cost	Bed	46,700	52,000	111,000				
32	0010	**DORMITORIES Mid Rise (4 to 8 story)**	S.F.	135	176	218				32
	0020	Total project costs	C.F.	14.90	16.35	19.60				
	2900	Electrical	S.F.	14.35	16.30	22	8.20%	10.20%	11.95%	
	3100	Total: Mechanical & Electrical	"	40	41.50	81.50	25.50%	34.50%	37.50%	
	9000	Per bed, total cost	Bed	19,200	43,900	111,000				
34	0010	**FACTORIES**	S.F.	51	76.50	117				34
	0020	Total project costs	C.F.	3.28	4.89	8.10				
	0100	Site work	S.F.	5.85	10.65	16.85	6.95%	11.45%	17.95%	
	2720	Plumbing		2.76	5.15	8.50	3.73%	6.05%	8.10%	
	2770	Heating, ventilating, air conditioning		5.35	7.70	10.40	5.25%	8.45%	11.35%	
	2900	Electrical		6.35	10.05	15.35	8.10%	10.50%	14.20%	
	3100	Total: Mechanical & Electrical	↓	18.15	24.50	37	21%	28.50%	35.50%	
36	0010	**FIRE STATIONS**	S.F.	102	139	182				36
	0020	Total project costs	C.F.	5.95	8.15	10.85				
	0500	Masonry	S.F.	15	27	34.50	8.60%	11.65%	16.45%	
	1140	Roofing		3.31	8.95	10.20	1.90%	4.94%	5.05%	
	1580	Painting		2.57	3.84	3.93	1.37%	1.57%	2.07%	
	1800	Equipment		1.31	2.57	4.47	.74%	1.98%	3.54%	
	2720	Plumbing		5.70	9.15	13.25	5.85%	7.35%	9.45%	
	2770	Heating, ventilating, air conditioning		5.60	9.10	14.10	5.15%	7.40%	9.40%	
	2900	Electrical		7.20	12.55	16.95	6.85%	8.65%	10.60%	
	3100	Total: Mechanical & Electrical	↓	35.50	45.50	52.50	18.75%	23%	27%	
37	0010	**FRATERNITY HOUSES & Sorority Houses**	S.F.	102	131	179				37
	0020	Total project costs	C.F.	10.10	10.55	12.70				
	2720	Plumbing	S.F.	7.65	8.80	16.10	6.80%	8%	10.85%	
	2900	Electrical		6.70	14.45	17.70	6.60%	9.90%	10.65%	
	3100	Total: Mechanical & Electrical	↓	17.90	26	31		15.10%	15.90%	
38	0010	**FUNERAL HOMES**	S.F.	107	146	265				38
	0020	Total project costs	C.F.	10.95	12.15	23.50				
	2900	Electrical	S.F.	4.73	8.65	9.50	3.58%	4.44%	5.95%	
	3100	Total: Mechanical & Electrical	↓	16.75	24.50	34	12.90%	12.90%	12.90%	
39	0010	**GARAGES, COMMERCIAL (Service)**	S.F.	60.50	93.50	129				39
	0020	Total project costs	C.F.	3.99	5.90	8.55				
	1800	Equipment	S.F.	3.42	7.70	12	2.21%	4.62%	6.80%	
	2720	Plumbing		4.20	6.45	11.80	5.45%	7.85%	10.65%	
	2730	Heating & ventilating		5.50	7.30	9.90	5.25%	6.85%	8.20%	
	2900	Electrical		5.75	8.80	12.65	7.15%	9.25%	10.85%	
	3100	Total: Mechanical & Electrical	↓	12.70	24.50	36	12.35%	17.40%	26%	
40	0010	**GARAGES, MUNICIPAL (Repair)**	S.F.	89	119	167				40
	0020	Total project costs	C.F.	5.55	7.05	12.15				
	0500	Masonry	S.F.	8.35	16.35	25.50	5.60%	9.15%	12.50%	
	2720	Plumbing		4	7.65	14.45	3.59%	6.70%	7.95%	
	2730	Heating & ventilating		6.85	9.90	19.10	6.15%	7.45%	13.50%	
	2900	Electrical		6.60	10.35	14.90	6.65%	8.15%	11.15%	
	3100	Total: Mechanical & Electrical	↓	30.50	42	61	21.50%	25.50%	28.50%	

50 17 00 | S.F. Costs		UNIT	UNIT COSTS			% OF TOTAL		
			1/4	MEDIAN	3/4	1/4	MEDIAN	3/4
41	**0010 GARAGES, PARKING**	S.F.	34.50	50.50	87			**41**
	0020 Total project costs	C.F.	3.26	4.42	6.45			
	2720 Plumbing	S.F.	.98	1.52	2.35	1.72%	2.70%	3.85%
	2900 Electrical		1.90	2.33	3.67	4.33%	5.20%	6.30%
	3100 Total: Mechanical & Electrical	↓	3.89	5.40	6.75	7%	8.90%	11.05%
	3200							
	9000 Per car, total cost	Car	14,600	18,400	23,400			
43	**0010 GYMNASIUMS**	S.F.	96.50	129	174			**43**
	0020 Total project costs	C.F.	4.81	6.50	8			
	1800 Equipment	S.F.	2.29	4.30	8.25	1.81%	3.30%	6.70%
	2720 Plumbing		6.10	7.25	9.35	4.65%	6.40%	7.75%
	2770 Heating, ventilating, air conditioning		6.55	10	20	5.15%	9.05%	11.10%
	2900 Electrical		7.40	10.10	13.30	6.60%	8.30%	10.30%
	3100 Total: Mechanical & Electrical	↓	27	36.50	43.50	19.75%	23.50%	29%
	3500 See also Division 11480 (MF2004 11 67 00)							
46	**0010 HOSPITALS**	S.F.	185	228	315			**46**
	0020 Total project costs	C.F.	14.05	17.45	25			
	1800 Equipment	S.F.	4.70	9.05	15.60	.96%	2.63%	5%
	2720 Plumbing		15.95	22.50	28.50	7.60%	9.10%	10.85%
	2770 Heating, ventilating, air conditioning		23.50	30	40.50	7.80%	12.95%	16.65%
	2900 Electrical		20	26.50	39	9.85%	11.55%	13.90%
	3100 Total: Mechanical & Electrical	↓	57	81.50	123	27%	33.50%	36.50%
	9000 Per bed or person, total cost	Bed	214,500	295,500	340,500			
	9900 See also Division 11700 (MF2004 11 71 00)							
48	**0010 HOUSING For the Elderly**	S.F.	91	115	142			**48**
	0020 Total project costs	C.F.	6.50	9	11.50			
	0100 Site work	S.F.	6.35	9.85	14.45	5.05%	7.90%	12.10%
	0500 Masonry		2.77	10.35	15.15	1.30%	6.05%	11%
	1800 Equipment		2.20	3.03	4.82	1.88%	3.23%	4.43%
	2510 Conveying systems		2.21	2.98	4.03	1.78%	2.20%	2.81%
	2720 Plumbing		6.75	8.60	10.85	8.15%	9.55%	10.50%
	2730 Heating, ventilating, air conditioning		3.47	4.91	7.35	3.30%	5.60%	7.25%
	2900 Electrical		6.80	9.20	11.80	7.30%	8.50%	10.25%
	3100 Total: Mechanical & Electrical	↓	23.50	28	37	18.10%	22.50%	29%
	9000 Per rental unit, total cost	Unit	84,500	99,000	110,500			
	9500 Total: Mechanical & Electrical	"	18,800	21,700	25,300			
50	**0010 HOUSING Public (Low Rise)**	S.F.	76.50	106	138			**50**
	0020 Total project costs	C.F.	6.80	8.50	10.55			
	0100 Site work	S.F.	9.75	14.05	22.50	8.35%	11.75%	16.50%
	1800 Equipment		2.08	3.39	5.15	2.26%	3.03%	4.24%
	2720 Plumbing		5.50	7.30	9.25	7.15%	9.05%	11.60%
	2730 Heating, ventilating, air conditioning		2.77	5.40	5.90	4.26%	6.05%	6.45%
	2900 Electrical		4.63	6.90	9.55	5.10%	6.55%	8.25%
	3100 Total: Mechanical & Electrical	↓	22	28.50	31.50	14.50%	17.55%	26.50%
	9000 Per apartment, total cost	Apt.	84,000	95,500	120,000			
	9500 Total: Mechanical & Electrical	"	17,900	22,100	24,500			
51	**0010 ICE SKATING RINKS**	S.F.	65.50	153	168			**51**
	0020 Total project costs	C.F.	4.81	4.92	5.65			
	2720 Plumbing	S.F.	2.44	4.58	4.69	3.12%	3.23%	5.65%
	2900 Electrical		7	10.75	11.35	6.30%	10.15%	15.05%
	3100 Total: Mechanical & Electrical	↓	11.65	16.50	20.50	18.95%	18.95%	18.95%
52	**0010 JAILS**	S.F.	199	257	330			**52**
	0020 Total project costs	C.F.	17.95	25	29.50			
	1800 Equipment	S.F.	7.75	23	39	2.80%	5.55%	10.35%
	2720 Plumbing		19.15	25.50	34	7%	8.90%	13.35%
	2770 Heating, ventilating, air conditioning		17.95	24	46.50	7.50%	9.45%	17.75%
	2900 Electrical	↓	20.50	27.50	35	8.20%	11.55%	14.95%

			UNIT COSTS			% OF TOTAL			
50 17 00 | S.F. Costs		UNIT	1/4	MEDIAN	3/4	1/4	MEDIAN	3/4	
52	3100	Total: Mechanical & Electrical	S.F.	54	99	117	28%	30%	34%	52
53	0010	**LIBRARIES**	S.F.	126	161	207				53
	0020	Total project costs	C.F.	8.60	10.80	13.75				
	0500	Masonry	S.F.	9.85	17.40	29	5.80%	7.60%	11.80%	
	1800	Equipment		1.72	4.63	7	.37%	1.50%	4.16%	
	2720	Plumbing		4.65	6.75	9.15	3.38%	4.60%	5.70%	
	2770	Heating, ventilating, air conditioning		10.30	17.40	22.50	7.80%	10.95%	12.80%	
	2900	Electrical		12.85	16.65	21	8.30%	10.30%	11.95%	
	3100	Total: Mechanical & Electrical	↓	38	48	58.50	19.65%	23%	26.50%	
54	0010	**LIVING, ASSISTED**	S.F.	116	137	162				54
	0020	Total project costs	C.F.	9.80	11.45	13				
	0500	Masonry	S.F.	3.42	4.08	4.79	2.37%	3.16%	3.86%	
	1800	Equipment		2.65	3.08	3.94	2.12%	2.45%	2.66%	
	2720	Plumbing		9.75	13.05	13.55	6.05%	8.15%	10.60%	
	2770	Heating, ventilating, air conditioning		11.60	12.10	13.25	7.95%	9.35%	9.70%	
	2900	Electrical		11.40	12.60	14.55	9%	10%	10.70%	
	3100	Total: Mechanical & Electrical	↓	32	37.50	43	26%	29%	31.50%	
55	0010	**MEDICAL CLINICS**	S.F.	118	146	185				55
	0020	Total project costs	C.F.	8.70	11.25	14.95				
	1800	Equipment	S.F.	3.20	6.70	10.45	1.05%	2.94%	6.35%	
	2720	Plumbing		7.85	11.10	14.80	6.15%	8.40%	10.10%	
	2770	Heating, ventilating, air conditioning		9.40	12.30	18.10	6.65%	8.85%	11.35%	
	2900	Electrical		10.20	14.45	18.90	8.10%	10%	12.25%	
	3100	Total: Mechanical & Electrical	↓	32.50	44	60.50	22.50%	27%	33.50%	
	3500	See also Division 11700 (MF2004 11 71 00)								
57	0010	**MEDICAL OFFICES**	S.F.	112	138	169				57
	0020	Total project costs	C.F.	8.30	11.25	15.25				
	1800	Equipment	S.F.	3.68	7.25	10.35	.70%	5.10%	7.05%	
	2720	Plumbing		6.15	9.50	12.80	5.60%	6.80%	8.50%	
	2770	Heating, ventilating, air conditioning		7.45	10.75	14.20	6.10%	8%	9.70%	
	2900	Electrical		9.10	13.05	18.10	7.65%	9.80%	11.70%	
	3100	Total: Mechanical & Electrical	↓	26	35.50	51.50	19.35%	23%	29%	
59	0010	**MOTELS**	S.F.	70.50	102	133				59
	0020	Total project costs	C.F.	6.25	8.40	13.70				
	2720	Plumbing	S.F.	7.15	9.05	10.80	9.45%	10.60%	12.55%	
	2770	Heating, ventilating, air conditioning		4.34	6.50	11.60	5.60%	5.60%	10%	
	2900	Electrical		6.65	8.40	10.45	7.45%	9.05%	10.45%	
	3100	Total: Mechanical & Electrical	↓	22.50	28.50	48.50	18.50%	24%	25.50%	
	5000									
	9000	Per rental unit, total cost	Unit	35,800	68,000	73,500				
	9500	Total: Mechanical & Electrical	"	6,975	10,600	12,300				
60	0010	**NURSING HOMES**	S.F.	110	142	174				60
	0020	Total project costs	C.F.	8.65	10.80	14.80				
	1800	Equipment	S.F.	3.47	4.60	7.70	2.02%	3.62%	4.99%	
	2720	Plumbing		9.45	14.30	17.25	8.75%	10.10%	12.70%	
	2770	Heating, ventilating, air conditioning		9.95	15.10	20	9.70%	11.45%	11.80%	
	2900	Electrical		10.90	13.60	18.55	9.40%	10.55%	12.45%	
	3100	Total: Mechanical & Electrical	↓	26	36.50	61	26%	29.50%	30.50%	
	9000	Per bed or person, total cost	Bed	48,900	61,000	79,000				
61	0010	**OFFICES Low Rise (1 to 4 story)**	S.F.	93	120	155				61
	0020	Total project costs	C.F.	6.65	9.15	12.05				
	0100	Site work	S.F.	7.20	12.60	18.85	5.90%	9.70%	13.60%	
	0500	Masonry		3.46	7.25	13.15	2.62%	5.50%	8.45%	
	1800	Equipment		.99	1.93	5.25	.69%	1.50%	3.63%	
	2720	Plumbing		3.32	5.15	7.50	3.66%	4.50%	6.10%	
	2770	Heating, ventilating, air conditioning		7.35	10.25	15	7.20%	10.30%	11.70%	
	2900	Electrical	↓	7.55	10.75	15.20	7.45%	9.60%	11.35%	

	50 17 00 | S.F. Costs	UNIT	UNIT COSTS			% OF TOTAL		
			1/4	MEDIAN	3/4	1/4	MEDIAN	3/4
61 3100	Total: Mechanical & Electrical	S.F.	20.50	27.50	41	18%	22%	27%	61
62 0010	**OFFICES Mid Rise (5 to 10 story)**	S.F.	98.50	119	157				62
0020	Total project costs	C.F.	6.95	8.90	12.60			
2720	Plumbing	S.F.	2.97	4.61	6.60	2.83%	3.74%	4.50%
2770	Heating, ventilating, air conditioning		7.45	10.65	17.05	7.65%	9.40%	11%
2900	Electrical		7.30	9.35	12.95	6.35%	7.80%	10%
3100	Total: Mechanical & Electrical		18.90	24.50	46.50	19.15%	21.50%	27.50%
63 0010	**OFFICES High Rise (11 to 20 story)**	S.F.	121	152	187				63
0020	Total project costs	C.F.	8.45	10.55	15.15			
2900	Electrical	S.F.	7.35	8.95	13.30	5.80%	7.85%	10.50%
3100	Total: Mechanical & Electrical		23.50	31.50	53.50	16.90%	23.50%	34%
64 0010	**POLICE STATIONS**	S.F.	145	190	240				64
0020	Total project costs	C.F.	11.55	14.15	19.35			
0500	Masonry	S.F.	13.60	24	30	6.70%	9.10%	11.35%
1800	Equipment		2.30	10.10	16.05	.98%	3.35%	6.70%
2720	Plumbing		8.10	16.15	20	5.65%	6.90%	10.75%
2770	Heating, ventilating, air conditioning		12.65	16.80	23	5.85%	10.55%	11.70%
2900	Electrical		15.85	22.50	30	9.80%	11.85%	14.80%
3100	Total: Mechanical & Electrical		52	62.50	84.50	25%	31.50%	32.50%
65 0010	**POST OFFICES**	S.F.	114	141	180				65
0020	Total project costs	C.F.	6.90	8.70	9.90			
2720	Plumbing	S.F.	5.15	6.40	8.05	4.24%	5.30%	5.60%
2770	Heating, ventilating, air conditioning		8.05	9.95	11.05	6.65%	7.15%	9.35%
2900	Electrical		9.45	13.30	15.75	7.25%	9%	11%
3100	Total: Mechanical & Electrical		27.50	35.50	40.50	16.25%	18.80%	22%
66 0010	**POWER PLANTS**	S.F.	795	1,000	1,925				66
0020	Total project costs	C.F.	22	47.50	102			
2900	Electrical	S.F.	56	119	177	9.30%	12.75%	21.50%
8100	Total: Mechanical & Electrical		140	455	1,025	32.50%	32.50%	52.50%
67 0010	**RELIGIOUS EDUCATION**	S.F.	92	120	148				67
0020	Total project costs	C.F.	5.10	7.35	9.15			
2720	Plumbing	S.F.	3.85	5.45	7.70	4.40%	5.30%	7.10%
2770	Heating, ventilating, air conditioning		9.75	11	15.55	10.05%	11.45%	12.35%
2900	Electrical		7.30	10.30	13.65	7.60%	9.05%	10.35%
3100	Total: Mechanical & Electrical		29.50	39	47.50	22%	23%	27%
69 0010	**RESEARCH Laboratories & Facilities**	S.F.	138	198	288				69
0020	Total project costs	C.F.	10.40	20	24			
1800	Equipment	S.F.	6.10	12.05	29.50	.94%	4.58%	8.80%
2720	Plumbing		13.85	17.70	28.50	6.15%	8.30%	10.80%
2770	Heating, ventilating, air conditioning		12.40	41.50	49.50	7.25%	16.50%	17.50%
2900	Electrical		16.15	26.50	43.50	9.55%	11.50%	15.40%
3100	Total: Mechanical & Electrical		49	92	132	29.50%	37%	42%
70 0010	**RESTAURANTS**	S.F.	134	172	224				70
0020	Total project costs	C.F.	11.25	14.75	19.35			
1800	Equipment	S.F.	8.65	21	32	6.10%	13%	15.65%
2720	Plumbing		10.60	12.85	16.85	6.10%	8.15%	9%
2770	Heating, ventilating, air conditioning		13.45	18.60	24	9.20%	12%	12.40%
2900	Electrical		14.15	17.45	22.50	8.35%	10.55%	11.55%
3100	Total: Mechanical & Electrical		44	47	60.50	21%	24.50%	29.50%
9000	Per seat unit, total cost	Seat	4,900	6,550	7,750			
9500	Total: Mechanical & Electrical	"	1,225	1,625	1,925			
72 0010	**RETAIL STORES**	S.F.	62.50	84	111				72
0020	Total project costs	C.F.	4.23	6.05	8.40			
2720	Plumbing	S.F.	2.26	3.78	6.45	3.26%	4.60%	6.80%
2770	Heating, ventilating, air conditioning		4.89	6.70	10.05	6.75%	8.75%	10.15%
2900	Electrical		5.65	7.70	11.10	7.25%	9.90%	11.65%
3100	Total: Mechanical & Electrical		14.95	19.20	26.50	17.05%	21%	23.50%

50 17 00 | S.F. Costs			UNIT COSTS			% OF TOTAL		
		UNIT	1/4	MEDIAN	3/4	1/4	MEDIAN	3/4
74 0010	**SCHOOLS Elementary**	S.F.	101	125	155				**74**
0020	Total project costs	C.F.	6.65	8.55	11.05			
0500	Masonry	S.F.	9.10	15.70	23.50	5.80%	11%	14.95%
1800	Equipment		2.78	4.70	8.75	1.89%	3.32%	4.71%
2720	Plumbing		5.85	8.30	11.05	5.70%	7.15%	9.35%
2730	Heating, ventilating, air conditioning		8.80	14	19.55	8.15%	10.80%	14.90%
2900	Electrical		9.60	12.70	15.95	8.40%	10.05%	11.70%
3100	Total: Mechanical & Electrical	↓	34	43	52.50	25%	27.50%	30%
9000	Per pupil, total cost	Ea.	11,700	17,400	50,000			
9500	Total: Mechanical & Electrical	"	3,300	4,175	14,500			
76 0010	**SCHOOLS Junior High & Middle**	S.F.	103	129	157				**76**
0020	Total project costs	C.F.	6.65	8.65	9.70			
0500	Masonry	S.F.	13.10	16.90	19.75	8%	11.40%	14.30%
1800	Equipment		3.36	5.40	8.15	1.81%	3.26%	4.86%
2720	Plumbing		6.10	7.55	9.35	5.30%	6.80%	7.25%
2770	Heating, ventilating, air conditioning		12.20	14.85	26	8.90%	11.55%	14.20%
2900	Electrical		10.30	12.40	15.95	7.90%	9.35%	10.60%
3100	Total: Mechanical & Electrical	↓	33.50	43	53	23.50%	27%	29.50%
9000	Per pupil, total cost	Ea.	13,300	17,500	23,500			
78 0010	**SCHOOLS Senior High**	S.F.	108	133	167				**78**
0020	Total project costs	C.F.	6.60	9.70	15.65			
1800	Equipment	S.F.	2.87	6.75	9.50	1.88%	2.98%	4.80%
2720	Plumbing		6.15	9.20	16.80	5.60%	6.90%	8.30%
2770	Heating, ventilating, air conditioning		12.50	14.35	27.50	8.95%	11.60%	15%
2900	Electrical		10.95	14.25	21	8.65%	10.15%	11.95%
3100	Total: Mechanical & Electrical	↓	36.50	42.50	71.50	23.50%	26.50%	28.50%
9000	Per pupil, total cost	Ea.	10,300	21,000	26,200			
80 0010	**SCHOOLS Vocational**	S.F.	88.50	128	159				**80**
0020	Total project costs	C.F.	5.50	7.90	10.90			
0500	Masonry	S.F.	5.20	12.80	19.60	3.53%	6.70%	10.95%
1800	Equipment		2.77	6.90	9.60	1.24%	3.10%	4.26%
2720	Plumbing		5.65	8.45	12.40	5.40%	6.90%	8.55%
2770	Heating, ventilating, air conditioning		7.90	14.75	24.50	8.60%	11.90%	14.65%
2900	Electrical		9.20	12.05	16.60	8.45%	10.95%	13.20%
3100	Total: Mechanical & Electrical	↓	32	35.50	61	23.50%	27.50%	31%
9000	Per pupil, total cost	Ea.	12,300	33,000	49,200			
83 0010	**SPORTS ARENAS**	S.F.	77.50	103	159				**83**
0020	Total project costs	C.F.	4.20	7.50	9.70			
2720	Plumbing	S.F.	4.49	6.80	14.35	4.35%	6.35%	9.40%
2770	Heating, ventilating, air conditioning		9.65	11.40	15.85	8.80%	10.20%	13.55%
2900	Electrical		8.05	10.95	14.15	8.60%	9.90%	12.25%
3100	Total: Mechanical & Electrical	↓	20	35	47	21.50%	25%	27.50%
85 0010	**SUPERMARKETS**	S.F.	71.50	83	97				**85**
0020	Total project costs	C.F.	3.98	4.81	7.30			
2720	Plumbing	S.F.	3.99	5.05	5.85	5.40%	6%	7.45%
2770	Heating, ventilating, air conditioning		5.85	7.80	9.50	8.60%	8.65%	9.60%
2900	Electrical		8.95	10.30	12.15	10.40%	12.45%	13.60%
3100	Total: Mechanical & Electrical	↓	23	25	32.50	20.50%	26.50%	31%
86 0010	**SWIMMING POOLS**	S.F.	116	194	415				**86**
0020	Total project costs	C.F.	9.25	11.55	12.60			
2720	Plumbing	S.F.	10.70	12.20	16.60	4.80%	9.70%	20.50%
2900	Electrical		8.70	14.10	20.50	5.75%	6.95%	7.60%
3100	Total: Mechanical & Electrical	↓	21	55	73.50	11.15%	14.10%	23.50%
87 0010	**TELEPHONE EXCHANGES**	S.F.	154	225	285				**87**
0020	Total project costs	C.F.	9.55	15.35	21			
2720	Plumbing	S.F.	6.50	10	14.65	4.52%	5.80%	6.90%
2770	Heating, ventilating, air conditioning	↓	15.05	30.50	37.50	11.80%	16.05%	18.40%

		50 17 00 | S.F. Costs	UNIT	UNIT COSTS			% OF TOTAL			
				1/4	MEDIAN	3/4	1/4	MEDIAN	3/4	
87	2900	Electrical	S.F.	15.65	25	44	10.90%	14%	17.85%	87
	3100	Total: Mechanical & Electrical	↓	46	87.50	124	29.50%	33.50%	44.50%	
91	0010	**THEATERS**	S.F.	96.50	120	183				91
	0020	Total project costs	C.F.	4.46	6.60	9.70				
	2720	Plumbing	S.F.	3.22	3.49	14.25	2.92%	4.70%	6.80%	
	2770	Heating, ventilating, air conditioning		9.40	11.35	14.05	8%	12.25%	13.40%	
	2900	Electrical		8.45	11.40	23	8.05%	9.95%	12.25%	
	3100	Total: Mechanical & Electrical	↓	21.50	32.50	66.50	23%	26.50%	27.50%	
94	0010	**TOWN HALLS City Halls & Municipal Buildings**	S.F.	108	137	179				94
	0020	Total project costs	C.F.	9.85	11.80	16.60				
	2720	Plumbing	S.F.	4.50	8.40	15.50	4.31%	5.95%	7.95%	
	2770	Heating, ventilating, air conditioning		8.15	16.15	23.50	7.05%	9.05%	13.45%	
	2900	Electrical		10.25	14.60	19.95	8.05%	9.45%	11.65%	
	3100	Total: Mechanical & Electrical	↓	35.50	45	69	22%	26.50%	31%	
97	0010	**WAREHOUSES & Storage Buildings**	S.F.	40.50	60	86				97
	0020	Total project costs	C.F.	2.11	3.30	5.45				
	0100	Site work	S.F.	4.15	8.25	12.45	6.05%	12.95%	19.85%	
	0500	Masonry		2.42	5.70	12.35	3.73%	7.40%	12.30%	
	1800	Equipment		.65	1.39	7.80	.91%	1.82%	5.55%	
	2720	Plumbing		1.34	2.41	4.50	2.90%	4.80%	6.55%	
	2730	Heating, ventilating, air conditioning		1.53	4.32	5.80	2.41%	5%	8.90%	
	2900	Electrical		2.38	4.48	7.40	5.15%	7.20%	10.10%	
	3100	Total: Mechanical & Electrical	↓	6.65	10.20	20	12.75%	18.90%	26%	
99	0010	**WAREHOUSE & OFFICES Combination**	S.F.	49.50	66	90.50				99
	0020	Total project costs	C.F.	2.53	3.67	5.45				
	1800	Equipment	S.F.	.86	1.66	2.47	.52%	1.20%	2.40%	
	2720	Plumbing		1.91	3.39	4.96	3.74%	4.76%	6.30%	
	2770	Heating, ventilating, air conditioning		3.02	4.72	6.60	5%	5.65%	10.05%	
	2900	Electrical		3.31	4.93	7.75	5.75%	8%	10%	
	3100	Total: Mechanical & Electrical	↓	9.25	14.25	22.50	14.40%	19.95%	24.50%	

Square Foot Project Size Modifier

One factor that affects the S.F. cost of a particular building is the size. In general, for buildings built to the same specifications in the same locality, the larger building will have the lower S.F. cost. This is due mainly to the decreasing contribution of the exterior walls plus the economy of scale usually achievable in larger buildings. The Area Conversion Scale shown below will give a factor to convert costs for the typical size building to an adjusted cost for the particular project.

The Square Foot Base Size lists the median costs, most typical project size in our accumulated data, and the range in size of the projects.

The Size Factor for your project is determined by dividing your project area in S.F. by the typical project size for the particular Building Type. With this factor, enter the Area Conversion Scale at the appropriate Size Factor and determine the appropriate cost multiplier for your building size.

Example: Determine the cost per S.F. for a 100,000 S.F. Mid-rise apartment building.

$$\frac{\text{Proposed building area} = 100,000 \text{ S.F.}}{\text{Typical size from below} = 50,000 \text{ S.F.}} = 2.00$$

Enter Area Conversion scale at 2.0, intersect curve, read horizontally the appropriate cost multiplier of .94. Size adjusted cost becomes .94 x $107.00 = $101.00 based on national average costs.

Note: For Size Factors less than .50, the Cost Multiplier is 1.1
For Size Factors greater than 3.5, the Cost Multiplier is .90

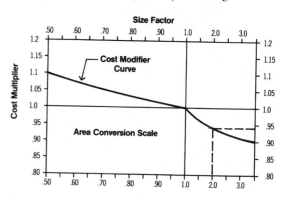

Square Foot Base Size							
Building Type	Median Cost per S.F.	Typical Size Gross S.F.	Typical Range Gross S.F.	Building Type	Median Cost per S.F.	Typical Size Gross S.F.	Typical Range Gross S.F.
Apartments, Low Rise	$ 84.50	21,000	9,700 - 37,200	Jails	$ 257.00	40,000	5,500 - 145,000
Apartments, Mid Rise	107.00	50,000	32,000 - 100,000	Libraries	161.00	12,000	7,000 - 31,000
Apartments, High Rise	116.00	145,000	95,000 - 600,000	Living, Assisted	137.00	32,300	23,500 - 50,300
Auditoriums	141.00	25,000	7,600 - 39,000	Medical Clinics	146.00	7,200	4,200 - 15,700
Auto Sales	105.00	20,000	10,800 - 28,600	Medical Offices	138.00	6,000	4,000 - 15,000
Banks	189.00	4,200	2,500 - 7,500	Motels	102.00	40,000	15,800 - 120,000
Churches	130.00	17,000	2,000 - 42,000	Nursing Homes	142.00	23,000	15,000 - 37,000
Clubs, Country	132.00	6,500	4,500 - 15,000	Offices, Low Rise	120.00	20,000	5,000 - 80,000
Clubs, Social	126.00	10,000	6,000 - 13,500	Offices, Mid Rise	119.00	120,000	20,000 - 300,000
Clubs, YMCA	143.00	28,300	12,800 - 39,400	Offices, High Rise	152.00	260,000	120,000 - 800,000
Colleges (Class)	152.00	50,000	15,000 - 150,000	Police Stations	190.00	10,500	4,000 - 19,000
Colleges (Science Lab)	242.00	45,600	16,600 - 80,000	Post Offices	141.00	12,400	6,800 - 30,000
College (Student Union)	179.00	33,400	16,000 - 85,000	Power Plants	1000.00	7,500	1,000 - 20,000
Community Center	134.00	9,400	5,300 - 16,700	Religious Education	120.00	9,000	6,000 - 12,000
Court Houses	180.00	32,400	17,800 - 106,000	Research	198.00	19,000	6,300 - 45,000
Dept. Stores	78.50	90,000	44,000 - 122,000	Restaurants	172.00	4,400	2,800 - 6,000
Dormitories, Low Rise	143.00	25,000	10,000 - 95,000	Retail Stores	84.00	7,200	4,000 - 17,600
Dormitories, Mid Rise	176.00	85,000	20,000 - 200,000	Schools, Elementary	125.00	41,000	24,500 - 55,000
Factories	76.50	26,400	12,900 - 50,000	Schools, Jr. High	129.00	92,000	52,000 - 119,000
Fire Stations	139.00	5,800	4,000 - 8,700	Schools, Sr. High	133.00	101,000	50,500 - 175,000
Fraternity Houses	131.00	12,500	8,200 - 14,800	Schools, Vocational	128.00	37,000	20,500 - 82,000
Funeral Homes	146.00	10,000	4,000 - 20,000	Sports Arenas	103.00	15,000	5,000 - 40,000
Garages, Commercial	93.50	9,300	5,000 - 13,600	Supermarkets	83.00	44,000	12,000 - 60,000
Garages, Municipal	119.00	8,300	4,500 - 12,600	Swimming Pools	194.00	20,000	10,000 - 32,000
Garages, Parking	50.50	163,000	76,400 - 225,300	Telephone Exchange	225.00	4,500	1,200 - 10,600
Gymnasiums	129.00	19,200	11,600 - 41,000	Theaters	120.00	10,500	8,800 - 17,500
Hospitals	228.00	55,000	27,200 - 125,000	Town Halls	137.00	10,800	4,800 - 23,400
House (Elderly)	115.00	37,000	21,000 - 66,000	Warehouses	60.00	25,000	8,000 - 72,000
Housing (Public)	106.00	36,000	14,400 - 74,400	Warehouse & Office	66.00	25,000	8,000 - 72,000
Ice Rinks	153.00	29,000	27,200 - 33,600				

Abbreviations

A	Area Square Feet; Ampere	BTUH	BTU per Hour	Cwt.	100 Pounds
ABS	Acrylonitrile Butadiene Stryrene;	B.U.R.	Built-up Roofing	C.W.X.	Cool White Deluxe
	Asbestos Bonded Steel	BX	Interlocked Armored Cable	C.Y.	Cubic Yard (27 cubic feet)
A.C.	Alternating Current;	°C	degree centegrade	C.Y./Hr.	Cubic Yard per Hour
	Air-Conditioning;	c	Conductivity, Copper Sweat	Cyl.	Cylinder
	Asbestos Cement;	C	Hundred; Centigrade	d	Penny (nail size)
	Plywood Grade A & C	C/C	Center to Center, Cedar on Cedar	D	Deep; Depth; Discharge
A.C.I.	American Concrete Institute	C-C	Center to Center	Dis., Disch.	Discharge
AD	Plywood, Grade A & D	Cab.	Cabinet	Db.	Decibel
Addit.	Additional	Cair.	Air Tool Laborer	Dbl.	Double
Adj.	Adjustable	Calc	Calculated	DC	Direct Current
af	Audio-frequency	Cap.	Capacity	DDC	Direct Digital Control
A.G.A.	American Gas Association	Carp.	Carpenter	Demob.	Demobilization
Agg.	Aggregate	C.B.	Circuit Breaker	d.f.u.	Drainage Fixture Units
A.H.	Ampere Hours	C.C.A.	Chromate Copper Arsenate	D.H.	Double Hung
A hr.	Ampere-hour	C.C.F.	Hundred Cubic Feet	DHW	Domestic Hot Water
A.H.U.	Air Handling Unit	cd	Candela	DI	Ductile Iron
A.I.A.	American Institute of Architects	cd/sf	Candela per Square Foot	Diag.	Diagonal
AIC	Ampere Interrupting Capacity	CD	Grade of Plywood Face & Back	Diam., Dia	Diameter
Allow.	Allowance	CDX	Plywood, Grade C & D, exterior	Distrib.	Distribution
alt.	Altitude		glue	Div.	Division
Alum.	Aluminum	Cefi.	Cement Finisher	Dk.	Deck
a.m.	Ante Meridiem	Cem.	Cement	D.L.	Dead Load; Diesel
Amp.	Ampere	CF	Hundred Feet	DLH	Deep Long Span Bar Joist
Anod.	Anodized	C.F.	Cubic Feet	Do.	Ditto
Approx.	Approximate	CFM	Cubic Feet per Minute	Dp.	Depth
Apt.	Apartment	c.g.	Center of Gravity	D.P.S.T.	Double Pole, Single Throw
Asb.	Asbestos	CHW	Chilled Water;	Dr.	Drive
A.S.B.C.	American Standard Building Code		Commercial Hot Water	Drink.	Drinking
Asbe.	Asbestos Worker	C.I.	Cast Iron	D.S.	Double Strength
ASCE.	American Society of Civil Engineers	C.I.P.	Cast in Place	D.S.A.	Double Strength A Grade
A.S.H.R.A.E.	American Society of Heating,	Circ.	Circuit	D.S.B.	Double Strength B Grade
	Refrig. & AC Engineers	C.L.	Carload Lot	Dty.	Duty
A.S.M.E.	American Society of Mechanical	Clab.	Common Laborer	DWV	Drain Waste Vent
	Engineers	Clam	Common maintenance laborer	DX	Deluxe White, Direct Expansion
A.S.T.M.	American Society for Testing and	C.L.F.	Hundred Linear Feet	dyn	Dyne
	Materials	CLF	Current Limiting Fuse	e	Eccentricity
Attchmt.	Attachment	CLP	Cross Linked Polyethylene	E	Equipment Only; East
Avg.,Ave.	Average	cm	Centimeter	Ea.	Each
A.W.G.	American Wire Gauge	CMP	Corr. Metal Pipe	E.B.	Encased Burial
AWWA	American Water Works Assoc.	C.M.U.	Concrete Masonry Unit	Econ.	Economy
Bbl.	Barrel	CN	Change Notice	E.C.Y	Embankment Cubic Yards
B&B	Grade B and Better;	Col.	Column	EDP	Electronic Data Processing
	Balled & Burlapped	CO₂	Carbon Dioxide	EIFS	Exterior Insulation Finish System
B.&S.	Bell and Spigot	Comb.	Combination	E.D.R.	Equiv. Direct Radiation
B.&W.	Black and White	Compr.	Compressor	Eq.	Equation
b.c.c.	Body-centered Cubic	Conc.	Concrete	EL	elevation
B.C.Y.	Bank Cubic Yards	Cont.	Continuous; Continued, Container	Elec.	Electrician; Electrical
BE	Bevel End	Corr.	Corrugated	Elev.	Elevator; Elevating
B.F.	Board Feet	Cos	Cosine	EMT	Electrical Metallic Conduit;
Bg. cem.	Bag of Cement	Cot	Cotangent		Thin Wall Conduit
BHP	Boiler Horsepower;	Cov.	Cover	Eng.	Engine, Engineered
	Brake Horsepower	C/P	Cedar on Paneling	EPDM	Ethylene Propylene Diene
B.I.	Black Iron	CPA	Control Point Adjustment		Monomer
Bit., Bitum.	Bituminous	Cplg.	Coupling	EPS	Expanded Polystyrene
Bit., Conc.	Bituminous Concrete	C.P.M.	Critical Path Method	Eqhv.	Equip. Oper., Heavy
Bk.	Backed	CPVC	Chlorinated Polyvinyl Chloride	Eqlt.	Equip. Oper., Light
Bkrs.	Breakers	C.Pr.	Hundred Pair	Eqmd.	Equip. Oper., Medium
Bldg.	Building	CRC	Cold Rolled Channel	Eqmm.	Equip. Oper., Master Mechanic
Blk.	Block	Creos.	Creosote	Eqol.	Equip. Oper., Oilers
Bm.	Beam	Crpt.	Carpet & Linoleum Layer	Equip.	Equipment
Boil.	Boilermaker	CRT	Cathode-ray Tube	ERW	Electric Resistance Welded
B.P.M.	Blows per Minute	CS	Carbon Steel, Constant Shear Bar	E.S.	Energy Saver
BR	Bedroom		Joist	Est.	Estimated
Brg.	Bearing	Csc	Cosecant	esu	Electrostatic Units
Brhe.	Bricklayer Helper	C.S.F.	Hundred Square Feet	E.W.	Each Way
Bric.	Bricklayer	CSI	Construction Specifications	EWT	Entering Water Temperature
Brk.	Brick		Institute	Excav.	Excavation
Brng.	Bearing	C.T.	Current Transformer	Exp.	Expansion, Exposure
Brs.	Brass	CTS	Copper Tube Size	Ext.	Exterior
Brz.	Bronze	Cu	Copper, Cubic	Extru.	Extrusion
Bsn.	Basin	Cu. Ft.	Cubic Foot	f.	Fiber stress
Btr.	Better	cw	Continuous Wave	F	Fahrenheit; Female; Fill
Btu	British Thermal Unit	C.W.	Cool White; Cold Water	Fab.	Fabricated

FBGS	Fiberglass	H.P.	Horsepower; High Pressure	LE	Lead Equivalent
F.C.	Footcandles	H.P.F.	High Power Factor	LED	Light Emitting Diode
f.c.c.	Face-centered Cubic	Hr.	Hour	L.F.	Linear Foot
f'c.	Compressive Stress in Concrete; Extreme Compressive Stress	Hrs./Day	Hours per Day	L.F. Nose	Linear Foot of Stair Nosing
		HSC	High Short Circuit	L.F. Rsr	Linear Foot of Stair Riser
F.E.	Front End	Ht.	Height	Lg.	Long; Length; Large
FEP	Fluorinated Ethylene Propylene (Teflon)	Htg.	Heating	L & H	Light and Heat
		Htrs.	Heaters	LH	Long Span Bar Joist
F.G.	Flat Grain	HVAC	Heating, Ventilation & Air-Conditioning	L.H.	Labor Hours
F.H.A.	Federal Housing Administration			L.L.	Live Load
Fig.	Figure	Hvy.	Heavy	L.L.D.	Lamp Lumen Depreciation
Fin.	Finished	HW	Hot Water	lm	Lumen
Fixt.	Fixture	Hyd.;Hydr.	Hydraulic	lm/sf	Lumen per Square Foot
Fl. Oz.	Fluid Ounces	Hz.	Hertz (cycles)	lm/W	Lumen per Watt
Flr.	Floor	I.	Moment of Inertia	L.O.A.	Length Over All
F.M.	Frequency Modulation; Factory Mutual	IBC	International Building Code	log	Logarithm
		I.C.	Interrupting Capacity	L-O-L	Lateralolet
Fmg.	Framing	ID	Inside Diameter	long.	longitude
Fdn.	Foundation	I.D.	Inside Dimension; Identification	L.P.	Liquefied Petroleum; Low Pressure
Fori.	Foreman, Inside	I.F.	Inside Frosted	L.P.F.	Low Power Factor
Foro.	Foreman, Outside	I.M.C.	Intermediate Metal Conduit	LR	Long Radius
Fount.	Fountain	In.	Inch	L.S.	Lump Sum
fpm	Feet per Minute	Incan.	Incandescent	Lt.	Light
FPT	Female Pipe Thread	Incl.	Included; Including	Lt. Ga.	Light Gauge
Fr.	Frame	Int.	Interior	L.T.L.	Less than Truckload Lot
F.R.	Fire Rating	Inst.	Installation	Lt. Wt.	Lightweight
FRK	Foil Reinforced Kraft	Insul.	Insulation/Insulated	L.V.	Low Voltage
FRP	Fiberglass Reinforced Plastic	I.P.	Iron Pipe	M	Thousand; Material; Male; Light Wall Copper Tubing
FS	Forged Steel	I.P.S.	Iron Pipe Size		
FSC	Cast Body; Cast Switch Box	I.P.T.	Iron Pipe Threaded	M²CA	Meters Squared Contact Area
Ft.	Foot; Feet	I.W.	Indirect Waste	m/hr.; M.H.	Man-hour
Ftng.	Fitting	J	Joule	mA	Milliampere
Ftg.	Footing	J.I.C.	Joint Industrial Council	Mach.	Machine
Ft lb.	Foot Pound	K	Thousand; Thousand Pounds; Heavy Wall Copper Tubing, Kelvin	Mag. Str.	Magnetic Starter
Furn.	Furniture			Maint.	Maintenance
FVNR	Full Voltage Non-Reversing	K.A.H.	Thousand Amp. Hours	Marb.	Marble Setter
FXM	Female by Male	KCMIL	Thousand Circular Mils	Mat; Mat'l.	Material
Fy.	Minimum Yield Stress of Steel	KD	Knock Down	Max.	Maximum
g	Gram	K.D.A.T.	Kiln Dried After Treatment	MBF	Thousand Board Feet
G	Gauss	kg	Kilogram	MBH	Thousand BTU's per hr.
Ga.	Gauge	kG	Kilogauss	MC	Metal Clad Cable
Gal., gal.	Gallon	kgf	Kilogram Force	M.C.F.	Thousand Cubic Feet
gpm, GPM	Gallon per Minute	kHz	Kilohertz	M.C.F.M.	Thousand Cubic Feet per Minute
Galv.	Galvanized	Kip.	1000 Pounds	M.C.M.	Thousand Circular Mils
Gen.	General	KJ	Kiljoule	M.C.P.	Motor Circuit Protector
G.F.I.	Ground Fault Interrupter	K.L.	Effective Length Factor	MD	Medium Duty
Glaz.	Glazier	K.L.F.	Kips per Linear Foot	M.D.O.	Medium Density Overlaid
GPD	Gallons per Day	Km	Kilometer	Med.	Medium
GPH	Gallons per Hour	K.S.F.	Kips per Square Foot	MF	Thousand Feet
GPM	Gallons per Minute	K.S.I.	Kips per Square Inch	M.F.B.M.	Thousand Feet Board Measure
GR	Grade	kV	Kilovolt	Mfg.	Manufacturing
Gran.	Granular	kVA	Kilovolt Ampere	Mfrs.	Manufacturers
Grnd.	Ground	K.V.A.R.	Kilovar (Reactance)	mg	Milligram
H	High Henry	KW	Kilowatt	MGD	Million Gallons per Day
H.C.	High Capacity	KWh	Kilowatt-hour	MGPH	Thousand Gallons per Hour
H.D.	Heavy Duty; High Density	L	Labor Only; Length; Long; Medium Wall Copper Tubing	MH, M.H.	Manhole; Metal Halide; Man-Hour
H.D.O.	High Density Overlaid			MHz	Megahertz
H.D.P.E.	high density polyethelene	Lab.	Labor	Mi.	Mile
Hdr.	Header	lat	Latitude	MI	Malleable Iron; Mineral Insulated
Hdwe.	Hardware	Lath.	Lather	mm	Millimeter
Help.	Helper Average	Lav.	Lavatory	Mill.	Millwright
HEPA	High Efficiency Particulate Air Filter	lb.; #	Pound	Min., min.	Minimum, minute
		L.B.	Load Bearing; L Conduit Body	Misc.	Miscellaneous
Hg	Mercury	L. & E.	Labor & Equipment	ml	Milliliter, Mainline
HIC	High Interrupting Capacity	lb./hr.	Pounds per Hour	M.L.F.	Thousand Linear Feet
HM	Hollow Metal	lb./L.F.	Pounds per Linear Foot	Mo.	Month
HMWPE	high molecular weight polyethylene	lbf/sq.in.	Pound-force per Square Inch	Mobil.	Mobilization
		L.C.L.	Less than Carload Lot	Mog.	Mogul Base
H.O.	High Output	L.C.Y.	Loose Cubic Yard	MPH	Miles per Hour
Horiz.	Horizontal	Ld.	Load	MPT	Male Pipe Thread

Abbreviations

MRT	Mile Round Trip	Pl.	Plate	S.F.C.A.	Square Foot Contact Area
ms	Millisecond	Plah.	Plasterer Helper	S.F. Flr.	Square Foot of Floor
M.S.F.	Thousand Square Feet	Plas.	Plasterer	S.F.G.	Square Foot of Ground
Mstz.	Mosaic & Terrazzo Worker	Pluh.	Plumbers Helper	S.F. Hor.	Square Foot Horizontal
M.S.Y.	Thousand Square Yards	Plum.	Plumber	S.F.R.	Square Feet of Radiation
Mtd., mtd.	Mounted	Ply.	Plywood	S.F. Shlf.	Square Foot of Shelf
Mthe.	Mosaic & Terrazzo Helper	p.m.	Post Meridiem	S4S	Surface 4 Sides
Mtng.	Mounting	Pntd.	Painted	Shee.	Sheet Metal Worker
Mult.	Multi; Multiply	Pord.	Painter, Ordinary	Sin.	Sine
M.V.A.	Million Volt Amperes	pp	Pages	Skwk.	Skilled Worker
M.V.A.R.	Million Volt Amperes Reactance	PP, PPL	Polypropylene	SL	Saran Lined
MV	Megavolt	P.P.M.	Parts per Million	S.L.	Slimline
MW	Megawatt	Pr.	Pair	Sldr.	Solder
MXM	Male by Male	P.E.S.B.	Pre-engineered Steel Building	SLH	Super Long Span Bar Joist
MYD	Thousand Yards	Prefab.	Prefabricated	S.N.	Solid Neutral
N	Natural; North	Prefin.	Prefinished	S-O-L	Socketolet
nA	Nanoampere	Prop.	Propelled	sp	Standpipe
NA	Not Available; Not Applicable	PSF, psf	Pounds per Square Foot	S.P.	Static Pressure; Single Pole; Self-Propelled
N.B.C.	National Building Code	PSI, psi	Pounds per Square Inch		
NC	Normally Closed	PSIG	Pounds per Square Inch Gauge	Spri.	Sprinkler Installer
N.E.M.A.	National Electrical Manufacturers Assoc.	PSP	Plastic Sewer Pipe	spwg	Static Pressure Water Gauge
		Pspr.	Painter, Spray	S.P.D.T.	Single Pole, Double Throw
NEHB	Bolted Circuit Breaker to 600V.	Psst.	Painter, Structural Steel	SPF	Spruce Pine Fir
N.L.B.	Non-Load-Bearing	P.T.	Potential Transformer	S.P.S.T.	Single Pole, Single Throw
NM	Non-Metallic Cable	P. & T.	Pressure & Temperature	SPT	Standard Pipe Thread
nm	Nanometer	Ptd.	Painted	Sq.	Square; 100 Square Feet
No.	Number	Ptns.	Partitions	Sq. Hd.	Square Head
NO	Normally Open	Pu	Ultimate Load	Sq. In.	Square Inch
N.O.C.	Not Otherwise Classified	PVC	Polyvinyl Chloride	S.S.	Single Strength; Stainless Steel
Nose.	Nosing	Pvmt.	Pavement	S.S.B.	Single Strength B Grade
N.P.T.	National Pipe Thread	Pwr.	Power	sst, ss	Stainless Steel
NQOD	Combination Plug-on/Bolt on Circuit Breaker to 240V.	Q	Quantity Heat Flow	Sswk.	Structural Steel Worker
		Qt.	Quart	Sswl.	Structural Steel Welder
N.R.C.	Noise Reduction Coefficient/ Nuclear Regulator Commission	Quan., Qty.	Quantity	St.;Stl.	Steel
		Q.C.	Quick Coupling	S.T.C.	Sound Transmission Coefficient
N.R.S.	Non Rising Stem	r	Radius of Gyration	Std.	Standard
ns	Nanosecond	R	Resistance	Stg.	Staging
nW	Nanowatt	R.C.P.	Reinforced Concrete Pipe	STK	Select Tight Knot
OB	Opposing Blade	Rect.	Rectangle	STP	Standard Temperature & Pressure
OC	On Center	Reg.	Regular	Stpi.	Steamfitter, Pipefitter
OD	Outside Diameter	Reinf.	Reinforced	Str.	Strength; Starter; Straight
O.D.	Outside Dimension	Req'd.	Required	Strd.	Stranded
ODS	Overhead Distribution System	Res.	Resistant	Struct.	Structural
O.G.	Ogee	Resi.	Residential	Sty.	Story
O.H.	Overhead	Rgh.	Rough	Subj.	Subject
O&P	Overhead and Profit	RGS	Rigid Galvanized Steel	Subs.	Subcontractors
Oper.	Operator	R.H.W.	Rubber, Heat & Water Resistant; Residential Hot Water	Surf.	Surface
Opng.	Opening			Sw.	Switch
Orna.	Ornamental	rms	Root Mean Square	Swbd.	Switchboard
OSB	Oriented Strand Board	Rnd.	Round	S.Y.	Square Yard
O.S.&Y.	Outside Screw and Yoke	Rodm.	Rodman	Syn.	Synthetic
Ovhd.	Overhead	Rofc.	Roofer, Composition	S.Y.P.	Southern Yellow Pine
OWG	Oil, Water or Gas	Rofp.	Roofer, Precast	Sys.	System
Oz.	Ounce	Rohe.	Roofer Helpers (Composition)	t.	Thickness
P.	Pole; Applied Load; Projection	Rots.	Roofer, Tile & Slate	T	Temperature; Ton
p.	Page	R.O.W.	Right of Way	Tan	Tangent
Pape.	Paperhanger	RPM	Revolutions per Minute	T.C.	Terra Cotta
P.A.P.R.	Powered Air Purifying Respirator	R.S.	Rapid Start	T & C	Threaded and Coupled
PAR	Parabolic Reflector	Rsr	Riser	T.D.	Temperature Difference
Pc., Pcs.	Piece, Pieces	RT	Round Trip	Tdd	Telecommunications Device for the Deaf
P.C.	Portland Cement; Power Connector	S.	Suction; Single Entrance; South		
P.C.F.	Pounds per Cubic Foot	SC	Screw Cover	T.E.M.	Transmission Electron Microscopy
P.C.M.	Phase Contrast Microscopy	SCFM	Standard Cubic Feet per Minute	TFE	Tetrafluoroethylene (Teflon)
P.E.	Professional Engineer; Porcelain Enamel; Polyethylene; Plain End	Scaf.	Scaffold	T. & G.	Tongue & Groove; Tar & Gravel
		Sch., Sched.	Schedule		
		S.C.R.	Modular Brick	Th., Thk.	Thick
Perf.	Perforated	S.D.	Sound Deadening	Thn.	Thin
PEX	Cross linked polyethylene	S.D.R.	Standard Dimension Ratio	Thrded	Threaded
Ph.	Phase	S.E.	Surfaced Edge	Tilf.	Tile Layer, Floor
P.I.	Pressure Injected	Sel.	Select	Tilh.	Tile Layer, Helper
Pile.	Pile Driver	S.E.R., S.E.U.	Service Entrance Cable	THHN	Nylon Jacketed Wire
Pkg.	Package	S.F.	Square Foot	THW.	Insulated Strand Wire

751

THWN	Nylon Jacketed Wire	USP	United States Primed
T.L.	Truckload	UTP	Unshielded Twisted Pair
T.M.	Track Mounted	V	Volt
Tot.	Total	V.A.	Volt Amperes
T-O-L	Threadolet	V.C.T.	Vinyl Composition Tile
T.S.	Trigger Start	VAV	Variable Air Volume
Tr.	Trade	VC	Veneer Core
Transf.	Transformer	Vent.	Ventilation
Trhv.	Truck Driver, Heavy	Vert.	Vertical
Trlr	Trailer	V.F.	Vinyl Faced
Trlt.	Truck Driver, Light	V.G.	Vertical Grain
TTY	Teletypewriter	V.H.F.	Very High Frequency
TV	Television	VHO	Very High Output
T.W.	Thermoplastic Water Resistant Wire	Vib.	Vibrating
		V.L.F.	Vertical Linear Foot
UCI	Uniform Construction Index	Vol.	Volume
UF	Underground Feeder	VRP	Vinyl Reinforced Polyester
UGND	Underground Feeder	W	Wire; Watt; Wide; West
U.H.F.	Ultra High Frequency	w/	With
U.I.	United Inch	W.C.	Water Column; Water Closet
U.L.	Underwriters Laboratory	W.F.	Wide Flange
Uld.	unloading	W.G.	Water Gauge
Unfin.	Unfinished	Wldg.	Welding
URD	Underground Residential Distribution	W. Mile	Wire Mile
		W-O-L	Weldolet
US	United States	W.R.	Water Resistant

Wrck.	Wrecker
W.S.P.	Water, Steam, Petroleum
WT., Wt.	Weight
WWF	Welded Wire Fabric
XFER	Transfer
XFMR	Transformer
XHD	Extra Heavy Duty
XHHW, XLPE	Cross-Linked Polyethylene Wire Insulation
XLP	Cross-linked Polyethylene
Y	Wye
yd	Yard
yr	Year
Δ	Delta
%	Percent
~	Approximately
Ø	Phase; diameter
@	At
#	Pound; Number
<	Less Than
>	Greater Than
Z	zone

Index

Index

Index

Index

Index

Index

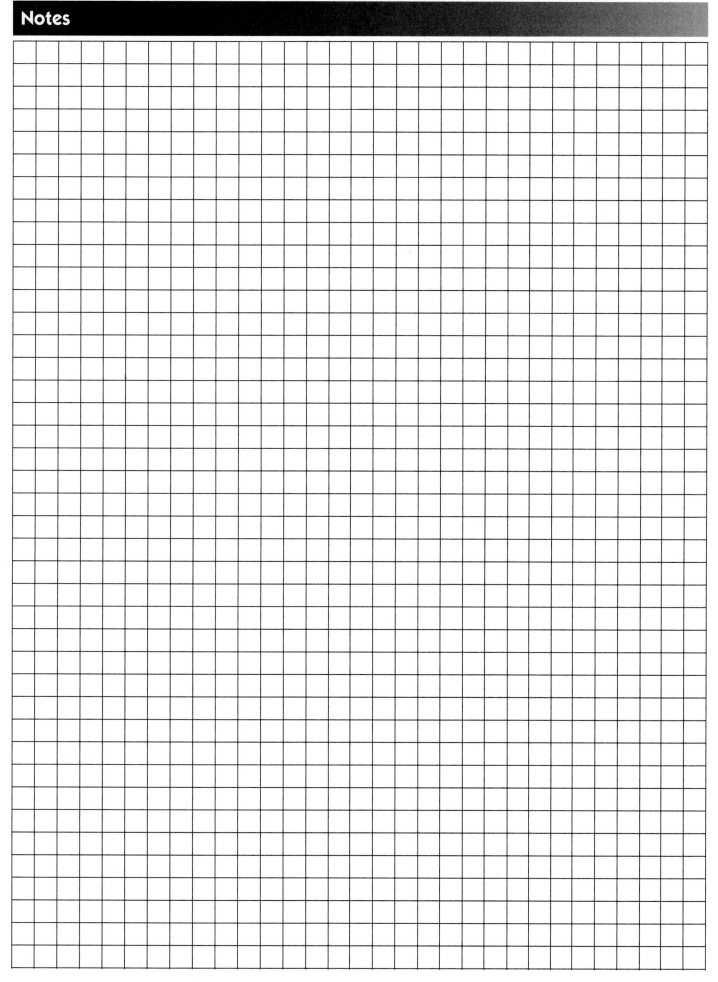

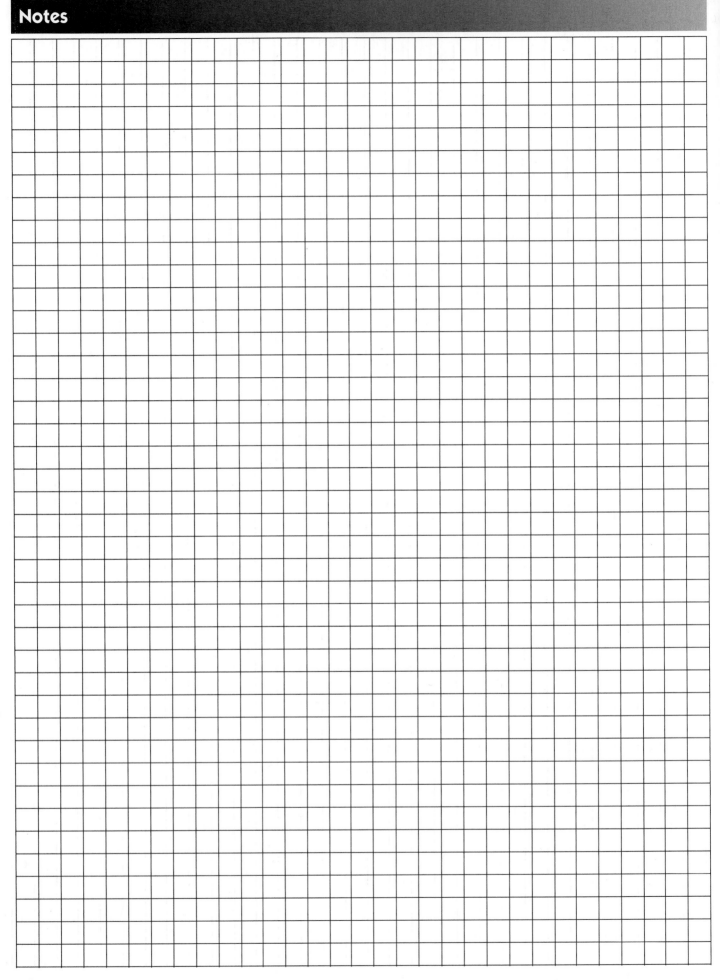

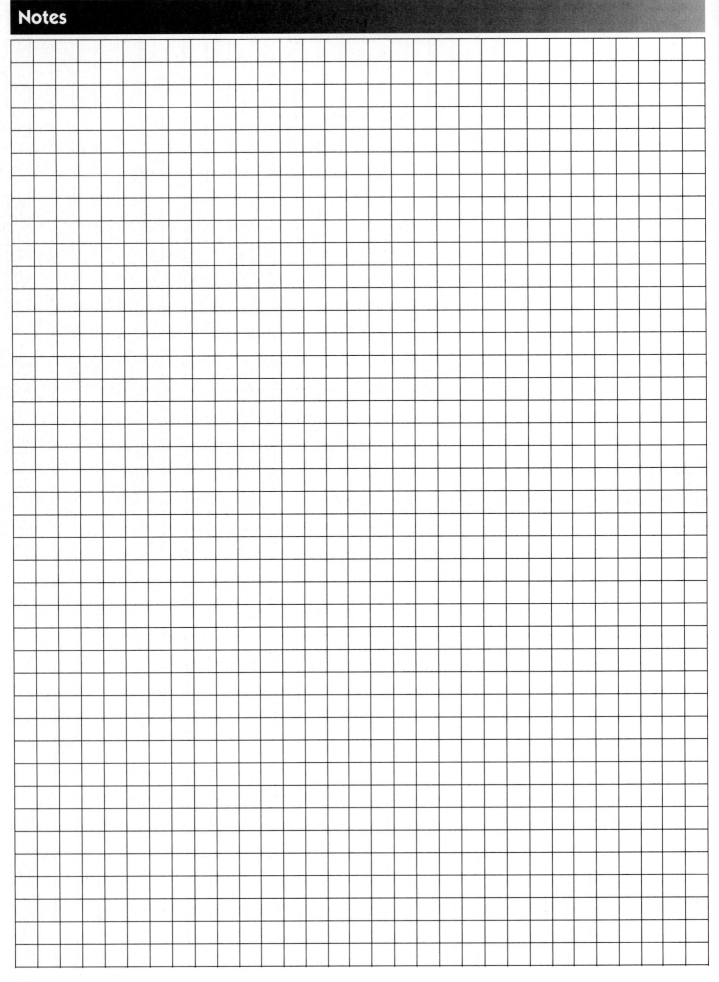

	CREW	DAILY OUTPUT	LABOR-HOURS	UNIT	2000 BARE COSTS				TOTAL INCL O&P
					MAT.	LABOR	EQUIP.	TOTAL	

		CREW	DAILY OUTPUT	LABOR-HOURS	UNIT	2000 BARE COSTS				TOTAL INCL O&P
						MAT.	LABOR	EQUIP.	TOTAL	
			CREW	DAILY OUTPUT	LABOR-HOURS	UNIT	2000 BARE COSTS				TOTAL INCL O&P